Geodynamics of a Cordilleran Orogenic System: The Central Andes of Argentina and Northern Chile

edited by

Peter G. DeCelles
Department of Geosciences
University of Arizona
Tucson, Arizona, USA

Mihai N. Ducea
Department of Geosciences
University of Arizona
Tucson, Arizona, USA, and
Universitatea Bucuresti
Facultatea de Geologie Geofizica
Strada N. Balcescu Nr 1
Bucuresti, Romania

Barbara Carrapa
Department of Geosciences
University of Arizona
Tucson, Arizona, USA

Paul A. Kapp
Department of Geosciences
University of Arizona
Tucson, Arizona, USA

THE
GEOLOGICAL
SOCIETY
OF AMERICA®

Memoir 212

3300 Penrose Place, P.O. Box 9140 ▪ Boulder, Colorado 80301-9140, USA

2015

Published by The Geological Society of America, Inc.
3300 Penrose Place, P.O. Box 9140, Boulder, Colorado 80301-9140, USA
www.geosociety.org

Printed in U.S.A.

GSA Books Science Editors: Kent Condie and F. Edwin Harvey

Library of Congress Cataloging-in-Publication Data

Geodynamics of a cordilleran orogenic system : the central Andes of Argentina and northern
Chile / edited by Peter G. DeCelles, Department of Geosciences, University of Arizona, Tucson,
Arizona, USA [and 3 others].
    pages cm.— (Memoir ; 212)
  Includes bibliographical references.
   ISBN 978-0-8137-1212-3 (cloth)
 1. Orogenic belts—Argentina. 2. Orogenic belts—Chile. 3. Orogenic belts—Andes.
4. Geodynamics—Andes. I. DeCelles, Peter G., 1958– editor.
 QE606.5.A7G46 2015
 551.8′209824—dc23
                                                                        2014031560

Cover, front: Nieves penitentes (~1.5 m tall) on north flank of Monte Pissis (6795 m, 22,293 ft), the
second highest volcano on Earth and third highest peak in the western hemisphere. The peak is an
inactive volcano in the Andean magmatic arc near the Argentina-Chile border. It bears the name of
French geologist Pierre Joseph Aimé Pissis (1812–1889). Photograph by P.G. DeCelles.
Back: The Salinas Grandes evaporite basin located in the eastern Puna Plateau at an elevation of
3412 m (11,194 ft), northwestern Argentina. Broad summit in background is Nevado de Chañi
(5930 m, 19,455 ft). Photograph by Martin Pepper.

10 9 8 7 6 5 4 3 2 1

Contents

iii

Preface and Acknowledgments

This volume represents many of the results of a six-year, multidisciplinary study of the central part of the Andean orogenic belt and its eastward adjacent foreland basin system, mainly between the latitudes of 22°S and 28°S, in northern Argentina and Chile. The study was designed to test the idea that diverse processes operating in cordilleran-style orogenic belts may be linked in feedback and feed-forward relationships. Such processes include crustal shortening, magmatism, upper mantle dynamics (especially delamination and dripping of lithospheric bodies), basin dynamics, and changes in surface elevation, among others. The central Andes provide an excellent place to study potential linkages among such processes, because the region has experienced a large amount of crustal shortening and thickening over the last 60 m.y. in response to approximately orthogonal convergence between the Nazca and South American plates. The region contains excellent examples of all features that are typical of cordilleran orogenic belts, including an active trench, forearc, and magmatic arc; a high hinterland plateau; a sizeable retroarc fold-thrust belt; and a long-lived foreland basin. The chapters in the volume provide new data from the upper mantle to the surface, from the forearc to the active foreland basin, and employ a wide array of techniques and methods, including geophysics, petrology, geochronology and thermochronology, sedimentology, structural geology, geomorphology, paleoaltimetry, and geodynamic modeling.

Most of the work represented in this volume was funded by a generous grant from the ExxonMobil Corporation, and much of the work involved collaborations among scientists from several universities and ExxonMobil. Without the efforts and vision of Carlos Dengo at ExxonMobil, Joaquin Ruiz at the University of Arizona, and Ricardo N. Alonso at the University of Salta, this project would not have proceeded beyond the conceptual stage. Additional funding was provided by the U.S. National Science Foundation and the Natural Sciences and Engineering Research Council of Canada. The volume benefited immensely from critical, insightful reviews provided by the following individuals, to whom the authors extend collective gratitude:

César Arriagada, Heinrich Bahlburg, Jason Barnes, Chris Beaumont, Cathy Busby, Peter Cawood, Robinson Cecil, Page Chamberlain, Mark Cooper, Brian S. Currie, Federico Dávila, Cecile Gautheron, E. Gierlowski-Kordesch, Ben Heit, Michael Hren, Ray Ingersoll, Cari Johnson, Teresa Jordan, Jonas Kley, Timothy Lawton, Cin-Ty Lee, Xiaoli Liu, Friedrich Lucassen, M. Mamani, Calvin Miller, Andreas Mulch, Onno Oncken, Chris Poulsen, Matt Pritchard, Keith Putirka, Victor Ramos, James Reynolds, M. Rosen, François Roure, Brandon Schmandt, M. Smith, Daniel Starck, Marlies ter Voorde, Roland van Heune, Jolante van Wijk, Giovanni Vezzoli, and Adolph Yonkee.

Peter G. DeCelles, Mihai N. Ducea*, Barbara Carrapa, and Paul A. Kapp
Department of Geosciences, University of Arizona, Tucson, Arizona, USA

*Also at Universitatea Bucuresti, Facultatea de Geologie Geofizica, Strada N. Balcescu Nr 1, Bucuresti, Romania.

The Geological Society of America
Memoir 212
2015

Geodynamic models of Cordilleran orogens: Gravitational instability of magmatic arc roots

Claire A. Currie
Department of Physics, University of Alberta, Edmonton, AB T6G 2E1, Canada

Mihai N. Ducea
Department of Geosciences, University of Arizona, Tucson, Arizona 85721, USA, and
Universitatea Bucuresti, Facultatea de Geologie Geofizica, Strada N. Balcescu Nr 1, Bucuresti, Romania

Peter G. DeCelles
Department of Geosciences, University of Arizona, Tucson, Arizona 85721, USA

Christopher Beaumont
Department of Oceanography, Dalhousie University, Halifax, NS B3H 3J5, Canada

ABSTRACT

Cordilleran orogens, such as the central Andes, form above subduction zones, and their evolution depends on both continental shortening and oceanic plate subduction processes, including arc magmatism and granitoid batholith formation. Arc and batholith magma compositions are consistent with partial melting of continental lithosphere and magmatic differentiation, whereby felsic melts rise upward through the crust, leaving a high-density pyroxenite root in the deep lithosphere. We study gravitational removal of this root using two-dimensional thermal-mechanical numerical models of subduction below a continent. The volcanic arc position is determined dynamically based on thermal structure, and formation of a batholith-root complex is simulated by changing the density of the arc lithosphere over time. For the model lithosphere structure, magmatic roots with even a small density increase are readily removed for a wide range of root strengths and subduction rates. The dynamics of removal depend on the relative rates of downward gravitational growth and lateral shearing by subduction-induced mantle flow. Gravitational growth dominates for high root densification rates, high root viscosities, and low subduction rates, resulting in drip-like removal as a single downwelling over 1–2.5 m.y. At lower growth rates, the root is removed over >3 m.y. through shear entrainment as it is carried sideways by mantle flow and then subducted. In all models, >80% of the root is removed, making this an effective way to thin orogenic mantle lithosphere. This can help resolve the mass problem in the central Andes, where observations indicate a thin mantle lithosphere, despite significant crustal shortening and thickening.

Currie, C.A., Ducea, M.N., DeCelles, P.G., and Beaumont, C., 2015, Geodynamic models of Cordilleran orogens: Gravitational instability of magmatic arc roots, *in* DeCelles, P.G., Ducea, M.N., Carrapa, B., and Kapp, P.A., eds., Geodynamics of a Cordilleran Orogenic System: The Central Andes of Argentina and Northern Chile: Geological Society of America Memoir 212, p. 1–22, doi:10.1130/2015.1212(01). For permission to copy, contact editing@geosociety.org. © 2014 The Geological Society of America. All rights reserved.

INTRODUCTION

The central Andes represents the type example of a Cordilleran orogen, in which crustal shortening has produced a continental mountain belt above an active subduction zone. Geological studies indicate that the central Andes was primarily built through 200 to >500 km of shortening of the western South American plate during Cenozoic subduction of the oceanic Nazca plate (e.g., McQuarrie et al., 2005; Oncken et al., 2006; Barnes and Ehlers, 2009). This resulted in the formation of the Altiplano-Puna Plateau, an internally drained plateau with an average elevation of ~4 km. Seismic receiver functions show present-day crustal thicknesses of 50–80 km below much of the plateau (Beck et al., 1996; Yuan et al., 2002; Beck and Zandt, 2002; McGlashan et al., 2008; Bianchi et al., 2013). At present, subduction continues at a plate convergence rate of 7–8 cm/yr, and shortening is localized on the eastern side of the plateau (e.g., Brooks et al., 2003; Oncken et al., 2006).

During orogenesis, shortening of the upper crust should be accompanied by thickening of the deeper lithosphere. However, several observations indicate that the mantle lithosphere is not anomalously thick beneath most of the central Andes. Seismic tomography studies show that many parts of the orogen have anomalously low velocities in the shallow mantle (<100 km depth; e.g., Myers et al., 1998; Beck and Zandt, 2002; Schurr et al., 2006; Bianchi et al., 2013). The regions of lowest velocities underlie regions of recent volcanism (e.g., northern Puna and eastern Altiplano). In addition, receiver functions show that the lithosphere-asthenosphere boundary is at 100–130 km depth below the central Altiplano, indicating a 30–50-km-thick mantle lithosphere (Heit et al., 2008). Other indicators of a thin, hot mantle lithosphere include elevated ^{3}He in groundwater of the Altiplano (Hoke et al., 1994), the occurrence of magmatism across the width of the plateau (e.g., Trumbull et al., 2006; Kay and Coira, 2009), and high crustal temperatures from surface heat flow (Springer and Forster, 1998; Springer, 1999) and seismic velocities (e.g., ANCORP Working Group, 2003).

One explanation for the lack of a thick lithosphere, despite significant upper-crustal shortening, is that the mantle lithosphere was anomalously thin prior to orogenesis. Geological evidence indicates that this region was close to sea level during the early Cenozoic, at the start of orogenesis (Sempere et al., 1997). If the lithosphere was hot and thin at that time, a surface elevation greater than 1 km would be expected, owing to thermal isostasy (e.g., Hyndman and Currie, 2011). The alternate possibility is that the mantle lithosphere has undergone thinning during orogen development. Observations of pulses of basaltic and ignimbritic magmatism (e.g., Kay et al., 1994; Kay and Coira, 2009; Ducea et al., 2013) and periods of abrupt surface uplift from paleoelevation data (e.g., Garzione et al., 2006) have been interpreted to reflect episodic removal of mantle lithosphere, and possibly lower crust, during orogenic shortening. In addition, seismic tomography images show small-scale (50–100-km-wide) high-velocity bodies at ~100 km depth below the Puna region (Schurr et al., 2006; Bianchi et al., 2013), which have been interpreted as fragments of detached continental lithosphere.

In this contribution, we investigate the dynamics of lithosphere removal in Cordilleran orogens. Various mechanisms for removing lithosphere have been proposed for the central Andes (Fig. 1). Continuous removal may occur through ablation (Fig. 1A), as continental mantle lithosphere in the mantle wedge corner is entrained by the oceanic plate through viscous drag at a rate that balances orogen shortening (Tao and O'Connell, 1992; Pope and Willett, 1998). However, ablative removal is a continuous process that does not result in an overall thin lithosphere. This is at odds with observational evidence for episodic removal events in the central Andes. These observations appear to require gravitational foundering of the lithosphere. Removal can be driven by the negative buoyancy of the mantle lithosphere, as it is cooler and therefore denser than the underlying material. Eclogitization and densification of the lower crust during orogenic shortening, and especially during magmatism, may provide an additional driving force (e.g., Kay and Kay, 1993; Ducea and Saleeby, 1998; Jull and Kelemen, 2001; Sobolev and Babeyko, 2005; Krystopowicz and Currie, 2013; Wang et al., this volume). Two distinct modes of gravitational foundering are: (1) Rayleigh-Taylor–type (RT) instability ("drip"; Fig. 1B), possibly induced by lithospheric shortening combined with magmatic extraction at deep levels under the arcs (e.g., Houseman et al., 1981; Houseman and Molnar, 1997; Molnar et al., 1998) and (2) delamination (Fig. 1C), in which mantle lithosphere peels along a weak crustal detachment layer (e.g., Bird, 1979). The two modes can be differentiated through patterns of surface deformation, uplift, and magmatism (Göğüş and Pysklywec, 2008), and both have been applied to explain observations of magmatism and surface uplift for the Altiplano-Puna Plateau (e.g., Kay and Kay, 1993; Garzione et al., 2006; Molnar and Garzione, 2007; Kay and Coira, 2009; Ducea et al., 2013).

In Cordilleran orogens, subduction-related magmatism at the volcanic arc or in the back-arc region may also induce lithosphere removal (Fig. 1D). Arc volcanism is produced through melting of the mantle wedge above a subducting plate at a depth of 100–150 km. A key problem is that arc magmas extruded at the surface have an andesitic and dacitic, not basaltic, composition (e.g., Lee et al., 2006, and references therein). In addition, Cordilleran arcs are characterized by thick (>25 km) crustal batholiths with a granitoid composition; such batholiths cannot be formed through direct melting of the mantle (Ducea, 2001). Melting of subducting oceanic crust (e.g., Defant and Drummond, 1990) or buoyant diapirs of subducted sediments and continental crust (e.g., Castro et al., 2010) have been proposed to explain this paradox, but the isotopic record of all major Cordilleran arcs rules out such an origin (e.g., Ducea and Barton, 2007). Mantle-derived magmas undergo differentiation during ascent (e.g., Hildreth and Moorbath, 1988; Ducea and Barton, 2007), especially in the lowermost parts of the crust, in what is referred to as the MASH zone (mixing-assimilation-storage and homogenization zone; Hildreth and Moorbath, 1988). Upwelling

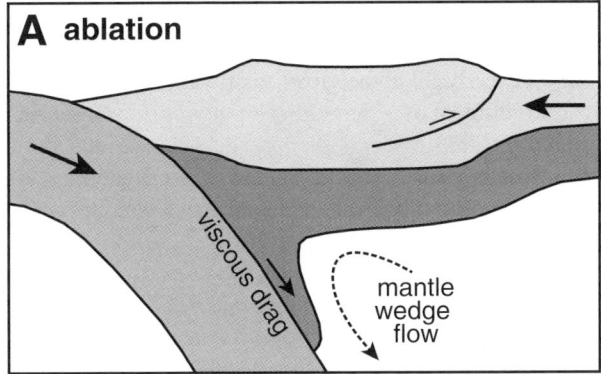

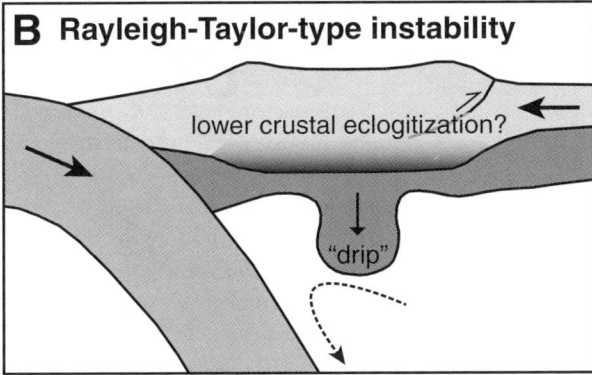

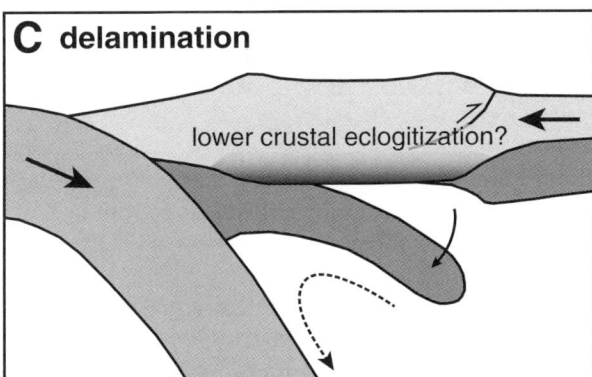

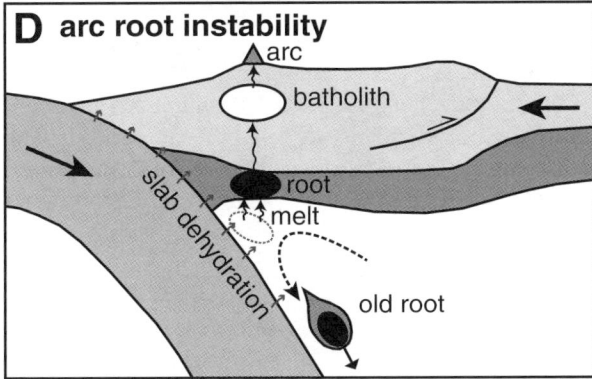

Figure 1. Schematic diagrams of proposed mechanisms of lithosphere removal in Cordilleran orogens (see text for discussion): (A) ablation, (B) Rayleigh-Taylor–type instability, (C) delamination, and (D) arc root instability. In this study, we use numerical models to examine the dynamics of gravitational removal of a volcanic arc root.

basaltic magmas stagnate at the base of the crust, where melt fractionation and assimilation of upper-plate lithosphere result in felsic magmas that are the source of granitoid batholiths and andesitic-dacitic volcanism. This leaves a high-density garnet-bearing or garnet-free pyroxenite (the garnet-bearing type being an eclogite rock sensu lato) residue in the deep lithosphere below the arc, which is then prone to gravitational removal (e.g., Kay and Kay, 1993; Lee et al., 2006).

Isotopic studies show that Cordilleran batholith formation is not a steady-state process but instead occurs during 5–15 m.y. high-flux events during which the magmatic flux is 3–4 times higher than in the intervening periods (e.g., Ducea, 2001; DeCelles et al., 2009, and references therein). Isotopic data also indicate continental lithosphere within the magma source during a high-flux event (e.g., Ducea, 2001; Ducea and Barton, 2007). These data have been explained using a model in which upper-plate shortening introduces fertile continental mantle lithosphere into the volcanic arc region, fueling a magmatic flare-up (e.g., Ducea, 2001; DeCelles et al., 2009). This lithosphere is partially melted and differentiated into a felsic melt, which rises up to form the batholith and a residual "root," which is primarily pyroxenitic/eclogitic (arclogitic) with lesser amounts of granulite (feldspathic) material. Subsequent foundering of the pyroxenitic part of the root is thus an important process for removing continental lithosphere.

At both island arcs and Cordilleran arcs, removal of dense roots has been inferred from geological field evidence, xenolith data, and seismic tomography studies (e.g., Ducea, 2002; Saleeby et al., 2003; Zandt et al., 2004; Behn and Kelemen, 2006). However, there have only been limited geodynamic studies of the removal process. Jull and Kelemen (2001) modeled removal for island arcs as an Rayleigh-Taylor drip and concluded that this can occur over time scales of less than 10 m.y. if temperatures are relatively high (>500 °C at the Moho for shortening lithosphere). Their models did not include sublithospheric mantle flow, which, as we show herein, has an important effect on the dynamics of gravitational instabilities. Behn et al. (2007) considered the foundering arc roots as falling spheres. Their models demonstrated that large, dense spheres strongly perturb slab-induced corner flow, leading to complex three-dimensional (3-D) flow that may explain observations of trench-parallel seismic anisotropy in the arc region.

In this study, we use numerical models to study the formation and removal of a dense mafic-ultramafic root at a Cordilleran volcanic arc. Using a simplified parameterization as a proxy for the development of the root, we examine the dynamics of its removal within an active subduction environment. We then assess the implications for Cordilleran orogen evolution.

NUMERICAL MODELS OF SUBDUCTION

Model Geometry and Governing Equations

The thermal-mechanical numerical models are regional-scale two-dimensional vertical cross sections through a subduction zone. The model domain has a width of 2000 km and extends from Earth's surface to a depth of 660 km. The initial geometry of the numerical models is shown in Figure 2A. The oceanic lithosphere is 90 km thick, with a 9 km crust. The continental plate is also 90 km thick, and crustal thickness varies from 48 km to 42 km. In this study, we examine the volcanic arc dynamics for a case without simultaneous shortening and formation of the

orogen. Therefore, we use a prethickened crust (48 km) in the vicinity of the subduction zone, which results in pressures of ~1.4 GPa at the base of the crust, well within the stability field for the development of a garnet pyroxenite residue (Ducea, 2002; Saleeby et al., 2003).

The finite-element method is used to calculate the coupled thermal-mechanical evolution of the lithosphere–upper-mantle

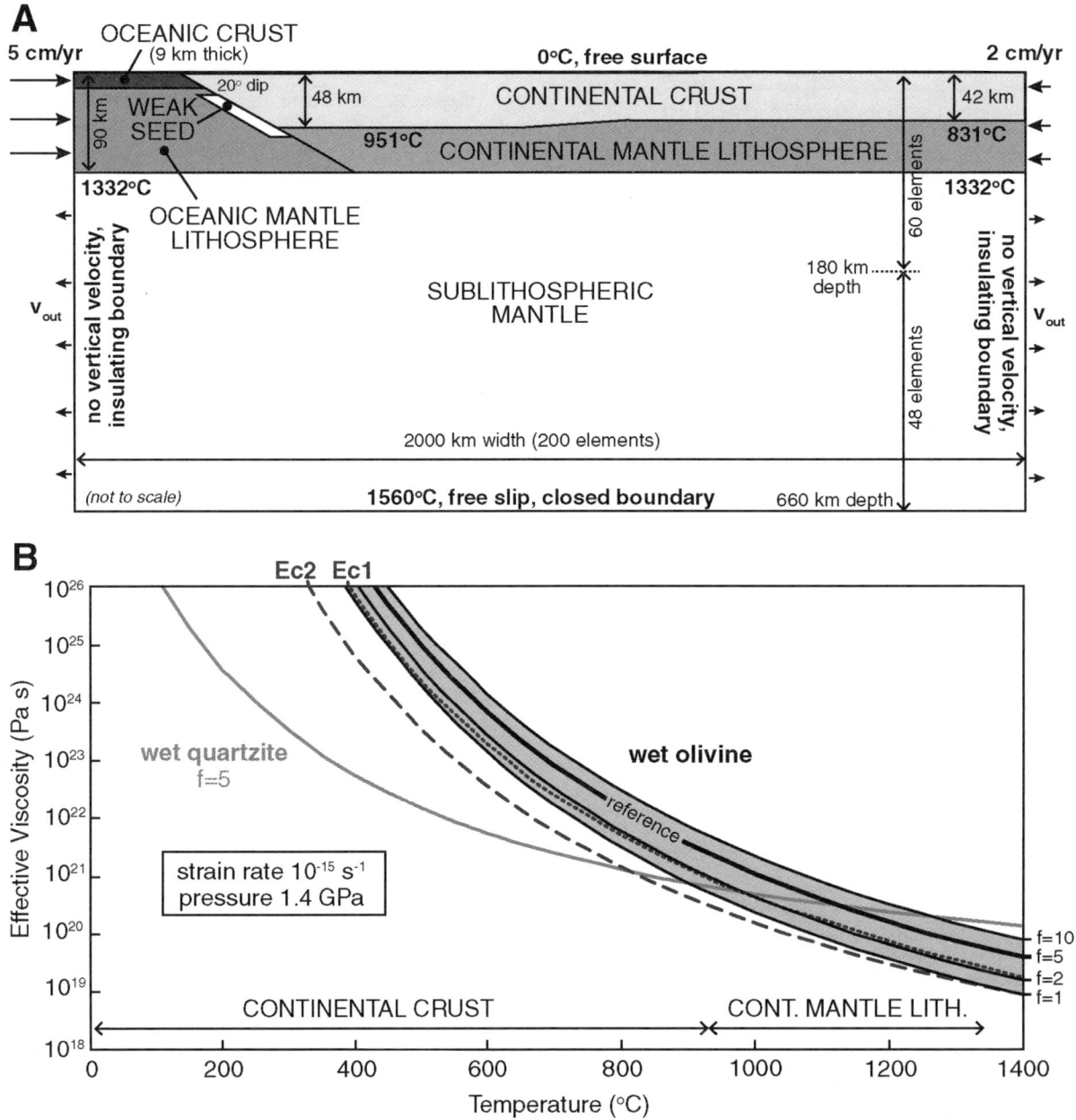

Figure 2. (A) Geometry and thermal-mechanical boundary conditions used in the numerical models. (B) Effective viscosity as a function of temperature for the rheologies used for continental crust (wet quartzite; Gleason and Tullis, 1995) and continental mantle lithosphere (wet olivine; Karato and Wu, 1993). See Table 1 for rheological parameters. Viscosities are calculated for a strain rate of 10^{-15} s^{-1}. The shaded gray region shows the range of root viscosities tested in the models, based on a linear scaling of the base wet olivine rheology (WO) by factor f. The dashed lines show the effective viscosity of dry eclogite from Jin et al. (2001; Ec1) and Zhang and Green (2007; Ec2).

system, under the assumptions of plane strain, incompressibility, and zero Reynolds number. The governing equations are: (1) conservation of volume when incompressible, (2) force balance, and (3) energy balance:

$$\frac{\partial v_j}{\partial x_j} = 0, \tag{1}$$

$$\frac{\partial \sigma_{ij}}{\partial x_i} + \rho g = 0, \tag{2}$$

$$\rho c_p \left(\frac{\partial T_K}{\partial t} + v_i \frac{\partial T_K}{\partial x_i} \right) = k \frac{\partial}{\partial x_i} \frac{\partial T_K}{\partial x_i} + A_T + \sigma'_{ij} \dot{\varepsilon}_{ij} + v_2 \alpha g T_K \rho, \tag{3}$$

where $x_{i,j}$ are spatial coordinates ($i,j = 1,2$), $v_{i,j}$ are components of velocity, ρ is density, g is (vertical) gravitational acceleration, c_p is specific heat, T_K is absolute temperature, t is time, k is thermal conductivity, A_T is volumetric radioactive heat production, and α is the volumetric thermal expansion coefficient. Repeated indices imply summation. In Equation 3, the last two terms on the right-hand side correspond to shear heating, assuming that all dissipated mechanical energy associated with deformation is converted to heat (term 3), and the temperature correction for adiabatic heating for vertical velocity v_2 (term 4).

The associated stress tensor is:

$$\sigma_{ij} = -P\delta_{ij} + \sigma'_{ij} = -P\delta_{ij} + 2\eta_{eff}\dot{\varepsilon}_{ij}, \tag{4}$$

where P is pressure (mean stress), σ'_{ij} is the deviatoric stress tensor, η_{eff} is effective viscosity, δ_{ij} is the Kronecker delta (1 for $i = j$ and 0 otherwise), and the strain rate tensor is:

$$\dot{\varepsilon}_{ij} = \frac{1}{2} \left(\frac{\partial v_i}{\partial x_j} + \frac{\partial v_j}{\partial x_i} \right). \tag{5}$$

These equations are solved using arbitrary Lagrangian-Eulerian (ALE) finite-element techniques (Fullsack, 1995), subject to the boundary conditions described in the following discussion and internal buoyancy forces. Mechanical and thermal calculations are carried out on an Eulerian mesh that stretches vertically to conform to the upper model surface. The Eulerian mesh has 200 elements in the horizontal direction (10 km width) and 108 elements vertically (3 km height in the upper 180 km, and 10 km height below). Material properties are tracked on a Lagrangian mesh and additional Lagrangian tracer particles that are advected with the model velocity field. The Lagrangian particles are used to update the Eulerian model material distribution at each time step. In the calculations, the thermal and mechanical fields are coupled through the temperature dependence of material densities and viscous rheologies, the shear and adiabatic heating terms in Equation 3, and redistribution of radioactive heat–producing materials by material flow.

The numerical modeling code (SOPALE) has been fully benchmarked for studies of gravitational instabilities (e.g., Pysk-

lywec et al., 2002). Our own tests show that the finite-element mesh used here can resolve the growth rate of Rayleigh-Taylor instabilities to within 6% of the analytic values (e.g., Houseman and Molnar, 1997). Additional tests with a viscous Stokes cylinder show that there is less than 7% difference in the sinking velocity computed using the preferred mesh and one with 2 km square elements.

Material Properties

Table 1 lists the thermal-mechanical properties of each model material in the reference model shown below. Materials have a viscous-plastic rheology and a temperature-dependent density. Frictional-plastic deformation follows a Drucker-Prager yield criterion:

$$J'_2 = P\sin\phi_{eff} + c_0\cos\phi_{eff}, \tag{6}$$

where J'_2 is the square root of the second invariant of the deviatoric stress tensor ($J'^2_2 = \frac{1}{2}\sigma'_{ij}\sigma'_{ij}$), c_0 is the cohesion, and ϕ_{eff} is the effective internal angle of friction, which includes the effects of pore-fluid pressure (e.g., Huismans and Beaumont, 2003; Beaumont et al., 2006). Plastic deformation is modeled by defining an effective viscosity that places the state of stress on yield (Fullsack, 1995; Willett, 1999). All materials undergo frictional-plastic strain softening through a decrease in ϕ_{eff} from 15° to 2° over accumulated strain (I'_2) of 0.5–1.5, as an approximation of material softening or an increase in pore-fluid pressure during deformation (Huismans and Beaumont, 2003).

At stresses below frictional-plastic yield, deformation follows a viscous power-law rheology, with effective viscosity (η^V_{eff}) given by:

$$\eta^V_{eff} = f(B^*)(\dot{I}'_2)^{(1-n)/n} \exp\left(\frac{Q + PV^*}{nRT_K} \right), \tag{7}$$

where f is a scaling factor (see following), $\dot{I}'_2$ is the square root of the second invariant of the strain rate tensor ($\dot{I}'^2_2 = \frac{1}{2}\dot{\varepsilon}_{ij}\dot{\varepsilon}_{ij}$), R is the gas constant (8.3145 J mol^{-1} K^{-1}), and B^*, n, Q, and V^* are the pre-exponential viscosity parameter, stress exponent, activation energy, and activation volume from laboratory data. The parameter B^* includes a conversion from the uniaxial laboratory experiments to the tensor invariant state of stress used in the models (Table 1).

The materials in our models are based on several well-constrained laboratory-derived viscous rheologies, and we use the scaling factor f (Eq. 7) to linearly scale the effective viscosity of the model materials relative to these base rheologies to approximate changes in strength owing to minor changes in composition or degree of hydration (Beaumont et al., 2006). For example, experimental data for olivine show a nearly linear decrease in effective viscosity with increasing water content (Hirth and Kohlstedt, 2003, and references therein), such that dry olivine is 5–10 times stronger than water-saturated

TABLE 1. MATERIAL PARAMETERS USED IN THE REFERENCE MODEL

	Continental crust	Continental mantle lithosphere	Oceanic crust	Oceanic mantle lithosphere	Sublithospheric mantle
Plastic rheology					
c_0 (MPa)	2	0	0	0	0
ϕ_{eff}[†]	15° to 2°	15° to 2°	15° to 2°	15° to 2°	15° to 2°
Viscous rheology					
f	5	5	0.1	10	1
A (Pa^{-n} s^{-1})	1.10×10^{-28}	3.91×10^{-15}	5.05×10^{-28}	3.91×10^{-15}	3.91×10^{-15}
B^* (Pa s$^{1/n}$)[§]	2.92×10^{6}	1.92×10^{4}	1.91×10^{5}	1.92×10^{4}	1.92×10^{4}
n	4.0	3.0	4.7	3.0	3.0
Q (kJ mol^{-1})	223	430	485	430	430
V^* (cm^3 mol^{-1})	0	10	0	10	10
Thermal parameters					
k (W m^{-1} K^{-1})[#]	2.25	2.25	2.25	2.25	2.25
A_T (W m^{-3})	1.0	0	0	0	0
c_p (J kg^{-1} K^{-1})	750	1250	750	1250	1250
Density[**]					
ρ_0 (kg m^{-3})	2900	3250	3000	3250	3250
T_0 (°C)	400	1332	0	1332	1332
Eclogite ρ_0 (kg m^{-3})	–	–	3350	–	–
Eclogite T_0 (°C)	–	–	500	–	–
α (K^{-1})	3.0×10^{-5}	3.0×10^{-5}	3.0×10^{-5}	3.0×10^{-5}	3.0×10^{-5}

[†]Frictional-plastic strain softening is included through a linear decrease in ϕ_{eff} over accumulated stain of 0.5 to 1.5.

[§]$B^* = \left(2^{(1-n)/n} 3^{-(n+1)/2n}\right) A^{-1/n}$. The term in brackets converts the pre-exponential viscosity parameter from uniaxial laboratory experiments (A) to the tensor invariant state of stress of the numerical models.

[#]Thermal conductivity at temperatures less than 1332 °C (i.e., within the lithosphere); at higher temperatures, thermal conductivity increases linearly from 2.25 W m^{-1} K^{-1} at 1332 °C to 58.75 W m^{-1} K^{-1} at 1372 °C. The high conductivity in the sublithospheric mantle maintains a nearly constant basal temperature for the lithosphere and an adiabatic temperature gradient of 0.4 °C/km in the sublithospheric mantle, without the need to explicitly model upper-mantle convection (Pysklywec and Beaumont, 2004).

[**]Density varies with temperature: $\rho(T) = \rho_0 [1 - \alpha(T - T_0)]$, where ρ_0 is the reference density at temperature T_0, and α is the volumetric thermal expansion coefficient.

olivine at a depth of 50–100 km (Karato and Wu, 1993; Hirth and Kohlstedt, 2003). The chosen rheologies and scaling factors follow those used in earlier studies (e.g., Beaumont et al., 2006; Krystopowicz and Currie, 2013). The oceanic crust uses the parameters of dry Maryland diabase (Mackwell et al., 1998), with $f = 0.1$, assuming that the crust is hydrated, and therefore weaker than the base rheology. For simplicity, the entire continental crust has a wet quartzite viscous rheology (Gleason and Tullis, 1995), with $f = 5$, to approximate a stronger composition than pure wet quartzite (Beaumont et al., 2006). Mantle materials follow a wet olivine rheology (Karato and Wu, 1993), with $f = 1$ in the sublithospheric mantle and $f = 5$ and $f = 10$ in the continental and oceanic mantle lithosphere, respectively. The higher values of f reflect dehydration and melt depletion of the mantle lithosphere during formation. Figure 2B shows the variation in effective viscosity with temperature for the chosen rheologies.

During subduction, the oceanic crust undergoes a metamorphic phase change as it reaches pressure and temperature conditions within the eclogite stability field (Hacker et al., 2003). At this time, its density increases by ~13% (Table 1), following the approach of Warren et al. (2008); all other parameters are unchanged. No other phase changes are included in the models.

Modeling Approach and Boundary Conditions

Models are run in three phases. In phase 1, the initial two-dimensional (2-D) thermal structure of the models is computed using the material thermal properties (Table 1), temperatures of 0 °C and 1560 °C for the top and bottom boundaries of the model domain, and a no-heat-flux (insulating) condition for the side boundaries. This yields temperatures of 830–950 °C for the continental Moho (Fig. 2A) and a continental surface heat flux of 65.5–71.5 mW/m², with the higher values associated with the region of thicker crust. The models then undergo isostatic adjustment, to allow the oceanic and continental plates to come into equilibrium. As a result, the oceanic plate sinks, such that its surface is at a depth of ~5.2 km.

In phases 2 and 3, plate convergence and subduction occur. The thermal and mechanical boundary conditions are given in Figure 2A. The thermal boundary conditions consist of prescribed temperatures along the top boundary (0 °C) and bottom boundary (1560 °C), and a no-heat-flux (insulating) condition for the side boundaries of the continental lithosphere and sublithospheric mantle. The side boundary of the oceanic lithosphere has prescribed temperatures that are given by a geotherm for a 50-m.y.-old oceanic plate (Stein and Stein, 1992). The mechanical boundary conditions include a stress-free top boundary and a free-slip bottom boundary. Plate convergence at 7 cm/yr is imposed through assigned velocities for the oceanic (5 cm/yr) and continental (2 cm/yr) plates at the side boundaries of the model. To maintain mass balance in the model domain, a small uniform outflux (V_{out}) through the side boundaries of the sublithospheric mantle occurs, equally distributed on each boundary. Models are solved in the continental reference frame by adding 2 cm/yr to all side boundaries. No surface processes (e.g., erosion) are included in the models.

In phase 2, subduction is initiated. Subduction is aided by a narrow, inclined zone of weak material between the oceanic and continental plates (Fig. 2A). This material is subducted with the oceanic plate and does not affect later model evolution. In addition, a high viscous strength is assigned to the continental crust ($f = 50$; Eq. 7) and continental mantle lithosphere ($f = 10$). This phase is run for a total of 10 m.y. (total convergence of 560 km). All the model experiments shown herein start at this point (phase 3), and times are reported as the time since the start of phase 3. During this phase, the viscous strength of the continental crust and mantle lithosphere is set to the reference

values ($f = 5$; Table 1), and magmatic processes at the volcanic arc are imposed.

SUBDUCTION ZONE BASE MODEL

We first present a model in which magmatic processes are not included. Figure 3A shows the model at the end of phase 2, after subduction has been established. The oceanic plate descends into the mantle along a well-defined shear zone, with little deformation of the overlying continent. At this time, the tip of the subducted plate is at the bottom of the model domain and is deflected horizontally along this impermeable boundary. Over the next 3–4 m.y., the oceanic plate undergoes retreat as a consequence of the reduction in the upper-plate strength at the start of phase 3. The trench shifts seaward by ~90 km (Fig. 3B), and there is minor distributed extension in the upper plate. In the mantle wedge corner (above a slab depth of 60–150 km), there is a slight decrease in the dip angle of the slab from ~40° to ~35°. After this, the subduction zone stabilizes, and there is steady subduction with little change in geometry.

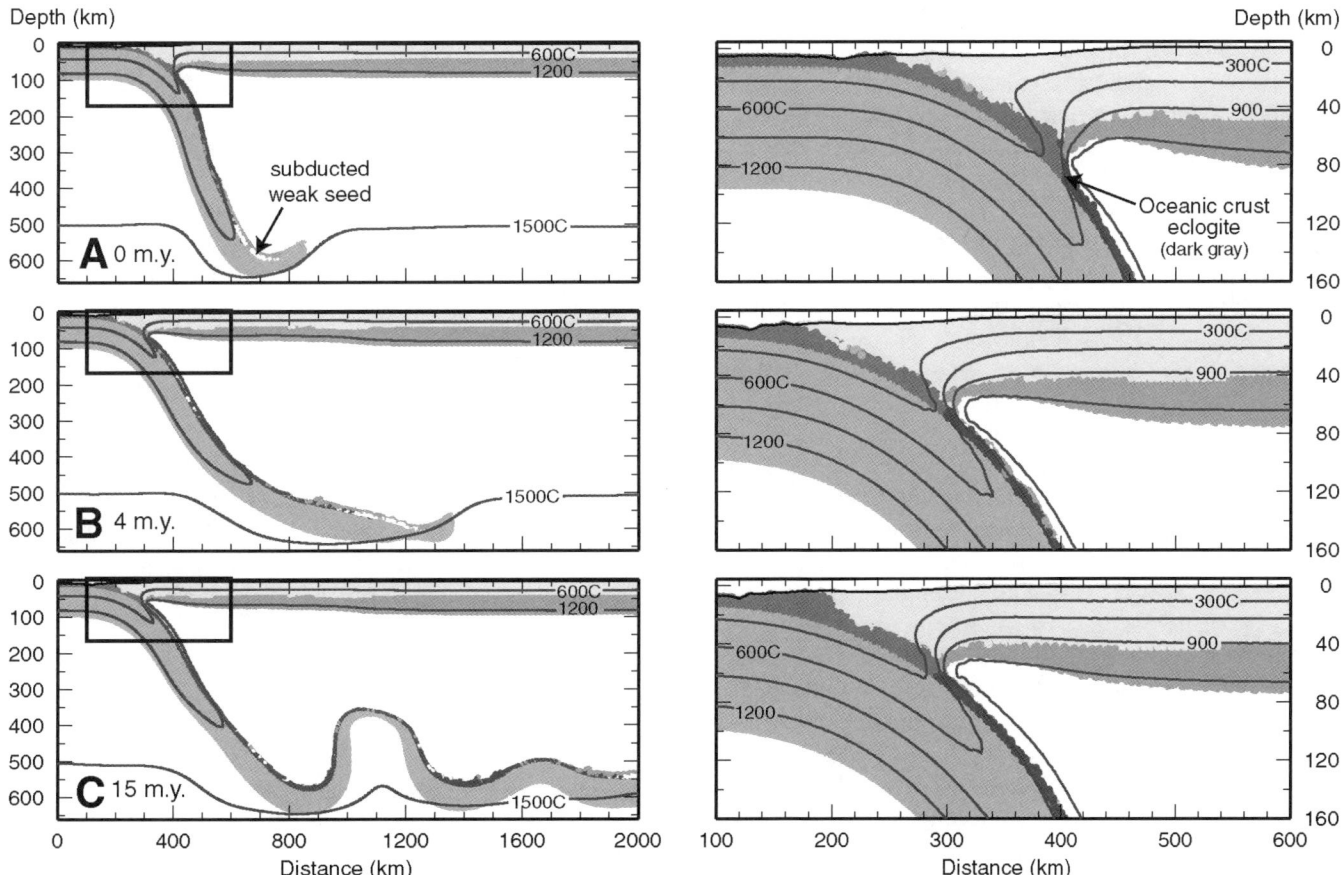

Figure 3. Evolution of a model with no volcanic arc. The entire model domain is shown on the left, and a close-up of the mantle wedge corner is shown on the right. Model parameters are those used in the reference model (Table 1). Material shading is same as in Figure 2. Times are relative to the start of phase 3 of the models (10 m.y. after subduction initiation).

Figure 3C shows the model at 15 m.y., which is the end of the model run. At this point, 1750 km of plate convergence has occurred, and the subducted slab has reached the side boundary of the model domain. Interactions with this boundary produce buckling of the slab in the deep part of the model domain, but there is little effect on the mantle wedge corner.

During model evolution, subduction and the associated mantle wedge flow cause shearing of the continental mantle lithosphere, resulting in minor ablation and thinning in the mantle wedge corner. Overall, the mantle lithosphere is relatively stable, with a thickness of ~15 km in this area.

IMPLEMENTATION OF MAGMATIC PROCESSES

Arc magmatism and the formation of a dense pyroxenite root are the result of a series of complex processes within the subduction zone. As the oceanic plate subducts, metamorphic phase changes in the subducting oceanic crust and mantle lithosphere lead to the release of aqueous fluids into the overlying mantle wedge, reducing its solidus temperature and promoting partial melting (e.g., Schmidt and Poli, 2003; Tatsumi, 2005). The basaltic melts then intrude the continental lithosphere, where fractional crystallization of the melt and localized partial melting of the host rocks result in a silicic melt, which rises to form crustal batholiths and surface magmatism, and a pyroxenite residue root, formed as either a cumulate or restite. Phase equilibria calculations of Jull and Kelemen (2001) predict that root densities are 50–250 kg/m^3 greater than that of mantle, depending on composition. Direct samples of pyroxenite xenoliths from the Sierra Nevada arc have densities of 3500–3600 kg/m^3 (300 kg/m^3 more dense than mantle), owing to their high garnet content (Ducea, 2002).

We do not attempt to model the details of arc magmatism but instead focus on the development of a high-density eclogitic pyroxenite root during a high-flux event at a Cordilleran arc (Fig. 1D). High-flux events are inferred to reflect times of enhanced magmatism related to the emplacement of melt-fertile continental lithosphere below the arc (Ducea, 2001; DeCelles et al., 2009), and therefore root formation through differentiation of continental lithosphere should be greatest during these events. We use a simplified model to simulate the development of the root. This is implemented at each time step in model calculations using a two-step approach: (1) determination of the location of the active volcanic arc (this is the region where localized basaltic melts are assumed to intrude the deep continental lithosphere), and (2) densification of the continental lithosphere in the arc region.

In the models, the arc is assumed to overlie the area where both: (1) the upper 30 km of the subducting plate is at a temperature less than 800 °C and (2) the mantle wedge has a temperature greater than 1200 °C (Fig. 4). The first condition accounts for the maximum stability temperature of hydrous phases in the oceanic mantle (e.g., chlorite) and allows for kinetic delay in dehydration reactions in oceanic crust and subducted sediments (e.g.,

Schmidt and Poli, 2003; Hacker et al., 2003). The latter condition is based on the solidus temperature for partially hydrated mantle (e.g., Schmidt and Poli, 2003) and constraints on arc magma source temperatures from geochemical analyses (e.g., Kelemen et al., 2003). By tying the definition of the volcanic arc to the thermal structure of the subduction zone, the location of the arc can migrate during model evolution. As shown in Figure 4 and subsequent figures, the arc region is located above a slab depth of ~80 to ~150 km, as observed in nature (e.g., Tatsumi, 2005; Syracuse and Abers, 2006).

The second step is to change the density of the lithosphere in the volcanic arc region to simulate magmatic differentiation, i.e., the formation of a high-density eclogite root and the emplacement of low-density felsic magmas within the arc crust. It is assumed that the eclogitic pyroxenite root resides in the continental mantle lithosphere below the arc (~15 km thick in our models) and that this entire region progressively becomes denser owing to differentiation through partial melting. The simplified model assumes that infiltration of small-volume basaltic mantle-derived melts produces sporadic, localized zones of partial melting in the continental lithosphere. The time scales of arc formation are tens of millions of years or more for large Cordilleran arcs. The Sierra Nevada arc, for example, consists of plutons with ages ranging from 230 Ma to ca. 80 Ma, and all of these plutons are confined to a surface area that is ~120 km wide. While the arc did form in a non-steady-state fashion, with short-lived 10–15 m.y. high-flux events separated by longer periods of lower magmatic flux (magmatic lulls), it is clear that the arc was built through incremental addition of mantle-derived melts and small batches of felsic differentiates in the upper crust (Coleman et al., 2004). The lifetime of a plutonic suite such as the classic Tuolumne Suite in the Yosemite region, central Sierra Nevada, is in the range of 10 m.y. (Coleman et al., 2004), and the composite pluton was assembled

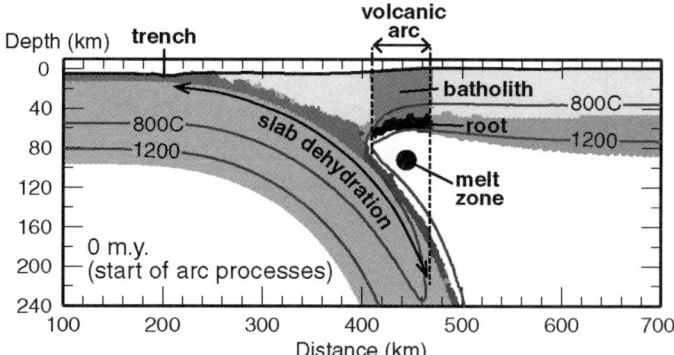

Figure 4. Diagram of how magmatic processes are included in the models. The location of the arc is determined by the region overlying an oceanic plate temperature of 800 °C (maximum temperature of slab dehydration) and a minimum mantle wedge temperature of 1200 °C (temperature for hydrous mantle melting). The continental mantle lithosphere in the arc region forms the "root" (region that undergoes density increase), while the crust forms the "batholith" (region of density decrease). Material shading is same as in Figure 2.

via small batches of melt at any given time. That observation implies that the root itself, while hot, was for the most part solid at any given time, and contained only small areas of partial melt. Melt fractions in excess of 20%, and in some cases up to 50%, are documented in studies of xenoliths representing the pyroxenitic part of the root (Lee et al., 2006), but they represent local areas of high-percent melt at any given time. Similarly, one-dimensional (1-D) thermal models show that basaltic intrusion may produce only limited melting of the arc complex (Annen and Sparks, 2002). Thus, at the million-year time scale or more, the root can be considered solid during the evolution of the arc.

In nature, the rate of root densification depends on the rates of basaltic magma intrusion, partial melting of the surrounding rock, melt differentiation, and extraction of the felsic melt component. Constraints on this come from the apparent intrusive flux of plutonic rocks and arc magmas. Typical island arcs have a flux rate of 20–40 km^3/m.y. per kilometer along strike (Reymer and Schubert, 1984); this range is commonly referred to as 1 Armstrong Unit (AU) (DeCelles et al., 2009). At Cordilleran arcs, the rate during a high-flux event is 3–4 times greater than this (Ducea, 2001). Assuming that the high-flux event corresponds to the sum of the background mantle-derived magmatic flux (1 AU) and melt productivity associated with the incorporation of fertile continental lithosphere into the magma (2–3 AU), the volume of felsic intrusives derived from continental lithosphere is 40–120 km^3/m.y./km. Given a melt to residue ratio of 1:1–1:3 for a Cordilleran batholith (Ducea, 2001; DeCelles et al., 2009), it is estimated that the root will form at a rate of 40–360 km^3/m.y. (per km along strike), as felsic components are extracted.

This is modeled by increasing the average (bulk) density of the continental mantle lithosphere within the arc root region (width of ~80 km, thickness of ~15 km), assuming that the entire region undergoes homogeneous melt extraction and densification. The root zone initially has the density of continental mantle lithosphere, 3250 kg/m^3, and the rate of densification is:

$$\frac{\delta\rho}{\delta t} = \frac{(\rho_{py} - \rho_m)\, R_r}{A_r}, \qquad (8)$$

where $(\rho_{py} - \rho_m)$ is the density contrast between pyroxenite and mantle, A_r is the cross-sectional area of the root, and R_r is the rate of root formation. For a 150–300 kg/m^3 density contrast (Jull and Kelemen, 2001; Ducea, 2002) and a root area of 1200 km^2, the rate of densification is 5–90 kg/m^3/m.y. In the models, the densification rate is a free parameter. The reference model shown next uses a value of 20 kg/m^3/m.y., and later we present models with densification rates of 10–80 kg/m^3/m.y. During each time step, the density of the root region is increased by the prescribed amount. At the same time, the average density of the volcanic arc crust ("batholith") is decreased at a rate that maintains mass balance in the models. This approximates the emplacement of felsic magmas in the shallow crust.

Magmatic processes are implemented at the start of phase 3 of the models, i.e., after a well-developed subduction zone is established. The start of this phase is taken to be the time at which basaltic melts start to intrude the continental mantle for this cross section through a subduction zone. This may represent either the along-strike migration of arc volcanism to this location or the reestablishment of magmatism in a region that underwent prior root removal, followed by rapid upper-plate shortening and the emplacement of fertile lithosphere below the arc. Note that the models do not include magmatic addition to the arc lithosphere. Rather, we assume that the main contribution to the root density comes from partial melting and differentiation of in situ continental lithosphere, as required by isotopic data for Cordilleran arcs (Ducea, 2001; DeCelles et al., 2009). Presumably, the heat source for partial melting comes from magmas derived from the mantle wedge, which are not explicitly included in our models.

RESULTS: MODELS WITH A HIGH-DENSITY ARC ROOT

Reference Model

The reference model uses the material properties in Table 1, and the density of the arc root region increases at 20 kg/m^3/m.y. The rheologies of the root and batholith are the same as those of the original material (mantle lithosphere and crust, respectively). The assumption is that the process of root formation only involves short-lived small-volume partial melts (e.g., Annen and Sparks, 2002; Coleman et al., 2004), and therefore the average root region can be considered as a solid. Figure 5 shows the evolution of this model. With rollback and slight shallowing of the slab during the first ~4 m.y. of phase 3, the arc location migrates seaward, and its width increases to 80–90 km. After this, it remains relatively stationary. As the model progresses, the base of the root is sheared by subduction-induced mantle flow, which causes a slight perturbation and initiates gravitational instability of the dense arc root. In the early stages of instability, the downward velocity is fairly slow, and therefore the perturbation is entrained by mantle flow and carried toward the wedge corner. The growth rate is initially exponential, but once the strain rate associated with instability exceeds the strain rate of shearing, the growth rate becomes superexponential, owing to the power-law rheology of this material (Molnar et al., 1998; Currie et al., 2008). As a result, the downward velocity of the root increases, and the root rapidly descends through the mantle wedge. It detaches from the upper plate and is carried downward with the subducting plate.

Overall, the removal process is rapid and efficient. The root is removed within 8 m.y. of the start of magmatic processes, and the dripping event (i.e., when the downward velocity exceeds the rate of lateral entrainment) occurs within 3 m.y. of the onset of instability. The instability involves nearly the entire root region. After removal, only a thin layer (<5 km) remains below the arc crust; some of this material is back-arc mantle lithosphere that was carried into the arc region by mantle flow.

10 *Currie et al.*

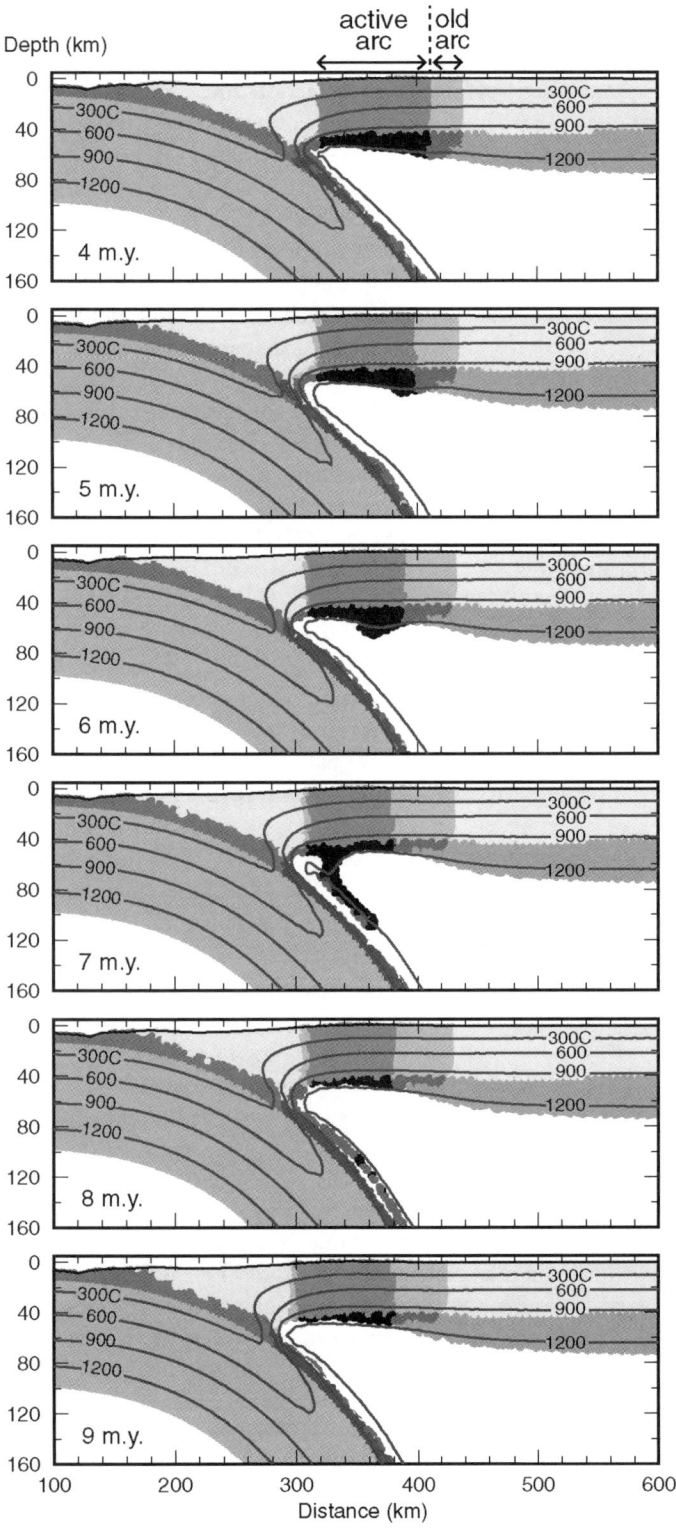

Figure 5. Evolution of the reference model with magmatic processes, using a densification rate of 20 kg/m³/m.y. in the volcanic arc root. The active arc is the region that forms the volcanic arc at each time step; this is where density changes in the continental lithosphere occur. The former arc tracks areas that were in the arc region earlier in the model evolution but are no longer undergoing density changes. Material shading is same as in Figure 2 and Figure 4.

Variations in Root Densification Rate

The root densification rate is a free parameter in the models, and it is estimated that it may vary from 5 to 90 kg/m³/m.y. in nature (see previous). Figure 6 shows models with densification rates of 10 and 40 kg/m³/m.y., keeping all other parameters the same as in the reference model. In both cases, the base of the root is sheared by mantle flow, causing an initial perturbation. The growth rate of the perturbation is proportional to the density contrast between the root and underlying material (Houseman and Molnar, 1997). With a densification rate of 10 kg/m³/m.y. (Fig. 6A), the growth rate is sufficiently slow that it does not exceed the rate of lateral entrainment. In this case, the root is swept into the mantle wedge corner and is removed through an ablation-like process as it is mechanically subducted with the subducting plate. This occurs over ~4 m.y., and approximately half of the thickness of the root is removed. The shallower part of the root remains intact and is not entrained in the early stages of the model, likely because it has a high viscosity due to lower temperatures. After the initial removal event, the remaining root is gradually thinned through shearing and ablation.

In contrast, a higher densification rate of 40 kg/m³/m.y. leads to a greater instability growth rate, enabling the initial perturbation to rapidly amplify (Fig. 6B). The root falls nearly vertically through the wedge and has a drip-like appearance, with a width of ~20 km. In comparison to the reference model, the removal event occurs over a shorter time and involves a greater area of root. The residual material in the tail of the instability is carried into the mantle wedge corner and is subducted, leaving very little root in the arc region. As the dense root is removed, the crust within the arc is slightly thickened (5 m.y. panel) and then relaxes after removal.

Lower-Viscosity Root

In these models, the root has a rheology that is 5 times stronger than wet olivine, which approximates the rheology of relatively dry olivine (Karato and Wu, 1993). However, this material may be weaker, due to factors such as higher temperatures related to magma emplacement in the arc root, the presence of melt, or a change in composition as the root region becomes garnet pyroxenite. Two laboratory studies have examined the rheology of dry eclogite (Jin et al., 2001; Zhang and Green, 2007). Their flow laws predict effective viscosities that are 2.5–5 times lower than that used in the reference model (Fig. 2B).

Figure 7 shows two models in which the root region is taken to have a wet olivine rheology with $f = 2$, which has a similar viscosity to that of the Jin et al. (2001) eclogite. The unaltered continental mantle lithosphere has the same rheology as the reference model (wet olivine with $f = 5$). With the weaker rheology, the root is easily sheared and perturbed by mantle wedge flow. With a densification rate of 20 kg/m³/m.y. (Fig. 7A), the perturbation grows at a relatively slow rate, and the root is carried into the wedge corner by mantle flow as it detaches. This removes

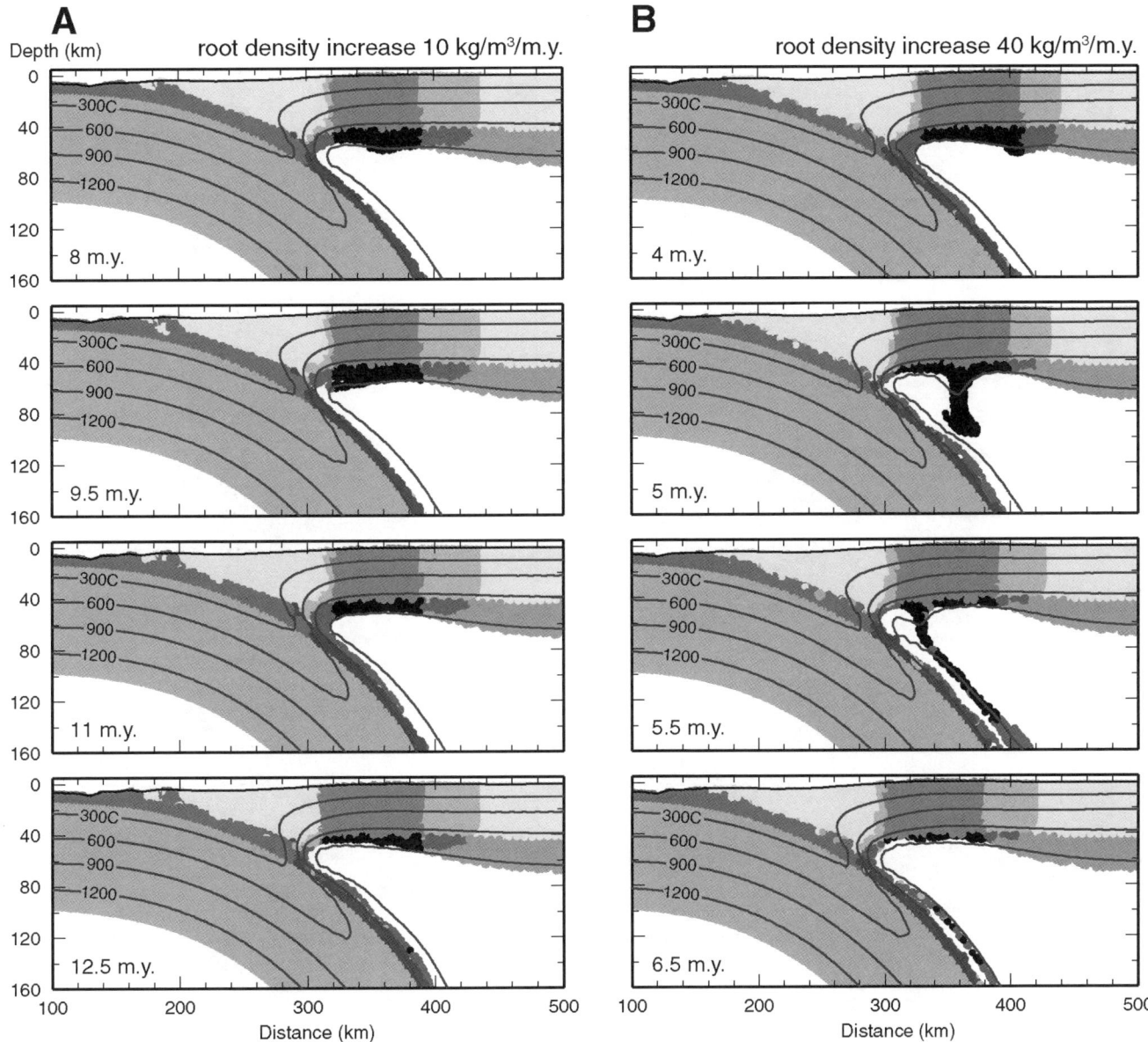

Figure 6. Model evolution for models with a root densification rate of: (A) 10 kg/m³/m.y. and (B) 40 kg/m³/m.y. Other model parameters are those of the reference model. Material shading is same as in Figure 2 and Figure 4.

approximately two-thirds of the root area, and the remaining root is removed through later shearing. By increasing the densification rate to 40 kg/m³/m.y. (Fig. 7B), the growth rate of the instability is increased, enabling a drip-like instability to form. This occurs earlier than in the comparable model with a stronger root (Fig. 6B).

Models with a weaker rheology (wet olivine with $f = 1$) experience even greater shearing of the root by mantle flow, with shearing velocities of ~4.2 cm/yr, i.e., ~0.4 cm/yr more than for $f = 2$. For a densification rate of 20 kg/m³/m.y., root removal occurs primarily through sideways entrainment (similar to Fig. 7A), whereas higher densification rates result in removal with a greater vertical velocity (similar to Fig. 7B).

Higher-Viscosity Root

We now test models in which the root viscosity is twice that of the reference model (wet olivine with $f = 10$). Rheological studies of eclogite show that the strength of eclogite may be controlled by the presence of omphacite, which is much weaker than garnet (Jin et al., 2001; Zhang and Green, 2007). As there is no omphacite in samples from the Sierra Nevada root (e.g., Ducea, 2002), pyroxenite roots may be stronger than predicted by the eclogite flow laws. Alternatively, a stronger rheology may correspond to a root that is dehydrated or where root temperatures are 50–100 °C cooler than in our models (Karato and

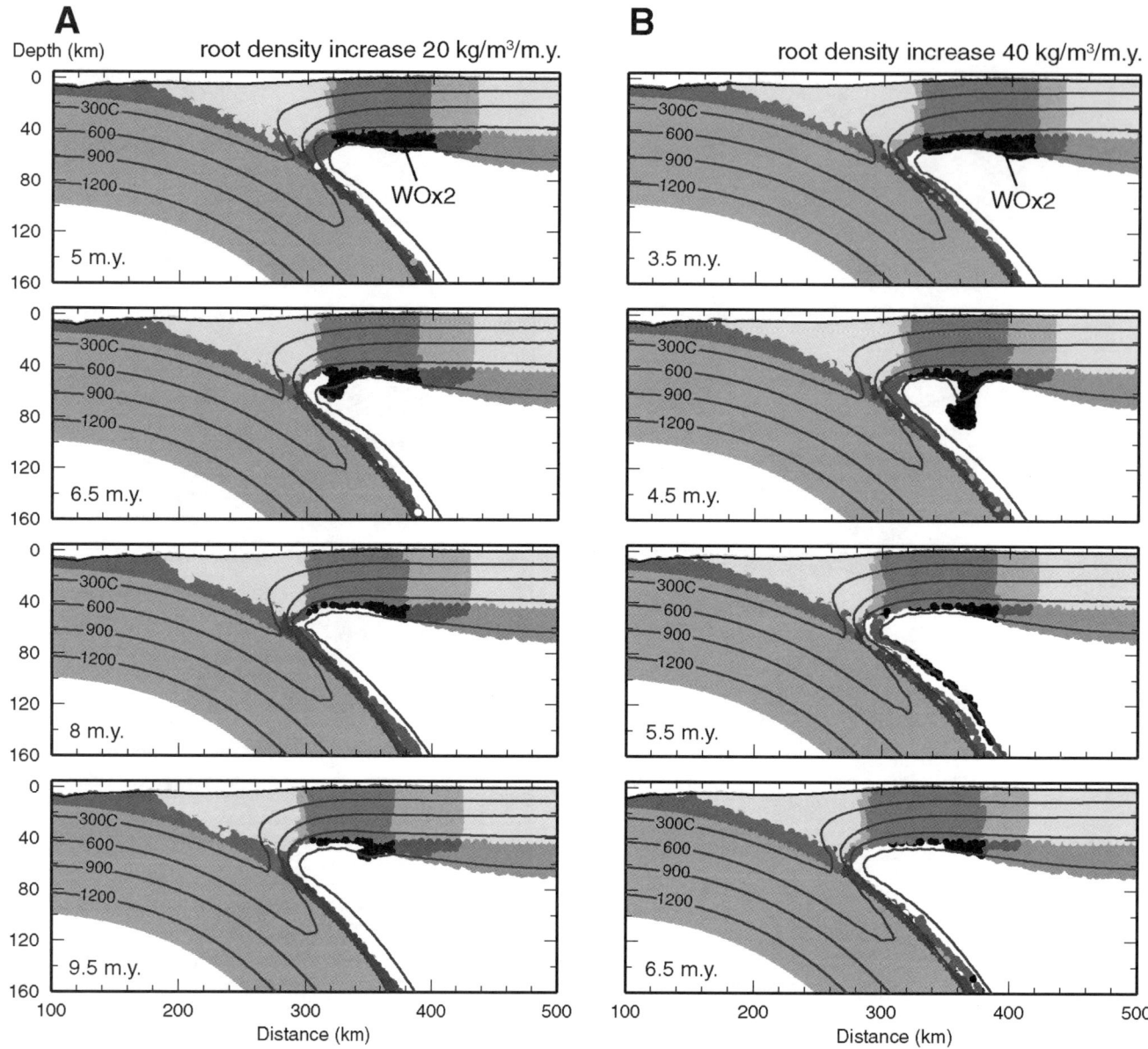

Figure 7. Evolution of a model with a weak magmatic arc root (wet olivine with $f = 2$; 2.5 times weaker than the reference model) and a densification rate of: (A) 20 kg/m³/m.y. and (B) 40 kg/m³/m.y. Material shading is same as in Figure 2 and Figure 4.

Wu, 1993). With the greater strength, the lithosphere is less easily perturbed by mantle flow, and the growth rate of the ensuing instability is decreased (Fig. 8). For densification rates of 20 and 40 kg/m³/m.y., the root removal time is increased relative to models with a weaker rheology. In both cases, the perturbation is initially swept sideways by mantle flow but then falls along a near-vertical trajectory once its growth rate is high enough. This removes almost all root material. In addition, the stronger root rheology leads to greater viscous coupling between the root and overlying arc crust, leading to slight crustal thickening above the detaching root.

Variations in Subduction Rate

The previous models demonstrate that mantle wedge corner flow plays an important role in the gravitational removal of the arc root region. We now investigate variations in the subduction rate, as this is what drives mantle wedge flow. For these models, the continental plate velocity is fixed at 2 cm/yr, as in the reference model. The oceanic plate velocities of 3 cm/yr and 7 cm/yr are examined, giving subduction rates of 5 cm/yr and 9 cm/yr, respectively; the reference model has a subduction rate of 7 cm/yr. In these models, the root parameters are those

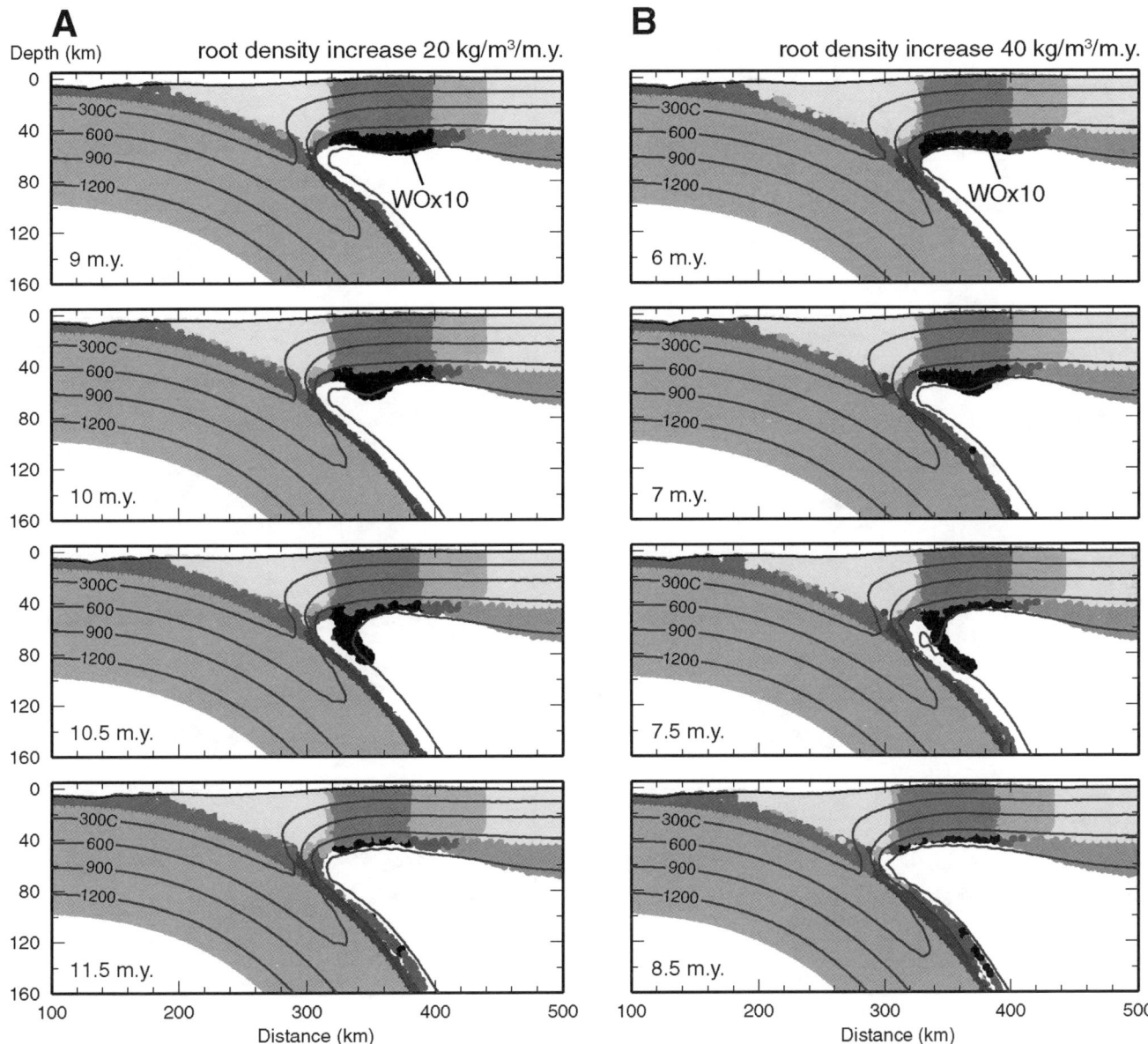

Figure 8. Evolution of a model with a strong magmatic arc root (wet olivine with $f = 10$; 2 times stronger than the reference model) and a densification rate of: (A) 20 kg/m³/m.y. and (B) 40 kg/m³/m.y. Material shading is same as in Figure 2 and Figure 4.

in the reference model (wet olivine with $f = 5$; densification rate of 20 kg/m³/m.y.).

With a lower subduction rate (Fig. 9A), the initial lithosphere perturbation and development of the instability are slightly delayed. However, once instability occurs, the root falls nearly vertically through the wedge and has a more drip-like appearance than in the reference model (Fig. 5). In this case, the downward velocity exceeds the mantle wedge flow velocity. In contrast, root instability develops at an earlier time with a higher subduction rate (Fig. 9B). This is a consequence of both greater shearing on the base of the root, as well as a minor decrease in the root viscosity, as the root has a non-Newtonian rheology, such that

its viscosity decreases with increased strain rate (Currie et al., 2008). However, the initial growth rate of the instability is not large enough to exceed the rate of mantle flow, and the perturbation is carried into the wedge corner before detaching.

DISCUSSION

Dynamics of Arc Root Removal

The models presented here highlight the range of dynamics that may occur as a high-density root forms below the frontal arc of a Cordilleran subduction system. We have tested a range of root

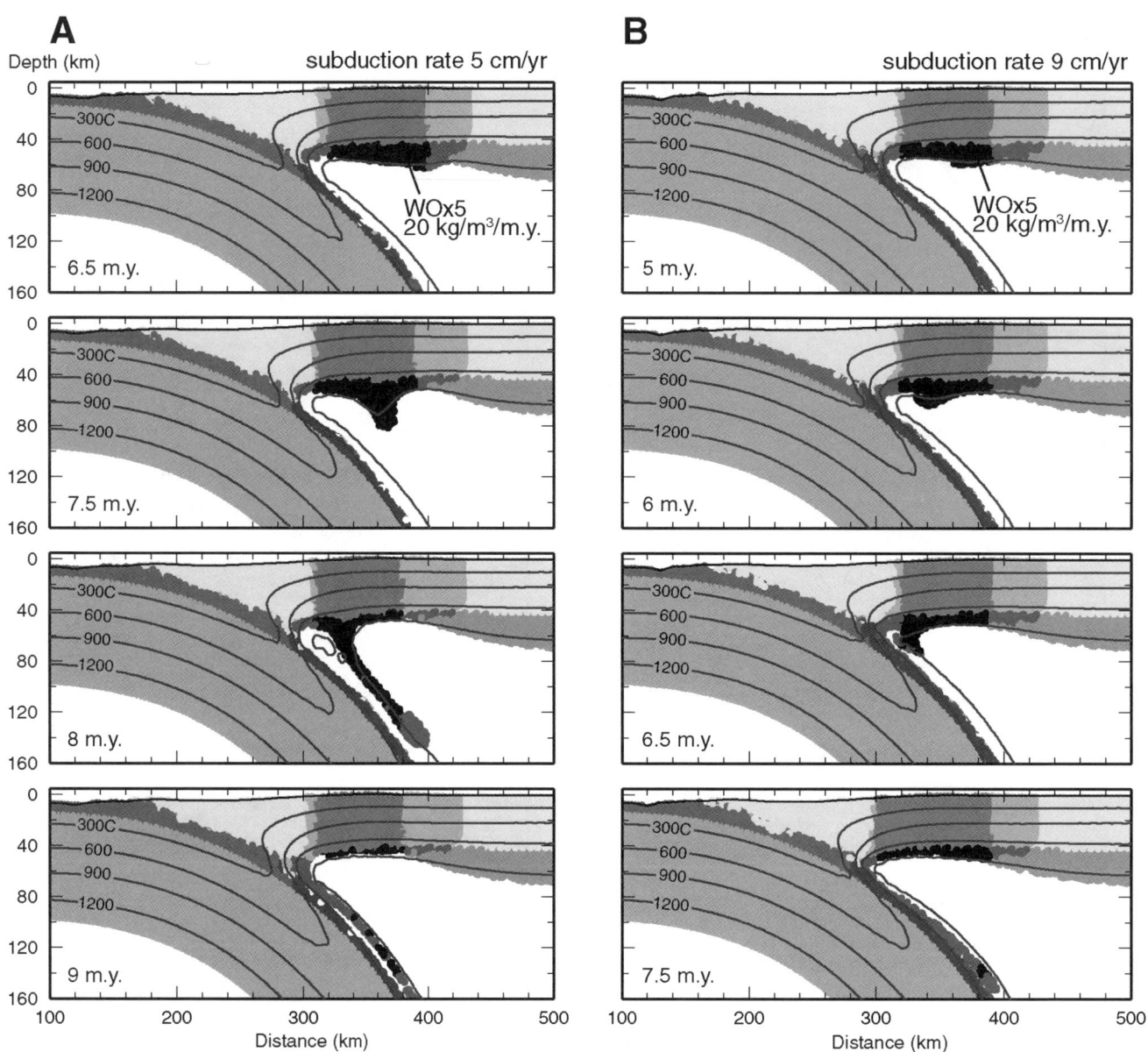

Figure 9. Model evolution for a subduction rate of: (A) 5 cm/yr and (B) 9 cm/yr. All other parameters are those of the reference model, with a root densification rate of 20 kg/m³/m.y. and root rheology of wet olivine (WO) ×5. Material shading is same as in Figure 2 and Figure 4.

rheologies, densification rates, and subduction rates, and in all cases, we find that >80% of the root region can be removed within 15 m.y. after the start of root densification. Table 2 and Figure 10 summarize the characteristics of root removal in the models.

The initiation of root removal is defined as the time at which the root region has a negative (downward) velocity at a depth of 60–63 km (i.e., the base of the root). As shown in Figure 10A, initiation of removal occurs earlier in models with a high densification rate. Removal is initiated within 2–3 m.y. after the start of densification for rates of 60–80 kg/m³/m.y. The rheology of the root also controls the onset of instability; stronger roots tend to have a

later initiation time. The effect of rheology is most significant at low densification rates. In all models, initiation occurs when the density of the root is 50–250 kg/m³ greater than that of the mantle (corresponding to absolute densities of 3300–3500 kg/m³), within the range of densities for arc-related pyroxenites (Jull and Kelemen, 2001; Ducea, 2002). Once initiated, root removal occurs within 5 m.y., with more rapid removal for roots with a greater density contrast relative to the mantle and for roots with a weaker rheology (Fig. 10B).

The removal of the dense arc root occurs primarily as a gravitational instability, driven by the density contrast between

TABLE 2. SUMMARY OF MODEL RESULTS

Root rheology	Rate of densification (kg/m³/m.y.)	Initiation of removal (m.y.)	Root density at initiation (kg/m³)	Duration of removal (m.y.)	Removal style*	Figure number
Subduction velocity = 7 cm/yr						
WO × 1	20	3.7	3323	2.1	Shear	
	40	3.8	3401	1.1	Drip	
	60	2.5	3398	1.0	Drip	
	80	2.3	3433	0.9	Drip	
WO × 2	20	5.2	3354	2.4	Shear	7A
	40	3.8	3402	1.5	Drip	7B
	60	3.1	3436	1.2	Drip	
	80	2.6	3458	0.9	Drip	
WO × 5	10	7.7	3327	3.9	Shear	6A
	20	4.8	3346	2.4	Drip	5
	40	3.6	3394	2.1	Drip	6B
	60	3.3	3448	1.3	Drip	
	80	3.2	3502	1.1	Drip	
WO × 10	10	10.2	3353	4.7	Shear	
	20	9.1	3433	1.8	Drip	8A
	40	6.2	3498	1.6	Drip	8B
	60	3.3	3449	1.7	Drip	
	80	2.5	3450	1.9	Drip	
Subduction velocity = 5 cm/yr						
WO × 5	20	5.9	3374	2.6	Drip	9A
Subduction velocity = 9 cm/yr						
WO × 5	20	4.4	3338	2.3	Shear	9B

Note: WO × *f* = wet olivine rheology (Karato and Wu, 1993), scaled by factor *f* (Eq. 7).
*Determined by comparing the horizontal velocity (V_h) and vertical velocity (V_v) at the base of the arc root lithosphere during root removal. Shear entrainment has $V_h > V_v$ throughout removal; drip-style removal has $V_v > V_h$ in the latter stages of removal.

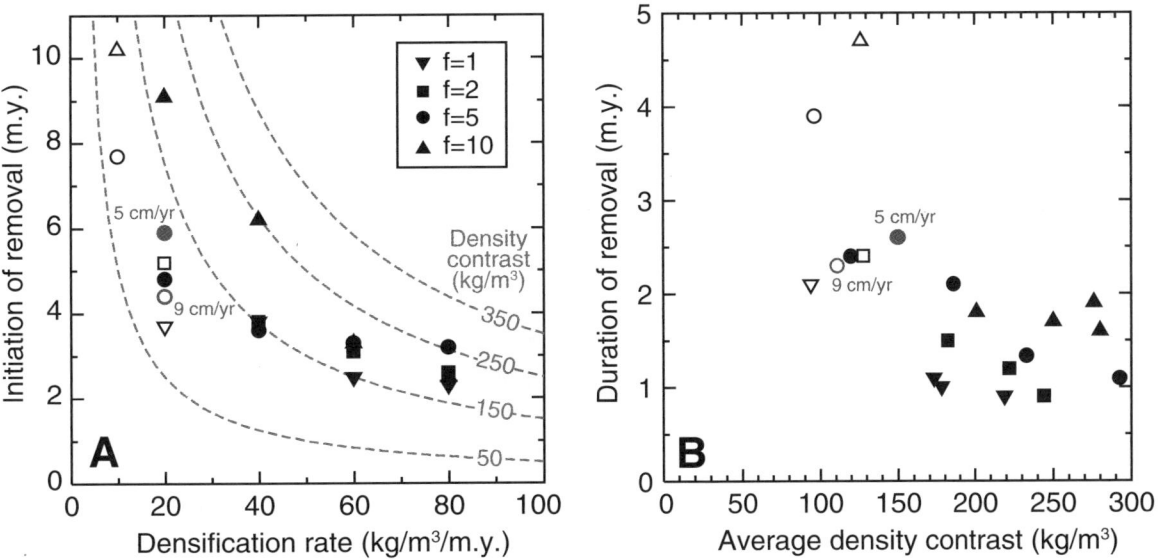

Figure 10. (A) Time of initiation of removal of the arc root for different densification rates (horizontal axis) and root strengths (symbols). Dashed gray lines are the density contrast of the root relative to a mantle density of 3250 kg/m³. (B) The duration of removal for each model, as a function of the average density contrast between the root and mantle during removal. Black symbols use a subduction rate of 7 cm/yr; gray symbols use the rate given on the figure. Solid symbols are models that were removed as a drip-like instability; open symbols are those that underwent shear entrainment. Model results are listed in Table 2.

Currie et al.

the root and underlying mantle. However, subduction-induced mantle wedge flow plays a key role in both the onset and dynamics of root removal (Fig. 11A). Mantle flow shears the base of the root, which causes a perturbation that initiates instability (Currie et al., 2008). The perturbation then grows in amplitude, at a rate determined by the density contrast between the root and underlying mantle and the viscosity of the root (Houseman and Molnar, 1997). As it grows, the perturbation is entrained by mantle flow and carried toward the wedge corner. The effect of mantle flow is greatest for weak roots, as they are more easily perturbed and readily entrained by mantle flow.

Two styles of removal are observed: (1) drip-like removal, in which the root falls subvertically through the mantle wedge, and (2) shear entrainment, in which the root is swept into the wedge corner by mantle flow. The style of removal is determined by the relative rates of gravitational growth of the instability and shearing by mantle flow. Drip-like removal occurs if gravitational growth dominates. These two styles are similar those of Behn et al. (2007), who approximated a detached arc root as a falling sphere. In their analysis, spheres with a large negative buoyancy (due to high density and/or large diameter) had a vertical trajectory, whereas spheres with a lower negative buoyancy were entrained by mantle wedge flow.

Figure 11B summarizes the style of removal observed for different root densification rates and strengths (viscosities) in our models. Drip-like removal is favored for roots with a high den-sification rate and high viscosity. Removal in this manner occurs within 1–2.5 m.y. of the onset of instability (Fig. 10B), which is consistent with the time scale observed for gravitational removal of a 1–10-km-thick eclogite layer in the models of Jull and Kelemen (2001). In most of these models, the average density contrast between the root and mantle is >150 kg/m³ during removal. Shear removal takes up to 5 m.y., with longer times associated with stronger roots. These models have a lower density contrast relative to the mantle. It should be noted that our models are based on a wet olivine rheology. A comparison of olivine and eclogite rheologies indicates that eclogite may have a rheology similar to the weaker models in our study (Fig. 2B). In this case, we predict that drip-like removal will occur for root densification rates of 40 kg/m³/m.y. or more (Fig. 11B).

Scaling Analysis for Root Removal

These results can be quantified by comparing the time scales for gravitational growth and shear entrainment. The main factor driving removal is gravitational instability of the high-density root. At the same time, mantle flow shears the base of the arc lithosphere, carrying the gravitational instability toward the wedge corner. In order for removal to occur as a drip, gravitational growth of the instability must result in a downward velocity that exceeds the horizontal velocity of the underlying mantle before the perturbation reaches the mantle wedge corner.

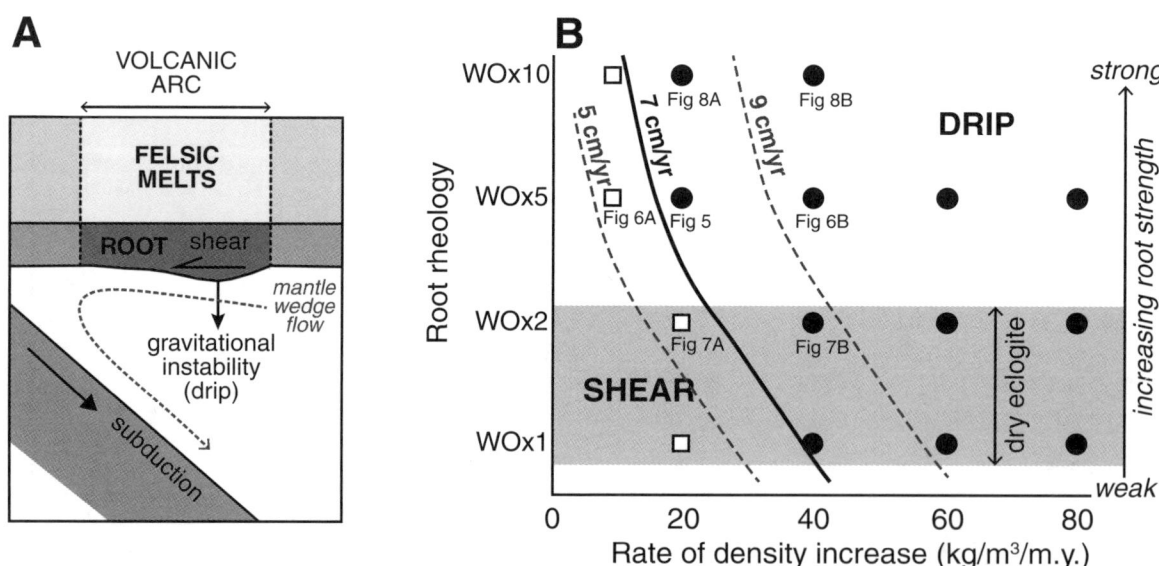

Figure 11. (A) Schematic diagram showing the relationship between root dynamics and subduction-induced mantle wedge flow. Flow perturbs the base of the root, initiating instability. The trajectory of the root as it destabilizes is determined by the relative rates of downward gravity-driven instability growth (V_v) and horizontal entrainment by mantle flow (V_h). (B) Diagram showing the style of removal for variations in root densification rate and viscosity (as a linear scaling of the base wet olivine [WO] rheology) for a subduction rate of 7 cm/yr. Drip-like removal (filled circles) occurs where V_v is greater than V_h within the volcanic arc region. Shear entrainment (white squares) occurs for $V_h > V_v$. Dashed lines show the boundary between style of removal for variations in subduction rate. The shaded gray region is the strength of dry eclogite from laboratory studies.

For a layer sheared by mantle flow, gravitational instability initially has an exponential growth rate (Molnar et al., 1998; Currie et al., 2008) given by:

$$q = \frac{c\Delta\rho g h}{2\eta_L}, \qquad (9)$$

where c is a nondimensional factor, $\Delta\rho$ is the density contrast between the root layer and underlying material, g is gravitational acceleration, h is the thickness of the layer undergoing instability, and η_L is the layer viscosity (e.g., Houseman and Molnar, 1997). At a time t after initiation, the downward velocity of the instability is:

$$V_v(t) = (Z_0 q)e^{qt}, \qquad (10)$$

where Z_0 is the amplitude of the perturbation that initiates instability (Houseman and Molnar, 1997).

We define the gravitational threshold time (t_g) as the time at which $V_v = 7$ cm/yr, as this corresponds to the maximum horizontal velocity of mantle wedge flow for a 7 cm/yr subduction rate. The arc root is approximated as a layer with constant density contrast and viscosity, and we assume an initial perturbation of 10% of the layer thickness. As the arc root layer that undergoes instability is 5–15 km thick, variations in density and viscosity related to temperature are relatively small. In this case, $c = 0.32$ (Houseman and Molnar, 1997). We neglect the transition from exponential to superexponential growth that occurs due to the non-Newtonian rheology of the arc root (Molnar et al., 1998). We also assume that Equations 9 and 10 apply throughout the evolution of the instability, although numerical experiments by Houseman and Molnar (1997) show that the growth rate of a Newtonian material rapidly increases once the perturbation amplitude equals the original layer thickness. These two factors result in enhanced instability growth, and therefore our calculated t_g values are overestimates.

Figure 12 shows t_g as a function of density contrast for different combinations of layer thickness and viscosity. The root viscosity is the product $f \times 10^{19}$ Pa s. This is consistent with the average viscosity in the lowermost root of the numerical models. In all cases, t_g decreases with increasing density contrast and layer thickness and decreasing root viscosity, as expected based on Equation 9.

The second critical time scale is that associated with lateral shearing of the root. As the perturbation grows gravitationally, it is swept sideways by mantle flow. We define the threshold time for shearing (t_s) as the time for a perturbation to be carried a horizontal distance of 80 km, the nominal width of the volcanic arc region. The rate of shearing depends on both the velocity of mantle wedge flow (set by the subduction rate) and the viscosity of the arc root, with a greater shearing for weaker (lower f) material (Currie et al., 2008). In our models, the observed shearing velocity is ~4.2 cm/yr for a root with $f = 1$, resulting in $t_s = 1.9$ m.y. (horizontal gray line on Fig. 12). As f increases, the

shearing velocity decreases and t_s increases; values of 2.1 m.y., 3.7 m.y., and 4.7 m.y. correspond to $f = 2$, 5, and 10, respectively.

To determine the style of root removal, t_g is compared to t_s. For $t_g < t_s$, the gravitational growth rate is sufficiently high that the root material is removed as a drip before it reaches the mantle wedge corner. Where $t_s < t_g$, shear entrainment occurs. Our calculations (Fig. 12) predict that a 5-km-thick root with $f = 1$ rheology requires a density contrast of at least 90 kg/m³ in order to be removed gravitationally. For $f = 10$, drip-like removal requires greater layer thicknesses and densities.

This analysis can be used to understand the styles of removal observed in the models. In models with a high densification rate, drip-like removal was observed for all root viscosities that we tested (Fig. 11B). In these models, the rapid increase in root density produces a high growth rate and low t_g, allowing gravitational removal of the root (Fig. 12). For a modest root densification rate (20–40 kg/m³/m.y.), the key factor controlling the style of removal is the strength of the root. For a weak root (low f), the greater shearing by mantle flow leads to low t_s. The initial

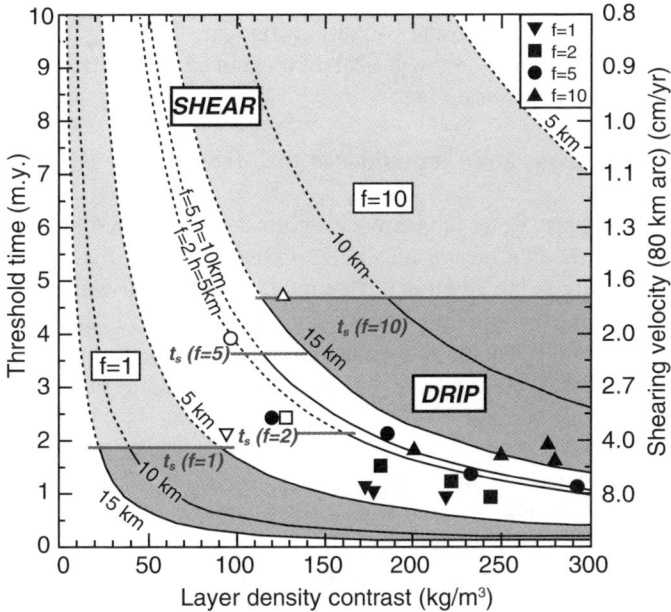

Figure 12. Threshold times for arc root removal. Lines show the gravitational threshold time (t_g) as a function of layer density, thickness, and viscosity ($f \times 10^{19}$ Pa s). This is defined as the time for a gravitational instability to develop a downward velocity of 7 cm/yr. Shaded regions encompass the range of thicknesses examined for $f = 1$ and $f = 10$. Horizontal gray lines are the shearing threshold time (t_s), set by the rate of shearing of the base of the arc root; weaker roots experience greater shearing (lower t_s). The scale on the right shows the shearing velocity that yields the time on the left axis for an 80-km-wide arc. Solid lines and darker-gray shading show where drip-like removal is favored ($t_g < t_s$). Dashed lines and light-gray shading show where removal through shear entrainment is favored ($t_g > t_s$). Model results for a subduction rate of 7 cm/yr are also shown, with solid symbols for models that exhibited drip-like removal and open symbols for models that exhibited shear entrainment (Table 2).

perturbation of the root and onset of instability occur earlier, when the root has a relatively low density (Fig. 10A), and a thinner layer that participates in the instability (Currie et al., 2008). As a result, the drip has a low gravitational growth rate, despite its low strength, and removal occurs through shear entrainment ($t_s < t_g$). In contrast, a stronger root is less susceptible to perturbation by mantle flow, resulting in a later perturbation, and thus higher density and growth rate and a reduced rate of shearing, and thus drip-like removal ($t_g < t_s$).

Effect of Subduction Rate

As observed in the models, another important parameter controlling root dynamics is the rate of subduction-induced mantle flow. As shown on Figure 10B, a lower subduction rate results in drip-like removal over a larger range of root densities and viscosities. In this case, the initial perturbation of the root is delayed due to low flow velocities (cf. Figs. 5 and 9A), and therefore the initiation of removal occurs at a higher density, producing higher gravitational growth rates. More importantly, the reduced shearing leads to a large t_s, and the root has sufficient time to be removed as a drip. With a higher velocity, t_s is reduced, and the root is rapidly carried into the wedge corner before it can grow gravitationally, except in cases where its density rapidly increases.

Assumptions and Limitations of the Models

The root zone in our models corresponds to the full thickness of the continental mantle lithosphere (~15 km thick), and its density is increased using a simplified parameterization that simulates the formation of a garnet pyroxenite residue. The assumption is that the entire region undergoes homogeneous densification at a constant rate, and that the root contains only small volumes of partial melt that are rapidly extracted. In nature, this is a complex process, and differentiation occurs heterogeneously in both space and time. Clearly, magmatism is focused under arcs in centers that are spaced typically at 50–80 km along the strike of the arc. This is seen both in active arcs where stratovolcanoes are aligned at this spacing, as well as in old arcs, where subsurface batholiths are exposed. These factors should result in small-scale variations in density and viscosity within the root region. These variations are below the resolution of our models, and it is unclear how they will affect the dynamics of the root.

Furthermore, our models do not include addition of mantle-derived magmas to the arc lithosphere, although there is an implicit assumption that magma intrusion provides the necessary heat for partial melting that leads to lithosphere differentiation. For Cordilleran arcs, up to 50% of the mass of the arc may be derived from mantle wedge magmatism (Ducea, 2002; DeCelles et al., 2009, and references therein). Using a basaltic magmatic flux of 20–40 km³/m.y. per kilometer along strike and a melt to residue ratio of 1:1–1:3 (DeCelles et al., 2009), the mantle-derived magmas would produce an eclogite pyroxenite

root that thickens at 0.25–1.5 km/m.y. for an 80-km-wide arc. This should enhance the potential for gravitational removal of the entire root complex, but the removal dynamics will depend on the distribution of basaltic melts in the root zone (i.e., whether they underplate or intrude). Behn et al. (2007) argued that a 2–4 km layer of underplated material should be removed within 1–10 m.y., which is on the same time scale as the instabilities observed in this study. Models that include magmatic addition are the focus of ongoing work. Such models will also be important for understanding dynamics of island arcs, where there is less involvement of upper-plate materials in arc magmatism.

Two other factors that may affect the behavior observed in the models are the geometry of the subducting plate and the structure of the continental lithosphere. In our models, the subduction geometry and velocity are nearly constant during model evolution. Temporal variations in these parameters are expected to affect the dynamics of the root. In particular, an episode of low-angle subduction could enable a dense arc root to remain in place until the slab is removed. Flat-slab and ultrashallow subduction under the arc, corresponding to the Laramide orogeny, has been proposed to explain the ~70 m.y. delay between the end of magmatism and inferred root removal for the Sierra Nevada arc in California (Saleeby et al., 2003; Zandt et al., 2004).

The structure of the continental lithosphere will also affect the dynamics of root removal. In the volcanic arc region, our models have a hot, prethickened orogenic crust, and the root forms within the underlying ~15-km-thick continental mantle lithosphere. The overall thermal structure of the orogen lithosphere (>900 °C at the arc Moho) is consistent with petrological and seismic constraints from modern arcs (Kelemen et al., 2003), and the crustal rheology is based on a wet quartzite flow law, consistent with seismic studies indicating that the central Andes has a dominantly felsic crust (e.g., Beck and Zandt, 2002; ANCORP Working Group, 2003). If the arc crust were weaker than in our models, possibly due to local heating by intruding melts and the presence of the melts themselves, it may be easier for the arc root to detach, as the viscous coupling between the root and crust is decreased. A weaker crust may also be more easily entrained during root removal, leading to thickening of the deep crust (e.g., Pysklywec and Beaumont, 2004; Wang et al., this volume). In contrast, the root removal process may be less efficient for regions where the continental crust is thinner or more mafic, or the lithosphere is cooler (and therefore thicker) than assumed here. A thinner, mafic crust and/or lower Moho temperature would result in an increase of both the root viscosity and the coupling between the crust and root, making it more difficult for the root to be removed. In addition, if the lithosphere were thicker, the root may occupy only the uppermost mantle lithosphere (top 10–20 km), and its removal may be further hindered by the underlying lithosphere. Additional models that explore a range of lithosphere structures are needed to determine how the removal style and amount of root removed are affected.

A major limitation of the models is that they are two dimensional. This may have several important consequences. For example, in 2-D models, a gravitational instability grows by pulling in material from within the model plane. In contrast, a 3-D instability involves material from the volume surrounding the initial perturbation, allowing instabilities to grow faster (e.g., Kaus and Podladchikov, 2001; Hasenclever et al., 2011). Furthermore, in 2-D models, mantle wedge flow is restricted to the model plane, and therefore, it can easily shear the base of the root. In 3-D models, flow may be diverted around any perturbation in the overriding lithosphere (e.g., Hasenclever et al., 2011). With the enhanced growth rates and reduced shearing, 3-D arc roots may favor a drip-like style of removal for a wider parameter space than predicted by the 2-D models (Fig. 11B).

Implications for Cordilleran Orogens

The formation and removal of dense eclogitic residues at volcanic arcs have been argued to be a fundamental processes in the evolution of Cordilleran orogens (DeCelles et al., 2009, and references therein). The dense root appears to occupy much of the lower lithosphere under arcs, where it is inferred that fertile continental mantle lithosphere undergoes melting and differentiation during high-flux events that have a duration of 5–15 m.y. (Ducea, 2001; DeCelles et al., 2009). Our models demonstrate that even a small increase in density of continental mantle lithosphere can result in its destabilization and removal for a wide range of root strengths and subduction rates (Fig. 11B). In the models, removal occurs within 15 m.y. of the start of densification, consistent with the duration of high-flux events. Over 80% of arc root material can be removed, making this an effective way to dispose of continental mantle lithosphere from a Cordilleran orogen. This may be an important mechanism to resolve the mass problem in Cordilleran orogens, where the mantle lithosphere is much thinner than expected on the basis of crustal shortening rates (e.g., Wernicke et al., 1996).

The models provide insight into surface observables that may be associated with this process. As root material is removed relatively quickly, it does not thermally equilibrate with the surrounding mantle wedge, and, therefore, it may be detected in seismic tomography studies as a high-velocity region (e.g., Beck and Zandt, 2002; Schurr et al., 2006; Bianchi et al., 2013). In our models, this material has different geometries, from 20- to 30-km-wide drips to small-scale (<10 km) features that are stretched and distorted by mantle flow (e.g., Fig. 6). If removal is primarily through shear entrainment, the size of the features is likely below the resolution of most seismic tomography studies.

The removal of a dense arc root may also be reflected in the surface topography. Figure 13 shows the surface elevation at the volcanic arc for models with different root densification rates. These models have the reference rheology structure (Table 1) and a convergence rate of 7 cm/yr. In the early stages of the models, there is little variation in surface topography. As the volcanic arc root detaches, the surface undergoes uplift of 300–900 m, as an isostatic response to the removal of the dense root. Surface uplift is greater for models with a high root densification

rate, as it is assumed that this is associated with a greater rate of crustal emplacement of low-density felsic magmas. The formation and drip-style removal of magmatic roots may be reflected by the anomalous surface deflections that are recorded in local "bobber basins" in the central Andes, such as the Arizaro Basin (DeCelles et al., this volume). One interesting observation is that uplift occurs during drip-style removal, but if the root is removed through shear entrainment, there is little surface response until after removal. In drip-style removal, the greater vertical velocity of the root may lead to stronger entrainment of the ductile lower crust, resulting in crustal thickening above the descending root, which may explain the earlier onset of uplift (e.g., Fig. 6B; Pysklywec and Beaumont, 2004; Wang et al., this volume).

Two additional surface observables are variations in surface magmatism and orogenic shortening. Magmatism is directly associated with the formation and removal of the eclogite pyroxenite root. Isostopic signatures and magma flux rates have been used to argue for the formation of arc roots during high-flux events (e.g., Ducea, 2001; DeCelles et al., 2009). Root removal is thus constrained to take place after major high-flux events. Recent numerical models suggest that the root material itself may undergo partial melting as it descends (Elkins-Tanton, 2007). Ducea et al. (2013) demonstrated that this may explain the upper-plate chemical signature that is observed in mafic magmas from central Andes. This requires that the edge of the descending root is conductively heated above its solidus within the time scale that it takes for the material to be removed. Elkins-Tanton (2007) showed that sufficient heating can occur within 1–2 m.y. for a descending lithosphere drip, similar to the time scale of removal in our models. However, the details of melting depend on the

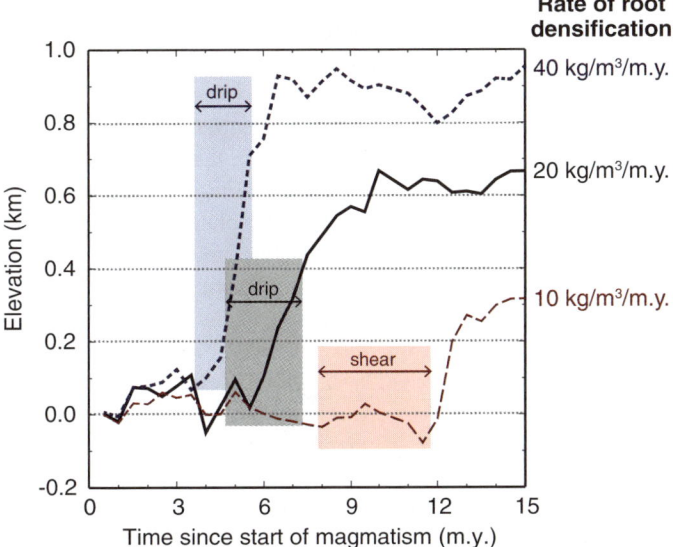

Figure 13. Relative surface elevation of the volcanic arc region for models with the reference material parameters (Table 1), subduction rate of 7 cm/yr, and different densification rates for the arc root. The shaded regions show the time scale for removal (Table 2).

initial temperature structure and composition of the removed lithosphere. A quantitative assessment of the relationship between the model dynamics and the composition of arc root material is needed to address whether significant melting could occur during arc root removal.

DeCelles et al. (2009) argued that the overall shortening rate of the orogen may be modulated by the formation and removal of arc roots. In their conceptual model, shortening rates decrease as continental mantle lithosphere fills the mantle wedge corner. As melt-fertile lithosphere enters the volcanic arc region, it undergoes partial melting and differentiation during a high-flux event. Subsequent gravitational removal of the root clears the wedge corner, enabling renewed shortening of the orogen and the introduction of new fertile continental material to the arc region. In addition, surface uplift associated with root removal (Fig. 13) may cause an outward propagation in the locus of orogenic shortening. In this manner, Cordilleran orogens may develop through a cyclical pattern of eclogite root buildup and removal, which may be recorded in the magmatic and shortening history of the orogen. High-flux events are observed to occur every 25–50 m.y. for Cordilleran orogens, which is interpreted as the time needed for continental mantle lithosphere to be thrust below the arc during orogen shortening (DeCelles et al., 2009). Our models assume that continental mantle lithosphere is already in place at the start of a high-flux event, which corresponds to a time of 0 m.y. in our models. They do not include upper-plate shortening, and therefore they do not address the relationships among orogen shortening, arc root removal, and the temporal spacing of high-flux events. This will be explored in the next generation of models.

CONCLUSIONS

The numerical models in this study have addressed one aspect of the proposed Cordilleran orogenic cycle (DeCelles et al., 2009), namely, the dynamic removal of a high-density eclogitic root under the frontal arc. We used a simplified parameterization to simulate chemical differentiation of continental lowermost crust and mantle lithosphere in the arc region, resulting in the formation of a dense, gravitationally unstable, garnet pyroxenite root. For the continental lithosphere structure used in this study, foundering of this material occurs readily for a wide range of densification rates, root viscosities, and subduction rates. Removal occurs within 15 m.y. of the start of densification, which is consistent with the 5–15 m.y. duration of high-flux events at Cordilleran arcs (Ducea, 2001; DeCelles et al., 2009). The dynamics of removal are strongly affected by subduction-driven flow in the underlying mantle wedge. This flow creates the initial perturbation that induces instability, and it can entrain the dense root as it destabilizes. Two styles of removal are observed: (1) drip-like removal, in which the root is removed as a single downwelling over 1–2.5 m.y.—this is observed for high densification rates, strong roots, and low subduction rates; and (2) shear entrainment, where the lower part of the root is swept sideways by mantle flow and then subducted with the oceanic plate—this is

a more gradual process that can take >3 m.y. In both cases, nearly the entire root region can be removed (>80% for the parameters tested here). In general, our models demonstrate that with minor densification, the root region of a volcanic arc is susceptible to gravitational foundering, making this an efficient way to remove mantle lithosphere from a Cordilleran orogen. By clearing the lithosphere from the mantle wedge corner region, this process may modulate the overall shortening rate of the orogen and may be observable in magmatic, paleoelevation, and seismic data (DeCelles et al., 2009; Ducea et al., 2013).

ACKNOWLEDGMENTS

We thank Cin-Ty Lee and an anonymous reviewer for helpful comments on an earlier version of this paper. The finite-element code (SOPALE) was developed by Philippe Fullsack, with additional modifications by Douglas Guptill (Dalhousie University, Halifax, Canada). Numerical models in this study were run using Compute Canada (WestGrid) computational resources. Research was supported by grants from ExxonMobil Upstream Research Company, Natural Sciences and Engineering Research Council of Canada (NSERC), and Romanian National Science Foundation grant PN-II-ID-PCE-2011-3-0217.

REFERENCES CITED

ANCORP Working Group, 2003, Seismic imaging of a convergent continental margin and plateau in the Central Andes (Andean Continental Research Project 1996 [ANCORP'96]): Journal of Geophysical Research, v. 108, no. B7, 2328, doi:10.1029/2002JB001771.
Annen, C., and Sparks, R.S.J., 2002, Effects of repetitive emplacement of basaltic intrusions on thermal evolution and melt generation in the crust: Earth and Planetary Science Letters, v. 203, p. 937–955, doi:10.1016/S0012-821X(02)00929-9.
Barnes, J.B., and Ehlers, T.A., 2009, End member models for Andean Plateau uplift: Earth-Science Reviews, v. 97, p. 105–132, doi:10.1016/j.earscirev.2009.08.003.
Beaumont, C., Nguyen, M.H., Jamieson, R.A., and Ellis, S., 2006, Crustal flow modes in large hot orogens, in Law, R.D., Searle, M.P., and Godin, L., eds., Channel Flow, Ductile Extrusion, and Exhumation of Lower-Mid Crust in Continental Collision Zones: Geological Society, London, Special Publication 268, p. 91–145.
Beck, S.L., and Zandt, G., 2002, The nature of orogenic crust in the Central Andes: Journal of Geophysical Research, v. 107, no. B10, 2230, doi:10.1029/2000JB000124.
Beck, S.L., Zandt, G., Myers, S.C., Wallace, T.C., Silver, P.G., and Drake, L., 1996, Crustal-thickness variations in the Central Andes: Geology, v. 24, p. 407–410, doi:10.1130/0091-7613(1996)024<0407:CTVITC>2.3.CO;2.
Behn, M.D., and Kelemen, P.B., 2006, Stability of arc lower crust: Insights from the Talkeetna arc section, south central Alaska, and the seismic structure of modern arcs: Journal of Geophysical Research, v. 111, B11207, doi:10.1029/2006JB004327.
Behn, M.D., Hirth, G., and Kelemen, P.B., 2007, Trench-parallel anisotropy produced by foundering of arc lower crust: Science, v. 317, p. 108–111, doi:10.1126/science.1141269.
Bianchi, M., Heit, B., Jakovlev, A., Yuan, X., Kay, S.M., Sandvol, E., Alonso, R.N., Coira, B., Brown, L., Kind, R., and Comte, D., 2013, Teleseismic tomography of the southern Puna Plateau in Argentina and adjacent regions: Tectonophysics, v. 586, p. 65–83, doi:10.1016/j.tecto.2012.11.016.
Bird, P., 1979, Continental delamination and the Colorado Plateau: Journal of Geophysical Research, v. 84, p. 7561–7571, doi:10.1029/JB084iB13p07561.

Brooks, B.A., Bevis, M., Smalley, R., Jr., Kendrick, E., Manceda, R., Lauria, E., Maturana, R., and Araujo, M., 2003, Crustal motion in the Southern Andes (36°–36°S): Do the Andes behave like a microplate?: Geochemistry Geophysics Geosystems, v. 4, 1085, doi:10.1029/2003GC000505.

Castro, A., Gerya, T., Garcia-Casco, A., Fernandez, C., Diaz-Alvarado, J., Moreno-Ventas, I., and Low, I., 2010, Melting relations of MORB-sediment mélanges in underplated mantle wedge plumes; implications for the origin of Cordilleran-type batholiths: Journal of Petrology, v. 51, p. 1267–1295, doi:10.1093/petrology/egq019.

Coleman, D.S., Gray, W., and Glazner, A.F., 2004, Rethinking the emplacement and evolution of zoned plutons: Geochronologic evidence for incremental assembly of the Tuolumne Intrusive Suite, California: Geology, v. 32, p. 433–436, doi:10.1130/G20220.1.

Currie, C.A., Huismans, R.S., and Beaumont, C., 2008, Thinning of continental backarc lithosphere by flow-induced gravitational instability: Earth and Planetary Science Letters, v. 269, p. 436–447, doi:10.1016/j.epsl.2008.02.037.

DeCelles, P.G., Ducea, M.N., Kapp, P., and Zandt, G., 2009, Cyclicity in Cordilleran orogenic systems: Nature Geoscience, v. 2, p. 251–257, doi:10.1038/ngeo469.

DeCelles, P.G., Carrapa, B., Horton, B.K., McNabb, J., Gehrels, G.E., and Boyd, J., 2015, this volume, The Miocene Arizaro basin, central Andean hinterland: Response to partial lithosphere removal? in DeCelles, P.G., Ducea, M.N., Carrapa, B., and Kapp, P.A., eds., Geodynamics of a Cordilleran Orogenic System: The Central Andes of Argentina and Northern Chile: Geological Society of America Memoir 212, doi:10.1130/2015.1212(18).

Defant, M.J., and Drummond, M.S., 1990, Derivation of some modern arc magmas by melting of young subducted lithosphere: Nature, v. 347, p. 662–665, doi:10.1038/347662a0.

Ducea, M., 2001, The California arc: Thick granitic batholiths, eclogitic residues, lithospheric-scale thrusting, and magmatic flare-ups: GSA Today, v. 11, no. 11, p. 4–10, doi:10.1130/1052-5173(2001)011<0004:TCATGB>2.0.CO;2.

Ducea, M.N., 2002, Constraints on the bulk composition and root foundering rates of continental arcs: A California arc perspective: Journal of Geophysical Research, v. 107, no. B11, 2304, doi:10.1029/2001JB000643.

Ducea, M.N., and Barton, M.D., 2007, Igniting flare-up events in Cordilleran arcs: Geology, v. 35, p. 1047–1050, doi:10.1130/G23898A.1.

Ducea, M.N., and Saleeby, J., 1998, A case for delamination of the deep batholithic crust beneath the Sierra Nevada, California: International Geology Review, v. 40, p. 78–93, doi:10.1080/00206819809465199.

Ducea, M.N., Seclaman, A.C., Murray, K.E., Jianu, D., and Schoenbohm, L.M., 2013, Mantle-drip magmatism beneath the Altiplano-Puna Plateau, Central Andes: Geology, v. 41, p. 915–918, doi:10.1130/G34509.1.

Elkins-Tanton, L.T., 2007, Continental magmatism, volatile recycling, and a heterogeneous mantle caused by lithospheric gravitational instabilities: Journal of Geophysical Research, v. 112, B03405, doi:10.1029/2005JB004072.

Fullsack, P., 1995, An arbitrary Lagrangian-Eulerian formulation for creeping flows and applications in tectonic models: Geophysical Journal International, v. 120, p. 1–23, doi:10.1111/j.1365-246X.1995.tb05908.x.

Garzione, C.N., Molnar, P., Libarkin, J.C., and MacFadden, B.J., 2006, Rapid late Miocene rise of the Bolivian Altiplano: Evidence for removal of mantle lithosphere: Earth and Planetary Science Letters, v. 241, p. 543–556, doi:10.1016/j.epsl.2005.11.026.

Gleason, G.C., and Tullis, J., 1995, A flow law for dislocation creep of quartz aggregates determined with the molten salt cell: Tectonophysics, v. 247, p. 1–23, doi:10.1016/0040-1951(95)00011-B.

Göğüş, O.H., and Pysklywec, R.N., 2008, Near surface diagnostics of dripping and delaminating lithosphere: Journal of Geophysical Research, v. 113, B11404, doi:10.1029/2007JB005123.

Hacker, B.R., Abers, G.A., and Peacock, S.M., 2003, Subduction factory 1. Theoretical mineralogy, densities, seismic wave speeds, and H_2O contents: Journal of Geophysical Research, v. 108, no. B1, 2029, doi:10.1029/2001JB001127.

Hasenclever, J., Morgan, J.P., Hort, M., and Rupke, L.H., 2011, 2D and 3D numerical models on compositionally buoyant diapirs in the mantle wedge: Earth and Planetary Science Letters, v. 311, p. 53–68, doi:10.1016/j.epsl.2011.08.043.

Heit, B., Koulakov, I., Asch, G., Yuan, X., Kind, R., Alcocer-Rodriguez, I., Tawackoli, S., and Wilke, H., 2008, More constraints to determine the seismic structure beneath the Central Andes at 21°S using teleseismic tomography analysis: Journal of South American Earth Sciences, v. 25, p. 22–36, doi:10.1016/j.jsames.2007.08.009.

Hildreth, W., and Moorbath, S., 1988, Crustal contributions to arc magmatism in the Andes of central Chile: Contributions to Mineralogy and Petrology, v. 98, p. 455–489, doi:10.1007/BF00372365.

Hirth, G., and Kohlstedt, D., 2003, Rheology of the upper mantle and the mantle wedge: A view from the experimentalists, in Eiler, J.M., ed., Inside the Subduction Factory: American Geophysical Union Geophysical Monograph 138, p. 83–105.

Hoke, L., Hilton, D.R., Lamb, S.H., Hammerschmidt, K., and Friedrichsen, H., 1994, ³He evidence for a wide zone of active mantle melting beneath the Central Andes: Earth and Planetary Science Letters, v. 128, p. 341–355, doi:10.1016/0012-821X(94)90155-4.

Houseman, G.A., and Molnar, P., 1997, Gravitational (Rayleigh-Taylor) instability of a layer with non-linear viscosity and convective thinning of continental lithosphere: Geophysical Journal International, v. 128, p. 125–150, doi:10.1111/j.1365-246X.1997.tb04075.x.

Houseman, G.A., McKenzie, D.P., and Molnar, P., 1981, Convective instability of a thickened boundary layer and its relevance for the thermal evolution of continental convergence belts: Journal of Geophysical Research, v. 86, p. 6115–6132, doi:10.1029/JB086iB07p06115.

Huismans, R.S., and Beaumont, C., 2003, Symmetric and asymmetric lithospheric extension: Relative effects of frictional-plastic and viscous strain softening: Journal of Geophysical Research, v. 108, no. B10, 2496, doi:10.1029/2002JB002026.

Hyndman, R.D., and Currie, C.A., 2011, Why is the North America Cordillera high? Hot backarcs, thermal isostasy and mountain belts: Geology, v. 39, p. 783–786, doi:10.1130/G31998.1.

Jin, Z.M., Zhang, J., Green, H.W., II, and Jin, S., 2001, Eclogite rheology: Implications for subducted lithosphere: Geology, v. 29, p. 667–670, doi:10.1130/0091-7613(2001)029<0667:ERIFSL>2.0.CO;2.

Jull, M., and Kelemen, P.B., 2001, On the conditions for lower crustal convective instability: Journal of Geophysical Research, v. 106, p. 6423–6446, doi:10.1029/2000JB900357.

Karato, S.I., and Wu, P., 1993, Rheology of the upper mantle: A synthesis: Science, v. 260, p. 771–778, doi:10.1126/science.260.5109.771.

Kaus, B.J.P., and Podladchikov, Y.Y., 2001, Forward and reverse modeling of the three-dimensional viscous Rayleigh-Taylor instability: Geophysical Research Letters, v. 28, p. 1095–1098, doi:10.1029/2000GL011789.

Kay, R.W., and Kay, S.M., 1993, Delamination and delamination magmatism: Tectonophysics, v. 219, p. 177–189, doi:10.1016/0040-1951(93)90295-U.

Kay, S.M., and Coira, B.L., 2009, Shallowing and steepening subduction zones, continental lithospheric loss, magmatism, and crustal flow under the Central Andean Altiplano–Puna Plateau, in Kay, S.M., Ramos, V.A., and Dickinson, W.R., eds., Backbone of the Americas: Shallow Subduction, Plateau Uplift, and Ridge and Terrane Collision: Geological Society of America Memoir 204, p. 229–259.

Kay, S.M., Coira, B., and Viramonte, J., 1994, Young mafic back arc volcanic rocks as indicators of continental lithospheric delamination beneath the Argentine Puna Plateau, Central Andes: Journal of Geophysical Research, v. 99, p. 24,323–24,339, doi:10.1029/94JB00896.

Kelemen, P.B., Rilling, J.L., Parmentier, E.M., Mehl, L., and Hacker, B.R., 2003, Thermal structure due to solid-state flow in the mantle wedge beneath arcs, in Eiler, J., ed., Inside the Subduction Factory: American Geophysical Union Geophysical Monograph 138, p. 293–311.

Krystopowicz, N.J., and Currie, C.A., 2013, Crustal eclogitization and lithosphere delamination in orogens: Earth and Planetary Science Letters, v. 361, p. 195–207, doi:10.1016/j.epsl.2012.09.056.

Lee, C.T.A., Cheng, Z., and Horodyskyj, U., 2006, The development and refinement of continental arcs by primary basaltic magmatism, garnet pyroxenite accumulation, basaltic recharge and delamination: Insights from the Sierra Nevada, California: Contributions to Mineralogy and Petrology, v. 151, p. 222–242, doi:10.1007/s00410-005-0056-1.

Mackwell, S.J., Zimmerman, M.E., and Kohlstedt, D.L., 1998, High temperature deformation of dry diabase with application to tectonics on Venus: Journal of Geophysical Research, v. 103, p. 975–984, doi:10.1029/97JB02671.

McGlashan, N., Brown, L., and Kay, S., 2008, Crustal thickness in the Central Andes from teleseismically recorded depth phase precursors: Geophysical Journal International, v. 175, p. 1013–1022, doi:10.1111/j.1365-246X.2008.03897.x.

McQuarrie, N., Horton, B.K., Zandt, G., Beck, S., and DeCelles, P.G., 2005, Lithospheric evolution of the Andean fold-thrust belt, Bolivia, and the origin of the central Andean plateau: Tectonophysics, v. 399, p. 15–37, doi:10.1016/j.tecto.2004.12.013.

Molnar, P., and Garzione, C.N., 2007, Bounds on the viscosity coefficient of continental lithosphere from removal of mantle lithosphere beneath the Altiplano and Eastern Cordillera: Tectonics, v. 26, TC2013, doi:10.1029/2006TC001964.

Molnar, P., Houseman, G.A., and Conrad, C.P., 1998, Rayleigh-Taylor instability and convective thinning of mechanically thickened lithosphere: Effects of non-linear viscosity decreasing exponentially with depth and of horizontal shortening of the layer: Geophysical Journal International, v. 133, p. 568–584, doi:10.1046/j.1365-246X.1998.00510.x.

Myers, S.C., Beck, S., Zandt, G., and Wallace, T., 1998, Lithospheric-scale structure across the Bolivian Andes from tomographic images of velocity and attenuation for P and S waves: Journal of Geophysical Research, v. 103, p. 21,233–21,252, doi:10.1029/98JB00956.

Oncken, O., Hindle, D., Kley, J., Elger, K., Victor, P., and Schemmann, K., 2006, Deformation of the Central Andean upper plate system—Facts, fiction and constraints for plateau models, *in* Oncken, O., Chong, G., Franz, G., Giese, P., Gotze, H.J., Ramos, V.A., Strecker, M.R., and Wigger, P., eds., The Andes: Active Subduction Orogeny: Berlin, Springer-Verlag, Frontiers in Earth Sciences, v. 1, p. 3–27.

Pope, D.C., and Willett, S.D., 1998, Thermal-mechanical model for crustal thickening in the Central Andes driven by ablative subduction: Geology, v. 26, p. 511–514, doi:10.1130/0091-7613(1998)026<0511:TMMFCT >2.3.CO;2.

Pysklywec, R.N., and Beaumont, C., 2004, Intraplate tectonics: Feedback between radioactive thermal weakening and crustal deformation driven by mantle lithosphere instabilities: Earth and Planetary Science Letters, v. 221, p. 275–292, doi:10.1016/S0012-821X(04)00098-6.

Pysklywec, R.N., Beaumont, C., and Fullsack, P., 2002, Lithospheric deformation during the early stages of continental collision: Numerical experiments and comparison with South Island, New Zealand: Journal of Geophysical Research, v. 107, no. B7, p. ETG-3-1–ETG-3-19, doi:10.1029/2001JB000252.

Reymer, A., and Schubert, G., 1984, Phanerozoic addition rates to the continental crust and crustal growth: Tectonics, v. 3, p. 63–77, doi:10.1029/ TC003i001p00063.

Saleeby, J., Ducea, M.N., and Clemens-Knott, D., 2003, Production and loss of high-density batholithic root, southern Sierra Nevada, California: Tectonics, v. 22, 1064, doi:10.1029/2002TC001374.

Schmidt, M.W., and Poli, S., 2003, Generation of mobile components during subduction of oceanic crust, *in* Holland, H.D., and Turekian, K.K., eds., Treatise on Geochemistry: The Crust, Volume 3: Oxford, Elsevier, p. 567–591, doi:10.1016/B0-08-043751-6/03034-6.

Schurr, B., Rietbrock, A., Asch, G., Kind, R., and Oncken, O., 2006, Evidence for lithospheric detachment in the Central Andes from local earthquake tomography: Tectonophysics, v. 415, p. 203–223, doi:10.1016/j .tecto.2005.12.007.

Sempere, T., Butler, R.F., Richards, D.R., Marshall, L.G., Sharp, W., and Swisher, C.C., III, 1997, Stratigraphy and chronology of Upper Cretaceous–Lower Paleogene strata in Bolivia and northwest Argentina: Geological Society of America Bulletin, v. 109, p. 709–727, doi:10.1130/0016-7606(1997)109<0709:SACOUC>2.3.CO;2.

Sobolev, S.V., and Babeyko, A.Y., 2005, What drives orogeny in the Andes?: Geology, v. 33, p. 617–620, doi:10.1130/G21557.1.

Springer, M., 1999, Interpretation of heat-flow density in the Central Andes: Tectonophysics, v. 306, p. 377–395, doi:10.1016/S0040-1951(99)00067-0.

Springer, M., and Forster, A., 1998, Heat-flow density across the Central Andean subduction zone: Tectonophysics, v. 291, p. 123–139, doi:10.1016/ S0040-1951(98)00035-3.

Stein, C.A., and Stein, S., 1992, A model for the global variation in oceanic depth and heat flow with lithospheric age: Nature, v. 359, p. 123–129, doi:10.1038/359123a0.

Syracuse, E.M., and Abers, G.A., 2006, Global compilation of variations in slab depth beneath arc volcanoes and implications: Geochemistry Geophysics Geosystems, v. 7, doi:10.1029/2005GC001045.

Tao, W.C., and O'Connell, R.J., 1992, Ablative subduction: A two-sided alternative to the conventional subduction model: Journal of Geophysical Research, v. 97, p. 8877–8904, doi:10.1029/91JB02422.

Tatsumi, Y., 2005, The subduction factory: How it operates in the evolving Earth: GSA Today, v. 15, no. 7, p. 4–10, doi:10.1130/1052-5173 (2005)015[4:TSFHIO]2.0.CO;2.

Trumbull, R.B., Riller, U., Oncken, O., Schueber, E., Munier, K., and Hongn, F., 2006, The time-space distribution of Cenozoic arc volcanism in the Central Andes: A new data compilation and some tectonic considerations, *in* Oncken, O., Chong, G., Franz, G., Giese, P., Gotze, H.J., Ramos, V.A., Strecker, M.R., and Wigger, P., eds., The Andes: Active Subduction Orogeny: Berlin, Springer-Verlag, Frontiers in Earth Sciences, v. 1, p. 29–44.

Wang, H., Currie, C.A., and DeCelles, P.G., 2015, this volume, Hinterland basin formation and gravitational instabilities in the Central Andes: Constraints from gravity data and geodynamic models, *in* DeCelles, P.G., Ducea, M.N., Carrapa, B., and Kapp, P.A., eds., Geodynamics of a Cordilleran Orogenic System: The Central Andes of Argentina and Northern Chile: Geological Society of America Memoir 212, doi:10.1130/2015.1212(19).

Warren, C.J., Beaumont, C., and Jamieson, R.A., 2008, Formation and exhumation of ultra-high-pressure rocks during continental collision: Role of detachment in the subduction channel: Geochemistry Geophysics Geosystems, v. 9, Q04019, doi:10.1029/2007GC001839.

Wernicke, B., Clayton, R., Ducea, M., Jones, C.H., Park, S., Ruppert, S., Saleeby, J., Snow, J.K., Squires, L., Fliedner, M., Jiracek, G., Keller, R., Klemperer, S., Luetgert, J., Malin, P., Miller, K., Mooney, W., Oliver, H., and Phinney, R., 1996, Origin of high mountains in the continents: The southern Sierra Nevada: Science, v. 271, p. 190–193, doi:10.1126/ science.271.5246.190.

Willett, S.D., 1999, Rheological dependence of extension in wedge models of convergent orogens: Tectonophysics, v. 305, p. 419–435, doi:10.1016/ S0040-1951(99)00034-7.

Yuan, X., Sobolev, S.V., and Kind, R., 2002, Moho topography in the Central Andes and its geodynamic implications: Earth and Planetary Science Letters, v. 199, p. 389–402, doi:10.1016/S0012-821X(02)00589-7.

Zandt, G., Gilbert, H., Owens, T.J., Ducea, M., Saleeby, J., and Jones, C.H., 2004, Active foundering of a continental arc root beneath the southern Sierra Nevada in California: Nature, v. 431, p. 41–46, doi:10.1038/ nature02847.

Zhang, J., and Green, H.W., II, 2007, Experimental investigation of eclogite rheology and its fabrics at high temperature and pressure: Journal of Metamorphic Geology, v. 25, p. 97–115, doi:10.1111/j.1525 -1314.2006.00684.x.

MANUSCRIPT ACCEPTED BY THE SOCIETY 3 JUNE 2014
MANUSCRIPT PUBLISHED ONLINE 9 SEPTEMBER 2014

The Geological Society of America
Memoir 212
2015

Imaging the Nazca slab and surrounding mantle to 700 km depth beneath the central Andes (18°S to 28°S)

Alissa Scire
C. Berk Biryol*
George Zandt
Susan Beck
Department of Geosciences, University of Arizona, 1040 E. 4th Street, Tucson, Arizona 85721, USA

ABSTRACT

The central Andes in South America is an ideal location to investigate the interaction between a subducting slab and the surrounding mantle to the base of the mantle transition zone. We used finite-frequency teleseismic P-wave tomography to image velocity anomalies in the mantle from 100 to 700 km depth between 18°S and 28°S in the central Andes by combining data from 11 separate networks deployed in the region between 1994 and 2009. Deformation of the subducting Nazca slab is observed in the mantle transition zone, with regions of both thinning and thickening of the slab that we suggest are related to a temporary stagnation of the slab in the mantle transition zone. Our study also images a strong low-velocity anomaly beneath the Nazca slab in the mantle transition zone, which is consistent with either a local thermal anomaly or a region of hydrated material. The shallow mantle (<165 km) under the Eastern Cordillera is generally fast, consistent with proposed underthrusting of the Brazilian cratonic lithosphere or a string of localized lithospheric foundering. Several discontinuous low-velocity anomalies are observed beneath parts of the Altiplano and Puna Plateau, including two strong low-velocity anomalies in the upper mantle under the Los Frailes volcanic field and the southern Puna Plateau, consistent with proposed asthenospheric influx following lithospheric delamination.

INTRODUCTION

The mantle beneath the central Andes of South America has been investigated using both local (e.g., Myers et al., 1998; Graeber and Asch, 1999; Schurr et al., 2006; Koulakov et al., 2006) and teleseismic (e.g., Dorbath et al., 1993; Engdahl et al., 1995; Heit et al., 2008) tomography, but each study was focused on a region defined by the individual footprint of the seismic array used. This has limited the depth of resolution of these teleseismic tomography studies in this region. More recently, Bianchi et al. (2013) combined teleseismic and local data to image part of the central Andes. Global tomography studies (e.g., Bijwaard et al., 1998; Zhao, 2004; Li et al., 2008; Fukao et al., 2001, 2009) provide images beneath South America down to lower-mantle depths; however, global tomography studies cannot provide detailed images of upper-mantle structure. Multiple temporary

*Current address: Department of Geological Sciences, University of North Carolina, Chapel Hill, 104 South Road, Mitchell Hall, Campus Box #3315, Chapel Hill, North Carolina 27599-3315, USA.

Scire, A., Biryol, C.B., Zandt, G., and Beck, S., 2015, Imaging the Nazca slab and surrounding mantle to 700 km depth beneath the central Andes (18°S to 28°S), *in* DeCelles, P.G., Ducea, M.N., Carrapa, B., and Kapp, P.A., eds., Geodynamics of a Cordilleran Orogenic System: The Central Andes of Argentina and Northern Chile: Geological Society of America Memoir 212, p. 23–41, doi:10.1130/2015.1212(02). For permission to copy, contact editing@geosociety.org. © 2014 The Geological Society of America. All rights reserved.

seismic networks and several permanent stations have been deployed in the central Andes, which allow us to combine data from these temporally and spatially distinct networks in a single study and make it possible to obtain a more detailed regional-scale model of the mantle. We used finite-frequency teleseismic P-wave tomography to image velocity anomalies in the mantle down to the base of the mantle transition zone at 660 km between 18°S and 28°S in the central Andes. Nonuniform distribution and spacing of denser networks in the broad span of the studied region result in coarser resolution in the shallower parts of the model but allow for regional-scale anomalies to be imaged and interpreted down to greater depths than smaller local studies.

CENTRAL SOUTH AMERICAN ANDES

The central Andes are characterized by significant along-strike variations in magmatism, upper-crustal shortening, crustal thickness, and lithospheric structure (Allmendinger et al., 1997; Kay and Coira, 2009). The Central Andean Plateau is generally defined as the area above 3 km elevation and is one of the largest high plateaus in the world, second only to the Tibetan Plateau (Fig. 1A). The central portion of the Central Andean Plateau includes two internally drained segments, the Altiplano to the north, a high, relatively flat plateau, and the Puna Plateau to the south, a region with a higher elevation and more topographic relief (Isacks, 1988; Whitman et al., 1996). The Altiplano has an average elevation of ~3.5 km, while the elevation of the Puna Plateau averages ~4.5 km. The edges of the plateau are defined to the west by the modern volcanic arc, the Western Cordillera, and to the east by the Eastern Cordillera, an inactive fold-and-thrust belt where shortening occurred from ca. 40 Ma to 15 Ma (McQuarrie et al., 2005; Oncken et al., 2006). Deformation in the Subandean zone east of the Eastern Cordillera began after ca. 10 Ma in Bolivia and is characterized by a thin-skinned fold-and-thrust belt north of 24°S (Allmendinger et al., 1997; Oncken et al., 2006). South of 24°S, deformation in the back arc shifts from the thin-skinned deformation of the Subandean zone to thick-skinned deformation in the Santa Barbara system and finally to the basement-cored uplifts of the Sierras Pampeanas (Kley and Monaldi, 2002). Receiver function studies (Yuan et al., 2000, 2002; Beck and Zandt, 2002; Wölbern et al., 2009) show a variable crustal thickness across the central Andes, with ~35-km-thick crust in the forearc, thickening to 70 km in the Western Cordillera. The crust under the Altiplano varies in thickness from 60 to 70 km, with an average of ~65 km, which then thickens locally to 75 km in the Eastern Cordillera. The crust under the Puna Plateau to the south appears to be slightly thinner, with an average crustal thickness of ~60 km, in spite of the higher average elevation of the Puna Plateau relative to the Altiplano. Under the Subandean zone, the crust is ~40 km thick, thinning to 30–35 km out on the Chaco Plain beyond the eastern limits of Andean deformation.

The subducting Nazca plate age is ca. 45 Ma at the trench as it enters the subduction zone beneath the central Andes (Müller et al., 2008), and it is converging with the South American plate at 61 mm/yr relative to a stable South American reference frame (Norabuena et al., 1999). Subduction of the Nazca plate under the central Andes occurs at an angle of ~30° to a depth of ~300 km as defined by the slab seismicity (Fig. 1B; Cahill and Isacks, 1992; Hayes et al., 2012) and is bracketed by the Peruvian flat slab region to the north and the Pampean flat slab region to the south. While there are few earthquakes between 300 and 500 km, some earthquakes do occur near the base of the mantle transition zone between 500 and 660 km (Fig. 1B). Local earthquake tomography studies between 21°S and 24°S indicate that seismic velocities in the forearc are fast, possibly resulting from the presence of plutonic remnants of earlier volcanic arcs (Graeber and Asch, 1999; Koulakov et al., 2006; Schurr et al., 2006). Generally, the thick crust of the Altiplano and Puna Plateaus has been imaged with low average crustal P- and S-wave velocities (Swenson et al., 2000; Ward et al., 2013). The presence of a low-velocity zone in the upper crust of the Altiplano, the Altiplano low-velocity zone, has been observed in several studies, and it is attributed to the presence of a small amount of partial melting of the crust at ~20 km depth (Yuan et al., 2000; Beck and Zandt, 2002; Ward et al., 2013). Further south in the Puna Plateau, there is a major low-velocity zone (~15–20 km) associated with the Altiplano-Puna volcanic complex known as the Altiplano-Puna magma body which is thought to be the source region for the extensive 11–1 Ma ignimbrite deposits at the surface (Chmielowski et al., 1999; Zandt et al., 2003; De Silva et al., 2006; Ward et al., 2013). Anomalously low velocities have been observed in the upper mantle under the southern Puna Plateau between 25°S and 27°S (e.g., Bianchi et al., 2013; Wölbern et al., 2009; Schurr et al., 2006; Koulakov et al., 2006) and, combined with geochemical analyses of the large ignimbrite deposits of the southern Puna volcanic field (e.g., Kay and Kay, 1993; Kay et al., 1994; Kay and Coira, 2009), have been used as evidence for the influx of asthenosphere following lithospheric delamination in the region. Underthrusting of Brazilian cratonic lithosphere has been proposed to explain anomalously high velocities under the Subandean zone (Myers et al., 1998; Beck and Zandt, 2002; Phillips et al., 2012). Similar observations of high-velocity anomalies near 16°S (Dorbath et al., 1993; Dorbath and Granet, 1996) and analyses of focal mechanisms further north at ~11°S (Dorbath et al., 1991) are also interpreted as evidence for cratonic underthrusting along much of the central Andes. Continuous removal of mantle lithosphere under the Altiplano/Eastern Cordillera boundary is suggested to accommodate the underthrusting of the Brazilian craton between 18°S and 20°S (McQuarrie et al., 2005; Myers et al., 1998; Beck and Zandt, 2002).

Tectonics of the Upper Plate—Development of the Altiplano-Puna Plateau

The tectonic development of the plateau is debated, with different studies arguing for different uplift histories and mechanisms. Original estimates of trench-perpendicular shortening appeared to be unable to fully explain the thickness of the

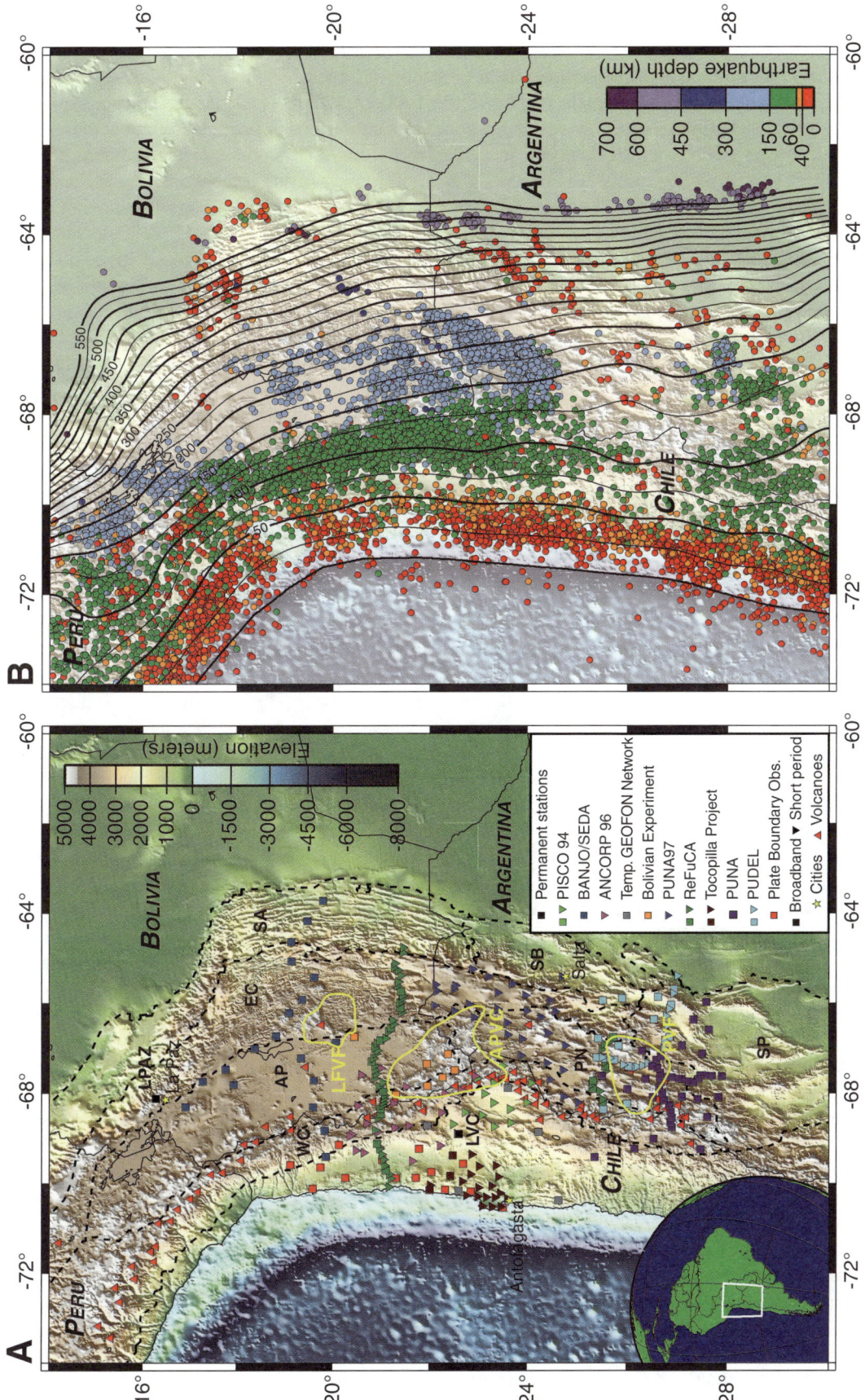

Figure 1. (A) Map showing seismic station locations for individual networks used in the study and topography of the central Andes. Yellow outlines mark locations of Los Frailes volcanic field (LFVF), Altiplano-Puna volcanic complex (APVC), and Puna volcanic field (PVF) (modified from Kay et al., 2010). Boundaries of geomorphic provinces (dashed lines) are modified from Tassara (2005). WC—Western Cordillera; AP—Altiplano; EC—Eastern Cordillera; SA—Subandean zone; PN—Puna; SB—Santa Barbara system; SP—Sierras Pampeanas. Red triangles mark location of Holocene volcanoes (Siebert and Simkin, 2002). (B) Map of seismicity (magnitude > 4) and Nazca slab depth contours. Slab contours are from the Slab1.0 global subduction zone model (Hayes et al., 2012). Earthquake data are from 1973 to 2012, U.S. Geological Survey–National Earthquake Information Center (NEIC) catalog (http://earthquake.usgs.gov/regional/neic/).

Altiplano crust (Kley and Monaldi, 1998) without along-strike contributions to crustal thickening, although more recent studies dispute this claim (e.g., Lamb, 2011). The timing of plateau uplift is uncertain, with structural reconstructions implying that plateau development was controlled by gradual shortening and thickening of the crust between 40 and 30 and 10 Ma (e.g., McQuarrie et al., 2005; Elger et al., 2005; Oncken et al., 2006; Lamb, 2011), while paleoelevation studies (e.g., Ghosh et al., 2006; Garzione et al., 2008) argue for a more recent (<10 Ma), rapid uplift of the plateau resulting from large-scale delamination of the lithosphere. Based on the geochemistry of the volcanic rocks, delamination of the lithosphere has been proposed as an explanation for the existence of high plateaus, particularly the Puna Plateau (Kay and Kay, 1993; Kay et al., 1994; Kay and Coira, 2009). Evidence for delamination from seismic studies exists in several regions of the central Andes. Seismic tomography studies have imaged high-velocity anomalies in the mantle, hypothesized to be delaminated blocks of lithosphere (Schurr et al., 2006; Asch et al., 2006; Koulakov et al., 2006; Bianchi et al., 2013), around 23°S, while other studies hypothesize localized lithospheric removal based on observations of anomalously low velocities below the crust, which are interpreted as asthenospheric mantle at shallow depths (Myers et al., 1998; Heit et al., 2008).

Tectonics of the Lower Plate—Slab Structure and the Fate of the Slab

Although the subducting Nazca slab has been imaged in small-scale teleseismic tomography studies and in global tomography studies, few studies have imaged the slab on a regional scale until recently (e.g., Pesicek et al., 2012; Bianchi et al., 2013). Global tomography studies show the Nazca slab in the central Andes descending into the lower mantle, although some studies indicate at least temporary stagnation in the mantle transition zone (e.g., Bijwaard et al., 1998; Zhao, 2004; Li et al., 2008; Fukao et al., 2001, 2009; Zhao et al., 2013). Unlike the stagnation of the Pacific slab under Japan, where the slab is imaged extending horizontally for several hundred kilometers in the transition zone with little evidence for it resuming subduction into the lower mantle (Fukao et al., 2009), the Nazca slab appears to resume subduction into the lower mantle after a period of thickening and stagnation in the mantle transition zone. Receiver function studies by Liu et al. (2003) and modeling studies by Quinteros and Sobolev (2013) indicate a correlation between abnormal mantle transition zone thicknesses and probable resistance to subduction of the Nazca plate into the lower mantle, which agrees with the images of the temporarily stagnated Nazca slab from global tomography studies, discussed earlier. In the lower mantle, the high-velocity anomaly associated with the slab becomes more amorphous (Bijwaard et al., 1998; Fukao et al., 2001; Li et al., 2008; Zhao et al., 2013). South of our study area, a nested regional-global tomography study by Pesicek et al. (2012) imaged a discontinuous Nazca slab at ~38°S, providing evidence for vertical tearing of the slab along the zone of weak-

ness provided by the subduction of the Mocha fracture zone. Engdahl et al. (1995) used teleseismic and regional P phases to image the Andean subduction zone between 5°S and 25°S down to ~1400 km depth. Their results indicate that the Nazca slab penetrates into the lower mantle throughout their study area, in agreement with global models. In the southern part of their study area, thickening of the slab in the mantle transition zone is observed, indicating at least temporary stagnation of the slab before it penetrates into the lower mantle. A teleseismic tomography study by Heit et al. (2008) along a line at 21°S had difficulties resolving the high-velocity slab anomaly without it being shifted vertically relative to the predicted location of the slab based on the observed slab seismicity from the EHB catalog (Engdahl et al., 1998) and used the earthquake location data to constrain the location of the slab in their final model. Similarly, Bianchi et al. (2013) used regional and global tomography data to constrain the location of the Nazca slab in their study, which imaged the lower crust and upper mantle down to depths of ~300 km under the Puna Plateau between 24°S and 29°S. They observed some variations in the amplitude of the slab anomaly, with a distinct low-velocity anomaly seen dividing the slab at ~26°S between 100 and 200 km depth. They noted that this feature is not well constrained and may be the result of vertical smearing of low-velocity anomalies in the crust down into the mantle, resulting in the discontinuous slab anomaly.

DATA

The data used in this study were collected by 284 short-period and broadband seismic stations belonging to 11 separate networks deployed at various time periods between 1994 and 2009 (Fig. 1A; Table 1). The networks include nine temporary deployments with varying numbers of stations as well as five stations from the GeoForschungsZentrum's (GFZ) global temporary GEOFON network. The data set also includes data from permanent stations of the Plate Boundary Observatory network, which were deployed before late 2009. Two permanent stations from the Global Seismograph Network and the Global Telemetered Seismograph Network, LVC and LPAZ, respectively, are also included.

Arrival time data were used for direct P phase arrivals for 250 teleseismic events with magnitudes greater than 5.0 between 30° and 90° away from the study region. PKIKP phase arrival time data from an additional 58 global events with a similar magnitude limit between 155° and 180° away from the study region were also used (Fig. 2A). Arrivals were picked using the multichannel cross-correlation technique described by VanDecar and Crosson (1990) and modified by Pavlis and Vernon (2010). In total, 12,126 direct P and 2671 PKIKP arrivals were picked in three frequency bands for the broadband stations with corner frequencies of 0.2–0.8 Hz, 0.1–0.4 Hz, and 0.04–0.16 Hz. Due to the limited frequency content of data from short-period stations, arrival times were picked in a single higher-frequency band with corner frequencies of 0.5–1.5 Hz.

TABLE 1. SEISMIC NETWORKS INCORPORATED INTO THIS STUDY

Network name	Sensor type	No. of stations	Reference
Global Seismograph Network	Broadband	1 (LVC)	Butler et al. (2004)
Global Telemetered Seismograph Network	Broadband	1 (LPAZ)	USGS (2013)
PISCO 94	Short period	19	Graeber and Asch (1999)
BANJO/SEDA	Broadband	20	Beck et al. (1996)
ANCORP 96	Short period	22	ANCORP Working Group (1999)
Temp. GEOFON Network	Broadband	5	GFZ (2013)
Bolivian Experiment (APVC)	Broadband	7	Chmielowski et al. (1999)
PUNA 97	Short period	32	Schurr et al. (1999)
ReFuCA	Short period/broadband	70	Heit et al. (2008)
Tocopilla Project	Short period/broadband	21	Sobiesiak et al. (2008)
PUNA	Broadband	44	Bianchi et al. (2013)
PUDEL	Short period/broadband	28	Bianchi et al. (2013)
Integrated Plate Boundary Observatory Chile	Broadband	14	Sodoudi et al. (2011)

The distribution of the number of rays per frequency band is dominated by the 0.2–0.8 Hz band, with 38% of total rays occurring in that frequency band. Rays are well distributed between the other three frequency bands, with 18% of total rays in the 0.1–0.4 Hz band, 17% in the 0.04–0.16 Hz, and 27% in the 0.5–1.5 Hz band. The azimuthal distribution of the arrivals used in this study shows a bimodal distribution of earthquakes, with the majority of events occurring in either the Middle America subduction zone or the South Sandwich subduction zone (Fig. 2B). Residual traveltimes were calculated relative to IASP91 (Kennett and Engdahl, 1991) and then demeaned for each event to calculate the relative traveltime residuals.

Crustal thickness estimates from Tassara and Echaurren (2012) were incorporated to calculate traveltime corrections for the relative residuals in order to compensate for crustal heterogeneity (Schmandt and Humphreys, 2010). Velocity estimates for the Central Andean crust are sparse, with only general averages existing for the entire crust, so a homogeneous layered crustal velocity model corresponding to the average velocity of the Altiplano crust was used to calculate the crustal corrections for stations in the Altiplano and Puna Plateau rather than a more elaborate crustal velocity model (Zandt et al., 1996; Swenson et al., 2000; Dorbath and Masson, 2000). Crustal corrections for stations in the forearc were calculated using a faster velocity model

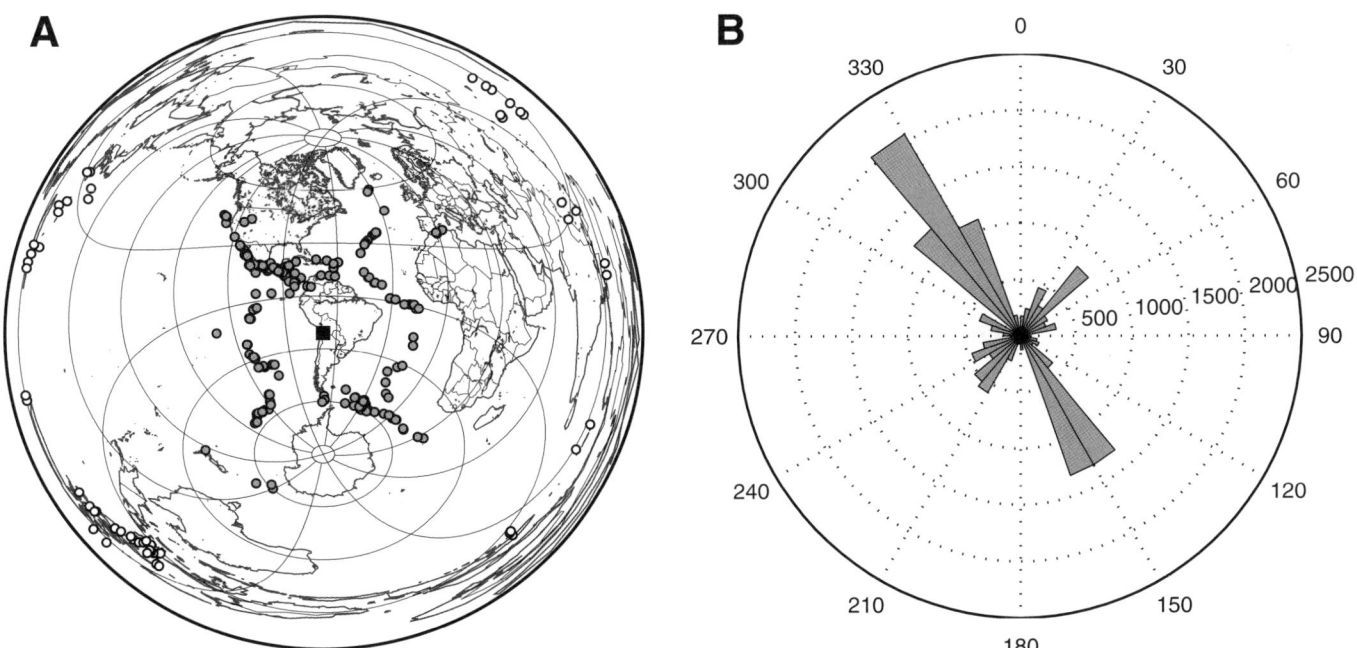

Figure 2. (A) Global map centered on our study region (black square) showing the location of events used in this study. Darker-gray circles mark events used for direct P arrivals, while open circles mark events used for PKIKP arrivals. (B) Plot showing the azimuthal distribution for all rays used in the study. Ray distribution is strongly controlled by the location of plate boundaries where the earthquakes are generated.

based on local tomography studies (Graeber and Asch, 1999; Koulakov et al., 2006; Schurr et al., 2006).

METHOD

The original tomography technique of Aki et al. (1977) assumes that seismic waves sample only along the infinitesimally thin ray path representing arrivals of infinite frequency. While this is convenient for practical purposes when formulating the tomographic problem, the data we use and the arrivals we pick have limited frequency content, leading to the need for a more realistic approximation. This is achieved by the finite-frequency approximation, which allows us to define frequency-dependent volumes around the geometrical ray path that are sampled by each arrival (Hung et al., 2000; Dahlen et al., 2000). This zone of sampling around the theoretical ray path is defined by the first Fresnel zone, the width of which is dependent on both frequency and distance along the ray path. The radius of the Fresnel zone increases with distance from the source or receiver, resulting in theoretical three-dimensional (3-D) "banana-doughnut" sensitivity kernels (Dahlen et al., 2000). The details of the algorithm used in this study to approximate the sensitivity kernels are discussed in Schmandt and Humphreys (2010). Due to differential sensitivity within the Fresnel zone, with zero sensitivity right along the theoretical ray path, sampling at each model layer is determined by the radius of the Fresnel zone at that depth and frequency and the "doughnut"-shaped sensitivity kernel that defines where the greatest sensitivity is within the Fresnel zone. Since the zone of sampling for each model layer and for each arrival is determined by the sensitivity kernels, multiple nodes can be sampled in each layer by a single arrival rather than the single node that lies along the geometrical ray path. The use of finite-frequency tomography improves sampling of the model space and recovery of the amplitude of velocity anomalies in the modeled region over ray theory–based techniques. In a global tomography study, Montelli et al. (2004) noted a 30%–50% increase in the amplitude of velocity perturbations recovered by a finite-frequency tomographic inversion over a ray-theory–based tomographic inversion.

The frequency-dependent relative traveltime residuals are inverted using the LSQR algorithm of Paige and Saunders (1982) in order to obtain velocity perturbations within the modeled volume. This algorithm aims to obtain the minimum energy/length model that satisfactorily explains the observed data. In the inversion process, we also incorporated station and event terms in order to compensate for receiver-side perturbations that are not taken into account in the crustal corrections and event-side velocity perturbations that remain outside the modeled volume. Since a homogeneous layered velocity model is used to calculate the crustal corrections, due to the lack of detailed information about the velocity structure in the study region, the station terms address local variations in the velocity structure from the homogeneous velocity model as well as errors in the a priori crustal thicknesses. Event terms

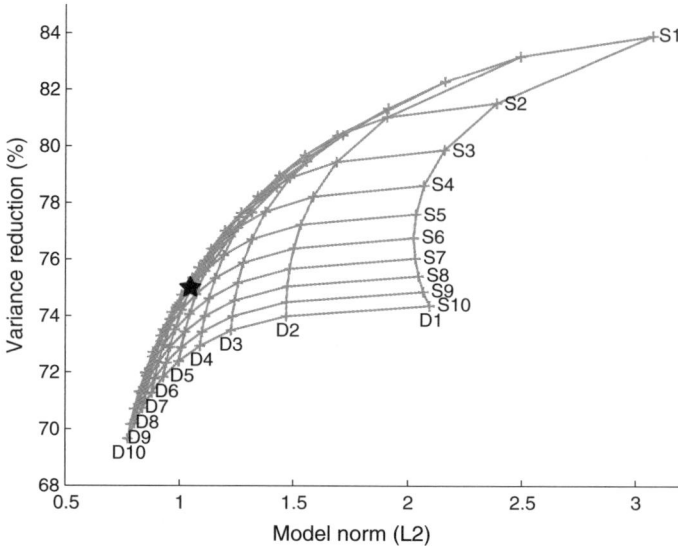

Figure 3. Plot showing the tradeoff analysis between the variance reduction and the Euclidean model norm (L2) performed to choose preferred overall damping (D1–D10) and smoothing (S1–S10) weights. The black star shows the damping (D6) and smoothing (S5) parameters used in this study.

are calculated to account for differences in the mean arrival time related to variations in mean velocity structure under the specific subset of stations that record a given event. Since this study incorporates data from different arrays, which were not all deployed contemporaneously, different subsets of stations record different events. This potentially introduces errors in calculated mean traveltime residuals related to variations in the velocity structure in different parts of the study region, which are also addressed by the calculation of the event terms. The inverse problem is regularized using norm and gradient damping (to account for depth-dependent theoretical ray path location uncertainty) as well as model smoothing, details of which are discussed in Schmandt and Humphreys (2010). A tradeoff analysis (Fig. 3) was performed between the variance reduction and the Euclidean model norm to choose the overall damping and smoothing weights. The chosen damping (6) and smoothing (5) parameters result in a variance reduction of 75.0%.

The model space was parameterized into a series of nodes in a nonuniform grid that increases in size with depth and distance from the center of the model. The nodes are spaced 40 km apart in the densest sampled portions of the model, with node spacing increasing to 55 km at the edges of the model space. The vertical distribution of nodes increases from 35 km in the shallowest parts of the model to 55 km in the deepest parts of the model. Node spacing within each horizontal layer also dilates with depth, with the nodes in the center part of the model increasing from 40 km spacing in the uppermost model layers to 56 km spacing in the lowermost model layers. Node spacing on the outer edges of the model increases from 55 km in the uppermost model layers to 78 km in the lowermost model layers.

Sampling of the model space was determined by calculating normalized hit quality maps (Fig. DR1[1]). The calculation of the hit quality maps relies on the idea that better sampling of a node is achieved with the intersection of rays from multiple azimuths (Schmandt and Humphreys, 2010). A node that has been sampled by multiple rays from all four of the geographical quadrants is assigned a hit quality of 1, while a node that is not sampled at all has a hit quality of 0. The hit quality maps for our data set indicate that most nodes are reasonably well sampled down to ~660 km depth, although in deeper layers, overall hit quality decreases as spreading of the rays with depth limits the number of crossing rays in the model space. Sampling of the uppermost layer (60 km depth) is strongly dependent on the distribution of stations, as noted by Biryol et al. (2011), and absorbs any errors in crustal corrections and has therefore been ignored in any interpretations. Similarly, the lowermost layer in the model (715 km) has also been removed from any interpretations because variations in lower-mantle structure not accounted for in the global model used to calculate traveltimes outside of the model are absorbed into the lowermost layer. The well-sampled parts of the forearc in the uppermost model layers are limited to between 18°S and 24°S because of a lack of stations in the forearc beyond those latitudes that were deployed contemporaneously with the networks used in this study. The extent of the well-sampled region in the eastern part of the study area is highly variable because of the distribution of stations and does not extend out into the stable foreland beyond the easternmost zone of Andean-related deformation.

Synthetic Resolution Tests

In addition to calculating the hit quality maps, we also performed a series of synthetic tests on our data set to investigate our resolution in our model. Initial synthetic tests used a "checkerboard" defined by alternating fast- and slow-velocity anomalies defined in eight node cubes that spanned two model layers (Fig. 4). Since node spacing dilates with depth, the location and size of the anomalies in the synthetic input also change with depth. Fast and slow anomalies for the synthetic input are defined as +5% and −5% anomalies, respectively. These anomalies are separated vertically by neutral background layers, which have no velocity anomalies. About 60% of the input amplitude of the anomalies is recovered by the inversion. The checkerboard tests show good lateral resolution throughout much of the model space (Fig. 4). Lateral smearing increases and amplitude recovery decreases toward the edges of the model space. Neutral layers (60, 95, 200, 240, 365, 410, 555, and 605 km depth) show evidence of vertical smearing as low-amplitude velocity anomalies are being incorrectly resolved in the neutral layers.

Additional synthetic recovery tests were performed to check on the ability of our inversion to resolve the subducting slab, since the Nazca slab is expected to be a prominent feature in our study region. A synthetic slab based on the Slab1.0 global subduction zone model (Hayes et al., 2012) with a +5% velocity anomaly was input into the inversion to test how well the model was able to resolve a continuous, dipping, high-velocity anomaly corresponding to the Nazca slab (Fig. 5). The synthetic slab anomaly was ~70–100 km thick, with variations resulting from the dilation of the grid spacing with depth. Recovery of the slab anomaly was affected by the vertical smearing along the ray paths, particularly in the shallower parts of the model. Cross sections through the synthetic model show the slab being shifted vertically along the ray paths in the shallower parts of the model, resulting in the shift of the resolved slab anomaly to the east in the shallowest layers. The resolved slab anomaly is more diffuse in the uppermost parts of the model, with the vertical smearing resulting in lower-amplitude (+1%–2%) positive velocity anomalies spread out in the model space rather than a distinctly resolved slab (Fig. 5). The resolved slab anomaly becomes more distinct with depth. In deeper parts of the model, ~60% of the input amplitude (+3%) is recovered, and little distortion of the shape of the input slab anomaly is observed, indicating that the slab should be well resolved below ~200 km.

Incorporating a Priori Slab Data into the Model

In order to improve our image of the uppermost mantle above the slab where the synthetic slab recovery tests (Fig. 5) indicate that the slab is not always well imaged, we constrain the location of the slab in our model using a priori information. Using the same method as Heit et al. (2008), we subtract the effect of the Nazca slab from the observed data using a priori information about the slab location to calculate theoretical residuals for rays traveling through a +4% fast synthetic slab model with the same geometry as the synthetic model shown in Figure 5, which is based on the Slab1.0 global subduction zone model (Hayes et al., 2012). The synthetic slab model used by Heit et al. (2008) had the same amplitude (+4%) as our synthetic slab but was based entirely on EHB earthquake locations. Because our model images the mantle under the Andes to a depth beyond the intermediate-depth seismicity, the use of the global subduction zone model was necessary since it offers some constraints on the location of the slab below the deepest intermediate-depth earthquakes. The ability to remove the effect of the slab from the observed data is a result of the linear nature of the tomographic inversion, and is discussed in detail by Heit et al. (2008). The final model superimposes the results of the "no-slab" inversion with the model of the slab, resulting in the tomograms seen in Figures 6 and 7. This model is used for interpretations of anomalies above 200 km, where synthetic tests indicated that the vertical smearing of the slab anomaly is the most prominent. We recognize that there is some uncertainty about the strength of the slab anomaly that could impact our final results above the slab in the upper 150 km.

---

[1]GSA Data Repository Item 2015001, Supplemental figures including additional resolution tests, is available at www.geosociety.org/pubs/ft2015.htm, or on request from editing@geosociety.org or Documents Secretary, GSA, P.O. Box 9140, Boulder, CO 80301-9140, USA.

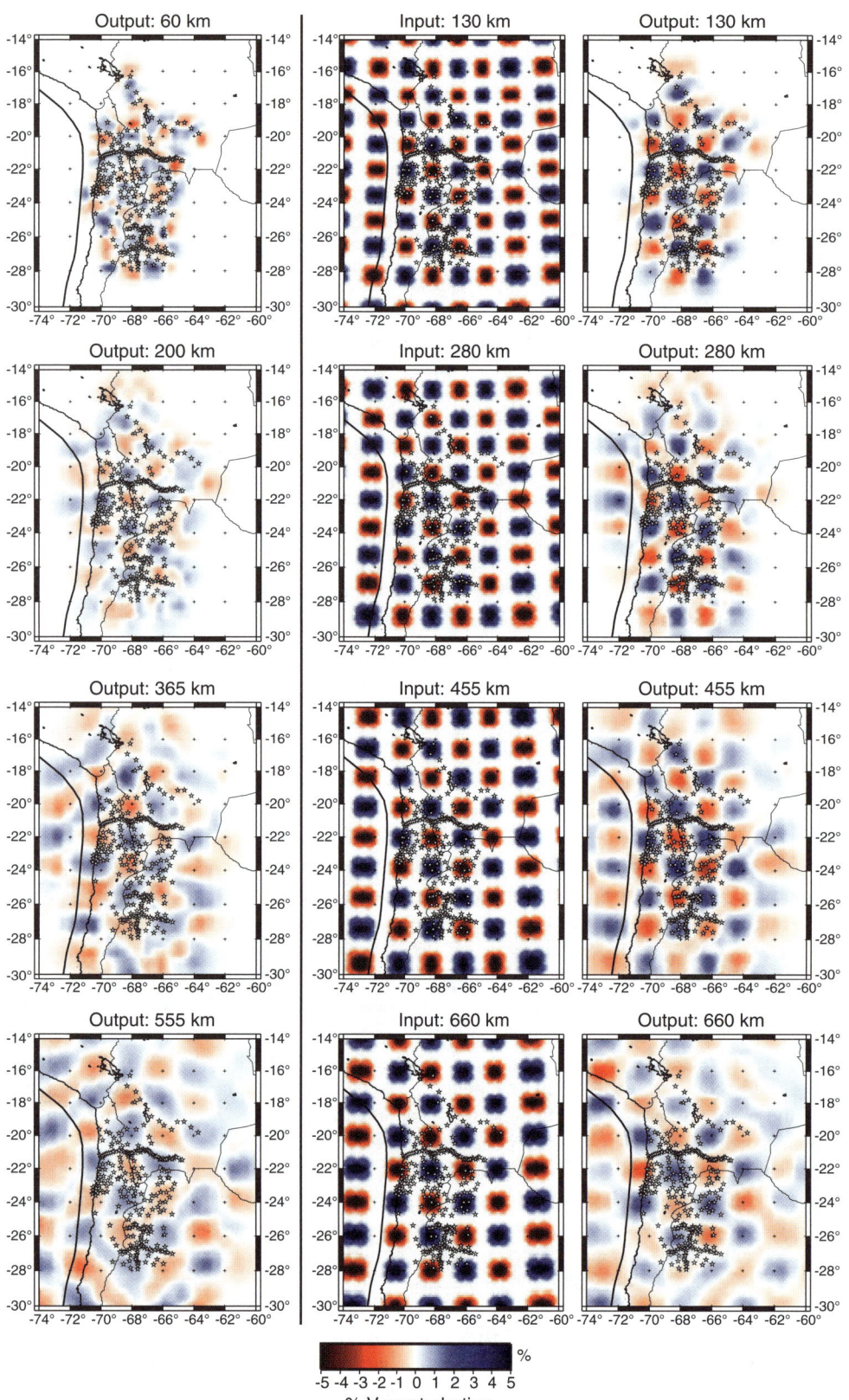

Figure 4. Horizontal depth slices for the checkerboard tests for every other model layer. Input for neutral layers (0% velocity deviation) is not shown. Output for neutral layers (60, 200, 365, 555 km depth) is shown in the left column. Resolution of velocity anomalies in neutral layers shown here indicates that some vertical smearing is occurring. The input and output for the layers with the checkerboard anomalies are shown in the middle and right columns, respectively. The checkerboard tests show that for shallower layers, resolution is controlled by station distribution, as expected. Deeper model layers indicate that while the input amplitude cannot be completely resolved, lateral changes in anomalous velocity resolve with little horizontal smearing. Resolution is lost toward the edges of the model region. Checkerboard tests for additional model layers and cross sections through the model are available in Figures DR2 and DR3 (see text footnote 1).

% Vp perturbation

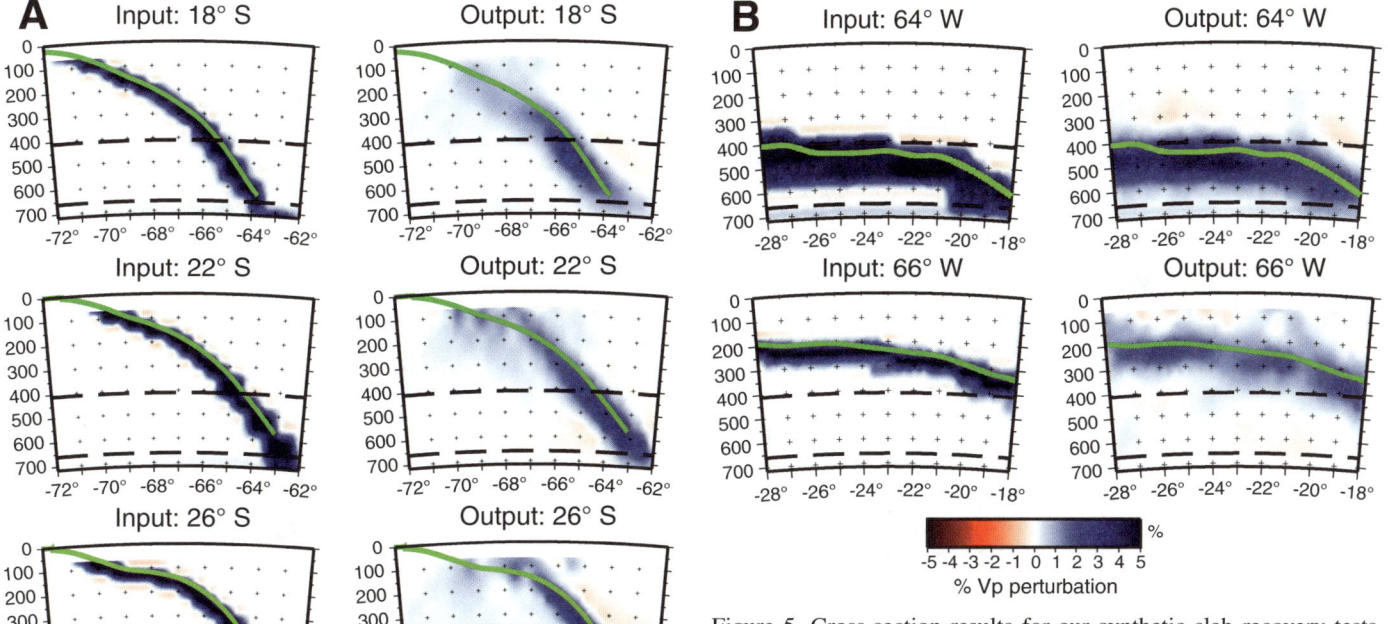

Figure 5. Cross-section results for our synthetic slab recovery tests. The geometry of our input slab model is based on the Slab1.0 contours (green line; Hayes et al., 2012). (A) E-W–oriented cross sections through the synthetic input (left) and output (right) model. Decreased amplitude recovery and vertical smearing of the recovered slab anomaly are observed in the upper 200 km of the model. In general, the amplitude recovery increases with depth. (B) N-S–oriented cross sections through the synthetic input (left) and output (right) model. The recovered slab anomaly at shallower depths (cross section at 66°W) is more diffuse than at deeper depths (cross section at 64°W).

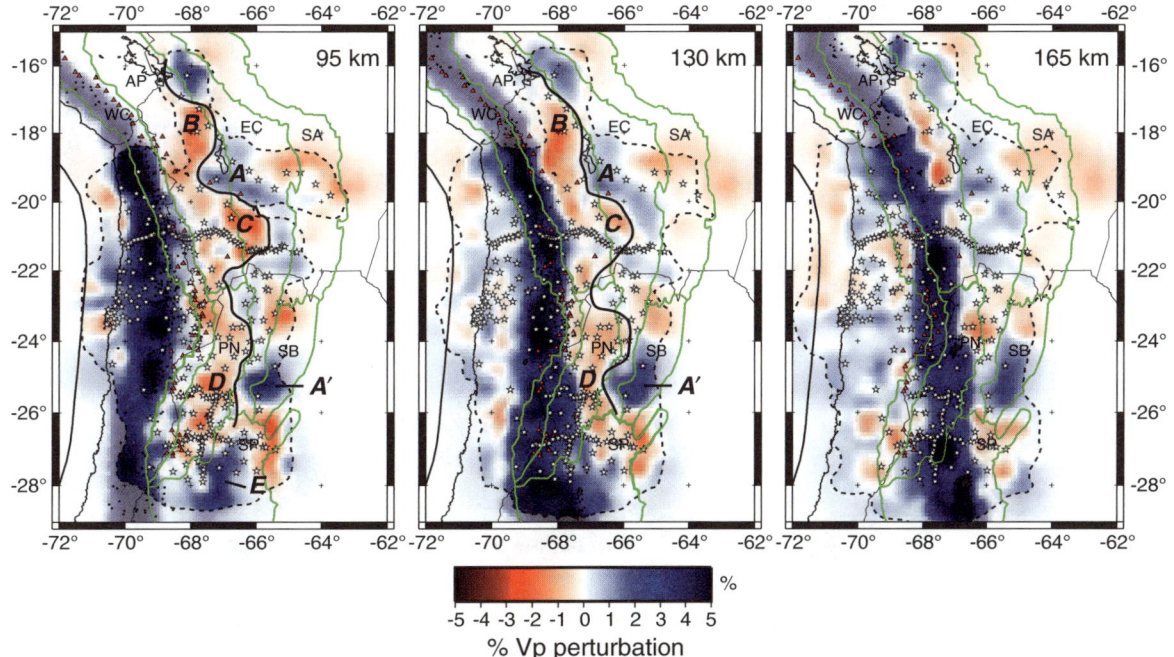

Figure 6. Horizontal depth slices for 95, 130, and 165 km from the tomography model using a priori information to constrain the slab as discussed in the text. The short dashed line represents the edge of the well-resolved region of the model, defined as regions with hit quality greater than 0.2 (Biryol et al., 2011). Geomorphic provinces (green lines) are the same as in Figure 1A. Stars mark locations of seismic stations used in the study. Red triangles mark locations of Holocene volcanoes (Siebert and Simkin, 2002). Black dots are earthquake locations from the EHB catalog (Engdahl et al., 1998). Solid black line marks edge of fast velocity anomaly (anomaly A and A′) interpreted as possible edge of cratonic lithosphere or delaminated lithosphere. Labeled anomalies (A, A′, B, C, D, and E) are discussed in the text. WC—Western Cordillera; AP—Altiplano; EC—Eastern Cordillera; SA—Subandean zone; PN—Puna; SB—Santa Barbara system; SP—Sierras Pampeanas. Additional depth slices (280–660 km depth) from this model are given in Figure DR4 (see text footnote 1).

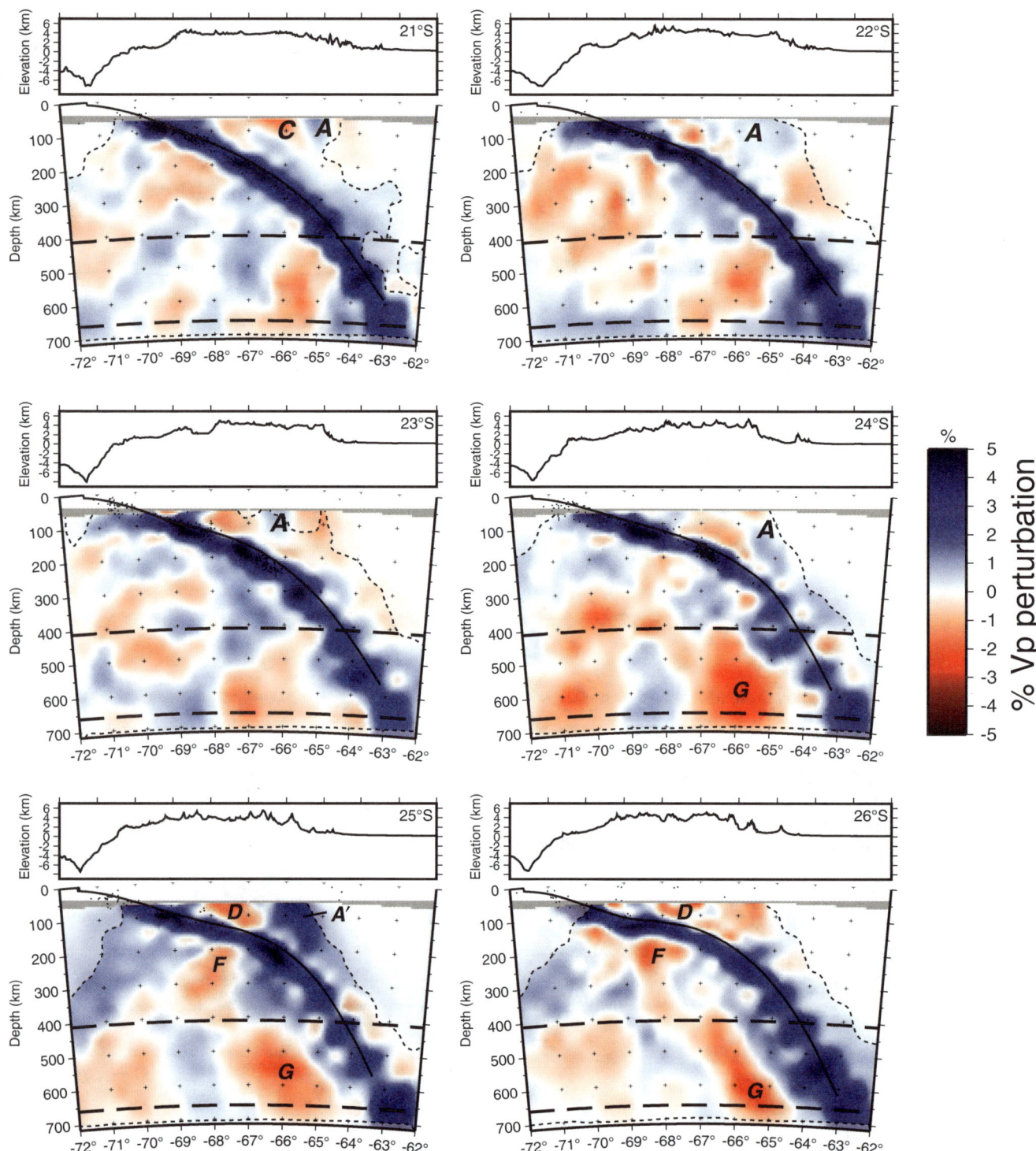

Figure 7. East-west–oriented cross sections through the tomography model shown in Figure 6. Location of the Nazca slab is constrained using a priori information as in Figure 6. Dashed lines are the same as in Figure 6. Black dots are earthquake locations from the EHB catalog (Engdahl et al., 1998). Solid black line marks the top of the Nazca slab from Slab1.0 model (Hayes et al., 2012). Labeled anomalies (A, A′, C, D, F, and G) are discussed in the text.

The results of a second tomographic inversion (Figs. 8 and 9), which does not use a priori data to constrain the location of the slab, are used to interpret anomalies below ~200 km, where the synthetic slab recovery tests (Fig. 5) indicate that vertical smearing of the slab is less dominant.

RESULTS

The final results from the teleseismic tomography inversions of the two models are displayed as horizontal depth slices (Figs. 6 and 8) and vertical cross sections (Figs. 7 and 9). The tomograms displayed in Figures 6 and 7 use the a priori slab information to constrain the location of the subducting slab and are used to interpret anomalies in the upper 200 km, while Figures 8 and 9 show tomograms that do not constrain the location of the slab and are used to interpret anomalies below 200 km. Additional depth slices from both models are displayed in Figures DR4 and DR5 (see footnote 1). Model resolution is best in the central part of the model and decreases toward the edges of the model space. As discussed previously, resolution in our model is reasonably good down to roughly 660 km depth. Below that depth, hit quality decreases, and anomalies are more poorly resolved. Resolution in the shallowest depth slices is strongly controlled by the location of the stations, as the ray paths become near vertical beneath the stations, and vertical smearing has a more pronounced effect and must be taken into account when interpreting anomalies in the shallowest parts of the model space. Horizontal resolution is good throughout the model, allowing us to interpret lateral differences in mantle structure with more confidence. We note that all major labeled anomalies discussed in the following sections (anomalies A through G) are present in both sets of inversions (Figs. 7, 8, 9, and 10), which attests to their robustness.

Imaging the Nazca Slab

The most prominent anomaly observed in our tomograms without incorporating any constraints on the location of the slab (Figs. 8 and 9) is the trench-parallel fast anomaly that appears to migrate east with depth, which is interpreted to be the subducting Nazca slab. The amplitude of this fast anomaly varies between +1% and +3%. The slab anomaly is discontinuous and has lower amplitudes at shallower depths but becomes more prominent below ~200 km. The fast anomaly that corresponds to the slab is discontinuous between 100 and 165 km between 24°S and 28°S (Fig. 9). Slow anomalies are resolved in place of the subducting slab, even though the synthetic slab resolution tests indicate that at least partial resolution of the slab should occur in this part of the model. Synthetic tests discussed previously (Fig. 5) indicate that the slab-shaped anomaly is being shifted vertically along the ray paths, resulting in a diffuse, low-amplitude, discontinuous slab anomaly in the shallowest part of the model. Similar difficulties in resolving the subducting slab were observed by Heit et al. (2008) for a teleseismic tomography study along a dense line of stations across the Andes at 21°S and were attributed to limitations in the ray path configuration. The vertical smearing of the Nazca slab anomaly along the ray paths complicates our ability to interpret anomalies in the uppermost parts of our model, hence our additional efforts to constrain the subducting slab in the upper mantle as shown in Figures 6 and 7.

In the deeper parts of the model, the Nazca slab anomaly is well resolved in our synthetic tests without needing to use a priori information to define the location of the subducting slab. For interpretations in the deeper part of our model, we use the inversion of the observed data without the imposition of an a priori slab (Figs. 8 and 9), allowing us to make observations about the shape of the subducting Nazca slab below 200 km.

Shallow Mantle (90–200 km)

The final results from the teleseismic tomography inversion using the a priori slab data to constrain the location of the Nazca slab are shown in Figures 6 and 7. Fast anomalies with amplitudes of between +1% and +2% are observed in the shallowest depth layers in the northern part of the study area (18°S to 22°S) under the eastern edge of the Altiplano and Eastern Cordillera (anomaly A and A'; Figs. 6 and 7). The fast anomaly is observed to extend under the Subandean zone at 20°S, but any determination of the extent of this fast anomaly under the Subandean zone is hindered by the lack of stations in the easternmost part of the study area to the north or south of this latitude. The fast anomalies continue under the Eastern Cordillera between 22°S and 26°S, although the amplitude of the anomalies is decreased, and the edges are less distinct (Fig. 7). An exception is the prominent, circular high-velocity feature located at 25°S between the Eastern Cordillera and the Santa Barbara system (anomaly A'; Figs. 6 and 7). This local anomaly remains strong to ~200 km depth, where it merges with the slab.

Two separate slow anomalies are observed in the northern part of the study area. A slow anomaly with amplitude of −2% is observed at 18°S under the Altiplano (anomaly B; Fig. 6). The lateral extent of this anomaly is unknown, since the limited distribution of stations in this part of the Andes results in a narrow region which is well resolved. A second, higher-amplitude (−2.5% to −3%) slow anomaly is observed at 21°S between 66°W and 67°W (anomaly C; Fig. 6). This anomaly is limited to the east by the presence of the fast anomalies discussed earlier and has a distinct western boundary at ~67°W, where the amplitude of the resolved anomaly decreases abruptly.

A distinct slow anomaly is observed at shallow depths under the Puna Plateau between ~24.5°S and 27.5°S (anomaly D; Figs. 6 and 7). While the amplitude of this anomaly varies between −1.0% and −1.5%, its lateral extent is well defined by the fast anomalies observed to the south and east. The western edge of this slow anomaly is defined by the location of the subducting slab. To the southeast of the Puna Plateau, fast velocity anomalies are observed with amplitudes between +1% and +2% (anomaly E; Fig. 6). The northwestern edges of these

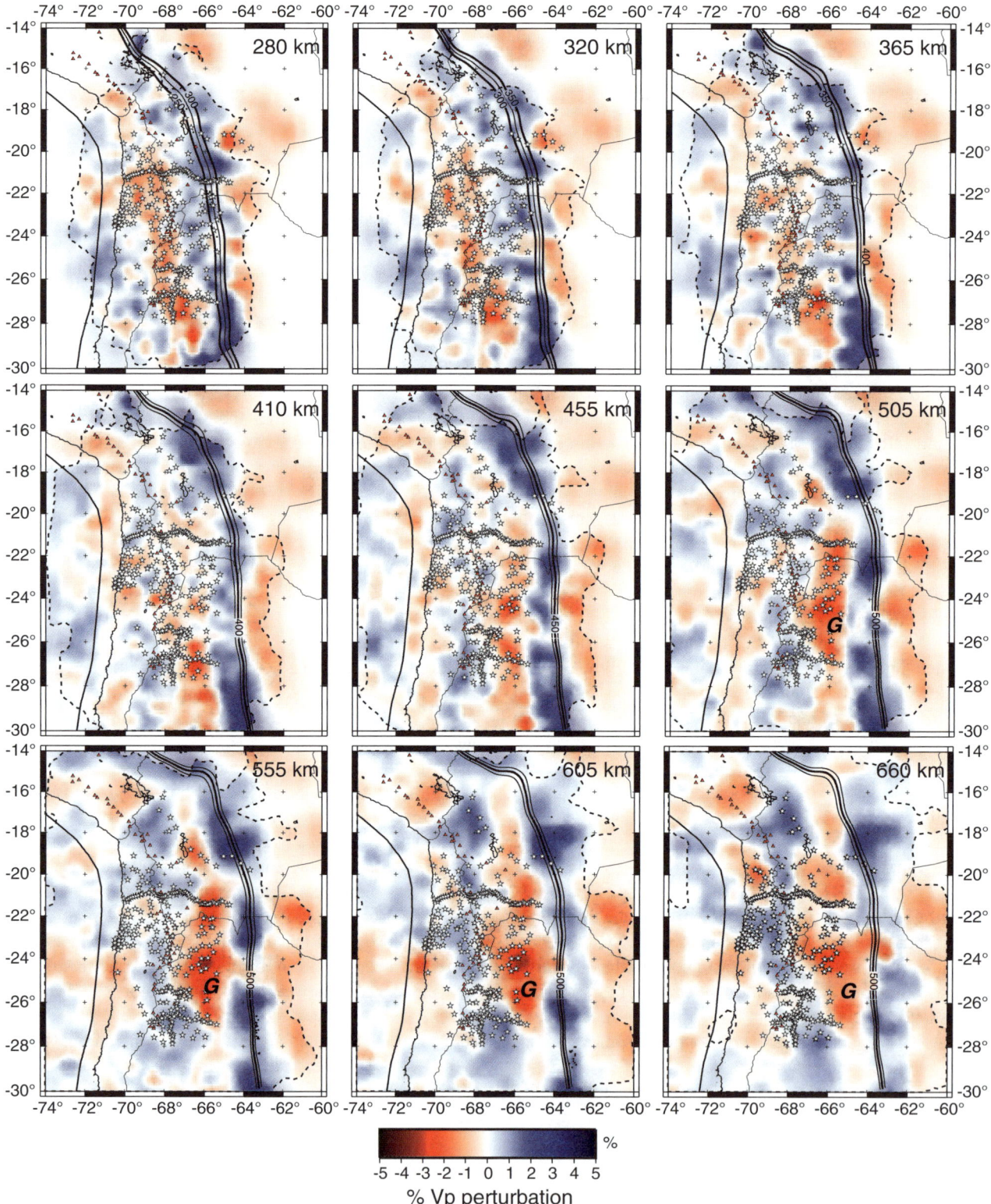

Figure 8. Horizontal depth slices from 280 to 660 km from the tomography model without a priori constraints on the location of the subducting slab. The short dashed line represents the edge of the well-resolved region of the model, defined as regions with hit quality greater than 0.2 (Biryol et al., 2011). Stars mark locations of seismic stations used in the study. Red triangles mark locations of Holocene volcanoes (Siebert and Simkin, 2002). Black dots are earthquake locations from the EHB catalog (Engdahl et al., 1998). Solid black lines are slab contours from Slab1.0 model (Hayes et al., 2012). Labeled anomaly G is discussed in the text. Additional depth slices (95–165 km depth) from this model are given in Figure DR5 (see text footnote 1).

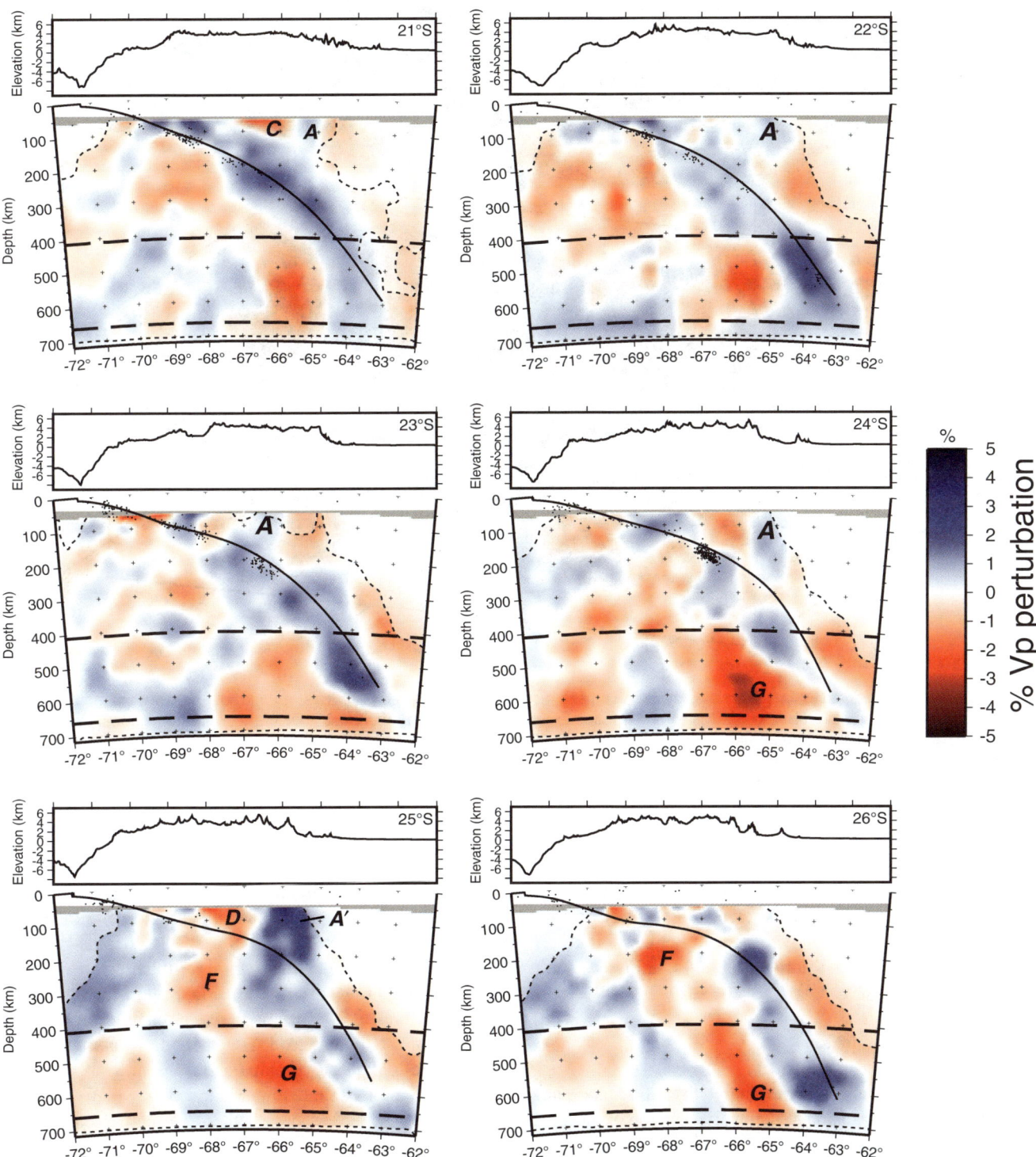

Figure 9. East-west–oriented cross sections through the tomography model without a priori constraints on the location of the subducting slab as in Figure 8. Dashed lines are the same as in Figure 8. Black dots are earthquake locations from the EHB catalog (Engdahl et al., 1998). Solid black line marks the top of the Nazca slab from the Slab1.0 model (Hayes et al., 2012). Labeled anomalies (A, A′, C, D, F, and G) are discussed in the text.

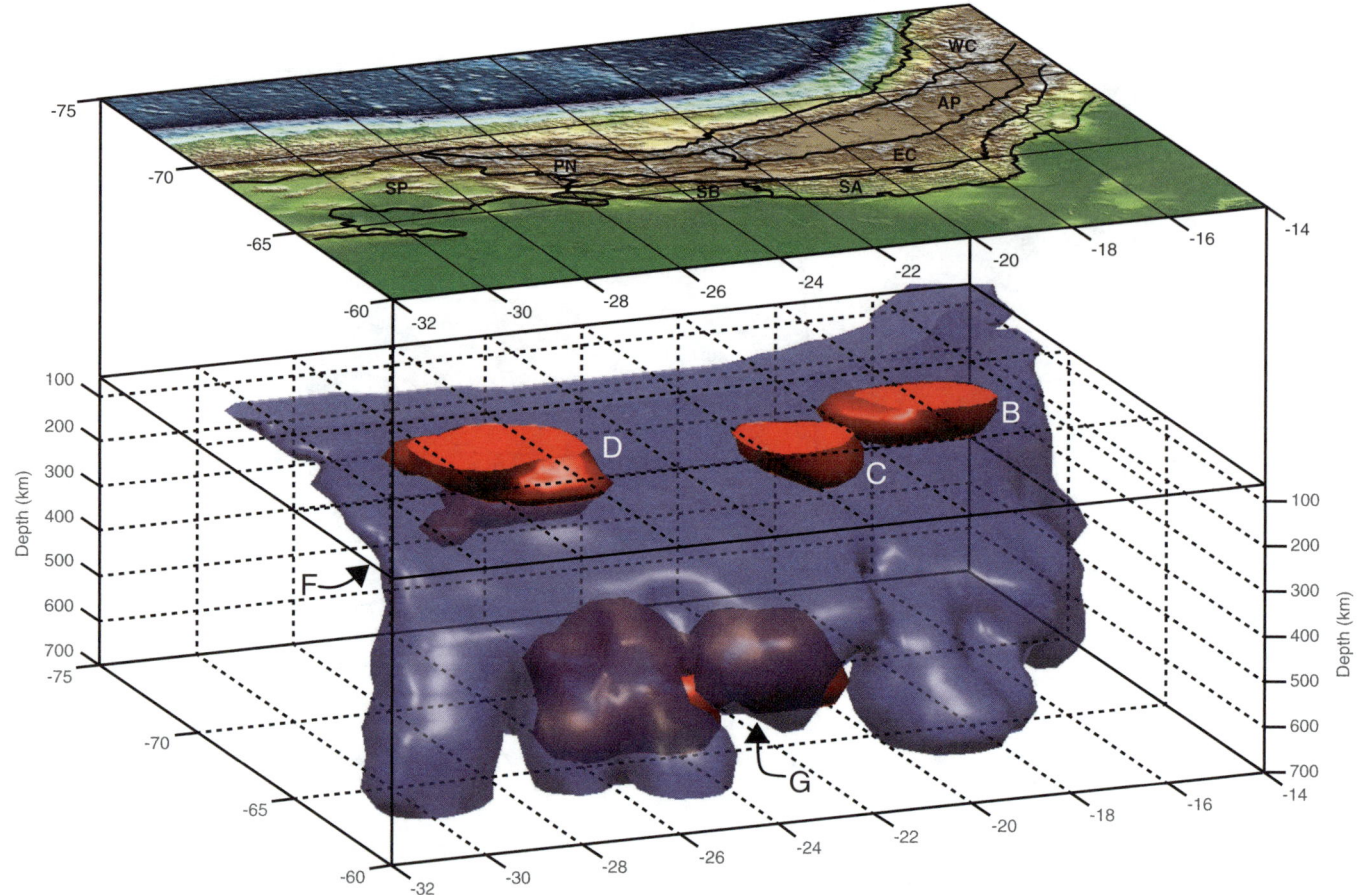

Figure 10. Three-dimensional (3-D) diagram of the resolved subducting Nazca slab and prominent mantle low-velocity anomalies inferred from our tomographic models. The isosurfaces for this model are obtained by tracing the most coherent low-velocity anomalies (less than negative 3%) and slab-related (greater than positive 3%), coherent fast anomalies in the tomographic model. Slab geometry above 200 km is determined entirely from the Slab1.0 model (Hayes et al., 2012). Anomalies B, C, D, F, and G are labeled as in previous figures. Boundaries of geomorphic provinces are as in Figure 1. WC—Western Cordillera; AP—Altiplano; EC—Eastern Cordillera; SA—Subandean zone; PN—Puna; SB—Santa Barbara system; SP—Sierras Pampeanas.

fast anomalies are distinct, although the eastern and southern extents of these anomalies are unknown, as they fall outside of our study area. A second low-velocity anomaly (anomaly F; Fig. 7) is observed below the slab under the Puna Plateau with amplitudes of −1% to −2%.

Deeper Mantle (~280–660 km)

Results for the teleseismic tomography inversion that do not use the a priori slab information to define the location of the subducting slab are shown in Figures 8 and 9. Since synthetic slab recovery tests (Fig. 5) indicate that the slab anomaly should be well resolved in the deeper parts of the model without the extensive vertical smearing effect noted in the shallow model layers, we can use the inversion of the observed data without constraining the location of the Nazca slab to make observations about the shape of the slab in the deeper parts of the model. The slab anomaly is resolved down to 660 km, with along-strike varia-

tions in amplitude and width being observed in the tomograms below 300 km depth (Fig. 8). A distinct region of thinning of the slab anomaly is observed at 24°S to 25°S at depths greater than 300 km (Figs. 8 and 9). Other regions of thinning are observed in the mantle transition zone at 20°S (Fig. 8). The fast slab anomaly appears to thicken in other parts of the mantle transition zone, particularly at 18°S and 26°S. Together, these anomalies give the slab a laterally periodic thinning and thickening appearance that is especially noticeable below 500 km (Fig. 8). We do not have the resolution to be certain whether this segmented appearance is the result of vertical tears or stress-induced deformation as the slab penetrates the mantle transition zone.

A high-amplitude (−3%) slow anomaly is observed below the subducting Nazca slab in the mantle transition zone (410–660 km depth) between 22°S and 28°S and is especially prominent south of 24°S (anomaly G; Figs. 8 and 9). Although other velocity anomalies are observed under the subducting Nazca slab, the amplitude of this anomaly makes it distinctive.

DISCUSSION

Shallow Mantle (90–200 km)

Figures 6 and 7 show results for our model where we incorporate the slab in order to better constrain the upper 200 km of the model. In general, we observe relatively slow velocities under the Altiplano and Puna Plateau at depths of 95 and 130 km and faster velocities to the east under parts of the Eastern Cordillera and Subandean zone (Fig. 6). We interpret the lateral change from slow to fast P-wave velocities as the edge of the continental craton underthrusting from the east. The presence of fast anomalies under the Eastern Cordillera (anomaly A; Fig. 6) suggests that the craton extends under parts of the Eastern Cordillera and is consistent with results from previous studies (Beck and Zandt, 2002; Dorbath et al., 1993). If correct, this suggests that the western extent of underthrusting of the craton is limited mostly to the Eastern Cordillera and does not underthrust the entire high plateau, as is suggested further north in southern Peru by Phillips et al. (2012). Another possible interpretation of the high-velocity anomalies under the Eastern Cordillera is that it represents cold downwellings of foundering lithospheric material. The cylindrical-shaped anomaly at 25°S (anomaly A′; Figs. 6 and 7) is especially suggestive of delaminating lithosphere and has been interpreted as such by Beck et al. (this volume). We observe a low-amplitude slow anomaly at 21°S that partially overlaps and continues to the south of the Los Frailes volcanic field in the Eastern Cordillera (anomaly C; Fig. 6). Heit et al. (2008) noted a low-velocity anomaly in the same location, although their anomaly was more laterally extensive. A similar low-velocity anomaly was observed by Myers et al. (1998) with regional P-wave tomography slightly north of the location of our anomaly and also correlated with the Los Frailes volcanic field. A fast P-wave anomaly under the central Altiplano (above ~130 km) in the northern part of the study area (north of 21°S) was observed previously by Myers et al. (1998) and interpreted as intact mantle lithosphere under this part of the Altiplano. This anomaly is not obvious in our results, probably due to the difference in resolution between our study and the local tomography study of Myers et al. (1998).

In general, we do not observe a prominent low-velocity anomaly under the Altiplano-Puna volcanic complex near 23°S. The anomalies observed under the Altiplano-Puna volcanic complex are low-amplitude, slow anomalies, which are not nearly as distinct as those observed under the Los Frailes and the Puna volcanic fields (anomalies C and D; Fig. 6). This is consistent with results from other studies (as discussed in Sobolev et al., 2006), which noted that the dominant low-velocity anomaly in the region under the Altiplano-Puna volcanic complex is in the crust rather than the upper mantle, unlike both the Los Frailes and the Puna volcanic fields.

In general, we observe low velocities at depths of 95 and 130 km beneath the Puna Plateau (anomaly D; Figs. 6 and 7). Slow P-wave velocities in the upper mantle beneath the southern

Puna Plateau have been observed in other seismological studies (e.g., Bianchi et al., 2013; Wölbern et al., 2009; Schurr et al., 2006; Koulakov et al., 2006) and interpreted as evidence for recent lithospheric removal. The local P-wave tomography studies show more heterogeneities in the uppermost mantle. Evidence for lithospheric thinning has been seen in shear wave attenuation studies (Whitman et al., 1992) and receiver function studies (Heit et al., 2007), and influx of asthenosphere to the base of the crust following delamination of the continental lithosphere has been proposed to explain both the geophysical observations and the geochemical characteristics of the large ignimbrite deposits in the southern Puna region (Kay and Kay, 1993; Kay et al., 1994; Kay and Coira, 2009). The location of the −1% to −2% velocity anomaly observed between 24°S and 28°S under the southern Puna Plateau in our results is consistent with asthenosphere in the upper mantle under the Puna Plateau. Local tomography studies (e.g., Bianchi et al., 2013; Schurr et al., 2006; Koulakov et al., 2006) have observed small fast anomalies in the upper mantle beneath the Puna Plateau, which have been interpreted as delaminated blocks of lithosphere. In general, our results are neither as detailed nor do they show as much small-scale heterogeneity as the local P-wave traveltime tomography models due to the regional scale of our study, and therefore we cannot robustly resolve the presence of such small-scale features. However, anomaly A′ (Fig. 6), which was not clearly resolved in previous studies, could represent the larger-scale delaminated material that was replaced by the influx of asthenosphere.

In the shallowest depth layer (95 km), fast velocity anomalies are observed to the southeast of the Puna Plateau (anomaly E; Fig. 6). The edges of this anomaly are distinct and correspond to the surface boundary of the Sierras Pampeanas. This corresponds to the surface location of the Precambrian Pampean terrane proposed by Ramos et al. (1986), which is thought to comprise the crystalline basement of the Sierras Pampeanas. Therefore, we could be imaging the edge of the thicker mantle lithosphere of the Pampean terrane. A fast anomaly in this area was also observed by Bianchi et al. (2013) and interpreted similarly. This is also the start of the transition from a normal dipping slab to a near-horizontal or flat slab. Hence, the fast velocities at a depth of 130 km are likely beginning to sample the downgoing slab as it flattens to the south. The slow anomaly under the slab (anomaly F; Fig. 7) is similar to an anomaly imaged by Bianchi et al. (2013), who interpreted it as a subslab region of hot asthenospheric mantle. However, this anomaly has a limited along-strike extent.

Deeper Mantle (~280–660 km)

Limited regional-scale imaging of the Nazca slab between 250 and 700 km depth had been done prior to this study. Engdahl et al. (1995) used teleseismic traveltime tomography to image the Nazca slab down to depths of 1400 km, concluding that the slab penetrates into the lower mantle after a period of temporary stagnation in the transition zone, where the slab thickens before resuming subduction into the lower mantle.

Local tomography studies in this region use earthquakes in the slab and usually only image the upper 150–200 km above the slab and often do not image the slab itself (e.g., Schurr et al., 2006; Koulakov et al., 2006).

The subducting Nazca slab is clearly imaged in our model between 280 and 660 km depth (Figs. 8 and 9). A few clusters of deep earthquakes between 500 and 600 km depth are observed in our study area. These deep earthquakes are located within the observed slab anomaly rather than toward the edges, which agrees with the idea that deep earthquakes occur within the cold cores of subducting slabs (Kirby et al., 1996). The amplitude of the observed deep slab in our tomograms varies with both latitude and depth. Some of the variation in amplitude of the deep slab is probably due to ray path distribution, as similar amplitude variations are observed in the synthetic slab recovery tests. Variations in slab thickness are also observed in our tomograms. However, similar variations in slab thickness are not reproduced in the output of the synthetic slab recovery tests (Fig. 5). This suggests that the observed variability in the thickness of the slab in our resulting model is not being controlled by ray path distribution or the geometry of our model, and thus probably represents real variations in the subducting slab. The subducting Nazca slab shows distinct variations along strike (Fig. 8), with localized zones of both thickening and thinning in the mantle transition zone.

Since we do not have resolution below the 660 km discontinuity, we cannot rule out the possibility that the slab stagnates in the transition zone east of our image solely based on our study. However, global tomography models show the Nazca slab penetrating into the lower mantle (e.g., Bijwaard et al., 1998; Li et al., 2008; Zhao, 2004; Fukao et al., 2001, 2009; Zhao et al., 2013). Our results show localized thickening of the Nazca slab in the mantle transition zone between 16°S and 18°S and south of 25°S (Figs. 8 and 9). Similar thickening of subducting slabs during penetration into the lower mantle has been observed in global models (e.g., Bijwaard et al., 1998; Li et al., 2008) and has been interpreted as evidence for temporary stagnation of a subducting slab in the mantle transition zone before the slab subducts into the lower mantle. Our images of the deeply subducted Nazca slab also show evidence of varying degrees of thinning in several places, possibly resulting from a changing stress state along strike as the slab deforms in the mantle transition zone before resuming subduction into the lower mantle (Fig. 10). At these depths, the resolution of the model is very uniform and does not show any lateral variations that could be perceived as lateral variations in slab thickness, indicating the robustness of these slab thickness variations. The most extreme areas of thinning occur at 20°S and 24°S and appear to correlate with along-strike changes in seismicity between 150 and 300 km depth (Fig. 1B). Assuming the National Earthquake Information Center (NEIC) earthquake catalogue is uniform at the magnitude 4 level, there is a distinct decrease in intermediate-depth seismicity south of ~24°S (Fig. 1B). While it is difficult to directly relate changes in the stress state of the slab in the mantle transition zone with changes in seismicity patterns between 150 and 300 km depth, it is pos-

sible that these regions of thinning represent tears in the subducting slab that separate different stress regimes, which could affect the pattern of intermediate-depth seismicity.

The high-amplitude, low-velocity anomaly below the slab between 22°S and 28°S in the mantle transition zone is well resolved in our model (anomaly G; Figs. 8 and 9). This anomaly can be explained either by a thermal anomaly or by volatiles in the mantle transition zone. An electrical conductivity study further to the south near 33°S also indicates the presence of a potentially water-rich or hot anomaly in the subslab mantle from transition zone depths (Burd et al., 2013). A thermal anomaly would locally increase temperature and therefore lower the seismic velocity. Similarly, the presence of increased water in the mantle transition zone could also lower the seismic velocity. From the seismic velocities alone, we cannot distinguish between these two causes. A similar low-velocity anomaly has been observed in the subslab mantle in other subduction zones (e.g., Obayashi et al., 2006; Zhao, 2004; Zhao et al., 2013). A low-velocity zone directly above the 410 km discontinuity under the subducting Pacific slab in the Honshu subduction zone has been interpreted as a local thermal anomaly (e.g., Obayashi et al., 2006; Bagley et al., 2009; Zhao, 2004; Zhao et al., 2013). Bagley et al. (2009) and Obayashi et al. (2006) calculated that the observed low-velocity anomalies can be explained by approximate temperature anomalies of 150 °C or 200 °C, respectively. The exact origin of these thermal anomalies is uncertain, although most studies suggest that they are associated with upwellings of hot lower-mantle material, either as a result of local-scale convection induced by the subduction of the nearby slab or as a result of the presence of small-scale plumes (Zhao, 2004; Bagley et al., 2009; Morishige et al., 2010). Wölbern et al. (2009) imaged a depressed 410 km discontinuity and an elevated 660 km discontinuity just to the west of our low-velocity anomaly at 67.5°W to 68°W. They argued that this thinning of the mantle transition zone is consistent with a local thermal anomaly in the subslab mantle but were unable to constrain the extent of the anomaly due to poor ray coverage at that depth.

It is also possible that water is the cause of the subslab anomaly observed in our tomograms. Schmerr and Garnero (2007) observed a depressed 410 km discontinuity east of the Nazca slab using SS phases and argued that it cannot be explained solely by a thermal origin. Several studies suggest that subduction zones can carry water into the mantle transition zone (e.g., Bercovici and Karato, 2003; Smyth and Jacobsen, 2006). The hydrated material is chemically buoyant and should remain on top of the mantle transition zone, potentially melting. A similar process was theorized by Bercovici and Karato (2003) as part of their transition zone water filter model of mantle convection, which argues for the hydration of the mantle transition zone. As the trench migrates west and the Nazca slab pushes the subslab mantle west, it will disrupt the 410 km discontinuity and could drag hydrated mantle material that has accumulated near the 410 km discontinuity into the mantle transition zone. A region of hydrated material in the mantle transition zone formed in this way could explain our low-velocity zone. The depressed 410 discontinuity observed by Schmerr and Garnero (2007) is much broader in the east-west direction than the width of the slab, suggesting that there is hydrated material both above and below the slab. In addition, the 410 discontinuity is significantly disrupted beneath the slab (Schmerr and Garnero, 2007). Hence, it is possible that our low-velocity anomaly is consistent with some component of hydrated material in the mantle transition zone beneath the slab.

CONCLUSION

Until recently, limited regional-scale seismic imaging has been done in the central Andes. Many studies of the central Andes have either looked at a small region under a single seismic network, limiting their maximum depth of resolution, or imaged the region as part of a global study, limiting their ability to resolve smaller-scale variations. Our regional-scale P-wave tomographic images of the central Andes show significant along-strike variation in upper-mantle structure. Fast anomalies under the Eastern Cordillera between 16°S and 26°S provide evidence for underthrusting of the Brazilian cratonic lithosphere under the Subandean zone and Eastern Cordillera along much of the central Andes or possibly a "curtain" of delaminating lithospheric material. Slow velocity anomalies underneath the Puna Plateau are related to the presence of partial melt in the shallow mantle, while fast velocity anomalies to the southeast correspond to the mantle lithosphere of the Precambrian Pampean terrane.

Detailed imaging of the structure of the Nazca slab in the deeper parts of the upper mantle had not been done in this part of the central Andes prior to this study. We have imaged previously unseen structure in the subducting Nazca slab at the base of the upper mantle and in the transition zone. Variations in the width of the slab anomaly in the deeper parts of the model show evidence for deformation of the slab between 300 and 660 km (Fig. 10). Thickening of the slab in the mantle transition zone is observed in several places, in agreement with the idea that the Nazca slab stagnates at least temporarily in the transition zone before resuming subduction into the lower mantle. A subslab low-velocity anomaly in the mantle transition zone between 22°S and 28°S is similar to those seen in other subduction zones and is interpreted as either a local thermal anomaly or a region of hydrated material in the mantle transition zone.

ACKNOWLEDGMENTS

Support for this research was provided by ExxonMobil as part of the COSA (Convergent Orogenic Systems Analysis) project. Additional support was provided by the National Science Foundation award EAR-0907880. We would like to acknowledge the GEOFON Program at GeoForschungsZentrum (GFZ) Potsdam as a source of much of the waveform data used in this study. Data were also obtained from the Incorporated Research Institutions for Seismology (IRIS) Data Management Center. The Global Seismographic Network is a cooperative scientific facility operated jointly by the Incorporated Research Institutions for Seismology (IRIS), the U.S. Geological Survey (USGS), and the National Science Foundation (NSF). We would like to thank the GEOFON Program at GFZ Potsdam and X. Yuan for access to the restricted data from the PUDEL network. We thank Brandon Schmandt, now at the University of New Mexico, and Gene Humphreys, of the University of Oregon, as the original authors of the teleseismic tomography code used in this study. Chevron-Texaco, Conoco-Phillips, and ExxonMobil provided additional support for Alissa Scire. We would also like to thank Ben Heit and Brandon Schmandt for their constructive and comprehensive reviews, which helped improve this manuscript.

REFERENCES CITED

Aki, K., Christoffersson, A., and Huseby, E.S., 1977, Determination of the three-dimensional seismic structure of the lithosphere: Journal of Geophysical Research, v. 82, no. 2, p. 277–296, doi:10.1029/JB082i002p00277.

Allmendinger, R.W., Jordan, T.E., Kay, S.M., and Isacks, B.L., 1997, The evolution of the Altiplano-Puna Plateau of the central Andes: Annual Review of Earth and Planetary Sciences, v. 25, p. 139–174, doi:10.1146/annurev.earth.25.1.139.

ANCORP Working Group, 1999, Seismic reflection image revealing offset of Andean subduction-zone earthquake locations into ocean mantle: Nature, v. 397, p. 341–344, doi:10.1038/16909.

Asch, G., Schurr, B., Bohm, M., Yuan, X., Haberland, C., Heit, B., Kind, R., Woelborn, I., Bataille, K., Comte, D., Pardo, M., Viramonte, J., Rietbrock, A., and Giese, P., 2006, Seismological studies of the central and southern Andes, *in* Oncken, O., Chong, G., Franz, G., Giese, P., Götze, H.J., Ramos, V.A., Strecker, M.R., and Wigger, P., eds., The Andes—Active Subduction Orogeny: Berlin, Springer-Verlag, Frontiers in Earth Sciences, v. 1, p. 443–457.

Bagley, B., Courtier, A.M., and Revenaugh, J., 2009, Melting in the deep upper mantle oceanward of the Honshu slab: Physics of the Earth and Planetary Interiors, v. 175, p. 137–144, doi:10.1016/j.pepi.2009.03.007.

Beck, S.L., and Zandt, G., 2002, The nature of orogenic crust in the central Andes: Journal of Geophysical Research, v. 107, no. B10, p. 2230, doi:10.1029/2000JB000124.

Beck, S.L., Zandt, G., Myers, S.C., Wallace, T.C., Silver, P.G., and Drake, L., 1996, Crustal-thickness variations in the central Andes: Geology, v. 24, p. 407–410, doi:10.1130/0091-7613(1996)024<0407:CTVITC>2.3.CO;2.

Beck, S.L., Zandt, G., Ward, K.M., and Scire, A., 2015, this volume, Multiple styles and scales of lithospheric foundering beneath the Puna plateau, Central Andes, *in* DeCelles, P.G., Ducea, M.N., Carrapa, B., and Kapp, P.A., eds., Geodynamics of a Cordilleran Orogenic System: The Central Andes of Argentina and Northern Chile: Geological Society of America Memoir 212, doi:10.1130/2015.1212(03).

Bercovici, D., and Karato, S., 2003, Whole-mantle convection and the transition-zone water filter: Nature, v. 425, p. 39–44, doi:10.1038/nature01918.

Bianchi, M., Heit, B., Jakovlev, A., Yuan, X., Kay, S.M., Sandvol, E., Alonso, R.N., Coira, B., Brown, L., Kind, R., and Comte, D., 2013, Teleseismic tomography of the southern Puna Plateau in Argentina and adjacent regions: Tectonophysics, v. 586, p. 65–83, doi:10.1016/j.tecto.2012.11.016.

Bijwaard, H., Spakman, W., and Engdahl, E.R., 1998, Closing the gap between regional and global travel time tomography: Journal of Geophysical Research, v. 103, no. B12, p. 30,055–30,078, doi:10.1029/98JB02467.

Biryol, C.B., Beck, S.L., Zandt, G., and Özacar, A.A., 2011, Segmented African lithosphere beneath the Anatolian region inferred from teleseismic P-wave tomography: Geophysical Journal International, v. 184, p. 1037–1057, doi:10.1111/j.1365-246X.2010.04910.x.

Burd, A.I., Booker, J.R., Mackie, R., Pomposiello, C., and Favetto, A., 2013, Electrical conductivity of the Pampean shallow subduction region of Argentina near 33°S: Evidence for a slab window: Geochemistry Geophysics Geosystems, v. 14, no. 8, doi:10.1002/ggge.20213.

Butler, R.F., Lay, T., Creager, K.C., Earl, P., Fischer, K.M., Gaherty, J.B., Laske, G., Leith, B., Park, J., Ritzwoller, M.H., Tromp, J., and Wen, L.,

2004, The Global Seismographic Network surpasses its design goal: Eos (Transactions, American Geophysical Union), v. 85, no. 23, p. 225–229, doi:10.1029/2004EO230001.

Cahill, T., and Isacks, B.L., 1992, Seismicity and shape of the subducted Nazca plate: Journal of Geophysical Research, v. 97, no. B12, p. 17,503–17,529, doi:10.1029/92JB00493.

Chmielowski, J., Zandt, G., and Haberland, C., 1999, The central Andean Altiplano-Puna magma body: Geophysical Research Letters, v. 26, no. 6, p. 783–786, doi:10.1029/1999GL900078.

Dahlen, F.A., Hung, S.-H., and Nolet, G., 2000, Fréchet kernels for finite-frequency traveltimes: 1. Theory: Geophysical Journal International, v. 141, p. 157–174.

De Silva, S., Zandt, G., Trumbull, R., Viramonte, J.G., Salas, G., and Jimenez, N., 2006, Large ignimbrite eruptions and volcano-tectonic depressions in the central Andes: A thermomechanical perspective, *in* Troise, C., De Natale, G., and Kilburn, C.R.J., eds., Mechanisms of Activity and Unrest at Large Calderas: Geological Society, London, Special Publication 269, p. 47–63, doi:10.1144/GSL.SP.2006.269.01.04.

Dorbath, C., and Granet, M., 1996, Local earthquake tomography of the Altiplano and the Eastern Cordillera of northern Bolivia: Tectonophysics, v. 259, p. 117–136, doi:10.1016/0040-1951(95)00052-6.

Dorbath, C., and Masson, F., 2000, Composition of the crust and upper-mantle in the central Andes (19°30′S) inferred from P wave velocity and Poisson's ratio: Tectonophysics, v. 327, p. 213–223, doi:10.1016/S0040-1951(00)00170-0.

Dorbath, C., Granet, M., Poupinet, G., and Martinez, C., 1993, A teleseismic study of the Altiplano and the Eastern Cordillera in northern Bolivia: New constraints on a lithospheric model: Journal of Geophysical Research, v. 98, no. B6, p. 9825–9844, doi:10.1029/92JB02406.

Dorbath, L., Dorbath, C., Jimenez, E., and Rivera, L., 1991, Seismicity and tectonic deformation in the Eastern Cordillera and the sub-Andean zone of central Peru: Journal of South American Earth Sciences, v. 4, p. 13–24, doi:10.1016/0895-9811(91)90015-D.

Elger, K., Oncken, O., and Glodny, J., 2005, Plateau-style accumulation of deformation: Southern Altiplano: Tectonics, v. 24, TC4020, doi:10.1029/2004TC001675.

Engdahl, E.R., van der Hilst, R.D., and Berrocal, J., 1995, Imaging of subducted lithosphere beneath South America: Geophysical Research Letters, v. 22, no. 16, p. 2317–2320, doi:10.1029/95GL02013.

Engdahl, E.R., van der Hilst, R.D., and Buland, R., 1998, Global teleseismic earthquake relocation with improved travel times and procedures for depth determination: Bulletin of the Seismological Society of America, v. 88, no. 3, p. 722–743.

Fukao, Y., Widiyantoro, S., and Obayashi, M., 2001, Stagnant slabs in the upper and lower mantle transition region: Reviews of Geophysics, v. 39, p. 291–323, doi:10.1029/1999RG000068.

Fukao, Y., Obayashi, M., Nakakuki, T., and the Deep Slab Project Group, 2009, Stagnant slab: A review: Annual Review of Earth and Planetary Sciences, v. 37, p. 19–46, doi:10.1146/annurev.earth.36.031207.124224.

Garzione, C., Hoke, G., Libarkin, J., Withers, S., McFadden, B., Eiler, J., Ghosh, P., and Mulch, A., 2008, Rise of the Andes: Science, v. 320, p. 1304–1307, doi:10.1126/science.1148615.

GeoForschungsZentrum (GFZ), 2013, GFZ Seismological Data Archive: http://geofon.gfz-potsdam.de/waveform/ (accessed October 2013).

Ghosh, P., Garzione, C., and Eiler, J., 2006, Rapid uplift of the Altiplano revealed through ^{13}C-^{18}O bonds in paleosol carbonates: Science, v. 311, p. 511–515, doi:10.1126/science.1119365.

Graeber, F.M., and Asch, G., 1999, Three-dimensional models of P wave velocity and P-to-S velocity ratio in the southern central Andes by simultaneous inversion of local earthquake data: Journal of Geophysical Research, v. 104, no. B9, p. 20,237–20,256, doi:10.1029/1999JB900037.

Hayes, G.P., Wald, D.J., and Johnson, R.J., 2012, Slab1.0: A three-dimensional model of global subduction zone geometries: Journal of Geophysical Research, v. 117, B01302, doi:10.1029/2011JB008524.

Heit, B., Sodoudi, F., Yuan, X., Bianchi, M., and Kind, R., 2007, An S receiver function analysis of the lithospheric structure in South America: Geophysical Research Letters, v. 34, L14307, doi:10.1029/2007GL030317.

Heit, B., Koulakov, I., Asch, G., Yuan, X., Kind, R., Alcocer-Rodriguez, I., Tawackoli, S., and Wilke, H., 2008, More constraints to determine the seismic structure beneath the central Andes at 21°S using teleseismic tomography analysis: Journal of South American Earth Sciences, v. 25, p. 22–36, doi:10.1016/j.jsames.2007.08.009.

Hung, S.-H., Dahlen, F.A., and Nolet, G., 2000, Fréchet kernels for finite-frequency traveltimes: II. Examples: Geophysical Journal International, v. 141, p. 175–203.

Isacks, B.L., 1988, Uplift of the central Andean Plateau and bending of the Bolivian orocline: Journal of Geophysical Research, v. 93, no. B4, p. 3211–3231, doi:10.1029/JB093iB04p03211.

Kay, R.W., and Kay, S.M., 1993, Delamination and delamination magmatism: Tectonophysics, v. 219, p. 177–189, doi:10.1016/0040-1951(93)90295-U.

Kay, S.M., and Coira, B., 2009, Shallowing and steepening subduction zones, continental lithosphere loss, magmatism and crustal flow under the central Andean Altiplano-Puna Plateau, *in* Kay, S.M., Ramos, V.A., and Dickinson, W.M., eds., Backbone of the Americas: Shallow Subduction, Plateau Uplift, and Ridge and Terrane Collisions: Geological Society of America Memoir 204, p. 229–260.

Kay, S.M., Coira, B., and Viramonte, J., 1994, Young mafic back arc volcanic rocks as indicators of continental lithospheric delamination beneath the Argentine Puna Plateau, central Andes: Journal of Geophysical Research, v. 99, no. B12, p. 24,323–24,339, doi:10.1029/94JB00896.

Kay, S.M., Coira, B., Caffe, P.J., and Chen, C.-H., 2010, Regional chemical diversity, crustal and mantle sources, and evolution of central Andean Puna Plateau ignimbrites: Journal of Volcanology and Geothermal Research, v. 198, p. 81–111, doi:10.1016/j.jvolgeores.2010.08.013.

Kennett, B.L., and Engdahl, E.R., 1991, Traveltimes for global earthquake locations and phase identification: Geophysical Journal International, v. 105, p. 429–465, doi:10.1111/j.1365-246X.1991.tb06724.x.

Kirby, S.H., Stein, S., Okal, E.A., and Rubie, D.C., 1996, Metastable mantle phase transformations and deep earthquakes in subducting oceanic lithosphere: Reviews of Geophysics, v. 34, p. 261–306, doi:10.1029/96RG01050.

Kley, J., and Monaldi, C., 1998, Tectonic shortening and crustal thickness in the central Andes: How good is the correlation?: Geology, v. 26, p. 723–726, doi:10.1130/0091-7613(1998)026<0723:TSACTI>2.3.CO;2.

Kley, J., and Monaldi, C., 2002, Tectonic inversion in the Santa Barbara system of the central Andean foreland thrust belt, northwestern Argentina: Tectonics, v. 21, no. 6, doi:10.1029/2002TC902003.

Koulakov, I., Sobolev, S.V., and Asch, G., 2006, P- and S-velocity images of the lithosphere-asthenosphere system in the central Andes from local-source tomographic inversion: Geophysical Journal International, v. 167, p. 106–126, doi:10.1111/j.1365-246X.2006.02949.x.

Lamb, S., 2011, Did shortening in thick crust cause rapid late Cenozoic uplift in the northern Bolivian Andes?: Journal of the Geological Society, London, v. 168, p. 1079–1092, doi:10.1144/0016-76492011-008.

Li, C., van der Hilst, R.D., Engdahl, E.R., and Burdick, S., 2008, A new global model for P wave speed variations in Earth's mantle: Geochemistry Geophysics Geosystems, v. 9, no. 5, Q05018, doi:10.1029/2007GC001806.

Liu, K.H., Gao, S., Silver, P.G., and Zhang, Y., 2003, Mantle layering across central South America: Journal of Geophysical Research, v. 108, no. B11, 2510, doi:10.1029/2002JB002208.

McQuarrie, N., Horton, B., Zandt, G., Beck, S., and DeCelles, P., 2005, Lithospheric evolution of the Andean fold-thrust belt, Bolivia, and the origin of the central Andean plateau: Tectonophysics, v. 399, p. 15–37, doi:10.1016/j.tecto.2004.12.013.

Montelli, R., Nolet, G., Masters, G., Dahlen, F.A., and Hung, S.-H., 2004, Global P and PP traveltime tomography: Rays versus waves: Geophysical Journal International, v. 158, p. 637–654, doi:10.1111/j.1365-246X.2004.02346.x.

Morishige, M., Honda, S., and Yoshida, M., 2010, Possibility of hot anomaly in the sub-slab mantle as an origin of low seismic velocity anomaly under the subducting Pacific plate: Physics of the Earth and Planetary Interiors, v. 183, p. 353–365, doi:10.1016/j.pepi.2010.04.002.

Müller, R.D., Sdrolias, M., Gaina, C., and Roest, W.R., 2008, Age, spreading rates, and spreading asymmetry of the world's ocean crust: Geochemistry Geophysics Geosystems, v. 9, no. 4, Q04006, doi:10.1029/2007GC001743.

Myers, S., Beck, S., Zandt, G., and Wallace, T., 1998, Lithospheric-scale structure across the Bolivian Andes from tomographic images of velocity and attenuation for P and S waves: Journal of Geophysical Research, v. 103, no. B9, p. 21,233–21,252, doi:10.1029/98JB00956.

Norabuena, E.O., Dixon, T.H., Stein, S., and Harrison, C., 1999, Decelerating Nazca–South America and Nazca-Pacific plate motions: Geophysical Research Letters, v. 26, no. 22, p. 3405–3408, doi:10.1029/1999GL005394.

Obayashi, M., Sugioka, H., Yoshimitsu, J., and Fukao, Y., 2006, High temperature anomalies oceanward of subducting slabs at the 410-km discontinuity: Earth and Planetary Science Letters, v. 243, p. 149–158, doi:10.1016/j.epsl.2005.12.032.

Oncken, O., Hindle, D., Kley, J., Elger, K., Victor, P., and Schemmann, K., 2006, Deformation of the central Andean upper plate system—Facts, fiction, and constraints for plateau models, *in* Oncken, O., Chong, G., Franz, G., Giese, P., Götze, H.J., Ramos, V.A., Strecker, M.R., and Wigger, P., eds., The Andes—Active Subduction Orogeny: Berlin, Springer-Verlag, Frontiers in Earth Sciences, v. 1, p. 3–27.

Paige, C.C., and Saunders, M.A., 1982, LSQR: An algorithm for sparse linear equations and sparse least squares: ACM Transactions on Mathematical Software, v. 8, no. 1, p. 43–71, doi:10.1145/355984.355989.

Pavlis, G., and Vernon, F., 2010, Array processing of teleseismic body waves with the USArray: Computers & Geosciences, v. 36, p. 910–920, doi:10.1016/j.cageo.2009.10.008.

Pesicek, J.D., Engdahl, E.R., Thurber, C.H., DeShon, H.R., and Lange, D., 2012, Mantle subducting slab structure in the region of the 2010 M8.8 Maule earthquake (30–40°S), Chile: Geophysical Journal International, v. 191, p. 317–324, doi:10.1111/j.1365-246X.2012.05624.x.

Phillips, K., Clayton, R.W., Davis, P., Tavera, H., Guy, R., Skinner, S., Stubailo, I., Audin, L., and Aguilar, V., 2012, Structure of the subduction system in southern Peru from seismic array data: Journal of Geophysical Research, v. 117, B11306, doi:10.1029/2012JB009540.

Quinteros, J., and Sobolev, S.V., 2013, Why has the Nazca plate slowed since the Neogene?: Geology, v. 41, no. 1, p. 31–34, doi:10.1130/G33497.1.

Ramos, V.A., Jordan, T.E., Allmendinger, R.W., Mpodozis, C., Kay, S.M., Cortes, J.M., and Palma, M., 1986, Paleozoic terranes of the central Argentine-Chilean Andes: Tectonics, v. 5, no. 6, p. 855–880, doi:10.1029/TC005i006p00855.

Schmandt, B., and Humphreys, E., 2010, Seismic heterogeneity and small-scale convection in the southern California upper mantle: Geochemistry Geophysics Geosystems, v. 11, Q05004, doi:10.1029/2010GC003042.

Schmerr, N., and Garnero, E.J., 2007, Upper mantle discontinuity and topography from thermal and chemical heterogeneity: Science, v. 318, p. 623–626, doi:10.1126/science.1145962.

Schurr, B., Asch, G., Rietbrock, A., Kind, R., Pardo, M., Heit, B., and Monfret, T., 1999, Seismicity and average velocities beneath the Argentine Puna Plateau: Geophysical Research Letters, v. 26, no. 19, p. 3025–3028, doi:10.1029/1999GL005385.

Schurr, B., Rietbrock, A., Asch, G., Kind, R., and Oncken, O., 2006, Evidence for lithospheric detachment in the central Andes from local earthquake tomography: Tectonophysics, v. 415, no. 1–4, p. 203–223.

Siebert, L., and Simkin, T., 2002, Volcanoes of the World: An Illustrated Catalog of Holocene Volcanoes and Their Eruptions: Smithsonian Institution, Global Volcanism Program Digital Information Series GVP-3: http://volcano.si.edu/search_volcano.cfm (accessed 4 June 2014).

Smyth, J.R., and Jacobsen, S.D., 2006, Nominally anhydrous minerals and Earth's deep water cycle, *in* Jacobsen, S.D., and van der Lee, S., eds., Earth's Deep Water Cycle: American Geophysical Union Geophysical Monograph 168, p. 1–11, doi:10.1029/168GM02.

Sobiesiak, M., Eggert, A., Woith, H., Grosser, H., Peyrat, S., Vilotte, J., Medina, E., Ruch, J., Walter, T., Victor, P., Barrientos, S., and Gonzalez, G., 2008, The M 7.7 Tocopilla earthquake and its aftershock sequence: Deployment of a Task Force local network: Eos (Transactions, American Geophysical Union), v. 89, Fall meeting supplement, abstract S24A-04.

Sobolev, S.V., Babeyko, A.Y., Koulakov, I., Oncken, O., and Vietor, T., 2006, Mechanism of the Andean orogeny: Insight from the numerical modeling, *in* Oncken, O., Chong, G., Franz, G., Giese, P., Götze, H.J., Ramos, V.A., Strecker, M.R., and Wigger, P., eds., The Andes—Active Subduction Orogeny: Berlin, Springer-Verlag, Frontiers in Earth Sciences, v. 1, p. 509–531.

Sodoudi, F., Yuan, X., Asch, G., and Kind, R., 2011, High-resolution image of the geometry and thickness of the subducting Nazca lithosphere beneath northern Chile: Journal of Geophysical Research, v. 116, B04302, doi:10.1029/2010JB007829.

Swenson, J.L., Beck, S.L., and Zandt, G., 2000, Crustal structure of the Altiplano from broadband regional waveform modeling: Implications for the composition of thick continental crust: Journal of Geophysical Research, v. 105, no. B1, p. 607–621, doi:10.1029/1999JB900327.

Tassara, A., 2005, Interaction between the Nazca and South American plates and formation of the Altiplano-Puna Plateau: Review of a flexural analysis along the Andean margin (15°–34°S): Tectonophysics, v. 399, p. 39–57, doi:10.1016/j.tecto.2004.12.014.

Tassara, A., and Echaurren, A., 2012, Anatomy of the Andean subduction zone: Three-dimensional density model upgraded and compared against global-scale models: Geophysical Journal International, v. 189, p. 161–168, doi:10.1111/j.1365-246X.2012.05397.x.

U.S. Geological Survey (USGS), 2013, Seismic Network Operations: GT LPAZ: http://earthquake.usgs.gov/monitoring/operations/station.php?network=GT&station=LPAZ (accessed October 2013).

VanDecar, J.C., and Crosson, R.S., 1990, Determination of teleseismic relative phase arrival times using multi-channel cross-correlation and least squares: Bulletin of the Seismological Society of America, v. 80, no. 1, p. 150–169.

Ward, K., Porter, R.C., Zandt, G., Beck, S.L., Wagner, L.S., Minaya, E., and Tavera, H., 2013, Ambient noise tomography across the central Andes: Geophysical Journal International, v. 194, p. 1559–1573, doi:10.1093/gji/ggt166.

Whitman, D., Isacks, B.L., Chatelain, J.-L., Chiu, J.-M., and Perez, A., 1992, Attenuation of high-frequency seismic waves beneath the central Andean plateau: Journal of Geophysical Research, v. 97, no. B13, p. 19,929–19,947, doi:10.1029/92JB01748.

Whitman, D., Isacks, B.L., and Kay, S.M., 1996, Lithospheric structure and along-strike segmentation of the central Andean Plateau: Seismic *Q*, magmatism, flexure, topography and tectonics: Tectonophysics, v. 259, p. 29–40, doi:10.1016/0040-1951(95)00130-1.

Wölbern, I., Heit, B., Yuan, X., Asch, G., Kind, R., Viramonte, J., Tawackoli, S., and Wilke, H., 2009, Receiver function images from the Moho and the slab beneath the Altiplano and Puna plateaus in the central Andes: Geophysical Journal International, v. 177, p. 296–308, doi:10.1111/j.1365-246X.2008.04075.x.

Yuan, X., Sobolev, S.V., Kind, R., Oncken, O., Bock, G., Asch, G., Schurr, B., Graeber, F., Rudloff, A., Hanka, W., Wylegalla, K., Tibi, R., Haberland, C., Rietbrock, A., Giese, P., Wigger, P., Rower, P., Zandt, G., Beck, S., Wallace, T., Pardo, M., and Comte, D., 2000, Subduction and collision processes in the central Andes constrained by converted seismic phases: Nature, v. 408, p. 958–961, doi:10.1038/35050073.

Yuan, X., Sobolev, S.V., and Kind, R., 2002, Moho topography in the central Andes and its geodynamic implications: Earth and Planetary Science Letters, v. 199, p. 389–402, doi:10.1016/S0012-821X(02)00589-7.

Zandt, G., Beck, S.L., Ruppert, S.R., Ammon, C.J., Rock, D., Minaya, E., Wallace, T.C., and Silver, P.G., 1996, Anomalous crust of the Bolivian Altiplano, central Andes: Constraints from broadband regional seismic waveforms: Geophysical Research Letters, v. 23, no. 10, p. 1159–1162, doi:10.1029/96GL00967.

Zandt, G., Leidig, M., Chmielowski, J., Baumont, D., and Yuan, X., 2003, Seismic detection and characterization of the Altiplano-Puna magma body, central Andes: Pure and Applied Geophysics, v. 160, p. 789–807, doi:10.1007/PL00012557.

Zhao, D., 2004, Global tomography images of mantle plumes and subducting slabs: Insights into deep Earth dynamics: Physics of the Earth and Planetary Interiors, v. 146, p. 3–34, doi:10.1016/j.pepi.2003.07.032.

Zhao, D., Yamamoto, Y., and Yanada, T., 2013, Global mantle heterogeneity and its influence on teleseismic regional tomography: Gondwana Research, v. 23, p. 595–616, doi:10.1016/j.gr.2012.08.004.

MANUSCRIPT ACCEPTED BY THE SOCIETY 3 JUNE 2014
MANUSCRIPT PUBLISHED ONLINE 9 SEPTEMBER 2014

The Geological Society of America
Memoir 212
2015

Multiple styles and scales of lithospheric foundering beneath the Puna Plateau, central Andes

Susan L. Beck*
George Zandt
Kevin M. Ward
Alissa Scire
Department of Geosciences, University of Arizona, 1040 E. 4th Street, Tucson, Arizona 85721, USA

ABSTRACT

Lithospheric foundering or delamination has been long recognized as an important process in the formation of the Andes, but the scale, timing, and surface uplift consequences remain controversial. We use recently completed ambient noise tomography and finite-frequency P-wave tomography results and other geologic and geophysical information to identify two ~200-km-diameter regions of piecemeal delamination in the Puna region between 21°S and 27°S. One location in the northern Puna Plateau is centered under the 11–1 Ma large-volume silicic Altiplano-Puna volcanic center, and the other in the southern Puna Plateau is centered approximately between the Arizaro Basin and 6–2 Ma Cerro Galan volcanic field. The foundering in the northern location has progressed to the point where the main thermal anomaly resides in the middle and upper crust, and the surface volcanic flare-up and mantle thermal anomalies are both in a waning stage. In the southern location, the main thermal anomaly is still in its waxing stage in the lower crust and upper mantle, and the foundering mantle material is imaged in the mantle wedge.

The differing patterns of back-arc volcanism in the two foundering centers suggest different styles and timing of delamination, with the foundering process coming to completion earlier in the north than in the south. Based on plate-motion reconstructions, the NE-SW–aligned Juan Fernandez Ridge swept southward through this area starting about ca. 14 Ma in the north and ca. 10 Ma in the south. Although we do not think the passage of the Juan Fernandez Ridge initiated foundering, it played an important role in facilitating delamination by increasing interplate coupling, and weakening and perhaps hydrating the upper plate, and its passage allowed the delaminated material to sink into the expanding space of the mantle wedge. Another important factor in this evolution is the upper-plate lithospheric strength variations

*slbeck@email.arizona.edu

Beck, S.L., Zandt, G., Ward, K.M., and Scire, A., 2015, Multiple styles and scales of lithospheric foundering beneath the Puna Plateau, central Andes, *in* DeCelles, P.G., Ducea, M.N., Carrapa, B., and Kapp, P.A., eds., Geodynamics of a Cordilleran Orogenic System: The Central Andes of Argentina and Northern Chile: Geological Society of America Memoir 212, p. 43–60, doi:10.1130/2015.1212(03). For permission to copy, contact editing@geosociety.org. © 2014 The Geological Society of America. All rights reserved.

inherited from the different geologic basements underlying the northern and southern Puna regions. As the larger-scale delamination progressed, leaving behind thin lithosphere and a mantle wedge with a mixture of continental lithospheric fragments and hot asthenosphere, smaller secondary Rayleigh-Taylor instabilities occurred beneath the southern Puna Plateau, influencing basin development, and subsequent melting of this "drip" material was the source of the ensuing low-volume mafic volcanism.

INTRODUCTION

The central Andean Plateau in the central Andes of South America is one of the largest plateaus on Earth, encompassing a region 1500 km along strike, with a maximum width of 400 km, and an average elevation of 4 km. It is the premier example of an active compressional orogenic system associated with an ocean-continent subduction plate boundary and, hence, an ideal location to study the lithospheric-scale structures of high plateaus and mechanisms of surface uplift and lithospheric removal. Most investigators agree that there are at least three processes that have important roles in the formation of high-elevation plateaus: (1) continuous isostatic uplift due to crustal shortening and thickening, (2) episodic lithospheric thinning or foundering, and (3) magmatic addition and/or crustal flow (for two recent reviews, see Barnes and Ehlers, 2009, and Kay and Coira, 2009). Seismology is an important tool with which to constrain the current structure and state of the upper-plate lithosphere and provide information on the relative roles of each of these processes.

Here, we will use the term "foundering" as a generic term, referring to any number of potential mechanisms in which the primary driving force is gravity, including the more fluid dynamic process of Rayleigh-Taylor instabilities (drips) and the more mechanical process of delamination. In our usage, the more general term of "lithospheric removal" includes processes where the primary force is not gravitational, for example, some types of subduction erosion.

The process of lithospheric foundering has long been invoked as an important mechanism in the development of the Andes (Kay and Kay, 1993; Kay et al., 1994; Beck and Zandt, 2002; McQuarrie et al., 2005; Asch et al., 2006; Garzione et al., 2006; Ghosh et al., 2006; DeCelles et al., 2009). Lithospheric foundering can be driven by at least two processes: (1) magmatic arc processes that produce an igneous eclogitic root and (2) crustal shortening and thickening that lead to a metamorphic eclogitization of a mafic lower crust. Therefore, foundering can include parts of the lower crust as well as mantle lithosphere. We should expect both processes to be important in the Andes because magmatic arcs have been present along much of the central Andes throughout the Mesozoic and Cenozoic, and well-documented upper-crustal shortening occurred during much of the Cenozoic. While many authors agree on the existence of foundering in the central Andes, there remains a vigorous debate on the causation, scale, timing, and surface expression.

In this paper, we focus on seismological constraints related to lithospheric foundering in the Puna region between 21°S and 27°S in the central Andes (Fig. 1). Following Kay and Coira (2009), we refer to the region between 21°S and 24°S as the northern Puna and the region between 24°S and 27°S as the southern Puna. Over the last two decades, numerous portable seismic deployments have been completed in the central Andes, resulting in an increased understanding of the crust and upper-mantle structure (Fig. 1). These portable seismic deployments included a range of instrumentation, from short-period to broadband sensors that recorded from 3 mo to 2 yr. Large numbers of seismological studies have been published for each area, but comparing results is often difficult due to the different methods used, particularly with respect to different types of tomography. Identifying locations where lithospheric removal is occurring or has occurred in the past is difficult, in part because it likely involves a range of processes occurring at different scales and at different times along strike of the Andes. In this study, we build on two recently completed studies, an ambient noise tomography (ANT) study that combined all the available broadband data in the Andes, producing a shear-wave (V_{sv}) velocity model for the crust down to ~50 km (Ward et al., 2013), and a teleseismic P-wave tomography study that combined data from many of the portable seismic deployments in the region to image the upper mantle (Scire et al., this volume). We use these new results to document two different styles and scales of lithospheric removal in the Puna region of the central Andes.

TECTONIC SETTING OF THE CENTRAL ANDES

The tectonics and volcanism of the central Andes have been reviewed by Isacks (1988), Allmendinger et al. (1997), Lamb and Hoke (1997), Oncken et al. (2006), Trumbull et al. (2006), Hoke and Lamb (2007), Ramos (2009), Kay and Coira (2009), and references therein, and so will only be briefly summarized here (Figs. 1 and 2). The central Andes are divided into five major tectonomorphic zones from west to east, the erosional forearc, Western Cordillera, Altiplano-Puna Plateau, Eastern Cordillera, and the Subandean zone (Fig. 1). The Western Cordillera is part of the Central volcanic zone, a linear series of active stratovolcanoes (Fig. 1). The Eastern Cordillera is a fold-and-thrust belt with deformed Paleozoic sedimentary rocks, Triassic to Miocene intrusions, and locally large ignimbrite deposits. Further east, there is the Subandean zone and Santa Barbara system, both active foreland thrust belts with along-strike variations in structural style and amounts of shortening (Kley, 1996).

Our focus in this paper is the central Andean Plateau, generally defined as the area with elevations above 3 km. The central

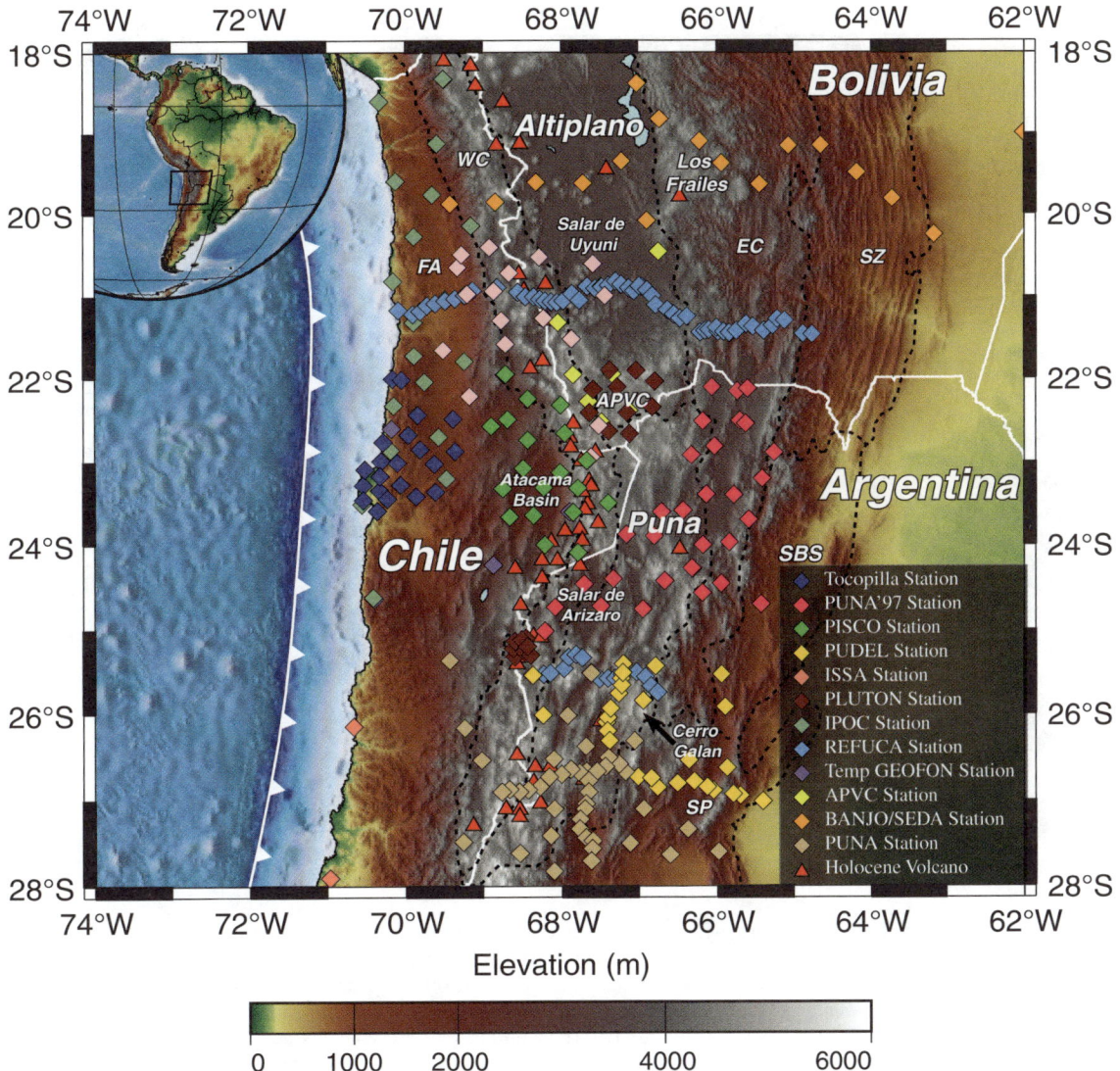

Figure 1. Map of the central Andes showing topography and seismic stations used in the ambient noise tomography and P-wave tomography (not all stations were used in both studies) analyses. Also shown are the morphotectonic boundaries (dashed lines) (FA—forearc, Altiplano; WC—Western Cordillera; EC—Eastern Cordillera; SZ—Subandean zone, Puna; APVC—Altiplano-Puna volcanic center; SBS—Santa Barbara system, SP—Sierras Pampeanas). The Holocene volcanic centers are shown as red triangles.

Andean Plateau includes both the Altiplano and Puna Plateaus, and parts of the adjacent Western and Eastern Cordilleras, areas with active volcanism, thick crust, and high elevations (Isacks, 1988; Allmendinger et al., 1997; Oncken et al., 2006; Trumbull et al., 2006; Kay and Coira, 2009). However, the Altiplano and Puna Plateaus are different in several important ways. First, the Puna is on average 0.5–1 km higher in elevation than the Altiplano (Isacks, 1988; Whitman et al., 1996). Much of the Altiplano is an internally drained, sediment-filled basin, while the Puna has higher-relief topography, smaller isolated basins, and more widespread volcanism, with many active volcanic centers, including several large ignimbrite centers. Whitman et al. (1992, 1996) showed that seismic waves propagate with higher frequen-

cies in the upper mantle beneath the Altiplano than beneath the Puna. These authors interpreted this in terms of seismic attenuation variations and concluded that the Altiplano has mantle lithosphere beneath the crust but that the lithospheric mantle was much thinner or virtually absent beneath the Puna.

Basement Geology

The basement geology of this region is poorly known due to the scarcity of outcrops and the volcanic cover. The Antofalla basement is thought to underlie much of the central Andes between 20°S and 26°S (Ramos, 2008). The Antofalla basement consists of Proterozoic to Early Cambrian metamorphic basement

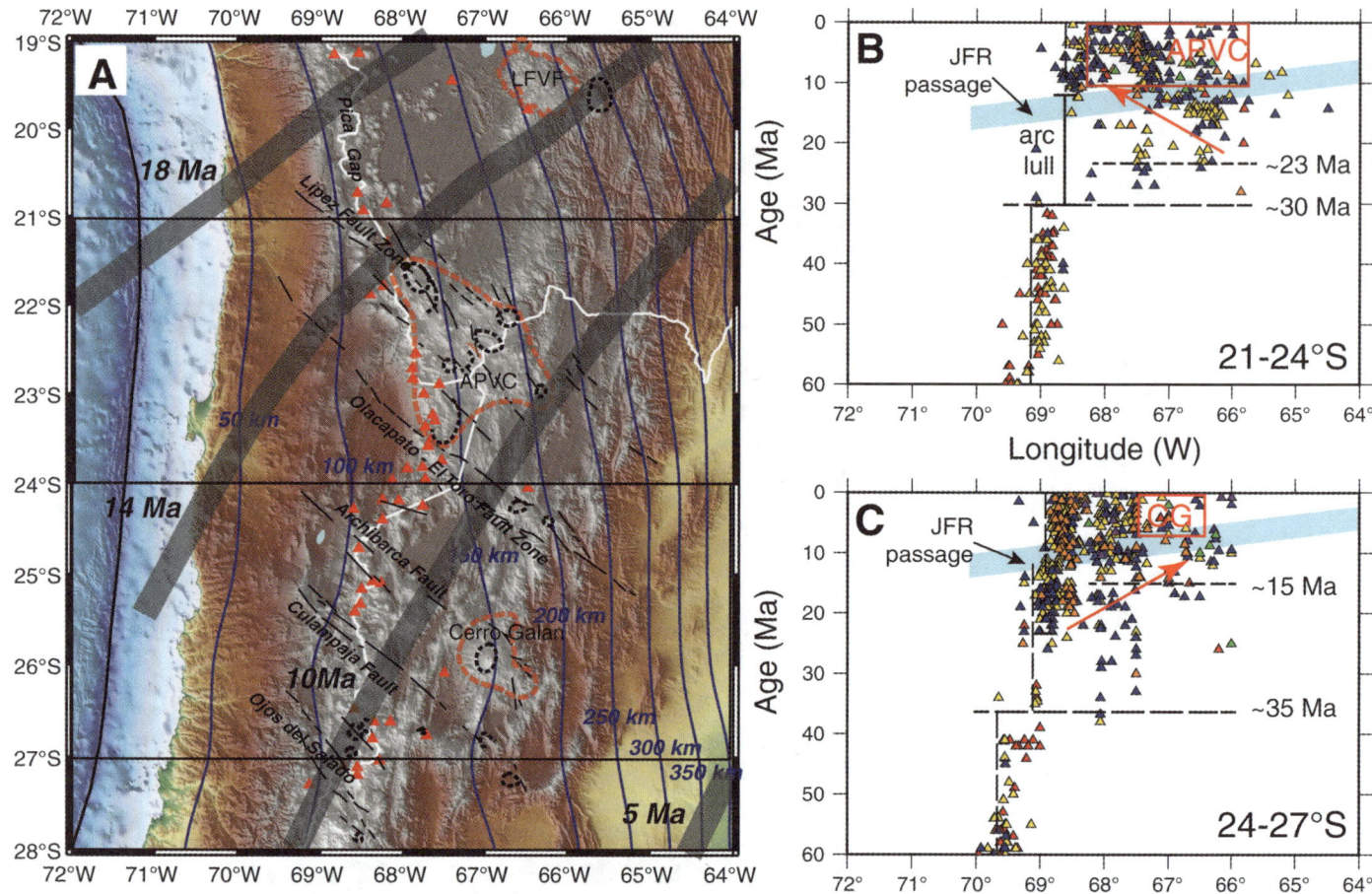

Figure 2. (A) Map of the study area showing topography, volcanic centers, faults, slab contours, and the migration of the subducting Juan Fernandez Ridge. The solid black lines are faults from Riller et al. (2001), solid blue lines are slab contours from Hayes et al. (2012), and thick transparent lines are past locations of the subducted Juan Fernandez Ridge based on the plate reconstruction of Yáñez et al. (2001). Dashed red lines outline the large ignimbrite volcanic fields: Los Frailes volcanic field (LFVF), Altiplano-Puna volcanic complex (APVC), and Cerro Galan. Small dashed areas are calderas from Petrinovic et al. (2010). (B) Summary of arc volcanism and ignimbrites in the central Andes as a function of age between 21°S and 24°S replotted using the database from Trumbull et al. (2006). Symbols are color coded for type of volcanic deposit; red—intrusives; gold—lava, breccia, and dome deposits; blue—pyroclastic, ash, tuff, and ignimbrite deposits; green—calderas; orange—stratovolcanoes and cones; purple—monogenic centers. Blue bar shows the passage of the Juan Fernandez Ridge (JFR). Note the gap in arc volcanism and the onset of widespread back-arc volcanism. The red box shows the Altiplano-Puna volcanic complex (APVC). (C) Summary of volcanism and ignimbrites in the central Andes as a function of age between 24°S and 27°S, replotted using the database from Trumbull et al. (2006). The red box shows Cerro Galan (CG). Symbols are the same as in B.

with Grenville-age affinities as well as early Paleozoic intrusives typical of a magmatic arc (Ramos, 2008). The Central Andean gravity high, as identified by Götze and Krause (2002), likely corresponds to the Antofalla basement. This suggests that the eastern extent of the Antofalla basement cuts obliquely across the central Puna Plateau, with most of the northern Puna region east of the Antofalla basement, while the Antofalla basement underlies much of the southern Puna region. Kay et al. (2010) suggested the possibility of different compositions of basement in the northern and southern Puna regions based on differences in the geochemistry of the ignimbrites. They suggested that the crustal contamination is more pelitic in the northern Puna region and more igneous gneiss-like in the southern Puna region. Hence, the northern and south-

ern Puna regions appear to have different basements, which may reflect differences in lithospheric rheology.

Deformation Patterns

The central Andes have some of the thickest crust on Earth, thought to be related to large magnitudes of crustal shortening (Fig. 3; Beck and Zandt, 2002; Yuan et al., 2002; Tassara and Echaurren, 2012; Oncken et al., 2006; McQuarrie, 2002; McQuarrie et al., 2005; Elger et al., 2005; Baby et al., 1997; Gotberg et al., 2010). The amount of upper-crustal shortening varies along strike, with as much as 250–330 km near ~18°S–22°S, decreasing southward to less than ~100 km of shortening at 30°S

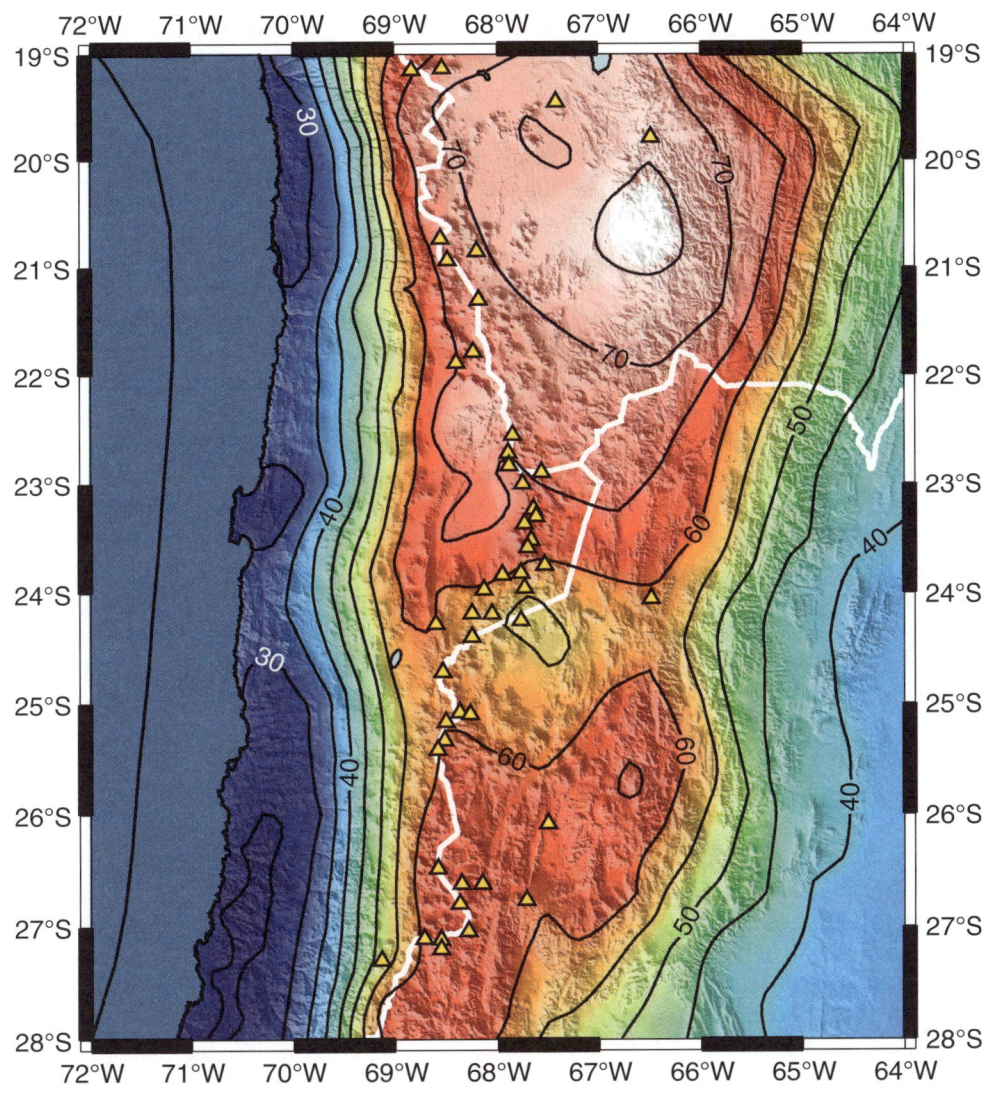

Figure 3. Map of the Puna region showing depth to Moho in km (contours) plotted using the database from Tassara and Echaurren (2012). The model values were interpolated using the natural neighbor algorithm (Sambridge et al., 1995). Also shown are the Holocene volcanic centers (yellow triangles).

(Baby et al., 1997; Kley and Monaldi, 1998, 2002; Müller et al., 2002; Elger et al., 2005; McQuarrie et al., 2005; Oncken et al., 2006). Total estimates of crustal shortening vary based on the assumptions made in constraining the deformation in the middle and lower crust, but all the studies agree that shortening is at its maximum at 19°S–21°S and decreases to the south. Although most of the central Andes formed from E-W shortening, there are series of NW-SE–striking strike-slip faults cutting across the Puna Plateau due in part to the differential shortening along strike (Fig. 2; Riller et al., 2001).

Oncken et al. (2006) summarized the timing and magnitude of the deformation between 21°S and 27°S, and we will only briefly review it here. There is documented deformation starting in the west by at least 46 Ma in the Precordillera and western volcanic arc (Elger et al., 2005; Oncken et al., 2006). The earliest deformation is likely even older, based on the foreland basin deposits that imply there was crustal shortening starting no later than Paleocene time in the Altiplano and the Puna as far south as

26°S (DeCelles and Horton, 2003; DeCelles et al., 2011). Quade et al. (this volume) used stable isotopic compositions of carbonates and volcanic glass to estimate paleoelevations between 22°S and 26°S. They found that the Western Cordillera and western Puna were high (~4 km) as early as 38 Ma, suggesting that shortening and magmatic processes had already thickened the crust by that time. Structural and thermochronology studies also suggest that deformation and uplift progressed from west to east, although not necessarily smoothly (Ramos, 1999; Carrapa and DeCelles, 2008; Carrapa and DeCelles, this volume; Reiners et al., this volume). Deformation in the Eastern Cordillera occurred between 37 and 46 Ma but increased between 36 and 30 Ma, with much less deformation in the center of the Puna region, suggesting there was a "jump" in deformation from the Western Cordillera to the Eastern Cordillera at this time (Oncken et al., 2006). The Western Cordillera, Altiplano, Puna, and Eastern Cordillera were actively deforming between 29 Ma and 20 Ma. Deformation continued between 19 and 8 Ma, with increasing deformation in the Puna. The final phase of deformation, starting ca. 7–10 Ma, shifted to the margins of the Altiplano and Puna and into the foreland thrust belts to the east (Oncken et al., 2006, and references therein). The crustal shortening to the east in the foreland is probably rooted in the lower crust under the Eastern Cordillera, continuing to thicken the crust to the west as the cratonic crust underthrusts from the east (Isacks, 1988).

There are also along-strike variations in the style of deformation in the foreland region of the central Andes. The Subandean zone from northern Bolivia to northern Argentina is a thin-skinned fold-and-thrust belt with stratigraphically controlled detachments in the thick Paleozoic section (Allmendinger et al., 1983; Kley, 1999; Oncken et al., 2006; DeCelles et al., 2011). In northern Argentina, at ~23°S, the structural style in the foreland changes to more of a tectonic inversion style in the Santa Barbara system, where many of the reverse faults are reactivated normal faults from previous Cretaceous rifting (Kley, 1996; Kley and Monaldi, 2002). Further south, the deformation transitions into basement-cored uplifts in the Sierras Pampeanas, similar to the Laramide uplifts in the western United States (Jordan and Allmendinger, 1986). Along-strike differences in the preexisting geology of the upper crust likely contribute to these differences in structural style (Allmendinger et al., 1997). Despite the changes in structural style, DeCelles et al. (2011) suggested that there was a flexural foreland basin that formed in the Paleocene and migrated east at least 600 km in northwestern Argentina.

It has been proposed that an important control on upper-plate deformation is the dip of the subducting Nazca plate, which has varied along strike and through time. In the central Andes, the Nazca plate is presently subducting at an angle of ~30° in the upper mantle beneath the South America plate, with a well-developed active volcanic arc (Hayes et al., 2012). However, this normal subduction dip may not have always been the case beneath the central Andean Plateau (e.g., Kay and Coira, 2009). The Juan Fernandez Ridge is a discontinuous chain of volcanic centers or seamounts, aligned roughly ENE-WSW, and has been attributed to a hotspot (von Huene et al., 1997), and the present location of the Juan Fernandez Ridge correlates with flat slab subduction in Chile and Argentina near 30°S. It has been argued that subduction of aseismic ridges (buoyant overthickened oceanic crust) is at least one of the contributing factors to flat slab subduction (Gutscher et al., 2000). Based on Yáñez et al. (2001) plate reconstructions, the Juan Fernandez Ridge was subducting beneath the northern Puna starting ca. 14–12 Ma and beneath the southern Puna starting ca. 12–8 Ma (Fig. 2).

Volcanism Patterns

The central Andes have a long history of volcanism, with both major stratovolcanoes and widespread ignimbrites and back-arc volcanism (Trumbull et al., 2006; Hoke and Lamb, 2007; Kay and Coira, 2009). The northern Puna region is the site of one of the largest silicic ignimbrite fields, the Altiplano-Puna volcanic complex, with multiple large-volume ignimbrite erup-

tions between 11 and 1 Ma (de Silva, 1989; Coira et al., 1993; de Silva et al., 2006a; Kay et al., 2010; Salisbury et al., 2011). Trumbull et al. (2006) summarized the spatial-temporal patterns of arc volcanism and ignimbrite eruptions between 20°S and 26°S and found that the arc volcanism was active in a fairly narrow belt between 60 and ca. 30 Ma. In Figure 2, we replot volcanic data from the database in Trumbull et al. (2006) using all of the volcanic types and slightly different along-strike segments, 21–24°S and 24–27°S, corresponding to the northern and southern Puna regions, respectively. The patterns of volcanism for these latitude ranges are very similar to the patterns Trumbull et al. (2006) showed using latitude ranges of 20–23°S and 23–26°S. We note that this database does not include any information about volumes of volcanic rocks. Kay and Coira (2009) showed time-space plots for parts of this area with a better representation of the associated volumes; nonetheless, Figures 2B and 2C provide a useful first-order view of space-time patterns in volcanism.

In the region between 21°S and 24°S, there was a lull in arc volcanism starting at ca. 30 Ma, with some ignimbrite activity in the back arc during the "gap" in arc volcanism (Fig. 2B). By 20–25 Ma, the volcanism occurred over a broad zone, reaching well into the back arc (65–66°W, depending slightly on latitude). By 17 Ma, back-arc volcanism started to increase, and by 11 Ma, there were caldera complexes and major ignimbrite activity, with the start of activity in the Altiplano-Puna volcanic center (see summary in Kay and Coira, 2009, and references therein). Numerous studies have attributed the gap in volcanism between ca. 26 Ma and 17 Ma as a possible indication of flat slab subduction (Kay et al., 1999; James and Sacks, 1999; Allmendinger et al., 1997; Ramos and Folguera, 2009). The broad zone of back-arc volcanism starting ca. 17–15 Ma has been interpreted as the result of the steepening of the slab following flat slab subduction and the inflow of warm asthenosphere into the mantle wedge. As pointed out by Trumbull et al. (2006), the timing of the gap in arc volcanism is prior to the passage of the subduction of the Juan Fernandez Ridge beneath the southern Altiplano and northern Puna, based on plate reconstructions of Yáñez et al. (2001) (Fig. 2B). Hence, Trumbull et al. (2006) argued that a gap in arc volcanism in the northern Puna region is not correlated with the subduction of the Juan Fernandez Ridge.

In the region between 24°S and 27°S, there is a decrease in arc volcanism but not an obvious cessation of volcanism (Fig. 2C). A more pronounced feature is inboard migration of the arc at ca. 35 Ma and another at ca. 12 Ma. Between ca. 27 Ma and 17 Ma, there were ignimbrite eruptions in the arc region and smaller volumes of back-arc volcanism (Kay and Coira, 2009). Between ca. 17 Ma and 11 Ma, the arc broadened, with several active stratovolcanoes (Kay and Coira, 2009). After ca. 7 Ma, volcanism increased across the arc and back arc, including the large-volume Cerro Galan ignimbrite eruptions ca. 6–2 Ma (Kay and Coira, 2009).

As noted by Trumbull et al. (2006) and Kay and Coira (2009), the pattern of volcanism was different between the northern Puna region, with the Altiplano-Puna volcanic center, and the southern Puna region, with Cerro Galan. In the northern segment, after the volcanic lull, back-arc activity started ca. 23 Ma, and there appears to have been a migration of westward younging volcanism, leading to reestablishment of the frontal arc ca. 10 Ma (Fig. 2B). The volcanic pattern in the southern Puna region is quite different, with the arc volcanism waning but not completely absent between ca. 30 Ma and 23 Ma. The back-arc volcanism broadening occurred between ca. 25 Ma and 10 Ma, and with a less pronounced eastward younging of the volcanism (Fig. 2C). Later, we argue that rather than being an effect only of slab dip changes, this broadening of volcanism was a manifestation of lithospheric foundering.

LITHOSPHERIC STRUCTURE OF THE CENTRAL ANDES

In this section, we first briefly review published geophysical studies on the Puna region. Then, we focus on summarizing recently completed ambient noise and P-wave tomography results for the Puna region.

Prior Geophysical Studies

Geophysical studies have shown that the central Andean Plateau has some of the thickest crust on Earth. The Altiplano has crustal thicknesses of 70–75 km (Beck and Zandt, 2002; Yuan et al., 2002; McGlashan et al., 2008; Wölbern et al., 2009). The crust has variable thicknesses under the Eastern Cordillera, ranging from 60 to 75 km, and then thinning to 40–45 km beneath the Subandean zone and to ~35–40 km under the foreland basin (Beck and Zandt, 2002; Yuan et al., 2002; Wölbern et al., 2009).

The depth to Moho in the Puna region is variable but generally thinner than the Altiplano. Recent studies have suggested the depth to Moho values range from 55 to 65 km (Yuan et al., 2002; Tassara, 2006; Wölbern et al., 2009; Tassara and Echaurren, 2012; Bianchi et al., 2013). Figure 3 shows a contour map of depth to Moho from the model of Tassara and Echaurren (2012) calculated using seismic constraints and gravity. We have used natural neighbor interpolation to contour the depth to Moho data (Sambridge et al., 1995). This produces a smooth model similar to that shown by Tassara and Echaurren (2012), consistent with both the seismic and gravity data. The northern Puna Plateau has a relatively smooth Moho depth at 65 km, and the southern Puna Plateau has a more variable Moho depth. At 24°S–25°S, the depth to the Moho locally thins to ~55 km in an area centered on the Arizaro Basin before deepening again to 60–65 km at 25°S–26°S in an area centered on the Cerro Galan volcanic field (Yuan et al., 2002; Tassara and Echaurren, 2012).

A key question is: How thick is the present-day lithosphere beneath the central Andean Plateau? The base of the lithosphere is difficult to measure because it is represented by a decrease in seismic velocity and is usually gradational rather than sharp. It is also difficult to interpret the seismic velocity images in the uppermost mantle because they are often heterogeneous, and it is difficult to

differentiate between the effects of thermal, compositional, and/or changes in physical state from seismic velocities alone.

Heit et al. (2007) used S-wave receiver functions to map out the base of the lithosphere across the central Andes. This technique is similar to P-wave receiver function analysis except that it isolates S-to-P conversions that arrive prior to the direct S-wave. It has the advantage of not being subject to interference from multiples but does have lower-frequency content and, hence, less resolution. Heit et al. (2007) migrated S-wave receiver functions using data from stations located between ~20°S and 21°S. They identified a conversion from the base of the crust (Moho) and a more diffuse conversion indicating a decrease in velocity that they interpreted as the lithosphere-asthenosphere boundary. The lithosphere-asthenosphere boundary depth is estimated at ~130 km under the Altiplano and western part of the Eastern Cordillera (66°W–67.5°W), shallowing to 100 km depth under the eastern part of the Eastern Cordillera near 65°W, and then steps down to 170 km beneath the Subandean zone (63°W). This would suggest that there is at least some lithospheric mantle beneath parts of the south-central Altiplano in the region of the Salar de Uyuni near 20°S–21°S.

Using gravity data, published values of crustal thickness, and the location of the slab, Tassara and Echaurren (2012) calculated an updated three-dimensional (3-D) density model for South America. They calculated a lithospheric thickness of ~80 km beneath the Altiplano, thinning to a localized minimum of ~60 km beneath the central Puna Plateau. The lithospheric thickness increases to the east, reaching ~120 km in the foreland near ~64°W–65°W. These values are slightly less than that observed by Heit et al. (2007) from the S-wave receiver functions but show a similar pattern.

Since earlier studies were published using sparse regional data (Whitman et al., 1992, 1996), there have been a number of portable seismic deployments carried out by multiple international groups in the central Andes to image the crust and upper mantle. One difficulty in comparing these studies is that different investigators used different methods and parameterizations, and different initial seismic velocity models, making it difficult to directly compare the tomographic images from different studies. Nevertheless, there are a number of important observations when we review all of these studies together, including some major along-strike variations.

Data from many of the portable seismic deployments over the past 20 yr in the central Andes have been used for local or regional earthquake traveltime tomography studies to image the lithosphere and mantle wedge (Myers et al., 1998; Graeber and Asch, 1999; Haberland and Rietbrock, 2001, 2003; Schurr et al., 2006; Koulakov et al., 2006). These studies used local earthquakes in the crust and slab to determine a 3-D velocity model in each study region. Other studies such as Heit et al. (2008) used teleseismic data to image the upper mantle in the vicinity of ~21°S, and Bianchi et al. (2013) used both teleseismic and regional distance earthquakes between 24°S and 28°S.

In the southern Altiplano and northern Puna regions, both Koulakov et al. (2006) and Schurr et al. (2006) used data from several portable seismic deployments between 1994 and 1997 and different tomography methods to image the lithosphere between 21°S and 24°S (see review in Kay and Coira, 2009). In general, their results are similar, although the magnitude of the anomalies and some details vary. The best station coverage is south of 22°S in northern Argentina. A large part of the upper mantle beneath the Puna Plateau shows low *Vp* values, consistent with the removal of much of the lithospheric mantle. The tomography studies of the Altiplano and Puna regions show significant heterogeneities in the uppermost mantle but in general suggest that the upper mantle under the central part of the Altiplano is generally faster (Myers et al., 1998) as compared to most of the Puna (Schurr et al., 2006; Koulakov et al., 2006). Between 22°S and 24°S, Schurr et al. (2006) and Asch et al. (2006) identified ~100-km-scale high-velocity anomalies in the upper mantle, generally close to the slab, that they interpreted as delaminated blocks.

Two new tomographic studies from the recently completed U.S.-Argentina-German PUNA-PUDEL experiment (Fig. 1) provide some of the first seismic images for this region. Bianchi et al. (2013) combined teleseismic traveltime data (International Seismological Center data) and regional earthquake data to image the upper mantle in the southern Puna (25°S–28°S). They found overall low *Vp* values between 25°S and 27°S in the eastern part of southern Puna Plateau above the slab. Below Salar Hombre Muerto (~25.5°S, 66.5°W), just north of Cerro Galan, Bianchi et al. (2013) identified an ~50-km-diameter fast anomaly at ~100 km depth that they interpreted as a delaminated block. Using surface wave tomography, Calixto et al. (2013) found a similar fast anomaly just north of Cerro Galan at 67°W at a depth of ~150 km.

Crustal Structure from Ambient Noise Tomography

A recently published study in the central Andes using ambient noise tomography (ANT) has produced a 3-D shear-wave velocity model for the Andes from 14°S to 40°S. Ward et al. (2013) cross correlated ambient seismic noise recorded at 330 permanent and portable broadband seismic stations deployed between 1994 and 2012 to determine phase velocity maps (8–40 s) and then inverted those for a 3-D shear-wave velocity (V_{sv}) model. A major advantage of ANT compared to other tomography techniques is that it is not based on earthquake sources, which are often not distributed optimally for the volume being imaged, but rather it is based on surface waves that sample the paths between contemporary stations so that a two-dimensional (2-D) array of stations provides an almost optimal resolution for the area within the perimeter of the outermost stations. Adjacent or overlapping arrays from different time periods can be combined to increase the area of coverage. Ward et al. (2013) used Rayleigh waves, so it is sensitive to the absolute velocities of the vertically polarized S-wave, or V_{sv}. The results for four depth layers are shown in Figure 4 for our study area. The ANT results show low shear-wave velocities in the crust through most of the southern Altiplano region (Fig. 4). For example, the V_{sv} 3.5 km/s contour under the central Andean Plateau is deeper than 30 km and locally extends

to 50 km. Previous studies modeling shear-coupled P-waves also showed low P-wave velocities (~5.8–6.2 km/s) in the southern Altiplano crust (Swenson et al., 1999, 2000). All these studies confirm that the Altiplano crust is thick and seismically slow, indicative of a felsic crust that is warm and weak (Beck and Zandt, 2002).

The ANT results show large variations in the shear-wave velocities beneath the Puna region (Fig. 4). The largest anomaly in the shear-wave velocity model is a large low-velocity body that coincides with the area of the Altiplano-Puna volcanic center that is visible throughout the crust in the northern Puna region (Fig. 4). Previous seismic studies of the Altiplano-Puna volcanic center had identified a very low-velocity layer at ~20 km depth and interpreted it as a partially crystallized magma body (Chmielowski et al., 1999; Zandt et al., 2003; de Silva et al., 2006b). Based on the new results in the upper crust (<30 km), the anomaly is outlined by the 2.9 km/s velocity contour and reaches a minimum velocity of <2.4 km/s and defines a slightly elliptical body ~200 km in diameter. The long axis of the elliptical-shaped body is oriented NW-SE, parallel to the alignment of the regional lineaments and faults that have been suggested to control the localization of eruptive centers in the region (de Silva, 1989; Riller et al., 2001). Ward et al. (2013) suggested that the upper-crustal low-velocity anomaly is a magma body and is related to ongoing silicic plutonic formation. In the lower crust (>30 km), the anomaly continues downward as a relatively low-velocity anomaly, although the absolute velocity increases, perhaps reflecting an increasingly more mafic composition of the crust. By 40–45 km depth, the central portion of the anomaly is relatively fast, indicating that the main portion of the thermal anomaly in this region is located in the upper crust, although it persists into the lower crust (Fig. 4).

In the southern Puna region (~24°S–27°S), the crustal shear-wave velocities are generally higher and much more heterogeneous and variable with depth (Fig. 4). In the northern part of this region, a NW-SE alignment of relatively high crustal velocities underlies the Atacama block of the forearc and connects to a zone of high velocities centered under the Arizaro Basin. This high-velocity zone correlates closely to the central Andean gravity high as first defined by Götze and Krause (2002) and bounded on the northeastern edge by the El Toro fault zone (Riller et al., 2001). South of and roughly parallel to this high-velocity zone, there is a low-velocity zone that starts at the arc and extends southeast to the location of the Cerro Galan volcanic field and beyond. In the upper crust, this low-velocity zone is outlined by the 3.2 km/s contour, has a minimum velocity of ~3 km/s, and is significantly smaller in magnitude compared to the Altiplano-Puna low-velocity anomaly. These velocity patterns are most prominent in the upper crust but extend into the lower crust with diminished amplitudes and changing shapes. In the lower crust (35–45 km depth), the low-velocity anomaly has a circular shape centered between the Arizaro Basin and Cerro Galan (Fig. 4). We will later suggest that this lower-crustal anomaly is the result of a recent lithospheric foundering event.

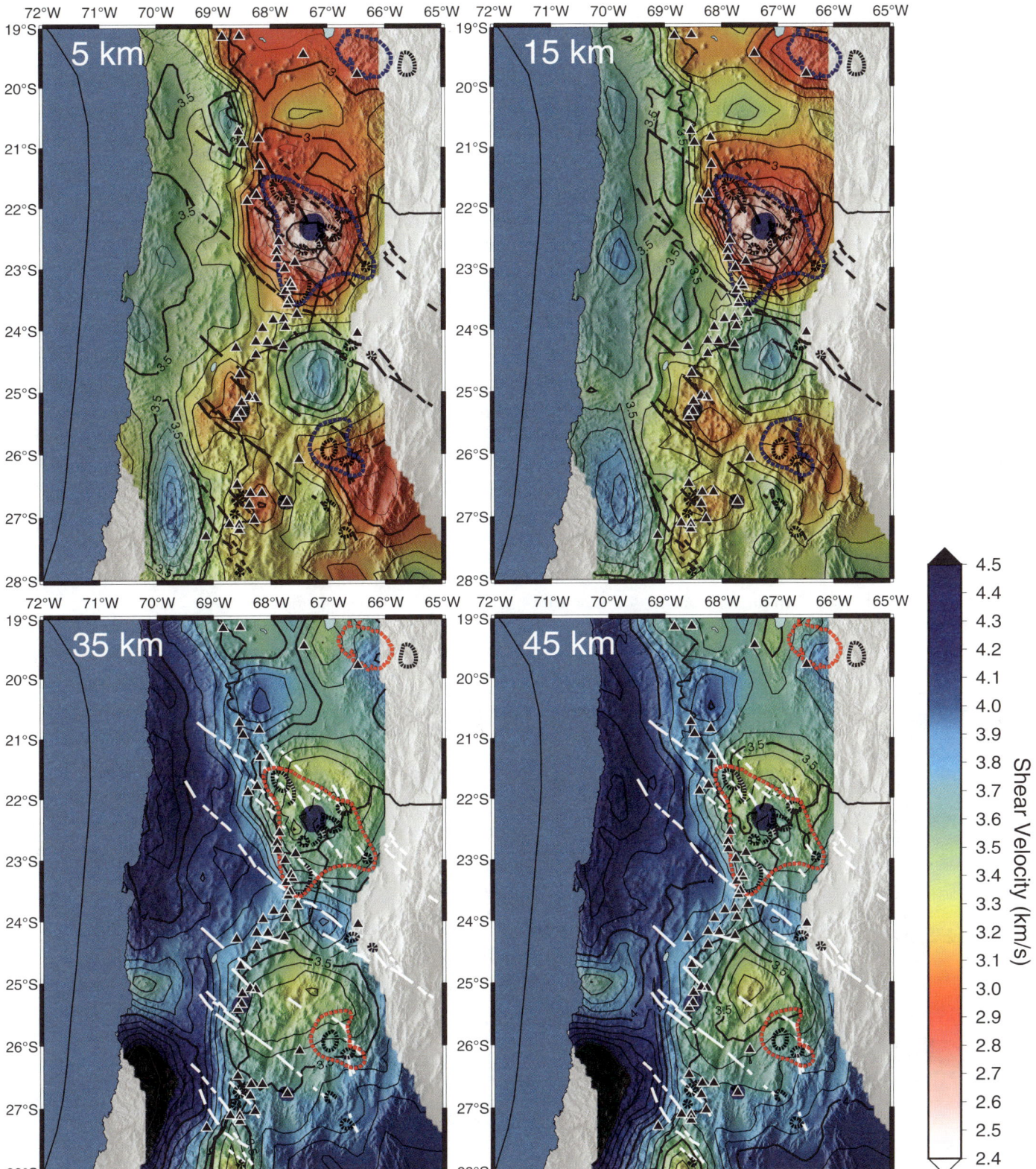

Figure 4. Map views of shear-wave velocity (V_{sv}) results for the upper crust (depths at 5 km and 15 km below sea level) and the lower crust (depths at 35 km and 45 km below sea level) from the ambient noise tomography (Ward et al., 2013). Also shown are NW-SE strike-slip faults (black and white lines), Holocene volcanic centers (black triangles), major ignimbrite regions (purple lines on 5 and 15 km depth slices and red lines on 35 and 45 km depth slices), and surface deformation locations (dark blue circles) from Pritchard and Simons (2004).

Across the entire region, most of active volcanic areas are also underlain by a low-velocity zone at ~15 km depth. The 15 km layer shows a close correlation between low-velocity areas and calderas, ignimbrite volcanic fields, and surface deformation locations (Fig. 4; Pritchard and Simons, 2004). However, the crust at ~24°S associated with the central Andean gravity high (Götze and Krause, 2002) is seismically fast, with shear-wave velocities of >3.6 km/s throughout the upper crust, suggesting that it is different than the crust to the north or south (Fig. 4).

P-Wave Tomography of the Upper Mantle

Scire et al. (this volume) used data collected from 284 seismic stations from 11 different networks operating between 1994 and 2009 for teleseismic P-wave tomography. P and PKIKP arrivals were picked (arrivals picked in four frequency bands) using a multichannel cross-correlation technique (Pavlis and Vernon, 2010). In this study, a finite-frequency P-wave teleseismic inversion scheme from Schmandt and Humphreys (2010) was used to image the mantle from the base of the crust to 700 km depth. Despite the fact that not all the stations were out at the same time, overlapping coverage and the use of permanent stations provided sufficient corrections for datum offsets, as evidenced by resolution tests and the continuity of the imaged subducted slab.

The resolution tests showed that the slab and surrounding mantle were well-imaged from ~150 km to 700 km, with some vertical smearing in the upper 100–150 km. Hence, using a method following Heit et al. (2008), a 4% high-velocity slab

based on the geometry of Slab1.0 model (Hayes et al., 2012) was prescribed to account for the slab in the starting model in order to sharpen the images in the upper 150 km. However, the major anomalies are present in both the "constrained slab" and "unconstrained slab" inversions. Figure 5 shows depth slices at 95 km, 130 km, and 165 km for the P-wave velocity model as percent perturbations from a starting model with the prescribed slab. The results resolve considerable P-wave velocity variations in the upper mantle above the slab beneath the Altiplano and Puna regions. Aside from the slab, the most prominent anomaly is an ~100-km-diameter high-velocity anomaly at ~25°S and 65.5°W that is roughly circular at 95 km depth and more elongate in the north-south direction at 165 km and surrounded by a broad halo of lower velocities. Anomaly "A" is approximately cylindrically shaped and similar in size and shape to the "Isabella anomaly" associated with the purported southern Sierra Nevada "delamination" (Zandt et al., 2004). Although parts of this anomaly were detected by investigators of the PUNA-PUDEL experiment (Bianchi et al., 2013; Calixto et al., 2013), much of the anomaly was on the edge of that deployment, and its precise lateral location and full size and shape were only resolved after combining data with previous deployments. Further south (27.5°S), another high-velocity anomaly, labeled "B," underlies the Sierras Pampeanas (also detected by Bianchi et al., 2013), where we might expect the presence of a thicker mantle lithosphere.

Figure 6 shows comparisons of cross sections of the topography, shear-wave velocities in the crust from ANT, and P-wave velocity perturbations in the mantle from teleseismic

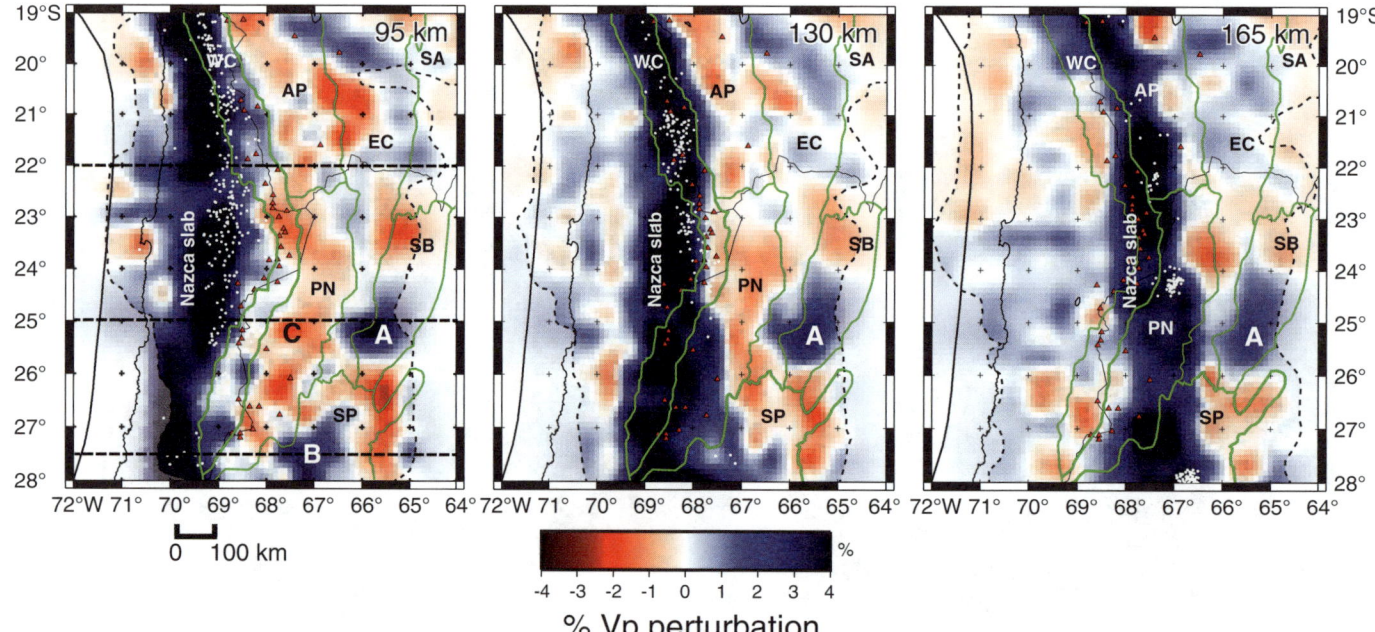

Figure 5. Teleseismic P-wave velocity model, plotted as % perturbation from a starting model with a prescribed slab for depth slices of 95 km, 130 km, and 165 km (Scire et al., this volume). A, B, and C are seismic anomalies discussed in the text. Holocene volcanoes are plotted as red triangles, and morphotectonic province boundaries are plotted as thin green lines. AP—Altiplano, PN—Puna, EC—Eastern Cordillera, SA—Subandean zone, SB—Santa Barbara system, SP—Sierras Pampeanas. White dots are earthquake locations within 17 km of the indicated depth.

tomography at three different latitudes. The three cross sections reveal very different lithospheric structures. At 22°S, the cross section cuts through the Altiplano-Puna volcanic complex and shows very low shear-wave velocities in the upper and middle crust, consistent with the previously identified Altiplano-Puna magma body. At this latitude, the upper mantle has generally low velocities and smaller variations in P-wave velocities across the mantle wedge. The seismically slowest mantle is located directly beneath the slowest crust. The forearc mantle wedge appears fast across a wide zone. The middle cross section at 25°S crosses the Arizaro Basin and the back arc ~100 km north of Cerro Galan. In the upper mantle, at this latitude, there is a very low-velocity anomaly (anomaly C; Fig. 5) centered under the Puna Plateau, and to the east, there is a high-velocity anomaly (anomaly A; Fig. 5), discussed earlier. Immediately above the lowest velocities in the mantle, there is a broad "V"-shaped low-velocity anomaly that connects westward to the arc and eastward under the seismically fast Arizaro Basin crust toward the eastern edge of the Puna Plateau. Cross sections just to the south

(not shown) show this eastern arm connects to an upper-crustal low-velocity zone located below Cerro Galan volcanic field. The southernmost cross section at 27.5°S crosses the Cordillera near the southern termination of the Central volcanic zone and the Puna and passes into the northern Sierras Pampeanas. The dip of the subducting slab is just starting to flatten at this location. Here, we still observe a thick, low-velocity crust under the high topography but a more "normal" higher-velocity crust under the lower average elevation of the block-faulted foreland. In the mantle, we observe low velocities under the arc and the previously described high-velocity anomaly (anomaly B; Fig. 5) directly beneath the crust of the Sierras Pampeanas. We believe the low velocities in the mantle east of 66°W are part of the broad halo of low velocities surrounding anomaly "A" (Fig. 5). In comparison to the cross section at 22°S, the forearc elevation profile and velocity structure appear distinctly different. Before we present our interpretation of these results, we briefly review recent numerical modeling studies of the lithospheric foundering process.

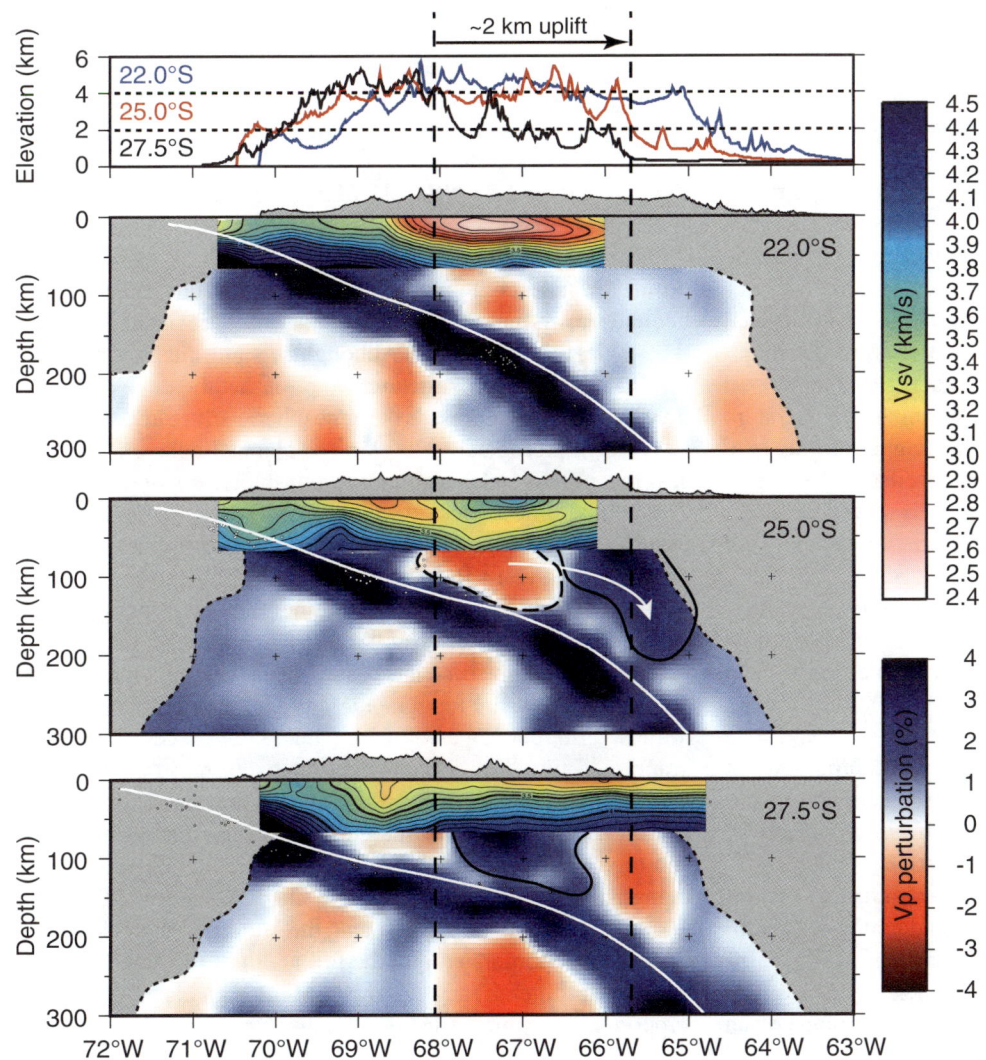

Figure 6. Cross sections of the topography (top panel), ambient noise tomography shear-wave velocity model (Ward et al., 2013), and P-wave tomography model (Scire et al., this volume) for latitudes 22°S (blue line), 25°S (red line), and 27.5°S (black line). The white line is the Slab1.0 model from Hayes et al. (2012). The three cross sections show the removal of lithospheric material at different stages from north to south. The lithosphere removal is more complete at 22°S, in progress at 25°S, and in the early stages with lithosphere still attached at 27.5°S. See text for further discussion.

LITHOSPHERIC FOUNDERING MODELING STUDIES

Lithospheric foundering has been the subject of many recent modeling studies, and a comprehensive review is outside the scope of this paper, and so we refer the interested reader to a concise review in Krystopowicz and Currie (2013). We focus on recent modeling studies that are directly relevant to the central Andes. Sobolev and Babeyko (2005), Babeyko et al. (2006), and Sobolev et al. (2006) published a series of related numerical modeling studies of the central Andes. They utilized a 2-D thermal-mechanical model to simulate the Nazca plate–South American plate convergence constrained by geologic and geophysical data. The upper plate is divided into a weak crust with regions of strong crust on either side in order to accommodate the observed crustal shortening. From these models, they inferred that the most important factors in controlling the deformation in the upper plate were the rate of westward motion of the South American plate, the crustal structure of the upper plate, and the shear coupling at the interplate interface. The models with upper-plate shortening at an average rate of ~10 mm/yr lead to eclogitization of the lower crust in ~15 m.y. and subsequent delamination (their term) of the dense region and adjacent mantle lithosphere. Once initiated, the actual foundering process requires less than ~5 m.y., and the foundered material is removed by the mantle corner flow. In the modeling time frame of 35 m.y., two delaminations occurred, the first near the arc, and the second ~200 km farther inland. These locations are controlled by the edges of the weaker orogenic crust. The combined effects of the two delaminations leave behind a doubled thickness of predominantly upper-crustal felsic crust and a very thin lithospheric mantle overlying hot asthenosphere, a situation conducive to the initiation of extensive crustal melting. The authors noted that during the process of delamination, the foundering material initially obstructs the shallow part of the mantle corner flow, temporarily lowering the temperature and increasing the interplate coupling, and suggested this may lead to an increase in shortening rate and a decrease in arc magmatic activity. Overall, delamination leads to a weaker upper plate, which is then more susceptible to increased tectonic and magmatic activity over a larger span of the orogen.

Krystopowicz and Currie (2013) investigated similar numerical models of delamination in a generic compressional orogen but examined a wider variation in eclogite density and rheology, and in crustal and mantle strength. Variations in these properties control the occurrence, onset time, duration, and style of delamination. In all these models, there is a central zone of relatively weak crust where deformation and thickening (orogen) are focused and where lower-crustal eclogitization can occur, bounded by stronger crust that remains relatively undeformed. The crustal strength difference between the two crustal domains is a key factor in controlling the occurrence and timing of delamination. Delamination occurs when the deforming crust is sufficiently weak and the eclogitized crust is sufficiently dense. The time required for delamination increases as the deforming crustal strength increases and the eclogite density decreases. Delamina-

tion always initiates at the boundary of the weak and strong crust, but mantle strength variations lead to two styles of delamination. In the case of a weak mantle, the delamination initiates at the edge and detaches inward as the weak mantle rolls back and breaks off in large (~200 km) pieces over a period of ~10 m.y. They called this "retreating delamination." Alternatively, in the case of a strong mantle, the delamination initiates at the same location but is too strong to roll back and, instead, sinks vertically beneath the site of initiation, forming a nearly horizontal detachment surface from the overlying crust. They called this "stationary delamination," and it results in the removal of nearly the entire mantle lithosphere from beneath the orogen very quickly (~3 m.y.), although much of the lower crust remains intact. Continued shortening and thickening induce smaller-scale Rayleigh-Taylor–type instabilities that drip into the hot asthenosphere that replaced the delaminated mantle (see figure 8 *in* Krystopowicz and Currie, 2013).

The link between lithospheric foundering and surface topography and uplift is not always straightforward because it depends on the density and rheology of the crust and mantle lithosphere. The pattern of surface uplift varies with the type of lithospheric delamination and changes through time as the delamination event progresses. Hence, surface uplift is affected by many factors, which can lead to a wide range of uplift magnitudes and histories (Krystopowicz and Currie, 2013; Wang et al., this volume).

FOUNDERING BENEATH THE PUNA PLATEAU

Our interpretation of delamination in the Puna region is based primarily on the combined seismic imaging results presented earlier and shown in cross section format in Figure 6 and illustrated schematically in Figure 7. Although seismic imaging can only delineate present-day structure, we posit that the three cross sections in Figure 6 represent the lithospheric structure at different stages of a piecemeal (~200-km-scale areas) delamination process that progressed in a nonuniform fashion from north to south. The southernmost cross section (27.5°S) shows the structure prior to or at an incipient delamination time (predelamination stage), the southern Puna cross section (25°S) shows the structure associated with recent or ongoing removal (waxing stage), and the northern Puna cross section (22°S) shows the evolved structure after some time has passed since delamination (waning stage). We relate the topographic profiles for each cross section to these different stages of the delamination process. The topographic profile at 27.5°S shows a relatively narrow (~200 km) high cordillera (~4 km) transitioning to the much lower average elevation (~2 km) of the "normal" Sierra Pampeanas crust and high-velocity lithospheric upper mantle. In comparison, the topographic profile at 25°S shows a much wider area (~350 km) of high elevations (~4 km), and the previously block-faulted foreland is incorporated into the full width of the high Puna and Eastern Cordillera. The central part of the high elevations is now supported by a thick crust and a low-velocity mantle (asthenosphere), and the eastern edge of the high plateau is underlain by

55

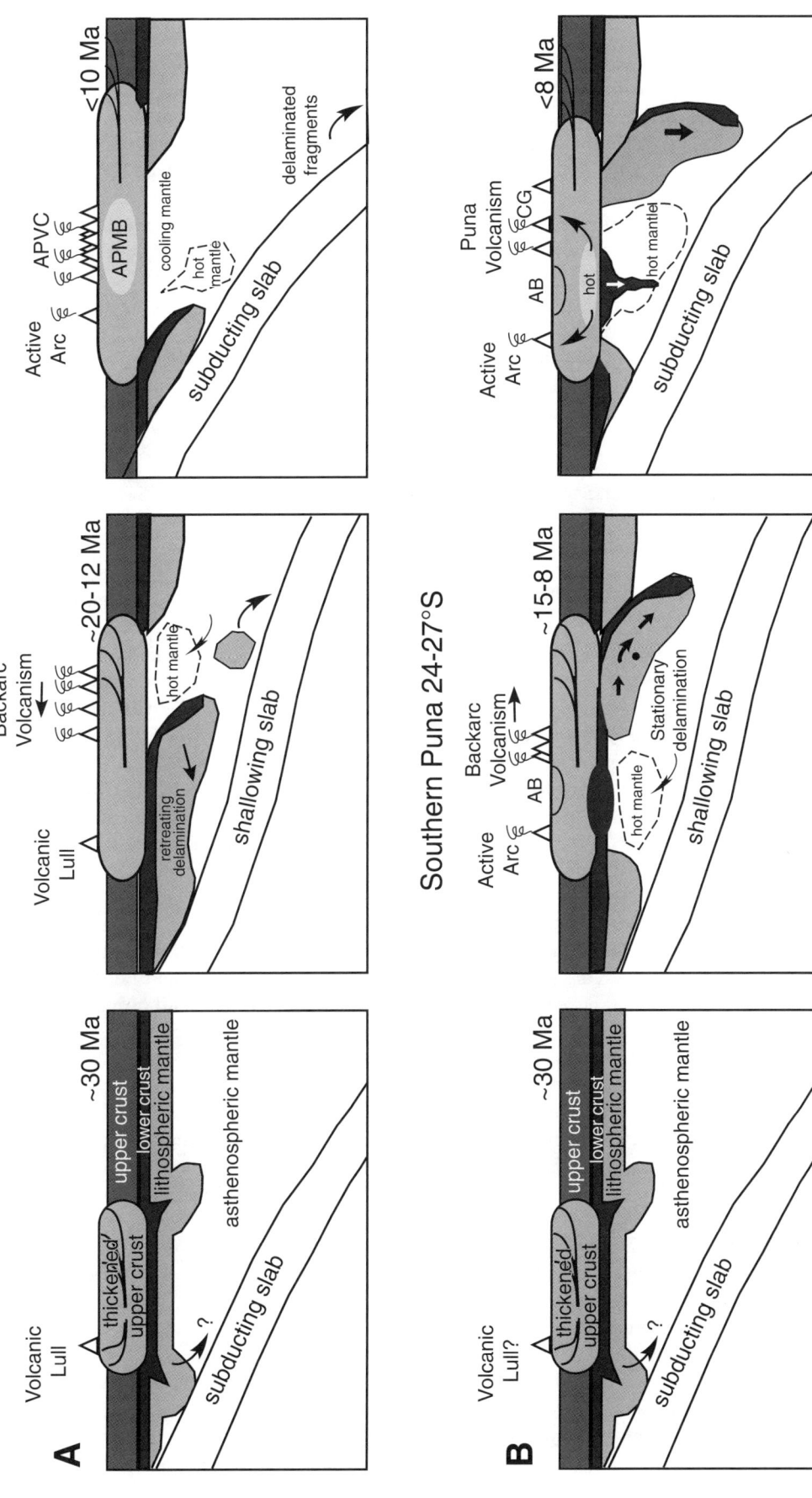

Figure 7. (A) Schematic cross sections showing the evolution of the northern Puna region (top row). Panel on the left starts at ca. 30 Ma. The middle panel shows the start of retreating delamination and associated back-arc volcanism. Right panel shows the completion of the delamination and formation of the Altiplano–Puna magma body (APMB) below the Altiplano–Puna volcanic complex (APVC). (B) Schematic cross sections showing the evolution of the southern Puna region (bottom row). Panel on left starts at ca. 30 Ma. Middle panel shows start of stationary delamination, start of back-arc volcanism, and formation of the Arizaro Basin. Right panel shows completion of stationary delamination and a Raleigh–Taylor–type drip beneath the Arizaro Basin (AB) and the Cerro Galan (CG) ignimbrite field, with asthenosphere beneath.

the delaminating lithospheric mantle. Modeling studies reviewed in the previous section show that as the delaminating mantle is nearing detachment from the crust, the downward forces locally thicken the crust by lower-crustal flow and cause isostatic surface uplift prior to final detachment. The seismic resolution near the crust-mantle boundary at this location is insufficient to confirm this detail. At face value, the elevation difference between 27.5°S and 25°S suggests that the uplift from the lithospheric mantle removal combined with any crustal thickening that occurred during or prior to the removal event can account for ~2 km of uplift, which is roughly consistent with modeling predictions. On the northernmost topographic profile (22°S), the high elevations are somewhat wider, and the relief is more subdued. We attribute this to a thermal relaxation of the heat input from the delamination process. The heat progressively rises into the crust from the mantle, causing crustal melting and crustal flow, widening and flattening the high topography, and leaving a cooling mantle. In the following sections, we describe some additional details and supportive evidence of our interpretation that involve geologic constraints, principally from the volcanic record, which is described in greater detail by Trumbull et al. (2006), Schnurr et al. (2007), Kay and Coira (2009), and Ducea et al. (2013).

Different Styles of Delamination in the Puna Plateau

The two principal arguments for the piecemeal nature of delamination, rather than a wholesale continuous removal under the entire Puna Plateau, are the seismic signatures left in the lower crust and upper mantle (Figs. 4 and 5) and the differing along-strike back-arc volcanic age patterns (Fig. 2). As we described earlier, the two roughly 200-km-wide, circular, low-velocity zones in the lower crust are separated by a ridge of high velocity underlying the location of the central Andean gravity high (Fig. 4). We suggest that these two low-velocity zones mark the locations of the two main delamination centers under the Puna Plateau. While the lower-crustal low-velocity zones are similar, the upper-crustal manifestations are quite different between the northern and southern Puna regions. In the northern Puna region, nearly the entire crust is occupied by a very low-velocity zone; in contrast, the southern Puna upper-crustal structure has generally higher velocities, and the low velocities are more aligned along the NW-SE faults that transect the plateau and are thought to be zones of weakness for magma transport (Riller et al., 2001). The upper-crustal low velocity is slow under the Cerro Galan region and the volcanic arc, but neither is as slow as the slowest velocities associated with the Altiplano-Puna magma body (Figs. 4 and 6). This difference in crustal structure is consistent with the difference in geochemical makeup of the Altiplano-Puna volcanic center ignimbrites and the southern Puna ignimbrites, which indicate much higher levels of crustal melting in the Altiplano-Puna volcanic center (Schnurr et al., 2007). An exception to this difference is the Cerro Galan ignimbrites, which are geochemically similar to the Altiplano-Puna volcanic center ignimbrites. Nonetheless, Schnurr et al. (2007) interpreted this comparison

as indicating that the southern Puna region may be in the early stages of an Altiplano-Puna volcanic center–like volcanic flare-up. Why are the two upper-crustal manifestations of this process so different? One possibility is that the southern area is in an earlier stage of development. Another possibility is that the compositions and rheology of the crust are different. As we reviewed earlier, existing data indicate that the central Andean gravity high region might be associated with either a Paleozoic arc (Götze and Krause, 2002) or the Proterozoic cratonic Antofalla terrane (Ramos, 2008). Perhaps the crust in the southern Puna region is already melt depleted and hence stronger and more resistant to extensive remelting.

We postulate that the change in Cenozoic volcanism in the central Andes that occurred in the time frame of ca. 30–17 Ma (Allmendinger et al., 1997; Kay et al., 1999; Trumbull et al., 2006) was in large part the surface expression of the lithospheric foundering process under the Puna Plateau. As we reviewed earlier, the pattern of volcanism was different in the 21°S–24°S segment (northern Puna region, including the Altiplano-Puna volcanic center) and the 24°S–27°S segment (southern Puna region, including Cerro Galan). In the northern segment, after the volcanic lull, there appears to be a westward-younging migration of volcanism, leading to a relatively recent reestablishment of the frontal arc. The volcanic pattern in the southern segment is quite different. The arc volcanism stepped eastward at ca. 35 Ma and decreased, but the arc was active again by ca. 25 Ma, with the major back-arc broadening starting at ca. 15 Ma. There is a suggestion of eastward younging of the widespread back-arc volcanic activity (Fig. 2C).

The differing volcanic patterns, in conjunction with our new lower-crustal and upper-mantle seismic images, suggest two separate delamination centers with different styles, and perhaps timing of foundering, as illustrated in Figure 7. The differences in patterns of volcanism can be explained by varying styles of foundering related to differences in the strength of the lithosphere (Krystopowicz and Currie, 2013). The northern pattern can be explained by a westward-directed retreating delamination, characteristic of weaker lithosphere that creates a westward-opening lithospheric window while clogging the mantle wedge under the arc (Fig. 7A). The southern pattern can be explained by an eastward-directed stationary delamination, characteristic of stronger lithosphere (Fig. 7B; Krystopowicz and Currie, 2013). In both cases, we attribute the initial arc lull in the northern Puna region and the decrease in arc magmatism in the southern Puna region to tectonically or magmatically induced lithospheric thickening that clogged the shallowest portion of the mantle wedge under the arc. This transferred shortening eastward toward the Eastern Cordillera, first in the northern Puna and later in the southern Puna. This jump in deformation also initiated the foundering process in the Eastern Cordillera, which started the widespread increase in back-arc volcanism ca. 20–23 Ma in the north and ca. 15–17 Ma in the south. Based on the plate-motion reconstructions of Yáñez et al. (2001), the NE-SW–aligned Juan Fernandez Ridge swept southward through this area starting about ca. 14–12 Ma in the north and ca. 12–8 Ma in the south. If the initiation of back-arc volcanism signals the beginning of foundering, and if the plate reconstructions are correct, the onset of flat subduction postdated the initiation of foundering and could not have been the direct causative agent.

On the other hand, a migrating Juan Fernandez Ridge could significantly shallow the slab dip and increase the interplate coupling, leading to more forearc erosion (and sediment subduction), larger stresses, and greater deformation in the upper plate. The shallow slab sweeping southeast would likely cause interactions with foundering pieces, mixing smaller pieces in the mantle wedge and sweeping larger fragments to the south and southeast. After delamination and passage of the Juan Fernandez Ridge, the upper-plate lithosphere would have a thick and weak crust, thin and weakened lithospheric mantle, and be underlain by a mixture of hot asthenosphere and forearc sediments and lithospheric fragments. This situation would be favorable for onset of high melt production in the mantle, and, in conjunction with crustal underthrusting in the fold-and-thrust belt, lead to increased crustal melting, and a silicic volcanic "flare-up." The Juan Fernandez Ridge had completed its sweep past the northern Puna area by ca. 12–10 Ma, and the first ignimbrite eruptions marking the onset of the Altiplano-Puna volcanic center flare-up occurred at ca. 11 Ma. In the northern Puna region, the delaminated material is not visible in the teleseismic P-wave tomography (Fig. 6), but locally higher-resolution tomography using local earthquakes has imaged "delaminated blocks" near the slab (Schurr et al., 2006).

In the southern Puna (22°S–27°S), this process was delayed, perhaps due to the southward migration of the Juan Fernandez Ridge (it completed its sweep past the Cerro Galan area only ca. 8 Ma) or due to the different composition of the crust, or a combination of these two factors. In either case, we suggest a large-scale delamination event initiated ca. 15 Ma but was interrupted by the passage of the Juan Fernandez Ridge through the area between ca. 10 Ma and 8 Ma. The large-scale stationary delamination progressed rapidly to the east, perhaps in part driven by the migration of the Juan Fernandez Ridge, and it only recently started to approach completion. Seismic tomography shows high velocities in the uppermost mantle under the Eastern Cordillera that we interpret as the delaminated material (anomaly A; Figs. 5 and 6). After passage of the Juan Fernandez Ridge, the conditions were set for starting a crustal melting event similar to the Altiplano-Puna volcanic center, but the process is still in its early stages (Fig. 7B). In the north, the melting event is in its waning stages, with a highly developed upper-crustal magmatic system and a waning thermal anomaly in the underlying mantle. In contrast, in the south, the melting event is still in its waxing stage, with prominent low-velocity anomalies in the mantle and lower crust but less melting in the upper crust, except locally, e.g., at Cerro Galan (Schnurr et al., 2007, Sobolev et al., 2006).

Small-Scale Rayleigh-Taylor Instabilities: Arizaro Basin

There is also evidence of a smaller local drip beneath the Arizaro Basin. The Arizaro Basin is located just east of the active arc at a latitude of ~25°S (Fig. 1). The basin is roughly circular,

100 km in diameter, and filled with ~3 km of sediments deposited during Miocene time (ca. 21–9 Ma; DeCelles et al., this volume). The geologic data from this saucer-shaped basin do not fit with most of the common origins of orogenic basins, such as topographic-load forearc basins, intra-arc basins, postcontraction extensional basins, or piggyback basins. Its formation has been attributed instead to the removal of an eclogitized lower crust in a Rayleigh-Taylor gravitational instability (DeCelles et al., this volume).

Wang et al. (this volume) used numerical models to specifically assess the Rayleigh-Taylor instability drip origin hypothesis for the Arizaro Basin. They used 2-D, finite-element, thermal-mechanical modeling to test a range of physical parameters and boundary conditions that are capable of fitting the main features of the basin formation. Their modeling demonstrated that a Rayleigh-Taylor–type gravitational instability could produce a basin of the correct size over a time scale that is consistent with the observations from Arizaro Basin. The model that fit the data best required a lower-crustal high-density anomaly embedded in a strong crust within a thin lithosphere. Over a model time span of ~15 m.y., the eclogitized portion of the ductile lower crust contracts, thickens, and "drips" through the thin lithosphere into the underlying mantle via a narrow, near-vertical conduit, entraining a thin layer of lithospheric mantle (Fig. 7B). This scale of "drip" is too small to be seismically imaged using currently available data and techniques.

Several aspects of the geophysics and geochemistry of the volcanics of the region are consistent with features of the localized instability model. As we have shown, the Arizaro Basin and its salar are underlain by crust with relatively fast and dense upper- to middle-crustal seismic velocities. Correlations with basement geology, as reviewed earlier, agree with the modeling requirement of stronger craton-like crust in this location. Also, there are multiple studies that indicate the lithospheric mantle is thin and the upper mantle is hot and probably melt laden (Schurr et al., 2006; Heit et al., 2008; Scire et al., this volume). Our ANT images show low-velocity zones in the lower crust that are consistent with lower-crustal flow toward the base of the crust under the basin (Fig. 6). Finally, the geochemistry of young mafic volcanics in the region indicates a shallow mantle source produced by melting of small sinking fragments of eclogitic materials like the Rayleigh-Taylor instability drip of lower-crustal material (Ducea et al., 2013).

CONCLUSIONS

Lithospheric foundering or delamination has been recognized as an important process in the formation of the Andes. We use recently published ambient noise tomography results for the crust and new P-wave tomography results for the upper mantle and other geologic and geophysical information to identify two regions of delamination in the Puna region between 21°S and 27°S. Based on the distribution of low-velocity zones in the lower crust (35–45 km), we postulate two separate zones of delamination in the Puna region, one in the northern Puna region centered under the Altiplano-Puna volcanic center and the other in the southern Puna region centered approximately between the Arizaro Basin and Cerro Galan. While the lower-crustal low-velocity zones are similar in both regions, the upper-crustal manifestations are quite different. In the northern Puna region, nearly the entire crust is occupied by a very low-velocity zone. In contrast, under the southern Puna region, much of the upper crust is seismically fast, and the lower crust and uppermost mantle are slow. The upper mantle is heterogeneous beneath the northern Puna but overall has low to moderate P-wave velocities, suggesting that the foundered lithosphere has already been mostly recycled into the mantle. In contrast, the upper mantle beneath the southern Puna has very low velocities adjacent to a very high-velocity anomaly that we interpret as asthenosphere replacing recently foundered lithospheric material (Fig. 6). The differing patterns suggest two separate delamination centers with different styles and timing of foundering (Fig. 7). The differences in patterns of volcanism can be explained by varying styles of foundering related to lithospheric strength (Krystopowicz and Currie, 2013). We summarize the main conclusions as follows:

1. The northern Puna volcanic pattern can be explained by a westward-propagating retreating delamination that created a westward-opening lithospheric window while clogging the mantle wedge under the arc.

2. The southern Puna volcanic pattern can be explained by an eastward-directed stationary delamination that completed more recently.

3. The different styles of delamination may be due to different strengths of the lithosphere, which is reflected in the different basement geology.

4. In both cases, we attribute the initial arc lull in magmatism in the northern Puna region and the decrease in arc magmatism in the southern Puna region to tectonically or magmatically induced lithospheric thickening, which clogged the shallowest portion of the mantle wedge.

5. This also transferred thickening to the east, where the foundering initiated, and started the widespread increase in back-arc volcanism ca. 20–23 Ma in the north and ca. 15 Ma in the south.

6. Based on the plate-motion reconstructions of Yáñez et al. (2001), the NE-SW–aligned Juan Fernandez Ridge swept southward through this area starting ca. 14–12 Ma in the north and ca. 12–8 Ma in the south. The Juan Fernandez Ridge may have decreased the slab dip (but not necessarily with completely horizontal subduction), causing increased interplate stress, and further weakening the upper plate and facilitating delamination.

7. As the larger-scale delamination progressed, leaving behind thin lithosphere and hot asthenosphere, a smaller secondary Rayleigh-Taylor–type instability occurred beneath the Arizaro Basin.

ACKNOWLEDGMENTS

This study was supported by the Convergent Orogenic Systems Analysis (COSA) project, funded by ExxonMobil and National Science Foundation grant EAR-0907880. We thank all of our COSA colleagues and other Andes researchers for discussions over the past several years. The paper was substantially improved by reviews from Matt Pritchard and Victor Ramos.

REFERENCES CITED

Allmendinger, R.W., Ramos, V.A., Jordan, T.E., Palma, M., and Isacks, B.L., 1983, Paleogeography and Andean structural geometry, northwest Argentina: Tectonics, v. 2, p. 1–16, doi:10.1029/TC002i001p00001.

Allmendinger, R., Jordan, T., Kay, S.M., and Isacks, B., 1997, The evolution of the Altiplano-Puna Plateau of the central Andes: Annual Review of Earth and Planetary Sciences, v. 25, p. 139–174, doi:10.1146/annurev.earth.25.1.139.

Asch, G., Schurr, B., Bohm, M.C., Haberland, C., Heit, B., Kind, R., Wolbern, I., Yuan, X., Bataille, K., Comte, D., Pardo, M., Viramonte, J., Rietbrock, A., and Giese, P., 2006, Seismological studies of the central and southern Andes, *in* Oncken, O., Chong, G., Franz, G., Giese, P., Goetze, H.-J., Ramos, V., Strecker, M., and Wigger, P., eds., The Andes: Active Subduction Orogeny: Berlin, Springer, Frontiers in Earth Sciences, p. 443–457.

Babeyko, A.Y., Sobolev, S.V., Vietor, T., Oncken, O., and Trumbull, R.B., 2006, Numerical study of weakening processes in the central Andean back-arc, *in* Oncken, O., Chong, G., Franz, G., Giese, P., Goetze, H.-J., Ramos, V., Strecker, M., and Wigger, P., eds., The Andes: Active Subduction Orogeny: Berlin, Springer, Frontiers in Earth Sciences, p. 495–512.

Baby, P., Rochat, P., Mascle, G., and Herail, G., 1997, Neogene shortening contribution to crustal thickening in the back arc of the central Andes: Geology, v. 25, p. 883–886, doi:10.1130/0091-7613(1997)025<0883:NSCTCT>2.3.CO;2.

Barnes, J.B., and Ehlers, T.A., 2009, End member models for Andean Plateau uplift: Earth-Science Reviews, v. 97, p. 105–132, doi:10.1016/j.earscirev.2009.08.003.

Beck, S.L., and Zandt, G., 2002, Nature of orogenic crust in the central Andes: Journal of Geophysical Research, v. 107, 2230, doi:10.1029/2000JB000124.

Bianchi, M., Heit, B., Jakovlev, A., Yuan, X., Kay, S.M., Sandvol, E., Alonso, R.N., Coira, B., Brown, L., Kind, R., and Comte, D., 2013, Teleseismic tomography of the southern Puna Plateau in Argentina and adjacent regions: Tectonophysics, v. 586, p. 65–83, doi:10.1016/j.tecto.2012.11.016.

Calixto, F.J., Sandvol, E., Kay, S., Comte, D., Alvarado, P., Heit, B., and Yuan, X., 2013, Velocity structure beneath the southern Puna Plateau: Evidence for delamination: Geochemistry Geophysics Geosystems, v. 14, p. 4292–4305, doi:10.1002/ggge.20266.

Carrapa, B., and DeCelles, P.G., 2008, Eocene exhumation and basin development in the Puna of northwestern Argentina: Tectonics, v. 27, TC1015, doi:101029/2007/TC002127.

Carrapa, B., and DeCelles, P.G., 2015, this volume, Regional exhumation and kinematic history of the central Andes in response to cyclical orogenic processes, *in* DeCelles, P.G., Ducea, M.N., Carrapa, B., and Kapp, P.A., eds., Geodynamics of a Cordilleran Orogenic System: The Central Andes of Argentina and Northern Chile: Geological Society of America Memoir 212, doi:10.1130/2015.1212(11).

Chmielowski, J., Zandt, G., and Haberland, C., 1999, The central Andean Altiplano-Puna magma body: Geophysical Research Letters, v. 26, p. 783–786, doi:10.1029/1999GL900078.

Coira, B., Kay, S.M., and Viramonte, J., 1993, Upper Cenozoic magmatic evolution of the Argentine Puna—A model for changing subduction geometry: International Geology Review, v. 35, p. 677–720, doi:10.1080/00206819309465552.

DeCelles, P.G., and Horton, B.K., 2003, Early to Middle Tertiary foreland basin development and the history of Andean crustal shortening in Bolivia: Geological Society of America Bulletin, v. 115, p. 58–77, doi:10.1130/0016-7606(2003)115<0058:ETMTFB>2.0.CO;2.

DeCelles, P.G., Ducea, M.N., Kapp, P., and Zandt, G., 2009, Cyclicity in Cordilleran orogenic systems: Nature Geoscience, v. 2, p. 251–257, doi:10.1038/ngeo469.

DeCelles, P.G., Carrapa, B., Horton, B.K., and Gehrels, G.E., 2011, Cenozoic foreland basin system in the central Andes of northwestern Argentina: Implications for Andean geodynamics and modes of deformation: Tectonics, v. 30, p. 1–30, doi:10.1029/2011TC002948.

DeCelles, P.G., Carrapa, B., Horton, B.K., McNabb, J., Gehrels, G.E., and Boyd, J., 2015, this volume, The Miocene Arizaro basin, central Andean hinterland: Response to partial lithosphere removal? *in* DeCelles, P.G., Ducea, M.N., Carrapa, B., and Kapp, P.A., eds., Geodynamics of a Cordilleran Orogenic System: The Central Andes of Argentina and Northern Chile: Geological Society of America Memoir 212, doi:10.1130/2015.1212(18).

de Silva, S., 1989, Altiplano-Puna volcanic complex of the central Andes: Geology, v. 17, p. 1102–1106, doi:10.1130/0091-7613(1989)017<1102 :APVCOT>2.3.CO;2.

de Silva, S., Zandt, G., Trumble, R.T., Viramonte, J.G., Salas, G., and Jimenez, N., 2006a, Large ignimbrite eruptions and volcano-tectonic depressions in the central Andes: A thermomechanical perspective, *in* Troise, C., De Natale, G., and Kilburn, C.R.J., eds., Mechanisms of Activity and Unrest at Large Calderas: Geological Society, London, Special Publication 269, p. 47–63.

de Silva, S., Zandt, G., Trumbull, R., and Viramonte, J., 2006b, Large-scale silicic volcanism—The result of thermal maturation of the crust: Advances in Geosciences, v. 1, p. 215–230, doi:10.1142/9789812707178_0021.

Ducea, M.N., Seclaman, A.C., Murray, K.E., Jianu, D., and Schoenbohm, L.M., 2013, Mantle-drip magmatism beneath the Altiplano-Puna Plateau, central Andes: Geology, v. 41, no. 8, p. 915–918, doi:10.1130/G34509.1.

Elger, K., Oncken, O., and Glodny, J., 2005, Plateau-style accumulation of deformation—The Southern Altiplano: Tectonics, v. 24, TC4020, doi:10.1029/2004TC001675.

Garzione, C.N., Molnar, P., Libarkin, J.C., and MacFadden, B.C., 2006, Rapid late Miocene rise of the Bolivian Altiplano: Evidence for removal of mantle lithosphere: Earth and Planetary Science Letters, v. 241, p. 543–556, doi:10.1016/j.epsl.2005.11.026.

Ghosh, P., Garzione, C.N., and Eiler, J.M., 2006, Rapid uplift of the Altiplano revealed through ^{13}C–^{18}O bonds in paleosol carbonates: Science, v. 311, no. 5760, p. 511–515, doi:10.1126/science.1119365.

Gotberg, N., McQuarrie, N., and Carlotto Caillaux, V., 2010, Comparison of crustal thickening budget and shortening estimates in southern Peru (12–14°S): Implications for mass balance and rotations in the "Bolivian orocline": Geological Society of America Bulletin, v. 122, p. 727–742, doi:10.1130/B26477.1.

Götze, H.-J., and Krause, S., 2002, The central Andean gravity high, a relic of an old subduction complex?: Journal of South American Earth Sciences, v. 14, p. 799–811, doi:10.1016/S0895-9811(01)00077-3.

Graeber, F., and Asch, G., 1999, Three-dimensional models of P-wave velocity and P-to-S-velocity ratio in the southern central Andes by simultaneous inversion of local earthquake data: Journal of Geophysical Research, v. 104, p. 20,237–20,256, doi:10.1029/1999JB900037.

Gutscher, M., Spakman, W., Bijwaard, H., and Engdahl, E.R., 2000, Geodynamics of flat subduction: Seismicity and tomographic constraints from the Andean margin: Tectonics, v. 19, no. 5, p. 814–833, doi:10.1029/1999TC001152.

Haberland, C., and Rietbrock, A., 2001, Attenuation tomography in the western central Andes: A detailed insight into the structure of a magmatic arc: Journal of Geophysical Research, v. 106, p. 11,151–11,167, doi:10.1029/2000JB900472.

Haberland, C., Rietbrock, A., Schurr, B., and Brasse, H., 2003, Coincident anomalies of seismic attenuation and electrical resistivity beneath the southern Bolivian Altiplano Plateau: Geophysical Research Letters, v. 30, 1923, doi:10.1029/2003GL017492.

Hayes, G.P., Wald, D.J., and Johnson, R.J., 2012, Slab1.0: A three-dimensional model of global subduction zone geometries: Journal of Geophysical Research, v. 117, B01302, doi:10.1029/2011JB008524.

Heit, B., Sodoudi, F., Yuan, X., Bianchi, M., and Kind, R., 2007, An S-receiver function analysis of the lithospheric structure in South America: Geophysical Research Letters, v. 34, L14307, doi:10.1029/2007GL030317.

Heit, B., Koulakov, I., Asch, G., Yuan, X., Kind, R., Alcozer, I., Tawackoli, S., and Wilke, H., 2008, More constraints to determine the seismic structure beneath the central Andes at 21°S using teleseismic tomography analysis: Journal of South American Earth Sciences, v. 25, p. 22–36, doi:10.1016/j .jsames.2007.08.009.

Hoke, L., and Lamb, S., 2007, Cenozoic behind-arc volcanism in the Bolivian Andes, South America: Implications for mantle melt generation and lithospheric structure: Journal of the Geological Society, London, v. 164, p. 795–814, doi:10.1144/0016-76492006-092.

Isacks, B.L., 1988, Uplift of the central Andean Plateau and bending of the Bolivian orocline: Journal of Geophysical Research–Solid Earth, v. 93, p. 3211–3231, doi:10.1029/JB093iB04p03211.

James, D.E., and Sacks, I.S., 1999, Cenozoic formation of the central Andes: A geophysical perspective, *in* Skinner, B., ed., Geology and Ore Deposits of the Central Andes: Society of Economic Geology Special Publication 7, p. 1–25.

Jordan, T.E., and Allmendinger, R.W., 1986, The Sierras Pampeanas of Argentina: A modern analogue of Rocky Mountain foreland deformation: American Journal of Science, v. 286, p. 737–764, doi:10.2475/ajs.286.10.737.

Kay, R.W., and Kay, S.M., 1993, Delamination and delamination magmatism: Tectonophysics, v. 219, p. 177–189, doi:10.1016/0040-1951(93)90295-U.

Kay, S.M., and Coira, B.L., 2009, Shallowing and steepening subduction zones, continental lithospheric loss, magmatism, and crustal flow under the central Andean Altiplano-Puna Plateau, *in* Kay, S.M., Ramos, V.A., and Dickinson, W.R., eds., Backbone of the Americas: Shallow Subduction, Plateau Uplift, and Ridge and Terrane Collision: Geological Society of America Memoir 204, p. 229–259, doi:10.1130/2009.1204(11).

Kay, S.M., Coira, B., and Viramonte, J., 1994, Young mafic back-arc volcanic rocks as indicators of continental lithospheric delamination beneath the Argentine Puna Plateau, central Andes: Journal of Geophysical Research, v. 99, p. 24,323–24,339, doi:10.1029/94JB00896.

Kay, S.M., Mpodozis, C., and Coira, B., 1999, Magmatism, tectonism and mineral deposits of the central Andes (22°–33°S latitude), *in* Skinner, B.J., ed., Geology and Ore Deposits of the Central Andes: Society of Economic Geology Special Publication 7, p. 27–59.

Kay, S.M., Coira, B.L., Caffe, P.J., and Chen, C.H., 2010, Regional chemical diversity, crustal and mantle sources and evolution of central Andean Puna Plateau ignimbrites: Journal of Volcanology and Geothermal Research, v. 198, p. 81–111, doi:10.1016/j.jvolgeores.2010.08.013.

Kley, J., 1996, Transition from basement-involved to thin-skinned thrusting in the Cordillera Oriental of southern Bolivia: Tectonics, v. 15, no. 4, p. 763–775, doi:10.1029/95TC03868.

Kley, J., 1999, Geologic and geometric constraints on a kinematic model of the Bolivian orocline: Journal of South American Earth Sciences, v. 12, p. 221–235, doi:10.1016/S0895-9811(99)00015-2.

Kley, J., and Monaldi, C.R., 1998, Tectonic shortening and crustal thickness in the central Andes: How good is the correlation?: Geology, v. 26, no. 8, p. 723–726, doi:10.1130/0091-7613(1998)026<0723:TSACTI>2.3.CO;2.

Kley, J., and Monaldi, C.R., 2002, Tectonic inversion in the Santa Barbara system of the central Andean foreland thrust belt, northwestern Argentina: Tectonics, v. 21, no. 6, doi:10.1029/2002TC902003.

Koulakov, I., Sobolev, S., and Asch, G., 2006, P- and S-velocity images of the lithosphere–asthenosphere system in the central Andes from local source tomographic inversion: Geophysical Journal International, v. 167, p. 106–126, doi:10.1111/j.1365-246X.2006.02949.x.

Krystopowicz, N.J., and Currie, C.A., 2013, Crustal eclogitization and lithosphere delamination in orogens: Earth and Planetary Science Letters, v. 361, p. 195–207, doi:10.1016/j.epsl.2012.09.056.

Lamb, S., and Hoke, L., 1997, Origin of the high plateau in the central Andes, Bolivia, South America: Tectonics, v. 16, p. 623–649, doi:10.1029/97TC00495.

McGlashan, N., Brown, L.D., and Kay, S.M., 2008, Crustal thicknesses in the central Andes from teleseismically recorded depth phase precursors: Geophysical Journal International, v. 175, p. 1013–1022, doi:10.1111/j.1365 -246X.2008.03897.x.

McQuarrie, N., 2002, The kinematic history of the central Andean fold-thrust belt, Bolivia: Implications for building a high plateau: Geological Society of America Bulletin, v. 114, p. 950–963, doi:10.1130/0016 -7606(2002)114<0950:TKHOTC>2.0.CO;2.

McQuarrie, N., Horton, B.K., Zandt, G., Beck, S.L., and DeCelles, P.G., 2005, Lithospheric evolution of the Andean fold-thrust belt, Bolivia, and the origin of the central Andean plateau: Tectonophysics, v. 399, p. 15–37, doi:10.1016/j.tecto.2004.12.013.

Müller, J.P., Kley, J., and Jacobshagen, V., 2002, Structure and Cenozoic kinematics of the Eastern Cordillera, southern Bolivia (21°S): Tectonics, v. 21, no. 5, doi:10.1029/2001TC001340.

Myers, S., Beck, S., Zandt, G., and Wallace, T., 1998, Lithospheric-scale structure across the Bolivian Andes from tomographic images of velocity and attenuation for P and S waves: Journal of Geophysical Research, v. 103, p. 21,233–21,252, doi:10.1029/98JB00956.

Oncken, O., Hindle, D., Kley, J., Elger, K., Victor, P., and Schemmann, K., 2006, Deformation of the central Andean upper plate system—Facts, fiction, and constraints for plateau models, *in* Oncken, O., Chong, G., Franz, G., Giese, P., Goetze, H.-J., Ramos, V., Strecker, M., and Wigger, P., eds., The Andes: Active Subduction Orogeny: Berlin, Springer, Frontiers in Earth Sciences, p. 3–27.

Pavlis, G., and Vernon, F., 2010, Array processing of teleseismic body waves with the USArray: Computers & Geosciences, v. 36, p. 910–920, doi:10.1016/j.cageo.2009.10.008.

Petrinovic, I.A., Martí, J., Aguirre-Díaz, G.J., Guzmán, S., Geyer, A., and Salado Paz, N., 2010, The Cerro Aguas Calientes caldera, NW Argentina: An example of a tectonically controlled, polygenetic collapse caldera, and its regional significance: Journal of Volcanology and Geothermal Research, v. 194, p. 15–26, doi:10.1016/j.jvolgeores.2010.04.012.

Pritchard, M.E., and Simons, M., 2004, An InSAR-based survey of volcanic deformation in the central Andes: Geochemistry Geophysics Geosystems, v. 5, Q02002, doi:10.1029/2003GC000610.

Quade, J., Dettinger, M.P., Carrapa, B., DeCelles, P., Murray, K.E., Huntington, K.A., Cartwright, A., Canavan, R.R., Gehrels, G., and Clementz, M., 2015, this volume, The Growth of the Central Andes 22–26°S, *in* DeCelles, P.G., Ducea, M.N., Carrapa, B., and Kapp, P.A., eds., Geodynamics of a Cordilleran Orogenic System: The Central Andes of Argentina and Northern Chile: Geological Society of America Memoir 212, doi:10.1130/2015.1212(15).

Ramos, V., 1999, Las Provincias geologicas del territorio Argentino, *in* Caminos, R., ed., Geologia Argentina: Servicio Geologico Minero Argentino, Anales, v. 29, p. 41–96.

Ramos, V., 2008, The Arequipa and related terranes: Annual Review of Earth and Planetary Sciences, v. 36, p. 289–324, doi:10.1146/annurev.earth.36.031207.124304.

Ramos, V., 2009, Anatomy and global context of the Andes: Main geologic features and the Andean orogenic cycle, *in* Kay, S.M., Ramos, V.A., and Dickinson, W.R., eds., Backbone of the Americas: Shallow Subduction, Plateau Uplift, and Ridge and Terrane Collision: Geological Society of America Memoir, 204, p. 31–65, doi:10.1130/2009.1204(02).

Ramos, V., and Folguera, A., 2009, Mesozoic-Cenozoic orogens, *in* Murphy, J.B., Keppie, J.D., and Hynes, A.J., eds., Ancient Orogens and Modern Analogues: Geological Society, London, Special Publication 327, p. 31–54, doi:10.1144/SP327.3.

Reiners, P.W., Thomson, S.N., Vernon, A., Willett, S.D., Zattin, M., Einhorn, J., Gehrels, G., Quade, J., Pearson, D., Murray, K.E., and Cavazza, W., 2015, this volume, Low-temperature thermochronologic trends across the central Andes, 21–28° S, *in* DeCelles, P.G., Ducea, M.N., Carrapa, B., and Kapp, P.A., eds., Geodynamics of a Cordilleran Orogenic System: The Central Andes of Argentina and Northern Chile: Geological Society of America Memoir 212, doi:10.1130/2015.1212(12).

Riller, U., Petrinovic, I., Ramelow, J., Strecker, M., and Oncken, O., 2001, Late Cenozoic tectonism, collapse caldera and plateau formation in the central Andes: Earth and Planetary Science Letters, v. 188, p. 299–311, doi:10.1016/S0012-821X(01)00333-8.

Salisbury, M.J., Jicha, B.R., de Silva, S.L., Singer, B.S., Jimenez, N.C., and Ort, M.H., 2011, Ar-40/Ar-39 chronostratigraphy of Altiplano-Puna volcanic complex ignimbrites reveals the development of a major magmatic province: Geological Society of America Bulletin, v. 123, p. 821–840, doi:10.1130/B30280.1.

Sambridge, M., Braun, J., and McQueen, H., 1995, Geophysical parameterization and interpolation of irregular data using natural neighbours: Geophysical Journal International, v. 122, no. 3, p. 837–857, doi:10.1111/j.1365-246X.1995.tb06841.x.

Schmandt, B., and Humphreys, E., 2010, Seismic heterogeneity and small-scale convection in the southern California upper mantle: Geochemistry Geophysics Geosystems, v. 11, Q05004, doi:10.1029/2010GC003042.

Schnurr, W.B.W., Trumbull, R.B., Clavero, J., Hahne, K., Siebel, W., and Gardeweg, M., 2007, Twenty-five million years of felsic magmatism in the southern Central volcanic zone of the Andes: Geochemistry and magma genesis of ignimbrites from 25–27°S, 67–72°W: Journal of Volcanology and Geothermal Research, v. 166, no. 1, p. 17–46, doi:10.1016/j.jvolgeores.2007.06.005.

Schurr, B., Rietbrock, A., Asch, G., Kind, R., and Oncken, O., 2006, Evidence for lithospheric detachment in the central Andes from local earthquake tomography: Tectonophysics, v. 415, p. 203–223, doi:10.1016/j.tecto.2005.12.007.

Scire, A., Biryol, C.B., Zandt, G., and Beck, S., 2015, this volume, Imaging the Nazca slab and surrounding mantle to 700 km depth beneath the central Andes (18°S to 28°S), *in* DeCelles, P.G., Ducea, M.N., Carrapa, B., and Kapp, P.A., eds., Geodynamics of a Cordilleran Orogenic System: The Central Andes of Argentina and Northern Chile: Geological Society of America Memoir 212, doi:10.1130/2015.1212(02).

Sobolev, S.V., and Babeyko, A.Y., 2005, What drives orogeny in the Andes?: Geology, v. 33, p. 617–620, doi:10.1130/G21557.1.

Sobolev, S.V., Babeyko, A.Y., Koulakov, I., Oncken, O., and Vietor, T., 2006, Mechanism of the Andean orogeny: Insight from the numerical modeling,

in Oncken, O., Chong, G., Franz, G., Giese, P., Goetze, H.-J., Ramos, V., Strecker, M., and Wigger, P., eds., The Andes: Active Subduction Orogeny: Berlin, Springer-Verlag, Frontiers in Earth Sciences, p. 509–531.

Swenson, J.L., Beck, S.L., and Zandt, G., 1999, Regional distance propagation of shear-coupled *P*-waves within the Northern Altiplano: Geophysical Journal International, v. 139, no. 3, p. 743–753, doi:10.1046/j.1365-246x.1999.00968.x.

Swenson, J.L., Beck, S.L., and Zandt, G., 2000, Crustal structure of the Altiplano from broadband regional waveform modeling: Implications for the composition of thick continental crust: Journal of Geophysical Research, v. 105, p. 607–621, doi:10.1029/1999JB900327.

Tassara, A., 2006, Factors controlling the crustal density structure underneath active continental margins with implications for their evolution: Geochemistry Geophysics Geosystems, v. 7, Q01001, doi:10.1029/2005GC001040.

Tassara, A., and Echaurren, A., 2012, Anatomy of the Andean subduction zone: Three-dimensional density model upgraded and compared against global-scale models: Geophysical Journal International, v. 189, p. 161–168, doi:10.1111/j.1365-246X.2012.05397.x.

Trumbull, R.B., Riller, U., Oncken, O., Scheuber, E., Munier, K., and Hongn, F., 2006, The time-space distribution of Cenozoic arc volcanism in the south-central Andes: A new data compilation and some tectonic implications, *in* Oncken, O., Chong, G., Franz, G., Giese, P., Götze, H-J., Ramos, V.A., Strecker, M.R., and Wigger, P., eds., The Andes—Active Subduction Orogeny: Berlin, Springer-Verlag, Frontiers in Earth Science Series, p. 29–44.

von Huene, R., Corvalan, J., Flueh, E.R., Hinz, K., Korstgard, J., Ranero, C.R., Weinrebe, W., and the CONDOR Scientists, 1997, Tectonic control of the subducting Juan Fernandez Ridge on the Andean margin near Valparaiso, Chile: Tectonics, v. 16, p. 474–488, doi:10.1029/96TC03703.

Wang, H., Currie, C.A., and DeCelles, P.G., 2015, this volume, Hinterland basin formation and gravitational instabilities in the central Andes: Constraints from gravity data and geodynamic models, *in* DeCelles, P.G., Ducea, M.N., Carrapa, B., and Kapp, P.A., eds., Geodynamics of a Cordilleran Orogenic System: The Central Andes of Argentina and Northern Chile: Geological Society of America Memoir 212, doi:10.1130/2015.1212(19).

Ward, K., Porter, R., Zandt, G., Beck, S., Wagner, L., Minaya, E., and Tavera, H., 2013, Ambient noise tomography across the central Andes: Geophysical Journal International, doi:10.1093/gji/ggt166.

Whitman, D., Isacks, B., Chatelain, J., Chiu, J., and Perez, A., 1992, Attenuation of high frequency seismic waves beneath the central Andean Plateau: Journal of Geophysical Research, v. 97, p. 19,929–19,947, doi:10.1029/92JB01748.

Whitman, D., Isacks, B., and Kay, S., 1996, Lithospheric structure and along-strike segmentation of the central Andean Plateau: Topography, tectonics and timing: Tectonophysics, v. 259, p. 29–40, doi:10.1016/0040-1951(95)00130-1.

Wölbern, I., Heit, B., Yuan, X., Asch, G., Kind, R., Viramonte, J., Tawackoli, S., and Wilke, H., 2009, Receiver function images from the Moho and the slab beneath the Altiplano and Puna Plateaus in the central Andes: Geophysical Journal International, v. 177, p. 296–308, doi:10.1111/j.1365-246X.2008.04075.x.

Yáñez, G.A., Ranero, C.R., von Huene, R., and Diaz, J., 2001, Magnetic anomaly interpretation across the southern central Andes (32°–34° S): The role of the Juan Fernandez Ridge in the Late Tertiary evolution of the margin: Journal of Geophysical Research, v. 106, p. 6325–6345, doi:10.1029/2000JB900337.

Yuan, X., Sobolev, S.V., and Kind, R., 2002, Moho topography in the central Andes and its geodynamic implications: Earth and Planetary Science Letters, v. 199, p. 389–402, doi:10.1016/S0012-821X(02)00589-7.

Zandt, G., Leidig, M., Chmielowski, J., Baumont, D., and Yuan, X., 2003, Seismic detection and characterization of the Altiplano-Puna magma body, central Andes: Pure and Applied Geophysics, v. 160, p. 789–807, doi:10.1007/PL00012557.

Zandt, G., Gilbert, H., Owens, T.J., Ducea, M., Saleeby, J., and Jones, C.H., 2004, Active foundering of a continental arc root beneath the southern Sierra Nevada, California: Nature, v. 431, p. 41–46, doi:10.1038/nature02847.

MANUSCRIPT ACCEPTED BY THE SOCIETY 3 JUNE 2014
MANUSCRIPT PUBLISHED ONLINE 9 SEPTEMBER 2014

Printed in the USA

The Geological Society of America
Memoir 212
2015

Along-strike variations in crustal seismicity and modern lithospheric structure of the central Andean forearc

Kathryn Metcalf
Paul Kapp
Department of Geosciences, University of Arizona, Tucson, Arizona 85721-0077, USA

ABSTRACT

The dynamics of the erosive central Andean forearc vary significantly along strike. In northern Chile at 20°S–27°S, and particularly at 22°S–25°S, the forearc in the Coastal Cordillera has been undergoing extension since at least the Pliocene, reactivating steep E-dipping faults of the Mesozoic Atacama fault system. This has been explained by forearc uplift driven by underplating, shallow slab dip, subduction of bathymetric features, and elastic rebound during the earthquake cycle. These processes, however, are active over a much wider area of the central Andean forearc than Coastal Cordillera extension and therefore cannot explain why extension is localized to the northern Chilean onshore outer forearc. We compiled crustal seismicity and the depth of lithospheric boundaries from existing studies to investigate other possible explanations for onshore forearc extension. At 22°S–25°S, seismicity increases above the background subduction-related level present to the north and south. Extensional focal mechanisms, consistent with steep E-dipping faults active at depths up to ~40 km, are also present onshore within this latitude range, but absent to the north and south; this is consistent with the distribution of mapped active fault scarps. The Salar de Atacama crust is seismically active at depths up to ~40 km. Thick lithosphere is present beneath the forearc, and the longitudinal axis of thickest lithosphere is deflected to the east at the latitude of the Salar de Atacama. To the east, the Puna Plateau lithosphere has been thinned by lithospheric removal events. The most robust correlation with onshore forearc extension is the thick, cold, strong crust and mantle lithosphere beneath the anomalous Salar de Atacama in the inner forearc of northern Chile. The combination of underplating-driven outer forearc uplift, the presence of the preexisting structure of the Atacama fault system favorable for reactivation, and the negative buoyancy beneath the Salar de Atacama is inferred to drive Coastal Cordillera normal faulting at this latitude. Recent (<10 Ma) lithospheric removal beneath the Puna Plateau to the east may have enhanced the effect of the negatively buoyant Salar de Atacama lithosphere on the forearc. This implies that both preexisting lithospheric structure and lithospheric processes in the hinterland may influence forearc dynamics.

Metcalf, K., and Kapp, P., 2015, Along-strike variations in crustal seismicity and modern lithospheric structure of the central Andean forearc, *in* DeCelles, P.G., Ducea, M.N., Carrapa, B., and Kapp, P.A., eds., Geodynamics of a Cordilleran Orogenic System: The Central Andes of Argentina and Northern Chile: Geological Society of America Memoir 212, p. 61–78, doi:10.1130/2015.1212(04). For permission to copy, contact editing@geosociety.org. © 2014 The Geological Society of America. All rights reserved.

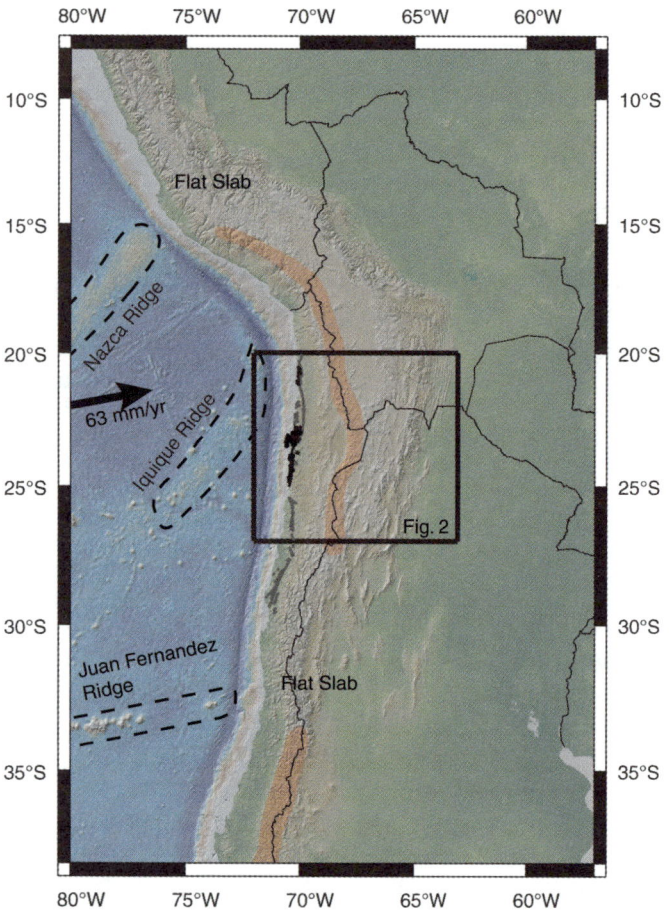

INTRODUCTION

The central Andes mark the region of greatest Cenozoic crustal shortening in the South American cordillera (e.g., Allmendinger et al., 1997; Kley and Monaldi, 1998). This region is distinguished from the northern and southern Andes by the Peruvian flat slab to the north and the Pampean flat slab to the south, with their respective volcanic gaps in the upper plate (Fig. 1). The retroarc thrust belt in Bolivia is a classic thin-skinned system, with the high (~3.7 km), low-relief Altiplano Plateau in the hinterland. Farther south in Argentina, the retroarc thrust belt is thick skinned, involving mechanical basement, and its hinterland region is occupied by the higher-elevation (~4.1 km) and higher-relief Puna Plateau (Fig. 2).

The central Andes are in overall approximately E-W compression as a consequence of the continental South American plate overriding the oceanic Nazca plate. Upper-plate advance tends to cause an increase in horizontal compressive stress in the upper plate (Russo and Silver, 1996; Oncken et al., 2006; Schellart, 2008) and increases the likelihood of flat or shallow

Figure 1. Map of the central Andes showing extent of the Mesozoic and modern Atacama fault system (active faults—black and inactive faults—gray) and the main magmatic arc (shaded orange). The flat slabs and volcanic gaps are the result of the subduction of aseismic ridges. Fault locations are from Scheuber and Gonzalez (1999) and Allmendinger and Gonzalez (2010). The Nazca–South America convergence vector is based on Kendrick et al. (2003).

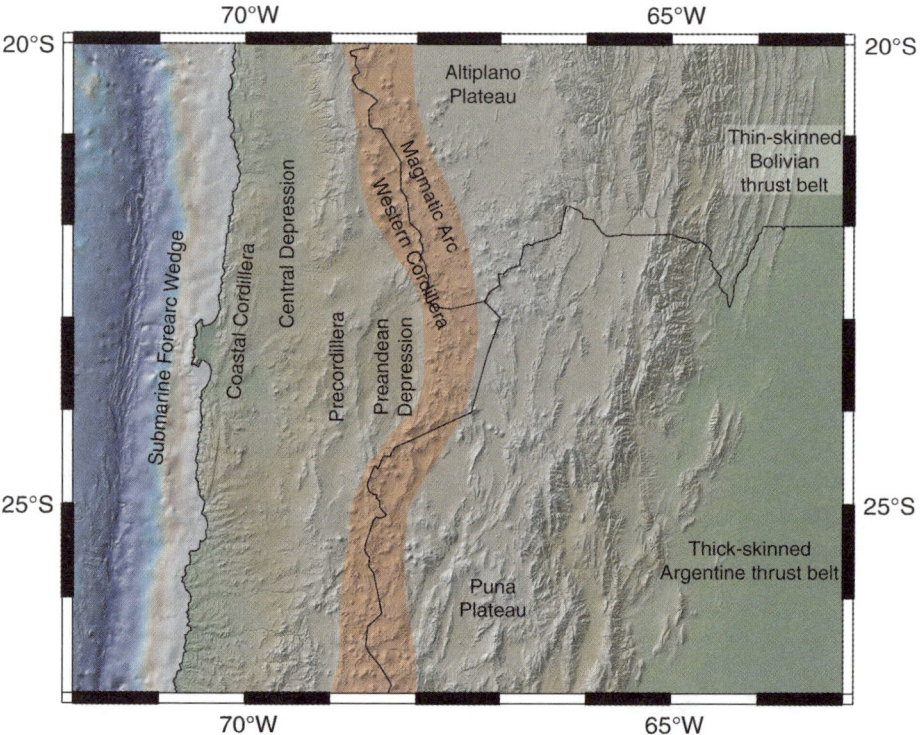

Figure 2. Morphotectonic provinces of the central Andes in the study area. The location of the active magmatic arc is shaded in orange.

subduction (e.g., Manea and Gurnis, 2007) and subduction erosion. The forearc along the central Andes is a type example of a tectonically erosive margin. Exposure of Jurassic arc rocks along the coast, 100 km east of the modern trench, indicates that the entire Jurassic forearc (an ~250 km E-W length of upper-plate lithosphere, assuming a typical width for continental forearcs) has undergone subduction erosion since at least 150 Ma (Scheuber and Reutter, 1992; Stern, 2004).

The modern central Andean forearc is divided into five longitudinal morphotectonic provinces: the (1) trench and submarine forearc wedge, (2) Coastal Cordillera, (3) Central Depression, (4) Precordillera (Cordillera de Domeyko and Sierra Moreno in northern Chile), and (5) Preandean Depression (Calama Basin and Salar de Atacama) (Fig. 2). The Western Cordillera marks the position of the modern volcanic arc and comprises the western margin of the Altiplano-Puna Plateau. In the hyperarid climate of the Atacama Desert, little sediment is transported to the trench, contributing to the erosive nature of the continental margin (Lamb and Davis, 2003). Along most of the central Andes, elevation decreases gradationally westward from the Western Cordillera to the Central Depression within the forearc (Fig. 2). An exception, however, is in northern Chile, where the anomalously low-elevation (2.3 km) Salar de Ata-cama and the bounding Precordillera to the west separate the Western Cordillera from the Central Depression (Fig. 2).

Despite the overall highly compressive tectonic setting, steeply dipping, N-S–striking normal faults are active in the region of Antofagasta, Chile (20°S–27°S) in the Coastal Cordillera (Fig. 3). Extension in the forearcs of broader compressive systems has been explained by a number of mechanisms, almost all calling on processes occurring along the subduction interface. These include anomalously shallow slab dip, subduction of bathymetric features such as oceanic seamounts and plateaus, subduction erosion and underplating, and elastic rebound following large subduction earthquake events (e.g., Loveless et al., 2010, and references therein). In the central Andes, however, there are no obvious along-strike variations in the geometry, kinematics, or dynamics of subduction nor magnitude of subduction erosion or underplating that can be called upon to explain why Coastal Cordillera normal faulting is spatially localized between the latitudes of 20°S and 27°S but most pervasive, laterally continuous, and active since the Pleistocene between 22°S and 25°S. This motivated us to explore potential orogenic processes or preexisting lithospheric heterogeneities farther inboard as potential explanations for the unique behavior of the forearc at this latitude.

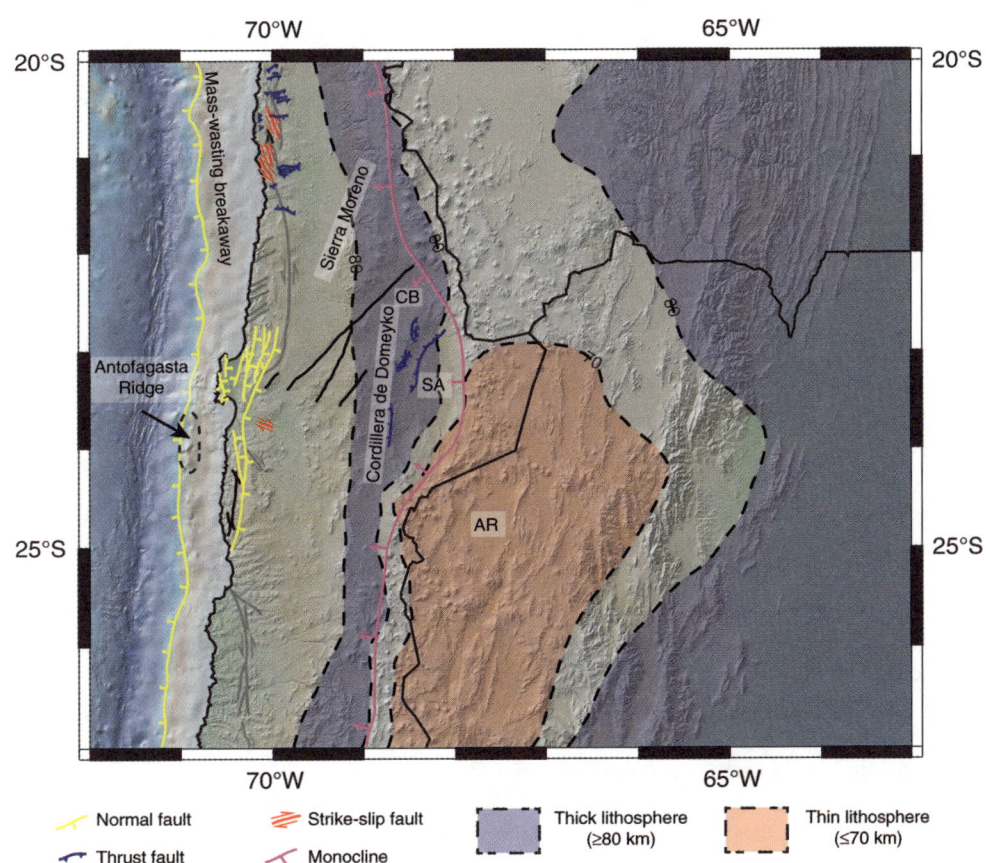

Figure 3. Neotectonic map of northern Chile. Faults in gray are inactive. Active extension is most prominent at 22°S–25°S. The main volcanic arc is bounded by a monocline on the west. Blue and red shadings represent areas of thick and thin lithosphere, respectively (Tassara and Echaurren, 2012). Fault locations are from Scheuber and Gonzalez (1999), Arriagada et al. (2006), Amilibia et al. (2008), and Allmendinger and Gonzalez (2010). CB—Calama Basin; SA—Salar de Atacama; AR—Salar de Arizaro.

DeCelles et al. (2009) presented a conceptual model of cyclicity in cordilleran orogenic systems. This model makes several predictions about forearc dynamics during different stages of the orogenic cycle. During growth of an eclogite root and associated subsidence in the hinterland of the orogen, the forearc is inferred to be placed in a subcritical state and to deform internally (e.g., by underplating or duplexing) in an attempt to attain critical taper. After the dense root detaches and the hinterland of the orogen isostatically rebounds, the forearc is inferred to be driven into a supercritical state. Accretionary forearcs may elongate, and thereby reduce taper, by incorporating more material into the accretionary prism, but erosive forearcs are limited to internal extension and mass wasting in order to lengthen and thereby decrease taper. In some cases, delamination may even promote shallow or flat subduction to fill the region of vacated lithosphere. This is the first model that attributes forearc dynamics to processes in the hinterland and implies that along-strike variations in hinterland processes or the stage in the cycle might cause along-strike variations in forearc dynamics.

Here, we use published geophysical data combined with published geological constraints to investigate along-strike variations in seismicity, lithospheric architecture, and the geologic framework that may correlate spatially with forearc normal faulting. Modern crustal seismicity corresponds well with the latitude of extension, and the Salar de Atacama has retained a lithospheric root. We infer that the combination of forearc uplift resulting from underplating, the favorable orientation of Mesozoic Atacama fault system structures, and the isolation of the Salar de Atacama lithospheric root by lithosphere removal beneath the Puna Plateau to the east has resulted in Coastal Cordillera normal faulting between 22°S and 25°S.

GEOLOGICAL AND GEOPHYSICAL BACKGROUND

Arc Migration and Subduction Erosion

The forearc of northern Chile is underlain by Mesozoic and older continental basement (Loewy et al., 2004; Sallares and Ranero, 2005; Ramos, 2008). Oceanic lithosphere has been subducting eastward beneath the western margin of South America since at least Jurassic time. The remains of the magmatic arc from that time are exposed along the coast. Younger arcs are located progressively to the east in the modern forearc; arc activity was established in the Western Cordillera beginning during Oligocene–Miocene time and has persisted to the present (Scheuber and Reutter, 1992; Stern, 2004). This pattern implies that the arc has migrated east in a series of jumps and that the forearcs of older magmatic arcs have been subducted (Rutland, 1971; Coira et al., 1982). Subduction erosion provides an explanation for arc migration and associated loss of lithosphere. The arc has migrated ~250 km since ca. 150 Ma, yielding a time-averaged landward trench migration rate of ~1.7 km/m.y. (Stern, 2004).

As the arc was being built during the Mesozoic, the Atacama fault system formed in a sinistral transtensional to transpressional

setting (Fig. 1). At the time of its formation, it was in an intra-arc to retroarc region, while the volcanic arc was located at the modern location of the Coastal Cordillera (Scheuber and Andriessen, 1990; Brown et al., 1993; Cembrano et al., 2005). Today, the Atacama fault system extends from 21°S to 30°S, and sections of it are reactivated as normal faults (Fig. 1).

Whereas records of the Jurassic and older forearc have been subducted, reflection seismology and marine deposits indicate that the entire central Andean forearc was undergoing both frontal and basal erosion during the mid-Miocene (Clift and Hartley, 2007). This led to subsidence of the forearc, similar to that which is occurring in the modern forearcs of Japan (von Huene et al., 1994; Wells et al., 2003) and Guatemala (Vannucchi et al., 2004) as the result of basal erosion. Evidence of mid-Miocene subsidence is found all along the central Andean forearc in both onshore basins (Clift and Hartley, 2007) and the midslope offshore (von Huene and Ranero, 2003).

Modern Geomorphology and Active Tectonics

Trench and Submarine Forearc Wedge

Due in part to the hyperarid climate in the central Andean forearc, the trench is underfilled (<1 km of sediment; von Huene and Scholl, 1991; Heuret et al., 2012). Because of this and the compressive nature of the margin, the forearc is erosive (Fig. 4). The depth of the trench below sea level averages ~7.5 km and varies from ~6.8 km to 8 km, with no systematic variations along strike. The submarine forearc wedge is dominated by slumping and mass wasting toward the trench (von Huene and Ranero, 2003; Sallares and Ranero, 2005). At the plate interface, pressurized fluids may fracture the basement, which is then carried down in grabens that developed on the subducting oceanic lithosphere in response to flexure (von Huene and Ranero, 2003). Based on their geometry and spacing in the Nazca plate, subducting grabens can transport as much as 28–36 km³/m.y. of upper-plate material into the subduction channel (von Huene and Ranero, 2003). The middle slope of the wedge is experiencing the greatest subsidence, suggesting that this is the present locus of maximum basal erosion (von Huene and Ranero, 2003).

Coastal Cordillera

The Coastal Cordillera averages ~1500 m in elevation and reaches elevations up to ~2750 m at 24.5°S. Rocks at these high elevations are undergoing normal faulting and surface uplift. Segments of the Mesozoic Atacama fault system from 20°S to 27°S have been reactivated as steep normal faults and strike-slip transfer zones since the mid- to late Miocene (e.g., Delouis et al., 1998; Scheuber and Gonzalez, 1999; González et al., 2003; Cembrano et al., 2005; Allmendinger and Gonzalez, 2010). The most pervasive, laterally continuous, and recently active normal faults have produced prominent fault scarps and occur from 22°S to 25°S (e.g., Delouis et al., 1998; González et al., 2003; Loveless et al., 2005; Allmendinger and Gonzalez, 2010). In this study, we focus on Coastal Cordillera faults, which have experienced mostly normal displacement since at least the Pleistocene. The normal faults generally strike N-S and dip steeply (~70° to subvertical) eastward. Estimates of slip on these faults range from ~400 m to a few meters since the Pleistocene (González et al., 2003). The Mejillones Peninsula is an exception. Normal faults on the peninsula dip both east and west and are generally listric (Allmendinger and Gonzalez, 2010). The Antofagasta ridge offshore (Fig. 3) might be an example of such a peninsula that has migrated toward the trench as subduction erosion progressed (von Huene and Ranero, 2003). In addition to faults, the forearc also exhibits pervasive tensional cracks (Delouis et al., 1998; González et al., 2003; Loveless et al., 2005; Allmendinger and Gonzalez, 2010; Baker et al., 2013).

Field observations have documented new slip on Coastal Cordillera normal faults a few weeks following large subduction zone earthquakes (e.g., Delouis et al., 1998). In places, normal faults have been reactivated by a few meters of reverse slip (Delouis et al., 1998; González et al., 2003; Loveless et al., 2010), which has been explained as coseismic extension followed by interseismic compression (Delouis et al., 1998). This pattern has been explained by low-magnitude extension in the weak, fractured forearc as the upper plate moves rapidly over the slab during a large seismic event, followed by compression during the interseismic period (Delouis et al., 1998). Models of Coulomb stress changes, however, suggest the opposite pattern of rapid shortening during the event and slow extension during the locked phase (Loveless et al., 2010). Observationally, interferometric synthetic aperture radar (InSAR) data following the 1995 Antofagasta earthquake did not conclusively show any coseismic slip on forearc normal faults (Loveless and Pritchard, 2008), while InSAR data following the 2007 Tocopilla earthquake found interseismic to postseismic slip reversal in two faults on the Mejillones Peninsula (Shirzaei et al., 2012). The seismic recurrence interval for the subduction interface is at least an order of magnitude less than that of the normal faults in the upper plate, suggesting that several subduction earthquakes are necessary to build up enough stress to generate an upper-plate earthquake (Cortés A., et al., 2012). It is not disputed that large subduction zone earthquakes may provide a mechanism that facilitates normal slip on individual faults; this has been recognized in several postseismic studies following, for example, the Maule and Tohoku earthquakes. However, this process does not explain the latitudinally localized distribution of prominent Coastal Cordillera normal faults that have developed over many seismic cycles.

Uplift of the coastal forearc likely occurred during the Oligocene and Miocene (Hartley et al., 2000), followed by a period

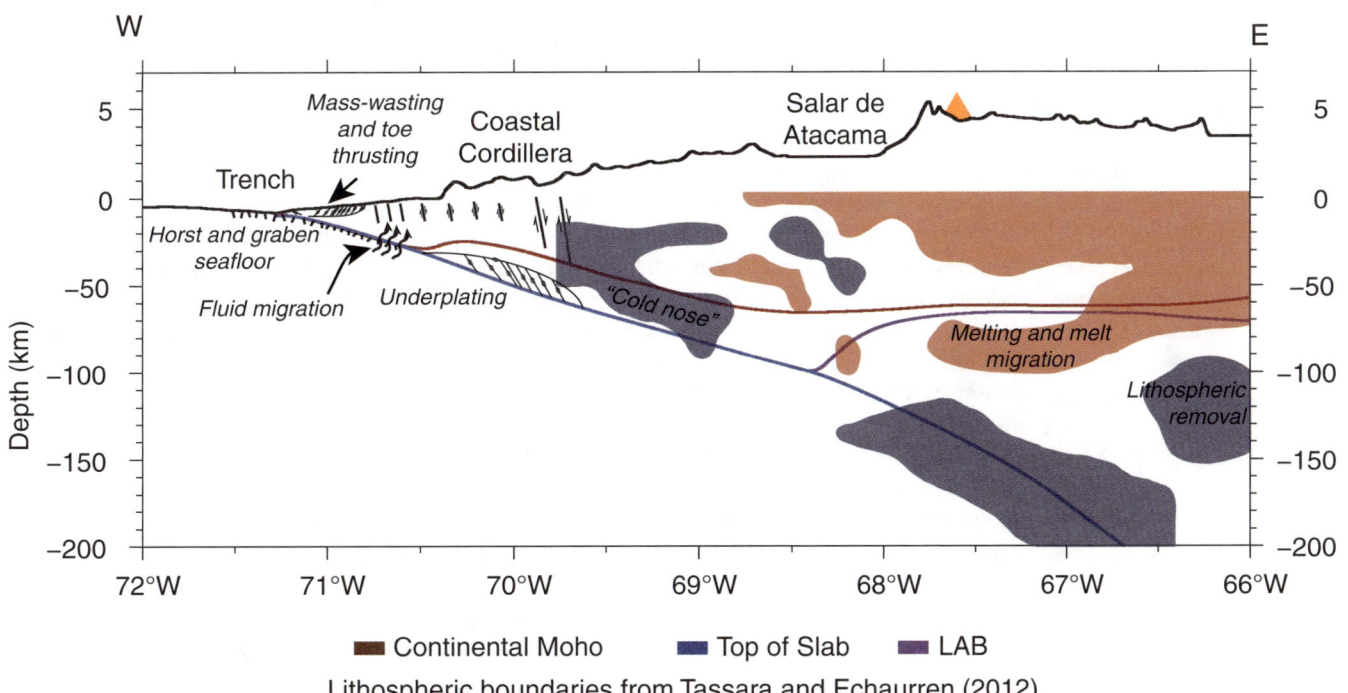

Figure 4. A cartoon model of forearc dynamics at 23.5°S. Horst and graben seafloor topography is subducted. The toe of the forearc wedge is undergoing thrusting, while the remainder to the east is undergoing extensional collapse. Offshore faults dip trenchward, and onshore normal faults dip arcward. Underplating is inferred beneath the Coastal Cordillera. The forearc crust and mantle are cold, sometimes referred to as a "cold nose," but below the magmatic arc, the asthenosphere circulates and arc melts are generated. Red and blue represent Vp anomalies ≤−2% and ≥ 2%, respectively (Schurr et al., 2006). Colored lines in the cross sections denote lithospheric boundaries (Tassara and Echaurren, 2012). Topography below sea level (0 km) is shown with no vertical exaggeration, and topography above sea level is shown with 10× vertical exaggeration. All geophysical data and interpretations are shown without vertical exaggeration. LAB—lithosphere-asthenosphere boundary.

of subsidence in the Pliocene (Ortlieb et al., 1996; Buchbinder et al., 2005; Clift and Hartley, 2007). At ca. 2 Ma, the central Andean forearc basins began to be uplifted (Clift and Hartley, 2007), and the oldest marine terraces were exposed (Regard et al., 2010). Quaternary marine terraces record interglacial highstands that have been preserved by tectonic uplift (Marquardt et al., 2004; Regard et al., 2010). The Mejillones Peninsula has Pliocene–Pleistocene marine deposits at elevations up to ~700 m (González et al., 2003), providing direct evidence that it was uplifted coeval with extension. Marine terraces in the graben indicate that the whole forearc has been uplifted, rather than just the exhumed footwalls (González et al., 2003). This shift from subsidence to uplift has commonly been attributed to the initiation of underplating beneath the forearc wedge (Delouis et al., 1998; Adam and Reuther, 2000), although the anomalous magnitude of uplift on the Mejillones Peninsula could also be a result of segmentation along the seismogenic subduction zone (Victor et al., 2011). Coastal Cordillera uplift is often invoked to explain Coastal Cordillera normal faulting at 22°S to 25°S; however, the uplift was relatively synchronous and of similar magnitude all along the coast of the erosive forearc (Clift and Hartley, 2007; Regard et al., 2010), and the observed scatter is inconsistent with the latitudes of active Coastal Cordillera normal faults.

Central Depression, Precordillera, and Preandean Depression

The Central Depression lies at lower elevations in the forearc between the Coastal Cordillera and the Western Cordillera. In the study area from 20°S to 27°S, the western boundary of the Central Depression is defined by normal faults. To the east, the Precordillera is the result of west-verging thrusting to the north and south of the Preandean Depression, whereas the Cordillera de Domeyko is characterized by east-verging thrusting and folding (e.g., Victor et al., 2004; Arriagada et al., 2006). A portion of the Cordillera de la Sal thrust system, on the western edge of the Salar de Atacama, offsets Quaternary gravels and is inferred to be active to this day (Arriagada et al., 2006). Farther inboard, the Preandean Depression, containing the Calama Basin and Salar de Atacama, deflects the volcanic front and plateau margin eastward (Fig. 2). The spatial distribution of Cenozoic arc rocks through time shows that this deflection did not exist until the late Oligocene (Haschke et al., 2002), and a magmatic gap existed at the latitude of the Salar de Atacama (~22°S–25°S) from ca. 28 Ma to 17 Ma, when the arc was reestablished to the east (Reutter et al., 2006).

Geophysical Background

Several seismic deployments aimed at illuminating lithospheric structure have been conducted in the central Andes. Receiver function (e.g., Beck and Zandt, 2002; Yuan et al., 2002; McGlashan et al., 2008; Wölbern et al., 2009; Lloyd et al., 2010) and seismic-refraction (Schmitz et al., 1999) studies have constrained crustal thicknesses that are inconsistent with surface elevations under the assumption of simple Airy isostasy. This implies that surface elevation is compensated by lateral variations in the density structure of the upper mantle. For example, assuming a felsic upper crust and mafic lower crust, the Arizaro Basin, containing the Salar de Arizaro (Fig. 3), has a thinner crust than would be expected for its elevation, whereas the Salar de Atacama has an elevation that is too low for its thick crust (Yuan et al., 2002). These disparities cannot be accommodated by realistic variations in crustal composition and density and therefore require lateral density variations in the continental upper mantle. The variable lithospheric thicknesses combined with seismic tomography images of low velocities near the base of the crust (e.g., Schurr and Rietbrock, 2004; Koulakov et al., 2006; Schurr et al., 2006; Heit et al., 2008) imply lithospheric removal beneath the southern Altiplano, Salar de Arizaro, and southern Puna (Fig. 3). Lithospheric removal is supported by the geochemistry and spatial distribution of basalts and ignimbrites, which range in age from ca. 30 Ma to <1 Ma (e.g., Kay and Kay, 1993; Kay et al., 1994; Kay and Coira, 2009), and by gravity data (Tassara et al., 2006; Prezzi et al., 2009; Tassara and Echaurren, 2012). The onshore forearc is interpreted to be strong and cold based on high Vp and Vs values and low attenuation (Schurr and Rietbrock, 2004; Schurr et al., 2006; Koulakov et al., 2006; Wölbern et al., 2009). The Salar de Atacama is underlain by lithosphere with moderately high Vp and very low attenuation, and by crust with deep (~40 km) seismicity; together these data suggest a cold, strong rheology (Schurr and Rietbrock, 2004). Given the thick crust (up to 65 km), low geothermal gradient, and likelihood of significant fluid migration from the subducting slab, the lower crust or mantle lithosphere in the inner forearc could be in the serpentinite or eclogite stability field. The Salar de Atacama is also located within the central Andean gravity high (Götze and Krause, 2002). The lithosphere beneath the forearc (up to ~100 km) is much thicker than in the hinterland to the east, where recent (<10 Ma) lithospheric removal events have produced a thin lithosphere (~65 km) and hot upper mantle.

Detailed forearc structure, showing offshore mass wasting, seafloor horst and graben topography, and regions of possible fluid migration, has been imaged using marine seismic-reflection studies (Oncken et al., 2003; von Huene and Ranero, 2003; Sick et al., 2006) and high-resolution bathymetry (von Huene and Ranero, 2003). In addition, relocation of microseismicity following large subduction zone earthquakes has provided a clear, well-resolved subduction interface and information about seismic locking depth (ranging from ~20 to ~70 km; e.g., Comte et al., 1994; Husen et al., 2000; Nippress and Rietbrock, 2007; Motagh et al., 2010; Schurr et al., 2012). This information affords the opportunity to clearly distinguish upper-plate seismicity from interface seismicity. Global positioning system (GPS) studies help to constrain current seismic and aseismic motion at the subduction interface and across the orogen (e.g., Khazaradze and Klotz, 2003; Béjar-Pizarro et al., 2010; Brooks et al., 2011), providing a control for deformation not represented by earthquakes.

With these constraints from the literature, we investigate the along-strike variations in seismicity and modern lithospheric structure of the forearc.

DATA COMPILATION

In this study, we compiled previously relocated earthquakes, focal mechanisms, and lithospheric boundaries based on gravity constrained by numerous seismic studies (Tassara and Echaurren, 2012). The data are shown in map view in Figure 5. The interpreted data are summarized in map view in Figure 6A and cross-section views in Figures 6B–6E.

Seismicity

Relocated hypocenters were compiled from the EHB Bulletin (International Seismological Center, 2009), Koulakov et al. (2006), and Schurr et al. (2006) (Table 1; Fig. 5). Uncertain-

ties on earthquake locations are greatest in the vertical direction. In order to correlate seismicity with surface features and to determine fault structure at depth, it is necessary to use only relocated earthquakes with minimum location error. Areas with denser station concentration record more lower-magnitude earthquakes than those with lower station concentration. Seismic networks also have different cutoffs for the magnitudes of earthquakes reported. To reduce the bias of station concentration and reporting, multiple data sets of varying duration and distribution were included. Regional patterns of seismicity were determined based on earthquakes with magnitude ≥4.5. However, all earthquakes were used in detailed local analyses. Error estimates were available for only the *EHB Bulletin* hypocenters, which are shown in cross section herein (Figs. 6B–6E). Many of the Koulakov et al. (2006) and Schurr et al. (2006) hypocenters were independently calculated for local tomography from the same raw data. Because error estimates and magnitudes were unavailable, both data sets were plotted to show

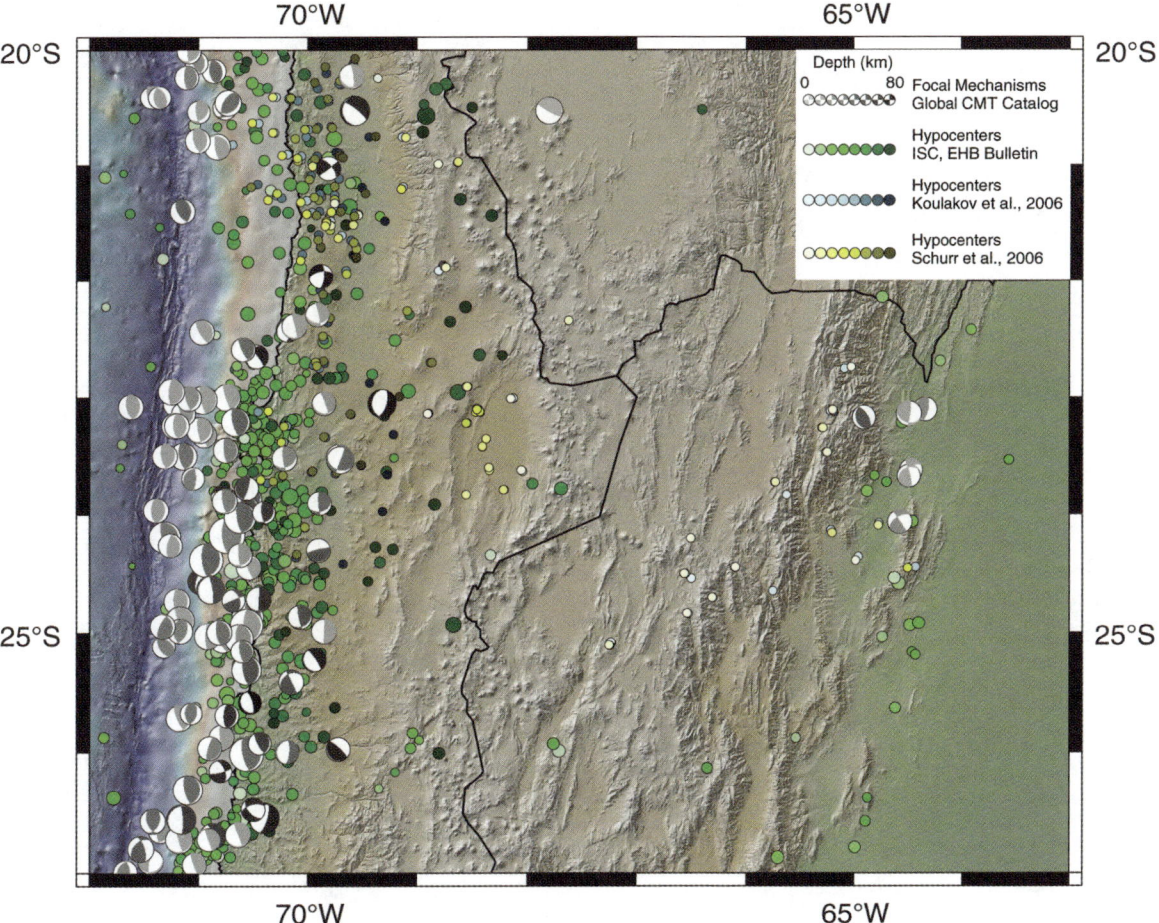

Figure 5. Map of seismicity at depths ≤80 km to more clearly show crustal seismicity. Focal mechanisms are lower-hemisphere projections with the compressional quadrants colored. Focal mechanisms and hypocenters are scaled by magnitude where available. Data from Koulakov et al. (2006) and Schurr et al. (2006) are scaled to magnitude 3, since no magnitudes were available. CMT—Centroid-Moment-Tensor; ISC—International Seismological Center.

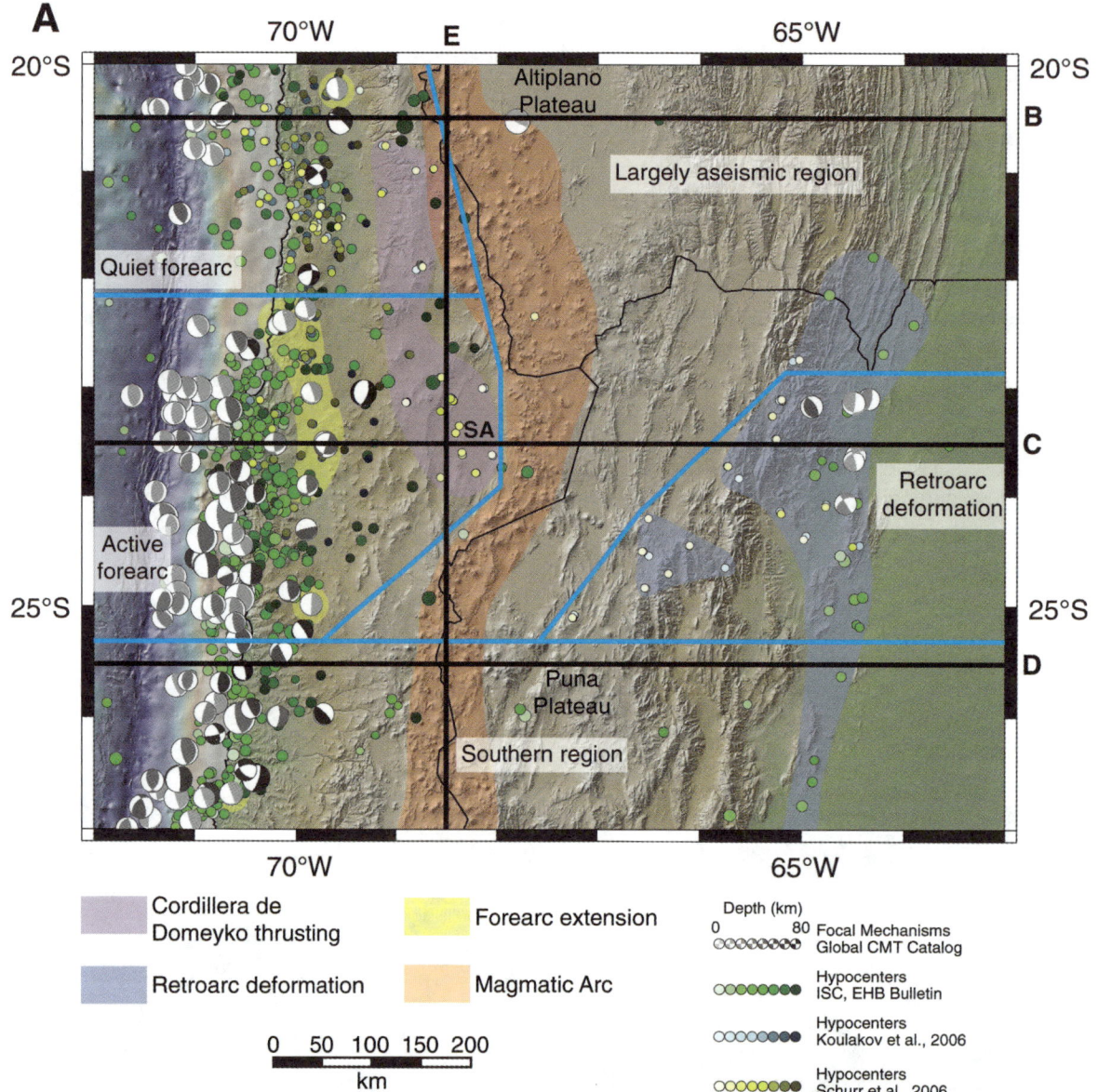

Figure 6 (*Continued on facing page*). (A) Seismicity map and interpretations. Seismicity is as in Figure 5. Blue lines divide labeled regions of seismicity. Shaded regions mark types of seismicity. Black lines are the locations of cross sections B–E. CMT—Centroid-Moment-Tensor; ISC—International Seismological Center; SA—Salar de Atacama.

TABLE 1. COMPILED HYPOCENTERS AND FOCAL MECHANISMS

Source	Latitudinal coverage	No. events	Temporal coverage	Magnitudes	Error bars	Map color*
Hypocenters						
EHB Bulletin	Global	697	1960–2007	Yes	Yes	Green
Koulakov et al. (2006)	20°S–25°S	238	1994–1997	No[†]	No	Light blue
Schurr et al. (2006)	20°S–25°S	172	1994–1997	No[†]	No	Yellow
Focal mechanisms						
Global CMT catalog[§]	Global	109	1976–2011	Yes	No	Gray

*Refers to the color in Figures 5 and 6, as well as the focal mechanisms in Figure 7.
[†]Where magnitudes were unavailable, they are scaled to 3.
[§]Global Centroid-Moment-Tensor (CMT) catalog.

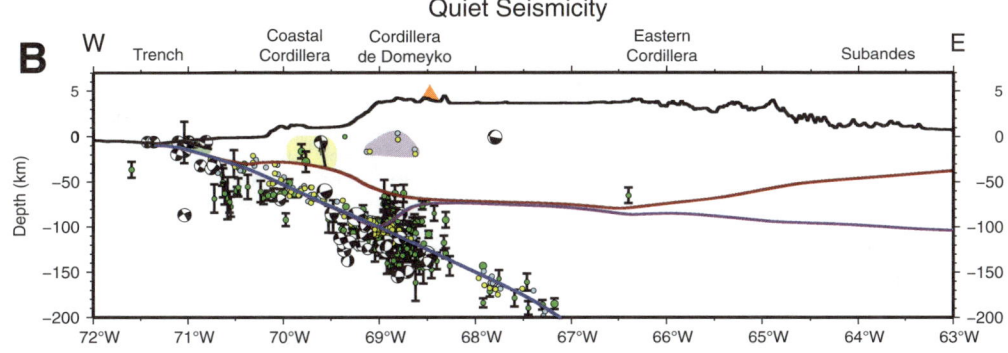

Quiet Seismicity

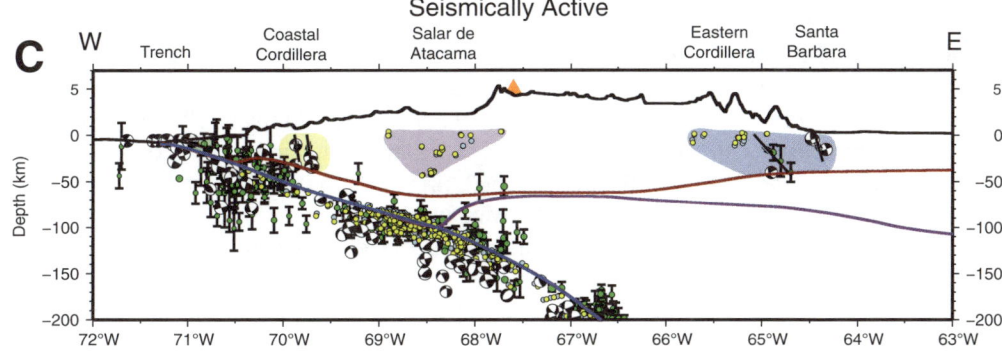

Seismically Active

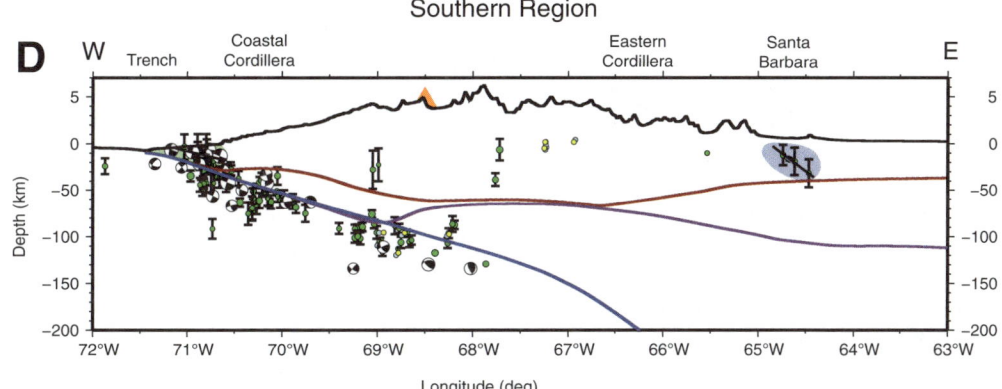

Southern Region

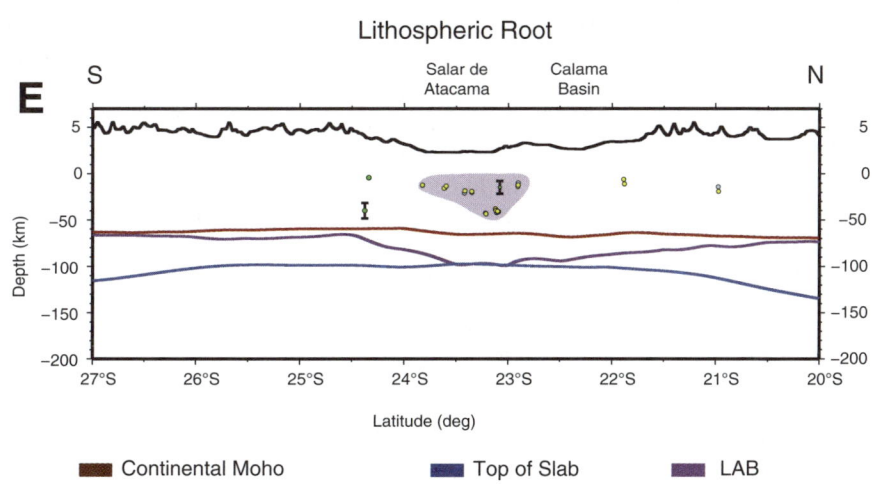

Lithospheric Root

| Continental Moho | Top of Slab | LAB |

Figure 6 (*Continued*). (B–E) Seismicity cross sections and interpretations. Seismicity from 0.5° to either side is projected onto each cross section. Focal mechanisms are shown in cross-section view (Northern Hemisphere projection). Colored lines in the cross sections denote lithospheric boundaries (Tassara and Echaurren, 2012). Fault plane interpretations on focal mechanisms are from published field observations. Topography below sea level (0 km) is shown with no vertical exaggeration, and topography above sea level is shown with 10× vertical exaggeration. All geophysical data and interpretations are shown without exaggeration. Orange triangles mark the volcanic front. Slab seismicity in D was removed to show lithospheric boundaries. LAB—lithosphere-asthenosphere boundary. See text for explanation.

the slight discrepancies in relocations of the same event, giving a sense of the location error.

We compiled focal mechanisms from the Global Centroid-Moment-Tensor (CMT) catalog (Table 1; Fig. 6). Focal mechanisms can only be calculated for earthquakes with a large enough magnitude to distinguish first motion from background noise, ~5 or higher for the global network. Therefore, we were able to use fault plane data for larger earthquakes only. Focal mechanisms provide two equally possible planes of slip. Without other data, it is impossible to identify the correct plane of slip, but with published field observations, we were able to determine the most likely plane.

All data sets cover almost the entire study area. Although there are several smaller, more local studies of varying durations and at varying times, we chose to use only the global data and larger, regional data sets. The distribution of seismicity shows active seismic deformation but does not take into account aseismic deformation or longer-term patterns of seismicity. Most local studies provide detailed information of only a few years, and the total instrumented seismic record extends for a few decades at most. This snapshot may not represent structures that are locked over longer time scales and may resolve structures that are only transiently active. Thus, it is important to compare the seismic record with the geologic record to distinguish between abandoned and locked faults and areas of persistent or transient deformation.

Lithospheric Boundaries

Tassara and Echaurren (2012) incorporated constraints on continental Moho depth, lithospheric thicknesses, and depth to slab from several geophysical studies into a gravity model. With these independently determined constraints, gravity data were used to extrapolate the depth to the Moho, slab, and lithosphere-asthenosphere boundary for all of South America. Here, the lithosphere-asthenosphere boundary is thermally defined and consistent with heat-flow maps and seismic attenuation (Tassara et al., 2006). The uncertainties in each of these boundaries are related to the uncertainties in the seismic data incorporated in this model. These are ±5 km for the Moho and ±10 km for the slab and lithosphere-asthenosphere boundary (Tassara and Echaurren, 2012). While a certain amount of smoothing is to be expected between seismic deployments, the lithospheric boundaries are best constrained in our study area where the density of published seismology studies is greatest (Tassara et al., 2006; Tassara and Echaurren, 2012).

RESULTS

Seismicity

The pattern of upper-plate seismicity across the orogen varies along strike. It is divided from north to south into regions of quiet seismicity, active seismicity, and the southern region (Figs. 6B–6D).

Quiet Seismicity: 20°S to 22°S

In the region of the study area north of ~22°S (Fig. 6B), the forearc expresses background subduction zone–related seismicity observed across the majority of the Nazca subduction zone. At 69.5°W and 20.25°S, there is one extensional focal mechanism at ~5 km depth consistent with steep E-dipping normal faulting. At 22°S, extensional focal mechanisms mark the northern boundary of the region with significant normal fault scarps (Fig. 6A). The northern latitudes (20°S to 22°S), with only moderate Coastal Cordillera normal fault scarps, show only moderate extensional seismicity. A concentration of seismicity occurs at depths of ~0–20 km along the western boundary of the Altiplano within the Sierra de Moreno. Although no focal mechanisms are available, field observations and analysis of satellite imagery suggest that this seismicity is most likely related to thrust faulting and growth of the monocline along the Sierra de Moreno (e.g., Amilibia et al., 2008; Jordan et al., 2010). The Altiplano Plateau and Bolivian retroarc thrust belt are largely aseismic. The geology suggests that the thin-skinned thrust belt in the Bolivian Andes has been recently active (McQuarrie et al., 2005), and GPS data show strain accumulation on a locked basal décollement zone (Brooks et al., 2011). Therefore, the observed aseismicity in the Bolivian thrust belt is interpreted to indicate an interseismic cycle longer than the seismic record.

Active Seismicity: 22°S to 25°S

Between latitudes ~22°S to ~25°S (Fig. 6C), the entire orogen is more seismically active than farther north and south. The subduction zone seismicity increases, and forearc extensional focal mechanisms are much more prevalent, consistent with the more extensive mapped forearc normal faults. The latitude of greatest active Coastal Cordillera normal faults (Fig. 3) corresponds with the greatest amounts of forearc seismicity and extensional focal mechanisms. The map view trend in the positions of the extensional focal mechanisms mimics the eastward-convex deflection of the volcanic front and Western Cordillera around the Salar de Atacama, and extensional focal mechanisms are ~20 km deeper toward the center of the deflection (Figs. 6A and 6C). Seismicity within the Salar de Atacama itself is shown by field studies and reflection seismic profiles to be related to active E-W contraction (e.g., Arriagada et al., 2006). In addition, the depths of seismicity in the crust below the Salar de Atacama (down to ~40 km) indicate a cold, rheologically strong crust. This is consistent with the very low seismic attenuation described by Schurr and Rietbrock (2004). The Puna Plateau is aseismic and presumably not undergoing significant upper-crustal strain. The thick-skinned retroarc thrust belt at this latitude, however, is seismically active. The inverted Salta Rift system experiences both dip-slip and oblique-slip earthquakes, consistent with steep E-dipping thrusting and sinistral strike-slip faulting. This seismicity is consistent with the active kinematics of the frontal part of the thrust belt as revealed by GPS data and field studies of faults (e.g., Cahill et al., 1992; Schurr et al., 1999; Kley and Monaldi, 2002; Echavarria et al., 2003; Meigs and Nabelek, 2010; Pearson et al., 2012).

Southern Region: 25°S to 27°S

South of ~25°S (Fig. 6D), the subduction zone seismicity returns to its background level. Extensional focal mechanisms are absent within the forearc, and only a few seismic events are located beneath the Western Cordillera at a depth of up to 40 km. As the normal fault scarps decrease in prevalence southwards, the active seismicity resumes its normal pattern. At this latitude, the retroarc remains seismically active, but less so than in the middle latitudes of the study area. Retroarc seismicity also steps west into the hinterland and broadens as the Puna Plateau narrows to the south (Fig. 6A).

Additional Observations

With the exception of the Bolivian retroarc thrust belt, the regional seismic patterns correspond well with the regional subaerial geologic record of fault activity. The forearc in the north of the study region contains systems of small strike-slip and thrust faults with low magnitudes of slip (Allmendinger and Gonzalez, 2010). To the south, the subduction interface has experienced more recent seismic activity, but the geologic record shows that the most pervasive and laterally continuous active Coastal Cordillera normal faults occur from 22°S to 25°S (e.g., Delouis et al., 1998; González et al., 2003; Loveless et al., 2010; Allmendinger and Gonzalez, 2010), with slip estimates ranging from ~400 m to a few meters since the Pleistocene (González et al., 2003). Although the instrumental record is only a short increment in the geological seismic cycle, the locations and densities of crustal seismicity correlate well with the regional pattern of mapped faults in the study area. The Bolivian retroarc thrust belt records an active recent geologic history but is aseismic during the instrumental record. GPS velocities in the aseismic Bolivian retroarc thrust belt are similar to those in the seismically active Argentinian system, suggesting that the faults may be slipping aseismically at depth while the near-surface segments of the faults remain locked (Brooks et al., 2011). In this case, the apparent discrepancy between the geological and seismic records is likely an artifact of the shorter seismic record.

At latitude 23°S–25°S, both forearc extensional and retroarc contractional hypocenters define linear, dipping arrays that extend to up to 40 km depth, near the base of the ~40-km-thick crust. This suggests that the scale of deformation along the surface faults may be crustal rather than restricted to the upper and middle crust. Additionally, these observations suggest that the crust in the forearc and frontal part of the retroarc is sufficiently cold to deform in a brittle manner at near-Moho depths.

The focal mechanisms in the subducting slab are consistent with outer-arc extension before entering the trench, thrust faulting at the subduction interface, and downdip extension at depths of ~50 km and greater (Figs. 6B–6D). Seismic activity is concentrated at the subduction interface, the intersection between the slab and the continental lithosphere-asthenosphere boundary, and below ~200 km. The cluster of seismicity at the intersection with the lithosphere-asthenosphere boundary also closely correlates in space with the location of the volcanic front (Figs. 6B–6D). More earthquakes are present in the cluster beneath the volcanic arc in the north (20°S to 22°S). As the volcanic arc is deflected around the Salar de Atacama (22°S to 25°S), the cluster of seismicity broadens to extend beyond the lithosphere-asthenosphere boundary contact (Fig. 6C). In the south (25°S to 27°S), the cluster is closer to the trench again, and the amount of seismicity decreases.

Moho and Lithosphere-Asthenosphere Boundaries

The slab geometry from Tassara and Echaurren (2012) corresponds well with slab seismicity. The crust is thickest beneath the Eastern Cordillera, up to 75 km at the latitude of the Altiplano, and up to 65 km at the latitude of the Puna, and thins toward the coast and foreland, consistent with thickening due to crustal shortening during the growth of the thrust belt and plateau. Beneath the Puna Plateau, the lithosphere is thin (~65 km), consistent with previous suggestions of recent lithospheric removal and asthenospheric upwelling (e.g., Koulakov et al., 2006; Schurr et al., 2006; Heit et al., 2008). Toward the foreland, the lithosphere thickens to ~100 km away from the high-heat-flow regions of the plateau. The forearc also has a thermal lithosphere-asthenosphere boundary that is thicker (~100 km) than that beneath the hinterland. This has been referred to as the "cold nose" of the forearc (van Keken et al., 2011), and it is observed globally in subduction systems. Thermal modeling requires a region of mantle decoupled from the slab above the region of corner flow where the slab and mantle are coupled (Wada and Wang, 2009). A cold forearc mantle is also correlated with high Vp/Vs ratios, most likely due to the formation of hydrous minerals such as serpentine and chlorite in the cold mantle wedge (Christensen, 2004). The location of the magmatic arc seems linked to this transition from the cold nose to the coupled slab and continental mantle, where fluids are able to flux out of the slab and into the mantle wedge more easily (van Keken et al., 2011).

Subduction Erosion

Roughly half of modern subduction zones are erosive (von Huene and Scholl, 1991). However, the mechanisms of subduction erosion (or tectonic erosion) are poorly understood. The first problem is determining the processes that remove material from the upper plate and transport that eroded material into the subduction zone. This has not been conclusively observed except in the case of subducting seamounts (e.g., Ranero and von Huene, 2000). A second outstanding issue is the fate of the tectonically eroded material. How much is subducted into the mantle, and how much is underplated or re-accreted to the upper plate? These questions could be interrogated with proper seismic imaging, but the fracture- and fluid-rich subduction zone is often not seismically imaged with the necessary detail.

Subduction erosion may be either frontal, removing material from the toe of the wedge, or basal, removing material from beneath the wedge (von Huene and Lallemand, 1990). Accretionary margins may also experience basal erosion (Cloos and

Shreve, 1988) but require frontal accretion to build the accretionary prism. Conversely, erosive margins require frontal erosion to remove forearc material but may also undergo basal erosion or accretion.

Submarine mass wasting and subsidence along with coastal uplift suggest that the forearc in northern Chile is undergoing frontal erosion, middle-slope basal erosion, and basal accretion or underplating beneath the Coastal Cordillera (Fig. 4). Frontal erosion could be accomplished by mass wasting into the trench and subducting grabens (Fig. 4; von Huene and Scholl, 1991; von Huene and Ranero, 2003), subduction of positive topographic features such as seamounts or ridges (von Huene and Lallemand, 1990; Ranero and von Huene, 2000), underthrusting in the toe of the wedge (Vannucchi et al., 2012), or some combination of these. Basal accretion may occur through dewatering, which increases density and viscosity (Götze and Krause, 2002; Shreve and Cloos, 1986; Cloos and Shreve, 1988), duplexing as the basal décollement steps down (Collot et al., 2008, and references therein), and/or by downward migration of the roof décollement (Vannucchi et al., 2012). With the present quality of seismic imaging, these cannot be discriminated. Forearc uplift or subsidence is commonly attributed to basal underplating or erosion, respectively, but independent evidence is scarce.

DISCUSSION: MECHANISMS FOR FOREARC EXTENSION

The onshore, E-dipping, forearc normal faults in the Coastal Cordillera have been attributed to reactivation of preexisting faults, underplating and coastal uplift, and slab dynamics. We examine these and other possible controlling factors in the upper plate to explore the possible conditions responsible for the latitudinal extent of these faults.

Reactivation of Preexisting Faults

The location of onshore, E-dipping, forearc normal faulting is largely coincident with the location of the Mesozoic Atacama fault system. The reactivation of these strike-slip faults is likely a large control on the steep dip of the Neogene normal faults. It could be argued that the presence or absence of preexisting strike-slip faults explains the presence or absence of N-S–striking normal faulting, respectively. However, the Atacama fault system extends southward to 30°S (Cembrano et al., 2005), whereas extension is concentrated at 22–25°S.

Underplating and Coastal Uplift

Coastal uplift is well documented along strike of the central Andes. Most studies attribute this to basal underplating of tectonically eroded material from the trench (Marquardt et al., 2004; Clift and Hartley, 2007; Regard et al., 2010). Studies also attribute normal faulting to a supercritical wedge due to uplift driven by underplating (Delouis et al., 1998; Adam and Reuther,

2000). When considered in isolation, increased gravitational potential energy from underplating beneath the Coastal Cordillera is a viable cause of the observed normal faulting. Coastal uplift, however, occurs over far greater latitudinal distances than does Coastal Cordillera normal faulting.

Slab Dynamics

Forearc tectonics are largely interpreted in the context of the behavior of the subducting slab. Retreat of either the slab or the upper plate should cause extension in the upper plate, but the opening of the South Atlantic advances the upper plate, which should lead to a more compressive state (Schellart, 2008). Slab dip and topography on the slab also affect forearc tectonics (e.g., Cloos and Shreve, 1996; van Hunen et al., 2002; Collot et al., 2008). In Costa Rica and Nicaragua, subducting seamounts imaged with reflection seismology cause both frontal and basal subduction erosion and extension in the forearc directly above them (Ranero and von Huene, 2000). No subducting seamounts have been imaged, however, at the latitude of the Coastal Cordillera normal faulting in the central Andes (e.g., Buske et al., 2002; Oncken et al., 2003; von Huene and Ranero, 2003), and the lateral extent of normal faulting is far greater than that expected from subduction of seamounts, which are generally spatially scattered or define narrow, linear chains. The Iquique Ridge is one of the more prominent bathymetric features on the Nazca plate entering the subduction zone of northern Chile (Fig. 1; Contreras-Reyes and Carrizo, 2011); however, the subduction angle remains normal through this region and does not shallow until the Nazca Ridge to the north and the Juan Fernandez Ridge to the south (Fig. 1). Since none of the observed features—the reactivation of preexisting faults, underplating and coastal uplift, and slab dynamics—is spatially limited to the latitude of Coastal Cordillera normal faulting, we look inboard for a controlling mechanism.

Retroarc

It should be noted that at the same latitude of greatest active onshore normal faulting and the Salar de Atacama, the retroarc transitions from the aseismic, thin-skinned thrust belt in the north to the seismic, thick-skinned thrust belt in the south (Figs. 3 and 6A). This is most likely controlled by preexisting along-strike heterogeneities in Paleozoic stratigraphic thickness and structure, such as the presence of Cretaceous rift structures in Argentina (e.g., Cahill et al., 1992; Allmendinger and Gubbels, 1996; Kley and Monaldi, 2002), and the significance, if any, in relation to forearc extension is cryptic at best.

Salar de Atacama

The only feature that is completely anomalous and strongly spatially correlated with onshore forearc normal faulting is the Salar de Atacama in the Preandean Depression. The greatest active onshore normal faulting is located at the latitude of the Mejillones Peninsula (Allmendinger and Gonzalez, 2010), which latitudinally corresponds with the center of the depression (Fig. 3). The normal-fault focal mechanisms follow this pattern, being more concentrated at the latitude of the depression, and their trends in map view follow the eastward deflection of the Western Cordillera and volcanic front around the depression (Fig. 7). The Salar de Atacama exhibits high crustal seismic velocities and low attenuation (Schurr and Rietbrock, 2004), a pronounced gravity high (Götze and Krause, 2002; Prezzi et al., 2009), and a crustal thickness greater than that expected based on assumptions of isostasy from elevation (Yuan et al., 2002). These observations indicate that the crust is cold, dense, and strong. The lithosphere beneath the Salar de Atacama is not anomalously thick for the forearc (~100 km), but it is anomalous for its longitude (Fig. 7). The thickest part of the lithosphere is near 69°W for most of the forearc, but it is shifted eastward to ~68.5°W at the latitude of the Salar de Atacama (Fig. 7). The Salar de Atacama appears to be underlain by a thick lithospheric root that is directly juxtaposed against the underlying subducting slab (Figs. 6C and 6E).

The origin of the anomalous crust and mantle lithosphere beneath the Salar de Atacama is enigmatic. Based on the fast Vp but moderate Vp/Vs values, the lower-crustal rocks must be of intermediate composition (consistent with a mix of gneiss and metabasite) and at high-pressure and low-temperature eclogite-facies conditions (Schurr and Rietbrock, 2004). The central Andean gravity high, which includes the Salar de Atacama, the Salar de Arizaro, and the Calama Basin, has been suggested to be the result of mafic intrusions related to a proposed Ordovician magmatic arc, or possibly to more recent Triassic magmatic activity (Götze and Krause, 2002). Alternatively, Reutter et al. (2006) suggested that the extension of the Salar de Atacama in the Atacama-Sey arm of the Salta Rift led to basaltic underplating, and the Cretaceous arc intruded dense magmatic bodies. During a presumed flat or shallow subduction event 28–17 Ma, the Salar de Atacama crust cooled slowly with only weak hydration, forming a strong, dense crustal block (Reutter et al., 2006).

The surface of the Salar de Atacama is not as high as the plateaus that it deflects, but it nevertheless rises to ~2.3 km. Seismic and borehole data suggest that the basin has been tilted westward during the Cenozoic, related to the growth of a monocline system on the western plateau boundary (Jordan et al., 2010). Large low-velocity anomalies beneath the Altiplano-Puna volcanic complex to the east and high-velocity anomalies to the west may have also

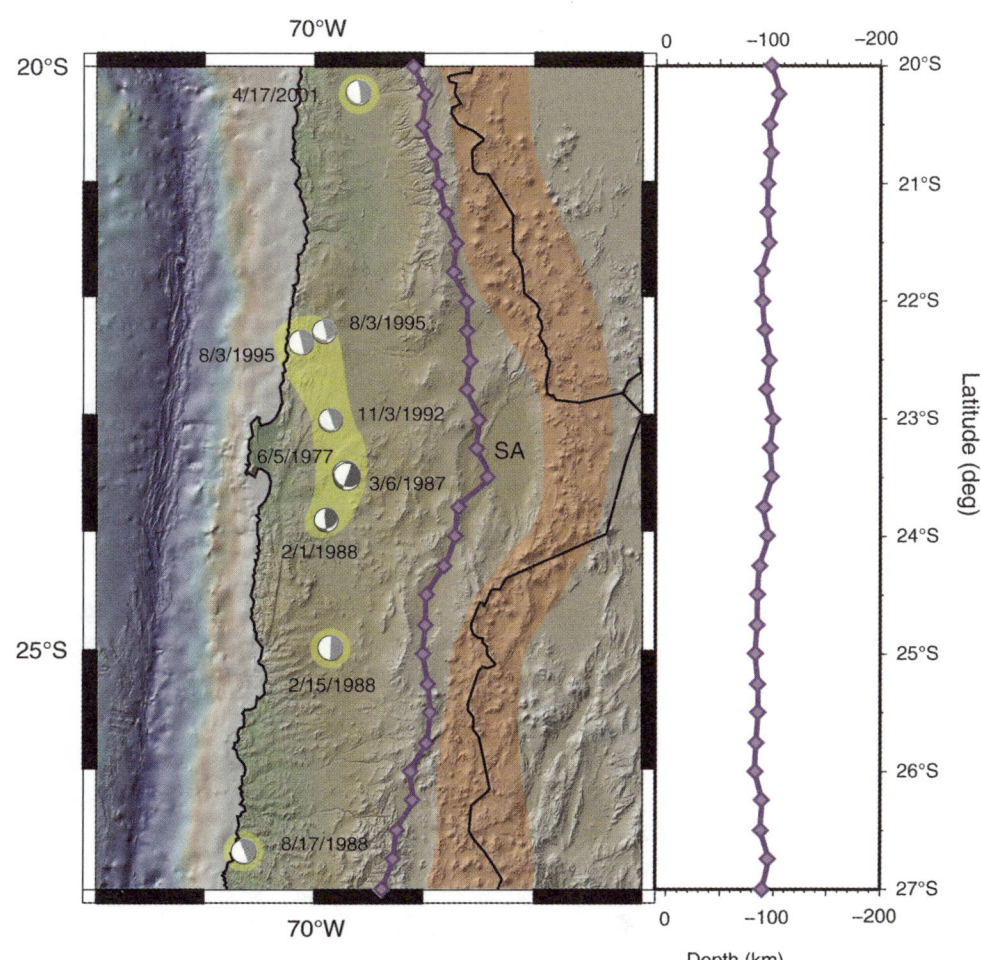

Figure 7. Deflections at the latitude of the Salar de Atacama (SA). Extensional focal mechanisms are shaded with yellow. The purple line, shown in map view and longitudinal section, marks the greatest depth to the lithosphere-asthenosphere boundary at each latitude (Tassara and Echaurren, 2012). The magmatic arc is shaded orange. All three are deflected eastward at the latitude of the Salar de Atacama. Extensional focal mechanisms are also deeper toward the central latitude of the basin.

contributed to the westward tilt of the block at a deeper level. While the sedimentary basin is subsiding, the Atacama block as a whole has been uplifted relative to the Central Depression, though not so high as the Altiplano and Puna Plateaus to the east. The elevation difference between the Atacama block and the plateaus is likely the surface expression of a negatively buoyant root attached to a heterogeneous block (Reutter et al., 2006).

Negative buoyancy must not be confused with surface subsidence. In the case of the Atacama block, the active thrust faults indicate thickening or bending of the lithosphere. If these thrust faults are rooted in the middle crust, the associated shortening may create a thicker, denser, more negatively buoyant root while simultaneously driving rock uplift at the surface, essentially equal to surface uplift in the hyperarid climate of the Atacama Desert. However, the thick root suppresses the elevation gain of the block. In the absence of faults rooted in the middle crust, these faults could be related to inner arc shortening as the lithosphere bends down in response to the negatively buoyant root. This also promotes minor thickening in the hinge zone. Rather than driving surface subsidence, a negatively buoyant root may simply prevent a block from uplifting as high as it might in the absence of such a root.

The thin lithosphere and slow-velocity anomalies beneath the Puna Plateau (Figs. 4 and 6C–6D) are strong evidence for lithospheric delamination. The lithospheric root beneath the Salar de Atacama is juxtaposed directly against the subducting slab. It is not clear whether the root is more or less likely to detach from this position. The slab may help support the root, making it less likely to detach, or shear stress from corner flow may destabilize it. If it were to detach, the asthenospheric wedge would flow into the vacancy, the volcanic front and Western Cordillera would likely straighten out, and the Salar de Atacama basin would isostatically and dynamically rebound. This would presumably decrease the negative vertical load of the Salar de Atacama in the inner forearc and thereby decrease a force that would contribute to extension within the outer forearc.

Hinterland Root Removal

The Cordilleran orogenic cycle model predicts forearc extension in the postdelamination stage (DeCelles et al., 2009). Removing a dense root would lead to isostatic and dynamic rebound, increasing the gravitational potential energy of the area above the delamination. This predicts that the forearc would be driven into a supercritical state at latitudes of the plateau that have experienced recent lithospheric removal. The observed Coastal Cordillera extension, however, seems related to the persistence of a dense crust and lithosphere beneath the Salar de Atacama. Lithospheric removal appears to be piecemeal and has been inferred at latitudes with only moderate extensional faulting (e.g., ~26°S [Kay and Kay, 1993; Kay et al., 1994; Drew et al., 2009] and 20°S [Beck and Zandt, 2002]), whereas areas of greater extensional faulting have experienced partial (24°S [Koulakov et al., 2006; Schurr et al., 2006]) or no recent (<10 Ma)

lithospheric removal. These spatial patterns of lithospheric removal and observed forearc extension are inconsistent with isostatic rebound driving the forearc wedge into a supercritical state, leading to extension. Lithospheric removal in the hinterland, however, may have increased the impact of the dense Salar de Atacama lithospheric root on forearc dynamics. Lithospheric root removal beneath the Puna Plateau would have increased the lateral gravitational potential energy gradient across the inner forearc and may help explain the active E-W shortening that characterizes the Salar de Atacama. The negative buoyancy of the Salar de Atacama may also have been enhanced as it was juxtaposed against the less dense asthenosphere beneath the hinterland to the east; a greater downward pull deep below the inner forearc may have contributed to the down-to-the-east normal faulting in the Coastal Cordillera (Fig. 8). Although the orogenic cycle model predicts extension in an erosional forearc as a result of delamination, and delamination has played a role in driving forearc extension, the dynamics of partial removal and preexisting structures can lead to localized forearc extension without the entire forearc being in a supercritical state.

The Right Combination of Factors

We propose that onshore, E-dipping, active forearc normal faulting in the Coastal Cordillera in northern Chile is caused by the negatively buoyant lithospheric root beneath the Salar de Atacama pulling down the inner forearc relative to the outer forearc, which is being uplifted in response to underplating (Fig. 8). The negative buoyancy of the Salar de Atacama may have been further enhanced by recent (<10 Ma) lithospheric delamination to the east of the forearc, but it was not until forearc uplift began at ca. 2 Ma and the development of a positive vertical force (due to underplating) to the west that the forearc began to extend. The two opposite forces, in combination with the presence of preexisting faults of suitable orientation, may have caused the forearc to fail in a normal sense with the east side down toward the lithospheric root (Fig. 8). The entire length of the central Andean Coastal Cordillera contains preexisting faults and is actively uplifting; however, these factors are not enough to drive significant extension. The contribution of stress from the Salar de Atacama root may have been sufficient to increase the differential stress just enough to cause low-magnitude extensional strain on the order of 0.02–0.2 mm/yr (Allmendinger and Gonzalez, 2010). However, modeling is necessary to investigate the validity of this proposed model for driving Coastal Cordillera normal faulting.

CONCLUSIONS

At the latitude of E-dipping, active Coastal Cordillera extensional faulting (22°S to 25°S), seismicity increases from background subduction-related seismicity (Fig. 6A). Extensional focal mechanisms represent steep E-dipping, N-S–striking faults that extend to a depth of 40 km (Fig. 6C). The thin-skinned Bolivian retroarc thrust belt is seismically inactive, but the thick-skinned

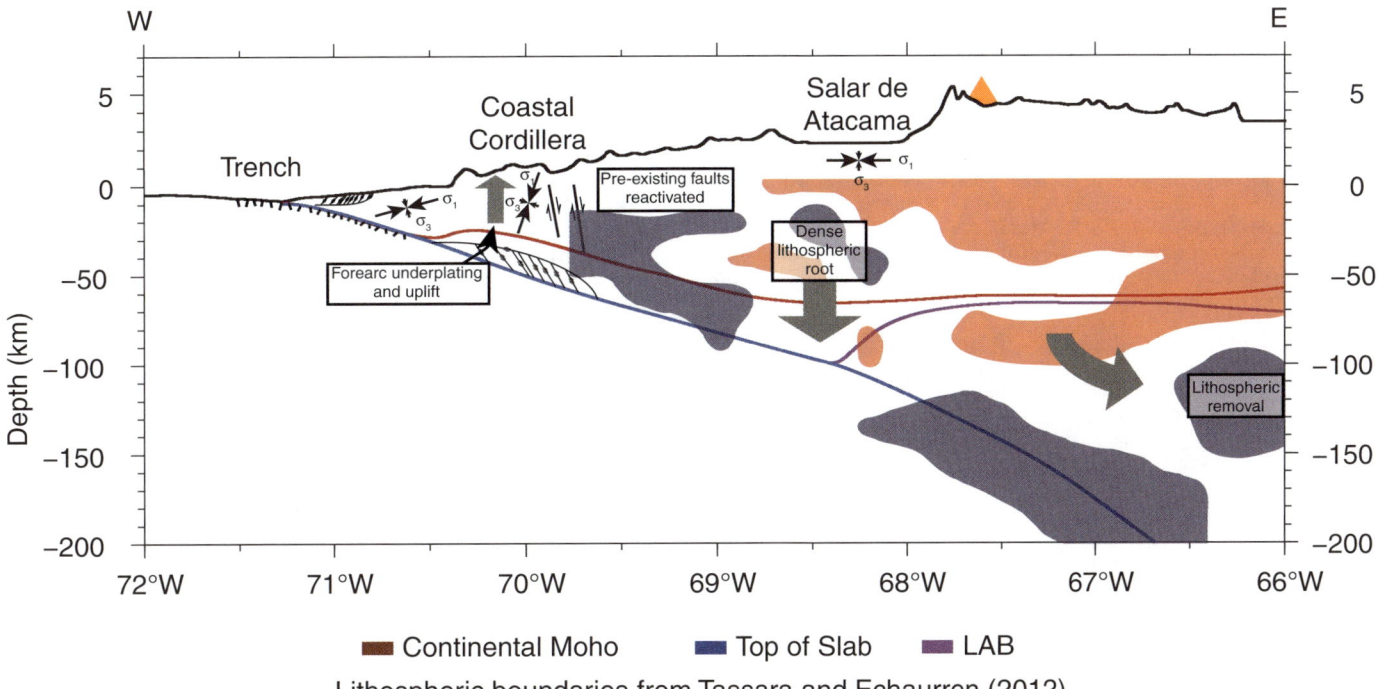

Figure 8. Cartoon of proposed forces driving onshore forearc normal faulting. Gray arrows indicate forces due to underplating and uplift and to the negative buoyancy of the dense, cold Atacama block and lithospheric root. Inferred principal stress directions are shown in the toe of the wedge, the Coastal Cordillera, and the Salar de Atacama. Topography below sea level (0 km) is shown with no vertical exaggeration, and topography above sea level is shown with 10× vertical exaggeration. All interpretations are shown without vertical exaggeration. Colored lines are as in Figure 6. Orange triangle marks the volcanic front. LAB—lithosphere-asthenosphere boundary.

Argentine retroarc thrust belt to the south is seismically active (Fig. 6A). The Sierra de Moreno (Fig. 3) in the northern Chilean Precordillera is seismically active to depths of ~20 km (Fig. 6B), while the Salar de Atacama in the central Precordillera contains seismicity at depths down to 40 km (Fig. 6B). The Puna Plateau lithospheric mantle, compared to the Altiplano Plateau, has been thinned by delamination events (Fig. 3). In contrast, the retroarc and forearc maintain thick (>80 km) mantle lithosphere (Fig. 3). The axis of thickest forearc mantle lithosphere is shifted eastward at the latitude of the Salar de Atacama (Fig. 7), owing to the presence of a lithospheric root (Fig. 6B). Outer forearc normal faulting is localized to the same latitude band as the negatively buoyant, dense lithospheric root beneath the Salar de Atacama in the inner forearc (Figs. 3 and 7).

In the absence of large changes in plate motion, subduction angle, slab age, slab topography, and forearc uplift, along-strike variations in forearc dynamics in northern Chile are most likely controlled by variations in the lithospheric architecture of the upper plate. The Salar de Atacama, the surface expression of the seismically defined Atacama block (Schurr and Rietbrock, 2004), corresponds with the latitude of greatest active Coastal Cordillera normal faults and deflects the volcanic arc, Western Cordillera, pattern of normal focal mechanisms, and longitudinal axis of thick lithosphere to the east. The forearc at latitudes 22°S–25°S was preconditioned for extension by the Mesozoic

Atacama fault system. Coastal Cordilleran uplift in response to underplating causes an increase in gravitational potential energy, but not enough to cause extension in areas to the north and south. Combined with the negative buoyancy of the cold, dense Atacama block, possibly enhanced by lithospheric removal beneath the Puna Plateau to the east, there may be enough differential stress to explain the reactivation of the Atacama fault system in a normal sense. Further modeling work is needed to substantiate this proposed model.

ACKNOWLEDGMENTS

This research was supported by ExxonMobil. Additional funding was provided by ConocoPhillips, ChevronTexaco, and the Peter J. Coney Scholarship Fund. Discussion with David Pearson improved ideas in this paper. We acknowledge Peter DeCelles and George Zandt for their helpful comments on previous drafts. This work benefited from constructive reviews by O. Oncken and R. von Huene.

REFERENCES CITED

Adam, J., and Reuther, C.D., 2000, Crustal dynamics and active fault mechanics during subduction erosion: Application of frictional wedge analysis on to the north Chilean forearc: Tectonophysics, v. 321, no. 3, p. 297–325, doi:10.1016/S0040-1951(00)00074-3.

Allmendinger, R.W., and Gonzalez, G., 2010, Invited review paper: Neogene to Quaternary tectonics of the Coastal Cordillera, northern Chile: Tectonophysics, v. 495, no. 1–2, p. 93–110, doi:10.1016/j.tecto.2009.04.019.

Allmendinger, R.W., and Gubbels, T., 1996, Pure and simple shear plateau uplift, Altiplano-Puna, Argentina and Bolivia: Tectonophysics, v. 259, no. 1–3, p. 1–13, doi:10.1016/0040-1951(96)00024-8.

Allmendinger, R.W., Jordan, T.E., and Kay, S.M., 1997, The evolution of the Altiplano-Puna Plateau of the central Andes: Annual Review of Earth and Planetary Sciences, v. 25, no. 1, p. 139–174, doi:10.1146/annurev.earth.25.1.139.

Amilibia, A., Sabat, F., McClay, K.R., Munoz, J.A., Roca, E., and Chong, G., 2008, The role of inherited tectono-sedimentary architecture in the development of the central Andean mountain belt: Insights from the Cordillera de Domeyko: Journal of Structural Geology, v. 30, no. 12, p. 1520–1539, doi:10.1016/j.jsg.2008.08.005.

Arriagada, C., Cobbold, P.R., and Roperch, P., 2006, Salar de Atacama basin: A record of compressional tectonics in the central Andes since the mid-Cretaceous: Tectonics, v. 25, no. 1, TC1008, doi:10.1029/2004TC001770.

Baker, A., Allmendinger, R.W., Owen, L.A., and Rech, J.A., 2013, Permanent deformation caused by subduction earthquakes in northern Chile: Nature Geoscience, v. 6, no. 6, p. 492–496, doi:10.1038/ngeo1789.

Beck, S.L., and Zandt, G., 2002, The nature of orogenic crust in the central Andes: Journal of Geophysical Research–Solid Earth, v. 107, no. B10, p. 1–16, doi:10.1029/2000jb000124.

Béjar-Pizarro, M., Carrizo, D., Socquet, A., Armijo, R., Barrientos, S., Bondoux, F., Bonvalot, S., Campos, J., Comte, D., de Chabalier, J.B., Charade, O., Delorme, A., Gabalda, G., Galetzka, J., and 10 others, 2010, Asperities and barriers on the seismogenic zone in north Chile: State-of-the-art after the 2007 Mw 7.7 Tocopilla earthquake inferred by GPS and InSAR data: Geophysical Journal International, v. 183, no. 1, p. 390–406, doi:10.1111/j.1365-246X.2010.04748.x.

Brooks, B.A., Bevis, M., Whipple, K., Arrowsmith, J.R., Foster, J., Zapata, T., Kendrick, E., Minaya, E., Echalar, A., Blanco, M., Euillades, P., Sandoval, M., and Smalley, R.J., 2011, Orogenic-wedge deformation and potential for great earthquakes in the central Andean backarc: Nature Geoscience, v. 4, no. 6, p. 380–383, doi:10.1038/ngeo1143.

Brown, M., Diaz, F., and Grocott, J., 1993, Displacement history of the Atacama fault system 25°00′S–27°00′S, northern Chile: Geological Society of America Bulletin, v. 105, no. 9, p. 1165–1174, doi:10.1130/0016-7606(1993)105<1165:DHOTAF>2.3.CO;2.

Buske, S., Luth, S., Meyer, H., Patzig, R., Reichert, C., Shapiro, S., Wigger, P., and Yoon, M., 2002, Broad depth range seismic imaging of the subducted Nazca slab, north Chile: Tectonophysics, v. 350, no. 4, p. 273–282, doi:10.1016/S0040-1951(02)00117-8.

Cahill, T., Isacks, B.L., Whitman, D., Chatelain, J.L., Perez, A., and Chiu, J.M., 1992, Seismicity and tectonics in Jujuy province, northwestern Argentina: Tectonics, v. 11, no. 5, p. 944–959, doi:10.1029/92TC00215.

Cembrano, J., Gonzalez, G., Arancibia, G., Ahumada, I., Olivares, V., and Herrera, V., 2005, Fault zone development and strain partitioning in an extensional strike-slip duplex: A case study from the Mesozoic Atacama fault system, northern Chile: Tectonophysics, v. 400, no. 1–4, p. 105–125, doi:10.1016/j.tecto.2005.02.012.

Christensen, N.I., 2004, Serpentinites, peridotites, and seismology: International Geology Review, v. 46, no. 9, p. 795–816, doi:10.2747/0020-6814.46.9.795.

Clift, P.D., and Hartley, A.J., 2007, Slow rates of subduction erosion and coastal underplating along the Andean margin of Chile and Peru: Geology, v. 35, no. 6, p. 503–506, doi:10.1130/G23584A.1.

Cloos, M., and Shreve, R.L., 1988, Subduction-channel model of prism accretion, mélange formation, sediment subduction, and subduction erosion at convergent plate margins: 1. Background and description: Pure and Applied Geophysics, v. 128, no. 3–4, p. 455–500.

Cloos, M., and Shreve, R.L., 1996, Shear-zone thickness and the seismicity of Chilean- and Marianas-type subduction zones: Geology, v. 24, no. 2, p. 107–110, doi:10.1130/0091-7613(1996)024<0107:SZTATS>2.3.CO;2.

Coira, B., Davidson, J., Mpodozis, C., and Ramos, V., 1982, Tectonic and magmatic evolution of the Andes of northern Argentina and Chile: Earth-Science Reviews, v. 18, no. 3–4, p. 303–332, doi:10.1016/0012-8252(82)90042-3.

Collot, J.Y., Agudelo, W., Ribodetti, A., and Marcaillou, B., 2008, Origin of a crustal splay fault and its relation to the seismogenic zone and underplating at the erosional north Ecuador–south Colombia oceanic margin: Journal of Geophysical Research–Solid Earth, v. 113, no. B12, p. 1–19, doi:10.1029/2008jb005691.

Comte, D., Pardo, M., Dorbath, L., Dorbath, C., Haessler, H., Rivera, L., Cisternas, A., and Ponce, L., 1994, Determination of seismogenic interplate contact zone and crustal seismicity around Antofagasta, northern Chile, using local data: Geophysical Journal International, v. 116, no. 3, p. 553–561, doi:10.1111/j.1365-246X.1994.tb03279.x.

Contreras-Reyes, E., and Carrizo, D., 2011, Control of high oceanic features and subduction channel on earthquake ruptures along the Chile-Peru subduction zone: Physics of the Earth and Planetary Interiors, v. 186, no. 1–2, p. 49–58, doi:10.1016/j.pepi.2011.03.002.

Cortés A., J., González L., G., Binnie, S.A., Robinson, R., Freeman, S.P.H.T., and Vargas E., G., 2012, Paleoseismology of the Mejillones fault, northern Chile: Insights from cosmogenic ^{10}Be and optically stimulated luminescence determinations: Tectonics, v. 31, no. 2, TC2017, doi:10.1029/2011TC002877.

DeCelles, P.G., Ducea, M.N., Kapp, P., and Zandt, G., 2009, Cyclicity in Cordilleran orogenic systems: Nature Geoscience, v. 2, no. 4, p. 251–257, doi:10.1038/ngeo469.

Delouis, B., Philip, H., Dorbath, L., and Cisternas, A., 1998, Recent crustal deformation in the Antofagasta region (northern Chile) and the subduction process: Geophysical Journal International, v. 132, no. 2, p. 302–338, doi:10.1046/j.1365-246x.1998.00439.x.

Drew, S.T., Ducea, M.N., and Schoenbohm, L.M., 2009, Mafic volcanism on the Puna Plateau, NW Argentina: Implications for lithospheric composition and evolution with an emphasis on lithospheric foundering: Lithosphere, v. 1, no. 5, p. 305–318, doi:10.1130/L54.1.

Echavarria, L., Hernández, R., Allmendinger, R., and Reynolds, J., 2003, Subandean thrust and fold belt of northwestern Argentina: Geometry and timing of the Andean evolution: American Association of Petroleum Geologists Bulletin, v. 87, no. 6, p. 965–985, doi:10.1306/01200300196.

González, G., Cembrano, J., Carrizo, D., Macci, A., and Schneider, H., 2003, The link between forearc tectonics and Pliocene–Quaternary deformation of the Coastal Cordillera, northern Chile: Journal of South American Earth Sciences, v. 16, no. 5, p. 321–342, doi:10.1016/S0895-9811(03)00100-7.

Götze, H.J., and Krause, S., 2002, The central Andean gravity high, a relic of an old subduction complex?: Journal of South American Earth Sciences, v. 14, no. 8, p. 799–811, doi:10.1016/S0895-9811(01)00077-3.

Hartley, A.J., May, G., Chong, G., Turner, P., Kape, S.J., and Jolley, E.J., 2000, Development of a continental forearc: A Cenozoic example from the central Andes, northern Chile: Geology, v. 28, no. 4, p. 331–334, doi:10.1130/0091-7613(2000)28<331:DOACFA>2.0.CO;2.

Haschke, M., Siebel, W., Gunther, A., and Scheuber, E., 2002, Repeated crustal thickening and recycling during the Andean orogeny in north Chile (21°–26°S): Journal of Geophysical Research–Solid Earth, v. 107, no. B1, p. 1–18, doi:10.1029/2001jb000328.

Heit, B., Koulakov, I., Asch, G., Yuan, X., Kind, R., Alcocer-Rodriguez, I., Tawackoli, S., and Wilke, H., 2008, More constraints to determine the seismic structure beneath the central Andes at 21°S using teleseismic tomography analysis: Journal of South American Earth Sciences, v. 25, no. 1, p. 22–36, doi:10.1016/j.jsames.2007.08.009.

Heuret, A., Conrad, C.P., Funiciello, F., Lallemand, S., and Sandri, L., 2012, Relation between subduction megathrust earthquakes, trench sediment thickness and upper plate strain: Geophysical Research Letters, v. 39, no. 5, L05304, doi:10.1029/2011GL050712.

Husen, S., Kissling, E., and Flueh, E.R., 2000, Local earthquake tomography of shallow subduction in north Chile: A combined onshore and offshore study: Journal of Geophysical Research–Solid Earth, v. 105, no. B12, p. 28,183–28,198, doi:10.1029/2000JB900229.

International Seismological Centre, 2009, EHB Bulletin: Thatcham, UK, International Seismological Centre, http://www.isc.ac.uk (accessed July 2014).

Jordan, T.E., Nester, P.L., Blanco, N., Hoke, G.D., Davila, F., and Tomlinson, A.J., 2010, Uplift of the Altiplano-Puna Plateau: A view from the west: Tectonics, v. 29, TC5007, doi:10.1029/2010TC002661.

Kay, R.W., and Kay, S.M., 1993, Delamination and delamination magmatism: Tectonophysics, v. 219, no. 1–3, p. 177–189, doi:10.1016/0040-1951(93)90295-U.

Kay, S.M., and Coira, B.L., 2009, Shallowing and steepening subduction zones, continental lithospheric loss, magmatism, and crustal flow under the central Andean Altiplano-Puna Plateau, *in* Kay, S.M., Ramos, V.A., and Dickinson, W.R., eds., Backbone of the Americas: Shallow Subduction, Plateau Uplift, and Ridge and Terrane Collision: Geological Society of America Memoir 204, p. 229–259, doi:10.1130/2009.1204(11).

Kay, S.M., Coira, B., and Viramonte, J., 1994, Young mafic back-arc volcanic rocks as indicators of continental lithospheric delamination beneath the

Argentine Puna Plateau, central Andes: Journal of Geophysical Research–Solid Earth, v. 99, no. B12, p. 24,323–24,339, doi:10.1029/94JB00896.

Kendrick, E., Bevis, M., Smalley, R., Brooks, B., Vargas, R., Lauriia, E., and Fortes, L., 2003, The Nazca South America Euler vector and its rate of change: Journal of South American Earth Sciences, v. 16, no. 2, p. 125–131, doi:10.1016/S0895-9811(03)00028-2.

Khazaradze, G., and Klotz, J., 2003, Short- and long-term effects of GPS measured crustal deformation rates along the south central Andes: Journal of Geophysical Research–Solid Earth, v. 108, no. B6, p. 1–15, doi:10.1029/2002jb001879.

Kley, J., and Monaldi, C.R., 1998, Tectonic shortening and crustal thickness in the central Andes: How good is the correlation?: Geology, v. 26, no. 8, p. 723–726, doi:10.1130/0091-7613(1998)026<0723:TSACTI>2.3.CO;2.

Kley, J., and Monaldi, C.R., 2002, Tectonic inversion in the Santa Barbara system of the central Andean foreland thrust belt, northwestern Argentina: Tectonics, v. 21, no. 6, p. 11-1–11-18, doi:10.1029/2002TC902003.

Koulakov, I., Sobolev, S.V., and Asch, G., 2006, P- and S-velocity images of the lithosphere-asthenosphere system in the central Andes from local-source tomographic inversion: Geophysical Journal International, v. 167, no. 1, p. 106–126, doi:10.1111/j.1365-246X.2006.02949.x.

Lamb, S., and Davis, P., 2003, Cenozoic climate change as a possible cause for the rise of the Andes: Nature, v. 425, no. 6960, p. 792–797, doi:10.1038/nature02049.

Lloyd, S., van der Lee, S., Franca, G.S., Assumpcao, M., and Feng, M., 2010, Moho map of South America from receiver functions and surface waves: Journal of Geophysical Research–Solid Earth, v. 115, p. 1–12, doi:10.1029/2009jb006829.

Loewy, S.L., Connelly, J.N., and Dalziel, I.W.D., 2004, An orphaned basement block: The Arequipa-Antofalla basement of the central Andean margin of South America: Geological Society of America Bulletin, v. 116, no. 1, p. 171–187, doi:10.1130/B25226.1.

Loveless, J.P., and Pritchard, M.E., 2008, Motion on upper-plate faults during subduction zone earthquakes: Case of the Atacama fault system, northern Chile: Geochemistry Geophysics Geosystems, v. 9, no. 12, p. 1–23, doi:10.1029/2008GC002155.

Loveless, J.P., Hoke, G.D., Allmendinger, R.W., Gonzalez, G., Isacks, B.L., and Carrizo, D.A., 2005, Pervasive cracking of the northern Chilean Coastal Cordillera: New evidence for forearc extension: Geology, v. 33, no. 12, p. 973–976, doi:10.1130/G22004.1.

Loveless, J.P., Allmendinger, R.W., Pritchard, M.E., and Gonzalez, G., 2010, Normal and reverse faulting driven by the subduction zone earthquake cycle in the northern Chilean forearc: Tectonics, v. 29, p. 1–16, doi:10.1029/2009TC002465.

Manea, V., and Gurnis, M., 2007, Subduction zone evolution and low viscosity wedges and channels: Earth and Planetary Science Letters, v. 264, no. 1–2, p. 22–45, doi:10.1016/j.epsl.2007.08.030.

Marquardt, C., Lavenu, A., Ortlieb, L., Godoy, E., and Comte, D., 2004, Coastal neotectonics in southern central Andes: Uplift and deformation of marine terraces in northern Chile (27°S): Tectonophysics, v. 394, no. 3–4, p. 193–219, doi:10.1016/j.tecto.2004.07.059.

McGlashan, N., Brown, L., and Kay, S., 2008, Crustal thickness in the central Andes from teleseismically recorded depth phase precursors: Geophysical Journal International, v. 175, no. 3, p. 1013–1022, doi:10.1111/j.1365-246X.2008.03897.x.

McQuarrie, N., Horton, B.K., Zandt, G., Beck, S., and DeCelles, P.G., 2005, Lithospheric evolution of the Andean fold-thrust belt, Bolivia, and the origin of the central Andean plateau: Tectonophysics, v. 399, no. 1–4, p. 15–37, doi:10.1016/j.tecto.2004.12.013.

Meigs, A.J., and Nabelek, J., 2010, Crustal-scale pure shear foreland deformation of western Argentina: Geophysical Research Letters, v. 37, p. 1–5, doi:10.1029/2010GL043220.

Motagh, M., Schurr, B., Anderssohn, J., Cailleau, B., Walter, T.R., Wang, R.J., and Villotte, J.P., 2010, Subduction earthquake deformation associated with 14 November 2007, Mw 7.8 Tocopilla earthquake in Chile: Results from InSAR and aftershocks: Tectonophysics, v. 490, no. 1–2, p. 60–68, doi:10.1016/j.tecto.2010.04.033.

Nippress, S.E.J., and Rietbrock, A., 2007, Seismogenic zone high permeability in the central Andes inferred from relocations of micro-earthquakes: Earth and Planetary Science Letters, v. 263, no. 3–4, p. 235–245, doi:10.1016/j.epsl.2007.08.032.

Oncken, O., Sobolev, S., Stiller, M., Asch, G., Haberland, C., Mechie, J., Yuan, X.H., Luschen, E., Giese, P., Wigger, P., Lueth, S., Scheuber, E., Götze,

H.-J., Brasse, H., and 18 others, 2003, Seismic imaging of a convergent continental margin and plateau in the central Andes (Andean Continental Research Project 1996 (ANCORP'96)): Journal of Geophysical Research–Solid Earth, v. 108, no. B7, p. 1–25, doi:10.1029/2002jb001771.

Oncken, O., Hindle, D., Kley, J., Elger, K., Victor, P., and Schemmann, K., 2006, Deformation of the central Andean upper plate system—Facts, fiction, and constraints for plateau models, *in* Oncken, O., Chong, G., Franz, G., Giese, P., Götze, H.-J., Ramos, V.A., Strecker, M.R., and Wigger, P., eds., The Andes: Active Subduction Orogeny: Berlin, Springer, Frontiers in Earth Sciences, p. 3–27.

Ortlieb, L., Zazo, C., Goy, J., Hillaire-Marcel, C., Ghaleb, B., and Cournoyer, L., 1996, Coastal deformation and sea-level changes in the northern Chile subduction area (23°S) during the last 330 ky: Quaternary Science Reviews, v. 15, no. 8–9, p. 819–831, doi:10.1016/S0277-3791(96)00066-2.

Pearson, D.M., Kapp, P., Reiners, P.W., Gehrels, G.E., Ducea, M.N., Pullen, A., Otamendi, J.E., and Alonso, R.N., 2012, Major Miocene exhumation by fault-propagation folding within a metamorphosed, early Paleozoic thrust belt: Northwestern Argentina: Tectonics, v. 31, no. 4, TC4023, doi:10.1029/2011TC003043.

Prezzi, C.B., Gotze, H.J., and Schmidt, S., 2009, 3D density model of the central Andes: Physics of the Earth and Planetary Interiors, v. 177, no. 3–4, p. 217–234, doi:10.1016/j.pepi.2009.09.004.

Ramos, V.A., 2008, The basement of the central Andes: The Arequipa and related terranes: Annual Review of Earth and Planetary Sciences, v. 36, no. 1, p. 289–324, doi:10.1146/annurev.earth.36.031207.124304.

Ranero, C.R., and von Huene, R., 2000, Subduction erosion along the Middle America convergent margin: Nature, v. 404, no. 6779, p. 748–752, doi:10.1038/35008046.

Regard, V., Saillard, M., Martinod, J., Audin, L., Carretier, S., Pedoja, K., Riquelme, R., Paredes, P., and Herail, G., 2010, Renewed uplift of the central Andes forearc revealed by coastal evolution during the Quaternary: Earth and Planetary Science Letters, v. 297, no. 1–2, p. 199–210, doi:10.1016/j.epsl.2010.06.020.

Reutter, K.J., Charrier, R., Götze, H.-J., Schurr, B., Wigger, P., Scheuber, E., Giese, P., Reuther, C.-D., Schmidt, S., Rietbrock, A., Chong, G., and Belmonte-Pool, A., 2006, The Salar de Atacama Basin: A subsiding block within the western edge of the Altiplano-Puna Plateau, *in* Oncken, O., Chong, G., Franz, G., Giese, P., Götze, H.-J., Ramos, V.A., Strecker, M.R., and Wigger, P., eds., The Andes: Active Subduction Orogeny: Berlin, Springer, Frontiers in Earth Sciences, p. 303–325.

Le Roux, J.P., Gomez, C., Venegas, C., Fenner, C., Middleton, H., Buchbinder, B., Frassinetti, D., Marquardt, C., Marchant, M., Gregory-Wodzicki, K.M., and Lavenu, A., 2005, Neogene-Quaternary coastal and offshore sedimentation in north central Chile: Record of sea-level changes and implications for Andean tectonism: Journal of South American Earth Sciences, v. 19, no. 1, p. 83–98, doi:10.1016/j.jsames.2003.11.003.

Russo, R.M., and Silver, P.G., 1996, Cordillera formation, mantle dynamics, and the Wilson cycle: Geology, v. 24, no. 6, p. 511–514, doi:10.1130/0091-7613(1996)024<0511:CFMDAT>2.3.CO;2.

Rutland, R.W., 1971, Andean orogeny and ocean floor spreading: Nature, v. 233, p. 252–255, doi:10.1038/233252a0.

Sallares, V., and Ranero, C.R., 2005, Structure and tectonics of the erosional convergent margin off Antofagasta, north Chile (23°30′S): Journal of Geophysical Research–Solid Earth, v. 110, no. B6, p. 1–19, doi:10.1029/2004jb003418.

Schellart, W.P., 2008, Subduction zone trench migration: Slab driven or overriding-plate-driven?: Physics of the Earth and Planetary Interiors, v. 170, no. 1–2, p. 73–88, doi:10.1016/j.pepi.2008.07.040.

Scheuber, E., and Andriessen, P., 1990, The kinematic and geodynamic significance of the Atacama fault zone, northern Chile: Journal of Structural Geology, v. 12, no. 2, p. 243–257, doi:10.1016/0191-8141(90)90008-M.

Scheuber, E., and Gonzalez, G., 1999, Tectonics of the Jurassic–Early Cretaceous magmatic arc of the north Chilean Coastal Cordillera (22°–26°S): A story of crustal deformation along a convergent plate boundary: Tectonics, v. 18, no. 5, p. 895–910, doi:10.1029/1999TC900024.

Scheuber, E., and Reutter, K.J., 1992, Magmatic arc tectonics in the central Andes between 21° and 25°S: Tectonophysics, v. 205, no. 1–3, p. 127–140, doi:10.1016/0040-1951(92)90422-3.

Schmitz, M., Lessel, K., Giese, P., Wigger, P., Araneda, M., Bribach, J., Graeber, F., Grunewald, S., Haberland, C., Luth, S., Rower, P., Ryberg, T., and Schulze, A., 1999, The crustal structure beneath the central Andean forearc and magmatic arc as derived from seismic studies—The PISCO

94 experiment in northern Chile (21°–23°S): Journal of South American Earth Sciences, v. 12, no. 3, p. 237–260, doi:10.1016/S0895-9811(99)00017-6.

Schurr, B., and Rietbrock, A., 2004, Deep seismic structure of the Atacama Basin, northern Chile: Geophysical Research Letters, v. 31, no. 12, L12601, doi:10.1029/2004GL019796.

Schurr, B., Asch, G., Rietbrock, A., Kind, R., Pardo, M., Heit, B., and Monfret, T., 1999, Seismicity and average velocities beneath the Argentine Puna Plateau: Geophysical Research Letters, v. 26, no. 19, p. 3025–3028, doi:10.1029/1999GL005385.

Schurr, B., Rietbrock, A., Asch, G., Kind, R., and Oncken, O., 2006, Evidence for lithospheric detachment in the central Andes from local earthquake tomography: Tectonophysics, v. 415, no. 1–4, p. 203–223, doi:10.1016/j.tecto.2005.12.007.

Schurr, B., Asch, G., Rosenau, M., Wang, R., Oncken, O., Barrientos, S., Salazar, P., and Vilotte, J.P., 2012, The 2007 M7.7 Tocopilla northern Chile earthquake sequence: Implications for along-strike and downdip rupture segmentation and megathrust frictional behavior: Journal of Geophysical Research–Solid Earth, v. 117, no. B5, p. 1–19, doi:10.1029/2011JB009030.

Shirzaei, M., Burgmann, R., Oncken, O., Walter, T.R., Victor, P., and Ewiak, O., 2012, Response of forearc crustal faults to the megathrust earthquake cycle: InSAR evidence from Mejillones Peninsula, northern Chile: Earth and Planetary Science Letters, v. 333–334, no. C, p. 157–164, doi:10.1016/j.epsl.2012.04.001.

Shreve, R.L., and Cloos, M., 1986, Dynamics of sediment subduction, mélange formation, and prism accretion: Journal of Geophysical Research–Solid Earth, v. 91, no. B10, p. 229–245, doi:10.1029/JB091iB10p10229.

Sick, C., Yoon, M.-K., Rauch, K., Buske, S., Lüth, S., Araneda, M., Bataille, K., Chong, G., Giese, P., Krawczyk, C., Mechie, J., Meyer, H., Oncken, O., Reichert, C., Schmitz, M., Shapiro, S., Stiller, M., and Wigger, P., 2006, Seismic images of accretive and erosive subduction zones from the Chilean margin, *in* Oncken, O., Chong, G., Franz, G., Giese, P., Götze, H.-J., Ramos, V.A., Strecker, M.R., and Wigger, P., eds., The Andes: Active Subduction Orogeny: Berlin, Springer, Frontiers in Earth Sciences, p. 147–169.

Stern, C.R., 2004, Active Andean volcanism: Its geologic and tectonic setting: Revista Geologica de Chile, v. 31, no. 2, p. 161–206.

Tassara, A., and Echaurren, A., 2012, Anatomy of the Andean subduction zone: Three-dimensional density model upgraded and compared against global-scale models: Geophysical Journal International, v. 189, no. 1, p. 161–168, doi:10.1111/j.1365-246X.2012.05397.x.

Tassara, A., Götze, H.-J., Schmidt, S., and Hackney, R., 2006, Three-dimensional density model of the Nazca plate and the Andean continental margin: Journal of Geophysical Research–Solid Earth, v. 111, no. B9, p. 1–26, doi:10.1029/2005jb003976.

van Hunen, J., van den Berg, A.P., and Vlaar, N.J., 2002, On the role of subducting oceanic plateaus in the development of shallow flat subduction: Tectonophysics, v. 352, no. 3–4, p. 317–333, doi:10.1016/S0040-1951(02)00263-9.

van Keken, P.E., Hacker, B.R., Syracuse, E.M., and Abers, G.A., 2011, Subduction factory: 4. Depth-dependent flux of H_2O from subducting slabs worldwide: Journal of Geophysical Research–Solid Earth, v. 116, no. B1, p. 1–15, doi:10.1029/2010JB007922.

Vannucchi, P., Galeotti, S., Clift, P.D., Ranero, C.R., and von Huene, R., 2004, Long-term subduction-erosion along the Guatemalan margin of the Middle America Trench: Geology, v. 32, no. 7, p. 617–620, doi:10.1130/G20422.1.

Vannucchi, P., Sage, F., Phipps Morgan, J., Remitti, F., and Collot, J.-Y., 2012, Toward a dynamic concept of the subduction channel at erosive convergent margins with implications for interplate material transfer: Geochemistry Geophysics Geosystems, v. 13, p. 1–24, doi:10.1029/2011GC003846.

Victor, P., Oncken, O., and Glodny, J., 2004, Uplift of the western Altiplano Plateau: Evidence from the Precordillera between 20° and 21°S (northern Chile): Tectonics, v. 23, no. 4, p. 1–24, doi:10.1029/2003TC001519.

Victor, P., Sobiesiak, M., Glodny, J., Nielsen, S.N., and Oncken, O., 2011, Long-term persistence of subduction earthquake segment boundaries: Evidence from Mejillones Peninsula, northern Chile: Journal of Geophysical Research–Solid Earth, v. 116, no. B2, p. 1–22, doi:10.1029/2010jb007771.

von Huene, R., and Lallemand, S., 1990, Tectonic erosion along the Japan and Peru convergent margins: Geological Society of America Bulletin, v. 102, no. 6, p. 704–720, doi:10.1130/0016-7606(1990)102<0704:TEATJA>2.3.CO;2.

von Huene, R., and Ranero, C.R., 2003, Subduction erosion and basal friction along the sediment-starved convergent margin off Antofagasta, Chile: Journal of Geophysical Research–Solid Earth, v. 108, no. B2, p. 1–17, doi:10.1029/2001jb001569.

von Huene, R., and Scholl, D.W., 1991, Observations at convergent margins concerning sediment subduction, subduction erosion, and the growth of continental-crust: Reviews of Geophysics, v. 29, no. 3, p. 279–316, doi:10.1029/91RG00969.

von Huene, R., Klaeschen, D., Cropp, B., and Miller, J., 1994, Tectonic structure across the accretionary and erosional parts of the Japan Trench margin: Journal of Geophysical Research–Solid Earth, v. 99, no. B11, p. 22,349–22,361, doi:10.1029/94JB01198.

Wada, I., and Wang, K., 2009, Common depth of slab-mantle decoupling: Reconciling diversity and uniformity of subduction zones: Geochemistry Geophysics Geosystems, v. 10, no. 10, p. 1–36, doi:10.1029/2009GC002570.

Wells, R.E., Blakely, R.J., Sugiyama, Y., Scholl, D.W., and Dinterman, P.A., 2003, Basin-centered asperities in great subduction zone earthquakes: A link between slip, subsidence, and subduction erosion?: Journal of Geophysical Research–Solid Earth, v. 108, no. B12, p. 1–30, doi:10.1029/2003jb002880.

Wölbern, I., Heit, B., Yuan, X., Asch, G., Kind, R., Viramonte, J., Tawackoli, S., and Wilke, H., 2009, Receiver function images from the Moho and the slab beneath the Altiplano and Puna Plateaus in the central Andes: Geophysical Journal International, v. 177, no. 1, p. 296–308, doi:10.1111/j.1365-246X.2008.04075.x.

Yuan, X., Sobolev, S.V., and Kind, R., 2002, Moho topography in the central Andes and its geodynamic implications: Earth and Planetary Science Letters, v. 199, no. 3–4, p. 389–402, doi:10.1016/S0012-821X(02)00589-7.

Manuscript Accepted by the Society 3 June 2014
Manuscript Published Online 9 September 2014

The Geological Society of America
Memoir 212
2015

Along-strike variation in structural styles and hydrocarbon occurrences, Subandean fold-and-thrust belt and inner foreland, Colombia to Argentina

Michael F. McGroder
Richard O. Lease*
David M. Pearson[†]
ExxonMobil Upstream Research Company, Houston, Texas 77252, USA

ABSTRACT

The approximately N-S–trending Andean retroarc fold-and-thrust belt is the locus of up to 300 km of Cenozoic shortening at the convergent plate boundary where the Nazca plate subducts beneath South America. Inherited pre-Cenozoic differences in the overriding plate are largely responsible for the highly segmented distribution of hydrocarbon resources in the fold-and-thrust belt. We use an ~7500-km-long, orogen-parallel ("strike") structural cross section drawn near the eastern terminus of the fold belt between the Colombia-Venezuela border and the south end of the Neuquén Basin, Argentina, to illustrate the control these inherited crustal elements have on structural styles and the distribution of petroleum resources.

Three pre-Andean tectonic events are chiefly responsible for segmentation of sub-basins along the trend. First, the Late Ordovician "Ocloyic" tectonic event, recording terrane accretion from the southwest onto the margin of South America (present-day northern Argentina and Chile), resulted in the formation of a NNW-trending crustal welt oriented obliquely to the modern-day Andes. This paleohigh influenced the distribution of multiple petroleum system elements in post-Ordovician time. Second, the mid-Carboniferous "Chañic" event was a less profound event that created modest structural relief. Basin segmentation and localized structural collapse during this period set the stage for deposition of important Carboniferous and Permian source rocks in the Madre de Dios and Ucayali Basins in Peru. Third, protracted rifting that lasted throughout the Mesozoic provided the framework for deposition of many of the source rocks in the Subandean belt, but most are not as widely distributed as the Paleozoic sources in Bolivia and Peru.

We attribute variations in the style of Andean deformation and distribution of oil versus gas in the Subandes largely to differences in pre-Cenozoic structure along

*Current address: U.S. Geological Survey, Anchorage, Alaska 99508, USA.
[†]Current address: Department of Geosciences, Idaho State University, Pocatello, Idaho 83209, USA.

McGroder, M.F., Lease, R.O., and Pearson, D.M., 2015, Along-strike variation in structural styles and hydrocarbon occurrences, Subandean fold-and-thrust belt and inner foreland, Colombia to Argentina, *in* DeCelles, P.G., Ducea, M.N., Carrapa, B., and Kapp, P.A., eds., Geodynamics of a Cordilleran Orogenic System: The Central Andes of Argentina and Northern Chile: Geological Society of America Memoir 212, p. 79–113, doi:10.1130/2015.1212(05). For permission to copy, contact editing@geosociety.org. © 2014 The Geological Society of America. All rights reserved.

the fold belt. The petroleum occurrences and remaining potential can be understood in the context of three major geographic subdivisions of the Subandes. Between Colombia and central Peru, rich, late postrift Cretaceous source rocks occur beneath Upper Cretaceous–Cenozoic strata that vary significantly in thickness, yielding large accumulations around the Cusiana field in Colombia and within the Oriente Basin in Ecuador, but weak or nonviable petroleum systems elsewhere, where the cover is too thin or too thick. The central Subandes of southern Peru and Bolivia evaded significant Mesozoic rifting, which kept Paleozoic sources shallow enough to delay primary or secondary hydrocarbon generation. After post-Oligocene propagation of the fold belt, however, numerous accumulations were formed around Camisea, Peru, and in the Santa Cruz–Tarija fold belt of central Bolivia. In the southern Subandes of Argentina, Paleozoic crustal thickening caused by terrane accretion and associated arc magmatism created a tectonic highland that precluded deposition and/or destroyed the effectiveness of any Paleozoic source rocks. Here, all accumulations in the foreland and in the fold belt rely on Triassic and younger source rocks deposited in both lacustrine and marine environments within narrow rifts. The trap styles and structurally restricted source rocks in the southern segment have yielded a much smaller discovered conventional resource volume than in the northern and central Subandean segments.

In comparing the Subandean system to other global fold belt petroleum systems, it is undoubtedly the rule rather than the exception that the robustness of hydrocarbon systems varies on a scale of a few hundred kilometers in the strike direction. Our work in the Andes of South America suggests that this is because most continents possess heterogeneous basement, superposed deformation, and subbasin stratigraphy that vary on roughly this length scale.

INTRODUCTION

Innumerable descriptions have been published on the geometry and kinematic evolution of fold-and-thrust belts around the globe, utilizing outcrop observations, industry seismic-reflection data, structural analysis techniques, and various other data types (e.g., Rich, 1934; Bally et al., 1966; Dahlstrom, 1970; Boyer and Elliott, 1982; Pfiffner, 1986; DeCelles et al., 2001). To a lesser degree, attempts have been made to relate those structural styles and attributes to the distribution of hydrocarbon resources in petroleum basins (e.g., Cooper, 2007). Both types of syntheses have been aimed in large part at two-dimensional, cross-sectional evaluations relating, for example, sedimentary wedge taper to wavelength of structural imbrication, mechanical stratigraphy to décollement preferences, influence of rift faults or other preexisting basement architecture on contraction styles, and the relation of source rock burial history and hydrocarbon charge to structural evolution. Some simple observations have been made about petroleum systems in these settings; for example, petroliferous fold belts commonly occur at the margins of petroliferous basins (e.g., the Zagros fold belt and the Arabian platform; Western Canada fold belt and basin). In other cases (e.g., Papua New Guinea), however, the hydrocarbon endowment is greater in the fold belt than in the adjacent basin, partly because the field sizes can be much larger in the fold belts than in the forelands.

In this paper, we build on the deep knowledge base of dip-oriented structural characteristics and extend our focus to the third (strike) dimension of the Subandean fold belt to address three questions: (1) Over what length scales do the structural styles of fold belts change and for what reasons? (2) How do the original sedimentary configuration and mechanical stratigraphy of a basin influence or control subsequent structural development of the contractional system? (3) How do these stratigraphic and structural changes in the strike direction conspire to yield boundaries between more and less viable petroleum systems? The first two questions have been addressed in South America in several historically important syntheses, including Allmendinger et al. (1983), Mpodozis and Ramos (1989), and Kley et al. (1999). This paper leverages those and other earlier works in order to generate new insights about the factors that control the distribution and quality of petroleum systems in retroarc fold belts.

We created a megaregional strike section near the eastern terminus of substantial Neogene deformation, near the boundary between the modern Andean wedge top and foreland basin (DeCelles and Giles, 1996). This location was chosen to traverse the traditional habitat for fold belt petroleum traps. Our compilation draws on roughly 100 dip-oriented two-dimensional (2-D) structural cross sections of the Subandes between latitudes 8°N and 40°S (Fig. 1). Our analysis relied most heavily on recently published sections that incorporated observations from industry seismic-reflection profiles, utilized ties to wells, and applied

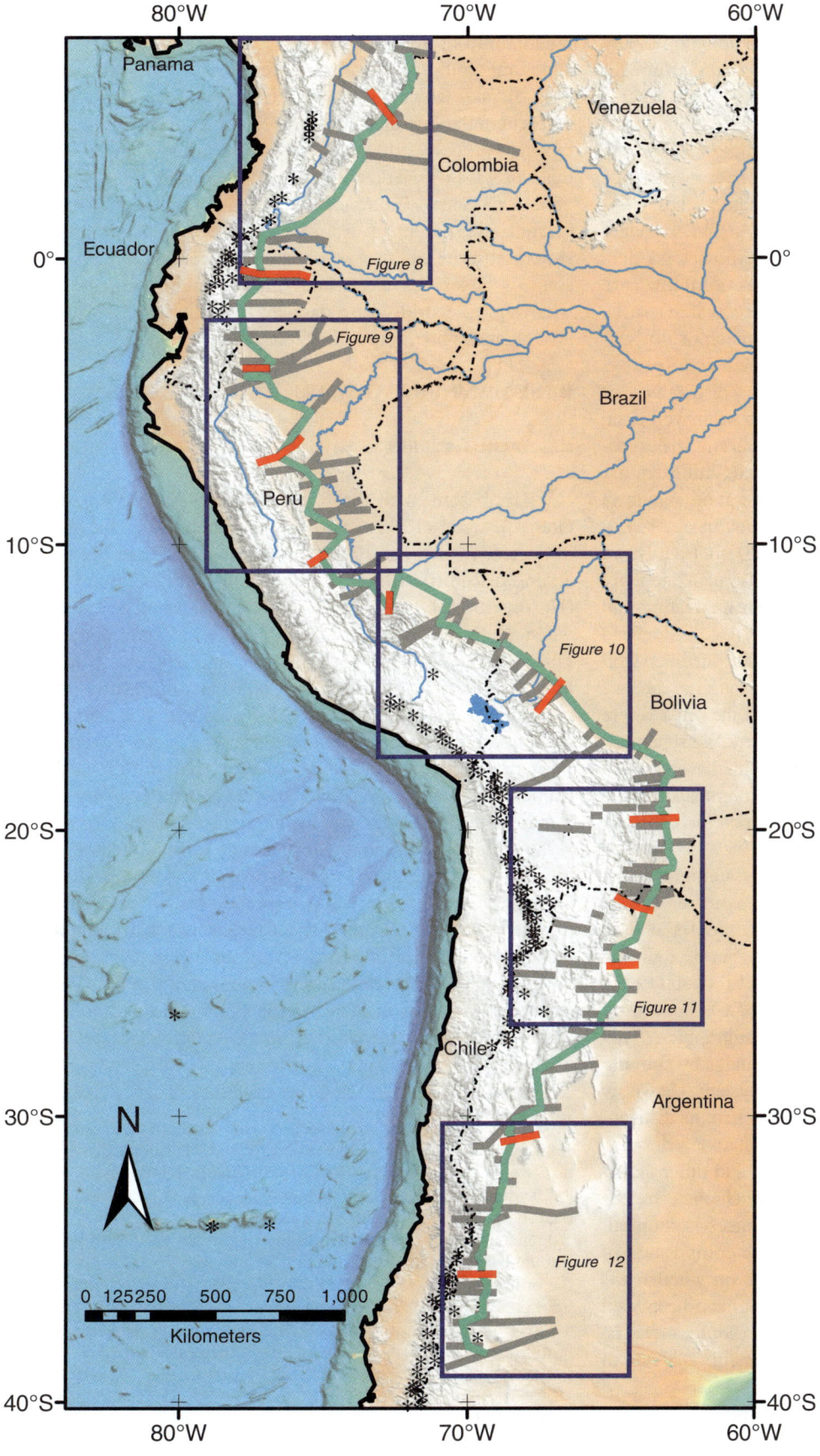

Figure 1. Map of South America showing the locations of cross sections discussed in this paper, and the location of geologic maps in Figures 8–12. Gray and red lines show locations of cross sections used in the construction of our along-strike cross section, the full extent of which is shown in teal. Red lines are locations of cross sections shown in Figure 7. Asterisks represent active volcanoes. Their absence in Peru and northern Argentina has been attributed to flat slab subduction.

modern structural balancing approaches in section construction. Other cross sections were not included in our compilation if they appeared overly schematic or were too close to other chosen sections. The resultant strike section presented in this paper extends from roughly the Colombia-Venezuela border to the south end of the Neuquén Basin in central Argentina, a distance of ~7500 km (Fig. 2). We chose to study this system over such a large distance because the length scale of stratigraphic and structural changes—especially the distribution of key source rocks—is much longer in the strike direction (typically hundreds of kilometers) than in the dip direction (typically kilometers to tens of kilometers). Additional description of the chosen location of the along-strike cross section is given in the Appendix. The spacing of dip sections along the strike section averages ~75 km but is, of course, irregular. Where a significant gap occurs in the distribution of dip sections, we attempted to fill in the section with well data, as in the Putumayo Basin of southern Colombia. The strike section in Figure 2 also incorporates Moho depth (Chulick et al., 2012), current depth of the oil- and gas-generation windows from published burial history models (e.g., Baby et al., 1995), and subducted slab depths from Tassara et al. (2006). Finally, we have not attempted to depict many details of "basement" geology along much of the section. Clearly, there is an interesting and important story about how those deep features have influenced evolution of overlying basins and structural styles, but since well penetrations and age control are limited along our section, we have focused our effort on the upper interval of the crust where viable petroleum systems and traps are preserved. We schematically illustrate the approximate level of "economic basement" in a gray color in our compilation of stratigraphic columns along the strike section (Fig. 3) and note that information on those deeper levels is limited in the literature.

At the beginning of this paper, we briefly summarize the well-established tectonic history of the Andes, focusing on pre-Andean events and their structural-stratigraphic consequences. More comprehensive treatments of this history can be found in França et al. (1995), Sempere (1995), Pindell and Tabbutt (1995), Tankard et al. (1995), Tectonic Analysis Ltd. (2006), and Ramos (2009). Second, we describe the structural-stratigraphic framework of the Neogene fold belt zone of the Subandes by summarizing the relationships in five subareas of the system, shown by boxes in Figure 1. Third, we relate this framework to the distribution of discovered oil and gas resources in the Subandes, drawing on Cooper (2007), C and C Reservoirs (2010), and other industry resource assessments contained therein. We characterize the genetics and quality of various petroleum systems based on our understanding of the structural and stratigraphic composition of the different oil and gas districts. We comment on whether the petroleum systems within individual subbasins have been well tested or not by drilling, and where remaining resources may be discovered based on our three-dimensional genetic analysis of these basins.

We conclude with some general observations about the quality of fold belt petroleum systems that emerge from our compara-

tive study of the anatomy of the 7500-km-long heterogeneous mountain front in South America. Our goal is to clarify how the structural-stratigraphic architecture of fold belts at the subbasin scale dictates why certain provinces are better settings for oil and gas accumulations than others. While we acknowledge a set of repeated processes that occur in most or all systems, a result of this study is that we do not believe there is such thing as a "standard" or "generic" retroarc fold belt; they possess tremendous diversity in basement framework, stratigraphic composition, structural style, and hydrocarbon endowment. It is our hope that some of the lessons of this study can be applied predictively elsewhere in the world where data are more limited or exploration investigations are more rudimentary.

BASEMENT FRAMEWORK AND ANDES HISTORY

Basement Terranes

The basement in the foreland of the Andes consists of a Grenville-aged (1.3–1.0 Ga) orogenic belt along the modern-day western margins of the Guyana and Brazilian Shields and a Pan-African (0.8–0.5 Ga) mobile belt south of central Bolivia (Hoffman, 1991). In Late Proterozoic to Early Cambrian time, when the Rodinian supercontinent broke apart, the western margin of these earlier orogenic belts became a passive continental margin (Hoffman, 1991; Blakey, 2008). Intraplate rifting prior to ocean floor spreading resulted in deposition of thick assemblages of Neoproterozoic clastic sediments known as the Puncoviscana Formation in Argentina and the Tucavaca belt in Bolivia (Fig. 3; Ramos, 2008). Similar-aged rocks were penetrated by two exploration wells in the Llanos Basin, suggesting rifting extended as far north as Colombia (Ramos, 2009). In the central Andes, these rocks currently lie inboard of the Arequipa-Antofalla basement terranes, which may represent the western margin of the Neoproterozoic–Cambrian basin that was fragmented from the continental margin during intraplate rifting (Fig. 4; Ramos, 2008). Following breakup, a fringing platform of carbonate overlain by clastic units of Middle Cambrian to Middle Ordovician age formed in northern South America (Ramos, 2009).

Figure 2. Strike-parallel cross section from northern Colombia to central Argentina. Vertical exaggeration (VE) 20:1. Sources of information are the ~100 dip cross sections that intersect this line (shown by "x" symbols; see Fig. 1 for locations), well data, and Moho depths from Chulick et al. (2012). Most along-strike changes record the heterogeneous tectonic and basin-filling history in the South American foreland, but note that some thickness changes reflect sharp doglegs in the profile, which were needed to capture key subsurface data points. Economic basement is shown (see Fig. 3), such that rocks classified as "basement" may vary in age. The hydrocarbon maturity window is not shown farther south due to the structural complexity and the limited presence of source rocks and burial history models at those latitudes. Some smaller Subandean structures, e.g., near Camisea and in the Huallaga Basin, are not shown due to the vertical exaggeration of the section. a.s.l.—above sea level.

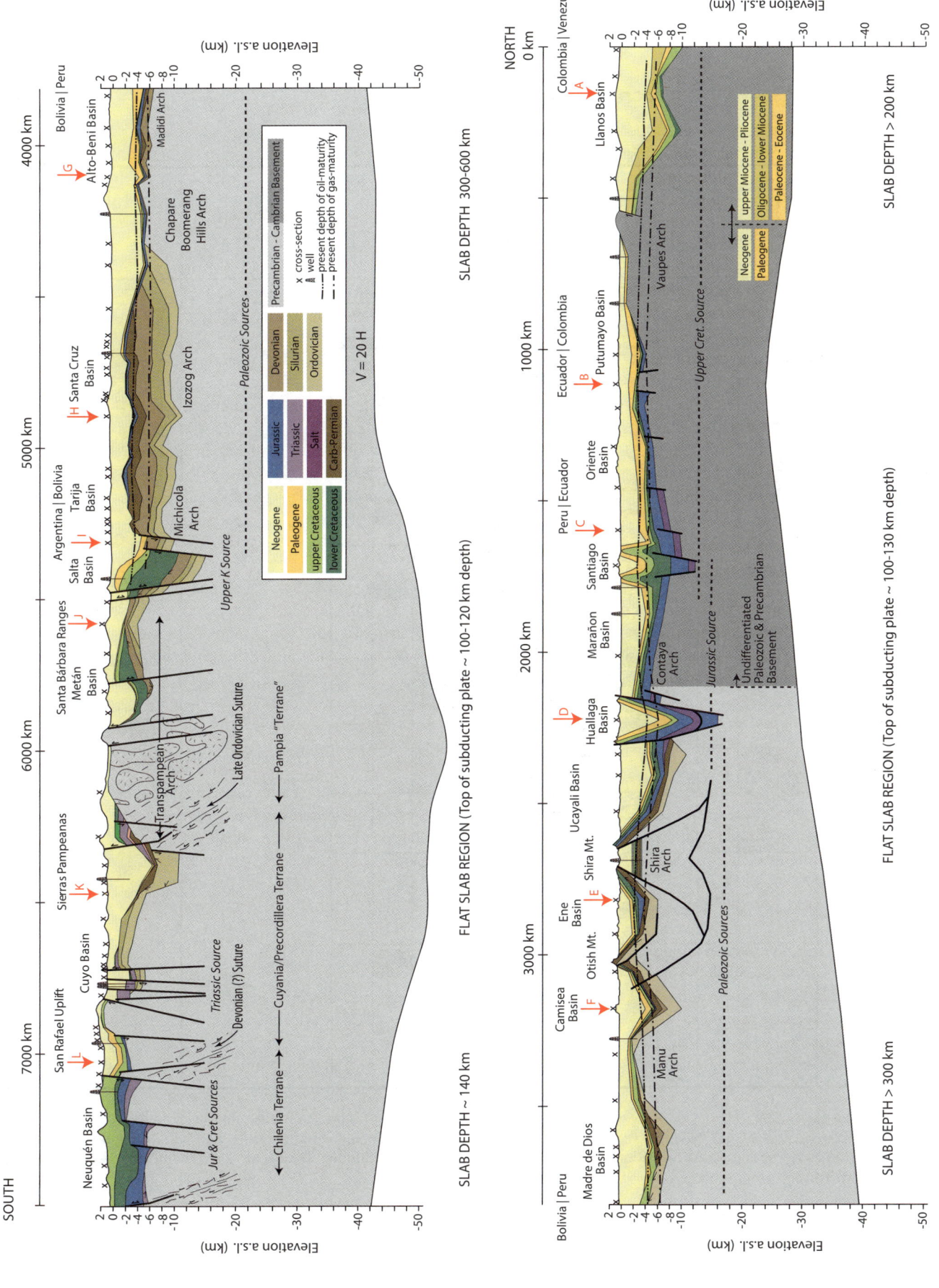

McGroder et al.

Hydrocarbon System

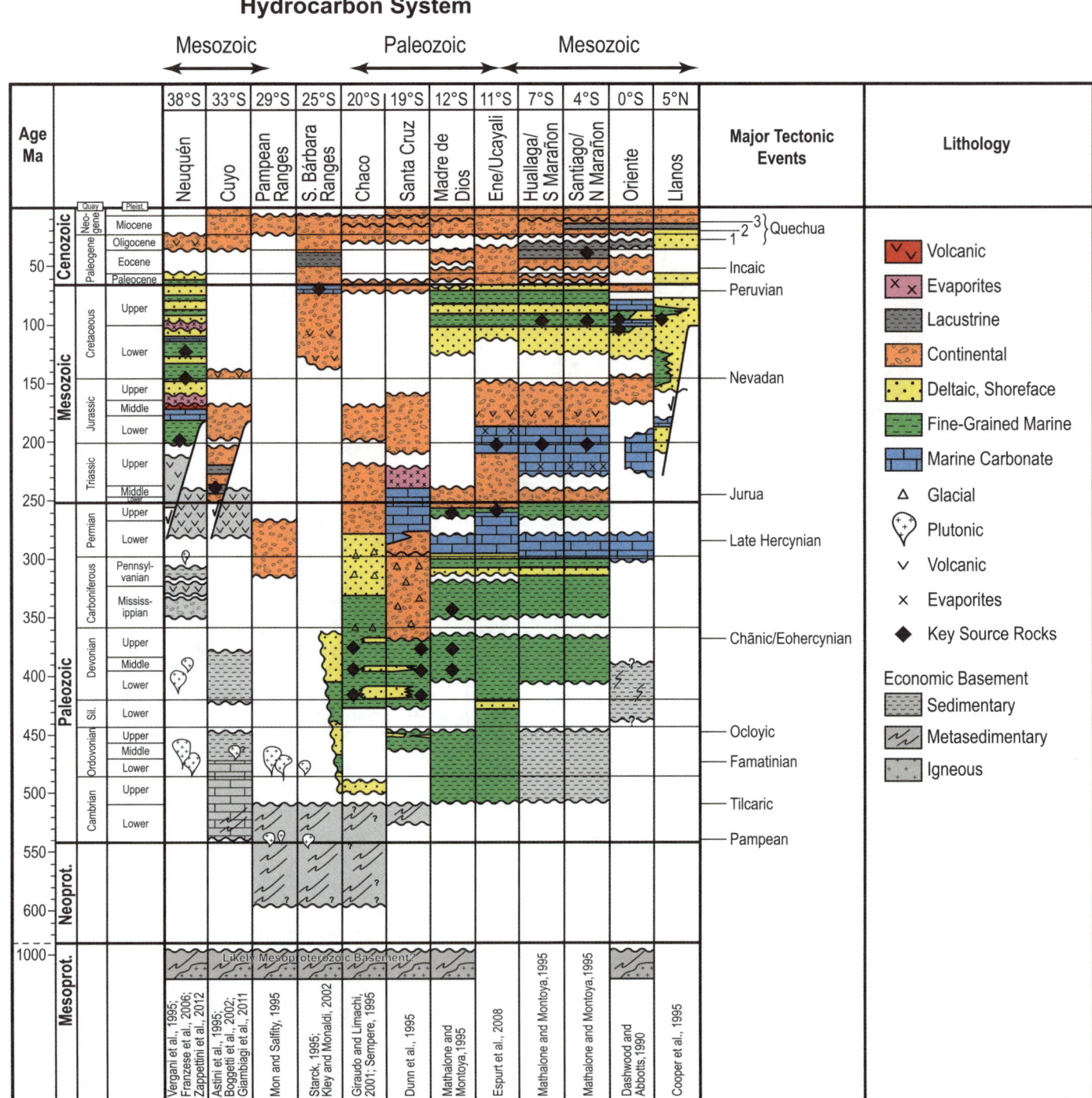

Figure 3. Chronostratigraphic chart showing ages and lithologies of strata encountered in Andean mountain front outcrops or subsurface wells in the Subandean zone or foreland. Note the regional extent of many of the key unconformities, which can be attributed to tectonic events listed to the right of the columns. Sources of information are shown beneath each column. Time scale is from International Commission on Stratigraphy (2013).

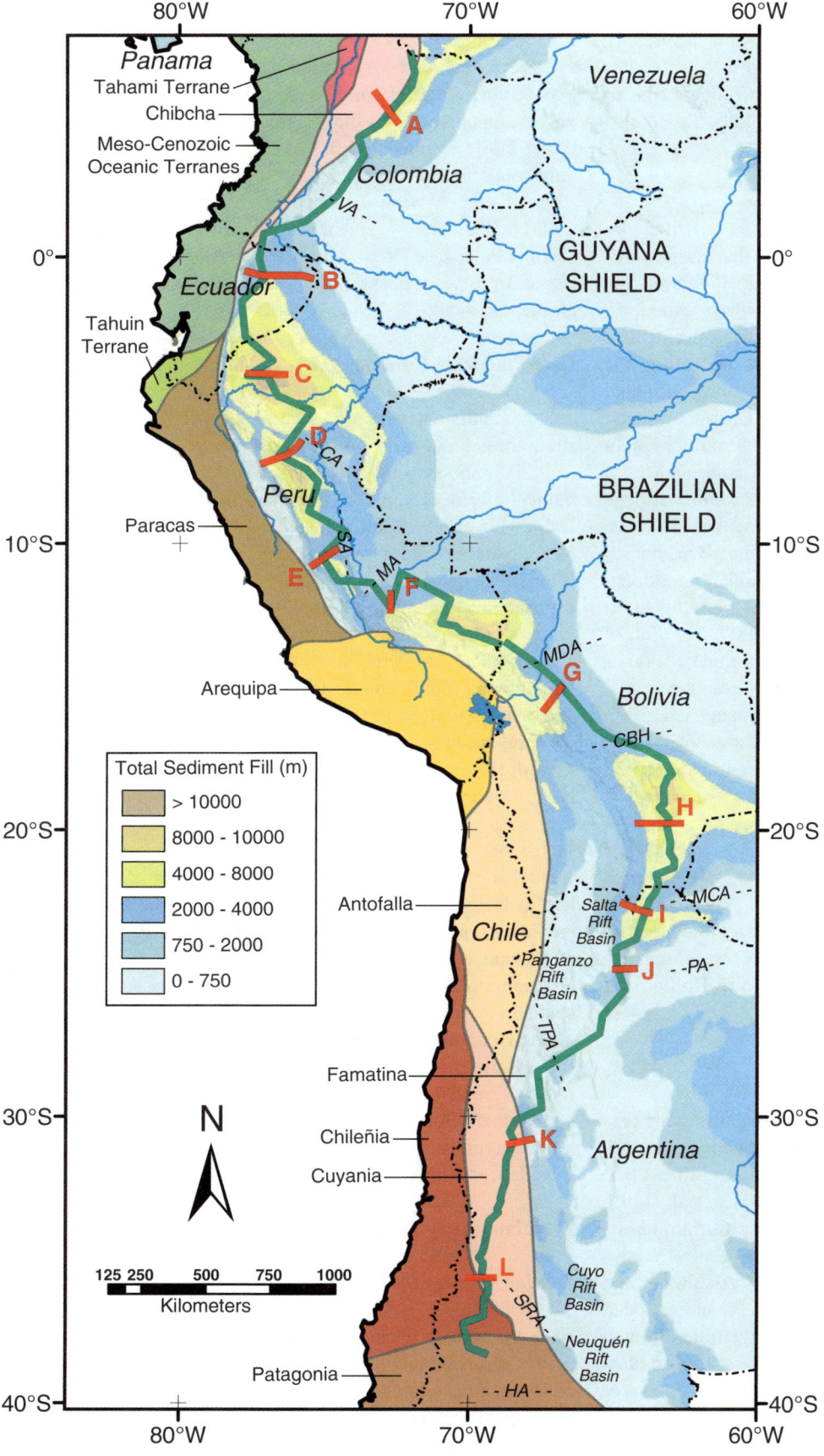

Figure 4. Tectonic map of South America showing sediment fill in blue-yellow shades and basement terranes in western South America, adapted from Ramos (2009). Red lines labeled A–L refer to cross sections shown in Figure 7. Basement "arch" abbreviations are as follows: VA—Vaupes arch; CA—Contaya high; SA—Shira arch; MA—Manu arch; MDA—Madidi arch; CBH—Chapare Boomerang Hills arch; MCA—Michicola arch; PA—Pampean arch; TPA—Transpampean arch; SRA—San Rafael arch; HA—Huincul arch. Sediment fill contours on this and succeeding maps are from Exxon Production Research Co. (1995).

At the southern limit of our transect, basement affinities of the North Patagonian Massif are less certain; however, it is generally established that the North Patagonian Massif formed the southern margin of the Neuquén Basin since at least Carboniferous to Permian time (Pankhurst et al., 2006; Ramos, 2009).

Early Paleozoic Orogenies and Crustal Welt

At the latitude of northern Argentina, northern Chile, and Bolivia, several contractional events affected the margin during early Paleozoic time, producing a tectonic highland (the "Transpampean arch") that persisted throughout the Paleozoic (Tankard et al., 1995). Following intraplate rifting during late Neoproterozoic time, subduction was reestablished along the margin at ca. 540 Ma (Pampean orogeny), culminating with final closure of the Puncoviscana-Tucavacan basin (Tilcaric orogeny; Aceñolaza et al., 1988) after ca. 510 Ma (Pearson et al., 2012). In Argentina, this event is recorded by the Upper Cambrian Mesón Quartzite, which unconformably overlies folded Puncoviscana Formation (Fig. 3). Following a brief (<20 Ma) period of tectonic quiescence, subduction was reinitiated in Early Ordovician time, resulting in emplacement of the "Famatinian" continental magmatic arc and associated regional low-pressure, high-temperature metamorphism (Rapela et al., 1998). Accretion of the Laurentian-affinity Cuyania-Precordillera terrane (Thomas and Astini, 1996) outboard of Argentina (Fig. 4) resulted in cessation of arc magmatism over much of the region at ca. 465 Ma and major shortening accommodated by predominantly N-S–striking, W-dipping thrust faults (Ocloyic tectonism; Astini and Dávila, 2004). A Late Ordovician cessation of magmatism at the modern latitude of Bolivia and Peru may have resulted from Ordovician reattachment of the Arequipa-Antofalla terrane (Bahlburg and Hervé, 1997), which was previously fragmented during Neoproterozoic rifting (Ramos, 2008). The record of these events is shown in Figure 3 by a series of regional unconformities. To the north and east of this evolving orogenic highland, the Chaco depocenter became established in Bolivia and Argentina during Ordovician time.

Mid-Paleozoic "Back Arc"

Through the course of the middle Paleozoic, following cessation of major Ordovician Famatinian arc magmatism, the area of the present-day central Andean retroarc fold-and-thrust belt was located east of an uplifted highland (Tankard et al., 1995; Starck, 1995). An absence of Silurian to early Carboniferous magmatic rocks at the modern latitudes of the central Andes may indicate a passive or strike-slip margin (Bahlburg and Hervé, 1997; Mišković et al., 2009). The Chaco Basin continued to be the site of clastic deposition in northern Argentina, Bolivia, and southern Peru (Fig. 5), where Devonian west-derived clastic sediments of the Huamampampa and Iquiri Formations prograded into a marine environment (Isaacson, 1975; Tankard et al., 1995). Source-prone facies of the Los Monos Formation

were deposited in a restricted marine setting in Middle Devonian time in central Bolivia.

Devonian- and Mississippian-aged rocks are more poorly exposed in northern Peru, Ecuador, and Colombia, but paleogeographic reconstructions suggest that the Devonian–Mississippian Chaco Basin extended north into Peru and east into the ancestral Amazonas trough between the Guyana and Brazilian Shields (Figs. 4 and 5; Isaacson and Díaz Martínez, 1995; Williams, 1995). These areas were more distant from the magmatic arc and any Paleozoic retroarc thrusting, and so may have been thinner and probably possessed a smaller volume of west-derived clastic sediments.

Late Paleozoic Contraction

A significant tectonic event is recorded in Carboniferous strata throughout much of the central South American Subandes (Fig. 3). This orogeny is variably termed the Chañic (Mon and Salfity, 1995) or the Eo-Hercynian event (Dalmayrac, 1978; Dalmayrac et al., 1980; Laubacher, 1978). Angular unconformities separating folded and slightly metamorphosed Lower and Middle Paleozoic strata from Lower Carboniferous units are well exposed in a large region N and NW of Lake Titicaca in southern Peru and in parts of western Bolivia. Uplift became accentuated along the Pampean arch in northern Argentina, Chile, and western Bolivia (Tankard et al., 1995; Tectonic Analysis Ltd., 2006), and Devonian to Carboniferous deformation propagated into the foreland in places, including the Shira arch in southern Peru (Fig. 4; Espurt et al., 2008). The Mississippian Ambo Formation in Peru was deposited at this time in a coastal plain environment (Fig. 5). It contains coals and tuffaceous, volcaniclastic, and lithic sandstones with metamorphic rock fragments derived from the west (Laubacher, 1978). The Ambo Formation has been identified as the source of gas and condensates in the giant Camisea gas field in southern Peru. Subsequent to the Mississippian Eo-Hercynian event, which is recognized widely across southern and central Peru, a more moderate folding event is recognized in southern Peru in the middle Permian, termed the Tardi-Hercynian or late Hercynian event (Fig. 3; Laubacher, 1978).

Elsewhere in the region, the Hercynian events are more difficult to identify. In the Eastern Cordillera in northern Argentina and southern Bolivia, an absence of Carboniferous strata probably reflects a remnant Ordovician highland and cessation of back-arc subsidence that postdated terrane accretion (Starck, 1995).

Permian–Early Jurassic Extension

Southern Andes

Following Chañic shortening, the late Paleozoic and Triassic were periods of extension and volcanism throughout much of the present-day central Andes, probably related to orogenic collapse of the composite Paleozoic contractional system (Kay et al., 1989; Figs. 5 and 6). Fernandez-Seveso and Tankard (1995) described some of the oldest rift-related sediments in the Paganzo

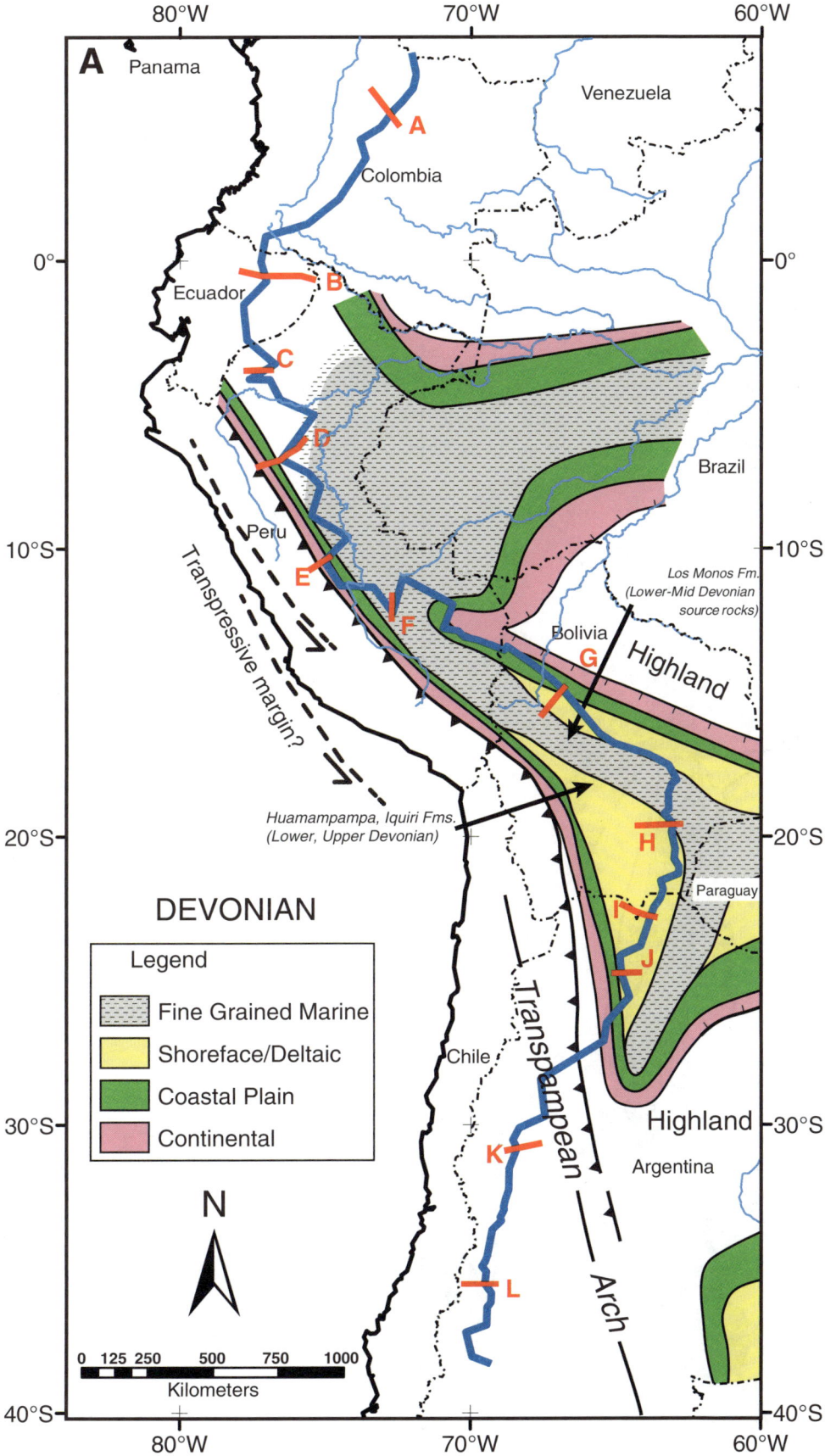

Figure 5 (*Continued on following page*). Paleogeographic maps showing inferred depositional limits of the Devonian (A) and Permian (B) basins in the modern retroarc of South America, as well as other domains traversed by the strike cross section (blue; Fig. 2). Figure is modified and simplified after Mpodozis and Ramos (1989), França et al. (1995), Tankard et al. (1995), and Tectonic Analysis Ltd. (2006). Note the fundamental difference between the northern Subandean zone, which was constructed within relatively stable basins and shelves of Paleozoic age, and the southern Subandean zone, which was built within a major tectonic highland formed during the Ocloyic and Chañic orogenies. In some areas (e.g., northern Argentina), subsequent erosion of formations of these ages is recorded by unconformities.

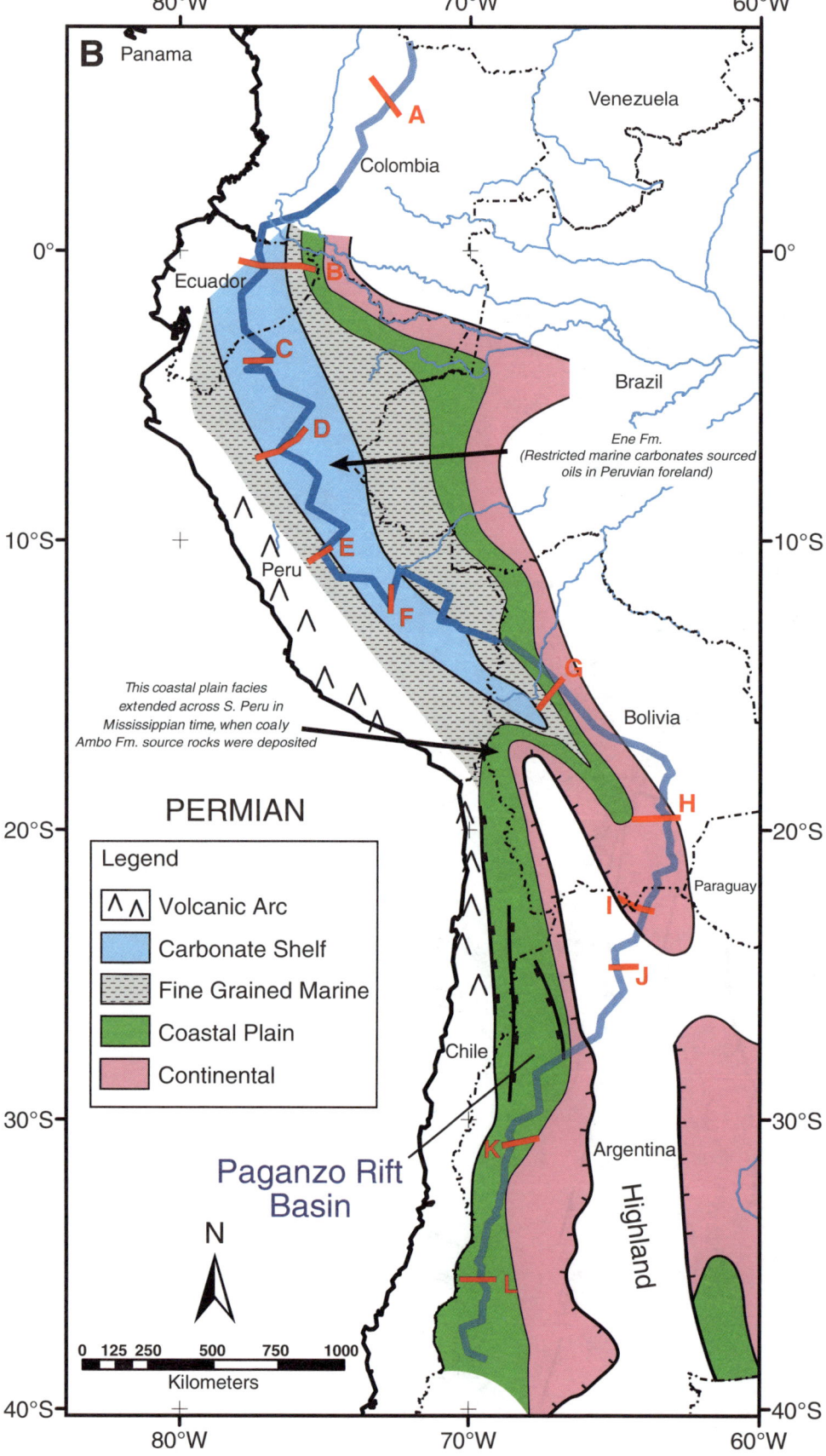

Figure 5 (Continued).

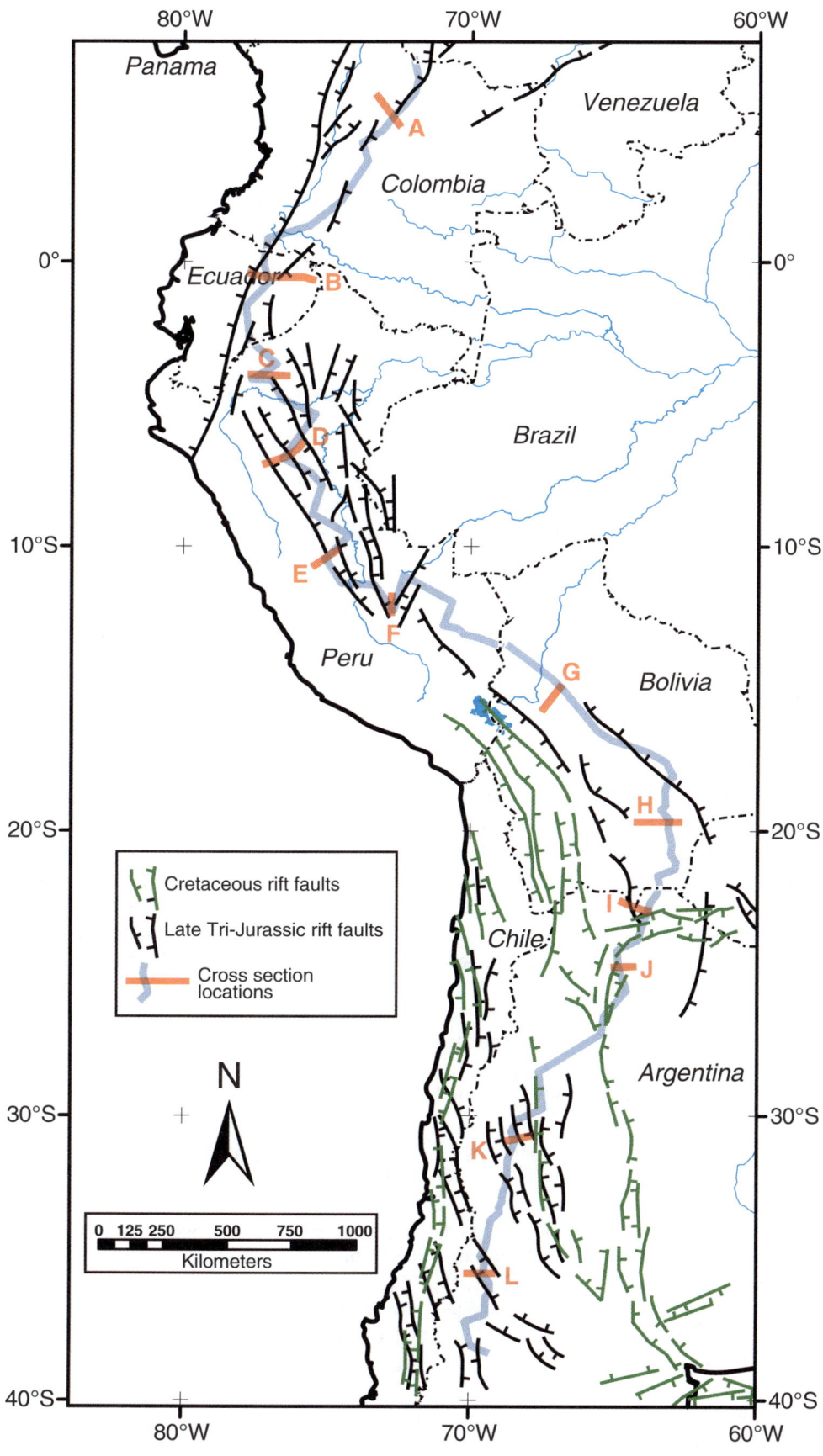

Figure 6. Map showing major Triassic–Jurassic and Cretaceous extensional faults in western South America, after Ramos (2009). Note paucity of normal faults in southeastern Peru and central Bolivia compared to central Peru and Argentina. Permian normal faults that are well developed in the Paganzo Basin in northern Argentina are not shown on this map.

Basin in northwestern Argentina (Fig. 5), which they suggested may have been influenced by strike-slip faults within the amalgamated Pampean, Precordillera, and Chileñia terranes beneath the Chañic unconformity. The oldest parts of the Paganzo Basin are structurally segmented, of Mississippian age, and largely nonmarine. However, marginal marine facies within the upper basin fill, together with increased connections between depocenters, suggest that the Transpampean arch had locally collapsed below sea level (Fernandez-Seveso and Tankard, 1995).

Triassic to Early Jurassic extension also had a profound impact on south-central South America. The onset of major extension in the Neuquén Basin is marked by the eruption of ignimbrites of the Lower to Middle Triassic Choiyoi Group and deposition of interbedded coarse clastic sediments (Fig. 3; Kay et al., 1989; Vergani et al., 1995). Basement-involved normal faults of Late Triassic to Early Jurassic age bounded the graben and created several internal depocenters. Extension continued into the Middle Jurassic, but depositional patterns within the Neuquén Basin become more integrated and marine, resulting in deposition of the Los Molles source rock of Pliensbachian to Toarcian age. During Middle Jurassic time, Neuquén deposition began to be dominated by thermal subsidence, which controlled facies as fluvial and deltaic systems prograded into a central marine axis of the west-facing triangular basin. The onset of basin inversion may have promoted restriction and deposition of evaporites during Callovian time, with more significant inversion strongly affecting Late Jurassic deposition.

The Cuyo Basin formed at roughly the same time as the Neuquén Basin and has a similar stratigraphy, with Choiyoi volcanic rocks giving way upward to fault-controlled lacustrine and alluvial-fluvial clastic rocks of Early Triassic age (Fig. 3). Unlike the Neuquén Basin, the Cuyo Basin does not contain marine strata and preserves only a thin Jurassic section. The thickness of Triassic and Jurassic rocks locally exceeds 2500 m (Giambiagi et al., 2003). The only significant source rock in the basin is the lacustrine Cacheuta Formation of Middle Triassic age (Dellapé and Hegedus, 1995). The Jurassic marine seaway documented throughout central and western Peru continued west of the present-day Altiplano, flooding the Neuquén Basin but not the Cuyo Basin due to the lesser amount of subsidence in that basin and/or the presence of a topographic barrier to the west (Pindell and Tabbutt, 1995).

Northern Andes

Farther north, Sempere et al. (2002) described southward-younging extension, in which Late Permian and Triassic rifting (Mitu and Pucara Formations) propagated southeastward from central Peru into western Bolivia, where the peak of extension was Late Triassic to Middle Jurassic in age. In central Peru, the main axis of rifting coincides broadly with the present-day mountain front, although industry seismic-reflection data demonstrate existence of Mitu-aged rifts east of the Andes in the Ucayali and southwestern Marañon Basins (Mathalone and Montoya, 1995). To the southeast, the rift axis follows the trend of the present-day

Eastern Cordillera in southern Peru, Bolivia, and northern Argentina (Fig. 6; Sempere et al., 2002). In Peru, the Triassic continental clastic rift deposits are overlain by the broadly distributed Jurassic Pucara Formation (Fig. 3). This shallow-marine carbonate unit gives way eastward and southeastward into eastern Peru and Bolivia to a widespread Jurassic fluvial and eolian assemblage that extends into the Paraná Basin and other continental basins to the east (França et al., 1995). The Jurassic units overlap the rift flanks of the Triassic Mitu basins and are assumed to have formed in a thermal sag phase following rifting, consistent with their broad distribution across western South America (Sempere et al., 2002).

Whereas the axis of the Triassic Mitu rift system extends south from Peru along the Eastern Cordillera in Bolivia, the overlying Jurassic marine sag basin is wide and straddles the rift in central Peru. To the south, the marine basin deviates west of the Altiplano into the Arequipa Basin, becoming a narrow back-arc basin where the Lower to Middle Jurassic (Toarcian to Bajocian) Socosani Formation interfingers with and overlies volcanic and volcaniclastic rocks of the Chocolate Formation (Sempere et al., 2002). Similar-aged volcanic rocks in coastal Peru and Chile yield geochemical signatures suggestive of a subduction-related arc genesis, indicating that the Jurassic carbonate shelf in Peru narrows to the south and likely interfingers with volcanic strata proximal to a west-facing Jurassic arc (Sempere et al., 2002; Romeuf et al., 1993, 1995). The narrowing and westward stepping of the Jurassic shelf around the west side of the Altiplano suggest that the Transpampean tectonic welt that had formed during Paleozoic time in northern Argentina, Chile, and southwestern Bolivia continued to exert an influence on deposition well into the Mesozoic (Tankard et al., 1995).

In Colombia, extensional faulting and basin formation were more protracted, with recognized pulses in the Late Triassic to Jurassic (Kammer and Sánchez, 2006) as well as in the Early Cretaceous (Horton et al., 2010, and references therein). Some of the Triassic–Jurassic fault systems in the Eastern Cordillera, which were reactivated as Cenozoic reverse faults, are considered to have originated as transtensional faults in Jurassic time (Sarmiento-Rojas et al., 2006).

Late Jurassic Inversion and Cretaceous Extension

Although Permian through mid-Cretaceous time in the Andes was generally characterized by extension, the margin evolved heterogeneously along strike. In the Neuquén Basin, for example, low-magnitude inversion of rift basins occurred during Middle to Late Jurassic time (Vergani et al., 1995; Grimaldi and Dorobek, 2011). Shortening of this age is also documented in Peru (Mathalone and Montoya, 1995; Jacques, 2003; Bump et al., 2008), but it is absent along much of the rest of the margin.

Whereas Permian to Jurassic extension is observed in central Argentina, Peru, and Bolivia, extension of this age has only been documented locally in northern Argentina, most notably within the Paganzo Basin, a late Paleozoic successor or collapse basin formed above sutures between the Pampean, Precordillera, and Chileñia terranes (Fig. 5; Fernandez-Seveso and Tankard, 1995). The more pronounced extensional events in northwestern Argentina are younger, postdating low-magnitude Jurassic inversion observed along strike. Extension in northern Argentina and southern Bolivia is primarily Cretaceous in age and resulted in deposition of up to 6 km of synrift clastic sediment in the Salta rift basin (Fig. 4; Salfity and Marquillas, 1994; Mon and Salfity, 1995). Like those in the Paganzo Basin, normal faults in the Salta rift may have reactivated older structures (Salfity, 1985). In contrast to the Neuquén and Cuyo Basins, where extension resulted in deposition of high-quality source rocks (Figs. 3 and 4), a lower-quality carbonate source rock within the Yacoraite Formation formed during postrift thermal relaxation in the Salta rift in latest Cretaceous time (Starck, 2011).

As described already, restorations of the Mesozoic depositional configuration in the Eastern Cordillera of Colombia recognize Late Triassic half grabens and a westward thickening of Jurassic and Lower Cretaceous strata into the Eastern Cordillera (Dengo and Covey, 1993; Toro et al., 2004; Kammer and Sánchez, 2006; Mora et al., 2010.). Extension climaxed in the Early Cretaceous, recording back-arc rifting and/or opening of the proto-Caribbean as North and South America began to separate (Pindell et al., 2005; Sarmiento-Rojas et al., 2006; Horton et al., 2010). There is some dispute about the polarity of the Early Cretaceous arc at the latitude of Colombia, but most suggest it was an east-facing fringing arc that docked in Albian or Cenomanian time (see Tectonic Analysis, Ltd., 2006; Villagómez et al., 2008).

Andean Retroarc Shortening

It is difficult to assign a regionally synchronous age for the onset of Andean shortening because western South America has been situated above an east-dipping subduction zone since the Paleozoic (see previous section on dispute about Early Cretaceous subduction polarity in Colombia), and there were contractional events that shortened crust in the retroarc during Carboniferous, Permian, and Jurassic time. Some descriptions consider the Andean orogeny to date from the Jurassic (Coira et al., 1982). However, many accounts of individual segments of the Subandean zone recognize a latest Cretaceous to Paleogene (commonly termed "pre-Andean") episode of back-arc flexural subsidence, followed by a Neogene "Andean" episode of crustal shortening reflecting eastward propagation of the orogenic wedge into its present position (e.g., Allmendinger et al., 1983; Cooper et al., 1995; Horton, 1998; DeCelles and Horton, 2003). This paper focuses primarily on the Neogene structures preserved near the modern mountain front because older Andean structures in the hinterland generally no longer have viable traps, seals, or live petroleum systems (e.g., Arriagada et al., 2006).

The styles and magnitudes of shortening differ along strike, as will be shown herein, but a general consensus exists that the area of greatest shortening is approximately at the latitude of central Bolivia, where several estimates propose that shorten-

ing across the entire Andes system exceeds 300 km (McQuarrie, 2002a). More discussion of the structural styles of retroarc shortening at the scale of individual subbasins will be addressed in subsequent paragraphs.

Numerous studies describe how the Nazca plate currently subducts beneath South America at various angles, with flat slabs present beneath much of Peru and also at the latitude of northern Chile and Argentina (e.g., Barazangi and Isacks, 1976, 1979; Gutscher et al., 2000; Martinod et al., 2010). These two regions lie above areas where the aseismic Nazca and Juan Fernandez Ridges are currently subducting beneath the South American margin (e.g., Jordan et al., 1983; Wagner et al., 2005; Gans et al., 2011). In northern Argentina, the seismically active "broken foreland" in the Sierras Pampeanas has been at least partly attributed to the interference of a shallow slab with the base of the South American crust (Jordan et al., 1983). Following our more detailed discussion of stratigraphy, structure, and petroleum systems at the subbasin scale, we will comment on how important these modern subduction elements appear to be for petroleum generation or preservation.

SUBANDEAN STRUCTURAL VARIATIONS IN THE STRIKE DIMENSION

Llanos

The Llanos Basin flanks the eastern edge of the Eastern Cordillera in Colombia (Figs. 7A and 8). The relatively narrow Subandean fold belt is a zone of thick-skinned structural inversion in this area (e.g., Dengo and Covey, 1993; Cooper et al., 1995; Toro et al., 2004). While overall shortening across the Eastern Cordillera and Subandes probably ranges between 100 and 150 km, estimates across the Subandes are modest—in the range of ~20 km (Colletta et al., 1990; Dengo and Covey, 1993; Mora et al., 2008). The narrow frontal zone of "thin-skinned" thrusting is interpreted to record footwall shortcut faults beneath thick-skinned inversion features localized along master Mesozoic normal faults (Fig. 7A; Mora et al., 2006, 2010). At depth, the major contractional faults carry basement rocks; previous authors show the normal faults to be cut by the contractional faults (e.g., Toro et al., 2004; Fig. 7A). Mora et al. (2010) and Parra et al. (2012) used apatite fission-track (AFT) thermochronology and field relations to infer that shortening across the Eastern Cordillera began in late Eocene time and progressed until the present, which is corroborated by detrital zircon U/Pb age spectra in the basin-fill sediments (Horton et al., 2010). The timing of trap formation at the leading edge of the Llanos fold-and-thrust belt is generally accepted to be late Miocene to Holocene, but the AFT data suggest that the earliest phase of structural growth in the external part of the system may have begun as early as late Oligocene time (Mora et al., 2010).

The 15–20-km-wide Llanos fold-and-thrust belt contains less than a dozen oil and gas condensate fields. Exploration began in the 1940s, but it was not until the late 1980s and early 1990s that the giant (>1 billion barrels of oil equivalent, BBOE) Cusiana

92

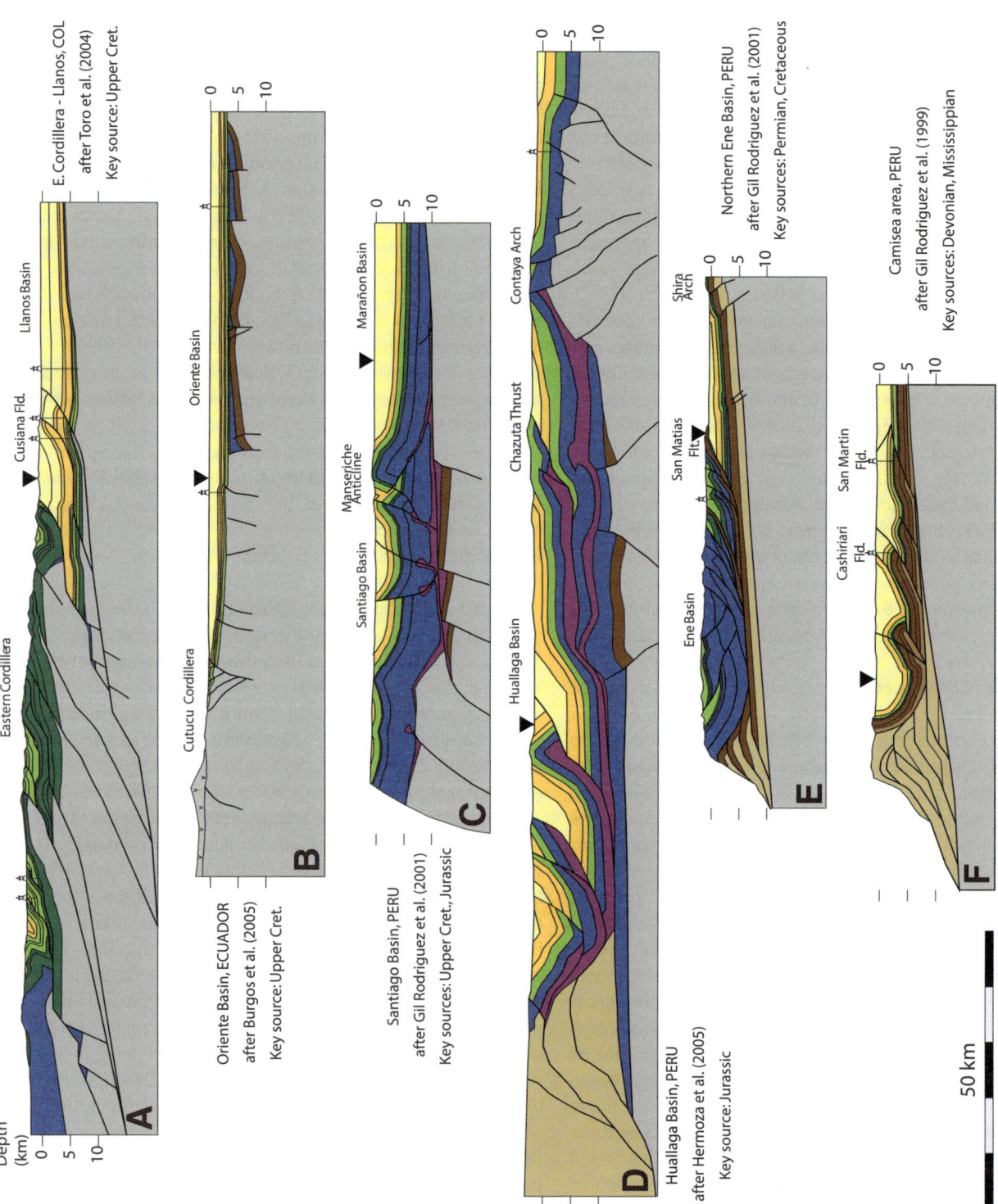

Figure 7 (*Continued on facing page*). Dip-oriented structural profiles illustrating structural styles along the Subandean mountain front between Colombia and Argentina, showing (A–F) Colombian, Ecuadorian, and Peruvian sections and (G–L) Bolivian and Argentinean sections. Letters refer to locations of sections shown in Figures 1, 4, and 8–12. All sections are shown at same scale with no vertical exaggeration. Intersections with the regional strike section are shown by inverted triangles. Sources are shown next to each section. The original authors interpreted many of the large culminations on the left sides of the sections as duplex structures, but it is also possible that some of them may contain basement or thick sections of rift fill.

93

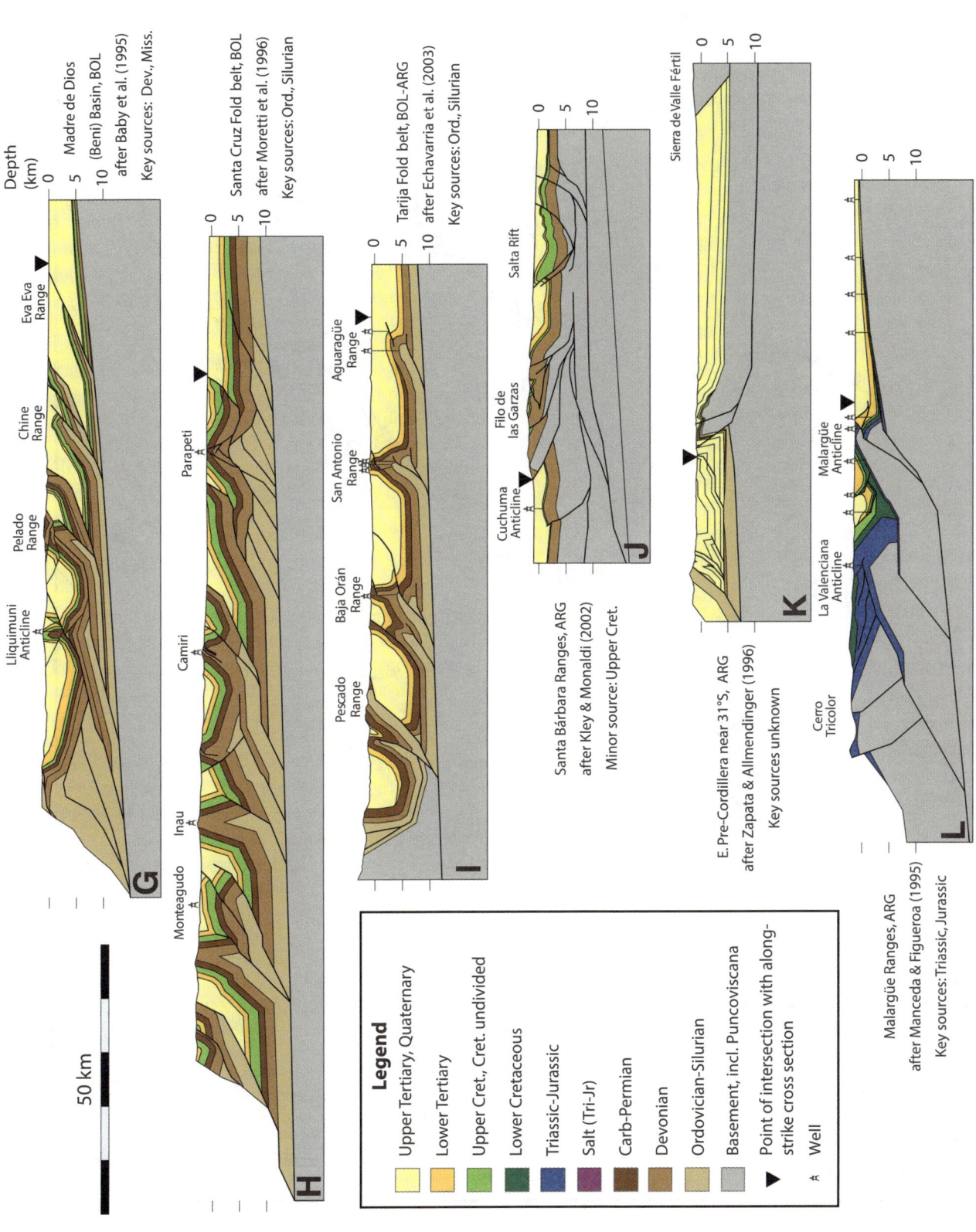

Figure 7 (*Continued*).

McGroder et al.

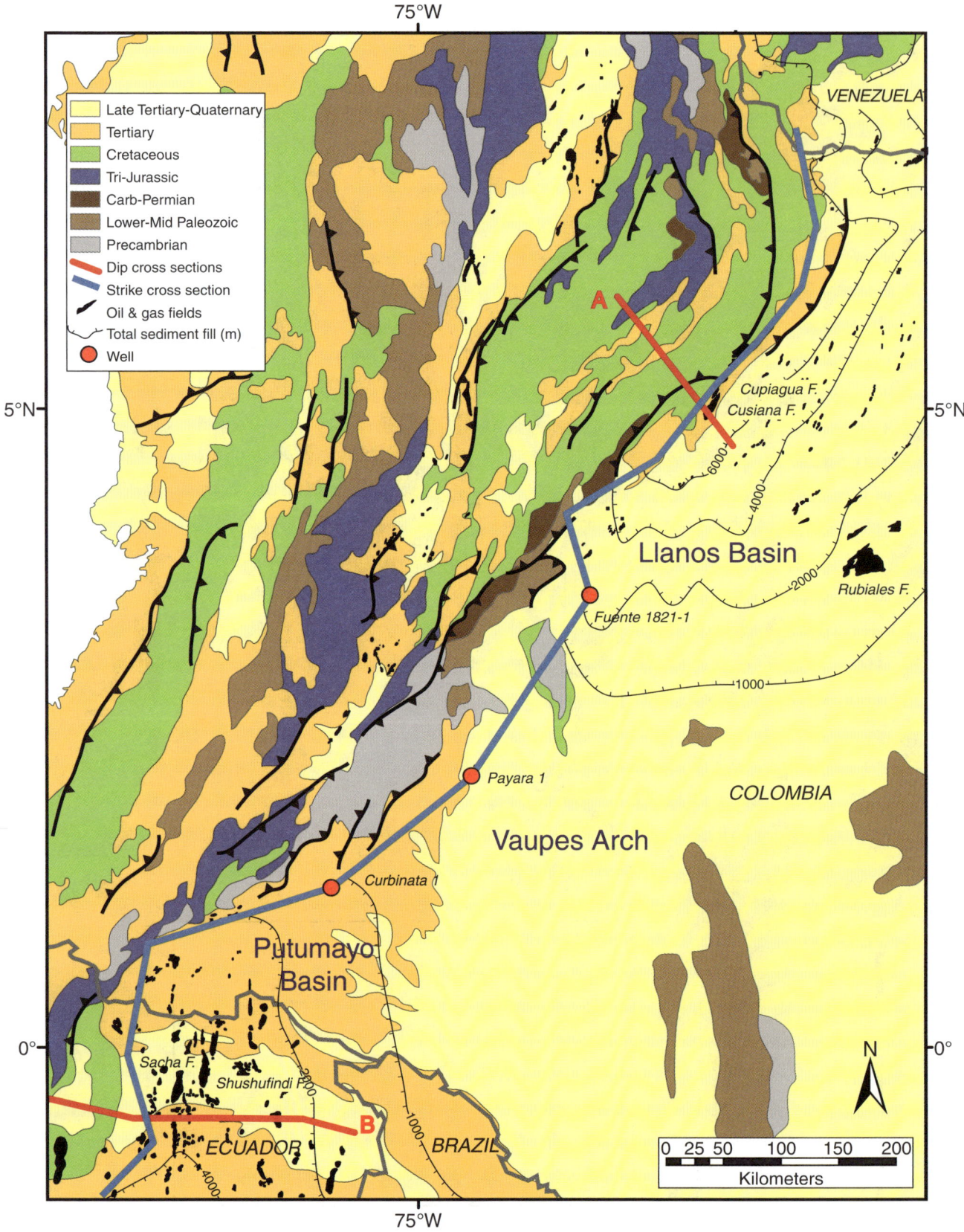

Figure 8. Simplified geologic map of the Llanos Basin and Eastern Cordillera, Colombia and northern Ecuador. Note thickening of the Llanos trough to the NE and thinning to the SW onto the Vaupes arch. Map information is simplified from Geologic Data Systems (1997).

and Cupiagua fields were discovered. The primary proven play in the trend is Upper Cretaceous to Eocene east-derived sandstone reservoirs in fault-propagation fold-type traps, sourced by the prolific marine shales of the Turonian-Coniacian Gacheta Formation (Figs. 3 and 7A). The narrow width of the productive belt around Cusiana is characteristic of inversion-type mountain fronts; the modest structural relief of the field helps maintain the trap and seal integrity of the faulted anticline (Fig. 7A). Most production in the trend is gas condensate, but Cusiana is primarily oil. Oil fields in the foreland to the east and south produce heavier crudes, indicative of the lower maturities and biodegradation in the external parts of the system where Cenozoic basin fill is thinner.

Vaupes, Putumayo, and Oriente

Southwest of the Cusiana area in the central Colombia Subandes, the foreland basin shallows onto the Vaupes arch, a broad, long-lived, low-relief basement feature where Cretaceous or Paleogene strata rest on basement at a depth as shallow as 1000 m below sea level (Figs. 2 and 8). No significant hydrocarbon accumulations have been discovered on the Vaupes arch (Fig. 8). Where the basement then deepens into the Oriente Basin in Ecuador, Upper Cretaceous source rocks become buried deeply enough to have produced hydrocarbons, filling Cretaceous sandstone reservoirs in several dozen basement fault-block and inversion traps in the foreland of eastern Ecuador and southwesternmost Colombia. Field sizes exceed one billion barrels of oil (BBO) for the largest fields, Sacha and Shushufindi, both discovered in 1969.

Large-displacement structures in the fold belt at this latitude are less common than in the Llanos Basin because the Mesozoic rift in the Cutucu Cordillera of Ecuador is narrower and shallower than the analogous rift basin in the Eastern Cordillera of Colombia (Fig. 7B; cf. Legrand et al., 2005; Toro et al., 2004). In contrast, modest shortening in Ecuador was distributed across numerous broad foreland structures, many of which originated as Permian to Jurassic normal fault blocks. Consequently, the only discovered hydrocarbons at the latitude of the Oriente Basin are in the foreland. West of the inverted rift in the Cutucu Cordillera, there is a sharp tectonic boundary across which allochthonous oceanic terranes are stitched against native South American basement (Fig. 4).

Santiago and Marañon

Continuing south into northern Peru, the Cenozoic foreland basin fill thickens to ~5 km in the western Marañon Basin of north-central Peru (Fig. 9). In northernmost Peru, the Subandean mountain front is located along the Manseriche anticline, where Cretaceous rocks are exposed at the surface (Fig. 7C). This anticline separates the Santiago Basin from the Marañon foreland basin to the east. In this location, the Santiago Basin is a synformal bathtub basin stranded between the frontal Manseriche

anticline and deeper-seated uplifts to the west. Mathalone and Montoya (1995) and Gil Rodriguez et al. (2001) inferred that bedding-parallel detachment in Triassic evaporites above deeper extensional fault blocks contributed to uplift of the relatively narrow anticlines in this basin (Fig. 7C). The presence of Cretaceous and Paleogene strata at the surface is noteworthy because it indicates that Cretaceous reservoirs are potentially breached or prone to seal failure over a large proportion of the Santiago Basin.

Between the late 1960s and late 1990s, seven exploration wells tested Cretaceous objectives on Neogene anticlines in the Santiago Basin. No commercial discoveries have been made, although one well (Tanguintza 1x) was declared a noncommercial gas and condensate discovery in the Upper Cretaceous Vivian Formation (Fig. 9). Minor hydrocarbon indications ("shows") reported from many of the tests suggest two causes of failure—trap breaching and reservoir cementation or overcompaction. The latter is supported by research by Mathalone and Montoya (1995), who demonstrated that the Upper Cretaceous Vivian Formation was overlain by up to 5 km of Upper Cretaceous and Cenozoic section prior to Neogene shortening (Fig. 9). These Cretaceous–Cenozoic thicknesses exceed those to the north and south, recording the position of the paleo-Amazon depocenter prior to mid-Cenozoic drainage reorganization in northern South America (Hoorn et al., 2010).

Huallaga, Ucayali, and Ene

At roughly 5°S latitude, the frontal fold belt changes trend from NNE-SSW in the southern Santiago Basin to NNW-SSE in the northern Huallaga Basin. Just south of there, the fold-and-thrust belt abuts the projection of the Contaya arch, an important foreland structural buttress with a protracted history and a number of small oil and gas traps localized along it (Figs. 4, 7D, and 9). The Contaya arch and other areas of central Peru were strongly influenced by rifting during the Triassic, leading to deposition of red beds of the Mitu Formation, which are overlain locally by evaporites and regionally by platform carbonates of the Pucara Group (Mathalone and Montoya, 1995; Rosas et al., 2007; Fig. 3). While there is some uncertainty about the extent to which Lower Mesozoic salt detachments influenced structural development of the Santiago anticlines, it is clear that evaporites controlled the way in which fault trajectories and detachments formed during shortening in central Peru. One of the largest-displacement thrusts in the Peruvian Andes, the Chazuta thrust, carried the Huallaga Basin ~40 km to the northeast based on good-quality seismic-reflection data (Hermoza et al., 2005; Fig. 7D).

Like the Santiago Basin to the north, the Huallaga Basin preserves folded Cretaceous and Cenozoic clastic rocks to the west of an uplifted frontal ridge where Jurassic strata are exposed. To the south, the western Ucayali segment of the fold-and-thrust belt crosses several transverse elements before extending south into the Pachitea and Ene Basins (Fig. 9). The frontal Subandean structure in this area is the San Matias fault. Gil Rodriguez et al. (2001) showed the San Matias structure as detaching in the

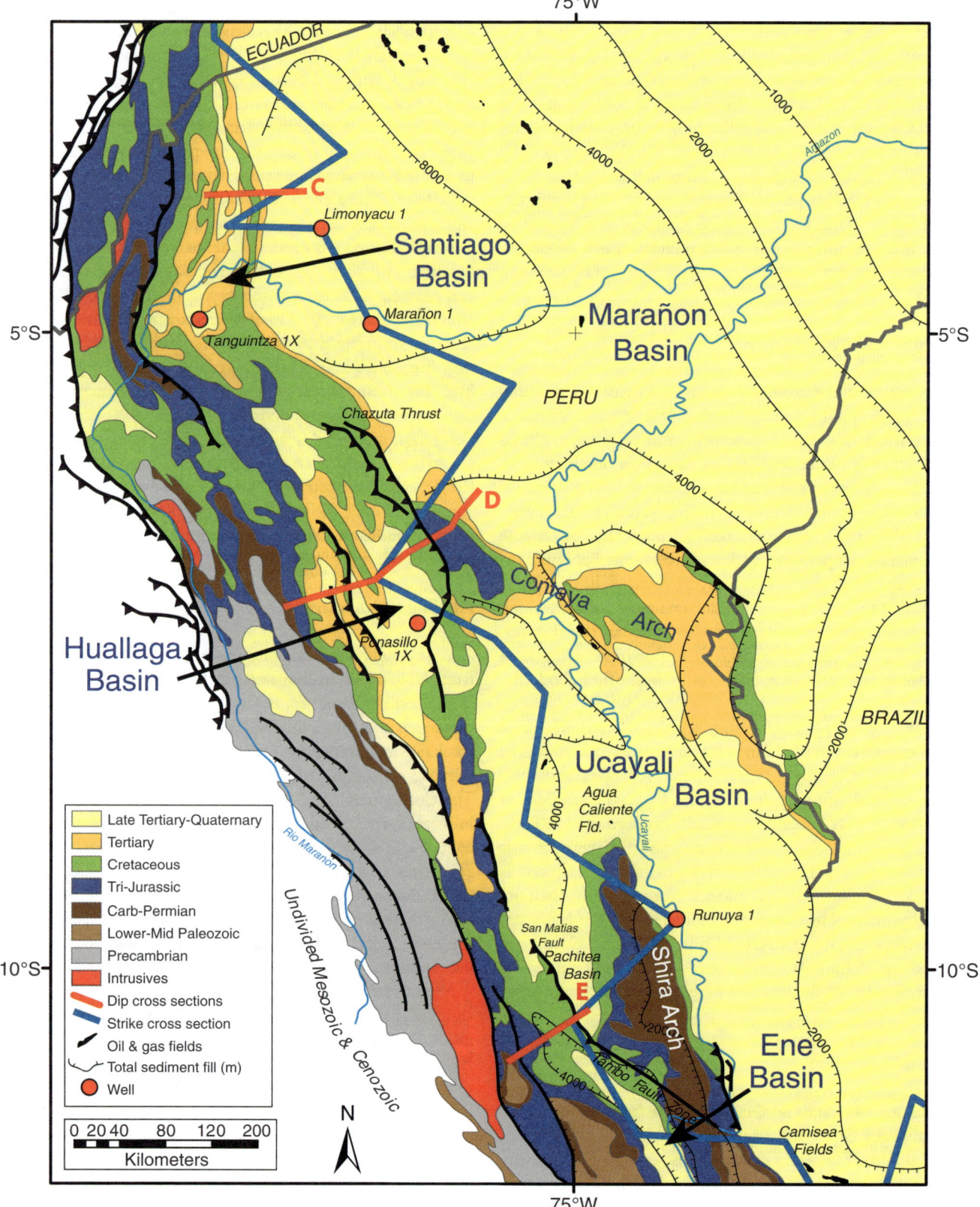

Figure 9. Simplified geologic map of the Santiago, Marañon, Huallaga, Ucayali, and Ene Basins, Peru. Sediment thickness in the western Marañon Basin is greater than most other areas of the Andean foreland. Map information is simplified from Geologic Data Systems (1997).

Permian and carrying a thick assemblage of Triassic and Jurassic units in its hanging wall (Fig. 7E). Restorations of the Pachitea and Ene fold belts by Gil Rodriguez et al. (2001) and Espurt et al. (2008) illustrate a hybrid structural style in which high-angle Permian to Triassic extensional faults and corresponding thick sediment packages are intersected and carried by bedding-parallel, low-angle Cenozoic detachments, which could be characterized as classic thin-skinned thrusts. However, the stratigraphic thickness changes between adjacent thrust sheets indicate the presence of preexisting rift architecture, giving the overall structure some similarity to an inversion feature, although basement does not appear to be involved in the shortening as far east as the mountain-front San Matias fault. Importantly, the Huallaga and Ene fold belts possess structural styles that differ markedly from the mountain-front structures in Ecuador and Colombia to the north. In the south, structures have undergone translation of several tens of kilometers on low-angle thrust faults detached on salt (Santiago, Huallaga Basins) and ultimately Paleozoic shale (western Huallaga and Ene Basins). This thin-skinned structural style and Paleozoic detachment level continue to the south into southeast Peru and Bolivia, where the greatest concentration of oil and gas fields exists in the entire Subandean fold-and-thrust belt between Colombia and Argentina.

Another important change occurs between the Santiago and Huallaga Basins from a petroleum systems standpoint. The Upper Cretaceous organic-rich interval, which is such a prolific source of hydrocarbons in Venezuela, Colombia, and Ecuador, becomes diluted by terrigenous clastic input such that total organic carbon content (TOC) decreases below 2%, and this unit ceases to be effective as a source south of the Contaya arch (Mathalone and Montoya, 1995). However, as the foreland and fold-and-thrust belt emerge southward from beneath the large sedimentary load of Amazonia in central Peru (Hoorn et al., 2010), older formations become potentially viable as petroleum source rocks. The foreland in central Peru has a number of small oil and gas accumulations in which oils have been correlated to a source in the Jurassic Pucara Formation, which occurs widely in the western foreland and the fold-and-thrust belt (Wine et al., 2003; ChemTerra International Consultants, 2000). However, no commercial hydrocarbons have been discovered anywhere in this segment of the Peruvian thrust belt (although only six wells have been drilled to date), potentially because the peak of maturity of Jurassic source rocks occurred prior to formation of the late Cenozoic fold-and-thrust belt. In Huallaga, the Ponasillo 1X well tested Upper Jurassic and Lower Cretaceous reservoirs on a simple anticlinal closure, but those reservoirs were tight. In the Ene segment, five unsuccessful exploration wells were drilled in the 1960s. These wells tested Cretaceous sandstones, and hydrocarbon shows were encountered in several of them. There are also oil seeps in the vicinity, indicating the presence of an active petroleum system. The Ene segment of the fold-and-thrust belt potentially has a viable source rock of Permian age—the Ene Formation (Fig. 3). This source rock is known to source the Agua Caliente and other oil fields in the foreland (Fig. 9) but has not been demonstrated as an effective source in the fold-and-thrust belt (Mathalone and Montoya, 1995).

Madre de Dios and Beni

The northwestern boundary of the Madre de Dios segment of the Subandean fold-and-thrust belt in southern Peru is defined by the Shira arch, a significant, thick-skinned contractional uplift with both late Paleozoic and late Cenozoic episodes of activity (Figs. 9 and 10; Espurt et al., 2008). A complex transverse fault zone, the Tambo fault zone, occupies an area where the detached system crosses the southern extension of the Shira arch. Unlike the Contaya arch to the north, Paleozoic rocks are at the surface over the entire length of the feature, and hence the late Paleozoic and Mesozoic hydrocarbon systems are not viable on the Shira arch. Southeast of the Tambo fault zone and the intersection of the thin-skinned belt with the Shira arch, the Madre de Dios segment of the frontal fold-and-thrust belt is a classical thin-skinned imbricate system with a basal detachment in the Lower Paleozoic, an outer foothills zone of simple fault-bend and fault-propagation folds, and an inner foothills zone of duplex structures (Gil Rodriguez et al., 1999; Espurt et al., 2011; Figs. 7 and 10).

The largest hydrocarbon discoveries in Peru are in the Camisea area in southern Peru. Between 1984 and 2012, 10 gas condensate discoveries, many with multi-TCF (trillion cubic feet) sizes, were made in Permian and Cretaceous sandstone reservoirs in relatively simple faulted anticline traps. The source of these hydrocarbons is the Mississippian Ambo Formation, a coaly source deposited in front of the late Paleozoic mountain front in central South America (Tectonic Analysis Ltd., 2006). It is notable that the Camisea discoveries occur in an area where several basement arches intersect the fold-and-thrust belt (Fig. 10). Near Camisea, the Manu arch records thinning of the Paleozoic and Mesozoic section atop it (House et al., 1999), which would have had the effect of maintaining Paleozoic source intervals at shallow levels over a long time period. In other words, the basement arch helped counteract the tendency of Paleozoic sources to prematurely yield hydrocarbons (compared with the late Cenozoic age of trap formation) in response to burial by (1) late Paleozoic retroarc basin fill, (2) Mesozoic rift fill, and (3) early to mid-Cenozoic foreland basin fill, thereby preserving and protecting the petroleum system so that young traps could be charged in late Cenozoic time. Places where the burial and timing were not so favorable may suffer from a fatal flaw of yield versus structural timing (Baby et al., 1995) or, if timing was favorable, may suffer from impaired (tight) reservoir quality due to excess burial, similar to the Candamo discovery in southeastern Peru (Fig. 10).

Baby et al. (1995) and McQuarrie et al. (2008a) described the Beni Subandes in northwestern Bolivia as a trend of narrow anticlinal areas separated by 10–20-km-wide piggyback synclinal basins with 6.5–7-km-thick Cenozoic basin fill. Detachment horizons occur in the Ordovician, Devonian, and Cretaceous. The ~1000-km-long segment of the fold belt in southeast Peru and northwest Bolivia differs from areas to the north and south in

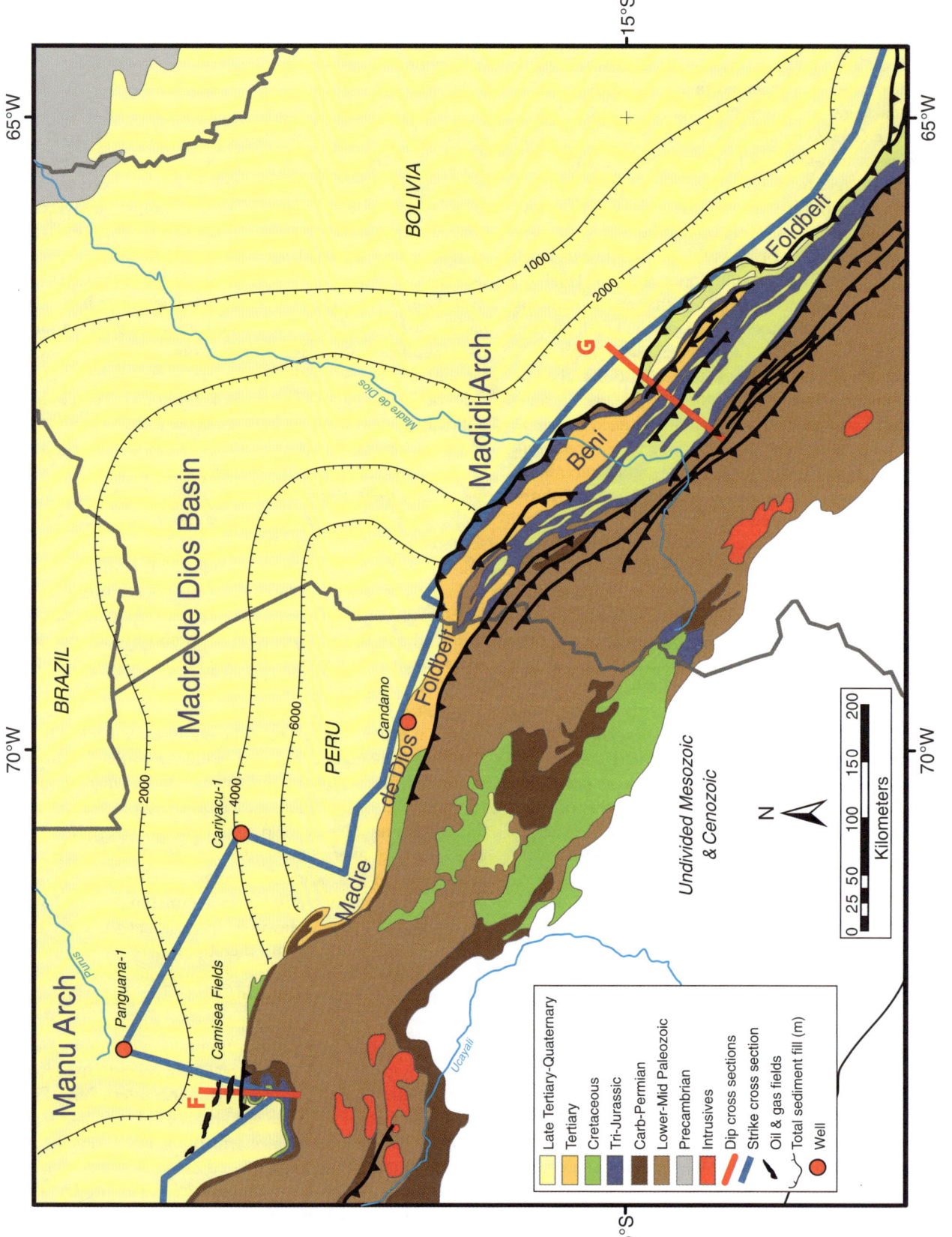

Figure 10. Simplified geologic map of the Madre de Dios Basin and fold belt, Bolivia and Peru. Excluding the Camisea area, the fold-and-thrust belt on this map is one of the most lightly explored trends of the entire Subandean system in South America. Map information is simplified from Geologic Data Systems (1997).

that there is little evidence of the pronounced Mesozoic rifting that occurred near the mountain front in Colombia, northern and central Peru, and Argentina. Triassic-aged rift-fill units are present in thrust sheets in the Eastern Cordillera in northwest Bolivia and southern Peru, but in the Subandes, the Lower Cretaceous strata typically sit on Permian strata (House et al., 1999), indicating that this area was on a structural high (i.e., Manu and Madidi arches) or perhaps a regional rift shoulder while areas to the west were extending in Permian and Mesozoic time (Figs. 2, 3, and 6).

The Beni Subandean fold belt segment in northwest Bolivia has been very lightly explored. Several wells in excess of 4000 m have tested Jurassic continental sandstone and Permian carbonate objectives, but none has proven successful. This segment of the fold belt is underlain by a rich Devonian source rock, but the great thickness of Cenozoic sediment is problematic for structural timing (Baby et al., 1995), and excess burial impaired sandstone reservoir quality. This may explain why the Candamo discovery in southern Peru was subeconomic (Fig. 10). Thermochronology suggests an overall eastward propagation of Cenozoic deformation (Barnes et al., 2006; McQuarrie et al., 2008a; Gillis et al., 2006) that was strongly influenced by the late Miocene onset of orographic rainfall and increased erosion (Poulsen et al., 2010; Norton and Schlunegger, 2011; Barnes et al., 2012). Late Miocene–Holocene tectonic-climatic coupling was probably not conducive for hydrocarbon systems; it led to deep erosion levels along many structures (e.g., breached anticlines of Fig. 7G), may be responsible for the exceptionally thick (7 km) Cenozoic sediment, and restricted propagation of the orogenic wedge (Isacks, 1988; Horton, 1999; McQuarrie et al., 2008b).

The Chapare basement high separates the Beni and Santa Cruz sectors of the Subandes (Fig. 4; Baby et al., 1994; Hérail et al., 1990). Prospectivity is limited atop the Chapare high due to the absence of Paleozoic source rocks (Baby et al., 1995). On the southern margin of the Chapare high, however, the north-tapering wedge of Paleozoic fill has localized deformation along the Boomerang Hills oblique ramp (Welsink et al., 1995; Hinsch et al., 2002). Here, a series of anticlinal traps together hold reserves of 50 million barrels of oil and condensate (MMBO) and 1.3 TCF of gas, primarily within Devonian reservoirs (Welsink et al., 1995).

Santa Cruz–Tarija

The retroarc fold-and-thrust belt at the latitude of the southern Bolivian Santa Cruz–Tarija segment represents the easternmost excursion of the Subandean system and probably also coincides with the area of greatest shortening anywhere in the Andes (326 ± 32 km; McQuarrie, 2002a, 2002b; Kley and Monaldi, 1998). Shortening just within the Subandean zone is estimated at ~100 km (Dunn et al., 1995) or 67 km (McQuarrie, 2002b). This maximum in shortening of the system coincides, not coincidentally, with the location of the thick, Paleozoic retroarc basin in the central Subandes and foreland (e.g., Allmendinger et al., 1983; Sempere, 1995; Allmendinger and Gubbels, 1996; Kley et al., 1999; McQuarrie, 2002a). Shale detachments in Lower Paleo-

zoic rocks are situated within a westward-thickening wedge of strata (Uba et al., 2009).

Numerous, parallel, detached anticlines in this part of the fold-and-thrust belt resemble a wide, Canadian-style mountain front more than the thicker-skinned inversion-style mountain front so prevalent in Colombia and Peru (Figs. 7H, 7I, and 11). Numerous gas fields are present in the Bolivian segment of the system, sourced by rich Devonian shales in the Los Monos Formation, as well as secondary Silurian and Ordovician organic-rich formations. Reservoirs are shallow-marine clastic units of Devonian and Carboniferous age. Ramos and Aguaragüe, two of the larger gas and condensate fields in the trend, have ultimate recoverable gas reserves that approach 4 and 1 TCF, respectively (C and C Reservoirs, 2010). Cretaceous sandstones, which are prevalent reservoirs in the northern half of the Subandean system, are generally uplifted and breached in this trend. The relative aridity of southern Bolivia (Bookhagen and Strecker, 2008; Hilley and Coutand, 2010) has resulted in lower erosion levels (Barnes et al., 2008), thinner Cenozoic sediment, and a wider Subandean zone (McQuarrie et al., 2008b), enhancing the prospectivity of this area compared to humid, more destructive, northern Bolivia.

Metán and Salta

Arguably the most dramatic change in structural style of the Subandean mountain front along the 7500-km-long transect described in this paper happens just south of the Bolivia-Argentina border, where shortening estimates decline significantly, and the basal detachment of the system descends from Lower Paleozoic levels to a level within low-grade metasedimentary or crystalline basement rocks (Mon and Salfity, 1995; Figs. 7J and 11). This transition has been the subject of much discussion and analysis (e.g., Allmendinger et al., 1983, 1997; Mon and Salfity, 1995; Kley and Monaldi, 1998; McQuarrie, 2002a). Previous authors concluded that this change in structural style relates in large part to the change in antecedent crustal configuration owing to the presence of a thick Paleozoic basin to the north and the remnant of the early Paleozoic Ocloyic orogeny—or the so-called Transpampean arch—to the south.

In detail, parallel anticlines detached in the Ordovician–Silurian succession in the southern Bolivia fold-and-thrust belt diminish in size and plunge southward into the subsurface (Fig. 7I). Just south of the Bolivia-Argentina border, at around 23°S, those structures give way to west-vergent Ordovician-detached or basement-involved structures in the Santa Bárbara Ranges (Fig. 7J; Kley et al., 1999; Kley and Monaldi, 2002). In addition to being west vergent, the structures in the Santa Bárbara Ranges differ in other ways from the Bolivian fold-and-thrust structures to the north: (1) They record less shortening; (2) they are more oblique to the main trend of the orogen; and (3) they carry a Cretaceous stratigraphic assemblage that formed in a major rift system in northern Argentina (Figs. 4 and 11; Mon and Salfity, 1995).

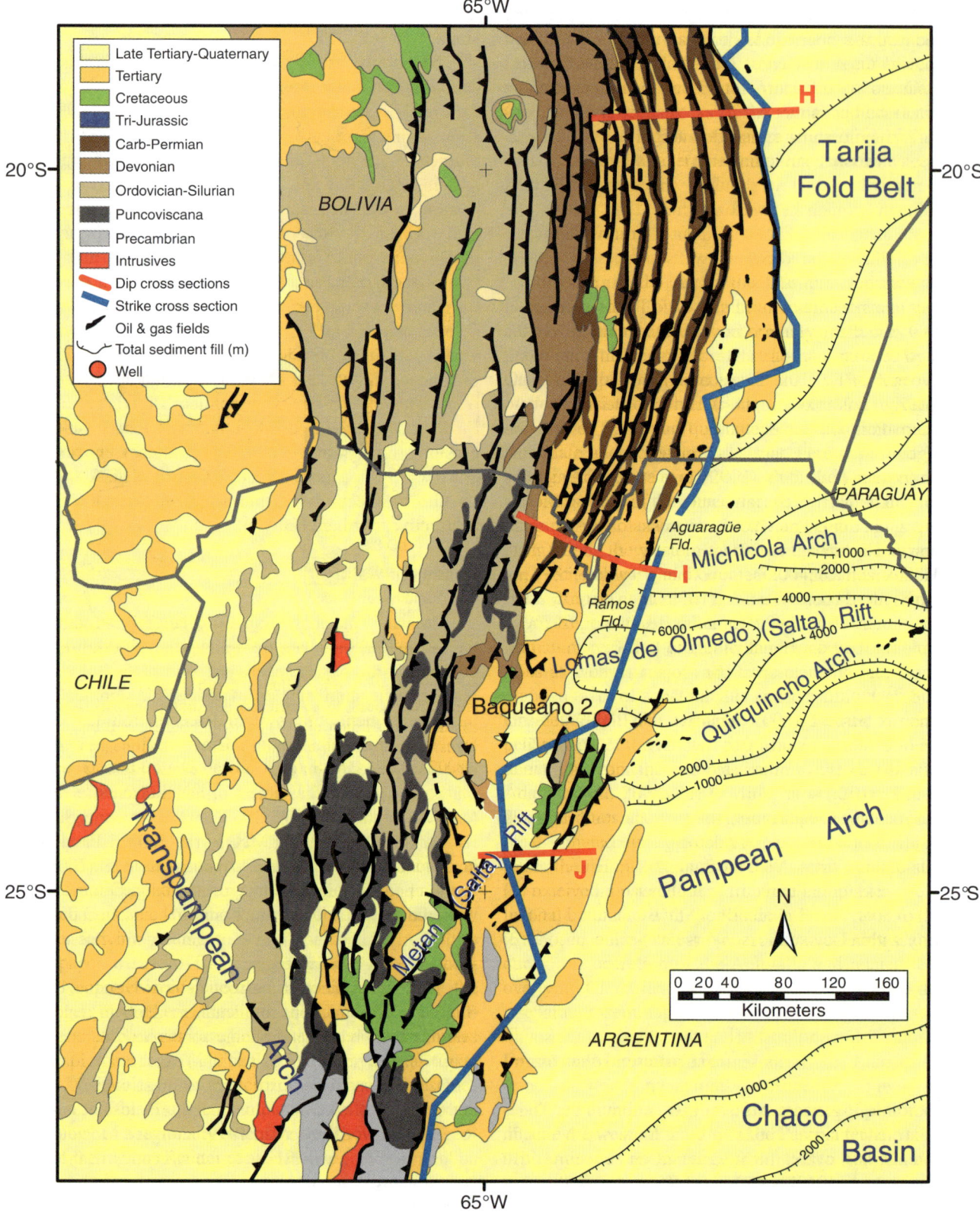

Figure 11. Simplified geologic map of the southern Santa Cruz–Tarija fold belt, Bolivia, and the complex mountain front in northwestern Argentina. Note the termination of closely spaced décollement-style thrusts near the Lomas de Olmedo rift. South of this intersection, the structural style of the Subandes changes dramatically, and the only viable petroleum systems are those associated with discrete Mesozoic rifts. Map information is simplified from Geologic Data Systems (1997).

The intersection of the fold-and-thrust belt with a Mesozoic rift system in northern Argentina fundamentally changes the character of the system to one in which faults are stratigraphically more deeply rooted and carry variable stratal thicknesses above them. Shortening in these rift-influenced parts of the system appears to be significantly lower in magnitude than in the thin-skinned Bolivian salient (McQuarrie, 2002a), although substantial structural relief has been built in much of the thicker-skinned Argentinian segments of the belt.

The Salta rift system at 23°S latitude profoundly influenced how Neogene contractional structures formed (Grier et al., 1991; Kley and Monaldi, 2002; Kley et al., 2005; Carrera et al., 2006; Carrera and Munoz, 2008; Iaffa et al., 2011). The Lomas de Olmedo, Metan, and related rifts (Figs. 2, 6, and 11) comprised an Early Cretaceous to early Cenozoic rift system that intersected the Subandean mountain front and abutted the northern margin of the Transpampean and Pampean arches in northern Argentina (Starck, 2011). The rift system also extended into the Altiplano and other areas of the interior Andes to the west (Mon and Salfity, 1995; Welsink et al., 1995). Up to 6 km of Lower Cretaceous and younger strata are present in the thickest Salta rift depocenters. Basal synrift units are continental red beds and alkaline volcanic rocks of the Pirgua subgroup. Upper Cretaceous to Paleogene postrift sandstones and lacustrine limestones of the Yacoraite Formation overlie these units (Fig. 3). Yacoraite limestones formed in a restricted carbonate basin and became a self-sourcing reservoir that produces in approximately a dozen small fields within and on the shoulders of the Lomas de Olmedo and Metan rifts. Cretaceous rift fill lies above a heterogeneous substrate ranging in age from Precambrian to Carboniferous. Stratigraphic relationships below the pre-Cretaceous unconformity indicate that the Lomas de Olmedo rift arm formed atop a NE-SW paleo-high termed the Michicola arch, which probably originated as a Chañic structure in Late Devonian to early Carboniferous time (Salfity et al., 1987; Starck, 2011). Presently, the Michicola and Quirquincho arches bound the Lomas de Olmedo rift, having evolved into Cretaceous–Cenozoic rift shoulders upon dismemberment of the ancestral Michicola arch (Fig. 11).

Cenozoic structures at 23°S–27°S latitude are a complex combination of inversions, detached thrusts and back thrusts, and basement-involved thrusts and back thrusts (Fig. 7). Contractional deformation began in Eocene time in the eastern Puna and western Eastern Cordillera and propagated eastward sporadically to the present seismically active Santa Bárbara system (Carrapa et al., 2011). The southernmost Subandean zone of northern Argentina was shortened by ca. 9–8 Ma (Echavarria et al., 2003). Ranges in the southern Eastern Cordillera developed within the southwestern arm of the Salta rift and are roughly continuous along-strike with the fold-and-thrust belt to the north, but they are clearly basement involved and bivergent. In contrast, the Sierras Pampeanas to the south are discontinuous, more obliquely oriented, basement-involved structures (Grier et al., 1991; Kley et al., 2005). Structural relief between Upper Cretaceous Salta rift rocks in the Eastern Cordillera and the

same units in the subsurface Lomas de Olmedo Basin exceeds 10 km (Pearson et al., 2012).

Petroleum systems in this segment of the Subandean belt rely principally on source rocks in the Campanian–Maastrichtian Cretaceous Yacoraite lacustrine limestone and overlying lacustrine black shales of the Olmedo Formation (Comínguez and Ramos, 1995). These sources matured upon burial by Eocene and younger foreland basin fill as Andean deformation encroached into the area. The limited distribution of fields probably reflects some combination of (1) sparse, discontinuous, or absent source, (2) low trap density in the less-deformed subsurface rift system to the east, and (3) problematic structural timing to the west where Andean thrusts or inversions may have uplifted and extinguished any hydrocarbon generation before those sources could charge any Andean-aged traps.

Cuyo-Bolsones and Neuquén

The southernmost segment of our megaregional cross section (Fig. 2) continues southward from the Salta area, where Cretaceous rifts formed within a heterogeneous substrate of Paleozoic sedimentary rocks, Ordovician metamorphic and igneous rocks, and the Grenville-aged Cuyania–Precordillera and Chileñia basement terranes (Fig. 4). Physiographically, the southern 2000 km section of our regional strike section crosses the Sierras Pampeanas, the Bolsones and Cuyo Basins, the San Rafael high, and the Neuquén Basin (Fig. 12). Structurally, it traverses the Transpampean arch and possesses a dramatically different structural and stratigraphic composition than southern Bolivia and areas to the north.

Southwest of the Salta rift, Carboniferous to Permian normal faults formed the Paganzo Basin (Figs. 4 and 6) atop the Transpampean arch (Fernandez-Seveso and Tankard, 1995). South of here, Mesozoic extension gets progressively younger, with mainly Early Triassic faulting in the Cuyo Basin (Dellapé and Hegedus, 1995) and Late Triassic extension in the Neuquén Basin (Vergani et al., 1995). The Cuyo Basin trends obliquely into the Precordillera at between 33°S and 34°S and contains 25 oil fields, discovered since 1932. The largest field, Vacas Muertas, has ultimate recoverable reserves estimated at 470 MMBO (Fig. 12; C and C Reservoirs, 2010). The Neuquén Basin trends at a high angle to the Precordillera mountain front between 37°S and 39°S and contains more than 70 oil and gas fields sourced primarily from marine shales of the Lower Jurassic Los Molles Formation and the uppermost Jurassic Vaca Muerta Formation (Fig. 3). Most of the production in this region is from inverted fault block traps in the foreland, except for a group of relatively small fields located in the Malargüe-Agrio thrust belt at the mountain front between the central Cuyo and Neuquén Basins (Fig. 7L). Cooper (2007) reported the ultimate recoverable reserves in the thrust belt fields to be ~700 MMBOE (million barrels of oil equivalent), with approximately two thirds of that volume being oil.

Directly south of Mendoza, Argentina, the Cuyo Basin occupies a lowland region in the foreland of the Precordillera of the

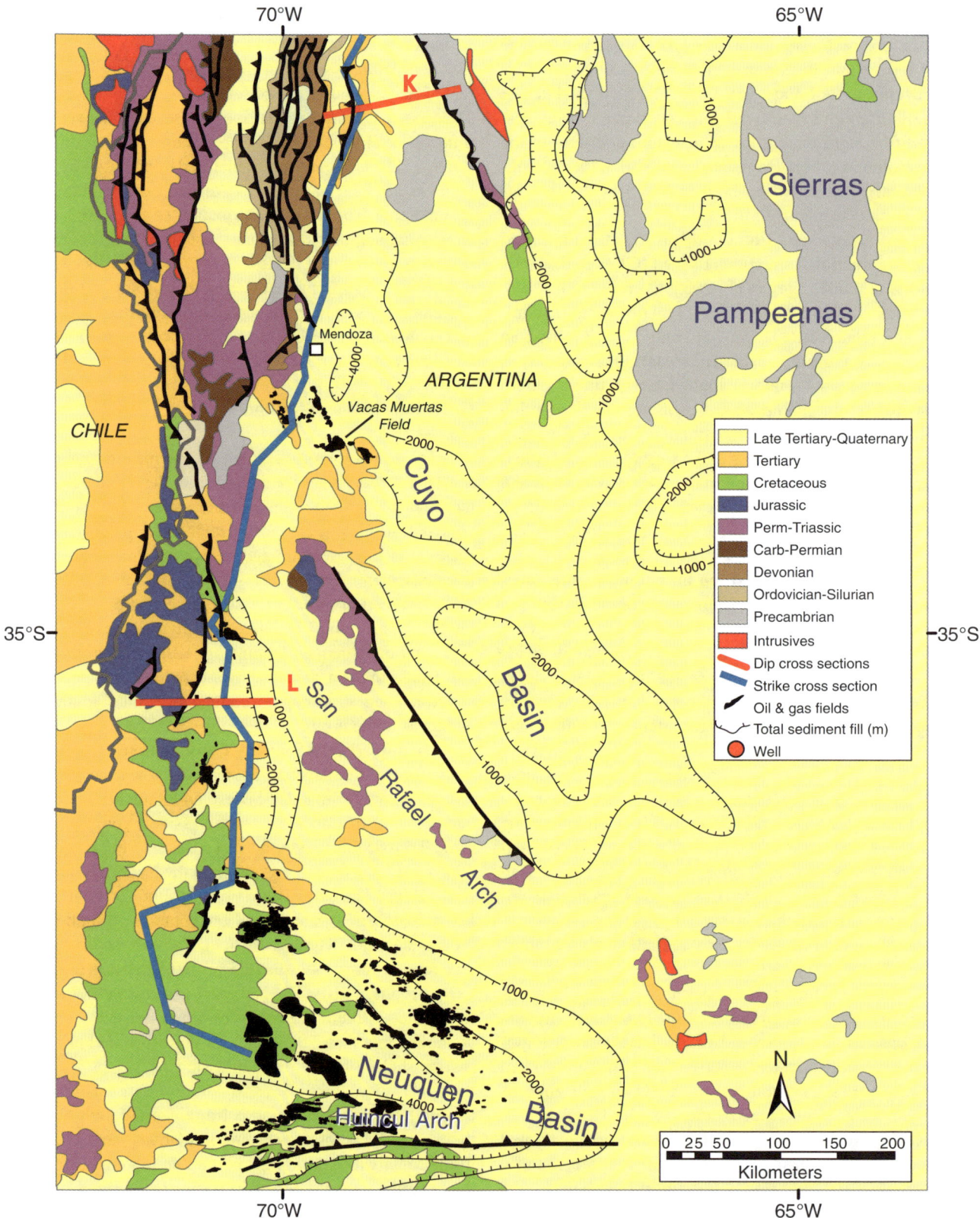

Figure 12. Simplified geologic map of the Neuquén-Cuyo segment of the Subandean fold belt, Argentina. Note the NW and the WNW trends of the Cuyo and Neuquén Basins, which intersect the Subandean zone at oblique angles. The only segment of the frontal system of the Andes that contains hydrocarbons is in the Malargüe-Agrio belt, at the northwestern end of the Neuquén Basin. Map information is simplified from Geologic Data Systems (1997).

Andes, west of the uplifted, crystalline-cored Pampean ranges (Fig. 12; Dellapé and Hegedus, 1995). The Cuyo Basin is smaller and contains a less complete section of Mesozoic strata, but otherwise it shares a common history with the Neuquén Basin. Rifting in the Cuyo began in Early Triassic time in half grabens when alluvial, fluvial, and lacustrine clastic sediments were deposited above Choiyoi Group volcanic rocks and a faulted low-grade metasedimentary basement of Paleozoic age. Extensional fault throws locally exceed 3000 m, which created isolated depocenters that generally coalesced during Jurassic sag phase subsidence. The key source rock in the basin is the Middle Triassic Cacheuta Formation. Jurassic, Cretaceous, and Paleogene strata generally have more regionally continuous facies; thicknesses of this interval range from 500 to 3000 m. Neogene nonmarine foreland basin clastic rocks were deposited above the older rift-sag sequence and locally exceed 2000 m in thickness. Inversion of the extensional fault blocks occurred in Miocene to Holocene time, aligning anticlinal fields above the older rift faults in three NNW trends of structures.

Following two episodes of rifting in the Late Triassic and Early Jurassic, the Neuquén Basin underwent a protracted depositional evolution extending into the Late Cretaceous, when rift-related thermal subsidence had decayed (Vergani et al., 1995; Ramos and Folguera, 2005; Zapata and Folguera, 2005). Most of the Mesozoic section in the basin thickens and deepens to the west, although west-derived clastic rocks are recognized in units as old as Middle Jurassic along the western margin of the basin. Multiple inversion events caused by contractional reactivation of normal faults created unconformities and influenced stratal architecture in the basin, in Callovian, Oxfordian-Kimmeridgian, early Valanginian, late Albian, and late Cenozoic time. Most of the hydrocarbon traps in this basin owe their origin to one or more of these inversion events.

This complex history of rift-sag subsidence and sedimentation in the Neuquén Basin was occurring east of the Mesozoic volcanic arc at 38°S, but the basin did not take on a foreland style of deposition until latest Cretaceous time (Ramos and Folguera, 2005; Zapata and Folguera, 2005). During the climax of Andean shortening in the Miocene and Pliocene, fold-and-thrust deformation was restricted to the western edge of the basin (Figs. 7L and 12). The structures in the Malargüe-Agrio fold-and-thrust belt root into a detachment below 10 km depth in the Paleozoic strata or basement (Manceda and Figueroa, 1995). Faults flatten upward in the direction of transport to the east into Jurassic shales or evaporites, or into Cretaceous shales, creating a large-scale, fault-bend fold style of deformation. Locally inverted normal faults and variable thicknesses of Triassic–Lower Jurassic units reside in the hanging walls of the thrust sheets. Shortening on individual east-vergent thrust faults is in the range of 10 km near the front of the external fold-and-thrust belt, and complex back thrusting, duplexes, and triangle zones accommodate the slip at the tips of the thrusts.

One conspicuous aspect of the Neuquén Basin, at the south end of our regional strike transect, is how little Cenozoic section is preserved atop the Mesozoic basin (Fig. 2), even east of the fold-and-thrust belt. The basin appears to be currently undergoing a regional uplift or inversion, potentially related to its proximity to the modern arc and trench to the west.

HYDROCARBON ENDOWMENT IN THE STRIKE DIMENSION

Geographic Summary

It is well known that the distribution of discovered hydrocarbons can be quite variable within individual foreland basins and fold belts. Clearly, the most fundamental control on viability of a petroleum system is whether a high-quality source rock is present in the section. Figure 13 shows the distribution of the major source intervals discussed in this paper, irrespective of their level of maturity, richness, or whether they are known to be locally effective in charging traps. Figure 13 also shows that large segments of the South American Subandean system have such sources yet have not produced significant oil and gas discoveries. In this section, we examine the entire system in the strike direction with the aim of explicitly defining how more and less productive segments of the system relate to one another and how the tectonic, depositional, and burial history of a given segment will control whether sources are able to charge reservoirs in traps that are generally of late Cenozoic age.

Beginning in the north, the Llanos Subandean belt has a relatively high resource density (3788 MMBOE recoverable in discovered fields; Cooper, 2007); however, these resources are localized within just a few fields in a narrow thin-skinned belt (Figs. 7A and 8). The productive area is bounded to the west by uplifted and breached reservoirs and sources, and to the east by the frontal fault of the system. To the southwest, Cenozoic thicknesses decrease, and the key source intervals yield only heavy oil, or they are immature. It is not surprising, therefore, that the only significant exploration success following up on the Cusiana-Cupiagua discoveries was along strike to the north.

The small amount of shortening transferred out of the Eastern Cordillera into the Subandes in the Llanos segment has limited the quantity of traps but has preserved a large fetch area containing source rocks at regional level in a synclinal maturation kitchen west of the field. The modest throws on the faults in the Llanos foothills mitigated against excess uplift, erosion, and breaching. Importantly, the shortcut fault pattern at the edge of the Eastern Cordillera allows the basal detachment fault trajectory to flatten upward rather than ascending the entirety of the stratigraphic section (Fig. 7A; Toro et al., 2004). This aspect of the Subandean system, as will be shown in subsequent descriptions of the belt to the south, is very basic but of greatest importance for preserving petroleum potential in contractional fold belt provinces. Structures that "keep their heads down," either because of modest shortening or because of favorable detachment geometries, are required for trap preservation in fold belt systems. Excess shortening and/or fault systems that lack multiple

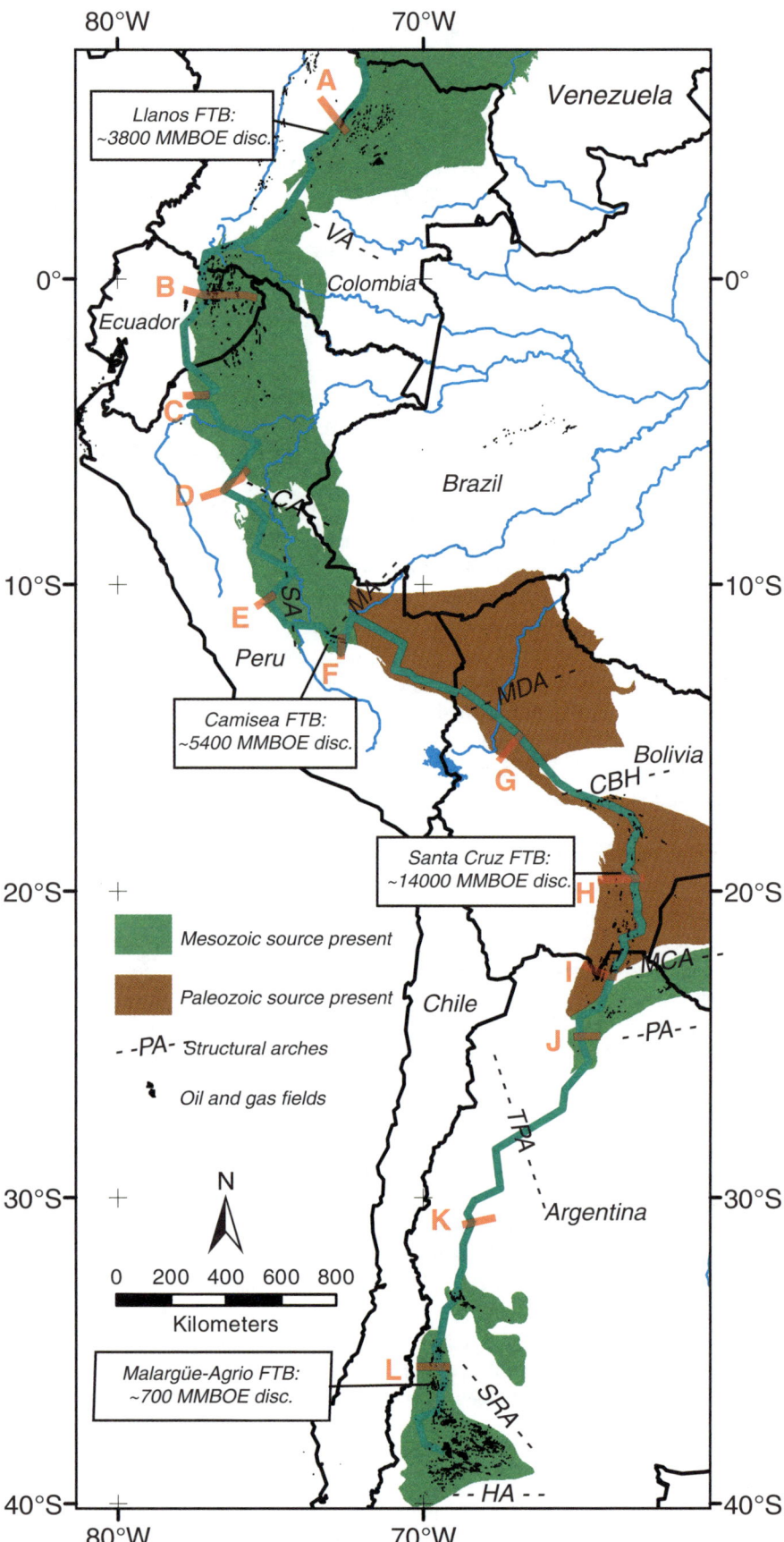

Figure 13. Source rock and resource distribution summary map. Brown polygons are areas with Paleozoic source rocks present, and green polygons are areas with known Mesozoic sources. Also plotted are discovered resources from Cooper (2007). Camisea estimate includes recent discoveries not included in Cooper (2007). Note the significant extent of Subandean areas with known source rocks but no discovered resources. Arch abbreviations are same as Figure 4. FTB—fold-and-thrust belt; MMBOE—million barrels of oil equivalent.

detachment levels can cause too much uplift and erosion, which destroy trap integrity and hydrocarbon retention.

Structural foreland arches also have an important impact on how petroleum systems evolve in fold belts. As shown in Figure 2, the level of maturity of the Upper Cretaceous source interval decreases to the south due to the long-lived structural high on the Vaupes arch. The structural relief on the Vaupes arch served to inhibit the Upper Cretaceous source interval from maturing during the Cenozoic, causing the Subandean system to the southwest of Cusiana to be high risk or nonprospective due to the probable lack of a mature Upper Cretaceous source.

The presence and quality of the Upper Cretaceous source remained intact to the south, and as the eastern terminus of the contractional system plunged southwest beyond the Vaupes arch, the source interval matured enough to charge several fields of BBO size in the foreland of Ecuador. The generally high structural elevation of basement in the Cutucu Cordillera precluded formation of a "Llanos-type" Subandean detached fold-and-thrust belt. The absence of a shallow section precluded the inversion faults from rolling into flat upper detachment levels, as in the Llanos (Fig. 7B), and so they surfaced and breached the Subandean traps. The proximity of the basement high in southwest Colombia and northern Ecuador to the western end of the Guyana Shield suggests that the Vaupes arch has a Precambrian ancestry and a genetic link to shield areas to the east.

In northern Peru, the fold belt system becomes even more deeply buried due to the depositional load of Amazonia (Fig. 9; Hoorn et al., 2010; Hoorn and Wesselingh, 2011). Mathalone and Montoya (1995) estimated the total thickness of Cretaceous and Cenozoic strata at >5 km, which caused Late Cretaceous sources to be gas prone and reservoir quality to be impaired by late Cenozoic time. In addition, the viable traps in the Santiago Basin tend to be subsidiary features on the flanks of larger, salt-detached, breached anticlines, so trap sizes are small by global fold belt standards.

In central Peru, the effectiveness (TOC) of the Upper Cretaceous source interval degrades such that no discovered hydrocarbons in the Marañon Basin south of 5°S can be correlated to those sources. Instead, several foreland discoveries are attributed to sources in the Jurassic Pucara Formation or older units (Mathalone and Montoya, 1995; ChemTerra International Consultants, 2000; Wine et al., 2003). However, in the northern Ucayali segment, the absence of significant hydrocarbon in the Ponasillo 1X well and three dry holes on young foreland structures just outboard (east) of the Chazuta thrust suggests that the Jurassic and older sources may have been overmature at the time of late Cenozoic trap formation, due to their location on the fringes of the thick, Amazonia depocenter. This depocenter, which is partly localized above the extensive Triassic rift system in Peru and also above the projection of an underlying Paleozoic rift that separates the Guyana and Brazilian Shields to the east, contributed in a positive way to maturing Cretaceous sources that charged foreland traps in Ecuador. However, this same depocenter was a destructive factor for older, deeper sources, causing them to

mature prior to trap formation in the late Cenozoic fold-and-thrust belt.

In the southern Ucayali and Ene fold belt segments, where Cretaceous and Cenozoic units thin, it is likely that the Jurassic and perhaps the Upper Paleozoic sections are locally still in a maturity window capable of generating hydrocarbons that could fill late Cenozoic traps. However, due to a paucity of drilling in this area, the ultimate potential of this segment of the fold-and-thrust belt remains uncertain.

In southern Peru, south of the Shira arch, the frontal fold-and-thrust belt deflects eastward beyond the margin of significant Mesozoic extension. In this area, detachment levels and the overall structural style become controlled by the thick Paleozoic Chaco clastic basin that exists throughout much of Bolivia. The Camisea gas condensate fields in Peru (Mississippian source, Permian and Cretaceous reservoirs), and numerous Bolivian fields (Silurian and Devonian source, Devonian and Carboniferous reservoirs) attest to the preservation of the Paleozoic sources at low maturity levels until traps formed in late Cenozoic time. In the vicinity of Camisea, the Manu arch was instrumental in maintaining Mississippian source rocks at shallow depths throughout the Mesozoic and early Cenozoic (Fig. 10). Along much of this central region of the Subandean fold-and-thrust belt, the Mesozoic section is thin or absent, suggestive of a rift shoulder or broad uplifted platform tectonic setting. Cooper (2007) estimated the reserve base of the Madre de Dios portion of the Peruvian fold belt at 3816 MMBOE. With the addition of recent discoveries around Camisea, we now estimate the resource base in that area at ~5400 MMBOE.

The Bolivian Paleozoic sources are among the richest in South America, and where trap timing and burial history have a favorable relationship, the quality of the petroleum system can be excellent, although fracture enhancement is required for the Paleozoic sandstones in some of the fields to produce commercially (C and C Reservoirs, 2010). The Bolivian thrust salient would likely contain even more resources if the level of erosion on some of the structures was not as great (Fig. 7). Many of the structures do not have a preserved Cenozoic cover, as they do in southeast Peru, and so the only viable targets are deeper in the Paleozoic beneath breached Carboniferous reservoirs at the surface. Discovered hydrocarbons in the Bolivian fold-and-thrust belt are mostly gas, and have been estimated at >45 TCF (~7.5 MMBOE) by C and C Reservoirs (2010) and at ~14,000 MMBOE by Cooper (2007; Fig. 13).

As described earlier (Fig. 11), the southern end of the Bolivian fold-and-thrust salient eventually deflects back to the SSW, and the character of the structural system and the petroleum system changes again. In addition to structural style changes, the Paleozoic source intervals here are deformed and/or eroded as the fold-and-thrust belt transects the Transpampean orogenic welt (Figs. 2 and 3). No known fold belt hydrocarbon accumulations can be tied to those older sources south of the Salta rift in northernmost Argentina. Instead, several isolated Mesozoic rift basins are superimposed on the Paleozoic orogen, and they have self-contained petroleum systems in the Salta, Cuyo, and Neuquén Basins. Although there are some hydrocarbon accumulations near the main topographic front of the Andes in classic "fold-and-thrust" systems (Fig. 12), the bulk of the resource associated with these basins resides in inverted rift systems. In the Salta Basin, roughly a dozen small fields exist in the foreland of the Andes, mostly within and on the shoulders of the Lomas de Olmedo rift arm. The Cuyo Basin has 25 fields sourced from the Triassic, and the Neuquén Basin has numerous fields sourced by Lower and Upper Jurassic and Lower Cretaceous intervals, and trapped in normal fault block or inverted fault block traps. On the west side of these inverted rifts in the Argentina foreland, the Malargüe-Agrio fold belt is estimated by Cooper (2007) to contain 723 MMBOE (Fig. 13).

Discussion

Two first-order paleogeographic controls on source presence affect the distribution of hydrocarbon resources in the Subandes structural belt. Between southern Bolivia and northern Argentina, the basement framework and structural style change abruptly where the Subandean system traverses the boundary between northern in situ South American basement and crust dominated by accreted terranes and Paleozoic orogenic fabrics to the south (Fig. 4). All hydrocarbon discoveries south of approximately latitude 22°S owe their origin to Mesozoic rifts superimposed on the older Paleozoic orogenic crust. The limits of Paleozoic source and reservoir distribution reflect the profound impact of the Ordovician Ocloyic orogeny and the Late Devonian–Mississippian Chañic compressional event, virtually isolating those older plays to areas north of the Argentina-Bolivia border. In the northern Subandes, the prolific Upper Cretaceous source rocks in Colombia and Ecuador become diluted and ineffective to the south. The southeastward shallowing of facies responsible for this dilution is observed in units as old as Permian (Fig. 5) and seems to be broadly related to the tectonic reorganizations in the late Paleozoic that affected the central Subandes. The Chaco Basin in Bolivia, which was a long-lived marine trough in the Paleozoic, became the site of continental deposition throughout most of the Mesozoic, isolating Jurassic and Cretaceous marine sources to central Peru, Ecuador, and Colombia (Tectonic Analysis Ltd., 2006).

Following the late Paleozoic tectonic reorganizations, Permian through Early Cretaceous rifts were superimposed on South American crust with variable geometry and distribution. In Argentina, rifts capitalized on basement features set up during contractional events and are therefore oriented oblique to the modern Andean deformation front. In Peru, Ecuador, and Colombia, rifts are generally parallel to the modern Subandean chain. Hydrocarbons in the South American Subandes and inner foreland owe their origins roughly equally to high-quality Paleozoic sources in Bolivia and southern Peru that evaded overprinting and burial by Mesozoic rifts, and to younger sources that tie directly to the rifts or collapse basins that formed elsewhere

in western South America beginning in the Permian. These rift basins generally had the positive effect of creating source-prone syn- to postrift sedimentary assemblages but probably also destroyed the viability of Paleozoic sources in large areas of Peru, Ecuador, and Colombia.

Because the Subandean mountain front traps are primarily late Cenozoic in age, the timing and magnitude of burial by either large craton-derived depositional systems or thick west-derived foreland fill are important. Central Peru seems to have suffered from too much east-derived sediment burial of Jurassic and older sources, such that they may have been overmature before the fold belt traps formed. Parts of the Madre de Dios, Beni, and Santa Cruz–Tarija segments of the system are gas prone due to the great thickness of Cenozoic foreland basin fill in those areas. The Llanos Basin has a variable thickness of Cenozoic fill, leading to hydrocarbon occurrences that range from absent, to abundant early-maturity heavy oil, to gas condensate and gas in fields like Cupiagua.

The trap styles themselves also dictate whether a given trend has preserved commercial accumulations. Numerous workers have recognized the importance of inherited fabrics in controlling or strongly influencing subsequent structural development along the Andes (Allmendinger et al., 1983; Grier et al., 1991; Mon and Salfity, 1995; Allmendinger and Gubbels, 1996; Kley et al., 1999, 2005; Jaillard et al., 2002; Sempere et al., 2002; Giambiagi et al., 2003; Carrera et al., 2006; Rosas et al., 2007; Iaffa et al., 2011). The thick-skinned inversion basins (Llanos, Cuyo, Neuquén) tend to have narrow structural corridors where traps are not breached and where direct migration from synclinal source kitchens is possible. The hybrid "thinner-skinned" inversion basins in Peru (e.g., Huallaga, Ene) have not yet been proven commercially successful, although this may be due more to issues with the petroleum systems than to the traps themselves. The expansive, truly thin-skinned system that extends from Camisea in southeast Peru to northernmost Argentina clearly has several favorable structural characteristics. Obviously, this style promotes development of a great quantity of traps (e.g., Figs. 7F, 7G, 7H, and 7I), but perhaps more important is the fact that subhorizontal basal and/or intermediate-level detachments allow tremendous amounts of shortening to occur without building the destructive structural relief so common in the thicker-skinned systems (e.g., Figs. 7A, 7J, 7K, and 7L). In simplest terms, long horizontal basal or intermediate detachments allow shortening and uplift to be distributed across numerous anticlines, thereby limiting the likelihood of breaching on any one structure. Where the fault ramps are steep and cross thick intervals of upper crust, even a modest amount of shortening can cause traps to be breached, as in the Malargüe-Agrio belt in Argentina. Systems with flat detachments also allow source rocks to remain at regional levels and continue to undergo prograde maturation in synclines overlying the tails of thrust sheets; this is less likely in thick-skinned systems. In the best thin-skinned systems, Cenozoic strata are preserved above all the frontal structures, and trap risks are minimal, as at Camisea. In the Santa Cruz–Tarija system in Bolivia, the fact that many structures have Carboniferous strata at the surface indicates that the fold belt traps are partly breached, and the only viable targets are Lower and Middle Paleozoic reservoirs, which tend to be tight.

Finally, it is worth briefly addressing the role of flat slab subduction on the retroarc petroleum system in the South American Subandes. We find no simple correlation between structural relief or thermal state, which can be attributed to flat slabs, and the effectiveness of the petroleum systems, except arguably in the Sierras Pampeanas foreland region of northern Argentina. There, young uplifts have clearly destroyed any preexisting petroleum system, although most of this region already had problematic petroleum systems due to Paleozoic orogenesis. Those events, which deformed the older Paleozoic substrate and created a long-lived highland that probably never had late Paleozoic source deposition atop it, precluded the creation and/or preservation of a Bolivian- or southeast Peru–style Paleozoic basin. So these significant Paleozoic tectonic events in Argentina had already impaired the petroleum potential of this area before any Cenozoic flat slabs arrived. If it were not for the widespread rifts of Mesozoic age that formed within the ancestral Transpampean high, this region would be completely devoid of oil and gas occurrences.

In southern Peru, we have difficulty reconciling recent flat-slab–related uplift with the preservation of Cenozoic strata above compressional folds in the Madre de Dios Basin. The Camisea gas fields are in structurally favorable traps because the crests of the anticlines are not eroded deeper than the Lower Cenozoic. The Camisea area also sits above two older structural arches (Manu and Madidi) that have condensed Mesozoic section atop them, which helped keep the Paleozoic source rocks from maturing prior to the late Cenozoic time of trap formation. The fact that the long-lived Manu arch was not subsequently uplifted enough to cause deeper erosion of the Camisea folds suggests that recent flat-slab–related uplift was not large enough to significantly impact the petroleum system in that area. So, in southern Peru as in Argentina, the main crustal elements that control the petroleum systems were in place prior to the Cenozoic.

PREDICTIVE CONCEPTS FOR GLOBAL EXPLORATION

Contractional mountain fronts contain some of the world's large hydrocarbon accumulations, for example, in the Zagros chain of Iran and Iraq, yet when viewed globally, more of these mountain fronts are devoid of reserves than are productive. Our megaregional strike section is an attempt to systematically analyze a large (~7500-km-long) retroarc continental system in its entirety with the aim of describing the primary and secondary controls on oil and gas prospectivity. This analysis complements that of Cooper (2007), who concluded that no single structural or stratigraphic factor alone has a high correlation with hydrocarbon reserves within global contractional mountain fronts.

Foremost in importance when evaluating these systems is to recognize that, excluding the constructive role of foreland basin burial of source rocks into the maturation window, fold-and-thrust belts tend to be destructive with respect to most elements of petroleum systems. Cooper (2007) pointed out that the vast majority of worldwide conventional reserves in these structural environments occur in Cenozoic deformation belts. The reason for this is that the shallow levels of compressional mountain belts, where petroleum resources reside, are ephemeral fixtures of the crust: Those above sea level erode, which causes isostatic rebound of the crust and a runaway cycle of uplift and denudation, which ultimately compromises seals for the traps and causes the reservoirs themselves to be unroofed. In addition, exhumation, driven by uplift and erosion, causes cooling of the upper crust and tends to shut off generation of hydrocarbons from source rocks except in large synclines or in areas where extensional collapse basins allow source kitchens to remain buried. An easy way to observe this problem is to look at any of the world's petroliferous mountain fronts and to note that the productive trends are commonly only a few tens of kilometers wide in the dip direction, beyond which the subsurface reservoirs are present at the surface. It is common for breached oil fields to be found in these hinterland areas, where trap closures that resided in the subsurface just 10–20 m.y. ago are now uplifted on the youngest thrust faults in the system.

In order for traps in a system that is generating hydrocarbons to maintain integrity and avoid uplift and breaching, it is obviously imperative that structural relief be modest in comparison to the thickness of sedimentary fill above the reservoir zones. The Camisea area in Peru (Fig. 7) is an excellent example of how traps that contain hydrocarbons tend to be modest in size and have minimal structural complexity. Duplex structures, as shown on the south end of the Camisea profile, can create large structural culminations but often have the undesirable effect of causing surface breaching or at least tilting of the roof thrust sheets, thereby reducing the height of shallower structural closures.

One of the most effective ways for traps to "keep their heads down" and avoid breaching is to form in a low structural taper (Davis et al., 1983) environment, which is promoted by the presence of preexisting sedimentary basin taper and widespread, weak décollement horizons. Such systems do not require extensive internal shortening in order for the basal detachment to fail, and so structural and topographic relief tends to be minimized, limiting the potential for breaching of traps. Both of these characteristics can be found in the Madre de Dios–Beni and Santa Cruz–Tarija segments of the Subandean zone in southeast Peru and Bolivia, where the highest concentration of fold-and-thrust belt oil and gas fields exists between Colombia and Argentina. The inversion mountain fronts, e.g., in the Llanos Basin (Fig. 8), are more abrupt and have narrower areas between uplifted internal zones and undeformed forelands. They are certainly capable of containing viable petroleum traps, especially where low-angle shortcut faults propagate out from beneath the higher-angle inversion faults, but the quantity of traps tends to be lower, and the fetch areas for source kitchens may be less extensive than in the fully detached, low-taper systems. Consistent with previous syntheses (Allmendinger et al., 1983; Grier et al., 1991; Mon and

Salfity, 1995; Allmendinger and Gubbels, 1996; Kley et al., 1999; Jaillard et al., 2002; Sempere et al., 2002; Giambiagi et al., 2003; Rosas et al., 2007), the South American examples in this paper testify to the importance of preexisting structural/stratigraphic fabrics—most importantly, whether the basins shortened in late Cenozoic time were previously affected by Mesozoic rifting—in setting up the fundamental control on structural styles in the modern mountain front.

Beyond the basic characteristic of structural relief, structural timing with respect to petroleum generation is a critically important limiting factor for filling of traps in contractional mountain fronts (Cooper, 2007). Because foreland basins flex under the load of encroaching retroarc thrust belts, they commonly contain thick successions of synorogenic clastic rocks derived from both the continent and mountain belt sides of the basins. In South America, the thickness of Cenozoic strata commonly exceeds 4 km in fold belt synclines and in the proximal foredeep just outboard (east) of the deformation front. These burial depths of Paleozoic and Mesozoic source rocks in the Subandean zone set up a delicate balance between the timing of hydrocarbon maturation and the timing of trap formation. Maturation sweet spots occur where relatively young source rocks, like the Upper Cretaceous formations in Colombia, are buried under a variable load of Cenozoic synorogenic clastic rocks. As demonstrated by the maturity window in Figure 2, there are areas of limited burial (e.g., over the Vaupes arch) where the source rock is submature or generates only early-maturity heavy oil, and also areas of "optimum" burial, where oil or condensate is the main commodity type. Elsewhere, where source rocks are older and deeper, and/or the Cenozoic loads are thicker, source rocks tend to be gas-mature or overmature, and young traps may be dry. An area where overmaturity appears to be a problem is in central Peru, where the Jurassic Pucara Formation is known to have charged traps with a complex, protracted history along the Contaya arch in the foreland, but is not known to have charged any traps in the fold-and-thrust belt. This conclusion could change as drilling increases in the lightly explored central Peruvian Subandean zone, but it appears that the thick sedimentary load of Amazonia (Hoorn et al., 2010; Hoorn and Wesselingh, 2011), where Cretaceous and Cenozoic depositional thicknesses locally exceed 6 km, resulted in maturation of Jurassic sources beyond the window where gas generation was viable before late Cenozoic traps formed in this trend (Fig. 2). Similarly, Cenozoic stratal thicknesses in southeast Peru and Bolivia locally exceed 4 km, and so Paleozoic source rocks in those parts of the Subandean zone are in the gas window or are overmature today.

With respect to structural arches that intersect the Subandean zone, whether they promote or impede petroleum generation and preservation depends on the nature of the structural arch and the age of the petroleum system. Over the Vaupes arch in southern Colombia, this long-lived structural high has impeded maturation of young, Upper Cretaceous sources. Conversely, the long-lived Manu and Madidi arches in southern Peru had the positive effect of keeping Mississippian source rocks in the

Camisea area from becoming buried too deeply between late Paleozoic and early Cenozoic time, such that liquid-rich condensate could fill those youthful traps in late Cenozoic time. The largest structural arch in our transect, the Paleozoic Transpampean arch, had a more profound effect than the arches to the north, completely precluding any significant Paleozoic source preservation and also setting up an orogenic basement fabric that gave rise to multiple rift/collapse basins that possess self-contained Mesozoic petroleum systems.

One aspect of the South American system that differs from other mountain front provinces (e.g., North America; the Middle East) is that virtually all the reservoirs in the Subandes are clastic. This fact is important because these rock types have a greater depth dependence on permeability and may not contain the quality of fracture systems that carbonate rock types contain. Tight sandstones have been encountered in Cretaceous reservoirs in the Santiago and Huallaga Basins in Peru and in Paleozoic reservoirs in Bolivia. This consideration is another reason why reserve sizes and reservoir quality tend to be greatest at the modestly buried frontal terminus of a fold belt system compared to the more deeply buried interior zones.

Finally, we note that relatively few of the sources and reservoirs in the Subandean and other global fold belt systems occur within the synorogenic assemblage of strata. In South America, Cenozoic reservoirs in Colombia are part of the foreland basin assemblage, as is some of the secondary (Cretaceous) production in the Camisea area in Peru. However, all of the major source rocks in the Subandes are pre-Cenozoic, and most of the key reservoirs in Argentina, Bolivia, and Peru are Paleozoic or Mesozoic. Cretaceous reservoirs are important in the foreland of Ecuador and Peru, but overall the Cenozoic sedimentary section of South America contains fewer reserves than some other foreland successions in the world (e.g., the Cretaceous of Canada), because the foreland was overfilled during most of this period, and it contains mostly nonmarine rocks with limited source and seal potential.

In summary, the Subandean system of South America is probably similar to other Cenozoic compressional mountain fronts in the world in terms of its tremendous diversity of structural styles, stratal types, and proven petroleum systems. Beyond the basic elements of a flexural foredeep, locally developed wedge-top depozones, and general rates and styles of orogenic wedge advance, we find the overall system to be remarkably nonuniform in the strike direction and challenge the notion that there is such a thing as a "generic" retroarc thrust belt. We concur with previous authors (e.g., Allmendinger et al., 1983) that the role of structural-tectonic inheritance cannot be overstated when trying to unravel the stratigraphic and petroleum systems characteristics of a Neogene contractional system.

While much of the ~7500-km-long mountain front contains high-quality source rocks, only a small proportion of the Subandes contains commercial hydrocarbons. In areas with viable source rocks, the chief factors that inhibited viable petroleum systems from being developed are (1) high-standing basement arches where source rocks were never buried enough to mature and (2) structural timing within thick foredeep depocenters where maturation preceded trap formation. In areas where a recently active petroleum system is present, but overall resource density is not high, the causes are (1) inversion mountain fronts where structural trap density is low and/or breaching is prevalent, and (2) high-relief anticlines in thin-skinned systems where shallow reservoirs are breached and production relies on tighter, Lower Paleozoic fractured sandstone reservoirs. In our view, most of the known petroleum occurrences in the Subandes are easily explained in the context of their subregional stratigraphic and structural framework, and we would expect future discoveries to be made offsetting those fields. We also suggest that one additional area could prove productive in the future—the district north and west of Camisea in Peru, where Paleozoic and/or Jurassic sources might be locally viable where the Cretaceous–Cenozoic depositional load is 4 km or less. The current approaches targeting unconventional reservoirs might also open up larger tracts of the Subandes to a new round of successful exploration since source rocks are abundant in these basins. These targets will likely be in more gently deformed areas than the conventional anticline traps, such as in wide bathtub synclines that are still in the oil- or gas-generating window.

ACKNOWLEDGMENTS

We thank ExxonMobil Upstream Research Company for permission to publish this paper. We thank Richard Chuchla, Mark Cooper, Pete DeCelles, and Jonas Kley for constructive reviews that significantly improved the manuscript. The analysis presented here was undertaken as part of the COSA (Convergent Orogenic Systems Analysis) research project, a collaboration between ExxonMobil and the University of Arizona; we thank Carlos Dengo and Joaquin Ruiz for their vision and sponsorship, and Pete DeCelles for his leadership of the project. Many others contributed to this summary through various dialogs and collaborations, including Leon Aden, Tom Becker, Gary Gray, Paul Kapp, Ted Keeling, Jerry Kendall, Carlos Lopez, and Lori Summa. Richard Lease acknowledges partial support from the National Science Foundation Continental Dynamics program (EAR-0907817).

APPENDIX. COMMENTS ON LOCATION OF ALONG-STRIKE CROSS SECTION

Given that a main goal of our analysis is to characterize the cause of along-strike segmentation of hydrocarbon systems, it was necessary to traverse an area close to the modern Subandean deformation front, in proximity to the traditional habitat for fold belt petroleum traps. However, in some locations, the section departs from that precise location in order to intersect more complete stratigraphic sections on the dip profiles. Those departures generated several zigzags in the strike section (e.g., into the Huallaga Basin, Peru), which lead to some odd geometries on the strike section that should not be misinterpreted as transverse structure elements. Secondly, there are areas such as northern Argentina, with significantly deformed forelands, where the choice

of location for the strike section was somewhat arbitrary. As with the zigzags, we attempted to focus discussion on the fundamental transverse elements rather than the artifacts caused by arbitrary locations of the strike section, and we encourage the reader to always view the vertically exaggerated (20×) strike section together with the intersecting dip sections (vertical exaggeration 1×) shown in Figure 7 in order to best comprehend the interplay of stratigraphy and structural elements in three dimensions.

REFERENCES CITED

Aceñolaza, F.G., Miller, H., and Toselli, A., 1988, The Puncoviscana Formation (Late Precambrian–Early Cambrian)—Sedimentology, tectonometamorphic history and age of the oldest rocks of NW Argentina, *in* Bahlburg, H., and Breitkreuz, Ch., eds., The Southern Central Andes: Berlin, Springer, Lecture Notes in Earth Sciences, v. 17, p. 25–37.

Allmendinger, R.W., and Gubbels, T., 1996, Pure and simple shear plateau uplift, Altiplano-Puna, Argentina and Bolivia: Tectonophysics, v. 259, no. 1–3, p. 1–13, doi:10.1016/0040-1951(96)00024-8.

Allmendinger, R.W., Ramos, V.A., Jordan, T.E., Palma, M., and Isacks, B.L., 1983, Paleogeography and Andean structural geometry, northwest Argentina: Tectonics, v. 2, no. 1, p. 1–16, doi:10.1029/TC002i001p00001.

Allmendinger, R.W., Jordan, T.E., Kay, S.M., and Isacks, B.L., 1997, The evolution of the Altiplano-Puna Plateau of the central Andes: Annual Review of Earth and Planetary Sciences, v. 25, p. 139.

Arriagada, C., Cobbold, P.R., and Roperch, P., 2006, Salar de Atacama Basin: A record of compressional tectonics in the central Andes since the mid-Cretaceous: Tectonics, v. 25, TC1008, doi:10.1029/2004TC001770.

Astini, R.A., and Dávila, F.M., 2004, Ordovician back arc foreland and Ocloyic thrust belt development on the western Gondwana margin as a response to Precordillera terrane accretion: Tectonics, v. 23, no. 4, TC4008, doi:10.1029/2003TC001620.

Astini, R.A., Benedetto, J.L., and Vaccari, N.E., 1995, The early Paleozoic evolution of the Argentine Precordillera as a Laurentian rifted, drifted, and collided terrane: A geodynamic model: Geological Society of America Bulletin, v. 107, no. 3, p. 253–273, doi:10.1130/0016-7606(1995)107<0253:TEPEOT>2.3.CO;2.

Baby, P., Specht, M., Oller, J., Montemurro, G., Colletta, B., and Letouzey, J., 1994, The Boomerang-Chapare transfer zone (recent oil discovery trend in Bolivia): Structural interpretation and experimental approach, *in* Roure, F., ed., Special Publication, EAPG Congress, Moscow: Paris, Editions Technip, p. 203–218.

Baby, P., Moretti, I., Guillier, B., Limachi, R., Mendez, E., Oller, J., and Specht, M., 1995, Petroleum system of the northern and central sub-Andean zone, *in* Tankard, A.J., Suárez Soruco, R., and Welsink, H.J., eds., Petroleum Basins of South America: American Association of Petroleum Geologists Memoir 62, p. 445–458.

Bahlburg, H., and Hervé, F., 1997, Geodynamic evolution and tectonostratigraphic terranes of northwestern Argentina and northern Chile: Geological Society of America Bulletin, v. 109, no. 7, p. 869–884, doi:10.1130/0016-7606(1997)109<0869:GEATTO>2.3.CO;2.

Bally, A.W., Gordy, P.L., and Stewart, G.A., 1966, Structure, seismic data, and orogenic evolution of southern Canadian Rocky Mountains: Bulletin of Canadian Petroleum Geology, v. 14, p. 337–381.

Barazangi, M., and Isacks, B., 1976, Spatial distribution of earthquakes and subduction of the Nazca plate beneath South America: Geology, v. 4, p. 686–692, doi:10.1130/0091-7613(1976)4<686:SDOEAS>2.0.CO;2.

Barazangi, M., and Isacks, B., 1979, Subduction of the Nazca plate beneath Perú; evidence from spatial distribution of earthquakes: Geophysical Journal of the Royal Astronomical Society, v. 57, p. 537–555, doi:10.1111/j.1365-246X.1979.tb06778.x.

Barnes, J.B., Ehlers, T.A., McQuarrie, N., O'Sullivan, P.B., and Pelletier, J.D., 2006, Variations in Eocene to Recent erosion across the central Andean fold-thrust belt, northern Bolivia: Implications for plateau evolution: Earth and Planetary Science Letters, v. 248, p. 118–133, doi:10.1016/j.epsl.2006.05.018.

Barnes, J.B., Ehlers, T.A., McQuarrie, N., O'Sullivan, P.B., and Tawackoli, S., 2008, Thermochronometer record of central Andean Plateau growth, Bolivia (19.5°S): Tectonics, v. 27, TC3003, doi:3010.1029/2007TC002174.

Barnes, J.B., Ehlers, T.A., Insel, N., McQuarrie, N., and Poulsen, C.J., 2012, Linking orography, climate, and exhumation across the central Andes: Geology, v. 40, p. 1135–1138, doi:10.1130/G33229.1.

Blakey, R.C., 2008, Gondwana paleogeography from assembly to breakup—A 500 m.y. odyssey, *in* Fielding, C.R., Frank, T.D., and Isbell, J.L., eds., Resolving the Late Paleozoic Ice Age in Time and Space: Geological Society of America Special Paper 441, p. 1–28, doi:10.1130/2008.2441(01).

Boggetti, D., Scolari, J., and Regazzoni, C., 2002, Cuenca Cuyana: Marco geologico y resena historica de la actividad petrolera, *in* Schiuma, M., Hinterwimmer, G., and Vergani, G., eds., Rocas Reservorio de las Cuencas Productivas de la Argentina; V Congreso de Exploracion y Desarrollo de Hidracarburos, Mar del Plata, Argentina: Buenos Aires, Instituto Argentino del Petróleo y del Gas, p. 585–604.

Bookhagen, B., and Strecker, M.R., 2008, Orographic barriers, high-resolution TRMM rainfall relief variations along the eastern Andes: Geophysical Research Letters, v. 35, L06403, doi:10.1029/2007GL032011.

Boyer, S.E., and Elliott, D., 1982, Thrust systems: American Association of Petroleum Geologists Bulletin, v. 66, no. 9, p. 1196–1230.

Bump, A., Kennan, L., and Fallon, J., 2008, Structural history of the Andean foreland, Peru, and its relation to subduction zone dynamics, *in* American Association of Petroleum Geologists Annual Convention and Exhibition: San Antonio, Texas, American Association of Petroleum Geologists, search and discovery article 30062, www.searchanddiscovery.com/pdfz/documents/2008/08030bump/images/bump.pdf.html.

Burgos, D.J., Christophoul, F., Baby, P., Antoine, P., Soula, J., Good, D., and Rivadeneira, M., 2005, Dynamic evolution of Oligocene–Neogene sedimentary series in a retroforeland basin setting: Oriente Basin, Ecuador, *in* Institut de recherche pour le développement, Universitat de Barcelona and Instituto Geológico y Minero de España, eds., 6th International Symposium on Andean Geodynamics (ISAG 2005, Barcelona), Extended Abstracts: Paris, IRD Editions, p. 127–130.

C and C Reservoirs, 2010, Exploration in Fold and Thrust Belts, Basin Analog Studies, DAKS: Houston, Texas, C and C Reservoirs.

Carrapa, B., Trimble, J.D., and Stockli, D.F., 2011, Patterns and timing of exhumation and deformation in the Eastern Cordillera of NW Argentina revealed by (U-Th)/He thermochronology: Tectonics, v. 30, TC3003, doi:10.1029/2010TC002707.

Carrera, N., and Munoz, J.A., 2008, Thrusting evolution in the southern Cordillera Oriental (northern Argentine Andes): Constraints from growth strata: Tectonophysics, v. 459, no. 1–4, p. 107–122, doi:10.1016/j.tecto.2007.11.068.

Carrera, N., Munoz, J.A., Sabat, F., Mon, R., and Roca, E., 2006, The role of inversion tectonics in the structure of the Cordillera Oriental (NW Argentinean Andes): Journal of Structural Geology, v. 28, no. 11, p. 1921–1932, doi:10.1016/j.jsg.2006.07.006.

ChemTerra International Consultants, 2000, Oil Generation in Sub-Andean Basins in Peru: Report for Parsep, Perupetro SA and Canadian Petroleum Institute, Perupetro Technical Archive, Calgary, Alberta, Canada, 45 p.

Chulick, G.S., Detweiler, S., and Mooney, W.D., 2012, Seismic structure of the crust and uppermost mantle of South America and surrounding oceanic basins: Journal of South American Earth Sciences, v. 42, p. 260–276, doi:10.1016/j.jsames.2012.06.002.

Coira, B., Davidson, J., Mpodozis, C., and Ramos, V., 1982, Tectonic and magmatic evolution of the Andes of northern Argentina and Chile, *in* Linares, E., ed., A Symposium on the Magmatic Evolution of the Andes: Earth-Sciences Review, v. 18, p. 303–332.

Colletta, B., Hebrard, F., Letouzey, J., Werner, P., and Rudkiewicz, J.L., 1990, Tectonic style and crustal structure of the Eastern Cordillera (Colombia) from a balanced cross-section, *in* Letouzey, J., ed., Petroleum and Tectonics in Mobile Belts: Paris, Editions Technip, p. 81–100.

Comínguez, A.H., and Ramos, V.A., 1995, Geometry and seismic expression of the Cretaceous Salta rift system, northwestern Argentina, *in* Tankard, A.J., Suárez Soruco, R., and Welsink, H.J., eds., Petroleum Basins of South America: American Association of Petroleum Geologists Memoir 62, p. 325–340.

Cooper, M., 2007, Structural style and hydrocarbon prospectivity in fold and thrust belts: A global review, *in* Ries, A.C., Butler, R.W.H., and Graham, R.H., eds., Deformation of the Continental Crust: The Legacy of Mike Coward: Geological Society [London] Special Publication 272, p. 447–472.

Cooper, M.A., Addison, F.T., Alvarez, R., Coral, M., Graham, R.H., Hayward, A.B., Howe, S., Martinez, J., Naar, J., Peñas, R., Pulham, A.J., and

Taborda, A., 1995, Basin development and tectonic history of the Llanos Basin, Eastern Cordillera, and Middle Magdalena Valley, Colombia: American Association of Petroleum Geologists Bulletin, v. 79, no. 10, p. 1421–1443.

Dahlstrom, C.D.A., 1970, Structural geology in the eastern margin of the Canadian Rocky Mountains: Bulletin of Canadian Petroleum Geology, v. 18, p. 332–406.

Dalmayrac, B., 1978, Géologie de la Cordillère Orientale de la region de Huanuco: Sa place dans une transversal des Andes du Pérou Central (9°S à 10°30′S): Travaux et Documents de l'Orstom, Paris, no. 93, 161 p.

Dalmayrac, B., Laubacher, B., and Marocco, R., 1980, Caractères généraux de l'évolution géologique des Andes Péruviennes: Travaux et Documents de l'Orstom, Paris, no. 122, 501 p.

Dashwood, M.F., and Abbotts, I.L., 1990, Aspects of the petroleum geology of the Oriente Basin, Ecuador, *in* Brooks, J., ed., Classic Petroleum Provinces: Geological Society [London] Special Publication 50, p. 89–117.

Davis, D., Suppe, J., and Dahlen, F.H., 1983, Mechanics of fold-and-thrust belts and accretionary wedges: Journal of Geophysical Research, v. 88, no. B2, p. 1153–1172, doi:10.1029/JB088iB02p01153.

DeCelles, P.G., and Giles, K.N., 1996, Foreland basin systems: Basin Research, v. 8, p. 105–123, doi:10.1046/j.1365-2117.1996.01491.x.

DeCelles, P.G., and Horton, B.K., 2003, Early to middle Tertiary foreland basin development and the history of Andean crustal shortening in Bolivia: Geological Society of America Bulletin, v. 115, no. 1, p. 58–77, doi:10.1130/0016-7606(2003)115<0058:ETMTFB>2.0.CO;2.

DeCelles, P.G., Robinson, D.M., Quade, J., Ojha, T.P., Garzione, C.N., Copeland, P., and Upreti, B.N., 2001, Stratigraphy, structure, and tectonic evolution of the Himalayan fold-thrust belt in western Nepal: Tectonics, v. 20, no. 4, p. 487–509, doi:10.1029/2000TC001226.

Dellapé, D., and Hegedus, A., 1995, Structural inversion and oil occurrence in the Cuyo Basin of Argentina, *in* Tankard, A.J., Suárez Soruco, R., and Welsink, H.J., eds., Petroleum Basins of South America: American Association of Petroleum Geologists Memoir 62, p. 359–367.

Dengo, C.A., and Covey, M.C., 1993, Structure of the Eastern Cordillera of Colombia: Implications for trap styles and regional tectonics: American Association of Petroleum Geologists Bulletin, v. 77, no. 8, p. 1315–1337.

Dunn, J.F., Hartshorn, K.G., and Hartshorn, P.W., 1995, Structural styles and hydrocarbon potential of the sub-Andean thrust belt of southern Bolivia, *in* Tankard, A.J., Suárez Soruco, R., and Welsink, H.J., eds., Petroleum Basins of South America: American Association of Petroleum Geologists Memoir 62, p. 523–543.

Echavarria, L., Hernandez, R., Allmendinger, R., and Reynolds, J., 2003, Subandean thrust and fold belt of northwest Argentina: Geometry and timing of the Andean evolution: American Association of Petroleum Geologists Bulletin, v. 87, p. 965–985, doi:10.1306/01200300196.

Espurt, N., Brusset, S., Baby, P., Hermoza, W., Bolaños, R., Uyen, D., and Déramond, J., 2008, Paleozoic structural controls on shortening transfer in the Subandean foreland thrust system, Ene and southern Ucayali basins, Peru: Tectonics, v. 27, TC3009, doi:10.1029/2007TC002238.

Espurt, N., Barbarand, J., Roddaz, M., Brusset, S., Baby, P., Saillard, M., and Hermoza, W., 2011, A scenario for late Neogene Andean shortening transfer in the Camisea Subandean Zone (Peru, 12°S): Implications for growth of the northern Andean Plateau: Geological Society of America Bulletin, v. 123, no. 9–10, p. 2050–2068, doi:10.1130/B30165.1.

Exxon Production Research Co., 1995, Tectonic Map of the World: Tulsa, Oklahoma, American Association of Petroleum Geologists Foundation, scale 1:10,000,000, 20 sheets.

Fernandez-Seveso, F., and Tankard, A.J., 1995, Tectonics and stratigraphy of the late Paleozoic Paganzo Basin of western Argentina and its regional implications, *in* Tankard, A.J., Suárez Soruco, R., and Welsink, H.J., eds., Petroleum Basins of South America: American Association of Petroleum Geologists Memoir 62, p. 285–301.

França, A.B., Milani, E.J., Schneider, R.L., López, P., O., López M., J., Suárez Soruco, R., Santa Ana, H., Wiens, F., Ferreiro, O., Rossello, E.A., Bianucci, H.A., Flores, R.F.A., Vistalli, M.C., Fernandez-Seveso, F., Fuenzalida, R.P., and Muñoz, N., 1995, Phanerozoic correlation in southern South America, *in* Tankard, A.J., Suárez Soruco, R., and Welsink, H.J., eds., Petroleum Basins of South America: American Association of Petroleum Geologists Memoir 62, p. 129–161.

Franzese, J.R., Veiga, G.D., Schwarz, E., and Gómez-Pérez, I., 2006, Tectonostratigraphic evolution of a Mesozoic graben border system: The Chachil depocentre, southern Neuquén Basin, Argentina: Journal of the

Geological Society [London], v. 163, no. 4, p. 707–721, doi:10.1144/0016 -764920-082.

Gans, C.R., Beck, S.L., Zandt, G., Hersh, G., Alvarado, P., Anderson, M., and Linkimer, L., 2011, Continental and oceanic crustal structure of the Pampean flat slab region, western Argentina, using receiver function analysis: New high-resolution results: Geophysical Journal International, v. 186, p. 45–58, doi:10.1111/j.1365-246X.2011.05023.x.

Geologic Data Systems, 1997, Digital Compilation of Geologic Maps of South America (unpublished): Denver, Colorado. Digital compilation by Geologic Data Systems (see citations included within, dated 1943–1996), scale range 1:2,000,000 to 1:100,000.

Giambiagi, L., Alvarez, P.P., Godoy, E., and Ramos, V.A., 2003, The control of pre-existing extensional structures on the evolution of the southern sector of the Aconcagua fold and thrust belt, southern Andes: Tectonophysics, v. 369, p. 1–19, doi:10.1016/S0040-1951(03)00171-9.

Giambiagi, L., Mescua, J., Bechis, F., Martinez, A., and Folguera, A., 2011, Pre-Andean deformation of the Precordillera southern sector, southern central Andes: Geosphere, v. 7, no. 1, p. 219–239, doi:10.1130/GES00572.1.

Gillis, R.J., Horton, B.K., and Grove, M., 2006, Thermochronology, geochronology, and upper crustal structure of the Cordillera Real: Implications for Cenozoic exhumation of the central Andean plateau: Tectonics, v. 25, TC6007, doi:10.1029/2005TC001887.

Gil Rodriguez, W., Baby, P., Marocco, R., and Ballard, J.F., 1999, North-south structural evolution of the Peruvian Subandean zone, *in* Fourth International Symposium on Andean Geodynamics, Goettingen (Germany): Paris, Institut de Recherche pour le Développement, p. 278–282.

Gil Rodriguez, W., Baby, P., and Ballard, J.-F., 2001, Structure et contrôle paléogéographique de la zone subandine péruvienne: Earth and Planetary Sciences, v. 333, p. 741–748.

Giraudo, R., and Limachi, R., 2001, Pre-Silurian control in the genesis of the central and southern Bolivian foldbelt: Journal of South American Earth Sciences, v. 14, p. 665–680, doi:10.1016/S0895-9811(01)00068-2.

Grier, M., Salfity, J.A., and Allmendinger, R.W., 1991, Andean reactivation of the Cretaceous Salta rift, northwestern Argentina: Journal of South American Earth Sciences, v. 4, p. 351–372, doi:10.1016/0895-9811(91)90007-8.

Grimaldi, G.O., and Dorobek, S.L., 2011, Fault framework and kinematic evolution of inversion structures: Natural examples from the Neuquén Basin, Argentina: American Association of Petroleum Geologists Bulletin, v. 95, no. 1, p. 27–60, doi:10.1306/06301009165.

Gutscher, M.-A., Spakman, W., Bijwaard, H., and Engdahl, E.R., 2000, Geodynamics of flat subduction: Seismicity and tomographic constraints from the Andean margin: Tectonics, v. 19, no. 5, p. 814–833, doi:10.1029/1999TC001152.

Hérail, G., Baby, P., Oller, J., López, M., López, O., Salinas, R., Sempere, T., Beccar, G., and Toledo, H., 1990, Structure and kinematic evolution of the sub-Andean thrust system of Bolivia, *in* Olivier, R.A., and Vatin-Perignon, N., eds., First International Symposium on Andean Geodynamics, Grenoble: Paris, Editions de l'Orstom, p. 179–182.

Hermoza, W., Brusseta, S., Baby, P., Gil, W., Roddaz, M., Guerrero, N., and Bolaños, M., 2005, The Huallaga foreland basin evolution: Thrust propagation in a deltaic environment, northern Peruvian Andes: Journal of South American Earth Sciences, v. 19, p. 21–34, doi:10.1016/j.jsames.2004.06.005.

Hilley, G.E., and Coutand, I., 2010, Links between topography, erosion, rheological heterogeneity, and deformation in contractional settings: Insights from the central Andes: Tectonophysics, v. 495, p. 78–92, doi:10.1016/j.tecto.2009.06.017.

Hinsch, R., Krawczyk, C.M., Gaedicke, C., Giraudo, R., and Demuro, D., 2002, Basement control on oblique thrust sheet evolution: Seismic imaging of the active deformation front of the central Andes in Bolivia: Tectonophysics, v. 355, no. 1–4, p. 23–39, doi:10.1016/S0040-1951(02)00132–4.

Hoffman, P.F., 1991, Did the breakout of Laurentia turn Gondwanaland inside-out: Science, v. 252, no. 5011, p. 1409–1412, doi:10.1126/science.252.5011.1409.

Hoorn, C., and Wesselingh, F., eds., 2011, Amazonia, Landscape and Species Evolution: A Look into the Past: Chichester, UK, Wiley-Blackwell.

Hoorn, C., Wesselingh, F.P., ter Steege, H., Bermudez, M.A., Mora, A., Sevink, J., Sanmartín, I., Sanchez-Meseguer, A., Anderson, C.L., Figueiredo, J.P., Jaramillo, C., Riff, D., Negri, F.R., Hooghiemstra, H., Lundberg, J., Stadler, T., Särkinen, T., and Antonelli, A., 2010, Amazonia through time: Andean uplift, climate change, landscape evolution, and biodiversity: Science, v. 330, p. 927–931.

Horton, B.K., 1998, Sediment accumulation on top of the Andean orogenic wedge: Oligocene to late Miocene basins of the Eastern Cordillera, southern Bolivia: Geological Society of America Bulletin, v. 110, no. 9, p. 1174–1192, doi:10.1130/0016-7606(1998)110<1174:SAOTOT>2.3.CO;2.

Horton, B.K., 1999, Erosional control on the geometry and kinematics of thrust belt development in the central Andes: Tectonics, v. 18, p. 1292–1304, doi:10.1029/1999TC900051.

Horton, B.K., Saylor, J.E., Nie, J.S., Mora, A., Parra, M., Reyes-Harker, A., and Stockli, D.F., 2010, Linking sedimentation in the northern Andes to basement configuration, Mesozoic extension, and Cenozoic shortening: Evidence from detrital zircon U-Pb ages, Eastern Cordillera, Colombia: Geological Society of America Bulletin, v. 122, p. 1423–1442, doi:10.1130/B30118.1.

House, N.J., Carpenter, D.G., Cunningham, P.S., and Berumen, M., 1999, Influence of Paleozoic arches on structural style and stratigraphy in the Madre de Dios basin in southern Peru and northern Bolivia, *in* INGEPET '99-Exploration and Exploration of Petroleum and Gas: Perupetro (on CD-ROM).

Iaffa, D.N., Sàbat, F., Bello, D., Ferrer, O., Mon, R., and Gutierrez, A.A., 2011, Tectonic inversion in a segmented foreland basin from extensional to piggy back settings: The Tucumán basin in NW Argentina: Journal of South American Earth Sciences, v. 31, p. 457–474, doi:10.1016/j.jsames.2011.02.009.

International Commission on Stratigraphy, 2013, International Chronostratigraphic Chart Version 2013/01: www.stratigraphy.org (accessed 1 June 2013).

Isaacson, P.E., 1975, Evidence for a western extracontinental land source during the Devonian Period in the central Andes: Geological Society of America Bulletin, v. 86, no. 1, p. 39–46, doi:10.1130/0016-7606(1975)86<39 :EFAWEL>2.0.CO;2.

Isaacson, P.E., and Díaz Martínez, E., 1995, Evidence for a middle–late Paleozoic foreland basin and significant paleolatitudinal shift, central Andes, *in* Tankard, A.J., Suárez Soruco, R., and Welsink, H.J., eds., Petroleum Basins of South America: American Association of Petroleum Geologists Memoir 62, p. 231–249.

Isacks, B.L., 1988, Uplift of the central Andean plateau and bending of the Bolivian orocline: Journal of Geophysical Research, v. 93, 3211, doi:10.1029/JB093iB04p03211.

Jacques, J., 2003, A tectonostratigraphic synthesis of the Sub-Andean basins: Implications for the geotectonic segmentation of the Andean Belt, Peru: Journal of the Geological Society [London], v. 160, p. 687–701, doi:10.1144/0016-764902-088.

Jaillard, E., Herail, G., Monfret, T., and Worner, G., 2002, Andean geodynamics: Main issues and contributions from the 4th ISAG, Gottingen: Tectonophysics, v. 345, p. 1–15, doi:10.1016/S0040-1951(01)00203-7.

Jordan, T., Isacks, B., Allmendinger, R., Brewer, J., Ando, C., and Ramos, V.A., 1983, Andean tectonics related to geometry of subducted plates: Geological Society of America Bulletin, v. 94, p. 341–361, doi:10.1130/0016 -7606(1983)94<341:ATRTGO>2.0.CO;2.

Kammer, A., and Sánchez, J., 2006, Early Jurassic rift structures associated with the Soapaga and Boyacá faults of the Eastern Cordillera, Colombia: Sedimentological inferences and regional implications: Journal of South American Earth Sciences, v. 21, p. 412–422, doi:10.1016/j.jsames.2006.07.006.

Kay, S.M., Ramos, V.A., Mpodozis, C., and Sruoga, P., 1989, Late Paleozoic to Jurassic silicic magmatism at the Gondwana margin—Analogy to the Middle Proterozoic in North America: Geology, v. 17, no. 4, p. 324–328, doi:10.1130/0091-7613(1989)017<0324:LPTJSM>2.3.CO;2.

Kley, J., and Monaldi, C.R., 1998, Tectonic shortening and crustal thickness in the central Andes: How good is the correlation?: Geology, v. 26, no. 8, p. 723–726, doi:10.1130/0091-7613(1998)026<0723:TSACTI>2.3.CO;2.

Kley, J., and Monaldi, C.R., 2002, Tectonic inversion in the Santa Bárbara system of the central Andean foreland thrust belt, northwestern Argentina: Tectonics, v. 21, no. 6, 1061, doi:10.1029/2002TC902003.

Kley, J., Monaldi, C.R., and Salfity, J.A., 1999, Along-strike segmentation of the Andean foreland: Causes and consequences: Tectonophysics, v. 301, no. 1–2, p. 75–94, doi:10.1016/S0040-1951(98)90223-2.

Kley, J., Rossello, E., Monaldi, C.R., and Habighorst, B., 2005, Seismic and field evidence for selective inversion of Cretaceous normal faults, Salta rift, northwest Argentina: Tectonophysics, v. 399, p. 155–172, doi:10.1016/j.tecto.2004.12.020.

Laubacher, G., 1978, Géologie de la Cordillère Orientale et de l'Altiplano au Nord et Nord-Ouest du lac Titicaca (Pérou): Travaux et Documents de l'Orstom, Paris, v. 95, 217 p.

Legrand, D., Baby, P., Bondoux, F., Dorbath, C., Bès de Berc, S., and Rivad-eneira, M., 2005, The 1999–2000 seismic experiment of Macas swarm (Ecuador) in relation with rift inversion in Subandean foothills: Tectono-physics, v. 395, p. 67–80, doi:10.1016/j.tecto.2004.09.008.

Manceda, R., and Figueroa, D., 1995, Inversion of the Mesozoic Neuquén rift in the Malargüe fold and thrust belt, Mendoza, Argentina, in Tankard, A.J., Suárez Soruco, R., and Welsink, H.J., eds., Petroleum Basins of South America: American Association of Petroleum Geologists Memoir 62, p. 369–382.

Martinod, J., Husson, L., Roperch, P., Guillaume, B., and Espurt, N., 2010, Horizontal subduction zones, convergence velocity and the building of the Andes: Earth and Planetary Science Letters, v. 299, p. 299–309, doi:10.1016/j.epsl.2010.09.010.

Mathalone, J.M.P., and Montoya, R.M., 1995, Petroleum geology of the sub-Andean basins of Peru, in Tankard, A.J., Suárez Soruco, R., and Welsink, H.J., eds., Petroleum Basins of South America: American Association of Petroleum Geologists Memoir 62, p. 423–444.

McQuarrie, N., 2002a, Initial plate geometry, shortening variations, and evo-lution of the Bolivian orocline: Geology, v. 30, no. 10, p. 867–870, doi:10.1130/0091-7613(2002)030<0867:IPGSVA>2.0.CO;2.

McQuarrie, N., 2002b, The kinematic history of the central Andean fold-thrust belt, Bolivia: Implications for building a high plateau: Geological Soci-ety of America Bulletin, v. 114, no. 8, p. 950–963, doi:10.1130/0016 -7606(2002)114<0950:TKHOTC>2.0.CO;2.

McQuarrie, N., Barnes, J.B., and Ehlers, T.A., 2008a, Geometric, kinematic, and erosional history of the central Andean Plateau, Bolivia (15–17°S): Tectonics, v. 27, TC3007, doi:10.1029/2006TC002054.

McQuarrie, N., Ehlers, T.A., Barnes, J.B., and Meade, B.J., 2008b, Temporal variation in climate and tectonic coupling in the central Andes: Geology, v. 36, p. 999–1002, doi:10.1130/G25124A.1.

Mišković, A., Spikings, R.A., Chew, D.M., Košler, J., Ulianov, A., and Schaltegger, U., 2009, Tectonomagmatic evolution of Western Ama-zonia: Geochemical characterization and zircon U-Pb geochronologic constraints from the Peruvian Eastern Cordilleran granitoids: Geo-logical Society of America Bulletin, v. 121, no. 9–10, p. 1298–1324, doi:10.1130/B26488.1.

Mon, R., and Salfity, J.A., 1995, Tectonic evolution of the Andes of northern Argentina, in Tankard, A.J., Suárez Soruco, R., and Welsink, H.J., eds., Petroleum Basins of South America: American Association of Petroleum Geologists Memoir 62, p. 269–283.

Mora, A., Parra, M., Strecker, M.R., Kammer, A., Dimaté, C., and Rodrí-guez, F., 2006, Cenozoic contractional reactivation of Mesozoic exten-sional structures in the Eastern Cordillera of Colombia: Tectonics, v. 25, TC2010, doi:10.1029/2005TC001854.

Mora, A., Parra, M., Strecker, M.R., Sobel, E.R., Hooghiemstra, H., Torres, V., and Vallejo-Jaramillo, J., 2008, Climatic forcing of asymmetric orogenic evolution in the Eastern Cordillera of Colombia: Geological Society of America Bulletin, v. 120, p. 930–949, doi:10.1130/B26186.1.

Mora, A., Horton, B.K., Mesa, A., Rubiano, J., Ketcham, R.A., Parra, M., Blanco, V., Garcia, D., and Stockli, D.F., 2010, Migration of Cenozoic deformation in the Eastern Cordillera of Colombia interpreted from fis-sion track results and structural relationships: Implications for petroleum systems: American Association of Petroleum Geologists Bulletin, v. 94, no. 10, p. 1543–1580, doi:10.1306/01051009111.

Moretti, I., Baby, P., Mendez, E., and Zubieta, D., 1996, Hydrocarbon gen-eration in relation to thrusting in the Sub Andean zone from 18 to 22°S, Bolivia: Petroleum Geoscience, v. 2, p. 17–28, doi:10.1144/petgeo.2.1.17.

Mpodozis, C., and Ramos, V.A., 1989, The Andes of Chile and Argentina, in Ericksen, G.E., Cañas Pinochet, M.T., and Reinemund, J.A., eds., Geology of the Andes and Its Relation to Hydrocarbon and Mineral Resources: Houston, Texas, Circum-Pacific Council for Energy and Min-eral Resources, p. 59–90.

Norton, K., and Schlunegger, F., 2011, Migrating deformation in the central Andes from enhanced orographic rainfall: Nature Communications, v. 2, no. 584, doi:10.1038/ncomms1590.

Pankhurst, R.J., Rapela, C.W., Fanning, C.M., and Márquez, M., 2006, Gond-wanide continental collision and the origin of Patagonia: Earth-Science Reviews, v. 76, no. 3–4, p. 235–257, doi:10.1016/j.earscirev.2006.02.001.

Parra, M., Mora, A., Lopez, C., Rojas, L.E., and Horton, B.K., 2012, Detect-ing earliest shortening and deformation advance in thrust belt hinter-lands: Example from the Colombian Andes: Geology, v. 40, p. 175–178, doi:10.1130/G32519.1.

Pearson, D.M., Kapp, P., Reiners, P.W., Gehrels, G.E., Ducea, M.N., Pullen, A., Otamendi, J.E., and Alonso, R.N., 2012, Major Miocene exhumation by fault-propagation folding within a metamorphosed, early Paleozoic thrust belt: Northwestern Argentina: Tectonics, v. 31, no. 4, TC4023, doi:10.1029/2011TC003043.

Pfiffner, O.A., 1986, Evolution of the North Alpine foreland basin in the central Alps, in Allen, P.A., and Homewood, P., eds., Foreland Basins: Interna-tional Association of Sedimentologists Special Publication 8, p. 219–228.

Pindell, J.L., and Tabbutt, K.D., 1995, Mesozoic–Cenozoic Andean paleogeogra-phy and regional controls on hydrocarbon systems, in Tankard, A.J., Suárez Soruco, R., and Welsink, H.J., eds., Petroleum Basins of South America: American Association of Petroleum Geologists Memoir 62, p. 101–128.

Pindell, J., Kennan, L., Maresch, W.V., Stanek, K.-P., Draper, G., and Higgs, R., 2005, Plate-kinematics and crustal dynamics of circum-Caribbean arc-continent interactions: Tectonic controls on basin development in proto–Caribbean margins, in Avé Lallemant, H.G., and Sis-son, V.B., eds., Caribbean–South American Plate Interactions, Ven-ezuela: Geological Society of America Special Paper 394, p. 7–52, doi:10.1130/2005.2394(01).

Poulsen, C.J., Ehlers, T.A., and Insel, N., 2010, Onset of convective rainfall during gradual late Miocene rise of the central Andes: Science, v. 328, p. 490–493, doi:10.1126/science.1185078.

Ramos, V.A., 2008, The basement of the central Andes: The Arequipa and related terranes: Annual Review of Earth and Planetary Sciences, v. 36, p. 289–324, doi:10.1146/annurev.earth.36.031207.124304.

Ramos, V.A., 2009, Anatomy and global context of the Andes: Main geologic features and the Andean orogenic cycle, in Kay, S.M., Ramos, V.A., and Dickinson, W.R., eds., Backbone of the Americas: Shallow Subduction, Plateau Uplift, and Ridge and Terrane Collision: Geological Society of America Memoir 204, p. 31–65, doi:10.1130/2009.1204(02).

Ramos, V.A., and Folguera, A., 2005, Tectonic evolution of the Andes of Neu-quén: Constraints derived from the magmatic arc and foreland deforma-tion, in Veiga, G.D., Spalletti, L.A., Howell, J.A., and Schwarz, E., eds., The Neuquén Basin, Argentina: A Case Study in Sequence Stratigraphy and Basin Dynamics: Geological Society [London] Special Publication 252, p. 15–35, doi:10.1144/GSL.SP.2005.252.01.02.

Rapela, C.W., Pankhurst, R.J., Casquet, C., Baldo, E., Saavedra, J., and Galindo, C., 1998, Early evolution of the proto-Andean margin of South America: Geology, v. 26, no. 8, p. 707–710, doi:10.1130/0091-7613 (1998)026<0707:EEOTPA>2.3.CO;2.

Rich, J.L., 1934, Mechanics of low-angle overthrust faulting illustrated by Cumberland thrust block, Virginia, Kentucky and Tennessee: American Association of Petroleum Geologists Bulletin, v. 18, p. 1584–1596.

Romeuf, N., Aguirre, L., Carlier, G., Soler, P., Bonhomme, M., Elmi, S., and Salas, G., 1993, Present knowledge of the Jurassic volcanogenic forma-tions of southern coastal Peru, in II International Symposium on Andean Geodynamics: Paris, Editions de l'Office de la Recherche Scientifique et Technique d'Outre-mer, p. 437–440.

Romeuf, N., Aguirre, L., Soler, P., Féraud, G., Jaillard, E., and Ruffet, G., 1995, Middle Jurassic volcanism in the northern and central Andes: Revista Geológica de Chile, v. 22, p. 245–259.

Rosas, S., Fontbote, L., and Tankard, A., 2007, Tectonic evolution and paleogeography of the Mesozoic Pucara Basin, central Peru: Jour-nal of South American Earth Sciences, v. 24, p. 1–24, doi:10.1016/j .jsames.2007.03.002.

Salfity, J.A., 1985, Lineamientos transversales al rumbo andino en el noroeste argentino: Cuarto Congreso Geológico Chileno, Actas, v. 2, p. 119–227.

Salfity, J.A., and Marquillas, R.A., 1994, Tectonic and sedimentary evolution of the Cretaceous–Eocene Salta Group basin, Argentina, in Salfity, J.A., ed., Cretaceous Tectonics of the Andes: Braunschweig, Vieweg Verlag, p. 266–315.

Salfity, J.A., Azcuy, C.L., López, G.O., Valencio, D.A., Vilas, J.F., Cuerda, A., and Laffitte, G., 1987, Cuenca Tarija, in Archangelsky, S., ed., El Sistema Carbonífero en la República Argentina: SCCS-Project PICG 211, Aca-demia Nacional de Ciencias de Córdoba: Córdoba, República Argentina, Academia Nacional de Ciencias, p. 15–39.

Sarmiento-Rojas, L.F., Van Wess, J.D., and Cloetingh, S., 2006, Mesozoic transtensional basin history of the Eastern Cordillera, Colombian Andes: Inferences from tectonic models: Journal of South American Earth Sci-ences, v. 21, p. 383–411, doi:10.1016/j.jsames.2006.07.003.

Sempere, T., 1995, Phanerozoic evolution of Bolivia and adjacent regions, in Tankard, A.J., Suárez Soruco, R., and Welsink, H.J., eds., Petroleum

Basins of South America: American Association of Petroleum Geologists Memoir 62, p. 207–230.

Sempere, T., Carlier, G., Soler, P., Fornari, M., Carlotto, V., Jacay, J., Arispe, O., Neraudeau, D., Cardenas, J., Rosas, S., and Jimenez, N., 2002, Late Permian–Middle Jurassic lithosphere thinning in Peru and Bolivia and its bearing on Andean-age tectonics: Tectonophysics, v. 345, p. 153–181, doi:10.1016/S0040-1951(01)00211-6.

Starck, D., 1995, Silurian–Jurassic stratigraphy and basin evolution of northwestern Argentina, *in* Tankard, A.J., Suárez Soruco, R., and Welsink, H.J., eds., Petroleum Basins of South America: American Association of Petroleum Geologists Memoir 62, p. 251–267.

Starck, D., 2011, Cuenca Cretácica-Paleógena del Noroeste Argentino, *in* Kozlowski, E., Legarreta, L., Boll, A., and Marshall, P., eds., Simposio Cuencas Argentinas: Visión actual, VIII Congreso de Exploración y Desarrollo de Hidrocarburos, Mar del Plata: Buenos Aires, Instituto Argentino del Petróleo y del Gas, p. 1–48.

Tankard, A.J., Uliana, M.A., Welsink, H.J., Ramos, V.A., Turic, M., França, A.B., Milani, E.J., de Brito Neves, B.B., Eyles, N., Skarmeta, J., Santa Ana, H., Wiens, F., Cirbián, M., López P., O., De Wit, G.J.B., Machacha, T., and Miller, R.McG., 1995, Tectonic controls of basin evolution in southwestern Gondwana, *in* Tankard, A.J., Suárez Soruco, R., and Welsink, H.J., eds., Petroleum Basins of South America: American Association of Petroleum Geologists Memoir 62, p. 5–52.

Tassara, A., Götze, H.-J., Schmidt, S., and Hackney, R., 2006, Three-dimensional density model of the Nazca plate and the Andean continental margin: Journal of Geophysical Research–Solid Earth, v. 111, no. B9, doi:10.1029/2005JB003976.

Tectonic Analysis, Ltd., 2006, Exploration Framework Atlas Series, Volume 3: Central Andes: West Sussex, UK, Tectonic Analysis Ltd., 67 p.

Thomas, W.A., and Astini, R.A., 1996, The Argentine Precordillera: A traveler from the Ouachita embayment of North American Laurentia: Science, v. 273, no. 5276, p. 752–757, doi:10.1126/science.273.5276.752.

Toro, J., Roure, F., Bordas-Le Floch, N., Le Cornec-Lance, S., and Sassi, W., 2004, Thermal and kinematic evolution of the Eastern Cordillera fold and thrust belt, Colombia: American Association of Petroleum Geologists Hedberg Series 1, p. 79–115.

Uba, C.E., Kley, J., Strecker, M.R., and Schmitt, A., 2009, Unsteady evolution of the Bolivian Subandean thrust belt: The role of enhanced erosion and clastic wedge progradation: Earth and Planetary Science Letters, v. 281, p. 134–146, doi:10.1016/j.epsl.2009.02.010.

Vergani, G.D., Tankard, A.J., Belotti, H.J., and Welsink, H.J., 1995, Tectonic evolution and paleogeography of the Neuquén Basin, Argentina, *in* Tankard, A.J., Suárez Soruco, R., and Welsink, H.J., eds., Petroleum Basins of South America: American Association of Petroleum Geologists Memoir 62, p. 383–402.

Villagómez, D., Spikings, R., Seward, D., Magna, T., Winkler, W., and Kammer, A., 2008, Thermotectonic history of the Northern Andes, *in* 7th International Symposium on Andean Geodynamics: Nice, France, p. 573–576.

Wagner, L., Beck, S., and Zandt, G., 2005, Upper mantle structure in the south central Chilean subduction zone (30° to 36°S): Journal of Geophysical Research, v. 110, B01308, doi:10.1029/2004JB003238.

Welsink, H.J., Franco, M.A., and Oviedo, G.C., 1995, Andean and pre-Andean deformation, Boomerang Hills area, Bolivia, *in* Tankard, A.J., Suárez Soruco, R., and Welsink, H.J., eds., Petroleum Basins of South America: American Association of Petroleum Geologists Memoir 62, p. 481–499.

Williams, K.E., 1995, Tectonic subsidence analysis and Paleozoic paleogeography of Gondwana, *in* Tankard, A.J., Suárez Soruco, R., and Welsink, H.J., eds., Petroleum Basins of South America: American Association of Petroleum Geologists Memoir 62, p. 79–100.

Wine, G., Martinez, E., Fernandez, J., Calderón, Y., and Galdos, C., 2003, Significant reserves may remain in parts of Peru's Marañon basin: Oil & Gas Journal, v. 101, no. 23, http://www.ogj.com/articles/print/volume-101/issue-23/exploration-development/significant-reserves-may-remain-in-parts-of-perus-maran-basin.html.

Zapata, T., and Allmendinger, R.W., 1996, The thrust front zone of the Precordillera, Argentina: A thick-skinned triangle zone: American Association of Petroleum Geologists Bulletin, v. 80, no. 3, p. 359–381.

Zapata, T., and Folguera, A., 2005, Tectonic evolution of the Andean fold and thrust belt of the southern Neuquén Basin, Argentina, *in* Veiga, G.D., Spalletti, L.A., Howell, J.A., and Schwarz, E., eds., The Neuquén Basin, Argentina: A Case Study in Sequence Stratigraphy and Basin Dynamics: Geological Society [London] Special Publication 252, p. 37–56, doi:10.1144/GSL.SP.2005.252.01.03.

Zappettini, E.O., Chernicoff, C.J., Santos, J.O.S., Dalponte, M., Belousova, E., and McNaughton, N., 2012, Retrowedge-related Carboniferous units and coeval magmatism in the northwestern Neuquén province, Argentina: International Journal of Earth Sciences, v. 101, no. 8, p. 2083–2104, doi:10.1007/s00531-012-0774-3.

MANUSCRIPT ACCEPTED BY THE SOCIETY 3 JUNE 2014
MANUSCRIPT PUBLISHED ONLINE 23 SEPTEMBER 2014

The Geological Society of America
Memoir 212
2015

U-Pb zircon geochronology of Neoproterozoic–Paleozoic sandstones and Paleozoic plutonic rocks in the Central Andes (21°S–26°S)

Jesse C. Einhorn*
George E. Gehrels[†]
Antoine Vernon[§]
Peter G. DeCelles
Department of Geosciences, University of Arizona, 1040 E. 4th Street, Tucson, Arizona 85721, USA

ABSTRACT

The crystalline basement of the Central Andes between 21°S and 26°S consists of a variety of Neoproterozoic–Paleozoic arc-type and basinal assemblages. We characterize these assemblages through analysis of U-Th-Pb ages of zircons sampled from 16 different plutonic suites and from 21 different sandstones in northern Argentina and Chile. The ages of igneous zircons show that magmatism occurred in three main phases: ca. 550 Ma (late Neoproterozoic); 490–464 Ma (Late Cambrian to Middle Ordovician); and 318–264 Ma (late Carboniferous–Late Permian). Detrital zircon ages are mainly 600–450 Ma from Paleozoic strata all across the orogen, reflecting derivation primarily from local Neoproterozoic, Cambrian, and Ordovician magmatic constructs. These relations suggest that crystalline basement in this portion of the Andes was assembled within a broad extensional convergent margin system during early Paleozoic time. Because similar arc and basinal assemblages characterize much of the Terra Australis orogen, it is not possible to constrain the degree of tectonic mobility within this convergent system.

INTRODUCTION

The Central Andes are underlain by a variety of Proterozoic through Mesozoic rocks that constitute the crystalline basement of the modern Andean orogen. These rocks occur outboard of the South American craton (Fig. 1), within the Terra Australis orogen (Coney, 1992; Cawood, 2005), which formed along the Gondwana margin during late Proterozoic and early Mesozoic

time. As with most segments of this orogen, the tectonic evolution of Central Andean crystalline basement is a matter of long-standing debate. Contesting hypotheses regard the western margin of Gondwana as either an assemblage of accreted terranes (Bahlburg and Herve, 1997; Rapalini, 2005; Ramos et al., 1986; Loewy et al., 2004; Ramos, 2008a, 2010), or a "mobile belt" of recycled but not significantly displaced continental fragments (Bock et al., 2000; Lucassen et al., 2000; Franz et al., 2006;

*Current address: ExxonMobil Exploration Company, 233 Benmar Drive, Houston, Texas 77060, USA.
[†]Corresponding author.
[§]Current address: Shell International Exploration and Production, Inc., 3333 Highway 6 South, Houston, Texas 77082-3101, USA.

Einhorn, J.C., Gehrels, G.E., Vernon, A., and DeCelles, P.G., 2015, U-Pb zircon geochronology of Neoproterozoic–Paleozoic sandstones and Paleozoic plutonic rocks in the Central Andes (21°S–26°S), *in* DeCelles, P.G., Ducea, M.N., Carrapa, B., and Kapp, P.A., eds., Geodynamics of a Cordilleran Orogenic System: The Central Andes of Argentina and Northern Chile: Geological Society of America Memoir 212, p. 115–124, doi:10.1130/2015.1212(06). For permission to copy, contact editing@geosociety.org. © 2014 The Geological Society of America. All rights reserved.

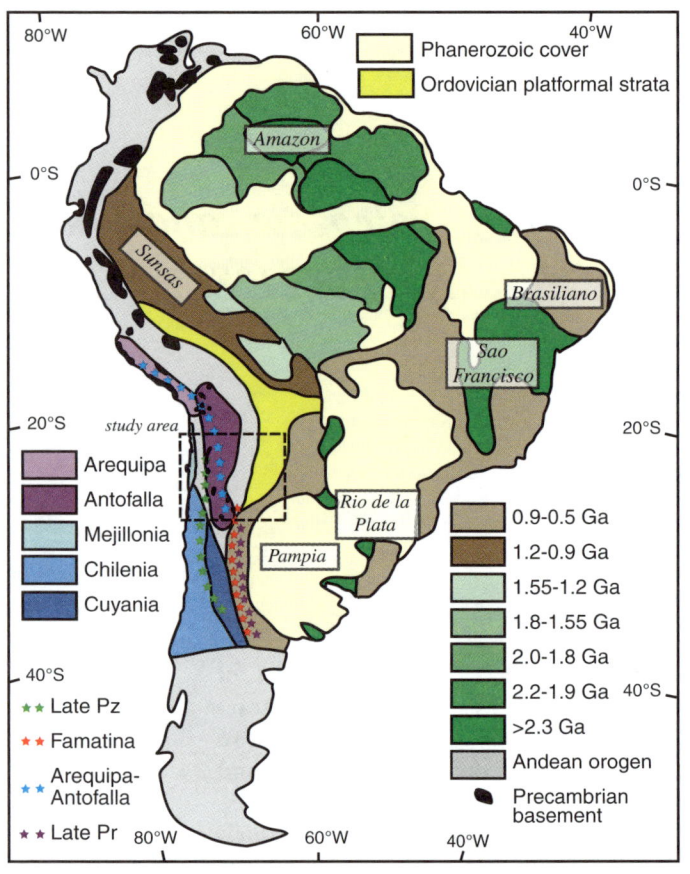

Figure 1. Sketch map showing the distribution of basement provinces of the South American craton, the occurrence of Precambrian (Pr) basement and undifferentiated Paleozoic (Pz) through Holocene rocks of the Andean orogen, and terranes within and adjacent to the study area. Figure is adapted from Goodwin (1996), Ramos (2008a, 2008b, 2010), and Bahlburg et al. (2009).

Cawood, 2005). Certainly, one of the complicating factors in resolving the pre-Cenozoic history of the region is the presence of widespread Cenozoic volcanic rocks that obscure the lithology and age of underlying basement.

In this paper, we use U-Th-Pb geochronology to help reconstruct the Neoproterozoic–Paleozoic tectonic history of the Andean orogen in northern Argentina and Chile (between latitudes 21°S and 26°S). Ages have been determined on 16 samples of Paleozoic plutonic rocks (440 analyses), and on detrital zircon grains from 21 samples of Neoproterozoic–Paleozoic sandstones (1517 analyses; Table 1). Each sample was collected from the typical lithology of a regionally extensive unit shown on the geologic map of the region (Reutter et al., 1994).

GEOLOGIC AND TECTONIC SETTING

Pre-Tertiary rocks in northern Argentina and adjacent regions of Chile have been divided into several main crustal blocks or terranes (Fig. 2). As described next, these include the Pampia, Antofalla, Cuyania, Chilenia, and Mejillonia terranes. Additional tectonic assemblages include a sequence of mainly Lower Paleozoic platformal strata that accumulated along the western margin of the South American craton, Cambrian and Ordovician igneous rocks of the Famatina and Tastil magmatic arcs, and broad regions of marine strata of early Paleozoic age east of the Antofalla terrane and of late Paleozoic age west of the Antofalla terrane.

Pampia Terrane

In the study area, the Pampia terrane consists of Neoproterozoic–Lower Cambrian marine strata of the Puncoviscana Formation that are unconformably overlain by Cambrian and Ordovician marine platformal strata and intruded by Cambrian–Ordovician plutons (Ramos et al., 2010). The Puncoviscana Formation consists of a thick (>3 km) succession of Neoproterozoic and Lower Cambrian quartzite, slate, conglomerate, and local carbonate and volcanic rocks that were deposited in a marine setting (Aceñolaza et al., 2002). Published interpretations of the tectonic setting of the Puncoviscana Formation include a passive margin (Aceñolaza et al., 2002), a N-S–trending rift (back-arc basin) along the western margin of the Pampia terrane (Ježek et al., 1985; Schwartz and Gromet, 2004; Ramos, 2008a), a forearc basin (Rapela et al., 2007), and a foreland basin (Kraemer et al., 1995; Keppie and Bahlburg, 1999; and Escayola et al., 2011). Immature recycled orogenic sandstone petrographic compositions as well as major- and trace-element data would seem to argue in favor of a tectonically active setting (e.g., Schwartz and Gromet, 2004; Zimmermann, 2005). Detrital zircon ages, with peaks at ca. 520 Ma, 660–610 Ma, and 1050–990 Ma, suggest derivation from the Pampia craton to the east (Ramos, 2008a; Adams et al., 2008, 2011; Miller et al., 2011; Escayola et al., 2011). Ramos (2008a) suggested that the Puncoviscana basin: (1) originated during Neoproterozoic time as a rift basin, separating the Pampia craton from the Antofalla and related terranes; (2) received most sediment from the Pampia terrane and its collisional boundary with the Rio de la Plata craton to the east; and (3) closed during latest Neoproterozoic–Early Cambrian time due to the accretion of the Antofalla and related terranes. The overlying Cambrian–Ordovician marine strata are described as a separate assemblage.

Famatina Magmatic Arc

The Famatina magmatic arc consists of plutonic rocks that intrude the western margin of the Pampia terrane (Fig. 2). Ages of the igneous rocks range from 490 to 464 Ma according to the compilation by Ramos et al. (2010) and more recent data from Ducea et al. (2010). The Famatina arc is widely interpreted to have formed in response to subduction along the western margin of Pampia prior to Ordovician arrival of the Cuyania and Antofalla terranes (e.g., Ramos, 2008a). In contrast, Omarini et al. (1999) and Ramos (2010) suggested that

TABLE 1. SAMPLE INFORMATION

IGNEOUS SAMPLES

Sample number	Sample name	Age (±2σ, Ma)	Latitude (°S)	Longitude (°W)
1	09AA19	311.7 ± 4.0	21.79821	69.11607
2	CL18	317.9 ± 12.2	22.36513	68.8383
3	CL19	292.4 ± 4.3	22.61917	68.81056
4	CL20	289.9 ± 4.6	23.04097	68.92259
5	CL29	293.3 ± 4.7	23.53725	69.03474
6	CL22	277.5 ± 4.3	23.66744	68.33968
7	08AA24	272.7 ± 4.1	24.95022	68.10528
8	08AA28	264.3 ± 4.3	24.57688	67.72931
9	08AA32	487.3 ± 6.3	24.62289	67.32837
10	08AA16	489.1 ± 9.5	24.25609	66.3814
11	08AA13	550.3 ± 9.0	24.48633	65.8911
12	09JA09	475.3 ± 6.5	25.2405	66.34455
13	09JA10	464.5 ± 6.5	25.38734	66.342015
14	08JA01	489.6 ± 6.8	25.44798	66.38937
15	09JA01	470.7 ± 6.7	25.47417	66.36774
16	09JA11	468.9 ± 9.5	25.49273	66.23559

DETRITAL SAMPLES

Sample number	Sample name	Latitude (°S)	Longitude (°W)	Depositional age	Maximum depositional age (Ma)
17	09AA41	24.77195	70.3809	Devonian	471
18	09AA40	23.77343	70.0372	Mid- to late Paleozoic	318
19	CL17-d	22.04404	69.78458	Carboniferous	461
20	09AA38	21.70621	69.62718	Devonian	467
21	09AA02	21.73096	69.12171	Upper Paleozoic	463
22	09AA25	23.98681	68.51179	Devonian	437
23	09AA26	23.76867	68.01766	Permian-Triassic	253
24	09AA31	23.61709	67.2921	Ordovician	466
25	08AA18	23.98373	66.88155	Ordovician	465
26	08AA37	24.41237	66.94588	Ordovician	453
27	PUNC1	24.34963	66.10309	Proterozoic	540
28	08DP23	25.0916	66.28395	Proterozoic	546
29	08DP17	25.11038	66.26045	Proterozoic	540
30	Meson	23.69771	65.72921	Cambrian	484
31	Victoria	23.69437	65.71798	Ordovician	539
32	08AA10	23.18207	65.20881	Cambrian	534
33	SM20070629-1	23.18323	65.05817	Carboniferous	379
34	08AA08	23.19126	65.05386	Carboniferous	447
35	08AA09	23.19126	65.05386	Carboniferous	500
36	08AA07	23.28024	65.01972	Devonian	405
37	08AA06	23.28682	65.01895	Silurian	462

the arc may have been associated with the outboard Antofalla terrane, rather than Pampia.

Antofalla Terrane

The Antofalla terrane consists of metasedimentary and metaplutonic rocks that are higher in metamorphic grade and more variable in protolith than the surrounding marine strata (Loewy et al., 2004; Ramos, 2008a). These rocks yield U-Pb ages as old as 1.88 Ga, with common ages of 1.5–1.4 Ga, 1.2–1.0 Ga, 0.7–0.6 Ga, and 0.5–0.4 Ga (Loewy et al., 2004). Exposures of these rocks are limited due to widespread Cenozoic volcanic and sedimentary cover: The outline of the terrane on Figure 2, adapted from Ramos (2008a), is based largely on gravity data from Götze et al. (1994) and Omarini et al. (1999). Keppie and Ramos (1999)

argued that the Antofalla and Arequipa terranes (Fig. 1) share a similar history and referred to the composite as the Arequipa-Antofalla terrane.

Cuyania Terrane

The Cuyania terrane is located outboard of the Famatina arc and the Pampia terrane in the south-central Andes (Astini et al., 1995; Thomas and Astini, 2003; Vujovich et al., 2004; Astini and Davila, 2004). The Cuyania terrane locally contains Mesoproterozoic igneous rocks that are interpreted as basement to extensive sequences of Lower Paleozoic (and possible Neoproterozoic) strata (Ramos, 2008b). Cuyania has been widely interpreted as an exotic tectonic fragment in the Andes, most likely derived from southern Laurentia, on the basis of stratigraphic,

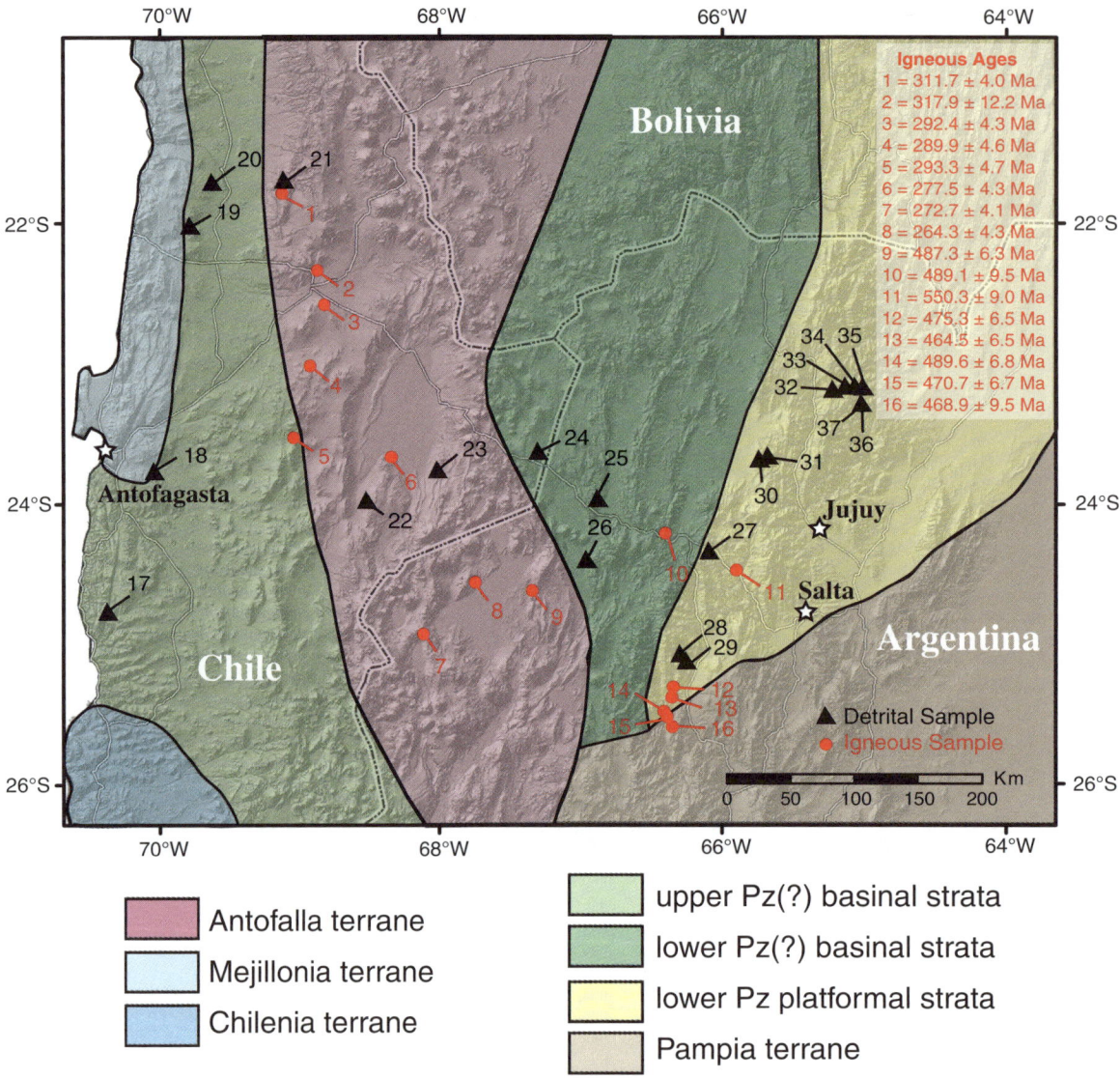

Figure 2. Sketch map showing the location of our U-Pb geochronologic samples in relation to the main terranes and tectonic assemblages in the study area (adapted from Ramos, 2008a). Weighted mean ages are reported for all igneous samples (uncertainties at 2σ). Pz—Paleozoic.

faunal, and paleomagnetic features (e.g., Astini and Thomas, 1999; Thomas and Astini, 2003; Ramos, 2010). Naipauer et al. (2010) reported U-Pb ages of detrital zircons that also support the interpretation that Cuyania has been displaced from its Laurentian site of origin.

Chilenia Terrane

The Chilenia terrane consists of rare exposures of metamorphic basement with Mesoproterozoic (ca. 1.06 Ga) ages, rare Lower Paleozoic strata, widespread Upper Paleozoic plutonic rocks, and both marine and continental strata (Ramos, 2008b; Bahlburg et al., 2009). Chilenia is suspected to have formed in proximity to Laurentia, given the presence of Grenville-age

metamorphic rocks and the lack of evidence for Brasiliano-age deformation (Keppie and Ramos, 1999). Accretion against Cuyania, and hence the margin of South America, is interpreted to have occurred during Devonian time, given the overlap of Carboniferous strata across both terranes (Ramos et al., 1986; Ramos, 2008b). Upper Paleozoic plutonic rocks and deformation record subduction along the Gondwana margin that continued into Mesozoic time (Bahlburg et al., 2009).

Mejillonia Terrane

Mejillonia was recognized as a separate tectonic fragment by Ramos (2008a) based on the occurrence of amphibolite-facies gneisses and schists, and the occurrence of ca. 525 Ma granitoids near Antofagasta (Fig. 2). Most of the terrane, however, consists of Upper Paleozoic marine strata that may have formed in a subduction complex (Bahlburg et al., 2009).

Paleozoic Platformal and Basinal Strata

Paleozoic marine strata are interpreted to occur in three separate belts: (1) an eastern sequence of Cambrian and Ordovician marine platformal strata that rests unconformably along the western margin of the Pampia terrane (e.g., Kumpa and Sanchez, 1988); (2) a thick sequence of marine strata of known and suspected Ordovician age that separates the Pampia margin from the Antofalla terrane (Bahlburg et al., 1987; Bahlburg, 1998); and (3) marine strata of known and suspected Carboniferous–Permian age that separate the Antofalla terrane from outboard Mejillonia and Chilenia terranes (Fig. 2; Bahlburg and Breitkreuz, 1993; Ramos, 2008a). These strata consist largely of quartz-arenitic turbidite sandstones and associated finer-grained lithofacies that formed in deep marine basins (Ramos, 2008a).

ANALYTICAL METHODS

The primary contribution of this study is determination of U-Pb ages from 16 igneous samples and 21 detrital samples, which were analyzed using the methods outlined next and by Gehrels et al. (2008). Sample locations are shown in Figure 2 and described in Table 1.

Zircon crystals were extracted from samples by traditional methods of crushing and grinding, followed by separation with a Wilfley table, heavy liquids, and a Frantz magnetic separator. Samples were processed such that as many zircons as possible were retained in the final heavy mineral fraction. For detrital samples, a large split of these grains (generally ~2000 grains) was incorporated into a 1″ (2.54 cm) epoxy mount together with fragments of our Sri Lanka zircon standard. For igneous samples, ~50 zircon grains were selected and mounted in a similar fashion. Mounts were polished down to a depth of ~20 μm, imaged optically and by cathodoluminescence (CL), and cleaned prior to isotopic analysis.

U-Pb geochronology of zircons was conducted by laser ablation–multicollector–inductively coupled plasma–mass spectrometry (LA-MC-ICP-MS) at the Arizona LaserChron Center. Two different mass spectrometers were used to analyze samples during the course of this study. In both cases, the analyses involved ablation of zircon with an excimer laser (wavelength of 193 nm) using a spot diameter of 30 μm. The ablated material was carried in helium into the plasma source of either a GVI Isoprobe or a Nu high-resolution ICP-MS. Both were equipped with flight tubes of sufficient width that U, Th, and Pb isotopes could be measured simultaneously. When using the GVI isoprobe, all measurements were made in static mode, using 10^{11} Ω Faraday detectors for ^{238}U, ^{232}Th, ^{208}Pb, and ^{206}Pb, a 10^{12} Ω Faraday collector for ^{207}Pb, and an ion-counting channel for ^{204}Pb. Ion yields were ~1.0 mV per ppm. Each analysis consisted of one 12 s integration on peaks with the

laser off (for backgrounds), twelve 1 s integrations with the laser firing, and a 30 s delay to purge the previous sample and prepare for the next analysis. The ablation pit was ~12 μm in depth. When using the Nu HR ICP-MS, measurements were made in static mode, using Faraday detectors with 3×10^{11} Ω resistors for ^{238}U, ^{232}Th, ^{208}Pb-^{206}Pb, and discrete dynode ion counters for ^{204}(Pb+Hg) and ^{202}Hg. Ion yields were ~0.8 mV per ppm. Each analysis consisted of one 15 s integration on peaks with the laser off (for backgrounds), fifteen 1 s integrations with the laser firing, and a 30 s delay to purge the previous sample and prepare for the next analysis. The ablation pit was ~15 μm in depth.

For each analysis, the errors in determining ^{206}Pb/^{238}U and ^{206}Pb/^{204}Pb result in a measurement error of ~1%–2% (at 2σ level) in the ^{206}Pb/^{238}U age. The errors in measurement of ^{206}Pb/^{207}Pb and ^{206}Pb/^{204}Pb also result in ~1%–2% (at 2σ level) uncertainty in age for grains that are older than 1.0 Ga, but they are substantially larger for younger grains due to the low intensity of the ^{207}Pb signal. For most analyses, the crossover in precision of ^{206}Pb/^{238}U and ^{206}Pb/^{207}Pb ages occurs at ca. 1.0 Ga.

Common Pb correction was accomplished by first subtracting ^{204}Hg from the measured ^{204}(Pb+Hg), based on measured ^{202}Hg and natural ^{202}Hg/^{204}Hg, and then using the resulting ^{204}Pb. The initial Pb composition was assumed from Stacey and Kramers (1975), with propagated uncertainties of 1.0 (e.g., 18.6 ± 1.0) for ^{206}Pb/^{204}Pb and 0.3 for ^{207}Pb/^{204}Pb. On both machines, an in-run analysis of fragments of a large zircon crystal (generally every fifth measurement) with known age of 563.5 ± 3.2 Ma (2σ error; Gehrels et al., 2008) was used to correct for this fractionation. The uncertainty resulting from the calibration correction is generally 1%–2% (2σ) for both ^{206}Pb/^{207}Pb and ^{206}Pb/^{238}U ages.

Interpreted ages are based on ^{206}Pb/^{238}U for younger than 1000 Ma grains and on ^{206}Pb/^{207}Pb for older than 1000 Ma grains. Analyses that are >20% discordant (by comparison of ^{206}Pb/^{238}U and ^{206}Pb/^{207}Pb ages) or >5% reverse discordant are not considered further. In addition, analyses in which the ^{206}Pb/^{238}U ratio changed in an unusual fashion during data acquisition were rejected, as such changes are generally due to ablation across an age boundary, fracture, or inclusion, or a domain with significant Pb loss.

The analytical data and calculated ages for each analysis are reported in Tables DR1 and DR2.[1] These files include concordia plots of all samples and age probability plots for the detrital samples. The reported ages of igneous samples were determined from the weighted mean (Ludwig, 2003) of the ^{206}Pb/^{238}U ages of concordant and overlapping analyses. Uncertainties of these ages are reported at the 2σ level and include all internal and external uncertainties. Highly discordant analyses were excluded from age calculations. These ages are reported in Table 1, on Figure 2, and in Table DR1 (see footnote 1).

---

[1]GSA Data Repository Item 2015002, Tables containing U-Pb data for igneous and detrital samples, is available at www.geosociety.org/pubs/ft2015.htm, or on request from editing@geosociety.org or Documents Secretary, GSA, P.O. Box 9140, Boulder, CO 80301-9140, USA.

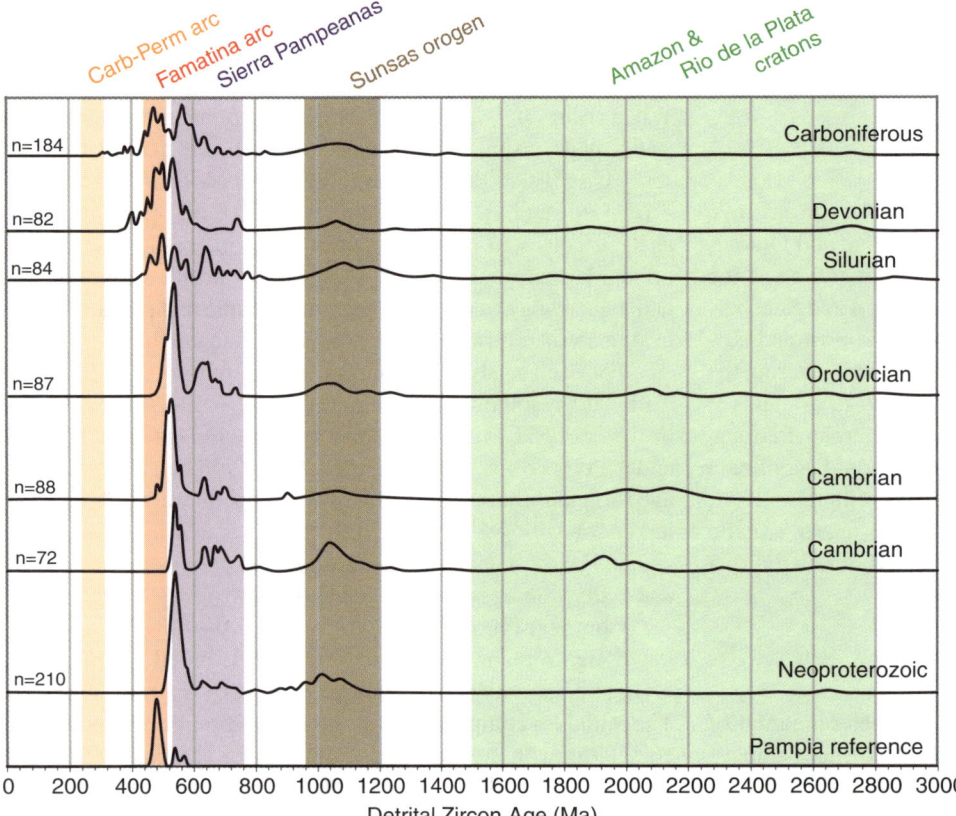

Figure 3. Relative age-probability curves for Neoproterozoic through Carboniferous strata of the passive margin sequence that accumulated on the Pampia craton. Shown for reference are 64 ages from igneous rocks of the Famatina arc and the western Pampia craton (ages from Ramos et al., 2010; Ducea et al., 2010; this study).

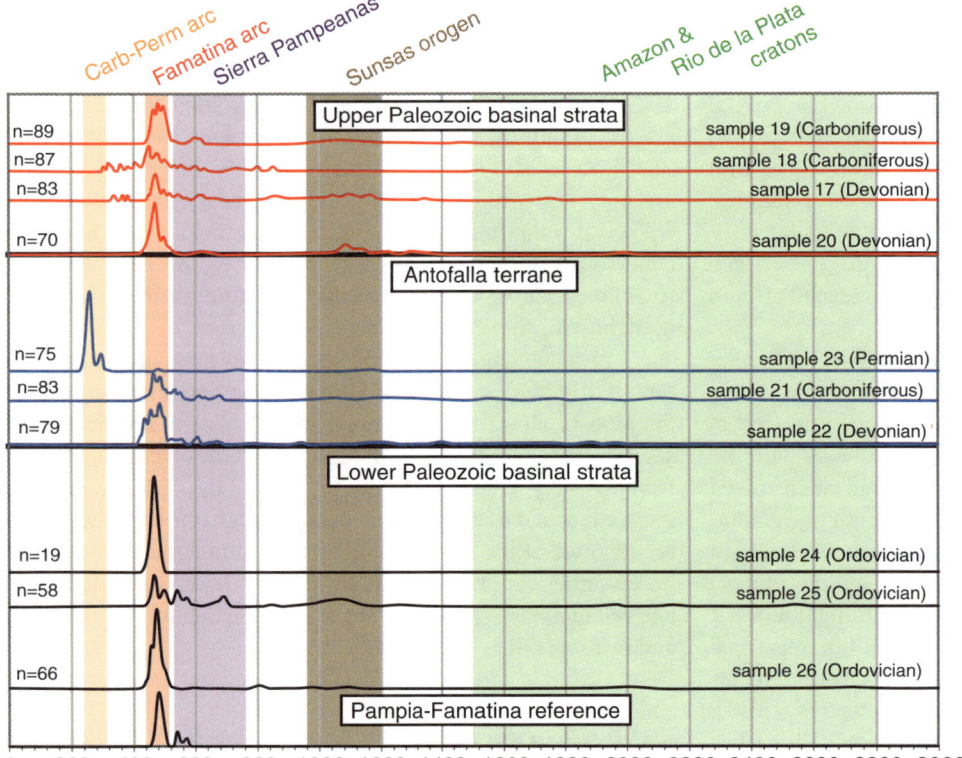

Figure 4. Relative age-probability curves for Paleozoic strata of the Antofalla terrane and adjacent basinal sequences, all located outboard of the Pampia craton. Shown for reference are 64 ages from igneous rocks of the Famatina arc and the western Pampia craton (ages from Ramos et al., 2010; Ducea et al., 2010; this study).

U-Pb ages from detrital zircons are shown on relative age-probability diagrams in Figures 3 and 4. These diagrams show each age and its uncertainty (for measurement error only) as a normal distribution and sum all ages from a sample into a single curve. Each curve is normalized by the number of constituent analyses so that they all contain the same area (Normalized Probability Plot program from www.laserchron.org). Figure 3 shows age distributions from strata belonging to the Lower Paleozoic platformal assemblage, and Figure 4 shows age distributions from all strata to the west. Figure 5 shows textures typical of detrital zircons analyzed in this study, with examples of grains with (Fig. 5E) and without (Fig. 5F) inherited cores. Analyses were conducted in the largest coherent domain present in each crystal.

For detrital samples, the ages of probability peaks are reported, along with the number of grains that make up each peak, using the "agepick" routine at www.laserchron.org. This routine determines the age of each probability peak that consists of at least three grains that overlap at 2σ. The youngest peak in age probability is interpreted as the maximum depositional age (Dickinson and Gehrels, 2009) and is reported in Table 1 and Table DR2 (see footnote 1).

ANALYTICAL RESULTS

Igneous Samples

Our samples of igneous rocks belong to three age groups: 318–264 Ma (late Carboniferous–Late Permian); 490–464 Ma (Late Cambrian to Middle Ordovician); and ca. 550 Ma (late Neoproterozoic; using the time scale of Ogg et al., 2008).

Samples 1 through 8 (Fig. 2; Table 1) are late Paleozoic in age and intrude mainly the Antofalla terrane. These samples are mostly devoid of inherited components, except for sample 1, which has numerous inherited cores ranging in age from 470 to 1250 Ma. Most samples of this suite reveal only igneous-type oscillatory zonation in CL images (Figs. 5A and 5B).

Cambrian–Ordovician ages were acquired from the Antofalla terrane (sample 9), the basinal assemblage to the east (sample 10), and from Pampia basement (sample 12–16; Fig. 2). All of these samples yield a cluster of relatively young ages that are interpreted as igneous components, and most samples contain abundant xenocrystic cores that yield ages ranging from ca. 500 Ma to >2.0 Ga. Most inherited components, however, yield ages of 700–500 Ma and 1200–900 Ma. Figures 5C and 5D show typical CL textures of grains with and without inherited cores.

The only sample that yielded a Neoproterozoic age, sample 11 (ca. 550 Ma), is clearly part of the basement to Pampia. No inherited components were observed in this sample.

Detrital Samples

All samples from the succession of Paleozoic platformal strata yielded a dominant cluster of 600–500 Ma ages and subordinate 1200–900 Ma ages. These grains were most likely derived from crystalline rocks of the Pampia terrane (Ramos, 2010), with possible input from more distant Sunsas (1.2–0.9 Ga) and Brasiliano (0.6–0.5 Ga) age provinces. These ages are similar to the set of ages reported previously from the Puncoviscana Formation (Adams et al., 2008, 2011). Silurian through Carboniferous strata in this assemblage also yield a significant proportion of 500–420 Ma grains, which could have been shed from any of the Ordovician igneous assemblages noted earlier. These grains are mostly devoid of inherited components (e.g., Fig. 5), suggesting that older basement is not extensive in whichever plutonic belt(s) contributed the grains.

Most of the samples from the Antofalla terrane and the adjacent basinal assemblages (Fig. 4) are dominated by 500–420 Ma grains, except one sample of Permian age (sample 23), which is dominated by 270–240 Ma detrital zircons. The predominance

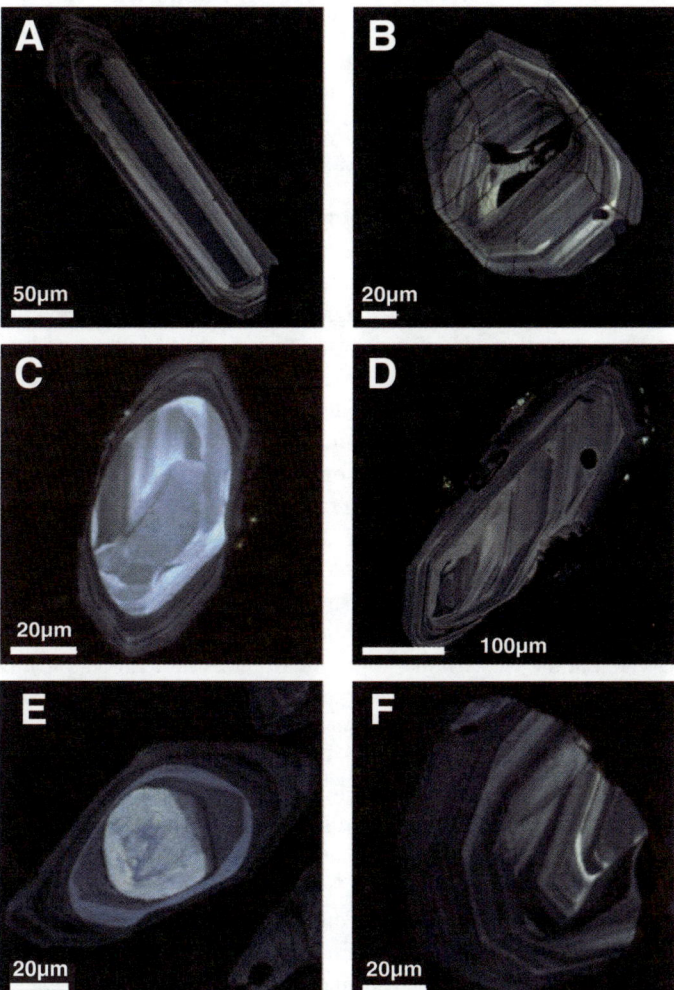

Figure 5. Cathodoluminescence (CL) images of zircons: (A, B) examples of the oscillatory zoning typical of most zircons in late Paleozoic samples (grains shown are from sample 4); (C, D) examples of zircon grains with and without inherited cores from sample 9, which are typical of the early Paleozoic plutons; (E, F) examples of detrital zircon grains with and without inherited cores (from sample 37).

Einhorn et al.

of 500–420 Ma grains relative to 600–500 Ma and older grains suggests that most of the detritus was shed from local arc sources within Antofalla and the adjacent basinal terranes, with little input from cratonal regions to the east. The scarcity of older detrital grains, together with the rare occurrence of xenocrystic inclusions in the early Paleozoic grains (e.g., Fig. 5), suggests that Proterozoic basement is not widespread within or beneath the Antofalla terrane and adjacent basinal assemblages. The lack of abundant detrital and magmatic ages in the 420–320 Ma range in Carboniferous and Permian samples supports the conclusion of Ramos (2008a, 2008b) that mid-Paleozoic magmatism was not widespread along the proto-Andean margin.

TECTONIC IMPLICATIONS

Our new geochronologic data, together with previously available U-Pb geochronology and geologic relations from the region (summarized by Damm et al., 1990, 1994; Lucassen et al., 2000, 2011; Pankhurst et al., 2000; Ramos, 2008a, 2008b, 2010; Bahlburg et al., 2009; Ramos et al., 2010; Casquet et al., 2012), suggest that three distinct magmatic episodes occurred along this portion of the proto-Andean margin, separated by two magmatic lulls. Following is a summary of these magmatic flare-ups and lulls, and their possible tectonic significance.

1. During Neoproterozoic–Cambrian time, plutons of 554–515 Ma age (Ramos et al., 2010; Ducea et al., 2010; this study) were emplaced along the western margin of the Pampia terrane. This magmatism presumably occurred in response to west-facing subduction (e.g., Cawood, 2005; Cawood and Buchan, 2007; Ramos, 2008a, 2010; Casquet et al., 2012). The Puncoviscana basin formed outboard of this magmatic arc and was filled primarily with sediment derived from the nearby magmatic arc as well as broader regions of the Pampia terrane (Fig. 3). These relations support the interpretation of Rapela et al. (2007) that the Puncoviscana basin formed as a forearc basin, rather than as a foreland basin (e.g., Ramos, 2008a).

2. A lull in magmatism occurred between ca. 515 Ma and ca. 490 Ma, marked in some areas by deformation associated with the Tilcaran-Pampean orogeny (Ramos, 2008a).

3. Magmatism resumed from ca. 490 Ma to ca. 460 Ma, with plutons of this age occurring along the western margin of Pampia, in the Famatina arc (which occurs along the western margin of the Pampia terrane), and in the Antofalla terrane. It is not clear whether these were separate magmatic arcs juxtaposed after Middle Ordovician time, or different exposures of a single very broad Ordovician arc system.

During the Ordovician magmatism, marine sediment was accumulating on the Pampia margin, in the adjacent marine basin, on the Antofalla terrane, and possibly in the marine basin outboard of the Antofalla terrane. All of these assemblages contain significant arc-derived detritus, and the off-shelf strata contain little else. The magmatic arc was surrounded by an apron of its own erosional detritus, perhaps in response to widespread crustal extension during slab rollback.

4. Magmatism waned after ca. 460 Ma, with little activity until ca. 320 Ma. Strata that accumulated during this time appear to have been shed primarily from the outboard Ordovician arcs (e.g., Antofalla), with no input from Pampia or other parts of the South American craton. The lack of 460–320 Ma magmatism is indicated by the dominance of Ordovician grains in strata of Ordovician, Silurian, and Devonian age (Table 1). Cessation of magmatism during Middle Ordovician time is interpreted to have coincided with arrival of the Cuyania terrane along the continental margin (Ramos, 2008a, 2008b).

5. The final magmatic episode represented in our data set extended from ca. 318 Ma to ca. 264 Ma. This magmatism is widely interpreted to have formed in a west-facing magmatic arc (e.g., Cawood, 2005; Cawood and Buchan, 2007; Ramos, 2008b; Bahlburg et al., 2009). The occurrence of Mejillonia within or outboard of this late Paleozoic convergent margin assemblage is enigmatic: Was it accreted along the margin during late Paleozoic time? Is it a western extension of Antofalla, serving as basement to the late Paleozoic arc system? Unfortunately, additional geochronologic data will be needed to answer these questions.

CONCLUSIONS

Our data contribute to the growing body of geochronologic data that shed light on the magmatic and tectonic evolution of this portion of the Andean orogen. These data provide strong support for the three phases of magmatism recognized by previous workers, and they also demonstrate that ca. 490 Ma magmatism occurred in western Pampia, the Famatina arc, and the Antofalla terrane. The detrital zircon data also demonstrate that these Ordovician magmatic arc systems dominated the provenance of Paleozoic marine strata all across the orogen—very little sediment was shed from the South American craton.

Collectively, these relations suggest that during early Paleozoic time, this portion of the proto-Andean margin consisted of a broad and low-lying magmatic arc that was blanketed by marine clastic strata derived almost entirely from local arc constructs. Given that the western margin of the South American craton was also blanketed by arc-derived clastic strata, at least the inboard margin of this magmatic arc system must have been in proximity to the continental margin.

Unfortunately, the available data are not yet sufficient to determine the degree of tectonic mobility within this broad early Paleozoic arc system. Complicating factors are the scarcity of exposures of Precambrian basement rocks outboard of the Pampia margin, the occurrence of Cambrian–Ordovician magmatism in several different regions within the orogen, and the widespread occurrence of locally derived marine strata within and between the magmatic constructs. Together, the available detrital geochronological data can be explained by a model incorporating only minimal allochthoneity of paleogeographic domains, and they do not require exotic provenance explanations. Such relations are one of the hallmarks of the Terra Australis orogen (Coney, 1992; Cawood, 2005).

ACKNOWLEDGMENTS

We thank Max Zattin, Dave Pearson, and Scott McBride for contributing unpublished data for use in this study. Thanks also go to Ricardo Alonso for consultations and logistical support. This project was funded by the Convergent Orogenic Systems Analysis collaboration between the Department of Geosciences at the University of Arizona and the ExxonMobil Exploration and Upstream Research Companies. Support for the Arizona LaserChron Center is provided by the National Science Foundation Instrumentation and Facilities Program (grant EAR-1032156). Thoughtful reviews by Friedrich Lucassen, Peter Cawood, Heinrich Bahlburg, and volume editor Paul Kapp helped us to substantially improve the manuscript.

REFERENCES CITED

Aceñolaza, F., Miller, H., and Toselli, A.J., 2002, Proterozoic–Early Paleozoic evolution in western South America—A discussion: Tectonophysics, v. 354, no. 1–2, p. 121–137, doi:10.1016/S0040-1951(02)00295-0.

Adams, C.J., Miller, H., Toselli, A.J., and Griffin, W.L., 2008, The Puncoviscana Formation of northwest Argentina: U-Pb geochronology of detrital zircons and Rb-Sr metamorphic ages and their bearing on its stratigraphic age, sediment provenance, and tectonic setting: Neues Jahrbuch für Geologie und Palaeontologie, Abhandlungen, v. 247, no. 3, p. 341–352, doi:10.1127/0077-7749/2008/0247-0341.

Adams, C.J., Miller, H., Acenolaza, F.G., Toselli, A.J., and Griffin, W.L., 2011, The Pacific Gondwana margin in the late Neoproterozoic–early Paleozoic: Detrital zircon U-Pb ages from metasediments in northwest Argentina reveal their maximum age, provenance and tectonic setting: Gondwana Research, v. 19, no. 1, p. 71–83, doi:10.1016/j.gr.2010.05.002.

Astini, R.A., and Davila, F.M., 2004, Ordovician back arc foreland and Ocloyic thrust belt development on the western Gondwana margin as a response to Precordillera terrane accretion: Tectonics, v. 23, TC4008, doi:10.1029/2003TC001620.

Astini, R.A., and Thomas, W.A., 1999, Origin and evolution of the Precordillera terrane of western Argentina: A drifted Laurentian orphan, in Ramos, V.A., and Keppie, J.D., eds., Laurentia-Gondwana Connections before Pangea: Geological Society of America Special Paper 336, p. 1–20.

Astini, R.A., Benedetto, J.L., and Vaccari, N.E., 1995, The early Paleozoic evolution of the Argentina Precordillera as a Laurentian rifted, drifted, and collided terrane: A geodynamic model: Geological Society of America Bulletin, v. 107, p. 253–273, doi:10.1130/0016-7606(1995)107<0253:TEPEOT>2.3.CO;2.

Bahlburg, H., 1998, The geochemistry and provenance of Ordovician turbidites in the Argentine Puna, in Pankhurst, R.J., and Rapela, C.W., eds., The Proto-Andean Margin of Gondwana: Geological Society, London, Special Publication 142, p. 127–142.

Bahlburg, H., and Breitkreuz, C., 1993, Differential response of a Devonian–Carboniferous platform–deeper basin system to sea-level change and tectonics, N. Chilean Andes: Basin Research, v. 5, p. 21–40, doi:10.1111/j.1365-2117.1993.tb00054.x.

Bahlburg, H., and Herve, F., 1997, Geodynamic evolution and tectonostratigraphic terranes of northwestern Argentina and northern Chile: Geological Society of America Bulletin, v. 109, p. 869–884, doi:10.1130/0016-7606(1997)109<0869:GEATTO>2.3.CO;2.

Bahlburg, H., Breitkreuz, C., and Zeil, W., 1987, Paleozoic basin development in northern Chile (21°–27°S): Geologische Rundschau, v. 76, p. 633–646, doi:10.1007/BF01821095.

Bahlburg, H., Vervoort, J.D., Du Frane, S.A., Bock, B., Augustsson, C., and Reimann, C., 2009, Timing and crust formation and recycling in accretionary orogens: Insights learned from the western margin of South America: Earth-Science Reviews, v. 97, p. 215–241, doi:10.1016/j.earscirev.2009.10.006.

Bock, B., Bahlburg, H., Worner, G., and Zimmermann, U., 2000, Tracing crustal evolution in the southern Central Andes from Late Precambrian to Permian with geochemical and Nd and Pb isotope data: The Journal of Geology, v. 108, p. 515–535, doi:10.1086/314422.

Casquet, C., Rapela, C.W., Pankhurst, R.J., Baldo, E.G., Galindo, C.M., Dahlquist, J.A., and Saavedra, J., 2012, A history of Proterozoic terranes in southern South America: From Rodinia to Gondwana: Geoscience Frontiers, v. 3, no. 2, p. 137–145, doi:10.1016/j.gsf.2011.11.004.

Cawood, P.A., 2005, The Terra Australis orogen: Rodinia break-up and development of the Pacific and Iapetus margins of Gondwana during the Neoproterozoic and Paleozoic: Earth-Science Reviews, v. 69, p. 249–279, doi:10.1016/j.earscirev.2004.09.001.

Cawood, P.A., and Buchan, C., 2007, Linking accretionary orogenesis with supercontinent assembly: Earth-Science Reviews, v. 82, p. 217–256, doi:10.1016/j.earscirev.2007.03.003.

Coney, P.J., 1992, The Lachlan belt of eastern Australia and circum-Pacific tectonic evolution: Tectonophysics, v. 214, no. 1–4, p. 1–25, doi:10.1016/0040-1951(92)90187-B.

Damm, K.W., Pichowiak, S., Harmon, R.S., Todt, W., Kelley, S., Omarini, R., and Niemeyer, H., 1990, Pre-Mesozoic evolution of the Andes; The basement revisited, in Kay, S.M., and Rapela, C.W., eds., Plutonism from Antarctica to Alaska: Geological Society of America Special Paper 241, p. 101–126.

Damm, K.W., Harmon, R.S., and Kelley, S., 1994, Some isotope and geochemical constraints on the origin and evolution of the Central Andean basement (19°–24°S), in Reutter, K.J., Scheuber, E., and Wigger, P.J., eds., Tectonics of the Southern Central Andes: Heidelberg, Germany, Springer, p. 263–275.

Dickinson, W.R., and Gehrels, G.E., 2009, Use of U-Pb ages of detrital zircons to infer maximum depositional ages of strata: A test against a Colorado Plateau Mesozoic database: Earth and Planetary Science Letters, v. 288, p. 115–125, doi:10.1016/j.epsl.2009.09.013.

Ducea, M.N., Otamendi, J., Stair, K.M., Bergantz, G., and Gehrels, G., 2010, Timing constraints on building an intermediate plutonic arc crustal section: U-Pb zircon geochronology of the Sierra Valle Fértil–La Huerta, Famatinian arc, Argentina: Tectonics, v. 29, TC4002, doi:10.1029/2009TC002615.

Escayola, M.P., van Staal, C.R., and Davis, W.J., 2011, The age and tectonic setting of the Puncoviscana Formation in northwestern Argentina: An accretionary complex related to Early Cambrian closure of the Puncoviscana Ocean and accretion of the Arequipa-Antofalla block: Journal of South American Earth Sciences, v. 32, p. 438–459, doi:10.1016/j.jsames.2011.04.013.

Franz, G., Lucassen, F., Kramer, W., Trumbull, R.B., Romer, R.L., Wilke, H.-G., Viramonte, J.G., Becchio, R., and Siebel, W., 2006, Crustal evolution at the Central Andean continental margin: A geochemical record of crustal growth, recycling and destruction, in Oncken, O., Chong, G., Franz, G., Giese, P., Götze, H.-J., Ramos, V.A., Strecker, M.R., and Wigger, P., eds., The Andes—Active Subduction Orogeny: Heidelberg, Germany, Springer-Verlag, p. 45–64.

Gehrels, G.E., Valencia, V., and Ruiz, J., 2008, Enhanced precision, accuracy, efficiency, and spatial resolution of U-Pb ages by laser ablation–multicollector–inductively coupled plasma–mass spectrometry: Geochemistry Geophysics Geosystems, v. 9, Q03017, doi:10.1029/2007GC001805.

Goodwin, A.M., 1996, Principles of Precambrian Geology: New York, Academic Press, 327 p.

Götze, H.J., Lahmeyer, B., Schmidt, S., and Strunk, S., 1994, The lithospheric structure of the Central Andes (20°–26°S) as inferred from interpretation of regional gravity, in Reutter, K.J., Scheuber, E., and Wigger, P.J., eds., Tectonics of the Southern Central Andes: Berlin, Springer, p. 7–21.

Ježek, P., Willner, A.P., Aceñolaza, F., and Miller, H., 1985, The Puncoviscana trough—A large basin of late Precambrian to Early Cambrian age on the Pacific edge of the Brasilian Shield: Geologische Rundschau, v. 74, p. 573–584, doi:10.1007/BF01821213.

Keppie, J.D., and Bahlburg, H., 1999, The Puncoviscana Formation of northwestern and central Argentina: Passive margin or foreland basin deposit?, in Ramos, V.A., and Keppie, J.D., eds., Laurentia-Gondwana Connections before Pangea: Geological Society of America Special Paper 336, p. 139–143.

Keppie, J.D., and Ramos, V.A., 1999, Odyssey of terranes in the Iapetus and Rheic Oceans during the Paleozoic, in Ramos, V.A., and Keppie, J.D., eds., Laurentia-Gondwana Connections before Pangea: Geological Society of America Special Paper 336, p. 267–276.

Kraemer, P.E., Escayola, M.P., and Martino, R.D., 1995, Hipotesis sobre la evolucion tectonica Neoproterozoica de las Sierras Pampeanas de Cordoba

(30°40′–32°40′), Argentina: Revista de la Asociación Geológica Argentina, v. 50, p. 47–59.

Kumpa, M., and Sanchez, M.C., 1988, Geology and sedimentology of the Cambrian Grupo Mesón (NW Argentina), *in* Bahlburg, H., Breitkreuz, C., and Giese P., eds., The Southern Central Andes: Contributions to Structure and Evolution of an Active Continental Margin: Heidelberg, Germany, Springer, Lecture Notes in Earth Sciences, v. 17, p. 39–54.

Loewy, S.L., Connelly, J.N., and Dalziel, I.W.D., 2004, An orphaned basement block: The Arequipa-Antofalla basement of the Central Andean margin of South America: Geological Society of America Bulletin, v. 116, p. 171–187, doi:10.1130/B25226.1.

Lucassen, F., Becchio, R., Wilke, H.G., Franz, G., Thirlwall, M.F., Viramonte, J., and Wemmer, K., 2000, Proterozoic–Paleozoic development of the basement of the Central Andes (18–26°S)—A mobile belt of the South American craton: Journal of South American Earth Sciences, v. 13, p. 697–715, doi:10.1016/S0895-9811(00)00057-2.

Lucassen, F., Becchio, R., and Franz, G., 2011, The early Palaeozoic high-grade metamorphism at the active continental margin of West Gondwana in the Andes (NW Argentina/N Chile): International Journal of Earth Sciences, v. 100, p. 445–463, doi:10.1007/s00531-010-0585-3.

Ludwig, K.R., 2003, Isoplot 3.00: Berkeley Geochronology Center Special Publication 4, 77 p.

Miller, H., Adams, C., Acenolaza, F.G., and Toselli, A.J., 2011, Evolution of exhumation and erosion in western West Gondwanaland as recorded by detrital zircons of late Neoproterozoic and Cambrian sedimentary rocks of NW and central Argentina: International Journal of Earth Sciences, v. 100, no. 2–3, p. 619–629, doi:10.1007/s00531-010-0559-5.

Naipauer, M., Vujovich, G.I., Cingolani, C.A., and McClelland, W.C., 2010, Detrital zircon analysis from the Neoproterozoic–Cambrian sedimentary cover (Cuyania terrane), Sierra de Pie de Palo, Argentina: Evidence of a rift and passive margin system?: Journal of South American Earth Sciences, v. 29, no. 2, p. 306–326, doi:10.1016/j.jsames.2009.10.001.

Ogg, J.G., Ogg, G., and Gradstein, F.M., 2008, The Concise Geologic Time Scale: Cambridge, UK, Cambridge University Press, 184 p.

Omarini, R.H., Sureda, R.J., Götze, J.-H., Seilacher, A., and Pfluger, F., 1999, Puncoviscana folded belt in northwestern Argentina: Testimony of late Proterozoic Rodinia fragmentation and pre-Gondwana collisional episodes: International Journal of Earth Sciences, v. 88, p. 76–97, doi:10.1007/s005310050247.

Pankhurst, R.J., Rapela, C.W., and Fanning, C.M., 2000, Age and origin of coeval TTG, I- and S-type granites in the Famatinian belt of NW Argentina: Transactions of the Royal Society of Edinburgh–Earth Sciences, v. 91, p. 151–168, doi:10.1017/S0263593300007343.

Ramos, V.A., 2008a, The basement of the Central Andes: The Arequipa and related terranes: Annual Review of Earth and Planetary Sciences, v. 36, p. 289–324, doi:10.1146/annurev.earth.36.031207.124304.

Ramos, V.A., 2008b, Field trip guide: Evolution of the Pampean flat slab region over the shallowly subducting Nazca plate, *in* Kay, S.M., and Ramos, V.,

eds., Field Guides to the Backbone of the Americas in the Southern and Central Andes: Geological Society of America Field Guide 13, p. 77–116.

Ramos, V.A., 2010, The Grenville-age basement of the Andes: Journal of South American Earth Sciences, v. 29, p. 77–91, doi:10.1016/j.jsames.2009.09.004.

Ramos, V.A., Jordan, T.E., Allmendinger, R.W., Mpodozis, C., Kay, S.M., Cortés, J.M., and Palma, M.A., 1986, Paleozoic terranes of the central Argentine-Chilean Andes: Tectonics, v. 5, p. 855–880, doi:10.1029/TC005i006p00855.

Ramos, V.A., Vujovich, G., Martino, R., and Otamendi, J., 2010, Pampia: A large cratonic block missing in the Rodinia supercontinent: Journal of Geodynamics, v. 50, p. 243–255, doi:10.1016/j.jog.2010.01.019.

Rapalini, A.E., 2005, The accretionary history of southern South America from the latest Proterozoic to the late Paleozoic: Some paleomagnetic constraints, *in* Vaughan, A.P.M., Leat, P.T., and Pankhurst, R.J., eds., Terrane Processes at the Margins of Gondwana: Geological Society, London, Special Publication 246, p. 305–328.

Rapela, C.W., Pankhurst, R.J., Casquet, C., Fanning, M., Baldo, E.G., Gonzalez-Casado, J.M., Galindo, C., and Dahlquist, J., 2007, The Rio de la Plata craton and the assembly of SW Gondwanaland: Earth-Science Reviews, v. 83, p. 49–82, doi:10.1016/j.earscirev.2007.03.004.

Reutter, K.J., Doebel, R., Bogdanic, T., and Kley, J., 1994, Geological map of the Central Andes between 20°S and 26°S, *in* Reutter, K.-J., Scheuber, E., and Wigger, P., eds., Tectonics of the Southern Central Andes: Berlin, Springer-Verlag, scale 1:1,000,000.

Schwartz, J.J., and Gromet, L.P., 2004, Provenance of Late Proterozoic–Early Cambrian basin, Sierra de Cordoba, Argentina: Precambrian Research, v. 129, p. 1–21, doi:10.1016/j.precamres.2003.08.011.

Stacey, J.S., and Kramers, J.D., 1975, Approximation of terrestrial lead isotope evolution by a two-stage model: Earth and Planetary Science Letters, v. 26, p. 207–221, doi:10.1016/0012-821X(75)90088-6.

Thomas, W.A., and Astini, R.A., 2003, Ordovician accretion of the Argentine Precordillera terrane to Gondwana: A review: Journal of South American Earth Sciences, v. 16, p. 67–79, doi:10.1016/S0895-9811(03)00019-1.

Vujovich, G.I., Fernandes, L., and Ramos, V.A., 2004, Cuyania: An exotic block to Gondwana: Gondwana Research, v. 7, p. 1005–1007, doi:10.1016/S1342-937X(05)71080-7.

Zimmermann, U., 2005, Provenance studies of very low- to low-grade metasedimentary rocks of the Puncoviscana complex, northwest Argentina, *in* Vaughan, A.P.M., Leat, P.T., and Pankhurst, R.J., eds., Terrane Processes at the Margins of Gondwana: Geological Society, London, Special Publication 246, p. 381–416.

MANUSCRIPT ACCEPTED BY THE SOCIETY 3 JUNE 2014
MANUSCRIPT PUBLISHED ONLINE 23 SEPTEMBER 2014

The Geological Society of America
Memoir 212
2015

The origin and petrologic evolution of
the Ordovician Famatinian-Puna arc

M.N. Ducea*

Department of Geosciences, University of Arizona, Tucson, Arizona 85721, USA, and
Universitatea Bucuresti, Facultatea de Geologie Geofizica, Strada N. Balcescu Nr 1, Bucuresti, Romania

J.E. Otamendi

Consejo Nacional de Investigaciones Científicas y Técnicas, Argentina, and
Departamento de Geología, Universidad Nacional de Río Cuarto, X5804BYA Río Cuarto, Argentina

G.W. Bergantz

Department of Earth and Space Sciences, University of Washington, Seattle, Washington 98195-1310, USA

D. Jianu
L. Petrescu
Universitatea Bucuresti, Facultatea de Geologie Geofizica, Strada N. Balcescu Nr 1, Bucuresti, Romania

ABSTRACT

Elemental chemistry, radiogenic isotopic data, and zircon U-Pb inheritance patterns for the Famatinian-Puna arc suggest that the primary petrogenetic process operating in the arc was mixing between subarc mantle-derived gabbroic magmas and metasedimentary materials without a substantial component of lower-crustal continental basement rocks. This mixing is observable in the field and evident in variations of chemical elemental parameters and isotopic ratios, revealing that hybridization coupled with fractionation of magmas took place in the upper 25 km of the crust. Intermediate and silicic plutonic rocks of the Famatinia-Puna arc formed in a subduction setting where the thermal and material input of mantle-derived magmas promoted fusion of fertile metasedimentary rocks and favored mixing of gabbroic and dioritic magmas with crustal granitic melts. Whole-rock geochemical and isotopic data for the Famatinian-Puna magmatic belt as a whole demonstrate that the petrologic model studied in detail in the Sierra Valle Fértil–La Huerta section has the potential to explain generation of plutonic and volcanic rocks across the Early Ordovician western Gondwanan proto-Pacific margin. This example further underscores the significance of passive-margin sedimentary accumulations in generating continental arcs.

*ducea@email.arizona.edu

Ducea, M.N., Otamendi, J.E., Bergantz, G.W., Jianu, D., and Petrescu, L., 2015, The origin and petrologic evolution of the Ordovician Famatinian-Puna arc, *in* DeCelles, P.G., Ducea, M.N., Carrapa, B., and Kapp, P.A., eds., Geodynamics of a Cordilleran Orogenic System: The Central Andes of Argentina and Northern Chile: Geological Society of America Memoir 212, p. 125–138, doi:10.1130/2015.1212(07). For permission to copy, contact editing@geosociety.org. © 2014 The Geological Society of America. All rights reserved.

INTRODUCTION

Interest in subduction-related magmatic belts centers on their importance to the study of: (1) the generation of continental-scale intermediate magmatism, (2) the causal mechanism of volcanic hazards and major earthquakes, (3) the nature of hydrothermal ore systems, and (4) the evolution and growth of the continental crust (Gill, 1981; Tatsumi and Eggins, 1995; Ducea and Barton, 2007). Ancient magmatic belts from destructive plate margins are dominated by Cordilleran-style intermediate and silicic plutonic rocks. The extent to which the dominant intermediate and silicic plutonic rocks from destructive plate margins reflect either addition of new material to the crust, recycling of preexisting crust, or a mixture between the two has been strongly debated (for a recent discussion, see Brown, 2013). The issue is central for deciphering the chemical evolution and net growth rate of the continental crust (Davidson and Arculus, 2006; Kemp et al., 2009; Cawood et al., 2012).

This study reviews isotopic data from one of Earth's largest magmatic arcs, the Ordovician Famatinian-Puna arc from NW Argentina and Bolivia. Our initial focus was the deepest known crustal section of the arc, which makes up the Sierra Valle Fértil–La Huerta, one of the Sierras Pampeanas (Fig. 1). The petrogenesis of Early Ordovician plutonism from Valle Fértil and La Huerta was deciphered through the combined study of petrography, whole-rock geochemistry, and radiogenic isotopes and U-Pb zircon geochronology (Otamendi et al., 2009a, 2012; Ducea et al., 2010). Subsequently, we extended our focus to various exposures of the Famatinian arc located to the north into the Puna Plateau. Preliminary observations from there, as well as previously published data, are used to interpret the petrogenetic and tectonic evolution of the arc. The principal conclusions drawn here are that the arc was active and in flare-up mode for a relatively short period of time, and it formed via a combination of subcraton (isotopically "enriched") mantle-derived magmas and melts derived from a thick and melt-fertile suite of passive-margin sedimentary rocks, regionally known as the Puncoviscana Formation and regionally correlative units. There is no evidence for the existence of a cratonal lower crust under the Famatinian arc.

GEOLOGICAL SETTING

Summary of the Geologic Evolution of the Late Neoproterozoic to Ordovician Western Gondwana Margin

The proto-Andean margin of western Gondwana has experienced fairly continuous subduction with relatively short interruptions during terrane accretions (Cawood, 2005) or periods when the margin was a transform fault since the latest Proterozoic–early Paleozoic. The Pampean magmatic arc was built on the once-passive margin of a western Gondwanan landmass when subduction began at ca. 550 Ma (Rapela et al., 1998; Schwartz et al., 2008). Subduction-related magmatic activity paused between ca. 515 and 495 Ma, stepped out to the west, and resumed on the western margin during the growth and evolution of the Famatinian magmatic arc (Pankhurst et al., 1998). The lack of subduction-related arc magmatism during the Late Cambrian was speculatively interpreted to have been caused by either accretion of the Pampean terrane to the proto-Pacific Gondwanan margin (Rapela et al., 1998) or a ridge-trench collision on the border of the Gondwanan landmasses (Schwartz et al., 2008). Current understanding shows that the Pampean thermo-tectonic orogeny was short-lived (ca. 530–515 Ma) and affected thick Neoproterozoic–Early Cambrian sedimentary sequences (Martino et al., 2009; Drobe et al., 2009). These thick, mostly marine sedimentary sequences, which are regionally referred to as the Puncoviscana Formation, were deposited in basins onto and outboard of landmasses from western Gondwana (e.g., Ježek et al., 1985; Pearson et al., 2012). The Pampean arc now comprises a N-S–trending belt from southern Córdoba (~33°S) into southern Bolivia (~22°S; Aceñolaza, 2003; Drobe et al., 2009; see also Fig. 1).

The Famatinian arc started at ca. 495 Ma, presumably when subduction was reestablished along the outboard boundary of the Pampean arc and "orogeny," leaving behind in its back arc the crystalline packages metamorphosed during the Early Cambrian (Fig. 1). The southern segment (28°S to 38°S, present-day coordinates) of the Famatinian arc was closed during the middle Ordovician (beginning at ca. 465 Ma), when a continental microplate that had rifted from North American Laurentian landmasses collided against the proto-Pacific Gondwana margin (Thomas and Astini, 1996; Ramos et al., 1996).

The Famatinian arc is exposed for ~2000 km along the strike of the modern central Andes, and the transition from plutonic to volcanic Famatinian rocks can be followed over large regions in northwestern Argentina (Rapela et al., 1992; Toselli et al., 1996; Pankhurst et al., 1998; Coira et al., 1999). The deepest plutonic levels of the arc are currently exposed along a roughly N-S–striking wide belt extending ~600 km between 28°S and 33°S (Fig. 1). Complementary Early Ordovician shallow-emplaced plutonic and eruptive igneous rocks interbedded with sedimentary rocks ("Faja Eruptiva") are found within the Puna-Altiplano region (Coira et al., 1999; Viramonte et al., 2007) and in the Sierra de Famatina (Mannheim and Miller, 1996), between 22°S and 28°S. The wall rocks of all the Famatinian plutonic rocks are supracrustal sedimentary packages consisting largely of siliciclastic sediments with subordinate interlayered carbonate beds—the Puncoviscana Formation and its metamorphic equivalents (Caminos, 1979; Ježek et al., 1985; Pearson et al., 2012). As shallower levels of the Famatinian paleo-arc crust are exposed northward along strike, the non- to weakly metamorphosed sedimentary stratigraphic units mapped in the Puna and northern Sierras Pampeanas, which are correlative to the metamorphosed strata that crop to the south (Aceñolaza et al., 2000), are uncovered. Late Neoproterozoic–Early Cambrian thick turbiditic packages and Late Cambrian shallow-marine sediments are the most likely protoliths to the metamorphic units hosting the Famatinian arc plutonic rocks (e.g., Aceñolaza, 2003; Collo et al., 2009), whereas epizonal plutons in Sierra de Famatina and

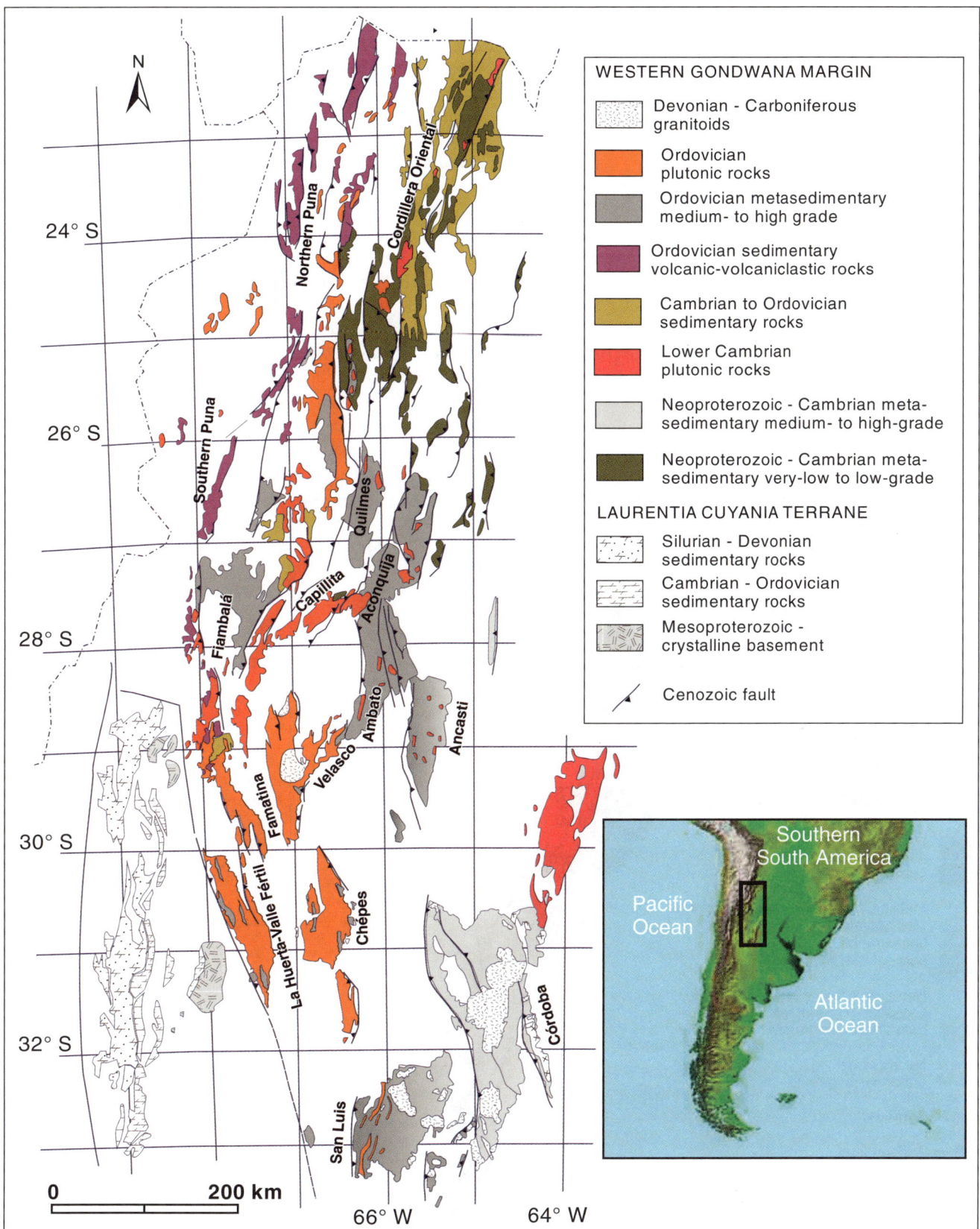

Figure 1. Map showing the distribution of pre-Carboniferous lithotectonic units in central-northwestern Argentina (modified after Pankhurst and Rapela, 1998; Coira et al., 1999; Hongn et al., 2010).

neighboring areas intruded into Early Ordovician volcanosedi-mentary cover sequences formed during the early magmatic arc stage of the Famatina (Toselli et al., 1996; Mángano and Buatois, 1996; Astini, 1998; Astini and Dávila, 2004).

Geology of the Sierra Valle Fértil–La Huerta Section

Within the western belt of the currently exposed Famatinian magmatic arc, the Sierra Valle Fértil–La Huerta section contains well-exposed sections showing the transition between lower- to upper-crustal levels (Fig. 2; Mirré, 1976; Vujovich et al., 1996; Otamendi et al., 2009a). In particular, cumulate textures in the mafic rocks are used as markers of paleohorizontal position (Ota-mendi et al., 2009a). The shallower part of the exposed section corresponds to its eastern boundary, whereas deeper levels of the crust are exposed to the west; the section is tilted almost 90° from its original position. From west to east, the lithologic units dis-play a progression from mafic to intermediate toward more silicic igneous compositions. The overall geometry of the lower part of the section is one of numerous sills of various magmatic units that invaded a preexisting crust in which only highly migmatized residual metasedimentary rocks are found. The upper part of the exposed section contains stock-like plutons of granodiorites and rare granites that are similar to the large plutonic masses found elsewhere in Cordilleran batholiths (e.g., the Sierra Nevada in California). Only minor faults exist in this section; there is no evidence that this arc was accompanied by structural (e.g., thick-ening) processes while it developed. The entire Sierra Valle Fér-til section was almost entirely exposed prior to the deposition of Permian–Triassic basalts. The collisional or transcollisional docking of the Cuyania-Precordillera microplate is inferred to have caused the exhumation of the deep crust in the area.

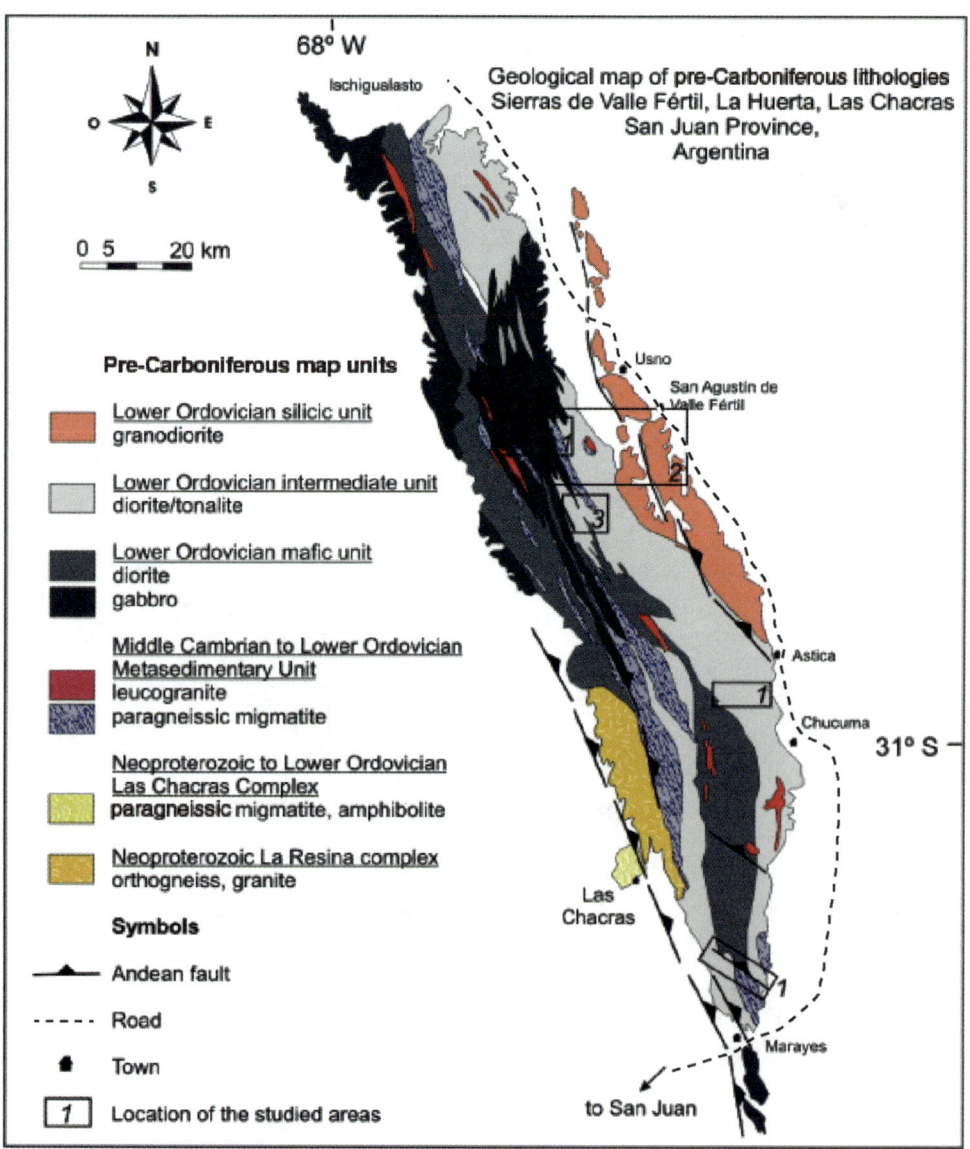

Figure 2. Simplified geologic map of the Sierras Valle Fertil–La Huerta taken af-ter the geological maps of Mirré (1976) and Vujovich et al. (1996). Map shows the location of areas chosen in our pre-vious studies and integrated here, these are: 1—Otamendi et al. (2009a), 2—Ducea et al. (2010) and Otamendi et al. (2012), 3—Otamendi et al. (2009b).

This deep-seated plutonic section of the Early Ordovician arc is almost entirely igneous, with minor framework of migmatitic metasedimentay rocks that were metamorphosed and partially melted during plutonism. There are no stratigraphic relationships, and mapping in the area is based entirely on magmatic-way-up indicators, metamorphic thermobarometry, and grouping of broad rocks types into rock units that predominate at various levels. The field relationships and petrographic observations and regional-scale geochemistry for every lithostratigraphic unit from the Sierra Valle Fertil has been presented in detail elsewhere (Otamendi et al., 2009a).

CHEMICAL, ISOTOPIC, AND GEOCHRONOLOGIC CONSTRAINTS ON THE PETROGENESIS OF PLUTONIC ROCKS FROM THE SIERRAS VALLE FÉRTIL–LA HUERTA

Limited Fractionation of Parental Magmas

The parental hydrous mafic magmas in the Famatinian paleo-arc have very low K_2O contents (0.2–0.4 wt%; DeBari, 1994; Otamendi et al., 2009a). Consequently, one of the most difficult compositional features of the igneous sequence to be explained by closed-system fractional crystallization is the relative covariation between K_2O and SiO_2 (Fig. 3A). In fact, this situation might be generally applicable to arcs because the great majority of primitive arc magmas have $K_2O < 0.9$ wt% (DeBari, 1994; Kelemen et al., 2003).

A simple calculation can be used to estimate the extent to which the major-element contents of the intermediate and silicic plutonic rocks are caused by closed-system fractional crystallization of primitive mafic magmas. We adopt an approach of examining K_2O and CaO versus SiO_2 trends, reflecting the fractionation of rock-forming minerals, specifically olivine, amphibole, and Ca-rich plagioclase. Pyroxenes have SiO_2 contents similar to, or even higher than, their host mafic magmas, and hence early fractional crystallization of pyroxenes alone would not generate the well-known SiO_2-enrichment trend of a subalkaline igneous series.

At pressures lower than 7 kbar, the sequence of crystallization as determined by petrographic studies is dominated by olivine + plagioclase (An > 90) ± orthopyroxene ± amphibole in primitive gabbroic magma chambers (Otamendi et al., 2010) and amphibole + plagioclase (An_{80}–An_{60}) in the gabbronoritic to dioritic magma bodies (Otamendi et al., 2009a). The increase

of SiO_2 in the derivative magma is caused by fractional crystallization of olivine, calcic plagioclase, and/or amphibole (Foden and Green, 1992; Eichelberger et al., 2006; Larocque and Canil, 2010). Olivine is the first ferromagnesian phase to crystallize in the most primitive rocks, but the appearance of olivine is limited to mafic-ultramafic cumulate layered bodies (Otamendi et al., 2010). This observation suggests that olivine cannot govern the crystallization sequence outside primitive magmatic chambers, and rules out olivine as part of the main assemblage controlling differentiation of arc magma from gabbro through diorite to tonalite and granodiorite.

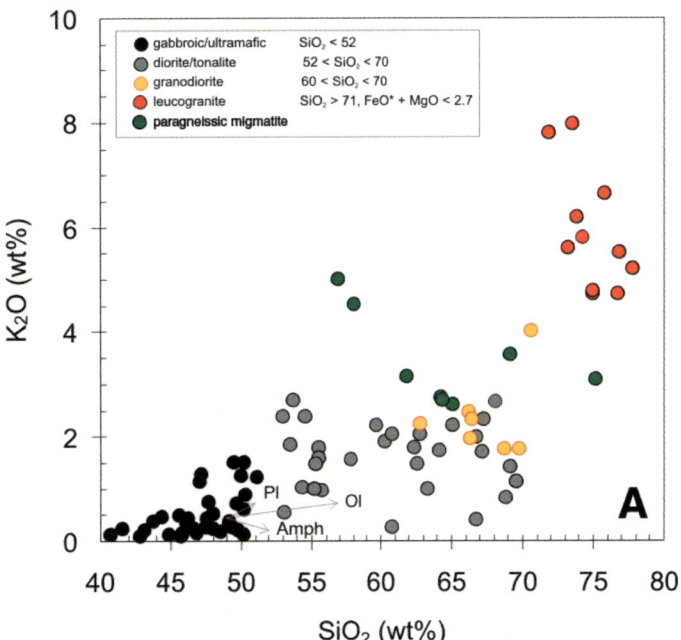

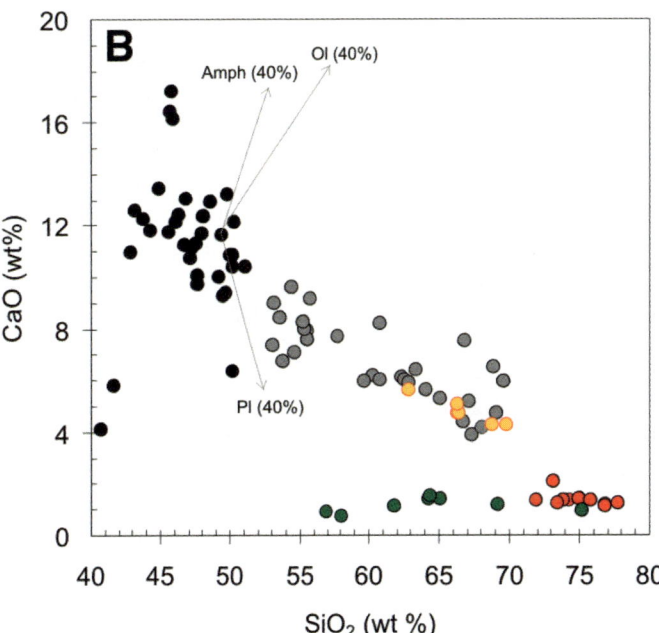

Figure 3. Representative whole-rock compositional variation for plutonic, metasedimentary, and anatectic granitic rocks from Sierras Valle Fertil–La Huerta shown in (A) K_2O and (B) CaO Harker-type diagrams. Vectors shows the compositional changes caused by fractional crystallization of distinct rock-forming minerals and the evolution of liquid after removing 40 wt% of a given mineral. Pl—plagioclase; Amph—amphibole; Ol—olivine.

In compositional terms, the effects of amphibole fractionation are broadly similar to those of olivine. The difference is that amphibole is found throughout the crystallization sequence, which makes amphibole the best candidate for driving the SiO_2-enrichment trend (Foden and Green, 1992; Larocque and Canil, 2010). Amphibole fractionation alone, however, cannot account for the well-defined decrease of CaO and increase of K_2O with increasing SiO_2 (Figs. 3A and 3B). To some extent, the combination of plagioclase and amphibole as early fractionating phases may replicate the igneous evolutionary trend, because the incorporation of K_2O in plagioclase is much lower than that of the magma (Fig. 3A). However, the CaO abundance of intermediate plutonic rocks is typically higher than 3.8 wt% and sets a limit to the proportion of calcic plagioclase involved in the process of fractional crystallization (Fig. 3B). A simple mass balance shows that if plagioclase fractionation were higher than 50%, the derivative magma would have CaO lower than those of the typical intermediate rocks, and even this proportion of plagioclase fractionation (50%) is not enough to yield the K_2O content of common intermediate rocks. In contrast, the linearity of geochemical data as seen in the Harker variation diagrams is commonly attributed to two-component magma mixing (Reid et al., 1983; Gray, 1984).

The K_2O contents of the Valle Fertil plutonic belt are not easily accounted for without involving a supracrustal (metasedimentary or its derivative granites) precursor in their ancestry. This is a common problem of calc-alkaline Cordilleran granites that was strikingly revealed by comparing various experimental petrology results with natural arc rocks (e.g., Patiño Douce, 1999).

Age Pattern of Inherited Zircon from the Plutonic Rocks

The age spectra of inherited zircon in the intermediate to silicic plutonic rocks provide an independent line of evidence for testing the ancestry of source materials of the igneous rocks from the Famatinian-Puna arc. The only area that has been investigated in detail for that purpose is the Valle Fertil section in the Sierra Pampeanas.

In a recent study (Ducea et al., 2010), we determined that inherited zircon cores within the plutonic rocks from Valle Fértil fingerprint several early magmatic events and cover the spectrum from Late Archean to early Paleozoic orogenic cycles. That paper provides a detailed study of zircon age populations for 15 plutonic rocks from the Sierra Valle Fértil; here, we evaluate the nature of inheritance in plutonic rocks by pooling the population of zircon ages older than 520 Ma from that study (Figs. 4A–4C). This cutoff age was chosen because it would date igneous or metamorphic events predating the first manifestation of the typical Famatinian magmatism, but it would still be able to record the late stage of the Pampean "orogeny" (e.g., Pankhurst and Rapela, 1998; Hongn et al., 2010). Figures 4A–4C show the histograms of age data and probability plots for representative individual plutonic rocks, which were then combined to construct the inherited zircon pattern for a composite of 12 tonalites and granodiorites.

The pattern of inherited zircon cores pooled from plutonic rocks from the Sierra Valle Fértil confirms at least three well-defined clusters of ages at around 1090, 600, and 530 Ma (see Fig. 4, lowest panel). Overall, this inheritance of zircon ages is typical and characteristic of three tectono-magmatic orogenic systems that were active from the Mesoproterozoic to the Early Cambrian in western Gondwana (de Brito Neves and Cordani, 1991; Trompette, 1997; Rapela et al., 2007; Adams et al., 2008; Drobe et al., 2009; Collo et al., 2009).

The Early Ordovician plutonism from the Sierra Valle Fértil was built up into a supracrustal sedimentary sequence that filled basins outboard of a western Gondwanan landmass (e.g., Ducea

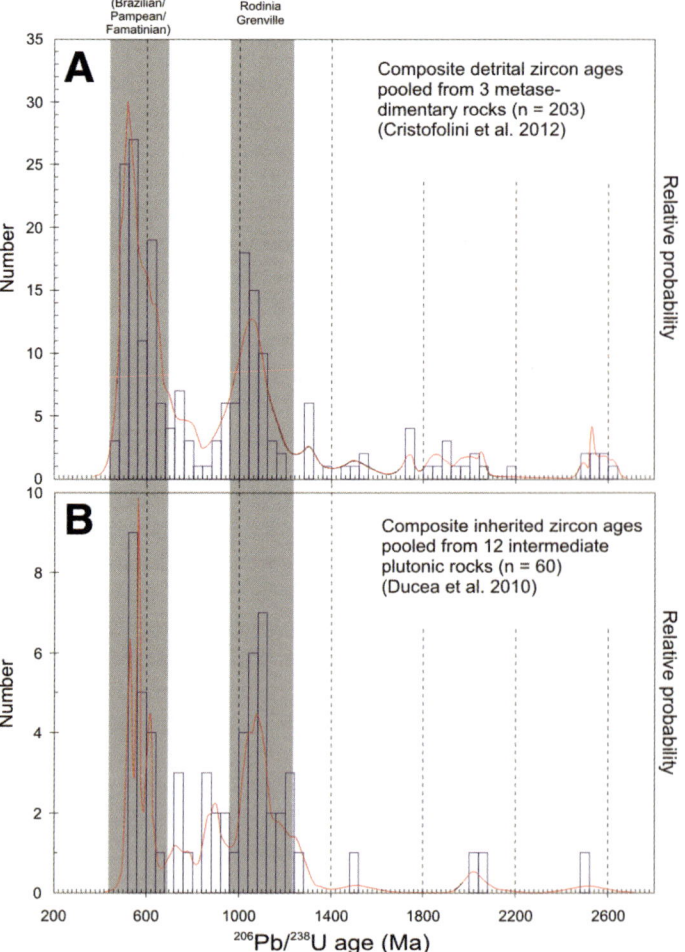

Figure 4. Combined histograms and probability plots illustrating the age data for metasedimentary and igneous rocks from the Sierra Valle Fertil. These plots were constructed with Isoplot 3.00 by Ludwig (2003). The number of analyses (*y* axis) gives the number of ages that fall in each histogram bin. Along the *x* axis, each plot covers a range in ages from 200 to 2800 Ma with 40 m.y. bin widths. These histograms are overlain by true probability plots for each of the age ranges. (A) Histogram of detrital zircon ages taken after Cristofolini et al. (2012). (B) Histogram of inherited zircon ages of plutonic rocks constructed using 14 specimens (Ducea et al., 2010).

et al., 2010). The spectra of inherited zircon ages preserved within the plutonic rocks reveal that the intermediate and silicic magmas had incorporated a significant amount of a (meta)sedimentary component, on average around 50% (±20%), based on a mass balance using Sr and Nd isotopes. The incorporation of inherited zircons must have resulted from widespread partial to nearly complete melting of pelitic and semipelitic host rocks and subsequent assimilation into the evolving magmas.

Modeling Radiogenic Isotopic Variations

A correlation exists between $\varepsilon_{Nd(i)}$ and $^{87}Sr/^{86}Sr_{(i)}$ among rock types and their isotopic composition, because a systematic trend of isotopic enrichment ranges from the igneous mafic to the metasedimentary migmatites and their anatectic leucogranitic complements (Fig. 5). Dioritic rocks define a cluster of isotopic data that appears nearly in the middle of the $\varepsilon_{Nd(i)}$ versus $^{87}Sr/^{86}Sr_{(i)}$ array between mafic and metasedimentary migmatites. Tonalitic rocks are isotopically more evolved than dioritic rocks in the Famatinian-Puna belt; however, the scatter of data for tonalites contrasts with a well-defined cluster displayed by diorites.

Despite the scatter of data, the initial ε_{Nd} and Sr isotopic compositions of rocks from the Famatinia-Puna arc display the hyperbolic trend characteristic of many arc plutonic suites in which the local upper plate contributes to the mass balance of the arc via assimilation (DePaolo and Wasserburg, 1979; McCulloch and Chappell, 1982; Gray, 1984). The most plausible interpretation for this data array is that the overall isotopic compositions reflect mixing between two end-member components. As a test for mixing, Figure 5 shows the projection of two mixing lines connecting the isotopically most primitive mafic rocks with either a metasedimentary migmatite or an anatectic leucogranite derived from partially melting metasedimentary rocks. The observation that several plutonic rocks fall within the band predicted by different two-component mixing hypotheses suggests that each plutonic rock is a hybrid product between a mantle-derived component and some (supra)crustal material, and that isotopic differences among rocks result from variable proportions of end-member components in the mixture (e.g., Gray, 1984).

Most primitive gabbroic rocks from Pocito and Valle Fértil–La Huerta suggest that the mafic component involved regionally in the Famatinian arc is isotopically enriched (Fig. 5; also see Kleine et al., 2004; Otamendi et al., 2012; Casquet et al., 2006). This reflects the existence of an ancient incompatible element–enriched subcontinental lithospheric mantle residing beneath the Early Ordovician arc. An alternate possibility is incorporation of melts and/or fluids released by the subducted sediments and oceanic crust into the mantle wedge. We do not favor the second hypothesis because, as shown elsewhere (e.g., Ducea and Barton, 2007), it would be unlikely to modify the isotopes (especially Nd isotopes) by slab-wedge interaction and still produce primitive mafic rocks. The primary material that makes up the mantle-derived component in this arc is continental lithosphere.

The most obvious candidate for a crustal end member is represented by the widespread Neoproterozoic to Early Cambrian Puncoviscana trough sedimentary sequences (Ježek et al., 1985; Mángano and Buatois, 2004; Zimmermann, 2005) or their metamorphic equivalents (Rapela et al., 1998; Becchio et al., 1999) and the Late Cambrian to Early Tremadocian formations, broadly known as Mesón Group, Negro Peinado, and La Aguadita (Aceñolaza, 2003; Collo et al., 2009). All of these sedimentary sequences were buried, and thus they hosted and interacted with Early Ordovician magmatism. In contrast, post-Tremadocian sedimentary sequences are excluded as potential sources of crustal granites because they are coeval with or even younger than the main plutonic arc activity (Mángano and Buatois, 1996; Bahlburg, 1998; Zimmermann and Bahlburg, 2003). Both Sr and Nd isotopes have been measured in metasedimentary rocks metamorphosed from greenschist- to granulite-facies conditions (Rapela et al., 1998; Becchio et al., 1999; Pankhurst et al., 1998). As the Sr- and Nd-isotopic compositions broadly overlap within the scatter of data, the sedimentary packages metamorphosed during the Early Cambrian (i.e., Pampean arc) and Early Ordovician (i.e., Famatinian-Puna arc) can be considered as a single isotopic component. These metasedimentary packages typically have initial $^{87}Sr/^{86}Sr$ higher than 0.713 and a wide range of $\varepsilon_{Nd(i)}$, from –3 to –10. The Late Cambrian low-grade metasedimentary sequence from Famatina has broadly the same $\varepsilon_{Nd(i)}$ values as those measured in schists and gneisses from the northern Sierras Pampeanas (Collo et al., 2009).

Purely granitic magmas generated after partially melting the fertile metasedimentary sequences, as suggested earlier, are a second crustal component for contaminating the evolving lineage of mantle-derived magmas. In terms of isotopic signature, strict granitic rocks seem to be divisible into two groups. One group of these granites clearly lies inside the isotopic compositional field of the metasedimentary sequences (Fig. 5). Therefore, as suggested by their whole-rock compositions, these granites crystallized from magmas solely generated after partially melting metasedimentary packages (Dahlquist et al., 2005). In contrast, some granite from the Sierra de Chepes studied by Pankhurst et al. (1998) has initial $^{87}Sr/^{86}Sr$ values much lower than those of typical metasedimentary rocks.

Regardless of the actual crustal contaminant, interaction between mafic magmas and upper-plate (local crust) supracrustal material provides the easiest explanation for the origin of intermediate and silicic plutonic rocks. The great majority of the intermediate and silicic plutonic rocks from the Famatinian-Puna arc appear nearly in the middle of the $^{87}Sr/^{86}Sr_{(i)}-\varepsilon_{Nd(i)}$ hyperbolic array between mafic primitive and metasedimentary/granitic components (Fig. 5). The lack of measurement of one isotopic system impedes projection of many other plutonic and volcanic rocks that either have $^{87}Sr/^{86}Sr_{(i)}$ between 0.706 and 0.710 (Mannheim, 1993; Saal et al., 1996; Saavedra et al., 1996) or $\varepsilon_{Nd(i)}$ between –2 and –6 (e.g., volcanosedimentary successions in Bock et al., 2000), but several of these Early Ordovician igneous rocks would most likely lie on the middle of the hyperbolic trend predicted

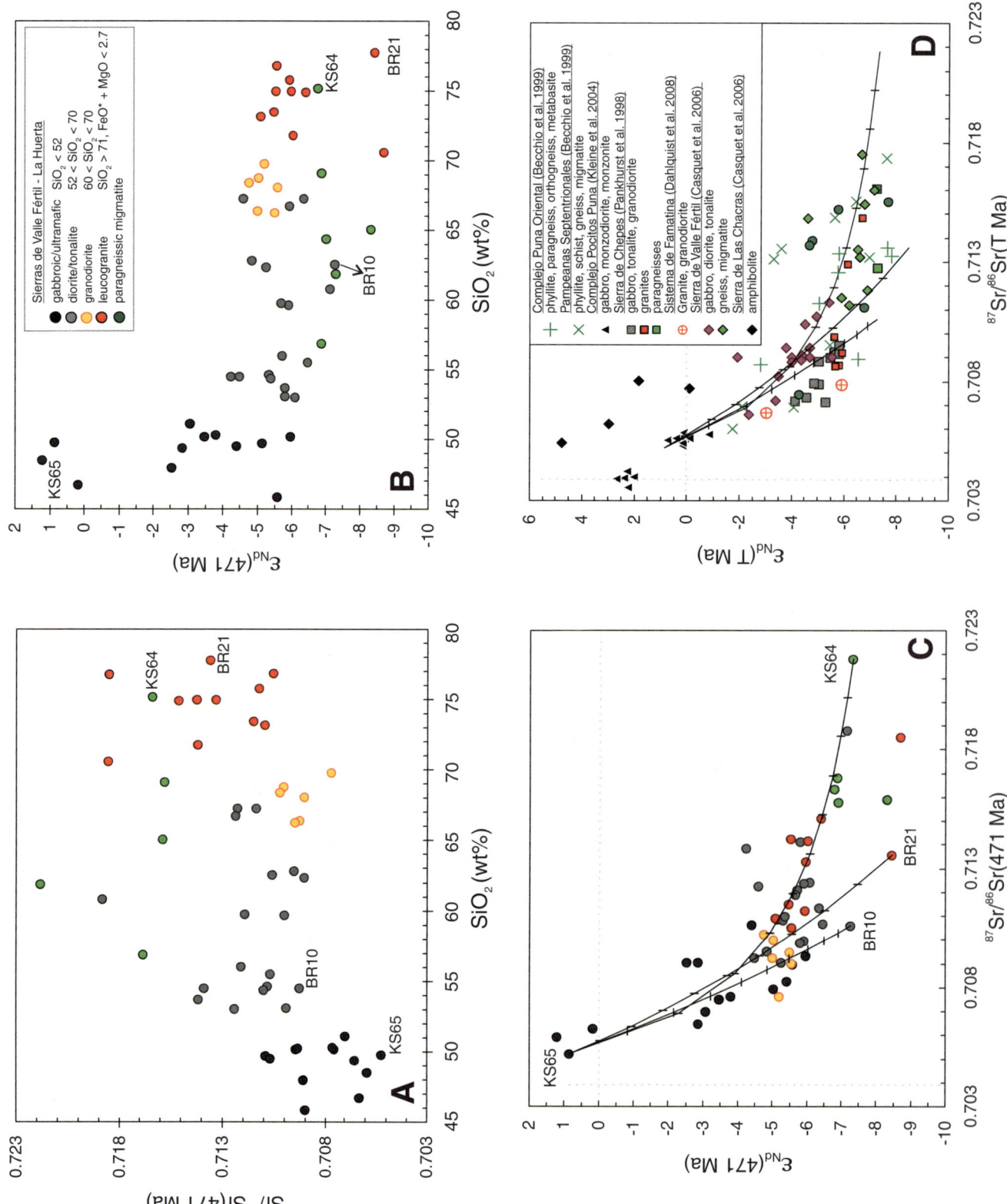

by the two-component mixing model. Our partial compilation of isotopic data gives evidence that most of the diorites, tonalites, granodiorites, and some monzogranites from the Famatinia-Puna arc fall on the hyperbola of two-component mixing where the end members are a primitive mafic igneous suite and a crustal material. However, we avoided modeling a particular case, because it has been shown that the mixing process is neither a single-step mechanism nor solely moved by magma mixing (Eichelberger et al., 2006). In effect, some middle member of the igneous suite may derive by hybridization in which one member is already a hybrid product between the two extreme end members (Beard, 2008; Otamendi et al., 2009b). Isotope variations would thereby reflect the end result of more complex petrogenetic mechanisms than two-component mixing, but in fact every single plutonic rock embodies the two components mixed at variable proportions. Moreover, the evidence extracted from isotopes is perfectly consistent with conclusions made from observing major-element variations. Also significant is that, at deep-seated levels from the Early Ordovician paleo-arc crust, there are observable field relationships supporting the occurrence of open-system processes (Otamendi et al., 2009b). Finally, the juvenile mafic and the crustal metasedimentary end members seem to be universal, as they have been found to play their role in most of the worldwide recognized Cordilleran-style plutonic chains, such as the Sierra Nevada and Peninsular Range (DePaolo, 1981; DePaolo et al., 1992; Pickett and Saleeby, 1994), and the Lachlan fold belt (Gray, 1984; Collins, 1996; Keay et al., 1997; Kemp et al., 2009).

PETROLOGIC IMPLICATIONS

Data for the high-grade partially melted metasedimentary rocks from Valle Fértil and La Huerta exhibit remarkably similar major-element contents to those of the medium- and low-grade metasedimentary rocks and sedimentary rocks elsewhere in the Early Ordovician magmatic belt (Fig. 6A). All these metasedimentary and sedimentary rocks define linear arrays for most major oxides against silica, reflecting the fact that they encompass pelites (SiO_2 ~59 wt%) to quartz-rich graywackes (SiO_2 ~80 wt%). Within the major-oxide covariant diagrams, the high-

---

Figure 5. Isotopic composition of rocks from Sierras de Valle Fertil–La Huerta compiled after Otamendi et al. (2009a, 2010, 2012). (A) Variation of $^{87}Sr/^{86}Sr$ at 471 Ma vs. SiO_2. (B) Variation of ε_{Nd} at 471 Ma vs. SiO_2. (C) Plot of ε_{Nd} vs. $^{87}Sr/^{86}Sr$ ratios (at 471 Ma) for plutonic rocks, metasedimentary migmatites, and anatectic and leucogranites. Hypothetical mixing models were computed using a primitive gabbroic rock and three potential crustal components. Mineralogy and whole-rock compositions of end-member rocks were provided in our previous studies. Tick marks on mixing lines are at 0.1 end-member fractions. (D) Plot of ε_{Nd} vs. $^{87}Sr/^{86}Sr$ ratios (at reported crystallization age) for rocks in type localities from the Famatinian-Puna arc. The geographic position of each locality is shown in Figure 2. Insets give the data sources for each locality. The three mixing lines computed for Valle Fertil and La Huerta rocks are shown for comparison.

grade metasedimentary rocks from Valle Fértil and La Huerta fall across the low-potassium limit, which is consistent with their having undergone partial melting and melt loss (Fig. 6A). Major-element whole-rock compositions do not provide solid constraints for correlating sedimentary rocks, because the chemistry of sediments reflects the mechanical unmixing of clay and sand that takes place in every sedimentary cycle (Taylor and McLennan, 1985). However, the correlation of major elements strongly suggests that the sedimentary packages that were stacked, buried, and metamorphosed between the Late Neoproterozoic and Early Ordovician had broadly similar sedimentary sequences of facies to those that remained in the upper crust (Fig. 6A). A sedimentary sequence made up by alternating beds of pelites and graywackes has the largest potential to produce granitic melts when experiencing granulite-facies temperatures (Thompson, 1996). By implication, the combination of fertility and volume of the Late Neoproterozoic and Early Ordovician sedimentary packages can account for all the felsic weakly and strongly peraluminous granites from the Ordovician system. Granitic batholiths entirely made up by peraluminous granitoids document that massive crustal anatexis occurred where a large-scale heat input acted upon widespread metasedimentary sequences (Rossi et al., 2002; Dahlquist et al., 2005). Our compilation of data also shows that metasediment-derived granites appear in almost all of the localities from the Early Ordovician magmatic system, and hence reflect crustal anatexis that took place pervasively at lower- to middle-crustal levels of the Famatinian-Puna paleo-arc (Fig. 6B). The importance of this interpretation, as inferred from field and compositional observations, is that metasediment-derived granites derived from partial melting of the upper plate are an essential component for driving the igneous evolutionary trend from gabbroic to monzogranitic rocks.

However, intracrustal melting of the metasedimentary sequence alone does not explain the vast volume of plutonic and volcanic rocks from the Early Ordovician magmatic belts. The plutonic suite spanning the range from gabbros to monzogranites has MgO, FeO, and CaO contents as well as Na_2O/K_2O ratios that are too high to be derived from any siliciclastic sedimentary protolith (e.g., Patiño Douce, 1999). Contrasting with the relative compositional homogeneity of the crustal granites, the igneous suite from gabbros to monzogranite spreads over a wide range of petrographic and chemical rock types (Figs. 6B and 6C). In effect, the nature of the latter suite of igneous (plutonic and volcanic) rocks needs to be further evaluated in the Early Ordovician magmatic belts.

Typical localities within the Famatinia-Puna magmatic arc consist of igneous rock suites with a compositional trend of increasing K_2O with increasing SiO_2 (Fig. 6B). This trend is also a distinctive characteristic of the Valle Fértil–La Huerta igneous suite and, as shown already, requires, to a large extent, a driving mechanism by crustal contamination of the evolving igneous magmas. As Figure 3A illustrates, at some point in the generation of the Early Ordovician igneous suites, every rock more evolved than a typical diorite must have incorporated a crustal component

134

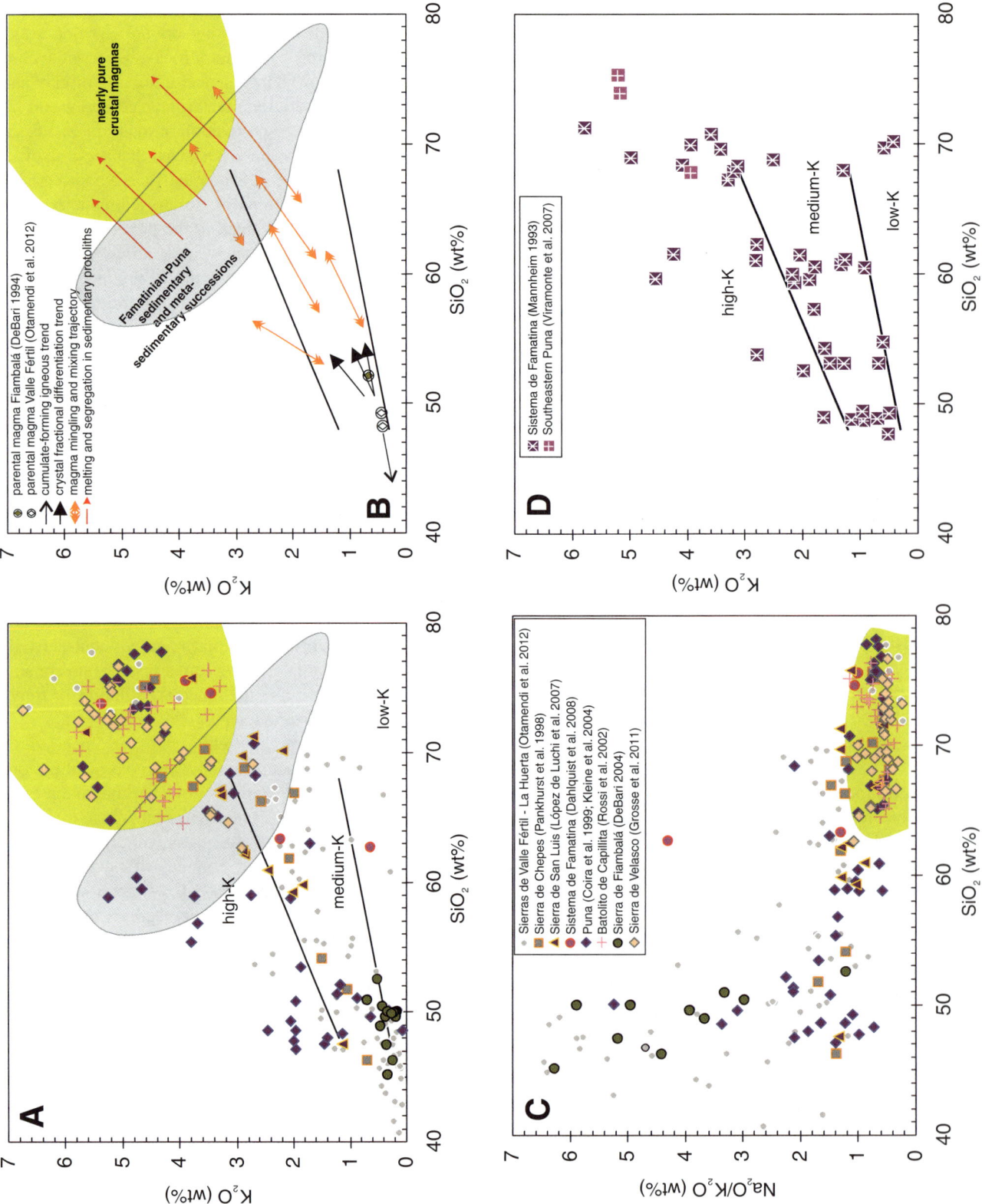

through either assimilation of metasedimentary rocks or interaction with metasediment-derived melts.

A significant volume of Early Ordovician magmatism erupted as either lava flows or pyroclastic rocks (Mannheim, 1993; Mannheim and Miller, 1996; Coira et al., 1999; Zimmermann and Bahlburg, 2003; Viramonte et al., 2007; and references therein). Field relationships are unequivocal about the complementary nature between Early Ordovician plutonic and volcanic rocks. In addition, the few available data for Famatinian-Puna volcanic rocks show compositional similarity with plutonic rocks (Fig. 6D). Thus, the same petrologic mechanisms as those observed for plutonic rocks were responsible for governing the whole-rock composition of eruptive volcanic sequences, with a slightly higher K_2O at a given content of SiO_2 for the volcanics.

IMPLICATIONS FOR CONTINENTAL ARC MAGMATISM

The Famatinian arc was clearly a continental arc, as it straddles the western margin of Gondwana in the modern coordinates of South America. There is no evidence that the South American cratonic basement crust was involved in magmatism in the Sierra Valle Fertil region; instead, the arc was emplaced exclusively into an extensive sedimentary assemblage that most likely constituted the accumulation of submarine passive-margin sediments along Gondwana's margin during the Neoproterozoic and early Paleozoic prior to subduction. However, all gabbros that have been analyzed in this section have radiogenic isotopic characteristics typical of old (Precambrian) continental lithosphere (Otamendi et al., 2009a, 2012), suggesting that South American lithosphere, perhaps thinned at a miogeoclinal margin, was in fact the framework of this arc. There is also no evidence that the crust became unusually thick during the Famatinian arc magmatism, as no rocks deeper than ~25–30 km are exposed at the surface to the west of the studied area, based on our preliminary field observations and mapping of index minerals within metamorphic framework rocks, or elsewhere within the exposed plutonic framework of the Famatinian arc in central South America. In addition, there is also no evidence for crustal magmas (felsic plutons or volcanics) that were derived from thicker parts of the crust. No shortening or extension disrupted the architecture of the section studied here during arc formation. Instead, the arc developed statically as a series of sills progressively emplaced into the existing crust, similar to the early stages (Jurassic) of Cordilleran magmatism in North America and to Cordilleran interior arcs that developed in

North America during periods of shallow subduction as magmatism migrated inland (Barton, 1996).

We show that magmatic input from the mantle at an average rate of mafic arc magmatism worldwide (Ducea and Barton, 2007) can provide enough heat and mass available for mixing with a preexisting metamorphic basement to generate a batholith-scale crustal section within a short period of time, some 15 m.y. or less. The architecture of the arc is one of multiple tens- to hundreds-of-meters-thick amphibole-rich gabbroic sills injected into a midcrustal section, where they mixed with partial melts derived from a metasedimentary framework. All mantle-derived melts intruded in the section were wet gabbros, and magmatic fractionation trends observable through field relationships suggest that some of these bodies transitioned to mafic diorites in their upper parts via closed-system fractionation. There is no evidence that gabbros in this section fractionated to intermediate (higher silica) rocks, nor is there evidence that they ever remelted to generate more felsic melts. A signature feature of the entire Sierra Valle Fertil area is that virtually every outcrop in which we observe transitions from the mafic to tonalitic/granodioritic rocks is in close proximity to a metasedimentary pendant or contains "ghosts" of it (identifiable rock enclaves, areas rich in cordierite and almandine garnet) within the more felsic units. Thus, we use our extensive field observations at the scale of this study and a moderate knowledge of the entire range to state that, with the exception of local closed-system fractionation to mafic diorite of gabbroic sills, the entire compositional diversity of the Famatinia-Puna arc—which includes the full compositional spectrum of Cordilleran calc-alkaline suites such as quartz diorites, monzonites, tonalites, granodiorites, and granites—was generated by various hybridization processes between mantle-derived gabbros and diorites and the Puncoviscana metasedimentary rocks and their high-grade equivalents.

We suggest that any incipient arc that developed on a continental upper plate in a subduction system may have similar characteristics to the Famatinian-Puna arc. They are static arcs emplaced as a series of mafic sills that ignite melting of and mixing with their framework rocks at magmatic rates that can be higher than 100 $km^3/km/m.y.$, depending on the extent to which the framework is melt fertile (e.g., Annen and Sparks, 2002). The late Paleozoic and early Mesozoic arcs of the North American Cordillera are equivalents to the Famatinian arc. In addition, island arcs emplaced into crust that experience long-lived subduction and have sizable trench and forearc accumulations (like modern Japan, the Caribbean, and parts of the Aleutians) where the upper plate is continental may have a similar crustal architecture.

CONCLUSIONS

1. Field relationships, the pattern of inherited zircon ages, and whole-rock compositional (elemental and isotopic) evidence clearly indicate that crustal contamination accompanying fractional crystallization explains the genesis of intermediate and silicic plutonic rocks of the Famatinian arc; on average a mixture

----

Figure 6. (A) Plot of K_2O vs. SiO_2 for plutonic and volcanic rocks in the Famatinian-Puna magmatic arc. The positions of lines separating low-, medium-, and high-K fields in the K_2O vs. SiO_2 diagram are taken after LeMaitre et al. (1989). (B) Schematic representation of petrologic process in the K_2O vs. SiO_2 covariation system. (C) Plot of Na_2O/K_2O ratios for same rocks as in panel A. (D) Plot of K_2O vs. SiO_2 for volcanic rocks in the Famatinian-Puna magmatic arc.

of ~50% mantle-derived gabbros and 50% crustal melts is suggested by field relationships in the Sierra Valle Fertil and is consistent with geochemical data.

2. Virtually every igneous rock more evolved than gabbros or basalts has been contaminated with a (supra)crustal component; model ages from all intermediate and silicic igneous rocks reflect the mixture between a Grenville-aged average crustal source and an underlying old continental mantle wedge.

3. The lithospheric architecture of the modern Cordilleran interior of the central Andes, including the Altiplano-Puna Plateaus, the Eastern Cordilleran region, and the western Sierras Pampeanas, has been profoundly influenced by the development of the Famatinian-Puna arc.

4. The Famatinian-Puna arc is an ancient equivalent of arcs formed on thin continental lithosphere covered by thick miogeoclinal sequences soon after subduction initiation.

ACKNOWLEDGMENTS

We acknowledge thorough and constructive reviews by Calvin Miller and Keith Putirka, whose detailed comments and criticism have significantly improved the quality of the manuscript, as well as guidance from volume editor Paul Kapp. This work was partly funded by COSA, an ExxonMobil–University of Arizona initiative. Otamendi was also funded by ANPCYT-Argentina through grants PICT 01904/07 and 00453/10, and SeCyT-UNRC, whereas the University of Bucharest group was funded by Romanian National Sciences CNCSIS grant PN-II-ID-PCE-2011-3-0217.

REFERENCES CITED

Aceñolaza, F.G., Miller, H., and Toselli, A.J., 2000, The Pampean and Famatinian cycles—Superposed orogenic events in West Gondwana: Zeitschrift für Angewandte Geologie: Sonderheft SH, v. 1, p. 337–344.

Aceñolaza, G.F., 2003, The Cambrian system in northwestern Argentina: Stratigraphical and palaeontological framework: Geologica Acta, v. 1, p. 23–39.

Adams, C.J., Miller, H., Toselli, A.J., and Griffin, W.L., 2008, The Puncoviscana Formation of northwest Argentina: U-Pb geochronology of detrital zircons and Rb-Sr metamorphic ages and their bearing on its stratigraphic age, sediment provenance and tectonic setting: Neues Jahrbuch für Geologie und Paläontologie, Abhandlungen, v. 247, p. 341–352.

Annen, C., and Sparks, R.S.J., 2002, Effects of repetitive emplacement of basaltic intrusions on thermal evolution and melt generation in the crust: Earth and Planetary Science Letters, v. 203, p. 937–955, doi:10.1016/S0012-821X(02)00929-9.

Astini, R.A., 1998, El Ordovícico en la región central del Famatina (provincia de La Rioja, Argentina): Aspectos estratigráficos, geológicos y geotectónicos: Revista de la Asociación Geológica Argentina, v. 53, p. 445–460.

Astini, R.A., and Dávila, F.M., 2004, Ordovician back arc foreland and Ocloyic thrust belt development on the western Gondwana margin as a response to Precordillera terrane accretion: Tectonics, v. 23, TC4008, doi:10.1029/2003TC001620.

Bahlburg, H., 1998, The geochemistry and provenance of Ordovician turbidites in the Argentine Puna, in Pankhurst, R.J., and Rapela, C.W., eds., The Proto-Andean Margin of Gondwana: Geological Society [London] Special Publication 142, p. 127–142.

Barton, M.D., 1996, Granitic magmatism and metallogeny of southwestern North America: Transactions of the Royal Society of Edinburgh–Earth Sciences, v. 87, p. 261–280, doi:10.1017/S0263593300006672.

Beard, J.S., 2008, Crystal–melt separation and the development of isotopic heterogeneities in hybrid magmas: Journal of Petrology, v. 49, p. 1027–1041, doi:10.1093/petrology/egn015.

Becchio, R., Lucassen, F., Kasemann, S., Franz, G., and Viramonte, J., 1999, Geoquímica y sistemática isotópica de rocas metamórficas del Paleozoico inferior. Noroeste de Argentina y Norte de Chile (21–27 S): Acta Geológica Hispánica, v. 34, p. 273–299.

Bock, B., Bahlburg, H., Wörner, G., and Zimmermann, U., 2000, Tracing crustal evolution in the southern central Andes from Late Precambrian to Permian with geochemical and Nd and Pb isotope data: The Journal of Geology, v. 108, p. 515–535, doi:10.1086/314422.

Brown, M., 2013, Granites: From genesis to emplacement: Geological Society of America Bulletin, v. 125, p. 1079–1113, doi:10.1130/B30877.1.

Caminos, R.J., 1979, Sierras Pampeanas Noroccidentales. Salta, Tucumán, Catamarca, La Rioja y San Juan, in Leanza, E.F., ed., Actas II Simposio de Geología Regional Argentina, v. 1: San Juan, University of San Juan, p. 225–291.

Casquet, C., Pankhurst, R.J., Fanning, C.M., Baldo, E., Galindo, C., Rapela, C.W., González-Casado, J.M., and Dahlquist, J.A., 2006, U-Pb SHRIMP zircon dating of Grenvillian metamorphism in western Sierras Pampeanas (Argentina): Correlation with the Arequipa-Antofalla craton and constraints on the extent of the Precordillera terrane: Gondwana Research, v. 9, p. 524–529, doi:10.1016/j.gr.2005.12.004.

Cawood, P.A., 2005, Terra Australis orogen: Rodinia breakup and development of the Pacific and Iapetus margins of Gondwana during the Neoproterozoic and Paleozoic: Earth-Science Reviews, v. 69, p. 249–279, doi:10.1016/j.earscirev.2004.09.001.

Cawood, P.A., Hawkesworth, C.J., and Dhuime, B., 2012, The continental record and the generation of continental crust: Geological Society of America Bulletin, v. 125, p. 14–32, doi:10.1130/B30722.1.

Coira, B., Kay, S.M., Pérez, B., Woll, B., Hanning, M., and Flores, P., 1999, Magmatic sources and tectonic setting of Gondwana margin Ordovician magmas, northern Puna of Argentina and Chile, in Ramos, V., and Keppie, J., eds., Laurentia-Gondwana Connections before Pangea: Geological Society of America Special Paper 336, p. 145–170.

Collins, W.J., 1996, Lachlan Fold Belt granitoids: Products of three-component mixing: Transactions of the Royal Society of Edinburgh–Earth Sciences, v. 87, p. 171–181, doi:10.1017/S0263593300006581.

Collo, G., Astini, R., Cawood, P.A., Buchan, C., and Pimentel, M., 2009, U-Pb detrital zircon ages and Sm-Nd isotopic features in low-grade metasedimentary rocks of the Famatina belt: Implications for late Neoproterozoic–early Palaeozoic evolution of the proto-Andean margin of Gondwana: Journal of the Geological Society [London], v. 166, p. 303–319, doi:10.1144/0016-76492008-051.

Cristofolini, E.A., Otamendi, J.E., Ducea, M.N., Pearson, D., Tibaldi, A.M., and Baliani, I., 2012, Detrital zircon ages from Sierra de Valle Fértil, entrapment of Middle and Late Cambrian marine successions in the deep roots of Early Ordovician arc: Journal of South American Earth Sciences, v. 37, p. 77–94, doi:10.1016/j.jsames.2012.02.001.

Dahlquist, J.A., Rapela, C.W., and Baldo, E.G., 2005, Petrogenesis of cordierite-bearing S-type granitoids in Sierra de Chepes, Famatinian orogen, Argentina: Journal of South American Earth Sciences, v. 20, p. 231–251, doi:10.1016/j.jsames.2005.05.014.

Dahlquist, J.A., Pankhurst, R.J., Rapela, C.W., Galindo, C., Alasino, P., Fanning, C.M., Saavedra, J., and Baldo, E., 2008, New SHRIMP U-Pb data from the Famatina complex: Constraining Early–Mid-Ordovician Famatinian magmatism in the Sierras Pampeanas, Argentina: Geologica Acta, v. 6, p. 319–333.

Davidson, J.P., and Arculus, R.J., 2006, The significance of Phanerozoic arc magmatism in generating continental crust, in Brown, M., and Rushmer, T., eds., Evolution and Differentiation of the Continental Crust: Cambridge, UK, Cambridge University Press, p. 135–172.

DeBari, S., 1994, Petrogenesis of the Fiambalá gabbroic intrusion, northwestern Argentina, a deep crustal syntectonic pluton in a continental magmatic arc: Journal of Petrology, v. 35, p. 679–713, doi:10.1093/petrology/35.3.679.

de Brito Neves, B.B., and Cordani, U.G., 1991, Tectonic evolution of South America during the Late Proterozoic: Precambrian Research, v. 53, p. 23–40, doi:10.1016/0301-9268(91)90004-T.

DePaolo, D.J., 1981, A neodymium and strontium isotopic study of the Mesozoic calc-alkaline granitic batholiths of the Sierra Nevada and Peninsular Ranges, California: Journal of Geophysical Research–Solid Earth, v. 86, no. B11, p. 10,470–10,488, doi:10.1029/JB086iB11p10470.

DePaolo, D.J., and Wasserburg, G.J., 1979, Petrogenetic mixing models and Nd-Sr isotopic patterns: Geochimica et Cosmochimica Acta, v. 43, p. 615–627, doi:10.1016/0016-7037(79)90169-8.

DePaolo, D.J., Perry, F.V., Scott, W., and Baldridge, W.S., 1992, Crustal versus mantle sources of granitic magmas: A two-parameter model based on Nd isotopic studies: Transactions of the Royal Society of Edinburgh–Earth Sciences, v. 83, p. 439–446, doi:10.1017/S0263593300008117.

Drobe, M., López de Luchi, M., Steenken, A., Frei, R., Naumann, R., Siegesmund, S., and Wemmer, K., 2009, Provenance of the Late Proterozoic to Early Cambrian metaclastic sediments of the Sierra de San Luis (eastern Sierras Pampeanas) and Cordillera Oriental, Argentina: Journal of South American Earth Sciences, v. 28, p. 239–262, doi:10.1016/j.jsames.2009.06.005.

Ducea, M.N., and Barton, D.M., 2007, Igniting flare-up events in Cordilleran arcs: Geology, v. 35, p. 1047–1050, doi:10.1130/G23898A.1.

Ducea, M.N., Otamendi, J.E., Bergantz, G., Stair, K., Valencia, V., and Gehrels, G., 2010, Timing constraints on building an intermediate plutonic arc crustal section: U-Pb zircon geochronology of the Sierra Valle Fértil, Famatinian arc, Argentina: Tectonics, v. 29, TC4002, doi:10.1029/2009TC002615.

Eichelberger, J.C., Izbekov, P.E., and Browne, B.L., 2006, Bulk chemical trends at arc volcanoes are not liquid lines of descent: Lithos, v. 87, p. 135–154, doi:10.1016/j.lithos.2005.05.006.

Foden, J.D., and Green, D.H., 1992, Possible role of amphibole in the origin of andesite: Some experimental and natural evidence: Contributions to Mineralogy and Petrology, v. 109, p. 479–493, doi:10.1007/BF00306551.

Gill, J.B., 1981, Orogenic Andesites and Plate Tectonics: Berlin, Springer-Verlag, 412 p.

Gray, C.M., 1984, An isotopic mixing model for the origin of granitic rocks in southeastern Australia: Earth and Planetary Science Letters, v. 70, p. 47–60, doi:10.1016/0012-821X(84)90208-5.

Grosse, P., Bellos, L.I., de los Hoyos, C.M., Larrovere, M.A., Rossi, J.N., and Toselli, A.J., 2011, Across-arc variation of the Famatinian magmatic arc (NW Argentina) exemplified by I-, S- and transitional I/S-type Early Ordovician granitoids of the Sierra de Velasco: Journal of South American Earth Sciences, v. 29, p. 289–305.

Hongn, F.D., Tubía, J.M., Aranguren, A., Vegas, N., Mon, R., and Dunning, G.R., 2010, Magmatism coeval with Lower Paleozoic shelf basins in NW-Argentina (Tastil batholith): Constraints on current stratigraphic and tectonic interpretations: Journal of South American Earth Sciences, v. 29, p. 289–305, doi:10.1016/j.jsames.2009.07.008.

Ježek, P., Willner, A.P., Aceñolaza, F.G., and Miller, H., 1985, The Puncoviscana trough—A large basin of Late Precambrian to Early Cambrian age on the Pacific edge of the Brazilian Shield: Geologische Rundschau, v. 74, p. 573–584, doi:10.1007/BF01821213.

Keay, S.M., Collins, W.J., and McCulloch, M.T., 1997, A three-component mixing model for granitoid genesis: Lachlan fold belt, eastern Australia: Geology, v. 25, p. 307–310, doi:10.1130/0091-7613(1997)025<0307:ATCSNI>2.3.CO;2.

Kelemen, P.B., Hanghøj, K., and Greene, A.R., 2003, One view of the geochemistry of subduction-related magmatic arcs, with emphasis on primitive andesite and lower crust, *in* Rudnick, R.L., and Gao, S., eds., The Crust: Treatise on Geochemistry Volume 3: Oxford, UK, Elsevier-Pergamon, p. 596–659.

Kemp, A.I.S., Hawkesworth, C.J., Collins, W.J., Gray, C.M., and Blevin, P.L., 2009, Isotopic evidence for rapid continental growth in an extensional accretionary orogen: The Tasmanides, eastern Australia: Earth and Planetary Science Letters, v. 284, p. 455–466, doi:10.1016/j.epsl.2009.05.011.

Kleine, T., Mezger, K., Zimmermann, U., Münker, C., and Bahlburg, H., 2004, Crustal evolution along the Early Ordovician proto-Andean margin of Gondwana: Trace element and isotope evidence from the Complejo Igneo Pocitos (northwestern Argentina): The Journal of Geology, v. 112, p. 503–520, doi:10.1086/422663.

Larocque, J., and Canil, D., 2010, The role of amphibole in the evolution of arc magmas and crust: The case from the Jurassic Bonanza arc section: Contributions to Mineralogy and Petrology, v. 159, p. 475–492, doi:10.1007/s00410-009-0436-z.

LeMaitre, R.W., Bateman, P., Dudek, A., Keller, J., Lameyre, J., Le Bas, M.J., Sabine, P.A., Schmid, R., Sørensen, S., Streckeisen, A., Woolley, A.R., and Zanettin, B., 1989, A Classification of Igneous Rocks and Glossary of Terms: Oxford, UK, Blackwell Scientific Publications, 193 p.

López de Luchi, M.G., Siegesmund, S., Wemmer, K., Steenken, A., and Naumann, R., 2007, Geochemical constraints on the petrogenesis of the Paleozoic granitoids of the Sierra de San Luis, Sierras Pampeanas, Argentina: Journal of South American Earth Sciences, v. 24, p. 138–166, doi:10.1016/j.jsames.2007.05.001.

Ludwig, K.R., 2003, Isoplot 3.00: Berkeley Geochronology Center Special Publication 4, 70 p.

Mángano, M.G., and Buatois, L.A., 1996, Shallow marine event sedimentation in a volcanic arc-related setting: The Ordovician Suri Formation, Famatina Range, northwest Argentina: Sedimentary Geology, v. 105, p. 63–90, doi:10.1016/0037-0738(95)00134-4.

Mángano, M.G., and Buatois, L.A., 2004, Integración de estratigrafía secuencial, sedimentología e icnología para un análisis cronoestratigráfico del Paleozoico inferior del noroeste argentino: Revista de la Asociación Geológica Argentina, v. 59, p. 273–280.

Mannheim, R., 1993, Genese der Vulkanite und Subvulkanite des altpaläozoischen Famatina-Systems, NW-Argentinien, und Seine geodynamische Entwicklung: Münchner Geologische Hefte, v. 7, p. 1–155.

Mannheim, R., and Miller, H., 1996, Las rocas volcánicas y subvolcánicas eopaleozoicas del Sistema de Famatina: Münchner Geologische Hefte, v. 19A, p. 159–186.

Martino, R.D., Guereschi, A.B., and Sfragula, J.A., 2009, Petrology, structure and tectonic significance of the Tuclame banded schists in the Sierras Pampeanas of Córdoba and its relationship with the metamorphic basement of northwestern Argentina: Journal of South American Earth Sciences, v. 27, p. 280–298, doi:10.1016/j.jsames.2009.01.003.

McCulloch, M.T, and Chappell, B.W., 1982, Nd isotopic characteristics of S- and I-type granites: Earth and Planetary Science Letters, v. 58, p. 51–64.

Mirré, J.C., 1976, Descripción Geológica de la Hoja 19e, Valle Fértil, Provincias de San Juan y La Rioja: Buenos Aires, Servicio Geológico Nacional Boletín 147, p. 1–70.

Otamendi, J.E., Vujovich, G.I., de la Rosa, J.D., Castro, A., Tibaldi, A., Martino, R., and Pinotti, L., 2009a, Geology and petrology of a deep crustal zone from the Famatinian paleo-arc, Sierras Valle Fértil–La Huerta, San Juan, Argentina: Journal of South American Earth Sciences, v. 27, p. 258–279, doi:10.1016/j.jsames.2008.11.007.

Otamendi, J.E., Ducea, M.N., Tibaldi, A.M., Bergantz, G., de la Rosa, J.D., and Vujovich, G.I., 2009b, Generation of tonalitic and dioritic magmas by coupled partial melting of gabbroic and metasedimentary rocks within the deep crust of the Famatinian magmatic arc, Argentina: Journal of Petrology, v. 50, p. 841–873, doi:10.1093/petrology/egp022.

Otamendi, J.E., Cristofolini, E., Tibaldi, A.M., Quevedo, F., and Baliani, I., 2010, Petrology of mafic and ultramafic layered rocks from the Jaboncillo Valley, Sierra de Valle Fértil, Argentina: Implications for the evolution of magmas in the lower crust of the Famatinian arc: Journal of South American Earth Sciences, v. 30, p. 29–45, doi:10.1016/j.jsames.2010.07.004.

Otamendi, J.E., Ducea, M.N., and Bergantz, G.W., 2012, Geological, petrological and geochemical evidence for progressive construction of an arc crustal section, Sierra de Valle Fértil, Famatinian arc, Argentina: Journal of Petrology, v. 53, p. 761–800, doi:10.1093/petrology/egr079.

Pankhurst, R.J., and Rapela, C.W., 1998, Introduction, *in* Pankhurst, R.J., and Rapela, C.W., eds., The Proto-Andean Margin of Gondwana: Geological Society [London] Special Publication 142, p. 1–9.

Pankhurst, R.J., Rapela, C.W., Saavedra, J., Baldo, E., Dahlquist, J., Pascua, I., and Fanning, C.M., 1998, The Famatinian magmatic arc in the central Sierras Pampeanas: An Early to Mid-Ordovician continental arc on the Gondwana margin, *in* Pankhurst, R.J., and Rapela, C.W., eds., The Proto-Andean Margin of Gondwana: Geological Society [London] Special Publication 142, p. 343–368.

Patiño Douce, A.E., 1999, What do experiments tell us about the relative contributions of crust and mantle to the origin of granitic magmas?, *in* Castro, A., Fernández, C., and Vigneresse, J.L., eds., Understanding Granites: Integrating New and Classical Techniques: Geological Society [London] Special Publication 168, p. 55–75.

Pearson, D.M., Kapp, P., Reiners, P.W., Gehrels, G.E., Ducea, M.N., Pullen, A., Otamendi, J.E., and Alonso, R.N., 2012, Major Miocene exhumation by fault-propagation folding within a metamorphosed, early Paleozoic thrust belt: Northwestern Argentina: Tectonics, v. 31, TC4023, doi:10.1029/2011TC003043.

Pickett, D.A., and Saleeby, J.B., 1994, Nd, Sr, and Pb isotopic characteristics of Cretaceous intrusive rocks from deep levels of the Sierra Nevada

Ducea et al.

batholith, Tehachapi Mountains, California: Contributions to Mineralogy and Petrology, v. 118, p. 198–215, doi:10.1007/BF01052869.

Ramos, V.A., Vujovich, G.I., and Dallmeyer, R.D., 1996, Los klippes y ventanas tectónicas de la estructura preándica de la Sierra de Pie de Palo (San Juan): Edad e implicanciones tectónicas: Proceedings XIII Congreso Geológico Argentino y III Congreso de Exploración de Hidrocarburos, v. 5, p. 377–392.

Rapela, C.W., Coira, B., Toselli, A., and Saavedra, J., 1992, The Lower Paleozoic magmatism of southwestern Gondwana and the evolution of Famatinian orogene: International Geology Review, v. 34, p. 1081–1142, doi:10.1080/00206819209465657.

Rapela, C.W., Pankhurst, R.J., Casquet, C., Baldo, E., Saavedra, J., Galindo, C., and Fanning, C.M., 1998, The Pampean orogeny of the southern proto-Andes: Cambrian continental collision in the Sierras de Córdoba, *in* Pankhurst, R.J., and Rapela, C.W., eds., The Proto-Andean Margin of Gondwana: Geological Society [London] Special Publication 142, p. 181–217.

Rapela, C.W., Pankhurst, R.J., Casquet, C., Baldo, E., Saavedra, J., Galindo, C., and Fanning, C.M., 2007, The Pampean orogeny of the southern proto-Andes: Cambrian continental collision in the Sierras de Córdoba: Earth-Science Reviews, v. 142, p. 181–217.

Reid, J.B., Evans, O.C., and Fates, D.G., 1983, Magma mixing in granitic rocks of the central Sierra Nevada, California: Earth and Planetary Science Letters, v. 66, p. 243–261, doi:10.1016/0012-821X(83)90139-5.

Rossi, J.N., Toselli, A.J., Saavedra, J., Sial, A.N., Pellitero, E., and Ferreira, V.P., 2002, Common crustal source for contrasting peraluminous facies in the early Paleozoic Capillitas Batholith, NW Argentina: Gondwana Research, v. 5, p. 325–337, doi:10.1016/S1342-937X(05)70726-7.

Saal, A., Toselli, A.J., and Rossi, J.N., 1996, Granitoides y rocas básicas de la Sierra de Paganzo: Münchner Geologische Hefte, v. 19A, p. 119–210.

Saavedra, J., Toselli, A.J., Rossi, J.N., and Pellitero, E., 1996, Granitoides y rocas básicas del Cerro Toro: Münchner Geologische Hefte, v. 19A, p. 229–240.

Schwartz, J.J., Gromet, L.P., and Miró, R., 2008, Timing and duration of calc-alkaline arc of the Pampean orogeny: Implications for the Late Neoproterozoic to Cambrian evolution of Western Gondwana: The Journal of Geology, v. 116, p. 39–61, doi:10.1086/524122.

Tatsumi, Y., and Eggins, S., 1995, Subduction Zone Magmatism: Cambridge, UK, Blackwell Science, 211 p.

Taylor, S.R., and McLennan, S.M., 1985, The Continental Crust: Its Composition and Evolution: Oxford, UK, Blackwell Scientific Publications, 312 p.

Thomas, W.A., and Astini, R.A., 1996, The Argentine Precordillera: A traveler from the Ouachita embayment of North American Laurentia: Science, v. 273, p. 752–757, doi:10.1126/science.273.5276.752.

Thompson, A.B., 1996, Fertility of crustal rocks during anatexis: Transactions of the Royal Society of Edinburgh–Earth Sciences, v. 87, p. 1–10, doi:10.1017/S0263593300006428.

Toselli, A.J., Durand, F.R., Rossi de Toselli, J.N., and Saavedra, J., 1996, Esquema de evolución geotectónica y magmática Eopaleozoica del sistema de Famatina y sectores de Sierras Pampeanas: Actas XIII Congreso Geológico Argentino, v. 5, p. 443–462.

Trompette, R., 1997, Neoproterozoic (~600 Ma) aggregation of Western Gondwana: A tentative scenario: Precambrian Research, v. 82, p. 101–112, doi:10.1016/S0301-9268(96)00045-9.

Viramonte, J.M., Becchio, R.A., Viramonte, J.G., Pimentel, M.M., and Martino, R.D., 2007, Ordovician igneous and metamorphic units in southeastern Puna: New U-Pb and Sm-Nd data and implications for the evolution of northwestern Argentina: Journal of South American Earth Sciences, v. 24, p. 167–183, doi:10.1016/j.jsames.2007.05.005.

Vujovich, G.I., Godeas, M., Marín, G., and Pezzutti, N., 1996, El complejo magmático de la Sierra de La Huerta, provincia de San Juan: Actas XIII Congreso Geológico Argentino y III Congreso de Exploración de Hidrocarburos, v. 3, p. 465–475.

Zimmermann, U., 2005, Provenance studies of very low- to low-grade metasedimentary rocks of the Puncoviscana complex, northwest Argentina, *in* Vaughan, A.P.M., Leat, P.T., and Pankhurst, R.J., eds., Terrane Processes at the Margins of Gondwana: Geological Society [London] Special Publication 246, p. 381–416.

Zimmermann, U., and Bahlburg, H., 2003, Provenance analysis and tectonic setting of the Ordovician clastic deposits in the southern Puna Basin, NW Argentina: Sedimentology, v. 50, p. 1079–1104, doi:10.1046/j.1365-3091.2003.00595.x.

MANUSCRIPT ACCEPTED BY THE SOCIETY 3 JUNE 2014
MANUSCRIPT PUBLISHED ONLINE 23 SEPTEMBER 2014

The Geological Society of America
Memoir 212
2015

Foundering-driven lithospheric melting: The source of central Andean mafic lavas on the Puna Plateau (22°S–27°S)

Kendra E. Murray
Department of Geosciences, University of Arizona, 1040 E. 4th Street, Tucson, Arizona 85721, USA

Mihai N. Ducea
Department of Geosciences, University of Arizona, 1040 E. 4th Street, Tucson, Arizona 85721, USA, and
Facultatea de Geologie Geofizica, Universitatea Bucuresti, Strada N. Balcescu Nr 1, Bucuresti, Romania

Lindsay Schoenbohm
Department of Chemical and Physical Sciences, University of Toronto Mississauga, Mississauga, ON L5L 1C6, Canada

ABSTRACT

Investigations of lithospheric foundering and related magmatism have long focused on the central Andes, where there are postulated links between the eruption of mantle-derived lavas and periodic loss of the lower lithosphere. Whole-rock elemental and Nd-Sr-Pb isotopic results from a suite of late Miocene–Quaternary mafic lavas erupted onto the Puna Plateau clarify the relationship between this hypothesized process and lava composition. Zinc and Fe provide a critical perspective because they are partitioned differently during the melting of asthenospheric and lithospheric mantle. All Puna lavas have Zn/Fe_T ($\times 10^4$) values >13, which requires clinopyroxene and perhaps garnet to be the dominant phase(s) in the melt source; this precludes a melt source of typical mantle asthenosphere. This result is contrary to classic models of delamination magmatism that suggest asthenospheric peridotite melts to generate these lavas. Pyroxenite (±garnet)–bearing lithospheric materials in the central Andes are likely common and heterogeneous in age, volatile content, and mineralogical composition, and if they are the melt source, this can explain the diversity in the elemental (La/Yb = 11–45; La/Ta = 22–40) and isotopic ($^{87}Sr/^{86}Sr$ = 0.7055–0.7080; ε_{Nd} = −1 to −7) compositions of these mafic magmas (MgO > 8%, Mg number > 60). We propose that compositionally diverse, gravitationally unstable pyroxenites both drive "dripping" of the lower lithosphere and are the source of the resulting melt. We also postulate that mantle-derived lavas erupted on the Puna Plateau were generated during localized foundering and melting of these materials. The cumulative effect of these drip events is a modern Puna Plateau with geodynamic anomalies including thin lithosphere and anomalously high surface elevation.

Murray, K.E., Ducea, M.N., and Schoenbohm, L., 2015, Foundering-driven lithospheric melting: The source of central Andean mafic lavas on the Puna Plateau (22°S–27°S), *in* DeCelles, P.G., Ducea, M.N., Carrapa, B., and Kapp, P.A., eds., Geodynamics of a Cordilleran Orogenic System: The Central Andes of Argentina and Northern Chile: Geological Society of America Memoir 212, p. 139–166, doi:10.1130/2015.1212(08). For permission to copy, contact editing@geosociety.org.

INTRODUCTION

In the central Andes, extensive differentiation and crustal mixing in Neogene arc magmas has obscured the composition of the regional mantle (Davidson et al., 1991), which restricts our understanding of Andean lithospheric evolution and arc petrogenesis (Thorpe et al., 1984; de Silva, 1989; Hildreth and Moorbath, 1988; Rogers and Hawkesworth, 1989; Davidson et al., 1991; Francis and Hawkesworth, 1994; Beck et al., 1996; Haschke et al., 2006; Lucassen et al., 2007; Kay and Coira, 2009; Mamani et al., 2010; Risse et al., 2013; Kay et al., 2013; Ducea et al., 2013). Mantle-derived mafic lavas erupted onto the Puna Plateau (Fig. 1) since 10 Ma are the most direct evidence for late Cenozoic upper-mantle composition. These lavas are thought to be the product of late Cenozoic loss of dense lower lithosphere (Kay et al., 1994), a process commonly referred to as lithospheric foundering or delamination (Kay and Kay, 1993; Göğüş and Pysklywec, 2008). Recycling of lithospheric material into the convecting upper mantle is required by chemical and mass balance in cordilleran orogens like the central Andes (Rudnick, 1995) and is hypothesized to play a key role in Andean geodynamic evolution (Kay et al., 1994; Beck and Zandt, 2002; Schurr et al., 2006; Garzione et al., 2006; DeCelles et al., 2009; Pelletier et al., 2010).

If lithospheric loss generated the Puna mafic lavas, then the scale of individual foundering events, the mantle material(s) that melted in response, and the relationship between these parameters and the composition of resulting melts remain equivocal. Kay et al. (1994) proposed a model for "delamination magmatism" based on trace-element compositions of the Puna mafic lavas, regional geophysical data, and structural observations. They used geographic trends in K and La/Ta values as an indicator of (1) the relative proportions of asthenospheric melt generated below different parts of the Puna Plateau, and/or (2) variability in the "arc-like" chemistry of the mantle source as a result of delamination. This trace-element interpretation required that the lavas were generated from a relatively homogeneous asthenospheric mantle peridotite that adiabatically upwelled and melted in the wake of a large foundering block of lower lithosphere (Kay and Kay, 1993). However, if K and La/Ta values fingerprint extensive replacement of the lower lithosphere by upwelling asthenosphere, then there should be other systematic trends in the elemental and isotopic values in Puna primitive lavas. This has been observed in other orogens (e.g., Iberian Massif—Gutierrez-Alonso et al., 2011; Canadian cordillera—Manthei et al., 2010). Instead, Kay et al. (1994) and subsequent studies of primitive Puna lavas (Kraemer et al., 1999; Drew et al., 2009; Risse et al., 2008, 2013) have

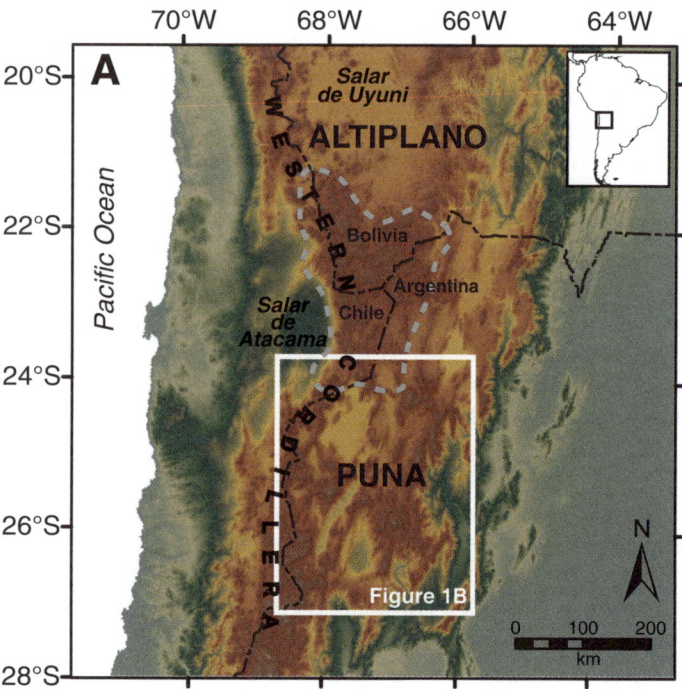

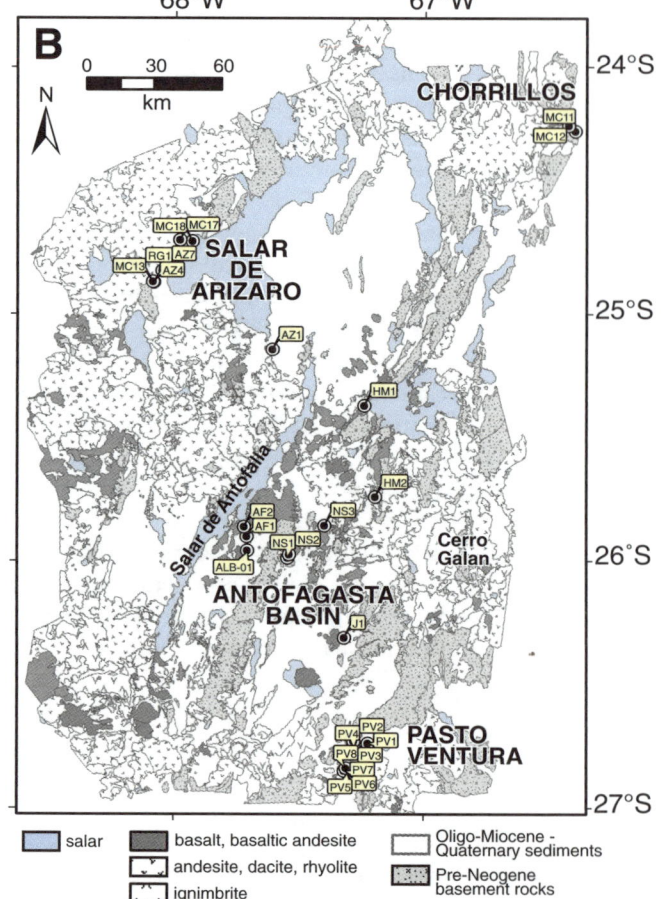

Figure 1. (A) Digital elevation map of the south-central Andes. The active arc stratovolcanoes are currently located in the Western Cordillera. Gray dotted line shows the regional extent of the Miocene Altiplano-Puna volcanic complex ignimbrites after de Silva et al. (2006). (B) Simplified geologic map of the Puna Plateau modified from Schnurr et al. (2006), with sample locations and regional names from this study.

reported a striking decoupling of whole-rock major-element, trace-element, and isotopic compositions. This compositional complexity has been attributed to variable re-enrichment of the mantle source by metasomatism or subduction erosion (Risse et al., 2013, and references therein), preserved pieces of enriched Paleozoic subcontinental lithospheric mantle (Drew et al., 2009), or minor to moderate amounts of crustal contamination (up to 20%–25%; Kay et al., 1994; Risse et al., 2013). If we consider that the lower crust and lithospheric mantle beneath long-lived cordilleran systems would tend toward and experience foundering because of profound heterogeneities in age, composition, and temperature (Elkins-Tanton, 2007), we evidently need better ways to clarify the links between lava composition and the materials driving foundering events.

Here, we find that the compositional complexity of Puna mafic lavas offers a critical insight into the processes driving melting in the Puna mantle. Considered together with high Zn/Fe_T values in the most primitive lavas (Le Roux et al., 2010), which require dominantly clinopyroxene (±garnet)–bearing source compositions, the trace-element and Nd-Sr isotopic data suggest that heterogeneous lithospheric material, and not asthenosphere, is the primary source of these lavas. This contribution expands upon Ducea et al. (2013) by presenting and interpreting the full trace-element and Nd-Sr-Pb isotopic data set for the 26 Puna lava samples used in that study. These results suggest that if lithospheric foundering is generating the Puna mafic lavas, then individual foundering "drips" must be small enough not to homogenize the Puna upper mantle or drive significant melting of the peridotitic asthenosphere. This likely precludes rapid, large-scale, and catastrophic loss of the Puna lower lithosphere. Additionally, we postulate that localized "dripping" of the lower lithosphere is a common result of localized, regular accumulation of dense lithospheric residues, and that the Puna lavas are evidence of a routine process in active central-Andean–type orogens over geologic time scales.

GEOLOGIC BACKGROUND

Composition of the Central Andean Upper Mantle

In the Andean Northern and Southern Volcanic Zones, there is a clear depleted mantle end member in the Nd and Sr isotopic compositions of arc volcanics (Davidson et al., 1991). In the Central Volcanic Zone, however, the depleted mantle is not evident, either because (1) it is obscured by prolonged differentiation and crustal mixing, or (2) it is not present in significant volumes. Additionally, there is probably no single upper-mantle composition in the Central Volcanic Zone because the region has been an active convergent margin throughout the Phanerozoic. Multiple phases of pre-Andean terrane accretion, arc magmatism, and deformation (Ramos, 2009, and references therein) likely generated lithosphere-scale chemical heterogeneities.

The only mantle xenoliths found in the Andes between 22°S and 28°S are from the Cretaceous Salta Rift volcanics,

most notably in Las Conchas valley (~26°S; Lucassen et al., 2005, 2002; Viramonte et al., 1999) located east of the Puna Plateau. Their compositions suggest significant mantle diversity beneath the central Andean margin during the Cretaceous. At that time, the active Andean magmatic arc was centered ~100 km west of its modern position, and this continental rift system was active in the back arc and foreland (Viramonte et al., 1999). The Las Conchas xenolith suite presented by Lucassen et al. (2005) is composed of peridotites (spinel lherzolites and harzburgites with 2%–18% calculated modal clinopyroxene) and minor pyroxenites; there are no garnet-bearing xenoliths. Although many of these xenoliths and the basanite flows that host them are clearly sourced from the depleted mantle ($\varepsilon_{Nd} > 0$, $^{87}Sr/^{86}Sr < 0.704$), there are also many peridotite and pyroxenite samples with elemental and isotopic variability suggesting that diverse mantle sources were tapped by the rift-induced melting (Lucassen et al., 2005). These heterogeneities are likely more profound in closer proximity to the continental margin to the west.

Lithospheric Foundering and Melt Generation

Foundering of dense lithosphere into the asthenosphere has been invoked to explain "missing" lower crust and lithospheric mantle in regions that have undergone significant crustal shortening and/or batholith-producing magmatism (Ducea, 2001, and references therein). Lithospheric foundering also has the potential to modulate the shape, volume, and surface elevation of orogenic wedges (DeCelles et al., 2009). There are two end-member models (Göğüş and Pysklywec, 2008) for the process in which gravitational instabilities lead to either peeling away (i.e., delamination; Bird, 1979; Kay and Kay, 1993) or convective dripping (i.e., Rayleigh-Taylor instabilities; Houseman et al., 1981; Jull and Keleman, 2001) of dense material in the lower lithosphere. Cordilleran magmatic arcs tend toward lithospheric foundering because ocean-continent convergence (1) generates lithosphere-scale structures and temperature inversions during crustal shortening; (2) produces large volumes of melt and dense residues (Ducea and Barton, 2007); and (3) brings crustal material to pressures and temperatures sufficient to produce dense phases and low-viscosity zones (Kay and Kay, 1993; Elkins-Tanton, 2005). Geophysical observations in the central Andes (Beck and Zandt, 2002; Yuan et al., 2002; Schurr et al., 2006; Bianchi et al., 2013) and the Sierra Nevada (Zandt et al., 2004; Frassetto et al., 2011) suggest that lithospheric removal is happening today. Numerical modeling has identified possible links and relative timing of orogen-scale responses to foundering events (Houseman and Molnar, 1997; Molnar and Garzione, 2007; Elkins-Tanton, 2005, 2007; Pelletier et al., 2010; Pysklywec and Cruden, 2004; Göğüş and Pysklywec, 2008). In the central Andes, proposed foundering events have been of particular interest to workers in the region who report changes in regional strain orientation and heat flow, and apparent fluctuations in the rates of crustal shortening, faulting, surface uplift, and sedimentation (England and Houseman,

1989; Kay et al., 1994; Garzione et al., 2008; DeCelles et al., 2009; Schoenbohm and Strecker, 2009).

The mafic volcanic record has been used to support other geological evidence for proposed lithosphere-removal events in a variety of tectonic settings, including the Colorado Plateau (Bird, 1979; Crow et al., 2011), Sierra Nevada (Farmer et al., 2002; Ducea, 2002), Iberian Massif (Gutierrez-Alonso et al., 2011), central Andes (Kay et al., 1994), Tibet (Chung et al., 2009), and Canadian cordillera (Manthei et al., 2010). Foundering is thought to generate mantle melts because foundering pieces of lithosphere make space for mantle asthenosphere to upwell, melt via decompression, and advect heat toward the base of the remaining lithosphere (Kay and Kay, 1993; Kay et al., 1994; Ducea and Saleeby, 1998). This convective overturn of the lower lithosphere should change the elemental and isotopic composition of the mantle source region and create a geochemical fingerprint (Ducea, 2011). This has been documented in the Iberian Massif (Gutierrez-Alonso et al., 2011), Sierra Nevada (Ducea and Saleeby, 1998), and Canadian cordillera (Manthei et al., 2010). Melts could also be generated directly from the descending lithosphere, which could expel fluids into the mantle asthenosphere and/or cross the melting solidus prior to reaching thermal equilibrium with the upper mantle (Elkins-Tanton, 2007). When conditions conspire to destabilize the lower part of the lithosphere, the style of detachment, volume and composition of foundering material, and the time scale of its removal should all contribute to the volume and chemistry of the resulting melt.

In active convergent margins like the Andes, subduction-zone melt production in the mantle wedge and subsequent assimilation, storage, homogenization, and fractional crystallization (Hildreth and Moorbath, 1988) during magma migration through the lithosphere modulate the composition and volume of arc magmas. It is critical to consider possible contributions from all these processes when interpreting trace-element and isotopic results in primitive cordilleran arc lavas.

South-Central Andean Arc

The southern part of the Central Volcanic Zone (~22°S–27°S) preserves a >150 m.y. record of arc magmatism, during which time there were periodic variations in magma chemistry over 10^5 to 10^8 m.y. time scales (Haschke et al., 2006; DeCelles et al., 2009). To explain these trends, workers have proposed models of repeated episodes of slab shallowing and breakoff (Haschke et al., 2006) in conjunction with other dynamic processes in the cordillera, including plateau formation, crustal thickening, and fold-and-thrust belt propagation (Ramos, 2009). DeCelles et al. (2009) proposed that this periodicity reflects cordilleran cyclicity controlled by retroarc thrusting, melt production, and lithospheric foundering. Perturbations in the timing, locus, and volume of melt production in the mantle wedge are related to the evolution of the orogen as a whole. Today in the south-central Andes, the overriding plate's great lithospheric and crustal thickness (>60 km thick crust in some regions; Yuan et al., 2002; Tassara and Echaurren,

2012) impedes magma migration and thus facilitates prolonged thermal and chemical interactions between the crust and magmas derived from melting in the mantle wedge (Thorpe et al., 1984; Hildreth and Moorbath, 1988; Davidson et al., 1991; Kay and Coira, 2009; Mamani et al., 2010). Since the Jurassic, the axis of the central Andean arc has migrated east ~250–350 km (Stern, 1991; Haschke et al., 2006). A corresponding amount of forearc crust has been removed, likely by subduction erosion. Some workers attributed the evolved compositions in central Andean lavas to this material being incorporated into arc magmas (Stern, 1991; Kay et al., 2013). The north-to-south transit of the Juan Fernández Ridge through this region during the Miocene (Yáñez et al., 2001) could have affected the dip of the subducting slab and related magmatism (Kay and Coira, 2009), though it does not appear to have generated a gap in volcanism as is observed today where the ridge subducts in the Chilean-Pampean flat slab region (Trumbull et al., 2006).

Neogene central Andean volcanics can be broadly divided into three categories based on volume, major-element composition, and eruptive style: (1) andesitic-dacitic stratovolcanoes along the modern arc front in the Western Cordillera (Fig. 1A), of which 44 are active today (de Silva and Francis, 1991); (2) intermediate to silicic ignimbrite flows erupted from large-volume (>1000 km³) caldera-forming eruptions, including the mid-Miocene to Pliocene Altiplano-Puna volcanic complex (APVC, Fig. 1A; de Silva, 1989); and (3) low-volume primitive lavas erupted east of the modern stratovolcanoes on both the Altiplano and Puna Plateaus (Fig. 1B), a subset of which were sampled for this study. These categories are not necessarily exclusive of each other (e.g., Volcán Ollagüe; Mattioli et al., 2006), but they do provide a useful framework for discussing the diversity of melt compositions and scenarios that produced them.

Volcanism on the Puna Plateau

The Puna Plateau (Fig. 1B) comprises a series of internally drained basins structurally bounded by uplifted blocks of Paleozoic plutonic and metasedimentary rocks (Allmendinger et al., 1997) through which Neogene–Quaternary volcanics erupted. The Puna is higher in mean elevation (~4000 m) and more structurally dissected than its northern continuation in Bolivia and Peru, the Altiplano (~3000 m; Whitman et al., 1996). All categories of Andean volcanics, as described previously, are common in the study area, including stratovolcanoes in the Western Cordillera, calderas and cones developed along NW-SE–trending lineaments (Matteini et al., 2002), a volcanic center at Cerro Tuzgle in the northern Puna (Coira and Kay, 1993), the Cerro Galan caldera complex (Francis et al., 1989), other silicic ignimbrites (Schnurr et al., 2007; Kay et al., 2010), and the mafic, small-volume lavas of interest to this study (Francis et al., 1989; Knox et al., 1989; Déruelle, 1991; Kay et al., 1994; Kraemer et al., 1999; Drew et al., 2009; Risse et al., 2013; Ducea et al., 2013).

Where dated, the small-volume mafic fissure flows and cinder cones on the Puna Plateau are late Miocene to Pleistocene or younger in age (7.3 to <0.1 Ma) and appear to have peaked in activity in the early Pliocene (Risse et al., 2008, 2013). The lavas are scattered across the plateau, occasionally grouped into volcanic fields, and are only present east of the modern arc stratovolcanoes. Some flows and cones are clearly associated with strike-slip and extensional faults (Marrett and Emerman, 1992; Zhou et al., 2013).

The original interpretation of these lavas as products of delamination magmatism (Kay et al., 1994) relied on geophysical, structural, and geochemical data from the Puna Plateau. These data fit a model (Kay and Kay, 1993) in which the asthenospheric mantle played a critical role in melt generation. The Puna lithosphere is anomalously thin, the plateau has a high surface elevation today, and the Puna mafic lavas are associated with normal and strike-slip faults that record a period of regional horizontal extension contemporaneous with eruption (Kay et al., 1994, and references therein). The presence of primitive lavas erupted in the midst of a cordilleran arc is itself an important piece of evidence for lithospheric foundering, as it requires some mechanism(s) for circumnavigating the prolonged residence time in the crust and resulting magma evolution that dominates all other Andean lavas. There is a roughly bull's-eye pattern in the Puna lava geochemistry in the major element K, and the trace elements Ba, Ta, and La centered ~26°S (Fig. 1B). At the center of this pattern (the Antofagasta Basin), the mafic lavas have relatively low K compositions and values of La/Ta < 25; on the perimeter of this pattern, the lavas are higher K and have values of La/Ta > 25. Kay et al. (1994) interpreted this pattern to be evidence of (1) differences in the relative percent melt in the asthenosphere beneath the Puna Plateau, because smaller volumes of melt have higher concentrations of incompatible elements (La, K; higher La/Ta), and (2) the strength of the subduction zone influence on the mantle source, because arc magmas tend to be enriched in large ion lithophile elements (Ba, K) and depleted in high field strength elements (Ta). The authors attributed this pattern to a delamination event: At the locus of lithospheric loss, there was greater melt generation and significant influx of non–slab-fluid-enriched asthenosphere, whereas on the edges of the event, less upwelling and melting occurred. Kay et al. (1994) labeled the low La/Ta lavas at the center of this pattern "OIB-like" (ocean-island basalt–like) or "intraplate" because of some chemical similarities to Pacific hotspot basalts (Knox et al., 1989). The higher La/Ta lavas were labeled "back-arc calc-alkaline," because their trace-element composition is typical of arc magmas.

Although the geographic patterns observed by Kay et al. (1994) are noteworthy, the nonsystematic variability in most aspects of the lavas' elemental and isotopic chemistry is arguably a more important characteristic of these flows. The Sr and Nd data from these lavas exemplify this variability. The Puna mafic lavas have the same range of Nd and Sr isotopic compositions as the regional stratovolcanoes, and no clear geographic pattern (Kay et al., 1994; Drew et al., 2009). In these lavas, increasing SiO_2 composition generally corresponds with more radiogenic Sr and Nd isotopic compositions, which is consistent with 20%–25%

crustal mixing (Kay et al., 1994; Risse et al., 2013). However, if crustal mixing is in part responsible for the observed isotopic variability in the more-evolved lavas, then a more subtle and nonsystematic contamination of the trace elements in even the most primitive (i.e., highest wt% MgO) lavas is possible (Glazner and Farmer, 1992). Drew et al. (2009) reported that even the least modified basaltic lavas (47.5 wt% SiO_2, 10.3 wt% MgO) have isotopic signatures requiring long-term enrichment of the source in incompatible elements relative to the depleted mantle ($^{87}Sr/^{86}Sr$ = 0.7055, ε_{Nd} = -0.8) and suggested that the mantle source region likely included some enriched, aged, subcontinental lithospheric mantle. Risse et al. (2013) preferred an alternative explanation in which the Puna upper mantle has experienced re-enrichment since the Miocene due to subduction erosion and foundering. Recently, Ducea et al. (2013) proposed that foundering pieces ("drips") of dense lithosphere melt as they descend into the asthenosphere and are the source of the Puna mafic lavas. Whereas attributing the trace-element and isotopic complexities of these lavas to melting drips of compositionally diverse lithospheric material does not resolve whether this enriched lithospheric material is remnant Paleozoic lithospheric mantle or younger mantle re-enriched by subduction-related processes, it does suggest a solution to the larger question of what is driving mantle melting beneath the Puna region and how that could be related to the evolution of the orogen.

In light of these complexities, we use an emerging approach for relating the elemental composition of mantle-derived lavas to the modal mineralogy of their primary source: ratios between first-row transition elements.

Zn/Fe$_T$ Values

Recent development of the first-row transition elements (the major elements Mn and Fe and trace elements Zn, Co, Ni) as tracers of the mineralogical composition of the sources of mid-ocean-ridge basalt (MORB) and ocean-island basaltic (OIB) lavas (Le Roux et al., 2010, 2011; Lee et al., 2010) offers a promising new approach to fingerprinting the sources of mantle-derived lavas erupted through the continental crust. Experimental results (Le Roux et al., 2010, 2011) suggest that if a mantle lithology is peridotitic (i.e., olivine and orthopyroxene are the dominant phases), then when melting occurs, there is minimal fractionation of Zn, Fe, and Mn. The melt will have the same Zn/Fe$_T$ and Mn/Fe$_T$ values as its source, and subsequent fractional crystallization of olivine will not modify these values in the resulting basaltic magma. In contrast, if clinopyroxene or garnet dominates the modal mineralogy in the mantle source, then these elements will fractionate during melting and generate magmas with higher Zn/Fe$_T$ values than the source.

We use the Zn/Fe$_T$ value (Le Roux et al., 2010) as a basis for our interpretation of the Puna mafic lavas because it offers a different approach than traditional trace-element ratios and connects Puna lava chemistry to the mantle materials hypothesized to play a key role in lithospheric foundering. Elemental ratios

TABLE 1. ELEMENTAL AND ISOTOPIC RESULTS

Locality:	Chorrillos					Salar de Arizaro	
Sample ID:	MC11*	MC12	MC13	MC17	MC18	AZ1	AZ4
Latitude (°S):	24.2592	24.2364	24.8684	24.6999	24.7055	25.1427	24.8249
Longitude (°W):	66.3967	66.4238	68.0886	67.9827	67.9331	67.6105	68.0525
Rock type:[†]	BTA	BTA	BTA	BA	BA	BA	BTA
SiO_2	54.43	55.96	54.35	54.38	54.66	56.04	52.80
TiO_2	1.339	1.164	1.207	1.200	1.074	1.065	1.101
Al_2O_3	13.46	14.88	16.30	17.28	15.96	16.42	16.64
FeO_T	6.32	5.90	6.86	7.91	6.88	6.82	7.72
MnO	0.106	0.094	0.111	0.133	0.121	0.115	0.136
MgO	7.88	6.31	5.52	4.91	6.38	5.48	5.27
CaO	6.23	6.60	7.76	7.96	7.91	7.40	8.30
Na_2O	2.52	2.92	3.65	3.00	3.30	3.51	3.45
K_2O	5.20	3.34	1.94	1.89	1.79	2.02	1.77
P_2O_5	0.862	0.529	0.377	0.315	0.297	0.287	0.358
Sum	98.34	97.69	98.08	98.99	98.39	99.17	97.55
LOI (%)	0.93	1.13	1.36	0.68	1.09	0.40	1.52
Ni	179	116	79	34	97	75	34
Cr	341	253	183	68	261	217	83
Sc	16	16	17	19	21	20	24
V	138	149	180	190	187	184	191
Ba	1923	910	541	440	448	495	462
Rb	181.7	113.3	46.1	33.1	46.7	60.0	40.4
Sr	966	730	698	789	620	639	756
Zr	475	236	201	199	173	182	186
Y	24.09	18.77	17.82	27.70	19.46	18.85	29.50
Nb	36.78	22.01	13.86	12.62	9.14	10.97	11.56
Ga	19	21	20	19	18	22	20
Cu	33	24	42	21	44	39	26
Zn	95	100	98	92	87	90	79
Pb	31.86	25.29	8.89	6.73	7.96	9.18	4.30
La	73.77	58.31	40.74	33.94	33.06	35.26	34.72
Ce	150.10	115.30	81.17	71.47	67.92	70.73	72.10
Th	15.85	14.03	6.16	5.01	5.22	7.98	5.20
Nd	68.83	51.28	37.85	36.39	32.09	32.64	35.38
U	3.75	4.18	1.29	0.93	1.17	1.42	1.04
Pr	18.00	13.62	9.79	8.99	8.27	8.54	9.03
Sm	13.30	9.56	7.47	7.61	6.38	6.30	7.35
Eu	3.08	2.17	1.92	1.94	1.64	1.61	1.83
Gd	9.15	6.90	5.85	6.51	5.19	5.01	6.32
Tb	1.15	0.88	0.79	0.99	0.75	0.71	0.97
Dy	5.58	4.33	3.94	5.73	4.16	3.82	5.69
Ho	0.92	0.73	0.70	1.09	0.76	0.70	1.10
Er	2.13	1.70	1.64	2.78	1.95	1.80	2.98
Tm	0.28	0.23	0.23	0.38	0.27	0.26	0.42
Yb	1.64	1.33	1.31	2.29	1.64	1.55	2.59
Lu	0.25	0.20	0.20	0.35	0.26	0.24	0.40
Hf	12.77	6.43	5.22	5.17	4.52	4.81	4.87
Ta	2.03	1.39	0.79	0.71	0.58	0.68	0.68
Cs	7.85	3.76	0.95	0.50	1.14	1.37	0.72
Sc	15.3	13.6	15.0	18.5	20.4	20.4	23.7
Mg#[§]	69	66	59	53	62	59	55
La/Yb	45.1	43.7	31.1	14.8	20.1	22.8	13.4
Zn/Fe**	19.3	21.8	18.4	15.0	16.3	17.0	13.2
$^{87}Sr/^{86}Sr_{(0)}$	0.708046	0.708269	0.70678	0.706787	0.707271	0.70643	0.70649
ε_{Nd}	−7.49	−7.96	−3.88	−5.07	−6.38	−2.13	−1.13
$^{206}Pb/^{204}Pb$	18.74099	18.73075	18.8796	18.92396	18.98098	18.891	18.9603
$^{207}Pb/^{204}Pb$	15.70293	15.68827	15.6403	15.63497	15.66287	15.6361	15.6447
$^{208}Pb/^{204}Pb$	39.20095	39.00297	38.8629	38.82893	38.89336	38.8583	38.8659

		Valle de Antofagasta					Locality:
AZ7*	RG1	HM1	HM2	AF1	AF2	ALB-01*	Sample ID:
24.5624	25.0757	25.3713	25.7414	25.9008	25.8614	25.9568	Latitude (°S):
67.9023	68.2070	67.2434	67.1981	67.7108	67.7211	67.7094	Longitude (°W):
B	B	A	D	TA	A	BTA	Rock type:[†]
51.87	51.25	62.92	66.69	61.50	57.85	52.31	SiO_2
1.082	1.121	0.905	0.768	1.066	1.306	1.416	TiO_2
16.08	15.80	15.90	15.50	15.67	15.84	15.71	Al_2O_3
8.09	7.42	4.71	3.60	4.97	6.56	7.84	FeO_T
0.142	0.127	0.065	0.041	0.069	0.097	0.129	MnO
8.06	6.14	2.41	1.48	2.86	4.36	8.19	MgO
8.68	8.90	4.40	3.01	4.75	6.09	7.73	CaO
3.16	3.10	3.84	3.72	3.83	3.75	3.31	Na_2O
1.35	1.76	2.94	3.71	3.07	2.10	1.88	K_2O
0.232	0.276	0.250	0.237	0.352	0.362	0.398	P_2O_5
98.75	95.90	98.34	98.75	98.13	98.32	98.92	Sum
0.82	2.07	−0.06	0.03	0.08	0.41	0.30	LOI (%)
127	83	23	14	38	65	191	Ni
361	223	46	22	64	108	323	Cr
27	22	11	7	11	17	21	Sc
222	199	116	87	124	171	179	V
277	400	721	860	755	596	486	Ba
34.4	39.4	110.8	154.4	103.5	65.9	47.5	Rb
487	503	496	503	659	577	680	Sr
144	174	216	249	239	229	189	Zr
21.38*	19.92	13.73	13.36	15.95	20.21	21.23	Y
7.78	8.92	10.96	10.22	12.38	14.40	27.44	Nb
20	18	23	24	23	23	19	Ga
52	48	21	23	24	32	42	Cu
89	101	94	80	90	106	84	Zn
6.04	8.15	17.06	19.08	15.67	12.99	8.27	Pb
20.20	26.10	44.38	53.66	54.04	45.48	39.38	La
43.87	55.68	87.45	103.88	105.86	92.89	75.84	Ce
3.05	3.38	13.23	17.65	14.53	8.86	6.43	Th
23.89	28.74	37.77	42.95	45.09	43.37	32.91	Nd
0.91	0.80	2.80	4.08	2.49	1.63	1.49	U
5.76	7.15	10.17	11.89	12.33	11.27	8.85	Pr
5.38	6.05	6.72	7.26	7.81	8.24	6.34	Sm
1.46	1.66	1.59	1.53	1.81	1.96	1.77	Eu
4.84	5.15	4.81	5.11	5.71	6.51	5.33	Gd
0.74	0.76	0.63	0.64	0.75	0.89	0.79	Tb
4.29	4.09	3.11	3.08	3.76	4.54	4.45	Dy
0.82	0.76	0.56	0.53	0.64	0.80	0.82	Ho
2.16	1.96	1.34	1.27	1.46	1.97	2.09	Er
0.30	0.27	0.18	0.17	0.19	0.26	0.29	Tm
1.85	1.63	1.07	0.97	1.08	1.52	1.74	Yb
0.29	0.25	0.16	0.15	0.16	0.23	0.26	Lu
3.89	4.62	5.79	6.74	6.38	6.10	4.58	Hf
0.51	0.55	0.76	0.76	0.77	0.96	1.67	Ta
0.96	1.00	3.67	6.08	2.19	1.43	1.47	Cs
27.5	21.5	10.2	7.2	10.4	15.2	21.0	Sc
64	60	48	42	51	54	65	Mg#[§]
10.9	16.1	41.5	55.2	49.9	29.8	22.6	La/Yb
14.2	17.5	25.7	28.6	23.3	20.8	13.8	Zn/Fe**
0.70549	0.706867	0.708797	0.709284	0.708036	0.70832	0.705749	$^{87}Sr/^{86}Sr_{(0)}$
−0.94	−1.99	−3.71	−5.36	−3.08	−3.92	−1.13	ε_{Nd}
18.8973	18.77795	18.87389	19.03484	18.96242	18.94922	18.9566	$^{206}Pb/^{204}Pb$
15.6372	15.63994	15.66733	15.68765	15.67573	15.66526	15.6546	$^{207}Pb/^{204}Pb$
38.7469	38.73998	39.04708	39.26717	39.13935	39.01588	38.9615	$^{208}Pb/^{204}Pb$

(Continued)

TABLE 1. ELEMENTAL AND ISOTOPIC RESULTS (*Continued*)

Locality:	Valle de Antofagasta					
Sample ID:	NS1*	NS2*	NS3	J1*	PV1	PV2
Latitude (°S):	25.9865	25.9732	25.8574	26.3110	26.7382	26.7286
Longitude (°W):	67.5483	67.5411	67.4004	67.3195	67.2248	67.2809
Rock type:[†]	BA	BA	A	B	BTA	BTA
SiO_2	53.60	53.98	59.23	51.40	53.15	52.23
TiO_2	1.131	1.039	1.113	1.572	1.094	1.149
Al_2O_3	14.83	14.43	16.24	15.46	15.07	15.70
FeO_T	7.80	7.65	6.53	8.94	7.29	7.77
MnO	0.137	0.138	0.090	0.143	0.123	0.131
MgO	9.25	9.17	3.41	8.43	7.00	7.40
CaO	7.39	7.57	6.16	7.99	8.10	8.37
Na_2O	2.94	2.85	3.45	3.29	3.29	3.46
K_2O	1.71	1.62	2.09	1.68	1.95	1.77
P_2O_5	0.266	0.244	0.275	0.343	0.344	0.346
Sum	99.06	98.68	98.58	99.24	97.41	98.32
LOI (%)	−0.25	−0.22	0.17	0.57	1.44	1.31
Ni	217	184	35	190	129	139
Cr	545	573	64	326	353	276
Sc	24	25	15	23	21	23
V	185	186	160	204	180	196
Ba	401	375	521	437	518	471
Rb	47.1	43.8	58.2	43.1	56.8	47.6
Sr	542	542	525	658	711	768
Zr	155	145	280	172	205	170
Y	20.87	20.98	20.32	22.68	20.76	20.39
Nb	16.98*	15.74*	11.57	22.63*	13.63	13.71
Ga	18	16	22	19	19	21
Cu	51	50	36	51	47	51
Zn	87	78	108	94	84	87
Pb	8.98	8.65	9.96	8.16	11.13	9.24
La	28.36	25.29	41.77	34.42	36.07	34.70
Ce	57.26	51.51	86.52	69.97	71.15	68.67
Th	5.34	4.61	9.34	5.21	7.29	6.81
Nd	26.67	24.36	41.46	33.35	33.78	32.81
U	1.25	1.09	1.41	1.15	2.19	1.50
Pr	6.92	6.24	10.67	8.53	8.72	8.37
Sm	5.26	4.91	8.39	6.45	6.57	6.44
Eu	1.42	1.39	1.87	1.80	1.70	1.74
Gd	4.61	4.46	6.58	5.54	5.44	5.38
Tb	0.72	0.69	0.88	0.82	0.77	0.75
Dy	4.15	4.05	4.52	4.62	4.34	4.19
Ho	0.82	0.82	0.81	0.89	0.81	0.77
Er	2.18	2.19	1.91	2.28	2.06	2.01
Tm	0.31	0.31	0.26	0.32	0.28	0.28
Yb	1.90	1.94	1.57	1.94	1.71	1.64
Lu	0.30	0.30	0.24	0.30	0.26	0.25
Hf	3.87	3.68	7.23	4.30	5.13	4.42
Ta	1.13	1.06	0.64	1.54	0.87	0.87
Cs	1.63	1.65	1.51	1.34	2.09	1.66
Sc	24.2	25.7	15.2	23.7	20.0	21.9
Mg#[§]	68	68	48	63	63	63
La/Yb	14.9	13.1	26.6	17.8	21.1	21.2
Zn/Fe**	14.3	13.1	21.3	13.5	14.8	14.4
$^{87}Sr/^{86}Sr_{(0)}$	0.705849	0.705543	0.709521	0.706108	0.70657	0.705465
ε_{Nd}	−2.38	−2.77	−7.00	−4.08	−5.31	−0.99
$^{206}Pb/^{204}Pb$	18.8312	18.83992	19.04474	18.97661	18.89649	18.98954
$^{207}Pb/^{204}Pb$	15.65526	15.6557	15.68221	15.66781	15.66558	15.6605
$^{208}Pb/^{204}Pb$	38.79381	38.78373	39.14805	38.96385	39.04852	39.10941

Note: LOI—loss on ignition. Major oxides are in wt% and trace elements are in ppm.
*High-Mg group.
[†]B—basalt; BA—basaltic andesite; BTA—basaltic trachyandesite; A—andesite; TA—trachyandesite; D—dacite.
[§]Mg# = [mol Mg/(mol Mg + mol Fe)] × 100, assuming all Fe^{2+}.
**Zn/Fe = (ppm Zn/ppm Fe) × 10^4, assuming all Fe^{2+}.

	PV3	PV4	PV5	PV6	PV7	PV8	
Pasto Ventura Basin							Locality:
	PV3	PV4	PV5	PV6	PV7	PV8	Sample ID:
	26.7592	26.7582	26.8589	26.8507	26.8495	26.8383	Latitude (°S):
	67.2878	67.2785	67.3035	67.3113	67.3188	67.3123	Longitude (°W):
	BTA	BTA	BTA	BA	BA	BTA	Rock type:[†]
	52.71	53.27	54.87	52.16	52.23	54.31	SiO_2
	1.126	1.066	0.950	1.105	1.086	1.114	TiO_2
	15.56	15.30	15.43	15.74	15.91	15.20	Al_2O_3
	7.77	7.43	6.93	8.13	8.10	7.61	FeO_T
	0.129	0.127	0.122	0.141	0.144	0.128	MnO
	7.96	7.82	6.45	7.92	7.80	7.53	MgO
	8.17	7.94	7.68	8.76	8.71	7.43	CaO
	3.58	3.55	3.26	3.29	3.34	3.51	Na_2O
	1.69	1.76	2.26	1.71	1.63	2.04	K_2O
	0.313	0.296	0.268	0.312	0.308	0.325	P_2O_5
	99.01	98.56	98.23	99.26	99.27	99.20	Sum
	0.65	0.80	0.63	0.44	−0.07	0.49	LOI (%)
	150	148	87	124	115	148	Ni
	392	386	287	397	378	345	Cr
	23	22	22	24	26	21	Sc
	202	194	191	214	214	188	V
	438	454	468	424	475	593	Ba
	43.3	47.6	75.4	42.0	43.4	50.3	Rb
	1482	815	562	665	734	684	Sr
	191	189	155	158	175	199	Zr
	18.19	18.09	18.54	19.87	23.60	18.74	Y
	10.34	10.16	10.34	11.22	12.44	13.34	Nb
	22	20	21	20	20	21	Ga
	55	53	51	58	41	50	Cu
	90	89	76	85	81	86	Zn
	8.03	8.65	12.86	8.98	7.19	9.13	Pb
	30.62	30.47	32.90	31.05	34.66	38.06	La
	62.42	62.05	63.87	62.34	68.99	74.84	Ce
	5.80	6.05	11.80	6.06	6.61	7.61	Th
	31.47	31.29	29.19	30.40	33.96	35.08	Nd
	1.29	1.44	2.40	1.26	1.40	1.27	U
	7.86	7.80	7.67	7.66	8.54	9.03	Pr
	6.31	6.23	5.82	6.10	6.86	6.66	Sm
	1.69	1.69	1.53	1.67	1.90	1.75	Eu
	5.06	4.93	4.82	5.16	5.66	5.25	Gd
	0.70	0.70	0.69	0.74	0.86	0.74	Tb
	3.84	3.80	3.83	4.08	4.82	4.03	Dy
	0.72	0.71	0.74	0.78	0.94	0.74	Ho
	1.79	1.79	1.94	2.03	2.41	1.85	Er
	0.25	0.25	0.26	0.28	0.34	0.26	Tm
	1.48	1.51	1.60	1.70	2.04	1.59	Yb
	0.22	0.23	0.25	0.26	0.31	0.24	Lu
	4.81	4.87	4.24	4.19	4.66	5.07	Hf
	0.63	0.63	0.75	0.71	0.77	0.79	Ta
	1.27	1.41	2.57	1.07	0.67	1.25	Cs
	22.2	21.0	23.0	26.0	29.0	19.7	Sc
	65	65	62	63	63	64	Mg#[§]
	20.7	20.2	20.5	18.2	17.0	24.0	La/Yb
	14.9	15.4	14.1	13.5	12.9	14.5	Zn/Fe**
	0.707932	0.708401	0.706561	0.706186	0.705400	0.705844	${}^{87}Sr/{}^{86}Sr_{(0)}$
	−3.10	−1.60	−4.41	−1.23	−0.80	−0.94	ε_{Nd}
	18.87577	18.84064	18.9247	18.99888	19.01626	18.90042	${}^{206}Pb/{}^{204}Pb$
	15.65271	15.65796	15.67863	15.65677	15.65247	15.65589	${}^{207}Pb/{}^{204}Pb$
	39.00766	38.96256	39.16369	39.05156	39.05077	39.02271	${}^{208}Pb/{}^{204}Pb$

such as Zn/Fe_T ($\times10^4$) are useful in two ways: First, they offer a tool for distinguishing between mantle-derived magmas sourced from peridotite (dominantly olivine and orthopyroxene; $Zn/Fe_T \times 10^4 = 9 \pm 1$) or pyroxenite (dominantly clinopyroxene with or without garnet; $Zn/FeT \times 10^4 = 13–20$; Le Roux et al., 2010, 2011). This is a critical consideration in the Puna region because it can help distinguish between asthenospheric and lithospheric melt sources. Second, this value is less susceptible to cryptic crustal contamination (Glazner and Farmer, 1992) than other commonly used elemental ratios because Fe is a major element.

SAMPLES AND METHODS

The 26 lavas described in this study are from the central and southern Puna Plateau and were targeted for sampling to supplement and expand upon the data set presented by Drew et al. (2009). Olivine thermometry analyses and interpretations from these samples are reported in Ducea et al. (2013). Geographically, the study area can be divided into four regions (Fig. 1B): (1) Chorrillos, in the northeastern corner of the Puna Plateau, (2) Salar de Arizaro in the northwestern Puna region, (3) Antofagasta Basin, just northeast of the Salar de Antofalla in the central Puna Plateau, and (4) Pasto Ventura, on the southern margin of the plateau.

Samples were prepared for analysis at the University of Arizona and analyzed at the University of Arizona and Washington State University. Fresh samples with no visible weathered surfaces and/or secondary minerals were powdered in an Al_2O_3-lined crusher. The GeoAnalytical Laboratory in the School of Earth and Environmental Sciences at Washington State University used aliquots for major- and trace-element analysis by X-ray fluorescence (XRF; Johnson, 1999) and inductively coupled plasma–mass spectrometry (ICP-MS; GeoAnalytical Lab Technical Notes, 2013). All isotopic analyses were performed at the University of Arizona, using thermal ionization mass spectrometry for Sr and Nd analysis, and multicollector (MC) ICP-MS in solution mode for Pb analysis (see Appendix).

In the XRF analyses for this study, all Fe was assumed to be Fe^{2+}. Total FeO (FeO_T) was calculated as $FeO_T = FeO + 0.8998 \times Fe_2O_3$. The Zn/Fe_T (Le Roux et al., 2010; Lee et al., 2010) was calculated by converting FeO_T wt% to ppm Fe_T. The Zn/Fe_T values are (ppm Zn)/(ppm Fe_T) and are multiplied by 10^4 to simplify notation in the text and figures.

RESULTS

Rock Types

Most lavas sampled are basaltic andesites, but five are andesitic and dacitic lavas, three are basaltic lavas, and the two samples from the Chorrillos region are basaltic trachyandesites that are shoshonitic (Table 1; Fig. 2). The samples in the suite form a high-K calc-alkaline trend typical of Neogene Central Volcanic Zone lavas (Fig. 2B).

For clarity in the following discussion, we divide the most primitive Puna lavas into two categories based on wt% MgO and magnesium number (Mg#; Table 1), which are useful benchmarks for identifying mantle-derived magmas. The most selective "high-Mg" group has lavas with MgO >8.0% ($n = 5$),

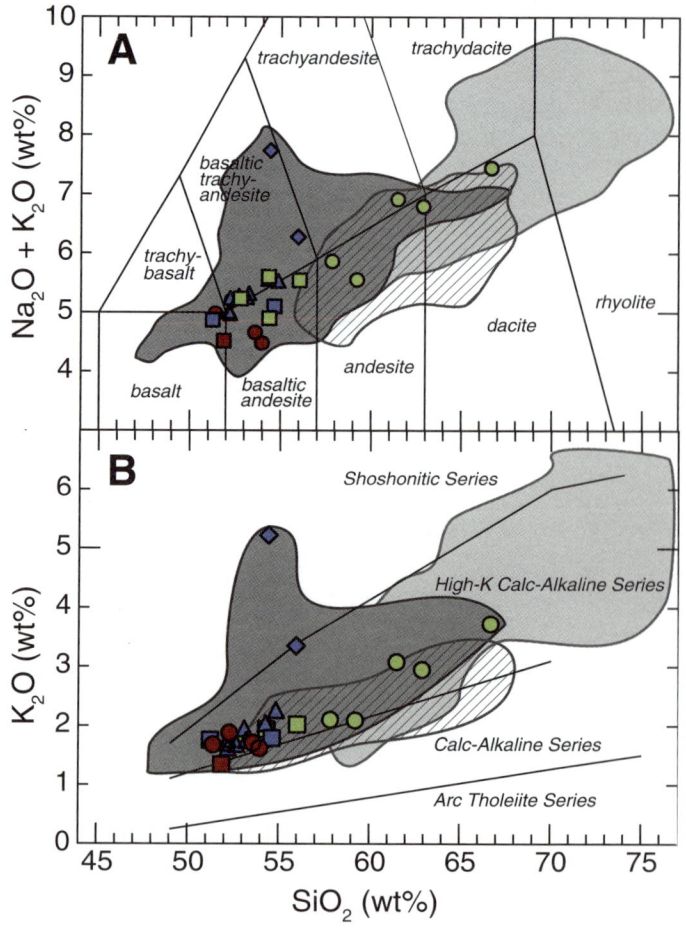

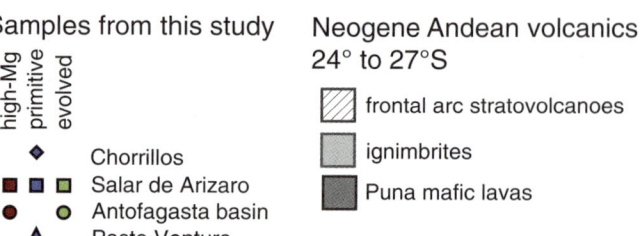

Figure 2. Major-element whole-rock data from Central Volcanic Zone lavas. Symbols are data from this study classified by region and Mg composition (see text for classification), and patterned fields are from a data compilation of Neogene volcanic rocks located 23°S to 27°S. This compilation includes other Puna mafic lavas (Knox et al., 1989; Francis et al., 1989; Déruelle, 1991; Kay et al., 1994; Kraemer et al., 1999; Drew et al., 2009), arc stratovolcanoes (Matthews et al., 1994; Trumbull et al., 1999; Rosner et al., 2003), and large-volume ignimbrites of the Puna region (Schnurr et al., 2007; Siebel et al., 2001) in addition to the Altiplano-Puna volcanic complex, which is centered between 21°S and 24°S (Hawkesworth et al., 1982; de Silva et al., 1994; Ort et al., 1996; Lindsay et al., 2001).

whereas the "primitive" group has lavas with MgO <8.0% but Mg# >60 ($n = 12$). The high-Mg-group lavas are most reliable for robust Zn/Fe_T interpretation because the compositions are closest to primary melt compositions (Le Roux et al., 2010). Trends in this select group help elucidate patterns evident in the larger suite that includes both the high-Mg and primitive-group lavas. All other lavas are in an "evolved" group ($n = 9$). We apply the same classifications to the compiled data from published Puna mafic lava studies (only those that report Zn values; Drew et al., 2009; Risse et al., 2013; Guzmán et al., 2006).

The high-Mg group from this study is composed of samples Az7 from Arizaro and Ns1, Ns2, J1, and ALB01 from Antofagasta. These lavas have MgO 8.06%–9.25% and Mg# from 63 to 68 (Table 1). They also have Ni concentrations of 127–217 ppm and Cr concentrations of 323–573 ppm, consistent with mantle-derived melts that have experienced only minor differentiation from primary melt compositions. The primitive-group lavas have MgO 6.14%–7.96%, 60–69 Mg#, Ni concentrations 83–179 ppm, and Cr concentrations 392–223 ppm (Table 1). There are primitive-group lavas from each geographic region.

Petrography

In thin section (Fig. 3), samples are commonly porphyritic and groundmass dominated; phenocrysts are dominantly olivine, clinopyroxene, and plagioclase, with minor orthopyroxene. Several of the evolved samples contain amphibole. Plagioclase is typically present only in the groundmass, which is dominated by plagioclase laths, 20–100 μm euhedral clinopyroxene crystals, and occasional broken olivine crystals. The groundmass commonly flows around phenocrysts. Quartz xenocrysts are present in two samples from Pasto Ventura (e.g., Fig. 3F) and two samples from Antofagasta; all visible xenocrysts were avoided during sample crushing for geochemical analysis. Olivine phenocryst abundance generally corresponds with the whole-rock MgO wt% composition. Cr-spinels are common as inclusions in olivine phenocrysts. Many samples are vesicular, and some vesicles contain secondary calcite (Fig. 3C).

Zn/Fe_T

The Zn/Fe_T ($\times 10^4$) values for the high-Mg-group and primitive-group lavas in this study are greater than 13 (Table 1; Fig. 4). The Chorrillos shoshonite samples MC11 and MC12 have the highest Zn/Fe_T ($\times 10^4$) values: 19.3 and 21.8, respectively. The high-Mg-group lavas have values between 13.1 and 14.3, and the primitive-group lavas have values have values between 12.9 and 17.5. All Puna mafic lavas, including those previously studied (Drew et al., 2009; Risse et al., 2013; Guzmán et al., 2006), have Zn/Fe_T ($\times 10^4$) values >12. These values are significantly higher than typical mantle peridotite and basaltic melts that would be generated from olivine-orthopyroxene–dominated sources (~9 ± 1; Le Roux et al., 2010; Lee et al., 2010) and are in the range expected of a source region dominated by clinopyroxene

± garnet. These values are not consistent with a peridotite mantle source (Fig. 4B).

Elemental and Isotopic Results

The lavas sampled in this study have trace-element patterns typical of calc-alkaline arc lavas (Fig. 5; Sun and McDonough, 1989; Hawkesworth et al., 1993). There are enrichments in large ion lithophile elements (LILEs; Ba, Rb, K) and other subduction-related elements (Th, Pb) and depletions in high field strength elements (HFSEs; Nb, Ta, Ti). The strength of this "arc signature" varies regionally and by lava type, as does the steepness of the rare earth element (REE) pattern (i.e., La/Yb; Table 1; Fig. 6), magnitude of the Eu anomaly, and the Nd-Sr isotopic compositions (Fig. 7; Table 2). Generally, Pb isotopes (Table 2) do not have significant systematic trends within this suite and are consistent with the range of previously reported values in the region (Mamani et al., 2008). In the following paragraphs, the elemental and isotopic results (Tables 1 and 2) are presented by region.

Pasto Ventura

The sampled flows in the Pasto Ventura region ($n = 8$) are primitive-group basaltic andesites. The samples have similar trace-element patterns (Fig. 5A), including a significant Ta-Nb depletion. The REE patterns of the Pasto Ventura lavas are also consistent: The La/Yb values are between 17 and 24 (Table 1), and Eu anomalies are of similar magnitude (Fig. 6A). In contrast, the Sr and Nd isotopic compositions of these lavas span almost the entire range of values in this study (Fig. 7A). Four samples have among the most primitive isotopic compositions reported from the Puna region ($\varepsilon_{Nd} = -1$; $^{87}Sr/^{86}Sr = 0.7054$–0.7058), whereas two others have more-evolved compositions ($\varepsilon_{Nd} = -4.6$, -5.3; $^{87}Sr/^{86}Sr = 0.7065, 0.7066$). The two additional lavas (PV3 and PV4) have Sr and Nd isotopic compositions that are not well correlated; the Sr isotopic values are high for the Nd values, so they plot to the right of the rest of the Pasto Ventura lavas in Figure 7A. They have ε_{Nd} values of -3.1 and -1.6 and $^{87}Sr/^{86}Sr$ values of 0.7079 and 0.7084. These samples are not higher in wt% SiO_2 than the other Pasto Ventura lavas (Fig. 7C) and are also not the samples with observed quartz xenoliths. The ranges of Pb isotope values in Pasto Ventura lavas are $^{206}Pb/^{204}Pb = 18.841$–$19.016$, $^{207}Pb/^{204}Pb = 15.652$–$15.679$, and $^{208}Pb/^{204}Pb = 38.96$–$39.16$ (Table 2).

Antofagasta Basin

Lavas sampled around the Valle de Antofagasta are basalts and basaltic andesites in the high-Mg group ($n = 4$), as well as evolved-group andesites and dacites ($n = 5$). The high-Mg-group lavas are less enriched in Ba, Rb, Th, and K, and more enriched in the HFSEs compared to the andesitic and dacitic flows in Antofagasta, and also in comparison to the lavas in other regions (Fig. 5B). They have La/Yb values from 13 to 22. The high-Mg-group lavas range in Nd isotopic composition from $\varepsilon_{Nd} = -1.1$ to -4.1. Their range of Sr isotopic compositions is narrower,

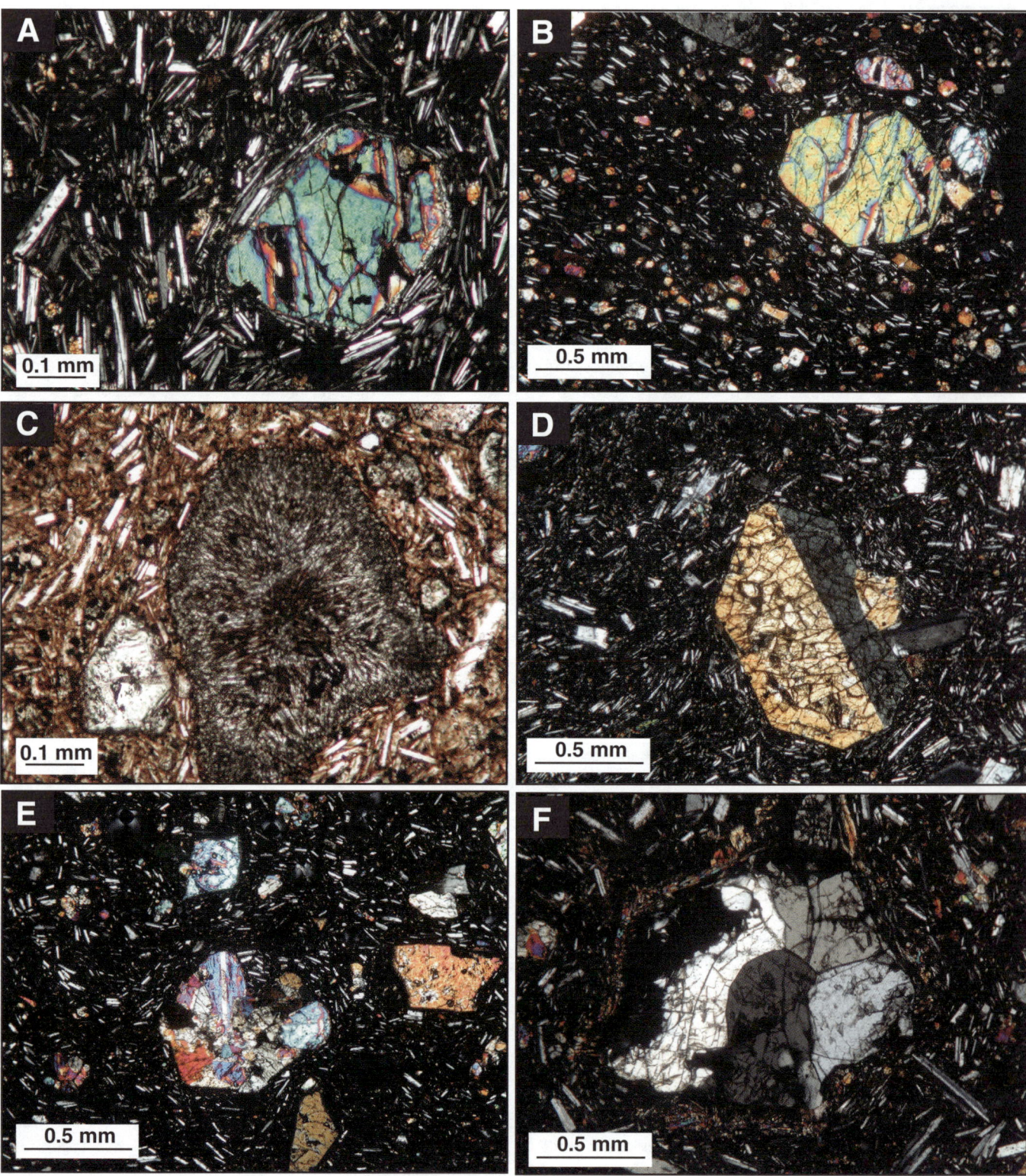

Figure 3. Photomicrographs of common textures in Puna lavas, in cross-polarized light unless otherwise noted. (A) Plagioclase laths aligned in a seritic texture around a large fractured olivine phenocryst in sample MC17. (B) Porphyritic texture typical of these lavas, with large olivine phenocrysts and small clinopyroxene, plagioclase, and olivine crystals in an aphanitic groundmass, as seen in sample PV5. (C) Sample PV8 has vugs filled with secondary phases, probably calcite. In plane-polarized light. (D) Large clinopyroxene phenocryst with polysynthetic twinning, sample PV6. (E) A polyphase enclave in sample Ns2. (F) Quartzite xenocryst in sample PV2, rimmed by a reaction texture.

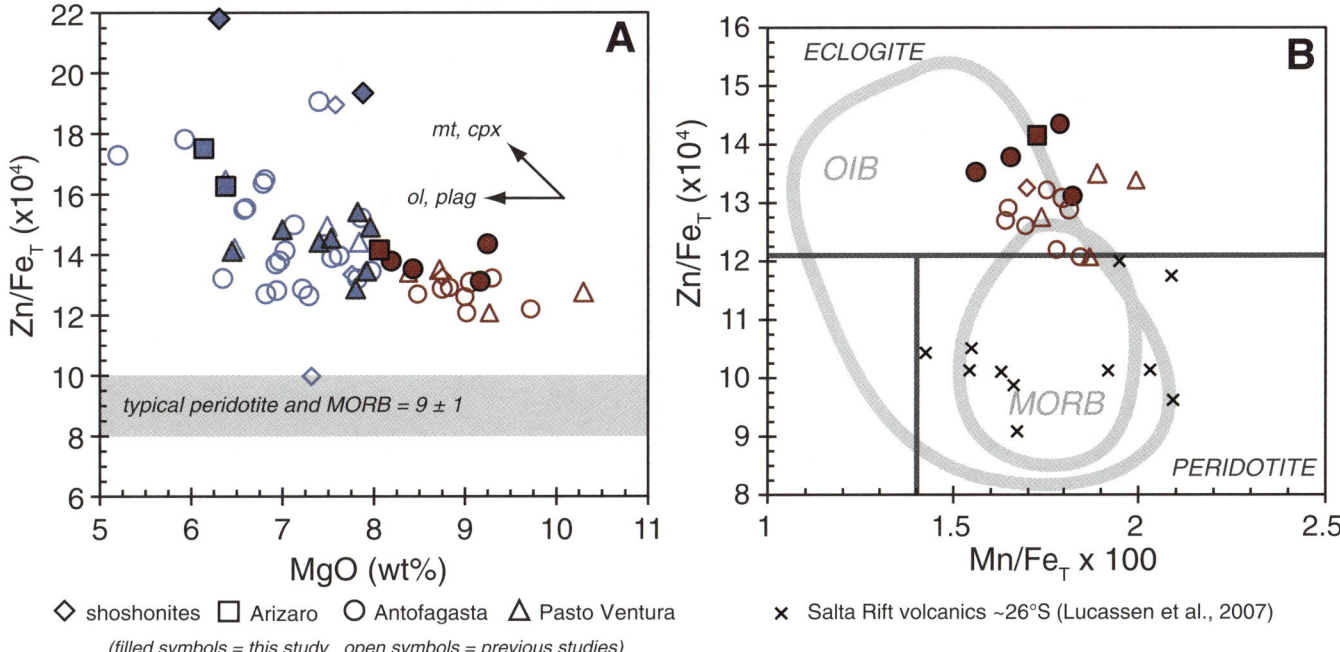

Figure 4. Zn/Fe$_T$ (×10^4) values for Puna mafic lavas in this study (filled symbols) and from other recent studies of Puna mafic lavas that include Zn compositional data (open symbols; Guzmán et al., 2006; Drew et al., 2009; Risse et al., 2013); symbol colors indicate classification by Mg compositions as in Figure 2. (A) Zn/Fe$_T$ values plotted again MgO wt% for the high-Mg-group and primitive-group lavas. Typical peridotite and mid-ocean-ridge basalt (MORB) values are from Lee et al. (2010). All Puna mafic lavas from this and other studies have higher Zn/Fe$_T$ than melts from peridotite. Vectors show melt evolution paths for fractional crystallization of olivine (ol) or plagioclase (plag; no change in magma Zn/Fe$_T$) and clinopyroxene (cpx) or magnetite (mt; increasing magma Zn/Fe$_T$ magma) (Lee et al., 2010), which could be responsible for the higher Zn/Fe$_T$ values in the less-primitive Puna lavas. (B) The high-Mg-group lavas plot in the eclogite field of Le Roux et al. (2011) in Zn/Fe$_T$ vs. Mn/Fe$_T$ space, whereas mafic flows from the Cretaceous Salta rift, ~26°S (Lucassen et al., 2007), have lower Zn/Fe$_T$ values and plot in the peridotite field. OIB—ocean-island basalt.

between 0.7055 and 0.7061. The andesitic and dacitic lavas in this region have steeper REE trends (Fig. 6B) than the Antofagasta high-Mg lavas, with La/Yb values from 30 to 55. The Sr isotopic composition of the evolved lavas ranges from 0.7080 to 0.7095 (Fig. 7). Although the ε_{Nd} values of the evolved-group lavas are more enriched as well (ε_{Nd} = –3.1 to –7.0), they are not as enriched as would be expected from the regional isotopic trend (Fig. 7). In these evolved Antofagasta samples, more radiogenic Sr compositions correlate with higher wt% SiO$_2$, which suggests assimilation-fractional-crystallization processes during magma migration through the crust. This is in contrast to the primitive-group Pasto Ventura samples PV3 and PV4, which also lie in an isotopic trend offset toward more radiogenic Sr but are not as evolved in composition. The ranges of Pb isotope values in Antofagasta lavas are ^{206}Pb/^{204}Pb = 18.831–19.045, ^{207}Pb/^{204}Pb = 15.655–15.688, and ^{208}Pb/^{204}Pb = 38.78–39.27 (Table 2).

Salar de Arizaro

Lavas from around the Salar de Arizaro are in the high-Mg group (Az7), the primitive group (MC18, RG1), and evolved group (n = 4). The evolved-group lavas from this region are basaltic andesites (Table 1; Fig. 2). The La/Yb values of Arizaro lavas range from 11 to 31 (Table 1). The high-Mg lava has the lowest

trace-element values (Fig. 5C) and a flatter REE pattern (Fig. 6C) than the lavas from this or the other regions. Whereas the high-Mg-group lava from Arizaro has among the most primitive Nd and Sr isotopic compositions in the suite (ε_{Nd} = –0.9; ^{87}Sr/^{86}Sr = 0.7055), these values are poorly correlated amongst the rest of the Arizaro lavas. They have a narrow range of ^{87}Sr/^{86}Sr values from 0.7065 to 0.7073, and those classified in the evolved group have lower Sr isotopic compositions than the two lavas in the primitive group. In contrast, the ε_{Nd} range is wider, from –1.1 to –6.4 (Table 2; Fig. 7A). Neither Nd nor Sr isotopic patterns correspond to SiO$_2$ wt% (Figs. 7B and 7C), suggesting that the variable isotopic composition is related to source variability rather than differentiation from a single source composition. The ranges of Pb isotope values in Arizaro lavas are small: ^{206}Pb/^{204}Pb = 18.778–18.960, ^{207}Pb/^{204}Pb = 15.635–15.663, and ^{208}Pb/^{204}Pb = 38.74–38.89 (Table 2).

Chorrillos

Samples MC11 and MC12 are basaltic trachyandesites, and their high-K compositions classify them as shoshonites (Fig. 2B). They are distinct from the rest of the Puna suite in their trace-element and isotopic composition (Figs. 5D and 6D) because they are more enriched overall in trace elements and have

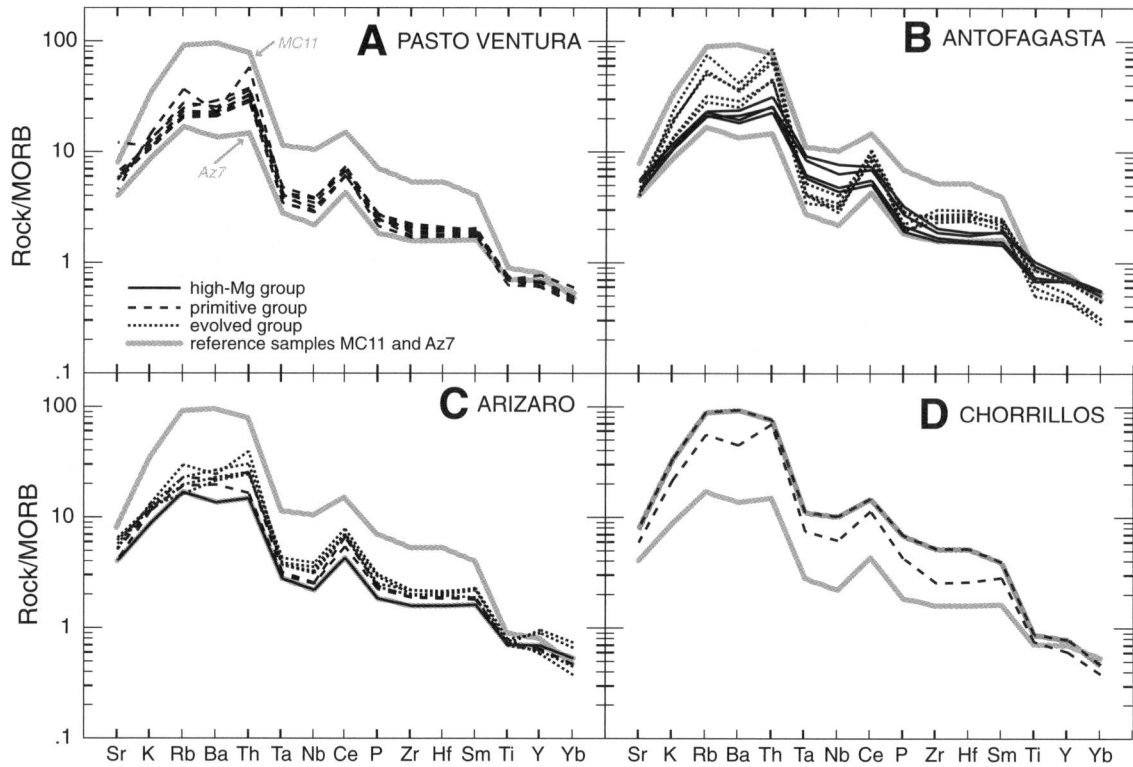

Figure 5. Trace-element diagram after Pearce and Parkinson (1993). In each panel, samples MC11 and Az7 are shown in thick gray lines for reference. High-Mg-, primitive-, and evolved-group lavas are represented by different patterned lines. (A) Pasto Ventura lavas. (B) Antofagasta region lavas. (C) Salar de Arizaro lavas. (D) Chorrillos region shoshonites. MORB—mid-ocean-ridge basalt.

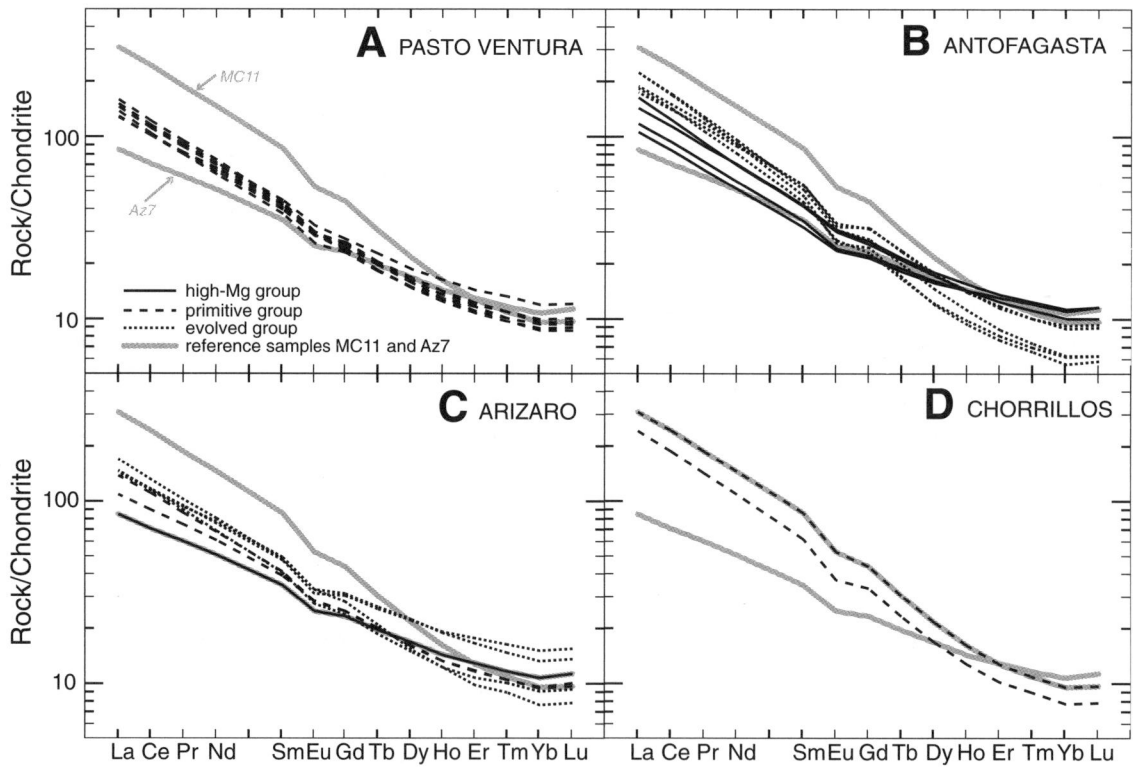

Figure 6. Rare earth element diagram after McDonough and Sun (1995). Symbols are the same as Figure 5. Reference samples on each panel are MC11 and Az7. (A) Pasto Ventura lavas. (B) Antofagasta region lavas. (C) Salar de Arizaro lavas. (D) Chorrillos region shoshonites.

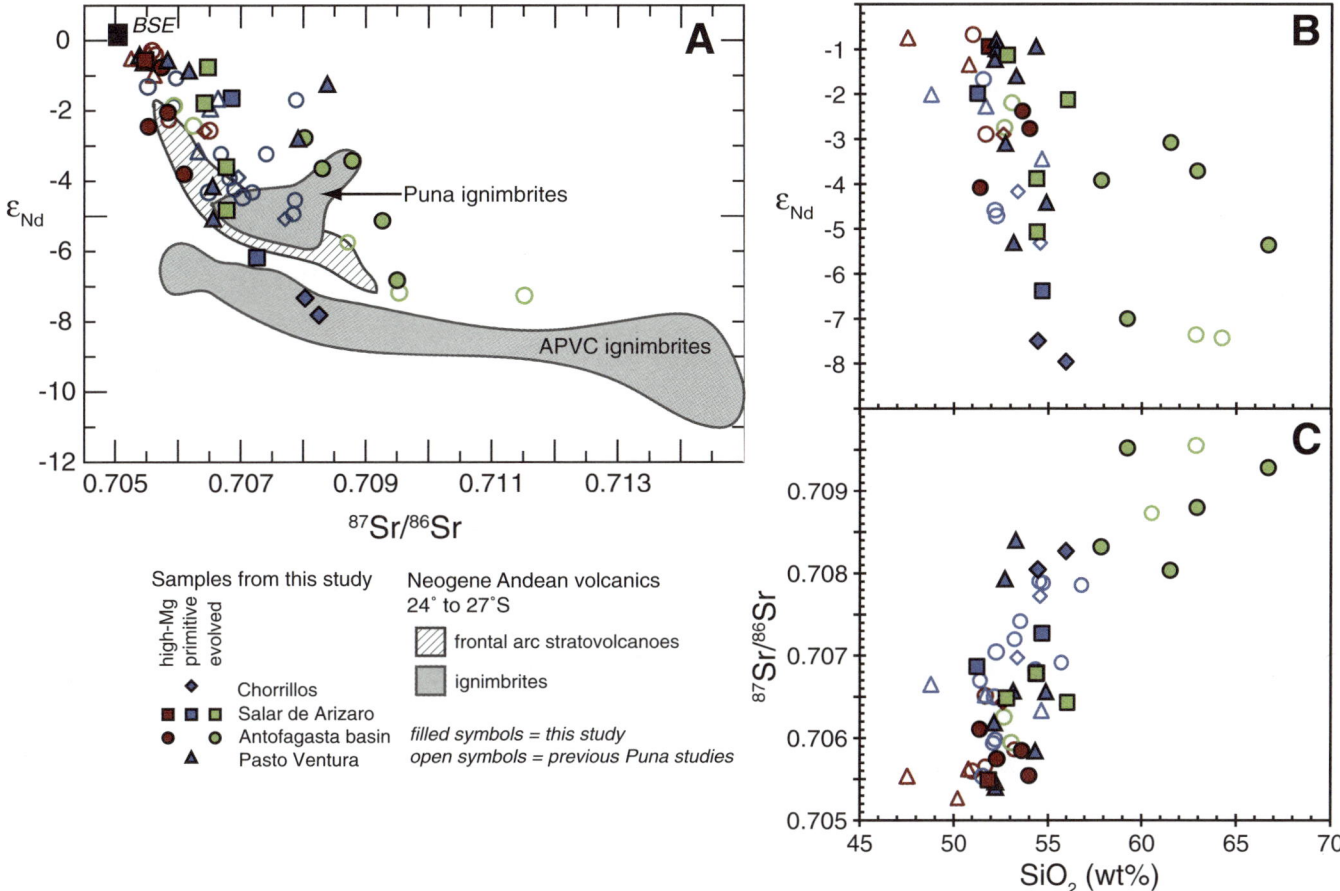

Figure 7. ε_{Nd} and $^{87}Sr/^{86}Sr$ data from this study (filled symbols) and other recent Puna lava studies (open symbols; Guzmán et al., 2006; Drew et al., 2009; Risse et al., 2013). (A) Fields identify isotopic variability in the regional compilation from Figure 2. *BSE*—bulk silicate earth; *APVC*—Altiplano-Puna volcanic complex. (B) ε_{Nd} values plotted against SiO_2 (wt%) show significant isotopic variability over a narrow range of compositions. (C) $^{87}Sr/^{86}Sr$ values plotted against SiO_2 also show significant isotopic variability over a narrow range of compositions, but lavas with SiO_2 compositions >56 wt% have increasing $^{87}Sr/^{86}Sr$ value with increasing SiO_2, which is expected from mixing with a radiogenic crustal source.

the steepest REE patterns among the primitive-group and high-Mg-group lavas (La/Yb = 45, 43; Table 1). Sample MC11 has the greatest enrichment of trace elements relative to MORB of all the samples in this study, consistent with other studies of high-K shoshonites on the Puna Plateau (Déruelle, 1991; Kay et al., 1994; Guzmán et al., 2006). These two samples have the most-evolved Nd isotopic values in the suite (ε_{Nd} = –7.5, –8.0; Table 2; Fig. 7), yet relatively unradiogenic Sr isotopic compositions ($^{87}Sr/^{86}Sr$ = 0.7080, 0.7082; Table 2; Fig. 7). The ranges of Pb isotope values in Chorrillos lavas are $^{206}Pb/^{204}Pb$ = 18.731–18.741, $^{207}Pb/^{204}Pb$ = 15.688–15.703, and $^{208}Pb/^{204}Pb$ = 39.00–39.20 (Table 2).

DISCUSSION

The elemental and isotopic data from Puna mafic lavas are variable and commonly poorly correlated, although the entire data set falls within the range of Central Volcanic Zone compositions (Figs. 2 and 7A). In the following, we discuss this variabil-

ity and how the high Zn/Fe_T values of the Puna lavas suggest both an explanation and a corresponding model for melt generation during lithospheric foundering.

Lithospheric Heterogeneities and Consequences for Melt Generation

The trace-element compositions of Puna mafic lavas have patterns typical of magmatic arcs, and several regional trends suggest the strength of this "arc signature" is the result of source variability. Trace-element ratios that track the magnitude of relative enrichments and depletions (i.e., Ba/Ta and La/Ta, which compare LILE to HFSE and light [L] REE to HFSE, respectively) have a geographic pattern. The lavas from the Antofagasta region have the lowest values and thereby smallest magnitude of enrichment in LILEs and LREEs relative to HFSEs (Fig. 8A; see Kay et al., 1994). Th/Nb values of Antofagasta lavas also suggest minimal subduction contributions. This

TABLE 2. ISOTOPIC RESULTS

Sample	Rb (ppm)	Sr (ppm)	Rb/Sr	$^{87}Rb/^{86}Sr$	$^{87}Sr/^{86}Sr$*	2σ	Sm (ppm)	Nd (ppm)
Chorrillos								
MC-11	169.6742	961.2434	0.1765	0.507595	0.708046	±8	12.593	69.553
MC-12	97.4946	737.6830	0.1322	0.380063	0.708269	±6	8.697	54.032
Salar de Arizaro								
MC-13	43.6132	698.2578	0.0625	0.179590	0.706781	±10	7.664	38.154
MC-17	30.9546	780.9488	0.0396	0.113968	0.706787	±9	6.851	34.818
MC-18	44.2608	621.0148	0.0713	0.204937	0.707271	±15	5.853	31.982
P09Az1	55.4249	450.0964	0.1231	0.354050	0.706432	±8	6.061	33.248
P09Az4	36.7573	523.4354	0.0702	0.201906	0.706485	±11	6.920	35.285
P09Az7	32.1364	346.5974	0.0927	0.266562	0.705490	±13	5.177	24.167
P09RG1	36.3305	354.4844	0.1025	0.294686	0.706867	±8	5.722	28.484
Valle de Antofagasta								
P09HM1	105.5934	360.1633	0.2932	0.843148	0.708797	±7	6.161	36.458
P09HM2	146.5127	368.2837	0.3978	1.144143	0.709284	±37	6.678	41.615
P09Af1	100.1269	482.1538	0.2077	0.597172	0.708036	±10	7.204	45.756
P09Af2	62.7785	411.7824	0.1525	0.438420	0.708320	±8	7.543	41.771
ALB-01	38.5361	810.7065	0.0475	0.136660	0.705749	±11	5.426	29.273
P09Ns1	45.3225	552.8262	0.0820	0.235704	0.705849	±8	5.225	27.291
P09Ns2	38.3972	682.1671	0.0563	0.161822	0.705543	±10	4.972	25.473
P09Ns3	54.7699	521.8122	0.1050	0.301873	0.709521	±8	7.995	41.722
P09J1	40.8399	664.3585	0.0615	0.176740	0.706108	±8	6.196	35.600
Pasto Ventura Basin								
P09PV1	52.5431	705.5590	0.0745	0.214119	0.706570	±9	6.140	33.257
P09PV2	46.7097	1015.7394	0.0460	0.132206	0.705465	±8	4.650	23.219
P09PV3	46.5352	1163.3329	0.0400	0.115029	0.707932	±7	5.470	27.462
P09PV4	41.1492	1011.9141	0.0407	0.116941	0.708401	±10	5.879	30.949
P09PV5	71.3857	136.4508	0.5232	1.504203	0.706561	±30	5.295	11.733
P09PV6	43.1501	982.7224	0.0439	0.126243	0.706186	±69	5.701	29.822
P09PV7	37.1057	874.8932	0.0424	0.121929	0.705400	±8	2.391	10.048
P09PV8	48.4734	489.3891	0.0990	0.284767	0.705844	±16	6.260	34.457

*Measured values.

ratio is useful for tracing sediment and other crustal contributions to arc magmas, and values less than 0.3 are typical for MORB and Pacific OIB lavas (Plank, 2005). The Antofagasta high-Mg-group lavas have the lowest Th/Nb values (0.23–0.31; Fig. 8C) compared to those from other regions (Arizaro: 0.39; Pasto Ventura primitive lavas: 0.51–1.14; Chorrillos shoshonites: 0.43–0.64), suggesting Antofagasta melts had minimal contributions from subducted sediments or continental crust. The distance from the modern arc front does not control the strength of the enrichments and depletions (Figs. 8B and 8D). Antofagasta is in the center of the study area and is the same distance from the arc front as Pasto Ventura, where high-Mg-group lavas have consistently more enriched trace-element compositions. Trace-element ratios that track the presence of garnet in the source, including La/Yb, Sm/Yb, and Hf/Sm (van Westrenen et al., 2011), also have regional differences between the Antofagasta and Pasto Ventura lavas (Fig. 9). All of these

trends are clearest among the high-Mg-group lavas (including those compiled from the literature; Fig. 8; Drew et al., 2009; Guzmán et al., 2006; Risse et al., 2013). The regional distinctions are also present in the more variable primitive-group data.

Kay et al. (1994) attributed the Ba/Ta and La/Ta geographic trends to postdelamination influx and melting of "OIB-like" asthenosphere beneath the Antofagasta region; however, the high Zn/Fe$_T$ values appear to preclude a dominantly asthenospheric mantle source for all the Puna lavas, including those from Antofagasta. There are no correlations between Zn/Fe$_T$ and La/Ta (Fig. 10A) or Th/Nb (Fig. 11A). Additionally, the high-Mg-group Antofagasta lavas have the same range of Nd, Sr, and Pb isotopic compositions as the high-Mg-group lavas from the other regions, and there are no correlations between these isotopic values and La/Ta values (Figs. 10B, 10C, 10D). Together, these observations suggest that while the lavas from the Antofagasta region are likely derived from a mantle source(s) depleted relative to

Sm/Nd	$^{147}Sm/^{144}Nd$	$^{143}Nd/^{144}Nd*$	2σ	$\varepsilon_{Nd}*$	$^{206}Pb/^{204}Pb$	$^{207}Pb/^{204}Pb$	$^{208}Pb/^{204}Pb$	Sample
								Chorrillos
0.181	0.109455	0.512254	±16	−7.49	18.741	15.703	39.201	MC-11
0.161	0.097308	0.512230	±10	−7.96	18.731	15.688	39.003	MC-12
								Salar de Arizaro
0.201	0.121526	0.512439	±11	−3.88	18.880	15.640	38.863	MC-13
0.197	0.118950	0.512378	±6	−5.07	18.924	15.635	38.829	MC-17
0.183	0.110628	0.512311	±13	−6.38	18.981	15.663	38.893	MC-18
0.182	0.110206	0.512529	±9	−2.13	18.891	15.636	38.858	P09Az1
0.196	0.118557	0.512580	±8	−1.13	18.960	15.645	38.866	P09Az4
0.214	0.129516	0.512590	±13	−0.94	18.897	15.637	38.747	P09Az7
0.201	0.121450	0.512536	±9	−1.99	18.778	15.640	38.740	P09RG1
								Valle de Antofagasta
0.169	0.102169	0.512448	±11	−3.71	18.874	15.667	39.047	P09HM1
0.160	0.097010	0.512363	±9	−5.36	19.035	15.688	39.267	P09HM2
0.157	0.095183	0.512480	±33	−3.08	18.962	15.676	39.139	P09Af1
0.181	0.109163	0.512437	±14	−3.92	18.949	15.665	39.016	P09Af2
0.185	0.112067	0.512580	±16	−1.13	18.957	15.655	38.962	ALB-01
0.191	0.115829	0.512516	±13	−2.38	18.831	15.655	38.794	P09Ns1
0.195	0.117997	0.512496	±13	−2.77	18.840	15.656	38.784	P09Ns2
0.192	0.115933	0.512279	±16	−7.00	19.045	15.682	39.148	P09Ns3
0.184	0.111486	0.512429	±8	−4.08	18.977	15.668	38.964	P09J1
								Pasto Ventura Basin
0.185	0.111607	0.512366	±14	−5.31	18.896	15.666	39.049	P09PV1
0.200	0.121070	0.512587	±10	−0.99	18.990	15.661	39.109	P09PV2
0.199	0.120407	0.512479	±16	−3.10	18.876	15.653	39.008	P09PV3
0.190	0.114837	0.512556	±12	−1.60	18.841	15.658	38.963	P09PV4
0.451	0.272836	0.512412	±11	−4.41	18.925	15.679	39.164	P09PV5
0.191	0.115584	0.512770	±16	2.57	18.999	15.657	39.052	P09PV6
0.238	0.143834	0.512597	±11	−0.80	19.016	15.652	39.051	P09PV7
0.182	0.109838	0.512590	±9	−0.94	18.900	15.656	39.023	P09PV8

those sourcing the other Puna lavas, this mantle material is likely not dominated by fresh upwelling asthenosphere, because such a source would be olivine-orthopyroxene–dominated peridotite, which would likely tap different isotopic reservoirs than the Central Volcanic Zone magmas.

The Zn/Fe_T values offer a useful perspective on these complex data because they suggest a mantle-lithosphere–dominated (i.e., pyroxenite-dominated) source for the Puna lavas (Fig. 4B). This is a key insight in two ways. First, it explains the nonsystematic compositional diversity of the Puna lavas because the lavas would be sourced from a central Andean mantle lithosphere that is a poorly mixed amalgamation of materials generated during Phanerozoic terrane accretion and orogenesis in the region. Second, if foundering of this lower lithosphere pyroxenite is the driver of Puna melt generation, then the pyroxenite itself would be more likely to melt than the asthenosphere upwelling to take its place, given the expected pressure-temperature (*P-T*) paths

(Fig. 12A). Before we explore the implications of this model for generating the Puna mafic lavas (Fig. 12), we first discuss the Zn/Fe_T results in more detail, assess the presence of garnet (and thereby eclogite) in the source, and consider the role of crustal contamination in generating compositional variability.

Possible Sources of High Zn/Fe_T Magmas

All Puna lavas in this study have Zn/Fe_T ($\times 10^4$) values >13 (Table 1; Figs. 4 and 10A), which, according to current understanding (Le Roux et al., 2011), is a consequence of the mineralogical composition of the mantle source region. Data from Puna lavas sampled by others who report Zn concentrations (Drew et al., 2009; Risse et al., 2013; Guzmán et al., 2006) are consistent with these results, and the lowest Zn/Fe_T ($\times 10^4$) value from a Puna lava is 12.1. There are, however, several other processes known or hypothesized to fractionate Zn and Fe during

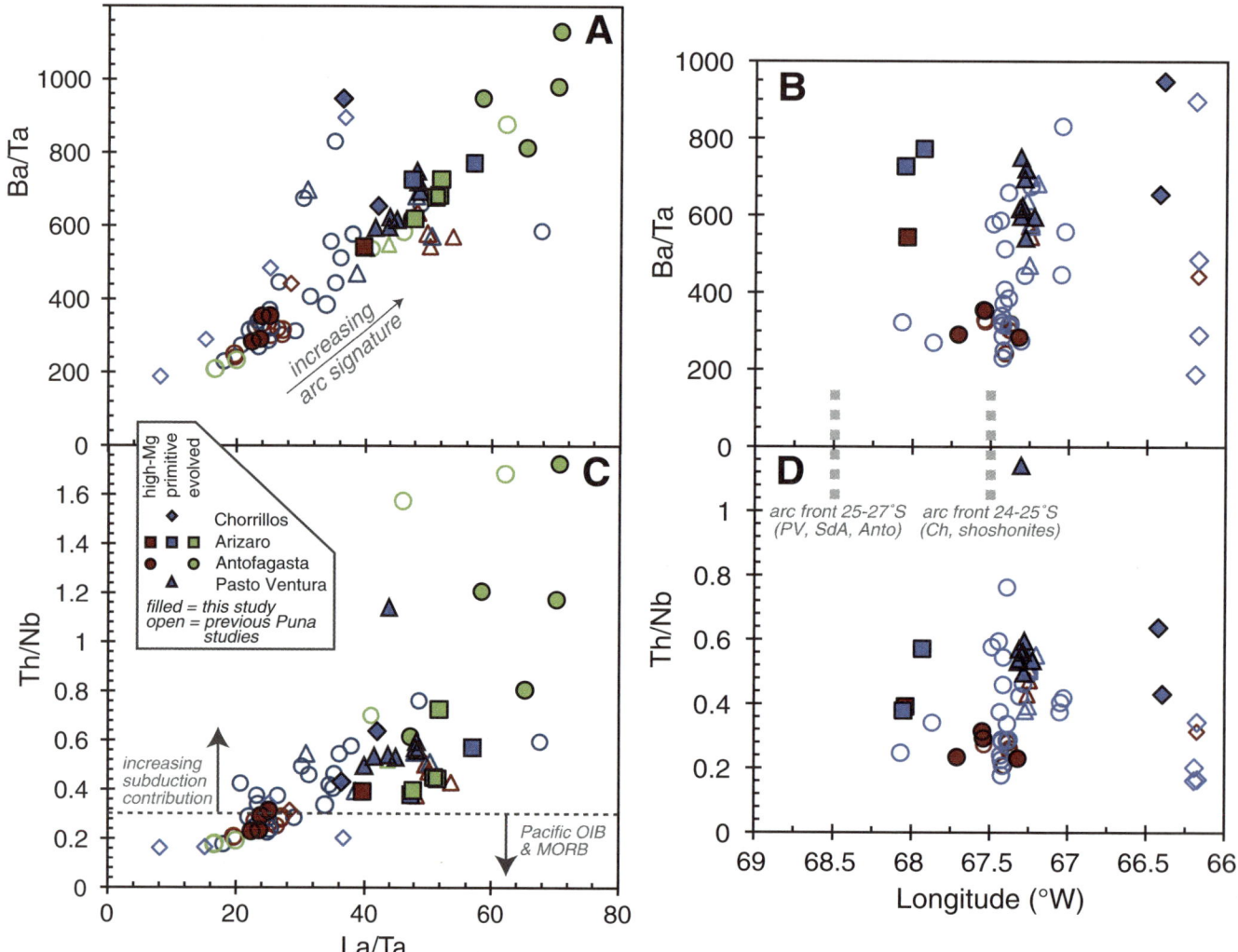

Figure 8. Geographic trends in trace-element ratios across the Puna Plateau. (A, C) Ba/Ta and Th/Nb values plotted vs. La/Ta values highlight the regional trend in elemental compositions that differentiates the high-Mg-group Antofagasta lavas from the rest of the Puna lavas (Kay et al., 1994) in this study (filled symbols) and other recent Puna lava studies (open symbols; Guzmán et al., 2006; Drew et al., 2009; Risse et al., 2013). Arrows interpreting Th/Nb fields separated by dotted gray line in C are after Plank (2005). (B, D) Trace-element ratios plotted against longitude, which is a good proxy for the relative sample distance from the active arc front. Approximate arc front longitude coordinates for 24.5°S to 27°S (SdA—Salar de Arizaro; PV—Pasto Ventura; Anto—Antofagasta basin) and 24.5°S to 23°S (Ch—Chorrillos) are indicated by the thick dotted gray lines. MORB—mid-ocean-ridge basalt; OIB—ocean-island basalt.

melt generation and differentiation: fractional crystallization of clinopyroxene and magnetite (Le Roux et al., 2010, 2011), the oxidation state of the melt source (Lee et al., 2010), and metasomatism by fluids fluxed from the subducting slab (Le Roux et al., 2010).

Clinopyroxene or magnetite fractional crystallization during magma ascent can increase the Zn/Fe_T values in magmas, but this cannot explain the high Zn/Fe_T values in the most mantle-derived samples from the Puna. Fractional crystallization of clinopyroxene or magnetite from a basaltic magma would fractionate Zn and Fe (Le Roux et al., 2011) and yield lavas with higher Zn/Fe_T values and lower wt% MgO and FeO_T (see vectors in Fig. 4A). This could be responsible for the Zn/Fe_T val-

ues of some of the primitive-group lavas, which have lower Mg and higher Zn/Fe_T than the highest-Mg-group lavas (Fig. 4A); these primitive-group lavas likely experienced some fractional crystallization of olivine and possibly clinopyroxene prior to eruption. However, clinopyroxene and magnetite fractionation could not have significantly changed the Zn/Fe_T of the high-Mg lavas from the Puna region without evolving their major-element composition. Even the most mantle-derived Puna lava sampled to date (PV07-SUR101, 10.3 wt% MgO; Drew et al., 2009) has a Zn/Fe_T (×10^4) value of 12.8.

During mantle melting, Fe^{3+} is more incompatible the Fe^{2+}; as a result, greater Fe^{3+}/Fe^{2+} in an oxidized mantle source would yield melts with low Zn/Fe_T values (Lee et al., 2010). Therefore,

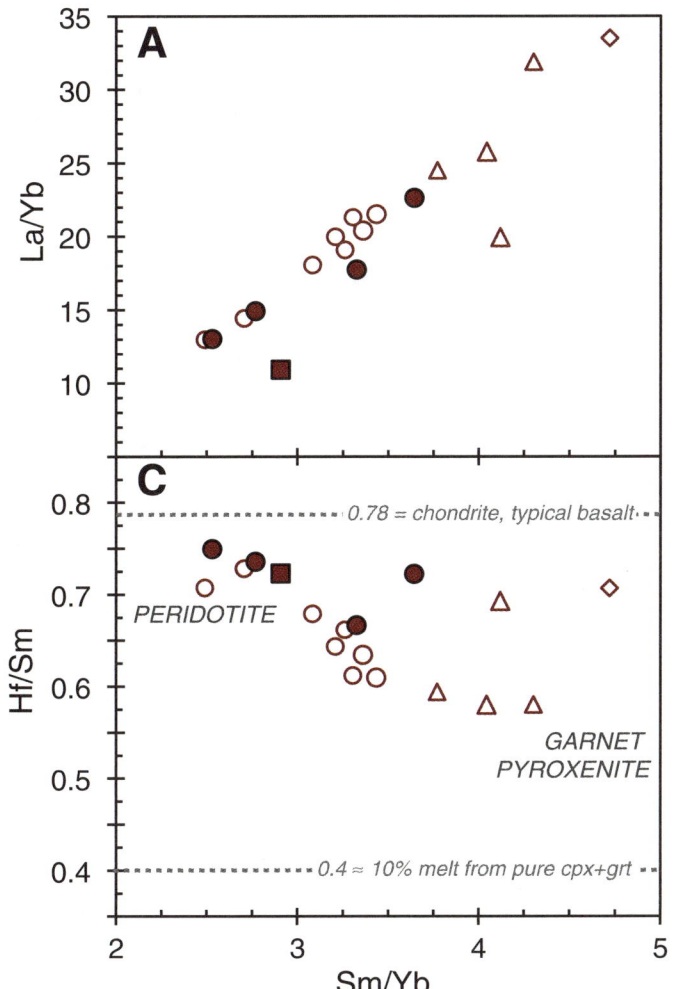

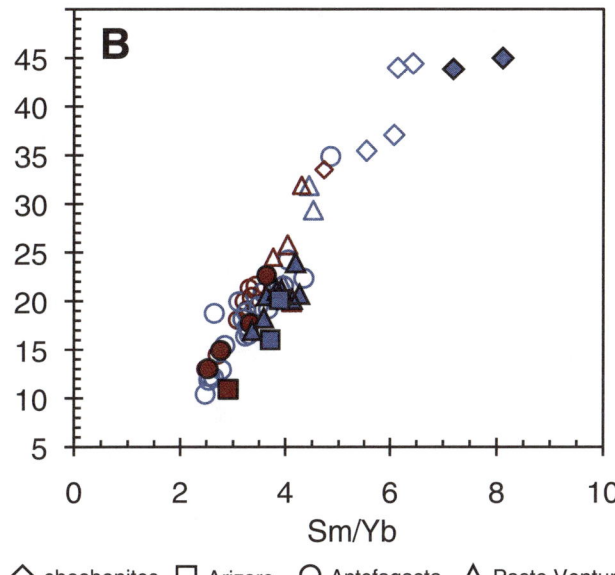

◇ shoshonites ☐ Arizaro ○ Antofagasta △ Pasto Ventura
(filled symbols = this study   open symbols = previous studies)

Figure 9. Trace-element ratios that commonly track garnet in melt source regions. (A) La/Yb plotted against Sm/Yb to compare the rare earth element (REE) slope to the heavy (H) REE slope. Greater values for both ratios indicate increasing likelihood of garnet in the source. (B) Note the Puna shoshonites have significantly higher values of both La/Yb and Sm/Yb. They are offset from the rest of the Puna lavas. (C) Hf/Sm values can be used to identify grossular garnet (common in eclogite) in the mantle source region (van Westrenen et al., 2011). Hf/Sm values of chondrite and 10% clinopyroxene (cpx) + garnet (grt) melt are after van Westrenen et al. (2011).

an oxidized Puna upper mantle cannot be the explanation for the observed high Zn/Fe$_T$ values in the Puna mafic lavas. This argument for a lack of upper-mantle oxidation in subduction settings is consistent with results from other arcs (Lee et al., 2010).

Our use of Zn/Fe$_T$ values as indicators of source mineralogy of arc basalts is a new application of this technique, and therefore it is possible that the high Zn/Fe$_T$ values reflect some Zn-rich source or partitioning effect in the metasomatized mantle wedge as yet unidentified. Le Roux et al. (2010) suggested that slab fluids could increase the Zn/Fe$_T$ value of peridotite in subduction systems, which would generate higher Zn/Fe$_T$ melts. Whereas it is beyond the scope of this paper to experimentally evaluate these potential effects, we suggest that the lack of correlation between Zn/Fe$_T$ values and La/Ta (Fig. 10A), Ba (Fig. 11B), Rb (Fig. 11C), Cs, Cu, Pb, and U (Table 1) in the Puna lavas from Pasto Ventura and Antofagasta indicates that there is no Zn-enriched source that corresponds to metasomatized and enriched mantle beneath the Puna Plateau. By this measure, however, the higher Zn/Fe$_T$ of the shoshonites from Chorrillos could be attributed to such as source.

In sum, the simplest explanation for the high Zn/Fe$_T$ values of the high-Mg-group Puna lavas, especially those from Antofa-

gasta and Pasto Ventura, is a source composition dominated by clinopyroxene and possibly garnet. Even with an extreme peridotite source with Zn/Fe$_T$ ($\times10^4$) = 11, it appears that there is no way to generate melts with Zn/Fe$_T$ ($\times10^4$) > 12.5 without the dominant presence of clinopyroxene or garnet at the source (Le Roux et al., 2010).

Garnet in the Source?

Compositional differences between the Antofagasta, Pasto Ventura, and shoshonite lavas could be explained in part by garnet-bearing eclogite in the source(s) of the shoshonites and Pasto Ventura high-Mg-group lavas, although Pasto Ventura heavy (H) REE depletions are not extreme. This is of interest because eclogite (i.e., garnet pyroxenite) is thought to be a key player in lithospheric foundering in the Sierra Nevada (Lee et al., 2006; Ducea, 2002), and also in the central Andes (Kay and Kay, 1991). The correlation between La/Yb and Sm/Yb values of the Puna lavas has a break in slope at La/Yb ~30 and Sm/Yb ~5 (Fig. 9B), which are values commonly cited as the transition to garnet-bearing sources (Haschke et al., 2006). Only the Puna shoshonites have La/Yb > 30 and Sm/Yb > 5 (Figs. 9A

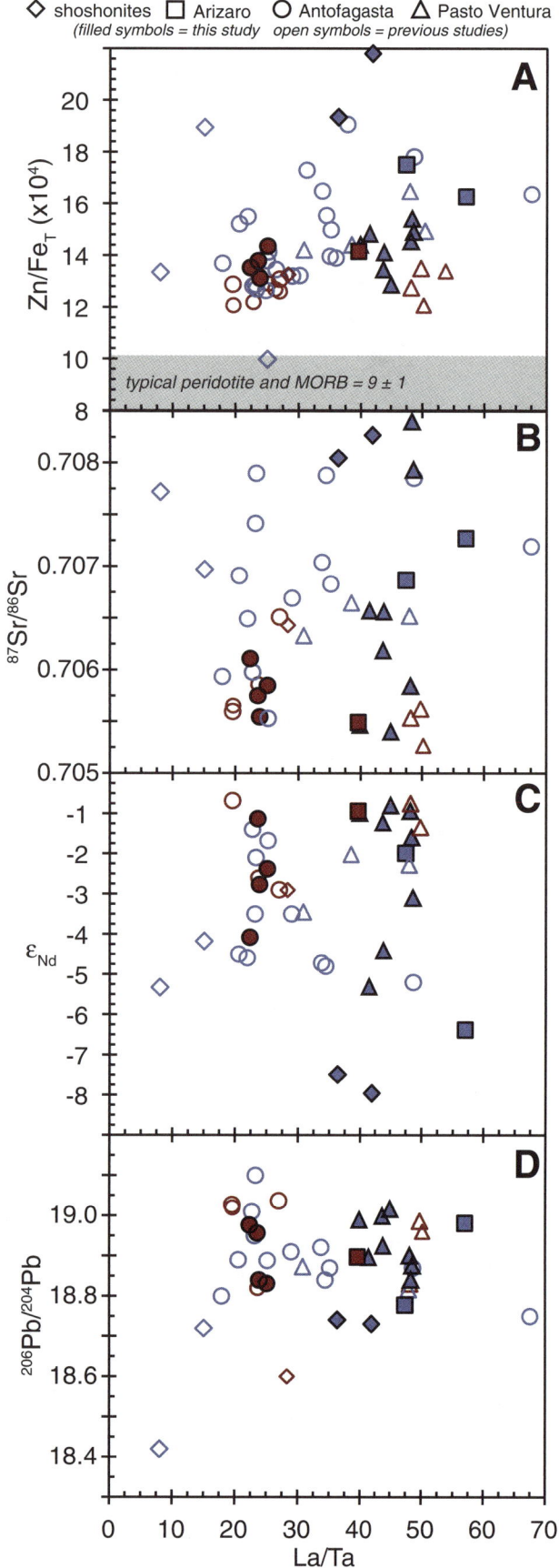

and 9B), but the Pasto Ventura high-Mg-group lavas (from Risse et al., 2013; Drew et al., 2009) have La/Yb and Sm/Yb values distinctly higher than the Antofagasta lavas. Another measure of the presence of eclogitic (i.e., grossular-rich) garnet, Hf/Sm (van Westrenen et al., 2011), varies with Sm/Yb, as would be expected if the Pasto Ventura lavas have some component of eclogite in their source (Fig. 9C).

Crustal Contributions to Isotopic Variability

Previous studies have attributed some of the variability in Puna lava whole-rock Sr-Nd isotopic compositions to magma mixing with radiogenic crustal material (up to 25 wt%; Kay et al., 1994; Drew et al., 2009; Risse et al., 2013). Additionally, oxygen isotope values in primary olivine crystals require that mafic melts in the Antofagasta region around Cerro Galán had $d^{18}O$ values of +6.7‰ to +7.2‰ during or prior to olivine crystallization (Risse et al., 2013; Kay et al., 2011). These values are above the typical bulk mantle and MORB $d^{18}O$ composition (+5.7‰ ± 0.2‰), and Risse et al. (2013) attributed this to early-stage crustal assimilation. This could explain some of the poorly correlated compositions exemplified by this study's lavas from Pasto Ventura. All are primitive-group lavas with identical trace-element patterns (Figs. 5 and 6), including a significant and similar magnitude of Ta-Nb depletion. However, the Nd isotopes from these lavas vary by more than four epsilon units and are decoupled from the other elemental trends (Figs. 7B and 10C). The absence of lower-crustal and mantle xenoliths in all Puna lavas suggests a relatively slow magma ascent rate, and an increased possibility for cryptic crustal contamination of the magmas (Glazner and Farmer, 1992).

A minor to moderate amount of crustal mixing is probably also reflected in the Zn/Fe_T trends of the primitive-group lavas from this and previous studies. Sr, Nd, and $^{206}Pb/^{204}Pb$ isotopic values all correlate with Zn/Fe_T values of the primitive-group lavas (Figs. 11C, 11D, and 11E). The high-Mg-group lavas have a small range of Zn/Fe_T values (12.1–14.3) over a moderate range of isotopic values with no regional trend or clear correlation. In contrast, the primitive-group lavas have a much larger range of Zn/Fe_T values (12.6–19.1, excluding the shoshonites), and more-evolved isotopic compositions are correlated with higher Zn/Fe_T (Figs. 11C, 11D, and 11E). Evolved magmatic rocks, including the evolved-group lavas from this study (Table 1), have high Zn/Fe_T values, due to fractional crystallization and significant

---

Figure 10. Elemental and isotopic data from the Puna Plateau that do not have the geographic trends highlighted in Figure 8. Only high-Mg-group and primitive-group lavas are plotted; closed symbols are from this study; open symbols are from recent studies of Puna lavas from Figure 4. Whereas the La/Ta values of Antofagasta lavas are significantly lower than lavas from the other regions on the Puna Plateau, high-Mg-group lavas (red symbols) from across the study region are not distinctive by the following measures: (A) Zn/Fe_T values; (B) $^{87}Sr/^{86}Sr$ values; (C) ε_{Nd} values; or (D) $^{206}Pb/^{204}Pb$ values. MORB—mid-ocean-ridge basalt.

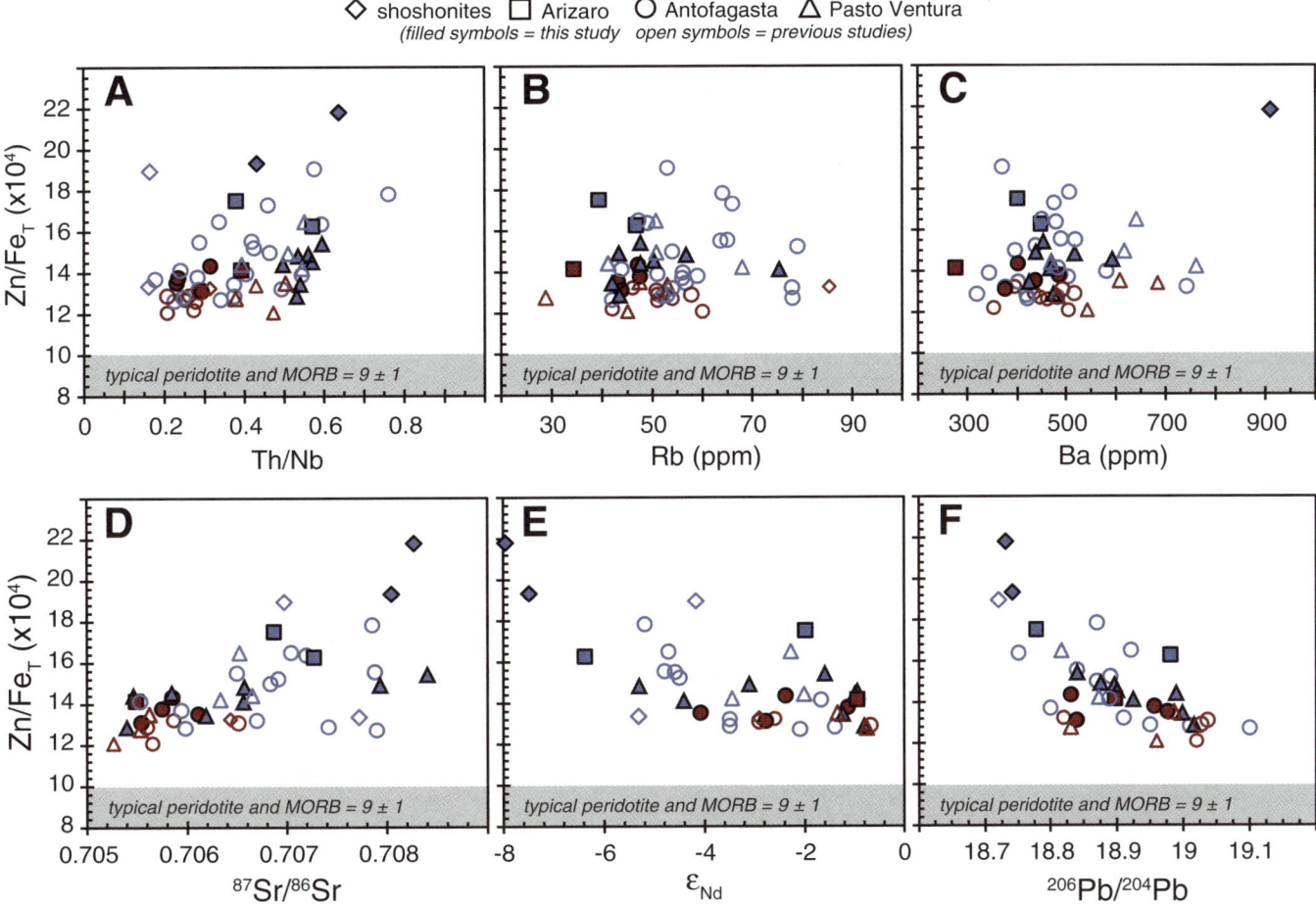

Figure 11. Zn/Fe$_T$ values plotted against key elemental and isotopic values in high-Mg-group and primitive-group lavas. (A–C) Zn/Fe$_T$ values plotted against values that track subduction-zone contributions (Th/Nb; Plank, 2005) and large ion lithophile elements (LILEs) associated with metasomatism in the mantle wedge (Rb, Ba). There are no patterns that suggest the high Zn/Fe$_T$ values in these lavas are associated with subduction-related enrichments or depletions. (D–F) Zn/Fe$_T$ values plotted against isotopic data. In contrast to values plotted in panels A, B, and C, these Zn/Fe$_T$ values trend with isotopic compositions of the primitive-group lavas. This suggests that the primitive-group lavas have more variability in elemental and isotopic compositions than the high-Mg-group lavas because of minor mixing with radiogenic (and high Zn/Fe$_T$) crustal material, probably in addition to variability in the mantle source region. MORB—mid-ocean-ridge basalt.

crustal mixing. Therefore, correlations between Zn/Fe$_T$ and Nd and Sr isotopic compositions are an expected result of crustal mixing in mantle-derived lavas with Mg compositions that suggest moderate evolution from primary melt compositions. Minor crustal mixing is also a possible reason for the high Zn/Fe$_T$ value of the primitive-group Puna shoshonites: Guzmán et al. (2006) attributed the poorly correlated compositions of the Puna shoshonites to crustal assimilation during rapid, turbulent ascent of those magmas.

The anomalously radiogenic Sr isotopic compositions of a few primitive-group and evolved-group lavas from this study (i.e., those offset to the right of the main Puna lava trend in Fig. 7A) are also likely the product of crustal contamination. The isotopic composition of Puna ignimbrites plots in the same region, suggesting there could be a high-Sr crustal component mixing with some of the Puna lavas (Fig. 7A).

As other studies have extensively explored Puna crustal mixing, we prefer instead to focus on the other source of compositional diversity in these lavas: variability in the composition of the mantle source region.

Array of Compositions beneath the Puna Plateau

In addition to crustal contamination, the compositional variability in the Puna mafic lavas is likely the result of variability in the source(s) of these melts. As previous studies have observed (Risse et al., 2013; Drew et al., 2009), there are likely at least two Nd-Sr isotopic end members that have relatively low $^{87}Sr/^{86}Sr$ compositions compared to the upper crust involved in the Altiplano-Puna volcanic complex ignimbrites. One is a relatively unradiogenic Sr and Nd source with a composition close to that of bulk silicate earth (BSE; Fig. 7A), which is represented by the highest-MgO

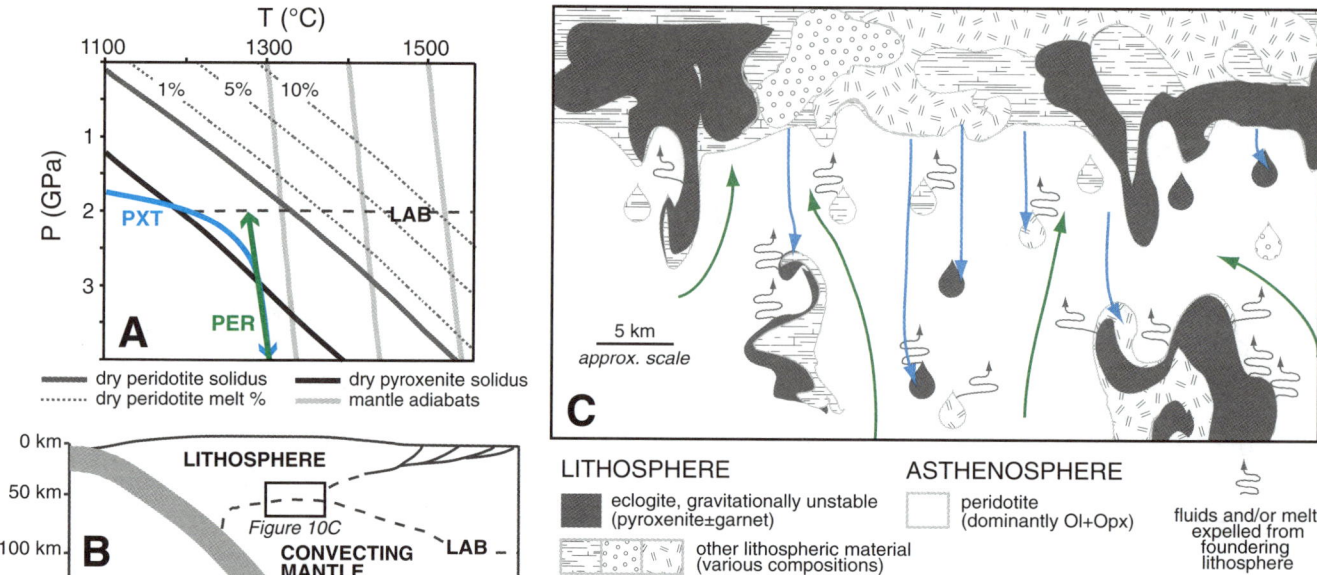

Figure 12. Summary of the mantle-drip hypothesis for generating primitive melts during lithospheric foundering: (A) Pressure-temperature (*P-T*) diagram after Ducea et al. (2013) showing hypothetical downward path for a descending pyroxenite drip (PXT, blue arrow) and adiabatic upward path for mantle peridotite upwelling to replace dripping lithospheric pyroxenite (PER, green arrows). The pyroxenite solidus is at lower *T* and higher *P* than the peridotite solidus, so lithospheric pyroxenites, including eclogites, would melt first and generate a greater percent melt along different *P-T* pathways when they exchange places during a foundering event. (B) Schematic cross section from west to east across the south-central Andes at ~26°S. Depth of lithosphere-asthenosphere boundary (LAB) is after Tassara and Echaurren (2012). (C) Cartoon cross-sectional view at the lithosphere-asthenosphere boundary beneath the Puna Plateau showing the proposed heterogeneity in composition, fertility, and density in the lower lithosphere and how that would play a key role in controlling the composition and volume of primary mantle melts generated in this setting. Lithospheric materials involved in foundering and melting could include accumulating residues from arc volcanism, remnant Famatinian crust and upper mantle, upwelled frozen asthenosphere, underthrust South American lithosphere, and material moved from the forearc during subduction erosion. Ol—olivine; Opx—orthopyroxene.

lava yet sampled on the Puna, P07-SUR101 (ε_{Nd} = –0.8, $^{87}Sr/^{86}Sr$ = 0.7055; Drew et al., 2009). The other end member is a source with more-evolved ε_{Nd} and similar $^{87}Sr/^{86}Sr$, which is represented by Antofagasta sample J1 (ε_{Nd} = –4.1, $^{87}Sr/^{86}Sr$ = 0.7061) and perhaps the shoshonites MC11 and MC12. The latter source could be in the lower crust, an isotopic reservoir that typically has a low Rb/Sr composition and a variety of crustal (i.e., negative) ε_{Nd} compositions (Risse et al., 2013). If the latter end member is lower crustal, however, it would be a crustal contaminant and not a mantle source rock. Given the elemental and isotopic variability amongst high-Mg-group lavas from this and other studies (*n* = 17, 14 with Sr-Nd isotopic data), we prefer to attribute the variability in the most mantle-derived lavas to a heterogeneous Puna upper mantle.

Orogenesis combined with continuous or punctuated convective overturn below the Puna Plateau would produce a lower lithosphere that is an amalgamation of accumulating residues from arc magmatism (Elkins-Tanton, 2007), material removed from the forearc during subduction erosion (Kay et al., 2013), underthrust continental lithospheric mantle from the retroarc transported toward the trench during shortening (DeCelles et al., 2009), residual pockets of unfoundered dense material, and upwelled asthenosphere subsequently frozen into the lithospheric

mantle (Figs. 12B and 12C). Which parts of this lithospheric mixture generate melts is a function of modal fertility, volatile content, and changes in pressure and/or temperature (Elkins-Tanton, 2007), which are not necessarily directly related to factors that would dictate the isotopic composition of those melting regions.

This heterogeneity might extend into the convecting upper mantle because it is likely isolated from fresh depleted mantle. The Puna Plateau is at the center of a continuous subduction zone that is ~7000 km in length. If depleted mantle is not present in significant volumes, it could be because the central Andes is a zone of upper-mantle stagnation (Schellart et al., 2007), in contrast to the northern and southern edge of the subduction system thousands of kilometers away. In that case, the convecting mantle in this region is eddy-like, filled with material shed by the local oceanic and continental plates for millions of years.

Model for Pyroxenite Melting during Lithospheric Foundering

The Zn/Fe$_T$ values of the high-Mg lavas on the Puna Plateau appear to require dominantly clinopyroxene ± garnet in materials melting to produce the Puna mafic lavas. This has implications for the mechanism(s) driving melting in the region. Given the abundant observations that point toward lithospheric loss from the Puna Plateau since the late Miocene, these results suggest an updated model for magmatism produced during convective removal of the lower lithosphere in cordilleran arc systems like the central Andes in which the lithospheric material foundering is itself the source of the primary melt. This links the materials driving foundering with the composition of the resulting melts, and it is contrary to the classic model (Kay and Kay, 1993) in which upwelling in the wake of delaminating blocks results in decompression melting of relatively homogeneous peridotite (i.e., olivine- and orthopyroxene-dominated mantle).

A principal role for clinopyroxene-bearing rocks (including pyroxenites and eclogites) in both foundering and melt generation fits well with current understanding of this process because pyroxenites, with or without garnet and/or amphibole and phlogopite, are common components of the subcontinental lithospheric mantle (Wilshire et al., 1988) and in arc settings occur as frozen veins of melt or as magmatic residues (Lee et al., 2006; Jull and Keleman, 2001). Furthermore, many pyroxenites, particularly eclogites, are denser than average peridotite mantle at upper-mantle *P-T* conditions (Jull and Keleman, 2001; Ducea, 2002; Hacker, 2004) and would therefore be present in gravitationally unstable parts of the lower lithosphere. Also, solidus temperatures for various pyroxenite compositions, particularly those with some volatile content (Elkins-Tanton, 2007), are commonly lower than for dry peridotite at upper-mantle conditions (Hirschmann and Stolper, 1996; Kogiso, 2004), and therefore pieces of foundering lithosphere are more likely to melt than the asthenosphere that adiabatically upwells in their wake (Fig. 12A).

In the "mantle-drip" (Ducea et al., 2013) or melting-drip model we propose (Fig. 12), the composition and size of the descending lithospheric drips, as well as their descent velocity, would control the amount of devolatilization and melting they experience (Elkins-Tanton, 2007). We hypothesize that smaller drips would heat up faster than larger drips and hit a solidus sooner after foundering. Additionally, smaller drips would also drive less convection in the asthenosphere and be less likely cause decompression melting (Elkins-Tanton, 2007). In contrast, during the foundering of large or rapidly descending pieces of lithosphere, the upwelling peridotite would be more likely to melt via decompression and contribute some melt to these magmas. We predict that if multiple batches of melts were generated from the same dripping event, the first should have the lowest melting temperatures, and the last should have the highest melting temperatures. If the lithospheric drip was large enough or descended fast enough into the convecting mantle to generate decompression melting in the asthenosphere, then the resulting peridotite-sourced melts should (1) have among the highest melt temperatures and (2) should be the youngest, because such melting would happen in the wake of the downgoing drip.

Ducea et al. (2013) tested these predictions using the silica-activity thermometer (Lee et al., 2009), olivine thermometer (Putirka, 2008), and Ar-Ar geochronology. They postulated that each regional volcanic field on the Puna Plateau (Pasto Ventura,

Antofagasta, Arizaro) was the locus of a lithospheric dripping event. As predicted, they found that the melt temperatures in each region were correlated with Zn/Fe_T values, and the older lavas had lower melting temperature than younger lavas. Additional thermometry and geochronology on the Puna lavas would further evaluate these findings. If indeed the regional volcanic fields on the Puna Plateau were each the locus of one (or more) related lithospheric dripping events, then each would have a compositional diversity inherited from the foundering material and modulated by the volume of the foundered piece(s) and perhaps some upwelling asthenosphere. Additionally, the geographic trend interpreted by Kay et al. (1994) could indeed reflect variations in the extent of lithospheric foundering.

For example, the high-Mg lavas from Antofagasta with relatively flat trace-element and REE patterns would have a lithospheric source with some component of peridotitic asthenosphere (i.e., the OIB-like signature of Kay et al., 1994). These lavas are the least enriched and arc-like of the Puna lavas and have some of the lowest Zn/Fe_T values. In our model, this reflects lithospheric removal that was frequent, rapid, or voluminous enough to not only melt the downgoing drips, but also perhaps to generate small amounts of peridotitic melt. It is also possible that the foundering lithosphere in this region is simply less enriched in incompatible elements than neighboring foundering lithosphere. In the Pasto Ventura region, the elemental compositions of the primitive lavas are nearly identical to each other, suggesting they were all sourced from a single piece of lithosphere or several drips from the same lithospheric composition. The isotopic variability in this region could be the result of minor mixing with asthenospheric melts or lower-crustal material. The Salar de Arizaro lavas are likely similar, with more crustal contributions accounting for the most-evolved lavas in the region. The Puna shoshonites also fit this model and represent the most enriched part of the Puna lower lithosphere involved in melting. Their high Mg numbers, high Zn/Fe_T ($\times10^4$) values (>19), evolved Sr-Nd isotopic ratios, steep REE patterns, significant Eu anomalies, extreme enrichment in LILEs, and high K compositions suggest the source was dominated by garnet-rich, old, enriched lithosphere, perhaps with some lower-crustal, plagioclase-rich component.

Implications for Lithospheric Foundering in the South-Central Andes

In the melting-drip model, foundering pieces of lithosphere do not need to be tens of kilometers in diameter in order to generate mantle melts, as is commonly suggested by schematic cartoons (see figure 11 *in* Kay et al., 1994; figure 13 *in* Kay et al., 2013). The evidence we present for dominantly pyroxenite melting is not a measure of the absolute volume of drips foundered from the Puna lithosphere. It does, however, offer some new insight: If the downgoing lithosphere (and not the upwelling asthenosphere) is generating these "delamination magmas," then (1) a much wider range of drip volumes could be responsible for melt generation than previously thought and (2) smaller lithospheric pieces are more likely to melt during lithospheric foundering than larger ones. Additionally, the apparent absence of a depleted-mantle end member in the first-row transition element ratios and the Nd-Sr isotopic values of these lavas suggests there was not voluminous enough (or rapid enough) convective overturn in the Puna upper mantle to generate dominantly peridotite-sourced melt. Geophysical imaging of the upper mantle in the central Andes resolves broad heterogeneities beneath the Puna region that are tens to hundreds of kilometers in diameter (Schurr et al., 2006; Bianchi et al., 2013). As suggested by our schematic cartoon in Figure 12C, we postulate that melt-generating foundering events could involve lithospheric drips that have kilometer-scale diameters. This is smaller than can currently be resolved by geophysical observations in the central Andes. We predict that as the resolution of these observations continues to improve, such heterogeneities will become more evident.

Catastrophic loss of large volumes of Andean lower lithosphere has been called upon to explain apparently rapid geodynamic changes in elevation of the Altiplano-Puna Plateau (cf. Garzione et al., 2008), and more broadly as a part of the orogenic cyclicity hypothesis of DeCelles et al. (2009). Our findings from the ~10 m.y. record of mantle melting archived in Puna mafic lavas are best explained by a piecemeal process of localized foundering events. The geodynamic response (i.e., change in surface topography, basin formation, faulting) to any single foundering event would likely be proportionally small, though a cluster of lithospheric drip events could conspire to generate a geodynamic response.

We postulate that in the same way that cordilleran arcs have a "background" rate of magma production punctuated by periodic high-flux events (DeCelles et al., 2009, and references therein), foundering-generated magmatism should have a complementary rate. That is to say, once an orogen reaches sufficient proportions (i.e., thick lithosphere with voluminous arc activity), small lithospheric loss events may occur frequently, producing small-volume mantle melts sourced from lithospheric drips. In contrast, catastrophic foundering events would be infrequent responses to significant shifts in the properties of the orogen that control the accumulation of dense lithospheric mantle and its ability to founder. The frequency of small lithospheric dripping events (and the net loss of lithosphere over a given time period) would be modulated over tens of millions of years by orogenic processes. For example, flare-ups in magmatic activity would rapidly generate large volumes of residual material that would tend toward eclogite compositions and gravitational instability (Ducea, 2001), while during periods of flat-slab subduction, the presence of the slab close to the base of the lithosphere would prevent foundering.

From this perspective, the ~10 m.y. record of mafic volcanism available on the Puna Plateau is just recording the "normal" noise of the cordilleran arc and its accumulation and evacuation of dense residual material. Preservation of such magmatic products in the geologic record would be sparse, because, as exemplified on the Puna Plateau mafic lavas, these melts are low in volume and often erupt in regions that also regularly experience voluminous volcanism, if they are erupted at all. Their elemental and isotopic variability reflects the heterogeneity of the lower lithosphere that is prone to foundering and melting, and this heterogeneity is itself a result of convergence, crustal thickening, arc magmatism, and lithospheric foundering (Elkins-Tanton, 2005).

CONCLUSIONS

In long-lived convergent orogens like the central Andes, the lithosphere is heterogeneous because its geochemistry and structure reflect prolonged accumulation and removal of crustal and mantle mass. Although our results broadly agree with previous workers (Kay et al., 1994) who proposed that Puna mafic lavas are the product of lithospheric foundering, we propose that lithospheric and not asthenosphere material is the source of these melts. The first-row transitional elements Zn and Fe provide this insight because their partitioning during mantle melting is sensitive to the difference between mantle sources with dominantly olivine + orthopyroxene (asthenospheric peridotite) and sources with dominantly clinopyroxene ± garnet (lithospheric pyroxenites and eclogites). The Zn/Fe_T evidence for pyroxenite (±garnet)–dominated lithospheric sources of these lavas is a new and invaluable perspective on their elemental and isotopic compositions, which are characterized by nonunique variability.

Pyroxenites (±garnet) are likely to be the gravitationally unstable components of the Andean lower lithosphere (Lee et al., 2006; Ducea, 2002) and would be more likely to generate melts during foundering than upwelling peridotitic mantle asthenosphere (Fig. 12). As a result, it is not necessary to catastrophically founder large volumes of the lithosphere to generate the Puna mafic lavas, as is commonly suggested in schematic diagrams of this process (Kay and Kay, 1993; Kay et al., 2013). Instead, our melting-drip model proposes that the fusible parts of small foundered pyroxenite drips melt during their thermal equilibration with the upper mantle. The resulting magmas are erupted as the Puna mafic lavas. Small-scale dripping of pyroxenite (±garnet)–bearing lithosphere would generate melts with heterogeneous and nonsystematic elemental and isotopic compositions because those pyroxenites are not only sourced from the modern arc cumulates and actively underthrusting continental lithosphere, but they are also inherited from earlier orogenic events in the region.

In sum, the classic model of delamination magmatism remains applicable to rapid foundering of very large pieces of the lower lithosphere, but our results suggest that piecemeal drizzle—and not catastrophic foundering—of dense garnet pyroxenites is a common mode of evacuating unstable lithosphere from beneath active cordilleran orogens. This is likely a common feature of the lithosphere in a long-lived and still-active convergent orogen like the central Andes, which has experienced repeated and prolonged orogenesis during the last 450 m.y.

APPENDIX

For whole-rock major- and trace-element and isotopic analysis, each sample was screened for weathered surfaces and secondary phases before powdering in an Al_2O_3-lined container. Powder aliquots were sent to the Geoanalytical Laboratory in the School of Earth and Environmental Sciences at Washington State University for major- and trace-element analysis by X-ray fluorescence (XRF) (Johnson, 1999) and inductively coupled plasma–mass spectrometry (ICP-MS). At the University of Arizona, 30–40 mg aliquots of whole-rock powder for each sample were dissolved in a hot mixture of concentrated HF and HNO_3 and spiked with Rb, Sr, and mixed Sm-Nd isotope spike solution (Otamendi et al., 2009). Rb, Sr, and the bulk of the REEs were separated in conventional cation columns filled with AG50W-X4 resin with 1 N to 4 N HCl. Cation column wash was subsequently used for Pb separation in Sr-Spec resin columns (Eichron, Darien, Illinois). Pb washes were loaded into the Sr-Spec columns in 8 N HNO_3, and the Pb was eluted using 8 N HCl (Drew et al., 2009). Nd and Sm separation from other REEs was achieved in anion columns filled with LN Spec resin using 0.1 N to 0.5 N HCl; half of the separations were performed in single-use 2 mL resin columns, and the other half in 6 mL resin columns.

Isotopic analyses were conducted at the University of Arizona, using thermal ionization mass spectrometry (TIMS) for Sr and Nd analysis, and multicollector (MC) ICP-MS in solution mode for Pb analysis. TIMS analyses were conducted on both an automated VG Sector MC instrument with adjustable 10^{11} Ω Faraday collectors and a Daly photomultiplier (Otamendi et al., 2009) and a VG Sector 54 instrument (Ducea, 2002). Sr and Rb cuts were loaded with a Ta_2O_5 slurry onto a single Ta filament. Nd and Sm cuts were loaded onto single Re filaments using resin beads. Concentrations of Rb, Sr, Sm, and Nd were determined by isotope dilution, with isotopic compositions determined on the same spiked runs. An off-line manipulation program was used for isotope dilution calculations. Typical runs consisted of acquisition of 100 isotopic ratios. The mean result of 10 analyses of the standard NRbAAA performed during the course of this study is: $^{85}Rb/^{87}Rb$ = 2.61199 ± 20. Fifteen analyses of standard Sr987 yielded mean ratios of: $^{87}Sr/^{86}Sr$ = 0.710285 ± 7 and $^{84}Sr/^{86}Sr$ = 0.056316 ± 12. The mean results of five analyses of the standard nSmb performed during the course of this study are: $^{148}Sm/^{147}Sm$ = 0.74880 ± 21, and $^{148}Sm/^{152}Sm$ = 0.42110 ± 6. Fifteen measurements of the La Jolla Nd standard were performed during the course of this study. The standard runs yielded the following isotopic ratios: $^{142}Nd/^{144}Nd$ = 1.14184 ± 2, $^{143}Nd/^{144}Nd$ = 511853 ± 2, $^{145}Nd/^{144}Nd$ = 0.348390 ± 2, and $^{150}Nd/^{144}Nd$ = 0.23638 ± 2. The Sr isotopic ratios of standards and samples were normalized to $^{86}Sr/^{88}Sr$ = 0.1194, whereas the Nd isotopic ratios were normalized to $^{146}Nd/^{144}Nd$ = 0.7219. The estimated analytical ±2σ uncertainties for samples analyzed in this study are: $^{87}Rb/^{86}Sr$ = 0.35%, $^{87}Sr/^{86}Sr$ = 0.0014%, $^{147}Sm/^{144}Nd$ = 0.4%, and $^{143}Nd/^{144}Nd$ = 0.0012%. Procedural blanks averaged from five determinations were: Rb = 10 pg, Sr = 150 pg, Sm = 2.7 pg, and Nd = 5.5 pg. Pb isotope analyses were conducted on a GV Instruments MC-ICP-MS (Thibodeau et al., 2007). All Pb samples were spiked with a Tl solution to acquire a Pb/Tl ratio of ~10 prior to analysis. Samples then entered the machine via free aspiration into a water-cooled chamber by way of a low-flow concentric nebulizer. A 2% HNO_3 blank was run before each sample, and the standard National Bureau of Standards (NBS)-981 was analyzed periodically during the sample run to ensure the stability of the instrument.

ACKNOWLEDGMENTS

This research was primarily supported by the Convergent Orogen System Analysis (COSA) collaboration between ExxonMobil and the Department of Geosciences at the University of Arizona, as well as National Science Foundation (NSF) grant EAR-0910941. MND acknowledges additional support from Romanian National Sciences Foundation (UEFISCDI) grant PN-II-ID-PCE-2011-3-0217. The Mark and Mary Lou Zoback graduate fellowship and a NSF Graduate Research Fellowship Program award provided additional support for KEM. S. Drew and C. Manthei collected the rock samples. E. Mortazavi and A. Thibodeau provided laboratory support. Thanks go to P. Reiners, P. DeCelles, A. Chapman, and M. Dettinger for useful discussion, suggestions, and comments on versions of this manuscript. Particular thanks go to M. Mamani and an anonymous reviewer for constructive comments that resulted in significant improvements to this manuscript.

REFERENCES CITED

Allmendinger, R.W., Jordan, T., Kay, S.M., and Isacks, B., 1997, The evolution of the Altiplano-Puna Plateau of the Central Andes: Annual Review of Earth and Planetary Sciences, v. 25, p. 139–174, doi:10.1146/annurev.earth.25.1.139.

Beck, S., and Zandt, G., 2002, The nature of orogenic crust in the central Andes: Journal of Geophysical Research–Solid Earth, v. 107, 2230, doi:10.1029/2000JB000124.

Beck, S., Zandt, G., Myers, S.C., Wallace, T., Silver, P., and Drake, L., 1996, Crustal-thickness variations in the central Andes: Geology, v. 24, p. 407–410, doi:10.1130/0091-7613(1996)024<0407:CTVITC>2.3.CO;2.

Bianchi, M., Heit, B., Jakovlev, A., Yuan, X., Kay, S.M., Sandvol, E., Alonso, R.N., Coira, B., Brown, L., Kind, R., and Comte, D., 2013, Teleseismic tomography of the southern Puna Plateau in Argentina and adjacent regions: Tectonophysics, v. 586, p. 65–83, doi:10.1016/j.tecto.2012.11.016.

Bird, P., 1979, Continental delamination and the Colorado Plateau: Journal of Geophysical Research, v. 84, p. 7561–7571, doi:10.1029/JB084iB13p07561.

Chung, S.-L., Chu, M.-F., Ji, J., O'Reilly, S.Y., Pearson, N.J., Liu, D., Lee, T.-Y., and Lo, C.-H., 2009, The nature and timing of crustal thickening in southern Tibet: Geochemical and zircon Hf isotopic constraints from postcollisional adakites: Tectonophysics, v. 477, p. 36–48, doi:10.1016/j.tecto.2009.08.008.

Coira, B., and Kay, S.M., 1993, Implications of Quaternary volcanism at Cerro Tuzgle for crustal and mantle evolution of the Puna Plateau, central Andes, Argentina: Contributions to Mineralogy and Petrology, v. 113, p. 40–58, doi:10.1007/BF00320830.

Crow, R., Karlstrom, K., Asmerom, Y., Schmandt, B., Polyak, V., and Dufrane, S.A., 2011, Shrinking of the Colorado Plateau via lithospheric mantle erosion: Evidence from Nd and Sr isotopes and geochronology of Neogene basalts: Geology, v. 39, p. 27–30, doi:10.1130/G31611.1.

Davidson, J.P., Harmon, R.S., and Worner, G., 1991, The source of central Andean magmas; some considerations, *in* Harmon, R.S., and Rapela, C.W., eds., Andean Magmatism and Its Tectonic Setting: Geological Society of America Special Paper 265, p. 233–243, doi:10.1130/SPE265-p233.

DeCelles, P.G., Ducea, M.N., Kapp, P., and Zandt, G., 2009, Cyclicity in Cordilleran orogenic systems: Nature Geoscience, v. 2, p. 251–257, doi:10.1038/ngeo469.

Déruelle, B., 1991, Petrology of Quaternary shoshonitic lavas of northwestern Argentina, *in* Harmon, R.S., and Rapela, C.W., eds., Andean Magmatism and Its Tectonic Setting: Geological Society of America Special Paper 265, p. 201–216, doi:10.1130/SPE265-p201.

de Silva, S.L., 1989, Altiplano-Puna volcanic complex of the central Andes: Geology, v. 17, p. 1102–1106, doi:10.1130/0091-7613(1989)017<1102:APVCOT>2.3.CO;2.

de Silva, S.L., and Francis, P., 1991, Volcanoes of the Central Andes: New York, Springer-Verlag, 216 p.

de Silva, S.L., Self, S., Francis, P., and Drake, R., 1994, Effusive silicic volcanism in the central Andes: The Chao dacite and other young lavas of the Altiplano-Puna volcanic complex: Journal of Geophysical Research, v. 99, p. 17,805–17,825, doi:10.1029/94JB00652.

de Silva, S.L., Zandt, G., Trumbull, R., Viramonte, J., Salas, G., and Jimenez, N., 2006, Large ignimbrite eruptions and volcano-tectonic depressions in the central Andes: A thermomechanical perspective, *in* Troise, C., De Natale, G., and Kilburn, C.R.J., eds., Mechanisms of Activity and Unrest at Large Calderas: Geological Society, London, Special Publication 269, p. 47–63, doi:10.1144/GSL.SP.2006.269.01.04.

Drew, S., Ducea, M.N., and Schoenbohm, L., 2009, Mafic volcanism on the Puna Plateau, NW Argentina: Implications for lithospheric composition and evolution with an emphasis on lithospheric foundering: Lithosphere, v. 1, p. 305–318, doi:10.1130/L54.1.

Ducea, M.N., 2001, The California arc: Thick granitic batholiths, eclogitic residues, lithospheric-scale thrusting, and magmatic flare-ups: GSA Today, v. 11, no. 1, p. 4–10, doi:10.1130/1052-5173(2001)011<0004:TCATGB>2.0.CO;2.

Ducea, M.N., 2002, Constraints on the bulk composition and root foundering rates of continental arcs: A California arc perspective: Journal of Geophysical Research–Solid Earth, v. 107, 2304, doi:10.1029/2001JB000643.

Ducea, M.N., 2011, Fingerprinting orogenic delamination: Geology, v. 39, p. 191–192, doi:10.1130/focus022011.1.

Ducea, M.N., and Barton, M.D., 2007, Igniting flare-up events in Cordilleran arcs: Geology, v. 35, p. 1047–1050, doi:10.1130/G23898A.1.

Ducea, M.N., and Saleeby, J., 1998, A case for delamination of the deep batholithic crust beneath the Sierra Nevada, California: International Geology Review, v. 40, p. 78–93, doi:10.1080/00206819809465199.

Ducea, M.N., Seclaman, A.C., Murray, K.E., Jianu, D., and Schoenbohm, L.M., 2013, Mantle-drip magmatism beneath the Altiplano-Puna Plateau, central Andes: Geology, v. 41, p. 915–918, doi:10.1130/G34509.1.

Elkins-Tanton, L.T., 2005, Continental magmatism caused by lithospheric delamination, *in* Foulger, G.R., Natland, J.H., Presnall, D.C., and Anderson, D.L., eds., Plates, Plumes, and Paradigms: Geological Society of America Special Paper 388, p. 449–461.

Elkins-Tanton, L.T., 2007, Continental magmatism, volatile recycling, and a heterogeneous mantle caused by lithospheric gravitational instabilities: Journal of Geophysical Research, v. 112, B03405, doi:10.1029/2005JB004072.

England, P., and Houseman, G., 1989, Extension during continental convergence, with application to the Tibetan Plateau: Journal of Geophysical Research–Solid Earth and Planets, v. 94, p. 17,561–17,579, doi:10.1029/JB094iB12p17561.

Farmer, G.L., Glazner, A., and Manley, C., 2002, Did lithospheric delamination trigger late Cenozoic potassic volcanism in the southern Sierra Nevada, California?: Geological Society of America Bulletin, v. 114, p. 754–768, doi:10.1130/0016-7606(2002)114<0754:DLDTLC>2.0.CO;2.

Francis, P., and Hawkesworth, C., 1994, Late Cenozoic rates of magmatic activity in the central Andes and their relationships to continental-crust formation and thickening: Journal of the Geological Society, London, v. 151, p. 845–854, doi:10.1144/gsjgs.151.5.0845.

Francis, P., Sparks, R.S.J., Hawkesworth, C., Thorpe, R., Pyle, D.M., Tait, S.R., Mantovani, M., and McDermott, F., 1989, Petrology and geochemistry of volcanic rocks of the Cerro Galan caldera, northwest Argentina: Geological Magazine, v. 126, p. 515–547, doi:10.1017/S0016756800022834.

Frassetto, A.M., Zandt, G., Gilbert, H., Owens, T.J., and Jones, C.H., 2011, Structure of the Sierra Nevada from receiver functions and implications for lithospheric foundering: Geosphere, v. 7, p. 898–921, doi:10.1130/GES00570.1.

Garzione, C., Molnar, P., Libarkin, J., and Macfadden, B., 2006, Rapid late Miocene rise of the Bolivian Altiplano: Evidence for removal of mantle lithosphere: Earth and Planetary Science Letters, v. 241, p. 543–556, doi:10.1016/j.epsl.2005.11.026.

Garzione, C., Hoke, G., Libarkin, J., and Withers, S., 2008, Rise of the Andes: Science, v. 320, p. 1304–1307, doi:10.1126/science.1148615.

GeoAnalytical Lab Technical Notes, 2013, Trace element analysis of rocks and minerals by ICP-MS: http://www.sees.wsu.edu/Geolab/note/icpms.html (accessed October 2013).

Glazner, A., and Farmer, G., 1992, Production of isotopic variability in continental basalts by cryptic crustal contamination: Science, v. 255, p. 72–74, doi:10.1126/science.255.5040.72.

Göğüş, O.H., and Pysklywec, R.N., 2008, Near-surface diagnostics of dripping or delaminating lithosphere: Journal of Geophysical Research, v. 113, B11404, doi:10.1029/2007JB005123.

Gutierrez-Alonso, G., Murphy, J.B., Fernandez-Suarez, J., Weil, A.B., Franco, M.P., and Gonzalo, J.C., 2011, Lithospheric delamination in the core of Pangea: Sm-Nd insights from the Iberian mantle: Geology, v. 39, p. 155–158, doi:10.1130/G31468.1.

Guzmán, S., Petrinovic, I., and Brod, J., 2006, Pleistocene mafic volcanoes in the Puna–Cordillera Oriental boundary, NW-Argentina: Journal of Volcanology and Geothermal Research, v. 158, p. 51–69, doi:10.1016/j.jvolgeores.2006.04.014.

Hacker, B.R., 2004, Subduction factory 3: An Excel worksheet and macro for calculating the densities, seismic wave speeds, and H_2O contents of minerals and rocks at pressure and temperature: Geochemistry Geophysics Geosystems, v. 5, Q01005, doi:10.1029/2003GC000614.

Haschke, M., Gunther, A., Melnick, D., Echtler, H., Reutter, K.-J., Scheuber, E., and Oncken, O., 2006, Central and Southern Andean tectonic evolution inferred from arc magmatism, *in* Oncken, O., Chong, G., Franz, G., Giese, P., Götze, H., Ramos, V., Strecker, M., and Wigger, P., eds., The Andes: Active Subduction Orogeny: Berlin, Springer, p. 337–353.

Hawkesworth, C.J., Hammill, M., Gledhill, A.R., Van Calsteren, P., and Rogers, G., 1982, Isotope and trace element evidence for late-stage intra-crustal melting in the High Andes: Earth and Planetary Science Letters, v. 58, p. 240–254, doi:10.1016/0012-821X(82)90197-2.

Hawkesworth, C.J., Gallagher, K., Hergt, J.M., and McDermott, F., 1993, Mantle and slab contribution in arc magmas: Annual Review of Earth and Planetary Sciences, v. 21, p. 175–204, doi:10.1146/annurev.ea.21.050193.001135.

Hildreth, W., and Moorbath, S., 1988, Crustal contributions to arc magmatism in the Andes of central Chile: Contributions to Mineralogy and Petrology, v. 98, p. 455–489, doi:10.1007/BF00372365.

Hirschmann, M.M., and Stolper, E.M., 1996, A possible role for garnet pyroxenite in the origin of the "'garnet signature'" in MORB: Contributions to Mineralogy and Petrology, v. 124, p. 185–208, doi:10.1007/s004100050184.

Houseman, G., and Molnar, P., 1997, Gravitational (Rayleigh-Taylor) instability of a layer with non-linear viscosity and convective thinning of continental lithosphere: Geophysical Journal International, v. 128, p. 125–150, doi:10.1111/j.1365-246X.1997.tb04075.x.

Houseman, G., McKenzie, D., and Molnar, P., 1981, Convective instability of a thickened boundary-layer and its relevance for the thermal evolution of continental convergent belts: Journal of Geophysical Research, v. 86, p. 6115–6132, doi:10.1029/JB086iB07p06115.

Johnson, D., 1999, XRF analysis of rocks and minerals for major and trace elements on a single low dilution Li-tetraborate fused bead: Advances in X-Ray Analysis, v. 41, p. 843–867.

Jull, M., and Keleman, P.B., 2001, On the conditions for lower crustal convective instability: Journal of Geophysical Research, v. 106, p. 6423–6446, doi:10.1029/2000JB900357.

Kay, R.W., and Kay, S.M., 1991, Creation and destruction of lower continental crust: Geologische Rundschau, v. 80, p. 259–278, doi:10.1007/BF01829365.

Kay, R.W., and Kay, S.M., 1993, Delamination and delamination magmatism: Tectonophysics, v. 219, p. 177–189, doi:10.1016/0040-1951(93)90295-U.

Kay, S.M., and Coira, B.L., 2009, Shallowing and steepening subduction zones, continental lithospheric loss, magmatism, and crustal flow under the central Andean Altiplano-Puna Plateau, *in* Kay, S.M., Ramos, V.A., and Dickinson, W.R., eds., Backbone of the Americas: Shallow Subduction, Plateau Uplift, and Ridge and Terrane Collision: Geological Society of America Memoir 204, p. 229–259.

Kay, S.M., Coira, B., and Viramonte, J., 1994, Young mafic back arc volcanic rocks as indicators of continental lithospheric delamination beneath the Argentine Puna Plateau, central Andes: Journal of Geophysical Research, v. 99, p. 24,323–24,339, doi:10.1029/94JB00896.

Kay, S.M., Coira, B.L., Caffe, P.J., and Chen, C.-H., 2010, Regional chemical diversity, crustal and mantle sources and evolution of central Andean Puna Plateau ignimbrites: Journal of Volcanology and Geothermal Research, v. 198, p. 81–111, doi:10.1016/j.jvolgeores.2010.08.013.

Kay, S.M., Coira, B., Wörner, G., Kay, R.W., and Singer, B.S., 2011, Geochemical, isotopic and single crystal $^{40}Ar/^{39}Ar$ age constraints on the evolution of the Cerro Galán ignimbrites: Bulletin of Volcanology, v. 73, p. 1487–1511, doi:10.1007/s00445-010-0410-7.

Kay, S.M., Mpodozis, C., and Gardeweg, M., 2013, Magma sources and tectonic setting of central Andean andesites (25.5–28°S) related to crustal thickening, forearc subduction erosion and delamination, *in* Gómez-Tuena, A., Straub, S.M., and Zellmer, G.F., eds., Orogenic Andesites and Crustal Growth: Geological Society, London, Special Publication 385, p. 303–334.

Knox, W., Coira, B., and Kay, S.M., 1989, Geochemical evidence on the origin of Quaternary basaltic andesites of the Puna, northwestern Argentina: Asociacion Geológica Argentina: Revista, v. XLIV, p. 194–206.

Kogiso, T., 2004, High-pressure partial melting of mafic lithologies in the mantle: Journal of Petrology, v. 45, p. 2407–2422, doi:10.1093/petrology/egh057.

Kraemer, B., Adelmann, D., Alten, M., Schnurr, W., Erpenstein, K., Kiefer, E., van den Bogaard, P., and Görler, K., 1999, Incorporation of the Paleogene foreland into the Neogene Puna Plateau: The Salar de Antofalla area, NW Argentina: Journal of South American Earth Sciences, v. 12, p. 157–182, doi:10.1016/S0895-9811(99)00012-7.

Lee, C.-T.A., Cheng, X., and Horodyskyj, U., 2006, The development and refinement of continental arcs by primary basaltic magmatism, garnet pyroxenite accumulation, basaltic recharge and delamination: Insights from the Sierra Nevada, California: Contributions to Mineralogy and Petrology, v. 151, p. 222–242, doi:10.1007/s00410-005-0056-1.

Lee, C.-T.A., Luffi, P., Plank, T., Dalton, H., and Leeman, W.P., 2009, Constraints on the depths and temperatures of basaltic magma generation on Earth and other terrestrial planets using new thermobarometers for mafic magmas: Earth and Planetary Science Letters, v. 279, p. 20–33, doi:10.1016/j.epsl.2008.12.020.

Lee, C.-T.A., Luffi, P., Le Roux, V., Dasgupta, R., Albarede, F., and Leeman, W.P., 2010, The redox state of arc mantle using Zn/Fe systematics: Nature, v. 468, p. 681–685, doi:10.1038/nature09617.

Le Roux, V., Lee, C.-T.A., and Turner, S.J., 2010, Zn/Fe systematics in mafic and ultramafic systems: Implications for detecting major element heterogeneities in the Earth's mantle: Geochimica et Cosmochimica Acta, v. 74, p. 2779–2796, doi:10.1016/j.gca.2010.02.004.

Le Roux, V., Dasgupta, R., and Lee, C.-T.A., 2011, Mineralogical heterogeneities in the Earth's mantle: Constraints from Mn, Co, Ni and Zn partitioning during partial melting: Earth and Planetary Science Letters, v. 307, p. 395–408, doi:10.1016/j.epsl.2011.05.014.

Lindsay, J., Schmitt, A., Trumbull, R., de Silva, S.L., Siebel, W., and Emmermann, R., 2001, Magmatic evolution of the La Pacana caldera system, central Andes, Chile: Compositional variation of two cogenetic, large-volume felsic ignimbrites: Journal of Petrology, v. 42, p. 459–486, doi:10.1093/petrology/42.3.459.

Lucassen, F., Escayola, M., Romer, R.L., Viramonte, J., Koch, K., and Franz, G., 2002, Isotopic composition of late Mesozoic basic and ultrabasic rocks from the Andes (23–32°S)—Implications for the Andean mantle: Contributions to Mineralogy and Petrology, v. 143, p. 336–349, doi:10.1007/s00410-001-0344-3.

Lucassen, F., Franz, G., Viramonte, J., Romer, R., Dulski, P., and Lang, A., 2005, The Late Cretaceous lithospheric mantle beneath the central Andes: Evidence from phase equilibria and composition of mantle xenoliths: Lithos, v. 82, p. 379–406, doi:10.1016/j.lithos.2004.08.002.

Lucassen, F., Franz, G., Romer, R., Schultz, F., Dulski, P., and Wemmer, K., 2007, Pre-Cenozoic intra-plate magmatism along the central Andes (17–34°S): Composition of the mantle at an active margin: Lithos, v. 99, p. 312–338, doi:10.1016/j.lithos.2007.06.007.

Mamani, M., Tassara, A., and Woerner, G., 2008, Composition and structural control of crustal domains in the central Andes: Geochemistry Geophysics Geosystems, v. 9, Q03006, doi:10.1029/2007GC001925.

Mamani, M., Wörner, G., and Sempere, T., 2010, Geochemical variations in igneous rocks of the central Andean orocline (13°S to 18°S): Tracing crustal thickening and magma generation through time and space: Geological Society of America Bulletin, v. 122, p. 162–182, doi:10.1130/B26538.1.

Manthei, C.D., Ducea, M.N., Girardi, J.D., Patchett, P.J., and Gehrels, G.E., 2010, Isotopic and geochemical evidence for a recent transition in mantle chemistry beneath the western Canadian Cordillera: Journal of Geophysical Research, v. 115, B02204, doi:10.1029/2009JB006562.

Marrett, R., and Emerman, S.H., 1992, The relations between faulting and mafic magmatism in the Altiplano Puna Plateau (central Andes): Earth and Planetary Science Letters, v. 112, p. 53–59, doi:10.1016/0012-821X(92)90006-H.

Matteini, M., Mazzuoli, R., Omarini, R., Cas, R., and Maas, R., 2002, Geodynamical evolution of central Andes at 24°S as inferred by magma composition along the Calama–Olacapato–El Toro transversal volcanic belt: Journal of Volcanology and Geothermal Research, v. 118, p. 205–228, doi:10.1016/S0377-0273(02)00257-3.

Matthews, S., Jones, A., and Gardeweg, M., 1994, Lascar volcano, northern Chile; evidence for steady-state disequilibrium: Journal of Petrology, v. 35, p. 401–432, doi:10.1093/petrology/35.2.401.

Mattioli, M., Renzulli, A., Menna, M., and Holm, P., 2006, Rapid ascent and contamination of magmas through the thick crust of the CVZ (Andes, Ollagüe region): Evidence from a nearly aphyric high-K andesite with skeletal olivines: Journal of Volcanology and Geothermal Research, v. 158, p. 87–105, doi:10.1016/j.jvolgeores.2006.04.019.

McDonough, W.F., and Sun, S.-s., 1995, The composition of the Earth: Chemical Geology, v. 120, p. 223–253, doi:10.1016/0009-2541(94)00140-4.

Molnar, P., and Garzione, C.N., 2007, Bounds on the viscosity coefficient of continental lithosphere from removal of mantle lithosphere beneath the Altiplano and Eastern Cordillera: Tectonics, v. 26, TC2013, doi:10.1029/2006TC001964.

Ort, M., Coira, B., and Mazzoni, M., 1996, Generation of a crust-mantle magma mixture: Magma sources and contamination at Cerro Panizos, central Andes: Contributions to Mineralogy and Petrology, v. 123, p. 308–322, doi:10.1007/s004100050158.

Otamendi, J., Ducea, M.N., and Tibaldi, A., 2009, Generation of tonalitic and dioritic magmas by coupled partial melting of gabbroic and metasedimentary rocks within the deep crust of the Famatinian magmatic arc, Argentina: Journal of Petrology, v. 50, no. 5, p. 841–873, doi:10.1093/petrology/egp022, doi:10.1093/petrology/egp022.

Pearce, J., and Parkinson, I., 1993, Trace element models for mantle melting: Application to volcanic arc petrogenesis, *in* Prichard, H.M., Alabaster, T., Harris, N.B.W., and Neary, C.R., eds., Magmatic Processes and Plate Tectonics: Geological Society, London, Special Publication 76, p. 373–403, doi:10.1144/GSL.SP.1993.076.01.19.

Pelletier, J.D., DeCelles, P.G., and Zandt, G., 2010, Relationships among climate, erosion, topography, and delamination in the Andes: A numerical modeling investigation: Geology, v. 38, p. 259–262, doi:10.1130/G30755.1.

Plank, T., 2005, Constraints from thorium/lanthanum on sediment recycling at subduction zones and the evolution of the continents: Journal of Petrology, v. 46, p. 921–944, doi:10.1093/petrology/egi005.

Putirka, K.D., 2008, Thermometers and barometers for volcanic systems: Reviews in Mineralogy and Geochemistry, v. 69, p. 61–120, doi:10.2138/rmg.2008.69.3.

Pysklywec, R.N., and Cruden, A.R., 2004, Coupled crust-mantle dynamics and intraplate tectonics: Two-dimensional numerical and three-dimensional analogue modeling: Geochemistry Geophysics Geosystems, v. 5, Q10003, doi:10.1029/2004GC000748.

Ramos, V.A., 2009, Anatomy and global context of the Andes: Main geologic features and the Andean orogenic cycle, *in* Kay, S.M., Ramos, V.A., and Dickinson, W.R., eds., Backbone of the Americas: Shallow Subduction, Plateau Uplift, and Ridge and Terrane Collision: Geological Society of America Memoir 204, p. 31–65.

Risse, A., Trumbull, R., Coira, B., Kay, S.M., and Bogaard, P., 2008, ^{40}Ar/^{39}Ar geochronology of mafic volcanism in the back-arc region of the southern Puna Plateau, Argentina: Journal of South American Earth Sciences, v. 26, p. 1–15, doi:10.1016/j.jsames.2008.03.002.

Risse, A., Trumbull, R.B., Kay, S.M., Coira, B., and Romer, R.L., 2013, Multistage evolution of late Neogene mantle-derived magmas from the central Andes back-arc in the southern Puna Plateau of Argentina: Journal of Petrology, v. 54, p. 1963–1995, doi:10.1093/petrology/egt038.

Rogers, N., and Hawkesworth, C., 1989, A geochemical traverse across the north Chilean Andes—Evidence for crust generation from the mantle wedge: Earth and Planetary Science Letters, v. 91, p. 271–285, doi:10.1016/0012-821X(89)90003-4.

Rosner, M., Erzinger, J., Franz, G., and Trumbull, R., 2003, Slab-derived boron isotope signatures in arc volcanic rocks from the central Andes and evidence for boron isotope fractionation during progressive slab dehydration: Geochemistry Geophysics Geosystems, v. 4, 9005, doi:10.1029/2002GC000438.

Rudnick, R.L., 1995, Making continental crust: Nature, v. 378, p. 571–578, doi:10.1038/378571a0.

Schellart, W.P., Freeman, J., Stegman, D.R., Moresi, L., and May, D., 2007, Evolution and diversity of subduction zones controlled by slab width: Nature, v. 446, p. 308–311, doi:10.1038/nature05615.

Schnurr, W., Risse, A., Trumbull, R., and Munier, K., 2006, Digital geological map of the southern and central Puna Plateau, NW Argentina, *in* Oncken, O., Chong, G., Franz, G., Giese, P., Götze, H., Ramos, V.A., Strecker, M.R., and Wigger, P., eds., The Andes: Active Subduction Orogeny: Frontiers in Earth Sciences: Berlin, Springer, p. 563–564, scale 1:250,000, available at http://link.springer.com/chapter/10.1007%2F978-3-540-48684-8_29.

Schnurr, W.B.W., Trumbull, R.B., Clavero, J., Hahne, K., Siebel, W., and Gardeweg, M., 2007, Twenty-five million years of silicic volcanism in the southern central volcanic zone of the Andes: Geochemistry and magma genesis of ignimbrites from 25 to 27°S, 67 to 72°W: Journal of Volcanology and Geothermal Research, v. 166, p. 17–46, doi:10.1016/j.jvolgeores.2007.06.005.

Schoenbohm, L.M., and Strecker, M.R., 2009, Normal faulting along the southern margin of the Puna Plateau, northwest Argentina: Tectonics, v. 28, TC5008, doi:10.1029/2008TC002341.

Schurr, B., Rietbrock, A., Asch, G., Kind, R., and Oncken, O., 2006, Evidence for lithospheric detachment in the central Andes from local earthquake tomography: Tectonophysics, v. 415, p. 203–223, doi:10.1016/j.tecto.2005.12.007.

Siebel, W., Schnurr, W., Hahne, K., Kraemer, B., Trumbull, R., van den Bogaard, P., and Emmermann, R., 2001, Geochemistry and isotope systematics of small- to medium-volume Neogene-Quaternary ignimbrites in the southern central Andes: Evidence for derivation from andesitic magma sources: Chemical Geology, v. 171, p. 213–237, doi:10.1016/S0009-2541(00)00249-7.

Stern, C., 1991, Role of subduction erosion in the generation of Andean magmas: Geology, v. 19, p. 78–81, doi:10.1130/0091-7613(1991)019<0078:ROSEIT>2.3.CO;2.

Sun, S.-s., and McDonough, W.F., 1989, Chemical and isotopic systematics of oceanic basalts: Implications for mantle composition and processes, *in* Saunders, A.D., and Norry, M.J., eds., Magmatism in the Ocean Basins: Geological Society, London, Special Publication 42, p. 313–345, doi:10.1144/GSL.SP.1989.042.01.19.

Tassara, A., and Echaurren, A., 2012, Anatomy of the Andean subduction zone: Three-dimensional density model upgraded and compared against global-scale models: Geophysical Journal International, v. 189, p. 161–168, doi:10.1111/j.1365-246X.2012.05397.x.

Thibodeau, A., Killick, D., Ruiz, J., Chesley, J., Deagan, K., Cruxent, J., and Lyman, W., 2007, The strange case of the earliest silver extraction by European colonists in the New World: Proceedings of the National Academy of Sciences of the United States of America, v. 104, no. 9, p. 3663–3666, doi:10.1073/pnas.0607297104.

Thorpe, R., Francis, P., O'Callaghan, L., Hutchison, R., and Turner, J., 1984, Relative roles of source composition, fractional crystallization and crustal contamination in the petrogenesis of Andean volcanic rocks: Philosophical Transactions of the Royal Society of London, ser. A, v. 310, p. 675–692, doi:10.1098/rsta.1984.0014.

Trumbull, R., Wittenbrink, R., Hahne, K., and Emmermann, R., 1999, Evidence for Miocene to Recent contamination of arc andesites by crustal melts in the Chilean Andes (25–26°S) and its geodynamic implications: Journal of South American Earth Sciences, v. 12, p. 135–155, doi:10.1016/S0895-9811(99)00011-5.

Trumbull, R.B., Riller, U., Oncken, O., Scheuber, E., Munier, K., and Hongn, F., 2006, The time-space distribution of Cenozoic volcanism in the south-central Andes: A new data compilation and some tectonic implications, *in* Oncken, O., Chong, G., Franz, G., Giese, P., Götze, H., Ramos, V.A., Strecker, M.R., and Wigger, P., eds., The Andes: Active Subduction Orogeny: Berlin, Springer, Frontiers in Earth Sciences, p. 29–43.

van Westrenen, W., Blundy, J.D., and Wood, B.J., 2011, High field strength element/rare earth element fractionation during partial melting in the presence of garnet: Implications for identification of mantle heterogeneities: Geochemistry Geophysics Geosystems, v. 2, 1039, doi:10.1029/2000GC000133.

Viramonte, J., Kay, S.M., Becchio, R., Escayola, M., and Novitski, I., 1999, Cretaceous rift related magmatism in central-western South America: Journal of South American Earth Sciences, v. 12, p. 109–121, doi:10.1016/S0895-9811(99)00009-7.

Whitman, D., Isacks, B., and Kay, S.M., 1996, Lithospheric structure and along-strike segmentation of the central Andean Plateau: Seismic Q, magmatism, flexure, topography and tectonics: Tectonophysics, v. 259, p. 29–40, doi:10.1016/0040-1951(95)00130-1.

Wilshire, H.G., Meyer, C.E., Nakata, J.K., Calk, L.C., Shervais, J.W., Nielson, J.E., and Schwarzman, E.C., 1988, Mafic and Ultramafic Xenoliths from Volcanic Rocks of the Western United States: U.S. Geological Survey Professional Paper 1443, 179 p.

Yáñez, G., Ranero, C., von Huene, R., and Dîaz, J., 2001, Magnetic anomaly interpretation across the southern central Andes (32–34°S): The role of the Juan Fernández Ridge in the Late Tertiary evolution of the margin: Journal of Geophysical Research, v. 106, p. 6325–6345, doi:10.1029/2000JB900337.

Yuan, X., Sobolev, S., and Kind, R., 2002, Moho topography in the central Andes and its geodynamic implications: Earth and Planetary Science Letters, v. 199, p. 389–402, doi:10.1016/S0012-821X(02)00589-7.

Zandt, G., Gilbert, H., Owens, T., Ducea, M.N., and Saleeby, J., 2004, Active foundering of a continental arc root beneath the southern Sierra Nevada in California: Nature, v. 431, p. 41–46, doi:10.1038/nature02847.

Zhou, R., Schoenbohm, L.M., and Cosca, M., 2013, Recent, slow normal and strike-slip faulting in the Pasto Ventura region of the southern Puna Plateau, NW Argentina: Tectonics, v. 32, p. 19–33, doi:10.1029/2012TC003189.

MANUSCRIPT ACCEPTED BY THE SOCIETY 3 JUNE 2014
MANUSCRIPT PUBLISHED ONLINE 15 OCTOBER 2014

The Geological Society of America
Memoir 212
2015

Miocene–Pliocene shortening, extension, and mafic magmatism support small-scale lithospheric foundering in the central Andes, NW Argentina

Lindsay M. Schoenbohm*

Department of Chemical and Physical Sciences, University of Toronto Mississauga, Mississauga, ON L5L 1C6, Canada

Barbara Carrapa

Department of Geosciences, University of Arizona, Tucson, Arizona 85721, USA

ABSTRACT

Lithospheric foundering may be a fundamental phenomenon in diverse tectonic settings and has been shown to affect surface deformation, subsidence, and uplift. In the central Andes, lithospheric removal has been proposed to have acted at the scale of the whole orogenic system, at a smaller scale, and cyclically. Although geophysical and geochemical data have led workers to infer lithospheric foundering beneath the central Andes, there is no consensus on the timing, magnitude, and location of such foundering events. New field mapping, sedimentology, and $^{40}Ar/^{39}Ar$ and U-Pb geochronology from the Puna Plateau in NW Argentina document the timing and spatial distribution of basaltic magmatism, contraction, and basin formation, and subsequent extension, all of which are predicted results of small-scale lithospheric drips. Our data are consistent with the formation of at least two small-scale (50–100 km) foundering events, alternating between the northern and southern Puna Plateau. Such "driplets" could be common in cordilleran systems and other plateaus and can contribute to significant recycling of lithosphere without causing extensive exhumation and surface uplift.

INTRODUCTION

There is growing recognition of the importance of lithospheric foundering, given that orogenic regions commonly host thinner mantle lithosphere than expected according to shortening estimates, and that orogenic continental crust is dominantly intermediate in composition, requiring recycling of dense, mafic residues into the mantle (Ducea, 2011). Although debated, litho-spheric removal as drip-like Rayleigh-Taylor instabilities has been proposed for several orogens around the world, including the Puna Plateau (Bianchi et al., 2013; Ducea et al., 2013; Kay and Kay, 1993; Kay et al., 2010, 2011) and the Altiplano Plateau (Garzione et al., 2008; Molnar and Garzione, 2007) of the central Andes, the Sierra Nevada (Ducea and Saleeby, 1998), Tibet (Turner et al., 1996), the Tien Shan (Molnar and Houseman, 2004), the Great Basin in the western United States (West

*lindsay.schoenbohm@utoronto.ca

Schoenbohm, L.M., and Carrapa, B., 2015, Miocene–Pliocene shortening, extension, and mafic magmatism support small-scale lithospheric foundering in the central Andes, NW Argentina, *in* DeCelles, P.G., Ducea, M.N., Carrapa, B., and Kapp, P.A., eds., Geodynamics of a Cordilleran Orogenic System: The Central Andes of Argentina and Northern Chile: Geological Society of America Memoir 212, p. 167–180, doi:10.1130/2015.1212(09). For permission to copy, contact editing@geosociety.org. © 2014 The Geological Society of America. All rights reserved.

et al., 2009), and the Colorado Plateau (Levander et al., 2011). In some cases, Rayleigh-Taylor instabilities have been imaged geophysically (e.g., Bianchi et al., 2013; Gilbert et al., 2007; Hales et al., 2005; Schurr et al., 2006; West et al., 2009) as dense features below or adjacent to regions of thin crust and lithosphere. Although immensely valuable in identifying recent foundering events, geophysical data are inherently limited in their ability to recognize older drips, which have descended deeply into the mantle. Geochemical data have also been used to identify Rayleigh-Taylor instabilities (e.g., DeCelles et al., 2009; Ducea et al., 2013; Kay and Kay, 1993; Manley et al., 2000) and can resolve older foundering events, but the magmatic response to foundering is highly variable, depending on the size and composition of foundering material (Elkins-Tanton, 2007), and surface erosion can remove evidence of the small-volume mafic volcanism often associated with lithospheric foundering. A possible recorder of the location and timing of past Rayleigh-Taylor instabilities, therefore, is surficial deformation. Although the pattern of deformation may vary depending on composition, size, and type of removal (e.g., single or double removal; Molnar and Houseman, 2004), analogue models have shown that in the case of a single drip, contraction and subsidence are predicted during drip formation, and uplift and extension should follow drip detachment (e.g., Göğüş and Pysklywec, 2008; Neil and Houseman, 1999; Pysklywec and Cruden, 2004). For example, surface subsidence and subsequent uplift have been successfully used to infer detachment of a dense plutonic root beneath the Wallowa Mountains in northeast Oregon in the mid-Miocene (Hales et al., 2005), which has since been confirmed by geophysical data (e.g., Darold and Humphreys, 2013).

In the Puna Plateau of northwest Argentina (Fig. 1), geochemical and geophysical data have been used to argue for removal of lithosphere through the formation and detachment of Rayleigh-Taylor instabilities (e.g., Kay and Kay, 1993; Kay et al., 1994; Schurr et al., 2006). Recent studies have further argued for small-scale (diameter of 50–100 km), repeated foundering events in the Puna Plateau (Bianchi et al., 2013; Calixto et al., 2013; Ducea et al., 2013; Kay et al., 2010, 2011). However, the precise timing and location of past drips are unknown, with inconsistencies between geophysical and magmatic indicators. We adopt a holistic approach, combining new structural, sedimentological, and geochronological data with existing geophysical, magmatic geochemical, and paleoelevation data to understand the history of the region.

We present new data from two key regions, Pasto Ventura in the southern Puna Plateau and the Arizaro region in the northwestern Puna Plateau (Fig. 1), which record evidence of shortening, followed by extension several million years after passage of the orogenic front through the region. This sequence of events is similar to model predictions and may be the result of past lithospheric foundering. These observations are corroborated by new $^{40}Ar/^{39}Ar$ geochronology of small-volume, mafic volcanism. Although other factors, such as wedge dynamics, gravitational spreading, and orogen-parallel extension, may play a role in surficial deformation in the Puna Plateau, when combined with existing geophysical and geochemical data, we will show that this history is best explained by the formation of multiple small drips, with the two most recent forming beneath the Arizaro region at 17–16 Ma and the southern Puna at 8–7 Ma. This interpretation contrasts with lithospheric foundering proposed for the Altiplano Plateau to the north, in which wholesale foundering and significant surface uplift are argued to have occurred at 10–7 Ma (Molnar and Garzione, 2007).

GEOLOGIC SETTING

Convergence between the Nazca and South American plates in the central Andes has resulted in formation of the broad, high-elevation (~3500 m) central Andean Plateau consisting of the internally drained Altiplano Plateau in Bolivia and the Puna Plateau in NW Argentina (Allmendinger et al., 1997). The Puna Plateau is distinguished from the Altiplano by higher elevations and relatively rugged topography with structurally bounded ranges and intervening basins (Fig. 1).

The mantle lithosphere is less than 10 km beneath most of the Altiplano and Puna Plateaus (Tassara and Echaurren, 2012). Moho depth is highly variable, ranging from 55 to 80 km, constrained by forward modeling of the Bouguer gravity anomaly (Tassara et al., 2006; Tassara and Echaurren, 2012), P-to-S teleseismic wave conversion (Yuan et al., 2000, 2002), and receiver-function analysis (Bianchi et al., 2013). However, the crust is thinned to <60 km in a quasi-circular region (Bianchi et al., 2013; Tassara et al., 2006; Tassara and Echaurren, 2012), with a diameter of 150–200 km, centered at 24°45′S, 67°45′W (Fig. 1B). High v_p/v_s ratios and low P-wave attenuation values have been used to argue for the presence of hot, asthenospheric mantle at shallow levels below the Puna Plateau and to infer the presence of sinking, detached lithosphere resting on the downgoing Nazca slab between 23°S and 24°S (Fig. 1B; Schurr et al., 2006). This feature is irregularly shaped, but it has a diameter of ~100 km in its longest dimension. In the southern Puna region, teleseismic tomography and Rayleigh-wave phase velocities indicate a region of dense material with a diameter of ~50 km resting or nearly resting on the subducting slab at a depth of 140–190 km (Fig. 1B; Bianchi et al., 2013; Calixto et al., 2013). In map view, this anomaly is located between 25°S and 26°S, just north of the Cerro Galan eruptive center (Fig. 1B), and it is also argued to be a detached lithospheric block (Bianchi et al., 2013; Calixto et al., 2013). Trench-parallel asthenospheric flow, based on shear-wave velocity anomalies, suggests that the anomalies may have migrated northward from their point of origin (Calixto et al., 2013).

Aspects of the magmatic history of the plateau are also consistent with lithospheric foundering. Small-volume, mafic, back-arc volcanism has developed in the southern Puna Plateau since ca. 7.3 Ma (Fig. 1B; Kay et al., 1994; Risse et al., 2008). Zn/Fe ratios of the most primitive of these flows indicate they are derived from melting of a peridotite source within 5–10 km below the base of the Moho (Ducea et al., 2013). These observations are consistent with modeling results (Elkins-Tanton, 2007), which suggest that magma is generated by heating and melting

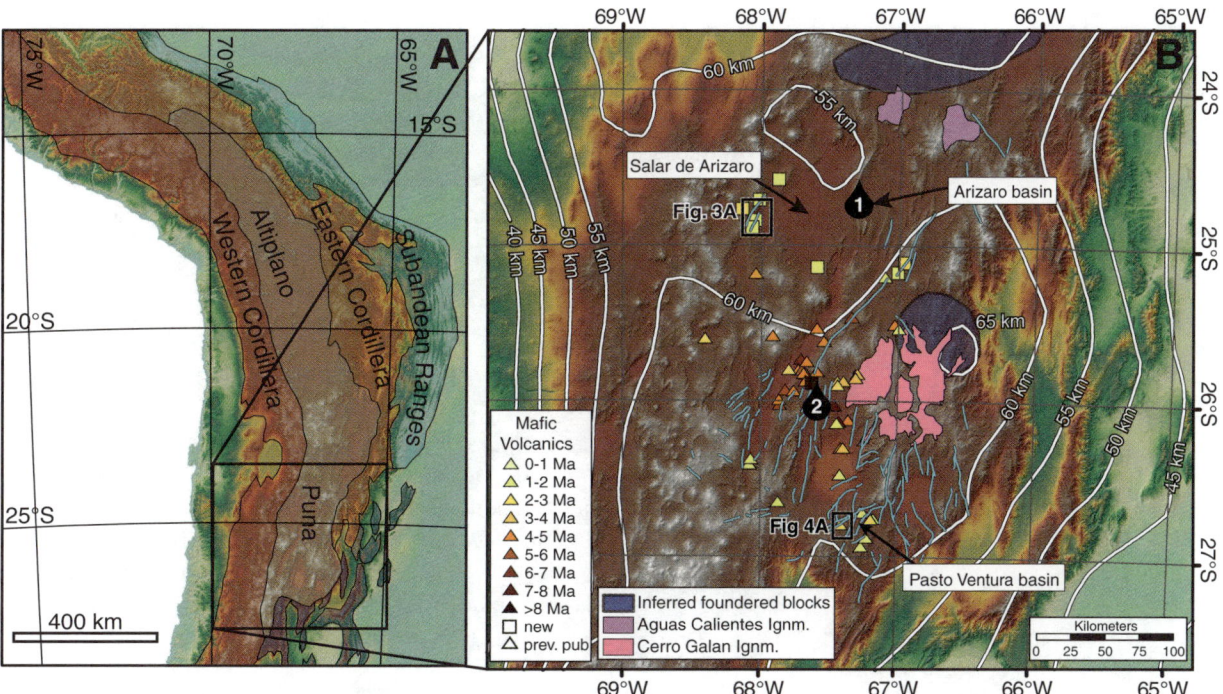

Figure 1. (A) Regional map on GTOPO30 DEM base (digital elevation model with horizontal grid spacing of 30 arc seconds), showing major tectonic units within the central Andes. (B) Shuttle Radar Topography Mission (SRTM) DEM showing age distribution of mafic volcanism (Risse et al., 2008; Zhou et al., 2011, personal commun.; this study), distribution of normal faulting (Schoenbohm and Strecker, 2009), contours of the Moho (Tassara and Echaurren, 2012, using a natural neighbor surface interpolation), the location of major ignimbrites (Ignm.; Kay et al., 2011), and the location of foundered lithospheric blocks inferred from v_p/v_s ratios, Qp values, teleseismic tomography, and Rayleigh-wave phase velocities (Bianchi et al., 2013; Calixto et al., 2013; Schurr et al., 2006). Proposed Rayleigh-Taylor instabilities are numbered sequentially.

of small-volume (<50 km) lithospheric drips. Small drips are also indicated by the remnants of stable mantle lithosphere beneath the Puna Plateau in some locations since the Ordovician (Drew et al., 2009), precluding wholesale lithospheric removal, as suggested for the Altiplano (Molnar and Garzione, 2007). Ignimbrites within the Puna Plateau have also been linked to repeated lithospheric foundering events (Kay et al., 2010, 2011) and could result from asthenospheric upwelling and melting (thermal conversion) of the mantle lithosphere after drip detachment.

Shortening began in the Chilean Cordillera in the early Cenozoic (Arriagada et al., 2006; Mpodozis et al., 2005), was within the Puna Plateau and its eastern margin at 38 Ma (Carrapa and DeCelles, 2008; Coutand et al., 2001), was within the Eastern Cordillera from ca. 21 Ma to ca. 4 Ma (Carrapa et al., 2011a, 2011b; Deeken et al., 2006), and migrated into the Santa Bárbara Subandes after ca. 4 Ma (Echavarria et al., 2003; Reynolds et al., 2000) (Fig. 2). Thus, although structural complexities exist, an in-sequence progression of the onset of deformation associated

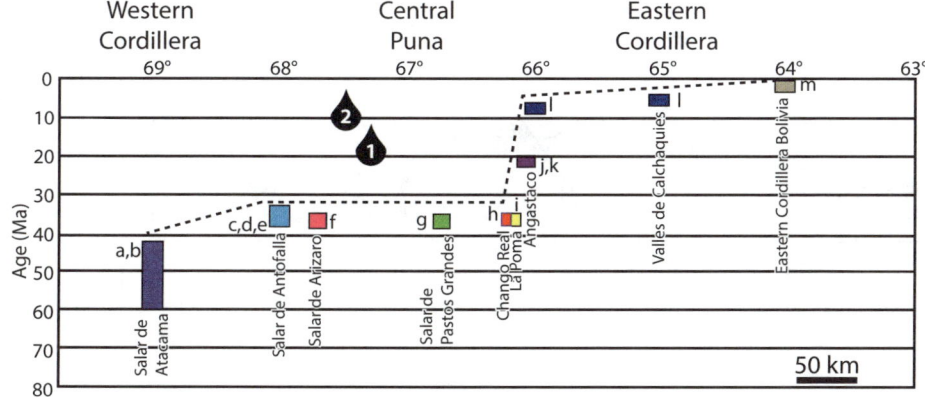

Figure 2. Square fields mark geographic area and age range of oldest Cenozoic deformation across the Puna Plateau by longitude. a—Arriagada et al. (2006); b—Mpodozis et al. (2005); c—Voss (2002); d—Kraemer et al. (1999); e—Carrapa et al. (2005); f—this study; g—Carrapa and DeCelles (2008); h—Hongn et al. (2007); i—Coutand et al. (2001); j—Deeken et al. (2006); k—Carrapa et al. (2011a); l—Carrapa et al. (2011b); m—Echavarria et al. (2003). Dashed lines mark limit of in-sequence deformation. Drip symbols mark timing and location of out-of-sequence shortening and extension as reported in this study.

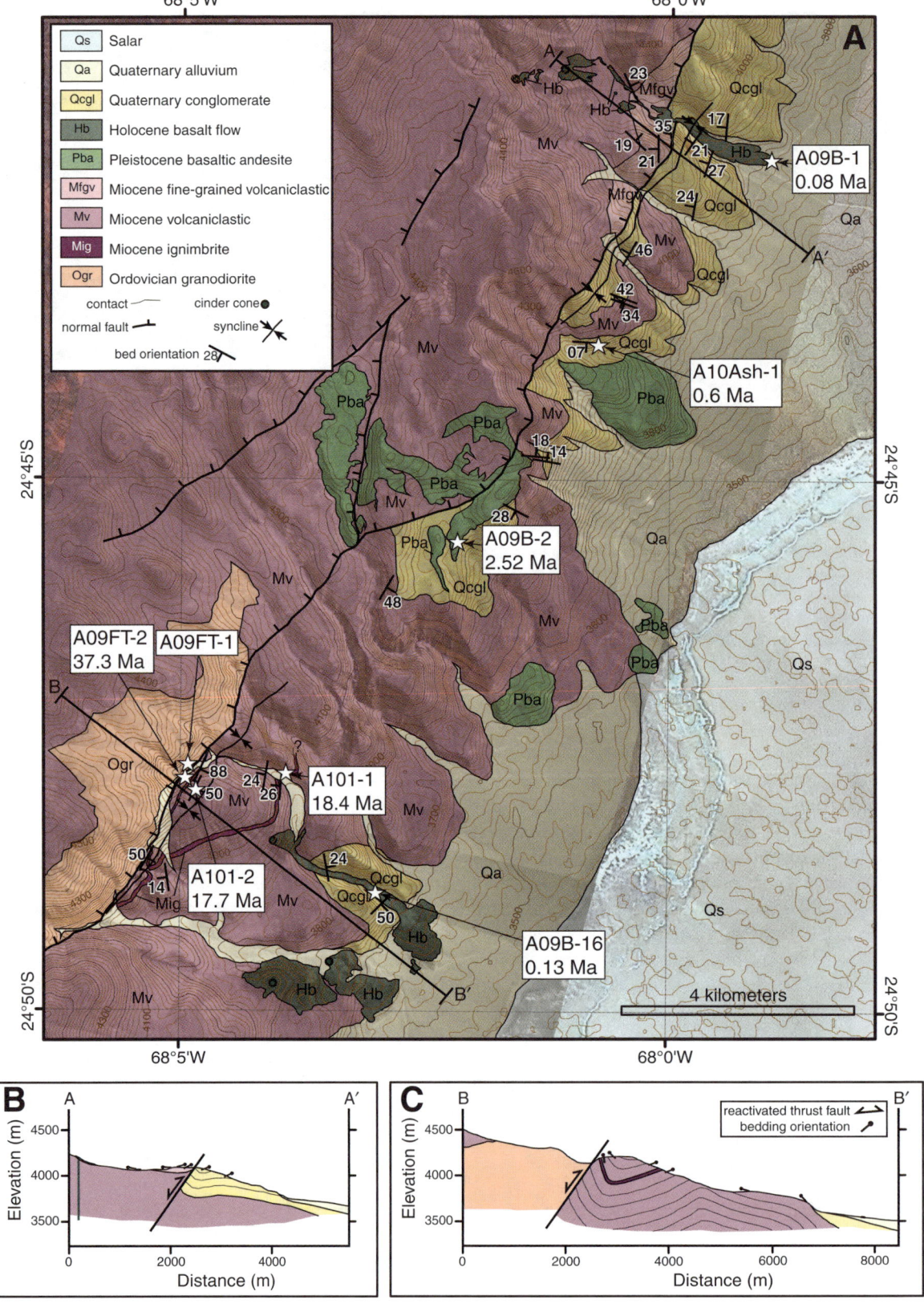

Figure 3. (A) Geologic map of the western margin of the Arizaro Basin on air photo base. Locations of samples dated in this study are indicated. (B) Cross section from southern part of western Arizaro Basin. (C) Cross section from northern part of western Arizaro Basin.

with west-to-east propagation of an orogenic wedge is present in the northern and central Puna Plateau (Carrapa et al., 2011a). In the Puna Plateau, widespread, low-magnitude extension has occurred since the late Miocene (Fig. 1B; Allmendinger, 1986; Montero Lopez et al., 2010; Risse et al., 2008; Schoenbohm and Strecker, 2009; Zhou et al., 2013). This extension likely reflects a combination of lithospheric foundering, gravitational spreading, and orogen-parallel extension (Riller et al., 2012; Schoenbohm and Strecker, 2009).

Deuterium ratios in volcanic glass suggest that elevations in the western Puna Plateau have been within 1 km of present values since at least 36 Ma (Canavan, 2012), suggesting uplift coincident with shortening and exhumation in the plateau (Arriagada et al., 2006; Carrapa and DeCelles, 2008; Coutand et al., 2001; Mpodozis et al., 2005). This is consistent with the gradual removal of excess lithospheric material through the detachment of small Rayleigh-Taylor instabilities, which would have a less significant impact on surface elevation that wholesale removal (Beiki-Ardakani et al., 2010). In contrast, paleoaltimetry proxies from the Altiplano suggest a rapid increase in surface elevation (~3 km) between 10 and 7 Ma, which has been used to argue for wholesale plateau delamination there (Garzione et al., 2008; Molnar and Garzione, 2007).

METHODS

We document the structural and sedimentological history of two key regions: Pasto Ventura along the southern margin of the plateau, where normal faulting and basaltic volcanism are abundant, and the Salar de Arizaro in the northwestern plateau, for which geophysical data suggest thin lithosphere (Fig. 1B). We completed geological maps at a scale of ~1:50,000 on an air photo base (Instituto Geografico Militar, Argentina) in a well-exposed section along the western margin of the Salar de Arizaro (Fig. 3) and within the northern Pasto Ventura basin (Fig. 4). We also measured a sedimentary section in the Pasto Ventura area (Fig. 4).

U-Pb zircon analyses on six intercalated ashes from both the Arizaro and Pasto Ventura regions were performed using the Cameca IMS 1270 ion microprobe at the University of California–Los Angeles (UCLA) National Ion Microprobe Facility (for pure ash layers in the Pasto Ventura section) and the inductively coupled plasma–mass spectrometer (ICP-MS) at the LaserChron Laboratory at the University of Arizona (for ignimbrite and impure ash layers in the Salar de Arizaro). Results are shown in Table 1, and additional details are available in Figure DR1 and Tables DR1 and DR2 of the GSA Data Repository.[1] Two ignimbrites in the Arizaro region have ages of 18.4 ± 0.7 Ma and 17.7 ± 0.5 Ma; an ash intercalated with Quaternary strata fur-

ther north is 0.6 ± 0.2 Ma. Ages in the Pasto Ventura sedimentary section range from 7.77 to 10.5 Ma. Additional methodological details are available in the GSA Data Repository (see footnote 1).

We complement our mapping data with $^{40}Ar/^{39}Ar$ dating of nine previously undated basaltic lava flows from the Arizaro, Calalaste, and Hombre Muerto regions (Fig. 1). These were dated at the U.S. Geological Survey (USGS) facility in Denver, Colorado, following standard methods. As summarized in Table 1 and Figures DR2–DR10 (see footnote 1), these range in age from –0.02 ± 0.03 Ma (A09B-2) to the oldest of the small-volume, mafic eruptions yet dated in the Puna Plateau at 8.7 ± 0.04 Ma (A10B-5). Additional methodological details are available in the GSA Data Repository (see footnote 1).

RESULTS

Arizaro Region

The Salar de Arizaro is a modern evaporite basin. It lies to the west of the Miocene Arizaro Basin, which contains ~3 km of eolian, alluvial, fluvial, and lacustrine sediment (Donato, 1987). We focused our mapping on the western margin of the modern salar. The major structure in this region is a NNE-striking, moderately WNW-dipping reverse fault (Fig. 3A). Its hanging wall consists of Ordovician Chuculaqui granodiorite and the volcaniclastic Lower Oligocene to Upper Miocene Quebrada de Agua complex (Zappetini et al., 2003). These units are carried over the upper Quebrada de Agua Formation in the south and Quaternary alluvial pebble-to-cobble conglomerate in the north (Fig. 3). In the southern footwall, volcaniclastic strata form a prominent asymmetric syncline, with an overturned western limb adjacent to the thrust fault and a shallowly dipping (20°–30°) eastern limb (Figs. 3A and 3C). Dips within the syncline progressively shallow up section, suggesting the presence of growth strata (Fig. 5), and therefore syndepositional deformation. In the northern footwall, Quaternary strata dip shallowly, mostly toward the northwest, toward the thrust fault (Figs. 3A and 3B).

The region is also cut by normal faults. The southern part of the major thrust is reactivated in a normal sense, as evidenced by a prominent bench where it crosses ridgelines, a result of extension along the fault. However, the major thrust is sealed in the north by basalt A09B-1, indicating no normal or thrust sense motion, at least along the northern portion of this fault, since this flow occurred (Fig. 3A). A normal fault in the interior of the range (Fig. 3A) offsets the Chuculaqui (Zappetini et al., 2003) basalt flow by ~120 m, with the hanging wall down to the west. Several small, monogenetic cinder cones and lava flows have erupted along the western margin of the Salar de Arizaro. These basalts are aligned along NNE trends, one along the edge of the modern salar, and one ~10 km to the west, close to the trace of the reverse fault, suggesting possible fault control.

Apatite fission-track (AFT) ages from the region are Oligocene and older (Carrapa and DeCelles, this volume). An age of 37.3 ± 4.8 Ma (A09FT-2; Fig. 3A) from the hanging-wall

[1]GSA Data Repository Item 2015003, Detailed analytical methods and results, is available at www.geosociety.org/pubs/ft2015.htm, or on request from editing@geosociety.org or Documents Secretary, GSA, P.O. Box 9140, Boulder, CO 80301-9140, USA.

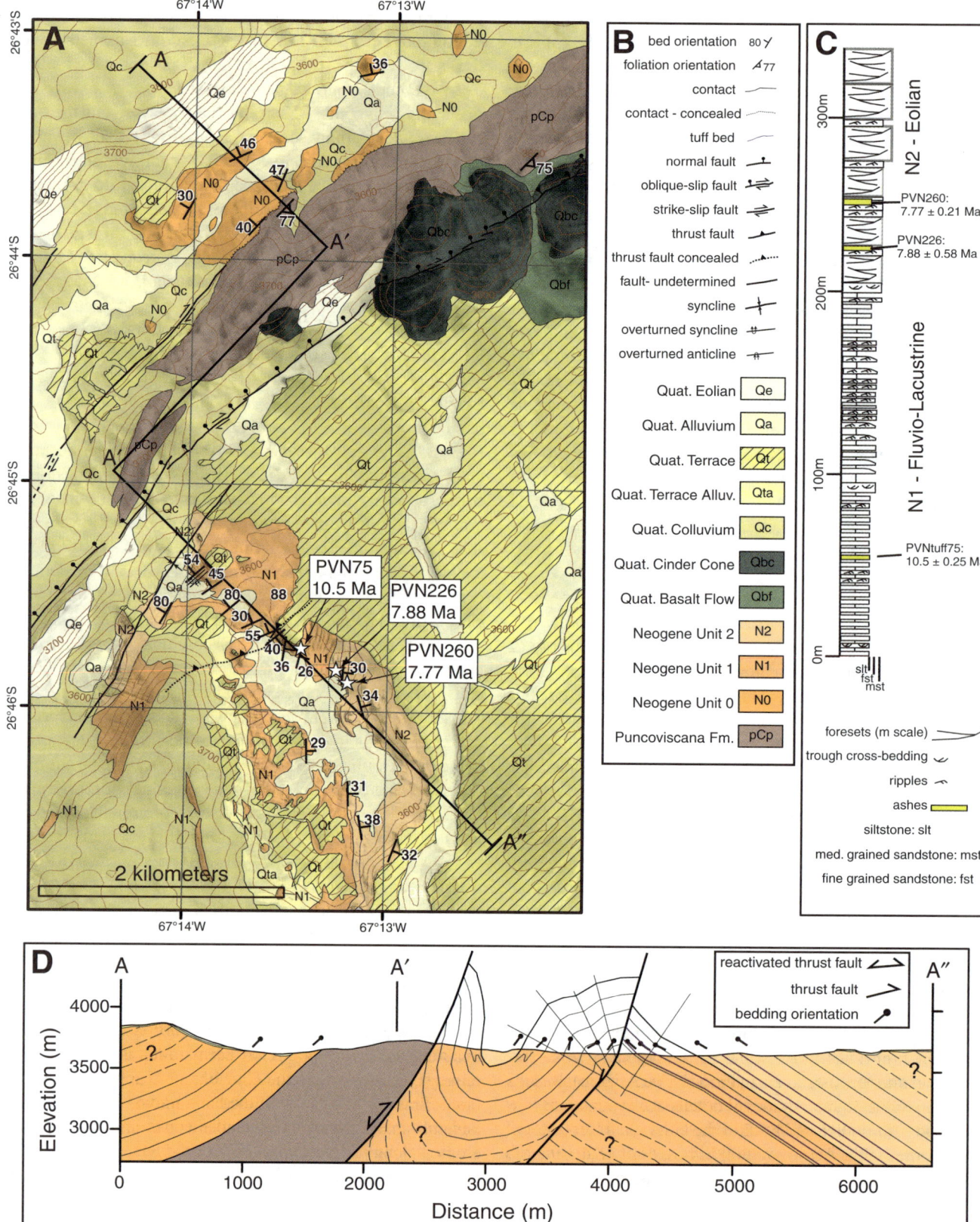

Figure 4. (A) Geologic map of the Pasto Ventura region on air photo base. (B) Legend for map. (C) Sedimentary log of units N1 and N2 in Pasto Ventura region. Ages shown on right are U-Pb dates on intercalated air-fall tuffs. Location of log is indicated on map. (D) Cross section with no vertical exaggeration.

TABLE 1. SUMMARY OF NEW GEOCHRONOLOGICAL DATA FOR SOUTHERN PUNA PLATEAU

Sample	Latitude (°S)	Longitude (°W)	Material	Method	Region	Age (Ma)	Error
A09B-1	24.7002	67.9826	Basalt – whole rock	$^{40}Ar/^{39}Ar$	Arizaro	0.08	0.06
A09B-2	25.1429	67.6099	Basalt – whole rock	$^{40}Ar/^{39}Ar$	Arizaro	−0.02	0.03
P09B-13	24.7055	67.9340	Basalt – whole rock	$^{40}Ar/^{39}Ar$	Arizaro	0.22	0.07
P09B-14	24.5628	67.9061	Basalt – whole rock	$^{40}Ar/^{39}Ar$	Arizaro	0.2	0.02
P09B-16	24.8136	68.0517	Basalt – whole rock	$^{40}Ar/^{39}Ar$	Arizaro	0.13	0.01
A10B-2	24.7599	68.0357	Basalt – whole rock	$^{40}Ar/^{39}Ar$	Arizaro	2.52	0.05
A10B-5	25.9239	67.7090	Basalt – whole rock	$^{40}Ar/^{39}Ar$	Calalaste	8.7	0.4
P09B-35	25.2643	67.0684	Basalt – whole rock	$^{40}Ar/^{39}Ar$	Hombre Muerto	0.87	0.06
P09B-35 (rep.)	25.2643	67.0684	Basalt – whole rock	$^{40}Ar/^{39}Ar$	Hombre Muerto	0.97	0.07
P09B-37	25.1978	67.0072	Basalt – whole rock	$^{40}Ar/^{39}Ar$	Hombre Muerto	2.99	0.13
P09B-35 (rep.)	25.1978	67.0072	Basalt – whole rock	$^{40}Ar/^{39}Ar$	Hombre Muerto	2.48	0.05
A10I-1	24.7980	68.0658	Ignimbrite - Zr	U/Pb by LA-MC-ICP-MS	Arizaro	18.4	0.7
A10I-2	24.7991	68.0811	Ignimbrite - Zr	U/Pb by LA-MC-ICP-MS	Arizaro	17.7	0.5
A10-ash1	24.7290	68.0144	Ash - Zr	U/Pb by LA-MC-ICP-MS	Arizaro	0.6	0.2
PVN75	26.7624	67.2235	Ash - Zr	U/Pb by SIMS	Pasto Ventura	10.5	0.1
PVN226	26.7641	67.2200	Ash - Zr	U/Pb by SIMS	Pasto Ventura	7.88	n/a
PVN260	26.7644	67.2194	Ash - Zr	U/Pb by SIMS	Pasto Ventura	7.77	0.21

Note: LA-MC-ICP-MS—laser ablation–multicollector–inductively coupled plasma–mass spectrometry; SIMS—secondary ion mass spectrometry.

granodiorite indicates limited exhumation (<6 km) of the basin margin in the Cenozoic. Inverse thermal modeling of a cobble from the footwall volcaniclastics (A09-FT1) supports cooling and exhumation in the Cretaceous and between ca. 30 Ma and 25 Ma, followed by heating between ca. 20 Ma and 18 Ma, most likely associated with sediment burial. Ignimbrites within the growing syncline are dated to 18.4 ± 0.7 Ma (A101-1) and 17.7 ± 0.5 Ma (A101-2) (Fig. 3A; Table 1; Fig. DR1; Table DR1 [see footnote 1]), supporting this interpretation. Relations among this syntectonic deformation, movement along the adjacent thrust fault, and cooling of sample A09-FT1 suggest active sedimentation and contraction across the western margin of the Salar de Arizaro at ca. 20–18 Ma.

No additional deformation was recorded until the late Quaternary. Conglomerate in the northern part of the study area contains a 0.6 ± 0.2 Ma (Ash10-1) intercalated air-fall tuff (Figs. 3A and 3C; Table 1; Fig. DR1; Table DR1 [see footnote 1]). Because these rocks were overthrust, compressional deformation must have occurred as recently as 0.6 Ma. Normal faulting in the area is at least younger than 2.52 ± 0.05 Ma (A10B-2), the $^{40}Ar/^{39}Ar$ age of the volcanic flow in the interior of the range offset by 120 m along a normal fault (Fig. 3A; Table 1; Fig. DR7 [see footnote 1]). The basalt flows along the NNE trend are considerably younger at, from north to south, 0.20 ± 0.02 Ma (P09B-14), 0.22 ± 0.07 Ma (P09B-13), 0.08 ± 0.06 Ma (A09B-1; Fig. 3A), and 0.13 ± 0.01 Ma (A09B-16; Fig. 3A; Table 1; Figs. DR2–DR5

[see footnote 1]). A flow to the southeast of the mapped area yields an even younger age: −0.02 ± 0.03 Ma (A09B-2) (Table 1; Fig. DR6 [see footnote 1]). If the onset of basaltic volcanism here reflects the onset of extension, as it does elsewhere in the Puna Plateau (Marrett and Emerman, 1992; Risse et al., 2008), then extension only began in the last ~200 k.y. in the study area.

Pasto Ventura Region

The Pasto Ventura region in the southern Puna Plateau lies in the zone of relatively thick crust and mantle lithosphere, with extensive normal faulting and volcanism (Fig. 1). Similar to the Arizaro region, a NE-striking, moderately NW-dipping thrust fault, later reactivated in a normal sense, dominates the structure of the Pasto Ventura study region (Fig. 4A). This fault carries the Early Cambrian (Escayola et al., 2011) Puncoviscana Formation in the hanging wall over deformed Cenozoic strata. To the west, undated Cenozoic strata in the hanging wall lie in depositional contact with Puncoviscana basement. Cenozoic strata in the footwall consist of ~200 m of lacustrine-fluvial deposits, which transition up section into eolian deposits (Fig. 4C). Paleo–wind directions in the eolian strata indicate a southward transport direction. An asymmetric syncline lies in the footwall of the thrust, with an overturned western limb and a more gently dipping 40°–50° eastern limb (Figs. 4A and 4F). Small folds and internal unconformities are present in the east limb of the syncline (Fig. 6), indicating

Figure 5. Growth strata in the volcaniclastic Quebrada de Agua complex on the western margin of the Salar de Arizaro. (A–B) Viewpoint is to the north and approximately corresponds to cross section in Figure 3B. Hill is ~300 m high. Black lines indicate bedding. Ignimbrite horizon (A101-1) is outlined by continuous black lines (Table 1; Fig. 3A), indicating deformation at this time. On left side, black arrow indicates thrust fault in which Ordovician Chuculaqui granodiorite is carried over the Quebrada de Agua complex; white arrow indicates reactivation of this plane in the late Quaternary as a normal fault. Note break in slope where normal fault intersects the ridgeline. An example of growth strata is pointed out. (C–D) Same growth strata, viewed from a position closer and more to the east. White lines indicate bedding. Growth strata are visible near the top of the outcrop.

syndepositional deformation. A second, subparallel thrust fault cuts the strata east of the major syncline. To the east of this fault, strata dip uniformly 30°–35° to the east. Several basalt flows and cinder cones erupted along the trace of the major thrust. Reactivation of the major thrust in an oblique right-lateral/normal sense is evidenced by offset Quaternary features in the south and cinder cones in the north (Fig. 4A).

Intercalated ashes are 10.5 ± 0.1 Ma in the lower fluvial-lacustrine unit and 7.88 ± 0.58 Ma and 7.77 ± 0.21 Ma in the upper, eolian unit (Fig. 4C; Table 1; Table DR2 [see footnote 1]). The internal unconformities (Fig. 6) are located at a stratigraphic location equivalent to the 10.5 Ma ash, indicating ongoing deformation at this time. Deformation must have continued until after ca. 7.8 Ma, the age of the youngest dated

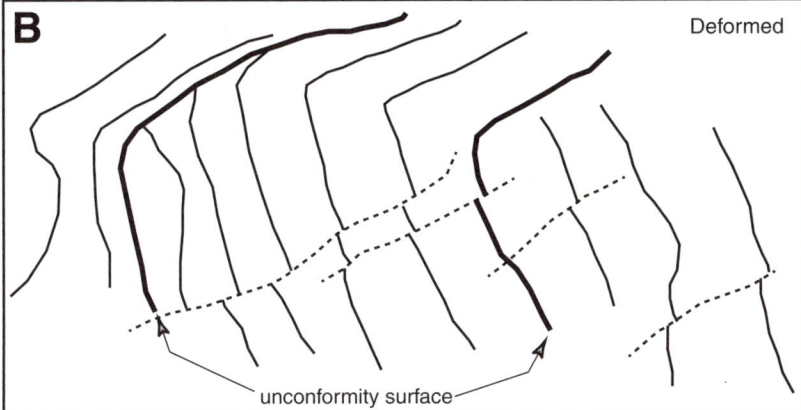

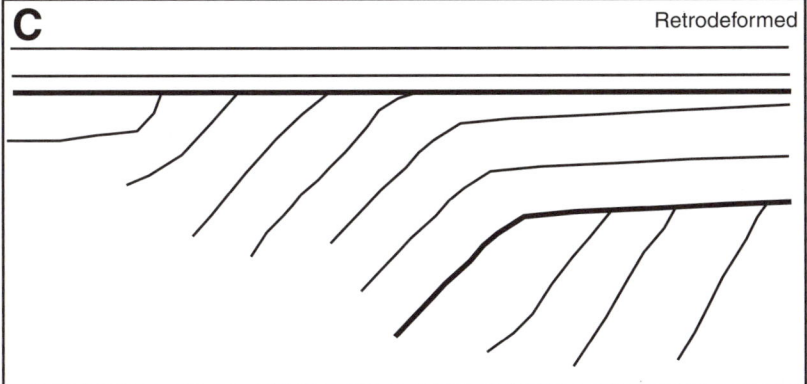

Figure 6. Syndepositional deformation in the Pasto Ventura region. (A) View to the N of T1 strata. The white lines highlight bedding contacts, while the heavy lines indicate minor thrust faults. (B) Sketch of the main features from the photo in A. (C) Strata retrodeformed such that the youngest strata in the photo are flat-lying. Two internal unconformities are present in this section, indicating syndepositional deformation. These strata contain ashes that are 9.9 ± 0.17 Ma and 10.5 ± 0.1 Ma (Table 1) in age, indicating contractional deformation at ca. 10 Ma.

tilted strata. Since ca. 7.8 Ma, the Pasto Ventura region has undergone a shift to extensional deformation, indicated by the reactivation of the major reverse fault as an oblique normal fault. Zhou et al. (2013) dated the basalt flow along the northern part of the fault (Fig. 4A) to 0.76 ± 0.16 Ma and mapped 16.1 m of normal sense and 34 m of right-lateral displacement. A 0.47 Ma flow is offset as well (Zhou et al., 2013). Therefore, the shift from shortening to extension occurred between ca. 7.8 and 0.47 Ma. Other faults that offset basalt

flows or Quaternary surfaces are present across the region as well, indicating NE-SW to NNE-SSW extension at a slow, time-integrated rate of 0.02–0.08 mm/yr since 0.8–0.5 Ma. Other data have been used to argue for extension as early as 7.3 Ma north of the Pasto Ventura area based on the presence of basaltic volcanism (Risse et al., 2008). Our dating indicates an even earlier onset of extension and volcanism, from an 8.7 ± 0.4 Ma ash collected in the Calalaste Range to the northwest (Table 1; Fig. DR8 [see footnote 1]).

Mafic Magmatism

A few hundred small-volume, monogenetic cinder cone eruptions are concentrated around the Calalaste Range, Salar de Antofogasta, and Salar de Antofalla, at ~26°S, 67°45′W, just west of the Cerro Galan ignimbrite complex (Drew et al., 2009; Ducea et al., 2013; Murray et al., this volume; Risse et al., 2008, 2013). Flows are found as far as 150 km to the NNW in the Arizaro region and 125 km to the SSE in the Pasto Ventura region. Age data are now available for 39 of these flows (Risse et al., 2008, and references therein; Zhou et al., 2013, personal commun; this study). The frequency of dated lava flows increases toward the present (Fig. 7), with no dated flows older than 8.7 Ma (this study). These data show an interesting spatial pattern as well (Fig. 1B). The oldest flows are located in the center of the eruptive region; over time, the eruptive region expanded, eventually reaching Arizaro and Pasto Ventura in the last million years. The pattern of increasing frequency with time could reflect an increase in eruptive activity toward the present or decreasing preservation and exposure with age. Alternatively, the eruption of mafic lavas at the surface could reflect the state of stress of the upper crust. Mafic magmas are dense enough that they may not erupt at the surface until the region undergoes extension (Marrett and Emerman, 1992). Significant residence time in the crust, as indicated by the geochemistry of these lava flows, is consistent with this latter scenario (Ducea et al., 2013; Murray et al., this volume; Risse et al., 2013).

DISCUSSION AND CONCLUSIONS

Even though the idea of lithospheric foundering in the Puna Plateau has been around for two decades (e.g., Kay and Kay,

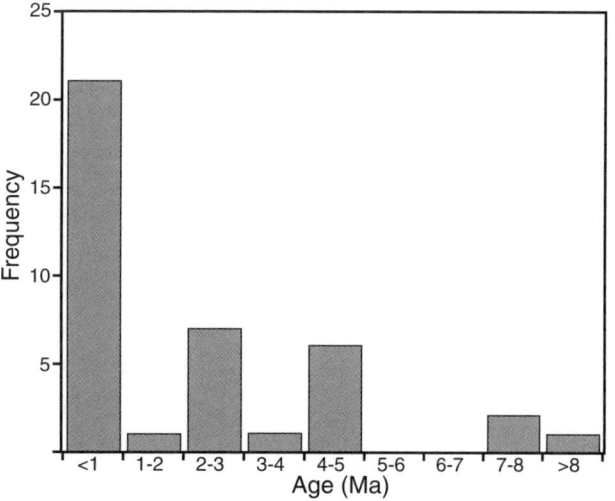

Figure 7. Frequency with time of dated small-volume mafic cinder cones and lava flows. Note increase in frequency with time, which could reflect a true increase or a poor preservation of earlier eruptions.

1993) and has gained considerable traction, particularly based on recent geophysical (Bianchi et al., 2013; Calixto et al., 2013; Tassara and Echaurren, 2012) and geochemical (Ducea et al., 2013; Murray et al., this volume) data, our structural data could reflect other processes, and they are limited to observations in only two locations. Our data are sparse on the Puna Plateau, given the poor exposure (slopes are mantled with debris) and the necessity of on-the-ground studies because of reactivation of older, shortening-related structures in recent extensional deformation (Schoenbohm and Strecker, 2009). There are, in fact, few other places in the southern Puna region in which such mapping is possible. We, thus, proceed with caution in interpreting our data in the larger geodynamic context.

Out-of-sequence deformation, such as we observe in the Arizaro and Pasto Ventura regions, could reflect normal wedge dynamics, in which the orogenic system is in subcritical state and deforming internally in order to build taper (e.g., Davis et al., 1983). Removal of material from the orogenic system, by erosion for example, is predicted to reduce taper and to drive the wedge into a subcritical state (e.g., DeCelles et al., 2009). Although out-of-sequence shortening within the plateau could be explained by wedge dynamics, the localized nature of deformation and the pattern of shortening followed by extension and associated mafic magmatism do not satisfy a wedge-dynamics explanation. Extension, alternatively, could be explained by gravitational spreading (e.g., Schoenbohm and Strecker, 2009), although this would not explain the shortening phase of deformation, which immediately preceded extension, and is, therefore, an incomplete explanation for our observations. Similarly, orogen-parallel extension (e.g., Riller et al., 2012), while allowing for simultaneous shortening and extension on the plateau, does not allow reactivation of structures of the same orientation over time, such as in Arizaro and Pasto Ventura.

Our favored mechanism to explain the observed geology, including out-of-sequence contraction followed by extension and mafic volcanism, is lithospheric foundering. Only a few modeling studies make specific predictions about the response of the surface to Rayleigh-Taylor instabilities, whether above marginal (Molnar and Houseman, 2004) or central (Göğüş and Pysklywec, 2008; Neil and Houseman, 1999) "drips." According to these models, during growth of a single Rayleigh-Taylor instability, material draws toward the locus of the drip, causing out-of-sequence shortening and thickening of overlying crust (Göğüş and Pysklywec, 2008; Neil and Houseman, 1999) in a spoke-like, radial pattern if the drip is cylindrical (Pysklywec and Cruden, 2004). The mantle lithosphere above the Rayleigh-Taylor instability may be substantially thinned if the event is large, lateral flow is insufficient to feed the drip, or the drip exhibits a nonlinear, temperature-independent rheology. Alternatively, thinning might occur more broadly with only a small local deflection in the lithosphere-asthenosphere boundary if the drip is small, the lithospheric viscosity is low enough to permit rapid lateral flow, or the drip rheology is temperature-dependent, allowing only the lower part of the lithosphere to be involved in dripping

(Conrad and Molnar, 1999; Elkins-Tanton, 2007; Göğüş and Pysklywec, 2008). In its initial stages, the negative buoyancy of the Rayleigh-Taylor instability overcomes the buoyant effect of crustal thickening, leading to subsidence. However, as the drip necks and detaches, the viscous traction of the drip decreases, and dense mantle lithosphere is simultaneously replaced by more buoyant asthenosphere, resulting in uplift (estimated from a few meters [Elkins-Tanton, 2007] to several kilometers [Molnar and Garzione, 2007]), and consequent extension over the locus of the drip (Elkins-Tanton, 2007; Göğüş and Pysklywec, 2008; Neil and Houseman, 1999). With smaller, multiple Rayleigh-Taylor instabilities ("driplets") at the scale of single basins (<100 km), surface subsidence, uplift, shortening, and extension are muted (Beiki-Ardakani et al., 2010). The effect of Rayleigh-Taylor instabilities on lithospheric topography may be short-lived as a result of lateral flow and conductive cooling, leading to a flat lithosphere-asthenosphere boundary a few million years after the event (Elkins-Tanton, 2007), particularly if the drip is small. Decompression melting of upwelling asthenosphere and dewatering of the sinking mantle lithosphere produce small-volume, high-potassium, mafic volcanism throughout the foundering event (Elkins-Tanton, 2005; Kay and Kay, 1993; Schott and Schmeling, 1998), but such dense magmas may not erupt at the surface until extension begins after drip detachment (Marrett and Emerman, 1992). Ignimbrite eruptions have been associated with Rayleigh-Taylor instabilities as well (Kay et al., 2010). In the case of small-scale lithospheric "driplets," low volumes of mafic magmatism are expected (Drew et al., 2009), with source material for melts dominated by pyroxenites derived from melting of downgoing lithosphere rather than by peridotites from an asthenospheric source (Ducea et al., 2013).

Based on these modeling predictions, we see evidence for two Rayleigh-Taylor instabilities in the Puna Plateau, which refine the location and timing of previously inferred foundering events (Fig. 8). We suggest that the first Rayleigh-Taylor instability reflected in our data began to form underneath the Arizaro sedimentary basin around 20 Ma. Subsidence in this basin is recorded by thick sediment accumulation between ca. 20 and ca. 11 Ma with rapid accumulation between ca. 19.5 and ca. 16 Ma (Boyd, 2010; DeCelles et al., this volume). The pattern of subsidence recorded by these strata led DeCelles et al. (this volume) to attribute basin formation to lithospheric foundering. To the west of the Arizaro Basin, along the western margin of the modern salar, we observe contraction simultaneous with sediment accumulation, recorded by syndepositional deformation of the ca. 18 Ma volcaniclastic rocks in the footwall of a major thrust fault (Fig. 3). Active deformation at this time is also indicated by folded strata of the middle Vizcachera Formation (ca. 19 to ca. 16 Ma), within the Arizaro Basin, in the footwall of a west-dipping reverse fault, placing the Macon Range crystalline basement rock on top of Miocene strata (Boyd, 2010). Detachment of the drip after 17–16 Ma is suggested by eruption of nearby ignimbrites of the Aguas Calientes caldera between 17.2 Ma and 10.5 Ma (Kay et al.,

2010). However, there is no evidence for mafic volcanism or normal faulting following Rayleigh-Taylor instability detachment. This may reflect the ephemeral nature of such evidence; normal faulting would be overprinted by subsequent shortening associated with a second drip (see following), and mafic volcanism was minor, and therefore subject to removal by erosion. Ages of dated mafic magmas decrease with time, supporting such a scenario (Fig. 7). Alternatively, the lack of mafic magmatism could reflect the size of the foundering material, with a large foundered block heating more slowly. Such a scenario would favor asthenospheric upwelling and eruption of large-volume ignimbrites (Kay et al., 2010, 2011). A relatively large foundered block is also favored by geophysical data, showing significant thinning of the crust and mantle lithosphere beneath the Arizaro region (Bianchi et al., 2013; Tassara et al., 2006; Tassara and Echaurren, 2012) and a relatively large anomaly (>100 km) located to the NE of the Arizaro region interpreted as foundered lithospheric material (Schurr et al., 2006). Shear-wave velocity anomalies suggest northward trench-parallel asthenospheric flow (Calixto et al., 2013), which could explain NE displacement of the imaged, foundered block from the region we propose for drip formation.

We also find evidence to support formation of a second Rayleigh-Taylor instability in the southern Puna Plateau, again refining the location and timing of a foundering event inferred

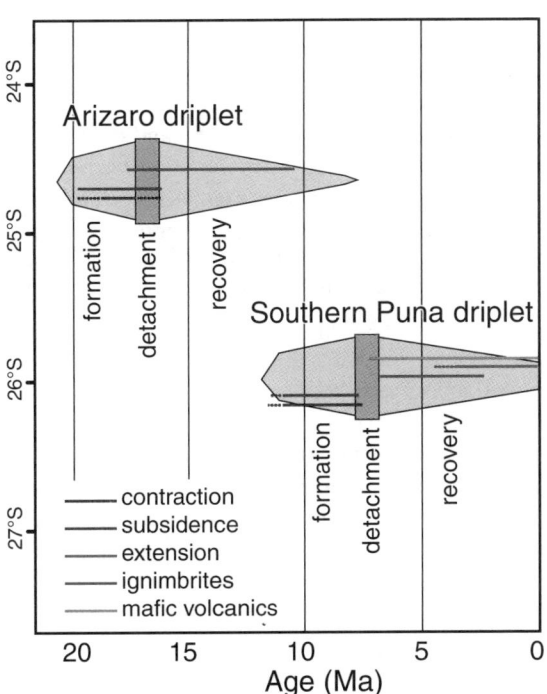

Figure 8. Summary of evidence for driplets in the Arizaro region and southern Puna Plateau. Colored bars show time ranges of contraction, subsidence, extension, ignimbrite eruption, and mafic volcanism. Dotted bars indicate uncertainties in timing. Gray polygons reflect stages and relative influence of the foundering process.

from geochemical and geophysical data (Fig. 8; e.g., Kay and Kay, 1993; Bianchi et al., 2013). The record of formation of this drip includes subsidence of the Pasto Ventura basin and synsedimentary contractional deformation from at least 10.5 Ma to 7.8 Ma. However, although deformation may have been more widespread (it is difficult to constrain in this poorly exposed region, which is dominated by reactivation of structures during multiple generations of deformation), it appears as though subsidence and deposition were less widespread than for the drip recorded in the Arizaro Basin. There is abundant support for the detachment of a Rayleigh-Taylor instability. Mafic volcanism began around 8.7 Ma in the Calalaste Range and migrated outward with time, reaching the Pasto Ventura and Arizaro regions in the last million years (Fig. 1B). This may indicate growth over time of the region affected by mafic magmatism, or it may reflect the growth of the region of horizontal extension in the crust, allowing an extant magma source to finally reach the surface. We favor the former interpretation, as there is evidence that extension reached the margins of the plateau as early as 6 Ma (Schoenbohm and Strecker, 2009). Shortening was followed by extension in both our study regions, with an earlier occurrence in Pasto Ventura (ca. 7–4 Ma) and a later occurrence in the Arizaro region (<0.6 Ma). The pattern of shortening followed by extension, although based on only two data points, thus also appears to migrate outward from the Calalaste region with time. Detachment of a Rayleigh-Taylor instability after 8 Ma is further suggested by eruption of the Cerro Galan ignimbrites between 6.6 Ma and 2.06 Ma (Kay et al., 2011). In the southern Puna Plateau, there is no evidence for extensive crustal thinning as there is in the northern Puna beneath the Arizaro region (Bianchi et al., 2013; Tassara et al., 2006; Tassara and Echaurren, 2012), suggesting that the foundered block may have been smaller, leaving less of a mark on crustal and mantle-lithospheric thickness (Elkins-Tanton, 2007). Indeed, a mantle anomaly interpreted as a foundered block slightly northeast of Cerro Galan has a diameter of ~50 km, about half that of the imaged anomaly northeast of Arizaro. Foundering of a smaller block could explain the lower degree of subsidence and the higher incidence of mafic magmatism in the Pasto Ventura region.

In summary, our study, combined with a growing body of additional evidence, supports partial, small-scale lithospheric removal in the Puna region, rather than wholesale plateau delamination as has been proposed for lithospheric removal in the Altiplano (e.g., Garzione et al., 2008; Molnar and Garzione, 2007). Evidence includes: (1) the irregular thinning of the lithospheric mantle and crust with a diameter of <100 km (Bianchi et al., 2013; Tassara et al., 2006; Tassara and Echaurren, 2012; Yuan et al., 2000, 2002); (2) the small volume and geochemical complexity of mafic magmatism, which suggest that drips could be <50 km in diameter (Ducea et al., 2013; Murray et al., this volume); (3) eruption of the Aguas Calientes and Cerro Galan ignimbrites (Kay et al., 2010, 2011); (4) low total horizontal extension (Schoenbohm and Strecker, 2009) at a low rate (Zhou et al., 2013); (5) geophysically imaged foundered lithosphere

with a diameter of ~50 km northeast of Cerro Galan (Bianchi et al., 2013; Calixto et al., 2013) and >100 km northeast of Arizaro (Schurr et al., 2006); and (6) the spatial and temporal pattern of out-of-sequence contraction and subsidence followed by extension, mafic volcanism, and basin incision that we document in this study. In addition, deuterium ratios in volcanic glass from the Puna Plateau suggest that the elevation has been within 1 km of modern values since at least ca. 36 Ma (Canavan, 2012): If lithospheric foundering is characterized by multiple, repeating, small-scale Rayleigh-Taylor instabilities, then the change in surface elevation as a result of each event is expected to be relatively small (<500 m; Beiki-Ardakani et al., 2010). We argue that a relatively large (~100 km) Rayleigh-Taylor instability formed and detached beneath the Arizaro Basin at ca. 17–16 Ma, and a second, smaller (~50 km) Rayleigh-Taylor instability detached beneath the southern Puna region at 8–7 Ma. It is likely that earlier driplets formed beneath the Puna Plateau, but their presence is not preserved in the geologic record due to overprinting by subsequent events.

ACKNOWLEDGMENTS

We acknowledge the support of National Science Foundation grant EAR-0911577 to Carrapa. We thank Peter DeCelles for useful discussion and comments. Robin Canavan, Glynis Jehle, John Boyd, Heather McPherson, Jonathan Pratt, and John Patrick Calhoun contributed to collecting and analyzing samples. We also thank collaborators George Gehrels at the University of Arizona and Axel Schmitt at the University of California–Los Angeles (UCLA) for U-Pb analyses, and Michael Cosca at the U.S. Geological Survey in Denver for $^{40}Ar/^{39}Ar$ analyses. The facilities at UCLA and the University of Arizona are partly supported by grants from the Instrumentation and Facilities Program, Division of Earth Sciences, National Science Foundation.

REFERENCES CITED

Allmendinger, R.W., 1986, Tectonic development, southeastern border of the Puna Plateau, northwestern Argentine Andes: Geological Society of America Bulletin, v. 97, p. 1070–1082, doi:10.1130/0016-7606 (1986)97<1070:TDSBOT>2.0.CO;2.

Allmendinger, R.W., Jordan, T.E., Kay, S.M., and Isacks, B.L., 1997, The evolution of the Altiplano-Puna Plateau of the central Andes: Annual Review of Earth and Planetary Sciences, v. 25, p. 139–174, doi:10.1146/annurev .earth.25.1.139.

Arriagada, C., Cobbold, P.R., and Roperch, P., 2006, Salar de Atacama Basin: A record of compressional tectonics in the central Andes since the mid-Cretaceous: Tectonics, v. 25, TC1008, doi:10.1029/2004TC001770.

Beiki-Ardakani, A., Pysklywec, R.P., and Schoenbohm, L.M., 2010, Surface topographic response to lithospheric instabilities and "driplets" beneath the central Andean Plateau [abs.]: Eos (Transactions, American Geophysical Union), v. 91, abstract T11A-2041.

Bianchi, M., Heit, B., Jakoviev, A., Yuan, X., Kay, S.M., Sandovol, E., Alonso, R.N., Coira, B., Brown, L., Kind, R., and Comte, D., 2013, Teleseismic tomography of the southern Puna Plateau in Argentina and adjacent regions: Tectonophysics, v. 586, p. 65–83, doi:10.1016/j .tecto.2012.11.016.

Boyd, J., 2010, Tectonic Evolution of the Arizaro Basin of the Puna Plateau, NW Argentina: Implications for Plateau-Scale Processes [M.S. thesis]: Laramie, Wyoming, University of Wyoming, 81 p.

Calixto, F.J., Sandvol, E., Kay, S., Mulcahy, P., Heit, B., Yuan, X.-H., Coira, B., Comte, D., and Alvarado, P., 2013, Velocity structure beneath the southern Puna plateau: Evidence for delamination: Geochemistry Geophysics Geosystems, v. 14, no. 10, p. 4292–4305, doi:10.1002/ggge20266.

Canavan, R., 2012, Cenozoic Paleoelevation Reconstructions of the Puna Plateau, NW Argentina [M.S. thesis]: Laramie, Wyoming, University of Wyoming, 126 p.

Carrapa, B., and DeCelles, P.G., 2008, Eocene exhumation and basin development in the Puna of northwestern Argentina: Tectonics, v. 27, TC1015, doi:10.1029/2007TC002127.

Carrapa, B., and DeCelles, P.G., 2015, this volume, Regional exhumation and kinematic history of the central Andes in response to cyclical orogenic processes, *in* DeCelles, P.G., Ducea, M.N., Carrapa, B., and Kapp, P.A., eds., Geodynamics of a Cordilleran Orogenic System: The Central Andes of Argentina and Northern Chile: Geological Society of America Memoir 212, doi:10.1130/2015.1212(11).

Carrapa, B., Adelmann, D., Hilley, G.E., Mortimer, E., Sobel, E.R., and Strecker, M.R., 2005, Oligocene range uplift and development of plateau morphology in the southern central Andes: Tectonics, v. 24, TC4011, doi:10.1029/2004TC001762.

Carrapa, B., Trimble, J.D., and Stockli, D.F., 2011a, Patterns and timing of exhumation and deformation in the Eastern Cordillera of NW Argentina revealed by (U-Th)/He thermochronology: Tectonics, v. 30, TC3003, doi:10.1029/2010TC002707.

Carrapa, B., Bywater-Reyes, S., DeCelles, P.G., Mortimer, E., and Gehrels, G.E., 2011b, Late Eocene–Pliocene basin evolution in the Eastern Cordillera of northwestern Argentina (25°–26°S): Regional implications for Andean orogenic wedge development: Basin Research, v. 24, no. 3, p. 249–268, doi:10.1111/j.1365-2117.2011.00519.x.

Conrad, C.P., and Molnar, P., 1999, Convective instability of a boundary layer with temperature- and strain-rate-dependent viscosity in terms of 'available buoyancy': Geophysical Journal International, v. 139, p. 51–68.

Coutand, I., Cobbold, P.R., de Urreiztieta, M., Gautier, P., Chauvin, A., Gapais, D., Rossello, E.A., and López-Gamundí, O., 2001, Style and history of Andean deformation, Puna Plateau, northwestern Argentina: Tectonics, v. 20, p. 210–234, doi:10.1029/2000TC900031.

Darold, A., and Humphreys, E., 2013, A plume-triggered delamination origin for the Columbia River flood basalt eruptions: Earth and Planetary Science Letters, v. 365, p. 232–242, doi:10.1016/j.epsl.2013.01.024.

Davis, D., Suppe, J., and Dahlen, F.A., 1983, Mechanics of fold-and-thrust belts and accretionary wedges: Journal of Geophysical Research: Solid Earth, v. 88, p. 1153–1172.

DeCelles, P.G., Ducea, M.N., Kapp, P., and Zandt, G., 2009, Cyclicity in Cordilleran orogenic systems: Nature Geoscience, v. 2, p. 251–257, doi:10.1038/NGEO469.

DeCelles, P.G., Carrapa, B., Horton, B.K., McNabb, J., Gehrels, G.E., and Boyd, J., 2015, this volume, The Miocene Arizaro Basin, central Andean hinterland: Response to partial lithosphere removal?, *in* DeCelles, P.G., Ducea, M.N., Carrapa, B., and Kapp, P., eds., Geodynamics of a Cordilleran Orogenic System: The Central Andes of Argentina and Northern Chile: Geological Society of America Memoir 212, doi:10.1130/2015.1212(18).

Deeken, A., Sobel, E.R., Coutand, I., Haschke, M., Riller, U., and Strecker, M.R., 2006, Development of the southern Eastern Cordillera, NW Argentina, constrained by apatite fission track thermochronology: From Early Cretaceous extension to middle Miocene shortening: Tectonics, v. 25, TC6003, doi:10.1029/2005TC001894.

Donato, E., 1987, Características estructurales del sector occidental de la Puna Salteña: Boletín de Informaciones Petroleras, v. 12, p. 89–97.

Drew, S.T., Ducea, M.N., and Schoenbohm, L.M., 2009, Mafic volcanism on the Puna Plateau, NW Argentina: Implications for the lithospheric composition and evolution with an emphasis on lithospheric foundering: Lithosphere, v. 1, p. 305–318, doi:10.1130/L54.1.

Ducea, M.N., 2011, Fingerprinting orogenic delamination: Geology, v. 39, p. 191–192, doi:10.1130/focus022011.1.

Ducea, M., and Saleeby, J., 1998, A case for delamination of the deep batholithic crust beneath the Sierra Nevada, California: International Geology Review, v. 40, p. 78–93, doi:10.1080/00206819809465199.

Ducea, M.N., Seclaman, A.C., Murray, K.E., Jianu, D., and Schoenbohm, L.M., 2013, Mantle-drip magmatism beneath the Antliplano-Puna Plateau, central Andes: Geology, v. 41, p. 915–918, doi:10.1130/G34509.1.

Echavarria, R., Hernandez, R., Allmendinger, R.W., and Reynolds, J.H., 2003, Sub-Andean thrust and fold belt of northwest Argentina: Geometry and

timing of the Andean evolution: American Association of Petroleum Geologists Bulletin, v. 87, p. 965–985, doi:10.1306/01200300196.

Elkins-Tanton, L.T., 2005, Continental magmatism caused by lithospheric delamination, *in* Foulger, G.R., Natland, J.H., Presnall, D.C., and Anderson, D.L., eds., Plates, Plumes, and Paradigms: Geological Society of America Special Paper 388, p. 449–461.

Elkins-Tanton, L.T., 2007, Continental magmatism, volatile recycling, and a heterogeneous mantle caused by lithospheric gravitational instabilities: Journal of Geophysical Research, v. 112, B03405, doi:10.1029/2005JB004072.

Escayola, M.P., van Staal, C.R., and Davis, W.J., 2011, The age and tectonic setting of the Puncoviscana Formation in northwestern Argentina: An accretionary complex related to Early Cambrian closure of the Puncoviscana Ocean and accretion of the Arequipa-Antofalla block: Journal of South American Earth Sciences, v. 32, p. 438–459, doi:10.1016/j.jsames.2011.04.013.

Garzione, C.N., Hoke, G.D., Libarkin, J.C., Withers, S., MacFadden, B., Eiler, J., Ghosh, P., and Mulch, A., 2008, Rise of the Andes: Science, v. 320, p. 1304–1307, doi:10.1126/science.1148615.

Göğüş, O., and Pysklywec, R., 2008, Near-surface diagnostics of dripping or delaminating lithosphere: Journal of Geophysical Research, v. 113, B11404, doi:10.1029/2007JB005123.

Hales, T.C., Abt, D.L., Humphreys, E.D., and Roering, J.J., 2005, A lithospheric instability origin for Columbia River flood basalts and Wallowa Mountains uplift in northeast Oregon: Nature, v. 438, p. 842–845, doi:10.1038/nature04313.

Kay, R.W., and Kay, S.M., 1993, Delamination and delamination magmatism: Tectonophysics, v. 219, p. 177–189, doi:10.1016/0040-1951(93)90295-U.

Kay, S.M., Coira, B., and Viramonte, J., 1994, Young mafic back arc volcanic rocks as indicators of continental lithospheric delamination beneath the Argentine Puna Plateau, central Andes: Journal of Geophysical Research, v. 99, p. 24,323–24,339, doi:10.1029/94JB00896.

Kay, S.M., Coira, B.L., Caffe, P.J., and Chen, C.-H., 2010, Regional chemical diversity, crustal and mantle sources and evolution of central Andean Puna Plateau ignimbrites: Journal of Volcanology and Geothermal Research, v. 198, p. 81–111, doi:10.1016/j.jvolgeores.2010.08.013.

Kay, S.M., Coira, B., Wörner, G., Kay, R.W., and Singer, B.W., 2011, Geochemical, isotopic and single crystal $^{40}Ar/^{39}Ar$ age constraints on the evolution of the Cerro Galán ignimbrites: Bulletin of Volcanology, v. 73, p. 1487–1511, doi:10.1007/s00445-010-0410-7.

Kraemer, B., Adelmann, D., Alten, M., Schnurr, W., Erpenstein, K., Kiefer, E., van den Bogaard, P., and Görler, K., 1999, Incorporation of the Paleogene foreland into the Neogene Puna plateau: The Salar de Antofalla area, NW Argentina: Journal of South American Earth Sciences, v. 12, p. 157–182.

Levander, A., Schmandt, B., Miller, M.S., Liu, K., Karlstrom, K.E., Crow, R.S., Lee, C.-T.A., and Humphreys, E.D., 2011, Continuing Colorado Plateau uplift by delamination-style convective lithospheric downwelling: Nature, v. 472, p. 461–465, doi:10.1038/nature10001.

Manley, C.R., Glazner, A.F., and Farmer, G.L., 2000, Timing of volcanism in the Sierra Nevada of California: Evidence for Pliocene delamination of the batholithic root?: Geology, v. 28, p. 811–814, doi:10.1130/0091-7613(2000)28<811:TOVITS>2.0.CO;2.

Marrett, R., and Emerman, S.H., 1992, The relations between faulting and mafic magmatism in the Altiplano-Puna Plateau (central Andes): Earth and Planetary Science Letters, v. 112, p. 53–59, doi:10.1016/0012-821X(92)90006-H.

Molnar, P., and Garzione, C.N., 2007, Bounds on the viscosity coefficient of continental lithosphere from removal of mantle lithosphere beneath the Altiplano and Eastern Cordillera: Tectonics, v. 26, TC2013, doi:10.1029/2006TC001964.

Molnar, P., and Houseman, G.A., 2004, The effects of buoyant crust on the gravitational instability of thickened mantle lithosphere at zones of intracontinental convergence: Geophysical Journal International, v. 158, p. 1134–1150, doi:10.1111/j.1365-246X.2004.02312.x.

Montero Lopez, M.C., Hongn, F.D., Strecker, M.R., Marrett, R., Seggiaro, R., and Sudo, M., 2010, Late Miocene–early Pliocene onset of N-S extension along the southern margin of the central Andean Puna Plateau: Evidence from magmatic, geochronological and structural observations: Tectonophysics, v. 494, p. 48–63, doi:10.1016/j.tecto.2010.08.010.

Mpodozis, C., Arriagada, C., Basso, M., Roperch, P., Cobbold, P., and Reich, M., 2005, Late Mesozoic to Paleogene stratigraphy of the Salar de Atacama Basin, Antofagasta, northern Chile: Implications for the tectonic evolution of the central Andes: Tectonophysics, v. 399, p. 125–154, doi:10.1016/j.tecto.2004.12.019.

Murray, K.E., Ducea, M.N., and Schoenbohm, L., 2015, this volume, Foundering-driven lithospheric melting: The source of central Andean mafic lavas on the Puna Plateau (22°S–27°S), *in* DeCelles, P.G., Ducea, M.N., Carrapa, B., and Kapp, P.A., eds., Geodynamics of a Cordilleran Orogenic System: The Central Andes of Argentina and Northern Chile: Geological Society of America Memoir 212, doi:10.1130/2015.1212(08).

Neil, E.A., and Houseman, G.A., 1999, Rayleigh-Taylor instability of the upper mantle and its role in intraplate orogeny: Geophysical Journal International, v. 138, p. 89–107, doi:10.1046/j.1365-246x.1999.00841.x.

Pysklywec, R.N., and Cruden, A.R., 2004, Coupled crust-mantle dynamics and intraplate tectonics: Two-dimensional numerical and three-dimensional analogue modeling: Geochemistry Geophysics Geosystems, v. 5, Q10003, doi:10.1029/2004GC000748.

Reynolds, J.H., Galli, C.I., Hernández, R.M., Idleman, B.D., Kotila, J.M., Hilliard, R.V., and Naeser, C.W., 2000, Middle Miocene tectonic development of the Transition zone, Salta Province, northwest Argentina: Magnetic stratigraphy from the Metán Subgroup, Sierra de González: Geological Society of America Bulletin, v. 112, p. 1736–1751, doi:10.1130/0016 -7606(2000)112<1736:MMTDOT>2.0.CO;2.

Riller, U., Cruden, A.R., Boutelier, D., and Schrank, C.E., 2012, The causes of sinuous crustal-scale deformation patterns in hot orogens: Evidence from scaled analogue experiments and the southern central Andes: Journal of Structural Geology, v. 37, p. 65–74, doi:10.1016/j.jsg.2012.02.002.

Risse, A., Trumbull, R.B., Coira, B., Kay, S.M., and van den Boggard, P., 2008, ^{40}Ar/^{39}Ar geochronology of mafic volcanism in the back-arc region of the southern Puna Plateau, Argentina: Journal of South American Earth Sciences, v. 26, p. 1–15, doi:10.1016/j.jsames.2008.03.002.

Risse, A., Trumbull, R.B., Kay, S.M., Coira, B., and Romer, R.L., 2013, Multistage evolution of Late Neogene mantle-derived magmas from the central Andes back-arc in the southern Puna Plateau of Argentina: Journal of Petrology, v. 54, p. 1963–1995.

Schoenbohm, L.M., and Strecker, M.R., 2009, Normal faulting along the southern margin of the Puna Plateau, northwest Argentina: Tectonics, v. 28, TC5008, doi:10.1029/2008TC002341.

Schott, B., and Schmeling, H., 1998, Delamination and detachment of a lithospheric root: Tectonophysics, v. 296, p. 225–247, doi:10.1016/S0040 -1951(98)00154-1.

Schurr, B., Rietbrock, A., Asch, G., Kind, R., and Oncken, O., 2006, Evidence for lithospheric detachment in the central Andes from local earth-

quake tomography: Tectonophysics, v. 415, p. 203–223, doi:10.1016/j .tecto.2005.12.007.

Tassara, A., and Echaurren, A., 2012, Anatomy of the Andean subduction zone: Three-dimensional density model upgraded and compared against global-scale models: Geophysical Journal International, v. 189, p. 161–168, doi:10.1111/j.1365-246X.2012.05397.x.

Tassara, A., Goetze, H.-J., Schmidt, S., and Hackney, R., 2006, Three-dimensional density model of the Nazca plate and the Andean continental margin: Journal of Geophysical Research, v. 111, B09404, doi:10.1029/2005JB003976.

Turner, S., Arnaud, N., Liu, J., Rogers, N., Haskesworth, C., Harris, N., Kelley, S., Van Calsternen, P., and Deng, W., 1996, Post-collision, shoshonitic volcanism on the Tibetan Plateau: Implications for convective thinning of the lithosphere and the source of ocean island basalts: Journal of Petrology, v. 37, p. 45–71, doi:10.1093/petrology/37.1.45.

Voss, R., 2002, Cenozoic stratigraphy of the southern Salar de Antofalla region, northwestern Argentina: Revista geológica de Chile, v. 29, p. 151–165.

West, J.D., Fouch, M.J., Roth, J.B., and Elkins-Tanton, L.T., 2009, Vertical mantle flow associated with a lithospheric drip beneath the Great Basin: Nature Geoscience, v. 2, p. 439–444, doi:10.1038/ngeo526.

Yuan, X., Sobolev, S.V., Kind, R., Onceen, O., Bock, G., Asch, G., Schurr, B., Graeber, F., Rudloff, A., Hanka, W., Wylegalla, K., Tibi, R., Haberland, C., Rietbrock, A., Giese, P., Wigger, P., Roewer, P., Zandt, G., Beck, S., Wallace, T., Pardo, M., and Comte, D., 2000, Subduction and collision processes in the central Andes constrained by converted seismic phases: Nature, v. 408, p. 958–961, doi:10.1038/35050073.

Yuan, X., Sobolev, S.V., and Kind, R., 2002, Moho topography in the central Andes and its geodynamical implications: Earth and Planetary Science Letters, v. 199, p. 389–402.

Zappetini, E.O., Blasco, G., Ramallo, E., and González, O., 2003, Socompa 2569-II Carta Geological de la Republica Argentina: Buenos Aires, Argentina, SEGEMAR, Instituto de Geologia y Recursos Minerales, scale 1:250,000.

Zhou, R.J., Schoenbohm, L.M., and Cosca, M., 2013, Recent, slow normal and strike-slip faulting in the Pasto Ventura region of the southern Puna Plateau, NW Argentina: Tectonics, v. 32, p. 19–33, doi:10.1029/2012TC003189.

MANUSCRIPT ACCEPTED BY THE SOCIETY 3 JUNE 2014
MANUSCRIPT PUBLISHED ONLINE 15 OCTOBER 2014

The Geological Society of America
Memoir 212
2015

Exhumation of the Precordillera and northern Sierras Pampeanas and along-strike correlation of the Andean orogenic front, northwestern Argentina

Roxana Safipour*
Barbara Carrapa
Peter G. DeCelles
Stuart N. Thomson
Department of Geosciences, University of Arizona, 1040 E. 4th Street, Tucson, Arizona 85721, USA

ABSTRACT

In NW Argentina (~26°S), adjacent to the Puna Plateau of the central Andes, exhumation and deformation propagated from west to east across the Puna Plateau and Eastern Cordillera during early–late Cenozoic time. Presently existing data do not indicate a clear eastward younging trend in exhumation in the Precordillera and northern Sierras Pampeanas at the southeastern flank of the Puna Plateau at ~28°S. In this study, we mapped an ~80-km-wide transect at a latitude of ~28°S in the Sierra de Las Planchadas and the Fiambalá Basin. Apatite fission-track (AFT) ages from six samples from thrust fault hanging walls are between 20 and 14 Ma, and apatite helium (U-Th)/He ages from five samples range from 21 Ma west of the Sierra de Las Planchadas to 2 Ma in the Fiambalá Basin. Several samples record mixed apatite (U-Th)/He ages and pre-Cenozoic AFT ages, indicating partial resetting through the 120–60 °C temperature window and suggesting a depth of exhumation between ~6 and 3 km. Thermal modeling of AFT and (U-Th)/He ages indicates cooling from ca. 22 Ma until 2 Ma and a younging trend to the east. These ages are consistent with previously published AFT ages in the Fiambalá region and AHe ages in the Eastern Cordillera to the northeast and suggest Miocene exhumation and deformation in the Precordillera at 28°S and the Eastern Cordillera at 26°S. From a kinematic standpoint, the region of the Precordillera at 28°S may be considered a continuation to the south of the Eastern Cordillera at 26°S, implying a single continuous SW-NE–striking deformation front along this portion of the central Andes. Detrital zircon U-Pb ages from Upper Miocene strata in the Fiambalá Basin suggest that Permian–Miocene rocks in the Sierra de Las Planchadas are the primary source of the Miocene basin fill. This is consistent with previously published eastward-directed paleocurrent data and other provenance proxies and indicates that uplift of the Sierra de Las Planchadas began no later than ca. 9 Ma.

*safipour@alum.mit.edu

Safipour, R., Carrapa, B., DeCelles, P.G., and Thomson, S.N., 2015, Exhumation of the Precordillera and northern Sierras Pampeanas and along-strike correlation of the Andean orogenic front, northwestern Argentina, *in* DeCelles, P.G., Ducea, M.N., Carrapa, B., and Kapp, P.A., eds., Geodynamics of a Cordilleran Orogenic System: The Central Andes of Argentina and Northern Chile: Geological Society of America Memoir 212, p. 181–199, doi:10.1130/2015.1212(10). For permission to copy, contact editing@geosociety.org. © 2014 The Geological Society of America. All rights reserved.

INTRODUCTION

The central Andes result from processes related to the convergence of the Nazca and South American plates since at least the Late Cretaceous (Allmendinger et al., 1997; Isacks, 1988). The topography of the central Andes is that of a classic ocean-continent convergent margin, with distinct longitudinal tectonomorphic zones (Fig. 1), including the Western Cordillera (the Miocene–modern arc), the Altiplano Plateau (a high-elevation internally drained region), the Eastern Cordillera (part of the retroarc fold-and-thrust belt), the Subandean zone (the frontal part of the fold-and-thrust belt where most of the modern deformation is taking place), and the Chaco Plain (a foreland basin above the Brazilian craton) (Allmendinger et al., 1997; Kley et al., 1997; McQuarrie et al., 2005; Strecker et al., 2007). South of 24°S, the Altiplano is replaced by the Puna, also a high-elevation, internally drained plateau disrupted by smaller basins and ranges, and the Subandean zone is replaced by the Santa Barbara system, a thick-skinned fold-and-thrust belt (Kley and Monaldi, 2002; Allmendinger et al., 1997; Strecker et al., 2007). South of 28°S latitude, the Altiplano and Puna Plateaus are replaced by the narrow Principal Cordillera and Precordillera and the intraforeland Laramide-style block uplifts of the Sierras Pampeanas system (Allmendinger et al., 1997). The magmatic arc is also extinguished at these latitudes (Cahill and Isacks, 1992).

The formation of the thick-skinned Sierras Pampeanas has been associated with flat slab subduction (Jordan et al., 1983; Ramos et al., 2002). The shape and position of the subducting Nazca slab can be determined by observing the positions of mantle earthquake hypocenters within the "Wadati-Benioff zone" in the slab. From 15°S to 27°S, the slab subducts at an angle of ~30° (Cahill and Isacks, 1992). The slab then shallows out to be nearly horizontal from 28°S to 33°S, extending inland several hundred kilometers with a dip of only 5° (Cahill and Isacks, 1992). The slab steepens once again south of 33°S (Cahill and Isacks, 1992). Whereas the steep slab region of the central Andes is associated with volcanism, a wide plateau, and thin-skinned to thick-skinned style deformation, the flat slab region lacks any modern volcanism and is characterized by a narrow or extinguished magmatic arc and basement-involved block uplifts of the Sierras Pampeanas system (Cahill and Isacks, 1992).

Both the deformation style and the total amount of crustal shortening in the Andes vary along strike (Allmendinger and Gubbels, 1996; Kley and Monaldi, 1998; Coutand et al., 2001; McGroder et al., this volume). In the northern central Andes, north of 24°S, the deformation is characterized by a thin-skinned fold-and-thrust belt with up to 340 km of documented shortening (McQuarrie and DeCelles, 2001). From 24°S to 28°S, the total crustal shortening decreases (Allmendinger et al., 1997; Kley and Monaldi, 1998), and deformation is accommodated by the thick-skinned Santa Barbara system. South of 28°S, in the flat slab region, shortening is expressed much farther inland in the Sierras Pampeanas system (Ramos et al., 2002).

North of the flat slab region, there is ample evidence to suggest that exhumation and deformation of the orogenic wedge propagated mostly in sequence from west to east in the Eastern Cordillera (McQuarrie et al., 2008; Carrapa et al., 2011; Barnes et al., 2012). Analyses of the Cenozoic sedimentary record show that a regional foreland basin system developed during the Late Cretaceous–early Cenozoic in Bolivia and Argentina and migrated eastward through time in response to eastward propagation of the orogenic strain front (Sempere et al., 1997; DeCelles and Horton, 2003; Horton, 2005; Carrapa et al., 2011; DeCelles et al., 2011). In northernmost Argentina, deformation propagated from the Eastern Cordillera eastward to the Subandean zone beginning at ca. 10–8 Ma, and then propagated in sequence from west to east across the Subandean zone until ca. 4.5 Ma, when out-of-sequence thrusting began to occur (Echavarria et al., 2003). The pattern of deformation and exhumation in the region remains controversial. For example, apatite (U-Th)/He data from synorogenic strata and basin-bounding structures together with growth strata indicate exhumation associated with an eastward propagation of deformation across the Eastern Cordillera between 25°S and 26°S from ca. 14 to ca. 4 Ma (Carrapa et al., 2011; DeCelles et al., 2011; Pearson et al., 2012). On the other hand, Hain et al. (2011) suggested an out-of-sequence pattern of deformation in the Lerma Valley. Apatite fission-track (AFT) data from the Angastaco Basin (Coutand et al., 2006) and from crystalline basement ranges within the Eastern Cordillera indicate

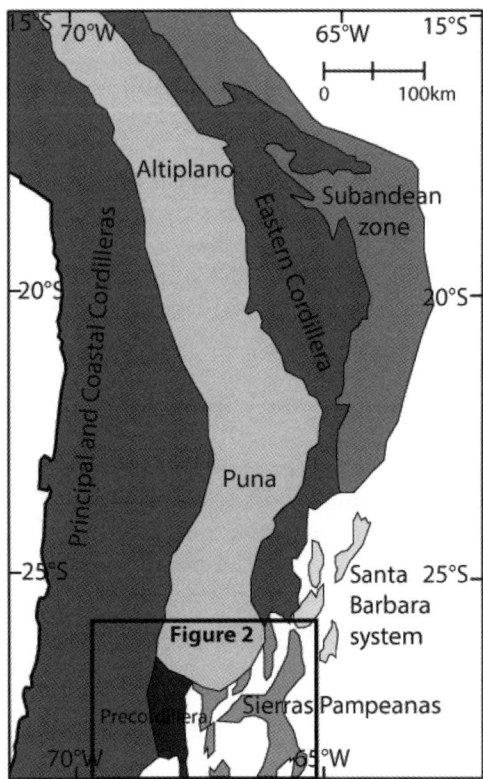

Figure 1. Major tectonomorphic zones of the central Andes, modified after Carrapa et al. (2011).

that the region was actively exhuming since at least ca. 21 Ma (Deeken et al., 2006). At the transition of the Eastern Cordillera, the eastern Puna Plateau, and the Sierras Pampeanas, the Sierra de Quilmes (~27°S) was exhumed during the Cretaceous and the late Cenozoic (Mortimer et al., 2007; Carrapa et al., 2014). This was coincident with, or possibly even subsequent to, the exhumation of the basin-bounding Sierra de Aconquija farther to the east (Sobel and Strecker, 2003; Mortimer et al., 2007).

Overall, thermochronologic studies from the Sierras Pampeanas suggest temporally nonuniform exhumation (Fig. 2; Coughlin et al., 1998; Carrapa et al., 2008, 2014; Dávila and Carter, 2013). Cooling ages in the Sierra de Aconquija are Late Cretaceous (Coughlin et al., 1998) and late Miocene–Pliocene (Sobel and Strecker, 2003), whereas in the Sierra de Chango Real to the west, AFT cooling ages are Paleocene (Mortimer et al.,

2007), Eocene–Oligocene (Coutand et al., 2001), and Late Cretaceous (Carrapa et al., 2014). Still farther to the west, in the southeastern part of the Puna Plateau, AFT cooling ages are early Miocene (Carrapa et al., 2006). Sedimentological proxies indicate a pulse of exhumation in the Sierras Pampeanas at ca. 6 Ma (Carrapa et al., 2008), which is consistent with the young exhumation ages in the Sierra de Aconquija (Sobel and Strecker, 2003). However, older cooling ages reported in other ranges of the northern Sierras Pampeanas and Eastern Cordillera (Carrapa et al., 2014; Coughlin et al., 1998; Coutand et al., 2001; Mortimer et al., 2007; Dávila and Carter, 2013) show temporally nonuniform exhumation and no clear directional trend for propagation of exhumation at latitudes south of 26°S.

Additional structural and thermochronologic analyses can provide insight into this problem by better resolving the timing of

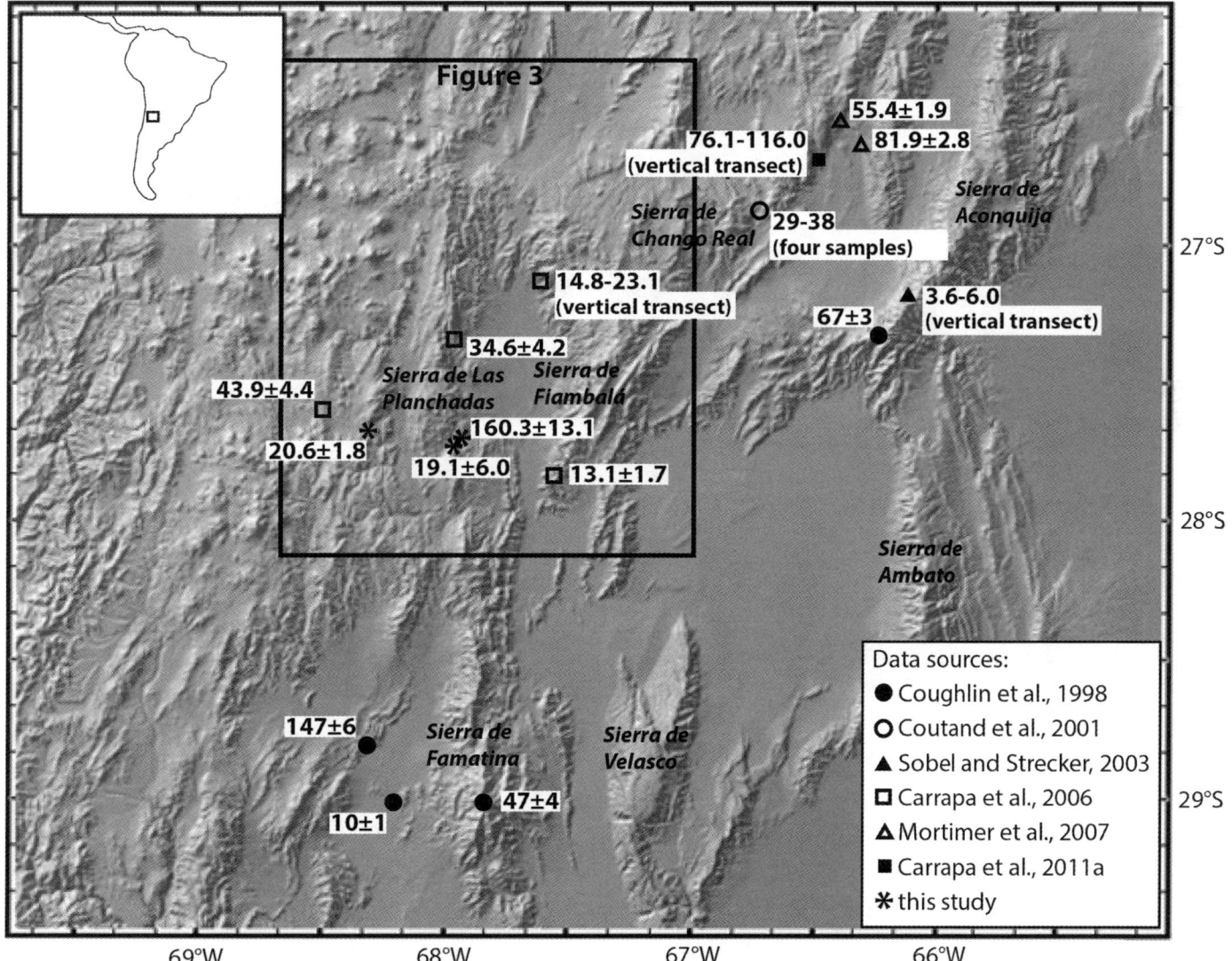

Figure 2. A regional summary of apatite fission-track (AFT) ages (Ma) in the Precordillera, northern Sierras Pampeanas, and southern Eastern Cordillera within latitudes 26°S–30°S.

exhumation in this region. In this study, we present a detailed map and thermochronologic and geochronologic data from the Fiambalá Basin and the Sierras de Las Planchadas within the Precordillera of NW Argentina. This section of the Precordillera is located at a transitional zone between the Puna Plateau to the north, the Eastern Cordillera to the northeast, and the Sierras Pampeanas to the east (Fig. 1). The tectonic evolution of the study area with respect to the Eastern Cordillera and Sierras Pampeanas is not well understood. The data presented in this study further constrain the provenance of Miocene basin strata and the timing of exhumation and deformation of the basin and of its bounding Precordillera at latitude ~28°S. We find that the Sierra de Las Planchadas is the source of the basin strata, and that exhumation of the Sierra de Las

Planchadas began prior to ca. 9 Ma and continued until ca. 2 Ma to the east. If this exhumation was driven by tectonic uplift, then the timing of uplift in the Sierra de Las Planchadas is consistent with a continuous deformation front connecting the Eastern Cordillera and the Precordillera between latitudes 26°S and 28°S.

GEOLOGIC SETTING

Geologic Overview

The Fiambalá Basin is located at the southern end of the Puna Plateau and to the east of the Precordillera between ~27°S and 28°S and ~68°W and 67°W (Fig. 3). The eastern edge of the

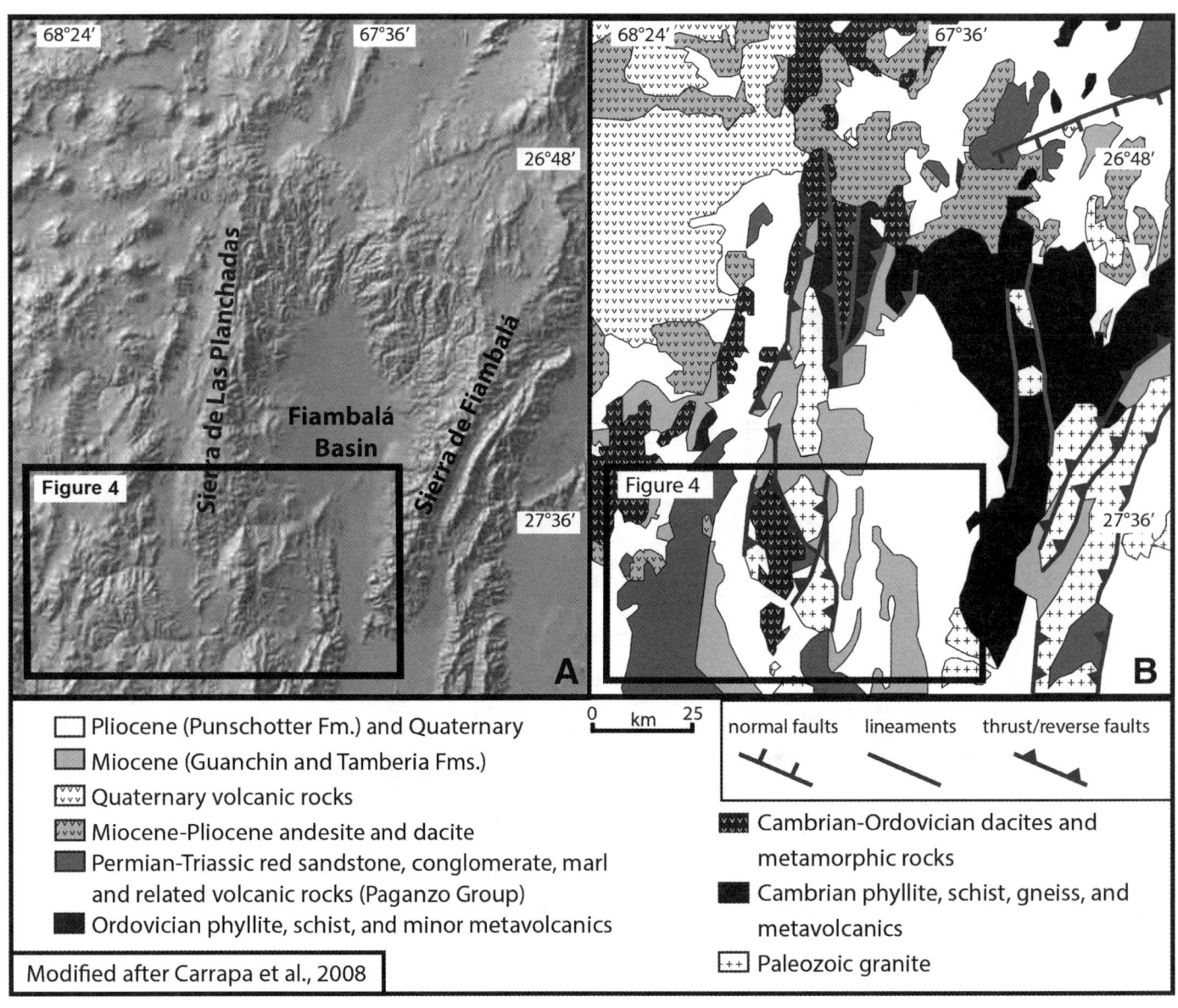

Figure 3. Topographic (A) and geologic (B) maps of the region around the Fiambalá Basin, modified after Carrapa et al. (2008). The field area mapped in this study (Fig. 4) is indicated by the black rectangle.

Fiambalá Basin is bounded by a buried west-verging, high-angle reverse fault with Cambrian and Precambrian granite and gneiss in the hanging wall (Fig. 3; Martínez et al., 1995; Rubiolo et al., 2009). To the north, the basin has a depositional contact with the Cambrian and Precambrian igneous basement (Fig. 3; Turner, 1967; Martínez et al., 1995; Rubiolo et al., 2009). To the south, the basin is open and drains into several smaller basins between intraforeland uplifts of the Sierras Pampeanas (Fig. 3; Martínez et al., 1995). The Sierra de Las Planchadas borders the Fiambalá Basin to the west and is composed mostly of a Paleozoic sedimentary package consisting of Ordovician siltstone, Carboniferous sandstone and shale, and Permian eolian sandstone (Fig. 3; Turner, 1967; Martínez et al., 1995; Rubiolo et al., 2009). The Sierra de Las Planchadas is part of the Precordillera and is characterized by both thin-skinned style thrusting and high-angle reverse faults (Rubiolo et al., 2009). The Fiambalá Basin is bounded by the Sierra de Fiambalá to the east (Martínez et al., 1995; Rubiolo et al., 2009), which is part of the Sierras Pampeanas Province (Jordan and Allmendinger, 1986) and is composed entirely of Cambrian and Precambrian igneous basement intruded by Carboniferous granites (Fig. 3; Martínez et al., 1995; Rubiolo et al., 2009). In the Sierra de Fiambalá and all major

ranges to the east of the Fiambalá Basin, thick-skinned block-uplift–style deformation prevails (Jordan et al., 1983; Jordan and Allmendinger, 1986). The Fiambalá Basin contains ~4 km of Neogene synorogenic deposits, which have been faulted and folded (Carrapa et al., 2006, 2008; Dávila and Astini, 2007).

Nine distinct rock units were observed in the field area. Descriptions of these units are summarized in Table 1. Figure 3 outlines the field area on regional topographic and geologic maps, and Figure 4 is the detailed geologic map produced from this study. Figure 5 is a schematic cross section, simplified from Rubiolo et al. (2009). The cross section is approximately located along the line A–A′ indicated on the map (Fig. 4). Sample locations are indicated on both the map and cross section.

THERMOCHRONOLOGICAL AND GEOCHRONOLOGICAL METHODS

AFT Thermochronology

AFT thermochronology is based on spontaneous fissioning of ^{238}U (e.g., Wagner, 1979; Gallagher et al., 1998). At temperatures above ~110 °C, the crystal lattice of apatite anneals, and

TABLE 1. GEOLOGIC UNITS IN NW ARGENTINA STUDY AREA

Formation	Description
Punaschotter Formation (Pliocene)	Poorly consolidated, poorly sorted, matrix-supported, cobble conglomerate. The Punaschotter Formation has been dated at ca. 3.7 Ma by zircon U-Pb dating of tuffs (Carrapa et al., 2006, 2008). Detrital apatite fission-track data from these deposits indicate that the southern Puna Plateau was the source for these deposits (Carrapa et al., 2006). Thickness is ~600 m.
Guanchín Formation (Upper Miocene)	Gray- and pink-colored fluvial sandstone containing channel conglomerates interbedded with volcanic tuffs (Carrapa et al., 2008). The Guanchín is planar bedded, trough cross-stratified, and poorly consolidated. The Guanchín Formation has been dated at 8.2–5.5 Ma by zircon U-Pb dating of tuffs (Carrapa et al., 2006, 2008). Thickness is ~1500 m.
Tambería Formation (Upper Miocene)	Lower member is ~1000 m of tan, medium-grained, tabular-bedded fluvial sandstones and mudstones; upper member is ~1000 m of tan sandstones with organized lenticular channel conglomerates interbedded with gray lacustrine beds (Carrapa et al., 2008). The midsection of the Tambería Formation has been dated at 9.0–8.2 Ma based on magnetostratigraphic data (Reynolds, 1987) and zircon U-Pb dating of interbedded volcanic tuffs (Carrapa et al., 2006, 2008). The basal section of the Tambería is not exposed in the Fiambalá Basin, but is exposed ~50 km to the south in the central Famatina region (Dávila and Astini, 2007).
La Cuesta Formation (Permian)	Dark red, medium-grained, quartzo-feldspathic sandstone capped by ~100 m of lacustrine facies (Turner, 1967; Rubiolo et al., 2009). Dominant sedimentary structure is large-scale (several meters) planar tangential and trough cross-stratification of eolian origin. Lacustrine deposits are lower medium- to upper fine-grained, finely laminated sandstone, interbedded with mudstone layers. Local thickness is ~1000 m.
Agua Colorada Formation (Carboniferous)	Shale, quartzite, and coarse-grained sandstone (Dávila and Astini, 2007; Rubiolo et al., 2009). Shales are light gray, blue-gray, pastel green, or tan. Quartzite is fine-grained, white, and is usually planar-bedded or rippled with burrows. Sandstones are composed of quartz and feldspar grains. Some matrix-supported beds of pebbly conglomerate are present, with well-rounded, poorly sorted quartzite clasts. Local thickness is ~200 m.
Punilla Formation (Devonian)	Fine-grained quartzite containing occasional matrix-supported conglomerate beds. Clasts are fine pebble-sized and are mostly quartzite with a few granites. Clasts are rounded to subrounded and well sorted. Thickness is at least 3500 m (Rubiolo et al., 2009).
Narvaez Granitoid (Ordovician)	Dark-green, coarse- to medium-grained granodiorite and white coarse- to medium-grained monzogranite containing pink K-feldspar and chlorite (Rubiolo et al., 2009). The granite is often heavily oxidized.
Las Planchadas Formation (Ordovician)	Fine-grained black- and red-colored basalts, dacites, and rhyolites (Rubiolo et al., 2009). Basalts contain plagioclase phenocrysts.
Suri Formation (Ordovician)	Fine-grained planar bedded or rippled (occasionally burrowed), dark-green sandstone and siltstone with some tan beds (Mángano and Buatois, 1997). The base of the unit is not exposed in the Sierra de Las Planchadas, so its thickness is not constrained, but is at minimum 1000 m.

186

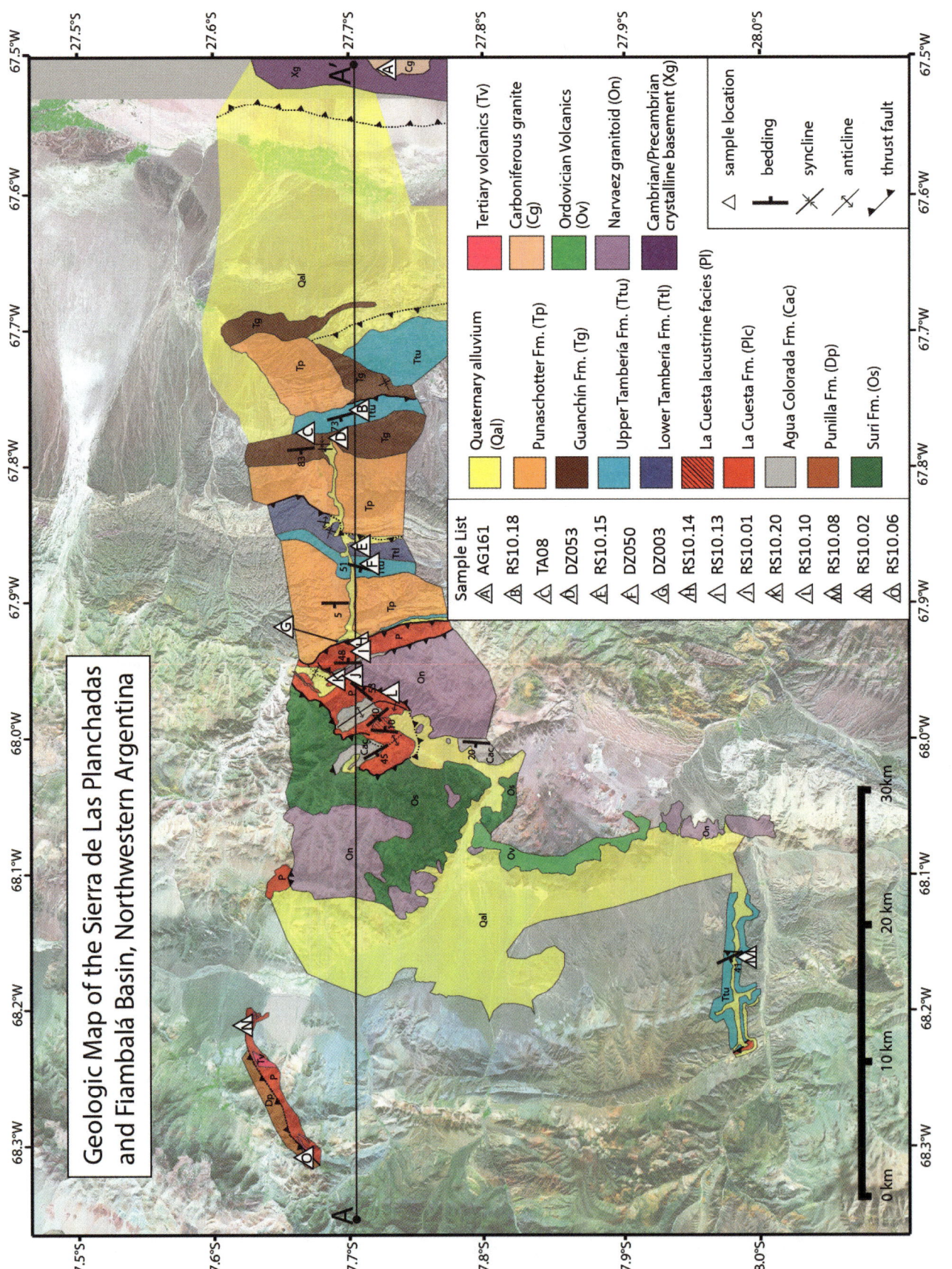

Figure 4. Geologic map of the Sierra de Las Planchadas and Fiambalá Basin. Sample locations are indicated with letters. Line A–A' is the approximate location of the cross section (Fig. 5) simplified from Rubiolo et al. (2009).

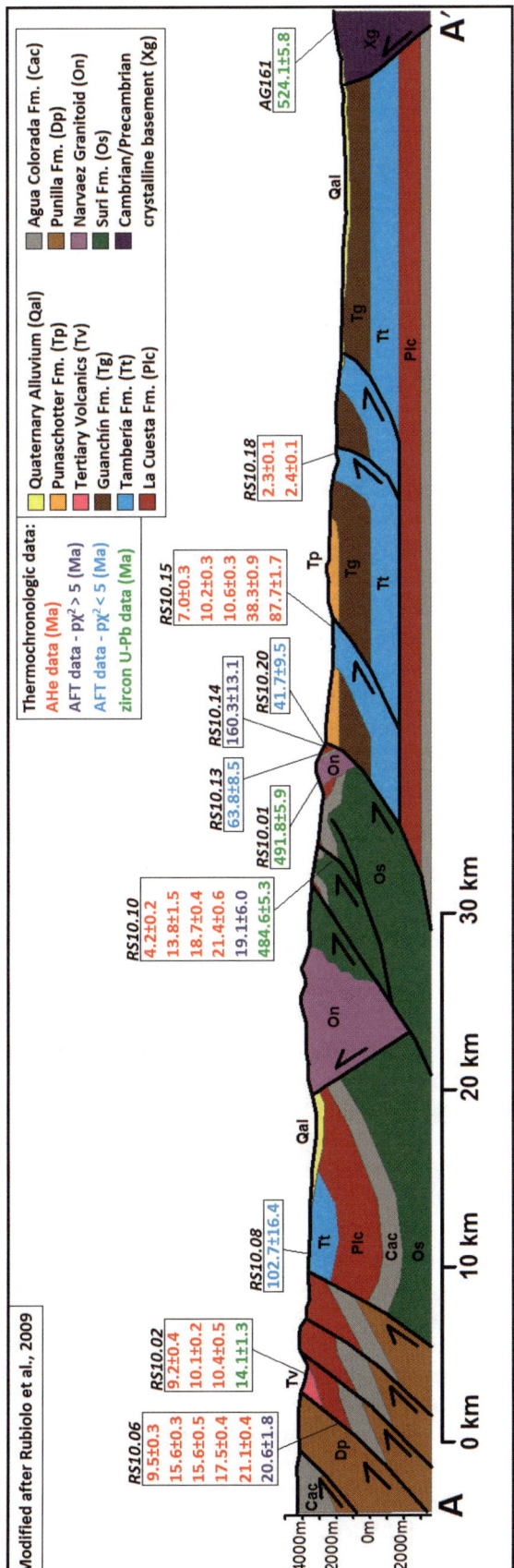

Figure 5. Approximate sample locations and analytical ages indicated on a schematic cross section simplified from Rubiolo et al. (2009). AFT—apatite fission track.

tracks are erased from the crystal over time periods of ~10^6 yr after they form. At temperatures below ~60 °C, the tracks are fully retained in the crystal (Gallagher et al., 1998). The temperature range 110–60 °C is known as the partial annealing zone (PAZ) (Wagner, 1979). In this temperature zone, tracks shorten over geologic time (>~10^6 yr) and are eventually annealed. The exact temperature limits of the PAZ depend on the cooling rate of the sample and the sample composition (Gallagher et al., 1998). The fission-track age calculated here represents a time when the crystal was within the 110–60 °C temperature range. If a typical geothermal gradient of 25 °C/km is assumed, such temperatures generally represent depths of ~4.5–2.5 km, and the fission-track age represents the time when the sample was exhumed through this depth range on its way to the surface. An AFT age thus typically represents the timing of exhumation of the sample, as opposed to higher-temperature geochronometers (e.g., U-Pb), which instead yield a crystallization age.

For this study, apatite grains were isolated from samples using standard mineral separation techniques. The grains were then etched in 5.5 *M* nitric acid at 21 °C for 20 s according to the protocol described by Donelick et al. (2005), which widened the tracks enough to be viewed and counted under a microscope. Approximately 20 grains were selected and counted for each sample. Uranium concentration was measured using the external detector method (Donelick et al., 2005). Grain ages were submitted to a χ^2 statistical test to quantify age dispersion for each sample (Galbraith, 2005). Samples that did not pass the χ^2 test are assumed to represent multiple populations of grains that have not fully annealed, and the age obtained is not assumed to provide any constraints on the timing of exhumation of the sample. Samples that passed the χ^2 test are assumed to represent a single population of grains that have been fully annealed. If these samples were exhumed quickly, then the AFT age represents a time when the samples were being exhumed through the PAZ. If the samples were exhumed slowly, some partial annealing may have occurred within the PAZ, causing the measured age to be younger than the true age of exhumation. In either case, the AFT age obtained for a sample that passes the χ^2 test represents a minimum age for the exhumation of the sample through the PAZ. Irradiation of the samples was carried out in the reactor at Oregon State University, and fission-track analysis was carried out at the University of Arizona Fission Track Laboratory.

Apatite (U-Th-[Sm])/He Dating

A second thermochronometer used in this study is the (U-Th-[Sm])/He system (for a review, see Farley et al., 1996). When applied to apatites, this is referred to as apatite helium (AHe) dating. Helium accumulates in apatites over time as the result of radioactive decay of ^{238}U, ^{235}U, and ^{232}Th to various isotopes of lead. All three of these systems produce α particles (helium nuclei) in the steps of their decay chains. The isotope ^{147}Sm also makes a contribution to radiogenic helium production in apatites, but its contribution is insignificantly small (Farley, 2002).

Like AFT dating, the AHe system is sensitive to low temperatures, because at temperatures above ~85 °C, helium diffuses out of the apatite grain (Wolf et al., 1998). Helium is partially retained within the apatite crystal at temperatures between ~85 °C and 40 °C, which is referred to as the partial retention zone (PRZ) (Wolf et al., 1998). Assuming a geothermal gradient of 25 °C/km, AHe analysis provides an age for exhumation through depths of ~3.5–1.5 km. Because AHe is sensitive to lower temperatures than AFT, AHe ages are expected to be younger than AFT ages for any given sample.

In this study, apatite grains were selected for analysis based on large size and a lack of inclusions. In the case of samples that had already undergone AFT analysis, apatite grains were plucked from the epoxy mounts. The grains were imaged with an optical microscope, and the three axial dimensions of each grain were measured from the images. Grains were then wrapped in Nb foil for analysis. Grains of sample 07WFS1, an internal laboratory standard that has an age of 65 ± 9 Ma, were also prepared and analyzed along with the unknown samples.

Grains were analyzed according to the methods described in Reiners (2002). The grains were heated with a laser to extract the ^{4}He gas. The ^{4}He gas was then cryogenically purified and spiked with ^{3}He. The $^{4}He/^{3}He$ ratio was measured with a quadrupole mass spectrometer. The grains were then retrieved, and the U and Th concentrations were measured by isotope-dilution inductively coupled plasma–mass spectrometry (ICP-MS). An age was calculated for each grain and corrected for α ejection based on the grain's dimensions. For those grains that were plucked from polished FT mounts, the α ejection correction factor was adjusted to account for the polished geometry of the grain, as describe in Reiners et al. (2007). Between two and five grains were analyzed for each sample, and from these individual grain ages, a weighted mean age was calculated for each sample using 1 ÷ standard error as the weights. All AHe analyses were carried out at the University of Arizona (U-Th-[Sm])/He Laboratory.

Thermal Modeling of AFT and (U-Th)/He Data

AFT and (U-Th)/He ages were modeled with HeFTy v1.8.0 (Ketcham et al., 2007) using the calibration of Flowers et al. (2009), and stopping distance and age calculation after Ketcham et al. (2011). Input parameters include age, error, track length, and Dpar (track etch pit diameter parallel to the crystallographic c-axis) data for AFT data and radius, uncorrected age, and U, Th, and Sm concentrations for apatite (U-Th)/He data. AFT and apatite (U-Th)/He ages were modeled together when possible. When an AFT age (with >5% χ^2) was not available, two apatite (U-Th)/He ages were selected and modeled together. No fits were obtained when more than two apatite (U-Th)/He ages were modeled. Several model runs using different permutations of ages were conducted. The best solutions are shown in Figure 6, and age envelopes for the good fits are presented in Figure 7. An initial constraint of temperature (T) > 120 °C was applied at

ca. 40 Ma, and near-surface temperatures were applied at 0 Ma; no additional constraints were applied.

Zircon U-Pb Dating

Uranium-lead (U-Pb) analysis was carried out on zircons. Over geologic time scales, uranium impurities in the zircon crystal lattice decay to lead: ^{238}U decays to ^{206}Pb with a half-life of 4.5 b.y., and ^{235}U decays to ^{207}Pb with a half-life of 700 m.y. (Gehrels, 2010). Both of these systems can independently be used to calculate an age. The zircon U-Pb system has a closure temperature of >800 °C (Gehrels, 2010), so unlike the AFT and AHe methods described earlier, zircon U-Pb analysis yields the crystallization age of the zircon and the igneous host rock from which it was derived.

For this study, zircons were isolated from the samples using standard mineral separation techniques and analyzed by the laser ablation (LA) ICP-MS method (Gehrels et al., 2008). For igneous samples, cathodoluminescence (CL) images were obtained of grains to identify possible inherited domains within the grains. During analysis, older inherited cores were analyzed separately from the younger domains. Approximately 30 grains were analyzed for igneous samples, and ~100 grains were analyzed for detrital samples. Results of igneous samples were plotted on Wetherill concordia diagrams (Wetherill, 1956), and results of detrital samples were reported as probability distribution plots to show the relative distribution of ages in the detrital samples.

RESULTS

AFT and AHe Data

Samples were collected from the hanging walls of major thrust faults in the Sierra de Las Planchadas. The geologic map (Fig. 4) indicates the locations of samples collected for thermochronologic analysis, and the cross section (Fig. 5) shows both the approximate sample locations (projected along-strike) and their structural context, as well as the analytical ages obtained. Samples from the Sierra de Las Planchadas were analyzed for AFT ages, and most samples were also analyzed using AHe. Samples from the Miocene strata in the Fiambalá Basin were analyzed only for AHe ages because prior work by Carrapa et al. (2006) showed that these strata were never buried deeply enough for AFT ages to be reset after deposition.

Table 2 summarizes AFT data. AFT ages range from 160.3 ± 13.1 Ma in the La Cuesta Formation in the hanging wall of the western basin-bounding thrust (Fig. 5, sample RS10.14), to 19.1 ± 6.0 Ma in the granodiorite in the hanging wall of the west-dipping thrust fault within the Sierra de Las Planchadas (Fig. 5, sample RS10.10). No clear directional trends exist in the AFT ages. Three AFT samples do not pass the χ^2 test: sample RS10.08 (102.7 ± 16.4 Ma), RS10.13 (63.8 ± 8.5 Ma), and RS10.20 (41.7 ± 9.5 Ma). The ages obtained for samples RS10.13 and RS10.20 are younger than the stratigraphic age of the rock (Permian La Cuesta Formation), indicating partial annealing of these samples.

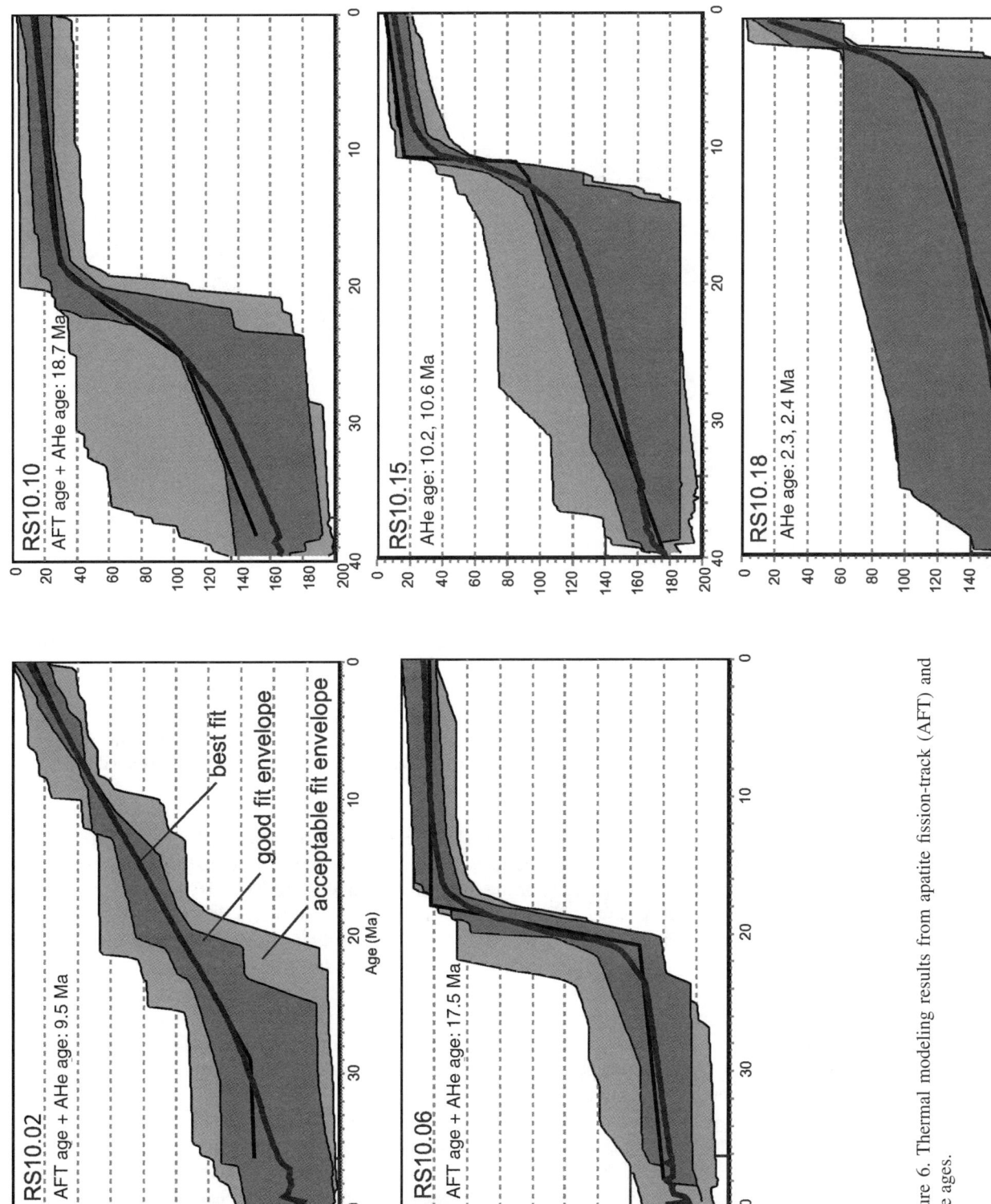

Figure 6. Thermal modeling results from apatite fission-track (AFT) and AHe ages.

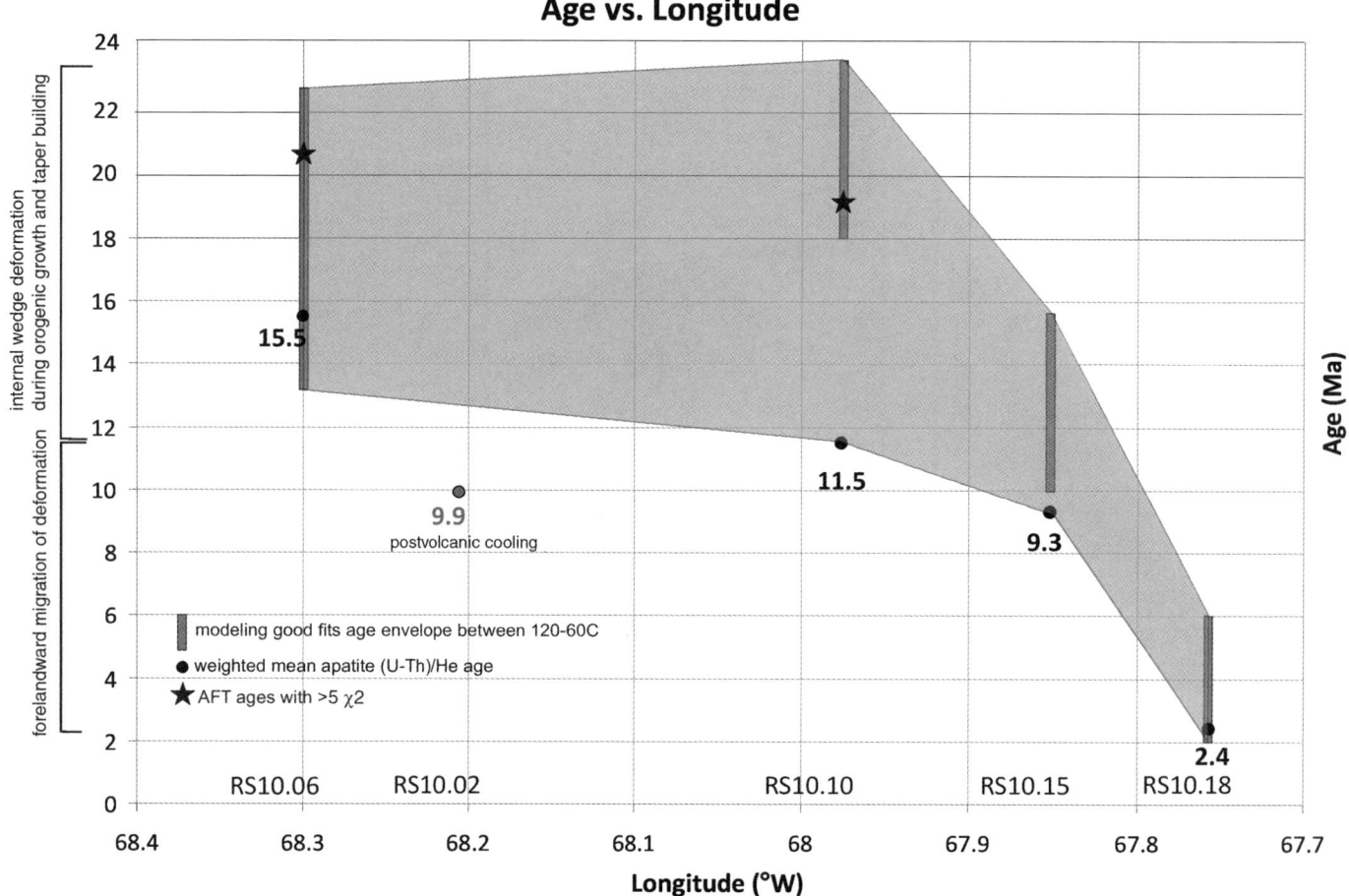

Figure 7. Apatite helium ages vs. longitude. The weighted mean age for each sample is calculated from the single-grain ages using $1/\sigma$ as the weights. A general west-to-east younging trend is seen in the weighted mean ages. The shaded gray area indicates the results of thermal modeling using both AHe and AFT ages. The modeling results corroborate the west-to-east younging trend seen in the weighted mean ages.

Three AFT samples do pass the χ^2 test: sample RS10.06 (20.6 ± 1.8 Ma), RS10.10 (19.1 ± 6.0 Ma), and RS10.14 (160.3 ± 13.1 Ma). These samples represent a single population of grains that have been fully annealed at $T > 120$ °C, corresponding to depth of ~4.5 km, assuming a 25 °C/km paleogeothermal gradient. These ages represent a minimum age for exhumation through the PAZ; if the samples were exhumed slowly, allowing partial annealing to occur in the PAZ, then the measured age will be younger than the true age of exhumation.

Table 3 summarizes AHe data. There is significant dispersion in the AHe ages for all samples except for the Miocene volcanic rocks (sample RS10.02). This dispersion may result from partial annealing of the grains, U-Th zonation within the grains, or compositional differences between grains (Reiners, 2002). For each sample, the weighted mean age was calculated using $1 \div$ standard error for the weights. In sample RS10.15, the two oldest ages of 38.3 ± 0.9 Ma and 87.7 ± 1.7 Ma were considered outliers and excluded from the weighted mean calculation. In Figure 7, the weighted mean age for each sample is plotted versus the longitude of the sample location. Moving from west to east, the weighted mean age is 15.5 Ma in the Devonian Punilla Formation in the high-elevation (>4000 m) western part of the field area (sample RS10.06), 9.9 Ma in the Miocene volcanic rocks just west of the Sierra de Las Planchadas (sample RS10.02), 11.5 Ma in the Ordovician granodiorite within the Sierra de Las Planchadas (sample RS10.10), 9.3 Ma in the lower Tambería Formation in the western part of the Fiambalá Basin (sample RS10.15), and 2.4 Ma in the upper Tambería Formation in the central Fiambalá Basin (sample RS10.18). Thermal modeling results (Fig. 6) confirm the younging trend to the east (Fig. 7).

Zircon U-Pb Data

Zircon U-Pb analysis was completed on five igneous samples from the ranges bounding the Fiambalá Basin to the west, north, and east. Zircon U-Pb analysis was also completed on four samples from the Upper Miocene strata in the Fiambalá Basin and one sample of Permian sandstone in the Sierra de Las Planchadas. Provenance of the basin strata can thus be compared to the sources in the surrounding ranges.

TABLE 2. APATITE FISSION TRACK DATA

Sample	Latitude (°S)	Longitude (°W)	Elevation (m)	Crystals	Spontaneous tracks ρs	Spontaneous tracks Ns	Induced tracks ρi	Induced tracks Ni	Dosimeter tracks ρd	Dosimeter tracks Nd	Dpar (μm)	Age dispersion $P\chi^2$%	Central age (Ma)	Error (1σ)
RS10.06	27.661889	68.300111	4436	40	159,046.4	182	2,265,975	2593	1,599,000	5118	2.3	100	20.62	±1.81
RS10.08	27.989917	68.160472	3370	20	483,100	269	1.306E+06	727	1,579,000	5051	2.3	0	102.70	±16.4
RS10.10	27.741222	67.976139	3206	10	37,445.53	11	565,087.1	166	1,565,000	5007	2.0	57	19.06	±5.99
RS10.13	27.702139	67.934917	2577	20	803,813.2	285	4,422,383	1568	1,551,000	4962	2.1	0	63.75	±8.54
RS10.14	27.704167	67.927222	2558	20	767,173.4	327	1,337,275	570	1,537,000	4918	2.1	65	160.27	±13.08
RS10.20	27.679972	67.949028	2795	20	435,524.4	194	216,1907	963	1,509,000	4828	2.1	0	41.71	±9.53

Note: ρ_s—spontaneous track density, N_s—number of spontaneous tracks, ρ_i—induced track density, N_i—number of induced tracks, ρ_d—dosimeter track density, N_d—number of dosimeter tracks, $P\chi^2$%—percent χ^2 probability.

TABLE 3. APATITE HELIUM DATA

Sample-grain	Latitude (°S)	Longitude (°W)	Mass (μg)	U (ppm)	eU (ppm)	Th (ppm)	Sm (ppm)	⁴He (nmol/g)	Length (μm)	Width (μm)	F_T	Age (Ma)	±1σ (Ma)
RS10.02-1	27.626639	68.205139	2.14	5.40	8.75	14.23	275.32	0.35	131	90	0.69	10.41	0.47
RS10.02-2			3.91	7.04	11.64	19.58	288.35	0.49	140	118	0.75	10.10	0.24
RS10.02-3			1.56	8.56	13.47	20.91	301.14	0.46	114	83	0.67	9.23	0.37
RS10.06-1	27.661889	68.300111	0.49	100.95	131.35	129.38	203.37	7.55	92	52	0.5	21.08	0.42
RS10.06-2			1.81	13.11	15.83	11.56	24.42	0.59	120	87	0.72	9.52	0.34
RS10.06-3			1.31	13.08	16.62	15.07	141.06	0.96	126	72	0.68	15.60	0.46
RS10.06-4			0.89	40.76	44.25	14.83	91.27	2.66	105	65	0.64	17.48	0.36
RS10.06-5			0.84	35.44	40.54	21.67	57.43	2.15	98	66	0.63	15.57	0.34
RS10.10-1	27.741222	67.976139	1.30	11.41	21.53	43.04	847.86	1.49	142	68	0.65	18.69	0.39
RS10.10-2			0.53	4.76	10.70	25.28	920.97	0.48	79	58	0.54	13.82	1.53
RS10.10-3			0.89	20.81	21.82	4.32	85.48	0.32	93	69	0.65	4.21	0.24
RS10.10-4			1.41	4.61	7.83	13.72	406.68	0.66	126	75	0.68	21.37	0.59
RS10.15-1	27.703556	67.851778	3.40	10.75	11.21	1.98	107.67	3.91	156	104	0.73	87.69	1.66
RS10.15-2			2.41	9.63	20.45	46.02	164.07	0.78	172	84	0.66	10.56	0.26
RS10.15-3			2.56	12.11	12.59	2.03	10.16	1.79	176	85	0.69	38.28	0.85
RS10.15-4			1.59	17.26	17.96	2.97	163.04	0.63	176	67	0.63	10.20	0.32
RS10.15-5			0.88	13.95	17.77	16.24	228.62	0.41	63	84	0.59	7.04	0.30
RS10.18-1	27.696188	67.756540	5.13	6.69	13.34	28.28	227.73	0.13	167	124	0.77	2.35	0.12
RS10.18-2			3.23	11.73	22.96	47.77	400.60	0.20	146	105	0.71	2.27	0.07

Note: F_T—α ejection correction factor.

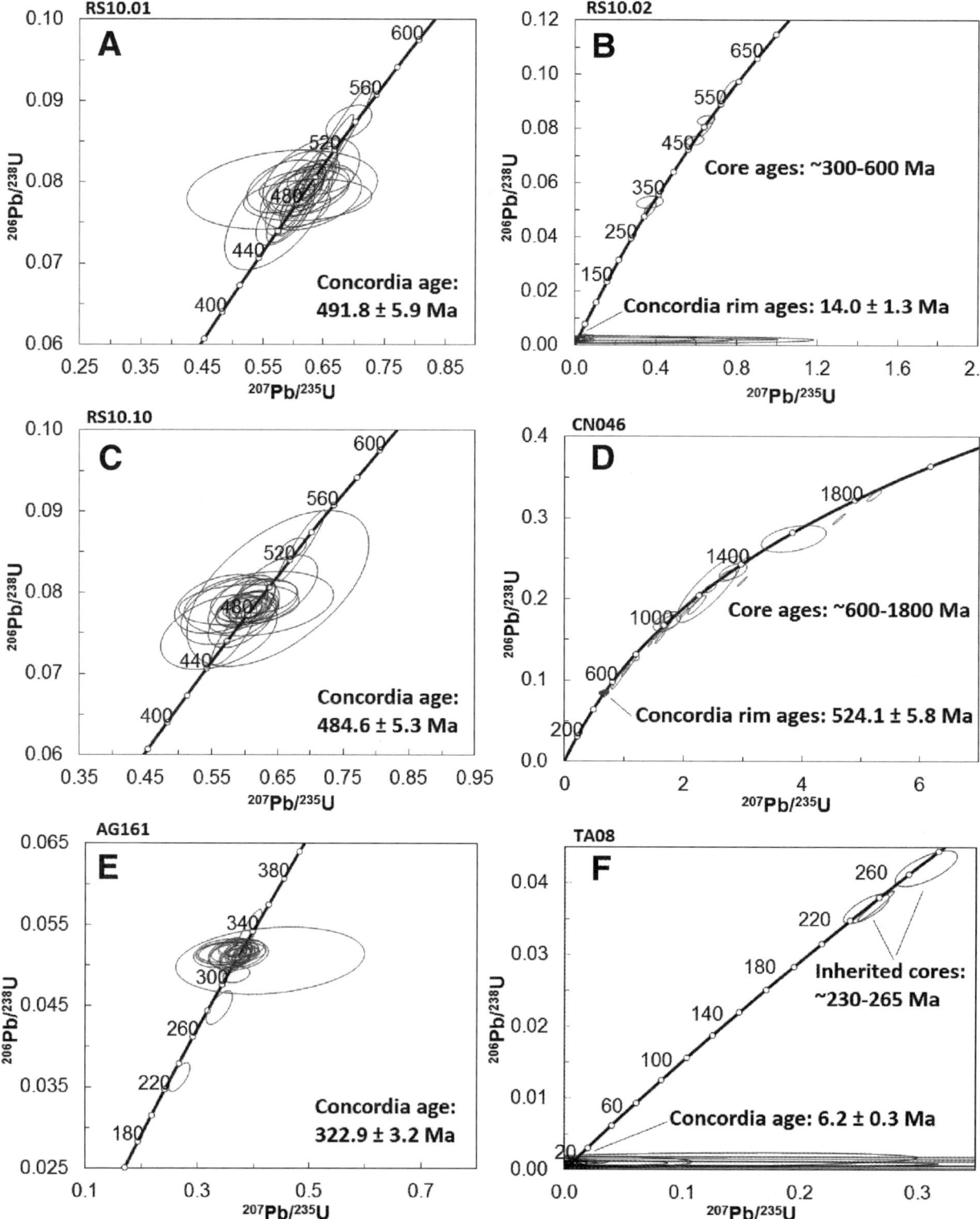

Figure 8. Results of zircon U-Pb analysis of igneous samples, plotted on concordia diagrams using the routines in Isoplot (Ludwig, 2008). Error ellipses are 68.3% confidence and include systematic errors. (A) Sample RS10.01, an Ordovician granodiorite in the Sierra de Las Planchadas. (B) Sample RS10.02, Miocene volcanics in the Valle de Chaschuil, west of the Sierra de Las Planchadas. (C) Sample RS10.10, an Ordovician granodiorite in the Sierra de Las Planchadas. (D) Sample CN046, a Cambrian-Precambrian granite near Cerro Negro on the northeastern flank of Fiambalá Basin. (E) Sample AG161, a Carboniferous leucogranite near Alto Grande, east of Fiambalá Basin. (F) Sample TA08, a volcanic ash from the top of the upper Tambería Formation in the Fiambalá Basin.

Approximately 30 zircon grains were analyzed by LA-ICP-MS for each igneous sample. Igneous sample locations are indicated on the map (Fig. 4) and cross section (Fig. 5), with the exception of sample CN046, which is a Cambrian-Precambrian granite north of the Fiambalá Basin and outside the mapped field area for this study (for a geologic map of the location of this sample, see Carrapa et al., 2006). Data from the igneous samples are shown on concordia diagrams in Figure 8 and in Table DR1 in the Data Repository (see footnote 1). Sample RS10.01 (Ordovician granodiorite from the eastern flank of the Sierra de Las Planchadas) has a concordia age of 491.8 ± 5.9 Ma. Sample RS10.02 (Miocene volcanic rocks in western part of field area) has a concordia rim age of 14.0 ± 1.3 Ma, and six grains contain older cores ranging from ca. 300 to 600 Ma. Sample RS10.10 (Ordovician granodiorite from the central Sierra de Las Planchadas) has a concordant age of 484.6 ± 5.3 Ma. Sample CN046 (Cambrian-Precambrian granite north of Fiambalá Basin) has a concordant rim age of 524.1 ± 5.8 Ma and older cores ranging from 1000 to 1900 Ma. Sample AG161 (Carboniferous granite east of Fiambalá Basin) has a concordant age of 322.9 ± 3.2 Ma.

Figure 9 is a stratigraphic column modified from Carrapa et al. (2006) with detrital zircon U-Pb sample locations marked. Figure 9 also indicates the location of sample TA08, a volcanic ash bed within a lacustrine-fluvial facies at the top of the upper Tambería Formation. Twenty zircon grains from sample TA08 were analyzed by LA-ICP-MS. Twelve grains had Cenozoic ages with a weighted mean age of 6.2 ± 0.3 Ma. A concordia plot of U-Pb analyses of sample TA08 is shown in Figure 8, and these data are also contained in Table DR1. This sample provides a new maximum stratigraphic age constraint: Since it is located at the transition from the upper Tambería Formation to the Guanchín Formation, it suggests that deposition of the Guanchín Formation commenced no earlier than ca. 6 Ma.

Approximately 100 zircon grains were analyzed by LA-ICP-MS for each detrital sample. Data from the detrital samples are shown on probability distribution diagrams in Figure 10 and in Table DR2 (see footnote 1). All five samples contain major populations of Grenville (Sunsas) (ca. 1100–1200 Ma), Cambrian (ca. 500 Ma), and Early Ordovician ages (ca. 480 Ma), as well as a minor population of Carboniferous age (ca. 350 Ma). The four samples from the Miocene section also contain Eocene and younger zircons attributable to Andean magmatism (Fig. 11). The youngest population ages were determined for the Cenozoic zircons using the Age Pick program (Gehrels, 2010). The Cenozoic zircons have a youngest population age of 23.9 ± 1.0 Ma in the lower Tambería Formation, 16.2 ± 4.6 Ma in the upper Tambería Formation, 5.0 ± 2.0 Ma in the Guanchín Formation, and 4.3 ± 1.1 Ma in the Punaschotter Formation.

DISCUSSION

AFT data indicate temporally differential exhumation. Three samples have fully annealed AFT ages that pass the χ^2 test, and therefore indicate a minimum age of exhumation through the PAZ: Samples RS10.06 and RS10.10 both have cooling ages of ca. 20 Ma, and sample RS10.14 has an age of ca. 160 Ma. The other three samples do not pass χ^2 and are assumed to be partially annealed. The AFT ages are older than the AHe ages, which is expected, since AFT is a slightly higher-temperature thermochronometer. The two fully annealed ca. 20 Ma ages corroborate the interpretation from the AHe ages, i.e., that exhumation was actively occurring during the Miocene.

The Miocene AFT ages determined in this study are shown within a regional context of previously published AFT ages in

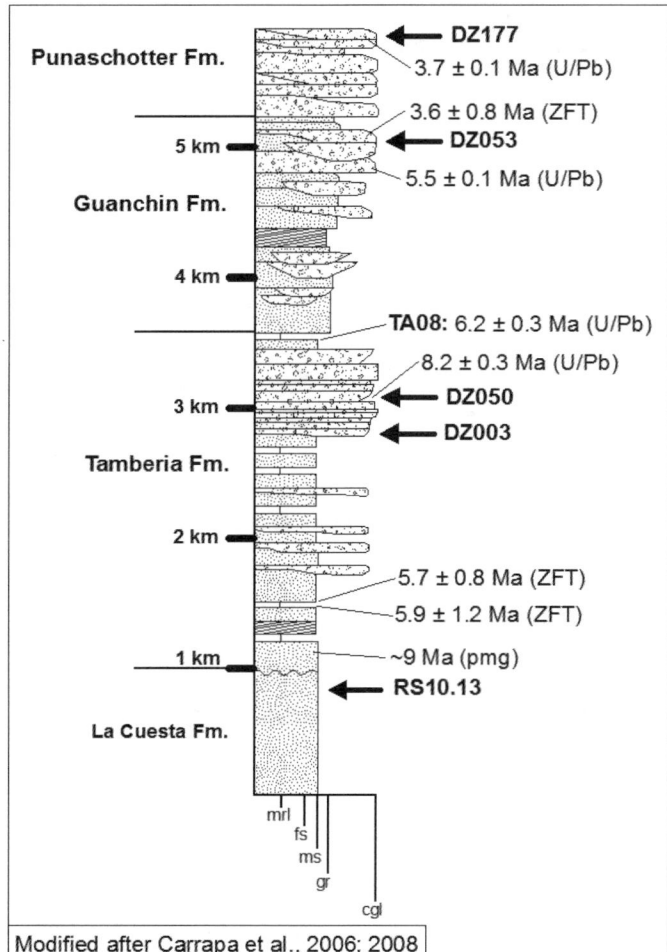

Figure 9. Stratigraphy of the Fiambalá Basin, modified after Carrapa et al. (2006, 2008). Locations of detrital zircon samples are indicated with arrows. In addition, sample TA08, a volcanic ash bed at the top of the upper Tambería Formation, is indicated. Previous stratigraphic age constraints from zircon U-Pb and zircon fission-track (ZFT) data (Carrapa et al., 2006, 2008) and paleomagnetic (pmg) data (Reynolds, 1987) are also indicated.

---

[1]GSA Data Repository Item 2015004, Zircon U-Pb data tables, is available at www.geosociety.org/pubs/ft2015.htm, or on request from editing@geosociety.org or Documents Secretary, GSA, P.O. Box 9140, Boulder, CO 80301-9140, USA.

Safipour et al.

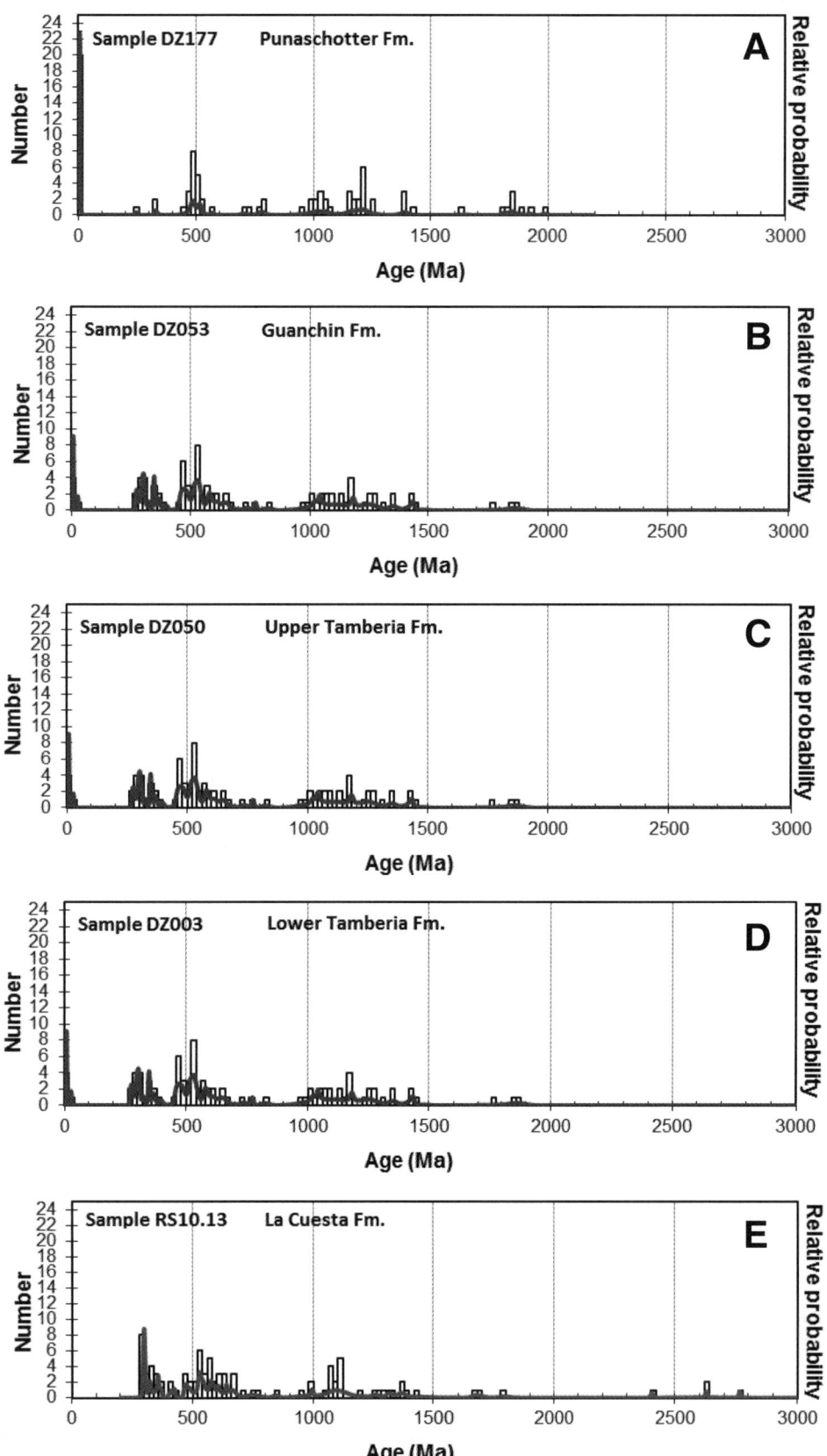

Figure 10. Results of zircon U-Pb analysis of detrital samples, plotted on probability distribution diagrams using the routines in Isoplot (Ludwig, 2008): (A) Sample DZ177, Punaschotter Formation. (B) Sample DZ053, Guanchín Formation. (C) Sample DZ050, upper Tambería Formation. (D) Sample DZ003, lower Tambería Formation. (E) Sample RS10.13, Permian La Cuesta Formation.

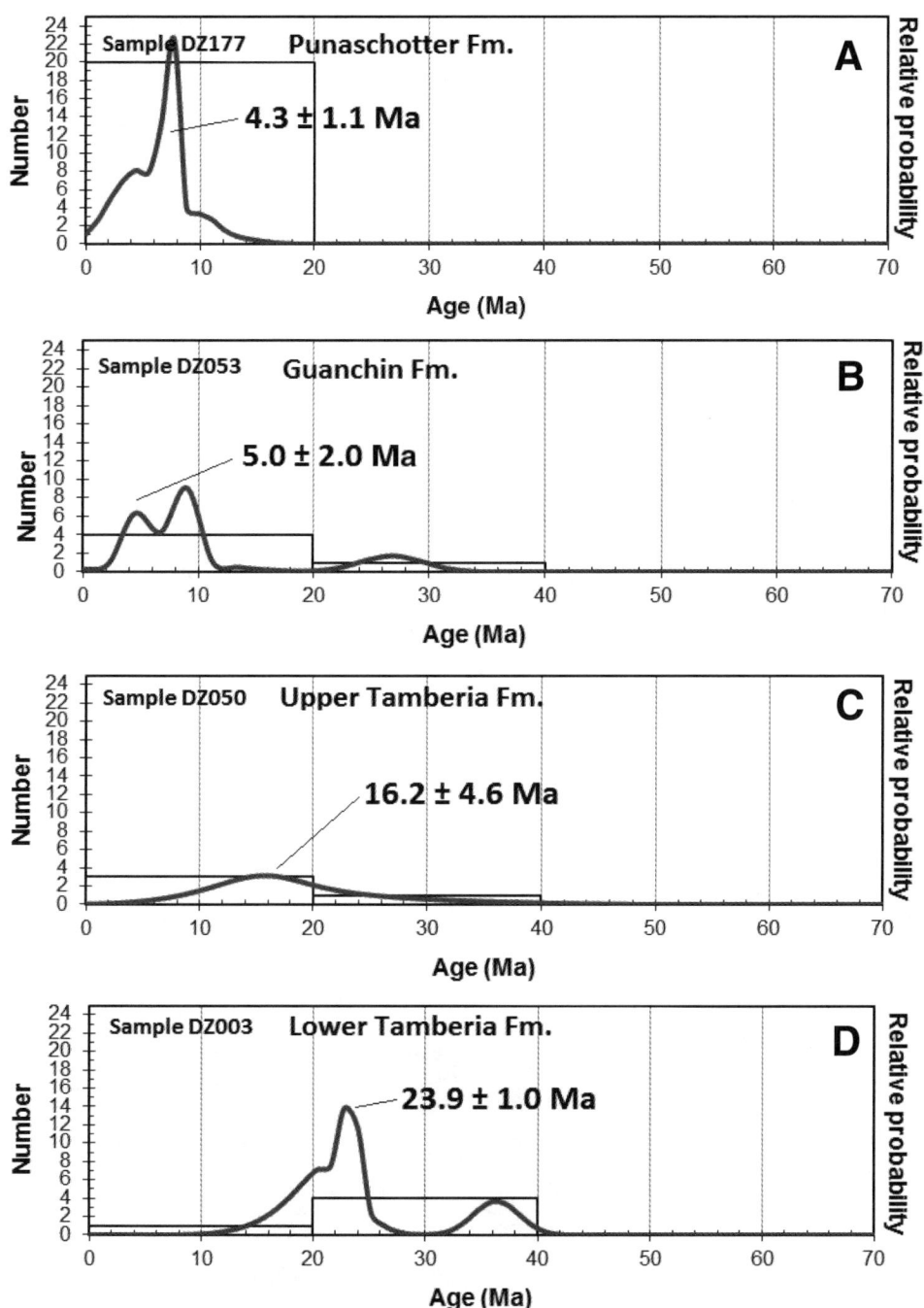

Figure 11. The Cenozoic portion of detrital zircon samples from Miocene–Pliocene strata within the Fiambalá Basin. The youngest population age peak is labeled for each sample. (A) Punaschotter Formation. (B) Guanchín Formation. (C) Upper Tambería Formation. (D) Lower Tambería Formation.

Figure 2. Carrapa et al. (2006) reported Eocene (to the west of our present transect) and middle Miocene (east of the transect) ages that, when combined with the ages reported herein, would appear to define a generally eastward-younging trend at latitude 27.6°S. However, AFT ages to the northeast of the Fiambalá Basin in the Sierra de Chango Real are Eocene–Oligocene (Coutand et al., 2001). Farther east in the Sierra de Aconquija, AFT ages are Late Cretaceous (Coughlin et al., 1998) and late Miocene–Pliocene (Sobel and Strecker, 2003). Thus, there is no apparent regional west to east younging trend recorded by the AFT ages at this latitude. Since the AFT system is sensitive to higher temperatures than the AHe system, it is possible that not enough erosion has occurred in this region to consistently exhume fully reset AFT ages. Furthermore, paleotopography can have a significant effect on exhumation ages; samples exhumed from inverted Cretaceous rift basins may have experienced burial to a great enough depth to reset the AFT age, whereas samples from the flanks of these rift basins may not have been fully annealed in the Cenozoic and instead still retain Cretaceous ages (Carrapa et al., 2014). Indeed, if the Cretaceous AFT ages are ignored on the criteria of possible partial annealing and/or differential burial, the Cenozoic ages regionally do decrease eastward. Sedimentologic evidence suggests that the Sierras Pampeanas were uplifted at ca. 6 Ma (Carrapa et al., 2008). This is consistent with the young ages found by Sobel and Strecker (2003), but it is not consistent with the old AFT ages found by Coutand et al. (2001) and Coughlin et al. (1998) and suggests that the AFT system does not consistently record the youngest exhumation of the Sierras Pampeanas. The presence of a younging trend in the AHe ages and lack of a clear trend in the AFT ages are further evidence that Cenozoic exhumation was of insufficient magnitude to be recorded by the AFT ages, indicating that total Cenozoic exhumation of the Sierra Pampeanas was limited to <~4 km. Another possible explanation for a lack of Cenozoic AFT ages is that the Andean foreland basin may have had a lower-than-average geothermal gradient during the late Miocene. Previous studies have suggested that the flattening of the Nazca slab and inboard migration of the mantle wedge during the late Miocene (Yáñez et al., 2001; Kay and Mpodozis, 2002) reduced the heat flow in the overlying crust, resulting in a geothermal gradient of only 15–18 °C/km (Collo et al., 2011; Dávila and Carter, 2013). With a geothermal gradient of 15 °C/km, the PAZ would be at ~7–4 km depth, and thus potentially up to 7 km of exhumation could occur without exposing reset AFT ages.

In all samples, the weighted mean AHe ages fall into the 2–16 Ma range. The AHe ages suggest that samples were being actively exhumed during the late Miocene–Pliocene. We interpret cooling and exhumation ages as an erosional response to active tectonic deformation within the Sierra de Las Planchadas during the late Miocene. This correlates well with the presence of Miocene syntectonic deposits in the Fiambalá Basin (Carrapa et al., 2006). Sample RS10.18 from the upper Tambería Formation has a weighted mean AHe age of 2.4 Ma, suggesting enhanced erosion associated with active deformation within the Fiambalá Basin as recently as the Pliocene; this observation is consistent with evidence for syndepositional local deformation observed in the Pliocene Punaschotter Formation (Carrapa et al., 2008).

Figure 7 is a plot of weighted mean AHe ages, AFT ages, and modeling age constraints versus longitude. The weighted mean AHe ages, AFT ages, and modeling results all show a general younging trend moving from west to east across the Sierra de Las Planchadas and Fiambalá Basin. Thermal modeling of AFT and AHe ages suggests that exhumation was concentrated within the Sierras de Las Planchadas between ca. 23 and 11.5 Ma and later migrated forelandward into the Fiambalá Basin. Exhumation within the Fiambalá Basin continued until 2.3 Ma.

There are several other possible explanations for lateral variations in thermochronologic cooling ages: The ages could be controlled by onset of erosion caused by lateral variations in the geometry of the basal décollement (e.g., Robert et al., 2011), variations in exhumation ages could be caused by uplift and erosion linked to duplexing within the thrust belt (e.g., Herman et al., 2010), or the eastward younging trend in the ages may be an

expression of the local eastward migration of exhumation and uplift. Field mapping has not identified any significant duplex structures within the Sierra de Las Planchadas (Figs. 4 and 5; Martínez et al., 1995; Rubiolo et al., 2009; this study), and thus duplexing is unlikely to be responsible for the variation in the AHe ages. Although the geometry of a basal décollement beneath the Sierra de Las Planchadas may be a control on the exhumation ages within the range, deformation in the central Andes has been migrating eastward since Late Cretaceous time (Jordan et al., 1993; Sempere et al., 1997; DeCelles and Horton, 2003; Horton, 2005; Gillis et al., 2006; McQuarrie et al., 2008; Carrapa et al., 2011; DeCelles et al., 2011; Barnes et al., 2012). Therefore, the west-to-east younging trend observed in the AHe ages more likely reflects the eastward migration of uplift-driven exhumation across the Sierra de Las Planchadas and into the Fiambalá Basin. The one exception to the west-to-east younging trend is the 11.5 Ma weighted mean age obtained for sample RS10.10. This age may be due to uplift and erosion in response to out-of-sequence thrusting, perhaps in response to subcritical taper conditions in the thrust belt.

The AHe and AFT ages and thermal modeling results obtained in this study are consistent with ages in the Eastern Cordillera at ~26°S, where Carrapa et al. (2011) found a west-to-east younging trend in AHe ages ranging from 14 to 3 Ma. If the two regions are correlated along strike, this indicates that exhumation in response to crustal deformation in the Precordillera at 28°S occurred synchronously with deformation in the Eastern Cordillera at 26°S and that the Eastern Cordillera and Precordillera may be a single continuous thrust front (Fig. 12).

The Permian La Cuesta Formation detrital zircon sample, RS10.13, contains zircons of Grenville (ca. 1100–1200 Ma),

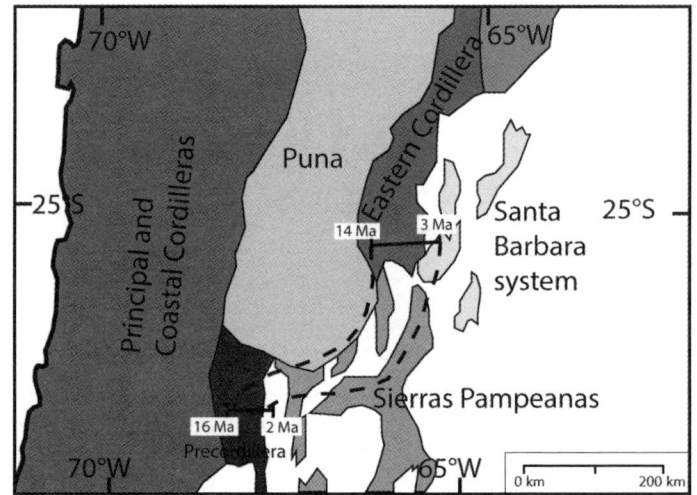

Figure 12. The eastward younging trend in AHe ages (Ma) found in this study at ~28°S is compared to the eastward younging trend in AHe ages observed by Carrapa et al. (2011) at ~26°S. Dashed lines indicate a proposed continuous thrust front connecting the Precordillera and the Eastern Cordillera during the Miocene.

Cambrian (ca. 500 Ma), and Early Ordovician (ca. 480 Ma) age populations, as well as a significant Carboniferous (ca. 350 Ma) population and a minor Early Permian (ca. 298 Ma) population. The four detrital zircon samples from the Miocene strata show little variation in provenance: All four samples contain major populations of Grenville (ca. 1100–1200 Ma), Cambrian, and Early Ordovician ages (ca. 480 Ma), as well as a minor population of Carboniferous age (ca. 350 Ma) and a Neogene population associated with Andean magmatism (<40 Ma). The Ordovician ages match the ages of the igneous intrusions (samples RS10.01 and RS10.10) that were dated in this study in the Sierra de Las Planchadas to the west, and the Cambrian and Carboniferous ages match the igneous samples that were dated in this study from the ranges to the north (sample CN046) and east (sample AG161) of the Fiambalá Basin. Major Ordovician plutonic bodies are not observed west of the Sierra de Las Planchadas or north or east of the basin (Rubiolo et al., 2009), so the Sierra de Las Planchadas is the only plausible source for these zircons. This is consistent with eastward paleocurrent indicators in the Miocene strata (Carrapa et al., 2006, 2008) and suggests that the Sierra de Las Planchadas to the west of the basin was the primary source of Fiambalá Basin strata. The Grenville signal likely comes from older inherited cores within the igneous zircons as well as zircons reworked from Paleozoic and Proterozoic strata. This age population is common in detrital zircon age spectra from Cenozoic strata throughout the central Andes (DeCelles et al., 2011). The Cambrian and Carboniferous zircons in the Miocene strata may be reworked from the Permian sandstones in the Sierra de Las Planchadas. If the Sierra de Las Planchadas is the source of the late Miocene strata in the Fiambalá Basin, then this indicates that uplift of the Sierra de Las Planchadas range occurred at the same time or before the late Miocene strata began to accumulate at ca. 9 Ma. Carrapa et al. (2008) have suggested the possibility of a growth structure in the Tambería Formation at the western margin of the Fiambalá Basin, which is consistent with our conclusion that the Sierra de Las Planchadas was being exhumed at the same time as deposition in the Fiambalá Basin.

The youngest populations of Cenozoic Andean-derived zircons show an up-section younging trend (Fig. 11). The youngest population age is 23.9 ± 1.0 Ma in the lower Tambería Formation, 16.2 ± 4.6 Ma in the upper Tambería Formation, 5.0 ± 2.0 Ma in the Guanchín Formation, and 4.3 ± 1.1 Ma in the Punaschotter Formation. This up-section younging trend indicates that local volcanism was occurring concurrent with deposition in the Fiambalá Basin.

CONCLUSIONS

Miocene AHe ages and thermal modeling of AHe and AFT ages show a younging trend in exhumation from west to east across the Sierra de Las Planchadas at latitude ~28°S, presumably indicative of the forelandward progression of erosion in response to uplift and deformation. This cooling history may reflect the exhumational response to deformation of the Precordillera at this latitude. In this case, deformation seemed to be concentrated in the Precordillera for over 10 m.y. This pattern is similar to the pattern observed in the Eastern Cordillera of NW Argentina, where deformation was concentrated in a narrow area between ca. 21 and 14 Ma (Carrapa et al., 2011) and later migrated eastward. The forelandward propagation of deformation is synchronous with the propagation of deformation inferred across the Eastern Cordillera at latitude 26°S (Carrapa et al., 2011). This suggests that a single continuous thrust front, striking NE-SW, connected the Eastern Cordillera and the Precordillera during the Miocene. No clear regional trend is present in the AFT ages at this latitude, which indicates that the AFT system is not sensitive to low enough temperatures to systematically record Cenozoic exhumation in the Sierras Pampeanas, and that the total amount of Cenozoic exhumation does not exceed ~7–4 km, dependent on the paleogeothermal gradient. Detrital zircon U-Pb ages of the Miocene and Pliocene strata in the Fiambalá Basin indicate provenance from Paleozoic igneous intrusions and sedimentary rocks in the Sierra de Las Planchadas to the west, as well as Miocene volcanic rocks that were erupted in the magmatic arc concurrent with deposition in the Fiambalá Basin. This indicates that uplift of the Sierra de Las Planchadas began prior to or at the same time as the initiation of deposition in the Fiambalá Basin at ca. 9 Ma.

ACKNOWLEDGMENTS

We wish to thank Nicky Giesler, for assistance in the field and in the geochronology laboratory; George Gehrels and Peter Reiners, for the use of their laboratories to process samples; Roswell Juan and Gayland Simpson, for assistance with mineral separation procedures; John Boyd, for analyzing sample TA08; Ricardo Alonso at the Universidad Nacional de Salta, for assistance and advice during the field season in Argentina; David Spatz, for the donation of field equipment; and Federico Dávila and an anonymous reviewer, for their helpful and constructive comments that helped us to improve the manuscript. Zircon U-Pb data for this project were acquired in the Arizona LaserChron Center, which is supported by National Science Foundation grant EAR-1032156. Funding for this project was provided by The ExxonMobil Corporation.

REFERENCES CITED

Allmendinger, R.W., and Gubbels, T., 1996, Pure and simple shear plateau uplift, Altiplano-Puna, Argentina and Bolivia: Tectonophysics, v. 259, p. 1–13, doi:10.1016/0040-1951(96)00024-8.

Allmendinger, R.W., Jordan, T.E., Kay, S.M., and Isacks, B.L., 1997, The evolution of the Altiplano-Puna Plateau of the central Andes: Annual Review of Earth and Planetary Sciences, v. 25, p. 139–174, doi:10.1146/annurev.earth.25.1.139.

Barnes, J.B., Ehlers, T.A., Insel, N., McQuarrie, N., and Poulson, C.J., 2012, Linking orography, climate, and exhumation across the central Andes: Geology, v. 40, no. 12, p. 1135–1138, doi:10.1130/G33229.1.

Cahill, T., and Isacks, B.L., 1992, Seismicity and shape of the subducted Nazca plate: Journal of Geophysical Research–Solid Earth, v. 97, p. 17,503–17,529, doi:10.1029/92JB00493.

Carrapa, B., Strecker, M.R., and Sobel, E.R., 2006, Cenozoic orogenic growth in the central Andes: Evidence from sedimentary rock provenance and apatite fission track thermochronology in the Fiambala Basin, southernmost Puna Plateau margin (NW Argentina): Earth and Planetary Science Letters, v. 247, p. 82–100, doi:10.1016/j.epsl.2006.04.010.

Carrapa, B., Hauer, J., Schoenbohm, L., Strecker, M.R., Schmitt, A.K., Villanueva, A., and Gomez, J.S., 2008, Dynamics of deformation and sedimentation in the northern Sierras Pampeanas: An integrated study of the Neogene Fiambala Basin, NW Argentina: Geological Society of America Bulletin, v. 120, p. 1518–1543, doi:10.1130/B26111.1.

Carrapa, B., Trimble, J.D., and Stockli, D.F., 2011, Patterns and timing of exhumation and deformation in the Eastern Cordillera of NW Argentina revealed by (U-Th)/He thermochronology: Tectonics, v. 30, TC3003, doi:10.1029/2010TC002707.

Carrapa, B., Reyes-Bywater, S., Safipour, R., Sobel, E.R., Schoenbohm, L.M., DeCelles, P.G., Reiners, P., and Stockli, D., 2014, The effect of inherited paleotopography on exhumation of the central Andes of NW Argentina: Geological Society of America Bulletin, v. 126, p. 66–77, doi:10.1130/B30844.1.

Collo, G., Dávila, F.M., Nóbile, J., Astini, R.A., and Gehrels, G., 2011, Clay mineralogy and thermal history of the Neogene Vinchina Basin, central Andes of Argentina: Analysis of factors controlling the heating conditions: Tectonics, v. 30, TC4012, doi:10.1029/2010TC002841.

Coughlin, T.J., O'Sullivan, P.B., Kohn, B.P., and Holcombe, R.J., 1998, Apatite fission-track thermochronology of the Sierras Pampeanas, central western Argentina: Implications for the mechanism of plateau uplift in the Andes: Geology, v. 26, p. 999–1002, doi:10.1130/0091-7613(1998)026<0999:AFTTOT>2.3.CO;2.

Coutand, I., Cobbold, P.R., de Urreiztieta, M., Gautier, P., Chauvin, A., Gapais, D., Rossello, E.A., and Lopez-Gamundi, O., 2001, Style and history of Andean deformation, Puna plateau, northwestern Argentina: Tectonics, v. 20, p. 210–234, doi:10.1029/2000TC900031.

Coutand, I., Carrapa, B., Deeken, A., Schmitt, A.K., Sobel, E.R., and Strecker, M.R., 2006, Propagation of orographic barriers along an active range front: Insights from sandstone petrography and detrital apatite fission-track thermochronology in the intramontane Angastaco basin, NW Argentina: Basin Research, v. 18, p. 1–26, doi:10.1111/j.1365-2117.2006.00283.x.

Dávila, F.M., and Astini, R.A., 2007, Cenozoic provenance history of synorogenic conglomerates in western Argentina (Famatina belt): Implications for central Andean foreland development: Geological Society of America Bulletin, v. 119, p. 609–622, doi:10.1130/B26007.1.

Dávila, F.M., and Carter, A., 2013, Exhumation history of the Andean broken foreland revisited: Geology, v. 41, p. 443–446, doi:10.1130/G33960.1.

DeCelles, P.G., and Horton, B.K., 2003, Early to middle Tertiary foreland basin development and the history of Andean crustal shortening in Bolivia: Geological Society of America Bulletin, v. 115, p. 58–77, doi:10.1130/0016-7606(2003)115<0058:ETMTFB>2.0.CO;2.

DeCelles, P.G., Carrapa, B., Horton, B.K., and Gehrels, G.E., 2011, Cenozoic foreland basin system in the central Andes of northwestern Argentina: Implications for Andean geodynamics and modes of deformation: Tectonics, v. 30, TC6013, doi:10.1029/2011TC002948.

Deeken, A., Sobel, E.R., Coutand, I., Haschke, M., Riller, U., and Strecker, M.R., 2006, Development of the southern Eastern Cordillera, NW Argentina, constrained by AFT-thermochronology: From Early Cretaceous extension to middle Miocene shortening: Tectonics, v. 25, TC6003, doi:10.1029/2005TC001894.

Donelick, R.A., O'Sullivan, P.B., and Ketcham, R.A., 2005, Apatite fission-track analysis: Reviews in Mineralogy and Geochemistry, v. 58, no. 1, p. 49–94, doi:10.2138/rmg.2005.58.3.

Echavarria, L., Hernández, R., Allmendinger, R.W., and Reynolds, J., 2003, Subandean thrust and fold belt of northwestern Argentina: Geometry and timing of the Andean evolution: American Association of Petroleum Geologists Bulletin, v. 87, p. 965–985, doi:10.1306/01200300196.

Farley, K.A., 2002, (U-Th)/He dating: Techniques, calibrations, and applications, *in* Porcelli, D.P., Ballentine, C.J., and Wieler, R., eds., Noble Gases: Reviews in Mineralogy and Geochemistry, v. 47, p. 819–844.

Farley, K.A., Wolf, R.A., and Silver, L.T., 1996, The effects of long alpha-stopping distances on (U-Th)/He ages: Geochimica et Cosmochimica Acta, v. 60, p. 4223–4229, doi:10.1016/S0016-7037(96)00193-7.

Flowers, R.M., Ketcham, R.A., Shuster, D.L., and Farley, K.A., 2009, Apatite (U-Th)/He thermochronometry using a radiation damage accumulation and annealing model: Geochimica et Cosmochimica Acta, v. 73, p. 2347–2365, doi:10.1016/j.gca.2009.01.015.

Galbraith, R.F., 2005, Statistics for Fission Track Analysis: Boca Raton, Florida, Chapman and Hall/CRC, 240 p.

Gallagher, K., Brown, R., and Johnson, C., 1998, Fission track analysis and its applications to geological problems: Annual Review of Earth and Planetary Sciences, v. 26, p. 519–572, doi:10.1146/annurev.earth.26.1.519.

Gehrels, G., 2010, Arizona LaserChron Center: https://sites.google.com/a/laserchron.org/laserchron/home/ (accessed September 2011).

Gehrels, G., Valencia, V., and Ruiz, J., 2008, Enhanced precision, accuracy, efficiency, and spatial resolution of U-Th-Pb ages by LA-ICPMS: Geochemistry Geophysics Geosystems, v. 9, Q03017, doi:10.1029/2007GC001805.

Gillis, R.J., Horton, B.K., and Grove, M., 2006, Thermochronology, geochronology, and upper crustal structure of the Cordillera Real: Implications for Cenozoic exhumation of the central Andean Plateau: Tectonics, v. 25, TC6007, doi:10.1029/2005TC001887.

Hain, M.P., Strecker, M.R., Bookhagen, B., Alonso, R.N., Pingel, H., and Schmitt, A.K., 2011, Neogene to Quaternary broken foreland formation and sedimentation dynamics in the Andes of NW Argentina (25°S): Tectonics, v. 30, TC2006, doi:10.1029/2010TC002703.

Herman, F., Copeland, P., Avouac, J., Bollinger, L., Mahéo, G., Le Fort, P., Rai, S., Foster, D., Pêcher, A., Stüwe, K., and Henry, P., 2010, Exhumation, crustal deformation, and thermal structure of the Nepal Himalaya derived from the inversion of thermochronological and thermobarometric data and modeling of the topography: Journal of Geophysical Research, v. 115, B06407, doi:10.1029/2008JB006126.

Horton, B.K., 2005, Revised deformation history of the central Andes: Inferences from Cenozoic foredeep and intermontane basins of the Eastern Cordillera, Bolivia: Tectonics, v. 24, TC3011, doi:10.1029/2003TC001619.

Isacks, B.L., 1988, Uplift of the central Andean Plateau and bending of the Bolivian orocline: Journal of Geophysical Research, v. 93, no. B4, p. 3211–3231, doi:10.1029/JB093iB04p03211.

Jordan, T.E., and Allmendinger, R.W., 1986, The Sierras Pampeanas of Argentina: A modern analogue of Laramide deformation: American Journal of Science, v. 286, p. 737–764, doi:10.2475/ajs.286.10.737.

Jordan, T.E., Isacks, B.L., Allmendinger, R.W., Brewer, J.A., Ramos, V.A., and Ando, C.J., 1983, Andean tectonics related to geometry of subducted Nazca plate: Geological Society of America Bulletin, v. 94, p. 341–361, doi:10.1130/0016-7606(1983)94<341:ATRTGO>2.0.CO;2.

Jordan, T.E., Allmendinger, R.W., Damanti, J.F., and Drake, R.E., 1993, Chronology of motion in a complete thrust belt: The Precordillera, 30–31°S, Andes Mountains: The Journal of Geology, v. 101, p. 135–156, doi:10.1086/648213.

Kay, S.M., and Mpodozis, C., 2002, Magmatism as a probe to the Neogene shallowing of the Nazca plate beneath the modern Chilean flat-slab: Journal of South American Earth Sciences, v. 15, p. 39–57, doi:10.1016/S0895-9811(02)00005-6.

Ketcham, R.A., Carter, A., Donelick, R.A., Barbarand, J., and Hurford, A.J., 2007, Improved modeling of fission-track annealing in apatite: The American Mineralogist, v. 92, no. 5–6, p. 799–810, doi:10.2138/am.2007.2281.

Ketcham, R.A., Gautheran, C., and Tassan-Got, L., 2011, Accounting for long alpha-particle stopping distances in (U-Th-Sm)/He geochronology: Refinement of the baseline case: Geochimica et Cosmochimica Acta, v. 75, p. 7779–7791, doi:10.1016/j.gca.2011.10.011.

Kley, J., and Monaldi, C.R., 1998, Tectonic shortening and crustal thickness in the central Andes: How good is the correlation?: Geology, v. 26, p. 723–726, doi:10.1130/0091-7613(1998)026<0723:TSACTI>2.3.CO;2.

Kley, J., and Monaldi, C.R., 2002, Tectonic inversion in the Santa Barbara system of the central Andean foreland thrust belt, northwestern Argentina: Tectonics, v. 21, no. 6, 1061, doi:10.1029/2002TC902003.

Kley, J., Müller, J., Tawackoli, S., Jacobshagen, V., and Manutsoglu, E., 1997, Pre-Andean and Andean-age deformation in the Eastern Cordillera of southern Bolivia: Journal of South American Earth Sciences, v. 10, p. 1–19, doi:10.1016/S0895-9811(97)00001-1.

Ludwig, K., 2008, Isoplot 3.6: Berkeley Geochronology Center Special Publication 4, 77 p.

Mángano, M.A., and Buatois, L.A., 1997, Slope-apron deposition in an Ordovician arc-related setting: The Vuelta de Las Tolas Member (Suri Formation), Famatina Basin, northwest Argentina: Sedimentary Geology, v. 109, p. 155–180, doi:10.1016/S0037-0738(96)00043-7.

Martínez, L. del V., Nullo, F.E., Caminos, R.L., Panza, J.L., and Chipulina, M.A., 1995, Mapa Geologica de la Provincia de Catamarca, República Argentina: Buenos Aires, Dirección Nacional del Servicio Geológico, Secretaría de Minería, scale 1:500,000.

McGroder, M.F., Lease, R.O., and Pearson, D.M., 2015, this volume, Along-strike variation in structural styles and hydrocarbon occurrences,

Subandean fold-and-thrust belt and inner foreland, Colombia to Argentina, *in* DeCelles, P.G., Ducea, M.N., Carrapa, B., and Kapp, P.A., eds., Geodynamics of a Cordilleran Orogenic System: The Central Andes of Argentina and Northern Chile: Geological Society of America Memoir 212, doi:10.1130/2015.1212(05).

McQuarrie, N., and DeCelles, P., 2001, Geometry and structural evolution of the central Andean backthrust belt, Bolivia: Tectonics, v. 20, p. 669–692, doi:10.1029/2000TC001232.

McQuarrie, N., Horton, B.K., Zandt, G., Beck, S., and DeCelles, P.G., 2005, Lithospheric evolution of the Andean fold-and-thrust belt, Bolivia, and the origin of the central Andean plateau: Tectonophysics, v. 399, p. 15–37, doi:10.1016/j.tecto.2004.12.013.

McQuarrie, N., Barnes, J.B., and Ehlers, T.A., 2008, Geometric, kinematic, and erosional history of the central Andean Plateau, Bolivia (15–17°S): Tectonics, v. 27, TC3007, doi:10.1029/2006TC002054.

Mortimer, E., Carrapa, B., Coutand, I., Schoenbohm, L., Sobel, E.R., and Gomez, J.S., 2007, Fragmentation of a foreland basin in response to out-of-sequence basement uplifts and structural reactivation: El Cajon–Campo del Arenal Basin, NW Argentina: Geological Society of America Bulletin, v. 119, p. 637–653, doi:10.1130/B25884.1.

Pearson, D.M., Kapp, P., DeCelles, P.G., Reiners, P.W., Gehrels, G.E., Ducea, M.N., Pullen, A., Otamendi, J.E., and Alonso, R.N., 2012, Major Miocene exhumation by fault-propagation folding within a metamorphosed, early Paleozoic thrust belt: Northwestern Argentina: Tectonics, v. 31, TC4023, doi:10.1029/2011TC003043.

Ramos, V.A., Cristallini, E.O., and Perez, D.J., 2002, The Pampean flat-slab of the central Andes: Journal of South American Earth Sciences, v. 15, p. 59–78, doi:10.1016/S0895-9811(02)00006-8.

Reiners, P.W., 2002, (U-Th)/He chronometry experiences a renaissance: Eos (Transactions, American Geophysical Union), v. 83, p. 21–26, doi:10.1029/2002EO000012.

Reiners, P.W., Thomson, S.N., McPhillips, D., Donelick, R.A., and Roering, J.J., 2007, Wildfire thermochronology and the fate and transport of apatite in hillslope and fluvial environments: Journal of Geophysical Research, v. 112, F04001, doi:10.1029/2007JF000759.

Reynolds, J., 1987, Chronology of Neogene Tectonics in Western Argentina (27°–33°S) Based on the Magnetic Polarity Stratigraphy of Foreland Basin Sediments [PhD thesis]: Ithaca, New York, Cornell University, 353 p.

Robert, X., van der Beek, P., Braun, J., Perry, C., and Mugnier, J., 2011, Control of detachment geometry on lateral variations in exhumation rates in the Himalaya: Insights from low-temperature thermochronology and numerical modeling: Journal of Geophysical Research, v. 116, B05202, doi:10.1029/2010JB007893.

Rubiolo, D., González, O., Seggario, R., Hongn, F., and Martínez, L., 2009, Carta Geológica de la República Argentina: Fiambalá, 2769-IV: Buenos Aires, Dirección Nacional del Servicio Geológico, Secretaría de Minería, scale 1:250,000.

Sempere, T., Butler, R.F., Richards, D.R., Marshall, L.G., Sharp, W., and Swisher, C.C., 1997, Stratigraphy and chronology of Upper Cretaceous–Lower Paleogene strata in Bolivia and northwest Argentina: Geological Society of America Bulletin, v. 109, p. 709–727, doi:10.1130/0016-7606(1997)109<0709:SACOUC>2.3.CO;2.

Sobel, E.R., and Strecker, M.R., 2003, Uplift, exhumation and precipitation: Tectonic and climatic control of late Cenozoic landscape evolution in the northern Sierras Pampeanas, Argentina: Basin Research, v. 15, p. 431–451, doi:10.1046/j.1365-2117.2003.00214.x.

Strecker, M.R., Alonso, R., Bookhagen, B., Carrapa, B., Hilley, G.E., Sobel, E.R., and Trauth, M.H., 2007, Tectonics and climate of the southern central Andes: Annual Review of Earth and Planetary Sciences, v. 35, p. 747–787, doi:10.1146/annurev.earth.35.031306.140158.

Turner, G., 1967, Descripción Geológica de la Hoja 13b, Chaschuil, Provincias de Catamarca y la Rioja: Buenos Aires, Argentina, Dirección Nacional de Geología y Minería, v. 106, p. 1–78.

Wagner, G.A., 1979, Correction and interpretation of fission track ages, *in* Jager, J.C., and Hunziker, J.C., eds., Lectures in Isotope Geology: Berlin, Springer-Verlag, p. 170–177.

Wetherill, G.W., 1956, Discordant uranium-lead ages: Transactions, American Geophysical Union, v. 37, p. 320–326, doi:10.1029/TR037i003p00320.

Wolf, R.A., Farley, K.A., and Kass, D.M., 1998, Modeling of the temperature sensitivity of the apatite (U-Th)/He thermochronometer: Chemical Geology, v. 148, p. 105–114, doi:10.1016/S0009-2541(98)00024-2.

Yáñez, G.A., Ranero, C.R., von Huene, R., and Díaz, J., 2001, Magnetic anomaly interpretation across the southern central Andes (32–32°S): The role of the Juan Fernández Ridge in the Late Tertiary evolution of the margin: Journal of Geophysical Research, v. 106, no. B4, p. 6325–6345, doi:10.1029/2000JB900337.

Manuscript Accepted by the Society 3 June 2014
Manuscript Published Online 15 October 2014

The Geological Society of America
Memoir 212
2015

Regional exhumation and kinematic history of the central Andes in response to cyclical orogenic processes

Barbara Carrapa
Peter G. DeCelles
Department of Geosciences, University of Arizona, 1040 E. 4th Street, Tucson, Arizona 85721, USA

ABSTRACT

Low-temperature thermochronological ages of samples from the central Andes correlate with major tectonic events during Late Cretaceous and Cenozoic times. Apatite fission-track (AFT) ages show prominent clusters during the Early–Late Cretaceous in the Coastal Cordillera and the Cordillera de Domeyko; Paleocene–Oligocene ages in the western Puna Plateau and Cordillera de Domeyko; and latest Eocene–Pliocene ages in the Eastern Cordillera. These ages track the expansion of the Andean orogenic edifice, the eastern front of which migrated rapidly eastward ~200 km and ~150 km during late Eocene and Pliocene times, respectively. During the intervening time interval, ca. 35–5 Ma, the orogenic strain front migrated slowly eastward through the Eastern Cordillera. A second cluster of Cretaceous ages in the Eastern Cordillera and Santa Bárbara Ranges documents exhumation related to extension in the Salta rift. The highly unsteady pace of orogenic wedge propagation suggests that kinematics controlled local climate, rather than vice versa. The frequency of AFT ages is anticorrelated with magmatic production in the central Andean arc and the rate of convergence between the Nazca and South American plates. We propose a link between AFT bedrock cooling ages in the central Andes and exhumation related to cyclical processes of shortening, wedge propagation, magmatism, and removal of dense roots from beneath the magmatic arc and thickened hinterland region. In particular, periods of sustained exhumation associated with local crustal shortening alternate with periods of rapid eastward wedge propagation during which exhumation was more spatially diffuse across the high-elevation hinterland. Episodes of spatially confined exhumation are correlated with periods of relatively low magmatic production in the central Andean arc and relatively slow or declining plate convergence rates. We speculate that shortening in the upper crust was contemporaneous with underthrusting of lower crust and mantle lithosphere beneath the magmatic arc. Because of thermal inertia, melting of these underthrusted rocks lagged behind the shortening events themselves, thus producing the observed temporal anticorrelation between rapid shortening-induced exhumation and arc magmatism.

Carrapa, B., and DeCelles, P.G., 2015, Regional exhumation and kinematic history of the central Andes in response to cyclical orogenic processes, *in* DeCelles, P.G., Ducea, M.N., Carrapa, B., and Kapp, P.A., eds., Geodynamics of a Cordilleran Orogenic System: The Central Andes of Argentina and Northern Chile: Geological Society of America Memoir 212, p. 201–213, doi:10.1130/2015.1212(11). For permission to copy, contact editing@geosociety.org.

INTRODUCTION

The central Andes contain the type example of an active continental margin magmatic arc and the world's largest noncollisional orogenic plateau (Fig. 1; Isacks, 1988). Processes responsible for the construction of the central Andes include crustal thickening in response to horizontal shortening associated with the formation of a retroarc orogenic system, magmatic processes associated with the volcanic arc, erosional processes, which are partially responsible for redistributing mass within the orogenic wedge, and geodynamic processes associated with the subducting Nazca plate oceanic slab and the overriding South American plate (e.g., Jordan et al., 1983; Isacks, 1988; Kay et al., 1994; Allmendinger et al., 1997; Kley and Monaldi, 1998; Beck and Zandt, 2002; Schurr et al., 2006; Oncken et al., 2006; Strecker et al., 2007; Pelletier et

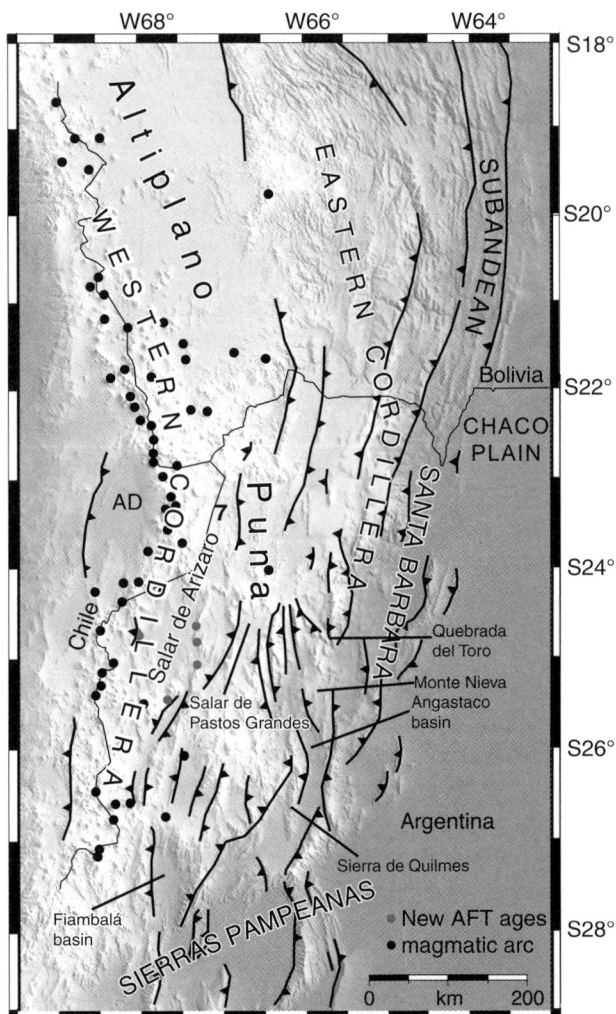

Figure 1. Shaded regional topography of the central Andes of northern Argentina and southern Bolivia, showing tectonomorphic zones (labeled), thrust faults (barbed lines), and volcanos (circles) (modified after Sobel and Strecker, 2003). AD—Atacama Desert; AFT—apatite fission track.

al., 2010). All of these processes have been proposed to be connected in a cyclical manner in cordilleran-type orogenic systems such as the North America Cordillera and the Andes (DeCelles et al., 2009). In this model, melt-fertile continental (foreland) lower crust and mantle lithosphere that are underthrusted beneath the arc melt to produce abnormally high magmatic flux, which differentiates into relatively felsic and mafic eclogitic components (Ducea and Barton, 2007). Isotopic and trace-element trends toward more-evolved magma compositions in the arc accompany the high-flux events and are interpreted to reflect melting of increasingly crustal components (Ducea, 2001). Metamorphism of overthickened crust also produces eclogite beneath the hinterland. Critical taper theory (e.g., Davis et al., 1983) suggests that during the formation of high-density roots, the orogenic wedge will be forced into a subcritical state in response to slightly decreased hinterland elevation; thus, the wedge will deform internally in order to increase taper. When the eclogitic roots reach critical mass, they may form convective instabilities that are prone to delamination or dripping into the asthenosphere (Kay and Kay, 1993; Ducea and Saleeby, 1998; Ducea, 2001; Zandt et al., 2004). This process of mass removal both relieves the accommodation space problem beneath the arc (e.g., McQuarrie et al., 2005) and causes isostatic uplift of the surface, potentially driving the orogenic wedge into a supercritical state (Garzione et al., 2006; DeCelles et al., 2009). Although the magnitude of such uplift is probably low (between 500 and 1000 m; Pelletier et al., 2010), and most of the elevation in the Puna Plateau region was achieved by Eocene–Oligocene time (Quade et al., this volume; Canavan et al., 2014), such small changes in surface uplift are enough to modify the state of stress within the orogenic wedge. In response to the excess gravitational potential energy in the uplifting orogenic wedge, the wedge will propagate forward in order to reduce its overall taper and regain a minimum-work state (Dahlen et al., 1984). Thus, the rate of magma production in the arc and the rate of crustal shortening of the orogenic wedge are anticorrelated, and the system oscillates between supercritical and subcritical states. Although critical taper theory has been formulated to explain thrust-belt–scale kinematics using a variety of different rheological parameterizations, the overall concepts of the original model are independent of scale and should apply at the orogenic belt scale (e.g., Willett et al., 1993).

In this paper, we use low-temperature thermochronology (primarily apatite fission-track [AFT] ages and, in places, apatite (U-Th)/He ages) as a proxy for the timing of exhumation and explore the spatio-temporal relationships among exhumation, deformation, and basin evolution with the goal of testing predictions made by the cordilleran cycle model (DeCelles et al., 2009). The study area in the central Andes of northwestern Argentina and Chile (Fig. 1) exhibits abundant evidence of lithospheric removal based on geophysical, petrological, and neotectonic studies (Kay et al., 1994; Schurr et al., 2006; Tassara et al., 2006; Drew et al., 2009; Ducea et al., 2013; Schoenbohm and Carrapa, this volume), and a rich record of arc magmatism has been documented in numerous studies (e.g., Haschke et al., 2002, 2006; Trumbull et al., 2006; Kay and Coira, 2009). A previous thermochronological compilation highlighted the relationships among tectonics, climate, erosion, and geomorphology (Strecker et al., 2007) but did not attempt correlations between exhumation and lithospheric-scale processes. Key questions that we address with this study are: (1) Was regional exhumation (and the kinematic activity for which we accept it as a proxy) steady or unsteady during the growth of the central Andes? (2) What

linkages exist among exhumation, kinematics, magmatic activity in the arc, and other potentially controlling processes? (3) Are trends in exhumation and deformation consistent with predictions of the cordilleran cycle model?

We compile previously published and new low-temperature thermochronological bedrock data from the Coastal Cordillera, Cordillera de Domeyko, Puna Plateau, Altiplano, and Eastern

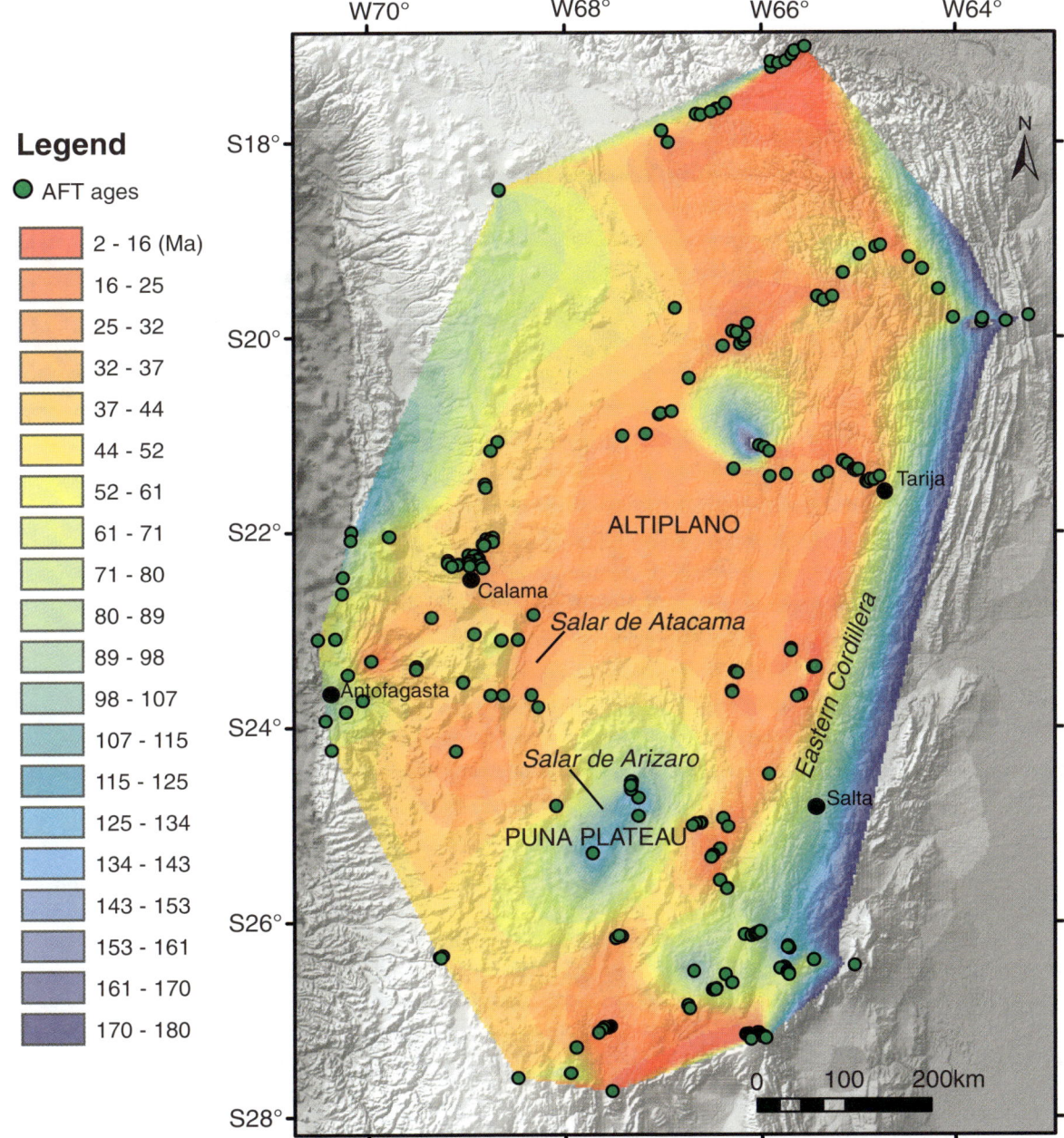

Legend

● AFT ages

■	2 - 16 (Ma)
■	16 - 25
■	25 - 32
■	32 - 37
■	37 - 44
■	44 - 52
■	52 - 61
■	61 - 71
■	71 - 80
■	80 - 89
■	89 - 98
■	98 - 107
■	107 - 115
■	115 - 125
■	125 - 134
■	134 - 143
■	143 - 153
■	153 - 161
■	161 - 170
■	170 - 180

Figure 2. Isochron map of the Altiplano, Puna, and northern Sierras Pampeanas based on data in Table DR1 (see text footnote 1). AFT—apatite fission track.

[1]GSA Data Repository Item 2015008, Analytical details of apatite fission-track thermochronology and thermal modeling, is available at www.geosociety.org/pubs/ft2015.htm, or on request from editing@geosociety.org or Documents Secretary, GSA, P.O. Box 9140, Boulder, CO 80301-9140, USA.

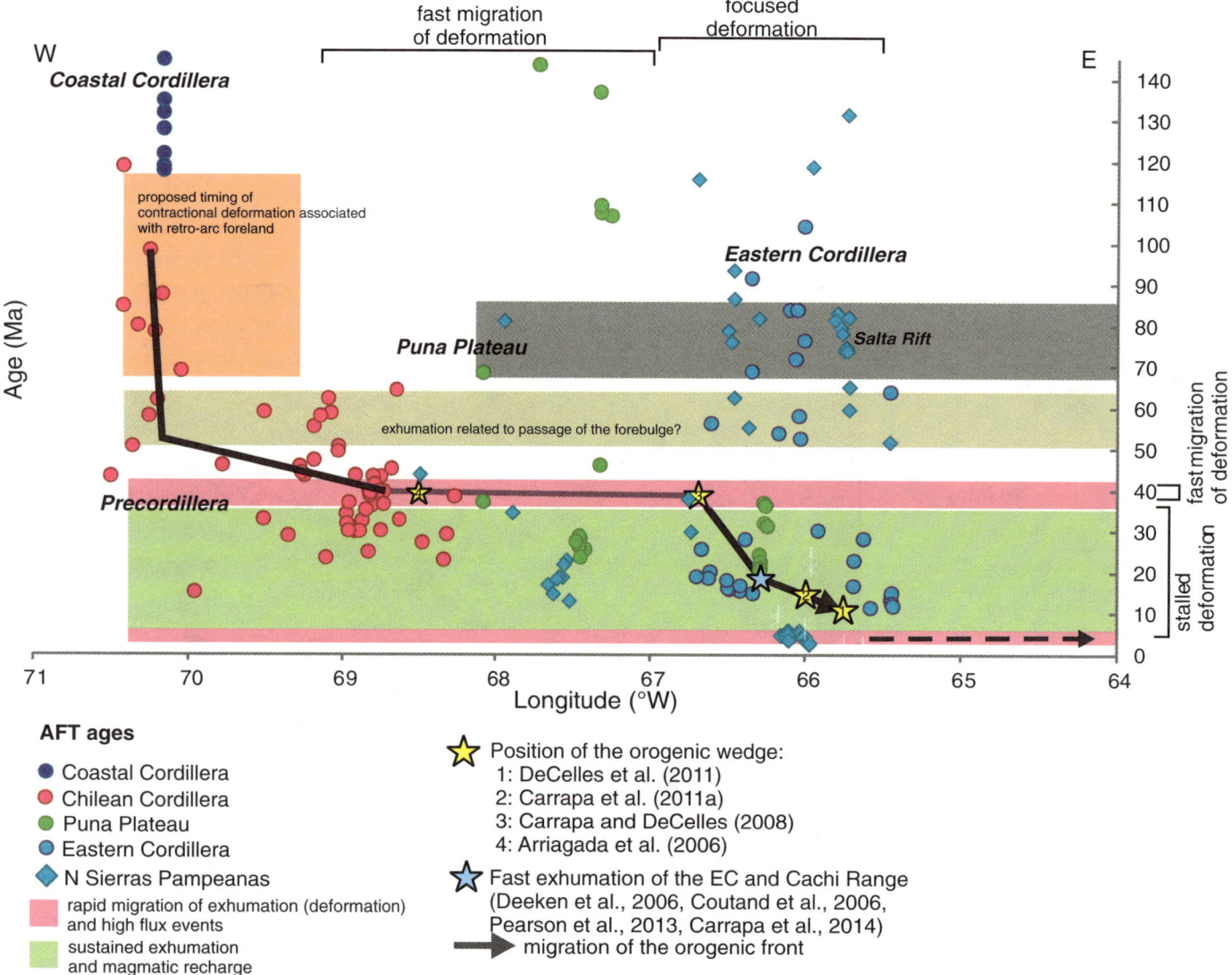

Figure 3. Spatio-temporal evolution of exhumation based on apatite fission-track (AFT) data presented in Table DR1 (see text footnote 1) and discussed in the text. EC—Eastern Cordillera.

Cordillera of northern Chile, northwestern Argentina, and Bolivia. Although we present data from the Altiplano as a general comparison (Fig. 2), we focus our discussion on the region in northwestern Argentina between ~21°S and 28°S.

Given the relatively low magnitude of Cenozoic maximum exhumation (6–8 km) in the Andes (Deeken et al., 2006; Strecker et al., 2007; Insel et al., 2012), AFT thermochronology is the ideal technique for recording Cenozoic and older exhumation signals. (U-Th)/He ages on zircons are Mesozoic and older, and (U-Th)/He ages on apatite are often partially reset, especially within the Puna Plateau (Carrapa et al., 2009; Reiners et al., this volume), and/or show large age scatter and low reproducibility in places such as the Salta rift province (e.g., Carrapa et al., 2014). On the other hand, apatite (U-Th)/He ages have been particularly useful in tracking the most recent exhumation signal within the

Eastern Cordillera associated with propagation of the deformation front through a region previously occupied by a thick regional foreland basin (Carrapa et al., 2012). In this paper, we include data from the northern Sierras Pampeanas, which we correlate with the Eastern Cordillera (Safipour et al., this volume), in our general compilation (Fig. 2), but we exclude them from the age versus longitude presentation in Figure 3 and from Figure 4.

EXHUMATION OF THE CENTRAL ANDES

Evidence of contractional deformation since the Cretaceous has been documented to the west of the Salar de Atacama in Chile (Mpodozis et al., 2005; Arriagada et al., 2006). Syndepositional deformation in Lower Cenozoic rocks of the Naranja

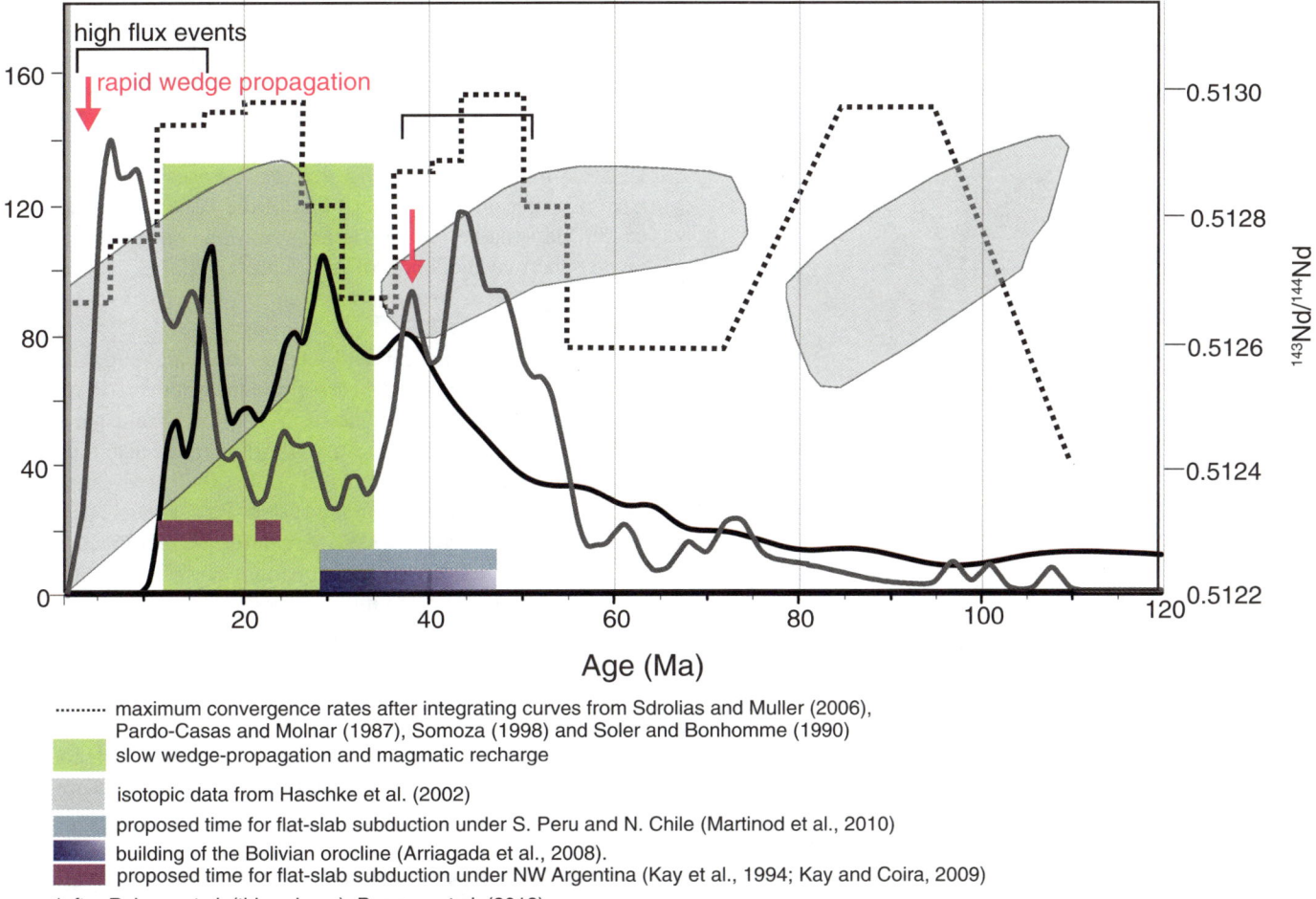

Puna Plateau-Eastern Cordillera, NW Argentina
AFT bedrock ages (n=124)
A He bedrock ages (n=245)*

.... maximum convergence rates after integrating curves from Sdrolias and Muller (2006), Pardo-Casas and Molnar (1987), Somoza (1998) and Soler and Bonhomme (1990)

slow wedge-propagation and magmatic recharge

isotopic data from Haschke et al. (2002)

proposed time for flat-slab subduction under S. Peru and N. Chile (Martinod et al., 2010)

building of the Bolivian orocline (Arriagada et al., 2008).

proposed time for flat-slab subduction under NW Argentina (Kay et al., 1994; Kay and Coira, 2009)

*after Reiners et al. (this volume), Pearson et al. (2013).
For detrital samples, only ages younger than depositional ages are here considered.

Figure 4. Probability density curves of apatite fission-track (AFT) and apatite (U-Th)/He ages (AHe; Tables DR1 and DR2 [see text footnote 1]) correlated with isotopic lulls (Haschke et al., 2002) and convergence rates (compiled from Sdrolias and Müller, 2006; Pardo-Casas and Molnar, 1987; Somoza, 1998; Soler and Bonhomme, 1990). High-flux events are constrained using the age of volcanic rocks after Trumbull et al. (2006). Only samples from NW Argentina are included here (Table DR1).

Formation suggests that this area was the site of active tectonic shortening during Cretaceous and Paleocene time (Arriagada et al., 2006). AFT data from the Cordillera de Domeyko (Maskaev and Zentilli, 1999; Andriessen and Reutter, 1994) and from the Coastal Cordillera (Juez-Larré et al., 2010) in Chile support active exhumation and deformation in the region since the Cretaceous (Fig. 2).

New AFT ages from six samples from the Salar de Arizaro region (Table 1) are Jurassic to Eocene, indicating limited exhumation in the region during the Cenozoic. An Eocene AFT age (37.3 ± 4.8 Ma) from the hanging wall of a thrust to the west of the Arizaro Basin (Fig. 2) suggests moderate local exhumation and deformation at that time. A granite cobble from Miocene synoro-

genic conglomerates (Schoenbohm and Carrapa, this volume) in the footwall of the same thrust has an AFT age of 68.7 ± 4.6 Ma (sample A09-FT1). This age, together with other Cretaceous and older ages (Table 1), suggests that although some exhumation occurred during the Cretaceous, most of the region was characterized by limited exhumation in the Cenozoic, in agreement with the tectono-thermal history of samples east of the Salar de Arizaro (DeCelles et al., this volume). This also indicates that localized lithospheric removal in the region (Schoenbohm and Carrapa, this volume) was associated with limited exhumation and uplift (Canavan et al., 2014). Limited exhumation of the Puna Plateau in the Cenozoic is in agreement with a high-elevation, dry plateau since ca. 36 Ma (Canavan et al., 2014).

TABLE 1. APATITE FISSION-TRACK (AFT) DATA FROM ARIZARO

Sample number	Lithology sample	Lat, long (degrees, decimal minutes)	No. Xls*	ρ_S (×10^5)†	$N_S^§$	ρ_i (×10^5)†	$N_i^§$	P(χ)^2 (%)#	ρ_D (×10^5)**	N_D††	Age (Ma)	±1σ	U (ppm)	Mean length (μm)
A09-FT1	Granitic clast in Miocene volcaniclastics	24°48.188'S, 68°5.228'W	25	8.661	676	36.47	2847	25.2	15.9	4082	68.7	4.6	28.97	13.7 ± 1
A09-FT2	Granitic basement	24°48.188'S, 68°5.228'W	20	1.274	81	9.77	621	65	15.6	4082	37.3	4.8	9.32	NA
ARB09-15	Granitic basement	24°43.430'S, 67°15.208'W	20	6.39	352	9.476	522	98.5	14.1	4082	173	14.7	8.19	NA
ARB09-13	Granitic basement	24°54.706'S, 67°15.238'W	20	3.231	108	7.928	265	82.9	14.4	4082	106	13.4	7.07	NA
A09-gr1	Granitic basement	25°17.522'S, 67°43.420'W	20	3.655	275	7.097	534	75.6	15.4	4082	143.7	13	5.8	NA
Mac 1	Granitic basement	24°38.197'S, 67°19.641'W	19	7.982	265	14.639	486	99.9	13.8	2320	137	12.2	13.25	NA

Note: Samples were analyzed with an Olympus microscope with drawing tube located above a digitizing tablet and a Kinetek computer-controlled stage driven by the FTStage program (Dumitru, 1993). Analysis performed with reflected and transmitted light at 1600× magnification. Samples were irradiated at Oregon State University. Samples where etched in 5.5 M nitric acid at 21 °C for 20 s. Following irradiation, the mica external detectors were etched with 21 °C in 40% hydrofluoric acid for 45 min. The pooled age is reported for all samples as they pass the χ^2 test, suggesting that they represent a single population. Error is 1σ, calculated using the zeta calibration method (Hurford and Green, 1983) with zeta of 366.4 ± 17.3 for apatite (Carrapa, 2011, personal observ.).

*No. Xls—the number of individual crystals dated.

†ρ_S and ρ_i are the spontaneous and induced track density measured, respectively (tracks/cm^2).

§N_S and N_i are the number of spontaneous and induced tracks counted, respectively.

#P(χ)^2 (%) is the χ^2 probability (Hurford and Green, 1983). Values greater than 5% are considered to pass this test and represent a single population of ages.

**ρ_D—the induced track density in external detector adjacent to CN5 dosimetry glass (tracks/cm^2).

††N_D—the number of tracks counted in determining ρ_D.

The Upper Eocene Geste Formation in the Salar de Pastos Grandes area of the eastern Puna (Fig. 1) contains proximal alluvial-fan deposits within a contractional growth structure (Carrapa and DeCelles, 2008). Detrital AFT ages from tripledated (U-Pb, AFT, and apatite [U-Th]/He) Paleozoic apatite grains in the Geste Formation indicate rapid Eocene exhumation of local western source terranes (Carrapa et al., 2009). Together, these observations indicate that the Geste Formation in this area was deposited in the proximal wedge-top portion of the foreland basin system, implying that the deformation front must have migrated into the eastern part of the Puna by no later than ca. 38 Ma. Apatite (U-Th)/He ages from the same deposits and inverse thermal modeling of AFT and (U-Th)/He data indicate Miocene–Pliocene exhumation related to subsequent basin incision (Carrapa et al., 2009).

Eocene deformation and exhumation are supported by ca. 38 Ma AFT ages recovered from the southern Puna margin (Coutand et al., 2001) and by 24–35 Ma K-feldspar multi–diffusion domain modeling of the Aguilar granite near the northeastern margin of the Puna (Insel et al., 2012). AFT ages from crystalline basement rocks in the eastern Puna and western Eastern Cordillera in northwestern Argentina (e.g., Luracatao and Cachi Ranges) indicate that the area was actively exhuming and deforming during early to middle Miocene time (Deeken, 2005; Pearson et al., 2012). Detrital AFT data from Miocene strata preserved in the Angastaco and Monte Nieva areas (Fig. 1) support rapid exhumation of the Eastern Cordillera during the Miocene (Coutand et al., 2006; DeCelles et al., 2011). An intraformational angular unconformity in ca. 14 Ma alluvial-fan deposits of the Angastaco Formation, derived from proximal Eastern Cordillera sources, indicates that this area was the site of wedge-top deposition and uplift by middle Miocene time (Carrapa et al., 2011, 2014b). This is corroborated by apatite (U-Th)/He data from the region between the Calchaquí Valley and the Rio de las Conchas to the east; combined with crosscutting relationships and the evidence of folding of Miocene–Pliocene strata, this indicates that the Eastern Cordillera was the site of active deformation between ca. 14 Ma and ca. 4 Ma (Carrera and Muñoz, 2008; Carrapa et al., 2014b; Pearson et al., 2013). Uplift of the Eastern Cordillera was associated with deformation and crustal thickening (Carrapa et al., 2014b). Growth strata in ca. 11–9 Ma alluvial-fan deposits in the Quebrada del Toro (Fig. 1) indicate active deformation and wedge-top deposition at this time (Mazzuoli et al., 2008; DeCelles et al., 2011). New paleoaltimetry data indicate that the Eastern Cordillera was uplifting between ca. 14 and 7 Ma (Carrapa et al., 2014a). Although the Eastern Cordillera was actively deforming during the Miocene and Pliocene, Cretaceous AFT and (U-Th)/He ages indicate limited Cenozoic exhumation of Cretaceous paleorift flanks and suggest significant paleotopography inherited from the Salta rift (Carrapa et al., 2014).

A few Cretaceous and older AFT ages, in addition to partially reset (U-Th)/He apatite ages within the Puna Plateau (Deeken et al., 2006; Carrapa et al., 2009; Reiners et al., this volume), indicate overall limited exhumation since the Cretaceous (<4 km, assuming a paleogeothermal gradient of 30 °C/km).

In particular, the presence of extensive Eocene evaporites in the Puna Plateau, along with stable isotopic evidence for coeval high paleoelevation, suggests that high and dry conditions limited the magnitude of erosion in the Puna during this time frame. These data also imply that the Puna region was never buried by a thick foreland basin, unlike parts of the Eastern Cordillera that were buried beneath thick Paleocene–Miocene foreland basin deposits (Starck and Vergani, 1996; DeCelles et al., 2011; Carrapa et al., 2012, 2014b) and subsequently incised during the late Miocene–Pliocene (Carrapa et al., 2012).

REGIONAL PATTERNS IN EXHUMATION AND KINEMATIC HISTORY

Figures 2 and 4 show the spatial-temporal patterns of cooling and exhumation based on AFT and apatite (U-Th)/He ages from the entire central Andean orogenic belt. We include data from the northern Sierras Pampeanas, which we correlate with the Eastern Cordillera (Safipour et al., this volume), in our general compilation (Fig. 2).

Cooling ages shown in Figure 3 are clustered into four major time-space groups: Cretaceous ages are abundant in the Chilean Coastal Cordillera and Precordillera (including the Cordillera de Domeyko) on the west, and in the Eastern Cordillera and frontal ranges to the east. Paleocene–Oligocene ages are abundant in the western Puna Plateau and Chilean Cordillera de Domeyko, and latest Eocene–Pliocene ages are abundant in the Eastern Cordillera. Each of these clusters represents a different tectonic-exhumation signal; correct interpretation depends strongly on local structural geological context and ancillary data sets that clarify each local situation. The gray line in Figure 3 tracks the orogenic strain front, which is based on the location of dated growth structures and wedge-top deposits discussed earlier and is essential for interpreting the meaning of the various cooling age signals.

The Cretaceous ages in the Chilean Precordillera may be reasonably interpreted to represent exhumation and cooling in response to shortening, crustal thickening, and development of orogenic topography during growth of the early Andes (Mpodozis et al., 2005; Arriagada et al., 2006). These ages are likely related to uplift and erosion in the nascent western thrust belt, which was confined to Chile. Contemporaneously, the Salta rift was active in Argentina to the east (Salfity and Marquillas, 1994). Uplift and exhumation of rift footwall blocks produced widespread Cretaceous cooling ages in the Eastern Cordillera and Santa Bárbara Ranges (Carrapa et al., 2014). By Paleocene time, the Salta rift was kinematically inactive and had been mostly buried by postrift thermal sag deposits of the Balbuena Subgroup (Salfity and Marquillas, 1994; Marquillas, 2005; DeCelles et al., 2011). Paleocene–Oligocene ages in the Puna Plateau reflect local uplift and exhumation in contractional structures associated with the eastward propagation of deformation into the eastern Puna region (Carrapa and DeCelles, 2008; Carrapa et al., 2009).

The Eocene–Pliocene ages are concentrated in the Eastern Cordillera and represent local structural growth; these ages track the eastward-migrating orogenic strain front, as determined mainly from growth structures in proximal wedge-top foreland basin deposits (Fig. 4). A fifth set of ages, in the ca. 68–47 Ma range and scattered over a broad region from the Eastern Cordillera to the Chilean Cordillera, with a few ages in the Puna Plateau, is not readily associated with the Salta rift or with local shortening in the thrust belt. These ages may result from partial annealing of apatites that experienced cooling during the Cretaceous rifting episode. Alternatively, these ages may be due to minor erosion associated with the passage through the region of the flexural forebulge, for which independent sedimentological evidence has been documented (DeCelles et al., 2011). We would expect no more than a few hundred meters of rock uplift in this case, so if this is indeed the cause of the Paleocene–early Eocene ages in the hinterland region, the rocks in which the ages are recorded must have been close to the top of the partial annealing zone before the flexural wave passed through the region. Together with evidence of wedge-top deposition in the same region at ca. 38 Ma (Carrapa and DeCelles, 2008), the forebulge cooling hypothesis would suggest an ~12–27 m.y. lag time between the migration of the forebulge and the arrival of the orogenic front. The forebulge was on the order of 300–400 km wide in this region (DeCelles et al., 2011). If the forebulge passed completely through this region between 65 Ma and 50 Ma, then its rate of migration would have been ~20–26 mm/yr.

We exclude Miocene–Pliocene AFT ages in the northern Sierras Pampeanas from our discussion of Figure 3 (Carrapa et al., 2006; Safipour et al., this volume) because the front of the orogenic belt swings southwestward in the transition zone to the Sierras Pampeanas such that a plot of age versus longitude becomes distorted. Nevertheless, these data are consistent with continued eastward propagation of the orogenic strain front (Fig. 4). Farther north at the latitude range of our main study area, orogenic strain migrated into the frontal Santa Bárbara Ranges during Pliocene time (Reynolds et al., 2000; Carrapa et al., 2011), and the present-day focus of seismic activity and neotectonic deformation is located in this region (Kendrick et al., 2006; Ramos et al., 2006; Metcalf and Kapp, this volume). Older (Cretaceous) cooling ages in the Sierras Pampeanas and the Bolivian Subandes (Fig. 2) may represent areas that have only recently (Miocene–Pliocene and younger) been uplifted and eroded, such that the present level of incision is insufficient to expose thermally reset rocks. Cretaceous ages in the Eastern Cordillera of NW Argentina record limited erosion of paleorift flanks that were only partially buried by Cenozoic foreland basin deposits and later exhumed (e.g., Carrapa et al., 2014).

Although the overall pattern of AFT and (U-Th)/He cooling ages has been approximated by a roughly 10 m.y. per degree of longitude trend of eastward-decreasing ages (Reiners et al., this volume), we note that an abrupt eastward shift in the timing of rapid exhumation during late Eocene time (ca. 40 Ma) is more consistent with structural and sedimentological constraints. In particular, Figure 3 highlights the rapid migration of exhumation mainly based on well-dated growth structures, which demonstrate an abrupt eastward jump of the orogenic strain front from the Chilean Cordillera de Domeyko into the eastern Puna region (Carrapa and DeCelles, 2008; tracked by the gray line in Fig. 3).

The large scatter of ages between ca. 40 and 5 Ma in the Eastern Cordillera (Figs. 3 and 4) is the result of exhumation slowly migrating through the region from latest Eocene through Miocene–Pliocene time, as demonstrated by growth structures (Carrera and Muñoz, 2008; Carrapa et al., 2011; DeCelles et al., 2011). The position of the modern deformation front requires that deformation had to abruptly shift into the Santa Bárbara system after ca. 4 Ma. AFT ages from the Bolivian Eastern Cordillera and Subandean zone generally correlate with ages from northwestern Argentina and are consistent with this regional exhumation history (Barnes et al., 2013; Fig. 2).

NATURE OF THE CONTROL

Numerous papers have been published attempting to synthesize the many available data sets from the central Andes in order to discern the overarching controls on the history of crustal thickening and mountain building (e.g., Isacks, 1988; Allmendinger et al., 1997; Beck and Zandt, 2002; McQuarrie et al., 2005; Oncken et al., 2006; Schellart, 2008; Kay and Coira, 2009). This paper contributes to this evolving synthesis (e.g., Barnes et al., 2013) by providing a regional-scale thermochronological analysis of the central Andes within the context of known structural and sedimentological events. The summary in this volume by Reiners et al. provides additional data, especially apatite (U-Th)/He ages. When combined with data from geochronological/geochemical compilations of magmatic arc rocks (Trumbull et al., 2006; Haschke et al., 2002, 2006), the kinematic history outlined herein may be interpreted in terms of larger controlling factors such as climate and tectonics.

Exhumation and cooling require the removal of overburden, either by tectonic processes (e.g., normal faulting) or by erosional surficial processes. Inasmuch as large-scale normal faults in the central Andes are not documented outside of the Salta rift region, we set aside this process as the main control on Cenozoic regional exhumation. Erosional surficial processes generally require precipitation, and much has been written about the potential effects of climate-induced erosion (or the lack thereof) in the central Andes (e.g., Sobel and Strecker, 2003; Lamb and Davis, 2003; Strecker et al., 2007; Pelletier et al., 2010; Reiners et al., this volume). Much of this work has focused on the role of aridity in reducing the rate at which mass can be removed by flowing water in the chronically dry central Andes. Almost all (~80%) precipitation in the central Andes is derived from Atlantic easterlies; as moist easterly air masses are lifted against the ~3-km-high eastern rampart of the Andes during the summer monsoon, this moisture condenses and rains out (Zhou and Lau, 1998; Strecker et al., 2007). Easterly air masses are largely

depleted of moisture upon arrival in the Eastern Cordillera and Puna Plateau, and the western flank of the central Andes is one of Earth's driest regions. The modern climate-orography setup in the central Andes provides an attractive modern analog for interpreting past erosional processes in the orogen. In this view, expressed eloquently by Sobel and Strecker (2003), the steep eastern orogenic front both creates and is in turn controlled by the climatic pattern: Orographic growth sets up the monsoon front, which in turn erosionally advects rocks toward the surface, confines active tectonic uplift to the region of rapid erosion, and further enhances the rain shadow such that hinterland regions are unable to export mass. Over time, orogenic taper becomes supercritical in response to the inability of surficial processes to export mass from the hinterland; new thrust faults become active in the foreland (i.e., the orogenic wedge propagates forward), and the cycle of enhanced erosion and advection repeats, thereby enlarging the high-elevation hinterland plateau.

If the model proposed by Sobel and Strecker (2003) is operating in the central Andes, three observable phenomena—the zone of rapid erosion (as proxied by rapid bedrock cooling rates), the region of high elevation, and the region of arid climate—should track the orogenic front through time. In detail, the zone of rapid erosion should be coupled with the orogenic strain front through time, and the orogenic hinterland region should gain elevation and aridity progressively eastward through time. The relationships between cooling ages and the timing of growth structures and wedge-top deposition demonstrate a tight linkage among exhumation, active deformation, and the position of the orogenic strain front (Fig. 4).

Aridity was established by early Cenozoic time in the central Andes of northern Chile and northwestern Argentina (Strecker et al., 2007; Quade et al., this volume). Evaporites and alluvial-fan deposits have been widespread in Chile and northwestern Argentina since the Paleocene (Hartley et al., 1992) and Eocene (Jordan and Mpodozis, 2006; Adelmann, 2001; Quade et al., this volume). Eocene–Miocene alluvial-fan and eolian deposits of the Vizcachera, Quebrada de los Colorados, and Angastaco Formations (Starck and Anzótegui, 2001; Carrapa et al., 2012; DeCelles et al., 2011) in the Puna Plateau and Eastern Cordillera of northwestern Argentina also indicate at least episodically dry climate. However, the dune field deposits preserved in the Quebrada de los Colorados and Angastaco Formations developed at relatively low elevations in the regional foreland basin (Carrapa et al., 2012; DeCelles et al., 2011), whereas those in the Vizcachera Formation formed at high elevation in the Puna Plateau (DeCelles et al., this volume; Quade et al., this volume). This suggests that, at least during the early Miocene, aridification in the central Andes may have been controlled by larger-scale climatic processes, rather than local tectonic orography (Starck and Anzótegui, 2001). More humid climatic conditions have been proposed in basins around the eastern perimeter of the Puna Plateau during the late Miocene (since ca. 7 Ma; Kleinter and Strecker, 2001). Sustained aridity was established at ca. 5 Ma in the Angastaco area within the Eastern Cordillera

(Bywater-Reyes et al., 2010). Recent stable isotope paleoaltimetry studies in the central Andes suggest that high elevations were achieved in the Puna Plateau region by the late Eocene (Canavan et al., 2014; Quade et al., this volume). Dry climate and high elevation in the central Andes would predict limited exhumation since the Eocene within the Puna and since at least the late Miocene–Pliocene within the Eastern Cordillera; these observations are consistent with the AFT ages (Fig. 4) and with data from Sobel and Strecker (2003). However, the unsteady eastward advance of the orogenic strain front cannot be explained simply as a response to steady-state orographic, precipitation-induced erosion along the eastern flank of the orogenic belt, nor is the kinematic history consistent with erosional pacing of kinematics (Sobel and Strecker, 2003). Rather, a more profound tectonic process seems to be at work here. To be clear, we find that the linkage between rapid erosion and local tectonic activity along the orogenic strain front (as envisioned by Sobel and Strecker, 2003) is strong, but that rapid erosion probably did not control the position of the strain front. Instead, we suggest that orography was first controlled by tectonic processes, and in turn possibly controlled local rainout-driven rapid exhumation. The question becomes: What controlled the regional kinematic history? Alternative controls could include magmatism in the arc and back-arc regions (Haschke et al., 2002), changing angles of Nazca plate subduction (e.g., Kay and Coira, 2009), along-strike variations in horizontal shortening related to oroclinal bending (Arriagada et al., 2008), and changes in the rate of plate convergence (Pardo-Casas and Molnar, 1987). It is understood that these are not necessarily independent processes (e.g., Martinod et al., 2010). In particular, Martinod et al. (2010) suggested that periods of flat slab subduction under southern Peru and northern Chile during Eocene–Oligocene times resulted in a stronger interplate coupling and in a concentration of shortening within the upper plate away from the trench. In Bolivia and Argentina, flat slab subduction has been suggested for the early Miocene (Oncken et al., 2006; Coira et al., 1993; Kay et al., 1994).

Figure 4 illustrates the frequency distributions of apatite (U-Th)/He and AFT ages for NW Argentina. Apatite (U-Th)/He ages cluster in the time intervals 50–38 Ma and 20–5 Ma, consistent with the timing of arc high-flux events documented by trends in $^{143}Nd/^{144}Nd$ and magmatic age data from Haschke et al. (2002) and Trumbull et al. (2006) (gray shaded areas in Fig. 4). This correlation may indicate that (U-Th)/He ages reflect thermal perturbations associated with the high-flux magmatic events rather than cooling associated with erosion. The AFT ages are less clustered than the apatite (U-Th)/He ages, but they do show a broad peak during the interval 40–15 Ma. The AFT ages are generally anticorrelated with the times of magmatic high-flux events. A relative lull in magmatic activity took place between ca. 35 Ma and 15 Ma (Fig. 4; Haschke et al., 2002; Trumbull et al., 2006), broadly coincident with the peak in AFT ages. As discussed already, the late Eocene AFT ages are mainly associated with exhumation in the Chilean Coastal Cordillera and Precordillera (including the Cordillera de Domeyko) and the eastern part

of the Puna Plateau, whereas the Oligocene–Miocene ages are from samples in the Eastern Cordillera. In general, the AFT ages track the development and erosion of local structural features and indicate times of rapid erosion owing to local crustal shortening and thickening, which are also supported by structural, sedimentological, and provenance data sets (e.g., Carrapa and DeCelles, 2008; Carrapa et al., 2012, 2014b). Thus, crustal shortening was, in general, out of phase with magmatism. On the other hand, eastward jumps in the location of the orogenic strain front took place during late Eocene and Pliocene time, at the ends of the magmatic high-flux events.

Coira et al. (1993) and Kay et al. (1994) argued that a magmatic lull at the latitude of the Puna Plateau calls for flat subduction of the Nazca plate during the period 26–17 Ma. Yáñez et al. (2001) used plate reconstructions of the position of the Juan Fernandez aseismic ridge (which currently is associated with flat subduction outboard of the Sierras Pampeanas to the south of our study area) to argue for flat subduction during the interval ca. 20–9 Ma in the Puna region. Oncken et al. (2006) employed the flat subduction mechanism to explain rapid shortening in the Bolivian central Andes (19°S–22°S) as a response of the upper plate to increased coupling with the flat subducting lower plate. Although both timings (late Oligocene–early Miocene and early–middle Miocene) proposed for flat subduction are roughly coincident with the broad peak in AFT ages (Fig. 4), these ages are mainly confined to the Eastern Cordillera, which continues north for >500 km into Bolivia and has a similar kinematic (Oncken et al., 2006) and exhumation (Barnes et al., 2013) history over a much broader region than the proposed flat subduction region. We also note that the data assembled by Trumbull et al. (2006) do not show a deep magmatic gap over either of these time intervals; rather, the greatest gap in magmatic activity was ca. 32–26 Ma. Also, there is no evidence for a Sierras Pampeanas–style (Laramide-style) basement uplift event during either of the proposed flat subduction intervals. Although we do not discount the potential importance of flat subduction in the central Andes during middle Cenozoic time, we suggest that it was not the primary control on regional exhumation patterns.

The AFT ages are also not well correlated with the Nazca–South America plate convergence rate (Fig. 4), implying that the timing of crustal shortening (as proxied by the AFT ages) was not directly tied to rates of plate convergence. Shortening rates in the Bolivian sector of the central Andes show a similarly poor correlation with plate convergence rates (Elger et al., 2005; Oncken et al., 2006).

Given the shortcomings in alternative explanations for the regional exhumation signal in the central Andes, we turn to the Cordilleran cycle model as a means of integrating the history of magmatism, shortening, wedge propagation, and bedrock cooling ages. This model predicts that shortening and magmatism should be out of phase, with minimum rates of shortening correlating with periods of maximum magmatic productivity. This is postulated to result from the growing accommodation space problem as underthrusting foreland lithosphere continues to accumulate beneath the arc. Melting and eclogite production beneath the arc and thickened hinterland region are delayed by 10–20 m.y. (Ducea, 2001), such that the magmatic high-flux event lags behind crustal shortening and thickening. As the dense hinterland roots become unstable and begin to founder, isostatic rebound causes the topographic surface to rise and the orogenic strain front to propagate rapidly toward the foreland. Crustal shortening accelerates as the high-flux event comes to an end. Conceivably, the AFT ages would mainly record the portion of the cordilleran cycle during which crustal shortening was relatively rapid, during and after the waning phase of the Eocene high-flux event. We refer to this as the recharge stage of the cycle (DeCelles et al., this volume).

We distinguish between rates of shortening and orogenic wedge propagation as follows: Propagation is the rate at which the orogenic wedge grows lengthwise in the tectonic transport direction, whereas shortening is the rate at which crust within the orogenic wedge reduces in length (DeCelles and DeCelles, 2001). Propagation is tracked by the orogenic strain front. Shortening and propagation rates are not necessarily coupled, and they represent different behavioral aspects of orogenic wedges. The orogenic wedge may shorten internally in order to build taper, or the wedge may propagate forward in order to reduce taper (e.g., Willett et al., 1993; DeCelles and Mitra, 1995). Thus, wedge propagation and shortening may be out-of-phase responses to different stress conditions within the wedge. In the case of the central Andes, the major propagation events took place during late Eocene and latest Miocene–Pliocene time (Fig. 4), and the major episodes of shortening-related exhumation occurred during Paleocene–middle Eocene and latest Eocene–Miocene time (Fig. 4). We suggest that the major propagation events heralded episodes of orogenic supercriticality, whereas the periods of locally controlled exhumation and shortening represent periods of time during which the orogenic wedge was subcritically to critically tapered, resulting in concentrated (stalled) exhumation (Fig. 4). The late Eocene propagation event may have been related to oroclinal bending, which predicts changes in wedge behavior and shortening (Martinod et al., 2010; Arriagada et al., 2008).

The cordilleran cycle model links these different states of orogenic wedge behavior to the history of magmatism and the growth and removal of dense lower-crustal and mantle lithospheric instabilities. Magmatic high-flux events were fueled by crustal shortening after 10–20 m.y. lag times, and periods during which dense arc and hinterland roots delaminated or dripped into the mantle both set the stage for rapid eastward propagation of the orogenic strain front and reduced arc magmatic activity to low levels. In this hypothetical context, rapid exhumation recorded by AFT dates is largely coupled with shortening during periods of slow wedge propagation and magmatic recharge. Apatite (U-Th)/He ages seem instead to correlate with high-flux events and with rapid wedge propagation (Fig. 4).

SUMMARY AND CONCLUSIONS

Our compilation of AFT and apatite (U-Th)/He ages in the central Andes demonstrates that bedrock cooling across the entire orogenic belt took place in response to several different processes. Cretaceous–Paleocene cooling ages in a relatively narrow region in Chile represent exhumation in the nascent central Andean thrust belt. Cretaceous cooling ages are also widespread in the Eastern Cordillera and frontal Santa Bárbara Ranges and are explained as the effect of Salta rift–related exhumation (Carrapa et al., 2014). An abrupt eastward shift in the locus of rapid exhumation from the Chilean thrust belt into the Puna Plateau region during late Eocene time (ca. 40–38 Ma) accompanied the development of growth structures in the eastern Puna region and is consistent with the attainment of regionally high elevation (Canavan et al., 2014; Quade et al., this volume). From late Eocene through late Miocene time, the zone of rapid exhumation was confined almost exclusively to the Eastern Cordillera, where growth structures and the sedimentology and provenance of wedge-top deposits (Carrera and Muñoz, 2008; Bywater-Reyes et al., 2010; Carrapa et al., 2012, 2014b; DeCelles et al., 2011) demonstrate a close linkage between rapid exhumation and local crustal shortening. During latest Miocene–Pliocene time, the orogenic strain front once again migrated rapidly forward into the Santa Bárbara and Subandean Ranges (Reynolds et al., 2000; Echavarria et al., 2003), but the evidence for bedrock cooling associated with this kinematic event is not yet exposed at the surface.

Our analysis suggests a strong correlation between rapid bedrock cooling/exhumation and local crustal shortening confined to relatively narrow (<100-km-wide) zones in the orogenic wedge. On the other hand, periods of rapid orogenic wedge propagation (not to be confused with shortening) are associated with regionally diffuse AFT cooling patterns (Fig. 4). Although a close linkage between rapid erosion controlled by rainout from orographically lifted moisture-laden air masses and bedrock cooling ages is plausible, the pattern of eastward migration of clusters of cooling ages (along with structural and sedimentological indicators of local thrust faulting) is highly unsteady and indicates that orographic precipitation was likely controlled by the kinematic and uplift history of the orogen, rather than vice versa.

A comparison of the cooling age spectra with published data on magmatic activity and plate convergence rate shows strong temporal anticorrelations. Bedrock AFT cooling ages, associated with local shortening, are correlated with a lull in magmatic activity and overall reduced plate convergence rate (Fig. 4). We suggest that this pattern is consistent with a model for cordilleran orogenic systems in which crustal shortening feeds magmatic high-flux events, which in turn produce dense eclogitic residues and roots beneath the arc and the hinterland of the retroarc thrust belt. When these roots become unstable and begin to founder into the mantle, the orogenic wedge becomes supercritically tapered and propagates forward. Apatite (U-Th/He) ages seem to record cooling associated with deformation driven by this mechanism. Within a few million years after the propagation event, arc magmatism is strongly reduced as the lower crust and lithosphere underthrusting beneath the hinterland during the preceding period of shortening heats up in anticipation of the next high-flux magmatic event. This way of explaining the nonuniform, unsteady distribution of cooling ages in the central Andes is attractive because it does not require multiple episodes of flat subduction or unrealistic assumptions about the long-term kinematic behavior of the orogenic wedge.

ACKNOWLEDGMENTS

We thank Cesar Arriagada, Giovanni Vezzoli, and an anonymous reviewer for constructive criticisms. Funding for this project come from NSF and ExxonMobil.

REFERENCES CITED

Adelmann, D., 2001, Känozoische Beckenentwicklung in der südlichen Puna am Beispiel des Salar de Antofolla (NW-Argentinien): Berlin, Frei Universität Berlin, 195 p.

Allmendinger, R., Jordan, T., Kay, S., and Isacks, B., 1997, The evolution of the Altiplano-Puna Plateau of the central Andes: Annual Review of Earth and Planetary Sciences, v. 25, p. 139–174, doi:10.1146/annurev.earth.25.1.139.

Andriessen, P., and Reutter, K.J., 1994, K-Ar and fission track mineral age determination of igneous rocks related to multiple magmatic arc systems along 23°S latitude of Chile and NW Argentina, *in* Reutter, K.-J., Scheuber, E., and Wigger, P.J., eds., Tectonics of the Southern Central Andes: Berlin, Springer-Verlag, p. 141–153.

Arriagada, C., Cobbold, P.R., and Roperch, P., 2006, Salar de Atacama Basin: A record of compressional tectonics in the central Andes since the mid-Cretaceous: Tectonics, v. 25, TC1008, doi:10.1029/2004TC001770.

Arriagada, C., Roperch, P., Mpodozis, C., and Cobbold, P.R., 2008, Paleogene building of the Bolivian orocline: Tectonic restoration of the central Andes in 2-D map view: Tectonics, v. 27, TC6014, doi:10.1029/2008TC002269.

Barnes, J.B., Ehlers, T.A., Insel, N., McQuarrie, N., and Poulsen, C.J., 2013, Linking orography, climate, and exhumation across the central Andes: Geology, v. 40, no. 12, p. 1135–1138.

Beck, S.L., and Zandt, G., 2002, The nature of orogenic crust in the central Andes: Journal of Geophysical Research, v. 107, 2230, doi:10.1029/2000JB000124.

Bywater-Reyes, S., Carrapa, B., Clementz, M., and Schoenbohm, L., 2010, Effect of late Cenozoic aridification on sedimentation in the Eastern Cordillera of northwest Argentina (Angastaco basin): Geology, v. 38, p. 235–238, doi:10.1130/G30532.1.

Canavan, R., Carrapa, B., Clementz, M., Quade, J., DeCelles, P.G., and Schoenbohm, L., 2014, Early Cenozoic uplift of the Puna Plateau, central Andes, based on stable isotope paleoaltimetry of hydrated volcanic glass: Geology, v. 42, no. 5, p. 447–450, doi:10.1130/G35239.1.

Carrapa, B., and DeCelles, P.G., 2008, Eocene exhumation and basin development in the Puna of northwestern Argentina: Tectonics, v. 27, TC1015, doi:10.1029/2007TC002127.

Carrapa, B., Adelmann, D., Hilley, G., Mortimer, E., Strecker, M.R., and Sobel, E.R., 2005, Oligocene uplift, establishment of internal drainage and development of plateau morphology in the southern central Andes: Tectonics, v. 24, TC4011, doi:10.1029/2004TC001762.

Carrapa, B., DeCelles, P.G., Reiners, P.W., Gehrels, G.E., and Sudo, M., 2009, Apatite triple dating and white mica ^{40}Ar/^{39}Ar thermochronology of syntectonic detritus in the central Andes: A multiphase tectonothermal history: Geology, v. 37, p. 407–410, doi:10.1130/G25698A.1.

Carrapa, B., Trimble, J., and Stockli, D., 2011, Patterns and timing of exhumation and deformation in the Eastern Cordillera of NW Argentina revealed by (U-Th)/He thermochronology: Tectonics, v. 30, TC3003, doi:10.1029/2010TC002707.

Carrapa, B., Bywater-Reyes, S., DeCelles, P.G., Mortimer, E., and Gehrels, G., 2012, Eocene–Miocene synorogenic basin evolution in the Eastern Cordillera of northwestern Argentina (25°–26°S): Regional implications for Andean orogenic wedge development: Basin Research, v. 23, p. 1–20.

Carrapa, B., Bywater-Reyes, S., Safipour, R., Sobel, E., Schoenbohm, L., Reiners, P., and Stockli, D., 2014a, The effect of inherited paleotopography on exhumation of the central Andes of NW Argentina: Geological Society of America Bulletin, v. 126, no. 1–2, p. 66–77, doi:10.1130/B30844.1.

Carrapa, B., Huntington, K.H., Clementz, M., Bywater -Reyes, S., Quade, J., Schoenbohm, L., and Canavan, R., 2014b, Uplift of the Central Andes of NW Argentina associated with upper crustal shortening, revealed by multi-proxy isotopic analyses: Tectonics, v. 33, doi:10.1002/2013TC003461.

Carrera, N., and Muñoz, J.A., 2008, Thrusting evolution in the southern Cordillera Oriental (northern Argentine Andes): Constraints from growth strata: Tectonophysics, v. 459, p. 107–122, doi:10.1016/j.tecto.2007.11.068.

Coira, B., Kay, S.M., and Viramonte, J., 1993, Upper Cenozoic magmatic evolution of the Argentine Puna—A model for changing subduction geometry: International Geology Review, v. 35, p. 677–720, doi:10.1080/00206819309465552.

Coutand, I., Cobbold, P.R., de Urreiztieta, M., Gautier, P., Chauvin, A., Gapais, D., Rossello, E.A., and Lòpez-Gamundí, O., 2001, Style and history of Andean deformation, Puna Plateau, northwestern Argentina: Tectonics, v. 20, p. 210–234.

Coutand, I., Carrapa, B., Deeken, A., Schmitt, A.K., Sobel, E., and Strecker, M.R., 2006, Orogenic plateau formation and lateral growth of compressional basins and ranges: Insights from sandstone petrography and detrital apatite fission-track thermochronology in the Angastaco Basin, NW Argentina: Basin Research, v. 18, p. 1–26, doi:10.1111/j.1365-2117.2006.00283.x.

Dahlen, F., Suppe, J., and Davis, D., 1984, Mechanics of fold-and-thrust belts and accretionary wedges: Cohesive Coulomb theory: Journal of Geophysical Research–Solid Earth (1978–2012), v. 89, p. 10,087–10,101.

Davis, D., Suppe, J., and Dahlen, F., 1983, Mechanics of fold-and-thrust belts and accretionary wedges: Journal of Geophysical Research–Solid Earth (1978–2012), v. 88, p. 1153–1172.

DeCelles, P.G., and DeCelles, P.C., 2001, Rates of shortening, propagation, underthrusting, and flexural wave migration in continental orogenic systems: Geology, v. 29, no. 2, p. 135–138, doi:10.1130/0091-7613(2001)029<0135:ROSPUA>2.0.CO;2.

DeCelles, P.G., and Mitra, G., 1995, History of the Sevier orogenic wedge in terms of critical taper models, northeast Utah and southwest Wyoming: Geological Society of America Bulletin, v. 107, no. 4, p. 454–462, doi:10.1130/0016-7606(1995)107<0454:HOTSOW>2.3.CO;2.

DeCelles, P.G., Ducea, M., Zandt, G., and Kapp, P., 2009, Cyclicity in cordilleran orogenic systems: Nature Geoscience, v. 2, p. 251–257, doi:10.1038/ngeo469.

DeCelles, P.G., Carrapa, B., Horton, B., and Gehrels, G., 2011, Cenozoic foreland basin system in the central Andes of northwestern Argentina: Implications for Andean geodynamics and modes of deformation: Tectonics, v. 30, TC6013, doi:10.1029/2011TC002948.

DeCelles, P.G., Carrapa, B., Horton, B.K., McNabb, J., Gehrels, G.E., and Boyd, J., 2015, this volume, The Miocene Arizaro basin, central Andean hinterland: Response to partial lithosphere removal? *in* DeCelles, P.G., Ducea, M.N., Carrapa, B., and Kapp, P.A., eds., Geodynamics of a Cordilleran Orogenic System: The Central Andes of Argentina and Northern Chile: Geological Society of America Memoir 212, doi:10.1130/2015.1212(18).

Deeken, A., 2005, Construction of the Southern Eastern Cordillera, NW-Argentina: From Early Cretaceous Extension to Middle Miocene Shortening, Constrained by AFT-Thermochronometry [Diploma thesis]: Berlin, Freien Universität Berlin, Institut für Geologie, Geophysik und Geoinformatik, Fachrichtung Allgemeine Geologie, 84 p.

Deeken, A., Sobel, E.R., Coutand, I., Haschke, M., Riller, U., and Strecker, M.R., 2006, Development of the southern Eastern Cordillera, NW Argentina, constrained by apatite fission track thermochronology: From Early Cretaceous extension to middle Miocene shortening: Tectonics, v. 25, TC6003, doi:10.1029/2005TC001894.

Drew, S.T., Ducea, M.N., and Schoenbohm, L.M., 2009, Mafic volcanism on the Puna Plateau, NW Argentina: Implications for the lithospheric composition and evolution with an emphasis on lithospheric foundering: Lithosphere, v. 1, p. 305–318, doi:10.1130/L54.1.

Ducea, M.N., 2001, The California arc: Thick granitic batholiths, eclogitic residues, lithospheric-scale thrusting, and magmatic flare-ups: GSA Today, v. 11, p. 4–10, doi:10.1130/1052-5173(2001)011<0004:TCATGB>2.0.CO;2.

Ducea, M.N., and Barton, M.D., 2007, Igniting flare-up events in Cordilleran arcs: Geology, v. 35, p. 1047–1050, doi:10.1130/G23898A.1.

Ducea, M., and Saleeby, J., 1998, A case for delamination of the deep batho-lithic crust beneath the Sierra Nevada, California: International Geology Review, v. 40, p. 78–93, doi:10.1080/00206819809465199.

Ducea, M.N., Seclaman, A.C., Murray, K.E., Jianu, D., and Schoenbohm, L., 2013, Mantle-drip magmatism beneath the Antiplano-Puna Plateau, cen-tral Andes: Geology, v. 41, p. 915–918, doi:10.1130/G34509.1.

Dumitru, T., 1993, FT STAge Systems, described: Nuclear Tracks and Radia-tion Measurements, v. 21, p. 575–580.

Echavarria, L., Hernandez, R., Allmendinger, R., and Reynolds, J., 2003, Sub-andean thrust and fold belt of northwestern Argentina: Geometry and tim-ing of the Andean evolution: American Association of Petroleum Geology Bulletin, v. 87, no. 6, p. 965–985, doi:10.1306/01200300196.

Ege, H., Sobel, E.R., Scheuber, E., and Jacobshagen, V., 2007, Exhumation history of the southern Altiplano Plateau (southern Bolivia) constrained by apatite fission track thermochronology: Tectonics, v. 26, TC1004, doi:10.1029/2005TC001869.

Elger, K., Oncken, O., and Glodny, J., 2005, Plateau-style accumulation of deformation—The southern Altiplano: Tectonics, v. 24, TC4020, doi:10.1029/2004TC001675.

Garzione, C.N., Molnar, P., Libarkin, J.C., and MacFadden, B.J., 2006, Rapid late Miocene rise of the Bolivian Altiplano: Evidence for removal of man-tle lithosphere: Earth and Planetary Science Letters, v. 241, p. 543–556, doi:10.1016/j.epsl.2005.11.026.

Hartley, A.J., Flint, S., Turner, P., and Jolley, E.J., 1992, Tectonic controls on the development of a semi-arid, alluvial basin as reflected in the stra-tigraphy of the Purilactis Group (Upper Cretaceous–Eocene), north-ern Chile: Journal of South American Earth Sciences, v. 5, p. 275–296, doi:10.1016/0895-9811(92)90026-U.

Haschke, M., Siebel, W., Günther, A., and Scheuber, E., 2002, Repeated crustal thickening and recycling during the Andean orogeny in north Chile (21°–26°S): Journal of Geophysical Research, v. 107, no. B1, p. ECV 6-1–ECV 6-18, doi:10.1029/2001JB000328.

Haschke, M., Günther, A., Melnick, D., Echtler, H., Reutter, K.-J., Scheuber, E., and Oncken, O., 2006, Central and southern Andean tectonic evolution inferred from arc magmatism, *in* Oncken, O., Chong, G., Franz, G., Giese, P., Götze, H.-J., Ramos, V., Strecker, M., and Wigger, P., eds., The Andes—Active Sub-duction Orogeny: Berlin, Springer, Frontiers in Earth Sciences, p. 337–353.

Hurford, A.J., and Green, P.F., 1983, The zeta age calibration of fission track dating: Chemical Geology, v. 41, p. 285–317.

Insel, N., Grove, M., Haschke, M., Barnes, J., Schmitt, A., and Strecker, M., 2012, Paleozoic to early Cenozoic cooling and exhumation of the base-ment underlying the eastern Puna Plateau margin prior to plateau growth: Tectonics, v. 31, TC6006, doi:10.1029/2012TC003168.

Isacks, B., 1988, Uplift of the central Andean Plateau and bending of the Boliv-ian orokline: Journal of Geophysical Research, v. 93, p. 3211–3231, doi:10.1029/JB093iB04p03211.

Jordan, T.E., and Mpodozis, C., 2006, Estratigrafia y evolucion tectonics de la cuenca paleogena de Arizaro-Pocitos, Puna Occidental (24°–25°S), *in* XI Congreso Geologico Chileno, Volume 2: Antofagasta, Chile, p. 57–60.

Jordan, T.E., Isacks, B.L., Allmendinger, R.W., Brewer, J.A., Ramos, V.A., and Ando, C.J., 1983, Andean tectonics related to geometry of subducting Nazca plate: Geological Society of America Bulletin, v. 94, p. 341–361, doi:10.1130/0016-7606(1983)94<341:ATRTGO>2.0.CO;2.

Juez-Larré, J., Kukowski, N., Dunai, T.J., Hartley, A.J., and Andriessen, P.A., 2010, Thermal and exhumation history of the Coastal Cordillera arc of northern Chile revealed by thermochronological dating: Tectonophysics, v. 495, p. 48–66, doi:10.1016/j.tecto.2010.06.018.

Kay, R.W., and Kay, S.M., 1993, Delamination and delamination magma-tism: Tectonophysics, v. 219, no. 1–3, p. 177–189, doi:10.1016/0040-1951(93)90295-U.

Kay, S.M., and Coira, B.L., 2009, Shallowing and steepening subduction zones, continental lithospheric loss, magmatism, and crustal flow under the central Andean Altiplano-Puna Plateau, *in* Kay, S.M., Ramos, V.A., and Dickinson, W.R., eds., Backbone of the Americas: Shallow Subduction, Plateau Uplift, and Ridge and Terrane Collision: Geological Society of America Memoir 204, p. 229–259, doi:10.1130/2009.1204(11).

Kay, S.M., Coira, B., and Viramonte, J., 1994, Young mafic back-arc volcanic rocks as indicators of continental lithosphere delamination beneath the Argentine Puna Plateau, central Andes: Journal of Geophysical Research, v. 99, p. 24,323–24,339, doi:10.1029/94JB00896.

Kendrick, E., Brooks, B.A., Bevis, M., Smalley, R., Jr., Lauria, E., Araujo, M., and Parra, H., 2006, Active orogeny of the south-central Andes studied with GPS geodesy: Revista de la Asociación Geológica Argentina, v. 61, no. 4, p. 555–566.

Kleinter, K., and Strecker, M.R., 2001, Changes in moisture regime and ecology in response to late Cenozoic orographic barriers: The Santa Maria Valley, Argentina: Geological Society of America Bulletin, v. 113, p. 728–742.

Kley, J., and Monaldi, C.R., 1998, Tectonic shortening and crustal thickness in the central Andes: How good is the correlation?: Geology, v. 26, p. 723–726, doi:10.1130/0091-7613(1998)026<0723:TSACTI>2.3.CO;2.

Lamb, S., and Davis, P., 2003, Cenozoic climate change as a possible cause for the rise of the Andes: Nature, v. 425, no. 6960, p. 792–797, doi:10.1038/nature02049.

Löbens, S., Sobel, E.R., Bense, F.A., Wemmer, K., Dunkl, I., and Siegesmund, S., 2013, Refined exhumation history of the northern Sierras Pampeanas, Argentina: Tectonics, v. 32, p. 453–472, doi:10.1002/tect.20038.

Martinod, J., Husson, L., Roperch, P., Guillaume, B., and Espurt, N., 2010, Horizontal subduction zones, convergence velocity and the building of the Andes: Earth and Planetary Science Letters, v. 299, p. 299–309, doi:10.1016/j.epsl.2010.09.010.

Maskaev, V., and Zentilli, M., 1999, Fission track thermochronology of the Domeyko Cordillera, northern Chile: Implications for Andean tectonics and porphyry copper metallogenesis: Exploration and Mining Geology, v. 8, p. 65–89.

Mazzuoli, R., Vezzoli, L., Omarini, R., Acocella, V., Gioncada, A., Matteini, M., Dini, A., Guillou, H., Hauser, N., and Uttini, A., 2008, Miocene mag-matism and tectonics of the easternmost sector of the Calama–Olacapato–El Toro fault system in central Andes at ~24 S: Insights into the evolution of the Eastern Cordillera: Geological Society of America Bulletin, v. 120, p. 1493–1517, doi:10.1130/B26109.1.

McQuarrie, N., Horton, B.K., Zandt, G., Beck, S., and DeCelles, P.G., 2005, Lithospheric evolution of the Andean fold–thrust belt, Bolivia, and the origin of the central Andean plateau: Tectonophysics, v. 399, p. 15–37, doi:10.1016/j.tecto.2004.12.013.

Metcalf, K., and Kapp, P., 2015, this volume, Along-strike variations in crustal seismicity and modern lithospheric structure of the central Andean forearc, *in* DeCelles, P.G., Ducea, M.N., Carrapa, B., and Kapp, P.A., eds., Geodynamics of a Cordilleran Orogenic System: The Central Andes of Argentina and Northern Chile: Geological Society of America Memoir 212, doi:10.1130/2015.1212(04).

Mpodozis, C., Arriagada, C., Basso, M., Roperch, P., Cobbold, P.R., and Reich, M., 2005, Late Mesozoic to Paleogene stratigraphy of the Salar de Ata-cama Basin, Antofagasta, northern Chile: Implications for the tectonic evolution of the central Andes: Tectonophysics, v. 399, p. 125–154, doi:10.1016/j.tecto.2004.12.019.

Oncken, O., Hindle, D., Kley, J., Elger, K., Victor, P., and Schemmann, K., 2006, Deformation of the central Andean upper plate system—Facts, fic-tion, and constraints for plateau models, *in* Oncken, O., Chong, G., Franz, G., Giese, P., Götze, H.-J., Ramos, V., Strecker, M., and Wigger, P., eds., The Andes—Active Subduction Orogeny: Berlin, Springer, Frontiers in Earth Sciences, p. 3–27.

Pardo-Casas, F., and Molnar, P., 1987, Relative motion of the Nazca (Farallon) and South American plates since Late Cretaceous time: Tectonics, v. 6, p. 233–248, doi:10.1029/TC006i003p00233.

Pearson, D., Kapp, P., Reiners, P., Gehrels, G., Ducea, M., Pullen, A., Otamendi, J., and Alonso, R., 2012, Major Miocene exhumation by fault-propagation folding within a metamorphosed, early Paleozoic thrust belt: Northwest-ern Argentina: Tectonics, v. 31, TC4023, doi:10.1029/2011TC003043.

Pearson, D.M., Kapp, P., DeCelles, P.G., Reiners, P.W., Gehrels, G.E., Ducea, M.N., and Pullen, A., 2013, Influence of pre-Andean crustal structure on Cenozoic thrust belt kinematics and shortening magnitude: North-western Argentina: Geosphere, v. 9, no. 6, p. 1766–1782, doi:10.1130/GES00923.1.

Pelletier, J.D., DeCelles, P.G., and Zandt, G., 2010, Relationships among cli-mate, erosion, topography, and delamination in the Andes: A numeri-cal modeling investigation: Geology, v. 38, p. 259–262, doi:10.1130/G30755.1.

Quade, J., Dettinger, M.P., Carrapa, B., DeCelles, P., Murray, K.E., Hunting-ton, K.A., Cartwright, A., Canavan, R.R., Gehrels, G., and Clementz, M., 2015, this volume, The Growth of the Central Andes 22–26°S, *in* DeCelles, P.G., Ducea, M.N., Carrapa, B., and Kapp, P.A., eds., Geo-dynamics of a Cordilleran Orogenic System: The Central Andes of

Argentina and Northern Chile: Geological Society of America Memoir 212, doi:10.1130/2015.1212(15).

Ramos, V.A., Alonso, R.N., and Strecker, M., 2006, Estructura y neotectónica de Las Lomas de Olmedo, zona de transición entre los Sistemas Subandino y de Santa Bárbara, provincia de Salta: Revista de la Asociación Geológica Argentina, v. 61, no. 4, p. 579–588.

Reiners, P.W., Thomson, S.N., Vernon, A., Willett, S.D., Zattin, M., Einhorn, J., Gehrels, G., Quade, J., Pearson, D., Murray, K.E., and Cavazza, W., 2015, this volume, Low-temperature thermochronologic trends across the central Andes, 21°S–28°S, *in* DeCelles, P.G., Ducea, M.N., Carrapa, B., and Kapp, P.A., eds., Geodynamics of a Cordilleran Orogenic System: The Central Andes of Argentina and Northern Chile: Geological Society of America Memoir 212, doi:10.1130/2015.1212(12).

Reynolds, J.A., Galli, C.I., Hernández, R.M., Idleman, B.D., Kotila, J.M., Hilliard, R.V., and Naeser, C.W., 2000, Middle Miocene tectonic development of the Transition zone, Salta Province, northwest Argentina: Magnetic stratigraphy from the Metán Subgroup, Sierra de González: Geological Society of America Bulletin, v. 112, p. 1736–1751, doi:10.1130/0016-7606(2000)112<1736:MMTDOT>2.0.CO;2.

Safipour, R., Carrapa, C., DeCelles, P.G., and Thomson, S.N., 2015, this volume, Exhumation of the Precordillera and northern Sierras Pampeanas and along-strike correlation of the Andean orogenic front, northwestern Argentina, *in* DeCelles, P.G., Ducea, M.N., Carrapa, B., and Kapp, P.A., eds., Geodynamics of a Cordilleran Orogenic System: The Central Andes of Argentina and Northern Chile: Geological Society of America Memoir 212, doi:10.1130/2015.1212(10).

Salfity, J.A., and Marquillas, R.A., 1994, Tectonic and sedimentary evolution of the Cretaceous–Eocene Salta Group Basin, Argentina, *in* Salfity, J.A., ed., Cretaceous Tectonics of the Andes: Wiesbaden, Germany, Friedrich Vieweg & Sohn, Earth Evolution Sciences, p. 266–315.

Schellart, W.P., 2008, Overriding plate shortening and extension above subduction zones: A parametric study to explain formation of the Andes Mountains: Geological Society of America Bulletin, v. 120, no. 11–12, p. 1441–1454, doi:10.1130/B26360.1.

Schoenbohm, L.M., and Carrapa, B., 2015, this volume, Miocene–Pliocene shortening, extension and mafic magmatism support small-scale lithospheric foundering in the central Andes, NW Argentina, *in* DeCelles, P.G., Ducea, M.N., Carrapa, B., and Kapp, P.A., eds., Geodynamics of a Cordilleran Orogenic System: The Central Andes of Argentina and Northern Chile: Geological Society of America Memoir 212, doi:10.1130/2015.1212(09).

Schurr, B., Rietbrock, A., Asch, G., Kind, R., and Oncken, O., 2006, Evidence for lithospheric detachment in the central Andes from local earthquake tomography: Tectonophysics, v. 415, p. 203–223, doi:10.1016/j.tecto.2005.12.007.

Sdrolias, M., and Müller, R.D., 2006, Controls on back-arc basin formation: Geochemistry Geophysics Geosystems, v. 7, Q04016, doi:10.1029/2005GC001090.

Sobel, E., and Strecker, M.R., 2003, Uplift, exhumation and precipitation: Tectonic and climatic control of late Cenozoic landscape evolution in the northern Sierras Pampeanas, Argentina: Basin Research, v. 15, p. 431–451, doi:10.1046/j.1365-2117.2003.00214.x.

Soler, P., and Bonhomme, M.G., 1990, Relation of magmatic activity to plate dynamics in central Peru from Late Cretaceous to present, *in* Kay, S.M., and Rapela, C.W., eds., Plutonism from Antarctica to Alaska: Geological Society of America Special Paper 241, p. 173–192, doi:10.1130/SPE241-p173.

Somoza, R., 1998, Updated Nazca (Farallon)–South America relative motions during the last 40 My: Implications for mountain building in the central Andean region: Journal of South American Earth Sciences, v. 11, p. 211–215, doi:10.1016/S0895-9811(98)00012-1.

Starck, D., and Anzótegui, L.M., 2001, The late Miocene climatic change—Persistence of a climatic signal through the orogenic stratigraphic record in northwestern Argentina: Journal of South American Earth Sciences, v. 14, p. 763–774, doi:10.1016/S0895-9811(01)00066-9.

Starck, D., and Vergani, G., 1996, Desarrollo tecto-sedimentario del Cenozoico en el sur de la Provincia de Salta-Argentina, *in* XIII Congreso Geológico Argentino, Actas I: Buenos Aires, p. 433–452.

Strecker, M., Alonso, R., Bookhagen, B., Carrapa, B., Hilley, G., Sobel, E., and Trauth, M., 2007, Tectonics and climate of the southern central Andes: Annual Review of Earth and Planetary Sciences, v. 35, p. 747–787, doi:10.1146/annurev.earth.35.031306.140158.

Tassara, A., Götze, H.J., Schmidt, S., and Hackney, R., 2006, Three-dimensional density model of the Nazca plate and the Andean continental margin: Journal of Geophysical Research–Solid Earth (1978–2012), v. 111, B09404, doi:10.1029/2005JB003976.

Trumbull, R.B., Riller, U., Oncken, O., Scheuber, E., Munier, K., and Hongn, F., 2006, The time-space distribution of Cenozoic volcanism in the south-central Andes: A new data compilation and some tectonic implications, *in* Oncken, O., Chong, G., Franz, G., Giese, P., Götze, H., Ramos, V., Strecker, M., and Wigger, P., eds., The Andes: Active Subduction Orogeny: Berlin, Springer, Frontiers in Earth Science, p. 29–43.

Willett, S.D., Beaumont, C., and Fullsack, P., 1993, Mechanical model for the tectonics of doubly vergent compressional orogens: Geology, v. 21, p. 371–374, doi:10.1130/0091-7613(1993)021<0371:MMFTTO>2.3.CO;2.

Yáñez, G.A., Ranero, C.R., von Huene, R., and Díaz, J., 2001, Magnetic anomaly interpretation across the southern central Andes (32°–35°S): The role of the Juan Fernández Ridge in the Late Tertiary evolution of the margin: Journal of Geophysical Research, v. 106, no. B4, p. 6325–6345, doi:10.1029/2000JB900337.

Zandt, G., Gilbert, H., Owens, T.J., Ducea, M., Saleeby, J., and Jones, C.H., 2004, Active foundering of a continental arc root beneath the southern Sierra Nevada in California: Nature, v. 431, p. 41–46, doi:10.1038/nature02847.

Zhou, J., and Lau, K.M., 1998, Does a monsoon climate exist over South America?: Journal of Climate, v. 11, p. 1020–1040, doi:10.1175/1520-0442(1998)011<1020:DAMCEO>2.0.CO;2.

MANUSCRIPT ACCEPTED BY THE SOCIETY 3 JUNE 2014
MANUSCRIPT PUBLISHED ONLINE 23 SEPTEMBER 2014

The Geological Society of America
Memoir 212
2015

Low-temperature thermochronologic trends across the central Andes, 21°S–28°S

P.W. Reiners
S.N. Thomson
A. Vernon
Department of Geosciences, University of Arizona, 1040 E. 4th Street, Tucson, Arizona 85721, USA

S.D. Willett
Geologisches Institut, Eidgenössische Technische Hochschule, NO E 33, Sonneggstrasse 5, 8092 Zürich, Switzerland

M. Zattin
Department of Geosciences, University of Padova, 35137 Padova, Italy

J. Einhorn*
G. Gehrels
J. Quade
D. Pearson[†]
K.E. Murray
Department of Geosciences, University of Arizona, 1040 E. 4th Street, Tucson, Arizona 85721, USA

W. Cavazza
Dipartimento di Scienze, Biologiche, Geologiche e Ambientali, Università di Bologna, 40126 Bologna, Italy

ABSTRACT

In this paper, we merge more than 200 new apatite and zircon (U-Th)/He analyses and 21 apatite fission-track analyses from 71 new samples with previous published thermochronologic data using the same systems to understand the growth and large-scale kinematics of the central Andes between 21°S and 28°S. In general, minimum dates decrease and the total range of dates increases from west to east across the range. Large variations in thermochronometer dates on the east side reflect high spatial gradients in depth of recent erosional exhumation. Almost nowhere in this part of the Andes has Cenozoic erosion exceeded ~6–8 km, and in many places in the eastern half of the range, erosion has not exceeded 2–3 km, despite these regions now being

*Current address: ExxonMobil Exploration Company, 233 Benmar Drive, Houston, Texas 77060, USA.
[†]Current address: Department of Geosciences, Idaho State University, Pocatello, Idaho 83209, USA.

Reiners, P.W., Thomson, S.N., Vernon, A., Willett, S.D., Zattin, M., Einhorn, J., Gehrels, G., Quade, J., Pearson, D., Murray, K.E., and Cavazza, W., 2015, Low-temperature thermochronologic trends across the central Andes, 21°S–28°S, *in* DeCelles, P.G., Ducea, M.N., Carrapa, B., and Kapp, P.A., eds., Geodynamics of a Cordilleran Orogenic System: The Central Andes of Argentina and Northern Chile: Geological Society of America Memoir 212, p. 215–249, doi:10.1130/2015.1212(12). For permission to copy, contact editing@geosociety.org. © 2014 The Geological Society of America. All rights reserved.

5–6 km above sea level. This means that west of the rapidly deforming and eroding eastern range front, uplift and erosion are largely decoupled as a result of meager precipitation, relatively low relief, internal drainage, and volcanic burial. We interpret the west-to-east pattern of decreasing minimum dates across the range as recording the time-transgressive eastward migration of a focused zone of deformation, erosion, and convergence between the South American plate and the eastern edge of the Andean orogenic plateau. At this scale, the thermochronologic data do not suggest major changes in rates of plateau propagation or shortening/convergence with time. We use the thermochronometer date-distance trend and a simple kinematic model to infer a rate of eastward propagation of deformation and plateau growth of 6–10 km/m.y. This plateau propagation model balances horizontal convergence, erosion, and crustal thickening and predicts rates of shortening and convergence between the Andes block and South American plate that are consistent with geologic and geodetic observations.

INTRODUCTION AND GEOLOGIC SETTING

The central Andes between ~21°S and 28°S are part of the archetypal doubly vergent Cordilleran orogen, characterized by ocean-continent subduction on the west and retroarc thrusting of the continental plate beneath the orogenic wedge on the east (Isacks, 1988; Allmendinger et al., 1997; Dickinson, 2004; DeCelles et al., 2009; Fig. 1). Oceanic subduction beneath the South American plate has been roughly continuous since at least the Jurassic, as recorded by Jurassic plutonic rocks west of 70°W (e.g., Haschke et al., 2002, 2006). To first order, younger magmatism gradually transgressed from west to east with time, and since the Oligocene most magmatism has occurred in a relatively broad swath from ~66°W to 69°W (Allmendinger et al., 1997).

In detail, however, the spatial distribution of magmatic rocks shows several relatively rapid shifts in locus and width since ca. 150 Ma, some of which are correlated with quasi-periodic cycles in the geochemical and isotopic compositions of the magmas (Haschke et al., 2006).

The spatial-temporal pattern of deformation in the shallow crust of the central Andes since ca. 50 Ma also shows a first-order west-to-east migration, but it is more complex than that of magmatism. In the Bolivian Andes, the locus of thrusting has generally shifted from west to east, but shortening rates have varied by as much as an order of magnitude, and structurally active domains have not propagated uniformly (e.g., Kley and Monaldi, 1998; Oncken et al., 2006). Geodetic observations (Bevis et al., 2001; Klotz et al., 2001; Brooks et al., 2011) indicate that, at least

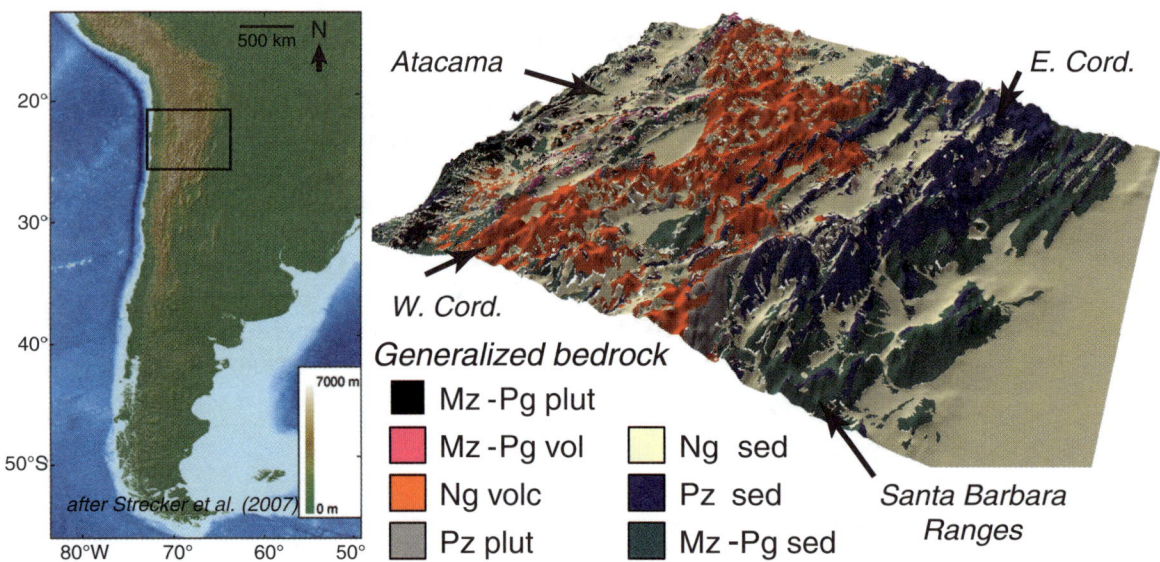

Figure 1. Generalized southern South America location map and perspective relief map, showing oblique shaded relief and primary bedrock types in transect between 21°S and 26°S, from coast (west of 70°W) to ~64°W. Arrows and labels show some of the important physiographic provinces referred to in this paper. Lithologic abbreviations as follows: Mz—Mesozoic; Pg—Paleogene; Ng—Neogene; Pz—Paleozoic; plut—plutonic; volc—volcanics; sed—sedimentary.

today, the vast majority of shortening occurs near the eastern front of the retroarc thrust belt, where material from the underthrusting South American plate is accreted to the range.

Evidence for nonuniform spatial-temporal patterns of deformation and arc magmatism in the central Andes has led to debate about the timing and evolution of the topographic form of the orogen (e.g., Strecker et al., 2007; Hain et al., 2011; Carrapa et al., 2011; DeCelles et al., 2011). Sedimentary rocks in many locations have been used to argue that at least some regions were low elevation and low relief until the Cenozoic (Lamb et al., 1997). Several studies, focusing mostly north of ~22°S, have suggested that surface uplift occurred in one or more discrete pulses affecting large regions as late as the late Miocene (Jordan et al., 1997; Garzione et al., 2008; Hoke and Garzione, 2008). However, the timing and rate at which different regions attained their current elevation are debated (e.g., Garzione et al., 2008; Barnes and Ehlers, 2009; Quade et al., this volume; Canavan et al., 2014).

The spatial-temporal pattern of uplift (both rock and surface) and the topographic development of the central Andes are largely controlled by the patterns of crustal shortening (e.g., Isacks, 1988; Kley and Monaldi, 1998; McQuarrie, 2002). The shortening-uplift relationship may be complicated by sublithospheric processes such as changes in magmatic flux (Haschke and Günther, 2003), frictional coupling at the subduction interface (Lamb and Davis, 2003), and convective instabilities that change the isostatic buoyancy of the range (DeCelles et al., 2009). It is possible that these mechanisms may act independently or in combination to produce punctuated episodes of rock uplift or surface elevation change. In addition, surface uplift and topographic growth are clearly modulated by erosional flux, further complicating inferences from shortening or accretionary constraints. Barnes and Ehlers (2009) provided a thorough introduction to considerations surrounding shortening, uplift, and topographic growth of the Andes through time.

As already noted in many studies (e.g., Strecker et al., 2007, and references therein), in the central Andes, the spatial distribution of erosion is strongly influenced by one of the largest continental precipitation gradients on Earth. Between ~21°S and 28°S, ranges of the eastern front of the retroarc receive mean annual precipitation higher than 2000 mm/yr, while the western flank receives ~20 mm/yr or less (e.g., Strecker et al., 2007; Bookhagen and Strecker, 2008), and there are regions where rainfall has never been recorded. Much of the central part of the Andes at this latitude, including the southern Puna Plateau, receives only ~200 mm/yr. Furthermore, much of the central part of the orogen is internally drained, has relatively low relief (compared with the Eastern and Western Cordilleras), and is covered in places by volcanic extrusive rocks. Low precipitation, low relief, internal drainage, and partial coverage by volcanics all limit rates of local erosion, as well as sediment transport and depths of erosion in the interior of the range. This, combined with the nearly complete lack of precipitation on the west flank of the range, means that the net erosional flux out of the Andes is likely highest on the retroarc side, and particularly near the active eastern thrusts. This is also the region where nearly all modern shortening is focused (Bevis et al., 2001; Klotz et al., 2001; Kendrick et al., 2006; Brooks et al., 2011).

In this paper, we use low-temperature thermochronometers to interpret the spatial-temporal patterns of deformation and uplift in the central Andes between ~21°S and 28°S. Cooling dates of low-temperature thermochronometers such as the fission-track and (U-Th)/He systems in apatite and zircon provide bounds on the possible thermal histories (primarily cooling histories) of rocks. Although cooling results from several processes, over large scales and with support of complementary geologic constraints, it can be reasonably interpreted as resulting from exhumation. In contractional orogens such as the central Andes, most exhumation is due to erosion rather than to normal faulting or ductile thinning. Episodes or varying rates of erosion may be attributable to a variety of causes, but in this setting, the most important variations in erosion rates in both space and time are reasonably associated with variations in rock uplift, through the development of topographic relief (and potentially enhanced precipitation) that drives erosion. Thus, to the extent that erosion can be related to rock uplift, and rock uplift can be related to contractional deformation, cooling dates can be used to elucidate spatial-temporal patterns of deformation. To the extent that deformation can be related to surface uplift, cooling dates also provide clues to patterns of topographic evolution through time.

In order to provide comparisons with and potentially tests of some of the orogen-scale ideas for the growth and evolution of the central Andes, this paper examines a large data set of thermochronometric dates to interpret the spatial-temporal pattern of erosion at the scale of the entire width of the orogen between 21°S and 28°S. We present new cooling dates for the apatite (U-Th)/He, apatite fission-track (AFT), and zircon (U-Th)/He systems for 71 new samples and combine them with previously published data for the same systems measured on other samples across the central Andes. We propose a simple model and interpretation of the significance of the spatial pattern of these dates at the scale of the entire width of the central Andes, focusing on the implications for the patterns of deformation and growth of the range through time.

SAMPLES AND METHODS

New thermochronologic dates from the apatite He, AFT, and zircon He systems, most of which were obtained from the same samples, are shown in Tables 1 and 2. Sample locations are shown in Figure 2. Nearly all these samples were collected between 22°S and 26°S, but we combine these with previously published data that cover a latitudinal range of 21°S–28°S. Sources of the other data north of 26°S include Coutand et al. (2001), Sobel and Strecker (2003), Maksaev and Zentilli (1999), Carrapa et al. (2005, 2009), Deeken et al. (2006), Juez-Larré et al. (2010), Carrapa et al. (2011), Pearson et al. (2012, 2013), and Mortimer et al. (2007). Also within this latitudinal range and shown in this compilation are data from subsurface mine

Reiners et al.

TABLE 1. NEW APATITE AND ZIRCON (U-Th)/He DATA

Sample name	Unit, lithologic, location, or age information	Aliquot name	⁴He (pmol)	⁴He (pmol ± 1σ)	U (ng)	U (ng ± 1σ)	Th (ng)	Th (ng ± 1σ)	Th/U (atomic)	Sm (ng)	Sm (ng ± 1σ)	Raw age (Ma)
Apatite												
08AA13	Quebrada del Toro, just above Tastil village	08AA13-A-1	0.05505	0.00032	0.2982	0.0086	0.0240	0.0007	0.08	0.840	0.017	33.43
		08AA13-A-3	0.06315	0.00036	0.3645	0.0105	0.0260	0.0008	0.07	0.758	0.015	31.47
08AA32	Macon east of Arizaro Salar	08AA32-A-1	0.01207	0.00010	0.0705	0.0026	0.0693	0.0019	1.01	0.200	0.004	25.66
		08AA32-A-2	0.01764	0.00014	0.0966	0.0032	0.1205	0.0033	1.28	0.355	0.007	26.03
		08AA32-A-3	0.01915	0.00016	0.0974	0.0032	0.1111	0.0031	1.17	0.410	0.008	28.55
08AA33	Macon east of Arizaro Salar	08AA33-A-2	0.01280	0.00009	0.0626	0.0025	0.0837	0.0023	1.37	0.342	0.007	28.63
		08AA33-A-3	0.01094	0.00007	0.0578	0.0023	0.0746	0.0021	1.33	0.367	0.007	26.70
08AA36	Macon east of Arizaro Salar	08AA36-A-1	0.00609	0.00008	0.0236	0.0018	0.0487	0.0014	2.12	0.277	0.006	31.78
		08AA36-A-2	0.00480	0.00013	0.0197	0.0018	0.0450	0.0013	2.35	0.203	0.004	29.07
08AA07	Devonian sandstone, Hornocal syncline	08AA07_A_1	0.00085	0.00004	0.0408	0.0010	0.0072	0.0002	0.18	0.437	0.009	3.68
		08AA07_A_2	0.00308	0.00005	0.0244	0.0006	0.0028	0.0001	0.12	0.427	0.009	22.36
		08AA07_A_3	0.00423	0.00004	0.0246	0.0006	0.0360	0.0008	1.50	0.304	0.006	23.36
		08AA07_A_4	0.00044	0.00002	0.0213	0.0006	0.0080	0.0007	0.39	0.232	0.005	3.44
		08AA07_A_5	0.00107	0.00005	0.0448	0.0011	0.0142	0.0004	0.32	0.572	0.011	4.07
08AA07a	Devonian sandstone, Hornocal syncline	08AA07a_1	0.00093	0.00003	0.0339	0.0017	0.0085	0.0008	0.26	0.355	0.007	4.73
		08AA07a_2	0.00098	0.00002	0.0344	0.0017	0.0249	0.0009	0.74	0.386	0.008	4.45
		08AA07a_3	0.00150	0.00010	0.0562	0.0019	0.0014	0.0008	0.03	0.668	0.013	4.87
		08AA07a_4	0.00367	0.00008	0.1330	0.0030	0.0852	0.0017	0.66	0.769	0.015	4.42
08AA11	Puncoviscana, Quebrada del Toro	08AA11_A_1	0.00014	0.00003	0.0044	0.0004	0.0162	0.0004	3.79	0.156	0.003	3.10
09AA31	Aguada de la Perdiz Fm, marine Ordovician sandstone	09AA31_A_1	0.00066	0.00004	0.0064	0.0003	0.0328	0.0007	5.24	0.753	0.015	8.05
		09AA31_A_2	0.00276	0.00005	0.0179	0.0005	0.1083	0.0022	6.22	0.564	0.011	11.55
CL09	Kg Jurassic granodiorite	CL09-A-1	0.00818	0.00008	0.0499	0.0022	0.0751	0.0021	1.54	0.152	0.003	22.32
		CL09-A-2	0.01552	0.00018	0.1007	0.0033	0.1322	0.0037	1.35	0.187	0.004	21.75
CL15	Kg Cretaceous granodiorite	CL15-A-1	0.03735	0.00033	0.1249	0.0039	0.2543	0.0070	2.09	2.125	0.043	36.81
		CL15-A-2	0.01157	0.00014	0.0586	0.0023	0.0723	0.0020	1.27	0.529	0.011	28.06
CL22	Kgd Late K–Early Pg granodiorite	CL22-A-1	0.02327	0.00017	0.1413	0.0043	0.2331	0.0065	1.69	2.805	0.056	21.57
		CL22-A-2	0.01217	0.00014	0.0638	0.0025	0.2165	0.0060	3.48	3.091	0.062	18.97
CL29	Ktig Paleocene granite	CL29-A-1	0.00244	0.00005	0.0080	0.0017	0.0291	0.0008	3.74	0.365	0.007	29.50
		CL29-A-2	0.00869	0.00011	0.0245	0.0018	0.0767	0.0021	3.21	0.985	0.020	36.66
09PRAD12	Kssb(a) El Buitre K-Pg granodiorite	09PRAD12aA	0.00515	0.00005	0.0231	0.0018	0.0327	0.0009	1.45	0.229	0.005	30.63
		09PRAD12aB	0.00715	0.00011	0.0308	0.0019	0.0432	0.0012	1.44	0.319	0.006	31.98
09PRAD13	Kssb(a) El Buitre K-Pg granodiorite	09PRAD13a1	0.01564	0.00014	0.0836	0.0029	0.0795	0.0022	0.97	0.303	0.006	28.18
09PRAD14	Kssb(a) El Buitre K-Pg granodiorite	09PRAD14aB	0.02869	0.00017	0.1331	0.0041	0.1868	0.0052	1.44	0.801	0.016	29.80
		09PRAD14aC	0.02012	0.00014	0.1070	0.0034	0.1495	0.0041	1.43	0.497	0.010	26.05
09PRAD15	Kssb(a) El Buitre K-Pg granodiorite	09PRAD15aA	0.01373	0.00015	0.0733	0.0027	0.0657	0.0019	0.92	0.283	0.006	28.48
		09PRAD15aB	0.03325	0.00023	0.1540	0.0047	0.1186	0.0033	0.79	0.497	0.010	33.68
		09PRAD15aC	0.01095	0.00014	0.0569	0.0024	0.0673	0.0019	1.21	0.313	0.006	27.71
09PRAD16	Kssb(a) El Buitre K-Pg granodiorite	09PRAD16a1	0.00773	0.00011	0.0363	0.0020	0.0583	0.0016	1.65	0.404	0.008	28.24
		09PRAD16a2	0.02452	0.00016	0.0920	0.0031	0.1808	0.0051	2.02	1.131	0.023	33.32
		09PRAD16a3	0.00440	0.00009	0.0192	0.0018	0.0369	0.0011	1.98	0.292	0.006	28.87
09PRAD17	Kssb(a) El Buitre K-Pg granodiorite	09PRAD17a-1	0.00677	0.00020	0.0294	0.0006	0.0476	0.0008	1.66	0.267	0.005	30.55
		09PRAD17a-2	0.01396	0.00043	0.0587	0.0011	0.0743	0.0013	1.30	0.398	0.008	33.67
09PRAD18	Kssb(a) El Buitre K-Pg granodiorite	09PRAD18aA	0.01367	0.00012	0.0685	0.0026	0.0921	0.0026	1.38	0.409	0.008	27.87
		09PRAD18aB	0.01131	0.00011	0.0627	0.0024	0.0943	0.0026	1.54	0.340	0.007	24.51

Raw age (Ma ± 1σ)	Mass (µg)	Half-width (µm)	U (ppm)	Th (ppm)	Sm (ppm)	⁴He (nmol/g)	F_T	F_T polished	Corrected age (Ma)	1s err anal. (Ma)	eU (ppm)	Longitude (°W)	Latitude (°S)	Elevation (m a.s.l.)	Distance from front (km)	Sample name
																Apatite
0.93	2.70	46.25	110.62	8.89	311.6	20.42	0.703	na	47.57	1.33	112.71	65.8911	24.4863	2825	−163	08AA13
0.88	2.28	44.75	159.73	11.41	332.1	27.67	0.690	na	45.62	1.27	162.41	65.8911	24.4863	2825	−163	
0.79	2.69	48.5	26.21	25.75	74.2	4.49	0.721	na	35.58	1.09	32.26	67.3284	24.6229	4631	−310	08AA32
0.69	2.55	45.5	37.86	47.24	139.2	6.92	0.709	na	36.72	0.97	48.96	67.3284	24.6229	4631	−310	
0.77	4.07	51.75	23.96	27.33	100.9	4.71	0.753	na	37.94	1.02	30.38	67.3284	24.6229	4631	−310	
0.86	1.81	38	34.59	46.29	189.0	7.08	0.647	na	44.27	1.33	45.47	67.3303	24.6337	4483	−310	08AA33
0.83	2.21	39.25	26.15	33.79	166.0	4.95	0.678	na	39.36	1.23	34.09	67.3303	24.6337	4483	−310	
1.68	1.58	34.5	14.91	30.76	174.8	3.85	0.637	na	49.89	2.63	22.14	67.3281	24.6350	4367	−309	08AA36
1.85	1.58	33.5	12.48	28.54	128.4	3.04	0.629	na	46.20	2.95	19.18	67.3281	24.6350	4367	−309	
0.17	1.50	40.25	27.18	4.81	291.4	0.57	0.651	na	5.65	0.27	28.32	65.0197	23.2802	4556	−113	08AA07
0.63	2.42	48.5	10.08	1.17	176.8	1.28	0.700	na	31.95	0.90	10.36	65.0197	23.2802	4556	−113	
0.49	1.16	36.5	21.14	30.96	261.4	3.63	0.612	na	38.16	0.80	28.42	65.0197	23.2802	4556	−113	
0.20	0.83	36	25.69	9.66	279.2	0.53	0.590	na	5.84	0.33	27.96	65.0197	23.2802	4556	−113	
0.21	1.64	44.5	27.30	8.63	348.3	0.65	0.664	na	6.13	0.32	29.33	65.0197	23.2802	4556	−113	
0.26	1.37	38	24.73	6.19	258.6	0.68	0.638	na	5.65	0.27	28.32	65.0197	23.2802	4556	−113	08AA07a
0.20	2.08	42.5	16.55	11.97	185.7	0.47	0.673	na	31.95	0.90	10.36	65.0197	23.2802	4556	−113	
0.36	2.61	47.5	21.56	0.54	256.3	0.58	0.704	na	38.16	0.80	28.42	65.0197	23.2802	4556	−113	
0.13	2.90	46.75	45.93	29.42	265.7	1.27	0.703	na	5.84	0.33	27.96	65.0197	23.2802	4556	−113	
0.76	0.79	29.25	5.53	20.43	196.4	0.18	0.542	na	5.71	1.40	10.34	65.6948	24.8908	1758	−128	08AA11
0.58	2.89	50.25	2.22	11.36	260.7	0.23	0.227	na	11.59	0.83	4.89	67.2921	23.6171	4547	−341	09AA31
0.29	2.03	47.75	8.80	53.36	278.1	1.36	1.358	na	16.71	0.41	21.34	67.2921	23.6171	4547	−341	
0.75	1.03	34.25	48.46	72.94	147.8	7.94	0.620	na	35.98	1.20	65.60	70.2603	22.6275	76	−688	CL09
0.59	1.65	36.25	61.21	80.32	113.7	9.43	0.634	na	34.27	0.94	80.09	70.2603	22.6275	76	−688	
0.88	5.96	64.5	20.95	42.64	356.3	6.26	0.780	na	47.18	1.12	30.97	69.3585	22.8731	1743	−585	CL15
0.91	1.83	45	31.94	39.39	288.1	6.31	0.693	na	40.51	1.32	41.20	69.3585	22.8731	1743	−585	
0.50	4.44	59.75	31.81	52.47	631.4	5.24	0.763	na	28.28	0.66	44.14	68.3397	23.6674	2376	−450	CL22
0.50	5.90	53.75	10.82	36.71	524.0	2.06	0.753	na	25.19	0.66	19.45	68.3397	23.6674	2376	−450	
3.32	1.19	36	6.72	24.50	306.9	2.05	0.628	na	47.01	5.29	12.47	69.0347	23.5373	2472	−527	CL29
1.61	3.69	48	6.64	20.77	266.8	2.35	0.723	na	50.70	2.23	11.52	69.0347	23.5373	2472	−527	
1.78	0.98	32.75	23.44	33.24	232.8	5.23	0.610	na	50.19	2.92	31.25	69.4657	23.3574	2232	−579	09PRAD12
1.52	1.09	31.5	28.12	39.45	291.6	6.53	0.587	na	54.47	2.59	37.39	69.4657	23.3574	2232	−579	
0.82	1.03	30	80.96	76.92	293.7	15.15	0.597	na	47.23	1.37	99.04	69.4629	23.3588	2165	−579	09PRAD13
0.71	1.74	36.25	76.45	107.29	459.9	16.48	0.637	na	46.79	1.12	101.67	69.4597	23.3583	2097	−578	09PRAD14
0.66	1.68	39	63.54	88.74	295.2	11.94	0.646	na	40.33	1.01	84.40	69.4597	23.3583	2097	−578	
0.89	1.04	35.25	70.46	63.15	272.3	13.19	0.630	na	45.22	1.41	85.30	69.4569	23.3574	2041	−578	09PRAD15
0.88	1.73	44.25	88.80	68.39	286.3	19.17	0.690	na	48.79	1.27	104.88	69.4569	23.3574	2041	−578	
0.95	1.21	35	46.88	55.44	258.1	9.03	0.646	na	42.93	1.47	59.91	69.4569	23.3574	2041	−578	
1.16	1.26	35.25	28.82	46.25	320.5	6.13	0.646	na	43.72	1.79	39.69	69.4657	23.3425	1937	−580	09PRAD16
0.81	3.63	38	25.31	49.76	311.3	6.75	0.671	na	49.64	1.21	37.01	69.4657	23.3425	1937	−580	
1.89	1.10	33	17.40	33.51	265.1	4.00	0.625	na	46.17	3.02	25.28	69.4657	23.3425	1937	−580	
1.00	0.97	31.25	30.31	49.00	274.6	6.97	0.577	na	52.96	1.73	41.83	69.4557	23.3538	2056	−578	09PRAD17
1.14	1.61	41.25	36.45	46.12	247.1	8.67	0.650	na	51.82	1.75	47.29	69.4557	23.3538	2056	−578	
0.82	1.85	38	37.07	49.79	221.4	7.39	0.648	na	43.02	1.26	48.77	69.4610	23.3863	1855	−578	09PRAD18
0.74	1.52	37.5	41.18	61.93	223.0	7.43	0.634	na	38.67	1.17	55.74	69.4610	23.3863	1855	−578	

(Continued)

Reiners et al.

TABLE 1. NEW APATITE AND ZIRCON (U-Th)/He DATA (*Continued*)

Sample name	Unit, lithologic, location, or age information	Aliquot name	⁴He (pmol)	⁴He (pmol ± 1σ)	U (ng)	U (ng ± 1σ)	Th (ng)	Th (ng ± 1σ)	Th/U (atomic)	Sm (ng)	Sm (ng ± 1σ)	Raw age (Ma)
09PRAD19	Kssb(a) El Buitre K-Pg granodiorite	09PRAD19aA	0.00707	0.00009	0.0405	0.0020	0.0442	0.0012	1.12	0.309	0.006	25.51
		09PRAD19aB	0.00429	0.00007	0.0263	0.0019	0.0341	0.0010	1.33	0.225	0.004	22.92
09PRAD20	Kssb(a) El Buitre K-Pg granodiorite	09PRAD20aA	0.01886	0.00011	0.0761	0.0027	0.0922	0.0026	1.24	0.515	0.010	35.42
09PRAD21	Kssb(a) El Buitre K-Pg granodiorite	09PRAD21a1	0.00737	0.00008	0.0452	0.0021	0.0512	0.0015	1.16	0.211	0.004	23.73
		09PRAD21a2	0.00700	0.00009	0.0338	0.0020	0.0568	0.0016	1.73	0.349	0.007	27.18
		09PRAD21a3	0.00477	0.00005	0.0296	0.0019	0.0381	0.0011	1.32	0.181	0.004	22.77
		09PRAD21a-1	0.00618	0.00018	0.0288	0.0005	0.0574	0.0009	2.05	0.295	0.006	26.81
		09PRAD21a-2	0.00707	0.00017	0.0396	0.0008	0.0558	0.0010	1.45	0.277	0.006	24.67
KMAD06	Loma Amarilla	KMAD06aA	0.00217	0.00003	0.0083	0.0003	0.0245	0.0009	3.03	0.186	0.004	28.12
		KMAD06aB	0.00061	0.00003	0.0042	0.0006	0.0059	0.0008	1.45	0.068	0.001	20.07
KMAD08	Loma Amarilla	KMAD08aA	0.00761	0.00006	0.0393	0.0008	0.0470	0.0011	1.23	0.174	0.003	27.81
		KMAD08aB	0.00487	0.00003	0.0226	0.0006	0.0302	0.0009	1.37	0.187	0.004	30.12
		KMAD08aD	0.00929	0.00006	0.0348	0.0007	0.0663	0.0013	1.96	0.147	0.003	33.93
KMAD09	Loma Amarilla	KMAD09aA	0.00225	0.00003	0.0128	0.0003	0.0183	0.0009	1.47	0.128	0.003	24.11
		KMAD09aB	0.00587	0.00009	0.0326	0.0007	0.0270	0.0009	0.85	0.058	0.001	27.86
		KMAD09aC	0.00123	0.00003	0.0102	0.0003	0.0083	0.0008	0.83	0.062	0.001	18.70
KMAD11	Loma Amarilla	KMAD11aA	0.00211	0.00005	0.0182	0.0004	0.0193	0.0008	1.09	0.063	0.001	17.15
		KMAD11aC	0.00387	0.00006	0.0157	0.0005	0.0401	0.0011	2.62	0.242	0.005	28.16
		KMAD11aD	0.00071	0.00002	0.0038	0.0002	0.0179	0.0008	4.85	0.108	0.002	16.11
KMAD13	Loma Amarilla	KMAD13aA	0.00213	0.00004	0.0141	0.0005	0.0546	0.0012	3.96	0.154	0.003	14.50
		KMAD13aC	0.00114	0.00005	0.0061	0.0002	0.0220	0.0009	3.68	0.206	0.004	18.28
KMAD14	Loma Amarilla	KMAD14aA	0.00038	0.00002	0.0029	0.0009	0.0098	0.0008	3.49	0.076	0.002	13.43
		KMAD14aB	0.00103	0.00003	0.0067	0.0003	0.0308	0.0009	4.74	0.288	0.006	13.35
09AA01	Paleozoic migmatite/ granitoid de Sierra Moreno	09AA01_A_1	0.05840	0.00020	0.2356	0.0057	0.0111	0.0003	0.05	0.651	0.013	45.18
		09AA01_A_2	0.12586	0.00038	0.4523	0.0110	0.0059	0.0002	0.01	1.528	0.031	51.07
		09AA01_A_3	0.12593	0.00032	0.4554	0.0111	0.0067	0.0002	0.02	1.246	0.025	50.77
09AA13	Cretaceous porphyry dike (Ksg)	09AA13_A_1	0.03854	0.00015	0.0985	0.0024	0.2531	0.0052	2.64	1.649	0.033	44.44
		09AA13_A_2	0.03453	0.00011	0.0944	0.0023	0.2660	0.0054	2.89	1.515	0.030	40.11
		09AA13_A_3	0.04067	0.00012	0.1138	0.0028	0.3094	0.0063	2.79	1.715	0.034	39.79
09AA14	Cretaceous porphyry dike (Ksg)	09AA14_A_2	0.02078	0.00010	0.0624	0.0015	0.1567	0.0032	2.58	1.211	0.024	38.08
		09AA14_A_3	0.05044	0.00016	0.1432	0.0035	0.3510	0.0072	2.51	2.690	0.054	40.64
09AA15	Cretaceous porphyry dike (Ksg)	09AA15A_A_1	0.01982	0.00011	0.0700	0.0017	0.2209	0.0045	3.24	0.907	0.018	29.72
		09AA15A_A_3	0.01424	0.00008	0.0521	0.0013	0.1170	0.0024	2.30	1.374	0.027	32.33
09AA17	Cretaceous porphyry dike (Ksg)	09AA17A_A_1	0.03302	0.00015	0.0989	0.0024	0.3424	0.0069	3.55	1.797	0.036	33.55
		09AA17A_A_2	0.02147	0.00012	0.0611	0.0015	0.1771	0.0036	2.98	1.181	0.024	38.04
		09AA17A_A_3	0.08025	0.00026	0.2427	0.0060	0.7160	0.0147	3.03	4.506	0.090	35.56
09AA18	Cretaceous porphyry dike (Ksg)	09AA18_A_1	0.01120	0.00008	0.0340	0.0008	0.1399	0.0029	4.22	0.813	0.016	30.45
		09AA18_A_2	0.03346	0.00010	0.1062	0.0026	0.4579	0.0093	4.42	1.612	0.032	28.60
		09AA18_A_3	0.04198	0.00018	0.1315	0.0032	0.6319	0.0129	4.93	2.994	0.060	27.29
Zircon												
2TT75	Santa Barbara subgroup of Salta Group	2TT75-86	0.21574	0.00163	0.1857	0.0039	0.0231	0.0005	0.13			205.61
		2TT75-90	1.59773	0.00596	2.2104	0.0465	0.2828	0.0050	0.13			128.75
		2TT75-92	3.48814	0.01676	3.1423	0.0642	1.0464	0.0176	0.34			187.87
		2TT75-100	4.50483	0.02384	2.7206	0.0568	0.8492	0.0144	0.32			279.17

Raw age (Ma ± 1σ)	Mass (μg)	Half-width (μm)	U (ppm)	Th (ppm)	Sm (ppm)	⁴He (nmol/g)	F_T	F_T polished	Corrected age (Ma)	1s err anal. (Ma)	eU (ppm)	Longitude (°W)	Latitude (°S)	Elevation (m a.s.l.)	Distance from front (km)	Sample name
1.04	1.23	29.25	32.84	35.80	250.8	5.73	0.580	na	43.96	1.79	41.26	69.4634	23.3873	1768	−578	09PRAD19
1.24	0.88	29.25	30.07	38.94	256.8	4.90	0.561	na	40.87	2.22	39.22	69.4634	23.3873	1768	−578	
1.00	2.06	37.25	36.91	44.71	249.7	9.15	0.666	na	53.15	1.50	47.41	69.4660	23.3893	1662	−578	09PRAD20
0.89	0.98	29	46.32	52.47	216.0	7.56	0.567	na	41.89	1.58	58.65	69.4672	23.3914	1598	−578	09PRAD21
1.15	1.41	30.5	23.92	40.23	247.5	4.95	0.614	na	44.29	1.87	33.37	69.4672	23.3914	1598	−578	
1.12	0.71	27.5	41.48	53.51	254.3	6.69	0.536	na	42.49	2.10	54.05	69.4672	23.3914	1598	−578	
0.87	1.26	28.25	22.90	45.64	234.5	4.92	0.581	na	46.15	1.49	33.62	69.4672	23.3914	1598	−578	
0.68	1.06	28.5	37.21	52.53	260.2	6.66	0.565	na	43.65	1.21	49.55	69.4672	23.3914	1598	−578	
0.82	1.62	38	5.13	15.13	115.0	1.34	0.633	na	44.44	1.30	8.68	68.6248	23.6904	2563	−479	KMAD06
2.44	0.68	31.5	6.11	8.64	100.7	0.90	0.583	na	34.45	4.18	8.14	68.6248	23.6904	2563	−479	
0.51	1.36	38.75	28.89	34.56	127.6	5.59	0.657	na	42.34	0.78	37.01	68.6714	23.6989	3095	−483	KMAD08
0.63	0.72	29.5	31.39	42.06	260.5	6.78	0.548	na	54.94	1.15	41.28	68.6714	23.6989	3095	−483	
0.55	1.11	32.75	31.36	59.76	132.8	8.37	0.613	na	55.35	0.89	45.41	68.6714	23.6989	3095	−483	
0.64	0.90	40.5	14.12	20.27	141.1	2.49	0.634	na	38.02	1.00	18.89	68.6503	23.6988	2859	−481	KMAD09
0.65	0.66	30.25	49.69	41.14	88.4	8.96	0.578	na	48.23	1.13	59.36	68.6503	23.6988	2859	−481	
0.71	0.50	31	20.21	16.41	124.1	2.45	0.563	na	33.19	1.26	24.06	68.6503	23.6988	2859	−481	
0.55	1.23	36.25	14.74	15.63	51.3	1.71	0.642	na	26.71	0.86	18.41	68.6424	23.6976	2738	−480	KMAD11
0.74	0.75	33.5	21.02	53.77	325.0	5.19	0.592	na	47.54	1.26	33.66	68.6424	23.6976	2738	−480	
0.79	0.52	27.25	7.30	34.53	208.3	1.37	0.539	na	29.88	1.46	15.41	68.6424	23.6976	2738	−480	
0.40	0.78	30	18.06	69.80	196.9	2.73	0.570	na	25.44	0.70	34.47	68.6336	23.6969	2650	−480	KMAD13
0.94	0.73	31.5	8.38	30.04	282.0	1.56	0.577	na	31.69	1.63	15.44	68.6336	23.6969	2650	−480	
2.30	1.60	39.5	1.79	6.09	47.3	0.24	0.658	na	20.39	3.50	3.22	68.6252	23.6023	2489	−482	KMAD14
0.53	0.75	32.75	8.93	41.28	385.9	1.38	0.598	na	22.31	0.88	18.63	68.6252	23.6023	2489	−482	
1.06	9.32	67.25	25.27	1.19	69.8	6.26	0.812	na	55.62	1.30	25.55	69.0812	21.7097	2193	−596	09AA01
1.20	14.58	85	31.03	0.40	104.8	8.63	0.839	na	60.84	1.43	31.12	69.0812	21.7097	2193	−596	
1.20	11.40	71.5	39.93	0.59	109.2	11.04	0.823	na	61.66	1.45	40.07	69.0812	21.7097	2193	−596	
0.75	11.59	84	8.50	21.83	142.3	3.32	0.832	na	53.43	0.90	13.62	68.6753	21.8420	4384	−548	09AA13
0.66	8.59	68.25	10.99	30.96	176.4	4.02	0.796	na	50.42	0.83	18.27	68.6753	21.8420	4384	−548	
0.66	13.39	75	8.50	23.11	128.1	3.04	0.823	na	48.37	0.80	13.93	68.6753	21.8420	4384	−548	
0.65	6.40	63.5	9.75	24.49	189.3	3.25	0.764	na	49.87	0.85	15.51	68.6786	21.8397	4274	−549	09AA14
0.68	13.47	71.75	10.63	26.05	199.6	3.74	0.813	na	50.00	0.84	16.75	68.6786	21.8397	4274	−549	
0.51	5.13	52	13.65	43.08	176.9	3.86	0.732	na	40.62	0.70	23.77	68.6855	21.8381	4069	−549	09AA15
0.58	2.92	44.75	17.85	40.09	470.6	4.88	0.715	na	45.21	0.81	27.28	68.6855	21.8381	4069	−549	
0.55	8.86	67	11.16	38.66	202.9	3.73	0.780	na	43.02	0.70	20.25	68.6926	21.8380	3784	−550	09AA17
0.65	4.21	54.25	14.51	42.09	280.7	5.10	0.746	na	51.01	0.87	24.40	68.6926	21.8380	3784	−550	
0.58	15.49	61.5	15.67	46.22	290.9	5.18	0.796	na	44.67	0.73	26.53	68.6926	21.8380	3784	−550	
0.53	2.21	43	15.36	63.25	367.5	5.06	0.664	na	45.83	0.79	30.23	68.6944	21.8338	3655	−551	09AA18
0.45	2.95	55	36.02	155.33	546.8	11.35	0.742	na	38.56	0.60	72.52	68.6944	21.8338	3655	−551	
0.43	3.45	51.5	38.11	183.18	867.8	12.17	0.737	na	37.03	0.59	81.16	68.6944	21.8338	3655	−551	
																Zircon
4.38	1.02	30.5	181.77	22.63		211.16	na	0.736	279.21	6.57	187.08	66.0791	25.2387	2740	−157	2TT75
2.60	5.62	56.75	393.59	50.35		284.50	na	0.846	152.27	3.23	405.42	66.0791	25.2387	2740	−157	
3.61	7.09	66	443.05	147.54		491.82	na	0.857	219.23	4.38	477.72	66.0791	25.2387	2740	−157	
5.57	13.68	76.75	198.92	62.09		329.38	na	0.884	315.80	6.53	213.51	66.0791	25.2387	2740	−157	

(Continued)

TABLE 1. NEW APATITE AND ZIRCON (U-Th)/He DATA (*Continued*)

Sample name	Unit, lithologic, location, or age information	Aliquot name	^{4}He (pmol)	^{4}He (pmol ± 1σ)	U (ng)	U (ng ± 1σ)	Th (ng)	Th (ng ± 1σ)	Th/U (atomic)	Sm (ng)	Sm (ng ± 1σ)	Raw age (Ma)
2TT142	Santa Barbara subgroup of Salta Group	2TT142-4	3.86877	0.01654	2.4452	0.0508	0.3315	0.0057	0.14			277.44
		2TT142-5	3.06866	0.01232	2.8730	0.0598	0.3882	0.0066	0.14			188.88
		2TT142-57	4.11101	0.01769	3.8314	0.0803	0.6496	0.0111	0.17			188.28
		2TT142-67	0.80425	0.00382	0.8604	0.0176	0.1177	0.0020	0.14			165.59
2TT948	Santa Barbara subgroup of Salta Group	2TT948-28	0.70611	0.00272	0.8228	0.0184	0.4587	0.0077	0.57			139.07
		2TT948-57	0.88909	0.00453	0.5319	0.0112	0.3856	0.0067	0.74			259.10
		2TT948-64	0.61440	0.00303	0.6949	0.0142	0.1301	0.0023	0.19			155.02
		2TT948-69	1.78869	0.00658	1.1986	0.0245	0.3010	0.0054	0.26			255.55
08AA06	Hornocal syncline	08AA06_Z_1	0.81599	0.00181	0.6339	0.0120	0.2971	0.0052	0.48			211.17
		08AA06_Z_2	0.29602	0.00071	0.4959	0.0096	0.1438	0.0026	0.30			102.78
		08AA06_Z_3	0.41226	0.00242	0.3919	0.0074	0.1432	0.0025	0.37			176.95
		08AA06_Z_4	0.21428	0.00055	0.6652	0.0127	0.2168	0.0037	0.33			55.26
08AA13	Quebrada del Toro, just above Tastil village	08AA13_Z_3	2.14690	0.00321	2.2191	0.0427	0.4837	0.0083	0.22			168.24
		08AA13_Z_1	4.37435	0.00858	2.9009	0.0546	0.9974	0.0176	0.35			253.10
		08AA13_Z_2	3.20984	0.00621	2.9804	0.0565	1.1317	0.0194	0.39			180.55
08AA18	West of Cauchari Salar	08AA18_Z_1	3.64341	0.00857	1.8228	0.0344	1.2481	0.0213	0.70			310.91
		08AA18_Z_2	2.28837	0.00486	1.2045	0.0228	0.5230	0.0093	0.45			311.13
		08AA18_Z_3	6.79289	0.01383	3.0021	0.0572	3.3768	0.0575	1.15			323.09
08AA32	East of Arizaro Salar, hanging wall of east-verging thrust, summit near antenna and hut	08AA32_Z_1	5.39469	0.03207	4.7105	0.0915	2.2117	0.0379	0.48			188.23
		08AA32_Z_2	3.29528	0.00703	2.8665	0.0538	1.3273	0.0222	0.48			189.18
		08AA32_Z_3	6.95801	0.02200	4.6629	0.0876	2.2681	0.0388	0.50			243.19
08AA33	East of Arizaro Salar, hanging wall of east-verging thrust, on ridge crest east from trail	08AA33_Z_1	3.18931	0.00717	2.1571	0.0410	1.1109	0.0186	0.53			239.60
		08AA33_Z_2	5.67640	0.01652	5.7416	0.1088	2.8173	0.0486	0.50			162.14
		08AA33_Z_3	3.52370	0.02810	3.3003	0.0648	1.3897	0.0239	0.43			177.47
08AA36	East of Arizaro Salar, just below gate	08AA36_Z_1	4.27113	0.01229	2.6400	0.0502	1.2947	0.0222	0.50			262.98
		08AA36_Z_2	1.32542	0.00336	1.2204	0.0230	0.2445	0.0042	0.21			189.25
		08AA36_Z_3	9.73763	0.02182	7.3062	0.1449	2.8696	0.0485	0.40			221.99
08AA41	West of Acay and north of 5000 m pass	08AA41_Z_3	0.30574	0.00056	0.6609	0.0130	0.3398	0.0058	0.53			76.07
09AA31	East of Zapaleri quadrangle, NNW of Passo Gualtiquina to Argentina, accessed from the WNW via the Pla ignimbrite Pampa, ridge S of #30	09AA31_Z_1	1.69780	0.00981	0.9187	0.0175	0.8420	0.0148	0.94			275.51
		09AA31_Z_2	1.14767	0.00236	0.6380	0.0122	0.5327	0.0089	0.86			272.49
		09AA31_Z_3	1.19550	0.00237	0.6512	0.0127	0.5931	0.0100	0.93			274.01
08AA24		08AA24-Z-1	1.51964	0.00830	1.1976	0.0295	1.8013	0.0278	1.54			171.37
		08AA24-Z-2	1.08465	0.00523	0.9308	0.0231	0.9752	0.0157	1.07			170.92
09PRAD04	Loma Amarilla	09PRAD04z1	0.23419	0.00138	0.4731	0.0119	0.6131	0.0100	1.33			69.91
		09PRAD04z2	0.83657	0.01254	1.6685	0.0425	1.5483	0.0248	0.95			75.83
09PRAD05	Loma Amarilla	09PRAD05z1	0.70442	0.01101	1.4304	0.0364	1.0594	0.0169	0.76			77.28
		09PRAD05z2	0.77915	0.01169	1.5841	0.0390	0.7469	0.0126	0.48			81.57
09PRAD012	Kssb(a) El Buitre K-Pg granodiorite	09PRAD012z2	1.41005	0.02092	3.8282	0.0963	6.8473	0.1145	1.84			47.84
09PRAD021	Kssb(a) El Buitre K-Pg granodiorite	09PRAD021z2	0.25862	0.00411	0.8858	0.0227	0.6617	0.0101	0.77			45.88
09AA01	Paleozoic migmatite/ granitoid de Sierra Moreno	09AA01_Z_2	0.40433	0.00105	1.4236	0.0271	0.1312	0.0023	0.09			51.38
		09AA01_Z_3	0.71248	0.00167	1.6396	0.0315	0.9562	0.0170	0.60			70.46
		09AA01_Z_1	3.19296	0.02566	6.6915	0.1253	1.0209	0.0174	0.16			84.86

Raw age (Ma ± 1σ)	Mass (μg)	Half-width (μm)	U (ppm)	Th (ppm)	Sm (ppm)	⁴He (nmol/g)	F_T	F_T polished	Corrected age (Ma)	1s err anal. (Ma)	eU (ppm)	Longitude (°W)	Latitude (°S)	Elevation (m a.s.l.)	Distance from front (km)	Sample name
5.65	8.12	49.75	300.98	40.81		476.22	na	0.843	329.17	7.07	310.57	66.0791	25.2387	2740	−157	2TT142
3.81	7.36	53	390.61	52.78		417.21	na	0.851	221.83	4.78	403.01	66.0791	25.2387	2740	−157	
3.80	5.47	44.25	700.12	118.70		751.23	na	0.830	226.87	4.98	728.01	66.0791	25.2387	2740	−157	
3.31	3.18	44.25	270.69	37.03		253.02	na	0.815	203.20	4.34	279.39	66.0791	25.2387	2740	−157	
2.74	4.35	48.75	189.04	105.38		162.23	na	0.822	169.16	3.56	213.80	66.0695	25.2431	2757	−156	2TT948
4.81	4.95	40.5	107.49	77.94		179.69	na	0.807	321.15	6.35	125.81	66.0695	25.2431	2757	−156	
3.06	2.60	31.25	267.51	50.07		236.54	na	0.769	201.50	4.38	279.28	66.0695	25.2431	2757	−156	
4.95	4.53	42.5	264.84	66.51		395.24	na	0.822	310.93	6.60	280.47	66.0695	25.2431	2757	−156	
3.59	2.53	33	250.94	117.61		323.01	0.706	na	299.15	5.08	278.58	65.0190	23.2868	4481	−113	08AA06
1.84	1.78	34	278.69	80.83		166.38	0.692	na	148.63	2.65	297.69	65.0190	23.2868	4481	−113	
3.19	1.65	30.25	237.64	86.85		249.97	0.673	na	262.83	4.74	258.05	65.0190	23.2868	4481	−113	
0.96	1.93	29.5	344.70	112.33		111.04	0.678	na	81.46	1.41	371.10	65.0190	23.2868	4481	−113	
3.02	4.36	41	508.84	110.90		492.29	0.758	na	222.05	3.98	534.90	65.8911	24.4863	2825	−163	08AA13
4.38	6.87	38.75	422.45	145.24		637.03	0.761	na	332.78	5.76	456.58	65.8911	24.4863	2825	−163	
3.10	6.36	38.5	468.32	177.83		504.37	0.760	na	237.48	4.08	510.11	65.8911	24.4863	2825	−163	
5.10	10.68	49.5	170.63	116.83		341.06	0.804	na	386.60	6.34	198.09	66.8816	23.9837	4085	−285	08AA18
5.34	4.03	41.25	298.77	129.72		567.62	0.752	na	386.60	6.34	198.09	66.8816	23.9837	4085	−285	
4.99	6.50	46.75	461.90	519.56		1045.16	0.780	na	413.80	7.10	329.25	66.8816	23.9837	4085	−285	
3.43	12.47	53	377.79	177.38		432.67	0.814	na	231.27	4.21	419.47	67.3284	24.6229	4631	−310	08AA32
3.17	7.25	50.75	395.39	183.09		454.54	0.792	na	238.80	4.00	438.42	67.3284	24.6229	4631	−310	
4.13	24.84	80	187.69	91.30		280.08	0.864	na	281.47	4.78	209.14	67.3284	24.6229	4631	−310	
4.05	7.54	38	285.99	147.28		422.85	0.762	na	314.60	5.31	320.60	67.3303	24.6337	4483	−310	08AA33
2.74	8.23	52.5	697.90	342.45		689.98	0.802	na	202.06	3.42	778.37	67.3303	24.6337	4483	−310	
3.43	5.48	46.5	601.73	253.39		642.47	0.778	na	228.10	4.40	661.28	67.3303	24.6337	4483	−310	
4.50	10.51	57.25	251.10	123.14		406.25	0.818	na	321.62	5.51	280.03	67.3281	24.6350	4367	−309	08AA36
3.36	2.60	34.5	469.80	94.11		510.25	0.715	na	264.57	4.70	491.92	67.3281	24.6350	4367	−309	
4.00	13.50	63.75	541.31	212.60		721.45	0.833	na	266.38	4.80	591.27	67.3281	24.6350	4367	−309	
1.30	1.59	36	416.75	214.32		192.81	0.707	na	107.56	1.84	467.12	66.2457	24.4302	4760	−202	08AA41
4.64	3.59	37	256.06	234.68		473.22	0.734	na	375.22	6.32	311.21	67.2921	23.6171	4547	−341	09AA31
4.38	3.34	44.25	191.26	159.69		344.03	0.747	na	364.79	5.87	228.79	67.2921	23.6171	4547	−341	
4.45	3.25	32.25	200.28	182.42		367.67	0.711	na	385.51	6.27	243.15	67.2921	23.6171	4547	−341	
3.25	4.16	40.25	287.82	432.89		365.20	0.744	na	230.29	4.37	389.55	68.1053	24.9502	4230	−380	08AA24
3.46	2.86	38.25	325.65	341.20		379.50	0.725	na	235.89	4.77	405.83	68.1053	24.9502	4230	−380	
1.39	2.28	35.5	207.13	268.42		102.53	0.697	na	230.29	4.37	270.21	70.0699	24.2247	1463	−613	09PRAD04
1.92	5.10	40.5	327.06	303.51		163.99	0.743	na	235.89	4.77	398.39	70.0699	24.2247	1463	−613	
2.03	3.06	35	467.58	346.31		230.26	0.710	na	100.35	1.99	548.96	70.0799	24.2057	2133	−614	09PRAD05
2.15	2.91	35.5	545.26	257.10		268.19	0.721	na	102.03	2.59	605.68	70.0799	24.2057	2133	−614	
1.11	4.37	40.75	875.64	1566.24		322.53	0.749	na	108.83	2.86	1243.71	69.4657	23.3574	2232	−579	09PRAD012
1.21	3.83	38	231.36	172.83		67.55	0.741	na	113.18	2.98	271.98	69.4672	23.3914	1598	−578	09PRAD021
0.93	2.95	36.5	482.88	44.51		137.14	0.720	na	71.38	1.30	493.34	69.0812	21.7097	2193	−596	09AA01
1.17	3.34	31.25	490.28	285.91		213.05	0.701	na	100.53	1.67	557.47	69.0812	21.7097	2193	−596	
1.64	17.89	59.5	374.02	57.06		178.47	0.834	na	101.81	1.97	387.43	69.0812	21.7097	2193	−596	

(Continued)

Reiners et al.

TABLE 1. NEW APATITE AND ZIRCON (U-Th)/He DATA (*Continued*)

Sample name	Unit, lithologic, location, or age information	Aliquot name	⁴He (pmol)	⁴He (pmol ± 1σ)	U (ng)	U (ng ± 1σ)	Th (ng)	Th (ng ± 1σ)	Th/U (atomic)	Sm (ng)	Sm (ng ± 1σ)	Raw age (Ma)
09AA14	Cretaceous porphyry dike (Ksg)	09AA14_Z_1	2.05904	0.00512	7.0677	0.1350	3.0272	0.0539	0.44			48.90
		09AA14_Z_2	1.91107	0.00309	6.4756	0.1234	2.8826	0.0489	0.46			49.36
		09AA14_Z_3	2.10052	0.01208	6.7805	0.1302	3.7135	0.0650	0.56			50.69
09AA18	Cretaceous porphyry dike (Ksg)	09AA18_Z_1	6.10382	0.03530	19.4813	0.3839	8.4878	0.1478	0.45			52.49
		09AA18_Z_2	2.86680	0.00531	2.8066	0.0526	0.5681	0.0101	0.21			178.09
		09AA18_Z_3	0.62109	0.00123	1.0433	0.0199	0.3016	0.0052	0.30			102.51
09AA38		09AA38_Z_1	2.86680	0.00531	2.8066	0.0526	0.5681	0.0101	0.21			178.09
		09AA38_Z_2	0.62109	0.00123	1.0433	0.0199	0.3016	0.0052	0.30			102.51
		09AA38_Z_3	0.61869	0.00357	1.2263	0.0232	0.4465	0.0076	0.37			85.59
09AA41		09AA41_Z_1	1.41537	0.00271	3.0482	0.0570	0.3273	0.0061	0.11			83.45
09KMAD06	Loma Amarilla	09KMAD06Z1	0.07759	0.00043	0.2322	0.0055	0.1240	0.0030	0.55			54.82
		09KMAD06Z2	0.88327	0.00587	0.9606	0.0215	1.0059	0.0227	1.07			135.28
		09KMAD06Z3	0.41362	0.00222	0.4739	0.0113	0.2703	0.0051	0.59			141.04
		09KMAD06Z4	1.83030	0.01051	0.7952	0.0177	0.7577	0.0173	0.98			338.95
		09KMAD06Z5	0.29266	0.00158	0.3691	0.0083	0.2759	0.0056	0.77			123.76
09KMAD13	Loma Amarilla	09KMAD13Z1	0.16026	0.00257	0.6363	0.0149	0.6439	0.0135	1.04			37.60
		09KMAD13Z2	0.31441	0.00115	0.3673	0.0083	0.2902	0.0053	0.81			132.38
		09KMAD13Z3	0.91870	0.00316	1.1595	0.0283	0.9458	0.0173	0.84			122.03
		09KMAD13Z4	0.49626	0.00171	0.7296	0.0161	0.6218	0.0131	0.87			104.17
		09KMAD13Z5	0.28369	0.00113	0.2877	0.0067	0.2008	0.0049	0.72			155.04
09KMAD11	Loma Amarilla	09KMAD11Z2	2.14596	0.01983	1.8016	0.0423	2.1032	0.0491	1.20			170.86
		09KMAD11Z3	0.40056	0.00365	0.4000	0.0119	0.3417	0.0069	0.88			152.67
		09KMAD11Z4	0.13894	0.00128	0.4832	0.0114	0.3809	0.0066	0.81			44.81
		09KMAD11Z5	0.50706	0.00461	0.5500	0.0129	0.3884	0.0067	0.72			144.85
		09KMAD09Z1	1.47389	0.00434	2.1046	0.0477	0.7074	0.0134	0.34			119.18
09KMAD09	Loma Amarilla	09KMAD09Z2	0.61466	0.00165	1.2692	0.0294	0.8778	0.0155	0.71			76.75
		09KMAD09Z3	1.03404	0.00239	1.3187	0.0331	0.6892	0.0121	0.54			128.12
		09KMAD09Z4	0.72683	0.00168	2.0021	0.0465	0.7818	0.0133	0.40			61.37
		09KMAD09Z5	0.49702	0.00132	0.7309	0.0175	0.2861	0.0057	0.40			114.39
		09KMAD08Z1	0.27018	0.00109	0.3213	0.0075	0.2297	0.0041	0.73			132.02
09KMAD08	Loma Amarilla	09KMAD08Z2	0.48768	0.00177	0.7601	0.0177	0.3633	0.0068	0.49			106.03
		09KMAD08Z3	0.59420	0.00228	0.6013	0.0138	0.5103	0.0087	0.87			150.85
		09KMAD08Z4	0.50754	0.00134	0.6035	0.0135	0.3711	0.0068	0.63			134.73
		09KMAD08Z5	0.74960	0.00197	0.7865	0.0188	0.6410	0.0117	0.84			146.50
		09KMAD14Z1	1.43148	0.00876	3.9545	0.0948	2.2272	0.0450	0.58			59.00
09KMAD14	Loma Amarilla	09KMAD14Z2	0.50021	0.00306	1.5357	0.0353	0.9604	0.0177	0.64			52.44
		09KMAD14Z3	0.39034	0.00240	0.9264	0.0226	0.7775	0.0143	0.86			64.90
		09KMAD14Z4	1.16166	0.00712	3.3024	0.0743	1.6946	0.0290	0.53			57.94
		09KMAD14Z5	0.61827	0.00381	1.7288	0.0438	1.0632	0.0203	0.63			57.67
		09JA01_2	1.08902	0.00812	1.8231	0.0336	0.0390	0.0009	0.02			109.26
09JA01	Ordovician granitoid near Quebrada Aguadita	09JA01_7	0.98626	0.00744	1.1875	0.0218	0.0270	0.0026	0.02			151.26
		09JA01_11	1.95974	0.01771	3.9208	0.0732	0.0584	0.0010	0.02			91.71
		09JA01_14	8.01413	0.05808	11.6989	0.2308	0.5309	0.0092	0.05			124.42
		09JA10_2	0.17467	0.00130	0.2534	0.0047	0.0504	0.0010	0.20			120.90
08JA01	Ordovician granitoid near Quebrada Aguadita	08JA01_1	1.83619	0.01313	1.5027	0.0283	0.2748	0.0049	0.19			213.28
		08JA01_11	2.10816	0.01676	5.7936	0.1095	0.3201	0.0054	0.06			66.30
		08JA01_25	1.15136	0.00925	1.6544	0.0319	0.2863	0.0053	0.18			122.77

Raw age (Ma ± 1σ)	Mass (μg)	Half-width (μm)	U (ppm)	Th (ppm)	Sm (ppm)	^{4}He (nmol/g)	F_T	F_T polished	Corrected age (Ma)	1s err anal. (Ma)	eU (ppm)	Longitude (°W)	Latitude (°S)	Elevation (m a.s.l.)	Distance from front (km)	Sample name
0.83	22.91	81.5	308.46	132.12		89.86	0.862	na	56.72	0.97	339.51	68.6786	21.8397	4274	−549	09AA14
0.83	21.19	61.75	305.67	136.07		90.21	0.843	na	58.56	0.98	337.64	68.6786	21.8397	4274	−549	
0.89	25.52	74.75	265.66	145.49		82.30	0.859	na	59.02	1.03	299.85	68.6786	21.8397	4274	−549	
0.96	38.61	87.25	504.53	219.82		158.08	0.879	na	59.69	1.09	556.19	68.6944	21.8338	3655	−551	09AA18
3.14	12.36	60.5	227.11	45.97		231.99	0.824	na	60.49	1.06	1062.06	68.6944	21.8338	3655	−551	
1.79	3.89	41.25	268.50	77.61		159.84	0.753	na	55.29	0.96	1072.13	68.6944	21.8338	3655	−551	
3.14	12.36	60.5	227.11	45.97		231.99	0.824	na	216.16	3.81	237.92	69.6272	21.7062	1219	−653	09AA38
1.79	3.89	41.25	268.50	77.61		159.84	0.753	na	136.07	2.38	286.74	69.6272	21.7062	1219	−653	
1.53	4.37	45.5	280.90	102.28		141.72	0.766	na	111.80	2.00	304.93	69.6272	21.7062	1219	−653	
1.48	3.71	38	820.74	88.13		381.10	0.743	na	112.26	2.00	841.45	70.3809	24.7720	2045	−627	09AA41
1.17	2.06	32	112.55	60.13		37.62	0.694	na	78.98	1.69	126.68	68.6248	23.6904	2563	−479	09KMAD06
2.60	5.25	47	182.92	191.55		168.21	0.771	na	175.40	3.37	227.94	68.6248	23.6904	2563	−479	
3.00	2.68	39.25	176.84	100.88		154.36	0.727	na	193.93	4.13	200.55	68.6248	23.6904	2563	−479	
6.58	1.90	27	418.11	398.41		962.39	0.660	na	513.39	9.96	511.74	68.6248	23.6904	2563	−479	
2.43	1.76	34.5	209.81	156.80		166.34	0.689	na	179.52	3.53	246.66	68.6248	23.6904	2563	−479	
0.93	4.76	43.5	133.72	135.33		33.68	0.762	na	49.35	1.22	165.52	68.6336	23.6969	2650	−480	09KMAD13
2.52	1.73	28	211.88	167.41		181.36	0.662	na	199.84	3.81	251.22	68.6336	23.6969	2650	−480	
2.49	5.17	51.75	224.30	182.95		177.72	0.779	na	156.73	3.19	267.29	68.6336	23.6969	2650	−480	
1.92	2.05	32	356.12	303.51		242.25	0.689	na	151.23	2.79	427.45	68.6336	23.6969	2650	−480	
3.13	1.96	37.25	146.67	102.39		144.63	0.703	na	220.55	4.46	170.74	68.6336	23.6969	2650	−480	
3.57	6.86	42.25	262.53	306.48		312.71	0.771	na	221.54	4.62	334.55	68.6424	23.6976	2738	−480	09KMAD11
3.97	4.08	36	97.98	83.70		98.12	0.735	na	207.73	5.40	117.65	68.6424	23.6976	2738	−480	
0.96	3.01	34.5	160.80	126.75		46.24	0.719	na	62.33	1.34	190.58	68.6424	23.6976	2738	−480	
3.15	2.03	30.25	270.50	191.00		249.39	0.684	na	211.65	4.61	315.38	68.6424	23.6976	2738	−480	
2.46	13.37	60	157.44	52.92		110.26	0.829	na	143.73	2.97	169.87	68.6503	23.6988	2859	−481	
1.51	5.21	45.5	243.83	168.63		118.08	0.773	na	99.35	1.95	283.46	68.6503	23.6988	2859	−481	09KMAD09
2.81	8.54	55.5	154.43	80.71		121.09	0.808	na	158.53	3.48	173.39	68.6503	23.6988	2859	−481	
1.27	3.99	39	502.33	196.16		182.37	0.748	na	82.10	1.70	548.42	68.6503	23.6988	2859	−481	
2.47	4.86	48.75	150.40	58.87		102.27	0.776	na	147.42	3.18	164.23	68.6503	23.6988	2859	−481	
2.66	1.14	28	282.77	202.12		237.76	0.642	na	205.78	4.14	330.27	68.6714	23.6989	3095	−483	
2.20	4.76	49.5	159.70	76.33		102.47	0.775	na	136.84	2.84	177.64	68.6714	23.6989	3095	−483	09KMAD08
2.90	4.21	50.5	142.68	121.09		141.00	0.764	na	197.49	3.80	171.13	68.6714	23.6989	3095	−483	
2.60	5.49	59	109.99	67.63		92.51	0.786	na	171.30	3.30	125.89	68.6714	23.6989	3095	−483	
2.92	5.59	54.5	140.60	114.59		134.00	0.784	na	186.88	3.73	167.52	68.6714	23.6989	3095	−483	
1.27	9.63	50.75	410.81	231.37		148.71	0.805	na	73.31	1.57	465.18	68.4000	23.5000	2302	−462	
1.07	3.95	34.5	388.92	243.22		126.68	0.730	na	71.85	1.47	446.08	68.4000	23.5000	2302	−462	09KMAD14
1.35	4.98	43.25	185.96	156.06		78.35	0.765	na	84.88	1.77	222.63	68.4000	23.5000	2302	−462	
1.18	9.25	47.75	356.93	183.16		125.56	0.797	na	72.66	1.48	399.97	68.4000	23.5000	2302	−462	
1.29	4.46	40.25	387.44	238.27		138.56	0.754	na	76.47	1.71	443.43	68.4000	23.5000	2302	−462	
2.11	0.35	24.5	5279.24	112.99		3153.57	na	0.678	152.76	3.52	1470.28	66.3886	25.4488	2315	−182	
2.93	0.23	37.25	5204.83	118.46		4322.79	na	0.735	191.23	4.15	468.48	66.3886	25.4488	2315	−182	09JA01
1.85	0.76	32	5186.69	77.19		2592.47	na	0.747	117.89	2.69	1438.69	66.3886	25.4488	2315	−182	
2.53	5.48	62.25	2135.53	96.91		1462.92	na	0.850	142.76	3.03	779.15	66.3886	25.4488	2315	−182	
2.29	2.26	35.5	112.17	22.30		77.31	na	0.780	154.91	3.24	117.41	66.3404	25.3908	2255	−179	
4.09	1.09	34.5	1379.12	252.21		1685.23	na	0.765	269.19	5.77	442.90	66.3886	25.4488	2315	−182	08JA01
1.31	3.74	37.25	441.97	76.49		307.59	na	0.742	154.30	3.28	459.95	66.3886	25.4488	2315	−182	
2.43	1.36	37.5	1214.02	210.12		844.89	na	0.766	112.02	2.33	414.74	66.3886	25.4488	2315	−182	

(Continued)

TABLE 1. NEW APATITE AND ZIRCON (U-Th)/He DATA (*Continued*)

Sample name	Unit, lithologic, location, or age information	Aliquot name	⁴He (pmol)	⁴He (pmol ± 1σ)	U (ng)	U (ng ± 1σ)	Th (ng)	Th (ng ± 1σ)	Th/U (atomic)	Sm (ng)	Sm (ng ± 1σ)	Raw age (Ma)
09JA10	Brealito pluton, Cerro Arcaguay	09JA10_10	1.07383	0.00597	0.5159	0.0098	0.1437	0.0025	0.29			351.18
		09JA10_20	0.55620	0.00298	0.5745	0.0107	0.0782	0.0014	0.14			171.43
		08JA06B_30	4.26397	0.02461	3.2126	0.0651	0.3336	0.0073	0.11			235.35
08JA06B	Pirgua Group sandstone	08JA06B_32	2.43687	0.01110	2.0542	0.0378	0.5432	0.0098	0.27			203.52
		08JA06B_38	3.20557	0.01833	2.7884	0.0520	0.3986	0.0071	0.15			202.67
		08JA06B_57	2.82918	0.01651	1.5930	0.0294	0.1858	0.0036	0.12			311.74
		08JA06_18	0.46882	0.00299	0.3268	0.0061	0.1365	0.0024	0.43			237.34
08JA06	Pirgua Group conglomerate	08JA06_36	0.21853	0.00126	0.1361	0.0044	0.1976	0.0034	1.49			218.01
		08JA06_46	0.29975	0.00165	0.2105	0.0040	0.1122	0.0019	0.55			230.08
		08JA06_66	0.13514	0.00084	0.0731	0.0015	0.1053	0.0018	1.48			250.99
		09JA09_16	1.40232	0.00796	0.9528	0.0175	0.1156	0.0020	0.12			259.35
09JA09	La Paya intrusive rock, Cachi Range	09JA09_41	1.34382	0.00861	1.4693	0.0281	0.6516	0.0106	0.45			151.64
		09JA09_42	0.62556	0.00457	0.5054	0.0095	0.1044	0.0018	0.21			214.86
		09JA11_30	0.08632	0.00063	0.2136	0.0042	0.0623	0.0011	0.30			69.75
09JA11	Angostura granite, Cerro Rumio	09JA11_6	0.94018	0.00608	1.6664	0.0308	0.4877	0.0083	0.30			97.13
		09JA11_19	0.68730	0.00391	1.2583	0.0233	0.1844	0.0031	0.15			97.17
CL1	JKv granodiorite	CL1Z1	3.01307	0.03246	4.7755	0.0837	3.6659	0.0894	0.79			98.30
		CL1Z2	3.73070	0.03884	6.7696	0.1201	3.1867	0.0779	0.48			91.33
CL7	Pzg late Paleozoic granite	CL7Z1	0.31564	0.00332	0.6079	0.0109	0.3279	0.0079	0.55			84.86
		CL7Z2	0.49594	0.00257	0.9421	0.0174	0.3558	0.0085	0.39			89.02
CL11	Kg Jurassic granodiorite	CL11Z1	1.78631	0.01056	2.7451	0.0481	2.2515	0.0539	0.84			100.31
		CL11Z2	1.93253	0.00977	2.9446	0.0521	2.5778	0.0627	0.90			100.08
CL14	Jig Early Jurassic granite	CL14Z1	0.33216	0.00172	1.7110	0.0308	0.8221	0.0195	0.49			32.27
		CL14Z2	0.31415	0.00167	1.2876	0.0247	0.9085	0.0218	0.72			38.68
CL15	Kg Cretaceous granodiorite	CL15Z1	0.73776	0.01586	2.4255	0.0425	2.1374	0.0517	0.90			46.54
		CL15Z2	0.61495	0.01353	2.3782	0.0418	1.3999	0.0335	0.60			41.98
CL22	Kgd Late K–Early Pg granodiorite	CL22Z1	0.34503	0.00173	0.4255	0.0084	0.2829	0.0072	0.68			128.64
		CL22-Z-3	0.16426	0.00102	0.3151	0.0078	0.1876	0.0029	0.61			84.21
		CL22-Z-4	1.17853	0.00596	1.7377	0.0426	1.2339	0.0197	0.73			106.82
CL23	Granite clast in clonglomerate of Orange (?) Fm.	CL23z1	1.81348	0.00975	1.5087	0.0271	1.0184	0.0242	0.69			189.32
		CL23z2	1.28421	0.00666	1.4669	0.0257	1.0219	0.0243	0.71			137.89
CL29	Ktig Paleocene granite	CL29-Z-3	0.87463	0.00482	1.0514	0.0259	0.6154	0.0095	0.60			134.09
		CL29-Z-4	1.25197	0.00668	1.1007	0.0271	0.9176	0.0143	0.86			173.80
		CL29z5	1.12429	0.00968	1.0625	0.0325	0.7671	0.0231	0.74			165.44
		CL29z6	0.95713	0.03154	1.0953	0.0336	0.6944	0.0208	0.65			139.38

Samples not included in transect because magmatic or hydrothermal cooling dates of severely radiation damaged zircon

Sample name	Unit, lithologic, location, or age information	Aliquot name	⁴He (pmol)	⁴He (pmol ± 1σ)	U (ng)	U (ng ± 1σ)	Th (ng)	Th (ng ± 1σ)	Th/U (atomic)	Sm (ng)	Sm (ng ± 1σ)	Raw age (Ma)
CL24	Volcanic tuff	CL24z1	0.19998	0.00112	7.3133	0.1308	3.5180	0.0858	0.49			4.56
		CL24nee22Z2	0.08296	0.00045	3.2303	0.0598	1.3428	0.0321	0.43			4.34
		CL24z3	0.28383	0.00247	10.8440	0.3285	4.9040	0.1479	0.46			4.39
		CL24z4	0.57196	0.00500	20.1348	0.6230	10.9742	0.3323	0.56			4.67
08AA28	Cenozoic granitoid	08AA28_Z_2	0.45328	0.00091	3.3074	0.0640	1.1547	0.0197	0.36			23.46
		08AA28_Z_3	0.12750	0.00026	0.9798	0.0188	0.6700	0.0117	0.70			20.75

F$_\tau$—alpha–ejection correction (see methods section and references therein); 1s err anal.—one–sigma analytical precision; eU—equivalent uranium concentration (U + 0.235 x Th); m a.s.l.—meters above sea level.

Raw age (Ma ± 1σ)	Mass (μg)	Half-width (μm)	U (ppm)	Th (ppm)	Sm (ppm)	⁴He (nmol/g)	F_T	F_T polished	Corrected age (Ma)	1s err anal. (Ma)	eU (ppm)	Longitude (°W)	Latitude (°S)	Elevation (m a.s.l.)	Distance from front (km)	Sample name
6.55	3.34	38.75	154.32	42.98		321.24	na	0.803	437.13	9.03	164.42	66.3404	25.3908	2255	−179	09JA10
3.17	2.39	36.75	240.55	32.73		232.87	na	0.786	218.07	4.43	248.24	66.3404	25.3908	2255	−179	
4.78	7.48	53	429.28	44.57		569.77	na	0.849	277.14	5.98	439.75	66.3140	25.4308	2058	−175	
3.59	8.11	52.5	253.15	66.95		300.32	na	0.851	239.03	4.49	268.89	66.3140	25.4308	2058	−175	08JA06B
3.78	7.67	57	363.60	51.97		418.00	na	0.857	236.41	4.63	375.81	66.3140	25.4308	2058	−175	
5.86	6.02	53.5	264.66	30.87		470.05	na	0.845	368.77	7.41	271.92	66.3140	25.4308	2058	−175	
4.28	1.03	26.25	318.30	132.89		456.61	na	0.711	333.68	6.84	349.53	66.3541	25.3124	2535	−183	
5.39	0.72	26.5	188.91	274.24		303.29	na	0.694	313.95	8.77	253.36	66.3541	25.3124	2535	−183	08JA06
4.01	1.26	26.5	167.15	89.09		238.02	na	0.718	320.49	6.28	188.09	66.3541	25.3124	2535	−183	
4.22	0.61	26.75	120.28	173.30		222.46	na	0.724	346.86	7.64	161.01	66.3541	25.3124	2535	−183	
4.81	3.11	39.5	306.33	37.16		450.85	na	0.818	317.17	6.50	315.06	66.3472	25.2484	2633	−185	
2.75	3.02	43.75	485.85	215.46		444.35	na	0.810	187.13	3.63	536.48	66.3472	25.2484	2633	−185	09JA09
4.10	1.56	31	323.59	66.82		400.51	na	0.770	279.09	6.08	339.29	66.3472	25.2484	2633	−185	
1.34	0.74	25	290.14	84.64		117.25	na	0.733	95.22	2.30	310.03	66.2355	25.4924	1987	−164	
1.75	4.22	31.75	395.32	115.69		223.05	na	0.795	122.22	2.52	422.51	66.2355	25.4924	1987	−164	09JA11
1.78	1.20	30.25	1046.71	153.42		571.74	na	0.753	129.01	3.03	1082.76	66.2355	25.4924	1987	−164	
1.81	25.17	52.25	189.69	145.62		119.69	0.826	na	119.04	2.19	223.91	70.3671	24.2346	1638	−644	CL1
1.72	23.92	68.5	283.01	133.22		155.97	0.853	na	107.01	2.02	314.32	70.3671	24.2346	1638	−644	
1.60	2.63	37	230.73	124.47		119.81	0.720	na	117.93	2.23	259.98	70.5070	23.0996	520	−698	CL7
1.55	4.11	37.25	229.29	86.59		120.70	0.743	na	119.73	2.08	249.64	70.5070	23.0996	520	−698	
1.60	3.31	39.75	828.65	679.65		539.23	0.735	na	136.40	2.17	988.36	70.1748	22.0827	224	−698	CL11
1.57	2.65	37.25	1111.14	972.72		729.24	0.719	na	139.16	2.18	1339.73	70.1748	22.0827	224	−698	
0.54	4.29	48.75	398.85	191.63		77.43	0.768	na	42.01	0.70	443.89	69.5168	23.4038	1519	−583	CL14
0.66	5.94	35.25	216.59	152.81		52.84	0.741	na	52.19	0.89	252.51	69.5168	23.4038	1519	−583	
1.21	10.91	60	222.34	195.93		67.63	0.814	na	57.15	1.49	268.39	69.3585	22.8731	1743	−585	CL15
1.12	11.09	54.5	214.44	126.23		55.45	0.813	na	51.65	1.38	244.11	69.3585	22.8731	1743	−585	
2.28	5.34	42	79.63	52.95		64.57	0.764	na	168.47	2.99	92.07	68.3397	23.6674	2376	−450	CL22
1.85	3.57	42	88.19	52.50		45.97	0.749	na	112.46	2.47	100.53	68.3397	23.6674	2376	−450	
2.26	6.96	41.5	249.58	177.23		169.27	0.772	na	138.36	2.92	291.23	68.3397	23.6674	2376	−450	
3.11	4.43	37.25	340.37	229.75		409.13	0.739	na	256.33	4.22	394.36	0.0000	0.0000	2554		CL23
2.20	5.53	43.75	265.15	184.72		232.13	0.770	na	179.06	2.85	308.56	0.0000	0.0000	2554		
2.93	4.81	48.75	218.61	127.97		181.86	0.774	na	173.26	3.79	248.68	69.0347	23.5373	2472	−527	CL29
3.64	5.91	46.5	186.18	155.21		211.77	0.777	na	223.64	4.68	222.66	69.0347	23.5373	2472	−527	
4.52	8.80	59.5	120.73	87.17		127.76	0.812	na	203.68	5.56	141.22	69.0347	23.5373	2472	−527	
5.92	4.36	45.25	251.37	159.35		219.65	0.764	na	182.39	7.75	288.82	69.0347	23.5373	2472	−527	
												Samples not included in transect because magmatic or hyrdrothermal cooling dates or severely radiation damaged zircon				
0.08	13.09	56.5	558.59	268.70		15.27	0.823	na	5.53	0.09	621.74					CL24
0.07	4.99	43.25	646.79	268.86		16.61	0.767	na	5.66	0.10	709.97					
0.12	28.24	68.25	383.98	173.65		10.05	0.857	na	5.12	0.14	424.79	69.0347	23.5373		−527	
0.13	55.14	71.75	365.13	199.01		10.37	0.870	na	5.36	0.15	411.90	69.0347	23.5373		−527	
0.41	13.40	58	246.89	86.20		33.84	0.823	na	28.51	0.50	267.15	67.7293	24.5769		−354	08AA28
0.34	5.07	35.25	193.15	132.08		25.13	0.740	na	28.06	0.46	224.18	67.7293	24.5769		−354	

TABLE 2. APATITE FISSION-TRACK DATA

Sample number	Unit, lithology, location, or age information*	No. of crystals	Spontaneous		Induced		$P(\chi)^2$	Dosimeter	
			ρ_s	N_s	ρ_s	N_s		ρ_d	N_d
CL1	JKv granodiorite	20	7.22	307	2.34	995	72.4	1.02	4852
CL6	J-K sandstone, Santa Ana Fm.	20	1.34	54	0.93	376	100.0	0.98	4667
CL7	Pzg late Paleozoic granite	20	2.97	224	1.22	917	78.7	0.97	4620
CL8	Jsg Jurassic granitoid	18	5.41	193	1.18	422	80.3	0.96	4574
CL9	Kg Jurassic granodiorite	20	3.66	182	1.09	544	90.8	0.95	4528
CL10	Kg Jurassic granodiorite	10	4.79	102	0.85	182	98.4	0.94	4482
CL11	Kg Jurassic granodiorite	20	6.34	300	1.23	582	98.3	0.93	4435
CL12	Jg Jurassic granite	14	6.82	194	1.15	328	0.1	0.92	4389
CL13	Kssb(a) El Buitre K-Pg granodiorite	13	0.91	36	1.00	396	71.2	0.91	4343
CL14	Jig Early Jurassic granite	20	2.12	107	1.06	534	84.4	0.90	4297
CL15	Kg Cretaceous granodiorite	20	2.45	137	1.41	788	0.2	0.89	4250
CL17	Pzs El Toco Fm. sandstone	14	4.51	142	1.40	442	0.0	0.88	4204
CL18	Pzgm granite	20	1.88	80	1.20	510	99.8	0.87	4158
CL20	Pzg granite	20	6.22	228	2.25	824	91.4	0.86	4112
CL21	Kgd Cretaceous granodiorite	6	0.98	18	0.51	93	81.8	0.86	4065
CL22	Kgd Late K–Early Pg granodiorite	20	1.77	123	1.18	824	100.0	0.85	4019
CL23	Granite clast in conglomerate of Orange (?) Fm.	20	1.06	65	0.55	338	93.6	0.84	3973
CL27	Cretaceous Purilactis sandstone	20	1.15	77	0.86	577	99.7	1.12	5316
CL28	Loma Amarilla	20	5.01	277	3.09	1711	69.7	1.11	5288
CL29	Ktig Paleocene granite	20	1.05	62	0.43	254	99.7	1.11	5260
Below: volcanic tuff sample; not included in transect.									
CL24	Volcanic tuff	20	0.26	19	0.85	619	87.5	0.83	3927

Note: ρ_s—spontaneous track density, N_s—number of spontaneous tracks, ρ_i—induced track density; N_i—number of induced tracks; ρ_d—dosimeter track density; N_d—number of dosimeter tracks; $P(\chi)^2$—χ^2 probability; m a.s.l.: meters above sea level.

samples of McInnes et al. (1999), but we only show samples within 0.5 km of the surface to be consistent with the other data sets. We also compiled AFT dates from samples between 26°S and 28°S, from Coughlin et al. (1998), Sobel and Strecker (2003), and Carrapa et al. (2006). We also report U/Pb dates determined on a subset of detrital zircon grains that were also dated by the (U-Th)/He method. These (U-Th)/He dates were analyzed by standard procedures described in Reiners et al. (2004). Alpha-ejection corrections for grains plucked from U/Pb grain mounts were made using the procedure in Reiners et al. (2007). U/Pb dates were measured by laser-ablation–inductively coupled plasma–mass spectrometry (ICP-MS) following methods in Gehrels et al. (2008). AFT dates were analyzed by standard procedures described in Zattin et al. (2003).

In the following section, we first present detrital zircon geochronologic and thermochronologic results from Paleogene sedimentary rocks in the Cordillera de Domeyko (Figs. 1 and 2). These are the only samples for which we discuss the geologic context and date interpretations in any detail, because this is a region with few previous geochronologic and thermochronologic data, and some of the results form important evidence for shallow range-internal erosion and ancient zircon He dates discussed later.

We then present cooling dates in the context of the large-scale date pattern across the entire orogeny by focusing on a cross-orogen transect extending from the Chilean coast at 71.5°W to the frontal Santa Barbara Ranges at 64°W–65°W (Fig. 2). Samples were collected across a range of elevations from 45 to 5670 m above sea level (Fig. 2). To emphasize cooling date patterns associated with erosional exhumation, rather than magmatic cooling, dates from samples of Cenozoic magmatic rocks with cooling dates that are indistinguishable within error from

Age (Ma ± 1σ)	Age for plot (Ma)	Mean confined track length (mm ± std. err.)	Std. dev.	No. of tracks measured	Loc X (°W)	Loc Y (°S)	Elevation (m a.s.l.)	Zm (m a.s.l.)	Distance from front (km)	Sample number
50.8 ± 3.4	50.8	13.60 ± 0.13	1.35	100	70.367134	24.23458	1638	1639	−644	CL1
23.7 ± 3.5	23.7	14.86 ± 0.11	1.08	100	69.1128	24.245817	3160	3036	−511	CL6
43.5 ± 3.3	43.5	13.24 ± 0.17	1.72	100	70.50699	23.099619	520	119	698	CL7
80.2 ± 7.1	80.2	14.05 ± 0.23	1.24	27	70.33236	23.09185	110	278	680	CL8
58.2 ± 5.1	58.2	13.38 ± 0.15	1.21	100	70.260254	22.627453	76	398	688	CL9
98.5 ± 12.3	98.5	13.58 ± 0.15	1.52	100	70.25423	22.458052	76	365	693	CL10
87.8 ± 6.4	87.8	13.43 ± 0.15	1.46	100	70.17482	22.082664	224	641	698	CL11
62.1 ± 14.0	62.1	–	–	–	70.205414	23.461185	725	880	−653	CL12
15.3 ± 2.7	15.3	13.75 ± 0.27	1.83	46	69.963715	23.319435	1211	1185	−633	CL13
33.1 ± 3.5	33.1	14.12 ± 0.13	1.28	100	69.51684	23.403782	1519	1733	−583	CL14
29.1 ± 3.8	29.1	–	–	–	69.35845	22.873102	1743	1642	−585	CL15
46.2 ± 8.2	46.2	–	–	–	69.784584	22.044035	1406	1514	−658	CL17
25.1 ± 3.0	25.1	13.06 ± 0.34	1.47	19	68.838295	22.365126	2439	2508	−547	CL18
43.7 ± 3.4	43.7	13.36 ± 0.15	1.54	100	68.922585	23.040968	2935	2918	−533	CL20
30.3 ± 7.8	30.3	–	–	–	68.7582	23.668657	2726	2824	−494	CL21
23.2 ± 2.3	23.2	13.46 ± 0.12	1.18	100	68.33968	23.667439	2376	2410	−450	CL22
29.4 ± 4.0	29.4	14.30 ± 0.39	1.24	10	68.320274	22.846033	2554	2696	−476	CL23
27.4 ± 3.4	27.4	14.46 ± 0.79	1.76	5	68.47902	23.099585	2336	2613	−484	CL27
32.9 ± 2.2	32.9	13.10 ± 0.17	1.65	100	68.63281	23.66815	2611	2664	−480	CL28
49.6 ± 7.1	49.6	12.28 ± 0.23	2.35	100	69.03474	23.537252	2472	2451	−527	CL29
									Below: volcanic tuff sample; not included in transect.	
4.7 ± 1.1		–	–	–	68.320274	22.846033			−476	CL24

known or inferred crystallization dates are not shown in figures or considered in interpretations. A limitation of this data set is a paucity of samples in the northeast and southwest corners of the cross-orogen swath. Although we see no reason why the main interpretations of this work would change with more complete sample coverage, this may be a consideration for future work attempting to interpret cooling dates at the cross-orogen scale.

In interpreting thermochronologic dates in the context of erosion rates, we adopt a single set of kinetic parameters for each thermochronometer as compiled in Willett and Brandon (2013). These yield apparent closure temperatures (assuming a cooling rate of 10 °C/m.y.) of ~67 °C, 116 °C, and 183 °C, for the apatite He, AFT, and zircon He systems, respectively. We recognize that kinetic models incorporating the effects of radiation damage (Flowers et al., 2009; Gautheron et al., 2009; Guenthner et al., 2013) and composition (e.g., Ketcham, 2005) would provide a

more robust inversion from date to model (steady) erosion rate. Kinetic differences among populations may also contribute to variation in the degree of partial resetting among older, "detrital" dates that are observed in some locations. However, incorporating these effects into the erosion-rate inversions in a reasonably realistic way would require more detailed knowledge of each sample's thermal history, and a more sophisticated inversion method than the current version of AGE2EDOT (Willett and Brandon, 2013), and the differences would not significantly affect the overall date-distance trends seen here or their interpretations.

CORDILLERA DE DOMEYKO AND EASTERN CORDILLERA DETRITAL SAMPLES

We first present detrital zircon U/Pb and (U-Th)/He cooling dates from samples collected from Paleogene sedimentary rocks

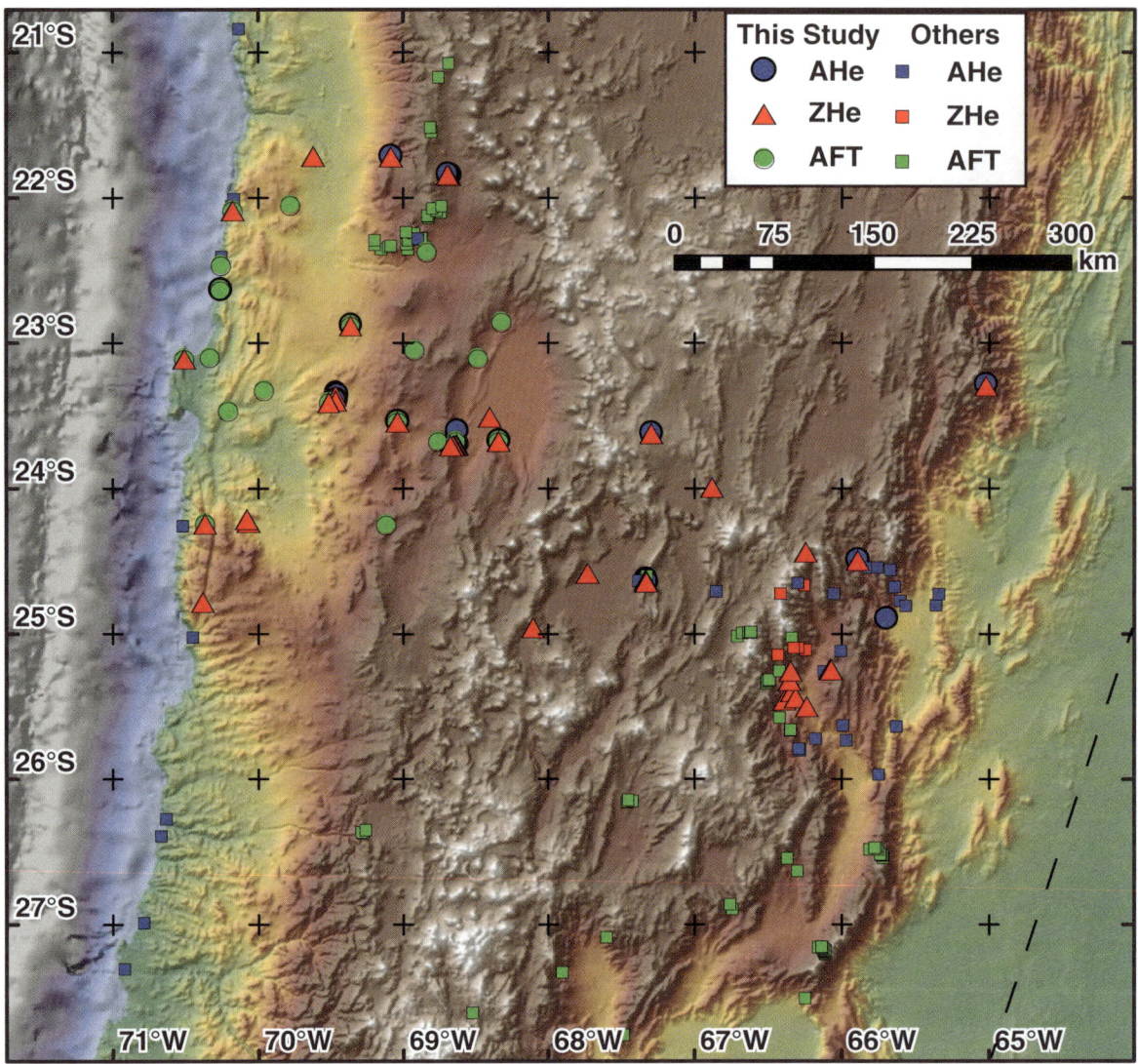

Figure 2. Plan view shaded relief and sample location map. Dashed trend is location of reference line used to define distance of samples from the "range front" with a northeast coordinate of 62.7°W and 21°S and southwest coordinate of 65°W and 28°S. Triangles and circles represent locations of new data presented in this study (red triangles—zircon (U-Th)/He; green circles—apatite fission track [AFT]; blue circles—apatite (U-Th)/He). Squares represent locations of samples from previous studies (see Samples and Methods section), with the same color scheme as the new samples.

on the eastern flank of the southern Cordillera de Domeyko, near the southwestern edge of the Salar de Atacama. These include 25 double-dated grains from five samples from the Paleogene Loma Amarilla Formation, a coarsening-upward clastic sequence reaching total exposed thicknesses greater than 2.5 km. This unit is interpreted as a syntectonic sedimentary package associated with Eocene deformation in this part of the Andes (Mpodozis et al., 2005). Its depositional age is not well constrained. Hammerschmidt et al. (1992) reported biotite ^{40}Ar/^{39}Ar plateau dates of 43.8 ± 0.5 Ma and 44.2 ± 0.9 Ma from the basal part of the unit. Mpodozis et al. (2005) reported a K/Ar date of 59.1 ± 2 Ma from a volcanic clast at the basal part of the unit, but this may represent reworked material. These authors also reported a much younger date of K/Ar date or 33.9 ± 0.3 Ma on hornblende from a volcanic clast only 100 m from the base of the unit, but they expressed doubt about the accuracy of the date due to the low-K content of the sample. They also noted that this date is similar to the total degassing dates on the same biotites studied by Hammerschmidt et al. (1992) and may represent partial Ar loss, possibly associated with a Miocene event that may be reflected in the 10–20 Ma dates of the earliest degassing steps (Mpodozis et al., 2005). Recognizing the uncertainties in these age constraints, we adopt a provisional estimate for the depositional age of the Loma Amarilla of somewhere between 34 and 44 Ma.

We also measured detrital zircon U/Pb and He double-dates from five grains from one sample of sandstone from the Naranja Formation, which directly underlies the Loma Amarilla Formation in parts of the Cordillera de Domeyko (Mpodozis et al., 2005; Arriagada et al., 2006). Seismic correlations in the subsurface of the Salar de Atacama suggest that the Naranja Formation is younger than Late Cretaceous, and similar units ~70 km to the south have been dated (whole-rock K/Ar) at 57.9 ± 1.9 Ma and 58.0 ± 3 Ma (Gardeweg et al., 1994). Mpodozis et al. (2005) interpreted the Naranja Formation as a post-tectonic sedimentary sequence following an episode of compressional deformation.

Loma Amarilla Zircon

Zircon grains from the Loma Amarilla Formation have U/Pb dates ranging from 55 to 308 Ma and (U-Th)/He dates ranging from 49 to 221 Ma (Tables 1 and 3). The U/Pb dates (Table 3) show a date gap between 100 and 200 Ma, with a secondary clustering of dates in the older group between 210 and 220 and between 250 and 260 Ma (Fig. 3). Grains with U/Pb dates younger than 100 Ma have crystallization (U/Pb) and cooling (He) dates that are concordant or nearly so, indicating derivation from first-cycle volcanic or hypabyssal sources and lack of burial or exhumation through depths where partial He loss from zircon occurs, typically 4–6 km (assuming geothermal gradients of ~20–30 °C per km). In contrast, most zircon grains with U/Pb dates older than 200 Ma have (U-Th)/He dates significantly younger (140–180 Ma) than crystallization dates. Three (four) of these older grains have cooling dates within 5% (10%) of the crystallization dates (Fig. 3), indicating that some of these older zircon grains have remained within a few kilometers of the surface since at least 200 Ma.

None of the dated zircons from the Loma Amarilla has a date within the presumed depositional age range of the unit (33–44 Ma; Fig. 3). However, only five zircon grains per sample were analyzed for either U/Pb or He dates, so younger grains approaching the presumed depositional ages could be found with more extensive analyses. Nonetheless, these results suggest that most of the grains in the Loma Amarilla Formation were formed significantly earlier than the depositional age of the unit. Notably, many of these grains have not been buried to depths where temperatures approached partial He loss from zircon (~140 °C) for as long as ~220 m.y.

Naranja Formation Zircon

Five zircon grains from one sample of the Naranja Formation have U/Pb dates of 72–81 Ma. (U-Th)/He dates on the same grains are the same within uncertainty, indicating that they are all first-cycle volcanic or hypabyssal grains that have remained within a few kilometers of the surface since the Late Cretaceous. As in the case of the Loma Amarilla, however, these dates are significantly older than the inferred depositional age of the unit (58 Ma).

Cordillera de Domeyko Provenance Changes?

Although there are too few double-dated grains to provide robust evidence, we note that both the Cretaceous Naranja Formation and the older (likely Eocene) section of the Loma Amarilla Formation contain abundant first-cycle zircon, whereas the younger Loma Amarilla Formation section contains exclusively zircons with He dates significantly younger than the U/Pb dates on the same grains. The younger Loma Amarilla unit also appears to contain zircons with a more restricted range of older U/Pb dates. We speculate that these features represent progressive unroofing of a magmatic crustal section, with volcanic and shallow igneous units eroded earlier, and deeper and older units eroded later within the stratigraphic sequence represented by the Loma Amarilla.

Comparison with Zircon Double-Dates from the Eastern Cordillera

Figure 3B shows (U-Th)/He and U/Pb double-dates of Cordillera de Domeyko zircons compared with those from several samples from the Eastern Cordillera. The latter include detrital zircons from the Eocene Santa Bárbara Subgroup of the Salta Group in the Tintin region, and the Cretaceous Pirgua sandstone of the Brealito subbasin of the Salta rift, as well as zircon grains from several samples of crystalline basement (primarily granitoids) surrounding the Brealito subbasin (Table 1). Both detrital and basement zircons from the Eastern Cordillera are generally much older than those from the Cordillera de Domeyko, with a primary U/Pb peak at ca. 460–560 Ma and secondary peaks around 1 Ga and older. Zircon He dates from these Eastern Cordillera grains are also generally older than those from the Cordillera de Domeyko. Only one zircon grain that we dated from this region has a He date within error of the U/Pb date. The majority have He dates ranging from 100 to 350 Ma, consistent with derivation from older rocks that cooled below ~150–200 °C between the mid-Cretaceous and Carboniferous. In detail, several of the early Paleozoic zircons show inverse He date–eU correlations, consistent with thermal histories involving residence at temperatures less than ~225–250 °C since formation (Guenthner et al., 2013). The main finding from the Eastern Cordillera zircons is that most of these rocks have not been hotter than the closure temperature of the zircon He system (~170–190 °C) since before the Cenozoic. Assuming a surface temperature of 10 °C and geothermal gradients of 20–30 °C per km, this means that they have been shallower than 6–8 km beneath the surface, and likely much shallower, since at least the mid-Cretaceous, and probably since the Carboniferous.

Cordillera de Domeyko Apatite

We measured (U-Th)/He dates in detrital apatite grains from six samples through the Loma Amarilla Formation covering an elevation range of ~450 m and a stratigraphic thickness of ~2.4 km. Apatite He dates near the top of this section are

Reiners et al.

TABLE 3. U-Pb GEOCHRONOLOGIC ANALYSES

Analysis	U (ppm)	$^{206}Pb/^{204}Pb$	U/Th	Isotope ratios					
				$^{206}Pb^*/^{207}Pb^*$	± (%)	$^{207}Pb^*/^{235}U^*$	± (%)	$^{206}Pb^*/^{238}U$	± (%)
09KMAD14Z-1	387.4	3780	2.1	23.0298	10.44	0.0672	10.56	0.0112	1.59
09KMAD14Z-2	310.9	968	2.0	14.5983	9.22	0.1113	9.37	0.0118	1.62
09KMAD14Z-3	200.8	2348	1.4	22.4797	9.92	0.0774	10.44	0.0126	3.25
09KMAD14Z-4	368.4	3380	2.1	21.7398	6.09	0.0757	6.17	0.0119	0.98
09KMAD14Z-5	319.9	2896	2.2	19.9767	11.22	0.0796	11.30	0.0115	1.29
09KMAD11Z-1	284.0	2664	1.7	19.1315	24.15	0.0863	24.17	0.0120	1.01
09KMAD11Z-2	208.3	4864	1.4	18.3904	8.90	0.2624	9.03	0.0350	1.53
09KMAD11Z-3	125.7	3808	1.3	18.3143	4.78	0.2705	5.09	0.0359	1.76
09KMAD11Z-4	155.4	652	1.3	11.3190	19.28	0.1242	19.43	0.0102	2.40
09KMAD11Z-5	242.9	5280	1.2	18.5690	4.58	0.3127	4.71	0.0421	1.11
09KMAD09Z-1	454.8	16,836	2.2	18.6006	1.92	0.3029	1.98	0.0409	0.50
09KMAD09Z-2	475.7	20,476	1.3	19.2012	2.26	0.3098	3.00	0.0431	1.97
09KMAD09Z-3	318.6	13,532	2.6	19.4528	2.21	0.2933	3.38	0.0414	2.55
09KMAD09Z-4	213.4	5356	3.0	18.6515	6.69	0.3060	6.77	0.0414	1.05
09KMAD09Z-5	116.2	6496	2.5	19.2177	6.29	0.2998	6.35	0.0418	0.85
09KMAD06Z-1	85.2	932	2.1	16.1367	19.56	0.1276	19.63	0.0149	1.75
09KMAD06Z-2	192.4	10,776	1.7	20.0863	7.30	0.2452	7.33	0.0357	0.74
09KMAD06Z-3	109.5	1876	2.5	16.8140	8.05	0.2846	8.11	0.0347	0.93
09KMAD06Z-4	339.2	3524	1.3	16.1145	13.83	0.2531	14.53	0.0296	4.45
09KMAD06Z-5	174.9	9372	1.8	18.6540	11.50	0.2657	11.92	0.0360	3.13
09KMAD13Z-1	98.3	1380	0.9	27.7667	46.92	0.0426	47.04	0.0086	3.34
09KMAD13Z-2	99.9	5504	1.3	19.6450	7.78	0.2675	8.14	0.0381	2.37
09KMAD13Z-4	212.0	10,516	2.5	19.7130	9.14	0.2321	9.17	0.0332	0.82
09KMAD13Z-5	119.0	6884	1.4	20.9595	19.26	0.2306	19.29	0.0351	1.04
09KMAD08Z-1	222.1	5448	2.0	15.6987	7.97	0.4305	8.59	0.0490	3.22
09KMAD08Z-2	229.7	11,564	2.2	20.0120	5.03	0.2917	5.05	0.0423	0.50
09KMAD08Z-3	214.1	9852	1.3	19.8949	6.66	0.2825	7.17	0.0408	2.66
09KMAD08Z-4	460.9	19,148	1.1	18.1382	8.60	0.3088	8.63	0.0406	0.80
09KMAD08Z-5	152.7	5620	2.1	19.6041	5.12	0.2858	5.25	0.0406	1.16
2TT75-86	70.5	7636	3.3	13.4681	2.51	1.7711	2.89	0.1730	1.42
2TT75-90	259.2	13,756	6.3	16.5018	4.62	0.6641	7.71	0.0795	6.18
2TT75-92	592.5	23,694	8.4	17.1697	1.11	0.7027	1.63	0.0875	1.20
2TT75-100	60.3	9260	2.9	13.6911	3.13	1.7711	3.44	0.1759	1.44
2TT142-4	509.1	18,506	6.1	16.5897	1.61	0.7602	2.76	0.0915	2.24
2TT142-5	463.6	27,568	3.0	12.8670	1.84	1.7603	3.01	0.1643	2.38
2TT142-57	498.7	16,388	2.2	7.3504	1.45	7.1391	3.24	0.3806	2.90
2TT142-67	247.6	10,530	23.4	17.4541	1.80	0.6346	2.76	0.0803	2.09
2TT948-28	308.3	10,046	1.3	16.9793	3.92	0.6643	4.24	0.0818	1.63

Error corr.	$^{206}Pb*/^{238}U*$	± (Ma)	$^{207}Pb*/^{235}U$	± (Ma)	$^{206}Pb*/^{207}Pb*$	± (Ma)	Best age (Ma)	± (Ma)	Analysis
					Apparent ages (Ma)				
0.151	72.0008	1.14	66.08	6.76	−143.34	259.30	72.00	1.14	09KMAD14Z-1
0.173	75.5045	1.22	107.13	9.52	883.78	191.12	75.50	1.22	09KMAD14Z-2
0.311	80.8841	2.61	75.74	7.62	−83.78	243.48	80.88	2.61	09KMAD14Z-3
0.159	76.5328	0.75	74.14	4.41	−2.43	147.10	76.53	0.75	09KMAD14Z-4
0.114	73.9293	0.95	77.78	8.46	197.71	261.46	73.93	0.95	09KMAD14Z-5
0.042	76.7560	0.77	84.07	19.50	297.22	558.20	76.76	0.77	09KMAD11Z-1
0.169	221.7204	3.33	236.56	19.07	386.64	200.30	221.72	3.33	09KMAD11Z-2
0.346	227.5195	3.93	243.06	11.01	395.94	107.19	227.52	3.93	09KMAD11Z-3
0.124	65.4052	1.56	118.89	21.80	1390.03	373.44	65.41	1.56	09KMAD11Z-4
0.236	265.9044	2.89	276.26	11.39	364.90	103.21	265.90	2.89	09KMAD11Z-5
0.252	258.1742	1.27	268.66	4.68	361.06	43.32	258.17	1.27	09KMAD09Z-1
0.656	272.2652	5.25	274.01	7.21	288.91	51.76	272.27	5.25	09KMAD09Z-2
0.755	261.3520	6.53	261.13	7.77	259.09	50.85	261.35	6.53	09KMAD09Z-3
0.155	261.5068	2.69	271.11	16.12	354.89	151.30	261.51	2.69	09KMAD09Z-4
0.134	263.9236	2.20	266.27	14.86	286.95	143.85	263.92	2.20	09KMAD09Z-5
0.089	95.5922	1.66	121.98	22.57	673.09	421.99	95.59	1.66	09KMAD06Z-1
0.101	226.2870	1.65	222.70	14.67	184.95	170.08	226.29	1.65	09KMAD06Z-2
0.115	219.9706	2.01	254.34	18.24	584.49	175.09	219.97	2.01	09KMAD06Z-3
0.306	187.9152	8.24	229.08	29.81	676.05	297.03	187.92	8.24	09KMAD06Z-4
0.263	227.6884	7.00	239.28	25.41	354.58	260.41	227.69	7.00	09KMAD06Z-5
0.071	55.0650	1.83	42.36	19.52	−629.81	1346.61	55.07	1.83	09KMAD13Z-1
0.291	241.1188	5.61	240.69	17.43	236.45	179.74	241.12	5.61	09KMAD13Z-2
0.089	210.4750	1.70	211.96	17.55	228.48	211.44	210.47	1.70	09KMAD13Z-4
0.054	222.1338	2.27	210.72	36.73	84.99	460.61	222.13	2.27	09KMAD13Z-5
0.375	308.4571	9.70	363.51	26.26	731.69	168.98	308.46	9.70	09KMAD08Z-1
0.099	267.3014	1.31	259.89	11.58	193.62	116.92	267.30	1.31	09KMAD08Z-2
0.371	257.5313	6.72	252.62	16.03	207.23	154.50	257.53	6.72	09KMAD08Z-3
0.093	256.6954	2.01	273.25	20.69	417.57	192.35	256.70	2.01	09KMAD08Z-4
0.221	256.7784	2.92	255.25	11.85	241.27	118.10	256.78	2.92	09KMAD08Z-5
0.492	1028.6187	13.50	1034.92	18.73	1048.23	50.65	1048.23	50.65	2TT75-86
0.801	493.0539	29.33	517.14	31.27	625.03	99.56	493.05	29.33	2TT75-90
0.735	540.7686	6.22	540.41	6.84	538.87	24.22	540.77	6.22	2TT75-92
0.418	1044.3327	13.88	1034.91	22.35	1015.04	63.41	1015.04	63.41	2TT75-100
0.812	564.2349	12.10	574.15	12.10	613.56	34.81	564.23	12.10	2TT142-4
0.791	980.4836	21.65	1030.97	19.49	1139.67	36.63	1139.67	36.63	2TT142-5
0.894	2079.0142	51.54	2128.93	28.89	2177.46	25.26	2177.46	25.26	2TT142-57
0.758	498.0961	10.02	498.94	10.87	502.81	39.60	498.10	10.02	2TT142-67
0.384	506.9117	7.95	517.26	17.19	563.21	85.33	506.91	7.95	2TT948-28

(Continued)

TABLE 3. U-Pb GEOCHRONOLOGIC ANALYSES (*Continued*)

Analysis	U (ppm)	$^{206}Pb/^{204}Pb$	U/Th	Isotope ratios					
				$^{206}Pb*/^{207}Pb*$	± (%)	$^{207}Pb*/^{235}U*$	± (%)	$^{206}Pb*/^{238}U$	± (%)
2TT948-57	206.8	4270	1.7	17.1781	3.28	0.6813	3.44	0.0849	1.01
2TT948-64	544.2	32,054	2.2	16.0814	2.01	0.9388	2.57	0.1095	1.60
2TT948-69	127.2	7976	1.7	17.8032	3.11	0.6671	3.87	0.0861	2.29
09JA01-2	1537.6	6933	22.3	17.3344	1.39	0.6037	2.87	0.0759	2.51
09JA01-7	991.0	36,958	77.5	17.4758	0.87	0.5931	2.58	0.0752	2.43
09JA01-11	825.2	3833	57.1	17.4354	1.14	0.5859	2.30	0.0741	2.00
09JA01-14	645.4	26,988	120.1	17.5173	1.05	0.5970	3.77	0.0759	3.62
09JA10-10	251.8	21,048	6.8	17.1925	4.11	0.5856	7.90	0.0730	6.75
09JA10-20	193.4	18,852	12.8	18.1198	5.11	0.5636	5.68	0.0741	2.48
09JA10-30	303.5	33,042	6.4	15.0015	3.02	0.9589	6.77	0.1043	6.06
09JA10-40	306.6	21,213	1.2	17.7518	2.95	0.5829	4.66	0.0750	3.61
09JA09-30	153.1	16,575	2.1	13.2237	2.63	1.9468	3.48	0.1867	2.27
09JA09-41	303.7	11,511	2.9	17.4272	3.14	0.5932	3.41	0.0750	1.32
09JA09-42	326.3	16,416	11.1	17.6386	1.89	0.5866	3.43	0.0750	2.86
09JA11-6	189.8	10,383	1.0	17.9129	3.43	0.5455	3.95	0.0709	1.97
09JA11-19	226.7	15,171	1.7	18.1431	2.77	0.5623	3.52	0.0740	2.18
09JA09_16	526.5	27,588	8.2	17.5970	3.50	0.5914	4.12	0.0755	2.17
08JA06B_32	158	39,508	1.5	8.8876	1.30	4.6597	5.90	0.3004	5.80
08JA06B_38	433	53,704	18.6	17.6088	2.50	0.6020	2.90	0.0769	1.60
08JA06B_57	452	27,488	12.0	16.9185	4.90	0.6376	6.30	0.0782	4.00
08JA06_18	358	14,004	1.2	9.1376	1.80	4.8138	2.90	0.3190	2.30
08JA06_36	103	6444	1.2	3.3820	1.10	27.2247	3.30	0.6678	3.10
08JA06_46	179	1200	1.8	18.5052	4.80	0.5553	5.90	0.0745	3.30
08JA06_66	169	3346	1.0	8.9488	1.30	3.9736	7.90	0.2579	7.80

Note: See Table 1 for lithology/unit/location information.

close to or slightly older than inferred ages of deposition for the Loma Amarilla (Fig. 4). However, dates generally decrease with lower elevations and stratigraphic depth, becoming as young as 20–25 Ma in the lowest samples. These dates are ~12–24 m.y. younger than the inferred depositional ages of 34–44 Ma, and the rough positive correlation with elevation and stratigraphic height is consistent with at least partial resetting of the predepositional (U-Th)/He dates by burial and exhumation through depths of at least the apatite He partial retention zone (~40–85 °C; roughly 1.5–3.0 km). The maximum date of exhumation through these depths is the minimum apatite He date at the base of the section: ca. 20–22 Ma.

CROSS-OROGEN–SCALE PATTERNS IN THERMOCHRONOMETER DATES

Numerous studies have focused on the detailed exhumational history of individual subranges or basins within the Andes. Here, we attempt to elucidate first-order orogen-scale trends in exhumation patterns at the latitude of 21°S–28°S as revealed by thermochronometer dates at this scale. To do this, we plot dates and topography of the range as a function of distance from the location of significant topographic relief at the eastern front of the Andes, represented by a line extending from 62.7°W, 21°S to 65°W, 28°S. Results for all three thermochronometers are shown

Error corr.	Apparent ages (Ma)						Best age (Ma)	± (Ma)	Analysis
	$^{206}Pb*/^{238}U*$	± (Ma)	$^{207}Pb*/^{235}U$	± (Ma)	$^{206}Pb*/^{207}Pb*$	± (Ma)			
0.294	525.1800	5.09	527.55	14.14	537.79	71.89	525.18	5.09	2TT948-57
0.623	669.8442	10.18	672.28	12.64	680.41	42.97	669.84	10.18	2TT948-64
0.592	532.6431	11.71	518.94	15.71	459.06	69.08	532.64	11.71	2TT948-69
0.875	471.5839	11.43	479.57	10.98	517.93	30.56	471.58	11.43	09JA01-2
0.942	467.2673	10.97	472.86	9.77	500.12	19.12	467.27	10.97	09JA01-7
0.869	460.7524	8.90	468.25	8.65	505.19	25.14	460.75	8.90	09JA01-11
0.960	471.3159	16.44	475.35	14.30	494.85	23.10	471.32	16.44	09JA01-14
0.854	454.3523	29.61	468.08	29.65	535.96	90.07	454.35	29.61	09JA10-10
0.437	460.6171	11.02	453.87	20.78	419.84	114.07	460.62	11.02	09JA10-20
0.895	639.7528	36.91	682.74	33.67	827.15	63.05	639.75	36.91	09JA10-30
0.774	466.4703	16.24	466.31	17.44	465.49	65.48	466.47	16.24	09JA10-40
0.653	1103.5587	23.02	1097.36	23.32	1085.10	52.78	1085.10	52.78	09JA09-30
0.387	466.0804	5.94	472.92	12.90	506.22	69.20	466.08	5.94	09JA09-41
0.835	466.4344	12.87	468.67	12.86	479.62	41.70	466.43	12.87	09JA09-42
0.498	441.4156	8.40	442.06	14.17	445.42	76.18	441.42	8.40	09JA11-6
0.619	460.1733	9.68	453.04	12.87	416.96	61.77	460.17	9.68	09JA11-19
0.527	469.0769	9.82	471.77	15.55	484.89	77.33	469.08	9.82	09JA09_16
0.980	1693.1000	85.90	1760.00	49.50	1840.40	23.20	1840.40	23.20	08JA06B_32
0.540	477.5000	7.20	478.50	11.10	483.40	54.20	477.50	7.20	08JA06B_38
0.630	485.6000	18.60	500.80	25.10	571.00	107.60	485.60	18.60	08JA06B_57
0.780	1784.9000	35.10	1787.30	24.20	1790.10	32.80	1790.10	32.80	08JA06_18
0.950	3297.3000	81.10	3391.60	32.50	3447.80	16.80	3447.80	16.80	08JA06_36
0.570	463.3000	14.80	448.40	21.30	372.60	108.80	463.30	14.80	08JA06_46
0.990	1479.1000	103.40	1628.80	64.40	1828.00	24.20	1828.00	24.20	08JA06_66

in Figure 5, and results for each system independently are shown in Figure 6.

Remarks on Date-Distance Relationships for All Systems Together

Figures 5 and 6 show the large-scale thermochronometer date variation. Data from samples spanning seven degrees of latitude have been collapsed to create a single transect in this figure, so undoubtedly some of the observed date variation at any position in the transect is due to along-strike variations in the timing and depth of exhumation within this part of the Andes. Another potential cause of date variation is relationships between thermochronometer date and local topographic relief, e.g., as in "vertical transects." However, very few data in this compilation were collected over sufficiently short horizontal distances (i.e., the critical wavelength of Braun, 2002) to justify simple interpretations of date-elevation relationships in this way. Examination of dates in this compilation as a function of sample elevation, as well as local deviation from mean elevation over 5–10 km scales, confirms that local elevation differences are not significantly positively correlated with date variation for nearly all data sets. With one possible exception, even when "vertical transects" are collected over relatively short distances, date-elevation correlations

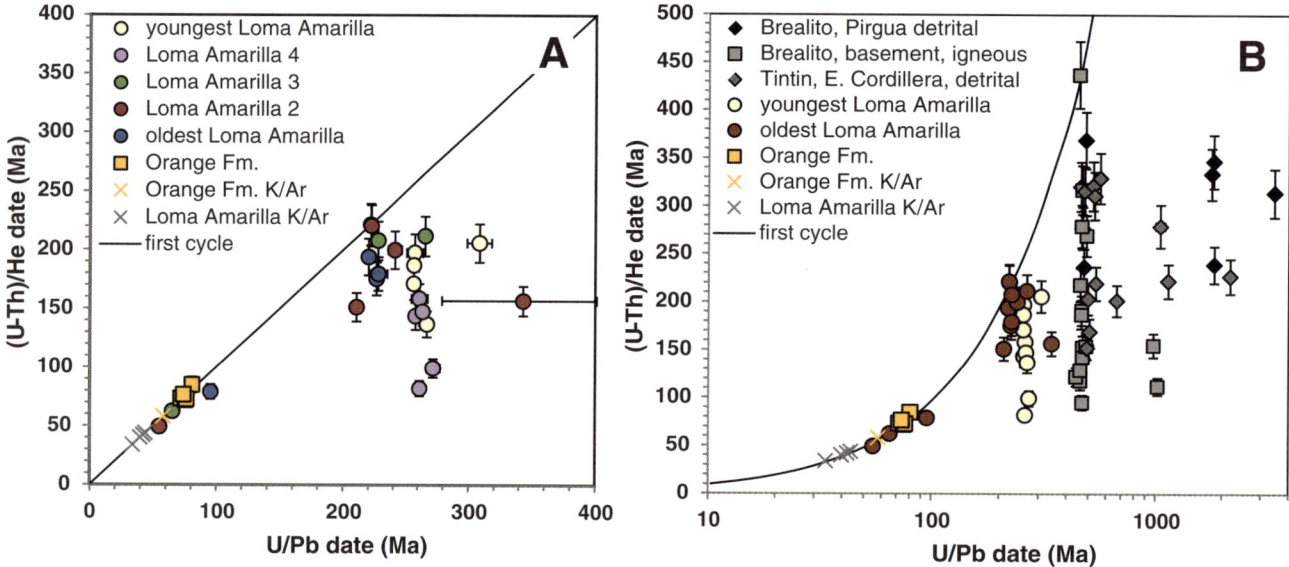

Figure 3. U/Pb and (U-Th)/He dates on single zircon grains from (A) the Cretaceous Orange Formation and Eocene Loma Amarilla Formation of the Cordillera de Domeyko, and (B) Eocene sediments of the Tintin Basin (gray diamonds), Cretaceous Pirgua Group of the Brealito subbasin (black diamonds), and crystalline basement rocks adjacent to the Brealito subbasin (gray squares). X-symbols represent K/Ar or $^{40}Ar/^{39}Ar$ dates on Orange and Loma Amarilla Formations in the Cordillera de Domeyko. Solid line is 1:1 trend of (U-Th)/He and U/Pb dates, marking position of hypothetical first-cycle (volcanic or rapidly cooled) detrital zircon. Both Cretaceous (Orange Formation) and early Eocene (oldest Loma Amarilla) zircon grains contain first-cycle zircon, likely derived from shallowly emplaced magmatic arc rocks to the west. Younger Loma Amarilla zircon grains, however, have He dates significantly younger than U/Pb dates, marking a source change with time, probably to more deeply eroded exposures with reset zircon (U-Th)/He dates.

are usually not evident in the data, and thermochronometer dates are either invariant with elevation, reflecting locally rapid exhumation, or show more complex patterns, such as inverse correlations that cannot be interpreted in the context of typical vertical transects. For example, Deeken et al. (2006) observed essentially invariant ca. 15–20 Ma AFT dates in two ~2 km transects ~30 km

apart in the Eastern Cordillera; a single sample at high elevation in one transect was 56 Ma. Sobel and Strecker (2003) also observed elevation-invariant dates ranging from 3.5 to 6.0 Ma in a 2 km transect in the southern Sierras Pampeanas. Another transect only ~100 km away showed an inverse date-elevation trend, with dates ranging from 60 to 85 Ma. Both the inverse

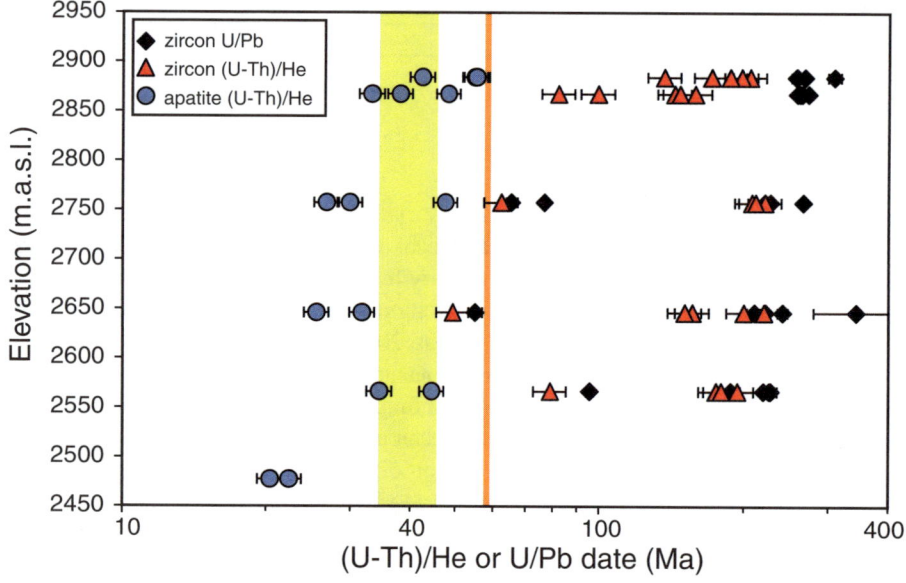

Figure 4. U/Pb and (U-Th)/He dates on detrital zircons and (U-Th)/He dates on detrital apatites from the Loma Amarilla Formation, eastern flank of the Cordillera de Domeyko. Yellow and orange bands represent presumed depositional age ranges of the Loma Amarilla and Orange Formations. Many of the apatite (U-Th)/He dates are younger than the depositional age of the formation and show a general trend to younger dates with increasing stratigraphic depth. This requires at least partial postdepositional resetting and exhumation sometime more recently than the minimum age of ca. 20–22 Ma.

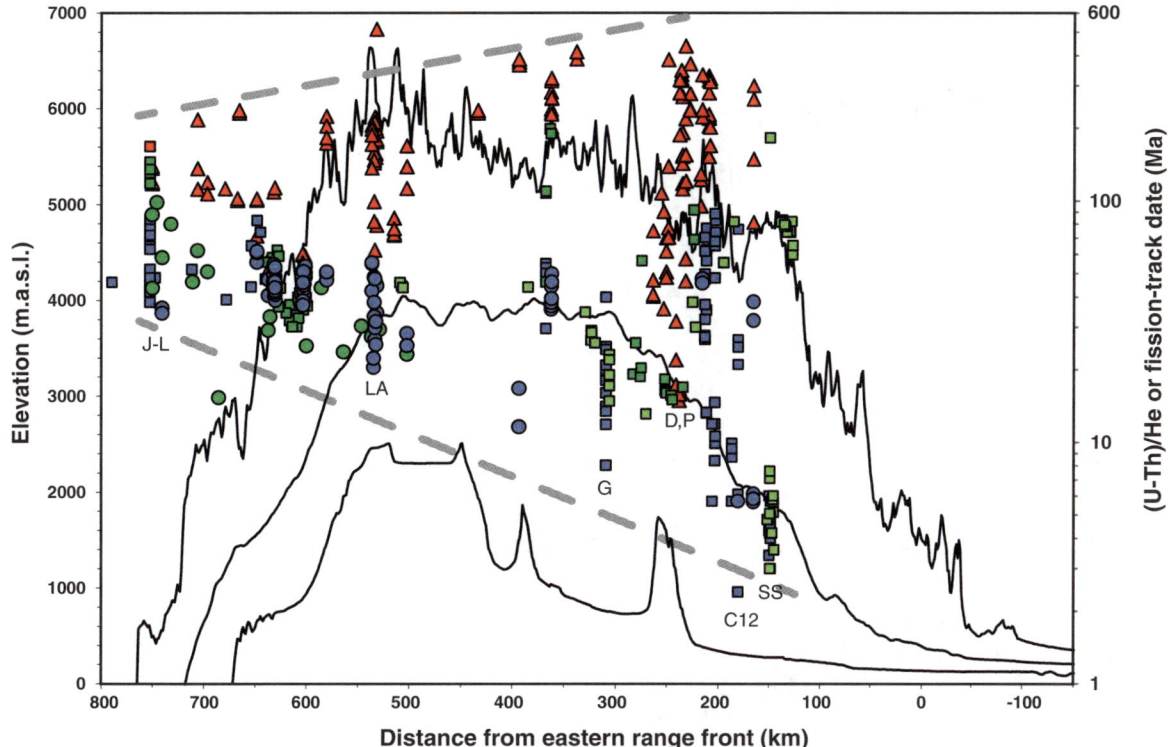

Figure 5. New and previously published low-temperature thermochronometer dates from the central Andes between 21°S and 28°S, superimposed on minimum, mean, and maximum topographic height (solid lines) in the transect, averaged over the same latitudinal swath. Green circles—apatite fission-track (AFT) data from this study; green squares—AFT data from other studies; blue circles—apatite (U-Th)/He data from this study; blue squares—apatite He data from other studies; red triangles—zircon (U-Th)/He data from this study; red squares—zircon He data from other studies. At the broadest scale, the envelope of maximum and minimum thermochronometer dates widens to the east. Approximate positions of data sets mentioned in text as collected over short distances or displaying large scatter due to exhumation from partial retention zone depths are marked with letters: J-L—Juez-Larré et al. (2010); LA—Loma Amarilla (this study); G—Geste Formation (Carrapa et al., 2009); D,P—Deeken et al. (2006) and Pearson et al. (2012); C12—Eastern Cordillera data of Carrapa et al. (2011); SS—Sobel and Strecker (2003).

date-elevation correlation and the large differences between these transects require that large horizontal gradients in exhumation depths are a much stronger control on thermochronometer dates than local elevation differences.

One exceptional case where a weak correlation between date and elevation is apparent is the apatite He data set of Juez-Larré et al. (2010), which was collected over a very short horizontal distance in the Coastal Cordillera. Here dates are roughly 40–60 Ma in the lower 1 km and 60–80 Ma in the upper 1 km. Aside from this example, most date variation cannot be explained by exhumation of rocks through steady topographic relief. Instead, date variations from place to place require differences in the depth or timing of exhumation, as described later herein.

Taking all the thermochronometer data together (Fig. 5), there is a broad correlation in the difference between minimum and maximum dates with distance across the range. This is also true for any thermochronometer separately—the widest range of dates is found on the east flank, and the smallest range of dates is found on the west (Fig. 6). In detail, dates of the apatite He

system may show a more abrupt increase in variation than the other thermochronometers, with the appearance of numerous ca. 83–26 Ma dates much older than the overall trend, at ~160–150 km from the range front in the Eastern Cordillera.

Zircon (U-Th)/He

Maximum zircon He dates generally increase from west to east, but several regions show dispersion to younger dates (Fig. 6). The 50–60 Ma dates at approximately ~650 km (Fig. 6) are from plutonic rocks with ca. 65 Ma crystallization dates, so their ca. 50–60 Ma zircon He dates likely reflect a large component of magmatic, rather than exhumational, cooling. Relatively young dates of ca. 50–70 Ma at ~550 km distance are detrital zircons from the Loma Amarilla Formation along the eastern flank of the Cordillera de Domeyko that have similar He and U/Pb dates (Fig. 4), so these dates likely represent magmatic (volcanic) cooling as well. A more important and obvious dispersion to younger zircon He dates is between ~230–260 km distance

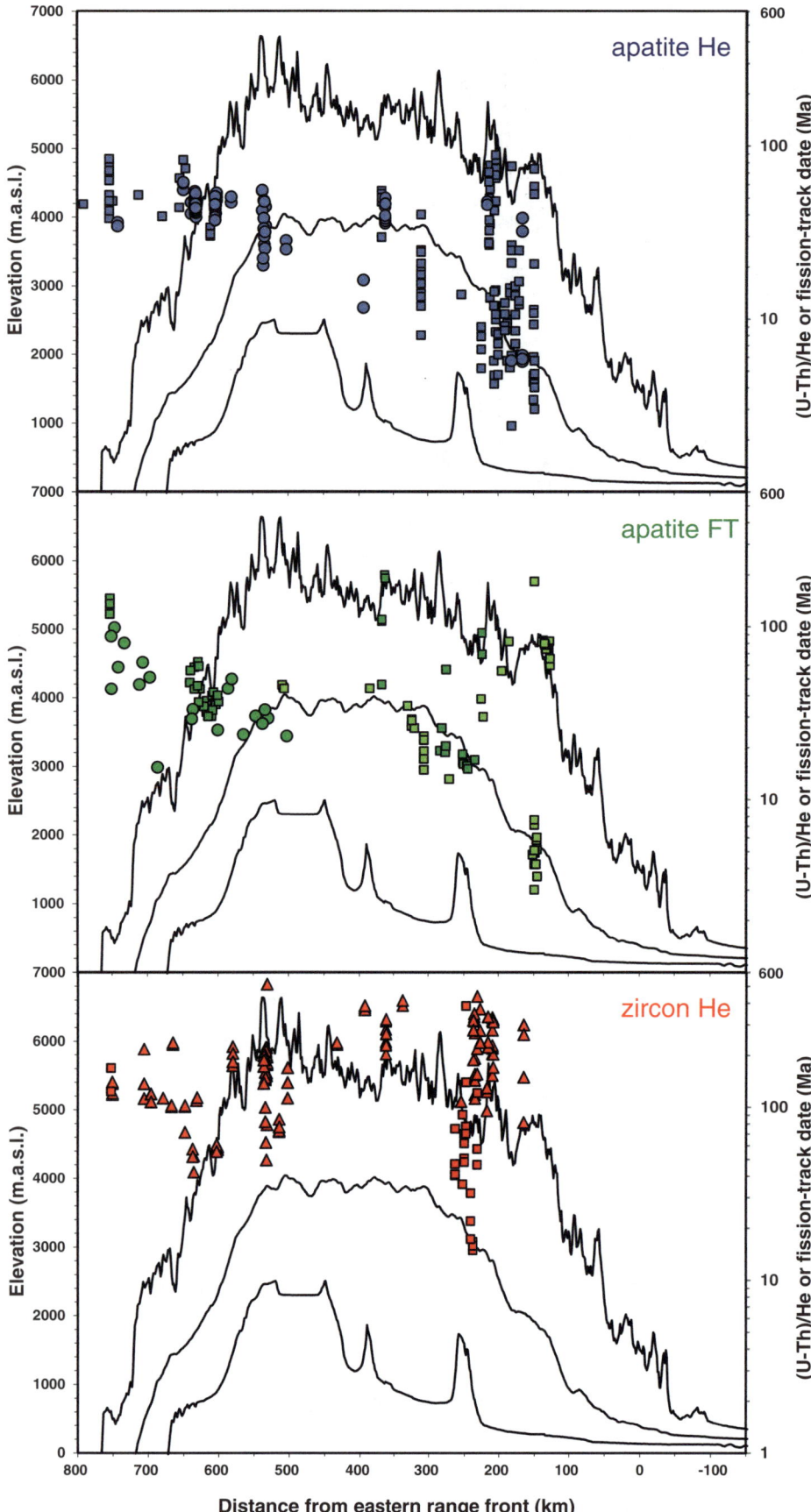

Figure 6. Thermochronometer dates as in Figure 5 but separated by system. Light-green squares in apatite fission-track (FT) panel denote samples collected south of 26°S.

(Fig. 6C), in the Eastern Cordillera near the Nevado de Cachi. As shown by Pearson et al. (2012), this is a region of exceptionally deep exhumation (>6–8 km) on a thrust fault hanging wall that appears to be a reactivated Cretaceous normal fault. This is also near the region where Deeken et al. (2006) found abundant ca. 16 Ma dates over a large elevation range, requiring >5 km of exhumation in the early-middle Miocene. The youngest of the zircon He dates in this region of the Eastern Cordillera are from a single sample yielding reproducible zircon He dates of 15 Ma, consistent with exhumation from below the zircon He closure depth (roughly 6–8 km) at this time. The wide date variation of other samples in this region is likely due to exhumation from depths within the zircon He partial retention zone.

The fact that most zircon He dates across the central Andes are older than ca. 90 Ma precludes Cenozoic exhumation deeper than roughly 6–8 km in most places in the range. The few Cenozoic dates in hanging walls of thrusts near the Cachi region are the only exception. Apparently, the several kilometers of rock and surface uplift inferred to have built the main topographic form of the Andes since ca. 40–60 Ma (e.g., Canavan et al., 2014; Quade et al., this volume) resulted in less than 6–8 km of erosion everywhere in the central Andes between 21°S and 28°S except in very localized regions, and possibly only one (Cachi). We also note that Ege et al. (2007) and Barnes et al. (2008) interpreted cumulative exhumation depths less than ~8.5 km, based on zircon fission-track dates, in the Eastern Cordillera of Bolivia, farther to the north.

Because most zircon He dates in this part of the central Andes do not record cooling associated with Cenozoic exhumation, the overall west-to-east increase in maximum zircon He dates may reflect the predominance of Mesozoic magmatic arc rocks in the west and either crystallization or cooling dates of much older continental crystalline rocks in the east.

Apatite Fission Track

AFT dates in this transect can be broadly divided into two groups (Fig. 6B). The first includes dates defining a broad ~10 km/m.y. trend of eastward-decreasing dates beginning at 60–70 Ma in the west (at ~750 km) and decreasing to 1–10 Ma in the east (at ~150 km). The second group includes samples in the eastern half of the orogen (150–300 km) in which AFT dates are significantly older (ca. 46–190 Ma) than at the same distance in the group displaying the date-distance trend. The AFT date differences at a given distance are most extreme in the easternmost samples of Sobel and Strecker (2003). Here, we interpret the large date differences in samples at about the same elevation only ~100 km apart as due to large differences in recent exhumation, possibly along reactivated normal faults (Pearson et al., 2012; Carrapa et al., 2013).

As seen in the zircon He dates, some AFT dates in the Cordillera de Domeyko and Atacama regions are slightly younger than predicted by a continuous 10 km/m.y. trend. Locally abundant Eocene magmatism and hydrothermal activity, rather than

exhumation, in the Atacama region may be responsible for these locally anomalous cooling ages. One anomalously young AFT date of 15 Ma is also seen at ~680 km from the deformation front. This comes from a tonalite with a preliminary U/Pb date of 138 Ma (Zattin, 2013, personal commun.). At this point, we can only speculate that the AFT date for this sample may be much younger than others at similar distances due to hydrothermal circulation or proximal magmatic heating.

Apatite (U-Th)/He

The pattern of apatite He dates in the cross-range transect is similar to that of the AFT dates. Most samples define a broad west-to-east decreasing trend with an approximate slope of 10 km/m.y. As with the AFT results, however, many samples in the eastern half of the range yield much older dates. The easternmost samples, for example, have apatite He dates as old as ca. 90 Ma and as young as 2.4 Ma in essentially the same longitudinal position.

Many of the samples in this suite do not have coexisting apatite He and AFT dates, making comparison between these systems at a given position difficult. Nevertheless, apatite He dates anywhere along the transect display a larger variation at a given distance than AFT dates. One example of this is the Loma Amarilla samples at ~550 km, where some apatite He dates are much younger than predicted by the 10 km/m.y. trends. Because many of the apatite He dates from the Loma Amarilla Formation are younger than the depositional age of their host rocks, this is likely a result of partial resetting by postdepositional burial and re-exhumation. However, in both the Loma Amarilla case, as well as the case of the Eocene Geste Formation (Carrapa et al., 2009), this burial and exhumation could not have exceeded ~2–3 km at the most, because apatite He dates appear to be only partially reset, and the AFT ages for the same samples show ages older than the depositional age of the hosting strata (Carrapa and DeCelles, 2008). Recent exhumation from variable and shallow depths is likely to also explain the large variations in apatite He dates in samples from the Eastern Cordillera and Santa Barbara Ranges between ~150 and 250 km in Figures 5 and 6 (Carrapa et al., 2011; Pearson et al., 2013; Carrapa et al., 2014).

SYNTHESIS OF THERMOCHRONOMETER TRENDS AND A KINEMATIC MODEL

In the following discussion, we focus on four primary observations of the date-distance trends in the cross-orogen transect in order to construct a simplified model for growth of the central Andes through time. The first observation is the abundance of pre-Cenozoic zircon He dates across the range. With one important exception, the Cenozoic orogenic processes that are thought to be responsible for rock and surface uplift of the central Andes (e.g., Oncken et al., 2006; Strecker et al., 2007) did not result in exhumation to depths greater than ~6–8 km (the typical zircon He closure depth). In fact, most of the samples have zircon

He dates older than Jurassic, and there are abundant Permian–Carboniferous zircon He dates, especially in the eastern part of the range, which limit exhumation in many areas to <6–8 km for hundreds of millions of years. The only known exception to this are the Miocene dates in the Cachi region of the Eastern Cordillera (Pearson et al., 2012), which require rapid and deep erosion of the hanging wall of a thrust fault from depths below the zircon He partial retention zone at 15–16 Ma. In principle, these relatively young dates might be explained by an anomalously high geothermal gradient in this region at this time, but there is no other evidence for this, and the date-elevation relationships for the AFT system also require at least 5 km of rapid exhumation here (Deeken et al., 2006).

A second important observation is that many regions within ~250 km (for apatite He) to 350 km (for AFT) of the eastern range front appear to have experienced less than 2–3 km of Cenozoic erosion. Interestingly, the oldest apatite He dates are similar to many of the AFT dates, at ca. 90 Ma, and both are found mostly in the eastern half of the range. These dates may record thermal relaxation, or possibly exhumation, associated with the Cretaceous rifting event that created the Salta rift and other structures in the eastern Andes (Salfity and Marquillas, 1994; Marquillas et al., 2005). Farther west, many of the oldest apatite thermochronometer dates are ca. 40–50 Ma, which may reflect a later exhumation event, possibly associated with Incaic deformation (Coira et al., 1982), which is also evident in paleoelevation proxies (Canavan et al., 2014; Quade et al., this volume), and which has been attributed to increased convergence between the Nazca and South American plates (Pardo-Casas and Molnar, 1987) or increased absolute westward motion of the South American plate (e.g., Coney and Evenchick, 1994; Silver et al., 1998).

The presence of mid-Cretaceous and older apatite He and AFT dates in rocks now at 3–5 km elevation in the central Andes requires that in some regions, erosion is largely decoupled from rock and surface uplift. West of the Eastern Cordillera, this can be at least partly attributed to low precipitation and internal drainage, at least since the mid-late Cenozoic. Reactivation of Cretaceous normal faults as thrusts may also explain the highly variable local exhumation depths and close juxtaposition of late Cenozoic and mid-Cretaceous dates in some locations (Sobel and Strecker, 2003; Pearson et al., 2012; Carrapa et al., 2013).

Third, with the exception of apatite He and AFT dates older than ca. 35 Ma in the eastern part of the orogen, dates from most samples decrease from west to east across the orogen. Given the variability of dates for these systems, as well as the abundance of much older samples that have escaped significant erosional exhumation in the eastern half of the range, robust regression of a distance-date trend is difficult. Nonetheless, simple linear regression of all the apatite He and AFT data in this transect yields apparent slopes of 5–8 km/m.y. A simpler fit that ignores most of the older dates in the eastern part of the range yields a trend with a slope of roughly 10 km/m.y. and apparent dates of 1–10 Ma within 100 km of the eastern range front (Fig. 6). The only zircon He dates that clearly reflect Cenozoic exhumational cooling (the

15 Ma dates in the Eastern Cordillera) are noteworthy in that they also fall near this ~10 km/m.y. trend (Fig. 6C).

The patterns in our data suggest that the eastern orogenic front is the location of the highest erosion rates, at least as averaged over million-year time scales; younger dates for the apatite systems are found farther to the east, and all dates younger than ca. 5 Ma are within 100–200 km of the eastern topographic front. Assuming that erosion rates have remained constant over the relevant time frame, we can use simple models relating dates to closure depths, calculating erosion rates from each date (Brandon et al., 1998; Reiners and Brandon, 2006; Willett and Brandon, 2013). The youngest AFT and apatite He dates in this region indicate average erosion rates of ~0.8–1.2 km/m.y. over the last ~2–3 m.y. In principle, the older thermochronometer dates farther west may also be converted into estimates of steady, time-averaged erosion rates. This would predict rates about an order of magnitude lower for samples on the west side of the range. However, we consider that rocks west of the Eastern Cordillera are unlikely to have experienced steady erosion rates over intervals of 40–90 m.y. Instead, we favor a model in which nearly all erosion (at least since creation of topography high enough to induce a strong rain shadow) occurs near the eastward-migrating eastern topographic front of the orogen, with very limited erosion to the west of it. Localization of almost all significant erosion within a few hundred kilometers of the eastern front is consistent with: (1) the location of most active shortening as indicated by structural, sedimentological, and global positioning system (GPS) evidence (Oncken et al., 2006; Bevis et al., 2001; Klotz et al., 2001; Echavarria et al., 2003; Kendrick et al., 2006; McQuarrie et al., 2008; Brooks et al., 2011); (2) relatively high precipitation at this location and very low precipitation west of the Eastern Cordillera (Hilley and Strecker, 2005; Strecker et al., 2007); (3) internal drainage west of the Eastern Cordillera, which severely limits base level and therefore stream power and the ability of drainages to export sediment from the interior of the range; and (4) highest topographic relief at the eastern front. We therefore envision the eastern locus of deformation and high erosion rates as migrating eastward with time. This means that the modern topographic, structural, and thermochronometric profile of this transect through the central Andes reflects the eastward growth of the edge of the Andean orogenic plateau, where localized convergence between it and the South American plate produce localized deformation, rock uplift, and erosion. If the majority of erosional exhumation anywhere in the range occurs at the eastern range front, then the westward increase in minimum thermochronometer dates reflects the rate of eastward plateau propagation as recorded by cessation of erosion in a given location as the plateau margin moves east of it.

A final observation is that apatite He dates of at least some samples in the interior of the orogen are younger than would be predicted by an end-member scenario of zero erosion in the interior, as outlined here. As noted earlier, examples of this are the Miocene dates in detrital apatite from Eocene sedimentary rocks in the Loma Amarilla and Geste Formations. These rocks were buried to, and subsequently exhumed from, partial retention zone depths for the apatite He system. This requires erosion to depths of less than 2–3 km as late as ca. 20 Ma in the Loma Amarilla, and as late as ca. 10 Ma in the Geste, as supported by thermal modeling of apatite He and AFT data (Carrapa et al., 2009). Erosion to maximum depths of ~2–3 km may have occurred elsewhere in the interior of the orogen as well, especially in locally high subranges and easily eroded lithologies such as the Eocene sedimentary rocks. However, because of the internal drainage and limited topographic relief, whatever erosion did occur likely resulted in relatively minor exhumation. In such an environment, eroded material is redeposited in nearby internal basins, further limiting topographic relief, raising base level, and protecting other regions from erosion. Subsequent minor deformation and internal drainage reconfigurations within the orogen then re-eroded these redeposited sediments, exposing partially reset apatite He dates, effectively recycling the same detritus multiple times without leading to exhumation deeper than a few kilometers. We propose the term recyclic erosion for this style of erosion within internally drained regions of subranges and basins.

A Simple Model for Thermochronometer Trends across the Central Andes

Here we present a simple thermo-kinematic model that attempts to reproduce basic features of the observations discussed herein. The basic form of this model is similar to the steady-state model of Thomson et al. (2010), balancing accretionary and erosional fluxes and predicting thermochronometer dates as a function of erosion rate throughout the orogen. However, here we assume progressive growth of the orogen, rather than flux steady state, with eastward propagation of the Andean Plateau driven by convergence and shortening focused at its eastern margin. We interpret the first-order pattern of west-to-east decreasing thermochronometer dates as the result of eastward propagation of a region of relatively high rates of erosional exhumation that follows this zone of convergence. Erosion is focused in this zone because of its relatively high precipitation, external drainage, and because this is also the zone of focused deformation and convergence between the Andean Plateau and the South American plate, as seen in GPS and the high concentration of active contractional structures (Fig. 7). Crust from the South American plate converges with the Andean Plateau on the eastern edge of this retroarc wedge with a material velocity V_c relative to the stable hinterland (Fig. 7). We assume that the South American crust thickens from an initial thickness h of 35 km to a final thickness H of 60 km in the plateau over a frontal erosion zone with width w_e of 150 km. The Andean Plateau crustal thickness estimate is based on the average depth to Moho in the region (Beck et al., this volume), and the undeformed South American crustal thickness is close to the average thickness of 37 km for regions east of 65°W (Yuan et al., 2002). Material velocity relative to the stable hinterland (V_c) decreases to zero at the eastern edge of frontal wedge, beneath which material is underplated and then exhumed

vertically at erosion rate $\dot{e}$. Crustal thickening to a thickness of 60 km (H) at the western edge of the frontal wedge results in eastward growth of the plateau at rate P.

The accretionary flux F_a to the plateau is therefore

$$F_a = (P + V_c)h, \tag{1}$$

and the erosional flux is

$$F_e = \dot{e}w_e. \tag{2}$$

The time-averaged or steady-state rate of growth of plateau crustal area, PH, is equal to the difference between the accretionary and erosional fluxes, so the rate of growth of excess crust of the plateau, $P(H - h)$, is

$$P(H - h) = V_c h - \dot{e}w_e. \tag{3}$$

Assuming all erosion occurs at the eastward-propagating frontal eroding wedge, then the slope of the date-distance relationship west of this zone corresponds to P. In the following discussion, we estimate P and $\dot{e}$ from the thermochronometer date-distance trends, and we assume reasonably well-constrained values for crustal thicknesses and an estimated width for the frontal erosion zone w_e. We then use Equation 3 to calculate V_c and compare it to GPS estimates of convergence rates between the Andes block and the foreland of the South American plate, and we compare $V_c + P$ to geologic estimates of shortening rates and flexural wave migration in the foreland. We also show that this model predicts growth of the observed width and crustal cross-sectional area of the orogenic plateau over a duration consistent with geologic evidence for growth history of the modern Andes.

To predict the distribution of thermochronometer dates in our cross-orogen transect, we start by assuming incoming, unreset dates in the South American plate of 40, 90, and 300 Ma for the apatite He, AFT, and zircon He systems, respectively (see horizontal dashed lines east of erosional front in Fig. 8). These values have no influence on model behavior west of the accretionary range front, but they simply represent unreset dates found in the eastern margin of the orogen (Figs. 6 and 8) and may be preserved west of there in cases where cumulative erosion is too low to produce reset dates (dashed colored lines in Fig. 8). We assume that material is incorporated into the orogenic plateau by underplating, followed by vertical exhumation within the 150-km-wide convergence/erosion zone (Fig. 7), allowing thermochronometer dates in this zone to be predicted using the model of Willett and Brandon (2013), accounting for transient thermal effects following the onset of erosion. Kinetic parameters assumed for the thermochronometers were discussed earlier herein. Erosion rates of 0.1, 0.4, 0.7, and 1.0 km/m.y. are used to predict dates. Once the 150-km-wide convergence/erosion zone has propagated east of a given position in the plateau, there is no further erosion, so the thermochronometer dates anywhere west of the erosion zone are simply $\tau + d/P$, where τ is the apparent

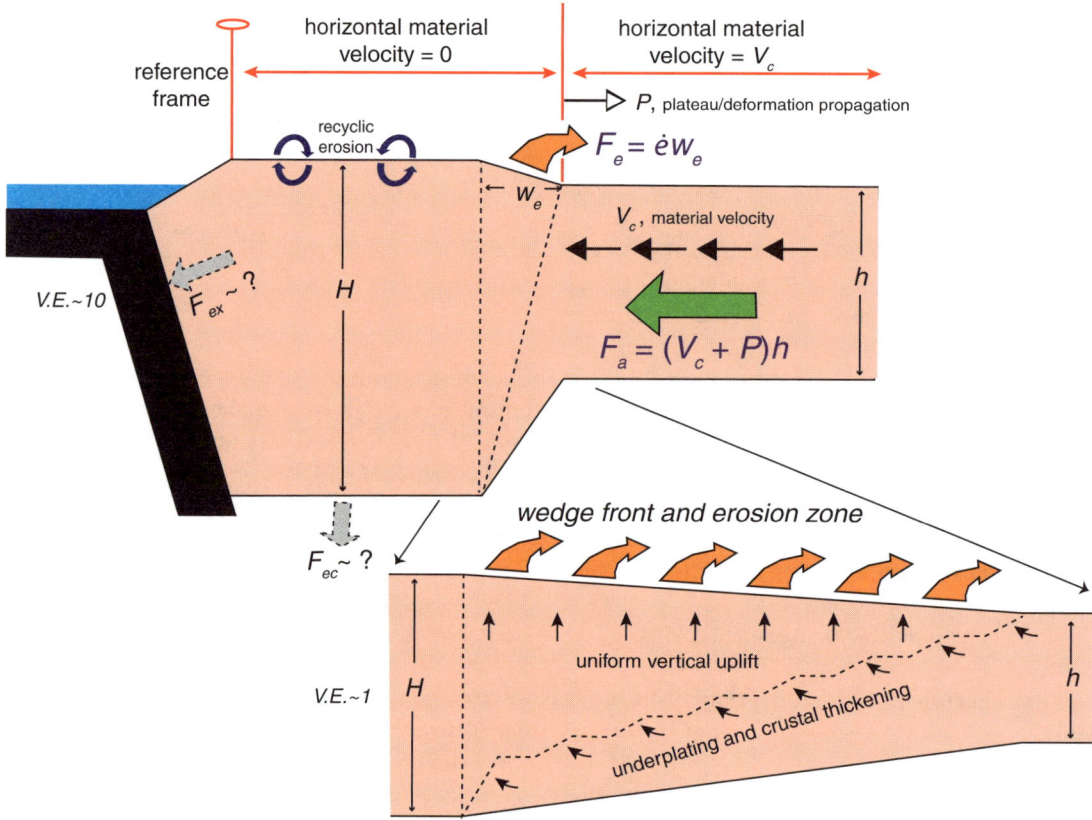

Figure 7. Schematic of thermokinematic model of the central Andes between 21°S and 28°S. Red lines show reference frame anchor on left side and boundaries and regions of distinct material velocity fields. Black arrow denotes horizontal material velocity V_c, which is the convergence rate between the stable Andes block and the stable foreland, and white arrow denotes P, the rate of propagation of the deformation front and east margin of the Andean orogenic plateau. Plateau crustal thickness is H, and incoming South American crust thickness is h. Erosion occurs within the 150-km-wide transition zone W_e of crustal thickening on the eastern front of the plateau. Material within this zone is assumed to be underplated within the zone of convergence and erosion and then experience purely vertical exhumation toward the surface at rate $\dot{e}$ before being overtaken by the eastward-propagating plateau, at which point its material velocity is zero. The accretionary influx (F_a) and erosional outflux (F_e) are shown by green and orange arrows, respectively, along with their values in terms of the other parameters. Also shown are possible outflux via subduction erosion "excretion" (F_{ex}) and gravitational instability (F_{ec}). Lower-right inset shows convergence/erosion zone on eastern margin of plateau where material is assumed to be underplated at depth and then experience purely vertical exhumation via erosion. Using $h = 35$ km, $H = 60$ km, plateau width of ~400–500 km, and west and east transition zones of 150 km each, the excess cross-sectional area of this part of the Andes is ~23,000 km², and this can be built within ~40–60 m.y. V.E.—vertical exaggeration.

date at the western margin of the erosion zone (set by the erosion rate), and d is the distance from this point.

Within the convergence/erosion zone, the thermal diffusivity of the crust is assumed to be 32 km/Ma², the initial (pre-erosional) geothermal gradient is 20 °C/km, and the surface temperature is assumed to be 10 °C. The relationship between date and erosion rate within the eroding zone is not very sensitive to any of the assumptions for the crustal thermal field within a wide range of values. In addition, although the nature of the thermal field may vary significantly across active orogens (especially those hosting active magmatic arcs), the primary effect of our thermal field assumptions here is simply to set the minimum ages of reset thermochronometers in the erosion zone. Once the orogenic front has propagated east of a given location (or a parcel of rock

has moved west of the convergence/erosion zone), the predicted dates simply steadily increase with distance at the propagation rate. The most important observation we are making here is the date-distance relationship west of the convergence/erosion zone, which we interpret as propagation rate, not the absolute date of a thermochronometer within this zone. Put another way, because we are concerned with the propagation rate of the orogenic front with time, cross-orogen variations in thermal properties do not play an important role in our interpretations. That said, transient increases in geothermal gradient in the interior and western parts of the range, for example, due to magmatism, may influence, and partially reset, some samples to the west of the orogenic front. This could explain at least part of the fact that many of the samples in the western part of the range show younger

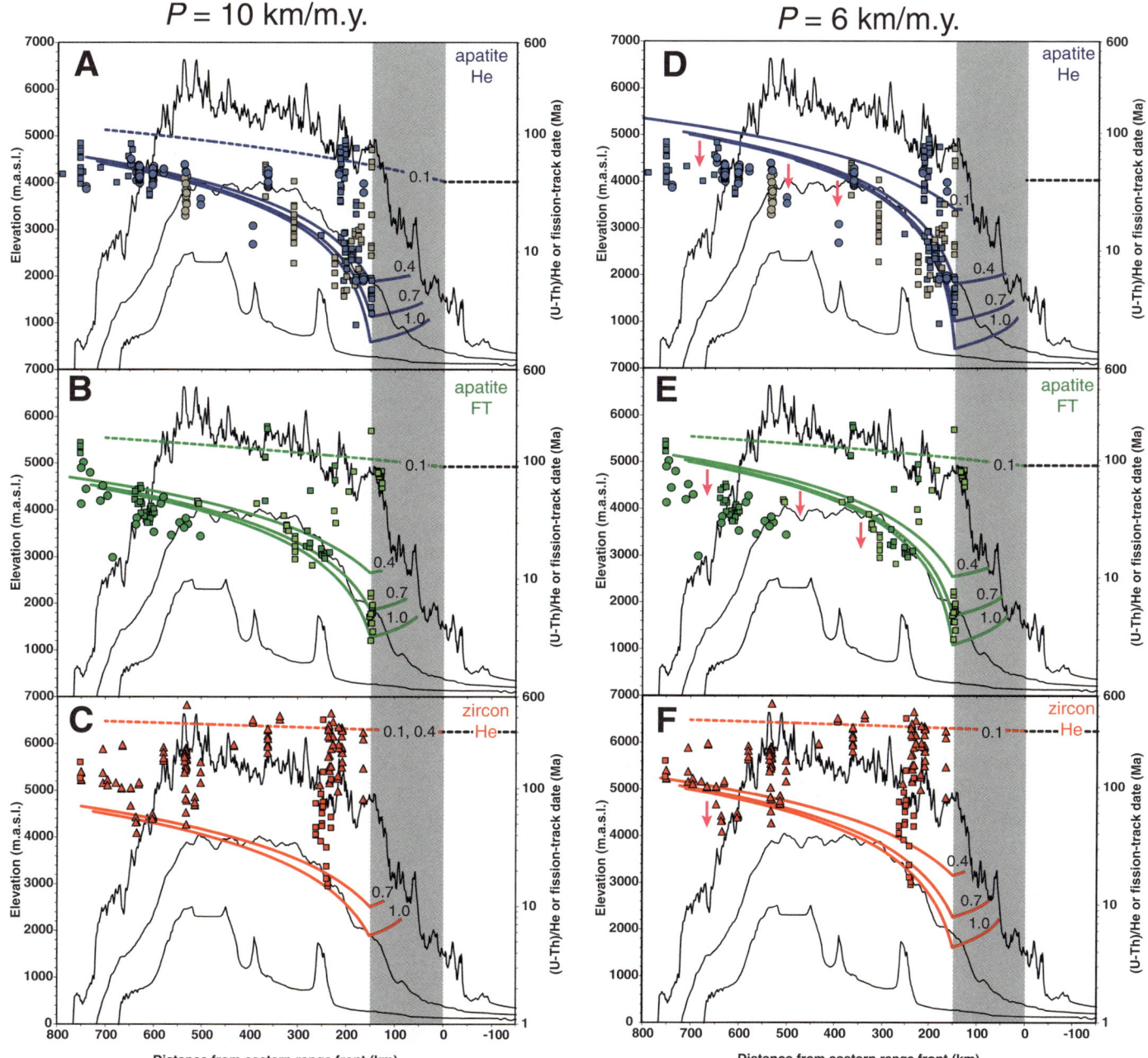

Figure 8. Predicted thermochronometer date-distance trends for the kinematic model described in the text involving eastward plateau growth by accretion of South American crust. (A–C) P (rate of eastward plateau growth) = 10 km/m.y.; (D–F) P = 6 km/m.y. Vertical gray bands represent location of 150-km-wide erosion zone. Horizontal dashed black line in eastern part of each panel represents the incoming pre-orogenic, unreset thermochronologic date assumed for each system prior to incorporation in the propagating wedge (40 Ma for apatite (U-Th)/He, 90 Ma for apatite fission track [AFT], and 300 Ma for zircon (U-Th)/He). Dashed colored lines for each system represent predicted thermochronometer dates for samples that were not exhumed from sufficiently great depths (closure depths) to expose reset dates during passage through the erosion zone. Solid colored trends represent predicted thermochronometer dates for systems that exposed partially reset dates during passage through the erosion zone, for erosion rates of 0.1, 0.4, 0.7, and 1.0 km/m.y., calculated using the transient AGE2EDOT model of Willett and Brandon (2013). The easternmost extensions of these colored solid lines within the erosion zone cannot be resolved using this transient model because of numerical stability, but the colored solid lines must connect in continuous curvature to the horizontal dashed black line east of the erosion zone. The inflection point in these dates at 150 km distance represents the location where erosion ceases and dates increase as a function of propagation rate P only. The slopes of the date-distance trends west of 150 km from the orogen front are inversely proportional to P. In general, model date trends for P = 10 km/m.y. provide better fits to the thermochronometer data in the western part of the orogen. However, if dates for all systems in the western part of the orogen are partially reset by small-magnitude, recyclic erosion, possibly combined with transient elevated geothermal gradients (shown schematically by pink arrows) west of the erosion zone as discussed in text, then P of 6.0 km/m.y. could also explain the general trend of dates across the orogen. Gray symbols in apatite He panels represent samples from sedimentary basins with clear evidence for postdepositional partial resetting.

thermochronometer dates than predicted by a steady orogenic propagation model. However, testing this would require more detailed investigations of relationships to the spatial-temporal patterns of magmatism.

Figure 8 shows the predicted date-distance trends for the three thermochronometers for P of 10 and 6 km/m.y., at erosion rates of 0.1–1.0 km/m.y. The general form of the date-distance trend for erosion rates greater than 0.1 km/m.y. is old, unreset dates east of the orogen front that then decrease to a minimum value within 150 km of the front as rocks are exhumed through their partial retention zone to expose reset dates. Within the erosion zone, the thermochronometer dates approach a minimum date, which is inversely proportional to the erosion rate. Dates are not shown in the easternmost part of the erosion zone because there has been insufficient exhumation to expose reset dates. However, the trends shown in Figure 8 can be considered to connect continuously with the horizontal dashed line delimiting the eastern edge of the convergence/erosion zone. West of this zone, thermochronometer dates increase linearly with distance to the west at rate P, reflecting increasing age since passage through the zone.

In most cases, erosion rates greater than 0.1 km/m.y. are required to produce sufficient exhumation to expose dates younger than the incoming unreset dates as the host rocks pass through the erosion zone. However, at relatively low P (6 km/m.y.), samples reside in the eroding zone for 150 km/6 km/m.y. = 25 m.y., so reset apatite He dates are just barely exposed at erosion rates of even 0.1 km/m.y. (total erosion of $\dot{e} \times w_e/P$ = 2.5 km). At fast P (10 km/m.y.), samples reside in the eroding zone for only 15 m.y. In this case, even erosion rates of 0.4 km/m.y. are just insufficient to exhume reset zircon He dates (total erosion of 6 km).

Except for the old, obviously unreset and pre-orogenic thermochronometer dates east of ~400 km from the front, the date-distance trends for propagation rates of ~6–10 km/m.y. reproduce well the general trend of decreasing thermochronometer dates seen between ~0 and 350 km from the front. West of this location, date-distance trends for P of 10 km/m.y. more closely match apatite He and AFT dates than the trends for P of 6 km/m.y., which tend to overpredict the dates because of slower accretion and eastward growth of the plateau. However, low-temperature thermochronometer dates in the western half of the orogen may be younger than predicted by the P = 6 km/m.y. trend due to later partial resetting by recyclic erosion, possibly combined with transient higher thermal gradients over time in this region. In any case, propagation rates greater than 10 km/m.y. and smaller than 6 km/m.y. do not provide a reasonable fit to the general trend of (obviously reset) thermochronometer dates within the orogen.

Throughout the orogen, unreset pre-Cenozoic dates for all thermochronometers coexist with much younger dates obviously related to Andean growth. This requires that some regions pass through the erosion/convergence zone with time-averaged erosion rates at least as low as 0.1 km/m.y. The wide range of dates seen in the eastern front, and in fact the whole eastern half of the orogen, is probably due to the wide range of erosion rates

experienced by different regions as they pass through the erosion/convergence zone. Obviously, this limits our ability to precisely constrain propagation rates and other parameters of our model, but it also underscores the fact that material that becomes part of the eastward-expanding Andean Plateau experiences a wide range of erosion rates, total exhumation, and rock uplift, some of which is probably due to differential motion on preexisting structures, as discussed earlier.

Predictions and Implications of the Eastward Plateau Propagation Model

Using the date-distance relationships for the apatite He and AFT systems in Figure 6, we adopt 10 km/m.y. as the best-fit rate of eastward propagation of plateau P (Fig. 8). Using the crustal thicknesses assumed above, w_e of 150 km, and $\dot{e}$ from 0 to 1 km/m.y., our model predicts V_c of 7.1–11 km/m.y. This agrees well with GPS estimates of convergence rates between the stable hinterland of the Andes block and the foreland from 23°S to 25°S (Bennett et al., 2013, personal commun.) as well as to the north at 20°S–22°S (Brooks et al., 2011). This model also allows us to predict the relative material velocity of material within the deforming zone (i.e., the shortening rate), as well as the apparent rate of migration of a fixed point in the foreland (e.g., a forebulge) as $V_c + P$, or 17–21 km/m.y. This is in reasonable agreement with the apparent rate of migration of the flexural profile in foreland basin sediments and apparent wave of forebulge migration, i.e., ~18–33 km/m.y. (DeCelles et al., 2011; Carrapa and DeCelles, this volume).

We can also check the self-consistency of this model by calculating the duration of time required to build the width or excess crustal thickness of the Andean orogenic plateau. These growth rates depend on P, which we determine from the thermochronometer date-distance trend, and our assumed crustal thickness values H and h. The parameters on the other side of Equation 3 (V_c, $\dot{e}$) are simply traded off against one another to balance $P(H - h)$. Thus, these duration calculations are not independent checks of the model, as comparisons to geodetic and shortening rates are. However, they do show that if the plateau and incoming South American crustal thicknesses can be reasonably constrained, the thermochronometer date-distance trends constrain both rates and durations of orogenic plateau growth, at least as well as the model trend fits the data all the way across the orogenic plateau.

For P of 10 km/m.y. (Fig. 8), building an Andean orogenic plateau of the observed width of 400–500 km would take ~40–50 m.y. At a rate of 8 km/m.y., the plateau would require ~50–62 m.y., and at the slowest rate shown in Figure 8, which does not fit the thermochronometer dates as well in the western part of the orogen, the plateau would require ~65–80 m.y. to attain the present width. Except for the greatest width (500 km) at the slowest rate (6 km/m.y.), these durations compare well with estimates of the timing of onset of major crustal shortening from structural and sedimentologic evidence in the central Andes (e.g.,

Sempere et al., 1997; DeCelles and Horton, 2003; McQuarrie et al., 2005; Elger et al., 2005; Arriagada et al., 2006; DeCelles et al., 2011; Barnes and Ehlers, 2009) and also are consistent with other mass-balance analyses (DeCelles and DeCelles, 2001; McQuarrie, 2002).

To check the implied duration for construction of the crustal cross-sectional area of the Andean Plateau, we assume excess crustal area of 11,250 km², corresponding to the area difference between crust of thickness H and h over a plateau width of 450 km. This implies that the west and east wedge zones (Fig. 7) were the same height and width prior to initiation of plateau growth. Building this crustal area using rate $P(H - h)$ and $P = 10$ km/m.y. (which can also be cast in terms of the parameters on the other side of Eq. 3) requires 45 m.y. (Fig. 9). Decreasing P to 6 km requires 75 m.y. Except for the lower range values for P, which are also less consistent with the thermochronometer date-distance trends, both this and the plateau width-duration calculation above are in good agreement with the geologic estimates of the duration of time since initiation of significant shortening and plate convergence in the central Andes, as cited above.

Our model of orogenic plateau growth uses only the thermochronologic date-distance trend across the orogen and observed crustal thicknesses of the Andes orogenic plateau and South American plate to predict convergence and shortening rates consistent with geodetic and geologic estimates and conclude that eastward plateau growth at a rate of 10 km/m.y. for ~45 m.y. balances convergence and the observed crustal thicknesses across

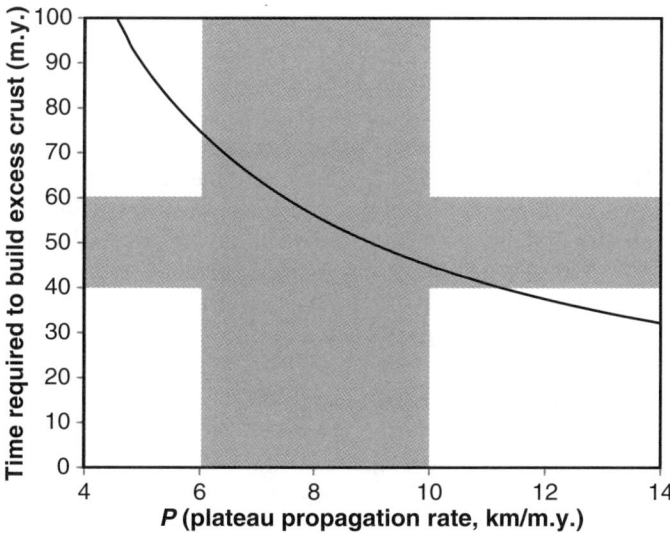

Figure 9. Durations required to accumulate the excess cross-sectional crustal area of the Andes orogenic plateau at 21°S–28°S, as depicted in Figure 7, and as a function of P. Horizontal gray band marks the approximate range of dates commonly cited for the onset of major crustal shortening that created the modern topographic expression of the central Andes. Vertical gray band marks the rates of plateau propagation most consistent with thermochronometer date-distance trends (Fig. 8). For the preferred value of $P = 10$ km/m.y., building the excess crustal area of the central Andes requires ~45 m.y.

the central Andes. This may appear to be in conflict with evidence for crustal loss in this part of the Andes by both subduction erosion on the Pacific side (e.g., von Huene et al., 1999) and eclogitic delamination by gravitational instability (e.g., Kay and Kay, 1993; Beck and Zandt, 2002; DeCelles et al., 2009; Pelletier et al., 2010; Krystopowicz and Currie, 2013; Beck et al., this volume). However, crust removed by subduction erosion on the Pacific side may simply be underplated beneath the orogen (e.g., Baby et al., 1997; Hartley et al., 2000), and while simple mass-balance constraints require that lithospheric mantle of the shortened South American plate must be lost beneath the central Andes, constraining likely volumes of *crustal* loss is more difficult. Dynamical models clearly predict loss of lower crust by gravitational instability over a wide range of conditions (Krystopowicz and Currie, 2013), which in turn predicts a number of magmatic and surficial manifestations that may be consistent with regional observations (Beck et al., this volume, and references therein). Seismic results also suggest that the lower crust above regions with mantle-like velocities in the central Andes is relatively felsic or intermediate in composition, consistent with loss of dense material typical of lower crust in other regions. Our model predicts a rough balance between accretion/shortening and crustal area over the last 40–60 m.y., which might mean that while most lithospheric mantle is removed, only relatively small volumes of crust are taken with it. Alternatively, this may mean that the lower crust that is removed is roughly balanced by magmatic contributions from the mantle (e.g., Arndt and Goldstein, 1989). Estimates of crustal addition rates are complicated by uncertainty about intrusive to extrusive ratios and juvenile versus recycled crustal proportions, but a typical estimate for mantle-derived arc magmatic addition to the crust is ~13–35 km²/m.y. (Francis and Hawkesworth, 1994; Haschke and Günther, 2003). This is only ~2%–6% of the plateau growth rate by convergence/shortening (Eq. 3; Fig. 6), but this may be important for felsification of overlying crust and refining the dense residue lost to gravitational instability.

Clearly, our simple model only explores a limited range of parameter values and makes several heuristic assumptions, including a simple erosion pattern, specific values of crustal thicknesses, and steady plateau propagation (and implied convergence/shortening rates) through time. However, the apparent balance of crustal shortening and cross-sectional area derived from the orogen-scale thermochronometer date-distance trend holds true whether or not plateau propagation and convergence rates were steady. Even if these rates varied strongly over the last 40–60 m.y., if their time-averaged values are close to those implied by the orogen-scale thermochronometer trends ($P = 6$–10 km/m.y.), then convergence and cross-sectional crustal area appear to roughly balance. This is surprising given the fact that shortening rates inferred from geologic observations are thought to have been significantly slower prior to ca. 10 Ma (e.g., figure 1.5 *in* Oncken et al., 2006; although this is for slightly north of our study area). This means that the inferred total convergence between the Andes block and South American plate from geologic evidence is insufficient to accrete enough material to generate the apparent excess crust of the Andean orogenic plateau, as noted in previous studies (e.g., Kley and Monaldi, 1998; DeCelles et al., 2011). Invoking significant net crustal loss by either subduction erosion or gravitational instability exacerbates this mass-balance problem.

One potential solution to this conundrum is that the plateau propagation and shortening rates inferred from the thermochronometer date-distance trends provide a more realistic estimate of long-term convergence and orogen growth rates through time, and that these rates were much higher than geologic estimates have inferred, especially prior to ca. 10 Ma, possibly due to volcanic burial or other factors that make total shortening difficult to resolve from surface observations. We do not claim that our cross-orogen-scale thermochronometer date-distance trends provide unambiguous evidence for steady and high rates of plateau propagation (and correspondingly high shortening and convergence rates) throughout the Cenozoic. Indeed, steady rates over the last 40–60 m.y. seem unlikely in the face of variations in rates and angles of convergence between the oceanic and South American plates (Pardo-Casas and Molnar, 1987; Somoza, 1998; Sdrolias and Müller, 2006), and as reflected in the spatial-temporal pattern of magmatism (e.g., Haschke et al., 2006). We also note that other authors have interpreted marked unsteadiness in the record of the migration of the strain front and flexural wave (Carrapa and DeCelles, 2008; Carrapa et al., 2011; DeCelles et al., 2011; Carrapa and DeCelles, this volume). Nevertheless, at least at the scale of this study, the thermochronologic data presented here do not support large variations in plateau growth rates or convergence rates and instead are broadly consistent with a simple model of constant growth through time at convergence and shortening rates similar to those of modern.

CONCLUSION

Low-temperature thermochronometer dates along a transect across the central Andes between 21°S and 28°S display broad west-to-east trends of decreasing minimum dates and increasing date variation. This pattern is consistent with focused erosion in a 100–200 km region on the eastern range front that has migrated eastward with time, with relatively little erosion (and essentially no net erosional outflux) in the range interior since ca. 40–60 Ma. The large variation in apatite He and AFT dates at the eastern range front is consistent with large spatial gradients in cumulative exhumation depth resulting from rapid but spatially variable recent erosion along the rapidly deforming front. This spatial variability may result from contrasting extents of rock uplift in adjacent walls of inverted Cretaceous normal faults.

Almost nowhere in this part of the central Andes has Cenozoic erosional exhumation exceeded depths of 6–8 km, and in many places in the eastern half of the range, erosion has not exceeded 2–3 km, despite the fact that these regions are now 5–6 km above sea level. This means that west of the rapidly deforming and eroding eastern range front, uplift and erosion

are largely decoupled, due to meager precipitation, relatively low relief, internal drainage, and ephemeral volcanic burial. This prevents simple interpretations of thermochronometer dates as indicators of *local* rock or surface uplift, as is often done in other ranges. Nevertheless, the broad west-to-east pattern of reset thermochronometer dates across the range can be used as a proxy for the eastward migration of the focused erosion and deformation zone, and therefore as a proxy for the growth of the eastern margin of the central Andes orogenic plateau through time. The orogenic plateau propagation rate of ~6–10 km/m.y. we determine is very similar to the deformation propagation rate determined from thermochronologic and other data by Carrapa et al. (2011) in the Eastern Cordillera (~8.3 km/m.y. over the last 14–3 m.y.). We suggest that a similar rate can be extended much farther west and into the past, and we present a simple kinematic model for eastward growth of the plateau with a thickness of ~60 km at a rate of ~10 km/m.y., as a result of convergence of the plateau with crust of the South American plate with a thickness of ~35 km at rates of ~11–17 km/m.y. At deformation/plateau propagation rates of ~8–10 km/m.y., the width and cross-sectional excess crust of the Andes at these latitudes is produced within ~45–55 m.y., without the need for crustal loss mechanisms other than surface erosion. The convergence/shortening rate inferred here is higher over a longer a duration than predicted by other geologic shortening rate estimates.

We acknowledge that at the scale of our sampling density in this transect, major abrupt variations through time in the inferred rate of plateau propagation and/or convergence/shortening cannot be elucidated. However, at least at this scale, the date-distance trends observed are consistent with a model of steady Cenozoic growth of the Andean orogenic plateau at a rate that reasonably reproduces geodetic and geologic convergence and shortening rates and balances crustal cross-section area. A corollary of this interpretation is that episodes of localized rapid erosion, presumably driven by episodes of localized deformation and rock and surface uplift, are not necessarily due to super-regional changes in erosion or uplift rates, but rather are due to the eastward propagation of localized zones of shortening, uplift, and erosion marking the eastward growth of the Andean orogenic plateau at a quasi-steady rate of 6–10 km/m.y.

ACKNOWLEDGMENTS

We are grateful to Stefan Nicolescu and Uttam Chowdhury for analytical support and Leandra Marshall for geographic information systems support. National Science Foundation (NSF) grant EAR-0732436 provided support of the Arizona LaserChron Center and NSF grant EAR-0732380 provided support of the Arizona Radiogenic Helium Dating Laboratory. William Cavazza and Massimilliano Zattin were supported by grants from the University of Bologna (Strategic Project 2/23/07) and the Carisbo Foundation (FS 2/27/03 Project). We acknowledge generous support from ExxonMobil through the Convergent Orogenic Systems Analysis collaboration. We appreciate helpful reviews from Pete DeCelles, Barbara Carrapa, Jason Barnes, and Cecile Gautheron, and helpful discussion with George Zandt, Nick Arndt, and Laurence Audin. Reiners also acknowledges support from a Marie Curie Incoming International Fellowship and the Centre de Recherches Pétrographiques et Géochimiques.

REFERENCES CITED

Allmendinger, R.W., Jordan, T.E., Kay, S.M., and Isacks, B.L., 1997, The evolution of the Altiplano-Puna plateau of the Central Andes: Annual Review of Earth and Planetary Sciences, v. 25, no.1, p. 139–174.

Arndt, N.T., and Goldstein, S.L., 1989, An open boundary between lower continental crust and mantle: Its role in crust formation and crustal recycling: Tectonophysics, v. 161, no. 3–4, p. 201–212, doi:10.1016/0040-1951(89)90154-6.

Arriagada, C., Cobbold, P.R., and Roperch, P., 2006, Salar de Atacama Basin: A record of compressional tectonics in the central Andes since the mid-Cretaceous: Tectonics, v. 25, TC1008, doi:10.1029/2004TC001770.

Baby, P., Rochat, P., Mascle, G., and Hérail, G., 1997, Neogene shortening contribution to crustal thickening in the back arc of the central Andes: Geology, v. 25, no. 10, p. 883–886, doi:10.1130/0091-7613(1997)025<0883:NSCTCT>2.3.CO;2.

Barnes, J.B., and Ehlers, T.A., 2009, End member models for Andean Plateau uplift: Earth-Science Reviews, v. 97, no. 1–4, p. 105–132, doi:10.1016/j.earscirev.2009.08.003.

Barnes, J.B., Ehlers, T.A., McQuarrie, N., O'Sullivan, P.B., and Tawackoli, S., 2008, Thermochronometer record of central Andean Plateau growth, Bolivia (19.5°S): Tectonics, TC3003, v. 27, doi:10.1029/2007TC002174.

Beck, S.L., and Zandt, G., 2002, The nature of orogenic crust in the central Andes: Journal of Geophysical Research, v. 107, no. B10, 2230, doi:10.1029/2000JB000124.

Beck, S.L., Zandt, G., Ward, K.M., and Scire, A., 2015, this volume, Multiple styles and scales of lithospheric foundering beneath the Puna Plateau, central Andes, *in* DeCelles, P.G., Ducea, M.N., Carrapa, B., and Kapp, P.A., eds., Geodynamics of a Cordilleran Orogenic System: The Central Andes of Argentina and Northern Chile: Geological Society of America Memoir 212, doi:10.1130/2015.1212(03).

Bevis, M., Kenrick, E., Smalley, R., Jr., Brooks, B., Allmendinger, R., and Isacks, B., 2001, On the strength of interplate coupling and the rate of back arc convergence in the central Andes: An analysis of the interseismic velocity field: Geochemistry Geophysics Geosystems, v. 2, 1067, doi:10.1029/2001GC000198.

Bookhagen, B., and Strecker, M.R., 2008, Orographic barriers, high-resolution TRMM rainfall, and relief variations along the eastern Andes: Geophysical Research Letters, v. 35, no. 6, L06403, doi:10.1029/2007GL032011.

Brandon, M.T., Roden-Tice, M.K., and Garver, J.I., 1998, Late Cenozoic exhumation of the Cascadia accretionary wedge in the Olympic Mountains, northwest Washington State: Geological Society of America Bulletin, v. 110, p. 985–1009, doi:10.1130/0016-7606(1998)110<0985:LCEOTC>2.3.CO;2.

Braun, J., 2002, Estimating exhumation rate and relief evolution by spectral analysis of age–elevation datasets: Terra Nova, v. 14, p. 210–214, doi:10.1046/j.1365-3121.2002.00409.x.

Brooks, B.A., Bevis, M., Whipple, K., Arrowsmith, J.R., Foster, J., Zapata, T., Kendrick, E., Minaya, E., Echalar, A., Blanco, M., Euillades, P., Sandoval, M., and Smalley, R.J., Jr., 2011, Orogenic-wedge deformation and potential for great earthquakes in the central Andean backarc: Nature Geoscience, v. 4, p. 380–383, doi:10.1038/ngeo1143.

Canavan, R., Carrapa, B., Clementz, M.T., Quade, J., DeCelles, P.G., and Schoenbohm, L.M., 2014, Early Cenozoic uplift of the Puna Plateau, central Andes, based on stable isotope paleoaltimetry of hydrated volcanic glass: Geology, v. 42, p. 447–450, doi:10.1130/G35239.1.

Carrapa, B., and DeCelles, P.G., 2008, Eocene exhumation and basin development in the Puna of northwestern Argentina: Tectonics, v. 27, TC1015, doi:10.1029/2007TC002127

Carrapa, B., and DeCelles, P.G., 2015, this volume, Regional exhumation and kinematic history of the central Andes in response to cyclical orogenic processes, *in* DeCelles, P.G., Ducea, M.N., Carrapa, B., and Kapp, P.A.,

eds., Geodynamics of a Cordilleran Orogenic System: The Central Andes of Argentina and Northern Chile: Geological Society of America Memoir 212, doi:10.1130/2015.1212(11).

Carrapa, B., Adelmann, D., Hilley, G.E., Mortimer, E., Sobel, E.R., and Strec, M.R., 2005, Oligocene range uplift and development of plateau morphology in the southern central Andes: Tectonics, v. 24, TC4011, doi:10.1029/2004TC001762.

Carrapa, B., Strecker, M.R., and Sobel, E.R., 2006, Cenozoic orogenic growth in the central Andes: Evidence from sedimentary rock provenance and apatite fission track thermochronology in the Fiambalá Basin, southernmost Puna Plateau margin (NW Argentina): Earth and Planetary Science Letters, v. 247, p. 82–100, doi:10.1016/j.epsl.2006.04.010.

Carrapa, B., DeCelles, P.G., Reiners, P.W., Gehrels, G.E., and Sudo, M., 2009, Apatite triple dating and white mica ^{40}Ar/^{39}Ar thermochronology of syntectonic detritus in the central Andes: A multiphase tectonothermal history: Geology, v. 37, p. 407–410, doi:10.1130/G25698A.1.

Carrapa, B., Trimble, J.D., and Stockli, D.F., 2011, Patterns and timing of exhumation and deformation in the Eastern Cordillera of NW Argentina revealed by (U-Th)/He thermochronology: Tectonics, v. 20, TC3003, doi:10.1029/2010TC002707.

Carrapa, B., Reyes-Bywater, S., Safipour, R., Sobel, E.R., Schoenbohm, L.M., DeCelles, P.G., Reiners, P.W., and Stockli, D., 2013, The effect of inherited paleotopography on exhumation of the Central Andes of NW Argentina: Geological Society of America Bulletin, v. 126, p. 66–77, doi:10.1130/B30844.1.

Coira, B., Davidson, J., Mpodozis, C., and Ramos, V., 1982, Tectonic and magmatic evolution of the Andes of northern Argentina and Chile: Earth-Science Reviews, v. 18, no. 3–4, p. 303–332.

Coney, P.J., and Evenchick, C.A., 1994, Consolidation of the American cordilleras: Journal of South American Earth Sciences, v. 7, p. 241–262, doi:10.1016/0895-9811(94)90011-6.

Coughlin, T.J., O'Sullivan, P.B., Kohn, B., and Holcombe, R.J., 1998, Apatite fission-track thermochronology of the Sierras Pampeanas, central western Argentina: Implications for the mechanism of plateau uplift in the Andes: Geology, v. 26, p. 999–1002, doi:10.1130/0091-7613(1998)026<0999:AFTTOT>2.3.CO;2.

Coutand, I., Cobbold, P.R., de Urreiztieta, M., Gautier, P., Chauvin, A., Gapais, D., Rossello, E.A., and López-Gamundí, O., 2001, Style and history of Andean deformation, Puna Plateau, northwestern Argentina: Tectonics, v. 20, p. 210–234, doi:10.1029/2000TC900031.

DeCelles, P.G., and DeCelles, P.C., 2001, Rates of shortening, propagation, underthrusting, and flexural wave migration in continental orogenic systems: Geology, v. 29, p. 135–138, doi:10.1130/0091-7613(2001)029<0135:ROSPUA>2.0.CO;2.

DeCelles, P.G., and Horton, B.K., 2003, Early to Middle Tertiary foreland basin development and the history of Andean crustal shortening in Bolivia: Geological Society of America Bulletin, v. 115, p. 58–77, doi:10.1130/0016-7606(2003)115<0058:ETMTFB>2.0.CO;2.

DeCelles, P.G., Ducea, M.N., Kapp, P., and Zandt, G., 2009, Cyclicity in Cordilleran orogenic systems: Nature Geoscience, v. 2, p. 251–257, doi:10.1038/ngeo469.

DeCelles, P.G., Carrapa, B., Horton, B.K., and Gehrels, G.E., 2011, Cenozoic foreland basin system in the central Andes of northwestern Argentina: Implications for Andean geodynamics and modes of deformation: Tectonics, v. 30, TC6013, doi:10.1029/2011TC002948.

Deeken, A., Sobel, E.R., Coutand, I., Haschke, M., Riller, U., and Strecker, M.R., 2006, Development of the southern Eastern Cordillera, NW Argentina, constrained by apatite fission track thermochronology: From Early Cretaceous extension to middle Miocene shortening: Tectonics, v. 25, TC6003, doi:10.1029/2005TC001894.

Dickinson, W.R., 2004, Evolution of the North American cordillera: Annual Review of Earth and Planetary Sciences, v. 32, p. 13–45, doi:10.1146/annurev.earth.32.101802.120257.

Echavarria, L., Hernández, R., Allmendinger, R., and Reynolds, J., 2003, Subandean thrust and fold belt of northwestern Argentina: Geometry and timing of the Andean evolution: American Association of Petroleum Geologists Bulletin, v. 87, p. 965–985, doi:10.1306/01200300196.

Ege, H., Sobel, E.R., Scheuber, E., and Jacobshagen, V., 2007, Exhumation history of the southern Altiplano Plateau (southern Bolivia) constrained by apatite fission track thermochronology: Tectonics, v. 26, no. 1, TC1004, doi:10.1029/2005TC001869.

Elger, K., Oncken, O., and Glodny, J., 2005, Plateau-style accumulation of deformation: Southern Altiplano: Tectonics, v. 24, TC4020, doi:10.1029/2004TC001675.

Flowers, R.M., Ketcham, R.A., Shuster, D.L., and Farley, K.A., 2009, Apatite (U-Th)/He thermochronometry using a radiation damage accumulation and annealing model: Geochimica et Cosmochimica Acta, v. 73, no. 8, p. 2347–2365, doi:10.1016/j.gca.2009.01.015.

Francis, P.W., and Hawkesworth, C.J., 1994, Late Cenozoic rates of magmatic activity in the central Andes and their relationships to continental crust formation and thickening: Journal of the Geological Society, London, v. 151, no. 5, p. 845–854, doi:10.1144/gsjgs.151.5.0845.

Gardeweg, M., Pino, H., Ramirez, C.F., and Davidson, J., 1994, Mapa Geologico del Area de Imilac y Sierra Almeida, Region de Antofagasta: Servicio Nacional de Geologia y Mineria Documento Trabajo 7, scale 1:100,000.

Garzione, C.N., Hoke, G.D., Libarkin, J.C., Withers, S., MacFadden, B., Eiler, J., Ghosh, P., and Mulch, A., 2008, Rise of the Andes: Science, v. 320, p. 1304–1307, doi:10.1126/science.1148615.

Gautheron, C., Tassan-Got, L., Barbarand, J., and Pagel, M., 2009, Effect of alpha-damage annealing on apatite (U-Th)/He thermochronology: Chemical Geology, v. 266, no. 3–4, p. 157–170, doi:10.1016/j.chemgeo.2009.06.001.

Gehrels, G.E., Valencia, V., and Ruiz, J., 2008, Enhanced precision, accuracy, efficiency, and spatial resolution of U-Pb ages by laser ablation–multicollector–inductively coupled plasma–mass spectrometry: Geochemistry Geophysics Geosystems, v. 9, Q03017, doi:10.1029/2007GC001805.

Guenthner, W.R., Reiners, P.W., Ketcham, R.A., Nasdala, L., and Geister, G., 2013, Helium diffusion in natural zircon: Radiation damage, anisotropy, and the interpretation of zircon (U-Th)/He thermochronology: American Journal of Science, v. 313, p. 145–198, doi:10.2475/03.2013.01.

Hain, M.P., Strecker, M.R., Bookhagen, B., Alonso, R.N., Pingell, H., and Schmitt, A.K., 2011, Neogene to Quaternary broken foreland formation and sedimentation dynamics in the Andes of NW Argentina (25°S): Tectonics, v. 30, TC2006, doi:10.1029/2010TC002703.

Hammerschmidt, K., Dobel, R., and Friedrichsen, H., 1992, Implication of dating of Early Tertiary volcanic rocks from the north-Chilean Precordillera: Tectonophysics, v. 202, p. 55–81, doi:10.1016/0040-1951(92)90455-F.

Hartley, A.J., May, G., Chong, G., Turner, P., Kape, S.J., and Jolley, E.J., 2000, Development of a continental forearc: A Cenozoic example from the central Andes, northern Chile: Geology, v. 28, no. 4, p. 331–334, doi:10.1130/0091-7613(2000)28<331:DOACFA>2.0.CO;2.

Haschke, M., and Günther, A., 2003, Balancing crustal thickening in arcs by tectonic vs. magmatic means: Geology, v. 31, p. 933–936, doi:10.1130/G19945.1.

Haschke, M., Scheuber, E., Gunthner, A., and Reutter, K.-J., 2002, Evolutionary cycles during the Andean orogeny: Repeated slab breakoff and flat subduction?: Terra Nova, v. 14, p. 49–55, doi:10.1046/j.1365-3121.2002.00387.x.

Haschke, M., Gunther, A., Melnick, D., Echtler, H., Reutter, K.-J., Scheuber, E., and Oncken, O., 2006, Central and southern Andean tectonic evolution inferred from arc magmatism, in Oncken, O., Chong, G., Franz, G., Giese, P., Götze, H.-J., Ramos, V., Strecker, M., and Wigger, P., eds., The Andes—Active Subduction Orogeny: Berlin, Springer, Frontiers in Earth Sciences, p. 337–353.

Hilley, G.E., and Strecker, M.R., 2005, Processes of oscillatory basin filling and excavation in a tectonically active orogen: Quebrada del Toro Basin, NW Argentina: Geological Society of America Bulletin, v. 117, p. 887–901, doi:10.1130/B25602.1.

Hoke, G.D., and Garzione, C.N., 2008, Paleosurfaces, paleoelevation, and the mechanisms for the late Miocene topographic development of the Altiplano Plateau: Earth and Planetary Science Letters, v. 271, p. 192–201, doi:10.1016/j.epsl.2008.04.008.

Isacks, B.L., 1988, Uplift of the central Andean Plateau and bending of the Bolivian orocline: Journal of Geophysical Research, v. 93, p. 3211–3231, doi:10.1029/JB093iB04p03211.

Jordan, T.E., Reynolds, J.H., and Erikson, J.P., 1997, Variability in age of initial shortening and uplift in the central Andes, 16–33°30′S, in Ruddiman, W.F., ed., Tectonic Uplift and Climate Change: New York, Plenum Press, p. 41–61.

Juez-Larré, J., Kubowskib, N., Dunai, T.J., Hartley, A.J., and Andriessen, P., 2010, Thermal and exhumation history of the Coastal Cordillera arc of northern Chile revealed by thermochronological dating: Tectonophysics, v. 495, p. 48–66.

Kay, R.W., and Kay, S.M., 1993, Delamination and delamination magmatism: Tectonophysics, v. 219, p. 177–189, doi:10.1016/0040-1951(93)90295-U.

Kendrick, E., Brooks, B.A., Bevis, M., Smalley, R., Lauria, E., Araujo, M., and Parra, H., 2006, Active orogeny of the south-central Andes studied with GPS geodesy: Revista de la Asociación Geológica Argentina, v. 61, no. 4, p. 555–566.

Ketcham, R.A., 2005, Forward and inverse modeling of low-temperature thermochronometry data: Reviews in Mineralogy and Geochemistry, v. 58, no. 1, p. 275–314, doi:10.2138/rmg.2005.58.11.

Kley, J., and Monaldi, C.R., 1998, Tectonic shortening and crustal thickness in the central Andes: How good is the correlation?: Geology, v. 26, p. 723–726, doi:10.1130/0091-7613(1998)026<0723:TSACTI>2.3.CO;2.

Klotz, J., Abolghasem, A., Khazaradze, G., Heinze, B., Vietor, T., Hackney, R., Bataille, K., Maurana, R., Viramonte, J., and Perdomo, R., 2001, Long-term signals in the present-day deformation field of the central and southern Andes and constraints on the viscosity of the Earth's upper mantle, in Oncken, O., Chong, G., Franz, G., Giese, P., Götze, H.-J., Ramos, V., Strecker, M., and Wigger, P., eds., The Andes—Active Subduction Orogeny: Berlin, Springer, Frontiers in Earth Sciences, p. 65–89.

Krystopowicz, N.J., and Currie, C.A., 2013, Crustal eclogitization and lithosphere delamination in orogens: Earth and Planetary Science Letters, v. 361, p. 195–207, doi:10.1016/j.epsl.2012.09.056.

Lamb, S., and Davis, P., 2003, Cenozoic climate change as a possible cause for the rise of the Andes: Nature, v. 425, p. 792–797, doi:10.1038/nature02049.

Lamb, S., Hoke, L., Kennan, L., and Dewey, J., 1997, Cenozoic evolution of the central Andes in Bolivia and northern Chile, in Burg, J.-P., and Ford, M., eds., Orogeny through Time: Geological Society, London, Special Publication 121, p. 237–264, doi:10.1144/GSL.SP.1997.121.01.10.

Maksaev, V., and Zentilli, M., 1999, Fission track thermochronology of the Domeyko Cordillera, northern Chile: Implications for Andean tectonics and porphyry copper metallogenesis: Exploration and Mining Geology, v. 8, p. 65–89.

Marquillas, R.A., Del Papa, C., and Sabino, I.F., 2005, Sedimentary aspects and paleoenvironmental evolution of a rift basin: Salta Group (Cretaceous–Paleogene), northwestern Argentina: International Journal of Earth Sciences, v. 94, p. 94–113, doi:10.1007/s00531-004-0443-2.

McInnes, B.I.A., Farley, K.A., Sillitoe, R.H., and Kohn, B.P., 1999, Application of apatite (U-Th)/He thermochronometry to the determination of the sense and amount of vertical fault displacement at the Chuquicamata porphyry copper deposit, Chile: Economic Geology and the Bulletin of the Society of Economic Geologists, v. 94, p. 937–947, doi:10.2113/gsecongeo.94.6.937.

McQuarrie, N., 2002, The kinematic history of the central Andean fold-thrust belt, Bolivia: Implications for building a high plateau: Geological Society of America Bulletin, v. 114, p. 950–963, doi:10.1130/0016-7606(2002)114<0950:TKHOTC>2.0.CO;2.

McQuarrie, N., Horton, B.K., Zandt, G., Beck, S.L., and DeCelles, P.G., 2005, Lithospheric evolution of the Andean fold-thrust belt, Bolivia, and the origin of the central Andean Plateau: Tectonophysics, v. 399, p. 15–37.

McQuarrie, N., Barnes, J.B., and Ehlers, T.A., 2008, Geometric, kinematic, and erosional history of the central Andean Plateau, Bolivia (15–17°S): Tectonics, v. 27, TC3007, doi:10.1029/2006TC002054.

Mortimer, E., Carrapa, B., Coutand, I., Schoenbohm, L., Sobel, E.R., Gomez, J.S., and Strecker, M.R., 2007, Fragmentation of a foreland basin in response to out-of-sequence basement uplifts and structural reactivation: El Cajón–Campo del Arenal Basin, NW Argentina: Geological Society of America Bulletin, v. 119, p. 637–653, doi:10.1130/B25884.1.

Mpodozis, C., Arriagada, C., Bassoc, M., Roperch, P., Cobbolde, P., and Reich, M., 2005, Late Mesozoic to Paleogene stratigraphy of the Salar de Atacama Basin, Antofagasta, northern Chile: Implications for the tectonic evolution of the central Andes: Tectonophysics, v. 399, p. 125–154, doi:10.1016/j.tecto.2004.12.019.

Oncken, O., Hindle, D., Kley, J., Elger, K., Victor, P., and Schemmann, K., 2006, Deformation of the central Andean upper plate system—Facts, fiction, and constraints for plateau models; central and southern Andean tectonic evolution inferred from arc magmatism, in Oncken, O., Chong, G., Franz, G., Giese, P., Götze, H.-J., Ramos, V., Strecker, M., and Wigger, P., eds., The Andes—Active Subduction Orogeny: Berlin, Springer, Frontiers in Earth Sciences, p. 3–27.

Pardo-Casas, F., and Molnar, P., 1987, Relative motion of the Nazca (Farallon) and South American plates since Late Cretaceous time: Tectonics, v. 6, no. 3, p. 233–248, doi:10.1029/TC006i003p00233.

Pearson, D.M., Kapp, P.A., Reiners, P.W., Gehrels, G., Ducea, M.N., Pullen, A., Otamendi, J., and Alonso, R., 2012, Major Miocene exhumation by fault-propagation folding within a metamorphosed, early Paleozoic thrust belt: Northwestern Argentina: Tectonics, v. 31, TC4023, doi:10.1029/2011TC003043.

Pearson, D.M., Kapp, P., DeCelles, P.G., Reiners, P.W., Gehrels, G.E., Ducea, M.N., and Pullen, A., 2013, Influence of pre-Andean crustal structure on Cenozoic thrust belt kinematics and shortening magnitude: Northwestern Argentina: Geosphere, v. 9, no. 6, p. 1766–1782, doi: 10.1130/GES00923.1.

Pelletier, J.D., DeCelles, P.G., and Zandt, G., 2010, Relationships among climate, erosion, topography, and delamination in the Andes: A numerical modeling investigation: Geology, v. 38, no. 3, p. 259–262, doi:10.1130/G30755.1.

Quade, J., Dettinger, M.P., Carrapa, B., DeCelles, P., Murray, K.E., Huntington, K.A., Cartwright, A., Canavan, R.R., Gehrels, G., and Clementz, M., 2015, this volume, The growth of the central Andes 22–26°S, *in* DeCelles, P.G., Ducea, M.N., Carrapa, B., and Kapp, P.A., eds., Geodynamics of a Cordilleran Orogenic System: The Central Andes of Argentina and Northern Chile: Geological Society of America Memoir 212, doi:10.1130/2015.1212(15).

Reiners, P.W., and Brandon, M.T., 2006, Using thermochronology to understand orogenic erosion: Annual Review of Earth and Planetary Sciences, v. 34, p. 419–466, doi:10.1146/annurev.earth.34.031405.125202.

Reiners, P.W., Spell, T.L., Nicolescu, S., and Zanetti, K.A., 2004, Zircon (U-Th)/He thermochronometry: He diffusion and comparisons with ^{40}Ar/^{39}Ar dating: Geochimica et Cosmochimica Acta, v. 68, p. 1857–1887, doi:10.1016/j.gca.2003.10.021.

Reiners, P.W., Thomson, S.N., McPhillips, D., Donelick, R.A., and Roering, J.J., 2007, Wildfire thermochronology and the fate and transport of apatite in hillslope and fluvial environments: Journal of Geophysical Research–Earth Surface, v. 112, F04001, doi:10.1029/2007JF000759.

Salfity, J.A., and Marquillas, R.A., 1994, Tectonic and sedimentary evolution of the Cretaceous–Eocene Salta Group basin, Argentina, *in* Salfity, J.A., ed., Cretaceous Tectonics of the Andes: Wiesbaden, Germany, Vieweg Verlag, p. 266–315.

Sdrolias, M., and Müller, R.D., 2006, Controls on back-arc basin formation: Geochemistry Geophysics Geosystems, v. 7, Q04016, doi:10.1029/2005GC001090.

Sempere, T., Butler, R.F., Richards, D.R., Marshall, L.G., Sharp, W., and Swisher, C.C., III, 1997, Stratigraphy and chronology of Upper Cretaceous–Lower Paleogene strata in Bolivia and northwest Argentina: Geological Society of America Bulletin, v. 109, p. 709–727, doi:10.1130/0016-7606(1997)109<0709:SACOUC>2.3.CO;2.

Silver, P.G., Russo, R.M., and Lithgow, B.C., 1998, Coupling of South American and African plate motion and plate deformation: Science, v. 279, p. 60–63, doi:10.1126/science.279.5347.60.

Sobel, E.R., and Strecker, M.R., 2003, Uplift, exhumation and precipitation: Tectonic and climatic control of late Cenozoic landscape evolution in the northern Sierras Pampeanas, Argentina: Basin Research, v. 15, p. 431–451, doi:10.1046/j.1365-2117.2003.00214.x.

Somoza, R., 1998, Updated Nazca (Farallon)–South America relative motions during the last 40 My: Implications for mountain building in the central Andean region: Journal of South American Earth Sciences, v. 11, p. 211–215.

Strecker, M.R., Alonso, R.N., Bookhagen, B., Carrapa, B., Hilley, G.E., Sobel, E.R., and Trauth, M.H., 2007, Tectonics and climate of the southern central Andes: Annual Review of Earth and Planetary Sciences, v. 35, p. 747–787, doi:10.1146/annurev.earth.35.031306.140158.

Thomson, S.N., Brandon, M.T., Reiners, P.W., Zattin, M., Isaacson, P.J., and Balestrieri, M.-L., 2010, Thermochronologic evidence for orogen-parallel variability in wedge kinematics during convergent orogenesis of the northern Apennines, Italy: Geological Society of America Bulletin, v. 122, no. 7–8, p. 1160–1179, doi:10.1130/B26573.1.

von Huene, R., Weinrebe, W., and Heeren, F., 1999, Subduction erosion along the north Chile margin: Journal of Geodynamics, v. 27, p. 345–358, doi:10.1016/S0264-3707(98)00002-7.

Yuan, X., Sobolev, S.V., and Kind, R., 2002, Moho topography in the central Andes and its geodynamic implications: Earth and Planetary Science Letters, v. 199, p. 389–402, doi:10.1016/S0012-821X(02)00589-7.

Willett, S.D., and Brandon, M.T., 2013, Some analytical methods for converting thermochronometric age to erosion rate: Geochemistry Geophysics Geosystems, v. 14, no. 1, p. 209–222.

Zattin, M., Stefani, C., and Martin, S., 2003, Detrital fission-track analysis and petrography as keys of Alpine exhumation: The example of the Veneto foreland (southern Alps, Italy): Journal of Sedimentary Research, v. 73, p. 1051–1061, doi:10.1306/051403731051.

MANUSCRIPT ACCEPTED BY THE SOCIETY 3 JUNE 2014
MANUSCRIPT PUBLISHED ONLINE 14 NOVEMBER 2014

The Geological Society of America
Memoir 212
2015

Climate and tectonics along the southern margin of the Puna Plateau, NW Argentina: Origin of the late Cenozoic Punaschotter conglomerates

Lindsay M. Schoenbohm
Department of Chemical and Physical Science, University of Toronto Mississauga, Mississauga, ON M6R 1L9, Canada

Barbara Carrapa
Department of Geosciences, University of Arizona, 1040 E. 4th Street, Tucson, Arizona 85721, USA

Heather M. McPherson*
Jonathan R. Pratt[†]
School of Earth Sciences, The Ohio State University, Columbus, Ohio 43210, USA

Sharon Bywater-Reyes
Department of Geosciences, University of Montana, Missoula, Montana 59812-1296, USA

Estelle Mortimer
School of Earth and Environment, University of Leeds, Leeds, LS2 9JT, UK

ABSTRACT

Basins around the margin of the Puna Plateau in NW Argentina each record the onset over a few hundred thousand years of alluvial-fan facies sedimentation with the deposition of the Punaschotter conglomerates in the late Cenozoic, despite differing tectonic histories. Their striking similarity suggests that these conglomerates might have a common origin, such as the climatically driven coarsening and increase in sedimentation rate proposed to have occurred worldwide at 4–2 Ma. With this contribution, we present new U-Pb geochronology of intercalated ashes that bracket the onset of Punaschotter deposition along the margin of the Puna Plateau in the Fiambalá and El Cajon Basins. Combining this with existing data, we explore the relative roles of climate and tectonics in shaping sedimentation in this region. Our analysis reveals that Punaschotter deposition was diachronous and only occurred if

*Current address: Chicago Bridge & Iron Company, Greenwood Village, Colorado 80111, USA.
[†]Current address: Department of Earth and Ocean Sciences, University of South Carolina, Columbia, South Carolina 29208, USA.

Schoenbohm, L.M., Carrapa, B., McPherson, H.M., Pratt, J.R., Bywater-Reyes, S., and Mortimer, E., 2015, Climate and tectonics along the southern margin of the Puna Plateau, NW Argentina: Origin of the late Cenozoic Punaschotter conglomerates, *in* DeCelles, P.G., Ducea, M.N., Carrapa, B., and Kapp, P.A., eds., Geodynamics of a Cordilleran Orogenic System: The Central Andes of Argentina and Northern Chile: Geological Society of America Memoir 212, p. 251–260, doi:10.1130/2015.1212(13). For permission to copy, contact editing@geosociety.org. © 2014 The Geological Society of America. All rights reserved.

both aridity and structural deformation were present. Development of orographic barriers and global climate change may have contributed to establishing regional aridity. In the eastern part of the study region (Santa Maria and Angastaco), onset of Punaschotter deposition tracks the establishment of aridity. In the already arid western basins (Fiambalá, Corral Quemado), a reactivation of structural deformation may have triggered the onset of Punaschotter deposition. Our study shows that initiation of coarse alluvial facies in this part of the Andes largely preceded the 4–2 Ma global climate change and emphasizes the coupled nature of local climate and tectonics in controlling sedimentation.

INTRODUCTION

Global terrigenous sediment accumulation has increased during the past few million years (Zhang et al., 2001). Because of the apparent lack of recent tectonic activity in many of these areas, the increase in sedimentation rates has been interpreted to reflect global climate change (Molnar, 2004). However, recent studies have argued against an increase in worldwide erosion (Willenbring and von Blanckenburg, 2009; Schumer and Jerolmak, 2009), and questions remain regarding the role of tectonics and climate on erosion and deposition of coarse clastics.

The intermontane basins along the margin of the Puna Plateau in NW Argentina (Fig. 1) offer an outstanding natural laboratory in which to test the competing influences of tectonics and climate on sedimentation. Here, the Punaschotter conglomerates (Penck 1920), deposited in basins bounding the Puna Plateau and locally within the plateau, are remarkably similar despite their different structural settings. First, they reflect an abrupt change from a variety of environmental settings (ephemeral braided streams, meandering channels, wetlands, or playa) to debris-flow–dominated, proximal alluvial fans (e.g., Bossi et al., 2000; Muruaga, 2001; Carrapa et al., 2008, 2011a; Bywater-Reyes et al., 2010). Second, they are preserved at the top of each stratigraphic section. Finally, previous studies suggested they were deposited simultaneously in late Pliocene–Quaternary time (Allmendinger, 1986; Strecker et al., 1989), approximately synchronous with globally observed changes in sedimentation. Global climate change at 4–2 Ma could thus be responsible for the appearance of these conglomerates in the stratigraphic record. However, local structural deformation, local climate forcing, or earlier climatic events may be wholly or partly responsible. With this contribution, we explore these relationships using new and existing geochronology, thermochronology, sedimentology, structural geology, and paleoclimate data. Our study shows that the Punaschotter conglomerates were deposited diachronously in different basins and are older than the 4–2 Ma global climate change. We attribute formation of the Punaschotter conglomerates to the occurrence of structural deformation under arid climate conditions—both conditions were necessary for the formation of the conglomerates and were achieved at different times in different basins.

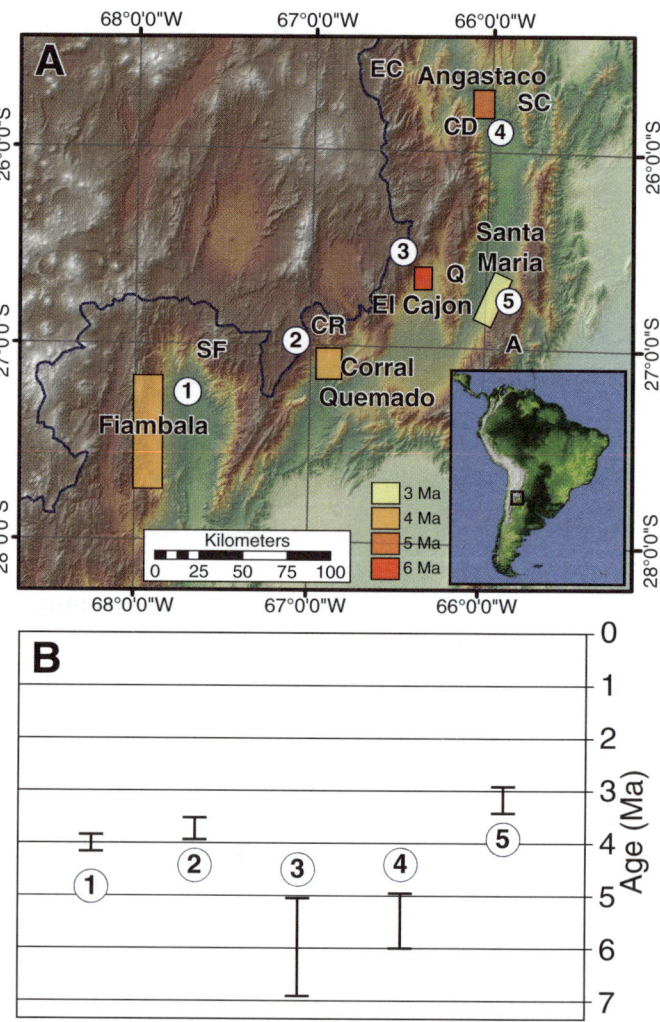

Figure 1. (A) Shaded digital elevation model (DEM) of the southeast margin of the Puna Plateau (inset shows location). Plateau boundary, defined by line of internal drainage, is marked with heavy dashed line. Numbers and colored boxes indicate approximate age and spatial extent of Punaschotter boundary. SF—Sierra de Fiambalá; CR—Chango Real; EC—Eastern Cordillera; CD—Cerro Durazno; SC—Sierra de los Colorados; Q—Quilmes; A—Aconquija. (B) Age of onset of Punaschotter conglomerate for locations shown in A. Bar marks possible age range. Note diachronous deposition.

GEOLOGIC SETTING

The basins of interest in this study are situated within and along the margin of the southern Puna Plateau in NW Argentina (Fig. 1). The plateau results from crustal thickening and the migration of an orogenic wedge (DeCelles et al., 2011) through the region beginning in the Late Cretaceous–early Cenozoic (Arriagada et al., 2006; Mpodozis et al., 2005), reaching the eastern ranges in the late Miocene–Pliocene (Coutand et al., 2001; Deeken et al., 2006; Carrapa and DeCelles, 2008; Carrapa et al., 2011a, 2011b; Echavarria et al., 2003; Reynolds et al., 2000). Deformation along the southern plateau boundary is more complicated and is reviewed in more detail in the following. In contrast to the Altiplano, which is argued to have experienced significant uplift since 10 Ma (e.g., Garzione et al., 2008), the Puna Plateau has likely been at an altitude similar to present day (~4000 m) since as early as 38 Ma, based on deuterium ratios in volcanic glass (Canavan et al., 2011). The plateau can be defined as the region of internal drainage, or the contiguous region above 3000 m elevation; these definitions yield broadly similar results (Fig. 1).

The edge of the modern southern Puna Plateau is marked by a number of intermontane, structurally bound, externally drained basins, which include, from southwest to northeast, the Fiambalá, Corral Quemado, El Cajon, Santa Maria, and Angastaco Basins (Fig. 1). These basins are generally filled with Neogene to Quaternary fluvial and lacustrine sedimentary strata, which have been variably deformed by thrusting along one or both basin-bounding ranges (e.g., Strecker et al., 1989; Coutand et al., 2001; Mortimer et al., 2007; Carrapa et al., 2008, 2011b). Strata within each basin indicate periods of deposition alternating with incision; this observation has been used to argue for lateral growth of the plateau through the incorporation of marginal basins (Sobel et al., 2003). In this model, orographically produced aridity and continued upwind range uplift lead to the defeat of drainage systems and the formation of internally drained basins, which fill with sediment over time, eventually becoming morphologically identical to the plateau. Recent work by the same group, however, has recognized nonsystematic basin filling with proximity to the plateau margin, which in addition to modeling of topographic data, indicates the importance of geodynamic processes in creating the modern plateau elevation (Strecker et al., 2009). The marginal basins record important tectonic and climatic information and therefore hold the key to understanding the interaction of these processes in the region.

The top of the sedimentary section in each basin is characterized by a distinctive unit, which regionally is called the Punaschotter conglomerate. The Punaschotter in each basin is characterized by medium- to coarse-grained, mainly clast- to locally matrix-supported, moderately to poorly sorted conglomerates with crude horizontal bedding and occasional imbrications (Fig. 2). These facies are interpreted to represent hyperconcentrated flood to debris-flow deposits associated with a proximal alluvial-fan environment (e.g., DeCelles et al., 1991). We take the first stratigraphic appearance of significantly thick (>10 m) alluvial-fan deposits to mark the onset of Punaschotter deposition. Other authors have defined the boundary differently based on qualitative geomorphic criteria, lithification, or the presence of an angular unconformity (e.g., Strecker et al., 1989; Muruaga, 2001); we have reinterpreted the formation names in these basins to be consistent with our facies definition. Collectively, we refer to the conglomerates as the "Punaschotter." When referring to individual basins, we use local formation names, indicating assignment to the Punaschotter (PS) conglomerates parenthetically (e.g., Tortoral Formation [PS]). From our study, we find that most of the contacts between the Punaschotter and underlying formations are transitional, whereas an erosional and/or angular unconformity is only locally observed.

ONSET OF DEPOSITION OF THE PUNASCHOTTER CONGLOMERATES

The timing of onset of deposition of the Punaschotter conglomerate, which is critical to evaluating the competing influences of climate and tectonics, has heretofore been poorly constrained. We remedy this by presenting 11 new U-Pb ages of volcanogenic zircons, which constrain this critical boundary in the Fiambalá and El Cajon Basins (Fig. 1; Table 1; Table DR1[1]). U-Pb geochronology was performed at the University of California at Los Angeles Cameca IMS 1270 ion microprobe facility, following the methods of Schmitt et al. (2003). We made field observations and constructed sedimentary logs across the base of the Punaschotter in the Fiambalá, El Cajon, Corral Quemado, and Angastaco Basins (Figs. 2 and 3), and we compiled geochronologic-stratigraphic data from new data and published sources (Fig. 4).

Fiambalá Basin

Six ashes and two ignimbrites were collected from the Fiambalá Basin (Fig. 3; Table 1). Five ashes and one ignimbrite were located within the Punaschotter Formation between 50 and 600 m above the base of the unit, ranging in age from 3.66 ± 0.05 Ma to 4.12 ± 0.09. A sixth ash collected from the underlying Guanchin formation is 4.08 ± 0.10 Ma; because the Punaschotter-Guanchin contact is structural in this location, the depth below the boundary cannot be determined. The final sample, an ignimbrite, was collected from near the base of the Guanchin section, which is 7.59 ± 0.20 Ma in age and significantly predates onset of Punaschotter deposition. The six ash samples range in age from 3.9 ± 0.12 (2σ) Ma to 4.12 ± 0.09 (2σ) Ma. It is possible within uncertainty that we have dated a single ash, perhaps reworked higher in the section. Alternatively, assuming a relatively uniform

---

[1]GSA Data Repository Item 2015005, Detailed analytical methods and results, is available at www.geosociety.org/pubs/ft2015.htm, or on request from editing@geosociety.org or Documents Secretary, GSA, P.O. Box 9140, Boulder, CO 80301-9140, USA.

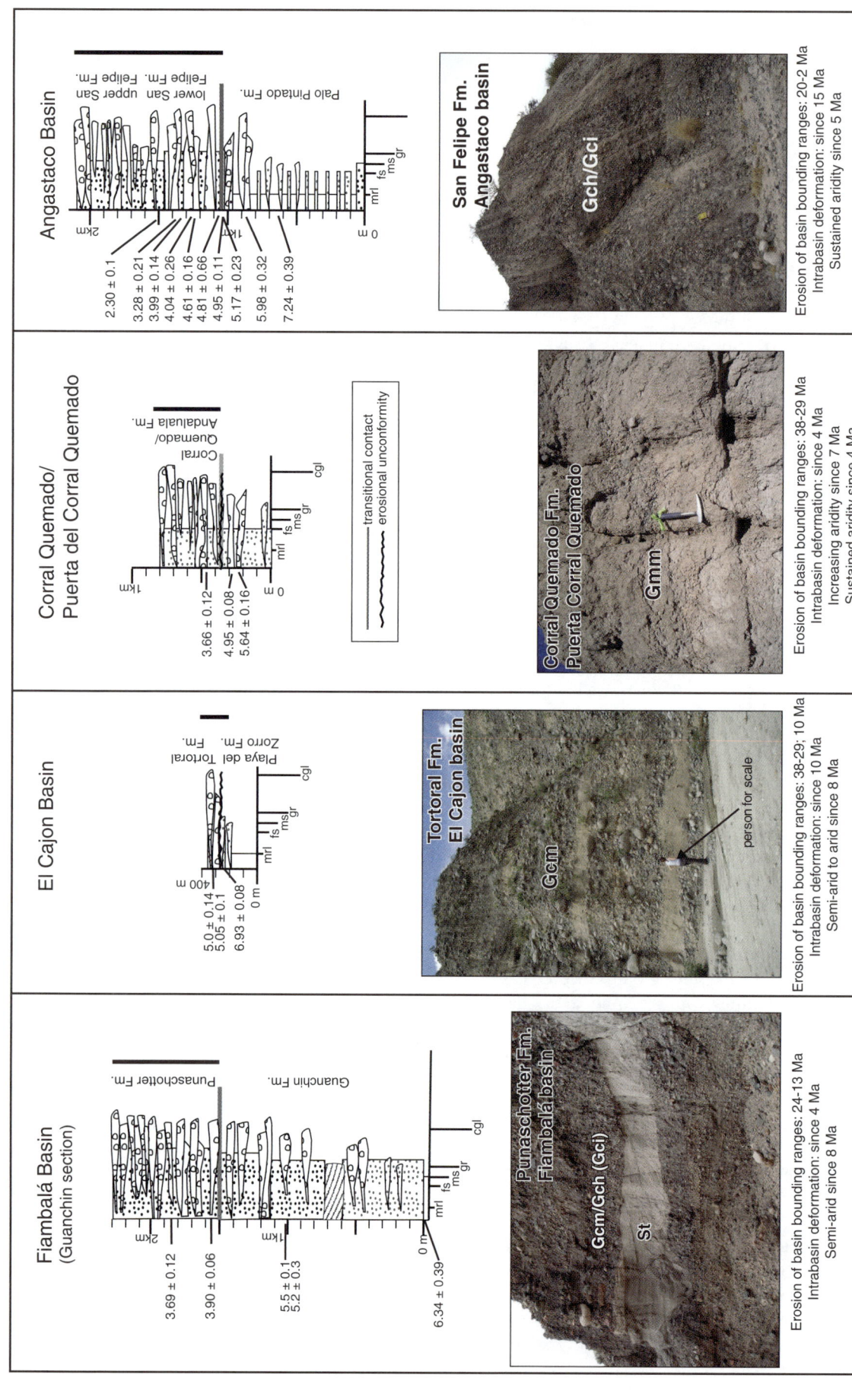

Figure 2. Schematic stratigraphic sections and facies of the Punaschotter conglomerate in the Fiambalá, El Cajon, Corral Quemado/Puerta del Corral Quemado, and Angastaco Basins. Logs are all at same scale. In photos, Punaschotter (or equivalent) is seen as clast- to matrix-supported (Gc, Gm), massive (m) to horizontally (h) stratified, occasionally imbricated (i), pebble to cobble to boulder conglomerates with occasional interbedded trough cross-stratified sandstone (St) (reworked ash in Fiambalá).

TABLE 1. ZIRCON U-Pb RESULTS

Sample	Weighted mean age (Ma)	±1σ (Ma)	MSWD*	Number of grains	Latitude (°S)	Longitude (°W)	Unit
FA3[†]	3.66	0.05	0.62	14	27°06′57.6″	67°49′03.7″	Punaschotter
FA4	3.99	0.06	1.67	4	27°08′45.4″	67°48′01.9″	Punaschotter
FA5	4.12	0.04	1.77	11	27°08′55.8″	67°47′47.0″	Punaschotter
FA7	3.93	0.05	1.02	13	27°18′39.6″	67°52′17.9″	Punaschotter
FA8	4.11	0.05	0.97	12	27°32′28.0″	67°50′02.1″	Punaschotter
FA9	3.90	0.06	1.06	9	27°42′31.2″	67°48′35.1″	Punaschotter
FA6	4.08	0.05	0.23	6	27°21′53.7″	67°52′24.1″	Guanchin
FA-ign[†]	7.59	0.10	1.23	10	27°06′16.9″	67°47′28.7″	Base of section
EC11	5.05	0.10	2.51	15	26°39′33.5″	66°23′28.0″	Punaschotter
EC12	5.00	0.14	1.69	9	26°39′34.2″	66°23′27.2″	Punaschotter
EC3	6.93	0.08	0.98	8	26°42′40.7″	66°21′07.2″	Penas Azules

*Mean square of weighted deviates.
[†]Ignimbrite.

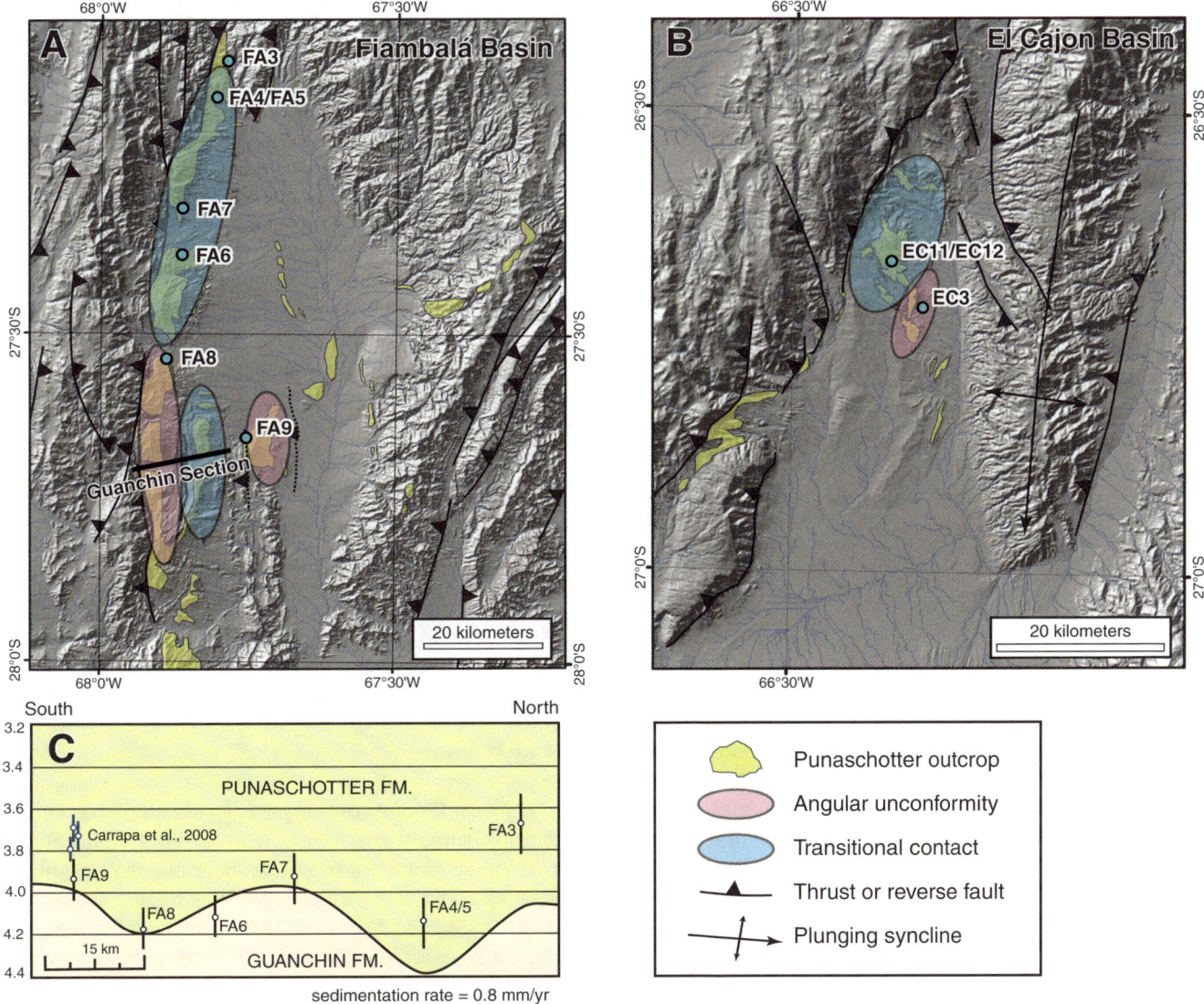

Figure 3. (A–B) Shaded relief maps of Fiambalá and El Cajon Basins. Yellow indicates Punaschotter deposits. Labels mark locations of ashes dated in this study. Blue-shaded regions indicate transitional contacts between Punaschotter-type deposits and underlying strata, whereas pink-shaded regions indicate unconformable contacts. Guanchin section (Fig. 2) is noted on Fiambalá map. (C) Illustration of the variation in age of the Punaschotter-Guanchin contact from north to south within the Fiambalá Basin. We use an arbitrary sedimentation rate of 0.8 mm/yr to calculate the age of the contact based on stratigraphic position and ash age for illustration purposes.

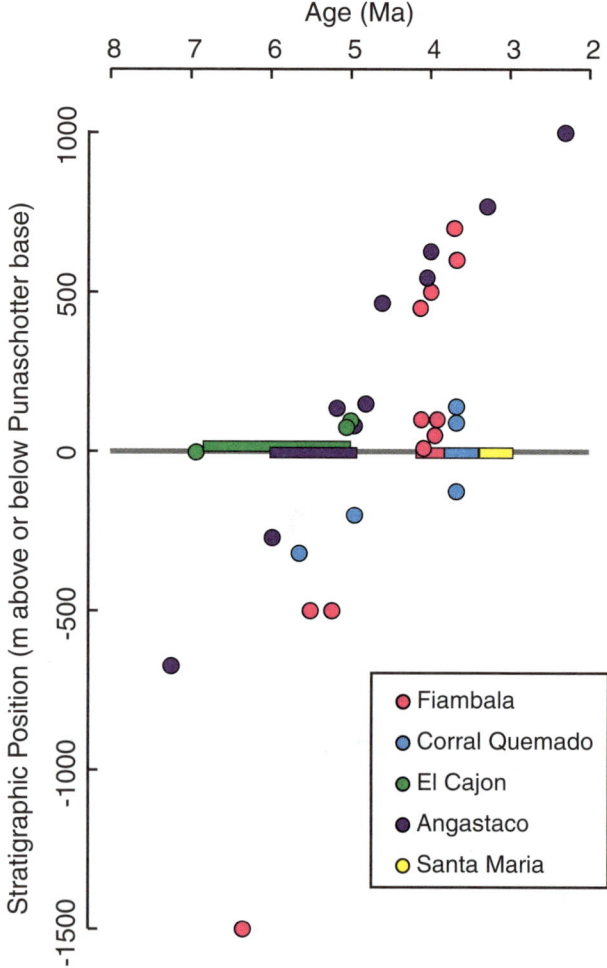

Figure 4. Stratigraphic position of dated ashes from across the transition to Punaschotter-type deposition based on data presented in Table DR2 (see text footnote 1). Rectangles mark bracketed age range of transition. Note general lack of change in sedimentation rate across boundary.

6.34 ± 0.29 Ma) provide corroboration (Figs. 2 and 4; Table DR2 [see footnote 1]; Carrapa et al., 2008).

Corral Quemado Basin

In the Corral Quemado Basin (Fig. 1), the key age control comes from a single, prominent, ~2-m-thick ash with remarkable uniformity in composition and thickness throughout the basin, dated using different methods in different locations (see Table DR2): 3.66 ± 0.12 Ma with zircon U-Pb (Coutand, 2009, personal commun.) and 3.66 ± 0.05 Ma using single crystal laser-fusion ^{40}Ar-^{39}Ar (Latorre et al., 1997). Although the contact appears to be conformable in the Corral Quemado region, the variation in age of the onset of Punaschotter deposition is similar to that in the Fiambalá Basin. Again, this could reflect difficulties in our criteria for picking the onset of Punaschotter deposition or a true difference in the timing of Punaschotter deposition in different parts of the basin. Butler et al. (1984) calculated a uniform sedimentation rate of 0.56 mm/yr across the boundary based on paleomagnetic data despite the change in facies (Fig. 2). As we do not have enough well-located samples to calculate a sedimentation rate independently, we use this rate in our analysis. In the eastern Puerta del Corral Quemado section, our interpretation of the stratigraphic descriptions indicates that the marker ash is located ~125 m below the base of the Punaschotter (Latorre et al., 1997). Using the rate of 0.56 mm/yr, the estimated age of the boundary is 3.44 Ma. In the Corral Quemado section in the central basin, according to our observations, the ash is instead located at least 140 m above the base of the Punaschotter, which is not exposed, yielding a preferred age for the boundary of >3.9 Ma. At the Villa Vil section in the northeast, the ash is within the Corral Quemado Formation, which we interpret as Punaschotter because of the proximal alluvial-fan facies, ~90 m above the transition (Muruaga, 2001). This translates to a preferred age of 3.8 Ma for the boundary. The age of the onset of Punaschotter deposition is therefore constrained to between 3.44 and >3.9 Ma. Two ashes deeper in the section dated using K-Ar (4.95 ± 2.0 Ma, 200 m below contact) and ^{40}Ar-^{39}Ar (5.64 ± 0.16 Ma, 320 m below contact) corroborate this age.

El Cajon Basin

In the El Cajon Basin (Figs. 1, 2, and 3B; Table 1), we collected two closely spaced ashes from the Tortoral Formation (Bossi et al., 2000), equivalent to the Punaschotter conglomerates, at least 90 m above the base of the unit. These yielded ages of 5.00 ± 0.28 Ma and 5.05 ± 0.20 Ma. In another location, we collected an ash within the underlying fluvial conglomerate (Bossi et al., 2000), within a few meters of the unconformable contact, giving an age of 6.93 ± 0.15 Ma. Just as in the Fiambalá Basin, the Tortoral (PS) Formation is conformable with the underlying strata in places, but it is marked by an unconformity in others, indicating syndepositional deformation (Fig. 3B). Because the base of the section below the Tortoral (PS) ashes is not exposed,

sedimentation rate, the age of the base of the Punaschotter conglomerate may vary nonsystematically from north to south within the basin by as much as several hundred thousand years, depending on sedimentation rate (Fig. 3C). A third possibility is that this variation reflects our criteria for defining the base of the Punaschotter conglomerate as the first appearance of conglomerate beds thicker than 10 m. The boundary is transitional in places (marked by blue regions in Fig. 3A) and marked by an unconformity in others (pink regions in Fig. 3A), indicating overall syndeformational deposition, with deformation localized in the western and central parts of the basin. The cluster of ages close to the Punaschotter contact brackets the onset of deposition to between 3.9 ± 0.06 Ma (youngest ash) and >4.12 ± 0.04 Ma (oldest ash). Additional samples collected from the top of the Punaschotter (3.05 ± 0.44 Ma and 3.69 ± 0.12 Ma) and from the middle to lower Guanchin Formation (5.23 ± 0.30 Ma, 5.50 ± 0.1 Ma, and

and because it is not known how much section was removed at the angular unconformity, the age of the boundary can only be constrained to lie between 5.05 and 6.93 Ma.

Angastaco Basin

The San Felipe Formation in the Angastaco Basin (Fig. 1) is equivalent to the Punaschotter conglomerates. A series of ashes sampled across the Palo Pintado–San Felipe (PS) boundary (Bywater-Reyes et al., 2010) brackets the age of the boundary to between 4.95 Ma and 5.98 Ma (Figs. 2 and 4; Table DR2 [see footnote 1]). We note that the boundary between the Palo Pintado and San Felipe formations is transitional at this location (Fig. 2). Based on a sedimentation rate of 0.14 mm/yr, we propose a preferred age for the boundary in the Angastaco Basin of 5.2 Ma (Table DR2).

Santa Maria Basin

In the Santa Maria Basin (Fig. 1), we consider the Corral Quemado Formation to be equivalent to the Punaschotter conglomerate on the basis of the proximal alluvial-fan facies descriptions. Zircon fission-track dating of intercalated ashes from above and below the contact bracket the age to between 2.96 Ma and 3.4 Ma (Table DR2; Strecker et al., 1989).

Summary

The Punaschotter conglomerates are both older than originally thought (late Miocene to Pliocene rather than Quaternary) and were deposited diachronously across the region between 2.96 Ma and 6.93 Ma. Variations in age of up to several hundred thousand years within a single basin (e.g., Fiambalá and Corral Quemado) exist as well (Fig. 3C; Table DR2). New evidence in the Fiambalá and El Cajon Basins suggests syndeformational deposition (Figs. 3A and 3B). Further, variations in thicknesses across the area suggest different magnitudes of erosion and deposition (Fig. 2). Also, our new data combined with existing data show no evidence for a significant change in sedimentation rate accompanying the change in sedimentary facies in either the Corral Quemado (Butler et al., 1984) or Angastaco (Bywater-Reyes et al., 2010) Basins, where sufficient data exist to make such an evaluation. These data combine to favor a more complex explanation than solely due to global climate change at 4–2 Ma for the onset of Punaschotter deposition in NW Argentina (e.g., Molnar, 2004).

CAUSE: LOCAL STRUCTURAL DEFORMATION?

Tectonically driven uplift of the basin-bounding ranges would increase relief and erosion rates in the source regions, possibly leading to changes in sedimentary facies. Exhumation and uplift can be inferred from bedrock or detrital thermochronology or from evidence for deformation within the basin such as angu-

lar unconformities or growth strata (Figs. 3 and 5; Table DR3 [see footnote 1]).

Exhumation of basin-bounding ranges mostly preceded deposition of the Punaschotter conglomerate in the region. Apatite fission-track (AFT) thermochronology from bedrock indicates relatively rapid exhumation of the Sierra de Fiambalá NE of the Fiambalá Basin between 24 Ma and 13 Ma (Carrapa et al., 2006). Intrabasin deformation and exhumation are also evident at the time of Punaschotter deposition, around 4 Ma (Carrapa et al., 2008, 2010; this study). The Chango Real range north of the Corral Quemado Basin and west of the El Cajon Basin was exhuming between 38 and 29 Ma (Coutand et al., 2001). Unconformities within the Corral Quemado section (below the Corral Quemado [PS] Formation) indicate syndepositional deformation there since ca. 4 Ma as well (Muruaga, 2001; this study). Uplift of the Quilmes Range between the El Cajon and Santa Maria Basins is constrained by detrital AFT data that record the first appearance of Quilmes-derived sediment in the El Cajon Basin at ca. 10 Ma (Mortimer et al., 2007; Pratt et al., 2008); this deformation could still be active. Growth strata within the Playa del Zorro Formation in the El Cajon Basin (Mortimer et al., 2007) and beneath the Tortoral (PS) Formation (this study) indicate intrabasin deformation since 10 Ma. West of the Angastaco Basin, AFT data indicate exhumation of the Eastern Cordillera by 18–20 Ma, progressing to Cerro Durazno by 7–12 Ma (Coutand et al., 2006; Deeken et al., 2006). Uplift and exhumation of the Sierra de los Colorados to the east of the Angastaco Basin began at 3.4–2.4 Ma based on paleocurrent and provenance data (Bywater-Reyes et al., 2010) and apatite (U-Th)/He thermochronology (Carrapa et al., 2011b). The presence of growth strata and unconformities throughout the

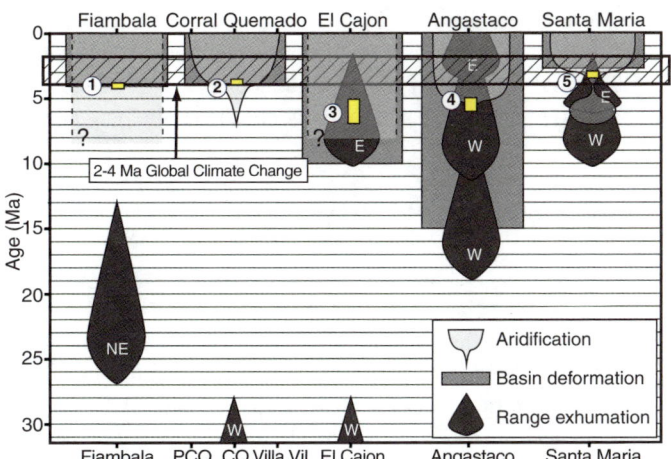

Figure 5. Compilation of thermochronologic, structural, and climate records. Age of onset of Punaschotter or other alluvial-fan deposition is marked with yellow bar. Teardrop shape of "range exhumation" field schematically reflects uncertain timing of onset and termination. Letters indicate direction of range from basin. Relative aridity is indicated by width of "aridification" field; uncertainty is indicated by dashed boundary. See Tables DR3 and DR4 for references (see text footnote 1). PCQ—Puerta del Corral Quemado; CQ—Corral Quemado.

section indicates intrabasin deformation since 15 Ma. AFT data indicate exhumation of the Aconquija Range east of the Santa Maria Basin between 6 and 3 Ma (Sobel and Strecker, 2003). These data are summarized in Table DR3 and presented graphically in Figure 5.

Deformation recorded within the basins locally corresponds to the onset of Punaschotter deposition. Although the main deformation recorded by thermochronometers preceded deposition of the Punaschotter conglomerate in the Fiambalá and Corral Quemado Basins, local angular unconformities indicate renewed syntectonic activity at the time of the transition (Muruaga, 2001; Carrapa et al., 2008, 2010; this study). However, there is no correspondence within the El Cajon or Angastaco Basins, where growth strata indicate deformation since ca. 10 Ma (Pratt et al., 2008) and ca. 15 Ma (Carrera and Muñoz, 2008; Carrapa et al., 2011b), respectively, and where the Punaschotter is transitional and conformable on the underlying Palo Pintado Formation (Carrapa et al., 2011a). Finally, the contact between the Corral Quemado (PS) and underlying strata is conformable (Strecker et al., 1989) in the Santa Maria Basin; unconformities higher in the section indicate protracted deformation.

West-to-east propagation of deformation is documented in the eastern basins, with uplift proceeding from the Eastern Cordillera to Cerro Durazno and the Quilmes, to the Aconquija and Sierra de los Colorados (Sobel and Strecker, 2003; Coutand et al., 2006; Deeken et al., 2006; Mortimer et al., 2007; Pratt et al., 2008). However, in the Fiambalá and Corral Quemado Basins in the west, the origin of some intrabasin deformation at ca. 4 Ma significantly postdates evidence for exhumation and major deformation in the adjacent ranges. It is not clear what caused this out-of-sequence deformation, although reactivation of structures within the wedge in order to maintain topographic taper, or changes in mass balance in the Puna Plateau related to lithospheric foundering are both possibilities. Strecker et al. (2009) noted the lack of correspondence between incision and deposition in marginal basins and distance from the plateau margin as well.

CAUSE: GLOBAL OR LOCAL ARIDIFICATION?

Alternatively, aridity and increased seasonality, driven either by global climate change or the growth of local orographic barriers, could explain the shift in sedimentary facies. Although alluvial fans can develop in a variety of climatic conditions (Dorn, 2009), they are often associated with dry climate. Alluvial fans are characteristic of modern arid environments (e.g., Owen et al., 1997; Clarke, 2006), possibly because arid regions are subject to more frequent large floods per small flood (Molnar et al., 2006), with highly variable discharge favoring sediment gravity flows such as those that make up the Punaschotter conglomerates (Schumm et al., 1987).

Evidence from climate proxies within the basins is shown in Figure 5 and compiled in Table DR4 (see footnote 1). Paleoclimatic interpretations based on sediment facies in the Fiambalá

and El Cajon Basins suggest that aridity preceded Punaschotter deposition. Evidence includes large mud cracks, silicified wood, and alluvial-fluvial facies in the Fiambalá Basin (Carrapa et al., 2006), and gypsum associated with lacustrine deposition in the El Cajon Basin (Mortimer et al., 2007). In these basins, aridity could have been the product of the development of orographic barriers as upwind ranges were uplifted, but phenomena such as expansion of the East Antarctic ice sheet invoked by Lamb and Davis (2003) to explain aridity in Chile at this time cannot be ruled out.

In the eastern basins, direct climate records demonstrate that Punaschotter deposition corresponds to the achievement of fully arid, seasonal conditions. The $\delta^{13}C$ and $\delta^{18}O$ data from the Puerta del Corral Quemado section indicate humid conditions prior to 8 Ma, with increasing aridity between 7 and 4 Ma and a sharp increase in aridity at ca. 4 Ma, with C_4 vegetation comprising up to 70% of plant biomass (Latorre et al., 1997). In the Santa Maria Basin, $\delta^{13}C$ data and the presence of calcic and silicic rhizoliths and authigenic clays suggest wet-dry seasonality between 5 and 7 Ma, whereas a period of relative humidity is recorded by an increase in C_3 plants between 3 and 5 Ma (Kleinert and Strecker, 2001). Arid conditions, evidenced by $\delta^{13}C$ and $\delta^{18}O$ data indicating the expansion of C_4 plants and intense soil-water evaporation, were reestablished by ca. 3 Ma (Kleinert and Strecker, 2001). More recent $\delta^{13}C$ and $\delta^{18}O$ data in the Angastaco Basin indicate an increase in aridification at ca. 6 Ma, becoming constant by ca. 5 Ma (Bywater-Reyes et al., 2010).

In the Corral Quemado, Angastaco, and Santa Maria Basins, aridification at ca. 6–7 Ma could correspond to concurrent worldwide changes in aridity and seasonality (Pagani et al., 1999). This is particularly clear in the Angastaco Basin, where there is no evidence of an orographic barrier until after the onset of aridity (Bywater-Reyes et al., 2010). However, orography could play a role in some basins, such as in Santa Maria, where the return to aridity at ca. 3 Ma has been attributed to uplift of the Aconquija to the east (Kleinert and Strecker, 2001). Punaschotter deposition in many of these basins clusters at ca. 4 Ma. This could reflect the time delay between the onset of aridity and the achievement of fully arid conditions, but it could also reflect a relationship with the increase in climate variability at 4–3 Ma expressed in ocean records (Zachos et al., 2001). These oscillations could cause landscape disequilibrium, enhancing erosion rates and potentially contributing to the change in sedimentary facies. This, combined with local deformation, can also explain more recent Quaternary deposits along the margin of the Puna Plateau (Hilley and Strecker, 2005).

SYNTHESIS: TECTONIC AND CLIMATIC CONTROLS ON SEDIMENTATION

Perhaps the most important finding of our study, based on new geochronologic data and compilation of regional structural and paleoclimatic data, is that both aridity/seasonality and structural deformation within the basin were necessary for

Punaschotter conglomerate deposition. If arid conditions were already established, such as in the Fiambalá Basin (Fig. 5), Punaschotter conglomerates were deposited whenever there was active deformation within the basin and presumed uplift of basin-bounding ranges. In the opposite scenario, in which local deformation was ongoing, such as in the Angastaco Basin (Fig. 5), Punaschotter conglomerate deposition did not begin until arid conditions were achieved. In the Corral Quemado and Santa Maria Basins, the onset of aridity, deformation, and Punaschotter deposition were approximately simultaneous (Fig. 5). The origin of the conglomerates in the El Cajon Basin remains elusive, but better climate data could resolve the question (Fig. 5). The simple explanation is therefore that climate and tectonics, working separately, are not enough; they must work in concert to provoke changes in sedimentary character. Although alluvial fans have been often associated with dry climate, fan deposition can develop in a variety of climatic conditions, and therefore alluvial fans are not diagnostic of aridity (Dorn, 2009). Our study shows, however, that aridity is an important and necessary factor controlling deposition of the Punaschotter conglomerate in NW Argentina. An interesting question is whether in some basins with a longer record of aridity (i.e., Fiambalá and El Cajon), earlier pulses of deformation could have or should have produced coarse, alluvial-fan, Punaschotter-type deposition. Our records of both deformation and paleoclimate, however, remain insufficient to answer that question at this time.

Not surprisingly, in exploring the question of whether climate or tectonics drives sedimentation and exhumation in and along the margin of the Puna Plateau, the answer is that both play a role and act together. Global climate oscillation and regional aridification coupled with local tectonics, driven either by propagation of deformation through the area or a later reactivation, are critical. The formation of local orographic barriers could play a role in some, but not all, of the basins.

ACKNOWLEDGMENTS

This research was supported by the National Science Foundation (EAR-0635584). The ion microprobe facility at the University of California–Los Angeles, used for U-Pb ash geochronology, is partly supported by a grant from the Instrumentation and Facilities Program, Division of Earth Sciences, National Science Foundation.

REFERENCES CITED

Allmendinger, R.W., 1986, Tectonic development, southeastern border of the Puna Plateau, northwestern Argentine Andes: Geological Society of America Bulletin, v. 97, p. 1070–1082, doi:10.1130/0016-7606 (1986)97<1070:TDSBOT>2.0.CO;2.

Arriagada, C., Cobbold, P.R., and Roperch, P., 2006, Salar de Atacama Basin: A record of compressional tectonics in the central Andes since the mid-Cretaceous: Tectonics, v. 25, TC1008, doi:10.1029/2004TC001770.

Bossi, G.E., Vides, M.E., Ahumada, A.L., Georgieff, S.M., Muruaga, C., and Ibanez, L.M., 2000, Analisis de las paleocorrientes y de la varianza de los componentes a tres niveles, Noegeno del Valle del Cajon, Catamar-

ca, Argentina: Asociación Argentina de Sedimentolgia Revista, v. 7, p. 23–47.

Butler, R.F., Marshall, L.G., Drake, R.E., and Curtis, G.H., 1984, Magnetic polarity stratigraphy and ^{40}K-^{40}Ar dating of late Miocene and early Pliocene continental deposits, Catamarca Province, NW Argentina: The Journal of Geology, v. 92, p. 623–636, doi:10.1086/628902.

Bywater-Reyes, S., Carrapa, B., Clementz, M., and Schoenbohm, L., 2010, Effect of late Cenozoic aridification on sedimentation in the Eastern Cordillera of northwest Argentina (Angastaco Basin): Geology, v. 38, p. 235–238, doi:10.1130/G30532.1.

Canavan, R.R., Clementz, M.T., Carrapa, B., Quade, J., DeCelles, P., and Schoenbohm, L.M., 2011, Paleoelevation of the Puna Plateau (northwestern Argentina) inferred from geochemical analyses of volcanic glass: Geological Society of America Abstracts with Programs, v. 43, no. 5, p. 539.

Carrapa, B., and DeCelles, P.G., 2008, Eocene exhumation and basin development in the Puna of northwestern Argentina: Tectonics, v. 27, TC1015, doi:10.1029/2007TC002127.

Carrapa, B., Adelmann, D., Hilley, G.E., Mortimer, E., Sobel, E.R., and Strecker, M.R., 2005, Oligocene range uplift and development of plateau morphology in the southern central Andes: Tectonics, v. 24, TC4011, doi:10.1029/2004TC001762.

Carrapa, B., Sobel, E.R., and Strecker, M.R., 2006, Cenozoic orogenic growth in the central Andes: Evidence from sedimentary rock provenance and apatite fission track thermochronology in the Fiambalá Basin, southernmost Puna Plateau margin (NW Argentina): Earth and Planetary Science Letters, v. 247, p. 82–100, doi:10.1016/j.epsl.2006.04.010.

Carrapa, B., Hauer, J., Schoenbohm, L., Strecker, M., Schmitt, A., Villaneva, A., and Sosa Gomez, J., 2008, Dynamics of deformation and sedimentation in the Sierras Pampeanas: An integrated study of the Neogene Fiambalá basin, NW Argentina: Geological Society of America Bulletin, v. 120, p. 1518–1543, doi:10.1130/B26111.1.

Carrapa, B., Hauer, J., Schoenbohm, L., Strecker, M., Schmitt, A., Villaneva, A., and Sosa Gomez, J., 2010, Dynamics of deformation and sedimentation in the northern Sierras Pampeanas: An integrated study of the Neogene Fiambalá Basin, NW Argentina: Reply: Geological Society of America Bulletin, v. 122, no. 5–6, p. 950–953.

Carrapa, B., Bywater-Reyes, S., DeCelles, P.G., Mortimer, E., and Gehrels, G., 2011a, Cenozoic synorogenic basin evolution in the Eastern Cordillera of northwestern Argentina (25°–26°S): Regional implications for Andean orogenic wedge development: Basin Research, v. 23, p. 1–20, doi:10.1111/j.1365-2117.2011.00519.x.

Carrapa, B., Trimble, J., and Stockli, D., 2011b, Timing and magnitude of deformation and exhumation of the Eastern Cordillera of NW Argentina revealed by (U-Th)/He thermochronology: Tectonics, v. 30, TC3003, doi:10.1029/2010TC002707.

Carrera, N., and Muñoz, J.A., 2008, Thrusting evolution in the southern Cordillera Oriental (northern Argentine Andes): Constraints from growth strata: Tectonophysics, v. 459, no. 1–4, p. 107–122, doi:10.1016/j.tecto .2007.11.068.

Clarke, J.D.A., 2006, Antiquity of aridity in the Chilean Atacama Desert: Geomorphology, v. 73, p. 101–114, doi:10.1016/j.geomorph.2005.06.008.

Coutand, I., Cobbold, P.R., de Urreiztieta, M., Gautier, P., Chauvin, A., Gapais, D., Rossello, E.A., and Lopèz-Gamundí, O., 2001, Style and history of Andean deformation, Puna Plateau, northwestern Argentina: Tectonics, v. 20, p. 210–234, doi:10.1029/2000TC900031.

Coutand, I., Carrapa, B., Deeken, A., Schmitt, A.K., Sobel, E.R., and Strecker, M.R., 2006, Orogenic plateau formation and lateral growth of compressional basins and ranges: Insights from sandstone petrography and detrital apatite fission-track thermochronology in the Angastaco Basin, NW Argentina: Basin Research, v. 18, p. 1–26, doi:10.1111/j.1365 -2117.2006.00283.x.

DeCelles, P.G., Gray, M.B., and Ridgway, I.K.D., 1991, Controls on synorogenic alluvial-fan architecture, Beartooth conglomerate (Paleocene), Wyoming and Montana: Sedimentology, v. 38, p. 567–590, doi:10.1111/j.1365-3091.1991.tb01009.x.

DeCelles, P.G., Carrapa, B., Horton, B.K., and Gehrels, G., 2011, Cenozoic foreland basin system in the central Andes of northwestern Argentina: Implications for Andean geodynamics and modes of deformation: Tectonics, v. 30, TC6013, doi:10.1029/2011TC002948.

Deeken, A., Sobel, E., Coutand, I., Haschke, M., Riller, U., and Strecker, M., 2006, Development of the southern Eastern Cordillera, NW Argentina, constrained by apatite fission track thermochronology: From Early Cretaceous extension to middle Miocene shortening: Tectonics, v. 25, TC6003, doi:10.1029/2005TC001894.

Dorn, R.I., 2009, The role of climatic change in alluvial fan development, *in* Parsons, A.J., and Abrahams, A.D., eds., Geomorphology of Desert Environments (2nd ed.): Dordrecht, Netherlands, Springer, p. 723–742, doi:10.1007/978-1-4020-5719-9 24.

Echavarria, R., Hernandez, R., Allmendinger, R.W., and Reynolds, J.H., 2003, Sub-Andean thrust and fold belt of northwest Argentina: Geometry and timing of the Andean evolution: American Association of Petroleum Geologists Bulletin, v. 87, p. 965–985, doi:10.1306/01200300196.

Garzione, C.N., Hoke, G.D., Libarkin, J.C., Withers, S., MacFadden, B., Eiler, J., Ghosh, P., and Mulch, A., 2008, Rise of the Andes: Science, v. 320, p. 1304–1307, doi:10.1126/science.1148615.

Hilley, G., and Strecker, M.R., 2005, Processes of oscillatory basin filling and excavation in a tectonically active orogen: Quebrada del Toro Basin, NW Argentina: Geological Society of America Bulletin, v. 117, no. 7–8, p. 887–901, doi:10.1130/B25602.1.

Kleinert, K., and Strecker, M.R., 2001, Changes in moisture regime and ecology in response to late Cenozoic orographic barriers: The Santa Maria Valley, Argentina: Geological Society of America Bulletin, v. 113, p. 728–742, doi:10.1130/0016-7606(2001)113<0728:CCIRTO>2.0.CO;2.

Lamb, S., and Davis, P., 2003, Cenozoic climate change as possible cause for the rise of the Andes: Nature, v. 425, p. 792–797, doi:10.1038/nature02049.

Latorre, C., Quade, J., and McIntosh, W.C., 1997, The expansion of C_4 grasses and global change in the late Miocene: Stable isotope evidence from the Americas: Earth and Planetary Science Letters, v. 146, p. 83–96, doi:10.1016/S0012-821X(96)00231-2.

Molnar, P., 2004, Late Cenozoic increase in accumulation rates of terrestrial sediment: How might climate change have affected erosion rates?: Annual Review of Earth and Planetary Sciences, v. 32, p. 67–89, doi:10.1146/annurev.earth.32.091003.143456.

Molnar, P., Anderson, R.S., Kier, G., and Rose, J., 2006, Relationships among probability distributions of stream discharges in floods, climate, bed load transport, and river incision: Journal of Geophysical Research, v. 111, F02001, doi:10.1029/2005JF000310.

Mortimer, E., Carrapa, B., Coutand, I., Schoenbohm, L., Sobel, E.R., Sosa-Gomez, J., and Strecker, M.R., 2007, Fragmentation of a foreland basin in response to out-of-sequence basement uplifts and structural reactivation: El Cajon–Campo del Arenal Basin, NW Argentina: Geological Society of America Bulletin, v. 119, p. 637–653, doi:10.1130/B25884.1.

Mpodozis, C., Arriagada, C., Basso, M., Roperch, P., Cobbold, P., and Reich, M., 2005, Late Mesozoic to Paleogene stratigraphy of the Salar de Atacama Basin, Antofagasta, northern Chile: Implications for the tectonic evolution of the central Andes: Tectonophysics, v. 399, p. 125–154, doi:10.1016/j.tecto.2004.12.019.

Muruaga, C.M., 2001, Estratigraphía y desarrollo tectosedimentario de los sedimentos terciarios en los alrededores de la Sierra de Hualfin, borde suroriental de la Puna, Catamarca, Argentina: Asociación Argentina de Sedimentológica Revista, v. 8, p. 27–50.

Owen, L.A., Windley, B.F., Cunningham, W.D., Badamgarav, J., and Dorjnamjaa, D., 1997, Quaternary alluvial fans in the Gobi of southern Mongolia: Evidence for neotectonics and climate change: Journal of Quaternary Science, v. 12, p. 239–252, doi:10.1002/(SICI)1099-1417(199705/06)12:3<239::AID-JQS293>3.0.CO;2-P.

Pagani, M., Freeman, K.H., and Arthur, M.A., 1999, Late Miocene atmospheric CO_2 concentrations and the expansion of C_4 grasses: Science, v. 285, p. 876–879, doi:10.1126/science.285.5429.876.

Penck, W., 1920, Der Südrand der Puna de Atacama (NW-Argentinien): Leipzig, Germany, Abhandlungen Mathematisch-Physiakalische Klasse der Sächsichen Akademie der Wissenschaften, v. 37, 420 p.

Pratt, J., Schoenbohm, L., Mortimer, E., and Strecker, M., 2008, Basin compartmentalization within a foreland: Structural and temporal analysis of the El Cajon Basin, NW Argentina: Eos (Transactions, American Geophysical Union), v. 89, no. 53, Fall Meeting supplement, abstract T53B-1927.

Reynolds, J.H., Galli, C.I., Hernández, R.M., Idleman, B.D., Kotila, J.M., Hilliard, R.V., and Naeser, C.W., 2000, Middle Miocene tectonic development of the Transition zone, Salta Province, northwest Argentina: Magnetic stratigraphy from the Metán Subgroup, Sierra de González: Geological Society of America Bulletin v. 112, p. 1736–1751, doi:10.1130/0016-7606(2000)112<1736:MMTDOT>2.0.CO;2.

Schmitt, A.K., Grove, M., Harrison, T.M., Lovera, O., Hulen, J., and Walters, M., 2003, The Geysers-Cobb Mountain magma system, California (Part 1): U-Pb zircon ages of volcanic rocks, conditions of zircon crystallization and magma residence times: Geochimica et Cosmochimica Acta, v. 67, p. 3423–3442, doi:10.1016/S0016-7037(03)00140-6.

Schumer, R., and Jerolmak, D.J., 2009, Real and apparent changes in sediment deposition rates through time: Journal of Geophysical Research, v. 114, F00A06, doi:10.1029/2009JF001266.

Schumm, S.A., Mosley, P.M., and Weaver, W.E., 1987, Experimental Fluvial Geomorphology: New York, John Wiley & Sons, 413 p.

Sobel, E.R., and Strecker, M.R., 2003, Uplift, exhumation and precipitation: Tectonic and climatic control of late Cenozoic landscape evolution in the northern Sierras Pampeanas, Argentina: Basin Research, v. 15, p. 431–451, doi:10.1046/j.1365-2117.2003.00214.x.

Sobel, E.R., Hilley, G.E., and Strecker, M.R., 2003, Formation of internally drained contractional basins by aridity-limited bedrock erosion: Journal of Geophysical Research, v. 108, no. B7, 2344, doi:10.1029/2002JB001883.

Strecker, M.R., Cereny, P., Bloom, A.L., and Malizia, D., 1989, Late Cenozoic tectonism and landscape development in the foreland of the Andes: Northern Sierras Pampeanas (26°–28°S), Argentina: Tectonics, v. 8, p. 517–534, doi:10.1029/TC008i003p00517.

Strecker, M.R., Alonso, R., Bookhagen, B., Carrapa, B., Coutand, I., Hain, M.P., Hilley, G.E., Mortimer, E., Schoenbohm, L., and Sobel, E.R., 2009, Does the topographic distribution of the central Andean Puna Plateau result from climatic or geodynamic processes?: Geology, v. 37, p. 643–646, doi:10.1130/G25545A.1.

Willenbring, J.K., and von Blanckenburg, F., 2009, Long-term stability of global erosion rates and weathering during late-Cenozoic cooling: Nature, v. 465, p. 211–214, doi:10.1038/nature09044.

Zachos, J.C., Pagani, M., Sloan, L., Thomas, E., and Billups, K., 2001, Trends, rhythms, and aberrations in global climate 65 Ma to present: Science, v. 292, p. 686–693, doi:10.1126/science.1059412.

Zhang, P., Molnar, P., and Downs, W.R., 2001, Increased sedimentation rates and grain sizes 2–4 Myr ago due to the influence of climate change on erosion rates: Nature, v. 410, p. 891–897, doi:10.1038/35069099.

Manuscript Accepted by the Society 3 June 2014
Manuscript Published Online 15 October 2014

The Geological Society of America
Memoir 212
2015

Testing the analytical protocols and calibration of volcanic glass for the reconstruction of hydrogen isotopes in paleoprecipitation

Matthew P. Dettinger*
Jay Quade
Department of Geosciences, University of Arizona, 1040 E. 4th Street, Tucson, Arizona 85721, USA

ABSTRACT

In this study, we investigate the foundations of the hydrated volcanic glass paleoelevation-paleoenvironment proxy by testing (1) the sensitivity of volcanic glass δD values to laboratory procedures, (2) the reliability of young glass reconstructions of observed meteoric water values, and (3) the paleoenvironmental reconstructions based on volcanic glass in comparison to those of other proxies. Volcanic glass δD values are insensitive to variations in laboratory pretreatments, displaying no significant shift in δD values with variations in duration of drying, rinsing, the temperature at which glass is dried, or with the grain size that is analyzed. However, clay adhering to the glass particles may skew δD results by ~10‰, and HF acid pretreatment can cause significant isotopic shifts. Parent-water δD values calculated from young (≤ 0.3 Ma) glass δD values are in close accord with modern surface-water δD values across ~3000 m of elevation in the central Andes. Combined glass δD values, pedogenic carbonate $\delta^{18}O$ values, and clumped isotope paleothermometry from the 16.3 Ma Los Cristales tuff provide a key test of resistance to resetting of δD values in ancient glass hydration layers. These proxies indicate hydrated volcanic glass preserves its stable isotopic information over geologic time and that the climatic patterns of the high and dry Puna Plateau were already in place by 16.3 Ma.

INTRODUCTION

High elevations are produced by major continental tectonic forces. Understanding the height and growth of ancient orogens can greatly increase our understanding of lithospheric-scale processes (Garzione et al., 2008; Pelletier et al., 2010; Quade et al., 2011), surficial processes (Clift et al., 2006; Strecker et al., 2007; Hoke and Garzione, 2008; Whipple, 2009; Mulch et al., 2010), and local and global climate (Insel et al., 2010; Molnar et al.,

2010; Rech et al., 2010; Eiler, 2011). Estimates of paleoelevation through time provide constraints on timing of surface uplift and therefore the degree of orogenic development at particular time periods. Various techniques have been used to constrain paleoelevations (e.g., Muñoz and Charrier, 1996; Wolfe et al., 1997; Gregory-Wodzicki et al., 1998; Poage and Chamberlain, 2001; Ghosh et al., 2006b; Barnes and Ehlers, 2009; Rech et al., 2010). The technique we explore in this study exploits the stable isotopic composition of the waters of hydration in volcanic glass for

*dettingm@email.arizona.edu

paleoenvironmental and paleoelevation reconstruction. The ratio of ^1H to ^2H (deuterium, D) is strongly influenced by the preferential loss of D as a body of moisture is forced upward by an orographic barrier (Poage and Chamberlain, 2001; Blisniuk and Stern, 2005). Therefore, the ratio of hydrogen isotopes (quantified as δD values in per mil [‰] relative to standard mean ocean water [SMOW]) in meteoric water is sensitive to elevation; by measuring the former, the latter can be estimated. The waters of hydration within volcanic glass are thought to archive this isotopic information (Friedman et al., 1993a) and so may be used for paleoelevation and paleoenvironmental reconstructions (e.g., Mulch et al., 2008).

"Volcanic glass" as used here refers to shards of shattered silicate glass bubbles or pumice in rhyolitic air-fall tuffs or ignimbrites. Our study examines late Oligocene– through Holocene-age tuffs from the central Andes of South America. These tuffs were erupted from andesitic-dacitic stratovolcanoes and from silicic large-volume caldera-forming eruptions of the Altiplano-Puna volcanic complex (Francis and Hawkesworth, 1994) within the modern Andean magmatic arc. This study scrutinizes these glasses in terms of laboratory methods, modern calibration, and other ancient contemporaneous proxies in order to dependably reconstruct hydrogen isotopes in paleoprecipitation.

Hydration of Glass

At zero age, glass contains 0.1–0.3 wt% magmatic water (Friedman et al., 1993b). Following eruption and deposition, volcanic glass begins to hydrate. Over 1–10 k.y., environmental water (i.e., marine, lacustrine, or near-surface soil and groundwater) diffuses into the glass structure, increasing the weight percent water to ~3%–8% wt% at full saturation (Onken, 1991; Friedman et al., 1993a, 1993b; Mulch et al., 2008), before the glass structure begins to break down and convert to secondary clays, zeolites, and other minerals. Hydration occurs through hydrolysis and condensation reactions and leaching of soluble elements, such as Na, Ca, Li, and Mo (Cailleteau et al., 2008; Valle et al., 2010). This process produces what the materials science community commonly refers to as a "gel" layer (Cailleteau et al., 2008). The rate of hydration varies considerably with tem-

perature; Friedman et al. (1966) proposed a range from 2 mm of hydration rind thickness per 10 k.y. at 5 °C to 10,000 mm per 10 k.y. at 100 °C. With further study, Onken (1991) suggested that the largely high-temperature–derived model of Friedman et al. (1966) underpredicts natural Earth surface hydration rates by a factor of ~5. However, neither of these studies takes into account more recent work on the structural and chemical changes that accompany hydration of glass. Due to hydrolysis and reorganization of silicate bonds, the gel layer, i.e., the hydrated portion of glass, is denser and slows further diffusion of water into and labile elements out of the glass (Cailleteau et al., 2008; Valle et al., 2010). The densification results in a drop in diffusion rates by three to five orders of magnitude (Crovisier et al., 2003; Cailleteau et al., 2008; Valle et al., 2010). Hydration of volcanic glass occurs early and decreases with time. Hence, current thinking suggests that the isotopic value of the hydration waters is overwhelmingly controlled by the isotopic composition of the water present early in the hydration process.

Isotopic Composition of Hydrated Glass

Friedman et al. (1993b) quantified the relationship between the δD value of the volcanic glass and the δD value of the original parent water. They used hollow rhyolitic spheres deposited in a lake in southwest Idaho to determine an empirical glass-water fractionation factor. Following eruption and deposition, the glass spheres absorbed environmental water from the lake. The glass walls acted as a semipermeable membrane, and salts trapped within the hollow sphere drove osmosis of water into the hollow center. As water diffused through the glass walls, H_2O (as opposed to HDO) was preferentially bound to the glass structure. Yet the diffusion of H_2O (lower mass) through the glass should be faster than HDO (higher mass). Friedman et al. (1993b) assumed that all competing forces balance each other, resulting in the δD value of the water inside the spheres after hydration matching the original parent water outside the spheres.

Friedman et al. (1993b) observed that measured δD values of the volcanic glass (expressed here as δD_{vg} in ‰, relative to SMOW; Table 1) and of the fluid phase within the spheres, a proxy for the original parent water, were –175‰ and –147‰,

TABLE 1. DEFINITIONS

$\delta^{18}O_{mw}$	Measured $\delta^{18}O$ value of meteoric water versus standard mean ocean water (SMOW)
δD_{mw}	Measured δD value of meteoric water versus SMOW
δD_{vg}	Measured δD value of waters of hydration within volcanic glass versus SMOW
δD_{cpw}	Calculated δD value of reconstructed parent water from volcanic glass versus SMOW
$\Delta_{HF\text{-}noHF}$	Measured difference, in per mil, between an HF acid–treated aliquot and nontreated aliquot(s) of a sample
$\delta^{18}O_{carb}$	Measured $\delta^{18}O$ value of carbonate versus Vienna Pee Dee Belemnite (VPDB)
$\delta^{18}O_{cpw}$	Calculated $\delta^{18}O$ value of reconstructed parent water from carbonate versus SMOW
$T(\Delta_{47})$	Temperature calculated from $\Delta 47$, the measured per mil difference in abundance of the mass 47 isotopologue derived from a sample and abundance of mass 47 isotopologue from the sample with stochastic (i.e., heated to 1000 °C) distribution

respectively. From these values, the authors determined the water-glass fractionation factor to be 1.0339 ± 0.0005. Similarly, a 20 ka New Zealand ash and local groundwater also analyzed by Friedman et al. (1993b) produced δD values of −82‰ and −45‰, respectively. After applying a −5‰ correction to the groundwater to account for compositional differences during the last glaciation, the authors calculated a fractionation factor of 1.0349. The calculated fractionation factors are close, and Friedman et al. (1993a) used an average of the two in the determination of calculated parent-water (cpw) compositions (δD_{cpw}) from δD_{vg} values:

$$\delta D_{cpw} = 1.0343 \times (\delta D_{vg} + 1000) - 1000. \quad (1)$$

Friedman et al. (1993a) went on to suggest that after initial hydration, glass does not experience further isotopic exchange with external water. There is no thermodynamic impetus for diffusion; not only does densification of the hydrated portion of the glass limit diffusion of water into the grain, but increasing hydration also reduces the diffusion gradient of water (Friedman et al., 1966; Cailleteau et al., 2008). In support of this, Friedman et al. (1993a) described two data sets, one of which varies in time and one of which varies in space. The first suite of data focused on ashes of different ages (6.8 ka to 2.1 Ma) collected at the same localities. These tuffs of different ages have different δD_{vg} values (differences range from −20‰ to +22‰) and are interpreted to show retention of primary δD values; diagenesis is presumed to reset values to some homogeneous value. Secondly, the 0.61 Ma Lava Creek ash, sampled across 13 western and central U.S. states and Saskatchewan, Canada, preserves δD_{vg} compositions both in and out of equilibrium with modern surface waters. Of the 25 samples, 19 display a >10‰ difference, and some samples differ by as much as 56‰. If exchange with surrounding water occurred continuously, such departures from equilibrium would not be expected. It is therefore unlikely that these glasses continued to exchange after initial hydration.

Using the fractionation factor developed by Friedman et al. (1993a, 1993b), several papers have employed the δD values of hydrated volcanic glass as a paleoaltimeter. For example, Mulch et al. (2008) and Cassel et al. (2009), later expanded in Cassel et al. (2012), used volcanic glass to estimate paleoelevation of the Sierra Nevada in the western United States. Both studies rely on an ~40‰ decrease in δD_{vg} values in ancient glasses from west to east, which strongly resembles the modern isotopic pattern, to infer that the Sierra Nevada has been a significant topographic barrier since Oligocene–Miocene time.

The results of Mulch et al. (2008) support the hypothesis of Friedman et al. (1993b), i.e., that δD_{vg} values experience little exchange following initial hydration. In the Great Basin, western United States, calculated Pleistocene parent-water values match modern water values, while calculated Miocene waters feature large departures from local modern waters. Ten of their 12 samples older than 6.5 Ma differed by ≥9‰ from modern local water compositions. If continuous isotopic exchange had occurred, the old glass samples should be in equilibrium with

modern water compositions. Because they are not, the authors suggested that the waters of hydration are resistant to diagenetic alteration. However, it should be noted that this probably applies only to relatively shallowly buried samples, as even the isotopic compositions of clay will begin to alter with sufficient burial temperatures, >80–100 °C (Yeh and Savin, 1977).

Central Andes

Our work focuses on a latitudinal band between ~23°S and 27°S in the southern central Andes (Fig. 1), which includes three generalized physiomorphic provinces: the Eastern Cordillera (including the Santa Barbara system and northernmost portion of the Sierras Pampeanas), Puna Plateau, and Western Cordillera (including the Precordillera). The study area is 500–600 km wide in the east-west direction. Precipitation across the range is Atlantic-sourced and falls primarily during the austral summer as a part of convective storms (Garreaud, 1999). Progressive rainout occurs when the moist westward-flowing winds meet the orographic barrier of the Eastern Cordillera, creating both a rain shadow and highly fractionated moisture over the Puna and Western Cordillera, which receive little input from the Pacific (Zhou and Lau, 1998). The aridity of the southern central Andes (Alpers and Brimhall, 1988) has ensured excellent preservation of glassy tuffs as old as 30–40 Ma. For this study, we used late Oligocene– through Holocene-age tuffs and modern natural surface waters to answer three primary questions: What is the effect of pretreatments on the measured δD_{vg} value of glass? What is the correlation of meteoric water δD values with elevation, and how well do young δD_{vg} and δD_{cpw} values mirror this correlation? Finally, what is the preservation potential of δD_{vg} values over geologic time?

Pretreatments

The motivation for the laboratory component of this study was not to test whether it is physically possible to change the δD_{vg} values in the laboratory. Anovitz et al. (2009) showed that water-obsidian exchange is indeed possible in sealed reaction vessels at 150 °C. Rather, these tests were intended to determine whether variations in sample preparation within the scope of normal laboratory work have an impact on the measured δD_{vg} value. If so, the preparation steps must rigorously follow a set protocol; if not, then there is more flexibility in the preparation of samples and ease of communication of results between laboratories. Potential confounding variables include grain size analyzed, temperature and/or duration of drying, presence of clay adhering to glass grains, or whether an HF treatment step is included. For example, at least one group, Cassel et al. (2012, p. 239), uses HF to remove "the exposed surface and near-surface rinds (outside of and the outermost gel layer)," while other groups, such as Friedman et al. (1993a), do not. With these tests, this study highlights those pretreatment steps that do and do not affect measured δD_{vg} values.

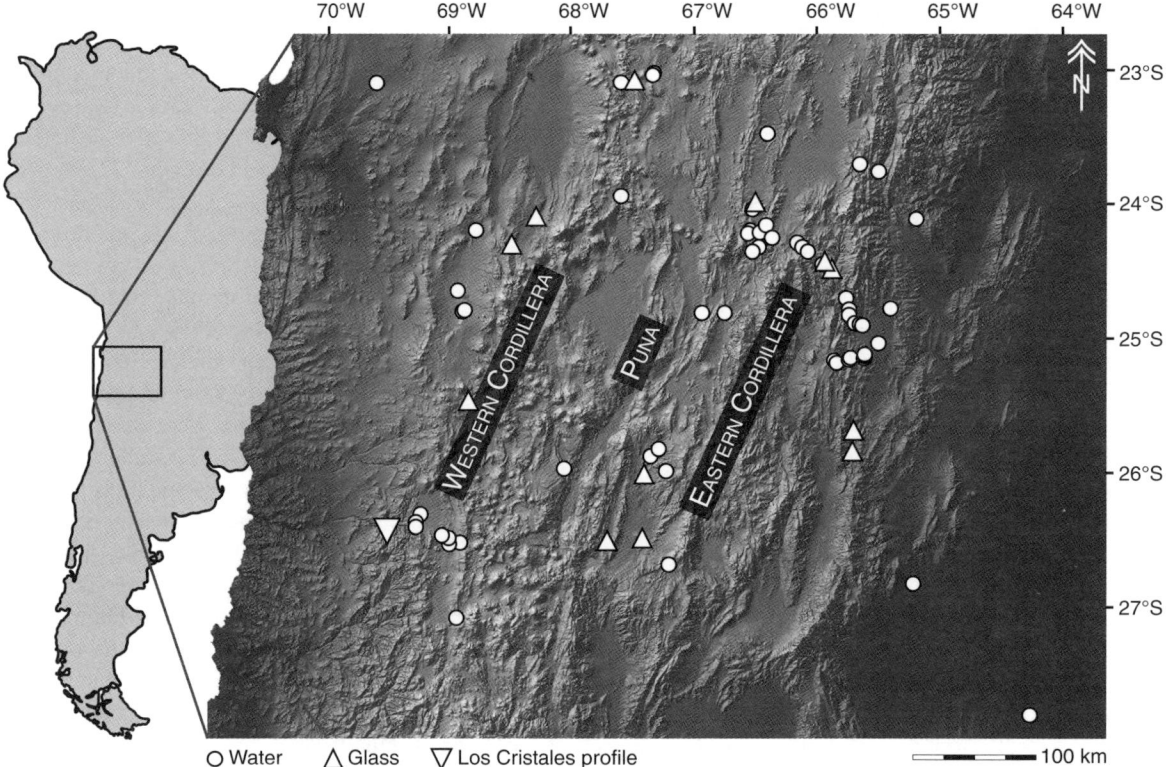

Figure 1. Southern central Andes with sampling sites of water (circles), glass (triangles), and the Los Cristales soil profile (upside-down triangle).

Modern Calibration

Extensive water and tuff sampling provides a calibration of the water-glass relationship. For volcanic glass to be useful as a paleoaltimetric-paleoenvironmental proxy, it has to archive isotopic compositions that faithfully reflect the composition of the parent water. Though initial hydration is rapid, a glass shard probably continues to hydrate for thousands of years, over which time the climate may vary. In the central Andes, we assume that little large-scale elevation change has occurred in the last 300,000 yr and that tuffs deposited since 0.3 Ma are at the same elevation as when they were originally deposited. These young δD_{vg} values should produce δD_{cpw} values that are similar to modern surface meteoric water (δD_{mw}) values, if glass is a faithful archive and if the fractionation factor in Equation 1 is accurate.

Test for Diagenesis

Preservation of original isotopic information is crucial to paleoaltimetry. Little is known about the diagenetic kinetics of volcanic glass. For example, at what depths and/or temperatures can one expect δD_{vg} values to no longer reflect original values? Are δD_{vg} values stable over geologic time? The Los Cristales soil profile from the Atacama Desert in northern Chile was selected to test the resistance to diagenesis of hydrated glass by comparing

glass-derived reconstructions to those of other paleoenvironmental proxies. The Los Cristales exposure preserves a buried calcareous paleosol developed on the glass-bearing 16.3 Ma Los Cristales ignimbrite at an elevation of 2450 m. The tuff is 4.5–5 m thick, and the paleosol developed in the uppermost meter. The carbonate is stage II–III (Gile et al., 1966) and features continuous clast coatings and some plugging in the upper 100 cm, with carbonate in fractures to a depth of ~140 cm below the top of the paleosol profile. The soil profile is capped with an undistorbed reddened clay zone. The profile is only shallowly buried by an unnamed and undated conglomerate that is at present only 50 m thick and is unlikely to have been much thicker (Nalpas et al., 2008).

The Los Cristales section in the Western Cordillera provides a key multiple proxy test of volcanic glass as a paleoenvironmental proxy. The measured $\delta^{18}O_{carb}$ value of a carbonate depends on the isotopic composition of its parent water, the temperature at which it precipitated, and its later diagenetic history. Clumped isotope paleothermometry, in turn, provides the paleotemperature of formation or, if in excess of Earth surface temperatures, a signal that the sample has experienced substantial burial. Because Δ_{47} values alter more readily than $\delta^{18}O_{carb}$ values (Huntington et al., 2010), calculated paleotemperatures, $T(\Delta_{47})$, within the expected range of soils at Earth's surface are indicative of little or no diagenesis. Therefore, in the carbonates of the Los Cristales paleosol, there are two independent systems that are sensitive to different degrees of diagenesis; $T(\Delta_{47})$ values are relatively sensitive, and $\delta^{18}O_{carb}$ values are relatively insensitive. These provide a reference frame for the diagenetic history of the Los Cristales paleosol and ignimbrite.

METHODS

The following methods were used in the analysis of the isotopic compositions of the water, volcanic glass, and carbonate samples and in the analysis of clumped isotope paleotemperatures. The techniques for the water, carbonate, and clumped isotope analyses followed standard procedures, whereas volcanic glass received experimental variations on the standard procedure, the so-called pretreatments. These methods are detailed below.

Water

Collected Waters

A range of natural waters (streams, rivers, springs) was collected, although the focus was on springs and small streams draining limited catchments. This was to ensure that water sampling was as elevation-specific as possible. At each sample site, 15 mL samples of unfiltered water or snow were sealed with Teflon and electrician's tape into a centrifuge tube and refrigerated in the laboratory. The $\delta^{18}O$ (SMOW) of water samples was measured using the CO_2 equilibration method on an automated sample preparation device attached directly to a Finnigan Delta S mass spectrometer at the University of Arizona. The δD values of water were measured using an automated chromium reduction device (H-Device) attached to the same mass spectrometer. The values were corrected based on internal laboratory standards, which were calibrated to SMOW and Standard Light Antarctic Precipitation (SLAP). The analytical precision for $\delta^{18}O$ and δD measurements was 0.08‰ and 0.6‰, respectively (1σ). Water isotopic results are reported using standard delta per mil notation relative to SMOW.

Volcanic Glass

Standard Procedure

A 100–200 g sample of tuff (air-fall or ash-flow) was crushed manually with mortar and pestle using a rocking motion to minimize reduction in mineral grain size. Samples were treated in a 6 N HCl acid solution overnight to dissolve any carbonate. After rinsing, the samples were wet-sieved using 850 mm, 250 mm, 125 mm, and 53 mm screens, as well as a pan to catch the <53 mm fraction. The separates were dried overnight at ~60 °C. The glass was separated from the desired size fraction (usually 250–125 mm) using a water-soluble solution of lithium heteropolytungstate (LST) with a specific gravity of 2.25. At this density, most glass floats, while most denser minerals sink. The resulting separates were rinsed and dried overnight. The glass separate was then examined for clay adhering to the outside of the grains and for mica flakes, which may also float in the LST. If clay was present, the sample was put in an ultrasound bath for ≥60 min, and again dried. Exploiting

mica's proclivity to adhere to a flat surface, a mica-bearing sample was poured back and forth between two sheets of weighing paper, shaking the sheets between pours to dislodge the mica, until no mica remained with the glass.

For analysis, 2–3 mg samples of glass were packed in silver foil. The δD_{vg} value was determined using an automated thermal conversion/element analyzer (TCEA) coupled to a Delta V Plus isotope ratio mass spectrometer with an analytical precision of ±2.5‰ (2σ). The isotope ratio was calibrated using an internal standard and the international standards NBS-30 biotite and IAEA-CH-7 polyethylene foil. All δD values are reported here in standard delta notation and referenced to SMOW.

Experimental Variations on the Standard Procedure

Subsamples of AD10-165, an air-fall tuff composed almost entirely of glass, were tested for effects of different preparation procedures on δD_{vg} values. Processed in four groups, group 1 minimized heat and tap water exposure, group 2 was prepared following standard procedure, and group 3 maximized heat exposure. The group number of a given sample is denoted by the number following the sample name but before other information such as grain size, e.g., AD10-165-2-125-250 denotes that it is sample AD10-165, group number 2, and contains grain sizes between 125 and 250 mm.

Group 1 samples minimized water exposure in the laboratory (15.5 h) while varying drying temperature and duration. Sample SLOW was dried for 1 wk at room temperature, while FAST was dried at 100 °C for 30 min. Sample NOUS received no ultrasound (clay is intact), but was otherwise identical to SLOW.

Group 2 samples varied in grain size and were treated to the standard procedure described earlier herein. Four grain-size fractions were analyzed: >250 mm, 125–250 mm, 53–125 mm, and <53 mm. An additional sample, AD10-165-2-125-250H, was composed of glass that sank during heavy liquid separation.

The group 3 sample was dried for 1 wk durations at ~80 °C to test effect of prolonged exposure to higher-than-average preparation temperatures.

Group 4 samples included additional clay-bearing samples prepared according to the standard procedure, except for the lack of ultrasound to remove the clay.

In addition to group 2, two other tuffs were tested for the effects of grain-size variation. AD10-168, a loose air-fall tuff, tested the size fractions >250 mm, 125–250 mm, 53–125 mm, and <53 mm. AD10-128, an indurate ignimbrite, featured fractions >850 mm, 250–850 mm, 125–250 mm, 53–125 mm, and <53 mm. An additional AD10-128 sample (AD10-128HOT) was dried in the oven for 2 wk.

Finally, aliquots from five samples were washed in 5% HF for 10 s and repeatedly rinsed. Each sample, prepared following the standard procedure as described already, had been analyzed at least two previous times. Sample ages ranged from 0.01 Ma to 24 Ma, and previously measured δD values varied between −62‰ and −142‰. Each treated subsample was analyzed alongside a nontreated sample for comparison.

Calcium Carbonate δ¹⁸O Analysis

Pedogenic carbonates were scraped from alluvial or bedrock clasts or sampled from nodules. Carbonate analyzed for δ¹⁸O (δ¹⁸O$_{carb}$) and δ¹³C values was heated at 250 °C for 3 h in vacuo before stable isotopic analysis using an automated sample preparation device (Kiel III) attached directly to a Finnigan MAT 252 mass spectrometer at the University of Arizona. Measured δ¹⁸O and δ¹³C values were corrected using internal laboratory standards calibrated to NBS-19 based on internal laboratory standards. Precision of repeated standards was ±0.11‰ for δ¹⁸O (1σ). Carbonate isotopic results are reported using standard delta per mil notation relative to Vienna Pee Dee Belemnite (VPDB).

Calcium Carbonate "Clumped Isotope" Analysis

Two carbonate powders produced for δ¹⁸O analysis were used for clumped isotope analysis, corresponding to the top and bottom of the Los Cristales soil profile. This carbonate was not heated prior to analysis, unlike carbonate used for δ¹⁸O and δ¹³C determinations. Clumped-isotope compositions of these carbonate powders were analyzed at the California Institute of Technology using protocols described by Ghosh et al. (2006a), Affek et al. (2008), and Huntington et al. (2009). Here, 7–10 mg aliquots were digested overnight at 25 °C in anhydrous orthophosphoric acid. After cryogenic trapping, the resulting CO_2 was purified of potential isobaric contaminants by passage through a dry ice–ethanol slush and a 30-m-long, 530 mm (interior diameter) Supelco GC column held at –10 °C. CO_2 was then analyzed using a Finnigan MAT 253 dual-inlet mass spectrometer configured for isotope ratio measurements of masses 44–49. Bulk composition (δ¹³C and δ¹⁸O) was computed from these measurements using a reference CO_2 tank of known isotopic composition, and Δ_{47} was derived from comparison with stochastic gases, i.e., CO_2 in a thermodynamic state of randomly distributed isotopes, with $\Delta_{47} = 0$, obtained by heating CO_2 aliquots at 1000 °C. Results are reported versus the CalTech laboratory standards.

PRETREATMENTS

The pretreatment tests were designed to investigate the effects of different preparation techniques during the analysis process of volcanic glass. The results of these tests are presented and discussed relative to the analytical precision of the analyses and the average results of samples that were prepared following the standard procedure.

Pretreatment Results

Analytical Precision

The analytical precision for δD$_{vg}$ values of the mass spectrometer is reported as 2.5‰ (2σ). By comparing results from multiple aliquots of 69 samples, we have found a slightly greater 2σ variability of ~2.9‰ on analysis of the same glass samples,

possibly due to slight compositional differences in the aliquots. Therefore, according to our data, 95% of aliquots of a given samples should fall within an ~5.7‰ range, and any results falling outside this range are likely to be significant. Within this framework, we evaluate the results of the following sensitivity tests of volcanic glass to different pretreatments (Table 2).

Heat and Water Exposure Sensitivity

Our pretreatment experiments show that within analytical precision, samples dried for 2 wk are indistinguishable from samples dried for shorter periods of time at lower temperatures. AD10-165-3, twice dried for 1 wk at 80 °C, has a δD$_{vg}$ value of –76‰ (Table 2). This is very similar to the average of group 2 values (–76‰, standard treatment: overnight at 60 °C) and AD10-165-1-SLOW (–77‰, room temperature for 1 wk).

Likewise, the δD$_{vg}$ value of AD10-165-1-FAST (–74‰), which experienced the shortest time in tap water (~15 h), is indistinguishable from group 2 values (~–76‰), which were in contact with tap water for ~30 h (Fig. 2).

Grain-Size Sensitivity

In all three of the grain-size variation tests, there is negligible systematic variation across grain sizes (Table 2). The differences between the averaged δD$_{vg}$ values of the coarsest (>250 mm) and finest (<53 mm) grain sizes for AD10-165 and AD10-168 are 3.6‰ and 2.6‰, respectively, i.e., well within the expected range of variability (Fig. 2). Although the values of AD10-128 appear more bimodal (~–110‰ or ~–120‰), this does not vary systematically with grain size.

Clay Content Sensitivity

Three subsamples that did not receive ultrasound treatment had "clay" adhering to the surface of the glass shards, often in folds or depressions. These clay-bearing samples, AD10-165-1-NOUS, AD10-165 "aliquot1," and AD10-165 "aliquot 2," returned similar values of –84‰, –85‰, and –83‰, respectively. However, subsamples in group 1, which were prepared like NOUS, except clay was removed via ultrasound, produced an average value of –75‰. The average of group 2, all of which also underwent ultrasound clay removal but were otherwise treated similarly to "aliquot 1" and "aliquot 2," is –73‰. A one-way analysis of variance (ANOVA) allows us to reject the "null" hypothesis that these two groups were sampled from the same population. At the 95% confidence level, the p-value is 0.00011, indicating that the clay-bearing subsamples clearly lie well outside of the expected range of variability (Fig. 2).

Effects of Hydrofluoric Acid Treatment

HF-treated subsamples produced measured values that varied considerably from nontreated subsamples. The largest differences between treated samples and nontreated sample averages ($\Delta_{HF-noHF}$) were +12‰ and –14‰ (Fig. 2). The smallest $\Delta_{HF-noHF}$ was –3‰ (compared to an eight nontreated sample average). Between treated and nontreated samples analyzed during the

TABLE 2. SUMMARY OF PRETREATMENT RESULTS

Sample	δD_{vg} (‰)	Grain size (µm)	H$_2$O duration (h)	Drying temp. (°C)	Drying duration*	Clay present?
Group 1: short water exposure, variations of drying duration and temperature						
AD10-165-1-NOUS	−84.4	125–250	15.5	20	1 wk	Yes
AD10-165-1-SLOW	−76.5	125–250	15.5	20	1 wk	No
AD10-165-1-FAST	−74.3	125–250	15.5	100	30 min	No
Group 2: standard treatment, multiple grain sizes						
AD10-165-2->250	−71.7	>250	31	60	15–20 h	No
AD10-165-2-125-250	−71.5	125–250	31	60	15–20 h	No
AD10-165-2-125-250H	−69.3	125–250	31	60	15–20 h	No
AD10-165-2-53-125	−75.2	53–125	31	60	15–20 h	No
AD10-165-2-<53	−75.3	<53	31	60	15–20 h	No
Group 3: long drying duration at relatively high temperatures						
AD10-165-3	−76.3	125–250	31	80	1 wk	No
Group 4: additional clay-bearing aliquots						
AD10-165 "aliquot 1"	−84.5	125–250	31	60	15–20 h	Yes
AD10-165 "aliquot 2"	−83.2	125–250	31	60	15–20 h	Yes
Additional grain-size tests on different tuffs						
AD10-168-1	−78.5	>250	31	60	15–20 h	No
AD10-168-1	−80.1	>250	31	60	15–20 h	No
AD10-168-2	−78.3	125–250	31	60	15–20 h	No
AD10-168-2	−81.4	125–250	31	60	15–20 h	No
AD10-168-3	−81.7	53–125	31	60	15–20 h	No
AD10-168-3	−80.8	53–125	31	60	15–20 h	No
AD10-168-4	−81.0	<53	31	60	15–20 h	No
AD10-168-4	−82.8	<53	31	60	15–20 h	No
AD10-128->850	−117.5	>850	31	60	15–20 h	No
AD10-128-250-850	−108.1	250–850	31	60	15–20 h	No
AD10-128-125-250	−121.4	125–250	31	60	15–20 h	No
AD10-128HOT	−118.0	125–250	31	60	2 wk (total)	No
AD10-128-53-125	−122.3	53–125	31	60	15–20 h	No
AD10-128-<53	−109.4	<53	31	60	15–20 h	No

HF treatment Sample	Age (Ma)	Prev. δD measurements (‰)		October 2012 (‰)	
		1	2	3	HF
AD10-165 125-250	0.01	Average of 8 aliquots: −74		−73	−77
AD-50	0.01	−62	−64	−70	−53
AD10-186	0.01	−134	−140	−136	−151
AD10-236	4	−141	−142	−150	−155
AD10-133	24	−68	−65	−60	−55

*Drying duration for each of the two drying steps in sample preparation.

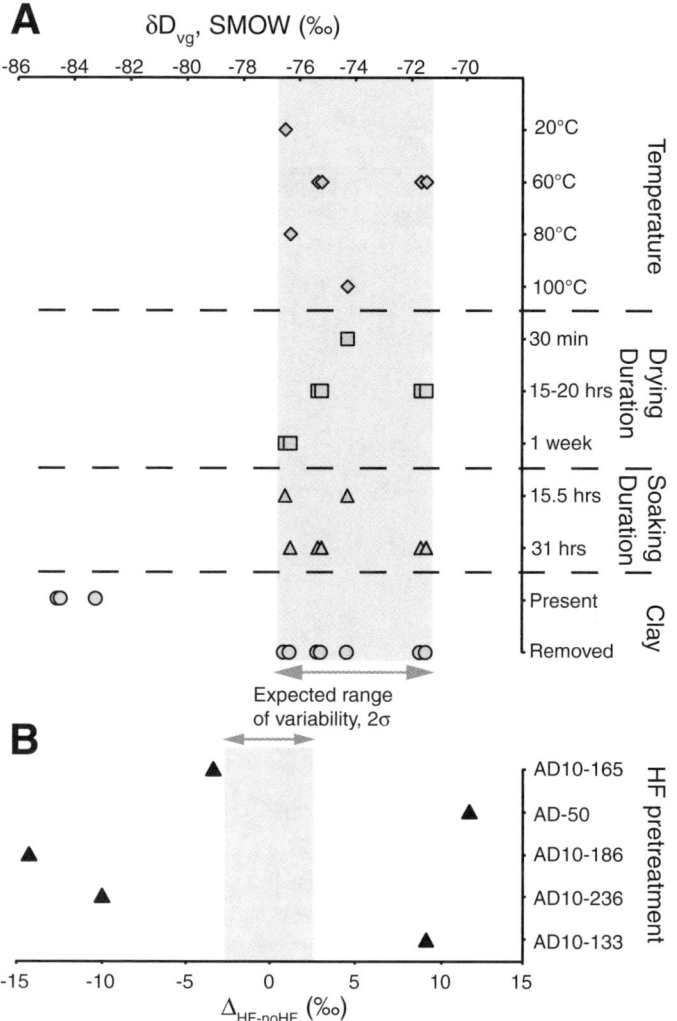

Figure 2. (A) δD_{vg} results of pretreatments involving different drying temperatures, duration of drying, duration of soaking, and presence/absence of clay on the same glass sample, AD10-165. SMOW—standard mean ocean water; vg—volcanic glass. (B) Effect of HF treatment versus no HF treatment ($\Delta_{HF\text{-}noHF}$) on five different glass samples (AD…). Only the presence of clay and the HF pretreatment significantly affected results, resulting in measured δD_{vg} values well outside the expected range of variation (2σ, gray bands), where $2\sigma = \pm2.9‰$ in both graphs and incorporates analytical and reproducibility error.

same run, i.e., not averaged with all previous analyses, the $\Delta_{HF\text{-}noHF}$ values were even more extreme: Two treated samples produced values that were +17‰ and −14‰ different from the non-treated samples, while the other three produced values only +5‰ and −4‰ different.

Two patterns, perhaps coincidental, are visible in our results. Two of the three 0.01 Ma tuffs possess the largest $\Delta_{HF\text{-}noHF}$ values, +17‰ and −14‰. The older 4 Ma and 24 Ma tuffs feature markedly less variation, only +5‰ and −4‰, respectively. Secondly, the lowest δD_{vg} values of the non-HF treated glasses decreased still further with HF treatment. Oppositely,

the δD_{vg} values of glasses with the highest original δD_{vg} values increased with HF treatment. It should be noted that there were even differences between the nontreated samples of this run and previous analyses of non-HF-treated aliquots. This inter-run variability ranged from +6‰ to −9‰, including two samples with 0‰ or 1‰ difference.

Pretreatment Discussion

Our data suggest that variations in drying temperature, drying duration, exposure to water during processing, and variations in grain size produce no systematic effects in the resultant δD_{vg} values. All variation observed in these tests falls within the 2σ range of expected values (Fig. 2). However, the presence of clay, adhering to the surface of grains and filling small depressions and crevices, does produce systematic and significant differences in δD_{vg} values. Despite representing a small portion of the sample mass, the presence of clay decreased the measured δD_{vg} value by ~10‰. The shift in δD_{vg} values produced by adhering clay is easily eliminated by simple ultrasound for ≥60 min.

The results from the HF experiments are less straightforward. HF treatment results in shifted δD values, but the cause of this change is unclear. HF acid may strip clay from the outside of glass shards more efficiently than ultrasound treatment. If so, δD_{vg} values for HF-treated AD10-165 should be higher than those that were ultrasonicated because removal of clays from sample AD10-165 increases the δD value of bulk glass. Instead, the HF-treated aliquot values were lower than the AD10-165 average. Another possibility is HF treatment results in greater influence of magmatic water preserved in the interior, unhydrated portion of the shard by removing some of the hydrated rind. Arguably, this would be most likely to affect younger grains that have had less time to fully hydrate, potentially causing the larger $\Delta_{HF\text{-}noHF}$ values seen in the 0.01 Ma samples. However, this explanation seems unlikely based on the direction of the isotopic shifts resulting from HF treatment. Estimates of the δD value of magmatic water are generally between −50‰ and −85‰ (Taylor, 1974; Kyser and O'Neil, 1984), including from a central Andean breccia pipe (Skewes et al., 2003). The lowest δD_{vg} values (~−140‰) have negative $\Delta_{HF\text{-}noHF}$ values, but the highest δD_{vg} values (~−60‰) have positive $\Delta_{HF\text{-}noHF}$ values. Both of these trends are opposite what would be expected if magmatic water exerted greater influence in HF-treated glasses.

Researchers are divided over the use of HF pretreatment. Friedman et al. (1993a, 1993b) do not mention the use of HF; Onken (1991) did not use it; but Mulch et al. (2008), Cassel et al. (2009), and Cassel et al. (2012) used a 5%–8% HF solution. Given the variability introduced by the HF treatment presented here and lack of a clear cause of this variability, this step in the preparation of samples requires further investigation.

These tests demonstrate that hydrated volcanic glass is mostly insensitive to basic laboratory preparation steps. Data can be compared across multiple laboratories and with published work without having to know exactly how the samples were prepared, with the exception of two factors: (1) The removal of clay adhering to the surface of the glass is paramount, and if not removed, clay could produce less accurate data. (2) When comparing samples that have had an HF treatment and those that have not, greater variation must be expected.

MODERN CALIBRATION

Following the establishment of basic laboratory procedures, the calibration of volcanic glass is the next step in assessing the reliability of glass as a paleoenvironmental archive. Here, we identify the isotopic compositions and trends of modern waters from the Central Andes and compare them to those of young volcanic glasses.

Modern Water Values

Eastern Cordillera

The natural surface waters collected for this study come from three distinct topographic regions of the central Andes: the Eastern Cordillera, the Puna, and the Western Cordillera (Table 3). Eastern Cordilleran water samples span elevations from 1190 m to 4548 m (Fig. 3A). The $\delta^{18}O$ values of these meteoric waters ($\delta^{18}O_{mw}$) range from −3.2‰ to −11.5‰, and δD values (δD_{mw}) are between −26.0‰ and −76.6‰ (Fig. 3B).

Additionally, isotopic compositions of precipitation are included from five Global Network of Isotopes in Precipitation (GNIP) monitoring sites in Argentina: Santiago del Estero, Tucuman, Salta, Los Molinos, and Purmamarca (IAEA/WMO, 2006). Weighted long-term averages provide an approximation of near-surface groundwater, which integrates yearly precipitation (Ingraham and Taylor, 1991). Shorter time periods (a combined 155 monthly observations) provide data with which to construct a regional local meteoric water line (Fig. 4). At altitudes of 187 m to 2400 m, these sites provide lower-elevation context to our suite of surface-water data. The $\delta^{18}O_{mw}$ values range from −4.1‰ to −7.9‰, and δD_{mw} values are between −9.8‰ and −55.3‰ (Table 3; IAEA/WMO, 2006).

The local meteoric water line (LMWL) of the Eastern Cordillera data is best described by the equation:

$$\delta D_{mw} = 8.0(\delta^{18}O_{mw}) + 12.6‰ \ (R^2 = 0.95). \qquad (2)$$

Equation 2 is close to that of the global meteoric water line (GWML; Fig. 4). The Eastern Cordillera LMWL uses 149 monthly GNIP observations, 28 sampled streams, and excludes 20 precipitation and stream samples that plot well to the right of the rest of the data, suggesting they have experienced evaporation.

Isotope-elevation relationships were calculated using a 30 sample subset of stream samples ($n = 26$) and weighted long-term GNIP averages ($n = 4$). Station and sampling elevations range from 187 m to 4548 m, and isotopic compositions plot close to the meteoric water line, minimizing the effects of evaporation. We also attempted to minimize the effects of confounding topog-

raphy by removing certain samples based on geography. For example, a sample taken at relatively low elevation in a deep valley downwind of a high-elevation mountain range would likely display anomalously low δD values relative to what its elevation would predict. Using this subset of Eastern Cordillera surface water, the isotope-elevation relationships are the following:

$$\text{H: } \delta D_{mw} = -13.7‰ \text{ km}^{-1} \times \text{elevation (in km)} - 9.0 \ (R^2 = 0.90); \quad (3)$$

$$\text{O: } \delta^{18}O_{mw} = -1.3‰ \text{ km}^{-1} \times \text{elevation (in km)} - 3.8 \ (R^2 = 0.83). \ (4)$$

A polynomial fit to sample points yields similar correlation coefficients.

Puna/Western Cordillera

The δD_{mw} values from the Puna and Western Cordillera are more variable for a given elevation than water collected from the Eastern Cordillera. Puna samples were collected from elevations between 2964 m and 4588 m (Fig. 3A). Due to the paucity of surface waters in the Atacama Desert, samples were not obtained lower than 2964 m or, for the most part, farther west than the western margin of the high-elevation Puna Plateau (Fig. 1). The $\delta^{18}O_{mw}$ values of Puna water range from +2.0‰ to −9.2‰, and δD_{mw} values range from −28.1‰ to −83.0‰ (Fig. 3). The LMWL described by these samples has the following equation:

$$\delta D_{mw} = 4.7(\delta^{18}O_{mw}) - 30.8‰ \ (R^2 = 0.75). \qquad (5)$$

Characteristics of Central Andean Surface Waters

Modern surface waters are important for calibrating isotope-elevation relationships and paleoreconstructions. The primary control on δD_{mw} values in the central Andes is elevation. Increasing elevation in the Eastern Cordillera is matched by decreasing δD_{mw} values. At higher (and drier) elevations, evaporation becomes a significant secondary control on observed δD_{mw} values. The shallow slope of the Puna LMWL (Fig. 4) appears similar to the characteristic effects of evaporation, but we cannot rule out other modifying factors, such as mixing of air masses, seasonality of precipitation, etc. Meteoric water is first highly fractionated through progressive rainout as it moves westward and upward over the orographic barrier, and surface waters reach δD_{mw} values of ~−80‰. At elevations higher than ~3500 m and/or farther west than ~66.1°W, water compositions are shifted (likely by evaporation) toward high δD_{mw} values (Fig. 3). The combined effects of elevation depletion in D and evaporative enrichment in D over the Puna give rise to observed values that differ by as much as 50‰–60‰ over less than 400 m of absolute elevation, although the samples are not necessarily close to one another. Few samples are available from the Western Cordillera due to the lack of small streams in the Atacama Desert, but the projected trend would be

<table>
<tr><th colspan="7">TABLE 3. WATER DATA</th></tr>
<tr><th>Sample</th><th>No.</th><th>$\delta^{18}O_{mw}$ (‰)</th><th>δD_{mw} (‰)</th><th>Elevation (m)</th><th>Longitude (°W)</th><th>Latitude (°S)</th></tr>
<tr><td colspan="7"><u>Eastern Cordillera</u></td></tr>
<tr><td>SL051508-</td><td>1</td><td>−6.5</td><td>−35.0</td><td>1190</td><td>65.496</td><td>25.041</td></tr>
<tr><td>SL051508-</td><td>2</td><td>−5.4</td><td>−26.0</td><td>1406</td><td>65.602</td><td>25.116</td></tr>
<tr><td>SL051508-</td><td>3</td><td>−5.3</td><td>−27.0</td><td>1409</td><td>65.603</td><td>25.117</td></tr>
<tr><td>SL051508-</td><td>4</td><td>−5.3</td><td>−26.0</td><td>1472</td><td>65.606</td><td>25.144</td></tr>
<tr><td>SL051508-</td><td>5</td><td>−5.7</td><td>−28.0</td><td>1488</td><td>65.609</td><td>25.151</td></tr>
<tr><td>SL051808-</td><td>6</td><td>−6.9</td><td>−38.3</td><td>1811</td><td>65.727</td><td>25.158</td></tr>
<tr><td>SL051808-</td><td>7</td><td>−5.5</td><td>−30.0</td><td>1843</td><td>65.735</td><td>25.165</td></tr>
<tr><td>SL051808-</td><td>8</td><td>−7.1</td><td>−37.4</td><td>2622</td><td>65.816</td><td>25.178</td></tr>
<tr><td>SL051508-</td><td>9</td><td>−5.6</td><td>−36.1</td><td>3102</td><td>65.847</td><td>25.184</td></tr>
<tr><td>SL051508-</td><td>10</td><td>−7.2</td><td>−46.8</td><td>3090</td><td>65.844</td><td>25.183</td></tr>
<tr><td>SL051508-</td><td>11</td><td>−7.3</td><td>−44.6</td><td>2952</td><td>65.850</td><td>25.167</td></tr>
<tr><td>SL051608-</td><td>13</td><td>−6.1</td><td>−32.1</td><td>1581</td><td>65.665</td><td>24.896</td></tr>
<tr><td>SL051608-</td><td>14</td><td>−6.0</td><td>−32.0</td><td>1607</td><td>65.674</td><td>24.892</td></tr>
<tr><td>SL051608-</td><td>15</td><td>−6.0</td><td>−31.8</td><td>1633</td><td>65.679</td><td>24.890</td></tr>
<tr><td>SL051608-</td><td>16</td><td>−6.3</td><td>−32.0</td><td>1631</td><td>65.683</td><td>24.889</td></tr>
<tr><td>SL051608-</td><td>17</td><td>−6.5</td><td>−37.5</td><td>1864</td><td>65.730</td><td>24.827</td></tr>
<tr><td>SL051608-</td><td>18</td><td>−6.6</td><td>−37.3</td><td>2010</td><td>65.731</td><td>24.791</td></tr>
<tr><td>SL051608-</td><td>20</td><td>−3.2</td><td>−30.5</td><td>2296</td><td>65.758</td><td>24.701</td></tr>
<tr><td>SL051608-</td><td>23</td><td>−7.6</td><td>−48.0</td><td>3096</td><td>65.950</td><td>24.451</td></tr>
<tr><td>SL051708-</td><td>25</td><td>−8.1</td><td>−51.0</td><td>3592</td><td>66.076</td><td>24.358</td></tr>
<tr><td>SL05??08-</td><td>26</td><td>−10.4</td><td>−74.5</td><td>3961</td><td>66.117</td><td>24.325</td></tr>
<tr><td>SL051708-</td><td>28</td><td>−6.3</td><td>−55.2</td><td>3940</td><td>66.156</td><td>24.290</td></tr>
<tr><td>SL051708-</td><td>32</td><td>−10.6</td><td>−76.6</td><td>4470</td><td>66.474</td><td>24.212</td></tr>
<tr><td>SL051708-</td><td>34</td><td>−8.5</td><td>−73.8</td><td>4367</td><td>66.548</td><td>24.205</td></tr>
<tr><td>SL051708-</td><td>35</td><td>−7.9</td><td>−67.8</td><td>4335</td><td>66.546</td><td>24.196</td></tr>
<tr><td>SL051908-</td><td>37</td><td>−11.5</td><td>−87.3</td><td>3670</td><td>66.407</td><td>23.468</td></tr>
<tr><td>SL051908-</td><td>39</td><td>−10.0</td><td>−73.1</td><td>4228</td><td>66.524</td><td>24.039</td></tr>
<tr><td>SL051908-</td><td>40</td><td>−5.2</td><td>−53.6</td><td>4349</td><td>66.415</td><td>24.151</td></tr>
<tr><td>SL060108-</td><td>76</td><td>−9.5</td><td>−68.0</td><td>4500</td><td>66.478</td><td>24.326</td></tr>
<tr><td>SL060108-</td><td>77</td><td>−7.7</td><td>−60.6</td><td>4321</td><td>66.518</td><td>24.354</td></tr>
<tr><td>SL060308-</td><td>99</td><td>−7.2</td><td>−72.1</td><td>3879</td><td>66.755</td><td>24.812</td></tr>
<tr><td>SL060408-</td><td>101</td><td>−9.3</td><td>−76.3</td><td>3836</td><td>66.363</td><td>24.248</td></tr>
<tr><td>SL060608-</td><td>108</td><td>−9.8</td><td>−69.8</td><td>4125</td><td>65.649</td><td>23.694</td></tr>
<tr><td>SL</td><td>127</td><td>−9.1</td><td>−72.9</td><td>3782</td><td>66.938</td><td>24.811</td></tr>
<tr><td>SL</td><td>128</td><td>−9.7</td><td>−72.3</td><td>4548</td><td>66.471</td><td>24.212</td></tr>
<tr><td colspan="7"><u>GNIP* weighted long-term precipitation averages</u></td></tr>
<tr><td>Santiago del Estero</td><td></td><td>−4.1</td><td>−9.8</td><td>187</td><td>64.270</td><td>27.780</td></tr>
<tr><td>Tucuman</td><td></td><td>−4.3</td><td>−16.7</td><td>430</td><td>65.217</td><td>26.817</td></tr>
<tr><td>Salta</td><td></td><td>−6.2</td><td>−34.8</td><td>1187</td><td>65.400</td><td>24.780</td></tr>
<tr><td>Los Molinos</td><td></td><td>−6.7</td><td>−27.1</td><td>1300</td><td>65.190</td><td>24.110</td></tr>
<tr><td>Purmamarca</td><td></td><td>−7.9</td><td>−55.3</td><td>2400</td><td>65.500</td><td>23.750</td></tr>
<tr><td colspan="7"><u>Puna−Western Cordillera</u></td></tr>
<tr><td>SL</td><td>111</td><td>−4.9</td><td>−40.3</td><td>3941</td><td>67.205</td><td>26.681</td></tr>
<tr><td>SL</td><td>123a</td><td>−5.7</td><td>−49.0</td><td>4321</td><td>67.231</td><td>25.989</td></tr>
<tr><td>SL</td><td>123b</td><td>−5.1</td><td>−51.5</td><td>4321</td><td>67.231</td><td>25.989</td></tr>
<tr><td>SL</td><td>125</td><td>−8.8</td><td>−58.4</td><td>3910</td><td>67.356</td><td>25.886</td></tr>
<tr><td>SL</td><td>126</td><td>−5.0</td><td>−46.8</td><td>4090</td><td>67.285</td><td>25.827</td></tr>
<tr><td>SL</td><td>132</td><td>−7.8</td><td>−75.1</td><td>3519</td><td>−68.058</td><td>25.972</td></tr>
<tr><td>AD09</td><td>88</td><td>2.0</td><td>−28.1</td><td>4048</td><td>67.599</td><td>23.938</td></tr>
<tr><td>AD09</td><td>93</td><td>−4.5</td><td>−54.3</td><td>4588</td><td>67.598</td><td>23.079</td></tr>
<tr><td>SPN</td><td>20B</td><td>−4.6</td><td>−48.0</td><td>3262</td><td>68.884</td><td>24.798</td></tr>
<tr><td>SPN</td><td>28 well</td><td>−3.7</td><td>−41.5</td><td>2924</td><td>68.930</td><td>24.650</td></tr>
<tr><td>SPN</td><td>20A</td><td>−5.9</td><td>−54.4</td><td>3262</td><td>68.884</td><td>24.798</td></tr>
<tr><td>Imilac spring</td><td>n/a†</td><td>−6.2</td><td>−50.8</td><td>2968</td><td>68.775</td><td>24.198</td></tr>
<tr><td>SPN</td><td>26</td><td>−2.3</td><td>−43.9</td><td>3297</td><td>68.890</td><td>24.807</td></tr>
<tr><td>AD</td><td>106</td><td>−2.9</td><td>−47.0</td><td>3278</td><td>68.889</td><td>24.807</td></tr>
<tr><td>AD</td><td>115</td><td>−1.9</td><td>−39.0</td><td>3262</td><td>68.884</td><td>24.798</td></tr>
<tr><td>AD</td><td>123</td><td>−5.3</td><td>−55.0</td><td>3270</td><td>68.887</td><td>24.803</td></tr>
<tr><td>AD10</td><td>138</td><td>−6.9</td><td>−65.0</td><td>3466</td><td>69.278</td><td>26.368</td></tr>
<tr><td>AD10</td><td>141</td><td>−3.4</td><td>−55.0</td><td>3354</td><td>69.244</td><td>26.311</td></tr>
<tr><td>AD10</td><td>145</td><td>−6.5</td><td>−65.0</td><td>3573</td><td>69.061</td><td>26.464</td></tr>
<tr><td>AD10</td><td>148</td><td>−6.5</td><td>−65.0</td><td>3854</td><td>68.913</td><td>26.520</td></tr>
<tr><td>AD10</td><td>151</td><td>−9.0</td><td>−69.0</td><td>3665</td><td>69.008</td><td>26.491</td></tr>
<tr><td>AD10</td><td>152</td><td>−7.9</td><td>−70.0</td><td>3640</td><td>69.004</td><td>26.530</td></tr>
<tr><td>AD10</td><td>163</td><td>−9.2</td><td>−77.0</td><td>4256</td><td>68.942</td><td>27.077</td></tr>
<tr><td>AD10</td><td>164</td><td>−5.6</td><td>−61.0</td><td>3541</td><td>69.283</td><td>26.409</td></tr>
<tr><td>AD10</td><td>222</td><td>−7.6</td><td>−77.0</td><td>4237</td><td>69.600</td><td>23.087</td></tr>
<tr><td>AD10</td><td>240</td><td>−7.6</td><td>−77.0</td><td>4329</td><td>67.333</td><td>23.010</td></tr>
<tr><td>AD10</td><td>243</td><td>−9.1</td><td>−83.0</td><td>4332</td><td>67.339</td><td>23.018</td></tr>
</table>

*GNIP—Global Network of Isotopes in Precipitation.
†n/a—not applicable.

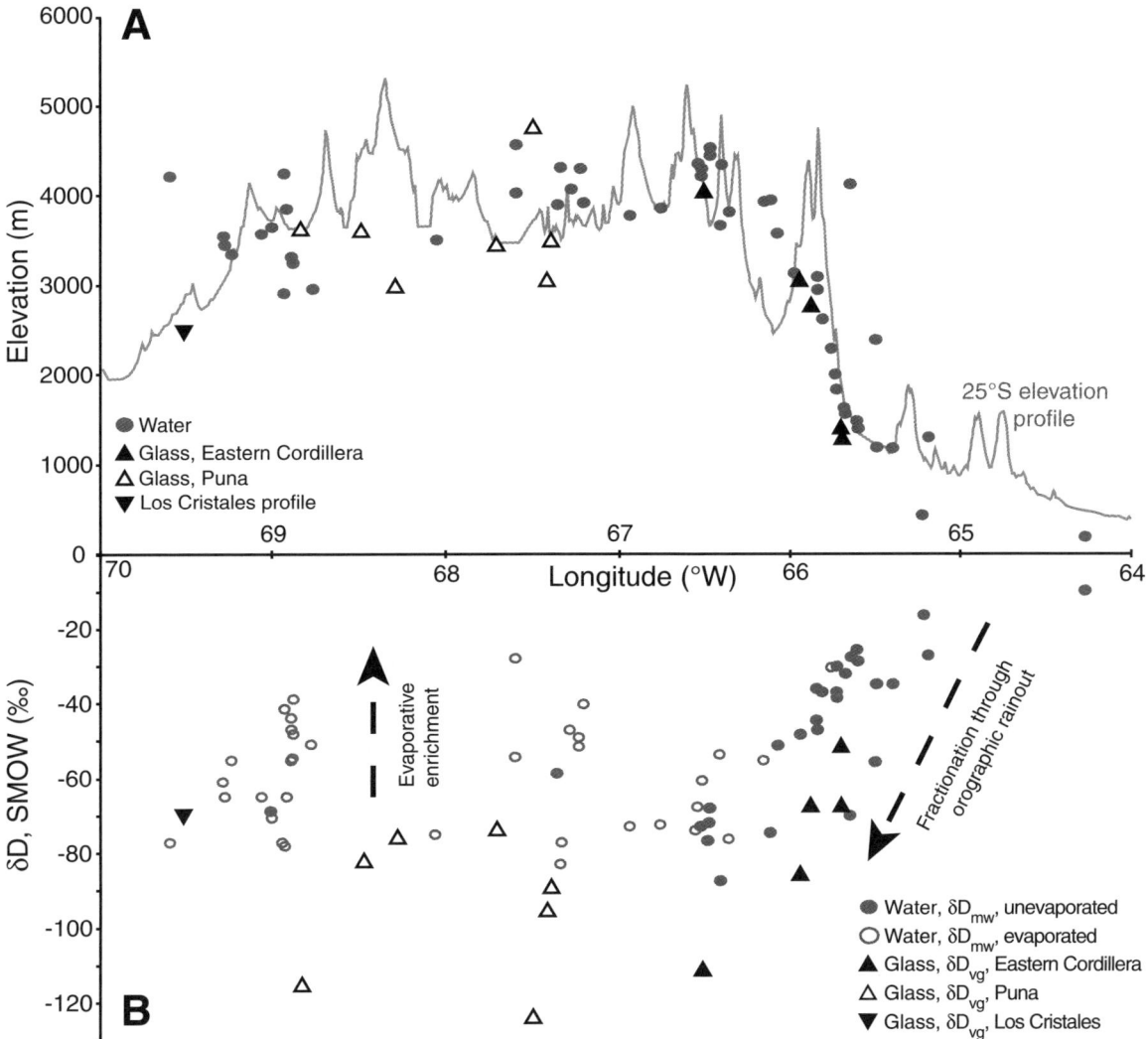

Figure 3. (A) Location of glass samples (triangles; Eastern Cordillera—solid, Puna—empty), water samples (circles), and Los Cristales samples (upside-down triangle) across the width of the central Andes with representative topographic profile (solid line) along 25°S. (B) The δD_{vg} values of glass samples (triangles: Eastern Cordillera—solid, Puna—empty; upside-down triangle—Los Cristales) and δD_{mw} values of unevaporated (solid circles) and evaporated (empty circles) surface waters observed across the central Andes. Elevation-controlled depletion of D causes the decrease in δD values in the Eastern Cordillera, and evaporation-related enrichment of D causes the scatter of higher values over the Puna and Western Cordillera. Solid triangles are used in δD_{vg}-δD_{mw}-elevation comparison (see Fig. 5). SMOW—standard mean ocean water; mw—meteoric water; vg—volcanic glass.

an increase in δD_{mw} values with decreasing elevation due to the influence of the Pacific Ocean.

Young Glass

Tuffs younger than 1 Ma were collected from an elevation range between 1311 m and 4757 m. Four of the five Eastern Cordilleran tuffs are probably younger than 0.01 Ma; the fifth is 0.3 Ma (Table 4). These glasses yield δD_{vg} values between −51‰ and −112‰. The calculated parent-water (δD_{cpw}) values, calculated using Equation 1, are between −19‰ and −81‰. Glasses

from the Puna, i.e., west of ~66.5°W, vary in age from younger than 0.01 Ma to 0.55 Ma. The δD_{vg} values of these seven glasses range between −74‰ and −125‰, producing reconstructed δD_{cpw} values of −42‰ to −95‰.

Comparison between Young Glass and Modern Meteoric Water

Our results from the Eastern Cordillera display a systematic relationship between the δD value of young glass and modern meteoric water (Fig. 5). Only the Eastern Cordillera data are

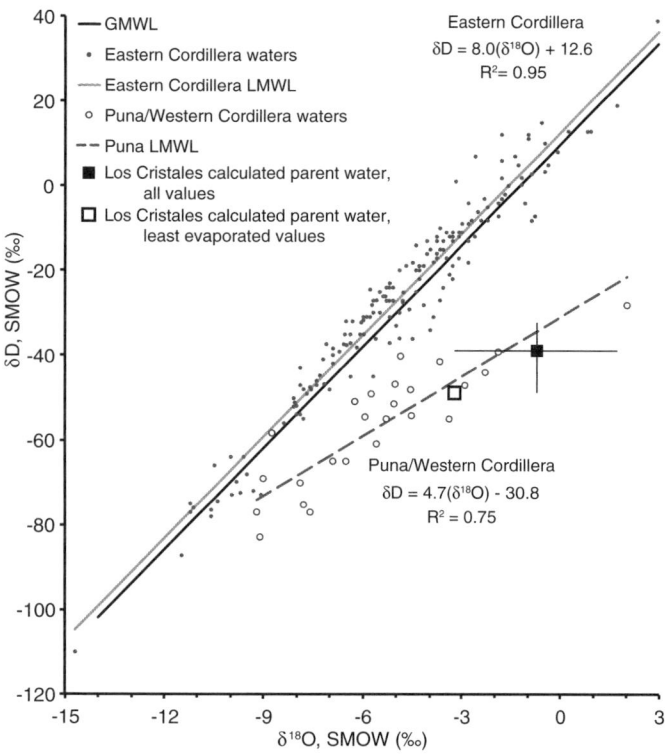

Figure 4. The global meteoric water line (GMWL, black), Eastern Cordillera local meteoric water line (LMWL, gray), and the Puna–Western Cordillera local meteoric water line (LMWL, dashed), with modern water data used to generate the latter two. The slope of the Puna–Western Cordillera LMWL follows an evaporation trend away from the negative extremes of the Eastern Cordillera LMWL. Also plotted: the isotopic composition of ca. 16 Ma parent water at the Los Cristales sampling site (squares), as calculated from δD_{vg}, $\delta^{18}O_{carb}$, and $T(\Delta_{47})$. Data point with error bars is the average of all values (solid square), with bars representing the range of measured values; the point with no error bars (empty square) is generated using only the least-evaporated data. Calculated isotopic values of paleowaters, both averaged and least evaporated, fall along modern Puna–Western Cordillera LMWL. SMOW—standard mean ocean water.

95% confidence interval (p level = 0.598 in a one-way ANOVA test). These results lead to three useful conclusions. First, our results verify that the Friedman et al. (1993b) fractionation factor, 1.0343, is a good approximation of the water-glass relationship for δD. Second, δD_{vg} values can be used to calculate δD_{mw} across a range of elevations and temperatures—an important first step toward estimating δD_{cpw}. Third, the slope of $-13.7‰$ km^{-1} (Eq. 3) for the δD_{mw}-elevation relationship in our area holds for the recent geologic past and is not an anomaly of the few years when the surface-water samples were collected.

TEST FOR DIAGENESIS

Los Cristales Ignimbrite and Paleosol: Glass and Carbonate Results

Volcanic glass and pedogenic carbonate were analyzed from the Los Cristales ignimbrite and associated calcic paleosol (Fig. 6) from north-central Chile. Glass was extracted from four samples of variably mixed soil and ignimbrite collected

used because this avoids the influence of evaporation, which, as observed in the δD_{mw} data, dominates the δD_{vg} values in the Puna and Western Cordillera (Fig. 3). The δD_{vg} values of the five young Eastern Cordillera tuffs, all ≤0.3 Ma, are offset from the observed δD_{mw}-elevation trend by an average of 33.3‰ ± 21.8‰ across an elevation range of ~3000 m. While the 2σ standard deviation is large, the mean is almost identical to the 34‰ difference between δD_{vg} and δD_{mw} values calculated by Friedman et al. (1993b). Additionally, δD_{cpw} values calculated from young tuff δD_{vg} values are statistically indistinguishable from δD_{mw} values at the

TABLE 4. GLASS DATA

Sample	δD_{vg}, average (‰)	δD_{cpw}*, average (‰)	Age (Ma)	Elevation (m)	Latitude (°S)	Longitude (°W)
Eastern Cordillera						
SL-42A	−67	−35	0.01	1408	25.855	65.703
SL-22	−86	−55	0.01	3051	24.459	65.939
SL-57	−51	−19	0.01	1311	25.715	65.700
SL-58	−68	−36	0.01	2773	24.494	65.882
SL-38	−112	−81	0.3	4041	23.998	66.508
Puna						
AD10-252	−81	−50	0.001	3594	24.290	68.475
SL-124	−89	−58	0.01	3492	25.998	67.390
AD10-128	−116	−85	0.01	3610	25.471	68.821
AD10-218	−125	−95	0.01	4757	23.098	67.496
AD10-251	−75	−43	0.01	2973	24.103	68.275
SL-117	−74	−42	0.2	3461	26.518	67.704
SL-112	−95	−64	0.55	3028	26.495	67.409

Note: vg—volcanic glass; cpw—calculated parent water.
*Calculated parent water uses an α(water-glass) of 1.0343 from Friedman et al. (1993a).

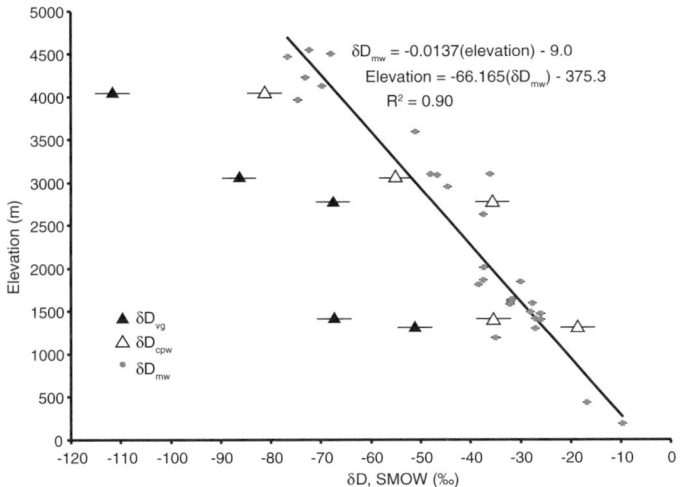

Figure 5. Isotope-elevation relationships of Eastern Cordillera waters (δD_{mw}, circles and trend line) and five ≤0.3 Ma glasses (δD_{vg}, solid triangles). Calculated δD_{cpw} values from glass (empty triangles) are statistically indistinguishable from modern observed surface-water δD_{mw} values. Error bars are 2σ. SMOW—standard mean ocean water; mw—meteoric water; vg—volcanic glass; cpw—calculated parent water.

at ~50 cm intervals, between 90 cm and 450 cm below the top of the paleosol. Two glass samples were run at each depth and averaged. The uppermost sampling level returned a δD_{vg} value of −78‰. The three deeper samples yielded δD_{vg} values between −66‰ and −70‰. From Equation 1, the calculated δD_{cpw} values come out between −34‰ and −47‰ and average −39‰.

Six soil carbonate samples were collected at depths of 20 cm to 140 cm at 20 cm intervals in the same profile as the glass samples. Three carbonate samples from each depth were analyzed, and the results are averaged in Table 5. The full range of $\delta^{18}O_{carb}$ values is between −0.2‰ and −5.1‰, with an average of −2.6‰ (VPDB). There is no systematic variation with depth (Fig. 6). The Δ_{47} results from soil carbonate samples at 20 cm and 140 cm yield $T(\Delta_{47})$ values of 24.3 °C and 21.0 °C, respectively. In order to estimate $\delta^{18}O_{cpw}$ based on $\delta^{18}O_{carb}$, we used the temperature-dependent calcite-water fractionation factor ($\alpha_{CaCO3-H2O}$) of Kim and O'Neil (1997):

$$1000\ln\alpha_{(CaCO3-H2O)} = 18.03(10^3 \times T^{-1}) - 32.42, \quad (6)$$

(where T is temperature, in Kelvin) and the relationship of α to δ-notation,

$$\alpha_{(CaCO3-H2O)} = (1000 + \delta^{18}O_{CaCO3})/(1000 + \delta^{18}O_{H2O}). \quad (7)$$

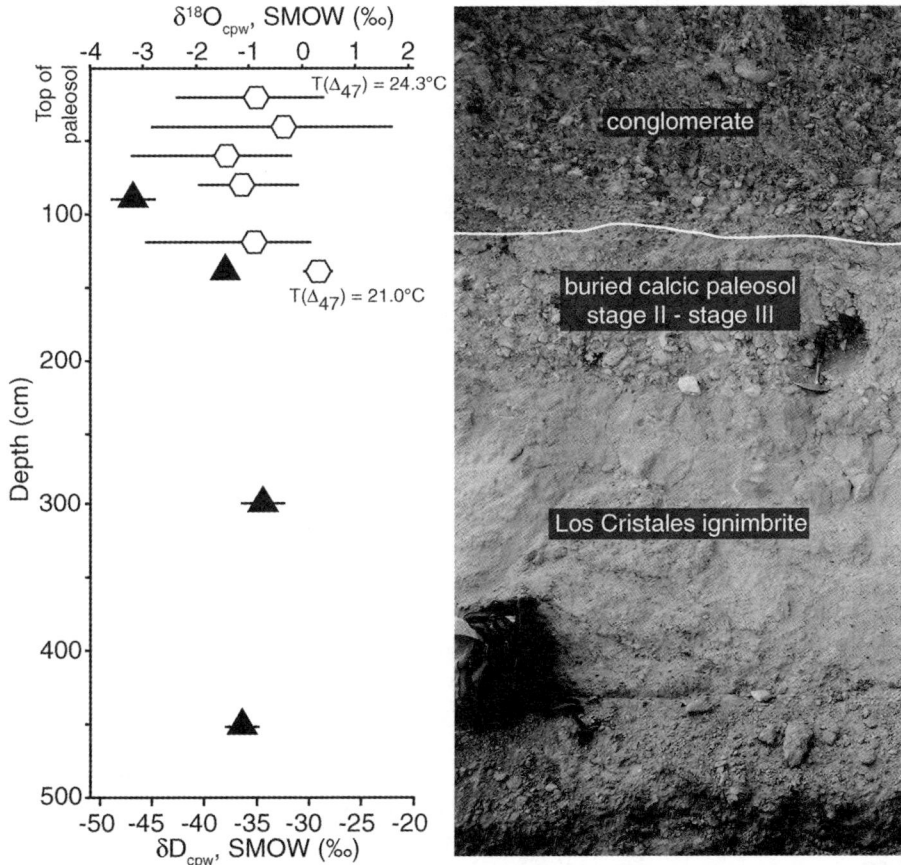

Figure 6. Los Cristales ignimbrite and soil profile: photograph of portion of section (backpack, 75 cm, and rock hammer, 33 cm, for scale). White line separates paleosol from overlying conglomerate. Isotopic profile presents the calculated average and range of calculated $\delta^{18}O_{cpw}$ values obtained from carbonate (hexagons) and calculated δD_{cpw} values obtained from glass (triangles), as well as clumped isotope paleotemperatures (noted next to the two analyzed soil carbonates). These systems appear to be in equilibrium and unaltered since formation. SMOW—standard mean ocean water; cpw—calculated parent water.

TABLE 5. LOS CRISTALES IGNIMBRITE AND BURIED CALCAREOUS PALEOSOL

	Latitude (°S)	Longitude (°W)	Elevation (m)	Tuff age (Ma)	Pedogenic carbonate development
Los Cristales section:	26.390	69.517	2450	Ca. 16.3*	Stage II–III

Pedogenic carbonate		Carbonate		Parent water[†]		
Sample	Depth (cm)	$\delta^{13}C$, average, VPDB (‰)	$\delta^{18}O$, average, VPDB (‰)	$\delta^{18}O$, average, SMOW (‰)	Δ_{47} (‰)	$T(\Delta_{47})$ (°C)
AD10 136-0.2	−20	−0.4	−2.8	−0.9	0.649	24.3
AD10 136-0.4	−40	−0.7	−2.2	−0.3		
AD10 136-0.6	−60	−0.7	−3.3	−1.4		
AD10 136-0.8	−80	−0.1	−3.0	−1.1		
AD10 136-1.2	−120	−1.4	−2.8	−0.9		
AD10 136-1.4	−140	−0.6	−1.6	0.3	0.664	21.0
Total average:			−2.6	−0.7		22.7

Volcanic glass		Glass	Parent water[§]
Sample	Depth (cm)	δD_{vg}, average, SMOW (‰)	δD_{cpw}, average, SMOW (‰)
AD10-136-0.9m	−90	−78.3	−46.7
AD10-136-1.4m	−140	−69.8	−37.9
AD10-136-3.0m	−300	−66.4	−34.4
AD10-136-4.5m	−450	−68.3	−36.4
Total average:		−70.7	−38.8

Note: VPDB—Vienna Pee Dee Belemnite; SMOW—standard mean ocean water.
*Nalpas et al. (2008).
[†]Using α(calcite–water) of Kim and O'Neil (1997) and the average $T(\Delta_{47})$ temperature.
[§]Using α(water–glass) of 1.0343 from Friedman et al. (1993a).

With an average $T(\Delta_{47})$ of 22.7 °C and an average $\delta^{18}O_{carb}$ (VPDB) value of −2.6‰, a $\delta^{18}O_{cpw}$ (SMOW) value of −0.7‰ was calculated. Using only the lowest $\delta^{18}O_{carb}$ values, interpreted to be the least evaporated, the $\delta^{18}O_{cpw}$ value becomes −3.0‰.

Combining the $\delta^{18}O_{cpw}$ and δD_{cpw} values produces a predicted range of parent-water values from −0.7‰ $\delta^{18}O$, −39‰ δD (average of all data) to −3.0‰ $\delta^{18}O$, −47‰ δD (lowest/least-evaporated results).

Preservation and Diagenesis Discussion

The consistency of our results from the Los Cristales section suggests that neither the volcanic glass nor the carbonate has been reset isotopically. $T(\Delta_{47})$ values of ~21–24 °C fall well within the range of those observed in modern soil carbonate (Huntington et al., 2010; Passey et al., 2010). Huntington et al. (2010) noted "cryptic resetting" in ancient mollusks; $T(\Delta_{47})$ values appeared reset, while $\delta^{18}O_{carb}$ values did not. This, and work by Quade et al. (2013), suggests that $\delta^{18}O_{carb}$ values are more resistant to resetting than $T(\Delta_{47})$. We therefore interpret the values of both the $T(\Delta_{47})$ and $\delta^{18}O_{carb}$ systems to be unaltered since burial following the 16.3 Ma deposition of the Los Cristales ignimbrite.

The $\delta^{18}O_{cpw}$ value estimated from carbonate and the δD_{cpw} value estimated from glass produce a calculated parent-water composition (−0.7‰ $\delta^{18}O$, −39‰ δD) that plots to the right of both the Eastern Cordillera LMWL and the GMWL in Figure 4, yet almost directly on the Puna–Western Cordillera LMWL. Using the lowest $\delta^{18}O_{carb}$ and δD_{vg} values (interpreted as the least evaporated) from the Los Cristales paleosol profile, the resulting reconstructed $\delta^{18}O_{cpw}/\delta D_{cpw}$ composition (−3.0‰ $\delta^{18}O$, −47‰ δD) still follows the trend of the Puna–Western Cordillera LMWL. Therefore, we conclude that both the volcanic glass and the soil carbonate formed in isotopic equilibrium with the same parent water, which, ca. 16.3 Ma, had similar δD_{mw} and $\delta^{18}O_{mw}$ values as surface waters found on the Puna today. Modern Puna waters derive their characteristic compositions from evaporative enrichment of highly fractionated easterly sourced moisture. This implies that orography and aridity similar to today were likely already in place at 16 Ma. Large mountain ranges must have been present to the east in order to create ancient $\delta D_{mw}/\delta^{18}O_{mw}$ values that fall on the modern fractionation- and evaporation-controlled LMWL.

Under favorable conditions, physical and isotopic preservation of glass is excellent. To be useful as a paleoelevation-paleoenvironment archive, two preservation requirements must be met: The glass itself cannot degrade to clay, and the waters of hydration must not experience isotopic exchange. Glass alters to clay much slower in arid or semiarid regions, such as the Great Basin, United States, or the central Andes of Argentina and Chile. Likewise, glass that is only modestly buried (<1–2 km) is much less likely to be altered, as in the case of the Los Cristales ignimbrite and of the Oligocene–Miocene tuffs investigated by Mulch et al. (2008) and Cassel et al. (2012).

However, the higher time and temperature limits of glass preservation are uncertain and untested. To our knowledge, no systematic survey of preservation potential of volcanic glass as a function of climate and age has been conducted. Such investigations are vital to testing the use of δD_{vg} in deeper geologic time and in wetter conditions. The north-south and east-west variations in climate, the diversity of structures and basins, and the abundance of tuffs make the Andes a prime natural laboratory for such testing.

CONCLUSIONS

Our research and that of other workers in this field (Friedman et al., 1993a; Mulch et al., 2008; Cassel et al., 2012) support the use of volcanic glass paleoenvironmental and paleoaltimetric reconstructions, especially within shallowly buried and dry settings such as in the central Andes. Except for the influence of adhering clay and potential complications of HF acid treatment, δD_{vg} values are insensitive to variations in preparation procedures. The δD_{cpw} values calculated from the youngest volcanic glasses match those expected for a given area, such as the Eastern Cordillera. The Los Cristales paleosol profile provides a key test of the resistance to alteration of volcanic glass after long-term (16 m.y.) burial. Isotopic compositions (δD_{vg}, $\delta^{18}O_{carb}$, and $T[\Delta_{47}]$) of the carbonate and glass are in equilibrium, arguing against isotopic resetting of any of these three systems. Our results confirm the fidelity of volcanic glass δD values in recording paleoenvironmental conditions shortly after burial.

ACKNOWLEDGMENTS

We would like to thank John Eiler, CalTech, for assistance with the clumped isotope paleothermometry, and Page Chamberlain and Andreas Mulch, for their helpful reviews of the manuscript. ExxonMobil Corporation provided funding through the Convergent Orogenic Systems Analysis (COSA) project. Dettinger would like to acknowledge the Peter J. Coney Award and Chevron for summer research support.

REFERENCES CITED

Affek, H.P., Bar-Matthews, M., Ayalon, A., Matthews, A., and Eiler, J.M., 2008, Glacial/interglacial temperature variations in Soreq cave speleothems as recorded by "clumped isotope" thermometry: Geochimica et Cosmochimica Acta, v. 72, no. 22, p. 5351–5360, doi:10.1016/j.gca.2008.06.031.

Alpers, C.N., and Brimhall, G.H., 1988, Middle Miocene climatic change in the Atacama Desert, northern Chile: Evidence from supergene mineralization at La Escondida: Geological Society of America Bulletin, v. 100, no. 10, p. 1640–1656, doi:10.1130/0016-7606(1988)100<1640:MMCCIT>2.3.CO;2.

Anovitz, L.M., Cole, D.R., and Riciputi, L.R., 2009, Low-temperature isotopic exchange in obsidian: Implications for diffusive mechanisms: Geochimica et Cosmochimica Acta, v. 73, no. 13, p. 3795–3806, doi:10.1016/j.gca.2009.02.035.

Barnes, J., and Ehlers, T.A., 2009, End member models for Andean Plateau uplift: Earth-Science Reviews, v. 97, no. 1–4, p. 105–132, doi:10.1016/j.earscirev.2009.08.003.

Blisniuk, P., and Stern, L., 2005, Stable isotope paleoaltimetry: A critical review: American Journal of Science, v. 305, no. 10, p. 1033–1074, doi:10.2475/ajs.305.10.1033.

Cailleteau, C., Angeli, F., Devreux, F., Gin, S., Jestin, J., Jollivet, P., and Spalla, O., 2008, Insight into silicate-glass corrosion mechanisms: Nature Materials, v. 7, no. 12, p. 978–983, doi:10.1038/nmat2301.

Cassel, E.J., Graham, S.A., and Chamberlain, C.P., 2009, Cenozoic tectonic and topographic evolution of the northern Sierra Nevada, California, through stable isotope paleoaltimetry in volcanic glass: Geology, v. 37, no. 6, p. 547–550, doi:10.1130/G25572A.1.

Cassel, E.J., Graham, S.A., Chamberlain, C.P., and Henry, C.D., 2012, Early Cenozoic topography, morphology, and tectonics of the northern Sierra Nevada and western Basin and Range: Geosphere, v. 8, no. 2, p. 229–249, doi:10.1130/GES00671.1.

Clift, P.D., Blusztajn, J., and Nguyen, A.D., 2006, Large-scale drainage capture and surface uplift in eastern Tibet–SW China before 24 Ma inferred from sediments of the Hanoi Basin, Vietnam: Geophysical Research Letters, v. 33, no. 19, L19403, doi:10.1029/2006GL027772.

Crovisier, J.-L., Advocat, T., and Dussossoy, J.-L., 2003, Nature and role of natural alteration gels formed on the surface of ancient volcanic glasses (natural analogs of waste containment glasses): Journal of Nuclear Materials, v. 321, no. 1, p. 91–109, doi:10.1016/S0022-3115(03)00206-X.

Eiler, J.M., 2011, Paleoclimate reconstruction using carbonate clumped isotope thermometry: Quaternary Science Reviews, v. 30, no. 25–26, p. 3575–3588, doi:10.1016/j.quascirev.2011.09.001.

Francis, P.W., and Hawkesworth, C.J., 1994, Late Cenozoic rates of magmatic activity in the central Andes and their relationships to continental-crust formation and thickening: Journal of the Geological Society, London, v. 151, p. 845–854, doi:10.1144/gsjgs.151.5.0845.

Friedman, I., Smith, R.L., and Long, W.D., 1966, Hydration of natural glass and formation of perlite: Geological Society of America Bulletin, v. 77, no. 3, p. 323, doi:10.1130/0016-7606(1966)77[323:HONGAF]2.0.CO;2.

Friedman, I., Gleason, J., and Warden, A., 1993a, Ancient climate from deuterium content of water in volcanic glass, *in* Swart, P.K., Lohmann, K.C., McKenzie, J., and Savin, S., eds., Climate Change in Continental Isotopic Records: American Geophysical Union Geophysical Monograph 78, p. 309–319.

Friedman, I., Gleason, J., Sheppard, R.A., and Gude, A.J., 1993b, Deuterium fractionation as water diffuses into silicic volcanic ash, *in* Swart, P.K., Lohmann, K.C., McKenzie, J., and Savin, S., eds., Climate Change in Continental Isotopic Records: American Geophysical Union Geophysical Monograph 78, p. 321–323.

Garreaud, R., 1999, Multiscale analysis of the summertime precipitation over the central Andes: Monthly Weather Review, v. 127, no. 5, p. 901–921, doi:10.1175/1520-0493(1999)127<0901:MAOTSP>2.0.CO;2.

Garzione, C., Hoke, G., Libarkin, J., Withers, S., MacFadden, B., Eiler, J., Ghosh, P., and Mulch, A., 2008, Rise of the Andes: Science, v. 320, no. 5881, p. 1304–1307, doi:10.1126/science.1148615.

Ghosh, P., Adkins, J., Affek, H., Balta, B., Guo, W., Schauble, E.A., Schrag, D., and Eiler, J.M., 2006a, ^{13}C–^{18}O bonds in carbonate minerals: A new kind of paleothermometer: Geochimica et Cosmochimica Acta, v. 70, no. 6, p. 1439–1456, doi:10.1016/j.gca.2005.11.014.

Ghosh, P., Garzione, C.N., and Eiler, J.M., 2006b, Rapid uplift of the Altiplano revealed through ^{13}C–^{18}O bonds in paleosol carbonates: Science, v. 311, no. 5760, p. 511–515, doi:10.1126/science.1119365.

Gile, L.H., Peterson, F.F., and Grossman, E., 1966, Morphological and genetic sequences of carbonate accumulation in desert soils: Soil Science, v. 101, no. 5, p. 347–360, doi:10.1097/00010694-196605000-00001.

Gregory-Wodzicki, K.M., Mcintosh, W.C., and Velasquez, K., 1998, Climatic and tectonic implications of the late Miocene Jakokkota flora, Bolivian Altiplano: Journal of South American Earth Sciences, v. 11, no. 6, p. 533–560, doi:10.1016/S0895-9811(98)00031-5.

Hoke, G., and Garzione, C., 2008, Paleosurfaces, paleoelevation, and the mechanisms for the late Miocene topographic development of the Altiplano Plateau: Earth and Planetary Science Letters, v. 271, no. 1–4, p. 192–201, doi:10.1016/j.epsl.2008.04.008.

Huntington, K.W., Eiler, J.M., Affek, H.P., Guo, W., Bonifacie, M., Yeung, L.Y., Thiagarajan, N., Passey, B., Tripati, A., Daëron, M., and Came, R., 2009, Methods and limitations of "clumped" CO_2 isotope δ_{47} analysis by gas-source isotope ratio mass spectrometry: Journal of Mass Spectrometry, v. 44, no. 9, p. 1318–1329, doi:10.1002/jms.1614.

Huntington, K.W., Wernicke, B.P., and Eiler, J.M., 2010, Influence of climate change and uplift on Colorado Plateau paleotemperatures from carbonate clumped isotope thermometry: Tectonics, v. 29, TC3005, doi:10.1029/2009TC002449.

Ingraham, N., and Taylor, B., 1991, Light stable isotope systematics of large-scale hydrologic regimes in California and Nevada: Water Resources Research, v. 27, no. 1, p. 77–90, doi:10.1029/90WR01708.

Insel, N., Poulsen, C.J., Ehlers, T.A., and Sturm, C., 2010, Response of meteoric $\delta^{18}O$ to surface uplift—Implications for Cenozoic Andean Plateau growth: Earth and Planetary Science Letters, v. 317–318, p. 1–11, doi:10.1016/j.epsl.2011.11.039.

International Atomic Energy Agency (IAEA)/World Meteorological Organization (WMO), 2006, Global Network of Isotopes in Precipitation: The GNIP Database: www.iaea.org/water (accessed 8 October 2011).

Kim, S., and O'Neil, J., 1997, Equilibrium and nonequilibrium oxygen isotope effects in synthetic carbonates: Geochimica et Cosmochimica Acta, v. 61, no. 16, p. 3461–3475, doi:10.1016/S0016-7037(97)00169-5.

Kyser, T.K., and O'Neil, J.R., 1984, Hydrogen isotope systematics of submarine basalts: Geochimica et Cosmochimica Acta, v. 48, no. 10, p. 2123–2133, doi:10.1016/0016-7037(84)90392-2.

Molnar, P., Boos, W.R., and Battisti, D.S., 2010, Orographic controls on climate and paleoclimate of Asia: Thermal and mechanical roles for the Tibetan Plateau: Annual Review of Earth and Planetary Sciences, v. 38, no. 1, p. 77–102, doi:10.1146/annurev-earth-040809-152456.

Mulch, A., Sarna-Wojcicki, A., Perkins, M., and Chamberlain, C., 2008, A Miocene to Pleistocene climate and elevation record of the Sierra Nevada (California): Proceedings of the National Academy of Sciences of the United States of America, v. 105, no. 19, p. 6819–6824, doi:10.1073/pnas.0708811105.

Mulch, A., Uba, C., Strecker, M., Schoenberg, R., and Chamberlain, C., 2010, Late Miocene climate variability and surface elevation in the central Andes: Earth and Planetary Science Letters, v. 290, no. 1–2, p. 173–182, doi:10.1016/j.epsl.2009.12.019.

Muñoz, N., and Charrier, R., 1996, Uplift of the western border of the Altiplano on a west-vergent thrust system, northern Chile: Journal of South American Earth Sciences, v. 9, p. 171–181, doi:10.1016/0895-9811(96)00004-1.

Nalpas, T., Dabard, M.P., Ruffet, G., Vernon, A., Mpodozis, C., Loi, A., and Hérail, G., 2008, Sedimentation and preservation of the Miocene Atacama Gravels in the Pedernales–Chañaral area, northern Chile: Climatic or tectonic control?: Tectonophysics, v. 459, no. 1–4, p. 161–173, doi:10.1016/j.tecto.2007.10.013.

Onken, J., 1991, The Effect of Microenvironmental Temperature Variation on the Hydration of Late Holocene Mono Craters Volcanic Ashes from East-

Central California [Master's thesis]: Tucson, Arizona, University of Arizona, 35 p.

Passey, B.H., Levin, N.E., Cerling, T.E., Brown, F.H., and Eiler, J.M., 2010, High-temperature environments of human evolution in East Africa based on bond ordering in paleosol carbonates: Proceedings of the National Academy of Sciences of the United States of America, v. 107, no. 25, p. 11,245–11,249, doi:10.1073/pnas.1001824107.

Pelletier, J.D., DeCelles, P.G., and Zandt, G., 2010, Relationships among climate, erosion, topography, and delamination in the Andes: A numerical modeling investigation: Geology, v. 38, no. 3, p. 259–262, doi:10.1130/G30755.1.

Poage, M., and Chamberlain, C., 2001, Empirical relationships between elevation and the stable isotope composition of precipitation and surface waters: Considerations for studies of paleoelevation change: American Journal of Science, v. 301, no. 1, p. 1, doi:10.2475/ajs.301.1.1.

Quade, J., Breecker, D.O., Daeron, M., and Eiler, J., 2011, The paleoaltimetry of Tibet: An isotopic perspective: American Journal of Science, v. 311, no. 2, p. 77–115, doi:10.2475/02.2011.01.

Quade, J., Eiler, J., Daëron, M., and Achyuthan, H., 2013, The clumped isotope geothermometer in soil and paleosol carbonate: Geochimica et Cosmochimica Acta, v. 105, p. 92–107, doi:10.1016/j.gca.2012.11.031.

Rech, J.A., Currie, B.S., Shullenberger, E.D., Dunagan, S.P., Jordan, T.E., Blanco, N., Tomlinson, A.J., Rowe, H.D., and Houston, J., 2010, Evidence for the development of the Andean rain shadow from a Neogene isotopic record in the Atacama Desert, Chile: Earth and Planetary Science Letters, v. 292, no. 3–4, p. 371–382, doi:10.1016/j.epsl.2010.02.004.

Skewes, A., Holmgren, C., and Stern, C., 2003, The Donoso copper-rich, tourmaline-bearing breccia pipe in central Chile: Petrologic, fluid inclusion and stable isotope evidence for an origin from magmatic fluids: Mineralium Deposita, v. 38, no. 1, p. 2–21, doi:10.1007/s00126-002-0264-9.

Strecker, M.R., Alonso, R.N., Bookhagen, B., Carrapa, B., Hilley, G.E., Sobel, E.R., and Trauth, M.H., 2007, Tectonics and climate of the southern central Andes: Annual Review of Earth and Planetary Sciences, v. 35, p. 747–787, doi:10.1146/annurev.earth.35.031306.140158.

Taylor, H.P., 1974, The application of oxygen and hydrogen isotope studies to problems of hydrothermal alteration and ore deposition: Economic Geology and the Bulletin of the Society of Economic Geologists, v. 69, no. 6, p. 843–883, doi:10.2113/gsecongeo.69.6.843.

Valle, N., Verney-Carron, A., Sterpenich, J., Libourel, G., Deloule, E., and Jollivet, P., 2010, Elemental and isotopic (^{29}Si and ^{18}O) tracing of glass alteration mechanisms: Geochimica et Cosmochimica Acta, v. 74, no. 12, p. 3412–3431, doi:10.1016/j.gca.2010.03.028.

Whipple, K.X., 2009, The influence of climate on the tectonic evolution of mountain belts: Nature Geoscience, v. 2, no. 2, p. 97–104, doi:10.1038/ngeo413.

Wolfe, J.A., Schorn, H.E., Forest, C.E., and Molnar, P., 1997, Paleobotanical evidence for high altitudes in Nevada during the Miocene: Science, v. 276, no. 5319, p. 1672–1675, doi:10.1126/science.276.5319.1672.

Yeh, H.W., and Savin, S.M., 1977, Mechanism of burial metamorphism of argillaceous sediments: 3. O-isotope evidence: Geological Society of America Bulletin, v. 88, no. 9, p. 1321–1330, doi:10.1130/0016-7606(1977)88<1321:MOBMOA>2.0.CO;2.

Zhou, J., and Lau, K.M., 1998, Does a monsoon climate exist over South America?: Journal of Climate, v. 11, no. 5, p. 1020–1040, doi:10.1175/1520-0442(1998)011<1020:DAMCEO>2.0.CO;2.

MANUSCRIPT ACCEPTED BY THE SOCIETY 3 JUNE 2014
MANUSCRIPT PUBLISHED ONLINE 14 NOVEMBER 2014

The Geological Society of America
Memoir 212
2015

The growth of the central Andes, 22°S–26°S

J. Quade*
M.P. Dettinger
B. Carrapa
P. DeCelles
K.E. Murray
Department of Geosciences, University of Arizona, Tucson, Arizona 85721, USA

K.W. Huntington
Department of Earth and Space Sciences, University of Washington, Seattle, Washington 98195-1310, USA

A. Cartwright
Mintec, Inc., 3544 East Fort Lowell Road, Tucson, Arizona 85716, USA

R.R. Canavan
Department of Geology, University of Wyoming, Laramie, Wyoming 82071, USA

G. Gehrels
Department of Geosciences, University of Arizona, Tucson, Arizona 85721, USA

M. Clementz
Department of Geology, University of Wyoming, Laramie, Wyoming 82071, USA

ABSTRACT

We synthesize geologic observations with new isotopic evidence for the timing and magnitude of uplift for the central Andes between 22°S and 26°S since the Paleocene. To estimate paleoelevations, we used the stable isotopic composition of carbonates and volcanic glass, combined with another paleoelevation indicator for the central Andes: the distribution of evaporites. Paleoelevation reconstruction using clumped isotope paleothermometry failed due to resetting during burial.

The Andes at this latitude rose and broadened eastward in three stages during the Cenozoic. The first, in what is broadly termed the "Incaic" orogeny, ended by the late Eocene, when magmatism and deformation had elevated to ≥4 km the bulk (~50%) of what is now the western and central Andes. The second stage witnessed the gradual building of the easternmost Puna and Eastern Cordillera, starting with deformation as early as 38 Ma, to >3 km by no later than 15 Ma. The proximal portions

*quadej@email.arizona.edu

Quade, J., Dettinger, M.P., Carrapa, B., DeCelles, P., Murray, K.E., Huntington, K.W., Cartwright, A., Canavan, R.R., Gehrels, G., and Clementz, M., 2015, The growth of the central Andes, 22°S–26°S, *in* DeCelles, P.G., Ducea, M.N., Carrapa, B., and Kapp, P.A., eds., Geodynamics of a Cordilleran Orogenic System: The Central Andes of Argentina and Northern Chile: Geological Society of America Memoir 212, p. 277–308, doi:10.1130/2015.1212(15). For permission to copy, contact editing@geosociety.org.

of the Paleogene foreland basin system were incorporated into the orogenic edifice, and basins internal to the orogen were enclosed and isolated from easterly moisture sources, promoting the precipitation of evaporites. In the third orogenic stage during the Pliocene–Pleistocene, Andean deformation accelerated and stepped eastward to form the modern Subandes, accounting for the final ~15%–20% of the current cross section of the Andes. About 0.5 km of elevation was added unevenly to the Western Cordillera and Puna from 10 to 2 Ma by voluminous volcanism.

The two largest episodes of uplift and eastward propagation of the orogenic front and of the foreland flexural wave, ca. 50 (?)–40 Ma and <5 Ma, overlap with or immediately postdate periods of very rapid plate convergence, high flux magmatism in the magmatic arc, and crustal thickening. Uplift does not correlate with a hypothesized mantle lithospheric foundering event in the early Oligocene. Development of hyperaridity in the Atacama Desert by the mid-Miocene postdates the two-step elevation gain to >3 km of most (~75%) of the Andes. Hence, the record suggests that hyperarid climate was a consequence, not major cause, of uplift through trench sediment starvation.

INTRODUCTION

The Puna-Altiplano is the highest-elevation region of the world not built by continent-continent collision. The development of this high region has preoccupied geologists for generations (e.g., Steinmann, 1929; Mégard, 1984), and as a result, much is known about the timing and extent of the deformation and depositional history. New developments in geochronology, thermochronology, and most recently paleoaltimetry are revolutionizing our view of Andean orogeny, but also stirring debate.

There are several different views of the chronology, spatial patterns, and causes of Andean uplift, based mostly on study of the geology of the Altiplano and adjacent areas. One view holds that deformation and volcanism had raised the western Andes and Altiplano by the Oligocene to mid-Miocene (Horton, 1999; Hoke and Garzione, 2008; Evenstar et al., 2009; Jordan et al., 2010), and subsequently deformation broadened eastward (Horton, 1999; Barnes et al., 2012). Others suggest that the whole of the Andes rose as a single mass, either gradually starting in the Miocene (Ehlers and Poulsen, 2009), or more abruptly in the late Miocene (Kay and Gordillo, 1994; Ghosh et al., 2006a; Garzione et al., 2006, 2008; Gregory-Wodzicki, 2000). One cause of abrupt uplift is thought to be removal of the mantle lithosphere beneath the Andes, either gradually by ablative removal (Pope and Willet, 1998), or abruptly by foundering of eclogitized mantle lithosphere (Garzione et al., 2006, 2008). Others explain late Miocene uplift by ductile lower-crustal flow and underthrusting (Allmendinger and Gubbels, 1996; Barke and Lamb, 2006). Sobel et al. (2003) and Strecker et al. (2007) stressed the role of Neogene orographic barrier development and climatic controls on sediment removal from the orogenic belt in expanding Andean topography. Yet another hypothesis suggests that global cooling in the mid- to late Miocene and attendant aridification of the Atacama Desert raised the Andes, due to the high shear stresses that developed as a result of trench starvation of sedi-

ment (Lamb and Davis, 2003). Studies have provided extensive geologic and thermochronologic evidence for deformation and exhumation, but only a few attempt to quantify paleoelevations achieved by this deformation. Evenstar et al. (2009) and Jordan et al. (2010) interpreted tilting of surfaces and ignimbrite flows to indicate that the western slope of the Andes stood ~2000 m above the adjacent forearc (now at ~1–1.5 km) by the early to mid-Miocene, and it has added a further ~1 km since then. Using isotopic evidence, Ghosh et al. (2006a) and Garzione et al. (2006, 2008) suggested that the whole of the Andes at the latitude of the Altiplano rose from ≤2 km to ~4 km between 10 and 5 Ma. Ehlers and Poulsen (2009) reinterpreted some of this same evidence to suggest the Andes also rose en masse, but more slowly, starting in the early Miocene. Recently, papers by many of the same authors show more convergence of views, with the recognition that uplift was probably time transgressive, west to east (e.g., Hoke and Garzione, 2008; Barnes et al., 2012).

Haschke et al. (2006) viewed the development of the Andes as a cyclical process involving gradual crustal thickening, deep lithospheric foundering, and slab shallowing, culminating in regional uplift. DeCelles et al. (2009) presented a holistic model in which crustal shortening, magmatism, upper-mantle dripping/delamination of dense eclogitic instabilities created by shortening and magmatism, and surface uplift are all linked. These processes operate on a cyclical schedule to build cordilleran-style orogenic belts such as the North and South American Cordilleras. The South American cycles are hypothesized to last 25–30 m.y. and are predicted to have culminated in Andean uplift during the early Oligocene and possibly the latest Neogene (Haschke et al., 2006).

In this paper, we journey south of the Altiplano to the Puna Plateau and the 22°S–26°S sector of the Andes (Fig. 1). For this sector, we merge the geologic and thermochronologic evidence for deformation with the gradual development of high elevations of the region. For paleoaltimetry, we rely on a large suite of new isotopic evidence from carbonates and volcanic glass, as well as from the new clumped isotope geothermometer. We use the modern distribution of salt lakes in South America for additional constraints on paleoelevation. Most reconstructions of surface elevation change have tended to focus on tectonic rather than volcanic contributions. Volcanic rocks have a conspicuous presence in the Andes, and to complete the picture of surface elevation changes, their contribution must be considered, especially during the last 10 m.y. In the final part of this paper, we turn to likely causes of Andean deformation and uplift during the Cenozoic.

GEOGRAPHIC AND GEOLOGICAL BACKGROUND

Physiographic and Climate Divisions

The modern Andes can be conveniently divided into five physio-tectonic zones from east to west between 22°S and 26°S: the Subandes, the Eastern Cordillera, and the Puna Plateau in the retroarc; the magmatic arc in the Western Cordillera; and the Atacama Desert in the forearc (Fig. 1). The Subandes (or more locally the Santa Bárbara system) at this latitude consist

of a series of variably elevated (<5 km) frontal ranges dividing externally drained, relatively low-elevation (<1 km) basins. The Subandes below 2 km are heavily vegetated, being fed by rains >1 m/yr carried by the prevailing easterlies (Zhou and Lau, 1998; Strecker et al., 2007). Above 2 km elevation, rainfall decreases rapidly. The Subandes ascend into the Eastern Cordillera, which attains elevations of >6 km. The Eastern Cordillera bounds the eastern edge of the Puna Plateau, hydrographically isolating the Puna basins and blocking the easterlies at this latitude. The Eastern Cordillera and Subandes are underlain by mainly Proterozoic and Lower Paleozoic clastic sedimentary rocks and Ordovician granitoids. These are locally capped by Cretaceous-age rift deposits of the Salta Group, and above them, by thick Cenozoic foreland basin deposits.

The Puna Plateau is the southern extension of the central Andean Plateau; the Altiplano Plateau in Bolivia forms the broader northern extension. The Puna Plateau stands at ~4 km average elevation, and the climate is arid to hyperarid due to partial interdiction of easterly moisture by the Eastern Cordillera. The Puna Plateau consists of N-S–trending basins bounded by thrust faults and filled by coarse clastic rocks and evaporites.

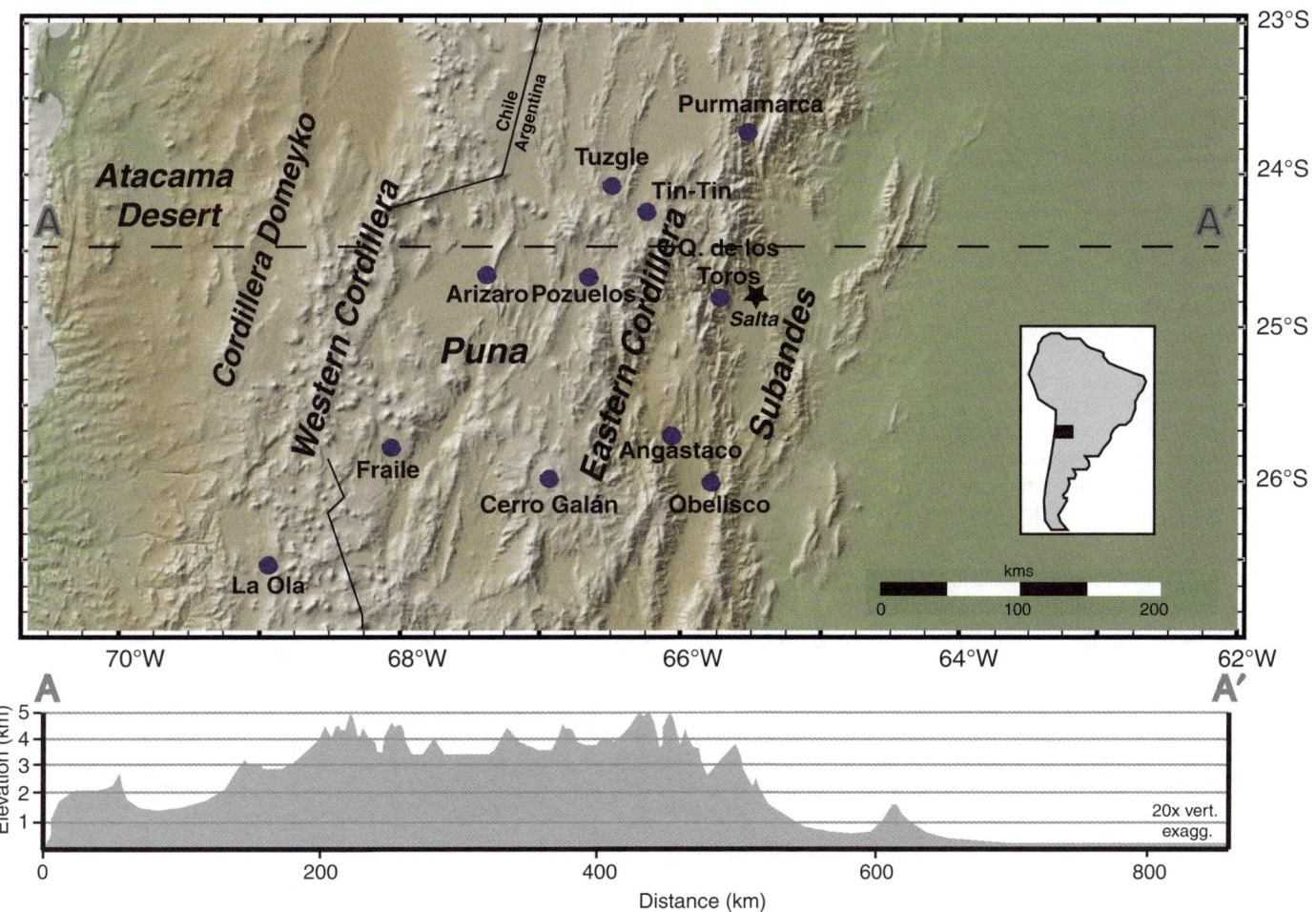

Figure 1. The major physiographic provinces of the central Andes and the main study locations.

The ranges reach >6 km elevations and are dominated by Lower Paleozoic sedimentary and igneous rocks; Cenozoic sedimentary rocks and volcanic rocks, including many voluminous late Neogene–age ignimbrites, are also widespread.

Bounding the western Puna region, the Western Cordillera is the main magmatic arc of the Andes. The arc consists of stratovolcanoes, flows, and ignimbrite sheets perched on top of the western edge of the otherwise moderate-relief Puna. Arid to hyperarid conditions prevail due to the nearly complete interdiction of the easterlies by the eastern Andes.

The Atacama Desert, a Martian-like region that is driest in this sector at 24°S–26°S, lies in the forearc west of the Andes. The forearc is underlain by Eocene and older intrusive, extrusive, and metasedimentary rocks, capped by mainly Neogene coarse clastic sedimentary rocks, evaporites, ignimbrites, and tephras.

Geologic Background and Sampling 22°S–26°S

The geologic history of this sector of the Andes has been substantially revised and expanded in the past few decades. The duration and locations of deformation and magmatism are sufficiently understood that we can compare this record to our evidence for the development of paleoelevation across the region. Here, we describe our sampling campaign in the context of a brief summary of the geologic evolution of the Andes at 22°S–26°S since the Late Cretaceous, where the geologic record is most complete.

The western margin of the central Andes between 22°S and 26°S has undergone some form of subduction and compression since the opening of the South Atlantic during the Early Cretaceous (Torsvik et al., 2009). In response, deformation (and uplift?) of the Andes at this and adjacent latitudes began in the west and gradually expanded eastward. This eastward progression is visible in a variety of structural (Mpodozis et al., 2005; Hongn et al., 2007; DeCelles et al., 2011; Carrapa et al., 2011b), paleotopographic and provenance (Hain et al., 2011; DeCelles et al., 2011), and thermochronologic (Deeken et al., 2006; Coutand et al., 2006; Carrapa and DeCelles, 2008; Carrapa et al., 2011b; Carrapa and DeCelles, this volume; Reiners et al., this volume) evidence. During the Eocene, a combination of arc magmatism and deformation created what we will term the "Incaic highlands," which stretched from the Cordillera de Domeyko in Chile eastward probably across much of the Puna Plateau (Fig. 1; Mpodozis et al., 2005; Trumbull et al., 2006; DeCelles et al., 2007; Carrapa et al., 2011b). To reconstruct absolute paleoelevations of the Incaic highlands, we used the δD value of waters of hydration in volcanic glass. This approach has been employed successfully in the western North American Cordillera to reconstruct Cenozoic paleoelevations (Mulch et al., 2008; Cassel et al., 2009). The method requires δD analysis of waters sealed in hydration rinds of volcanic glass to reconstruct the δD value of ancient water, which in turn is a function of elevation. We analyzed well-preserved glass from numerous ash-fall tuffs in basin deposits around Salar de Fraile and Salar de Arizaro in western

Argentina (Fig. 1). U-Pb dating of volcanic zircons in these tuffs provided the age control on our paleoelevation reconstructions from these tuffs.

During the Paleogene, a broad foreland basin stretched eastward of the Western Cordillera and Cordillera de Domeyko (the "Incaic highlands"), starting at about of 67°W. Thick deposits belonging the Santa Bárbara and Metán Subgroups were shed from the highlands eastward into the foreland basin (Fig. 2; DeCelles et al., 2011; Carrapa et al., 2011a). These deposits contain abundant surficial carbonate, the isotopic composition of

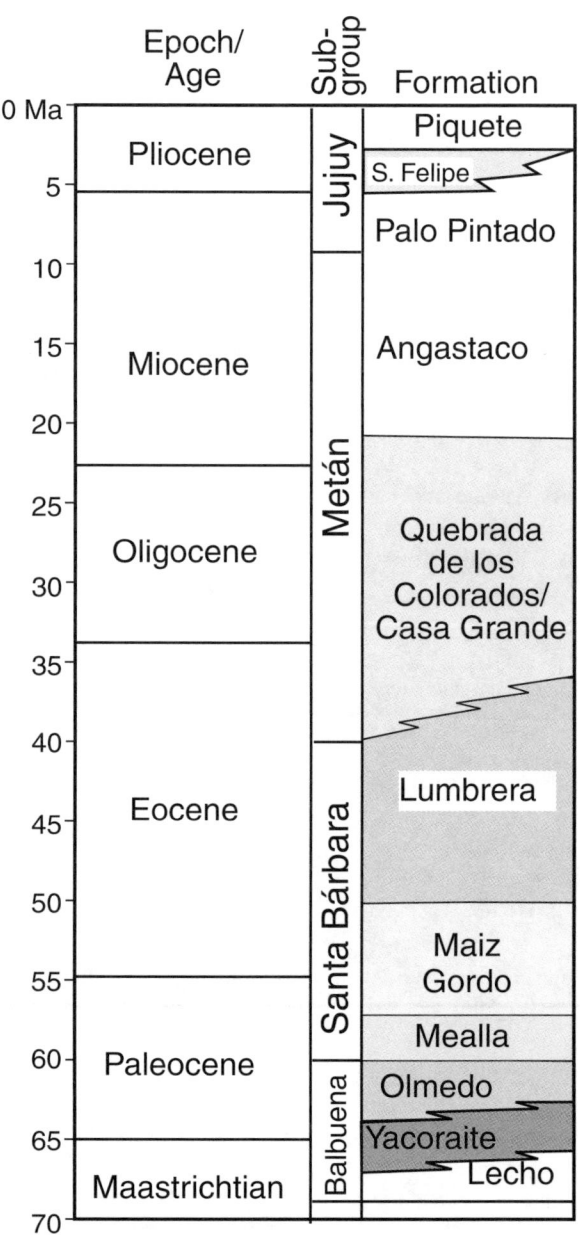

Figure 2. Stratigraphic chart of the eastern Puna and Cordillera, and Subandean zone of northwestern Argentina, modified from DeCelles et al. (2011).

which, if not reset during diagenesis, should record low-elevation paleoenvironments of the time. For this paper, we sampled paleosol carbonate nodules (Fig. 3B) from the back-bulge deposits of the Mealla and Maiz Gordo Formations (Santa Barbara Group) at the Obelisco (Fig. 3A), Tin-Tin, and San Antonio de los Cobres sections (Fig. 3C); and the foredeep deposits of the Quebrada de los Colorados Formation (Métan Subgroup) at the San Antonio de los Cobres and Tin-Tin sections (Fig. 1). Paleosol carbonate is abundant in these formations, providing a means of establishing the low-elevation isotopic composition of paleorainfall ($\delta^{18}O$), paleotemperature (Δ_{47}), and paleovegetation ($\delta^{13}C$). Together, these records allow us to correct our paleoelevation reconstructions of the Incaic highlands for any major changes in paleoclimate during the Paleogene.

Deformation and uplift spread eastward during the Neogene, raising the proximal portion of the former foreland up into what is now the Eastern Cordillera and Subandes. Again, volcanic glass is widely available to chronicle these changes and establish absolute paleoelevations, sampled from in and around Tuzgle, Cerro Galàn, La Ola, San Antonio de los Cobres, and other locations (Fig. 1). Geologic evidence for hydrographic isolation and aridification of montane basins, also compiled here, assists in reconstructing the eastward expansion and uplift of the eastern Andes during the Neogene.

METHODS

Isotopic Analysis of Carbonate

Pedogenic carbonates were scraped from alluvial or bedrock clasts or sampled from nodules. Carbonate analyzed for $\delta^{18}O$ and $\delta^{13}C$ values was heated at 250 °C for 3 h in vacuo before stable isotopic analysis using an automated sample preparation device (Kiel III) attached directly to a Finnigan MAT 252 mass spectrometer at the University of Arizona. Measured $\delta^{18}O$ and $\delta^{13}C$ values were corrected using internal laboratory standards calibrated to NBS-19. Precision of repeated standards is ±0.11‰ for $\delta^{18}O$ (1σ). Carbonate isotopic results are reported using standard delta per mil notation relative to Vienna Peedee belemnite (VPDB; Table 1).

Carbonate was analyzed for clumped isotope (Δ_{47}) thermometry at the California Institute of Technology. Samples were digested in phosphoric acid at 90 °C and purified using an automated sample preparation device coupled to a Thermo MAT 253 mass spectrometer configured to measure m/z 44–49 (Eiler and Schauble, 2004), following the methods of Passey et al. (2010). Measured Δ_{47} values were normalized and corrected for scale compression using heated gases (Huntington et al., 2009), corrected for 90 °C acid digestion (0.081‰; Passey et al., 2010), and converted to the absolute reference frame (Dennis et al., 2011) using heated CO_2 and carbonate standards (see GSA Data Repository for details[1]). Corrected Δ_{47} values were converted to temperature using both the Caltech (Ghosh at al., 2006b) and Harvard (Dennis and Schrag, 2010) calibration curves, translated to the absolute reference frame by Dennis et al. (2011). Temperature estimates (TΔ_{47}) and $\delta^{18}O$ and $\delta^{13}C$ values of carbonate (VPDB) from clumped isotope measurements are summarized in Table 2, along with estimates of the $\delta^{18}O$ values (relative to Vienna standard mean ocean water [VSMOW]) of water calculated from the measured carbonate (TΔ_{47}) and $\delta^{18}O$ values using the equation of Kim and O'Neil (1997). Temperatures discussed in the text were calculated using the Caltech calibration, with temperatures calculated using the Harvard calibration in parentheses.

Isotopic Analysis of Volcanic Glass

Each 100–200 g sample of tuff (air-fall or ash-flow) was purified for analysis following procedures described in Dettinger and Quade (this volume). For analysis, 2–3 mg of glass were packed in silver foil. The δD_{vg} values were determined using an automated thermal conversion/element analyzer (TCEA) coupled to a Delta V Plus isotope ratio mass spectrometer with an analytical precision of ±2.5‰ (2σ). The isotope ratios were calibrated using an internal standard and the international standards NBS-30 biotite and IAEA-CH-7 polyethylene foil. All δD values are reported here in standard delta notation and referenced to SMOW (Table 3).

Age Control

The ages of ash-fall and ash-flow tuffs were determined by U-Pb dating of volcanic zircons. Grain ages were determined by multicollector–laser ablation–inductively coupled plasma–mass spectrometry (MC-LA-ICP-MS) at the University of Arizona LaserChron Center. Individual zircon grains were ablated with a New Wave DUV193 Excimer laser (operating at a wavelength of 193 nm) using a spot diameter of 35 μm. The ablated material is carried in He gas into the plasma source of a Micromass Isoprobe, which is equipped with a flight tube of sufficient width that U, Th, and Pb isotopes are measured simultaneously. All measurements were made in static mode, using Faraday detectors for ^{238}U, ^{232}Th, ^{208}Pb–^{206}Pb, and an ion-counting channel for ^{204}Pb. Ion yields were ~1 mv per ppm. Each analysis consisted of one 20 s integration on peaks with the laser off (for backgrounds), twenty 1 s integrations with the laser firing, and a 30 s delay to purge the previous sample and prepare for the next analysis. The ablation pit was ~20 m deep. Common Pb correction was made by using the measured ^{204}Pb and assuming an initial Pb composition from Stacey and Kramers (1975) (with uncertainties of 1.0 for $^{206}Pb/^{204}Pb$ and 0.3 for $^{207}Pb/^{204}Pb$). Our measurement of ^{204}Pb was unaffected by the presence of ^{204}Hg because backgrounds were measured on peaks (thereby subtracting any background

---

[1]GSA Data Repository Item 2015007, Heated gas and carbonate standard data used to standardize clumped isotope data (Huntington et al., 2009) and convert data to the absolute reference frame (Dennis et al., 2011), is available at www.geosociety.org/pubs/ft2015.htm, or on request from editing@geosociety.org or Documents Secretary, GSA, P.O. Box 9140, Boulder, CO 80301-9140, USA.

Figure 3. Examples of materials and locations sampled for paleoelevation reconstruction. (A) Stacked vertisols from the Mealla-Maiz Gordo Formation at Obelisco (Fig. 1). These paleosols contain abundant soil carbonate. (B) Detail of soil carbonate nodules from the red beds in part C. (C) Exposure near San Antonio de los Cobres (Fig. 1) showing red beds of the Quebrada de los Colorados Formation at the bottom of the photo, and above that, Neogene tuffs resting on the red beds along an angular unconformity. The basal white tuff dates to 7.3 Ma (see Table 3, SL-41). (D) Stacked ignimbrites of Neogene age near Cerro Galán sampled for δD analysis of volcanic glass: Blanco (B), the lower Merihuaca (LM), the Real Grande (RG), and the Galán (G) ignimbrites (see Table 3). (E) View of the Salar de Fraile Basin (Fig. 1). The red beds in the background belong to the Quiñoas Formation and contain evaporites as far back as the late Eocene. (F) Detail of ash-fall tuff with well-preserved glass in the Eocene–Oligocene Quiñoas Formation at Salar de Fraile.

TABLE 1. STABLE ISOTOPIC RESULTS FROM CARBONATES

Samples	Formation	Location	Latitude (°S)	Longitude (°W)	Elevation (m)	Local thickness (m)	$\delta^{13}C$ (‰, PDB)	$\delta^{18}O$ (‰, PDB)	Age (Ma)	Material
SL44A	Mealla Fm.	Obelisco	25.98330	65.74860	1540	3	-6.24	-4.29	61	Stromatolite
SL44B	Mealla Fm.	Obelisco	25.98330	65.74860	1540	3	-3.55	-3.83	61	Stromatolite
SL44c	Mealla Fm.	Obelisco	25.98330	65.74860	1540	3	-2.14	-7.95	61	Stromatolite
SL44D	Mealla Fm.	Obelisco	25.98330	65.74860	1552	14	-3.43	-6.79	61	Intraclastic breccias
SL-45C(A)	Mealla Fm. ?	Obelisco	26.00064	65.78368	1558	272	-9.05	-5.63	63	Vertisol carbonate
SL-45C(B)	Mealla Fm. ?	Obelisco	26.00064	65.78368	1558	272	-9.10	-5.93	63	Vertisol carbonate
SL-45F(A)	Mealla Fm. ?	Obelisco	26.00064	65.78368	1558	280	-9.82	-6.27	63	Vertisol carbonate
SL-45F(B)	Mealla Fm. ?	Obelisco	26.00064	65.78368	1558	280	-9.11	-5.91	63	Vertisol carbonate
SL-45A(A)	Mealla Fm. ?	Obelisco	26.00064	65.78368	1558	267	-9.23	-5.67	63	Vertisol carbonate
SL-45A(B)	Mealla Fm. ?	Obelisco	26.00064	65.78368	1558	267	-9.04	-6.07	63	Vertisol carbonate
SL-45A(C)	Mealla Fm. ?	Obelisco	26.00064	65.78368	1558	267	-8.43	-6.03	63	Vertisol carbonate
SL-45D(A)	Mealla Fm. ?	Obelisco	26.00064	65.78368	1558	275	-9.36	-6.01	63	Vertisol carbonate
SL-45D(B)	Mealla Fm. ?	Obelisco	26.00064	65.78368	1558	275	-8.81	-5.63	63	Vertisol carbonate
SL-45B(A)	Mealla Fm. ?	Obelisco	26.00064	65.78368	1558	271	-8.72	-5.56	63	Vertisol carbonate
SL-45B(B)	Mealla Fm. ?	Obelisco	26.00064	65.78368	1558	271	-8.69	-5.38	63	Vertisol carbonate
SL-45E(A)	Mealla Fm. ?	Obelisco	26.00064	65.78368	1558	278	-8.58	-5.92	63	Vertisol carbonate
SL-45E(B)	Mealla Fm. ?	Obelisco	26.00064	65.78368	1558	278	-8.46	-5.85	63	Vertisol carbonate
SL-47A	Mealla Fm. ?	Obelisco	26.01087	65.78390	1563	110	-7.99	-5.59	63	Vertisol carbonate
SL-47B	Mealla Fm. ?	Obelisco	26.01087	65.78390	1563	110	-8.11	-5.81	63	Vertisol carbonate
SL-47C	Mealla Fm. ?	Obelisco	26.01087	65.78390	1563	110	-7.87	-7.46	63	Vertisol carbonate
SL-59A	Maiz Gordo Fm.	San Antonio	24.22901	66.25286	3850	87	-15.43	-2.99	55	Groundwater calcrete
SL-59B	Maiz Gordo Fm.	San Antonio	24.22901	66.25286	3850	87	-15.65	-2.56	55	Groundwater calcrete
SL-60B	Maiz Gordo Fm.	San Antonio	24.22906	66.25277	3840	97.5	-15.17	-4.37	55	Groundwater calcrete
SL-60C	Maiz Gordo Fm.	San Antonio	24.22906	66.25277	3840	97.5	-9.06	-2.48	55	Groundwater calcrete
SL-61A	Maiz Gordo Fm.	San Antonio	24.22906	66.25277	3840	97.5	-16.05	-2.34	55	Groundwater calcrete
SL-61B	Maiz Gordo Fm.	San Antonio	24.22906	66.25277	3840	97.5	-16.20	-3.36	55	Groundwater calcrete
SL-61C	Maiz Gordo Fm.	San Antonio	24.22906	66.25277	3840	97.5	-16.36	-2.09	55	Groundwater calcrete
SL62	Maiz Gordo Fm.	San Antonio	24.22924	66.25243	3844	124	-13.25	-4.76	55	Groundwater calcrete
SL63	Maiz Gordo Fm.	San Antonio	24.22928	66.25199	3848	154	-10.19	-0.99	55	Reworked soil carbonate

(Continued)

TABLE 1. STABLE ISOTOPIC RESULTS FROM CARBONATES (*Continued*)

Samples	Formation	Location	Latitude (°S)	Longitude (°W)	Elevation (m)	Local thickness (m)	δ¹³C (‰, PDB)	δ¹⁸O (‰, PDB)	Age (Ma)	Material
SL64	Maiz Gordo Fm.	San Antonio	24.22928	66.25199	3848	154	-3.78	-9.36	55	Reworked soil carbonate
SL65	Maiz Gordo Fm.	San Antonio	24.22936	66.25187	3849	161	-6.57	-2.62	55	Reworked soil carbonate
SL68	Maiz Gordo Fm.	San Antonio	24.22973	66.25157	3853	188	-8.66	-3.18	55	Soil carbonate
SL69	Maiz Gordo Fm.	San Antonio	24.22973	66.25157	3853	190	-9.83	-7.48	55	Soil carbonate
SL-70A	Maiz Gordo Fm.	San Antonio	24.22973	66.25157	3853	194	-7.41	-2.59	55	Soil carbonate
SL-70B	Maiz Gordo Fm.	San Antonio	24.22973	66.25157	3853	194	-7.79	-2.76	55	Soil carbonate
SL-70C	Maiz Gordo Fm.	San Antonio	24.22973	66.25157	3853	194	-7.27	-1.88	55	Soil carbonate
SL-71	Maiz Gordo Fm.	San Antonio	24.23009	66.25122	3864	229	-7.90	-3.50	55	Carbonate in supersol
SL-72	Maiz Gordo Fm.	San Antonio	24.23009	66.25122	3864	221	2.65	-2.72	55	Carbonate in supersol
SL-73A	Maiz Gordo Fm.	San Antonio	24.23009	66.25122	3864	221	-10.47	-8.45	55	Carbonate in supersol
SL-73B	Maiz Gordo Fm.	San Antonio	24.23009	66.25122	3864	221	-10.99	-7.00	55	Carbonate in supersol
SL-73C	Maiz Gordo Fm.	San Antonio	24.23009	66.25122	3864	300	-11.41	-7.46	37	Soil carbonate
SL-74A	Q. de los Colorados	San Antonio, lower	24.23262	66.2457	3847	300	-7.27	-6.92	37	Soil carbonate
SL-74B	Q. de los Colorados	San Antonio, lower	24.23262	66.2457	3847	300	-7.77	-6.73	37	Soil carbonate
SL-74C	Q. de los Colorados	San Antonio, lower	24.23262	66.2457	3847	302	-7.72	-6.74	37	Soil carbonate
SL75	Q. de los Colorados	San Antonio, lower	24.23262	66.2457	3847	400	-7.45	-7.82	37	Soil carbonate
SL-29A	Q. de los Colorados	San Antonio, lower	24.24091	66.24138	3836	400	-8.16	-5.97	37	Soil carbonate
SL-29B	Q. de los Colorados	San Antonio, lower	24.24091	66.24138	3836	400	-8.23	-5.69	37	Soil carbonate
SL-29C	Q. de los Colorados	San Antonio, lower	24.24091	66.24138	3836	410	-8.03	-5.18	37	Soil carbonate
SL-30A	Q. de los Colorados	San Antonio, lower	24.24091	66.24138	3836	410	-7.35	-5.63	37	Soil carbonate
SL-30B	Q. de los Colorados	San Antonio, lower	24.24091	66.24138	3836	410	-7.07	-4.56	37	Soil carbonate
SL-30C	Q. de los Colorados	San Antonio, lower	24.24091	66.24138	3836	410	-7.64	-5.57	37	Soil carbonate
SL-136A-1	Q. de los Colorados	San Antonio, upper	24.23589	66.24085	3837	430	-8.17	-4.11	33	Soil carbonate
SL-136A-2	Q. de los Colorados	San Antonio, upper	24.23589	66.24085	3837	430	-8.53	-4.78	33	Soil carbonate
SL-136A-3	Q. de los Colorados	San Antonio, upper	24.23589	66.24085	3837	430	-8.19	-4.15	33	Soil carbonate
SL-136A-4	Q. de los Colorados	San Antonio, upper	24.23589	66.24085	3837	430	-8.47	-4.58	33	Soil carbonate
SL-136B-1	Q. de los Colorados	San Antonio, upper	24.23589	66.24085	3837	460	-8.08	-4.30	31	Soil carbonate
SL-136B-2	Q. de los Colorados	San Antonio, upper	24.23589	66.24085	3837	460	-8.63	-5.15	31	Soil carbonate
SL-136B-3	Q. de los Colorados	San Antonio, upper	24.23589	66.24085	3837	460	-8.14	-4.57	31	Soil carbonate

(*Continued*)

TABLE 1. STABLE ISOTOPIC RESULTS FROM CARBONATES (*Continued*)

Samples	Formation	Location	Latitude (°S)	Longitude (°W)	Elevation (m)	Local thickness (m)	$\delta^{13}C$ (‰, PDB)	$\delta^{18}O$ (‰, PDB)	Age (Ma)	Material
SL-136B-4	Q. de los Colorados	San Antonio, upper	24.23589	66.24085	3837	460	−8.30	−4.81	31	Soil carbonate
SL-136C-1	Q. de los Colorados	San Antonio, upper	24.23589	66.24085	3837	560	−7.69	−5.47	27	Soil carbonate
SL-136C-2	Q. de los Colorados	San Antonio, upper	24.23589	66.24085	3837	560	−7.91	−4.15	27	Soil carbonate
SL-136C-3	Q. de los Colorados	San Antonio, upper	24.23589	66.24085	3837	560	−7.88	−4.32	27	Soil carbonate
SL-136C-4	Q. de los Colorados	San Antonio, upper	24.23589	66.24085	3837	560	−8.05	−4.36	27	Soil carbonate
SL-135-A1	Q. de los Colorados	Tin-Tin	25.24116	66.07684	2740		−9.02	−6.85	43	Soil carbonate
SL-135-A2	Q. de los Colorados	Tin-Tin	25.24116	66.07684	2740		−9.17	−7.72	43	Soil carbonate
SL-135-A3	Q. de los Colorados	Tin-Tin	25.24116	66.07684	2740		−8.88	−6.51	43	Soil carbonate
SL-135-B1	Q. de los Colorados	Tin-Tin	25.24116	66.07684	2740		−9.18	−6.93	43	Soil carbonate
SL-135-B2	Q. de los Colorados	Tin-Tin	25.24116	66.07684	2740		−9.23	−8.37	43	Soil carbonate
SL-135-B3	Q. de los Colorados	Tin-Tin	25.24116	66.07684	2740		−9.08	−8.02	43	Soil carbonate
SL-135-C1	Q. de los Colorados	Tin-Tin	25.24116	66.07684	2740		−9.52	−6.47	43	Soil carbonate
SL-135-C2	Q. de los Colorados	Tin-Tin	25.24116	66.07684	2740		−9.10	−7.15	43	Soil carbonate
SL-135-C3	Q. de los Colorados	Tin-Tin	25.24116	66.07684	2740		−9.03	−6.75	43	Soil carbonate
SL-135-D1	Q. de los Colorados	Tin-Tin	25.24116	66.07684	2740		−9.11	−6.30	43	Soil carbonate
SL-135-D2	Q. de los Colorados	Tin-Tin	25.24116	66.07684	2740		−9.19	−6.56	43	Soil carbonate
SL-135-D3	Q. de los Colorados	Tin-Tin	25.24116	66.07684	2740		−9.46	−6.63	43	Soil carbonate
SL-135-E1	Q. de los Colorados	Tin-Tin	25.24116	66.07684	2740		−9.39	−5.55	43	Soil carbonate
SL-135-E2	Q. de los Colorados	Tin-Tin	25.24116	66.07684	2740		−8.99	−5.86	43	Soil carbonate
SL-135-E3	Q. de los Colorados	Tin-Tin	25.24116	66.07684	2740		−8.88	−5.89	43	Paleosol nodules
SL-134A-2	Lumbrera supersol	Sierra de Hornocal	23.16314	65.08318	4043		−10.75	−7.55	50	Sandstone cement
SL-134A-3	Lumbrera supersol	Sierra de Hornocal	23.16314	65.08318	4043		−9.11	−14.68	50	Sandstone cement
SL-134A-4	Lumbrera supersol	Sierra de Hornocal	23.16314	65.08318	4043		−8.99	−12.25	50	Sandstone cement
SL-134B-1	Lumbrera supersol	Sierra de Hornocal	23.16314	65.08318	4043		−9.59	−9.56	50	Sandstone cement
SL-134B-2	Lumbrera supersol	Sierra de Hornocal	23.16314	65.08318	4043		−10.61	−8.78	50	Sandstone cement
SL-134B-3	Lumbrera supersol	Sierra de Hornocal	23.16314	65.08318	4043		−9.70	−9.55	50	Sandstone cement
SL-134B-4	Lumbrera supersol	Sierra de Hornocal	23.16314	65.08318	4043		−11.27	−6.54	50	Sandstone cement
SL-134C-1	Lumbrera supersol	Sierra de Hornocal	23.16314	65.08318	4043		−10.17	−9.80	50	Sandstone cement
SL-134C-2	Lumbrera supersol	Sierra de Hornocal	23.16314	65.08318	4043		−10.54	−5.94	50	Sandstone cement

(*Continued*)

TABLE 1. STABLE ISOTOPIC RESULTS FROM CARBONATES (Continued)

Samples	Formation	Location	Latitude (°S)	Longitude (°W)	Elevation (m)	Local thickness (m)	$\delta^{13}C$ (‰, PDB)	$\delta^{18}O$ (‰, PDB)	Age (Ma)	Material
SL-134C-3	Lumbrera supersol	Sierra de Hornocal	23.16314	65.08318	4043		-10.66	-5.83	50	Sandstone cement
SL-134C-4	Lumbrera supersol	Sierra de Hornocal	23.16314	65.08318	4043		-10.75	-6.02	50	Sandstone cement
SL-134D-1	Lumbrera supersol	Sierra de Hornocal	23.16314	65.08318	4043		-10.16	-8.31	50	Sandstone cement
SL-134D-2	Lumbrera supersol	Sierra de Hornocal	23.16314	65.08318	4043		-10.27	-9.31	50	Sandstone cement
SL-134D-3	Lumbrera supersol	Sierra de Hornocal	23.16314	65.08318	4043		-11.13	-6.92	50	Sandstone cement
SL-81A	Sijes Fm.	S. Pastos Grandes	24.67004	66.65445	3956		1.49	-1.13	5	Lacustrine siltstone
SL-81B	Sijes Fm.	S. Pastos Grandes	24.67004	66.65445	3956		-1.37	-5.38	5	Lacustrine siltstone
SL-81C	Sijes Fm.	S. Pastos Grandes	24.67004	66.65445	3956		-0.95	-1.52	5	Lacustrine siltstone
SL-89A	Sijes Fm.	S. Pastos Grandes	24.67581	66.65656	3963		9.31	4.76	5	Lacustrine siltstone
SL-89B	Sijes Fm.	S. Pastos Grandes	24.67581	66.65656	3963		7.46	4.44	5	Lacustrine siltstone
SL-90	Sijes Fm.	S. Pastos Grandes	24.67425	66.65434	3955		-3.42	3.70	5	Lacustrine siltstone
SL-91	Pozuelos Fm.	S. Pastos Grandes	24.69445	66.69913	3858		-4.57	-0.33	7	Sandstone cement
SL-92	Pozuelos Fm.	S. Pastos Grandes	24.69415	66.69955	3869		-3.02	1.38	7	Travertine
SL-93	Pozuelos Fm.	S. Pastos Grandes	24.69401	66.69985	3874		-5.12	1.81	7	Travertine
SL-95	Pozuelos Fm.	S. Pastos Grandes	24.69464	66.70111	3865		-3.98	-0.56	7	Travertine
SL-96	Pozuelos Fm.	S. Pastos Grandes	24.69464	66.70111	3865		-3.77	-3.49	7	Travertine
SL-98	Pozuelos Fm.	S. Pastos Grandes	24.69467	66.70199	3857		-3.15	-4.78	7	Travertine
SL-94A	Pozuelos Fm.	S. Pastos Grandes	24.69464	66.70111	3865		-1.98	-3.97	7	Travertine
SL-94B	Pozuelos Fm.	S. Pastos Grandes	24.69464	66.70111	3865		-1.77	-4.22	7	Travertine
SL-94C	Pozuelos Fm.	S. Pastos Grandes	24.69464	66.70111	3865		-1.01	-4.08	7	Travertine
SL-94D	Pozuelos Fm.	S. Pastos Grandes	24.69464	66.70111	3865		-0.93	-4.11	7	Travertine
SL-94E	Pozuelos Fm.	S. Pastos Grandes	24.69464	66.70111	3865		-2.87	-2.07	7	Travertine
SL-94BLACKA	Pozuelos Fm.	S. Pastos Grandes	24.69464	66.70111	3865		-5.23	-1.03	7	Travertine
SL103B	Modern carbonate	Pasto Chico	23.60313	66.43287	3751		3.39	-7.04	0	Biocarbonate
SL103D	Modern carbonate	Pasto Chico	23.60313	66.43287	3751		3.17	-4.45	0	Biocarbonate
SL103E	Modern carbonate	Pasto Chico	23.60313	66.43287	3751		-3.68	-4.15	0	Biocarbonate
SL53	Palo Pintado Fm.	Angastaco Basin	25.67815	66.09182	1835		-10.30	-6.21	7	Soil carbonate
SL54	Palo Pintado Fm.	Angastaco Basin	25.67815	66.09182	1835		-8.44	-5.39	7	Soil carbonate
SL100	Blanca Lila Fm.	Pasto Chico	24.52308	66.71008	3797		-5.80	-3.81	1	Marl

(Continued)

TABLE 1. STABLE ISOTOPIC RESULTS FROM CARBONATES (Continued)

Samples	Formation	Location	Latitude (°S)	Longitude (°W)	Elevation (m)	Local thickness (m)	δ13C (‰, PDB)	δ18O (‰, PDB)	Age (Ma)	Material
SL102	Blanca Lila Fm.	Pasto Chico	24.24771	66.36310	3836		2.44	−9.33	1	Travertine
SL109	Yacoraite Fm.	Purmamarca	23.70706	65.53575	2628		0.40	1.87	65	Stromatolite
SL 109 S1	Yacoraite Fm.	Purmamarca	23.70706	65.53575	2628		1.00	5.18	65	Stromatolite
SL-109-S2	Yacoraite Fm.	Purmamarca	23.70706	65.53575	2628		0.71	4.16	65	Stromatolite
SL 109 M1	Yacoraite Fm.	Purmamarca	23.70706	65.53575	2628		0.69	1.29	65	Micrite
SL 109 S3	Yacoraite Fm.	Purmamarca	23.70706	65.53575	2628		1.00	2.55	65	Stromatolite
SL 109 M2	Yacoraite Fm.	Purmamarca	23.70706	65.53575	2628		0.40	0.46	65	Micrite
SL 109 OOLITE1	Yacoraite Fm.	Purmamarca	23.70706	65.53575	2628		1.07	1.02	65	Oolite
SL 109 M3	Yacoraite Fm.	Purmamarca	23.70706	65.53575	2628		0.76	−4.40	65	Micrite
SL 109 OOLITE 2	Yacoraite Fm.	Purmamarca	23.70706	65.53575	2628		0.29	0.73	65	Oolite

Note: PDB—Peedee belemnite.

TABLE 2. SUMMARY OF CLUMPED ISOTOPIC DATA FOR YACORAITE (MAASTRICHTIAN) AND MEALLA (PALEOCENE) FORMATION SAMPLES

Sample	Δ47 (‰, Caltech)	±1 SE (‰, analytical)	Δ47 (‰, ARF)	T(Δ47) (°C, ARF, Ghosh)	T(Δ47) (°C, ARF, D&S)	δ13C (‰, measured)	δ18Ocarb (‰, measured)	δ18OH2O (‰), calc. (ARF, Ghosh)	δ18OH2O (‰), calc. (ARF, D&S)
Soil carbonate, Mealla Formation (Obelisco)									
SL45B	0.601	0.008	0.650	39	45	−8.79	−5.65	−0.6	0.5
SL47	0.510	0.011	0.556	64	97	−7.90	−5.81	3.6	8.3
SL45c	0.594	0.010	0.643	40	48	−8.81	−5.57	−0.3	1.1
SL46	0.415	0.009	0.459	97	193	−7.64	−5.58	8.6	18.8
SL44a	0.584	0.011	0.632	43	53	−5.85	−3.17	2.7	4.4
SL44c	0.505	0.010	0.551	65	101	−1.80	−8.35	1.1	6.3
Marine/lacustrine carbonate, Yacoraite Formation									
SL109M1	0.564	0.011	0.612	48	63	0.77	1.37	8.1	10.6
SL109M2	0.568	0.008	0.615	47	61	0.49	−0.53	6.0	8.4
SL109S	0.564	0.009	0.611	48	64	0.59	3.07	9.8	12.5

Note: All samples were reacted at 90 °C, with 0.081‰ acid digestion correction. δ18OH2O was calculated using values from Kim and O'Neil (1997) for calcite. Analytical precisions are reported for analysis of a single sample aliquot. Precision in δ18O and δ13C is 0.02‰ and 0.01‰ or better (1 s.d.). Transfer function from Caltech reference frame to absolute reference frame: Δ47,ARF = 1.0292 (D47,Caltech) + 0.0312 (R^2 = 0.9995). Long-term s.d. in Δ47 for standards over the analytical session is 0.009‰ to 0.012‰ (Carrara Marble), on par with the analytical precision of the samples. ARF—absolute reference frame; Ghosh—Ghosh et al., 2006b; D&S—Dennis and Shrag, 2010.

TABLE 3. DEUTERIUM ISOTOPE ANALYSES AND ASSOCIATED PALEOELEVATION RECONSTRUCTIONS FROM VOLCANIC GLASS

Sample no.	Modern elevation (m)	Latitude (°S)	Longitude (°W)	δD glass analysis 1 (‰)	δD glass analysis 2 (‰)	δD glass average (‰)	δD water average (‰)	Age (Ma)	Paleo-elevation* (m) linear	Paleo-elevation* (m) polynomial	Paleo-elevation† (m) polynomial	Tuff or formation
Eastern Andes												
SL-84	3945	24.67255	66.65714	-110	-113	-111	-81	5	4958	4512	4401	Tuff, lower Sijes Fm.
SL-36A	3621	23.40156	66.35854	-98	-98	-98	-67	8	4069	3986	4081	Tuff, Susques
SL-42A	1408	25.85466	65.70324	-69	-66	-67	-35	0.01	1965	2057	2118	Tuff, Holocene
SL-51	1805	25.69376	66.03663	-61	-61	-61	-29	7	1523	1529	1489	Tuff, Palo Pintado Fm.
SL-51	1805	25.69376	66.03663	-64	-61	-62	-29	7	1576	1595	1569	Tuff, Palo Pintado Fm.
SL-22	3051	24.45889	65.93869	-87	-86	-86	-55	0.01	3263	3362	3530	Tuff, Holocene
SL-38	4041	23.99760	66.50781	-111	-113	-112	-82	0.3	5027	4545	4413	Tuzgle ignimbrite
SL-57	1311	25.71498	65.70011	-49	-53	-51	-19	0.01	855	651	398	Tuff, Holocene
SL-58	2773	24.49417	65.88216	-66	-70	-68	-36	0.01	1983	2078	2142	Tuff, Holocene
SL-41	3866	24.22840	66.50781	-117	-116	-116	-86	7.3	5317	4675	4444	Tuff, San Antonio de los Cobres
SL-38	4041	23.99760	66.50781	-109	-114	-111	-81	0.3	4958	4512	4401	Tuzgle ignimbrite
SL-86	3975	24.67596	66.65868	-123	-125	-124	-94	5	5824	4857	4420	Tuff, lower Sijes Fm.
SL-84	3945	24.67255	66.65714	-110	-113	-111	-81	5	4958	4512	4401	Tuff, lower Sijes Fm.
SL-88	4013	24.67596	66.66057	-108	-110	-109	-79	5	4833	4448	4374	Tuff, lower Sijes Fm.
SL-55	1819	25.68088	66.07773	-78	-75	-76	-44	7	2565	2706	2851	Tuff, Palo Pintado Fm.
SL-97	3839	24.69528	66.70174	-99	-102	-101	-70	7	4258	4113	4174	Tuff, upper Pozuelos Fm.
SL-110	1872	27.22685	66.92015	-92	-94	-93	-62	6.7	3723	3736	3875	Tuff, Corral Quemado
SL-112	3028	26.49480	67.40940	-95		-95	-64	0.55	3857	3836	3960	Campo de Piedra Pumice
SL-115	3423	26.12967	67.41487	-59		-59	-27	0.001	1393	1366	1291	Young tuff, Laguna volcano
SL-117	3461	26.51837	67.70420	-75	-73	-74	-42	0.2	2394	2529	2655	Blanca Ignimbrite
SL-118	3417	26.05197	67.42115	-70	-68	-69	-37	2	2077	2185	2265	Galán Ignimbrite
SL-119	4501	25.97513	67.21388	-89	-91	-90	-59	2	3497	3558	3716	Galán Ignimbrite
SL-120	4473	25.97557	67.21407	-79	-83	-81	-49	5	2870	3006	3171	Tuff, Real Grande
SL-121	4339	25.97918	67.21622	-84	-84	-84	-53	5.6	3105	3222	3392	Tuff, lower Merihuaca
SL-122	4302	0.00000	0.00000	-110	-115	-112	-82	6	5054	4558	4417	Blanco Ignimbrite
SL-124	3492	25.99822	67.38950	-92	-86	-89	-58	0.01	3469	3535	3694	Young unnamed tuff
PVN226	3660			-84		-84	-53	7.88	3106	3224	3394	Unnamed tuff
PVN260	3663			-95		-95	-64	7.77	3856	3835	3960	Unnamed tuff
PVN123	3643			-104		-104	-73	9.90	4475	4248	4265	Unnamed tuff
AT2-007	1867			-73		-73	-42	4.81	2374	2508	2633	Unnamed tuff
VV-01	1400			-75		-75	-43	6.35	2463	2602	2736	Unnamed tuff
AT4-003	1847			-76		-76	-45	5.17	2589	2730	2877	Unnamed tuff
AT6-001	1890			-95		-95	-64	4.61	3843	3826	3952	Unnamed tuff
AT4-001	1820			-74		-74	-42	6.90	2390	2525	2651	Unnamed tuff
AT7-010	1857			-85		-85	-54	5.98	3188	3296	3466	Unnamed tuff
AT1-001a	1984			-75		-75	-43	7.24	2489	2628	2765	Unnamed tuff
AT3-009	1822			-75		-75	-43	4.04	2475	2614	2749	Unnamed tuff
LVT1-006	1200			-78		-78	-46		2660	2801	2954	Unnamed tuff

(Continued)

TABLE 3. DEUTERIUM ISOTOPE ANALYSES AND ASSOCIATED PALEOELEVATION RECONSTRUCTIONS FROM VOLCANIC GLASS (Continued)

Sample no.	Modern elevation (m)	Latitude (°S)	Longitude (°W)	δD glass analysis 1 (‰)	δD glass analysis 2 (‰)	δD glass average (‰)	δD water average (‰)	Age (Ma)	Paleo-elevation* (m) linear	Paleo-elevation* (m) polynomial	Paleo-elevation[†] (m) polynomial	Tuff or formation
Salar de Arizaro Basin												
A09-I-1	3737			−94	−91	−92	−61	5	3671	3696	3840	Unnamed tuff
A09-I-2	3559			−89	−95	−92	−61	34.8	3641	3673	3820	Unnamed tuff
ARB09-4	3807			−77	−72	−75	−43	1	2456	2594	2728	Unnamed tuff
ARB09-7	3782			−92	−86	−89	−57	1	3416	3491	3654	Unnamed tuff
ARB09-9	3833			−82	−82	−82	−50	0.4	2948	3079	3246	Unnamed tuff
ARB09-2	4140			−94	−95	−94	−63	18.8	3821	3809	3938	Unnamed tuff
ARB09-1	3827			−89	−90	−90	−59	19.5	3496	3557	3715	Unnamed tuff
ARB09-14	4185			−86	−102	−94	−63	17.9	3804	3797	3928	Unnamed tuff
ARB09-6	3763			−98	−97	−98	−67	13.9	4058	3979	4075	Unnamed tuff
ARB09-8	3821			−88	−90	−89	−57	15.8	3422	3496	3658	Unnamed tuff
Salar de Fraile Basin												
1SF49	3400	25.89020	68.09000	−102	—	−102	−71	38.5	4327	4157	4205	Tuff, Quinoas I Formation
5SF128	3522	25.89020	68.09000	−124	—	−124	−94	34.8	5841	4862	4418	Tuff, Quinoas I Formation
5SF257.5	3550	25.89020	68.09000	−103	—	−103	−72	33.2	4404	4205	4237	Tuff, Quinoas I Formation
5SF362	3564	25.89020	68.09000	−105	—	−105	−74	32.9	4541	4287	4288	Tuff, Quinoas I Formation
6SF128	3593	25.89020	68.09000	−124	—	−124	−94	31.1	5841	4862	4418	Tuff, Quinoas II Formation
6SF148	3600	25.89020	68.09000	−96	—	−96	−65	32.2	3952	3905	4017	Tuff, Quinoas II Formation
6SF284	3650	25.89020	68.09000	−116	—	−116	−86	31	5294	4665	4442	Tuff, Quinoas II Formation
6SF479	3675	25.89020	68.09000	−91	—	−91	−60	29.8	3576	3621	3773	Tuff, Quinoas II Formation
6SF664	3708	25.89020	68.09000	−102	—	−102	−71	24.2	4322	4154	4203	Tuff, Chacras Fm.
6SF753	3750	25.89020	68.09000	−142	—	−142	−113	23.3	7073	5067	3942	Tuff, Chacras Fm.
7SF9	3813	25.89020	68.09000	−122	−120	−121	−91	19.2	5629	4794	4441	Tuff, Portrero Grande Fm.

*This paper.
[†]Canavan et al. (2014).

[204]Hg and [204]Pb), and because very little Hg was present in the argon gas. Additional details about analytical procedures are described by Gehrels et al. (2008).

Interelement fractionation of Pb/U is generally ~15%, whereas fractionation of Pb isotopes is generally <2%. In-run analysis of fragments of a large zircon crystal (every sixth measurement) with known age of 563.5 ± 3.2 Ma (2σ error; Gehrels et al., 2008) was used to correct for this fractionation. Fractionation also increases with depth into the laser pit by up to 5%. This depth-related fractionation was accounted for by monitoring the fractionation observed in the standards. Analyses that displayed >10% change in ratio during the 20 s measurement were interpreted to be variable in age (or perhaps compromised by fractures or inclusions) and are excluded from further consideration. Also excluded are analyses that yielded >15% uncertainties in [206]Pb/[238]U ages or were >5% reverse discordant.

The measured ages of 12 tuff samples are reported in Table 4, with errors reported at the 2σ level. Ages for each sample are defined by analysis of 4–28 grains.

PALEOALTIMETRY

Isotopic Results and Implications from Carbonates

Carbonates were sampled from a variety of formations spanning most of the Cenozoic (Table 1; Fig. 4). The oldest of these come from the Maastrichtian–Paleocene (Danian) Yacoraite Formation of the Balbuena Group. Younger samples come from the Paleocene Mealla Formation and Paleocene–Eocene Maiz Gordo Formations of the Santa Bárbara Subgroup, and the overlying Eocene–Oligocene Quebrada de los Colorados Formation of the Metán Group. Here, we present $\delta^{18}O$, $\delta^{13}C$, and Δ_{47} results

TABLE 4. U-Pb (ZIRCON) GEOCHRONOLOGIC ANALYSES BY LASER-ABLATION–MULTICOLLECTOR–INDUCTIVELY COUPLED PLASMA–MASS SPECTROMETRY

Sample	No. of grains	Weighted mean age (Ma)	±2σ (%)	MSWD
1SF49	22	38.5	0.7 (1.8)	0.4
5SF128	18	34.7	0.7 (2.0)	0.2
5SF257.5	8	33.3	1.4 (4.3)	0.2
5SF362	11	32.9	2.9 (8.7)	0.1
6SF128	23	31.1	0.9 (2.8)	0.2
6SF148	4	32.3	2.5 (7.7)	0
6SF284	23	31	0.8 (2.6)	0.2
6SF479	20	29.8	0.3 (1.0)	6.7
6SF664	10	24.2	0.9 (3.5)	4.2
6SF753	11	23.3	1 (4.4)	0.1
7SF9	19	19.2	0.6 (3.1)	0.3
SL-122	28	5.8	0.1 (2.2)	9.7

Note: MSWD—mean square of weighted deviates.

from the carbonates, mostly from paleosols. Where unaltered by burial, $\delta^{18}O$ values from all carbonates provide information on paleoelevation and local climate; $\delta^{13}C$ values from paleosols relate directly to the nature of vegetation cover; and Δ_{47} results yield surface paleotemperatures, which in turn can provide estimates of paleoelevation.

Balbuena Subgroup

Micrites, stromatolites, and oolites ($n = 9$) from the Yacoraite Formation (Fig. 2) sampled near Purmamarca (Fig. 1) yielded $\delta^{13}C$ (VPDB) values of +0.7‰ ± 0.3‰ and $\delta^{18}O$ (VPDB) values of +1.4‰ ± 2.7‰ (Table 1; Fig. 4). These results are in line with most values previously obtained for the Yacoraite Formation from a broad area of northwestern Argentina (Marquillas et al., 2007).

The Yacoraite Formation is the oldest and most deeply buried of the formations we studied and is therefore the most likely to have been modified by diagenetic resetting. The positive isotopic values for both carbon and oxygen, however, strongly argue against resetting of either system. Results are consistent with a marine or low-elevation lacustrine setting for the Yacoraite for this time period, although the $\delta^{18}O$ (VPDB) values are on the high side of typical marine averages for this time period (Veizer et al., 1999). The origin of the Yacoraite has been the focus of much discussion (Palma, 2000; Marquillas et al., 2007), and it appears that the Yacoraite Formation is both marine and lacustrine. Even if entirely lacustrine in origin, the elevated $\delta^{18}O$ values of the Yacoraite are very difficult to explain by diagenetic resetting at higher temperatures, which tends to decrease $\delta^{18}O$ values.

The three Yacoraite Formation samples analyzed for Δ_{47} paleothermometry yielded indistinguishable, warmer than Earth-surface temperatures (T), of 47–48 °C (or 61–64 °C using Dennis et al., 2011; Table 2 herein). These warm temperatures suggest that the samples underwent some degree of diagenetic "resetting." Resetting could be due to diffusion of carbon and oxygen through the mineral lattice (Passey and Henkes, 2012) or to recrystallization, although carbonates from the Yacoraite and from paleosols in overlying foreland formations appear to remain largely unaltered and micritic, arguing against wholesale recrystallization and secondary introduction of cements. Quade et al. (2013) documented such resetting of Miocene-age carbonates from Nepal and Pakistan when buried >3–4 km (or >125 °C), and burial temperatures of ~100 °C or higher can cause solid-state C-O bond reordering (Henkes et al., 2014). The Yacoraite was overlain by 2–6 km of younger foreland deposits in the region (Carrapa et al., 2011a; DeCelles et al., 2011), and so deep burial could account for the resetting. Although resetting due to diffusive reordering of C-O bonds or cryptic recrystallization has obscured any primary paleotemperature information in these samples, the positive bulk $\delta^{13}C$ (VPDB) and $\delta^{18}O$ (VPDB) values do not appear to be altered by this process.

If our interpretation is correct, then higher than Earth-surface $T(\Delta_{47})$ values cannot necessarily be used to screen for diagenetic alteration of the bulk isotopic composition of samples if the $T(\Delta_{47})$ values reflect diffusive reordering or recrystallization in a

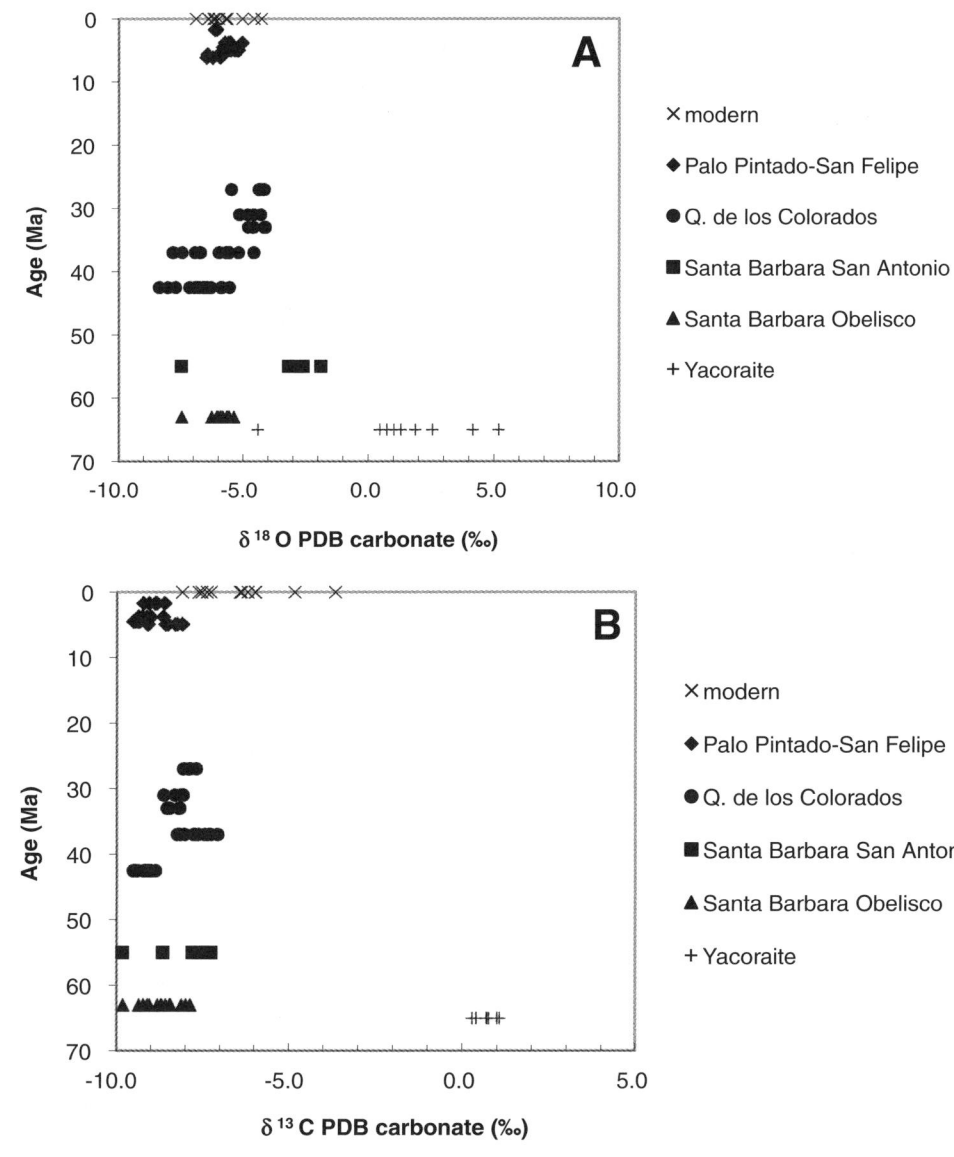

Figure 4. (A) $\delta^{18}O$ values and (B) $\delta^{13}C$ values of carbonates from Late Cretaceous to recent carbonates from areas now exposed in the easternmost Puna, Eastern Cordillera, and Subandes, 22°S–26°S (see Fig. 1 for locations). PDB—Peedee belemnite.

rock-buffered system (e.g., Huntington et al., in press). Moreover, although $\delta^{18}O$ values of water can be calculated using the $T(\Delta_{47})$ and $\delta^{18}O$ values of carbonate from the same sample (Table 2), such values will be difficult to interpret if the $T(\Delta_{47})$ value reflects diffusive reordering rather than the temperature of calcite crystallization from either Earth-surface or diagenetic waters.

Santa Bárbara and Metán Subgroups

Pedogenic carbonate isotopic values from Vertisols or reworked from Vertisols and Calcisols in the Mealla and Maiz Gordo Formations of the Santa Bárbara Subgroup (Fig. 2), and from the overlying Quebrada de los Colorados Formation of the Metán Subgroup are very similar and fall mostly in the −8‰ to −9‰ range for $\delta^{13}C$ (VPDB) and −4‰ to −6‰ for $\delta^{18}O$ (VPDB) (Fig. 4; Table 1).

These $\delta^{13}C$ (VPDB) values are consistent with soils that were covered by dominantly C_3 plants respiring at moderate to high rates (Fig. 4; Cerling and Quade, 1993; Quade et al., 2007). The C_3 cover is consistent with the absence or rarity of C_4 plants globally before the Miocene. The high soil respiration rates suggest semiarid to subhumid conditions, which typify elevations <1000 m in the region today. It is highly inconsistent with the desolate, nearly plantless setting of the >3000-m-high Andes. This holds for both the modern and the past, based on the extensive evaporites in partially age-equivalent deposits of the Quiñoas Formation at Salar de Fraile, discussed in the Evaporites and Paleoaltimetry section.

The consistently high $\delta^{18}O$ values of these formations also point to low paleoelevations of <1000 m (Fig. 5), entirely consistent with the foredeep to back-bulge foreland settings

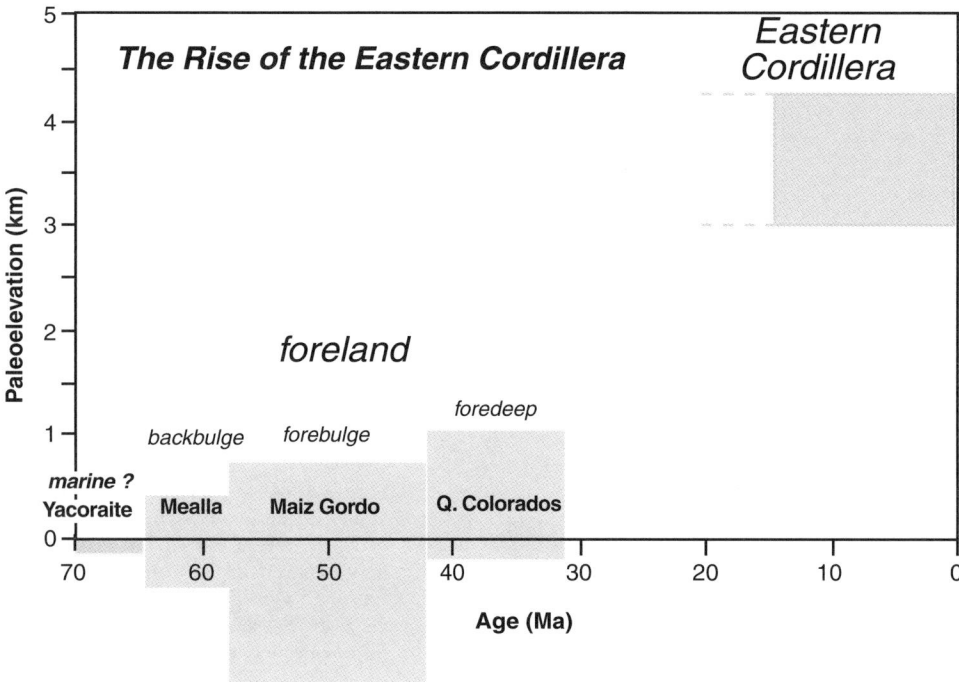

Figure 5. Uplift of the Eastern Cordillera region during the Cenozoic, based on oxygen isotopic results from carbonates, deuterium isotope values from volcanic glass, and the distribution of evaporites.

envisioned for the Santa Bárbara and Metán Subgroup deposits (DeCelles et al., 2011). For comparison, modern elevations in the Andean foreland basin system range from near sea level in the far east to ~400 m in the proximal foredeep and wedge-top depozones. They are also consistent with values from modern low-elevation carbonates sampled in Quebrada del Toro (Table 1; Fig. 1). We regard this comparison as valid even though Paleogene climate was warmer and the $\delta^{18}O$ value of the ocean was slightly lower. This region has probably always been embedded in the subtropical easterlies, which provides a probable reason why large departures from the present are not visible isotopically.

The Santa Bárbara and Metán Subgoup paleosol carbonates give a wide range of $T(\Delta_{47})$ values, most of which are too high for primary soil temperatures (Quade et al., 2013) and point to some diagenetic resetting (Table 2). Like the Yacoraite Formation carbonates, these samples were buried to 2–6 km. Four of the samples have similar $\delta^{18}O$ and $\delta^{13}C$ of carbonate values (around −6‰ and −8 ‰, respectively) but very different $T(\Delta_{47})$ values, ranging from 39 °C to 97 °C (or 45–193 °C using Dennis et al., 2011). The other two samples vary in their $\delta^{18}O$ and $\delta^{13}C$ carbonate values, but their temperatures fall within the large temperature range of the other paleosol carbonate samples. The fact that $\delta^{18}O$ and $\delta^{13}C$ values appear consistent among samples with highly variable $T(\Delta_{47})$ values strongly suggests that the bulk isotopic composition has remained primary while diffusive C-O bond reordering or cryptic recrystallization has occurred to varying extents among the samples. As a result, reconstruction of $\delta^{18}O$ of water values calculated from the $T(\Delta_{47})$ and $\delta^{18}O$ of carbonate data would be misleading.

Bywater-Reyes et al. (2010) studied a thick Neogene-age sequence of basin-fill deposits now in the eastern part of the Eastern Cordillera near the village of Angastaco. They reported $\delta^{18}O$ (VPDB) values from paleosol carbonates in the Palo Pintado (9–5 Ma) and San Felipe Formations (5–2 Ma) of −5.8‰ ± 0.5‰ (n = 35). This range excludes some results from the bottom of the section that appear not to be paleosol carbonate due to their low $\delta^{13}C$ values (<−14‰; figure 2 *in* Bywater-Reyes et al., 2010).

The Angastaco area deposits today are moderately elevated at ~2 km. Carrapa et al. (2011a) interpreted the Palo Pintado Formation as wedge-top deposits and the San Felipe Formation as intermontane paleovalley deposits (Fig. 2). The elevated $\delta^{18}O$ (VPDB) values in these formations point to a paleoelevation of ≤1 km. Paleosol carbonate $T(\Delta_{47})$ measurements averaging 24 ± 4 °C (n = 9; 1 s.d.) and as warm as 32 °C also suggest low paleoelevations (Carrapa et al., 2014b).

Carbonates are rare in younger deposits in the now high Puna Plateau. These are confined to a few carbonates in lake deposits otherwise dominated by clastic sediments or evaporites. The $\delta^{18}O$ values in these carbonates tend to be high and variable at −1.4‰ ± 3.2‰ (Table 1). These high values were likely produced by evaporation from shallow lakes and salt pans. The rare paleosols in these deposits are gypsiferous and lack carbonate, a characteristic of soils of the modern hyperarid Atacama (Quade et al., 2007). Relatively cool paleosol carbonate temperatures indicated by two $T(\Delta_{47})$ measurements are ~15 °C, suggesting higher elevations than the Angastaco area samples (Carrapa et al., 2014b). Taken together, conditions on the Puna Plateau were clearly arid to hyperarid and likely at high elevation since at least 10 Ma.

Isotopic Results and Implications from Volcanic Glass

More than 30 tuffs with preserved volcanic glass were analyzed for δD composition in this study (Table 3). Almost all come from >3000 m elevation in the Puna and Eastern Cordillera. Some tuffs were previously dated by other studies, whereas others were dated for this study using zircon U-Pb geochronology (Table 4). The samples that we report here range in age from 38 to 2 Ma. The δD (SMOW) values for the glass range from −73‰ to −142‰ (Table 3).

We reconstruct ancient water values (Table 2) using a water-glass fractionation factor ($\alpha_{water-glass}$) of 1.0342, which Friedman et al. (1993) established experimentally, and confirmed empirically. Dettinger and Quade (this volume) studied the δD composition of modern water and Quaternary-age glass in the eastern Andes. They observed a similar 30‰–35‰ fractionation between the δD values of water and glass across a range of elevations, supporting Friedman's conclusions.

Dettinger and Quade (this volume) sampled modern water across a wide elevation range and determined the following relationship for the Eastern Cordillera:

$$\text{elevation (m)} = -66.2 \times (\delta D_{water}) - 375.3 \quad (r^2 = 0.90, n = 25). \quad (1)$$

In the western part of the Puna, δD_{water} values increase less rapidly with elevation, and a polynomial fit to the combined Eastern Cordillera and Puna data is:

$$\text{elevation (m)} = -0.4775 \times (\delta D_{water})^2 - 109.63(\delta D_{water}) - \\ 1222.8 \ (r^2 = 0.90, n = 29). \quad (2)$$

The δD values of 2–0 Ma glass also from the Eastern Cordillera display the same slope as Equation 1 but are offset by 30‰–35‰, supporting the idea that our modern water sampling is representative of long-term $\delta D_{water-elevation}$ relationships.

Potential Complications

Considerable uncertainty attends any use of such modern relationships to reconstruct paleoelevation, chief among them the effects of climate change. This can be partly redressed by comparing ancient lowland samples to modern archives at low elevation, in order to discern net climate-driven changes in isotopic compositions of known elevation. Differences between ancient and modern archives of similar elevation equate to climate change, which presumably influences both low- and high-elevation records. We therefore correct for the effects of climate change by comparing carbonates from the Eocene–Oligocene Quebrada de los Colorados Formation to Quaternary carbonates. The $\delta^{18}O$ (VPDB) values of carbonate near sea level from this time period are in the −8‰ to −4‰ range, for an average of around −6‰, very similar to values of Quaternary-age carbonate (Fig. 4). The isotopic range probably arises from variable evaporation of soil water, perhaps coupled with some contribution to soil nodules composition by groundwater fed from higher elevations, since

many of the foreland soils show hydromorphic features. Evidently, climate change in this region during the Cenozoic did not appreciably alter the isotopic composition of lowland rainfall. Moreover, we can safely assume that changes in continentality in the Andes had little effect on the isotopic composition of rainfall over the Cenozoic. For example, the distance that moisture must travel over land via the NE trades, now and in the past, is 2000–3000 km, which is much larger than any changes due to tectonic shortening (~150 km) of the orogen. Therefore, climate change and tectonic shortening of Andes likely played a minimal role in altering the isotopic composition of meteoric water across all elevations.

Another key question is how the slope of Equations 1 and 2 may have differed in the past. Ehlers and Poulsen (2009) have suggested that slopes were less steep in the past when the Andes as a whole were lower, i.e., less change in isotopic compositions for a given change in elevation. In this view, the isotopic gradients only attained their present steepness once the bulk of the Andes, rising en masse, crossed a certain threshold. If correct, our reconstructed paleoelevations described in the next section using Equations 1 and 2 are minima.

However, we argue that isotopic gradients up the east face of the paleo-Andes and onto the Puna probably did not differ appreciably from today because the geologic evidence (see Ancient Evaporites and Paleoelevation section) points to high elevations already in place in the Puna by 38 Ma. Modeling predicts that the dry climate of the pre-uplift Andes should have produced high (>+3‰) $\delta^{18}O$ (SMOW) values for rainfall (Ehlers and Poulsen, 2009). Our results are uniformly more negative (<–5‰) for all formations back to the Paleocene, arguing for the presence of elevated uplands through nearly all the Cenozoic. In our view, the only changes have been that the east-bounding topographic front stepped eastward with time. If correct, isotopic gradients should not have appreciably changed.

Western Puna Plateau: Salars de Fraile and Arizaro

Using δD values from glass (Table 3), we can reconstruct the paleoaltimetric evolution of the region through time, bearing in mind the assumptions just described. Eocene- and Oligocene-age tuffs from two large hinterland basins in the western Puna (Fig. 1), the Fraile and Arizaro Basins, span the entire period roughly 40 Ma to present. Glass from 10 tuffs in the Fraile Basin yielded average δD values of −120‰ ± 14‰ (1σ), and water δD values of −90‰ ± 14‰ using the water-glass fractionation factor from Friedman et al. (1993). U-Pb dating of zircons from some of these tuffs reveals an age that ranges from 38.4 to 19.2 Ma (Table 4). Ten additional tuffs from the Arizaro Basin (21.5–1 Ma; Boyd, 2010; DeCelles et al., this volume) yielded average glass δD values of −89‰ ± 7‰ and average reconstructed water δD values of −58‰ ± 7‰ (Canavan et al., 2014).

Using Equation 2, these isotopic results translate into paleoelevations of 4100–5100 m for the Fraile glasses, and 2600–4000 m for the Arizaro glasses (Fig. 6). Use of Equation 1 yields even higher paleoelevations, but we regard Equation 2

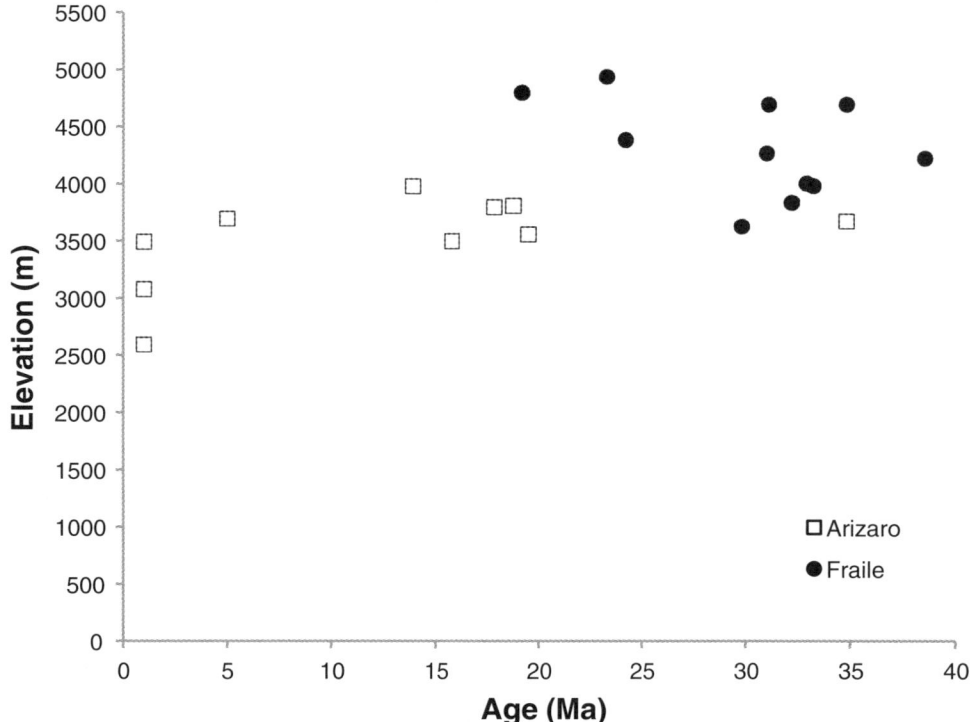

Figure 6. Paleoelevation reconstruction for 22°S–26°S based on the δD value of glass from the Fraile and Arizaro areas of the Puna Plateau (see Fig. 1). The modern elevation of these Puna basins is 3500–4000 m.

as more appropriate, because it includes natural waters from the Eastern Cordillera and Puna. As already noted, no correction for climate change was made, because the δ^{18}O value of lowland soil carbonates contemporaneous with the Fraile and Arizaro tuffs does not differ appreciably from the modern values. The modern elevation of the Fraile and Arizaro deposits is 3500–4200 m. Hence, paleoelevations since the late Eocene have been at least as high as the present. A few of the youngest (<5 Ma; Fig. 6) samples from Arizaro appear to mark a recent drop in elevation for that basin, not inconsistent with the evidence for late Neogene extension and normal faulting (Schoenbohm and Carrapa, 2011).

Our paleoelevation results, which are based on empirically derived models, can be compared to the results of Canavan et al. (2014), who used a theoretical model (Rowley and Garzione, 2007) to estimate paleoelevations of the Puna. Using this fairly different approach based on atmospheric thermodynamics and the process of Raleigh distillation during orographic uplift, Canavan et al. (2014) produced modeled paleoelevations for the Puna Plateau from 38 Ma to recent, also based on δD of volcanic glass samples (including samples from the Fraile and Arizaro Basins). The paleoelevation estimates made by Canavan et al. (2014) average ~1–0.5 km lower in elevation than ours but tell a similar story of an already elevated Puna Plateau (at least 2.5–3.5 km) by the late Eocene.

Eastern Andes

We analyzed δD values of glass from 22 tuffs and ignimbrites (Table 3) in the eastern portion of the Andes. They range in age from near-modern to 8 Ma and include glass from some of the Altiplano-Puna volcanic complex eruptions and other large ignimbrites, including Cerro Galán and Tuzgle (Fig. 1). The oldest of these samples yielded paleoelevations between 4000 and 4700 m, somewhat higher than elevations of the samples today at 3600–3900 m. This suggests that much of the Eastern Cordillera and eastern Puna Plateau had attained near-modern elevation by the late Miocene (Fig. 5).

The δD analyses of glass from two tuffs sampled by the first author (Quade) from the Palo Pintado Formation (9–5 Ma) in the Angastaco area (Fig. 1) indicate paleoelevations of ~1.5 km (Table 3). The modern sample elevation is 1.8 km, suggesting that the Angastaco area was modestly elevated while accommodating sediment during the late Miocene. As previously discussed in the section on carbonates, Carrapa et al. (2011a) placed the Palo Pintado Formation in the wedge-top depozone. These data are consistent with other paleoelevation proxies (δD, Δ$_{47}$) from the Eastern Cordillera, which, combined with geological evidence, indicate that the Angastaco Basin and areas to the east (La Vina) had attained their modern elevation during the mid-Miocene (Carrapa et al., this volume).

EVAPORITES AND PALEOALTIMETRY

Modern Salt Evaporite Distribution

Evaporites are common in the geologic record of the central Andes. They provide information not only about paleoclimate and orography, following Vandervoort et al. (1995), but we suggest also paleoaltimetry. The modern distribution of saline lakes in South America provides a vital perspective on their interpretation.

Saline lakes are found along almost the entire length of South America south of 10°S (Fig. 7). North of this, the climate is generally too wet for saline lakes to develop, the exception being far northeastern Brazil. South of 10°S, modern saline lakes occupy two basic positions, depending on prevailing wind direction. South of ~29°S, all saline lakes are east of the Andes. Prevailing

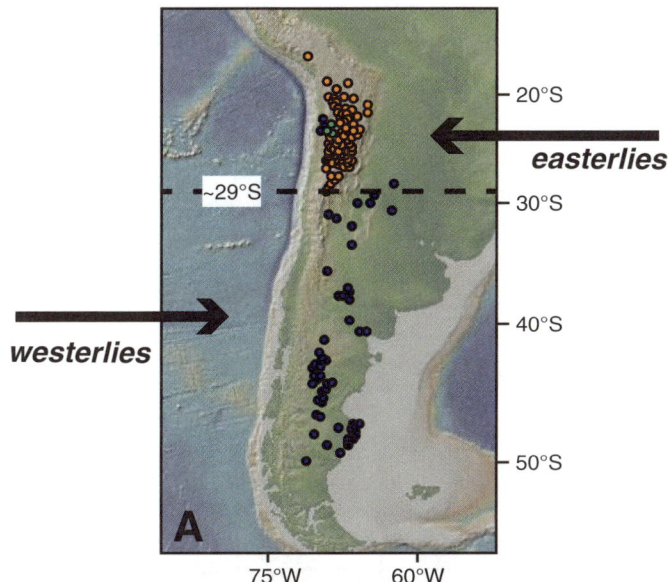

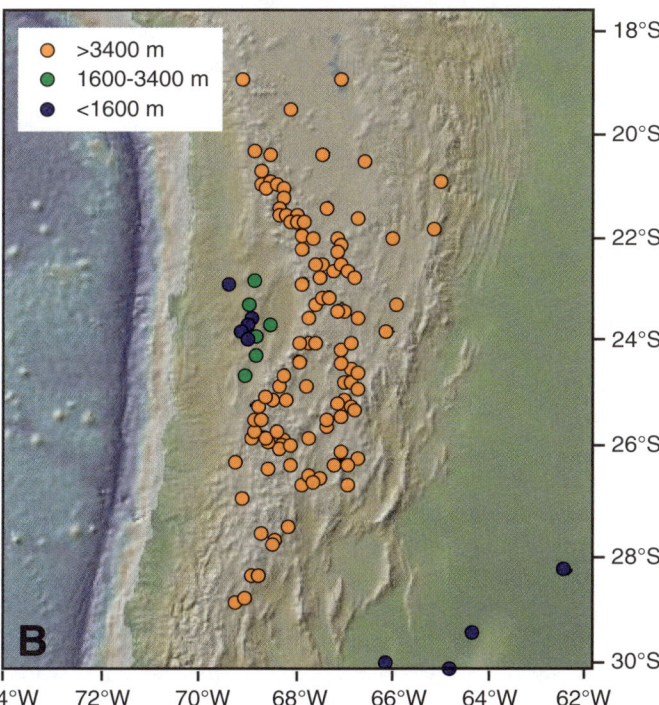

Figure 7. Distribution of salt pans in South America >1 km in diameter by elevation: (A) all of southern South America, with prevailing winds indicated, and (B) the subtropics of South America north of 30°S.

winds at these latitudes are westerly, and the moisture they bear is blocked by the Andes (Fig. 7A). What little rain penetrates the Andean rain shadow is carried by easterly intrusions of moist air from the Atlantic. Tectonically, the lakes lie mostly in the foredeep, back-bulge (e.g., Cohen et al., this volume), or in shallow local cratonic and volcanic depressions. They also are confined to elevations of <1600 m, and mostly <1000 m.

Between 29°S and ~10°S, saline lakes take on a completely different distribution and are confined to the high Andes and the Atacama Desert to the west. No dry/saline lakes are present east of the Andes in this latitudinal belt, although lakes are widespread in this region. This is because moisture at these latitudes is born from the Atlantic Ocean by the easterlies. Easterlies are blocked by the high Eastern Cordillera, creating a rain shadow to the west. In the Andes between 10°S and 29°S, the saline lakes are, without exception, at high elevation, >2900 m and mostly >3400 m (Fig. 8). However, in the Atacama Desert, saline/dry lakes range in elevation from 500 to 5500 m.

Between 10°S and 29°S (a latitude range that is the region of focus in this study), the strong relationship between saline lakes and high elevation is due to a combination of aridity and internal drainage on the Puna Plateau. Easterly air masses rise along the steep eastern front of the Andes, cool, and rain out most of their moisture between 0 and 2 km elevation. By 3 km, rainfall averages 20 cm/yr, compared to ~100–300 cm/yr at 500 m elevation at the eastern front of the Andes (Strecker et al., 2007). Evaporation far exceeds precipitation on the Puna Plateau, and a strongly negative moisture balance favors evaporite formation.

Internal basin drainage, of course, is also essential to evaporite development. Many basins on the high Puna Plateau >2900 m are hydrographically closed, whereas basins at lower elevations to the east in the Subandes and Santa Bárbara Ranges are all externally drained. Ranges in the Subandes can be very high (>5000 m) but are not continuous, and easterly moisture passes around and behind them. This effect diminishes with elevation, to the point where drainages in the higher Subandes lose stream power, leading to their eventual hydrographic isolation (Humphrey and Konrad, 2000; Sobel et al., 2003; Strecker et al., 2007). The point at which local drainage changes from external to internal depends on the competing factors of uplift rate, rainfall, and rock erodibility. The whole of the Subandes is undergoing rapid shortening and exhumation (Echavarria et al., 2003; Carrapa et al., 2011b). At <2 km elevation, these tectonic forces that would disrupt and potentially isolate drainage systems are exceeded by the higher rainfall, locally assisted by erodible rocks, thus promoting external drainage. At >2 km, tectonic processes gradually overwhelm diminishing rainfall, and basins cannot maintain external drainage. Internal drainage leading to saline lake formation develops very consistently at elevations >3000 m all along this sector of the eastern Andes.

At the latitudes examined in this study (22°S–26°S), we therefore take the presence of ancient evaporites in basin deposits of the Andes to indicate high paleoelevation (>2900 m, and likely >3400 m), provided certain conditions were met. For this to be

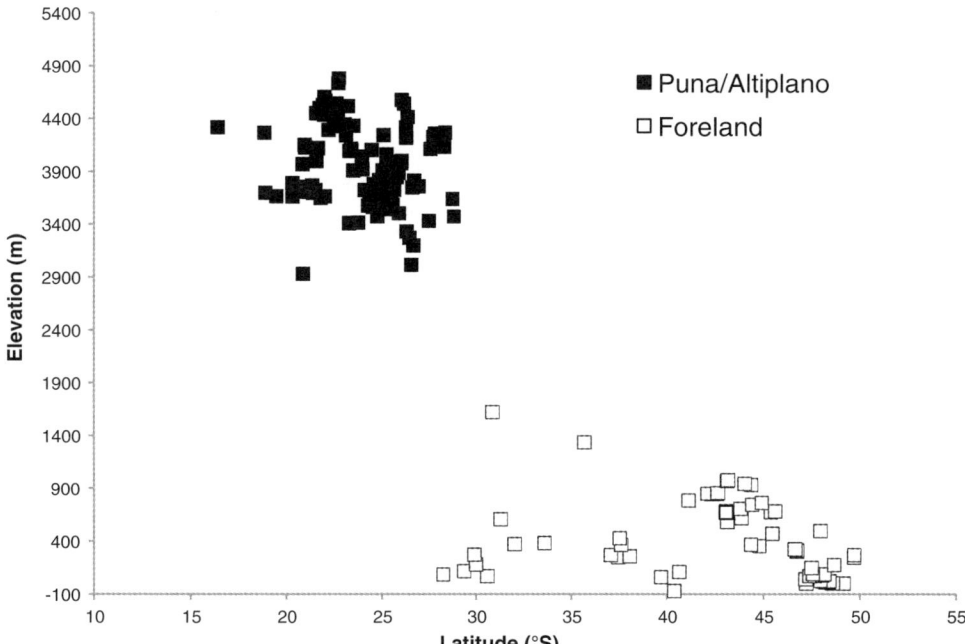

Figure 8. Elevation distribution of salt pans in South America >1 km in diameter, excluding the Atacama Desert. No salt pans are found between 1600 and 2900 m, and most salt pans are confined to >3400 m (all north of 29°S) and <1000 m (all south of <29°S).

true, the easterlies, not the westerlies must have dominated rainfall at this latitude. Had the westerlies dominated, the pattern of salt lakes would be like it is today south of 29°S, where salt lakes abound in the low-elevation foreland, and their presence would not imply high elevations. Second, the paleolatitude of the basin of interest and the transition between easterlies and westerlies must not have changed significantly during the Cenozoic. All indications are that the easterlies were as far south and possibly farther south throughout the Cenozoic, when global climate was warmer and the subtropics more expansive. Moreover, South America has not drifted in latitude appreciably during the Cenozoic.

It is important to be clear that we are making the saline lake–high elevation connection only for the midlatitude Andes between 10°S and 29 °S. Obviously, in other parts of the world, salt lakes occur in low-elevation basins, often in rain shadows, at this and other latitudes. Death Valley, USA, and other internally drained Basin and Range grabens are good examples. There, however, the tectonic setting is extensional, and the basins reside in the lee of the Sierra Nevada and other high ranges that block the moisture-bearing westerlies. The Andes currently are generally contractional, except for areas at the margin of the plateau and in the interior of the Puna, which are characterized by recent localized normal faulting (Schoenbohm and Strecker, 2009; Zhou et al., 2013). Hence, no low-elevation basins are found on the Puna, without exception. This latitude of the Andes has probably been under continuous compression since the cessation of Salta rifting in the Late Cretaceous to Paleocene (Pardo-Casas and Molnar, 1987; Soler and Bonhomme, 1990).

Saline and dry lakes are widespread at a range of elevations in the Atacama Desert west of the Western Cordillera of the Andes. As with Death Valley, salt pans in the Atacama have

no implications for paleoelevation, but merely indicate that the Andes to the east were sufficiently elevated to block the easterlies. The discussion that follows is confined to geologic sections that during deposition were east of the coeval Andean magmatic arc—the modern setting for the uniformly high-elevation saline lakes. If our hypothesis—that the present Andes is a key to understanding the paleo-Andes—is correct, then evaporites in the geologic record of the Andes at this latitude indicate paleoelevations >3000 m. This analysis expands upon the observations of Vandervoort et al. (1995), who used evaporite distribution to infer development of orographic blockage of moisture but not necessarily paleoelevation on the eastern Andes during the Neogene.

Ancient Evaporites and Paleoelevation

Evaporites are known from a number of Cenozoic sections in this sector of the Andes (Alonso et al., 1991; Vandervoort et al., 1995), but they are conspicuously absent from age-equivalent deposits in the modern Subandes (Strecker et al., 2007). This is consistent with our view that evaporites—past and present—are confined to elevations >3000 m. Here, we survey the occurrence of evaporites from west to east across the Andes. In general, the age of the oldest evaporites tends to decrease eastward (Fig. 9). Evaporites first appear in Upper Cretaceous strata in the forearc in Chile, in Upper Eocene rocks in the eastern forearc and Puna Plateau, and in mid-Miocene strata in the eastern Andes.

Starting our survey of evaporites in the modern forearc in Chile, the oldest evaporites are exposed in the Cordillera de Domeyko on the western margin of the Salar de Atacama (Hartley et al., 1992; Mpodozis et al., 2005) (Fig. 1). This includes thick evaporites of the Tonel Formation, which is very poorly dated but thought to be mid- to Late Cretaceous in age. The upper Orange (Naranja) Formation, of Paleocene to early Eocene age, also contains evaporites (Mpodozis et al., 2005; Arriagiada et al., 2006). All these evaporites were deposited to the east of the concurrent magmatic arc, before it stepped eastward after ca. 27 Ma. Again drawing on modern analogs, the presence of evaporites would place this region of the present forearc at high elevations in a retroarc position as far back as the Paleocene to early Eocene, and possibly as far back as the Tonel Formation, whatever its age turns out to be.

Evaporites are also abundant in the Oligocene–Miocene Paciencia Formation, the forerunner of the modern Salar de Atacama. By the mid-Oligocene, the main axis of the magmatic arc had shifted eastward to its current position in the Western Cordillera, and thus the Paciencia Formation was deposited in the forearc. The entire forearc then and today lies in the rain shadow of the lofty Western Cordillera, producing hyperarid conditions. In this setting, evaporites form at a broad range of elevations and provide no constraints on paleoelevation for the period <27 Ma. Since the Salar de Atacama area today lies at 2500 m, we suggest it must have lost 1–2 km of elevation since the Eocene (when it lay in a retroarc setting). Evenstar et al. (2009) and Jordan et al. (2010) documented tilting of surfaces and ignimbrite flows to show that the western slope of the Andes stood ~2000 m above the adjacent forearc (now at ~1–1.5 km) by the early to mid-Miocene, and it has added a further ~1 km since then. Their analysis focused on relief, not paleoelevation, so the surface and ignimbrite tilting evidence could support the idea that relief development was due to both the drop of the Atacama (the forearc) and the rise of the Andes.

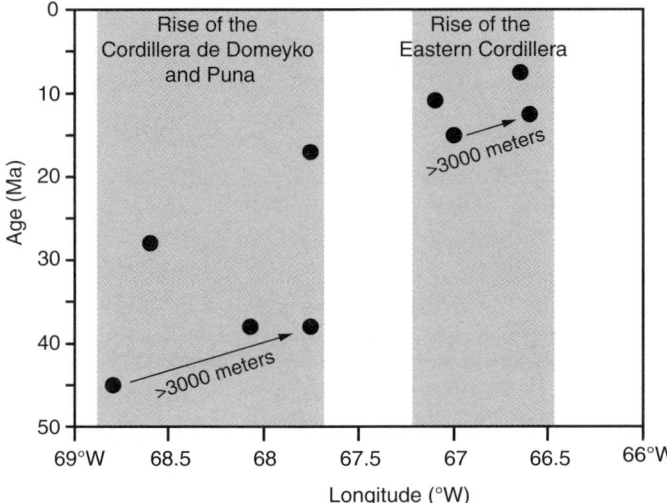

Figure 9. Distribution of evaporites in geologic records from the central Andes, 22°S–26°S, showing the distribution (solid black dots) of evaporite ages. The arrow traces the approximate age and longitude of the eastward rise of the Andes to >3000 m in two steps (shaded in gray), first the Cordillera de Domeyko and most of the Puna by the late Eocene, followed by the Eastern Cordillera by the mid-Miocene.

Evaporites are extensive in the Andes in sections exposed in the central Puna Plateau and the lee of the Western Cordilleran magmatic arc. At Salar de Fraile, they extend back to Member I of the Quinoas Formation (≥38 Ma; Kraemer et al., 1999; Carrapa et al., 2011b; Table 4). Farther north, on the southwest flank of the modern Salar de Arizaro, the oldest evaporites date to the late Eocene (DeCelles et al., this volume). All these evaporites occur just to the east of the main axis of the contemporaneous magmatic arc of the ancestral Western Cordillera (Trumbull et al., 2006). Hence, these basins resided east of the magmatic arc in the paleo-Puna, and not within the forearc (i.e., Atacama-like setting). As such, we interpret the presence of thick evaporites in these sections to indicate the attainment of >3000 m elevations by the late Eocene in this part of the west-central Andes.

Farther eastward in the Puna Plateau, evaporites have been forming since the middle Miocene and are actively forming today. According to the survey of Vandervoort et al. (1995), the oldest evaporites at ca. 15 Ma are found in the Salar de Hombre Muerto (Fig. 9). Prior to 15 Ma, fluvial red beds as old as 37–39 Ma (DeCelles et al., 2007; Carrapa and DeCelles, 2008), but no evaporites, characterize basin sedimentation. Vandervoort et al. (1995) interpreted ca. 15 Ma to mark blockage of easterly moisture and the development of internal drainage in the area. We agree but go further to say that this is also when paleoelevations in the eastern Puna rose above 3000 m, the lower elevation limit for evaporites today.

In contrast, moving eastward into the eastern part of the Eastern Cordillera, Subandes, and Santa Bárbara Ranges, portions of almost the entire Cenozoic record are exposed but do not contain evaporites (DeCelles et al., 2011). Depending on stratigraphic level and location, these strata consist of a broad range of fluvial conglomerate, sandstone, and siltstone; paleosols; eolianites; and local lacustrine siltstone and marl. Evaporites are conspicuously absent, according to our surveys and many other studies (e.g., Starck and Vergani, 1996; Reynolds et al., 2000; Uba et al., 2006; Strecker et al., 2007; Carrapa et al., 2011a; DeCelles et al., 2011; Siks and Horton, 2011). Their absence is consistent with isotopic evidence for basin paleoelevations <3 km.

Volcanic Contributions to Surface Elevation

In general, the role of magmatic addition in the Cenozoic crustal thickening of the central Andes is thought to be minor compared to tectonic shortening, with volcanic rocks contributing as little as 1.5% of the crustal thickness today (Isacks, 1988; Haschke and Gunther, 2003). However insignificant this contribution is to the evolution of crustal thickness in the central Andes, the isostatically neutral redistribution of mass from the midcrust to the surface during eruptions would add to local surface elevations.

Arc volcanism and voluminous eruptions of Miocene–Quaternary silicic volcanic fields (such as the Altiplano-Puna volcanic complex, 10–1 Ma) contributed to surface elevation change in this sector of the central Andes. The Western Cordillera is an andesite edifice upon which the modern active stratovolcanoes are built (e.g., Kay and Coira, 2009); the active volcanic centers rise 1–2 km above the average elevation of the Puna Plateau. In addition, ignimbrite lava flows can be tens to hundreds of meters thick and extend for tens of kilometers from source calderas, filling in topography and adding to surface elevations. Altiplano-Puna volcanic complex ignimbrite flows are distributed over an ~70,000 km² region of the central Andes (de Silva and Gosnold, 2007), although most flows are concentrated in areas around the volcanic centers and cover a total mapped area of ~17,000 km² (Francis and Hawkesworth, 1994). Using estimates of the productivity of andesitic arc volcanoes of the Central volcanic zone and previously compiled data on the age and estimated volumes of ignimbrite eruptions in the region (Francis and Hawkesworth, 1994; de Silva and Gosnold, 2007; Kay et al., 2010), we examined the timing and significance of the accumulation of volcanic material during the last 10 m.y.

The Cenozoic rate of andesitic volcanic addition in the Central volcanic zone is ~3 km³/m.y. per kilometer of arc length (Francis and Hawkesworth, 1994), and silicic supervolcanic eruptions periodically exceeded this rate. In this sector of the central Andes (22°S–26.5°S, ~500 km arc length), the andesitic arc has erupted ~1500 km³/m.y., and therefore ~18,000 km³ since 12 Ma (Fig. 10A). The silicic volcanic fields in the Central volcanic zone contributed significant additional material (Fig. 10A). The Altiplano-Puna volcanic complex extrusion rate exceeded the background rate of volcanism in the Central volcanic zone by at least an order of magnitude, with 200–2400 k.y. pulses of even higher rates of volcanism at ca. 10, 8, 6, and 4 Ma (de Silva and Gosnold, 2007). The total volume of material erupted during the Altiplano-Puna volcanic complex flare-up exceeded 10,000 km³ (Francis and Hawkesworth, 1994; de Silva and Gosnold, 2007). In the southern Puna (i.e., south of 24°S), Cerro Galán and other volcanic centers not considered part of the Altiplano-Puna volcanic complex erupted >1500 km³ (Kay et al., 2010) since the late Miocene (Fig. 10A).

To estimate the thickness of these accumulating volcanic rocks, we distribute the volumes over various conservatively large areas appropriate for the two voluminous types of arc volcanics: andesites and ignimbrites. Arc andesite volumes are distributed over swaths 100 km wide and as long as the arc length in question (Fig. 10B). Here, we make two Altiplano-Puna volcanic complex thickness estimates, one using an area of 17,000 km², and the other 35,000 km². Finally, the volumes of Puna ignimbrites from Kay et al. (2010) are distributed over areas estimated from maps provided by those authors. Note that the system is mass neutral and merely involves the redistribution of material from the lower or upper crust to the surface. Therefore, any change in density, say the change between a magma at kilometers depth to a frothy tuff at atmospheric pressure, will result in a volume change that is not compensated by isostasy.

The results of this simplified approach suggest that volcanic additions in the southern central Andes erupted >25,000 km³ of material since 12 Ma, which would result in 500–1000 m of sur-

face height gain across the entire region. The andesitic edifice in the Western Cordillera likely gained ~400 m of average surface height (Fig. 10B). Ignimbrites between 22°S–26.5°S contributed ~200–600 m of surface elevation in the last 10 m.y., with greater additions in the northern part of the study area that includes the Altiplano-Puna volcanic complex. At the center of the Altiplano-Puna volcanic complex activity, ignimbrites likely contributed

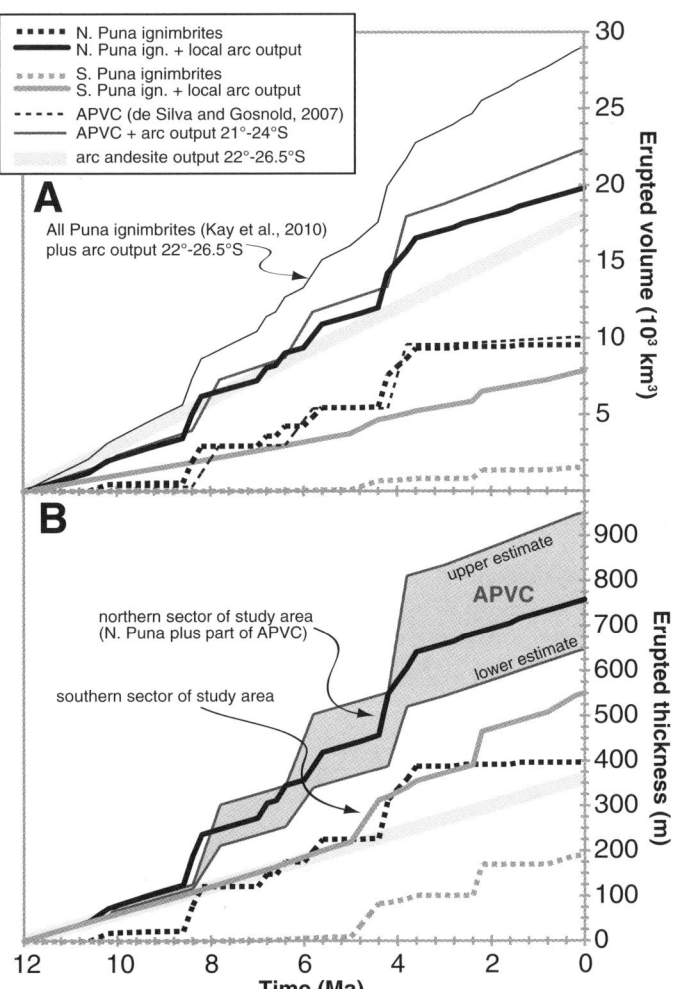

Figure 10. Estimates of the volume and thickness of volcanics erupted 21°S–26.5°S in the central Andes. Dotted lines indicate cumulative ignimbrite eruptions; solid lines combine ignimbrite results with the rate of andesite production, ~3 km³/m.y. per kilometer of arc length (Francis and Hawkesworth, 1994). (A) Cumulative erupted volumes of the Altiplano-Puna volcanic complex (APVC; 21°S–24°S; de Silva and Gosnold, 2007), Puna ignimbrites (22°S–26.5°S; northern Puna ignimbrites include some but not all of the Altiplano-Puna volcanic complex centers, notably excluding Pastos Grandes; Kay et al., 2010), and "background" andesite arc accumulation (22°S–26.5°S; Francis and Hawkesworth, 1994). (B) Cumulative volcanic thickness estimates for arc andesites and ignimbrites. Northern Puna ignimbrites were spread over 17,000 km² (the local andesites over 28,000 km²), and southern Puna ignimbrites were spread over 8000 km² (local andesites over 17,500 km²). Cumulative arc and ignimbrite curves (solid lines) from the same region are combined so they stack on each other.

~600–1000 m of surface elevation (Fig. 10B). These calculations are consistent with previous estimates, i.e., that the volcanics in the region compose the upper ~0.3–0.5 km of the Altiplano-Puna crust (Isacks, 1988), with locally thicker and thinner volcanics.

SYNTHETIC HISTORY AND BROADER IMPLICATIONS

The geologic and isotopic evidence converges on a coherent and detailed picture of the tectono-topographic evolution of the central Andes during the Cenozoic. The Andes grew upward and eastward in three main stages: The largest event occurred during the Eocene (the "Incaic orogeny" of older literature), when the western half the Andes attained >3–4 km elevation, followed by the gradual growth of the Eastern Cordillera in the Oligocene–Miocene, and the modern Subandes and Santa Bárbara Ranges after ca. 8 Ma (Fig. 11).

Late Cretaceous–Paleocene

The late Mesozoic to Paleocene magmatic arc was centered in what is now western Chile; a foredeep covered central Chile; and a rift system—the "Salta Rift"—occupied the back arc in westernmost Argentina (Fig. 11A; Salfity and Marquillas, 1994). East of the magmatic arc, most of what is now the Andes was covered by large lakes and coastal plains, represented by the Balbuena Subgroup, interspersed with residual paleotopographic highs of the Salta Rift and perhaps a more expansive Huaytiquina High (Mpodozis et al., 2005). The only paleoaltimetric or paleoclimatic information on the magmatic arc at this time is the presence of evaporites in the poorly dated (mid-Cretaceous?) Tonel Formation. Whatever its age, the presence of evaporites in the Tonel Formation demonstrates very dry conditions that require orographic blockage of rainfall to the east. However, we are hesitant to assign high elevations to the Tonel basin itself, in contrast to Cenozoic evaporites, because South America was not under strong compression during all of the Cretaceous. Since the opening of the South Atlantic during the Early Cretaceous, spreading rates have been variable (Torsvik et al., 2009), and slower rates may explain the Salta rift system during the Late Cretaceous.

Paleocene–Eocene

During the Eocene, the western part of the Andes—fully 50% of its modern cross section—rose to elevations of 3–4 km (Fig. 11B). This period of extensive Eocene deformation has

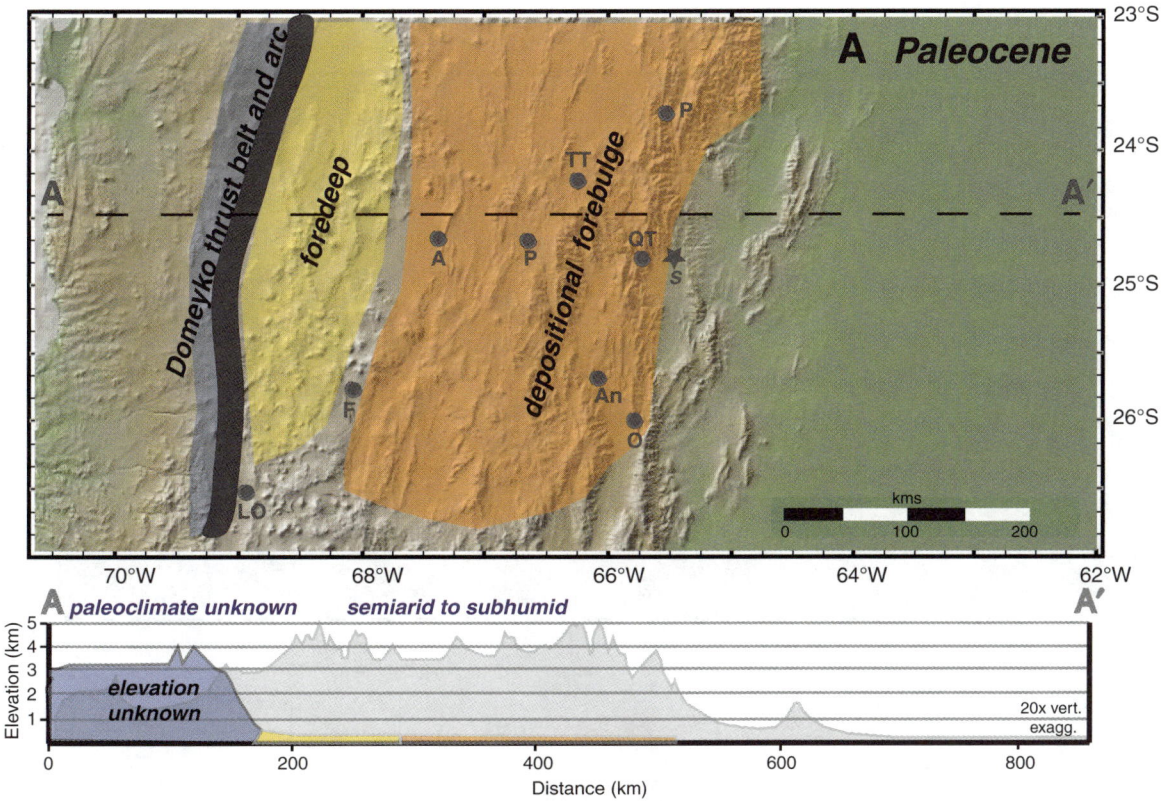

Figure 11 (*Continued on following pages*). Growth of the central Andes 23°S–27°S by the end of the (A) Paleocene, (B) late Eocene, (C) mid-Miocene, (D) late Miocene, (E) present, showing the approximate age intervals for accretion of each stage of the Andes from A to D. Topographic profile of the modern Andes is shown in background of cross sections. Letters are abbreviations of sampled sections shown on Figure 1. Thick gray line denotes the approximate position of the topographic/strain front.

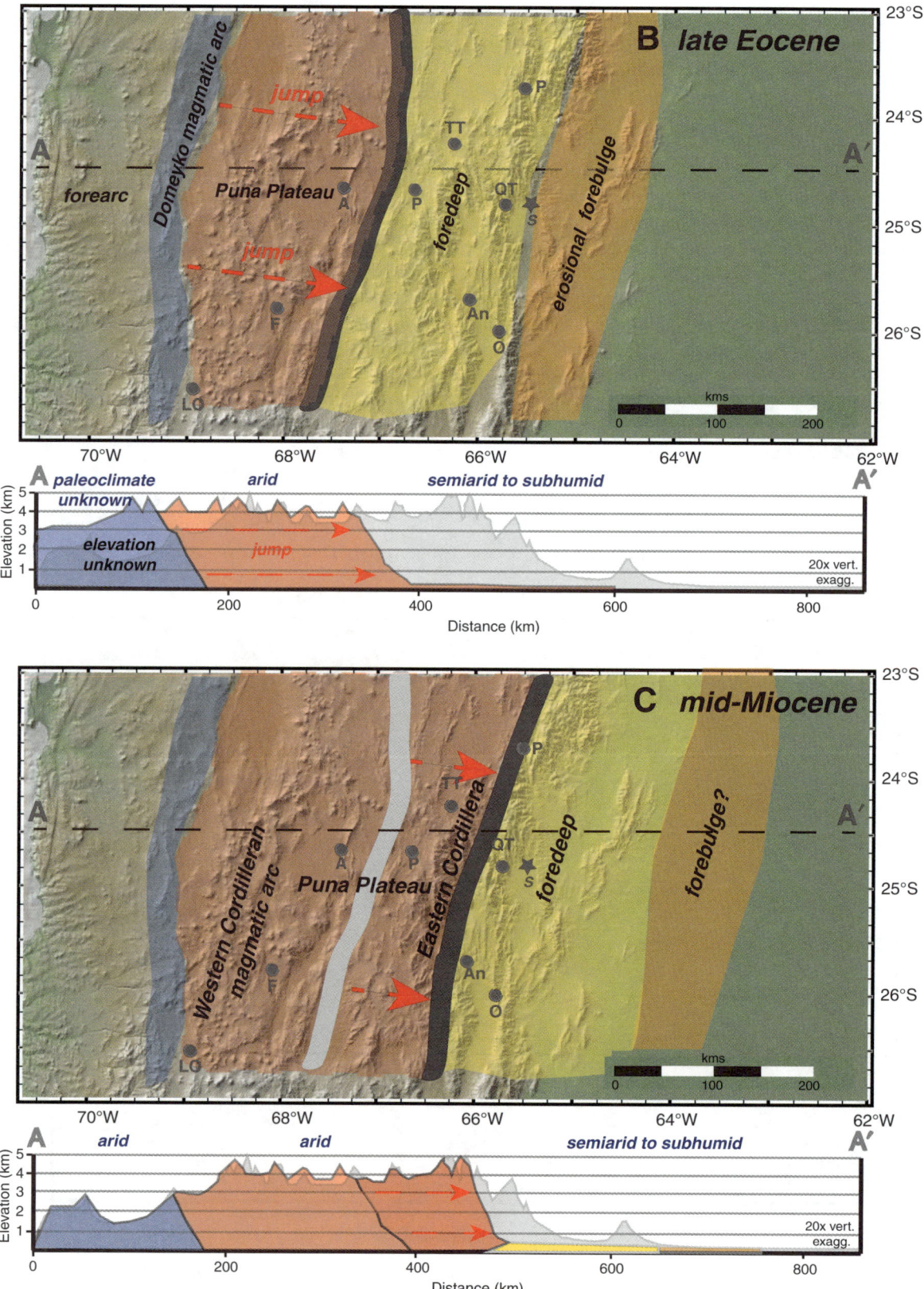

Figure 11 (*Continued*).

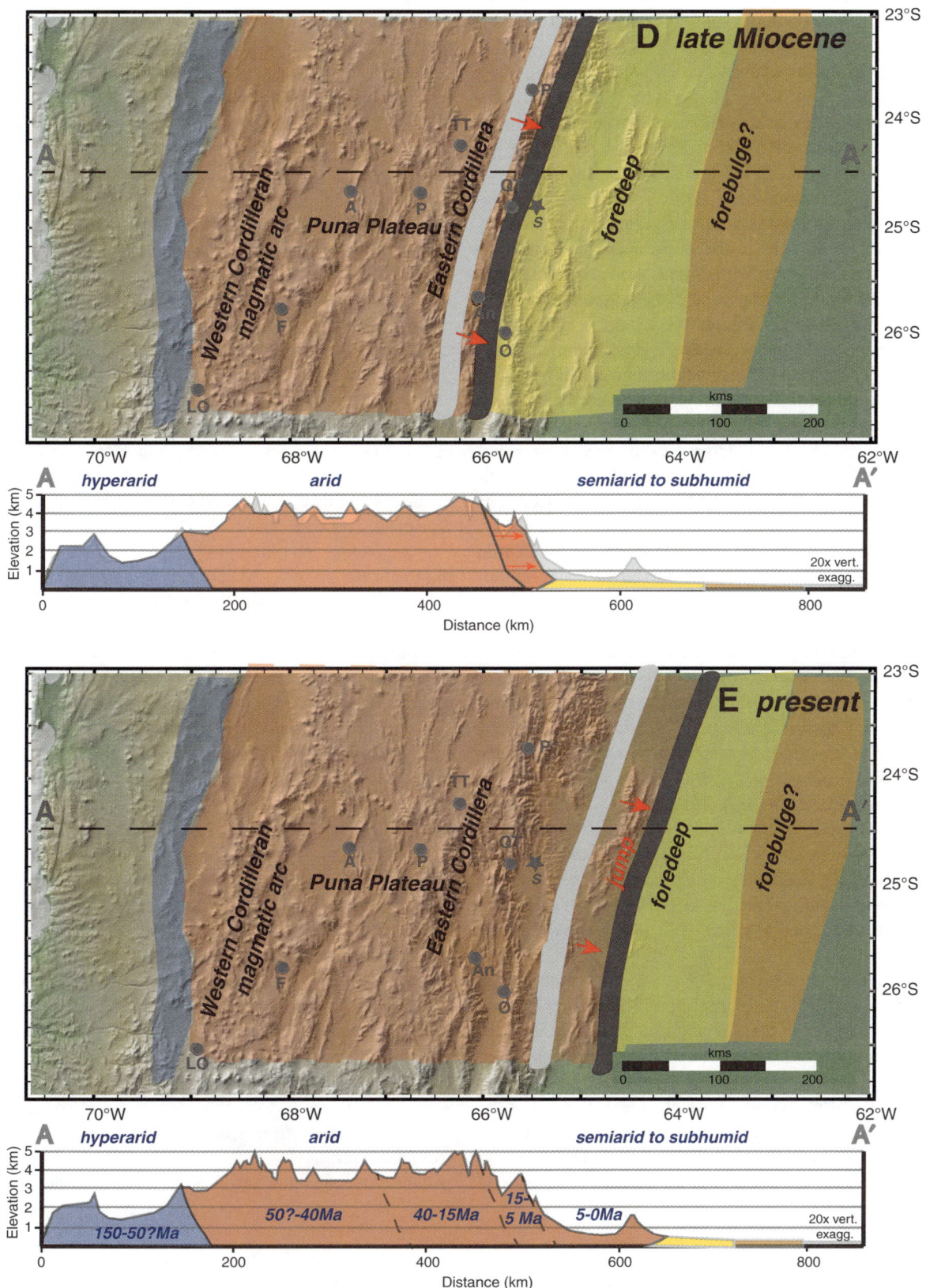

Figure 11 (*Continued*).

long been recognized in the Andes of Peru, Bolivia, and Argentina (Steinmann, 1929; Mégard, 1984; Mpodozis et al., 2005). A regional-scale flexural foreland basin developed to the east of the orogen, and the corresponding flexural wave migrated eastward at a rate >20 mm/yr during the period 50–40 Ma between 22°S and 26°S (DeCelles et al., 2011). The magmatic arc was narrowly focused along 69°W–69.5°W in the Cordillera de Domeyko in central Chile, 50–150 km to the west of its present location (Fig. 11A; Trumbull et al., 2006). We have no paleo-altimetric information on these volcanoes. Immediately east of the magmatic arc, evaporites in the upper Naranja Formation (Paleocene?; Arriagiada et al., 2006) signify arid conditions, at a minimum pointing to the orographic blockage of the easterlies by high elevations farther to the east, and very probably indicating high (>3 km) paleoelevations at this time. This inference, combined with low δD values from glass starting by ca. 38 Ma from the Salar de Fraile and Arizaro Basin, suggests that the Cordillera de Domeyko and most of the Puna Plateau had attained an elevation of ~4 km by ca. 38 Ma. Extensive evaporites interbedded with these tuffs also point to high elevations and, combined with the lack of paleosols, bioturbation, and fossils, attest to arid conditions since the Paleogene at this latitude.

Eastward, intermontane deposits such as in the Salares de Fraile and Arizaro give way to clastic fluvial and alluvial-fan deposits of the Upper Eocene Geste Formation, now cropping out at 67°W just west of Salar de Pastos Grandes (Fig. 1). The Geste Formation has been interpreted as wedge-top deposits marking the position of the late Eocene orogenic strain front (DeCelles et al., 2007; Carrapa and DeCelles, 2008) and therefore the eastern topographic front of the Eocene Andes. Thermochronologic evidence from this time period shows that exhumation, especially 50–30 Ma, was extensive across the width of the orogen, from the Cordillera de Domeyko in the west (Maksaev and Zentilli, 1999) to mountain ranges delivering sediment to the Geste Formation on the east (Carrapa and DeCelles, 2008). Despite the extensive deformation, however, exhumation was shallow, << 7–8 km everywhere across the Incaic Andes, as indicated by (U-Th)/He and apatite fission-track data (Reiners et al., this volume; Carrapa et al., this volume).

East of this topographic break at ~67°W, basin filling in response to flexural subsidence commenced during the Paleocene–early Eocene, in the area now comprising the Eastern Cordillera. Carbonates in these basin deposits, represented by the Santa Bárbara and Metán Subgroups, show that this foreland basin setting was <1 km elevation through at least the end of the Oligocene, and was climatically semiarid to subhumid, as it is today. Low elevations are consistent with the forebulge and backbulge setting of the Paleocene–early Eocene deposits envisioned by DeCelles et al. (2011).

Oligocene to Late Miocene (ca. 35–5 Ma)

During the Oligocene–early Miocene, the eastern margins of the Puna Plateau and the Eastern Cordillera began to rise and propagate slowly eastward (Fig. 11C), coeval with subduction erosion (Stern, 1991) and aridification (Strecker et al., 2007). The slow eastward propagation of the orogenic strain front from ca. 35 Ma to ca. 5 Ma is tracked by a slowly migrating (~4 mm/yr) flexural wave in the coeval foreland basin system and by exhumation (DeCelles et al., 2011; Carrapa et al., 2011b). Intermontane basins in a probable upland setting continued to receive coarse clastic sediments, eolianites (during the early Miocene), evaporites, and tephras in the western and central portions of the Andes. The uplands expanded to the east as the topographic front of the Andes migrated slowly eastward ~150 km, hydrographically isolating and desiccating basins by the mid-Miocene as far east as 66.5°W. The orogenic front was in the Angastaco area at 66°W by ca. 14 Ma, deforming and exhuming older foredeep deposits and depositing syndeformational sandstones and conglomerates of the Angastaco Formation (Carrapa et al., 2011a, 2011b).

Evidence of gradual eastward stepping (Carrapa et al., 2011b; Carrapa and DeCelles, this volume; Reiners et al., this volume) of exhumation across the eastern Puna Plateau and into the Eastern Cordillera is documented by a number of thermochronologic studies: Sierra de Calalaste and Cobres Granite at 24–29 Ma (Carrapa et al., 2005; Deeken et al., 2006); the Cumbres de Luracatao, Complejo Oire, and others at 17–20 Ma (Deeken et al., 2006); and the Cachi Range by 15 Ma (Fig. 1; Coutand et al., 2006; Deeken et al., 2006; Carrapa et al., 2014a). To some extent, volcanism also followed these patterns, being characterized by (1) a magmatic gap during much of the Oligocene, (2) post-Oligocene eastward stepping of eruptive centers, with the Western Cordillera arc centered at ~68.5°W, and (3) a gradual eastward broadening of volcanism and proliferation of ignimbrites.

We have no stable isotopic evidence from individual continuous stratigraphic sections that might illustrate the progression of uplift of the eastern Puna Plateau and Eastern Cordillera from near sea level during the Eocene to 3–4 km today. The region changed from a lowland foreland basin during early Miocene time to high elevation by late Miocene time (Carrapa and DeCelles, this volume), but sedimentary basins capable of archiving this progressive rise in elevation are not preserved. This is likely a result of the fact that once the region had become incorporated into the orogenic wedge and risen above the elevation of the active wedge-top depozone, it entered an erosive regime in which long-term sediment accommodation was precluded by rapid erosion in the rising mountains. Not until the orogenic plateau had expanded to include what now lies in the eastern Puna was the region able to again accommodate sediment. Another way to conceptualize this is that while the region was climbing from low elevation to high (i.e., plateau) elevation, it was incapable of storing sediment over the long term due to oscillatory filling and excavating processes like those described by Hilley and Strecker (2005).

Deposition and long-term preservation of sediment in the eastern Puna Plateau and Eastern Cordillera resumed during the middle to late Miocene with the accumulation of evaporites and tuffs at high elevations. The δD values from the tuffs show that the Eastern Cordillera was >3.5 km in elevation by 8 Ma (this study), and perhaps as early as 14 Ma (Carrapa et al., this volume), which is corroborated by the appearance of evaporites in the eastern Puna ca. 15 Ma (Vandervoort et al., 1995).

At the same time, to the east, a low-lying foreland region occupied what is now the Santa Bárbara tectonomorphic domain (DeCelles et al., 2011). Because the orogenic load was expanding eastward at a slow rate, the flexural wave migrated very slowly (~4 mm/yr) through the region (DeCelles et al., 2011). Our stable isotopic evidence from carbonates shows that this foreland basin lay at <1 km elevation.

Late Miocene to Present

The present form of the Andes began to take shape with a second rapid migration in the position of the orogenic strain front into the Subandes and Santa Bárbara Ranges, where nearly all major deformation was focused from latest Miocene onward (Figs. 11D–11E; e.g., Reynolds et al., 2000; Carrapa et al., 2011a; DeCelles et al., 2011; Metcalf and Kapp, this volume). Deposition in arid intermontane basins across the Puna Plateau continued through to the present. Deformation was minor in the late Neogene Puna Plateau, with extension locally replacing shortening (Schoenbohm and Strecker, 2009). Late Miocene to Pliocene volcanism, especially associated with voluminous ignimbrites of the so-called Altiplano-Puna volcanic complex flare-up (de Silva, 1989), raised the elevation of the Western Cordillera and Puna Plateau by an average of ~0.5 km.

The eastward migration of the orogenic strain front drove the foreland basin flexural wave rapidly (>45 mm/yr) through the region, stacking wedge-top deposits of the Palo Pintado Formation on top of foredeep facies in the Angastaco and Quebrada de los Colorados Formations, culminating with intermontane basin development during deposition of the San Felipe Formation (Carrapa et al., 2011a). Modest elevations of ~1.5 km appear to have developed during Palo Pintado deposition (9–5 Ma; Bywater-Reyes et al., 2010). The Santa Bárbara Ranges rose to their current maximum elevation of ~2 km mostly in the last 2 m.y. (Reynolds et al., 2000; Hain et al., 2011).

Causes and Broader Implications

Overall, our results break new ground in showing that the western half of the Andes rose to 3–4 km by the late Eocene. We agree with Evenstar et al. (2009) and Jordan et al. (2010) that the western slope of the Andes could have stood ~2 km above the adjacent forearc (now at ~1–1.5 km) by the early to mid-Miocene, and rose an additional ~1 km since then, partly due to ignimbrite additions. Our results also show that these high elevations were in place by the late Eocene and encompassed the Cordillera de Domeyko and most of the Puna Plateau. Much of the rest of the Andes out to the Eastern Cordillera was elevated ≥3 km by ca. 15 Ma, and the Subandes are the focus of most recent growth in the past 5 m.y.

This uplift history provides support for some existing models of central Andean uplift but contradicts others. For example, Lamb and Davis (2003) suggested that drying of the Atacama Desert due to global cooling in the Neogene caused uplift of the Andes. In this view, hyperaridification of the Atacama by the mid-Miocene (Rech et al., 2010) starved sediment supply to the adjacent Peru Trench, increasing shear stresses in the subduction zone. According to Lamb and Davis (2003), this led to accelerated uplift of the Andes in the last 10–15 m.y. Clearly, this explanation does not hold up if most of the Andes was elevated >3 km prior to 15 Ma, as we suggest here. Our evidence does not preclude the role of aridity in influencing tectonics, in that aridification of the Andes probably has promoted sediment storage and slowed erosion in the Andes, thus contributing to its height and eastward expansion (e.g., Sobel et al., 2003).

Others have suggested that the Andes, not at this latitude but further north in the Altiplano, rose en masse from ≤2 km to 4 km since 10 Ma (Ghosh et al., 2006a; Garzione et al., 2008). In our view, the isotopic evidence presented in these papers for late Miocene uplift is quite convincing. However, we would caution that the data reported are from the eastern Altiplano, and it is premature to speculate about young uplift for the entire Andean orogen. Additional paleoelevation data are needed to assess the uplift history of the western Altiplano and adjacent Western Cordillera.

In our opinion, the original idea that changes in convergence rates between South America and the Nazca plate control uplift of the Andes (Pardo-Casas and Molnar, 1987) still has significant explanatory power. Convergence rates have fluctuated between rapid (15 cm/yr) during 50–40 Ma and 20–10 Ma and slower (5–10 cm/yr) rates during 70–50, 40–20, and 10–0 Ma (Pardo-Casas and Molnar, 1987; Somoza, 1998). Peak convergence rates coincided in time with the rise of the Cordillera de Domeyko (at around 50–40 Ma) and Eastern Cordillera (by 15 Ma). If this idea is correct, the question remains open on how high convergence rates translate into uplift. Overall shortening has been estimated to be >100–150 km for the Cenozoic at this latitude and in Bolivia (for summary, see Kley and Monaldi, 1998), creating mass-balance problems and instabilities. The excess shallow crust not eroded away contributes to mountain building. Deeper crust and lithosphere can be removed or redistributed in a variety of ways beyond crustal thickening, including ablative removal (Pope and Willet, 1998), foundering of eclogitized mantle lithosphere (Kay and Gordillo, 1994; Ducea and Saleeby, 1998; Zandt et al., 2004; Garzione et al., 2006, 2008; Ducea et al., 2013), or ductile lower-crustal flow and underthrusting (Allmendinger and Gubbels, 1996; Barke and Lamb, 2006).

Haschke et al. (2006) made an intriguing attempt to unify the disposal of mantle lithosphere by delamination with plate convergence rates, igneous activity, slab breakoff, and ultimately uplift to explain cycles of Andean mountain building, much as DeCelles et al. (2009) proposed for western North America. Building on Kay and Kay (1991) and many others, they suggest that periods of rapid convergence lead to high-flux igneous events and attendant lithospheric thickening and densification (by eclogitization) of the mantle lithosphere and/or descending ocean slab. The eclogitic lithosphere is dense and therefore unstable, and it founders, causing uplift of the overlying crust as the dense eclogitic keel recycles into the asthenosphere. The sequence of events envisioned by Haschke et al. (2006) requires rapid con-

vergence followed by high-flux volcanic events, followed by lithospheric foundering and uplift. Haschke et al. (2006) summarized the geochemical, geophysical, and geologic evidence in support of two such cycles in the Cenozoic (Fig. 12). The first cycle spanned 70–40 Ma, and our data suggest it culminated

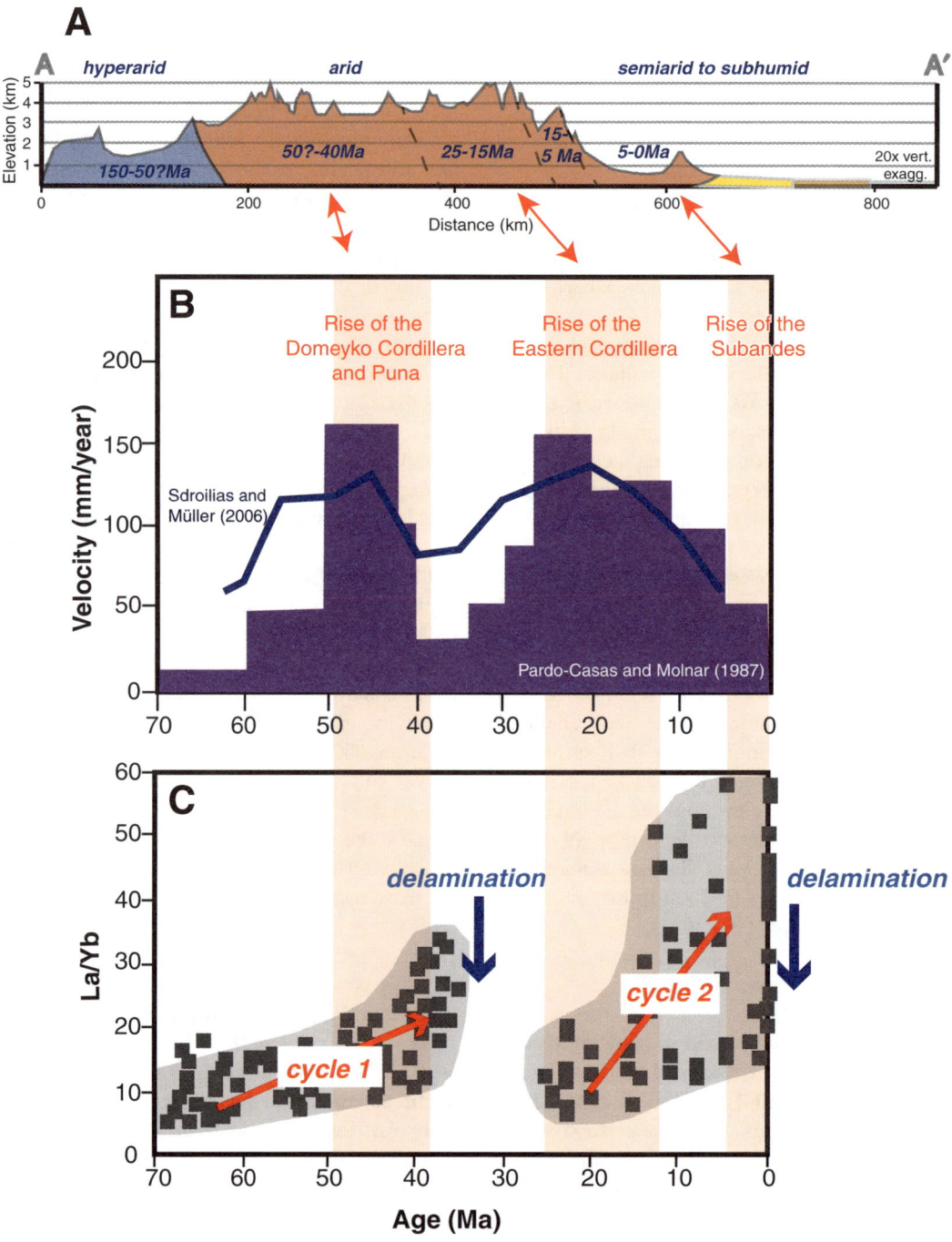

Figure 12. Summary of Andean uplift 22°S–26°S: (A) growth increments of the central Andes during the Cenozoic, (B) convergence rates between South America and the Nazca plate (modified from Pardo-Casas and Molnar, 1987; Somoza, 1998), and (C) cordilleran growth cycles as expressed by La/Yb ratios of Andean igneous rocks, modified from Haschke et al. (2006). La/Yb ratios are thought to positively correlate with crustal thickness. Blue arrows denote times of possible crustal foundering events.

in the rise of the western half of the Andes to high elevations by the late Eocene. The second cycle of increased convergence and volcanism occurred after the Oligocene magmatic gap starting ca. 25 Ma. This cycle plausibly culminated in the rise of the Eastern Cordillera by 15 Ma, completing construction of most of the modern Andes. Eruption of the voluminous Altiplano-Puna volcanic complex in the last 10 m.y. may be an expression of the ongoing status of this second "cordilleran" cycle (DeCelles et al., 2009), as the deformation front at this latitude stepped rapidly eastward, building the Subandes.

Despite these general correlations, the idea that lithospheric removal led to all Andean uplift does not appear consistent with our evidence. For one, according to isotopic evidence from Haschke et al. (2006), the first lithospheric removal event during the Cenozoic should have occurred early in the Oligocene ca. 35 Ma (Fig. 12). This would be too late to explain the Eocene uplift of the Cordillera de Domeyko and Puna Plateau. The 38 Ma date from tuffs and evaporites at base of the Fraile section provides the critical minimum age constraint for uplift to high elevation of the Puna Plateau. Therefore, we interpret the attainment of high elevations on the Puna Plateau by at least 38 Ma as a result of crustal thickening *preceding* lithospheric removal. Lastly, according to isotopic evidence from Haschke et al. (2006), the Andes is currently undergoing a second Cenozoic cycle of crustal thickening (not lithospheric removal), consistent with the second rapid phase of flexural wave propagation and the recent growth of the Subandes (Fig. 12; DeCelles et al., 2011).

New geochemical and geophysical evidence suggests that lithospheric removal has and currently is occurring beneath the Andes, but not on the sort of regional scale required to explain the three broad uplift events that we now recognize (Beck et al., this volume; Murray et al., this volume). Rather, geophysical data point to small-scale (<100 km^2), low-velocity zones, especially beneath recent large ignimbrite fields (Beck et al., this volume). If representative of the entire central Andes, our conclusions pose a challenge to our colleagues working to the north in the Altiplano, where studies suggest that lithospheric removal (Ghosh et al., 2006a; Garzione et al., 2006, 2008) is the cause of major uplift in that region. The Andes of Bolivia and southern Peru experienced similar convergence rates in the Cenozoic as our study area. The validity of our suggestion that crustal thickening, not lithospheric removal, plays a central role in uplift can be tested by an integrated, full-orogen reconstruction of uplift at that latitude. Special attention should be paid to the relatively unstudied western Andes, where we now know that uplift in the Puna was much earlier (Eocene) than previously recognized.

ACKNOWLEDGMENTS

We gratefully acknowledge the support of our entire Convergent Orogenic System Analysis research group by ExxonMobil. This paper benefited from numerous discussions with Mike Blum, Gary Johnson, Jane Lock, Mike McGroder, Kurt Rudolf, and especially Joellen Russell and Jerry Kendall. We are very grateful to John Eiler, Glynis Jehle, Nami Kitchen, and Ricardo Alonso for all their assistance. Finally, beware of waxy Argentinian diesel fuel: We thank two Argentinian geologists whose names we forgot for giving us a long ride to Susques after the diesel in our truck froze.

REFERENCES CITED

Allmendinger, R., and Gubbels, T., 1996, Pure and simple shear plateau uplift—Puna, Argentina and Bolivia: Tectonophysics, v. 259, p. 1–13, doi:10.1016/0040-1951(96)00024-8.

Alonso, R.N., Jordan, T.E., Tabutt, K.T., and Vandervoort, D.S., 1991, Giant evaporite beds of the Neogene central Andes: Geology, v. 19, p. 401–404, doi:10.1130/0091-7613(1991)019<0401:GEBOTN>2.3.CO;2.

Arriagiada, C., Cobbold, P., and Roperch, P., 2006, Salar de Atacama Basin; a record of compressional tectonics in the central Andes since the mid-Cretaceous: Tectonics, v. 25, TC1008, doi:10.1029/2004TC001770.

Barke, R., and Lamb, S., 2006, Late Cenozoic uplift of the Eastern Cordillera, Bolivia: Earth and Planetary Science Letters, v. 249, p. 350–367, doi:10.1016/j.epsl.2006.07.012.

Barnes, J.B., Ehlers, T.A., Insel, N., McQuarrie, N., and Poulsen, C.J., 2012, Linking orography, climate, and exhumation across the central Andes: Geology, v. 40, no. 12, p. 1135–1138, doi:10.1130/G33229.1.

Beck, S.L., Zandt, G., Ward, K.M., and Scire, A., 2015, this volume, Multiple styles and scales of lithospheric foundering beneath the Puna Plateau, central Andes, *in* DeCelles, P.G., Ducea, M.N., Carrapa, B., and Kapp, P.A., eds., Geodynamics of a Cordilleran Orogenic System: The Central Andes of Argentina and Northern Chile: Geological Society of America Memoir 212, doi:10.1130/2015.1212(03).

Boyd, J., 2010, Tectonic Evolution of the Arizaro Basin of the Puna Plateau, NW Argentina: Implications for Plateau-Scale Processes [M.S. thesis]: Laramie, Wyoming, University of Wyoming, 83 p.

Bywater-Reyes, S., Carrapa, B., Clementz, M., and Schoenbohm, L., 2010, Effect of late Cenozoic aridification on sedimentation in the Eastern Cordillera of northwest Argentina (Angastaco Basin): Geology, v. 38, no. 3, p. 235–238, doi:10.1130/G30532.1.

Canavan, R.R., Carrapa, B., Clementz, M.T., Quade, J., DeCelles, P., and Schoenbohm, L.M., 2014, Early Cenozoic uplift of the Puna Plateau, central Andes, based on stable isotope paleoaltimetry of hydrated volcanic glass: Geology, v. 42, no. 5, p. 447–450, doi: 10.1130/G35239.1

Carrapa, B., and DeCelles, P.G., 2008, Eocene exhumation and basin development in the Puna of northwestern Argentina: Tectonics, v. 27, TC1015, doi: 10.1029/2007TC002127.

Carrapa, B., and DeCelles, P.G., 2015, this volume, Regional exhumation and kinematic history of the central Andes in response to cyclical orogenic processes, *in* DeCelles, P.G., Ducea, M.N., Carrapa, B., and Kapp, P.A., eds., Geodynamics of a Cordilleran Orogenic System: The Central Andes of Argentina and Northern Chile: Geological Society of America Memoir 212, doi:10.1130/2015.1212(11).

Carrapa, B., Adelmann, D., Hilley, G.E., Mortimer, E., Sobel, E.R., and Strecker, M.R., 2005, Oligocene range uplift and development of plateau morphology in the southern central Andes: Tectonics, v. 24, TC4011, doi:10.1029/2004TC001762.

Carrapa, B., Bywater-Reyes, S., DeCelles, P.G., Mortimer, E., and Gehrels, G.E., 2011a, Late-Pliocene basin evolution in the Eastern Cordillera of northwestern Argentina (25–26°S): Regional implications for Andean orogenic wedge development: Basin Research, v. 23, p. 1–20.

Carrapa, B., Trimble, J.D., and Stockli, D.F., 2011b, Patterns and timing of exhumation and deformation in the Eastern Cordillera of NW Argentina revealed by (U-Th)/He thermochronology: Tectonics, v. 30, TC3003, doi:10.1029/2010TC002707.

Carrapa, B., Bywater-Reyes, S., Safipour, R., Sobel, E.R., Schoenbohm, L.M., DeCelles, P.G., Reiners, P.W., and Stockli, D., 2014a, The effect of inherited paleotopography on exhumation of the Central Andes of NW Argentina: Geological Society of America Bulletin, v. 126, no. 1–2, p. 66–77, doi:10.1130/B30844.1.

Carrapa, B., Huntington, K.H., Clementz, M., Bywater-Reyes, S., Quade, J., Schoenbohm, L., and Canavan, R., 2014b, Uplift of the Central Andes of NW Argentina associated with upper crustal shortening, revealed by multi-proxy isotopic analyses: Tectonics, p. 1039–1054, doi:10.1002/2013TC003461.

Cassel, E.J., Graham, S.A., and Chamberlain, C.P., 2009, Cenozoic tectonic and topographic evolution of the northern Sierra Nevada, California, through stable isotope paleoaltimetry of volcanic glass: Geology, v. 37, no. 6, p. 547–550, doi:10.1130/G25572A.1.

Cerling, T.E., and Quade, J., 1993, Stable carbon and oxygen isotopes in soil carbonates, *in* Swart, P., McKenzie, J.A., and Lohman, K.C., eds., Continental Indicators of Climate, Proceedings of Chapman Conference, Jackson Hole, Wyoming: American Geophysical Union Geophysical Monograph 78, p. 217–231.

Cohen, A., McGlue, M.M., Ellis, G.S., Zani, H., Swarzenski, P.W., Assine, M.L., and Silva, A., 2015, this volume, Lake formation, characteristics, and evolution in retroarc deposystems: A synthesis of the modern Andean orogen and its associated basins, *in* DeCelles, P.G., Ducea, M.N., Carrapa, B., and Kapp, P.A., eds., Geodynamics of a Cordilleran Orogenic System: The Central Andes of Argentina and Northern Chile: Geological Society of America Memoir 212, doi:10.1130/2015.1212(16).

Coutand, I., Carrapa, B., Deekin, A., Schmidt, A.K., Sobel, E.R., and Strecker, M.R., 2006, Propagation of orographic barriers along an active range front: Insights from sandstone petrography and detrital apatite fission-track thermochronology in the intramontane Angastaco Basin, NW Argentina: Basin Research, v. 18, p. 1–26, doi:10.1111/j.1365-2117.2006.00283.x.

DeCelles, P.G., Carrapa, B., and Gehrels, G.E., 2007, Detrital zircon U-Pb ages provide provenance and chronostratigraphic information from Eocene synorogenic deposits in northwestern Argentina: Geology, v. 35, no. 4, p. 323–326, doi:10.1130/G23322A.1.

DeCelles, P.G., Ducea, M.N., Kapp, P., and Zandt, G., 2009, Cyclicity in Cordilleran orogenic systems: Nature Geoscience, v. 2, p. 251–257, doi:10.1038/ngeo469.

DeCelles, P.G., Carrapa, B., Horton, B.K., and Gehrels, G.E., 2011, Cenozoic foreland basin system in the central Andes of northwestern Argentina: Implications for Andean geodynamics and modes of deformation: Tectonics, v. 30, TC6013, doi:10.1029/2011TC002948.

DeCelles, P.G., Zandt, G., Beck, S.L., Currie, C.A., Ducea, M.N., Kapp, P., Gehrels, G.E., Carrapa, B., Quade, J., and Schoenbohm, L.M., 2015, this volume, Cyclical orogenic processes in the Cenozoic central Andes, *in* DeCelles, P.G., Ducea, M.N., Carrapa, B., and Kapp, P.A., eds., Geodynamics of a Cordilleran Orogenic System: The Central Andes of Argentina and Northern Chile: Geological Society of America Memoir 212, doi:10.1130/2015.1212(22)

Deeken, A., Sobel, E.R., Coutand, I., Haschke, M., Riller, U., and Strecker, M.R., 2006, Development of the southern Eastern Cordillera, NW Argentina, constrained by apatite fission track thermochronology: From Early Cretaceous extension to middle Miocene shortening: Tectonics, v. 25, TC6003, doi:10.1029/2005TC001894.

Dennis, K.J., and Schrag, D.P., 2010, Clumped isotope thermometry of carbonatites as an indicator of diagenetic alteration: Geochimica et Cosmochimica Acta, v. 74, p. 4110–4122, doi:10.1016/j.gca.2010.04.005.

Dennis, K.J., Affeck, H.P., Passey, B.H., Schrag, D.P., and Eiler, J.M., 2011, Defining the absolute reference frame for clumped isotope studies of CO_2: Geochimica et Cosmochimica Acta, v. 75, p. 7117–7131, doi:10.1016/j.gca.2011.09.025.

de Silva, S., 1989, Altiplano-Puna volcanic complex of the central Andes: Geology, v. 17, no. 12, p. 1102–1106, doi:10.1130/0091-7613(1989)017<1102:APVCOT>2.3.CO;2.

de Silva, S.L., and Gosnold, W.D., 2007, Episodic construction of batholiths: Insights from the spatiotemporal development of an ignimbrite flare-up: Journal of Volcanology and Geothermal Research, v. 167, no. 1–4, p. 320–335, doi:10.1016/j.jvolgeores.2007.07.015.

Dettinger, M.P., and Quade, J., 2015, this volume, Testing the analytical protocols and calibration of volcanic glass for the reconstruction of hydrogen isotopes in paleoprecipitation, *in* DeCelles, P.G., Ducea, M.N., Carrapa, B., and Kapp, P.A., eds., Geodynamics of a Cordilleran Orogenic System: The Central Andes of Argentina and Northern Chile: Geological Society of America Memoir 212, doi:10.1130/2015.1212(14).

Ducea, M., and Saleeby, J., 1998, The age and origin of a thick-ultramafic keel from beneath the Sierra Nevada batholith: Contributions to Mineralogy and Petrology, v. 133, p. 169–185, doi:10.1007/s004100050445.

Ducea, M.N., Seclaman, A.C., Murray, K.E., Jianu, D., and Schoenbohm, L.M., 2013, Mantle-drip magmatism beneath the Altiplano-Puna Plateau, central Andes: Geology, v. 41, no. 8, p. 915–918, doi:10.1130/G34509.1.

Echavarria, L., Hernandez, R., Allmendinger, R., and Reynolds, J., 2003, Subandean thrust and fold belt of northwestern Argentina; geometry and timing of Andean evolution: American Association of Petroleum Geologists Bulletin, v. 87, no. 6, p. 965–985, doi:10.1306/01200300196.

Ehlers, T.A., and Poulsen, C.J., 2009, Influence of Andean uplift on climate and paleoaltimetry estimates: Earth and Planetary Science Letters, v. 281, p. 238–248, doi:10.1016/j.epsl.2009.02.026.

Eiler, J.M., and Schauble, E.A., 2004, $^{18}O^{13}C^{16}O$ in Earth's atmosphere: Geochimica et Cosmochimica Acta, v. 68, no. 23, p. 4767–4777, doi:10.1016/j.gca.2004.05.035.

Evenstar, L.A., Hartley, A.J., Stuart, F.M., Mather, A.E., Rice, C.M., and Chong, G., 2009, Multiphase development of the Atacama planation surface recorded by cosmogenic ^3He exposure ages: Implications for uplift and Cenozoic climate change in western South America: Geology, v. 37, no. 1, p. 27–30, doi:10.1130/G25437A.1.

Francis, P., and Hawkesworth, C., 1994, Late Cenozoic rates of magmatic activity in the central Andes and their relationships to continental-crust formation and thickening: Journal of the Geological Society, London, v. 151, p. 845–854, doi:10.1144/gsjgs.151.5.0845.

Friedman, I., Gleason, J., Sheppard, R.A., and Gude, A.J., 1993, Deuterium fractionation as water diffuses into silicic volcanic ash, *in* Swart, P., McKenzie, J.A., and Lohman, K.C., eds., Continental Indicators of Climate, Proceedings of Chapman Conference, Jackson Hole, Wyoming: American Geophysical Union Geophysical Monograph 78, p. 321–323.

Garzione, C., Molnar, P., Libarkin, J.C., and MacFadden, B.J., 2006, Rapid late Miocene rise of the Bolivian Altiplano: Evidence for removal of the mantle lithosphere: Earth and Planetary Science Letters, v. 241, p. 543–556, doi:10.1016/j.epsl.2005.11.026.

Garzione, C., Hoke, G., Libarkin, J., and Withers, S., 2008, Rise of the Andes: Science, v. 320, p. 1304–1307, doi:10.1126/science.1148615.

Gehrels, G.E., Valencia, V., and Ruiz, J., 2008, Enhanced precision, accuracy, efficiency, and spatial resolution of U-Pb ages by laser ablation–multicollector–inductively coupled plasma–mass spectrometry: Geochemistry Geophysics Geosystems, v. 9, Q03017, doi:10.1029/2007GC001805.

Ghosh, P., Eiler, J.M., and Garzione, C., 2006a, Rapid uplift of the Altiplano revealed in abundances of ^{13}C–^{18}O bonds in paleosol carbonate: Science, v. 311, p. 511–515, doi:10.1126/science.1119365.

Ghosh, P., Adkins, J., Affek, H., Balta, B., Guo, W., Schauble, E.A., Schrag, D., and Eiler, J.M., 2006b, ^{13}C–^{18}O bonds in carbonate minerals: A new kind of paleothermometer: Geochimica et Cosmochimica Acta, v. 70, p. 1439–1456, doi:10.1016/j.gca.2005.11.014.

Gregory-Wodzicki, K.M., 2000, Uplift history of the central northern Andes: Geological Society of America Bulletin, v. 112, no. 7, p. 1091–1105, doi:10.1130/0016-7606(2000)112<1091:UHOTCA>2.0.CO;2.

Hain, M.P., Stecjer, M.R., Bookhagen, B., Alonso, R.N., Pingel, H., and Schmitt, A.K., 2011, Neogene to Quaternary broken foreland formation and sedimentation dynamics in the Andes of NW Argentina (25°S): Tectonics, v. 30, TC2006, doi:10.1029/2010TC002703.

Hartley, A.J., Flint, S., Tuner, P., and Jolley, E.J., 1992, Tectonic controls on the development of a semi-arid, alluvial basin as reflected in the stratigraphy of the Purilactis Group (Upper Cretaceous–Eocene), northern Chile: Journal of South American Earth Sciences, v. 5, p. 275–296, doi:10.1016/0895-9811(92)90026-U.

Haschke, M., and Gunther, A., 2003, Balancing crustal thickening in arcs by tectonic vs. magmatic means: Geology, v. 31, no. 11, p. 933–936, doi:10.1130/G19945.1.

Haschke, M., Gunther, A., Melnick, D., Echtler, H., Reutter, K.J., Scheuber, E., and Oncken, O., 2006, Central and southern Andean tectonic evolution inferred from arc magmatism, *in* Oncken, O., Chong, G., Franz, G., Giese, P., Götze, H., Ramos, V.A., Strecker, M.R., and Wigger, P., eds., The Andes—Active Subduction Orogeny: Berlin, Springer-Verlag, Frontiers in Earth Sciences, v. 1, p. 337–353.

Henkes, G.A., Passey, B.H., Grossman, E.L., Shenton, B.J., Pérez-Huerta, A., and Yancey, T.E., 2014, Temperature limits for preservation of primary calcite clumped isotope paleotemperatures: Geochimica et Cosmochimica Acta, v. 139, p. 362–382, doi:10.1016/j.gca.2014.04.040.

Hilley, G.E., and Strecker, M.R., 2005, Processes of oscillatory basin filling and excavation in a tectonically active orogen: Quebrada del Toro Basin, NW Argentina: Geology, v. 117, no. 7–8, p. 887–901.

Hoke, G.D., and Garzione, C.N., 2008, Paleosurfaces, paleoelevation, and the mechanisms for the late Miocene topographic development of the

Altiplano Plateau: Earth and Planetary Science Letters, v. 271, p. 192–201, doi:10.1016/j.epsl.2008.04.008.

Hongn, F.D., del Papa, C., Powell, J., Petrinovic, I., Mon, R., and Deraco, V., 2007, Middle Eocene deformation and sedimentation in the Puna–Eastern Cordillera transition (23′26°S); control of pre-existing heterogeneities on the pattern of initial Andean shortening: Geology, v. 35, p. 271–274, doi:10.1130/G23189A.1.

Horton, B., 1999, Erosional control on the geometry and kinematics of thrust belt development in the central Andes: Tectonics, v. 18, no. 6, p. 1292–1304, doi:10.1029/1999TC900051.

Humphrey, N.F., and Konrad, S.K., 2000, River incision or diversion in response to bedrock uplift: Geology, v. 28, p. 43–46, doi:10.1130/0091-7613(2000)28<43:RIODIR>2.0.CO;2.

Huntington, K.W., Eiler, J.M., Affek, H.P., Guo, W., Bonifacie, M., Yeung, L.Y., Thiagarajan, N., Passey, B., Tripati, A., Daëron, M., and Came, R., 2009, Methods and limitations of "clumped" CO_2 isotope (Δ_{47}) analysis by gas-source isotope ratio mass spectrometry: Journal of Mass Spectrometry, v. 44, p. 1318–1329, doi:10.1002/jms.1614.

Huntington, K.W., Saylor, J., Quade, J., and Hudson, A.M., 2014, High Late Miocene-Pliocene elevation of the Zhada basin, SW Tibetan plateau, from clumped isotope thermometry: Geological Society of America Bulletin, doi:10.1130/B31000.1.

Isacks, B., 1988, Uplift of the central Andean Plateau and bending of the Bolivian orocline: Journal of Geophysical Research–Solid Earth and Planets, v. 93, no. B4, p. 3211–3231.

Jordan, T.E., Nester, P.L., Blanco, N., Hoke, G.D., Dávila, F., and Tomlinson, A.J., 2010, Uplift of the Altiplano-Puna Plateau: A view from the west: Tectonics, v. 29, TC5007, doi:10.1029/2010TC002661.

Kay, R.W., and Kay, S.M., 1991, Creation and destruction of lower continental crust: Geologische Rundschau, v. 80, p. 259–278, doi:10.1007/BF01829365.

Kay, S.M., and Gordillo, C.E., 1994, Pocho volcanic rocks and the melting of depleted continental lithosphere above a shallowly dipping subduction zone in the central Andes: Contributions to Mineralogy and Petrology, v. 117, p. 25–44, doi:10.1007/BF00307727.

Kay, S.M., and Coira, B., 2009, Shallowing and steepening subduction zones, continental lithospheric loss, magmatism, and crustal flow under the central Andes, *in* Kay, S.M., Ramos, V.A., and Dickinson, W.R., eds., Backbone of the Americas: Shallow Subduction, Plateau Uplift, and Ridge and Terrane Collision: Geological Society of America Memoir 204, p. 229–259.

Kay, S.M., Coira, B.L., Caffe, P.J., and Chen, C.-H., 2010, Regional chemical diversity, crustal and mantle sources and evolution of central Andean Puna Plateau ignimbrites: Journal of Volcanology and Geothermal Research, v. 198, no. 1–2, p. 81–111, doi:10.1016/j.jvolgeores.2010.08.013.

Kim, S.-T., and O'Neil, J.R., 1997, Equilibrium and nonequilibrium oxygen isotope effects in synthetic carbonates: Geochimica et Cosmochimica Acta, v. 61, no. 16, p. 3461–3475, doi:10.1016/S0016-7037(97)00169-5.

Kley, J., and Monaldi, C.R., 1998, Tectonic shortening and crustal thickness in the central Andes: How good is the correlation?: Geology, v. 26, no. 8, p. 723–726, doi:10.1130/0091-7613(1998)026<0723:TSACTI>2.3.CO;2.

Kraemer, B., Adelmann, D., Alten, M., Schurr, W., Erpenstein, K., Kiefer, E., van den Bogaard, P., and Görler, K., 1999, Incorporation of the Paleogene foreland into the Neogene Puna Plateau: The Salar de Antofalla area, NW Argentina: Journal of South American Earth Sciences, v. 12, p. 157–182, doi:10.1016/S0895-9811(99)00012-7.

Lamb, S., and Davis, S.D., 2003, Cenozoic climate change as a possible cause for the rise of the Andes: Nature, v. 425, p. 792–797, doi:10.1038/nature02049.

Maksaev, V., and Zentilli, M., 1999, Fission track thermochronology of the Domeyko Cordillera, northern Chile; implications for Andean tectonics and porphyry copper metallogenesis: Exploration and Mining Geology, v. 8, no. 1–2, p. 65–89.

Marquillas, R., Sabino, I., Sial, A.N., del Papa, C., Ferreira, V., and Mathews, S., 2007, Carbon and oxygen isotopes of Maastrichtian–Danian shallow marine carbonates: Yacoraite Formation, northwestern Argentina: Journal of South American Earth Sciences, v. 23, p. 304–320, doi:10.1016/j.jsames.2007.02.009.

Mégard, F., 1984, The Andean orogenic period and its major structures in central and northern Peru: Journal of the Geological Society, London, v. 141, p. 893–900, doi:10.1144/gsjgs.141.5.0893.

Metcalf, K., and Kapp, P., 2015, this volume, Along-strike variations in crustal seismicity and modern lithospheric structure of the central Andean forearc, *in* DeCelles, P.G., Ducea, M.N., Carrapa, B., and Kapp, P.A., eds., Geodynamics of a Cordilleran Orogenic System: The Central Andes of Argentina and Northern Chile: Geological Society of America Memoir 212, doi:10.1130/2015.1212(04).

Mpodozis, C., Arriagiada, C., Basso, M., Roperch, P., Cobbold, P., and Reich, M., 2005, Late Mesozoic to Paleogene stratigraphy of the Salar de Atacama Basin, Antofagasta, northern Chile: Implications for the tectonic evolution of the central Andes: Tectonophysics, v. 399, p. 125–154, doi:10.1016/j.tecto.2004.12.019.

Mulch, A., Sarna-Wodjicki, A.M., Perkins, M.E., and Chamberlain, C.P., 2008, A Miocene to Pleistocene climate and elevation record of the Sierra Nevada (California): Proceedings of the National Academy of Sciences of the United States of America, v. 105, no. 19, p. 6819–6824, doi:10.1073/pnas.0708811105.

Murray, K.E., Ducea, M.N., and Schoenbohm, L., 2015, this volume, Foundering-driven lithospheric melting: The source of central Andean mafic lavas on the Puna Plateau (22°S–27°S), *in* DeCelles, P.G., Ducea, M.N., Carrapa, B., and Kapp, P.A., eds., Geodynamics of a Cordilleran Orogenic System: The Central Andes of Argentina and Northern Chile: Geological Society of America Memoir 212, doi:10.1130/2015.1212(08).

Palma, R.M., 2000, Lacustrine facies in the upper Cretaceous Balbuena subgroup (Salta Group): Andina Basin, Argentina, *in* Gierolowski-Kordesch, E.H., and Kelts, K.R., eds., Lake Basins through Space and Time: American Association of Petroleum Geologists, Studies in Geology, v. 46, p. 323–328.

Pardo-Casas, F., and Molnar, P., 1987, Relative motion of the Nazca (Farallon) plate since Late Cretaceous time: Tectonics, v. 6, no. 3, p. 233–248, doi:10.1029/TC006i003p00233.

Passey, B.H., and Henkes, G.A., 2012, Carbonate clumped isotope bond reordering and geospeedometry: Earth and Planetary Science Letters, v. 351–352, p. 223–236, doi:10.1016/j.epsl.2012.07.021.

Passey, B.H., Levin, N.E., Cerling, T.E., Brown, F.H., and Eiler, J.M., 2010, High-temperature environments of human evolution in East Africa based on bond-ordering in paleosol carbonates: Proceedings of the National Academy of Sciences of the United States of America, v. 107, no. 25, p. 11,245–11,249, doi:10.1073/pnas.1001824107.

Pope, D.C., and Willet, S.D., 1998, Thermal-mechanical model for crustal thickening in the central Andes driven by ablative subduction: Geology, v. 26, p. 511–514, doi:10.1130/0091-7613(1998)026<0511:TMMFCT>2.3.CO;2.

Quade, J., Rech, J., Latorre, C., Betancourt, J., Gleason, E., and Kalin-Arroyo, M., 2007, Soils at the hyperarid margin: The isotopic composition of soil carbonate from the Atacama Desert: Geochimica et Cosmochimica Acta, v. 71, p. 3772–3795, doi:10.1016/j.gca.2007.02.016.

Quade, J., Eiler, J., Daëron, M., and Achuythan, H., 2013, The clumped isotope paleothermometer in soils and paleosol carbonate: Geochimica et Cosmochimica Acta, v. 105, p. 92–107, doi:10.1016/j.gca.2012.11.031.

Rech, J.A., Currie, B.S., Shullenberger, E.D., Dunagan, S.P., Jordan, T.E., Blanco, N., Tomlinson, A.J., Rowe, H.D., and Houston, J., 2010, Evidence for the development of the Andean rain shadow from a Neogene isotopic record in the Atacama Desert, Chile: Earth and Planetary Science Letters, v. 292, p. 371–382, doi:10.1016/j.epsl.2010.02.004.

Reiners, P.W., Thomson, S.N., Vernon, A., Willett, S.D., Zattin, M., Einhorn, J., Gehrels, G., Quade, J., Pearson, D., Murray, K.E., and Cavazza, W., 2015, this volume, Low-temperature thermochronologic trends across the central Andes, 21°S–28°S, *in* DeCelles, P.G., Ducea, M.N., Carrapa, B., and Kapp, P.A., eds., Geodynamics of a Cordilleran Orogenic System: The Central Andes of Argentina and Northern Chile: Geological Society of America Memoir 212, doi:10.1130/2015.1212(12).

Reynolds, J., Galli, C., Hernández, R., Idleman, B., Kotila, J., Hilliard, R., and Naeser, C., 2000, Middle Miocene tectonic development of the Transition zone, Salta Province, northwest Argentina: Magnetic stratigraphy from the Metán Subgroup, Sierra de González: Geological Society of America Bulletin, v. 112, p. 1736–1751, doi:10.1130/0016-7606(2000)112<1736:MMTDOT>2.0.CO;2.

Rowley, D.B., and Garzione, C.N., 2007, Stable isotope-based paleoaltimetry: Annual Review of Earth and Planetary Sciences, v. 35, p. 463–508, doi:10.1146/annurev.earth.35.031306.140155.

Salfity, J.A., and Marquillas, R.A., 1994, Tectonics and sedimentary evolution of the Cretaceous–Eocene Salta Group Basin, Argentina, *in* Salfity, J.A., ed., Cretaceous Tectonics of the Andes: Braunschweig-Wiesbaden, Germany, Friedrich Viewig and Sohn, Earth Evolution Sciences, p. 266–315.

Schoenbohm, L.M., and Carrapa, B., 2011, Evidence from timing of contraction, extension, sedimentation, and magmatism for small-scale lithospheric foundering in the Puna Plateau, NW Argentina: San Francisco, California, American Geophysical Union, Fall meeting, abstract T13I-07.

Schoenbohm, L.M., and Strecker, M.R., 2009, Normal faulting along the southern margin of the Puna Plateau, northwest Argentina: Tectonics, v. 28, no. 5, TC5008, doi:10.1029/2008TC002341.

Sdrolias, M., and Müller, R.D., 2006, Controls on back-arc basin formation: Geochemistry Geophysics Geosystems, v. 7, no. 4, doi: 101029/2005GC001090.

Siks, B.C., and Horton, B.K., 2011, Growth and fragmentation of the Andean foreland basin during eastward advance of fold-thrust deformation, Puna Plateau and Eastern Cordillera, northern Argentina: Tectonics, v. 30, no. 6, TC6017, doi:10.1029/2011TC002944.

Sobel, E.R., Hilley, G.E., and Strecker, M.R., 2003, Formation of internally drained contractional basins by aridity-limited bedrock incision: Journal of Geophysical Research, v. 108, 2344, doi:10.1029/2002JB001883.

Soler, P., and Bonhomme, M.G., 1990, Relation of magmatic activity to plate dynamics in central Peru from Late Cretaceous to present, *in* Kay, S.M., and Rapela, C.W., eds., Plutonism from Antarctica to Alaska: Geological Society of America Special Paper 241, p. 173–192.

Somoza, R., 1998, Updated Nazca (Farallon)–South America relative motions during the last 40 Ma: Implications for mountain building in the central Andean region: Journal of South American Earth Sciences, v. 11, no. 3, p. 211–215, doi:10.1016/S0895-9811(98)00012-1.

Stacey, J.S., and Kramers, J.D., 1975, Approximation of terrestrial lead isotope evolution by a two-stage model: Earth and Planetary Science Letters, v. 26, p. 207–221, doi:10.1016/0012-821X(75)90088-6.

Starck, D., and Vergani, G., 1996, Desarrollo tecto-sedimentario del Cenozoico en el sur de la Provincia de Salta, Argentina, *in* XIII Congreso Geológico Argentino y III Congreso de Exploración de Hidrocarburos: Buenos Aires, Argentina, Actas I, p. 433–452.

Steinmann, G., 1929, Geologie von Peru: Heidelberg, Germany, Karl Winter, 448 p.

Stern, C.R., 1991, Role of subduction erosion in the generation of Andean magmas: Geology, v. 19, p. 78–81, doi:10.1130/0091-7613 (1991)019<0078:ROSEIT>2.3.CO;2.

Strecker, M.R., Alonso, R.N., Bookhagen, B., Carrapa, B., Hilley, G.E., Sobel, E.R., and Trauth, M.H., 2007, Tectonics and climate of the southern central Andes: Annual Review of Earth and Planetary Sciences, v. 35, p. 747–787, doi:10.1146/annurev.earth.35.031306.140158.

Torsvik, T.H., Rousse, S., Labails, C., and Smerhurst, M.A., 2009, A new scheme for the opening of the South Atlantic Ocean and the dissection of Aptian salt basin: Geophysical Journal International, v. 177, no. 3, p. 1315–1333, doi:10.1111/j.1365-246X.2009.04137.x.

Trumbull, R.B., Riller, U., Oncken, O., Scheuber, E., Munier, K., and Hongn, F., 2006, The time-space distribution of Cenozoic volcanism in the south-central Andes: A new data compilation and some tectonic implications, *in* Oncken, O., Chong, G., Franz, G., Giese, P., Götze, H., Ramos, V.A., Strecker, M.R., and Wigger, P., eds., The Andes—Active Subduction Orogeny: Berlin, Springer-Verlag, Frontiers in Earth Sciences, v. 1, p. 29–43.

Uba, C.E., Heubeck, C., and Hulka, C., 2006, Evolution of the late Cenozoic Chaco foreland basin, southern Bolivia: Basin Research, v. 18, p. 145–170, doi:10.1111/j.1365-2117.2006.00291.x.

Vandervoort, D.S., Jordan, T.E., Zeitler, P.K., and Alonso, R.N., 1995, Chronology of internal drainage development and uplift, southern Puna Plateau, Argentine central Andes: Geology, v. 23, p. 145–148, doi:10.1130/0091 -7613(1995)023<0145:COIDDA>2.3.CO;2.

Veizer, J., Ala, D., Azmy, K., Bruckschen, P., Buhl, D., Bruhn, F., Carden, G.A.F., Diener, A., Ebneth, S., Godderis, Y., Jasper, T., Korte, C., Pawellek, F., Podlaha, O.G., and Strauss, H., 1999, Sr-87/Sr-86, δ^{13}C and δ^{18}O evolution of Phanerozoic seawater: Chemical Geology, v. 161, p. 59–88, doi:10.1016/S0009-2541(99)00081-9.

Zandt, G., Gilbert, H., Owens, T.J., Ducea, M., and Saleeby, J., 2004, Active foundering of a continental arc root beneath the southern Sierra Nevada in California: Nature, v. 431, p. 41–46, doi:10.1038/nature02847.

Zhou, J., and Lau, K.M., 1998, Does monsoon climate exist over South America?: Journal of Climate, v. 11, p. 1020–1040, doi:10.1175/1520 -0442(1998)011<1020:DAMCEO>2.0.CO;2.

Zhou, R., Shoenbohm, L.M., and Cosca, M., 2013, Recent, slow normal and strike-slip faulting in the Pasto Ventura region of the southern Puna Plateau, NW Argentina: Tectonics, v. 32, p. 19–33, doi:10.1029/2012TC003189.

MANUSCRIPT ACCEPTED BY THE SOCIETY 3 JUNE 2014
MANUSCRIPT PUBLISHED ONLINE 20 NOVEMBER 2014

The Geological Society of America
Memoir 212
2015

Lake formation, characteristics, and evolution in retroarc deposystems: A synthesis of the modern Andean orogen and its associated basins

Andrew Cohen*
Department of Geosciences, University of Arizona, Tucson, Arizona 85721, USA

Michael M. McGlue[†]
Geoffrey S. Ellis
Energy Resources Program, U.S. Geological Survey, Denver, Colorado 80225, USA

Hiran Zani
*Instituto Nacional de Pesquisas Espacias, Divisão de Sensoriamento Remoto,
São José dos Campos, São Paulo, Brazil 12227-010*

Peter W. Swarzenski
U.S. Geological Survey, 400 Natural Bridges Drive, Santa Cruz, California 95060, USA

Mario L. Assine
Department of Applied Geology, Universidade Estadual Paulista, Campus de Rio Claro, Rio Claro, São Paulo, Brazil 13506-900

Aguinaldo Silva
Campus do Pantanal/Geografia, Universidade Federal de Mato Grosso do Sul, Corumbá, Mato Grosso do Sul, Brazil 79304-020

ABSTRACT

Lake deposystems are commonly associated with retroarc mountain belts in the geological record. These deposystems are poorly characterized in modern retroarcs, placing limits on our ability to interpret environmental signals from ancient deposits. To address this problem, we have synthesized our existing knowledge about the distribution, morphometrics, and sedimentary geochemical characteristics of tectonically formed lakes in the central Andean retroarc. Large, active mountain belts such as the Andes frequently create an excess of sediment, to the point that modeling and observational data both suggest their adjacent retroarc basins will be rapidly overfilled by sediments. Lake formation, requiring topographic closure, demands special

*cohen@email.arizona.edu
[†]Current address: Department of Earth and Environmental Sciences, University of Kentucky, Lexington, Kentucky 40506-0053, USA.

Cohen, A., McGlue, M.M., Ellis, G.S., Zani, H., Swarzenski, P.W., Assine, M.L., and Silva, A., 2015, Lake formation, characteristics, and evolution in retroarc deposystems: A synthesis of the modern Andean orogen and its associated basins, *in* DeCelles, P.G., Ducea, M.N., Carrapa, B., and Kapp, P.A., eds., Geodynamics of a Cordilleran Orogenic System: The Central Andes of Argentina and Northern Chile: Geological Society of America Memoir 212, p. 309–335, doi:10.1130/2015.1212(16). For permission to copy, contact editing@geosociety.org. © 2014 The Geological Society of America. All rights reserved.

conditions such as topographic isolation and arid climatic conditions to reduce sediment generation, and bedrock lithologies that yield little siliciclastic sediment.

Lacustrine deposition in the modern Andean retroarc has different characteristics in the six major morphotectonic zones discussed. (1) High-elevation hinterland basins of the arid Puna-Altiplano Plateau frequently contain underfilled and balanced-filled lakes that are potentially long-lived and display relatively rapid sedimentation rates. (2) Lakes are rare in piggyback basins, although a transition zone exists where basins that originally formed as piggybacks are transferred to the hinterland through forward propagation of the thrust belt. Here, lakes are moderately abundant and long-lived and display somewhat lower sedimentation rates than in the hinterland. (3) Wedge-top and (4) foredeep deposystems of the Andean retroarc are generally overfilled, and lakes are small and ephemeral. (5) Semihumid Andean backbulge basins contain abundant small lakes, which are moderately long-lived because of underfilling by sediment and low sedimentation rates. (6) Broken foreland lakes are common, typically underfilled, large, and long-lived playa or shallow systems.

INTRODUCTION

Lake deposits within structural basins contain some of the most complete records of Earth history. In addition to their well-known use in providing long-term, high-resolution records of climate change (Olsen, 1986; Scholz et al., 2007; Melles et al., 2012), these lake deposits also retain useful records of tectonic activity, and a wide array of surficial and biological processes, including weathering and erosion rates. Our ability to extract these records from ancient lake deposits rests on a firm foundation of understanding the linkages among climatic, tectonic, hydrologic, and biologic processes and the resulting sedimentology and stratigraphy preserved on the lake floor. This can best be accomplished with thorough investigations of modern processes and lakes, guided by the theoretical underpinning of facies models, derived either conceptually or with the help of computer modeling.

Prior studies of rift-lake systems clearly illustrate the value of linking modern process studies in sedimentology and geomorphology with subsurface stratigraphy (from coring and shallow seismic-reflection data) in tectonically formed lake basins (e.g., Rosendahl, 1987; Scholz and Rosendahl, 1988; Cohen, 1990; Scholz et al., 1993, 1998; Tiercelin et al., 1994; Soreghan and Cohen, 1996; Cohen et al., 1997; Morley et al., 2000; Colman et al., 2003; McGlue et al., 2006; Karp et al., 2012). This synergy revolutionized our understanding of the ways in which rifts evolve as tectonostratigraphic systems and paved the way for a much more mature use of their deposits in paleoclimatology and basin analysis. In contrast with rift systems, we know relatively little about the sedimentology of modern lake-basin deposystems associated with retroarc mountain belts. Studies of modern orogenic-belt depozones have focused primarily on megafans, alluvial fans, and fluvial/overbank systems, where the linkages to active tectonics are perhaps more immediately evident (e.g., Räsänen et al., 1990, 1992; Mertes et al., 1996; Kronberg et al., 1998; Horton and DeCelles, 2001;

Singh et al., 2001; Aalto et al., 2003; Leier et al., 2005; Assine and Silva, 2009; Horton, 2011). Modern foreland-basin lakes and wetlands have received far less study from the point of view of basin analysis. Whereas a number of individual lake basins within orogenic-belt depozones have been subject to detailed study, the focus of the vast majority of these studies has been for Quaternary paleoclimatology and paleoecology (e.g., Seltzer et al., 2000; Fritz et al., 2004), rather than basin analysis linking climate, ecosystem, geomorphic, and tectonic drivers of lake evolution. This knowledge gap is remarkable given the extensive investigations that have been made on ancient foreland-basin lake deposits in areas such as natural resources (Castle, 1990; Cole, 1998; Rubble and Phillip, 1998; Bohacs et al., 2000), limnogeology (Smoot, 1983; Buchheim and Eugster, 1998), paleoclimatology (Roehler, 1993; Cole, 1998; Sloan and Morrill, 1998; Cabrera et al., 2002), sediment-delivery modeling, and paleohydrology in large lake basins (Pietras et al., 2003), not to mention the controversies that have surrounded many of these issues. More broadly, compressional orogenic-zone basin deposits have the potential to provide detailed, integrated records of tectonic and climate history of mountain belts, and therefore of plate-tectonic history (e.g., Heller et al., 1988; Jordan, 1995; DeCelles et al., 1998, 2011; Uba et al., 2007; Carrapa et al., 2011). Effectively interpreting the sedimentary record, however, demands a more thorough understanding of deposystems in modern foreland basins than we currently possess.

Previous studies by our research group have been directed at addressing this information gap by investigating specific modern basin examples in and around the Andean orogenic belt (Omarini, 2007; McGlue et al., 2011, 2012a). Here, we make a first attempt to synthesize our sedimentologic and geomorphic data, along with preexisting data from various published sources, on a range of modern Andean lakes of tectonic origin (Fig. 1A). Our goal is to provide a broad, comparative data set that can form the basis of comparative facies and limnogeologic models of modern lake formation, sedimentation, and geologic evolution within an active retroarc system. Tectonically formed lakes in the Andean retroarc have evolved in a wide range of tectonic, geomorphic, bedrock geologic, and climatic settings, providing an excellent basis for predictive-model development. The time is particularly opportune to take a synthetic approach to Andean limnogeology, as our understanding of the Andean retroarc system itself is in a phase of rapid change (DeCelles et al., 2009). Emerging tectonogeomorphic and geophysical models related to orogenic cycles and tectonic-climatic feedbacks in mountain belts provide a new framework for understanding the context of modern lake distribution. Thus, there is a strong potential that understanding controls on modern Andean lake deposystems can lead to a new and richer understanding of similar basins (for example, associated with the Laramide or Sevier orogenies) in the rock record.

GEOMORPHIC AND TECTONIC DISTRIBUTION OF LAKES IN THE ANDEAN RETROARC BELT

Modern lakes, wetlands, playas, and salars can be found in a wide variety of settings in the Andean region. The depressions that accommodate these bodies of water owe their origins to a variety of tectonic (fault-bounded) and nontectonic processes (e.g., eolian excavation, volcanism, glacial overdeepening, or glacial moraine and landslide damming). In most cases, individual lake origins and subsequent evolution are influenced by multiple external drivers that create or modify accommodation. It is useful to characterize the region's lakes by tectonogeomorphic setting, specifically because the tectonic drivers determine the likelihood of long-term preservation of lacustrine strata in the geologic record (Cohen, 2003). This in turn increases the value of actualistic studies for interpreting the ancient rock record.

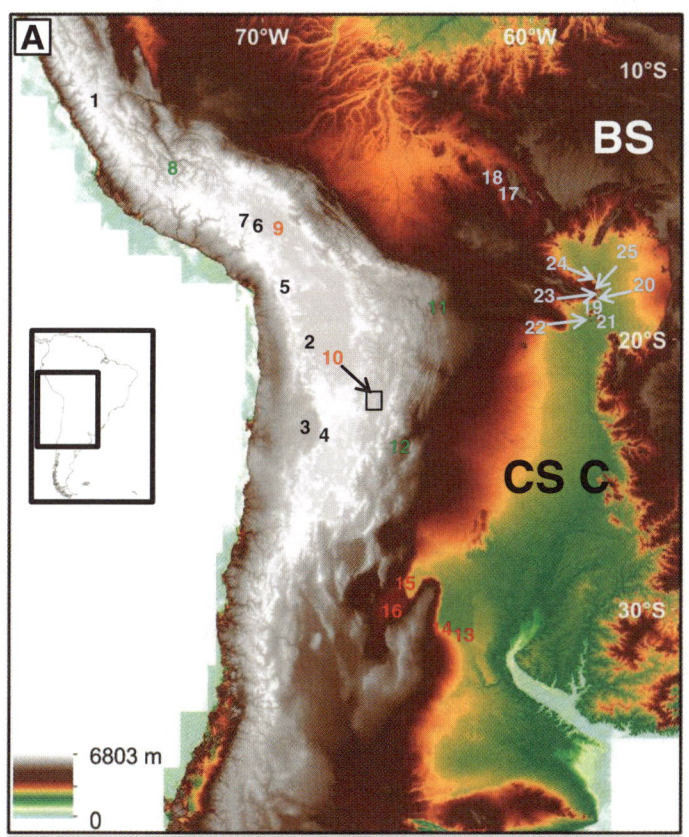
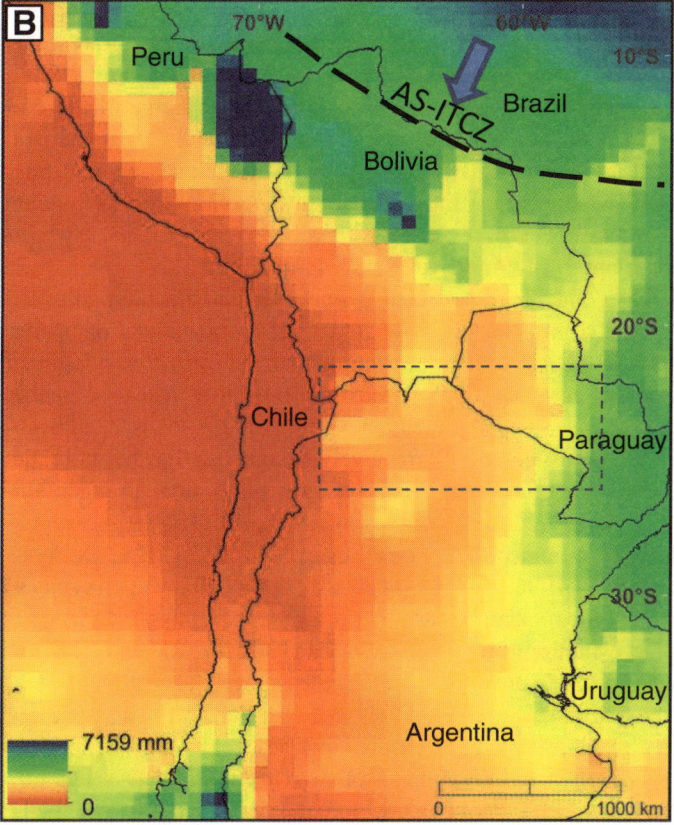

Figure 1. (A) Digital elevation model (DEM) of central South America showing lake locations mentioned in the text. Morphotectonic zones are color coded to lake localities as follows: hinterland (black, 1–7); hinterland/piggyback transition (orange, 9–10); piggyback (green, 8, 11–12); broken foreland (red, 13–16); back-bulge (light blue, 17–25). Lakes: 1—Lake Junin; 2—Salar de Uyuni; 3—Salar de Atacama; 4—Laguna Miscanti; 5—Lago Chungará; 6—Lake Umayo; 7—Lagunillas; 8—Lake Pacucha; 9—Lake Titicaca; 10—Laguna de los Pozuelos (note: box corresponds to map area of Fig. 6); 11—Lake Opabusu; 12—Laguna La Brea; 13—Laguna Mar Chiquita; 14—Laguna del Plata; 15—Salina de Ambargasta; 16—Salinas Grandes; 17—Lake Chaplin; 18—Lake Bella Vista; 19—Lagoa Caceres; 20—Lagoa Castelo; 21—Lagoa Negra; 22—Lagoa Jacadigo; 23—Baia Vermelha; 24—Lagoa Gaiva; 25—Lagoa Mandioré. DEM processed from Shuttle Radar Topography Mission (SRTM) 90 m data using RiverTools. BS—southwestern Brazilian Shield; C—Chaco; CS—Chaco Seco. (B) Mean annual precipitation (MAP) over the same region of South America. Data are from Legates and Willmott (1990). The southern extent (austral summer) of the Intertropical Convergence Zone (AS-ITCZ), extending across southern Amazonia, and associated NE (South Atlantic) moisture sources are also shown. Dashed rectangle marks the area used for the lake morphometric study discussed in Figure 12.

Lake Formation and the "Overfilling" Problem in Retroarc Deposystems

Regardless of the formation mechanism, the production of a lake on Earth's surface requires both a topographic depression and a source of water. Maintaining a topographic depression, in turn, involves a balance between sediment supply and whatever mechanisms are available to generate new accommodation for that sediment. Carroll and Bohacs (1999) developed a three-part classification scheme for lakes, which categorizes them based on the relative balance of formation rates of potential accommodation relative to sediment and water infill. In tectonically formed basins, such as within an orogenic belt, this largely, although not exclusively, involves some combination of faulting, flexure, or local downwarping of the crust.

In extensional or transtensional lake basins, for example, in continental rifts, relatively limited sediment supplies from areally small hinterlands are typically coupled with considerable normal-fault throws at the basin boundaries (Cohen, 1990). This leads to a common outcome of basins that are topographically (and often bathymetrically) deep, with sediments underfilling the basin over much of its history. The resulting geometry and accommodation volumes over time produce the characteristic rift-lake "sandwich" stratigraphy, with fluvial and deltaic deposits bracketing lacustrine ones in many evolving rift deposystems (Lambiase, 1990), often at rapid rates and over geologically brief intervals. In contrast, retroarc fold-and-thrust belts, with their large mountain fronts feeding sediment into flexure-controlled forelands, produce very different patterns of accumulation. Under such settings, sediment overfilling, with the formation of alluvial deposystems might be expected to be favored, and topographic depressions, which are required to produce lakes, intuitively would be expected to be relatively uncommon. Furthermore, the subsidence of foreland basins typically plays out over much longer periods (10^7–10^8 yr) than in rift basins (10^6–10^7 yr) and is probably a much more pulsed or cyclic process (e.g., DeCelles and Giles, 1996; DeCelles et al., 2009). Modeling experiments as well as the rock record itself generally support this idea. Geomorphic evolution models parameterized for realistic topographies and uplift/flexure rates for the Andes suggest that the formation of topographic depressions within the foreland of an orogenic belt is a difficult process (Pelletier, 2007; Engelder, 2012). Increasing accommodation volumes or decreasing sediment accumulation rates sufficient to create topographic depressions and a lake under such circumstances then require some combination of the following processes or conditions:

1. Localized topographic isolation. Topographic depressions can develop where alluvial and fluvial sediment sources are diverted from a basin by local bedrock highs. High-relief structures (e.g., steep-reverse or normal faults with significant throw) can be expected to provide particularly effective barriers to sediment accumulation. In retroarcs, the most common context for developing this type of topography and associated lacustrine basins is in association with thick-skinned, basement-

cored uplifts, such as the modern Sierras Pampeanas (Jordan and Almendinger, 1986), or the ranges of Laramide age in western North America (Pietras et al., 2003; Lawton, 2008). Localized basin isolation and topographic depressions can also occur when the regional flexural wave (which generates both structural highs and lows) interacts with preexisting geologic structures and in the process reactivates older faults. For example, reactivation of Salta Rift features within the Cordillera Oriental occurred in the Tertiary (Grier et al., 1991), and similar processes may have impacted the Andean back-bulge region in the Quaternary (Ussami et al., 1999).

2. Climate. A reduction in sediment yield by decreasing stream discharge (and variability and/or intensity of discharge) could result in the underfilling of sedimentary basins and allow topographic depressions to be maintained. This hypothesis is supported by both modeling (Pelletier, 2007; Engelder, 2012) and long-term observational data in the Andes (e.g., Hilley and Strecker, 2005; Uba et al., 2007; but see Bywater-Reyes et al. [2010] for a more equivocal climate–sedimentation rate relationship in the Andes). This contrast between sediment underfilling under long-term arid conditions versus sediment overfilling as climate becomes more humid results in an increased likelihood of lacustrine conditions over longer intervals and larger areas under drier conditions. This outcome is evident in the increased abundance of lakes in the modern Andean retroarc south of 35°S, where the westerlies take over as the principal climate system. As a result, the east side of the Andes is much drier, and this pattern is also borne out in retroarc basin systems elsewhere (e.g., Smith et al., 2008). This climatic effect can be expected to be most pronounced where erodible watersheds are largest and most integrated (i.e., away from pronounced topographic isolation; e.g., Jordan et al., 2001). In turn, this effect accounts for perhaps the largest single difference in lacustrine basin evolution between rifts and foreland basins. Rifts are structurally deep holes surrounded by small watersheds yielding little sediment, and thus they are prone to sediment underfilling. As a result, rift lakes over the long-term behave like natural rain gauges, increasing in size, depth, and lacustrine deposit extent as climate gets wetter. In contrast, foreland basin systems can experience immense sediment loads spilling into much broader receiving basin areas. This creates a tendency toward the overfilling of basins by sediment, and it will cause lacustrine deposition to disappear as climate gets wetter.

3. Bedrock lithology. Differences in erodibility can have a profound impact on the rate at which retroarc basins fill, and thus the likelihood that they can retain topographic closure and remain underfilled by sediment (Carroll et al., 2006). The erosion of highly resistant crystalline bedrock, for example, as exposed in many basement-cored uplifts, is more conducive to produce sediment underfilling in adjacent basins than easily eroded volcaniclastic or semi-unconsolidated sediments. Many, if not most, of the large lakes formed in modern or ancient foredeep settings (where sedimentation is most focused in retroarcs) occur adjacent to mountain ranges that are dominated by limestone

bedrock. Examples include the large lakes associated with the Sevier and Laramide foredeeps (Peterson and Draney Limestones, Flagstaff Limestone; e.g., Drummond et al., 1996; Zaleha, 2006; Gierlowski-Kordesch et al., 2008), or the Swiss Molasse Basin (Platt and Keller, 1992). This most likely results from the increased likelihood of chemical weathering of a large fraction of the carbonate-bearing watershed. A significant proportion of these weathered sediments, rather than being deposited within the foredeep as particulate material, is instead carried as dissolved load out of the deposystem. This would increase potential accommodation as the downstream basin continues to subside, enhancing the probability that a large topographic depression can be maintained, where a lake can form.

4. Short-term (nontectonic) interacting factors. Many of the lakes that occur within a retroarc system owe their existence to processes that are either unrelated to regional tectonics, such as competitive aggradation within large alluvial and fluvial systems or eolian deflation (e.g., Assine and Soares, 2004), or processes that are related in indirect ways, for example, volcanic impoundments, especially within the Altiplano region, where volcanoes are most abundant (e.g., Pueyo et al., 2011). It is important in discussing patterns of lake formation within the retroarc to understand when short-term lake-forming mechanisms have a long-term connection to the accumulation of tectonically controlled lacustrine basins. As discussed herein, there are strong feedback mechanisms between the climatic and tectonic conditions across a large orogenic belt like the Andes. Thus, even though a lake may have a short-lived mode of origin that is seemingly unrelated to tectonics, the probability of that mode producing large numbers of topographic depressions across the retroarc landscape may in fact be strongly regulated by the long-term evolution of the orogen.

Climatology of the Andean Retroarc and Its Relationship to Lake Formation

The climate, and in particular the effective moisture balance of the Andean retroarc region, plays a key role in determining both the nature of standing waterbodies in this region and their short term areal extent (i.e., expansions and contractions related to precipitation variability in basins that lack a surficial outlet, which in this paper we term "closed"). Our intention here is not to give a thorough review of the region's climatology and forcing variables, but rather to explain the broad patterns of effective moisture-balance variability in the study region.

The primary sources of available moisture within the central Andean retroarc (11°S–30°S tropical-subtropical belt; Fig. 1B) are regulated by the seasonal migration of the Intertropical Convergence Zone (ITCZ), which supplies moisture predominantly from the South Atlantic (Garreaud et al., 2009), supplemented in the southern region by the South American monsoon. Although wind fields strengthen and weaken seasonally, there are no complete reversals of wind vectors between the South Atlantic and South America, the hallmark of a fully monsoonal system. Pre-

cipitation in the study area reaches a maximum during austral summer, especially between southern Amazonia and northern Argentina. Except for the far southeastern portion of the study region (Fig. 1 map area), precipitation is minimal during the austral winter, when the ITCZ retreats north of the study area. Summer moisture in the Chaco region (northeastern Argentina and Paraguay) is also strongly influenced by Amazonian-sourced water vapor, which is transported south between the Brazilian Plateau and the Andes as far as 40°S (the southern limit of summer moisture) by the South American Low-Level Jet (Labraga et al., 2000; Doyle and Barros, 2002). Between ~20°S and 30°S, there is also a general gradient of decreasing precipitation from the east (Brazilian Shield) to the west, which reaches a minimum in northern Argentina in the arid Chaco (Chaco Seco). This trend may be related to southwesterly transport of the Amazonian-sourced moisture and rainout from east to west over the Chaco region, where large summer convective storms often develop. Interannual variation in the strength of these sources is modulated by variations in sea-surface temperature (SST) and El Niño–Southern Oscillation (ENSO) variability; high-latitude effects (Antarctic Oscillation and North Atlantic Oscillation) also appear to have some influence on precipitation variation.

The impacts of these interactions are highly influenced by the orographic effects imposed by the Andes. Easterly winds shed almost their entire remaining volume of precipitation within a geographically narrow belt upon reaching the abrupt eastern edge of the Andes. This is evident in the narrow belt of slightly higher mean annual precipitation (MAP) at the Andean front at both 23°S and 26°S (Figs. 2A and 2B); in some areas, extreme rainfall anomalies are associated with the lower-elevation mountain front (e.g., Trauth et al., 2000). This rainout leads to much lower precipitation levels at higher elevations and locations across the western side of the tropical and subtropical Andes, and eventually hyperaridity toward much of the Western Cordillera. Large-scale air mass subsidence affects all of the western flank of the Cordillera year-round south of ~15°S. In the northernmost part of the study area, Amazonian and Atlantic moisture reaches the high Cordillera as a result of middle- and upper-level easterlies associated with the formation of the Bolivian High (Lenters and Cook, 1997). This results in strong convection over the northern Altiplano region, which is therefore much wetter than the southern Altiplano and Argentine Puna Plateau. The southern portion of the South American Cordillera lies within the westerly belt, where Pacific moisture sources and winter rainfall dominate. Arid conditions prevail to the east of the Andes here, but this zone lies outside of the latitudinal belt of concern in this study.

The climatic gradients discussed here have clearly played an important role in the evolution of lakes in this region, but their effect on lakes is frequently counterintuitive (Carroll and Bohacs, 1999). Although topographic depressions will initially display an increase in lake level and an enlargement of lake area as precipitation increases in its watershed, this enlargement is continuously being offset by the increase in water and sediment discharge that normally accompanies a precipitation increase. Thus, over

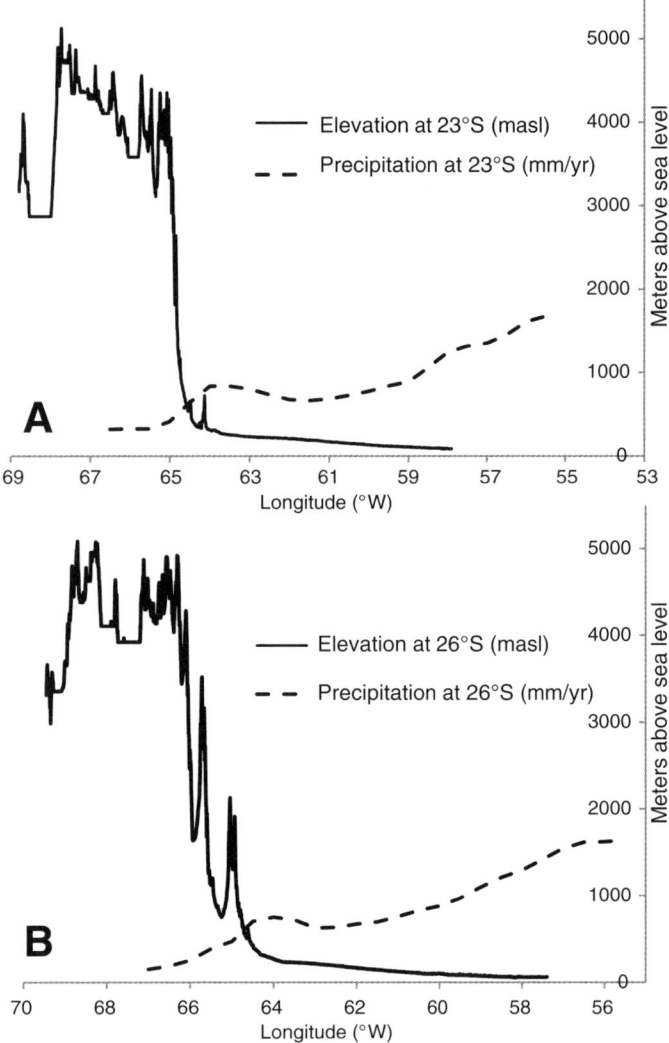

◄

Figure 2. Longitudinal transects of elevation and mean annual precipitation at two latitudes: (A) 23°S and (B) 26°S across the Andean retroarc.

in retroarcs because of the potentially enormous sediment loads that mountain belts can generate. In order to make sense of climate's role on lake formation in the Andean retroarc, we need to consider the relationship of the climate pattern to the depozones within which the lakes are forming.

Lake Depozones in the Andean Retroarc System

Lacustrine deposition in the modern Andean retroarc is focused in six major morphotectonic zones, each of which is characterized by a different set of dominant processes and environments (Figs. 3 and 4).

Hinterland Basin Lakes

The high-elevation plateaus (Altiplano or Puna areas) of western Argentina and Bolivia and parts of Chile and Peru are occupied in places by extensive hinterland basins (sensu Horton, 2011). These basins, typified by the numerous large modern salars of northwestern Argentina and southwestern Bolivia, are somewhat enigmatic in origin, but they likely owe their existence to irregularities in lithospheric thickness below the Andean plateaus. Areas of anomalously thick lithosphere in the Puna-Altiplano appear to be associated with the development of thick and dense eclogitic roots, which in turn are hypothesized to form during the high-flux magmatic events associated with the cordilleran orogenic cycle (DeCelles et al., 2009). At Earth's surface, these dense eclogitic roots are manifest by broad, saucer-shaped basins, lacking obvious border-fault boundaries, although they are often characterized by localized extension and normal faulting. DeCelles et al. (2009) hypothesized that the growth of such basins would terminate as the eclogitic root drips off the lithosphere, resulting in a topographic rebound and uplift of the crust below the prior depocenter (hence the colloquial term "bobber" basin has been applied to these basins). The lateral scale of modern topographic basins within the Altiplano is intriguingly similar to the lithospheric seismic-velocity heterogeneities observed

geologically longer time periods, a region of high precipitation can only support the continued existence of topographic depressions and lakes if a countervailing mechanism exists to maintain a sill, such as the examples of topographic diversion from localized uplifts or changes in watershed bedrock composition and sediment supply. These mechanisms are particularly critical

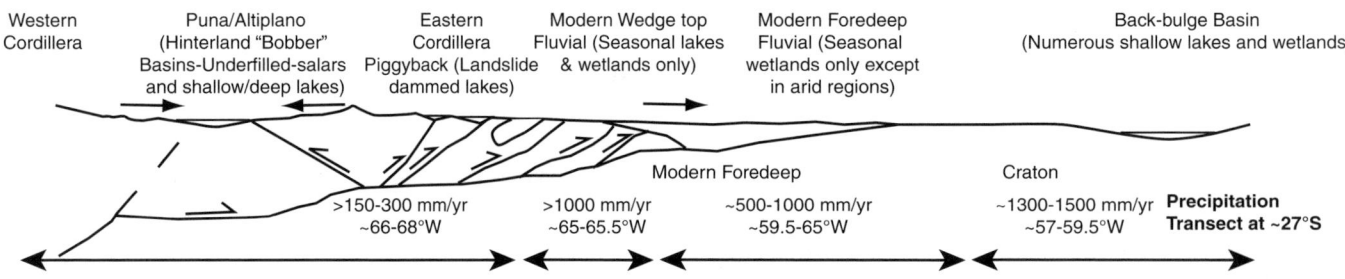

Figure 3. Schematic cross section of Andean tectonics and depozones in a region of thin-skinned deformation, modified after Horton (2011). Generalized precipitation gradient is representative of conditions at ~27°S.

Figure 4. Schematic cross section of Andean tectonics and depozones in a region of thick-skinned deformation, and basement-cored uplifts, modified after Costa and Vita-Finzi (1996) and Dávila et al. (2010). Generalized precipitation gradient is representative of conditions at ~32°S, across the Sierras Pampeanas.

through seismic tomography (Beck and Zandt, 2002). Seismic stratigraphy across several of the modern hinterland basins suggests they are currently undergoing broad subsidence, interrupted by only occasional normal faults (e.g., D'Agostino et al., 2002; Lowenstein et al., 2003). Individual topographic depressions owe their origin within the Andean hinterland to a variety of complex interactions of localized uplifts (nascent "bobbers") adjacent to preexisting topographic lows, which become cut off, producing the numerous medium-sized (tens of kilometers) basins of the Puna Plateau. In contrast, in the Altiplano region, topographic closure is manifest by fewer but larger structural basins integrated into a single large interior drainage system (Horton, 2011). Some lakes in this region also owe their origins to the abundant volcanism of the hinterland zone, through the damming of drainages (e.g., Pueyo et al., 2011).

As noted previously, the Puna and Altiplano Plateau regions where hinterland basins are developing are largely blocked from receiving easterly derived moisture. As a result, modern hinterland lake basins are dominated by semiarid to arid depositional processes and internal drainage systems (dry alluvial fans, saline lakes, and extensive salt flats or salars; Sobel et al., 2003; Alonso, 2006; Fig. 5A), and for the most part are underfilled (sensu Carroll and Bohacs, 1999). Several of these salars are commercially exploited for their high lithium (Li) and BO_3 contents. In the northern, semiarid parts of the Altiplano, higher precipitation on the eastern margin of the hinterland has developed interconnected drainages and the formation of freshwater lakes with surface outlets (e.g., Lake Titicaca).

Piggyback Basin Lakes

The thrust belt of the Eastern Cordillera is a region of very high relief and is almost entirely an erosional environment today. Relatively small piggyback basins, separated by thrust faults with surface expression, occupy isolated valleys lying atop the various thrust sheets that make up the Eastern Cordillera. We make a distinction in this study between piggyback basin lakes (sensu Ori and Friend, 1984) and lakes of the lower wedge-top zone (discussed in the following). This is because there are significant contrasting geomorphic and hydrologic consequences for lake development in those regions of the thrust belt where bedrock thrust sheets isolate currently existing deposystems (the piggyback zone in our terminology) versus lower-elevation areas

of the wedge top where active deposystems are not isolated by bedrock highs (i.e., blind thrusts terminate in the subsurface). High-altitude (2000–4000 m above sea level) piggyback basins that lie within the western part of the Eastern Cordillera experience semiarid to arid climates and have similar depositional characteristics to the hinterland basins (i.e., underfilled saline lakes and dry fans), although they are generally much smaller in surface area. Because of the ongoing eastward advancement of the Andean thrust belt, piggyback basins at the transition to the Andean hinterland appear to be undergoing an evolution from thrust-sheet (reverse fault)–bounded basins to the normal-fault boundaries more typical of the hinterland basins. Seismic stratigraphic, aeromagnetic, and some outcrop data from Laguna de los Pozuelos (Figs. 5B and 6) and Lake Titicaca (both on the eastern margin of the Puna-Altiplano) are in fact consistent with this evolutional- or hybrid-basin model. Older reverse-fault structures, still evident in surface-topographic controls of drainage, are now being superseded by either normal faulting (e.g., Caffe et al., 2002) or broadly subsiding depocenters with little evidence of faulting. Our research in the Pozuelos Basin highlights the potential influence of localized normal faults on basin-floor gradients and the distribution of depositional environments in Quaternary lake systems (McGlue et al., 2013).

Below ~2000 m, considerably higher precipitation causes most small intermontane piggybacks to be overfilled by sediment and dominated by fluvial and wet-fan deposystems, with only rare (and small) lakes and wetlands (e.g., Laguna La Brea; Fig. 5C). Most natural lakes within the thrust belt are formed by landslides, rather than by being structurally dammed (e.g., Trauth and Strecker, 1999). The former process of lake formation is particularly common in the lower-elevation portions of the Andean thrust belt because of the combination of extreme mass wasting and the existence of numerous, narrowly constricted river valleys. Given the high-precipitation characteristic of the region, such lakes are likely to be short-lived and catastrophically drained.

Wedge-Top and Foredeep Lakes

At the eastern transition from the Andean thrust belt to the modern wedge top (sensu DeCelles and Giles, 1996) and foreland basin, the depositional system is filled to capacity by a huge excess of sediment (Fig. 7). In the wedge top and foredeep of the modern Andes, very large distributary depositional fans

Figure 5. Photos of selected Andean retroarc lakes, wetlands, and fluvial deposystems. (A) Salar de Pocitos, an NaCl-dominated brine salar of the central Puna hinterland region, northern Argentina (A. Cohen photo). (B) Laguna de los Pozuelos, Argentina, a seasonally fluctuating (closed-basin), shallow saline lake at the hinterland-piggyback transition (M. McGlue photo). (C) Laguna La Brea, a northern Andean piggyback basin lake (M. McGlue photo). (D) Multiresolution seamless image database (MrSID) image of the Andean foredeep, upper Rio Bermejo megafan channel belt near Embarcacion, Argentina (image centered at 23°19′42″S, 63°51′07″W). Image field of view is approximately 25 km E-W. (E) Seasonal wetland associated with the Chaco Seco (foredeep) region of the Rio Bermejo megafan. (F) Laguna La Amarga (La Pampa Province Argentina), a permanent, closed-basin, saline lake in the arid foredeep (C. Whitlock photo).

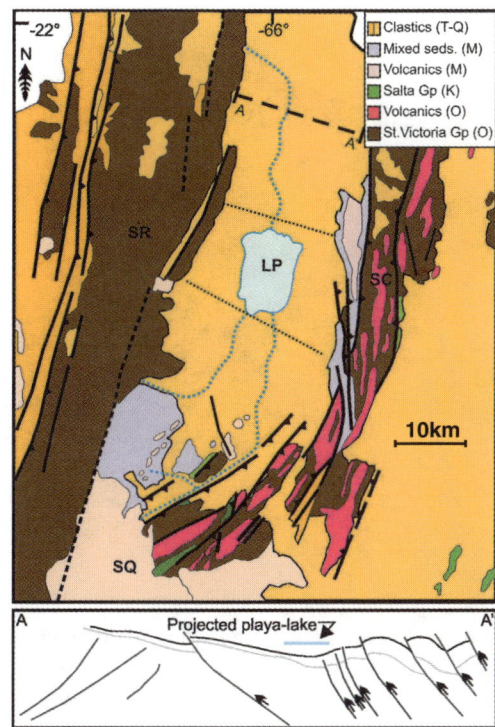

Figure 6. Laguna de los Pozuelos. Simplified geologic map (top) and structural cross section (bottom) of the Pozuelos Basin, a modern basin at the hinterland-to-piggyback transition (from McGlue et al., 2011). Map area is indicated on Figure 1A. LP—Laguna de los Pozuelos; SR—Sierras de Rinconada; SC—Sierras de Cochinoca; SQ—Sierras de Quichagua; T—Tertiary; Q—Quaternary; M—Miocene; K—Cretaceous; O—Ordovician.

Rio Bermejo megafan is typical, with a highly migratory channel belt depositing mostly fine sands and silts within the foredeep and spilling across the forebulge. Wetlands are highly localized within the foredeep (Fig. 5E), and lakes are rare in the northern Argentinian (Chaco Seco) foredeep. Further south, in the more arid portions of the foredeep, sediment underfilling and topographic closure become more common, leading to the formation of permanent, saline lakes lacking a surface outlet (Fig. 5F). Many of the lakes in this region are also highly modified by eolian processes (deflationary yardang lakes; Fig. 9A).

Forebulge and Back-Bulge Basin Lakes

The flexural forebulge of the Andean foreland deposystem lies ~200–250 km east of the western edge of the modern foredeep (Fig. 10). The forebulge is extensively overridden by megafans today and, as a result, has little or no surface topographic expression. Possible exceptions exist in the northern Andean foreland, where river-channel gradients (Beni region) and Neogene stratigraphy (Amazon region) suggest the presence of forebulge topography (Aalto et al., 2003; Roddaz et al., 2005). Furthermore, no clear relationship exists between the forebulge position and topographic residuals from an idealized

(megafans; sensu Leier et al., 2005) are being deposited by each of the major east-flowing river systems (Figs. 5D and 8). These low-gradient fan systems deposit relatively coarse-grained sediments close to the mountain fronts (e.g., regions of high residual elevation overlying blind frontal thrusts, thus within the wedge-top depozone), but their deposits rapidly fine in the foredeep. The

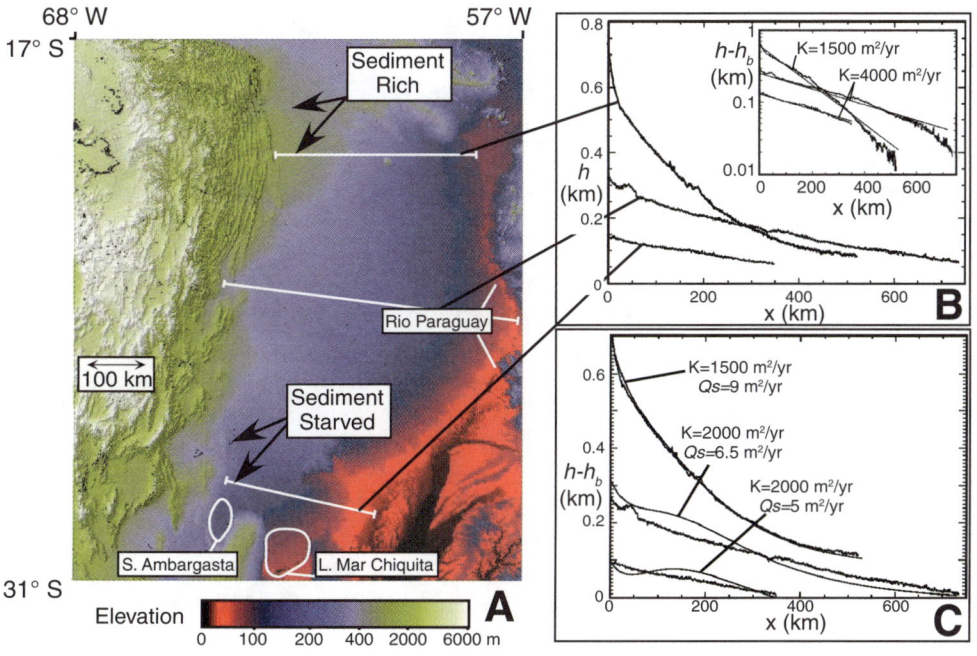

Figure 7. (A) Digital elevation model (DEM) of the Eastern Cordillera (northern Argentina–Bolivia) showing the transition from the thrust belt to the foreland basin and zones of sediment-overfilled and -underfilled (sediment-starved) regions. (B–C) Modeling profiles show the effects of variable sediment flux, Q (corresponding to series of profiles from higher precipitation [north] to lower precipitation [south]), on graded profiles across the foredeep. Reduced precipitation and (or) basin isolation (e.g., in broken foreland southern DEM area) generally yield lower sediment fill and the possibility of topographic closure (lakes). Figure is modified from Pelletier (2007).

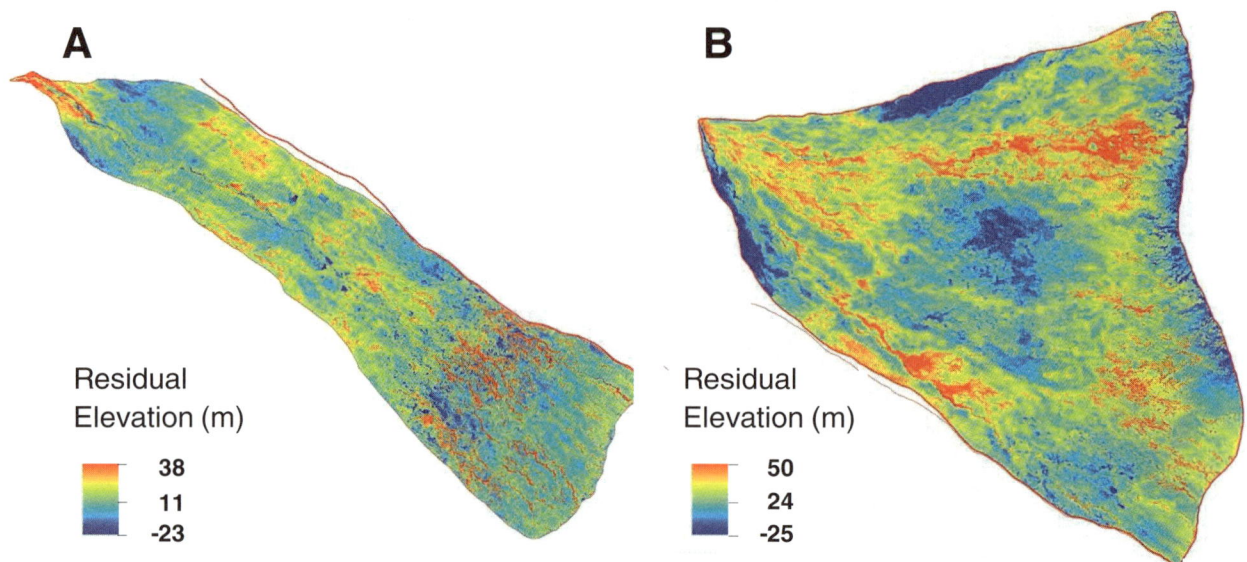

Figure 8. Digital elevation models of the (A) Bermejo and (B) Pilcomayo megafans showing residual elevations relative to a graded profile. See Zani et al. (2012) for methods explanation. Localized topographic lows ("holes") shown in blue represent sediment-underfilled areas, where wetlands and shallow lakes can occasionally develop. An uncertain amount of the topographic variability results from woodland treetop versus unforested area elevations, although this probably only accounts for 10–20 m maximum variation.

graded-stream profile along the Bermejo and Pilcomayo megafans (Fig. 8; developed using the methods described in Zani et al., 2012). The forebulge is associated with a relative residual high along the Rio Bermejo megafan, whereas the opposite appears to be the case for the Rio Pilcomayo, especially away from the active channel belt. Although the forebulge, where we have observed it, is primarily soil mantled, wetlands occur even on some parts of the forebulge crest (Fig. 9B). Much of the northern Argentine foredeep is blanketed by thick (up to tens of meters) Neotropical Quaternary loess (Sayago, 1995; Sayago et al., 2001), which further complicates attempts to correlate sedimentary patterns in the foredeep with tectonic controls. However, the position of the forebulge is marked by a broad transition in the megafans from a highly distributive system (multithread braided channels) to dominantly meandering channel belts with numerous oxbow lakes and wetlands, which continue into the back-bulge basin depozone (Figs. 9C and 9D; for a more extensive discussion of lakes and wetlands associated with the distal portions of these types of distributive systems, see Davidson et al., 2013). Because the position of the forebulge also coincides with a climatic transition (wetter to the east), and an extensive north-south spring line, it is unclear the extent to which underlying tectonics, hydrology, and/or climate control this transition in the deposystem.

In most areas of the Andean retroarc, the forebulge has few or no lakes, although there is one enigmatic exception to this generalization. The northern forebulge region within the study area includes parts of the Beni Plain of Bolivia from ~12.5°S to 15°S, where extraordinary numbers of shallow (generally <3 m), rectilinear lakes occur. The origin of these lakes has been

highly controversial. Some authors have suggested a tectonic origin (e.g., Plafker, 1964; Allenby, 1988) in alignment with local subsurface tensional features. However, as noted by Dumont and Fournier (1994), these lakes are also aligned with paleodune features. Even more remarkably, these rectilinear lakes (including some as much as 20 km long) appear to be connected to extensive mounded earthworks of pre-Columbian age (e.g., Lombardo and Prümers, 2010; Lombardo et al., 2012), suggesting their origins and orientation may be, at least in part, anthropogenic.

The back-bulge depozone (sensu DeCelles and Giles, 1996) forms a belt east of the forebulge in northeast Argentina, central Paraguay, southeast Bolivia, and southwest Brazil. Throughout this region, these basins experience considerably more humid climates than the Andean foredeep, with precipitation generally increasing toward the north. However, because of their distal setting relative to major sediment sources and thus extremely low sediment supply rates, they have sufficient accommodation to allow numerous lakes and wetlands to form, especially along the axis of the Paraguay River (Fig. 11A). Basin formation within the back-bulge depozone is probably also influenced by more local tectonic elements, for example, in the Pantanal Basin (Brazil-Bolivia), which is ponded adjacent to the relict topography of the Brazilian craton (Fig. 11B). Here, distinctly larger lakes occur (up to ~150 km^2, though generally <5 m deep), as well as craton-fed megafans and wetlands (Fig. 11C). As in the foredeep, megafan deposition has often been secondarily modified by eolian processes (lunette formation and shallow lakes), most recently during a mid-Holocene arid period (Fig. 11D; Assine and Soares, 2004; McGlue et al., 2012b).

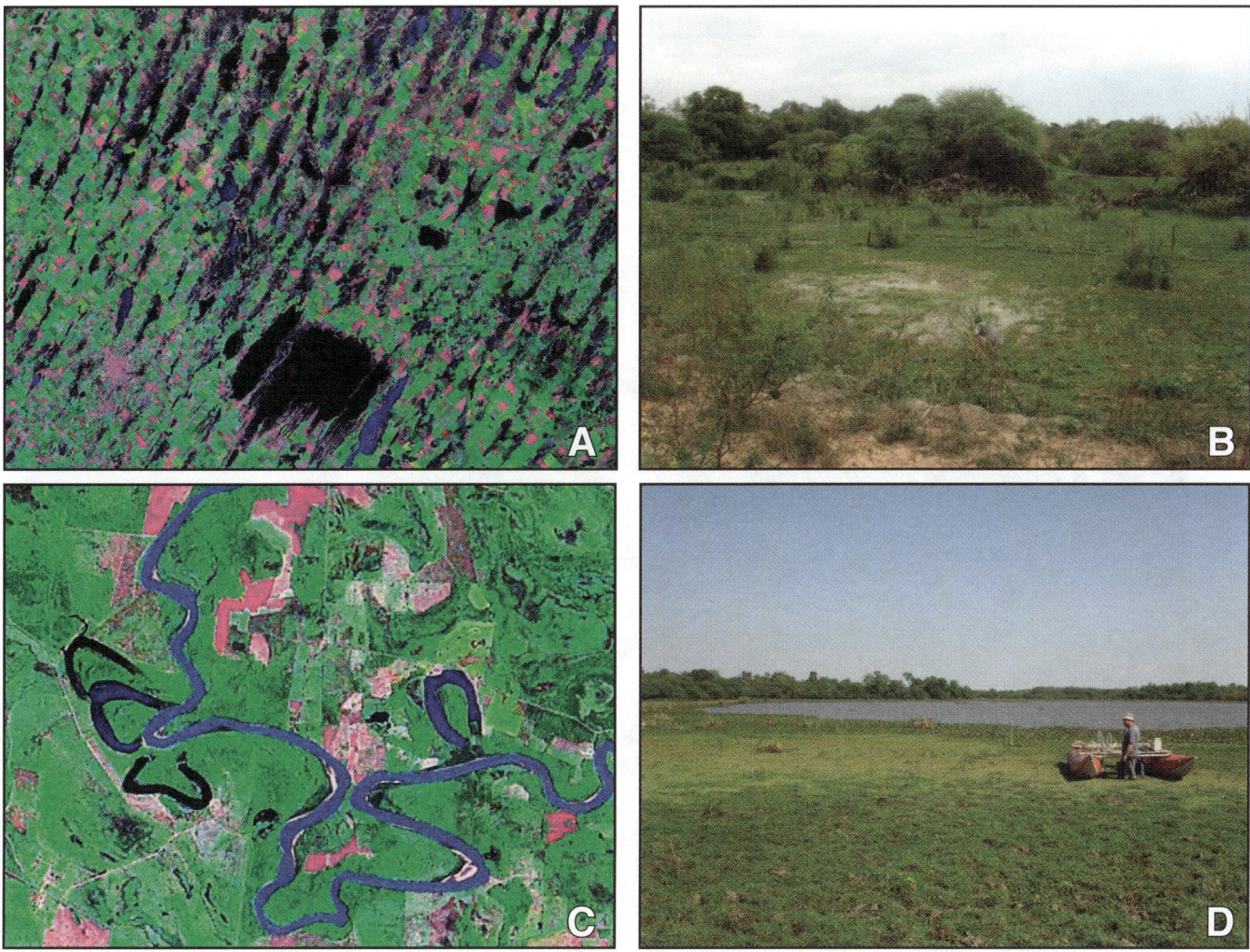

Figure 9. Lakes and wetlands of the Andean retroarc region. (A) Yardang-controlled lakes (erosion into Pliocene sediments) in W. Buenos Aires (MrSID image centered at 35°51′48″S, 62°22′27″W). Image field of view is 80 km E-W. (B) Wetlands on flexural forebulge east of Ingeniero Guillermo Juarez, Hwy 81, Formosa Province, Argentina (A. Cohen photo). (C) Meandering reach of lower Rio Bermejo. Note major oxbows and wetlands typical of the Chaco–Formosa Provinces back-bulge basin (MrSID image centered at 26°06′48″S, 59°38′58″W). Image field of view is 10 km E-W. (D) Laguna Vedado, eastern Chaco Province (back-bulge basin), a shallow lake (5 m) and seasonal wetland system ~85 km northwest of Resistencia (A. Cohen photo).

Broken Foreland Lakes

In the regions of the Andean retroarc experiencing flat-slab subduction, very different type of shallow crustal structure and topography occurs, comprising basement-cored uplifts (the Sierras Pampeanas) with steeply dipping, flanking reverse faults exposing crystalline basement rocks at higher elevations (Jordan and Almendinger, 1986; Rapela et al., 1998). The floors of the intervening basins are effectively isolated from much of the sediment load coming off the Eastern Cordillera, and this isolation, coupled with relatively resistant bedrock, has allowed topographic closure to develop between several of the ranges. Because the ranges also act as moisture traps (albeit lower-elevation ones than the main Andean front), the more westerly of these basins are relatively arid in comparison with the Pampean region to the east. Consequently, the depocenters of the western basins are currently occupied by hypersaline lakes and playas ("salinas"; e.g., Zanor et al., 2012; Fig. 11E). A major topographic depression is also found in the frontal foredeep east of the easternmost Pampean ranges, the Sierras de Cordoba, resulting in the Mar Chiquita topographic (and endorheic) basin, currently filled by Argentina's largest lake, Laguna Mar Chiquita (Fig. 11F; Martínez, 1995; Reati et al., 1997). Closure on the east of the Mar Chiquita depression may be linked to the position of the forebulge (i.e., the San Guillermo High; Dávila et al., 2010; see Fig. 10), although note that Chase et al.'s (2009) model places the bulge considerably more to the west of this position, and some authors have suggested that the San Guillermo High is a buried Pleistocene intraforeland uplift (Mon and Gutiérrez, 2009).

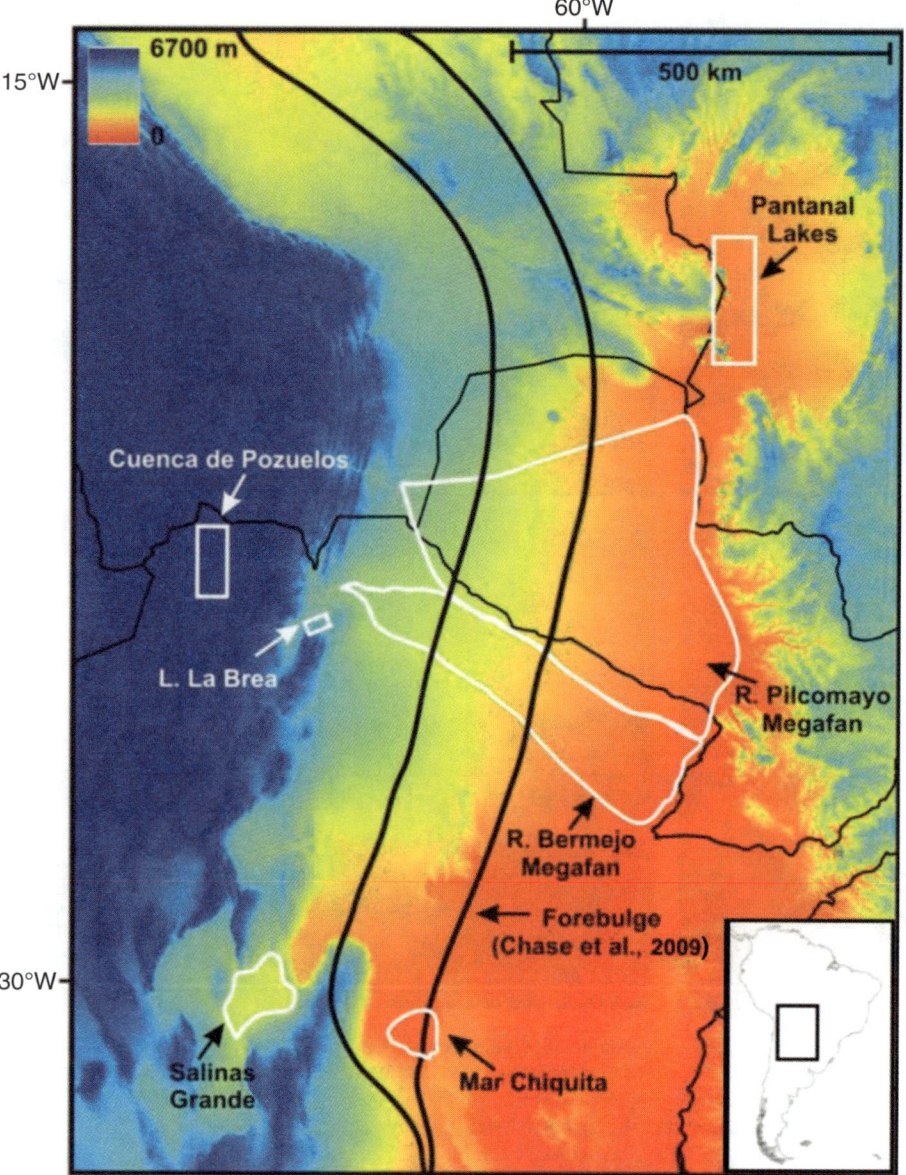

Figure 10. Digital elevation model of the southern and central portions of the study region, showing the position of the flexural forebulge. Position of the forebulge is inferred from geoid anomalies and flexural modeling (Chase et al., 2009).

COMPARATIVE GEOMORPHOLOGY, SEDIMENTOLOGY, AND SEDIMENTARY GEOCHEMISTRY OF ANDEAN RETROARC LAKE BASINS

As discussed in the previous section, lake deposystems within the Andean retroarc encompass a wide range of lake, wetland, and playa types formed under a wide range of climatic conditions. Comparison of the characteristics of lacustrine deposystems within a larger, tectonically controlled deposystem has previously proven useful, both by providing analogs for inferring similar settings in the geologic record, and as hydrocarbon and mineral exploration tools (e.g., Cohen, 1990). Here, we consider variations in lake morphometry, sediment accumulation rates, and sediment geochemistry among all Andean retroarc lakes

for which our own or published data are currently available. We restrict our study to those lakes for which origins, evolution, and topographic closure are linked to their tectonic setting within the retroarc. For this reason, we have excluded from our analysis, for example, glacial lakes of the high Andes. Although the sample size of lakes available for this first study of its kind is still limited, there are intriguing differences between lakes of different depozones, which appear to have predictable relationships to regional tectonoclimatic characteristics.

Lake Morphometry

We analyzed lake morphometry in the central Andes from ~21.5°S to 26°S and ~57°W to 67.5°W (Fig. 12). This region includes lakes from the hinterland (including piggyback basins

Figure 11. Images of Andean retroarc lakes and wetlands. (A) Lagoa Mandioré, Pantanal (southwest Brazil), a large but shallow, highly productive lake west of Rio Paraguay, modified back-bulge basin. Serras do Amolar of the Brazilian craton in background (M. McGlue photo). (B) Pantanal region, Brazil, showing Rio Taquiri megafan derived from Brazilian craton to the east, feeding the Paraguay River and large lakes (including Lagoa Mandioré) to the west. Inset map shows mean position of Intertropical Convergence Zone (ITCZ) for June–August (JJA) and December–February (DJF). (C) Nhecolândia deflation lakes of the southern Pantanal (M. Assine photo). (D) Pantanal wetlands east of the Paraguay River (A. Cohen photo). (E) Desiccated salt pan on the eastern margin of Salina de Ambargasta (A. Cohen photo). (F) Shoreline of Laguna Mar Chiquita, the large, closed-basin foredeep lake lying east of the Sierras Pampeanas (A. Cohen photo). Note effects of rising late twentieth-century lake levels flooding the coastal community of Miramar.

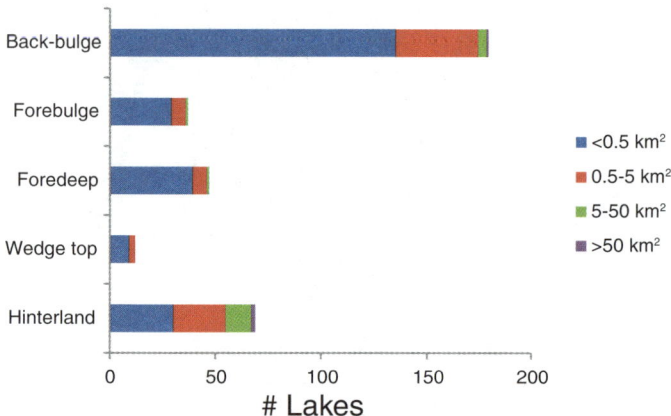

Figure 12. Distribution of lakes by area within different morphotectonic zones of the Andean retroarc. The region from which these lake data were obtained is shown in Figure 1B (dashed rectangle).

along the transition zone of the Puna and Eastern Cordillera), wedge-top, foredeep, forebulge, and back-bulge depozones. Lake boundaries were extracted manually using ArcGIS software and GeoCover 2000 data downloaded from http://nasa.ssc.zulu.gov/mrsid/. Surface areas and perimeters were calculated in ArcGIS. In this context, lake surface areas were based on the snapshot captured by Landsat sensors during the season of acquisition, which appears to have been during or shortly after the wet summer season.

Across our study area (*n* = 345 lakes total), the vast majority of the lakes have surface areas ≤0.5 km². The highest density of lakes analyzed was located in the back-bulge depozone (*n* = 180), where MAP ranged from ~800 to 1500 mm. The vast majority of these lakes are small (<0.5 km²). The largest lakes in the back-bulge (e.g., Lake Ypacaraí, Paraguay) were adjacent to minor topographic highs, and we hypothesize that their evolution may have been influenced by interactions of the migrating flexural profile with preexisting topography or neotectonics. A similar situation exists in the Pantanal, as some of the largest lakes in that basin are situated amongst the Neoproterozoic Serra do Amolar mountains, and their depocenters are adjacent to shoreline-coincident topography (McGlue et al., 2011). Lakes were least abundant in the wedge top, forebulge, and foredeep (*n* = 12, 37, and 47, respectively), which we attribute to limited preservation of topographic depressions in these depozones. By contrast, greater numbers of lakes were located in the hinterland (*n* = 69). The largest lakes in our survey were found in this depozone, with 14 lakes exhibiting a surface area >10 km². The presence of large lakes in the hinterland is consistent with the arid climate (MAP = 45–535 mm) limiting the rate of sediment delivery, as well as localized tectonic influences (e.g., normal faulting) on the creation of accommodation. Although it seems counterintuitive, shallow closed-basin lakes that are areally expansive are common features of arid environments, where sediment delivery is limited. In the long-term, wetter climates promote infilling of lake basins by sediment and ultimately the replacement of lakes by fluvial deposysy-

tems (e.g., Bohacs et al., 2000). In the Andean hinterland, water feeding these lakes is most likely sourced from local springs and runoff from the high Andes peaks surrounding the basins.

Sediment Accumulation Rates

Sediment accumulation rates have a potentially strong linkage with both tectonic and climatic forcing in lake basins. Characterizing systematic differences in accumulation rates between lakes or basin types may lead to fundamental insights into the probability that a given basin type is prone to over-, balanced-, or underfilling (sensu Carroll and Bohacs, 1999). Sediment accumulation rates ultimately affect other variables such as lake morphometry (discussed earlier herein) and sediment geochemistry (through clastic- or chemical-sediment dilution of other components).

However, comparing sediment accumulation rates among lake basins is not straightforward. As Sadler (1981) demonstrated, measured sediment accumulation rates are highly sensitive to the duration of the interval over which the measurement is inferred, as a result of an increasing probability of hiatuses and erosional episodes over longer durations. As a result, we present our comparative sedimentation rate data using a Sadler-type plot relating rates to measurement duration. We compiled all available rate information (constrained by a combination of ^{210}Pb, ^{14}C, U-series, Ar/Ar, and other dating methods) for nonglacial lakes within the retroarc depozone. Data were compiled from published literature and our own data. For all data points, we collected rate and interval information from all pairs of dated horizons within cored intervals, excluding obvious age-reversal data points from the analysis. Where only graphic data were available in the form of age-versus-depth age models (i.e., no published tables of ages were provided), we digitized those data using DigitizeIt.

A Sadler-type plot of sedimentation rate information indicates a clear separation between lake basins within different morphotectonic zones of the retroarc (Fig. 13A). When adjusted for measurement interval, the highest accumulation rates are observed in lakes from the Altiplano hinterland settings (e.g., ~0.2–3 mm/yr for observation durations of 10^4 yr; orange symbols on Fig. 13A), perhaps because of large watershed areas coupled with little vegetation stabilizing the landscape and (or) abundant loose volcanogenic debris in most of the hinterland settings. Somewhat lower sedimentation rates are observed in the lakes lying on the transition between hinterland and piggyback settings (orange edge and yellow center points on Fig. 13A), in piggyback basins (green points on Fig. 13A), and broken foreland basins (light-green points), all with 0.08–1 mm/yr for observation durations of 10^4 yr. The trend of these data conforms with the estimated sediment accumulation rate of 0.32–0.34 mm/yr over ~2 m.y. duration for the Miocene lake deposits of the Palo Pintado Formation, which formed in a wedge-top setting (Bywater-Reyes et al., 2010). In all cases, the intermontane settings of these lakes may isolate them from large watershed feeder areas. An exception is Laguna Mar Chiquita, which occupies an extramontane broken foreland setting and exhibits a large (>37,000 km²) watershed. Accumulation rates are relatively high for this basin, albeit over relatively short durations of measurement. The lack of rate data for lakes in the foredeep reflects an absence of such lakes in the modern Andean retroarc, although it is reasonable to predict that any lake forming in such a setting would be subject to very high accumulation rates. The lowest rates we observe are from lakes in the back-bulge settings (0.01–0.3 mm/yr), which are far removed from major montane sources. Furthermore, most of these back-bulge regions are partially or wholly bypassed by major sediment-distribution networks derived from the Andes, although some also receive sediment from the east (i.e., cratonic sources, such as in the case of some of the Pantanal lakes).

The range of sedimentation rate values observed in Andean retroarc lakes centers around global lacustrine means, but with a higher fall-off in rates with longer measurement duration (i.e., slope of data), most notably for the back-bulge deposits. This indicates a long-term lower preservation probability of these orogenic lacustrine deposystems (i.e., greater erosion or nondepositional hiatus likelihood) relative to the mean lacustrine slope. In this regard, the Andean lake sedimentation rate slopes are more similar to average fluvial deposystems (Sadler, 1981), albeit with lower mean accumulation rates for any given duration of measurement. This inference is clearly evident from the late Quaternary core records from shallow lakes (e.g., Laguna de los Pozuelos and various Pantanal lakes; McGlue et al., 2012b, 2013). Although some of this difference between retroarc lakes and all lakes might be attributable to greater dating uncertainty with greater duration, there is no reason to expect such an artifact would affect retroarc basins systematically.

Comparative Lacustrine Sediment Geochemistry

We have also compiled available sediment geochemistry data for Andean retroarc lakes (Fig. 13). The variables we consider, including total organic carbon (TOC), total inorganic carbon (TIC), carbon:nitrogen ratios in organic matter (C:N), biogenic silica concentrations (BiSi), and the $\delta^{13}C$ and $\delta^{18}O$ of organic matter, are ones that have been frequently assessed in limnogeology studies to interpret paleoenvironments (Cohen, 2003). However, even for these important variables, data are available for far fewer lakes ($n = 15$, not all variables available for all lakes) than for morphometry or sedimentation rates, and thus generalizations about lake-class characteristics remain problematic. Few of the depozone sample populations are distinguishable by univariate distributions around the mean at the 2σ level. Furthermore, most of the data sets are derived from subsurface core samples (surface sediment sample indicated on figure). Thus, some of the differences (and within-lake variability) observed may relate to varying climatic conditions at the time of deposition as much as depozone setting. In compiling these data, our intent is to stimulate additional research on comparative sedimentology in the region based on an admittedly limited range of observations, rather than to make definitive statements about the contrasts or

similarities among lakes. Our data set consists of both published data and our own, previously unpublished information.

In the plots that follow, TOC was determined either by coulometry, elemental analysis, or through loss on ignition (LOI) using standard techniques. The precision associated with TOC determined by coulometry or elemental analysis is typically better than 0.10 wt%, whereas the precision of LOI is usually more variable (Heiri et al., 2001). Similarly, TIC was determined coulometrically or through LOI methods. The carbon and nitrogen stable isotopic compositions of lacustrine organic matter are generally determined in tandem using a continuous-flow mass spectrometer, usually coupled with an elemental analyzer. Values reported here use the conventional δ notation relative to Vienna Peedee belemnite (VPDB) for carbon or atmospheric nitrogen (for nitrogen). Sample preparation is a known complicating factor in determining $\delta^{13}C$, particularly with the procedure for the removal of carbonates (e.g., Brodie et al., 2011), which can be accomplished through wet rinse or vapor-phase acidification in a desiccator. The precision of these analyses is on the order of 0.10‰ and 0.20‰ for $\delta^{13}C$ and $\delta^{15}N$, respectively. The ratio of carbon to nitrogen can be expressed either as mass or atomic values (Meyers and Teranes, 2001). These data are most reliable for the determination of organic matter source when TOC exceeds 1.0 wt% and when contributions from inorganic nitrogen are taken into account (e.g., Talbot, 2001). In most cases, biogenic silica was determined using a wet-chemical extraction technique that consists of exposing samples to hot sodium hydroxide over specific time steps (e.g., DeMaster, 1979). The precision of this analysis is ~1.0 wt%.

Total Organic Carbon (TOC) and Total Inorganic Carbon (TIC)

TOC versus TIC comparisons between lakes are complicated by the variable methods used to calculate these data among all studies. In particular, estimates of TOC and TIC by LOI have well-known uncertainties, particularly when carbonate is actively precipitating (Dean, 1974; Heiri et al., 2001). Our data suggest possible contrasts between lake depozones, although none is statistically significant at the $p < 0.05$ level (Fig. 13B). Analyzed hinterland lakes displayed moderate TIC and relatively high TOC values, with considerable within-lake variability in TOC. This variability is particularly evident over glacial-interglacial time scales, undoubtedly related to overall vast changes in lacustrine productivity associated with those changes. The overall high levels of TOC in a number of Altiplano lakes are noteworthy, because, as is evident from our sedimentation rate data, these are also areas of high overall sediment rates, primarily from siliciclastic inputs. This is because they all occur in arid settings with little terrestrial organic matter (OM) input (the role of aerosol OM contributions is unknown), and because ultraviolet B photodegradation can be expected to oxidize OM, and photoinhibition can be expected to constrain photosynthesis at these altitudes (Villafañe et al., 1999). Thus, a high concentration of OM in the face of high total sedimentation rates implies very high flux rates of OM.

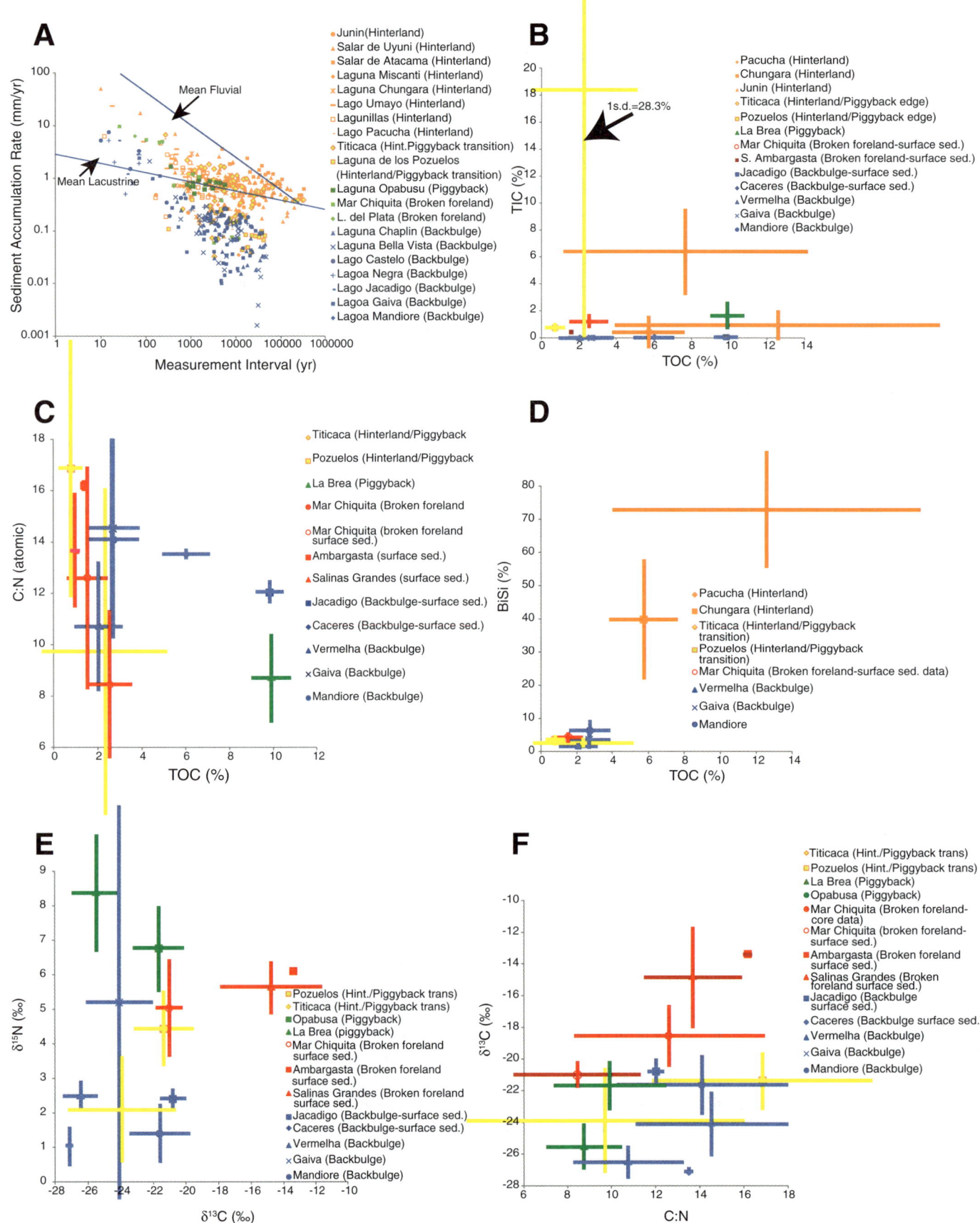

Figure 13. Sediment accumulation rate and sediment geochemistry for Andean retroarc lakes. For all plots: orange symbols—hinterland lakes; green—piggyback basin lakes; orange edge/yellow center—hinterland-to-piggyback transition; blue—back-bulge lakes. Geochemical data are from core samples unless indicated by "surface sed." on key. (A) Plot of sediment accumulation rates as a function of measurement interval for modern Andean retroarc lakes. Data sources: Junin (Seltzer et al., 2000); Uyuni (Fritz et al., 2004); Atacama (Bobst et al., 2001); Miscanti (Grosjean et al., 2001); Chungará (Moreno et al., 2007); Umayo (Ekdahl et al., 2008); Lagunillas (Ekdahl et al., 2008); Pacucha (Hillyer et al., 2009); Titicaca (Fritz et al., 2007); Pozuelos (McGlue et al., 2013); Opabusu (Carnes, 2011); Mar Chiquita (Piovano et al., 2007); del Plata (Larizzatti et al., 2001); Chaplin (Burbridge et al., 2004); Bella Vista (Burbridge et al., 2004); Castelo (Bonachea et al., 2010); Negra (Bonachea et al., 2010); Jacadigo (Bonachea et al., 2010); Gaíva (McGlue et al., 2012b); and Mandioré (McGlue et al., 2012b). (B) Total inorganic carbon (TIC) vs. total organic carbon (TOC) means (solid figures) and 1σ ranges for depocenter/near-depocenter cores, unless otherwise indicated below as surface sediments from modern lakes of the Andean retroarc region. Data sources, methods, and number of measurements: Pacucha (Hillyer et al., 2009; loss on ignition [LOI], $n = 122$ TOC, 84 TIC); Chungará (Moreno et al., 2007; coulometry, $n = 125$ TOC, 106 TIC); Junin (Seltzer et al., 2000; method not indicated, $n = 117$ TOC, 143 TIC); Titicaca (Ballantyne et al., 2011; TIC-AAS (atomic absorption spectroscopy), TOC-elemental analyzer, $n = 398$ TOC, 403 TIC); Pozuelos from depocenter core 6A (McGlue et al., 2013; elemental analyzer and coulometry, $n = 33$ TOC and TIC); La Brea (J. Omarini, 2010, personal commun.; LOI, $n = 10$ TOC and TIC); Mar Chiquita (McGlue et al., 2015; elemental analyzer and coulometry data, $n = 60$); Ambargasta (surface sediments, this paper; LOI, $n = 1$ TOC and TIC); Jacadigo (surface sediments, this paper; LOI, $n = 4$ TOC and TIC); Caceres (surface sediments, this paper; LOI, $n = 8$ TOC and TIC); Vermelha ($n = 42$ TOC and TIC), Gaíva ($n = 66$ TOC and TIC), and Mandioré ($n = 29$ TOC and 67 TIC) (Vermelha, Gaíva, and Mandioré all from McGlue et al., 2011; elemental analyzer and coulometry). (C) C:N (expressed as molar ratios) vs. TOC means (solid figures) and 1σ ranges for depocenter/near-depocenter cores from modern lakes of the Andean retroarc region. Samples are from near-depocenter cores unless otherwise noted as surface sediment samples TOC data, as in B. C and N data sources and number of measurements: Titicaca (Ballantyne et al., 2011; $n = 254$); Pozuelos (McGlue et al., 2012; core 6A, $n = 33$); La Brea (McGlue et al., 2012; $n = 6$); Mar Chiquita (Piovano et al., 2004; core data, $n = 36$); Ambargasta surface sediment (this report; $n = 2$); Salinas Grandes surface sediment (TOC and C:N, this report; $n = 3$); Jacadigo surface sediment (McGlue et al., 2011; this report; $n = 4$), Caceres surface sediment (this report; $n = 8$); Vermelha (McGlue et al., 2011; surface samples, $n = 42$); Gaiva (McGlue et al., 2012b; $n = 66$); and Mandioré (McGlue et al., 2012b; $n = 67$). (D) TOC vs. biogenic silica (BiSi) means (solid figures) and 1σ ranges for depocenter/near-depocenter cores from modern lakes of the Andean retroarc region. Samples are from near-depocenter cores unless otherwise noted as surface sediment samples. TOC data as in B. BiSi data sources and number of measurements: Pacucha ($n = 94$; Hillyer et al., 2009); Chungará ($n = 62$; Moreno et al., 2007); Titicaca ($n = 250$; Ballantyne et al., 2011); Pozuelos ($n = 33$; McGlue et al., 2012b); Mar Chiquita ($n = 61$; McGlue et al., 2012b); Vermelha (surface sediment, $n = 44$; McGlue et al., 2011); Gaíva ($n = 66$; McGlue et al., 2012b); and Mandioré ($n = 58$; McGlue et al., 2012b). (E) δ^{13}C vs. δ^{15}N means (solid figures) and 1σ ranges for depocenter/near-depocenter cores from modern lakes of the Andean retroarc region. Samples are from near-depocenter cores unless otherwise noted as surface sediment samples. Stable isotope data sources and number of measurements: Pozuelos ($n = 33$; M. McGlue, 2012, personal commun.); Titicaca (Ballantyne et al., 2011; $n = 420$); Opabusa ($n = 89$; Carnes, 2011); La Brea ($n = 6$; J. Omarini, 2010, personal commun.); Mar Chiquita surface sediments ($n = 58$ for δ^{13}C, $n = 55$ for δ^{15}N; McGlue et al., 2015); Ambargasta ($n = 1$, this study); Salinas Grandes ($n = 2$, this study); Jacadigo ($n = 5$, this study); Caceres ($n = 8$, this study); Vermelha surface sediments ($n = 42$; McGlue et al., 2011); Gaíva ($n = 66$; McGlue et al., 2012b); and Mandioré ($n = 29$; McGlue et al., 2012b). (F) C:N vs. δ^{13}C means (solid figures) and 1σ ranges for depocenter/near-depocenter cores and surface sediment samples from modern lakes of the Andean retroarc region. Sample sizes and data sources are as in C and E.

The two lakes on the transition between hinterland and piggyback zones, lying at similar altitudes and in similar precipitation regimes, had lower TOC values. Lake Titicaca has previously been documented to be mesotrophic, with primary productivity levels on par with other large tropical lakes (Richerson et al., 1992). They also had completely dissimilar TIC concentrations, with Lake Titicaca showing extremely variable (and bimodal) TIC concentrations, whereas Laguna de los Pozuelos was generally TIC poor. Carbonate-bedrock sources are more prevalent in the northern Altiplano and Lake Junin watersheds, which may explain the higher TIC values observed in these areas. Lakes in the wetter lowland settings (low-elevation piggyback, broken foreland, and back-bulge) below the Andean arid zone generally displayed very low TIC values and high but highly variable TOC.

Carbon:Nitrogen (C:N) Ratios

The carbon to nitrogen ratio in the organic fraction of lake sediments (expressed here as atomic or molar ratios) is a commonly measured variable that is often indicative of whether the dominant source of OM is autochthonous or allochthonous, the latter including atmospheric or bird-derived material from outside the immediate watershed (Meyers and Teranes, 2001). Because of the absence of cellulose in aquatic algae and the low proportions of N in cellulose and woody tissue, C:N ratios are typically much higher in lake sediments dominated by terrestrial sources of OM, although an exact threshold of autochthonous versus allochthonous OM cannot be specified (Cohen, 2003). Low values of C:N (<10) can reasonably be interpreted to indicate algal sources, but definitively identifying terrestrial sources is complicated by diagenesis (which tends to elevate C:N ratios over time) and autochthonous production in N-limited lakes (e.g., Hecky et al., 1993). In fact, diagenetic effects are probably responsible for much of the variation seen in this data set (Fig. 13C).

All depozones had C:N ratios that averaged between 9.8 and 12.6, with no statistically significant difference among depozones. Mesotrophic Lake Titicaca sediments displayed relatively low C:N ratios, which is somewhat surprising given the lake's size. However, the variability in ratios at Lake Titicaca suggests that potentially significant and variable N loss from diagenesis could affect these values. In contrast, C:N ratios in Laguna de los Pozuelos are both high and highly variable. Some of this pattern may result from low N proportional to OM delivered to the lake at its southern deltas. However, the high productivity of this lake and the large variability of the data both suggest that variable diagenetic loss of N may be the major factor responsible for these results and perhaps in the other lakes displaying large amounts of variability in C:N ratios. Highly variable C:N ratios are also evident in Lagoa Gaíva and Lagoa Mandioré in the back-bulge and Salina de Ambargasta in the broken foreland, albeit less variable than Pozuelos. Most of these back-bulge lakes lie within a wetland-dominated region of the Brazilian Pantanal, which is heavily influenced in terms of sedimentation by allochthonous materials derived from a seasonal flood pulse of the adjacent upper Paraguay River (McGlue et al., 2011). Laguna La Brea (piggyback zone) shows a lower C:N ratio, typical of more autochthonous production, which is not surprising considering that this lake has a very small watershed in comparison with those of the Pantanal, with their seasonal connection to the Paraguay River system. Sediments from Lake Opabusa, another piggyback lake not shown in Figure 13C because of a lack of TOC data, nonetheless also yields relatively low C:N ratios (mean = 9.95). In both cases, there may be a tectonogeomorphic control on these values, in that the spatially isolated watersheds within the lowland piggyback zone may be unfavorable for delivering large quantities of allochthonous OM to the lake. The broken foreland lake samples showed a very wide range of C:N ratios. Notably, modern Laguna Mar Chiquita sediment C:N ratios, typical of a predominantly autochthonous OM source, are much lower than those of core sediments from the same lake, suggesting the possible loss of N through diagenesis. This may also explain the high (and highly variable) ratios in the highly oxidizing arid lake environments of the other broken foreland lakes and the arid Laguna de los Pozuelos data.

Biogenic Silica (BiSi)

Biogenic silica is the proportion of Si produced by all Si-secreting organisms, including diatoms, other siliceous algae, and sponges, although in most lake depocenters (including those of the Andean retroarc,) the dominant contributors are diatoms. Therefore, BiSi is often interpreted as an indicator of diatom productivity (e.g., Schelske et al., 1986). However, it is clearly influenced by terrigenous sediment flux and diatom dissolution as well, and in lakes with strong flood pulse inputs, river-derived sponge spicules probably are major contributors to the total BiSi load. BiSi data show statistically significant differences between high values for the analyzed hinterland lakes versus all other depozone classes. The two hinterland basins for which data were available had extremely high (and highly variable) BiSi and TOC concentrations, possibly as a result of more autochthonous primary production over time, especially from diatoms (Fig. 13D). All other lakes measured to date in the retroarc show relatively

low BiSi values, with no significant differences among those lake classes. Internal production in back-bulge lakes may be more driven by green algae and other autotrophs than diatoms. Because many of these low-elevation lakes appear to be highly productive, it is possible some are Si- or N-limited (e.g., Lagoa Mandioré), which would encourage the growth of N-fixing cyanobacteria, which do not secrete silica. However, a dominance of N-fixing cyanobacteria is unlikely to explain the low BiSi values for Mar Chiquita (or other broken foreland lakes) given their high N-isotope values discussed in the following section.

$\delta^{13}C_{organic\ matter}$ versus $\delta^{15}N_{organic\ matter}$

Carbon and nitrogen stable isotope records from lacustrine OM have been commonly used by paleolimnologists to help understand lake and watershed ecosystem dynamics (Meyers and Teranes, 2001; Talbot, 2001). Our data set of $\delta^{13}C_{organic\ matter}$ and $\delta^{15}N_{organic\ matter}$ from the Andean retroarc shows a possible grouping of lakes by geographic (and perhaps tectonic) setting, although the differences are not statistically significant (Fig. 13E). The lakes of the relatively low-elevation piggyback basin zone (the only low-elevation lakes of the Andean front for which we have OM data) have OM characterized by relatively negative $\delta^{13}C$ values, suggesting a dominance of algal OM contributions or C_3 terrestrial OM. It is noteworthy, however, that Lake Opabusu, which lies within this zone, is surrounded by primarily C_4 vegetation (Carnes, 2011). As a group, these lakes also display relatively enriched $\delta^{15}N$ OM values, suggesting OM in these lakes is derived primarily from algae that are not N_2 assimilators, or, alternatively, indicating a significant soil contribution (the latter possibility being considered less likely given the depleted $\delta^{13}C$ data).

Lake sediments from the Pantanal back-bulge zone display a similar range of depleted $\delta^{13}C$ values as found in the piggyback lakes, consistent with a dominant contribution of autochthonous phytoplankton production. However, the nitrogen isotope field for the back-bulge lakes is less enriched in $\delta^{15}N$, suggesting either less soil input or more N_2-assimilating cyanobacteria in these lakes. Cyanobacterial blooms were frequently observed during our sampling of these lakes (McGlue et al., 2011). In this respect, the back-bulge lakes are similar to the semihumid Altiplano Lake Titicaca, where the C isotopic signature suggests a dominant algal contribution consistent with its large size, and where relatively depleted N isotopes are consistent with that lake's status as strongly N-limited, with strong contributions to its algal biomass from N-fixing cyanobacteria (Richerson et al., 1992; Ballantyne et al., 2011). The broken foreland basin lakes, and Laguna de los Pozuelos all occur in much more arid settings than the back-bulge or low-elevation piggybacks—clearly a function of orographic effects on precipitation in their respective settings. These lakes have sediment OM that is more enriched in $\delta^{13}C$, consistent with their closed hydrology, lower proportional autochthonous production of OM, and higher inputs of terrestrial C_4-derived vegetation. They also display intermediate $\delta^{15}N$ values, perhaps indicative of more soil enrichment or less N$_2$-assimilating cyanobacteria in the OM pool than in the back-bulge setting.

A plot of C:N versus δ^{13}C (Fig. 13F) shows a possible trend of more humid (and more frequently open) hydrologic systems in the relatively humid back-bulge and low-elevation piggyback settings versus the other lake types. There is a near-complete separation between the δ^{13}C of OM from hydrologically closed lakes (>−21.5‰) compared to open lakes (<−21.5‰). The only exception is the slightly more enriched mean value for Lagoa Mandioré (open) versus Lake Opabusa (closed). There is considerable overlap in C:N data across climates, probably as a result of diagenetic loss of N as previously discussed, as indicated by the generally much larger 1σ ranges associated with the C:N data.

DISCUSSION

Our investigations highlight the diversity of lake types present in the Andean retroarc. Based on the available literature and our own studies of the modern Andean retroarc lakes, we make some preliminary generalizations about the characteristics of lakes forming within each major depozone (Table 1). Not surprisingly, lakes are most common and probably have the greatest potential for leaving stratigraphic records in basins that are underfilled by sediments, most of which are in arid to semiarid environments. Predominantly closed-basin lakes of the hinterland, hinterland-to-piggyback transition, and the inter- and extramontane broken foreland probably represent the regions with the highest probability of lacustrine sediment preservation, given their combination of probable long duration in areas of active subsidence and relatively high sediment accumulation rates. This supposition is borne out in the Cenozoic rock record of the Andes, where significant lake deposits from Pliocene–Pleistocene piggyback and hinterland basins have been documented (e.g., Cladouhos et al., 1994; Sobel et al., 2003; Bywater-Reyes et al., 2010; González Bonorino and Abascal, 2012; McGlue et al., 2013). South of our primary study area, within the Patagonian climatic belt dominated by prevailing westerlies, the foredeep is considerably more arid, with lower Andean uplift rates and probably lower sediment yields. In this setting the foredeep is also a favorable environment for long-term accumulation of lake deposits. Evidence of foredeep lakes is present in the rock records of both the Andean and Sevier (western U.S. Mesozoic) foreland basins. A lake, represented by laminated siltstones and carbonates of the Upper Lumbrera Formation (DeCelles et al., 2011), appears to have occupied a foredeep depozone during part of the Eocene–Oligocene in Argentina, whereas saline-alkaline lake strata (marlstones and laminated siltstones) of the Upper Impora Formation mark the late Eocene distal foredeep in Bolivia (DeCelles and Horton, 2003). In both cases, total lake strata are relatively thin (10–20 m). The formation of topographic depressions may have been limited in the Andean foredeep by the erosion of the dominantly siliciclastic fold-and-thrust belt. In the rock record of the western United States, extensive Lower Cretaceous lacustrine strata (>75 m thick) are present in the fore-

deep depozones of Wyoming and Idaho (Drummond et al., 1996; Zaleha, 2006). Carbonate bedrock in the adjacent thrust belt and a dry prevailing climate associated with rain-shadow development may have assisted in the preservation of topographic depressions and lake formation in the Sevier foreland. Analogous rain-shadow effects and relationships to lake evolution are evident in modeling experiments for Laramide-age broken foreland basins of the western United States (Sewall and Sloan, 2006).

Back-bulge deposystems are also areas of abundant modern lakes. However, these lakes are likely to be much more ephemeral than the regions mentioned previously. This fact, coupled with the overall slow rates of lacustrine sediment accumulation in the back-bulge, makes the accumulation of thick sequences of back-bulge lacustrine rocks in the stratigraphic record less likely. Moreover, the total accommodation available for sediment and water in the back-bulge is limited by the minor flexural response cratonward of the forebulge (DeCelles and Giles, 1996). Back-bulge basins most likely develop where a strong viscous coupling exists between the base of the continental plate and downward circulating mantle-wedge material that becomes entrained by the subducting oceanic plate (i.e., dynamic subsidence; DeCelles, 2011). Lake systems in the back-bulge are likely to be broad, shallow, and characterized by low-relief margins. Examples of back-bulge lake systems in the Andean Cenozoic record are controversial, whereas much better records are preserved in the back-bulge of the Sevier foreland (e.g., Zaleha, 2006). In Argentina, Del Papa (1999) documented the Lower Paleocene Maíz Gordo Formation, a 230-m-thick succession of carbonate paleolake deposits that accumulated in a basin marked by low-gradient ramp margins. Marquillas et al. (2005) argued that these deposits, as well as the overlying "Faja Verde" lake beds, formed during the postextensional sag phase of the Salta Rift. More extensive tectonostratigraphic research indicates that the Maíz Gordo lake strata most likely mark deposition in a back-bulge depozone associated with early Andean orogenesis (DeCelles et al., 2011). In the western United States, deposits of numerous groundwater wetlands and floodplain lakes are present in the Upper Jurassic Morrison Formation of Colorado, Utah, and South Dakota (western United States; Dunagan and Turner, 2004; Turner and Peterson, 2004). These deposits are laterally discontinuous and range up to 20 m thick in Colorado (Dunagan and Turner, 2004) and South Dakota (M. McGlue, 2011, personal observ.). Zaleha (2006) described a relatively large Early Cretaceous lake, Lake Minnewaste, in a back-bulge position within the Sevier foreland basin system. Where it crops out in South Dakota, the Minnewaste Limestone Member of the Lakota Formation is a thin (up to 10 m) carbonate deposit with a mound-like geometry. These beds are interpreted as supralittoral deposits from a saline lake that produced bedded evaporites in its center (Trees, 2012).

Based on our Andean observations, the lowest-probability environment for lacustrine stratal preservation within the retroarc is the wedge-top environment. Although in principle, long-lived lake systems could form where thrust-sheet margins create topographic depressions, this would in most cases be counteracted by

TABLE 1. GENERALIZED CHARACTERISTICS OF MODERN LAKES FROM THE ANDEAN RETROARC (11°S–35°S) AND PROBABLE ANCIENT EXAMPLES

Depozone	Climatic/hydrologic features	Dominant local closure (i.e., lake-forming) mechanisms	Lake size and abundance	Likely duration as lacustrine depositsystems (potential for stratigraphic record preservation)	Sediment accumulation rates (~10^4 yr duration time frame)	Basin-fill characteristics (sensu Carroll and Bohacs, 1999)	Sediment organic matter (OM)/geochemical characteristics	Ancient Andean examples
Hinterland basins (associated with lithospheric removal)	Hyperarid-semiarid (0–400 mm/yr mean annual precipitation [MAP]), wetter in north (up to 900 mm/yr) and east. Closed basins, mostly shallow lakes. Precip from east blocked by orographic rainout; no westerly sources.	Sag basins with minor normal faulting.	Lakes common, intermediate average size.	Long-lived (10^6–10^7 yr), potentially correlated with duration of pull-down phase of eclogite accumulation in cordilleran cycle model.	0.2–3 mm/yr	Underfilled to balanced-filled.	Low to moderate total inorganic carbon (TIC), relatively high (and highly variable) total organic carbon (TOC), no data for C:N, high BiSi, no stable isotope data.	Vizcachera Fm., Arizaro Basin, Argentina (DeCelles et al., this volume)
Hinterland/piggyback transitional basins	Arid-semihumid (150–1000 mm/yr MAP, wetter in north and east). Closed basin except in north. Shallow to deep lakes. Precip from east reduced by orographic rainout; no westerly sources.	Fault-bounded (older structures reverse faults, younger normal faults) plus sags.	Lakes moderately abundant, larger in size.	Long-lived (10^6–10^7 yr), transitional phase for basins originally formed as piggybacks as they are transferred to hinterland through forward propagation of thrust belt.	0.08–1 mm/yr	Underfilled to balanced-filled.	Highly variable TIC, low TOC, highly variable C:N (diagenetic overprint?), low BiSi, depleted ^{15}N and ^{13}C (in OM).	None identified to date
Piggyback basins	Semiarid-semihumid (500–1000 mm/ yr MAP). Wetter to east/lower elevation. Open and closed basins. Mostly shallow lakes but landslide dammed can be deep.	Primarily landslide dammed on axial end and reverse faults on E and W margins.	Lakes rare and small.	Potentially long-lived (10^6–10^7 yr) for lakes with purely thrust sheet margin closure. Very short-lived (<10^3 yr) for landslide-dammed lakes.	~1 mm/yr (limited data)	Underfilled to balanced-filled.	At low elevations, low TIC, high TOC. No TOC/TIC data for high elevations, low C:N (? limited data), no BiSi data, enriched ^{15}N, depleted ^{13}C.	Lerma Valley Gp. (González Bonorino and Abascal, 2012); unnamed late Quaternary lake beds, Santa Maria Basin, etc., Argentina (Trauth et al., 2000)
Wedge top	Semiarid-semihumid (600–1000 mm/yr MAP). Lakes mostly open drainage, seasonal wetlands, very shallow.		Lakes rare and small.	Ephemeral. Little opportunity for sustained topographic closure because of rapid infilling. Open basin wetland/ abandoned channel duration scales to avulsion rates of megafan channels.	No modern lake data	Mostly fluvial-overfilled.	No modern lake data.	Miocene Palo Pintado Fm., (Bywater-Reyes et al., 2010)
Foredeep	Semiarid (south, 600–800 mm/yr) to humid (north, up to 2000 mm/yr). Lakes mostly open drainage, seasonal wetlands. Tectonically formed lakes mostly in arid foredeep south of study region. Mostly shallow lakes.	Eolian deflation ponds and competitive aggradation between megafan channels.	Lakes moderately abundant (more so in arid regions) and small.	Ephemeral. Little opportunity for sustained topographic closure because of infilling, except in areas/times with arid conditions (low sediment yield and underfilling). Open basin wetland and abandoned channel duration scales to avulsion rates of megafan channel systems.	No modern lake data	Mostly overfilled except under arid climates (underfilled or balanced-filled). Carbonate watershed bedrock (rare for modern Andean foredeep watersheds), could also be expected to be underfilled to balanced-filled.	No modern lake data.	Lumbrera Fm. (DeCelles et al., 2011); Yecua Fm. (distal foredeep; Uba et al., 2006)

(Continued)

TABLE 1. GENERALIZED CHARACTERISTICS OF MODERN LAKES FROM THE ANDEAN RETROARC (11°S–35°S) AND PROBABLE ANCIENT EXAMPLES (Continued)

Depozone	Climatic/hydrologic features	Dominant local closure (i.e., lake-forming) mechanisms	Lake size and abundance	Likely duration as lacustrine deposystems (potential for stratigraphic record preservation)	Sediment accumulation rates (~10^4 yr duration time frame)	Basin fill characteristics (sensu Carroll and Bohacs, 1999)	Sediment organic matter (OM)/geochemical characteristics	Ancient Andean examples
Back bulge	Semiarid to humid (600–2000 mm/yr), wetter to east and north. Shallow open lakes and wetlands.	Eolian deflation, abandoned fluvial channels and competitive aggradation along fluvial systems.	Lakes abundant and mostly small.	Moderately long-lived (10^4–10^6[?] yr) because of low sedimentation rates.	0.01–0.3 mm/yr	Fluvial	Very low (near 0) TIC, variable TOC, Intermediate to high C:N (fluvial inputs of terrestrial carbon), low BiSi; depleted ^{15}N and ^{13}C.	Maiz-Gordo Fm. (DeCelles et al., 2011)
Broken foreland basins between Sierras Pampeanas (intermontane)	Semiarid (200–700 mm/yr). Closed salt lakes and salinas (intermittent outlet from S. Ambargasta). Currently shallow but can be deep.	Localized reverse fault boundaries.	Lakes (playas) common and large.	Long-lived (10^6–10^7 yr) because of typical low sedimentation rates and underfilling. Individual lake phases potentially much shorter because of relative aridity.	No modern lake data in this duration range but probably slightly lower than extramontane based on shorter-duration comparisons.	Underfilled given modern climate conditions (arid-semiarid).	Low TIC and TOC? (minimal modern data). Relatively high C:N, no BiSi data, enriched ^{15}N and ^{13}C.	
Broken foreland (extramontane)	Semiarid (300–1100 mm/yr at L. Mar Chiquita historically). L. Mar Chiquita closed, relatively shallow.	Sag depression within foredeep, bound on E. by San Guillermo High.	Few examples but Mar Chiquita is very large.	Long-lived (10^6–10^7 yr), though potentially shorter than broken foreland intermontane because of higher sediment inputs and rates (greater probability of overfilling).	0.08–0.3 mm/yr	Underfilled given modern climate conditions (arid-semiarid).	Low TOC and TIC, highly variable C:N (diagenetic N loss?), low BiSi, enriched ^{15}N, intermediate ^{13}C.	No exposed modern examples

sediment overfilling from erosion of adjacent high-relief areas. Landslide-dammed lakes, although common in modern Andean piggyback settings, have very low long-term stratal preservational potential, because they fill and breach on time scales of decades to centuries (Table 1; Trauth and Strecker, 1999).

Ultimately, we see the potential for characteristic differences among lakes formed in differing morphotectonic zones to serve as a basis for correctly identifying zones of origin for retroarc lake deposits of unknown specific origin (e.g., back-bulge vs. foredeep?) based on these suites of facies characteristics. In Table 1, we identify some examples of ancient Andean retroarc lake and wetland deposits, previously pigeonholed by morphotectonic zone on the basis of other evidence. It is noteworthy that in each of these cases, the characteristics of the ancient examples closely correspond with the predicted gross lithologies and facies patterns observed in our modern lake examples. However, it is probably premature to establish a definitive range of facies characteristics for lake types in each morphotectonic zone at this point. First, the range of modern lakes we have studied, even within our limited study area, is far from exhaustive. Second, the range of combinations of lakes, climate, and bedrock expressed today in the Andes is far from a complete catalog of possibilities within a retroarc. For example, carbonate-bedrock watersheds feeding into a foredeep or broken foreland are relatively uncommon in the Andes. In retroarc basins of the Sevier and Laramide orogenic zones, where carbonates were common in paleowatersheds, large foredeep lakes were also common, possibly for reasons we have discussed earlier. A broader synthesis of retroarc lakes will need to incorporate modern studies of these basins from other modern orogenic systems besides the Andes as well as comparative systems in ancient examples from around the world.

CONCLUSIONS

In this study, we make a first attempt to compare and contrast sedimentologic and geomorphic data from a wide range of modern lakes of various tectonic origins formed within the Andean retroarc orogenic system. As in all lacustrine deposystems, maintenance of a topographic depression within retroarc basins (a prerequisite for lake formation) involves a balancing act between creation and destruction of accommodation. Large, actively uplifting mountain belts, such as the Andes, and associated orographic precipitation patterns create an excess of sediment to the point that topographic depressions in many, if not most, retroarc basins are rapidly filled by sediment. Lake formation in this context requires special conditions that deviate from this norm, driven by one or more of the following:

1. Topographic isolation. Alluvial and fluvial sediment sources can be diverted from a basin by local bedrock highs, or the regional flexural wave can interfere with preexisting topography, in the process reactivating older faults and diverting sediment from an individual basin.

2. Climate. A reduction in sediment yield by decreasing stream discharge (and variability or intensity of discharge) under more arid climatic conditions can result in the sediment starvation of retroarc basins, which allows topographic depressions to form. On geologically short time scales, increased precipitation can raise lake levels and enlarge lakes, but this is counterbalanced by increased sediment supply and eventually by the depositional infilling of the lake and formation of a through-flowing fluvial system,

3. Bedrock lithology. Differences in erodibility and solubility can have a profound impact on the rate at which retroarc basins infill with sediment, and thus the likelihood that they can persist as topographic depressions that are not infilled by sediment and can hold lakes.

4. Short-term (nontectonic) interactions. Many lakes within a retroarc owe their existence to processes that are partly or wholly unrelated to regional tectonics, such as competitive aggradation within fluvial systems, glacial excavation, or eolian deflation.

The climate, in particular, the effective moisture balance of the Andean retroarc region, plays a key role in determining the nature of standing waterbodies in this region, their susceptibility to becoming completely infilled by sediment, and their short-term spatial extent. The primary sources of available moisture within the central Andean retroarc are regulated by the seasonal migration of the ITCZ, which brings moisture predominantly from the South Atlantic, supplemented in the southern region by the South American monsoon. This results in relatively high precipitation in regions that are also far from primary sources of Andean sediments, allowing basins to remain sediment starved. Precipitation in the retroarc is also highly influenced by the orographic effects imposed by the Andes. After crossing eastern low-elevation regions of Argentina, Brazil, Paraguay, and Bolivia, easterly winds shed almost their entire remaining volume of precipitation within a geographically narrow belt upon reaching the abrupt eastern edge of the Andes, which in turn produces very high sediment yields into the foreland. This rainout leads, in turn, to much lower precipitation levels at higher elevations and more westerly locations across the tropical and subtropical Andes.

Lacustrine deposition in the modern Andean retroarc is focused in six major morphotectonic zones.

(1) The high-elevation plateaus (Altiplano or Puna areas) of western Argentina and Bolivia and parts of Chile and Peru are occupied in places by extensive hinterland basins. These basins are underfilled by sediment as a result of their combined tectonic and climatic setting, and they are typified by the numerous large, modern, internally drained salars of northwestern Argentina and southwestern Bolivia. The formation and evolution of these basins appear to be associated with the development of thick and dense eclogitic roots, which in turn are hypothesized to form during the high-flux magmatic events associated with the cordilleran orogenic cycle, which can allow these lakes to persist over geologically long time intervals.

(2) The thrust belt of the Eastern Cordillera is a region of very high relief and is almost entirely an erosional environment today. Relatively small piggyback basins occupy isolated valleys lying atop the various thrust sheets that make up the Eastern

Cordillera, but few hold lakes as a result of rapid sediment infilling. Because of the ongoing eastward advancement of the Andean thrust belt, piggyback basins at the transition to the Andean hinterland appear to be undergoing an evolution from thrust sheet (reverse fault)–bounded basins to the normal fault boundaries more typical of the hinterland basins, and some of these basins are quite large and contain big lakes (e.g., Lake Titicaca, Laguna de los Pozuelos), with sediment input limited by their high elevation, orographic moisture blockage, and arid to semiarid settings.

(3) At the eastern transition from the Andean thrust belt to the modern wedge-top and (4) foredeep depozones, the combination of adjacent relief and heavy precipitation causes the depositional system to be overloaded by a huge excess of fluvial sediments, and lakes are rare, represented by local and seasonal wetlands only. This pattern changes south of the subtropics, where westerlies prevail, causing the Argentine wedge-top and flexural foredeep to be much more arid (Patagonia) and sediment starved. Here, topographic closure and internally drained lakes are relatively common.

(5) The back-bulge depozone is quite distal to Andean sediment sources. This fact, plus the location of the back-bulge within the tropical-subtropical rain belt, causes wetlands and small, short-lived lakes to be quite common within the study area. However, many of these lakes also owe their existence to other factors, such as reactivation of preexisting faults on the craton and eolian deflation.

(6) In regions of the Andean retroarc affected by flat-slab subduction and the formation of basement-cored uplifts (e.g., Sierras Pampeanas), localized basin isolation, nonerosive crystalline basement in watersheds, and the formation of orographic rain shadows all lead to the formation of relatively low-elevation topographic depressions (salars in intramontane settings) and the Mar Chiquita (the largest lake in Argentina) in an extramontane setting.

Lake abundance and sediment accumulation rates are both strongly correlated with morphotectonic setting in the Andean retroarc (Table 1). In contrast, OM and other geochemical characteristics are highly variable among Andean retroarc lake depozones, and these contrasts may result from long-term climate variability or diagenesis unrelated to the specific depozone.

This first synthesis of retroarc lakes suggests lakes are most common and have the greatest potential for leaving stratigraphic records in arid to semiarid, sediment-underfilled environments. This would include the closed-basin lakes of the hinterland, hinterland-to-piggyback transition, and the inter- and extramontane broken foreland within the easterly belt and perhaps the Patagonian foredeep within the westerly belt. These depozones have the highest probability of lacustrine sediment preservation, given their combination of probable long duration in areas of active subsidence and relatively high sediment accumulation rates, and occurrence in a rain shadow. Back-bulge deposystems, while replete with modern lakes, are also areas where lakes are likely to be somewhat less persistent. This, coupled with their overall slow rates of lacustrine sediment accumulation, makes the accumulation of thick sequences of back-bulge lacustrine rocks in the stratigraphic record less likely, although lake deposits are certainly known from this environment in the stratigraphic record. The lowest-probability environments for lacustrine stratal preservation within the Andean retroarc are the piggyback and wedge-top depozones.

ACKNOWLEDGMENTS

We thank ExxonMobil for their support through the Convergent Orogenic Systems Analysis (COSA) grant to the University of Arizona for this project. We also thank the American Chemical Society (PRF grant 45910-AC8) and the National Science Foundation (NSF grant EAR-0542993), and grants from Sigma Xi and the University of Arizona–Chevron Corporation Summer Research Fund (both to McGlue). We thank the Limnological Research Center (University of Minnesota) for their considerable assistance with core preparation and access to their facilities. We thank the Federal University of Mato Grosso do Sul-Campus of the Pantanal, the Ecology in Action nongovernmental organization (ECOA), and the Saint Teresa Farm for their generous support during our fieldwork in the Pantanal. Swarzenski thanks the U.S. Geological Survey Coastal and Marine Geology Program for support in sediment provenance studies. Thanks go to E. Gierlowski-Kordesch, M. Smith, and M. Rosen for very helpful reviews of earlier versions of this manuscript. We also thank the following individuals for their assistance with various aspects of this project: J. Omarini, C. Landowski, J. Ash, E. Abel, E. Piovano, A. Kirschbaum, C. Gans, M. Trees, and J. Pelletier. Any use of trade, product, or firm names is for descriptive purposes only and does not imply endorsement by the U.S. government.

REFERENCES CITED

Aalto, R., Maurice-Bourgoin, L., Dunne, T., Montgomery, D.R., Nittrouer, C.A., and Guyot, J.L., 2003, Episodic sediment accumulation on Amazonian flood plains influenced by El Niño/Southern Oscillation: Nature, v. 425, p. 493–497, doi:10.1038/nature02002.

Allenby, R.J., 1988, Origin of rectangular and aligned lakes in the Beni Basin of Bolivia: Tectonophysics, v. 145, p. 1–20, doi:10.1016/0040-1951(88)90311-3.

Alonso, R.N., 2006, Ambientes evaporíticos continentales de Argentina. Temas de la Geología Argentina I (2): INSUGEO, Serie Correlación Geológica, v. 21, p. 155–170.

Assine, M.L., and Silva, A., 2009, Contrasting fluvial styles of the Paraguay River in the northwestern border of the Pantanal wetland, Brazil: Geomorphology, v. 113, p. 189–199, doi:10.1016/j.geomorph.2009.03.012.

Assine, M.L., and Soares, P.C., 2004, Quaternary of the Pantanal, west-central Brazil: Quaternary International, v. 114, p. 23–34, doi:10.1016/S1040-6182(03)00039-9.

Ballantyne, A.P., Baker, P.A., Fritz, S.C., and Poulter, B., 2011, Climate-mediated nitrogen and carbon dynamics in a tropical watershed: Journal of Geophysical Research, v. 116, G02013, doi:10.1029/2010JG001496.

Beck, S.L., and Zandt, G., 2002, The nature of orogenic crust in the central Andes: Journal of Geophysical Research, v. 107, 2230, doi:10.1029/2000JB000124.

Bobst, A.L., Lowenstein, T.K., Jordan, T.E., Godfrey, L.V., Ku, T.L., and Luo, S., 2001, A 106 ka paleoclimate record from drill core of the Salar de Atacama, northern Chile: Palaeogeography, Palaeoclimatology, Palaeoecology, v. 173, p. 21–42, doi:10.1016/S0031-0182(01)00308-X.

Bohacs, K.M., Carroll, A.R., Neal, J.E., and Mankiewicz, P.J., 2000, Lake-basin type, source potential, and hydrocarbon character: An integrated sequence-stratigraphic-geochemical framework, *in* Gierlowski-Kordesch, E.H., and Kelts, K.R., eds., Lake Basins through Space and Time: American Association of Petroleum Geologists, Studies in Geology, v. 46, p. 3–33.

Bonachea, J., Bruschi, V., Hurtado, M.A., Forte, L.M., da Silva, M., Etcheverry, R., Cavallotto, J.L., Dantas, M., Pejon, O., Zuquette, L.V., de O. Bezerra, M.A., Remondo, J., Rivas, V., Gomez-Arozamena, J., Fernandez, G., and Cendrero A., 2010, Natural- and human-forcing in recent geomorphic change: Case studies in the Rio de la Plata Basin: The Science of the Total Environment, v. 408, p. 2674–2695, doi:10.1016/j.scitotenv.2010.03.004.

Brodie, C.R., Casford, J.S.L., Lloyd, J.M., Leng, M.J., Heaton, T.H.E., Kendrick, C.P., and Yongqiang, Z., 2011, Evidence for bias in C/N, δ^{13}C and δ^{15}N values of bulk organic matter, and on environmental interpretation, from a lake sedimentary sequence by pre-analysis acid treatment methods: Quaternary Science Reviews, v. 30, no. 21–22, p. 3076–3087, doi:10.1016/j.quascirev.2011.07.003.

Buchheim, H.P., and Eugster, H.P., 1998, Eocene fossil lake: The Green River Formation of Fossil Basin, southwestern Wyoming, *in* Pittman, J., and Carroll, A., eds., Modern and Ancient Lacustrine Depositional Systems: Utah Geological Association Guidebook 26, p. 1–17.

Burbridge, R.E., Mayle, F.E., and Kileen, T.J., 2004, Fifty-thousand-year vegetation and climate history of Noel Kempff Mercado National Park, Bolivian Amazon: Quaternary Research, v. 61, p. 215–230, doi:10.1016/j.yqres.2003.12.004.

Bywater-Reyes, S., Carrapa, B., Clementz, M., and Schoenbohm, L., 2010, Effect of late Cenozoic aridification on sedimentation in the Eastern Cordillera of northwest Argentina (Angastaco Basin): Geology, v. 38, no. 3, p. 235–238, doi:10.1130/G30532.1.

Cabrera, L., Cabrera, M., Gorchs, R., and de las Heras, F.X.C., 2002, Lacustrine basin dynamics and organosulphur compound origin in a carbonate-rich lacustrine system (late Oligocene Mequinenza Formation, SE Ebro Basin, NE Spain): Sedimentary Geology, v. 148, p. 289–317, doi:10.1016/S0037-0738(01)00223-8.

Caffe, P.J., Trumbull, R.B., Coira, B.L., and Romer, R.L., 2002, Petrogenesis of early Neogene magmatism in the northern Puna; implications for magma genesis and crustal processes in the central Andean Plateau: Journal of Petrology, v. 43, p. 907–942, doi:10.1093/petrology/43.5.907.

Carnes, A., 2011 Hydrologic Variation and Lake Sediments: A Reconstruction of the Bolivian Lowlands over the Last 5,500 Years [M.S. thesis]: Durham, North Carolina, Duke University, 22 p.

Carrapa, B., Bywater-Reyes, S., DeCelles, P., Mortimer, E., and Gehrels, G., 2011, Late Eocene–Pliocene basin evolution in the Eastern Cordillera of northwestern Argentina (25–26°S): Regional implications for Andean orogenic wedge development: Basin Research, v. 24, p. 249–268, doi:10.1111/j.1365-2117.2011.00519.x.

Carroll, A.R., and Bohacs, K.M., 1999, Stratigraphic classification of ancient lakes: Balancing tectonic and climatic controls: Geology, v. 27, p. 99–102, doi:10.1130/0091-7613(1999)027<0099:SCOALB>2.3.CO;2.

Carroll, A.R., Chetel, L., and Smith, M.E., 2006, Feast to famine: Sediment supply control on Laramide basin fill: Geology, v. 34, p. 197–200, doi:10.1130/G22148.1.

Castle, J.W., 1990, Sedimentation in Eocene Lake Uinta (Lower Green River Formation), northeastern Uinta Basin, Utah, *in* Katz, B.J., ed., Lacustrine Basin Exploration—Case Studies and Modern Analogs: American Association of Petroleum Geologists Memoir 50, p. 243–263.

Chase, C.G., Sussman, A.J., and Coblentz, D.D., 2009, Curved Andes: Geoid, forebulge, and flexure: Lithosphere, v. 1, no. 6, p. 358–363, doi:10.1130/L67.1.

Cladouhos, T.T., Allmendinger, R.W., Coira, B., and Farrar, E., 1994, Late Cenozoic deformation in the central Andes: Fault kinematics from the northern Puna, northwestern Argentina and southwestern Bolivia: Journal of South American Earth Sciences, v. 7, p. 209–228, doi:10.1016/0895-9811(94)90008-6.

Cohen, A.S., 1990, A tectonostratigraphic model for sedimentation in Lake Tanganyika, Africa, *in* Katz, B.J., ed., Lacustrine Basin Exploration—Case Studies and Modern Analogs: American Association of Petroleum Geologists Memoir 50, p. 137–150.

Cohen, A.S., 2003, Paleolimnology: The History and Evolution of Lake Systems: Oxford, UK, Oxford University Press, 500 p.

Cohen, A.S., Lezzar, K.-E., Tiercelin, J.J., and Soreghan, M., 1997, New palaeogeographic and lake-level reconstructions of Lake Tanganyika: Implications for tectonic, climatic, and biological evolution in a rift lake: Basin Research, v. 9, p. 107–132, doi:10.1046/j.1365-2117.1997.00038.x.

Cole, R.D., 1998, Possible Milankovitch cycles in the lower Parachute Creek Member of Green River Formation (Eocene), north-central Piceance Creek Basin, Colorado: An analysis, *in* Pitman, J.K., and Carroll, A.R., eds., Modern and Ancient Lake Systems: Utah Geological Association Guidebook 26, p. 233–259.

Colman, S.M., Karabanov, E.B., and Nelson, C.H., 2003, Quaternary sedimentation and subsidence history of Lake Baikal, Siberia, based on seismic stratigraphy and coring: Journal of Sedimentary Research, v. 73, p. 941–956, doi:10.1306/041703730941.

Costa, C.H., and Vita-Finzi, C., 1996, Late Holocene faulting in the southeast Sierras Pampeanas of Argentina: Geology, v. 24, p. 1127–1130, doi:10.1130/0091-7613(1996)024<1127:LHFITS>2.3.CO;2.

D'Agostino, K.D., Seltzer, G., Baker, P., Fritz, S., and Dunbar, R., 2002, Late Quaternary lowstands of Lake Titicaca: Evidence from high-resolution seismic data: Palaeogeography, Palaeoclimatology, Palaeoecology, v. 179, p. 97–111, doi:10.1016/S0031-0182(01)00411-4.

Davidson, S.K., Hartley, A.J., Weissmann, G.S., Nichols, G.J., and Scuderi, L.A., 2013, Geomorphic elements on modern distributive fluvial systems: Geomorphology, v. 180–181, p. 82–95, doi:10.1016/j.geomorph.2012.09.008.

Dávila, F.M., Lithgow-Bertelloni, C., and Giménez, M., 2010, Tectonic and dynamic controls on the topography and subsidence of the Argentine Pampas: The role of the flat-slab: Earth and Planetary Science Letters, v. 295, p. 187–194, doi:10.1016/j.epsl.2010.03.039.

Dean, W., 1974, Determination of carbonate and organic matter in calcareous sediments and sedimentary rocks by loss on ignition: Comparison with other methods: Journal of Sedimentary Petrology, v. 44, p. 242–248.

DeCelles, P.G., 2011, Foreland basin systems revisited: Variations in response to tectonic settings, *in* Busby, C., and Azor, A., eds., Tectonics of Sedimentary Basins: Recent Advances: Chichester, UK, John Wiley & Sons, p. 405–426.

DeCelles, P.G., and Giles, K.A., 1996, Foreland basin systems: Basin Research, v. 8, p. 105–123, doi:10.1046/j.1365-2117.1996.01491.x.

DeCelles, P.G., and Horton, B.K., 2003, Early to Middle Tertiary foreland basin development and the history of Andean crustal shortening in Bolivia: Geological Society of America Bulletin, v. 115, p. 58–77, doi:10.1130/0016-7606(2003)115<0058:ETMTFB>2.0.CO;2.

DeCelles, P.G., Gehrels, G.E., Quade, J., Ojha, T.P., Kapp, P.A., and Upreti, B.N., 1998, Neogene foreland basin deposits, erosional unroofing, and the kinematic history of the Himalayan fold-thrust belt, western Nepal: Geological Society of America Bulletin, v. 110, p. 2–21, doi:10.1130/0016-7606(1998)110<0002:NFBDEU>2.3.CO;2.

DeCelles, P.G., Ducea, M.N., Kapp, P., and Zandt, G., 2009, Cyclicity in cordilleran orogenic systems: Nature Geoscience, v. 2, p. 251–257, doi:10.1038/ngeo469.

DeCelles, P.G., Carrapa, B., Horton, B.K., and Gehrels, G.E., 2011, Cenozoic foreland basin system in the central Andes of northwestern Argentina: Implications for Andean geodynamics and modes of deformation: Tectonics, v. 30, TC6013, doi:10.1029/2011TC002948.

DeCelles, P.G., Carrapa, B., Horton, B.K., McNabb, J., Gehrels, G.E., and Boyd, J., 2015, this volume, The Miocene Arizaro Basin, central Andean hinterland: Response to partial lithosphere removal?, *in* DeCelles, P.G., Ducea, M.N., Carrapa, B., and Kapp, P.A., eds., Geodynamics of a Cordilleran Orogenic System: The Central Andes of Argentina and Northern Chile: Geological Society of America Memoir 212, doi:10.1130/2015.1212(18).

Del Papa, C.E., 1999, Sedimentation on a ramp type lake margin: Paleocene-Eocene Maız Gordo Formation, northwestern Argentina: Journal of South American Earth Sciences, v. 12, p. 389–400, doi:10.1016/S0895-9811(99)00025-5.

DeMaster, D.J., 1979, The Marine Budgets of Silica and Si32 [Ph.D. dissertation]: New Haven, Connecticut, Yale University, 308 p.

Doyle, M., and Barros, V.R., 2002, Midsummer low-level circulation and precipitation in subtropical South America and related sea surface temperature anomalies in the South Atlantic: Journal of Climate, v. 15, p. 3394–3410, doi:10.1175/1520-0442(2002)015<3394:MLLCAP>2.0.CO;2.

Drummond, C.N., Wilkinson, B.H., and Lohmann, K.C., 1996, Climatic control of fluvial-lacustrine cyclicity in the Cretaceous Cordilleran foreland basin, western United States: Sedimentology, v. 43, p. 677–689, doi:10.1111/j.1365-3091.1996.tb02020.x.

Dumont, J.F., and Fournier, M., 1994, Geodynamic environment of Quaternary morphostructures of the Subandean foreland basins of Peru and Bolivia:

Characteristics and study methods: Quaternary International, v. 21, p. 129–142, doi:10.1016/1040-6182(94)90027-2.

Dunagan, S.P., and Turner, C.E., 2004, Regional paleohydrologic and paleoclimatic settings of wetland/lacustrine depositional systems in the Morrison Formation (Upper Jurassic), Western Interior, U.S.A.: Sedimentary Geology, v. 167, p. 269–296, doi:10.1016/j.sedgeo.2004.01.007.

Ekdahl, E.J., Friz, S.C., Baker, P., Rigsby, C.A., and Coley, K., 2008, Holocene multidecadal- to millennial-scale hydrologic variability on the South American Altiplano: The Holocene, v. 18, p. 867–876, doi:10.1177/0959683608093524.

Engelder, T.M., 2012, Investigating the Coupling between Tectonics, Climate and Sedimentary Basin Development [Ph.D. dissertation]: Tucson, Arizona, University of Arizona, 233 p.

Fritz, S.C., Baker, P.A., Lowenstein, T.K., Seltzer, G.O., Rigsby, C.A., Dwyer, G.S., Tapia, P., Arnold, K.K., Ku, T.L., and Luo, S., 2004, Hydrologic variation during the last 170,000 in the Southern Hemisphere Tropics of South America: Quaternary Research, v. 61, p. 95–104, doi:10.1016/j.yqres.2003.08.007.

Fritz, S.C., Baker, P.A., Seltzer, G.O., Ballantyne, A., Tapia, P., Cheng, H., and Edwards, R.L., 2007, Quaternary glaciation and hydrologic variation in the South American Tropics as reconstructed from the Lake Titicaca drilling project: Quaternary Research, v. 68, p. 410–420, doi:10.1016/j.yqres.2007.07.008.

Garreaud, R.D., Vuille, M., Compagnucci, R., and Marengo, J., 2009, Present-day South American climate: Palaeogeography, Palaeoclimatology, Palaeoecology, v. 281, p. 180–195, doi:10.1016/j.palaeo.2007.10.032.

Gierlowski-Kordesch, E.H., Jacobson, A.D., Blum, J.D., and Valero Garćes, B.L., 2008, Watershed reconstruction of a Paleocene–Eocene lake basin using Sr isotopes in carbonate rocks: Geological Society of America Bulletin, v. 120, p. 85–95, doi:10.1130/B26070.1.

González Bonorino, G., and Abascal, L. del Valle, 2012, Drainage and base-level adjustments during evolution of a late Pleistocene piggyback basin, Eastern Cordillera, central Andes of northwestern Argentina: Geological Society of America Bulletin, v. 124, p. 1858–1870, doi:10.1130/B30395.1.

Grier, M.E., Salfity, J.A., and Allmendinger, R.W., 1991, Andean reactivation of the Cretaceous Salta Rift, northwestern Argentina: Journal of South American Earth Sciences, v. 4, no. 4, p. 351–372, doi:10.1016/0895-9811(91)90007-8.

Grosjean, M., van Leeuwen, J.F.N., van der Knaap, W.O., Geyh, M., Ammann, B., Tanner, W., Messerli, B., and Nunez, L.A., 2001, A 22,000 ^{14}C year BP sediment and pollen record of climate change from Laguna Miscanti (23°S), northern Chile: Global and Planetary Change, v. 28, p. 35–51, doi:10.1016/S0921-8181(00)00063-1.

Hecky, R.E., Campbell, P., and Hendzel, L.L., 1993, The stoichiometry of carbon, nitrogen and phosphorus in particulate matter of lakes and oceans: Limnology and Oceanography, v. 38, p. 709–724, doi:10.4319/lo.1993.38.4.0709.

Heiri, O., Lotter, A., and Lemcke, G., 2001, Loss on ignition as a method for estimating organic and carbonate content in sediments: Reproducibility and comparability of results: Journal of Paleolimnology, v. 25, p. 101–110, doi:10.1023/A:1008119611481.

Heller, P.L., Agevine, C.L., Winslow, N.S., and Paola, C., 1988, Two-phase stratigraphic model of foreland basin development: Geology, v. 16, p. 501–504, doi:10.1130/0091-7613(1988)016<0501:TPSMOF>2.3.CO;2.

Hilley, G.E., and Strecker, M.R., 2005, Processes of oscillatory basin filling and excavation in a tectonically active orogeny: Quebrada del Toro Basin, NW Argentina: Geological Society of America Bulletin, v. 117, p. 887–901, doi:10.1130/B25602.1.

Hillyer, R., Valencia, B.G., Bush, M.B., Silman, M.R., and Steinetz-Kannan, M., 2009, A 24,700 yr paleolimnological history from the Peruvian Andes: Quaternary Research, v. 71, p. 71–82, doi:10.1016/j.yqres.2008.06.006.

Holdahl, S.R., Faucher, F., and Dragert, H., 1989, Recent vertical crustal motion in the Pacific Northwest: Eos (Transactions, American Geophysical Union), v. 68, p. 1240.

Horton, B.K., 2011, Cenozoic evolution of hinterland basins in the Andes and Tibet, *in* Busby, C., and Azor, A., eds., Tectonics of Sedimentary Basins: Recent Advances: Chichester, UK, John Wiley & Sons, doi:10.1002/9781444347166.ch21.

Horton, B.K., and DeCelles, P.G., 2001, Modern and ancient fluvial megafans in the foreland basin system of the central Andes, southern Bolivia: Implications for drainage network evolution fold-thrust belts: Basin Research, v. 13, p. 43–63, doi:10.1046/j.1365-2117.2001.00137.x.

Jordan, T.E., 1995, Retroarc foreland and related basins, *in* Busby, C.J., and Ingersoll, R.V., eds., Tectonics of Sedimentary Basins: Cambridge, Massachusetts, Blackwell Science Publishing, p. 331–362.

Jordan, T.E., and Almendinger, R.W., 1986, The Sierras Pampeanas of Argentina: A modern analogue of Rocky Mountain foreland deformation: American Journal of Science, v. 286, p. 737–764, doi:10.2475/ajs.286.10.737.

Jordan, T.E., Schlunegger, F., and Cardozo, N., 2001, Unsteady and spatially variable evolution of the Neogene Andean Bermejo foreland basin, Argentina: Journal of South American Earth Sciences, v. 14, p. 775–798, doi:10.1016/S0895-9811(01)00072-4.

Karp, T., Scholz, C.A., and McGlue, M.M., 2012, Structure and stratigraphy of the Lake Albert Rift, East Africa: Observations from seismic reflection and gravity data, *in* Baganz, O.W., Bartov, Y., Bohacs, K., and Nummedal, D., eds., Lacustrine Sandstone Reservoirs and Hydrocarbon Systems: American Association of Petroleum Geologists Memoir 95, p. 299–318.

Kronberg, B.I., Fralick, P.W., and Benchimol, R.E., 1998, Late Quaternary sedimentation and paleohydrology in the Acre foreland basin, SW Amazonia: Basin Research, v. 10, p. 311–323, doi:10.1046/j.1365-2117.1998.00067.x.

Labraga, J., Frumento, O., and López, M., 2000, The atmospheric water vapour cycle in South America and the tropospheric circulation: Journal of Climate, v. 13, p. 1899–1915, doi:10.1175/1520-0442(2000)013<1899:TAWVCI>2.0.CO;2.

Lambiase, J.J., 1990, A model for the tectonic control of lacustrine stratigraphic sequences in continental rift basins, *in* Katz, B.J., ed., Lacustrine Exploration: Case Studies and Modern Analogues: American Association of Petroleum Geologists Memoir 50, p. 265–276.

Larizzatti, F.E., Fávaro, D.I.T., Moreira, S.R.D., Mazzilli, B.P., and Piovano, E.L., 2001, Multielemental determination by instrumental neutron activation analysis and recent sedimentation rates using ^{210}Pb dating method at Laguna del Plata, Cordoba, Argentina: Journal of Radioanalytical and Nuclear Chemistry, v. 249, p. 263–268, doi:10.1023/A:1013271316393.

Lawton, T.F., 2008, Laramide sedimentary basins, *in* Miall, A.D., ed., The Sedimentary Basins of the United States and Canada: Amsterdam, Netherlands, Elsevier, p. 431–452.

Legates, D.R., and Willmott, C.J., 1990, Mean seasonal and spatial variability in gauge-corrected, global precipitation: International Journal of Climatology, v. 10, p. 111–127, doi:10.1002/joc.3370100202.

Leier, A.L., DeCelles, P.G., and Pelletier, J.D., 2005, Mountains, monsoons, and megafans: Geology, v. 33, p. 289–292, doi:10.1130/G21228.1.

Lenters, J., and Cook, K., 1997, On the origin of the Bolivian High and related circulation features of the South American climate: Journal of the Atmospheric Sciences, v. 54, p. 656–678, doi:10.1175/1520-0469(1997)054<0656:OTOOTB>2.0.CO;2.

Lombardo, U., and Prümers, H., 2010, Pre-Columbian human occupation patterns in the eastern plains of the Llanos de Moxos, Bolivian Amazonia: Journal of Archaeological Science, v. 37, p. 1875–1885, doi:10.1016/j.jas.2010.02.011.

Lombardo, U., May, J.H., and Veit, H., 2012, Mid- to late Holocene fluvial activity behind pre-Columbian social complexity in the southwestern Amazon Basin: The Holocene, v. 22, p. 1035–1045, doi:10.1177/0959683612437872.

Lowenstein, T.K., Hein, M.C., Bobst, A.L., Jordan, T.E., Ku, T.-L., and Luo, S., 2003, An assessment of stratigraphic completeness in climate-sensitive closed-basin lake sediments: Salar de Atacama, Chile: Journal of Sedimentary Research, v. 73, p. 91–104, doi:10.1306/061002730091.

Marquillas, R.A., del Papa, C., and Sabino, I.F., 2005, Sedimentary aspects and paleoenvironmental evolution of a rift basin: Salta Group (Cretaceous/Paleogene), northwestern Argentina: International Journal of Earth Sciences, v. 94, p. 94–113, doi:10.1007/s00531-004-0443-2.

Martínez, D.E., 1995, Changes in the ionic composition of a saline lake, Mar Chiquita, Province of Córdoba, Argentina: International Journal of Salt Lake Research, v. 4, p. 25–44.

McGlue, M.M., Scholz, C.A., Karp, T., Ongodia, B., and Lezzar, K.E., 2006, Facies architecture of flexural margin lowstand delta deposits in Lake Edward, East African Rift: Constraints from seismic reflection imaging: Journal of Sedimentary Research, v. 76, p. 942–958, doi:10.2110/jsr.2006.068.

McGlue, M.M., Silva, A., Corradini, F.A., Zani, H., Trees, M.A., Ellis, G.S., Parolin, M., Swarzenski, P.W., Cohen, A.S., and Assine, M.L., 2011, Limnogeology in Brazil's "Forgotten Wilderness": A synthesis from the floodplain lakes of the Pantanal: Journal of Paleolimnology, v. 46, p. 273–289, doi:10.1007/s10933-011-9538-5.

McGlue, M.M., Ellis, G.S., Cohen, A.S., and Swarzenski, P.W., 2012a, Playa-lake sedimentation and organic matter accumulation in an Andean piggyback basin: The recent record from the Cuenca de Pozuelos, northwest Argentina: Sedimentology, v. 59, p. 1237–1256, doi:10.1111/j.1365-3091.2011.01304.x.

McGlue, M.M., Silva, A., Zani, H., Corradini, F.A., Parolin, M., Abel, E.J., Cohen, A.S., Assine, M.L., Ellis, G.S., Trees, M.A., Kuerten, S., Gradella, F.S., and Rasbold, G.G., 2012b, Lacustrine records of Holocene flood pulse dynamics in the Upper Paraguay River watershed (Pantanal wetlands, Brazil): Quaternary Research, v. 78, p. 285–294, doi:10.1016/j.yqres.2012.05.015.

McGlue, M.M., Cohen, A.S., Ellis, G.S., and Kowler, A., 2013, Late Quaternary stratigraphy, sedimentology and geochemistry of an underfilled lake basin in the Puna (north-west Argentina): Basin Research, v. 25, p. 638–658, doi:10.1111/bre.12025.

McGlue, M.M., Ellis, G.S., and Cohen, A.S., 2015, The modern muds of Laguna Mar Chiquita (Argentina): Particle size and organic matter geochemical trends from a large saline lake in the "broken" Andean foreland, *in* Egenhoff, S., Fishman, N., and Larsen, D., eds., Paying Attention to Mudrocks—Priceless!: Geological Society of America Special Paper (in press).

Melles, M., Brigham-Grette, J., Minyuk, P., Nowaczyk, N.R., Wennrich, V., De Conto, R.M., Anderson, P.M., Andreev, A.A., Coletti, A., Cook, T.L., Haltia-Hovi, E., Kukkonen, M., Lozhkin, A.V., Rosen, P., Tarasov, P., Vogel, H., and Wagner, B., 2012, 2.8 million years of Arctic climate change from Lake El'gygytgyn, NE Russia: Science, v. 337, no. 6092, p. 315–320, doi:10.1126/science.1222135.

Mertes, L.A.K., Dunne, T., and Martinelli, L.A., 1996, Channel-floodplain geomorphology along the Solimoẽs-Amazon River, Brazil: Geological Society of America Bulletin, v. 108, p. 1089–1107, doi:10.1130/0016-7606(1996)108<1089:CFGATS>2.3.CO;2.

Meyers, P.A., and Teranes, J.L., 2001, Sediment organic matter, *in* Last, W.M., and Smol, J.P., eds., Tracking Environmental Change Using Lake Sediments: Volume 2: Physical and Geochemical Methods: New York, Springer, p. 239–269.

Mon, R., and Gutiérrez, A.A., 2009, The Mar Chiquita Lake: An indicator of intraplate deformation in the central plain of Argentina: Geomorphology, v. 111, p. 111–122, doi:10.1016/j.geomorph.2009.04.009.

Moreno, A., Giralt, S., Valero Garces, B., Saez, A., Bao, R., Prego, R., Pueyo, J.J., Gonzalez Sampiriz, P., and Taberner, C., 2007, A 14 kyr record of the tropical Andes: The Lago Chungara sequence (18S, northern Chilean Altiplano): Quaternary International, v. 161, p. 4–21, doi:10.1016/j.quaint.2006.10.020.

Morley, C.K., Vanhauwaert, P., and De Batist, M., 2000, Evidence for high-frequency cyclic fault activity from high-resolution seismic reflection survey, Rukwa Rift, Tanzania: Journal of the Geological Society of London, v. 157, p. 983–994, doi:10.1144/jgs.157.5.983.

Olsen, P.E., 1986, A 40-million-year lake record of early Mesozoic orbital climatic forcing: Science, v. 234, p. 842–848, doi:10.1126/science.234.4778.842.

Omarini, J., 2007, Estudio sedimentológico, geoquímico y ambiental de la Laguna La Brea, departamento de Santa Bárbara, Provincia de Jujuy [thesis]: Universidad Nacional de Salta, Argentina, 76 p.

Ori, G.G., and Friend, P.F., 1984, Sedimentary basins formed and carried piggyback on active thrust sheets: Geology, v. 12, p. 475–478, doi:10.1130/0091-7613(1984)12<475:SBFACP>2.0.CO;2.

Pelletier, J.D., 2007, Erosion-rate determination from foreland basin geometry: Geology, v. 35, p. 5–8, doi:10.1130/G22651A.1.

Pietras, J.T., Carroll, A.R., and Rhodes, M.K., 2003, Lake basin response to tectonic drainage diversion: Eocene Green River Formation, Wyoming: Journal of Paleolimnology, v. 30, p. 115–125, doi:10.1023/A:1025518015341.

Piovano, E.L., Ariztegui, D., Bernasconi, S.M., and McKenzie, J.A., 2004, Changes in subtropical Laguna Mar Chiquita (Argentina) over the last 230 years: The Holocene, v. 14, p. 525–535, doi:10.1191/0959683604hl729rp.

Piovano, E.L., Ariztegui, D., Cordoba, F., Cioccale, M. and Sylvestre, F., 2007, Hydrological variability in South America below the Tropic of Capricorn (Pampas and Patagonia, Argentina) during the last 13.0 ka, *in* Vimeux, F., Sylvestre, F., and Khodri, M., eds., Past Climate Variability in South America and Surrounding Regions: Berlin, Springer, Developments in Paleoenvironmental Research, v. 14, p. 323–351.

Plafker, G., 1964, Oriented lakes and lineaments of northeastern Bolivia: Geological Society of America Bulletin, v. 75, p. 503–522, doi:10.1130/0016-7606(1964)75[503:OLALON]2.0.CO;2.

Platt, N.H., and Keller, B., 1992, Distal alluvial deposits in a foreland basin setting—The Lower Freshwater Molasse (Lower Miocene), Switzerland: Sedimentology, architecture and paleosols: Sedimentology, v. 39, p. 545–565, doi:10.1111/j.1365-3091.1992.tb02136.x.

Pueyo, J.J., Sáez, A., Giralt, S., Valero-Garcés, B.L., Moreno, A., Bao, R., Schwalb, A., Herrera, C., Klosowska, B., and Taberner, C., 2011, Carbonate and organic matter sedimentation and isotopic signatures in Lake Chungará, Chilean Altiplano, during the last 12.3 kyr: Palaeogeography, Palaeoclimatology, Palaeoecology, v. 307, no. 1–4, p. 339–355, doi:10.1016/j.palaeo.2011.05.036.

Rapela, C.W., Pankhurst, R.J., Casquet, C., Baldo, E., Saavedra, J., and Galindo, C., 1998, Early evolution of the proto-Andean margin of South America: Geology, v. 26, p. 707–710, doi:10.1130/0091-7613(1998)026<0707:EEOTPA>2.3.CO;2.

Räsänen, M.E., Salo, J.S., Jungner, H., and Pittman, L.R., 1990, Evolution of the western Amazonian lowland relief: Impact of Andean foreland dynamics: Terra Nova, v. 2, p. 320–332, doi:10.1111/j.1365-3121.1990.tb00084.x.

Räsänen, M.E., Neller, R., Salo, J., and Junger, H., 1992, Recent and ancient fluvial deposition systems in the Amazonian foreland basin, Peru: Geological Magazine, v. 129, p. 293–306, doi:10.1017/S0016756800019233.

Reati, G.J., Florin, M., Fernandez, G.J., and Montes, C., 1997, The Laguna de Mar Chiquita (Córdoba, Argentina): A little known, secularly fluctuating, saline lake: International Journal of Salt Lake Research, v. 5, p. 187–219, doi:10.1007/BF01997137.

Richerson, P.J., Neale, P.J., Tapia, R.A., Carney, H.J., Lazzaro, X., Vincent, W., and Wurtsbaugh, W., 1992, Patterns of planktonic primary production and algal biomass, *in* Dejoux, C., and Iltis, A., eds., Lake Titicaca: A Synthesis of Limnological Knowledge: Dordrecht, Kluwer Academic Publishers, p. 196–222.

Roddaz, M., Baby, P., Brusset, S., Hermoza, W., and Darrozes, J.M., 2005, Forebulge dynamics and environmental control in western Amazonia: The case study of the Arch of Iquitos (Peru): Tectonophysics, v. 399, p. 87–108, doi:10.1016/j.tecto.2004.12.017.

Roehler, H.W., 1993, Eocene Climates, Depositional Environments, and Geography, Greater Green River Basin, Wyoming, Utah, and Colorado: U.S. Geological Survey Professional Paper 1506-F, 74 p.

Rosendahl, B.R., 1987, Architecture of continental rifts with special reference to East Africa: Annual Review of Earth and Planetary Sciences, v. 15, p. 445–503, doi:10.1146/annurev.ea.15.050187.002305.

Rubble, T.E., and Phillip, R.P., 1998, Stratigraphy, depositional environments and organic geochemistry of source rocks in the Green River petroleum system, Uinta Basin, Utah, *in* Pitman, J.K., and Carroll, A.R., eds., Modern and Ancient Lake Systems: Utah Geological Association Guidebook 26, p. 233–259.

Sadler, P., 1981, Sediment accumulation rates and the completeness of stratigraphic sections: The Journal of Geology, v. 89, p. 569–584, doi:10.1086/628623.

Sayago, J.M., 1995, The Argentine Neotropical loess: An overview: Quaternary Science Reviews, v. 14, p. 755–766, doi:10.1016/0277-3791(95)00050-X.

Sayago, J.M., Collantes, M.M., Karlson, A., and Sanabria, J., 2001, Genesis and distribution of the late Pleistocene and Holocene loess of Argentina: A regional approximation: Quaternary International, v. 76–77, p. 247–257, doi:10.1016/S1040-6182(00)00107-5.

Schelske, C.L., Conley, D.J., Soermer, E.F., Newberry, T.L., and Campbell, C.D., 1986, Biogenic silica and phosphorus accumulation in sediments as indices of eutrophication in the Laurentian Great Lakes: Hydrobiologia, v. 143, p. 79–86, doi:10.1007/BF00026648.

Scholz, C.A., and Rosendahl, B.R., 1988, Low lake stands in Lakes Malawi and Tanganyika, East Africa, delineated with multifold seismic data: Science, v. 240, p. 1645–1648, doi:10.1126/science.240.4859.1645.

Scholz, C.A., Johnson, T.C., and McGill, J., 1993, Deltaic sedimentation in a rift-valley lake: New seismic reflection data from Lake Malawi (Nyasa): East Africa: Geology, v. 21, p. 395–398, doi:10.1130/0091-7613(1993)021<0395:DSIARV>2.3.CO;2.

Scholz, C.A., Moore, T.C., Hutchinson, D.R., Klitgord, K.D., and Golmshtok, A.J., 1998, Comparative sequence stratigraphy of low-latitude versus high-latitude lacustrine rift basins: Seismic data examples from the East African and Baikal Rifts: Palaeogeography, Palaeoclimatology, Palaeoecology, v. 140, p. 401–420, doi:10.1016/S0031-0182(98)00022-4.

Scholz, C.A., Johnson, T.C., Cohen, A.S., King, J.W., Peck, J., Overpeck, J.T., Talbot, M.R., Brown, E.T., Kalindekafe, L., Amoako, P., Lyons, R.P., Shanahan, T.M., Castañeda, I., Heil, C.W., Forman, S.L., McHargue,

L.R., Beuning, K.R., Gomez, J., and Pierson, J., 2007, East African megadroughts between 135–75 kyr ago and implications for early human history: Proceedings of the National Academy of Sciences of the United States of America, v. 104, p. 16,416–16,421, doi:10.1073/pnas.0703874104.

Seltzer, G., Rodbell, D., and Burns, S., 2000, Isotopic evidence for late Quaternary climate change in tropical South America: Geology, v. 28, p. 35–38, doi:10.1130/0091-7613(2000)28<35:IEFLQC>2.0.CO;2.

Sewall, J.O., and Sloan, L.C., 2006, Come a little bit closer: A high-resolution climate study of the early Paleogene Laramide foreland: Geology, v. 34, p. 81–84, doi:10.1130/G22177.1.

Singh, A.K., Parkash, B., Mohindra, R., Thomas, J.V., and Singhvi, A.K., 2001, Quaternary alluvial fan sedimentation in the Dehradun Valley piggyback basin, NW Himalaya: Tectonic and palaeoclimatic implications: Basin Research, v. 13, p. 449–471, doi:10.1046/j.0950-091x.2001.00160.x.

Sloan, L.C., and Morrill, C., 1998, Orbital forcing and Eocene continental temperatures: Palaeogeography, Palaeoclimatology, Palaeoecology, v. 144, p. 21–35, doi:10.1016/S0031-0182(98)00091-1.

Smith, M.E., Carroll, A.R., and Singer, B.S., 2008, Synoptic reconstruction of a major ancient lake system: Eocene Green River Formation, western United States: Geological Society of America Bulletin, v. 120, p. 54–84, doi:10.1130/B26073.1.

Smoot, J., 1983, Depositional subenvironments in an arid closed basin: The Wilkins Peak Member of the Green River Formation (Eocene), Wyoming, USA: Sedimentology, v. 30, p. 801–827, doi:10.1111/j.1365-3091.1983.tb00712.x.

Sobel, E.R., Hilley, G.E., and Strecker, M.R., 2003, Formation of internally drained contractional basins by aridity-limited bedrock incision: Journal of Geophysical Research–Solid Earth, v. 108, no. B7, 2344, doi:10.1029/2002JB001883.

Soreghan, M.J., and Cohen, A.S., 1996, Textural and compositional variability across littoral segments of Lake Tanganyika: The effect of asymmetric basin structure on sedimentation in large rift lakes: American Association of Petroleum Geologists Bulletin, v. 80, p. 382–409.

Talbot, M.R., 2001, Nitrogen isotopes in palaeolimnology, *in* Last, W.M., and Smol, J.P., eds., Tracking Environmental Change Using Lake Sediments: Volume 2: Physical and Geochemical Methods: Dordrecht, Netherlands, Kluwer Academic Press, p. 401–439.

Tiercelin, J.J., Cohen, A.S., Soreghan, M.J., and Lezzar, K.E., 1994, Pleistocene–modern deposits of the Lake Tanganyika rift basin, East Africa: A modern analog for lacustrine source rocks and reservoirs, *in* Lomando, A.J., Schreiber, B.C., and Harris, P.M., eds., Lacustrine Reservoirs and Depositional Systems: Tulsa, Oklahoma, Society for Sedimentary Geology, Core Workshop 19, p. 37–60.

Trauth, M.H., and Strecker, M.R., 1999, Formation of landslide-dammed lakes during a wet period between 40,000 and 25,000 yr B.P. in northwestern Argentina: Palaeogeography, Palaeoclimatology, Palaeoecology, v. 153, p. 277–287, doi:10.1016/S0031-0182(99)00078-4.

Trauth, M.H., Alonso, R.A., Haselton, K.R., Hermanns, R.L., and Strecker, M.R., 2000, Climate change and mass movements in the NW Argentine Andes: Earth and Planetary Science Letters, v. 179, p. 243–256, doi:10.1016/S0012-821X(00)00127-8.

Trees, M., 2012, Lacustrine Sedimentation and Paleolimnology in an Early Cretaceous Backbulge-Basin Lake [M.Sc. thesis]: Tucson, Arizona, University of Arizona, 48 p.

Turner, C.E., and Peterson, F., 2004, Reconstruction of the Upper Jurassic Morrison Formation extinct ecosystem—A synthesis: Sedimentary Geology, v. 167, p. 309–355, doi:10.1016/j.sedgeo.2004.01.009.

Uba, C.E., Heubeck, C., and Hulka, C., 2006, Evolution of the late Cenozoic Chaco foreland basin, southern Bolivia: Basin Research, v. 18, p. 145–170, doi:10.1111/j.1365-2117.2006.00291.x.

Uba, C.E., Strecker, M.R., and Schmitt, A., 2007, Increased sediment accumulation rates and climatic forcing in the central Andes during the late Miocene: Geology, v. 35, p. 979–982, doi:10.1130/G224025A.1.

Ussami, N., Shiraiwa, S., and Dominguez, J.M.L., 1999, Basement reactivation in a sub-Andean foreland flexural bulge: The Pantanal wetland, SW Brazil: Tectonics, v. 18, p. 25–39, doi:10.1029/1998TC900004.

Villafañe, V.E., Andrade, M., Lairana, V., Zaratti, F., and Helbling, E.W., 1999, Inhibition of phytoplankton photosynthesis by solar ultraviolet radiation: Studies in Lake Titicaca, Bolivia: Freshwater Biology, v. 42, p. 215–224, doi:10.1046/j.1365-2427.1999.444453.x.

Zaleha, M.J., 2006, Sevier orogenesis and nonmarine basin filling: Implications of new stratigraphic correlations of Lower Cretaceous strata throughout Wyoming, USA: Geological Society of America Bulletin, v. 118, p. 886–896, doi:10.1130/B25715.1.

Zani, H., Assine, M.L., and McGlue, M.M., 2012, Revealing geoforms in Pantanal wetland (Brazil) with remote sensing: A method to enhance SRTM-DEM for megafan environments: Geomorphology, v. 161–162, p. 82–92, doi:10.1016/j.geomorph.2012.04.003.

Zanor, G.A., Piovano, E.L., Ariztegui, D., and Vallet-Coulomb, C., 2012, A modern subtropical playa complex: Salina de Ambargasta, central Argentina: Journal of South American Earth Sciences, v. 35, p. 10–26, doi:10.1016/j.jsames.2011.10.007.

MANUSCRIPT ACCEPTED BY THE SOCIETY 3 JUNE 2014
MANUSCRIPT PUBLISHED ONLINE 14 NOVEMBER 2014

The Geological Society of America
Memoir 212
2015

Simulating foreland basin response to mountain belt kinematics and climate change in the Eastern Cordillera and Subandes: An analysis of the Chaco foreland basin in southern Bolivia

Todd M. Engelder
Jon D. Pelletier*
Department of Geosciences, University of Arizona, 1040 E. 4th Street, Tucson, Arizona 85721, USA

ABSTRACT

The relative importance of crustal thickening, lithospheric delamination, and climate change in driving surface uplift and the associated changes in accommodation space and depositional facies in the adjacent foreland basin in the central Andes has been a topic of vigorous debate over the past decade. Interpretation of structural, geochemical, geomorphic, and geobiologic field data has led to two proposed end-member Tertiary surface uplift scenarios for the Eastern Cordillera and Subandes in the vicinity of the Bolivian orocline. A "gradual uplift" model proposes that the rate of surface uplift has been relatively steady since deformation propagated into the Eastern Cordillera during the late Eocene. In this scenario, the mean elevation of the region was >2 km above mean sea level (msl) by the late Miocene or earlier. Alternatively, a "rapid uplift" model suggests that the mean elevation of the Altiplano was <1 km above msl, and the peaks of the Eastern Cordillera were more than 2 km below their modern elevations until rapid uplift began ca. 10 Ma. Determining which of these uplift scenarios is most consistent with the stratigraphic record is complicated by the potentially confounding effects of global climate changes and lithospheric delamination in the stratigraphic record. In this study, we use a coupled mountain-belt–sediment-transport model to predict the foreland basin stratigraphic response to these end-member surface uplift scenarios. Our model results indicate that the location and height of the migrating deformation front play the dominant roles in controlling changes in accommodation space and grain size within the foreland basin. Changes in accommodation space and rates of sediment supply related to climate change and lithospheric delamination play secondary roles. Our results support the conclusion that the Eastern Cordillera likely gained most of its modern elevation prior to 10 Ma, in contrast with recent proposals that most of the modern elevation was obtained during the late Miocene. This conclusion is consistent with the most comprehensive paleoaltimetric analysis of the region to date.

*jdpellet@email.arizona.edu

Engelder, T.M., and Pelletier, J.D., 2015, Simulating foreland basin response to mountain belt kinematics and climate change in the Eastern Cordillera and Subandes: An analysis of the Chaco foreland basin in southern Bolivia, *in* DeCelles, P.G., Ducea, M.N., Carrapa, B., and Kapp, P.A., eds., Geodynamics of a Cordilleran Orogenic System: The Central Andes of Argentina and Northern Chile: Geological Society of America Memoir 212, p. 337–357, doi:10.1130/2015.1212(17).

INTRODUCTION

Sediments eroded from mountain belts are primarily deposited in foreland basins, which are elongate troughs created by flexural loading of the lithosphere adjacent to mountain belts (DeCelles and Giles, 1996). Foreland basin stratigraphy can provide useful data for reconstructing climate and crustal deformation histories within ancient mountain belts (e.g., Heller et al., 1988; DeCelles et al., 1998; Marzo and Steel, 2000; Uba et al., 2007). The fluvial systems that transport sediments from eroding mountain belts are sensitive to changes in sediment supply, discharge, and sediment accommodation rates, and thus, the foreland basin stratigraphy in these systems is a partial record of changes in climate and the geometry of the mountain belt load through time. The flexural profile of foreland basin accommodation space depends on the rigidity of the lithosphere that is underthrusted beneath the mountain belt as well as the width and elevation of the mountain belt load. If the foreland basin geometry and rigidity of the underthrusted lithosphere can be constrained, then it is possible to infer changes in the mountain belt load that occurred during the development of the foreland basin. As such, numerous studies have inferred tectonic histories for ancient mountain belts by fitting foreland basin geometries predicted by a coupled mountain belt and foreland basin numerical model to observed foreland basin isopach data (e.g., Toth et al., 1996; Ford et al., 1999; Garcia-Castellanos et al., 2002; Prezzi et al., 2009).

The eastern margin of the central Andes (i.e., the portion of the mountain belt that is to the east of the Altiplano Plateau) in southern Bolivia is an ideal region for constraining late Cenozoic changes in mountain belt geometry and climate through numerical modeling of foreland basin stratigraphy. The sediments preserved in the foreland basin have been well documented, and field and laboratory constraints have been reported in the literature for shortening and exhumation rates in the hinterland for the period of time over which the eastern margin of the central Andes has developed. A W-E transect of the Cenozoic foreland basin stratigraphy is exposed within the retroarc fold-and-thrust belt. Detailed isopach maps and stratigraphic sections for the Cenozoic foreland basin stratigraphy have been developed based on a combination of measured sections from the fold-and-thrust belt and correlations between well-log data and two-dimensional (2-D) seismic data (Sempere et al., 1997; DeCelles and Horton, 2003; Echavarria et al., 2003; Uba et al., 2005, 2006). In addition to data from the foreland basin, the timing and magnitude of shortening and exhumation within the mountain belt have been constrained by field mapping and thermochronology (McQuarrie, 2002; Muller et al., 2002; Oncken et al., 2006; Ege et al., 2007; Barnes et al., 2008, 2012). While most of the papers in this volume focus on the NW Argentina transect through the central Andes, we focus here on the Bolivian transect due to these unusually rich data sets. We further assume that the two transects (Bolivia and NW Argentina) behave in a broadly similar way. This assumption is supported by basin evolution and thermochronologic studies (e.g., DeCelles et al., 2011; Carrapa et al., 2011a,

2011b) as described further in the Discussion section. At a larger scale, the Andes as a whole have been an invaluable natural laboratory for exploring feedbacks between climate and rock uplift (Montgomery et al., 2001; Strecker et al., 2007).

An unresolved issue for the central Andes is: When did the topography of the central Andes rise to its modern elevation? One model for the topographic development of the central Andes near the latitude of the Bolivian orocline posits that the mean topography reached near-modern elevation following Neogene crustal thickening within the Subandean zone; thus, earlier crustal thickening within the Eastern Cordillera was not sufficient for the eastern margin of the central Andes to rise to near-modern elevations (Isacks, 1988; Gubbels et al., 1993). In addition to crustal thickening, continental lithospheric foundering has been proposed as a mechanism for the generating at least some of the rapid Neogene surface uplift posited by this model (Kay and Kay, 1993; Garzione et al., 2008; DeCelles et al., 2009). Continental lithospheric foundering involves removal of negatively buoyant lower crust (i.e., eclogite root) and mantle lithosphere by either delamination or Rayleigh-Taylor instability. Initial paleoelevation and geomorphic studies did support late Miocene rapid surface uplift of the region (Ghosh et al., 2006; Hoke et al., 2007; Garzione et al., 2008). Pedogenic carbonate and carbonate cement samples collected between 17°S and 18°S within the Altiplano and Eastern Cordillera Neogene stratigraphy contain decreasing $\delta^{18}O$ values in progressively younger units (Garzione et al., 2008). Garzione et al. (2008) interpreted this oxygen-isotope trend as evidence for an increase in elevation of the Eastern Cordillera of 2.5 ± 1 km from 10 to 7 Ma. Although these results are in agreement with fossil leaf and clumped isotope data collected from the Altiplano and Eastern Cordillera (e.g., Gregory-Wodzicki et al., 1998; Ghosh et al., 2006), the elevation gain predicted by all three methods can be complicated (i.e., biased toward larger values) by climate change due to the uplift of the Andes (Ehlers and Poulsen, 2009). Also, recent studies have interpreted the change in Miocene oxygen isotopes to be the result of regional climate change caused by uplift of an already significantly elevated (≥2 km) central Andes above an orographic threshold (Poulsen et al., 2010; Insel et al., 2012). Most recently, Quade et al. (this volume) showed that, while rapid uplift may have occurred locally in the Eastern Cordillera as documented by Garzione et al. (2008), the easternmost Puna and Eastern Cordillera rose gradually to >3 km by no later than 15 Ma. Evidence for late Miocene rock uplift is recorded by paleosurfaces that exist on both the eastern and western margins of the central Andes. Barke and Lamb (2006) estimated 1.7 ± 0.7 km of localized rock uplift for the San Juan Del Oro surface of the Eastern Cordillera and Interandean tectonomorphic regions since 12–9 Ma, when the surface was abandoned and incised. The Barke and Lamb (2006) results do not constrain the magnitude of surface uplift, however. The difference between rock and surface uplift is particularly significant because localized rock uplift can be much higher than mean regional surface uplift due to isostatic effects (England and Molnar, 1990). We will refer to this conceptual model for surface uplift as the rapid uplift end-member model. Although Quade et al. (this volume) appear to have disproven this model, it is still important to test this model using other lines of evidence (besides paleoaltimetry).

A second end-member model for the topographic evolution of the central Andes invokes gradual surface uplift since the late Eocene, when deformation propagated from the Western Cordillera into the Eastern Cordillera. Evidence for pre-Neogene deformation comes from Eocene exhumation ages within the Eastern Cordillera and changes in paleocurrent directions within Paleogene stratigraphy of the Altiplano and Eastern Cordillera (Horton et al., 2002; McQuarrie et al., 2005; Ege et al., 2007; Barnes et al., 2008). Although pre-Neogene shortening is considered low (<100–150 km) for building a high-elevation plateau, long-term shortening should have led to crustal thickening and isostatic rebound in the Eastern Cordillera. Therefore, unless erosion rates exceeded or were equal to rock uplift rates, the Eastern Cordillera should have been rising (i.e., surface uplift as defined by England and Molnar, 1990) since the late Eocene, barring some mechanism for keeping the elevation of the Andes low during an extended period of deformation. The gradual uplift end-member model posits that the Andes gained the majority of its modern elevation prior to ca. 10 Ma.

The Cenozoic stratigraphy of the Chaco foreland basin in Bolivia shows an increase in both grain size and sedimentation rates during the late Miocene (Uba et al., 2006, 2007). This stratigraphic change could have been the result of rapid surface uplift. However, previous studies have inferred that this depositional trend might instead be primarily controlled by distance from the approaching fold-and-thrust belt (DeCelles and Horton, 2003; Uba et al., 2006), which provides the topographic load to drive flexural subsidence and accommodation space creation. Thus, the primary mechanism that caused these changes in late Miocene stratigraphy is still uncertain. In addition to thrust belt migration, climate change and continental lithosphere foundering have also been emphasized as potential controls on the foreland basin stratigraphy of the central Andes in recent years. For example, Kleinert and Strecker (2001) documented a change from previously dry to wetter conditions in the Santa Maria Basin of northern Argentina between 9 and 7 Ma. Based on climate model studies, Poulsen et al. (2010) and Insel et al. (2012) have shown that surface uplift of a significantly elevated central Andes in the late Miocene would have caused an increase in precipitation along the eastern front of the central Andes. Uba et al. (2007) found a significant (i.e., factor of 5) increase in time-averaged depositional rates within the stratigraphy of the Subandean thrust belt in Bolivia at ca. 8–7 Ma based on U-Pb dating of tuffs within Miocene volcanic rocks. Coeval with changes in depositional rates and climate conditions within the foreland depositional facies of the Cenozoic fluvial units, a shift occurs from single-thread sinuous channels to more amalgamated alluvial megafan facies (Uba et al., 2006). Uba et al. (2007) interpreted these changes in depositional rates and facies to be a consequence of a change from semiarid to more humid climate conditions during the intensification of the South American monsoon. An increase in mean annual precipitation should increase sediment supply through enhanced erosion rates and increase the transport capacity of foreland basin fluvial systems via greater mean annual discharge.

Lower-crustal delamination is another mechanism that has been proposed for controlling the late Cenozoic foreland basin stratigraphy in the central Andes. The cordilleran cycle, a conceptual model for mountain belt development and cyclicity proposed by DeCelles et al. (2009), posits that as lower crust and mantle lithosphere are underthrust beneath a growing mountain belt, magmatic and petrologic processes lead to the formation of a dense eclogitic root that acts as a subsurface negatively buoyant load, lowering the elevation of the mountain belt relative to a state of isostatic equilibrium (DeCelles et al., 2009; Pelletier et al., 2010). If sufficient shortening takes place, the eclogite root reaches a critical thickness or volume and is delaminated or removed via a Rayleigh-Taylor instability. This cordilleran cycle, as posited by DeCelles et al. (2009), includes episodic periods of modest (i.e., 0.5–1 km) increases in surface uplift and shortening rates, both of which could have had a significant effect on the adjacent foreland basin through the modification of both rates of sediment supply and creation of accommodation space. Presently, the Puna Plateau is thought to be in a postdelamination state in this conceptual model (Schurr et al., 2006; DeCelles et al., 2009) and, as a consequence, has a mean elevation that is significantly higher than the Altiplano Plateau. The Altiplano, in turn, is thought to have had a delamination event at 10 Ma (Kay et al., 1994). DeCelles et al. (2009) proposed that the Altiplano may already be in an early stage of a new cycle, and its lower elevation is a consequence of newly forming eclogite loads. Evidence for delamination beneath the Altiplano comes from seismic velocity analysis of the eastern Altiplano lithosphere at 20°S (Beck and Zandt, 2002). A low-velocity zone occurs within the upper mantle beneath the thickened crust of the eastern Altiplano and western edge of the Eastern Cordillera and is interpreted as a location where the cold-fast upper mantle has been removed by a delamination event. Might stratigraphic trends in the late Miocene Chaco foreland basin deposits in Bolivia be a signature of lithospheric foundering instead of climate change or fold-and-thrust belt propagation?

Previous studies have numerically modeled the evolution of the central Andes as a coupled mountain-belt–foreland-basin system. Flemings and Jordan (1989) first simulated the rapid uplift end-member model for the last 5 m.y. of deformation in the central Andes with a two-dimensional model. They concluded that the foreland basin adjacent to the central Andes should have shifted from narrow and underfilled to broad and overfilled as sediment supply outpaced sediment accommodation. However, the results of their model are potentially limited by the fact the mountain belt component of their model was simplified by assuming a constant topographic slope and sediment supply through time. Prezzi et al. (2009) recently applied a more rigorous deformation and erosion model to test the effect of mountain load geometry and elastic thickness of the South American lithosphere on the sediment accommodation rates within the foreland basin of the central Andes since the middle Miocene. They found that by decreasing the elastic thicknesses beneath the Eastern Cordillera and Interandean zones between 14 and 6 Ma and by deforming a detailed structural cross section developed by McQuarrie (2002), their model results adequately fit modern gravity anomalies and isopach distributions for the late Cenozoic stratigraphy recently described by Uba et al. (2006) in the Subandean zone of Bolivia.

In this study, we explore the linkages among thrust-belt kinematics, climate change, continental lithosphere delamination, and the foreland basin stratigraphy in the central Andes using a coupled, two-dimensional numerical model. Flemings and Jordan (1989) simulated the foreland response to a fixed topographic slope and constant sediment supply and, thus, did not capture the effects of feedbacks among mountain belt erosion, sediment supply to the foreland basin, and sediment accommodation within the foreland basin. The Prezzi et al. (2009) study focused on the changes in foreland basin accommodation based on surface uplift due to a very specific history of crustal deformation and changes in elastic thickness under constant climate conditions. Both studies focused on crustal thickening as the dominant mechanism for driving foreland basin development. In contrast, in this paper, we aim to constrain the relationship of the Cenozoic foreland basin stratigraphy to end-member surface uplift models, climate change, and lower-crustal delamination in order to place firmer constraints on the paleoelevation history of central Andes. First, we determine which end-member surface uplift model is most consistent with the available stratigraphic and tectonic data for the development of the eastern margin of the central Andes. Second, we determine whether there is some

signal (c.g., unconformity, grain size change) of climate change or continental lithospheric delamination recorded in the upper Miocene Chaco foreland basin stratigraphy.

GEOLOGIC BACKGROUND

The central Andes of southern Bolivia contain a high-elevation, internally draining, low-relief plateau. This portion of the Andes also has the highest magnitude of total shortening (i.e., ~285 km of minimum crustal shortening within the Eastern Cordillera, Interandean, and Subandean zones; Isacks, 1988; McQuarrie, 2002; Oncken et al., 2006). The orogenic belt is generally broken up into tectonomorphic regions, which include (from west to east): the Western Cordillera, Altiplano, Eastern Cordillera, Interandean, Subandean, and Chaco foreland basin zones (Fig. 1). The modern topographic divide between westward internal drainage into the Altiplano basin and eastward drainage into the Chaco foreland basin resides within the Eastern Cordillera, a bivergent fold-and-thrust belt. Thrusts within the Eastern Cordillera detach in Ordovician-aged horizons and predominantly exhume Paleozoic through Mesozoic units (McQuarrie, 2002). Further east, the Interandean and Subandean zones are where deformation is currently active. These zones contain dominantly eastward-verging imbricate thrusts that generally exhume younger stratigraphic units compared to the Eastern Cordillera. The modern deformation front is located between 0.5 and 1 km above mean sea level (msl). Beyond the Subandean zone, the Chaco foreland basin extends an additional 250–600 km east into Bolivia, Paraguay, and the Pantanal wetlands of Brazil until it onlaps Precambrian

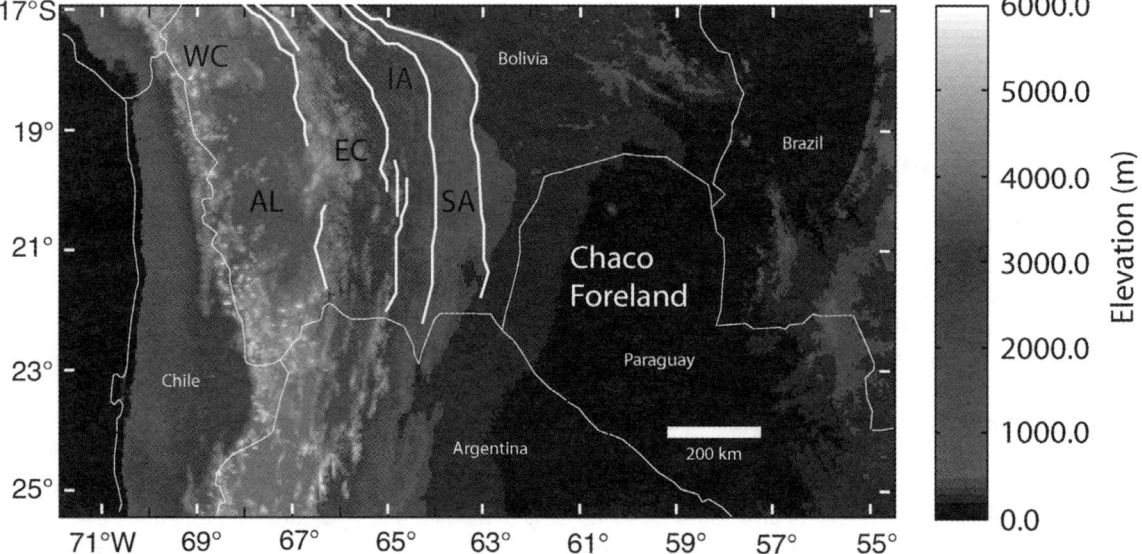

Figure 1. Digital elevation model for the central Andes displaying the tectonomorphic zones after Barnes et al. (2008). Topography is from the Shuttle Radar Topography Mission (SRTM) 90 m data set. The tectonomorphic regions are the following: WC—Western Cordillera; AL—Altiplano; EC—Eastern Cordillera; IA—Interandean zone; SA—Subandean zone. The thick white lines are thrust faults located on the boundary between major divisions, and the thin white lines mark political boundaries.

basement highs in eastern Bolivia and southwestern Brazil (Horton and DeCelles, 1997). An exception occurs between 19°S and 20°S, where the foredeep basin deposits pinch out onto the Alto de Izozog basement high over a distance that is less than 200 km from the deformation front (Uba et al., 2006).

The Chaco foreland basin stratigraphy has been documented within the exposed thrust sheets of the Eastern Cordillera, Interandean, and Subandean zones by DeCelles and Horton (2003), Echavarria et al. (2003), Horton (2005), and Uba et al. (2005). Isopach trends have also been developed for the buried modern foreland basin units based on correlations between well data and seismic cross sections (Uba et al., 2006). Starting at the base of the Cenozoic stratigraphy from the Subandean zone (Fig. 2), the Petaca and Yecua Formations contain fluvial deposits with well-developed paleosols, paleocurrent trends that dominantly show transport toward or along strike with the approaching mountain belt, and thicknesses that are low, considering the amount of time contained within the units. The Yecua Formation is unique to the Subandean zone Cenozoic units because it also contains shallow-marine facies in addition to fluvial and lacustrine facies in outcrops located north of 20°S. Moving up through the stratigraphy into the Tariquía, Guandacay, and Emborozú Formations, overall grain size increases from that of the Yecua Formation, paleocurrent data indicate a shift to transport from the west, and depositional rates increase. Based on dating volcanic ash layers in the Cenozoic foreland basin stratigraphy, there is a factor of five increase in depositional rates between the Yecua and Tariquía Formations (Uba et al., 2007). Higher in the stratigraphic section, depositional rates decrease by half upward into the Guandacay Formation. Although deposition is generally conformable within the Cenozoic stratigraphy, unconformities bound the Petaca Formation, and an angular unconformity separates the Guandacay

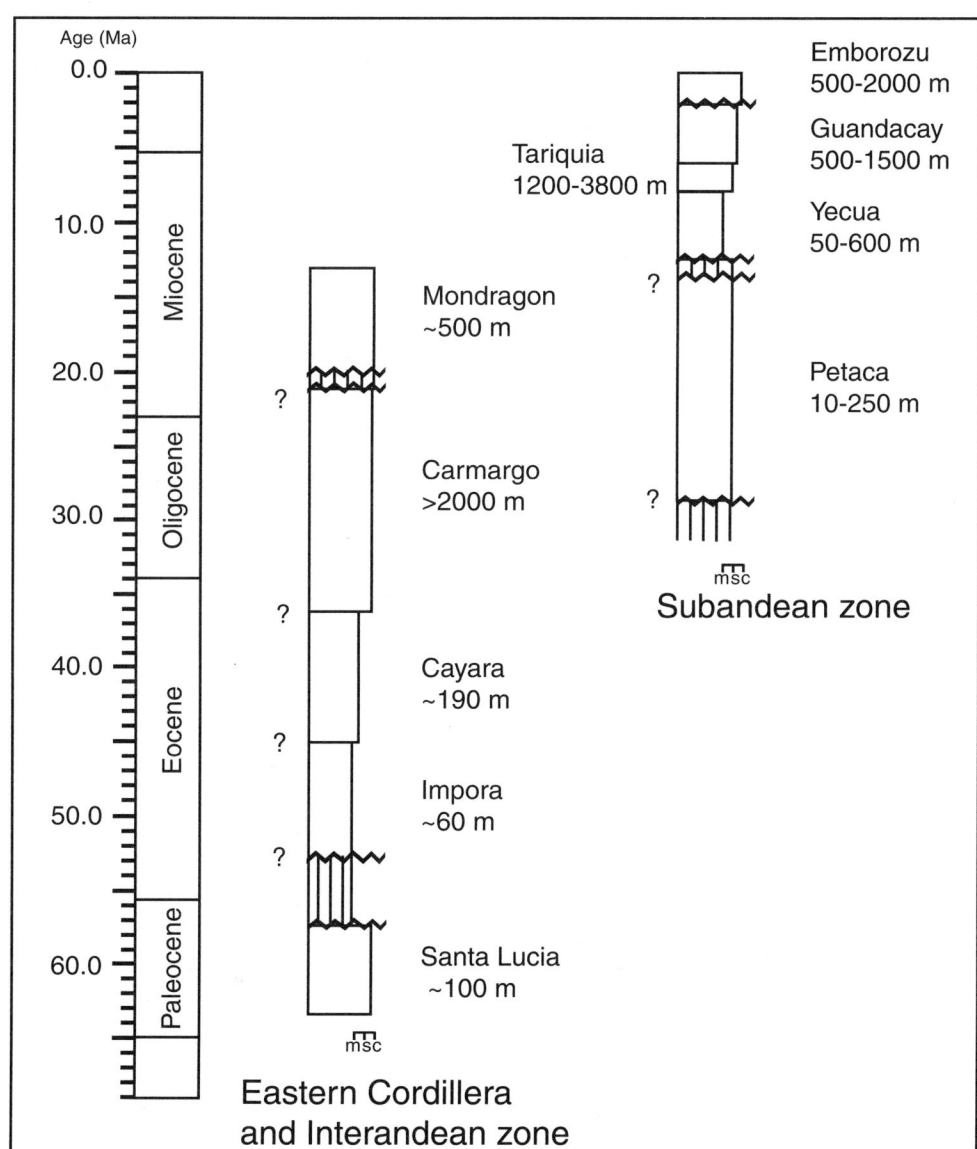

Figure 2. Generalized stratigraphic columns for the Cenozoic stratigraphy of the Chaco foreland basin after DeCelles and Horton (2003) and Uba et al. (2006, 2007). Formation thicknesses are displayed beneath the formation names. The question marks represent uncertainty in the boundaries between formations due to a lack of absolute age data. A detailed discussion of the age uncertainties for the Cenozoic stratigraphy of the Eastern Cordillera, Interandean, and Subandean zones can be found in DeCelles and Horton (2003) and Uba et al. (2006). m—mudstone; s—sandstone; c—conglomerate.

and Emborozú Formations. The upper unconformity of the Petaca Formation has been interpreted to record the uplift and passage of the forebulge as the Subandean zone passed from the back bulge into the foredeep basin (Uba et al., 2006). In contrast, the angular unconformity located between the Guandacay and Emborozú Formations has been interpreted as the initiation of wedge-top deposition in the Subandean zone. Low-angle unconformities have been documented at the base of coarsening-upward cycles within the age-equivalent stratigraphy of the Tariquía and Guandacay Formations in northern Argentina (Echavarria et al., 2003). However, cycles and significant unconformities were not documented in the Tariquía and Guandacay Formations of the southern Bolivia Subandean zone (Uba et al., 2006). Farther west and older in age, similar stratigraphic trends occur within the remnant foreland basin stratigraphy of the Eastern Cordillera and Interandean zones (Fig. 2). The transition from the Cayara into the Camargo Formation involves an overall increase in grain size and changes in paleocurrent direction and apparent depositional rates. One stratigraphic difference between this remnant foreland basin and the Neogene foreland basin of the Subandean zone is that an unconformity does not bound the lowest Cenozoic unit in the section. Overall, the most significant trends within the Cenozoic foreland basin stratigraphy of the eastern margin of the central Andes are the apparent increase in depositional rates and grain size with decreasing age, and thus, any numerical models for the foreland basin evolution of the eastern margin of the central Andes must honor these trends.

MODEL DESCRIPTION

Summary

The numerical model of this paper is a 2-D kinematic model that couples an actively deforming and eroding moun-

TABLE 1. TIME-AVERAGED SHORTENING RATES USED IN EACH MODEL SIMULATION

Zone	Shortening rate (m/yr)	Duration (m.y.)
Altiplano	0.0014	30
Eastern Cordillera back thrust	0.0015	38
Eastern Cordillera forethrust	0.0022	30
Interandean	0.0044	22
Western Subandes	0.0037	18
Eastern Subandes	0.0067	10

tain belt with sediment transport and flexure within a depositional foreland basin. A 2-D model is valid for this region of the Andes because the mountain front is nearly linear, and paleoflow directions for the majority of the Cenozoic foreland basin stratigraphy are perpendicular to the mountain front. The topography of the numerical model represents the mean topography of the eastern margin of the central Andes (i.e., western edge of the model begins in the Altiplano) located between 18°S and 21°S. In our kinematic model, shortening and rock uplift rates are prescribed. Although time-averaged shortening rates (Table 1) can be directly calculated from published values and used as preliminary input into the model, obtaining valid rock uplift rates is an iterative process of running simulations, checking the results against independent published data (e.g., modern topography and exhumation magnitudes), and adjusting the rock uplift rates within each tectonomorphic zone (Fig. 3). In this paper, we report model outcomes that are consistent with published data for the central Andes, including the modern topography, shortening rates from balance cross sections, exhumation magnitudes, Cenozoic isopach data, and modern basin geometries.

All model simulations begin at 43 Ma. Evidence from field mapping and thermochronology suggests that deformation did

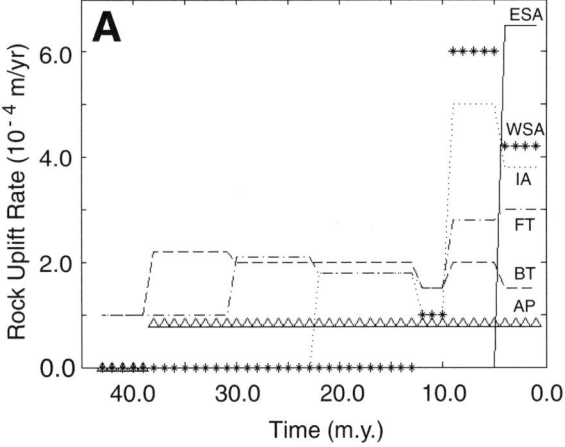
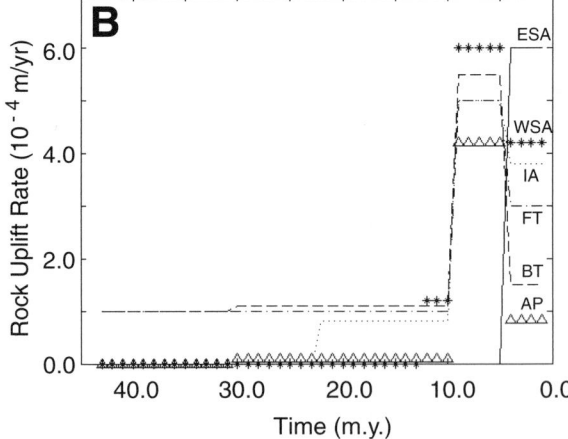

Figure 3. The distribution of prescribed zonal rock uplift rates within the eastern margin of the central Andes hinterland through time for both the gradual (A) and rapid (B) end-member uplift models. Symbols for each zone are the following: triangles—Altiplano (AP); dashed line—Eastern Cordillera back thrust (BT); dashed dotted-line—Eastern Cordillera forethrust (FT); dotted line—Interandean (IA); asterisk—western Subandean (WSA); solid line—eastern Subandean (ESA).

not propagate into the Eastern Cordillera until the late Eocene, and, therefore, starting the model before then is unnecessary and would have been poorly constrained with available data. The older foreland basin section in the region was interpreted by DeCelles and Horton (2003) to represent distal foredeep to fore-bulge deposits and only represents a partial foreland basin profile. Having a more complete foreland basin profile that includes the foredeep is necessary to constrain the topography of the adjacent mountain belt.

Rock Deformation

Dynamic models for crustal deformation (e.g., Simpson, 2004) that directly solve for visco-plastic deformation within a fold-and-thrust belt could be applied to the central Andes. This approach, however, comes with the drawbacks of longer computational times and the difficulty of calibrating the model to multiple types of calibration data. We therefore take a simpler approach in this paper. In the model, the mountain belt is partitioned into individual blocks for each tectonomorphic region (i.e., Altiplano, Eastern Cordillera, etc.). Exceptions are made for the Eastern Cordillera and Subandean zones. The Eastern Cordillera, a bivergent fold-and-thrust belt, was divided into separate back-thrust and fore-thrust blocks because these two regions were rapidly exhumed at different periods of time (Barnes et al., 2008). The Subandean zone was divided into a western and eastern block to allow the central and eastern Subandean zone to continue to subside until deformation propagated into the region during the late Miocene. Deformation is simulated by uniform rock uplift and shortening applied to each actively deforming block. Time-averaged shortening rates were determined for input to the model by dividing the total shortening determined from field-based balanced cross sections for each tectonomorphic zone by the duration of fault activity (McQuarrie et al., 2005; Muller et al., 2002). The positions of the model nodes located to the east of the deformation front are fixed, while the model nodes to the west of the deformation front are allowed to translate toward the foreland basin as if riding over a décollement. The total propagation of the deformation front (and hence the forebulge) is ~536 km, which is close to the ~600 km of total propagation of the forebulge since the late Eocene estimated by DeCelles and Horton (2003).

Bedrock Channel Erosion

A stream-power model is used to determine bedrock incision rates within the mountain belt (Howard and Kerby, 1983; Whipple and Tucker, 1999):

$$\frac{\partial h}{\partial t} = -k_e A^m \left|\frac{\partial h}{\partial x}\right|^n = -k_e A^{1/2} S = -k_e l S, \quad (1)$$

where h is elevation above msl (m), t is time (yr), k_e is the bedrock erodibility (yr^{-1} for $n = 1.0$ and $m = 0.5$), A is the

contributing drainage basin area (m^2), x is the lateral distance from the model origin (m), l is the distance along the principal channel from the headwaters of the drainage basin (m), S is the local channel slope (unitless), and m and n are constants that determine the dependence of local erosion on discharge and channel slope. In this standard 2-D implementation of the stream-power model, h represents the longitudinal profile of the main channel of a three-dimensional (3-D) river basin. The usual approach in such cases (e.g., Whipple and Tucker, 1999) is to assume that contributing area goes as the square of distance from the drainage divide. This assumption implicitly adds tributary area to the main channel in a manner that is statistically consistent with the way in which tributaries add area in real 3-D river basins. In Equation 1, we assume that the length of the principal channel is proportional to the square root of the contributing drainage area (Hack, 1957). We also assume that the area and slope exponents m and n have values of 0.5 and 1, respectively. Evidence from theory and field studies predicts that the ratio of m/n is near 0.5 (Whipple and Tucker, 1999). Although the slope exponent n can range between 0.66 and 2.0 depending on the relationship between slope and stream power (shear stress), we chose to make the stream power linearly proportional to the slope ($n = 1.0$ and $m = 0.5$), consistent with the assumption of many other studies (e.g., Kirby and Whipple, 2001; Snyder et al., 2000). Bedrock incision rates for active mountain belts are on the order of 0.1–1.0 mm/yr (Montgomery and Brandon, 2002). When m/n is equal to 1/2, we found that k_e must be on the order of 10^{-6} (m/yr) to appropriately reproduce the range of calculated exhumation rates (i.e., 0.1–0.6 mm/yr) for time scales of 10^2 to 10^7 yr for the central Andes (Safran et al., 2005; Ege et al., 2007; Barnes et al., 2008). The boundary between the bedrock portion of the model (where stream-power–driven erosion of bedrock is modeled) and the alluvial portion of the model (where grain-size–dependent sediment transport is modeled) is allowed to fluctuate in the model. If a given model node contains sediment or sediment is deposited (i.e., upstream sediment supply exceeds the transport capacity), then the node is treated as an alluvial channel, and no bedrock incision occurs. Otherwise, bedrock incision occurs, and the amount of newly created sediment transported downstream is controlled by the transport capacity.

Sediment Transport

Sediment transport of a bimodal grain-size distribution (i.e., gravel and sand) is simulated with a modified version of the diffusion model approach, which states that sediment flux is proportional to local channel slope. Linear slope-dependent sediment flux, when combined with conservation of mass, gives a diffusion equation for the evolution of the longitudinal profile of the foreland basin (Paola et al., 1992). In the model of this paper, we added a threshold slope term to the diffusion equation model because transport of gravel requires a threshold slope to initiate transport. The equation for alluvial deposition and erosion along a channel profile in our model is a combination of mass balance (i.e., the Exner equation) and the slope-dependent sediment-transport equation:

$$q_s = k_g (S - \varphi), S > \varphi \qquad (2)$$
$$q_s = 0 \qquad\qquad , S \leq \varphi$$

$$\frac{\partial h}{\partial t} = -\frac{\partial q_s}{\partial x} = \frac{q_{s,\text{upstream}} - q_{s,\text{downstream}}}{\partial x}, \qquad (3)$$

where k_g is the transport coefficient (m²/yr), φ is the threshold slope, which is a function of grain size, and q_s is the local sediment flux (m²/yr). The transport coefficients for gravel and sand are determined from a relationship that depends on discharge and river type (Paola et al., 1992). Paola et al. (1992) calculated the values for braided and meandering rivers to range between 1.0 and 7.0 × 10⁴ m²/yr for a drainage basin with a length of 100 km and a mean annual precipitation of 1 m. We chose a value of 1.0 × 10⁴ m²/yr for the gravel transport coefficient, which is a reasonable value for braided, gravel-dominated streams with relatively small drainage areas. The transport coefficient for sand was chosen to be ~6.5 × 10⁴ m²/yr, which is on the order of the transport coefficient used by Flemings and Jordan (1989) for their simulation of the central Andes. The threshold slope for gravel entrainment is ~0.001, which represents effective channel parameters of a median grain size of 0.02 m and bankfull flow depth of 1.85 m. The chosen median grain size is on order with the median grain size of bed-load material observed within the modern Rio Pilcomayo located in the eastern Subandean zone (Mugnier et al., 2006).

Channel armoring is an emergent property of the model obtained by incorporating a threshold slope for gravel transport. If discharge is insufficient to entrain gravel present on the channel bed during a time step, then the finer-grained sand deposited beneath the gravel is prevented from being transported during that time step. This process should reduce both alluvial and bedrock erosion in regions of lower regional slope where gravel is present. Active bed material layers in natural channels are on the order of 1–2 grains thick (Hassan et al., 2006). Although our active layer is on the order of a meter, it is valid to have a larger active layer for the purposes of this study because we are focusing on long-term (>10⁵ yr) grain-size trends that average over many individual flooding events.

Flexural Isostasy

The bedrock and alluvial surface dynamics models are coupled to a flexural foreland basin in order to quantitatively assess accommodation space creation and the migration of the forebulge through time in the model. The flexural model solves for the displacement of a thin elastic beam subjected to a spatially distributed vertical load (Turcotte and Schubert, 2002):

$$D \frac{\partial^4 w(x)}{\partial x^4} + (\rho_m - \rho_s) g w(x) = L(x), \qquad (4)$$

where w is the deflection of Earth's crust (m), D is the flexural rigidity (Nm), ρ_m is the density of the mantle (kg/m³), ρ_s is the density of the mountain crust or foreland sediment (kg/m³), g is the acceleration due to gravity (m/s²), and $L(x)$ is the topographic load (kg/ms²). For each simulation, except for those involving eclogite root delamination (discussed later), we solve for the deflection due to a topographic load every time step using a Fourier transform method (Watts, 2001). Viscous relaxation effects were not considered in the model because we interpret stratigraphic patterns over geologic time scales that are greater than relaxation time scales.

Several studies have calculated the flexural rigidity of the central Andes using 2-D methods (Horton and DeCelles, 1997; Stewart and Watts, 1997; Tassara, 2005). Their results suggested that flexural rigidities along the eastern margin of the central Andes range between 1.5 × 10²³ and 4.0 × 10²⁴ Nm. We chose to use a flexural parameter of 150 km because this value best fits the observed modern basin geometry between 18°S and 20°S. This value is consistent with the results of Chase et al. (2009), who found that the flexural parameter should be less than 220 km for the central Andes. The flexural parameter is defined as the following (Turcotte and Schubert, 2002):

$$\alpha = \left[\frac{4D}{(\rho_m - \rho_s)g} \right]^{1/4}, \qquad (5)$$

where α is the flexural parameter (km). Rearranging Equation 5 and applying values from Table 2 yields a flexural rigidity of ~6.8 × 10²³ Nm, which is well within the range of calculated effective flexural rigidities for the central Andes. Although previous studies have suggested that the elastic thickness varies in space and time (e.g., Toth et al., 1996; Prezzi et al., 2009), such variations are difficult to constrain. Therefore, our model implements a uniform and constant elastic thickness.

TABLE 2. PARAMETERS USED
IN END-MEMBER MODEL SIMULATIONS

k_e (m/yr)	1.0 × 10⁻⁶
k_g (m²/yr)	1.0 × 10⁴
k_s (m²/yr)	6.5 × 10⁴
ϕ	0.001
α (km)	150
ρ_s (kg/m³)	2750
ρ_m (kg/m³)	3300
ρ_e (kg/m³)	3600
g (m/s²)	9.8

Numerical Methods

We ran three types of experiments using our numerical model: (1) an end-member surface uplift model experiment, (2) an eclogite root foundering experiment, and (3) a climate change experiment. The end-member surface uplift model experiment simulates the last 43 m.y. when deformation is concentrated in the eastern margin of the central Andes. The purpose of this experiment is to contrast the foreland basin response (e.g., changes in depositional rates and grain size) to a mountain belt that is gradually uplifting through time against the foreland basin response to rapid surface uplift caused by Neogene crustal thickening. The parameters used in this experiment are found in Table 2. During the simulations of this experiment, mean annual precipitation rates were held constant, and time steps were ~700 yr. Topographic profiles were sampled at 22, 10, and 0 Ma for visual comparison.

Following the end-member surface uplift model experiment, we ran an eclogite root foundering experiment. The purpose of this experiment is to contrast the foreland basin response to rapid surface uplift in the mountain belt caused by crustal thickening alone against rapid surface uplift caused by a combination of foundering and crustal thickening. The model duration, time steps, and interval of topographic profile sampling in the foundering experiment were the same as in the end-member surface uplift experiment. We also applied the parameters in Table 2 and the same kinematic histories in the mountain belt (e.g., shortening rates and propagation of deformation) as in the rapid uplift model. An eclogite root grows in our model between 25 and 10 Ma beneath the Altiplano and Eastern Cordillera back-thrust zones, where significant crustal thickening has taken place such that lower-crustal rocks are subjected to pressures sufficient enough to produce eclogite. We assume that eclogite foundering is caused by a Rayleigh-Taylor instability (defined as the diapiric drip of a dense layer overlying a less dense layer) and that the time scales over which the growth of the eclogite drip occur can be calculated to within first order by the results of the linear-stability analysis of Turcotte and Schubert (2002):

$$\tau_a = \frac{13.04\mu}{(\rho_e - \rho_m)b}, \qquad (6)$$

where τ_a is the amount of time required for an instability to grow by a factor of e, μ is the viscosity of the upper and lower layers, and b is the original thickness of the eclogite layer. The Rayleigh-Taylor instability is not modeled explicitly, but instead we prescribe a rate of eclogite root growth consistent with the observed spacing between foundering events in the central Andes and the thickness of the root required to initiate foundering.

Equation 6 is rearranged to solve for b in order to prescribe the thickness of the eclogite root required to initiate foundering in our model. Based on geophysical models and paleoelevation proxies, the length of time over which the foundering event occurred beneath the Altiplano was ~3 m.y. (i.e., between 10 and

7 Ma; Molnar and Garzione, 2007; Garzione et al., 2008). We assume that the majority of the foundering time is spent growing the instability in the lower crust–mantle interface by the initial factor of e due to the high resistance to flow of the mantle when minimal perturbations in the lower crust–mantle interface exist. Later, when a significant pressure gradient exists at the base of the eclogite layer due to the formation of a drip, the pinching and foundering portion of the drip event occurs much more rapidly. Based on this argument, we inferred that τ_a is approximately less than or equal to 3 m.y. Using densities of 3300 and 3600 kg/m^3 for upper mantle and eclogite and a root growth period of 3 m.y., the range in maximum eclogite layer thickness is calculated to be between 1.87 and 46.88 km as the effective viscosity of the underlying mantle layer varies between 4.0 × 10^{19} and 1.0 × 10^{21} Pa s. This range in eclogite layer thickness is consistent with the 10–20 km eclogite root thickness that occurred prior to removal during a numerical simulation of mantle drip for the Puna Plateau (Quinteros et al., 2008). We used a similar scale (i.e., 12.5 km) for the thickness of the eclogite root required to initiate delamination.

Between 10 and 7 Ma, the root is removed at a linear rate until it is completely removed at 7 Ma. During the period of root foundering, we allowed our flexure algorithm to specify the magnitude of rock uplift due to foundering and superimposed that result on the results of the rapid uplift model. Our model is kinematic, so the viscous coupling between the sinking root and the overlying lithosphere is not simulated in our model. However, there is a possibility that viscous coupling may result in significant (i.e., on the order of hundreds of meters) subsidence and rebound (Göğüş and Pysklywec, 2008).

An experiment with climate change during the late Miocene was also conducted in order to determine its impact on the foreland basin stratigraphy. At 9 Ma, we simulated an increase in mean annual precipitation by doubling both the bedrock erodibility and sediment transport coefficients. Such increases could be triggered by intensification of the South American monsoon system and/or by an increase in orographic activity associated with the mountain belt attaining a threshold elevation. In this experiment, it is also necessary to increase bedrock uplift rates along with the bedrock erodibility and sediment transport coefficients because exhumation must keep pace with erosion in order for the model to reproduce the modern topography of the central Andes at the end of the simulation.

The bedrock erodibility (k_e) of the stream power model and the transport coefficients (k_s and k_g) of the diffusion model have both been shown theoretically to be proportional to discharge (Paola et al., 1992; Whipple and Tucker, 1999). If we assume that mean annual discharge scales proportionally with mean annual precipitation rate, then erosion rates and sediment transport rates scale proportionally with mean annual precipitation rate. This is reasonable assumption for erosion rates because long-term erosion rates have been shown to correlate with mean annual precipitation rates in the Andes. Sediment transport rates in alluvial rivers have also been proposed to correlate with mean annual

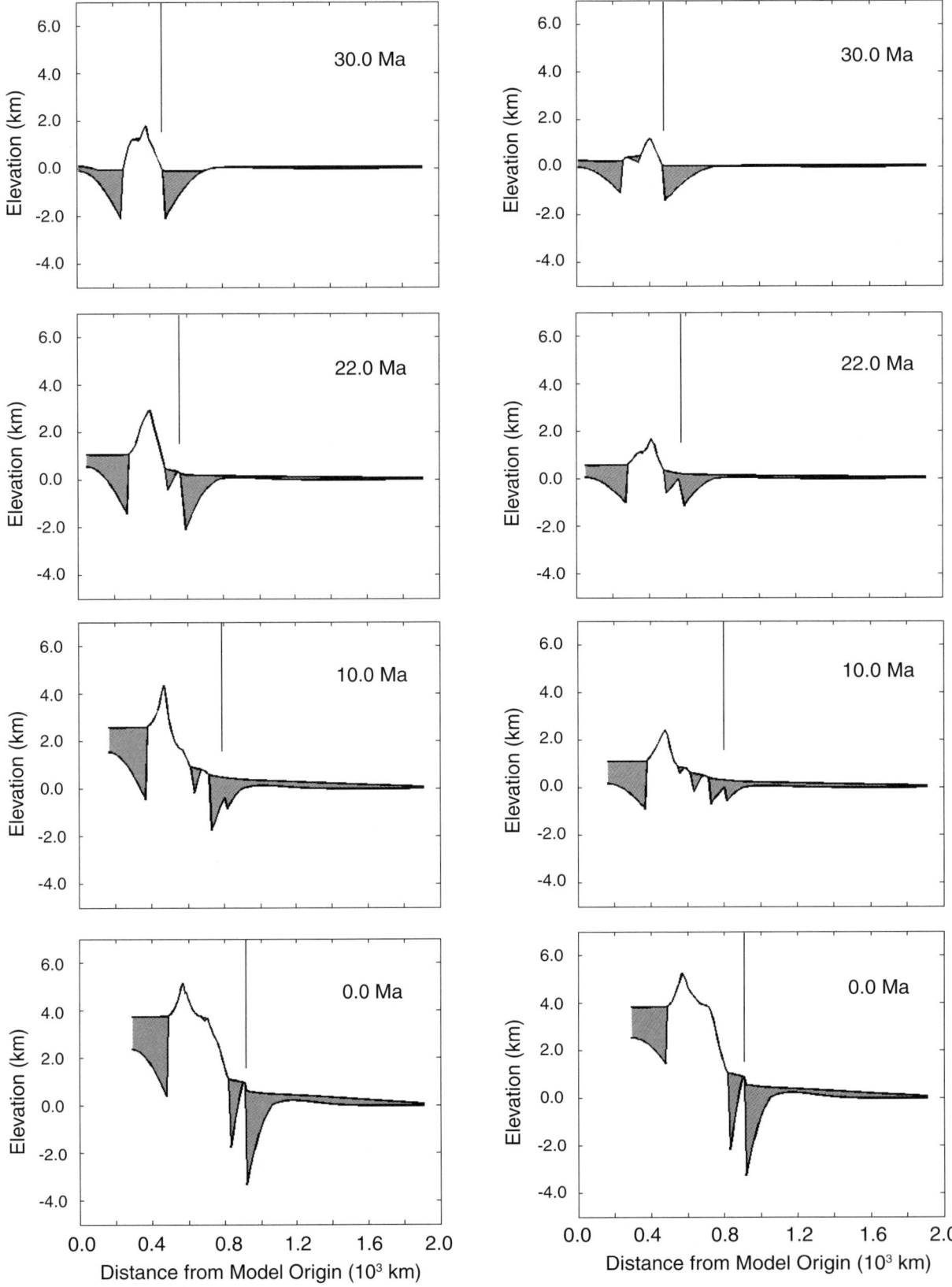

Figure 4. The evolution of the eastern margin of the central Andes as a series of cross sections in time for both the gradual (left column) and rapid (right column) end-member uplift models. Sedimentary deposits are shaded gray, and the location of the deformation front at the time represented by the snapshot is marked by the gray vertical line.

precipitation rates (Molnar et al., 2006). Although the magnitude of increase in mean annual precipitation rate at the onset of the South American monsoon is unknown, long-term erosion rates since the Eocene have not varied by more than a factor of two compared with modern rates (Safran et al., 2005; Ege et al., 2007; Barnes et al., 2008). Thus, both erosion rates and sediment transport rates increase by a factor of two at 10 Ma. The model duration, time steps, and interval of topographic profile sampling are the same as in the end-member uplift models.

RESULTS

Summary of Model Outputs

Topographic-profile and sediment-flux time-series plots show the development of the eastern margin of the central Andes in response to the gradual uplift end-member model (Figs. 4 and 5). Between 43 and 22 Ma, deformation is concentrated within the Eastern Cordillera. During this period of time, topographic slopes are low, and thus the sediment fluxes from the mountain belt into the Altiplano and foreland basin are also low. Sediment eroded off of the Eastern Cordillera is not transported beyond the foredeep, which is underfilled to completely filled. The sediment flux leaving the back-bulge basin toward the east is zero at this time. The Subandean zone in its predeformed state is located more than 400 km from the deformation front near the forebulge crest (Fig. 5A). Low sediment flux from the mountain belt is consistent with observed paleocurrent data from the Petaca Formation, located in the Subandean zone, showing westward paleotransport from the South American craton. Between 30 and 22 Ma, there is a period of higher-than-average sediment flux (Fig. 5B). Sedi-

ments with low erodibility contained within wedge-top basins are exhumed as the eastern portion of the Eastern Cordillera begins to uplift at 30 Ma. Exhumation of the wedge-top basin deposits causes a period of high sediment delivery to the foredeep.

By 22 Ma, rock uplift rates between 0.1 and 0.2 mm/yr cause the Eastern Cordillera to rise to a peak elevation of 3 km above msl. The increase in topographic relief causes the mean sediment flux from the mountain belt into the foreland basin to increase, such that sediment from the mountain belt outpaces foreland accommodation rates and is transported out of the foredeep. Subsidence rates in the back-bulge basin are an order of magnitude lower than in the foredeep. Consequently, the back-bulge basin is rapidly filled, and sediment from the mountain belt begins to bypass the filled back-bulge basin by 27 Ma (Fig. 5B). At this time, the deformation front propagates east into the Interandean zone and creates a step in topography located 500 km from the left end of the model domain in Figure 5A. Again, there is a period of higher-than-average sediment flux between 22 and 15 Ma caused by exhumation of wedge-top basins within the Interandean zone. The 22 Ma snapshot represents the time period when the Camargo Formation is transitioning into the overlying Mondragon Formation of the Eastern Cordillera and when the Petaca Formation is being deposited in the Subandean zone. These results are also consistent with available data from NW Argentina showing that the Eastern Cordillera was actively deforming and eroding at ca. 18–22 Ma (Deeken et al., 2006; Coutand et al., 2006). Similarly, thick sedimentary depocenters indicate high sedimentation rates in NW Argentina (Carrapa et al., 2011a).

By 10 Ma, the core of the Eastern Cordillera is within 1 km of its modern peak elevation, and the deformation front and

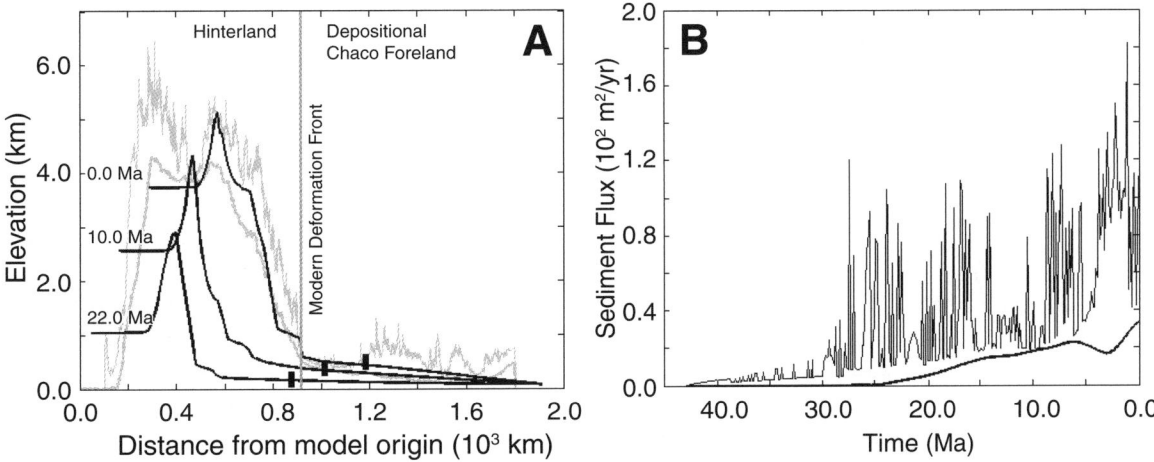

Figure 5. (A) Topographic evolution of the central Andes and (B) time series of sediment supply rates into and out of the foreland basin for the gradual uplift model. The thick black lines in A represent snapshots of topography throughout the simulation, and the thick gray lines represent the maximum and mean topography for a north-south sweep of modern topography between 18°S and 21°S. The black rectangles represent the locations of the forebulge crest during each snapshot in time, with the oldest forebulge location on the left. In B, the thick line represents the sediment flux leaving the back-bulge basin or right edge of the model, and the thin lines represent the sediment flux at the deformation front into the foredeep of the foreland basin.

forebulge migrate another 100 km toward the South American craton (Fig. 5A). As a result of the increase in relief and mountain front slopes, the average sediment flux into the foredeep increases from 20 to 30 m²/yr and overfills the foredeep. A plot of sediment flux leaving the back-bulge basin shows that at least half of the mean sediment supply to the foreland is leaving the back-bulge basin at this time (Fig. 5B). A significant decrease in the amount of sediment leaving the back-bulge basin and a significant increase in the amount of sediment entering the foredeep occur between 5 and 2 Ma. This is the period in time when the eastern Subandean zone is actively uplifting in both southern Bolivia and northwestern Argentina. Thermochronologic data and growth structures (Carrapa et al., 2011b; DeCelles et al., 2011) indicate that the deformation front was in the Eastern Cordillera until ca. 4 Ma and migrated out into the Subandes after that. This is broadly similar to what observed in Bolivia (Barnes et al., 2008, 2012). High rock uplift rates in the eastern Subandean zone lead to exhumation of wedge-top basins and the development of a topographic step. Although sediment flux from the mountain belt is high during this time, rapid creation of topographic relief in the front of the thrust belt causes high subsidence rates in the foredeep. A significant portion of the sediment supply to the foreland basin is deposited and stored in the foredeep. The final topography at 0 Ma fits the average and maximum topography of the modern central Andes well. However, the alluvial slopes of the depositional foreland basin are a few hundred meters higher than the average channel elevations today. It is unclear if this error is due to an overprediction of the sediment supply, an underprediction of transport coefficient, or if sediment fluxes at the downstream end of the model domain are too low.

Although rates of shortening and the propagation of the deformation front are identical for both the gradual and rapid uplift models, the surface uplift and sediment flux responses to the rapid uplift model are notably different (Figs. 6A and 6B). Between 43 and 22 Ma, low rock uplift rates (i.e., 0.1–0.2 mm/yr) in the Eastern Cordillera lead to only 1.7 km of peak surface uplift over this period of time. As such, mean sediment supply rates from the mountain belt remain low (i.e., <10 m²/yr) over the first 21 m.y. of simulation. Despite the fact that the mean sediment flux into the foreland basin is half the rate of the gradual uplift model, the foredeep and back-bulge basins in the rapid uplift model are able to fill and bypass sediment from the mountain belt by 25 Ma. Following an additional 12 m.y. of active uplift between 22 and 10 Ma, peak elevation of the Eastern Cordillera in the rapid uplift model rises to 2.3 km above msl, and deformation propagates east into the western Subandean zone. Mean sediment flux into the foreland basin and sediment fluxes leaving the back-bulge basin in the rapid uplift model remain low compared to the values for the gradual uplift model at this time. At 10 Ma, rock uplift rates across the mountain belt increase from 0.1 to 0.3–0.5 mm/yr as the central Andes rapidly uplift in response to deformation in the Subandean zone. As a result, mean sediment flux from the mountain belt increases from 20 to 60 m²/yr. However, the sediment flux leaving the back-bulge basin decreases shortly after the initiation of rapid uplift across the mountain belt because subsidence rates in the foredeep quickly respond to surface uplift within the mountain belt, trapping sediment within the foredeep near the edge of the mountain belt. The final topography in the rapid uplift model at 0 Ma equally reproduces the modern mean and maximum topographies of the central Andes mountain

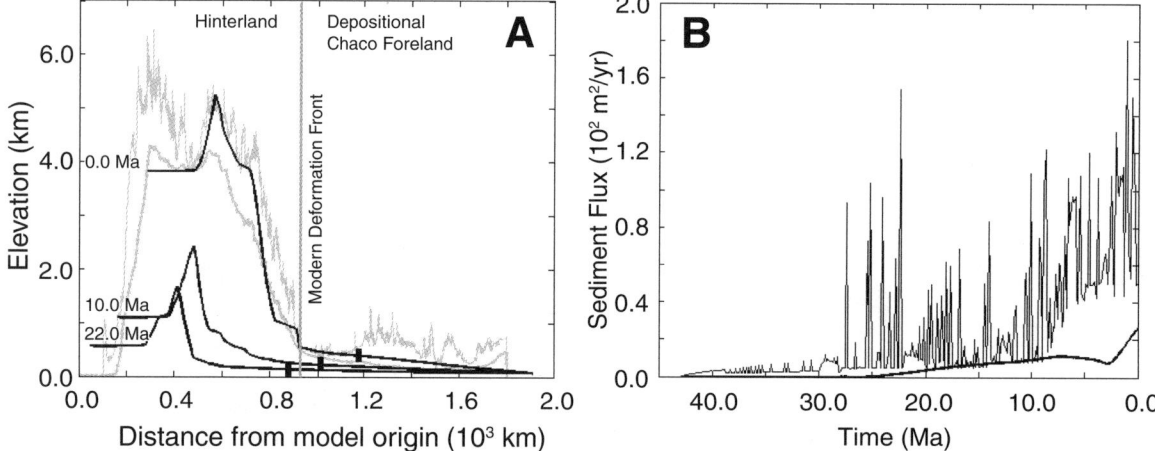

Figure 6. (A) Topographic evolution of the central Andes and (B) time series of sediment supply rates into and out of the foreland basin for the rapid uplift model. The thick black lines in A represent snapshots of topography throughout the simulation, and the thick gray lines represent the maximum and mean topography for a north-south sweep of modern topography between 18°S and 21°S. The black boxes represent the locations of the forebulge crest during each snapshot in time, with the oldest forebulge location on the left. In B, the thick line represents the sediment flux leaving the back-bulge basin or right edge of the model, and the thin lines represent the sediment flux at the deformation front into the foredeep of the foreland basin.

belt and foreland basin. Again, the predicted foreland topography is slightly higher than that of the observed topography.

Constraints on Foreland Basin Depositional Rates

In both the gradual and rapid uplift models, surface uplift of the eastern margin of the central Andes leads to an increase in sediment supply and subsidence within the foreland basin. Figures 7A and 7B show the sediment thickness curves for four depozones (i.e., the forethrust region of the Eastern Cordillera, western Interandean zone, western Subandean zone, and eastern Subandean zone, respectively) within the foreland basin for both the gradual and rapid uplift models. The solid lines represent the results for the simulations described in the previous section. Between 43 and 22 Ma, deformation and surface uplift are limited to the Eastern Cordillera. At this time, both the eastern region of the Eastern Cordillera and western Interandean zones are located within the foredeep basin of the gradual uplift model (Fig. 7A). Between 40 and 30 Ma, the deformation front remains in the core of the Eastern Cordillera. A spatially fixed load leads to constant subsidence rates and thus constant depositional rates in the Eastern Cordillera depozones over this period of time. The Interandean depozone, however, is located in the distal foredeep, and therefore it develops an acceleration in depositional rates between 31 and 25 Ma that is characteristic of an increase in subsidence rates due to an approaching mountain belt. Further to the east, the western and eastern Subandean depozones are located within the back-bulge basin. Low subsidence rates lead to low depositional rates between 43 and 36 Ma. Between 36 and 25 Ma, the forebulge crest migrates through the western Subandean depozone and leads to erosion and development

of an unconformity. Sediment deposited while the eastern Subandean depozone was located in the back bulge is completely eroded away during this time period. The shaded region between 36 and 25 Ma represents the approximate depositional age for the Camargo Formation. Predicted depositional thicknesses for the Camargo Formation range between 1 and 2 km in the Eastern Cordillera and Interandean depozones, but they underpredict the >2 km thickness of the Camargo Formation observed in the Camargo syncline (DeCelles and Horton, 2003). By 22 Ma, deformation propagates into the Interandean zone. Therefore, the Interandean depozone is uplifted and exhumed at this time. Between 22 and 10 Ma, depositional rates within the western Subandean zone accelerate as it reaches the foredeep. At 9 Ma, a sharp inflection point occurs that reflects a significant increase in depositional rates in the eastern Subandean zone. These high depositional rates are inferred to be the result of high rock uplift rates in the western Subandean zone. Gradual uplift model simulations were also conducted for constant and lower rock uplift rates for the western Subandean zone. However, a slow and gradual kinematic model for the deformation front could not reproduce the extreme increase in isopach thickness for the Tariquía Formation given the constraints on deformation propagation rates during the late Miocene. Reproducing accurate accommodation magnitudes for the foreland basin stratigraphy while also fitting the modern topography motivated us to apply an increase in rock uplift rates at the front of the mountain belt relative to the gradual uplift model. Rock uplift rates within the back thrust and forethrust of the Eastern Cordillera, however, remained gradual and constant. Increased uplift rates between 9 and 4 Ma caused an increase in sediment accommodation rates. As a result, a larger portion of the incoming sediment flux is stored in the foredeep depozone.

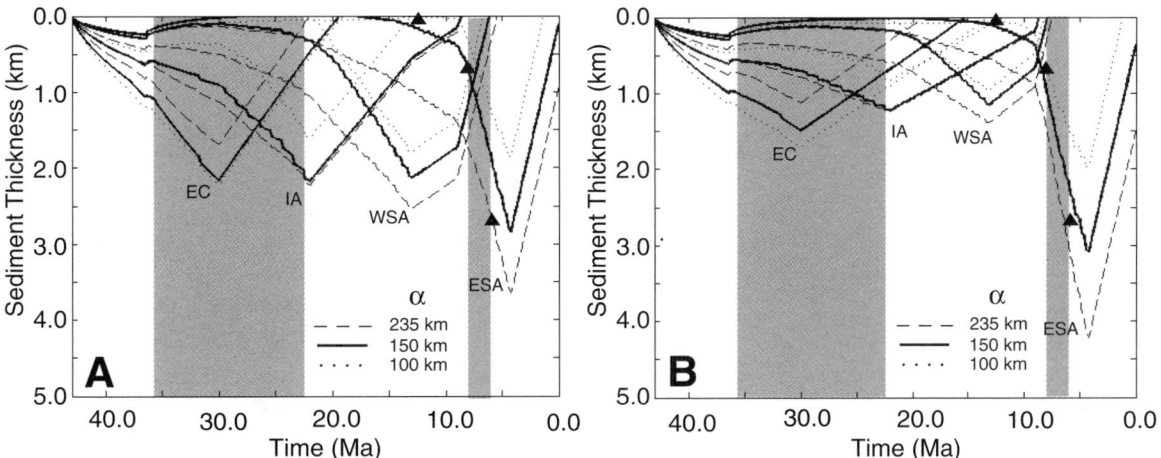

Figure 7. Uncompacted Tertiary sediment thickness for depozones located in the eastern Subandean zone (ESA), western Subandean zone (WSA), western Interandean zone (IA), and within the Eastern Cordillera forethrust region (EC) for the (A) gradual uplift and (B) rapid uplift models. The dotted, solid, and dashed lines represent simulations with flexural parameters of 100, 150, and 235 km. Shaded regions represent the time periods during which the Tariquía Formation of the Subandean zone and Camargo Formation of the Eastern Cordillera and Interandean zone were deposited. The subsidence curve from Uba et al. (2006) for the location in the Subandes that we sampled is represented by black triangles in both plots.

Extremely high depositional rates continue until ca. 4 Ma, when the basin begins to uplift and exhume. This period of time represents the deposition of the Guandacay Formation and the transition into the Emborozú Formation.

Although shortening rates and thrust belt propagation rates in the rapid uplift model are the same as in the gradual uplift model, the total thicknesses of basins for the rapid uplift model are significantly different than those for the gradual uplift model (Fig. 7B). Early in the model experiment, between 43 and 22 Ma, lower rock uplift rates within the Eastern Cordillera cause depositional thicknesses within the eastern forethrust region of the Eastern Cordillera and Interandean zones to be 500 m less compared to thicknesses predicted in the gradual uplift model. Both locations develop Camargo Formation thicknesses of ~500 m, which is much less than the observed >1 km thickness for the Camargo Formation in the Camargo syncline. By 12 Ma, low rock uplift rates in the Eastern Cordillera and Interandean zones lead to a factor of 2 decrease in the total sediment thickness of the western Subandean depozone compared to the thickness predicted for this depozone in the gradual uplift model. At 9 Ma, the entire mountain belt between the western Subandean zone and Altiplano rapidly uplifts. This rapid uplift causes an order of magnitude increase in depositional rates within the eastern Subandean depozone. Approximately 2 km of sediment are deposited between 8 and 6 Ma. However, this is still 1–1.8 km less than the maximum isopach values reported by Uba et al. (2006) for the Tariquía Formation.

Additional simulations were conducted for both surface uplift models to test the sensitivity of the sediment thickness curves to the rigidity of the South American plate. The dotted lines represent the results for simulations with the lowest rigidity calculated for the South American plate (i.e., 1.5×10^{23} Nm). Decreasing the rigidity in the gradual uplift model simulation

does not significantly affect the depositional thicknesses within the Eastern Cordillera (Fig. 7A). However, the maximum thicknesses of the other three depozones decrease by ~300–500 m. Conversely, increasing the rigidity of the South American plate (dashed lines) to the upper limit of rigidities previously calculated for the South American plate (i.e., 4.0×10^{24} Nm) causes the maximum depositional thickness of the three westernmost depozones to increase with respect to the initial results. Changing the rigidities of the South American plate in the rapid uplift model has similar effects as in the gradual uplift model (Fig. 7B). Although the rigidity of the South American plate can range over an order of magnitude, the predicted thickness of the Camargo Formation within Eastern Cordillera and Interandean depozones of the rapid uplift model is less than a kilometer.

Role of Eclogite Root Foundering on Surface Uplift and Foreland Development

The eclogite root foundering simulation involves the growth and removal of a subsurface load that modifies the flexural response of the foreland basin and rock uplift distribution within the mountain belt. Prior to 10 Ma, rock uplift rates and shortening rates are prescribed to be identical to the rapid uplift model for the region of the mountain belt that is deforming. Between 25 and 10 Ma, an eclogite load is allowed to uniformly grow beneath the Altiplano and back-thrust region of the Eastern Cordillera, where the deformed crust is sufficiently thick for lower-crustal rocks to undergo a phase transition to eclogite (Fig. 8A). The subsurface eclogite load, the location of which is represented by the black bar in Figure 8, sets up a flexural profile (identified by the thick black line) that superimposes onto the flexural profile caused by the topographic load. Although the total deflection due to the presence of the eclogitic root is small (i.e., ~1 m) at 25 Ma, by 10 Ma

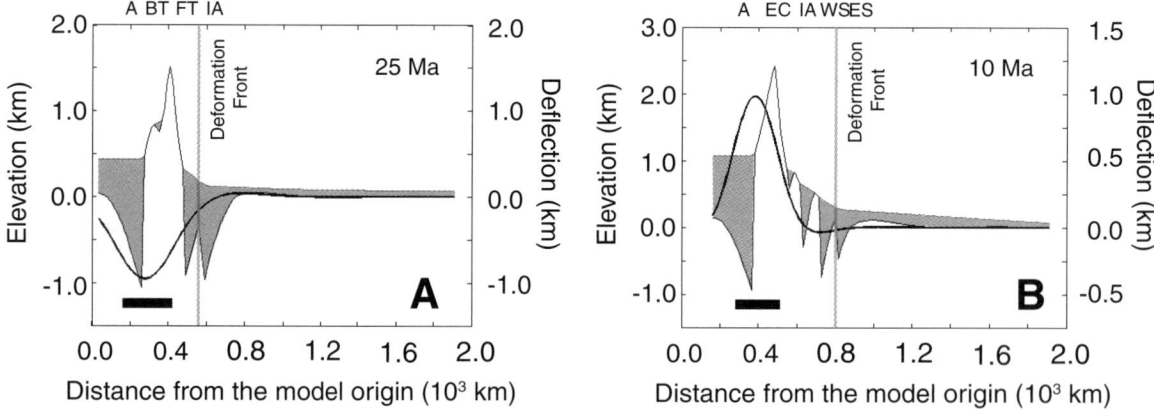

Figure 8 (A and B). Topographic and flexural profile snapshots in time for the eclogite delamination model. The shaded gray regions represent Tertiary sedimentary deposits, and the pre-Tertiary bedrock is white. Lithospheric deflection caused by accumulation and delamination of an eclogite load is represented by a thick black line, the scale of which is shown on the right vertical axis, and the position of the eclogite load is represented by the black horizontal bars. The locations of the tectonomorphic zones are represented by the following labels: A—Altiplano; EC—Eastern Cordillera; BT—Eastern Cordillera back thrust; FT—Eastern Cordillera forethrust; IA—Interandean; WS—western Subandean; ES—eastern Subandean.

the total deflection is on the order of 1 km at the center of the eclogite root. At 25 Ma, the edge of the eclogite load is located ~150 km from the current deformation front. As a result of the distance between the load and deformation front, subsidence caused by the eclogite load is small in the foredeep, and nearby to the east, subsidence changes to rock uplift in the distal foredeep. As time progresses, the Interandean and Subandean depozones are subject to an additional component of rock uplift because they are located on the flexural forebulge related to the eclogite root load. Following 10 Ma, the polarity of lithospheric deflection reverses as the eclogite root is removed (Fig. 8B). The solid black curve is the total amount of deflection that occurs over 3 m.y. as the subsurface load is removed. A significant amount of rock uplift occurs directly over the center of the load in the eastern Altiplano and western back-thrust region of the Eastern Cordillera and rapidly decays into the core of the Eastern Cordillera. Further east, the Interandean and western Subandean zones experience subsidence on the order of meters to tens of meters. Beyond the western Subandean zone, the deflection due the eclogite foundering is less than a meter. Rock uplift rates in the Altiplano and the core of the Eastern Cordillera due to the eclogite root foundering range between 3.0 and 6.0×10^{-4} m/yr. Maximum deflection rates in the eastern Interandean and western Subandean zones are on the order of 10^{-7} and 10^{-5} m/yr. Rock uplift rates near the mountain front and within the foreland basin (i.e., Interandean and Subandean zones) prescribed in the rapid uplift model at 10 Ma are on the order of 6.0–5.0×10^{-4} m/yr, and, therefore, the flexural signal from a delamination event is small when compared to the subsidence or rock uplift due to crustal thickening. Conversely, rock uplift rates due to the delamination are slightly greater in the Altiplano and the back-thrust region of the Eastern Cordillera than the prescribed uplift rates due to crustal thickening.

Sediment supply into the foredeep predicted by the eclogite root foundering model shows a similar behavior to that of the rapid uplift model. Sediment supply rates are low until ca. 10 Ma, when the eclogite drip event accompanied by the rapid uplift of the eastern Interandean and western Subandean zones lead to increased channel slopes in the front of the mountain belt (Fig. 9A). The amount of sediment bypassing the foreland basin in the eclogite root foundering model is also similar to the rates predicted by the rapid uplift model, with the exception of a greater magnitude of sediment flux between 15 and 5 Ma. This increase in sediment leaving the foreland per time period is either the result of increased erosion rates in the mountain belt or decreased sediment accommodation rates in the foreland basin. A comparison of sediment thicknesses predicted by the rapid uplift and eclogite root foundering models through time for the Interandean, western Subandean, and eastern Subandean zones shows that the drip event modified sediment accommodation by less than 10% of the maximum basin thicknesses before foundering (Fig. 9B). The western Interandean zone basin deposits are ~100 m thicker at 22 Ma for the eclogite root foundering model (solid line) compared to the rapid uplift model (dashed line). Between 25 and 22 Ma, the western Interandean zone is located close enough to the edge of the eclogite load to experience subsidence, which causes the change in sediment thickness between models. Conversely, the Subandean basins are located far enough from the eclogite load at that time to undergo rock uplift. The western Subandean basin is the least affected basin by the drip event because of its proximity to the inflection point between subsidence and rock uplift caused by accumulation of eclogite. The eastern Subandean zone also experiences rock uplift due to foundering, and, thus, lesser sediment thicknesses between 20 and 10 Ma. Decreased accommodation in the Subandean zones leads to an increase in sediment leaving the basin prior to 10 Ma. Following 10 Ma, the drip event leads to minor subsidence in the eastern Subandean zone, and the sediment thickness difference between the two models becomes negligible.

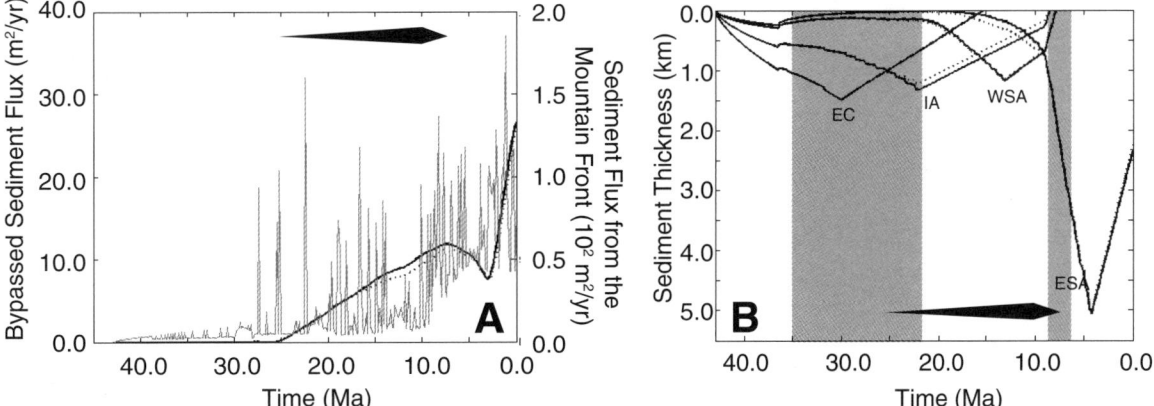

Figure 9. (A) Time-series data for sediment fluxes into and out of the foreland basin and (B) uncompacted sediment thickness for depozones located in the eastern Subandean (ESA), western Subandean (WSA), Interandean (IA), and Eastern Cordillera (EC) tectonomorphic zones for the eclogite delamination model. The black symbol at the top of A and the bottom of B represents the growth and delamination of the eclogite root between 25 and 10 Ma. In A and B, the thick lines represent the eclogite delamination model, and the thin dashed lines represent the rapid uplift model.

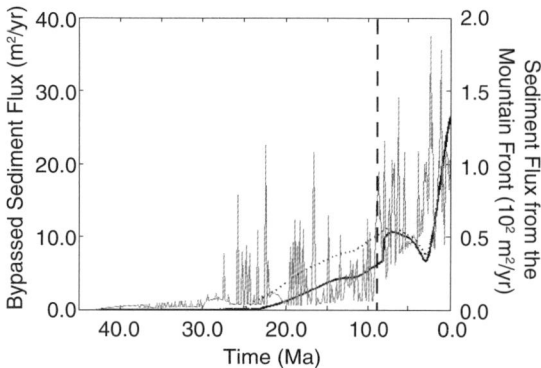

Figure 10. Sediment flux time-series data for the climate change model. The thick line represents the sediment flux leaving the back-bulge basin during the climate change model, and the dashed lines represent the sediment flux leaving the back-bulge basin during the rapid uplift model. The dashed vertical line represents the timing of the onset of the South American monsoon.

Role of Climate Change in Foreland Basin Development

The results for the climate change experiment (i.e., a doubling of the bedrock erodibility and sediment transport coefficients in late Miocene time) are reported here. Bedrock uplift rates during the late Miocene within the actively deforming mountain belt in this simulation are increased with respect to those of the rapid uplift model to maintain similar surface uplift histories, and, thus, the basin accommodation histories between the rapid uplift model and the climate change model are the same. We only report observations for sediment flux and grain size. Although erosion rates and sediment transport rates are a factor of two lower for the first 35 m.y., the first-order behavior of sediment flux entering the basin appears to be unchanged from that of the

rapid uplift model (Fig. 10). The effect of lowering both erosion and sediment transport coefficients by a factor of two early on in the simulation is a few-million-year delay in the time when sediments begin to exit the back-bulge basin. Decreasing both erosion rates and sediment transport rates also has a cumulative effect of decreasing sediment bypass rates (solid black line) by almost half the value of sediment bypass predicted by the rapid uplift model (dashed line). Following 9 Ma, the sediment bypass rates suddenly increase by a factor of 1.5 times the previous flux.

An abrupt change in erosion rates and transport efficiency due to climate change should be expected to have an effect on grain size within the depositional basin, and, therefore, we tracked the gravel-sand interface for each of the experiments (Fig. 11). A comparison of the gradual uplift, rapid uplift, and climate change model results (Fig. 11) reveals that in general the gravel-sand boundary closely tracks the location of the deformation front (shown as the dashed line). An exception to this observation occurs immediately following times when the deformation front propagates basinward, and the gravel-sand interface lags behind the deformation front for ~1–2 m.y. until the regional slope of the newly formed wedge-top basin increases above the threshold slope for gravel transport. Once gravel reaches the deformation front, gravel progradation into the foredeep appears to be limited within a narrow zone of ~50 km from the deformation front. Figure 11A compares the gravel-sand boundaries for the gradual and rapid uplift models. Prior to 10 Ma, the mean location of the gravel-sand boundaries for both models generally remain in front of the deformation front following the 2 m.y. lag periods. However, the initiation of rapid uplift in the Interandean and western Subandean zones at ca. 10 Ma causes the gravel-sand boundary to retrograde back into the wedge-top zone as accommodation rates exceed sediment supply rates. Both models appear to closely overlap each other, with two exceptions. Between 22 and 15 Ma, the gravel front in the gradual uplift model progrades

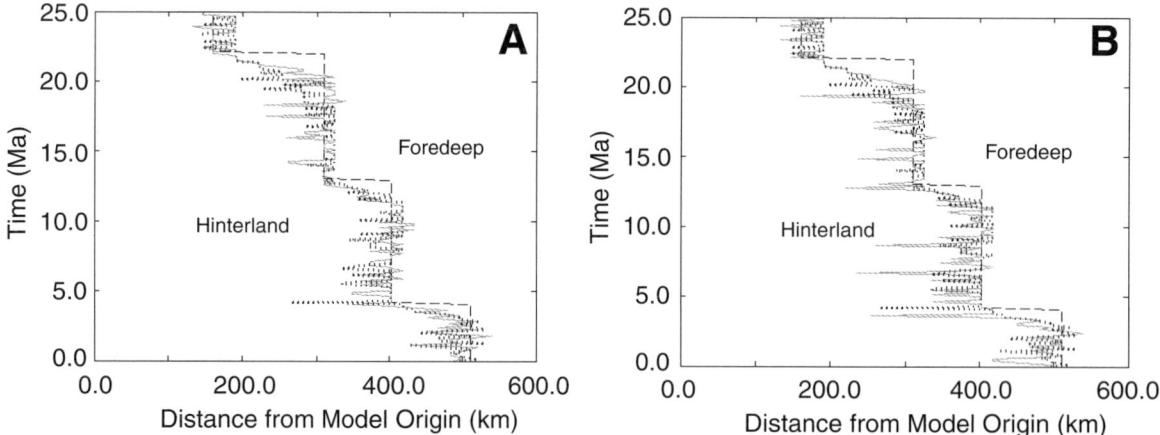

Figure 11. Gravel-sand interface time-series data for the (A) gradual and rapid uplift models and the (B) rapid uplift and climate change models. In A, the gradual and rapid uplift models are represented by the solid and dotted lines, respectively, and in B, the climate change and rapid models are represented by the solid and dotted lines, respectively. The location of the deformation front is shown by the dashed line in both plots.

more rapidly than the gravel front of the rapid uplift model due to a combination of greater uplift rates in the Interandean zone wedge-top basin and greater sediment supply rates. Between 8 and 5 Ma, the mean location of the gravel-sand interface of the rapid uplift model retrogrades more than the gravel-sand interface of the gradual uplift model due to greater subsidence rates. Figure 11B compares the results for the climate change model to the rapid uplift model and, thus, should highlight the effects of changes in precipitation on grain size. Although both the bedrock erodibilities and transport coefficients are a factor of two less in the climate change model compared to the rapid uplift model between 25 and 9 Ma, the gravel-sand boundaries generally track each other well through time. However, there are a few brief periods of time when the gravel-sand interface of the climate change model lags behind that of the rapid uplift mode. When the bedrock erodibilities and transport coefficients increase by a factor of root 2, the results of the climate change model do not vary significantly from the trends of the rapid uplift model.

DISCUSSION

Two end-member surface uplift models have been proposed for the central Andes (Garzione et al., 2008). One model involves a long history of constant deformation since the late Eocene, accompanied by a gradual increase in mean surface elevation prior to 10 Ma. The opposing end-member model suggests that the central Andes did not gain significant elevation until after 10 Ma. Our results show that if the elastic properties of the underthrusted South American plate remain effectively constant, then rapid rock uplift is likely to have occurred within the Interandean and western Subandean zones during the late Miocene in order to generate sufficient accommodation space for the Tariquía Formation observed in outcrops within the central to eastern Subandean zone (Fig. 7). Constraints on the initiation of rapid rock uplift in the Interandean and Subandean zones are based on the isopach data for the Yecua and Petaca Formations. If rock uplift rates were greater in the Interandean and Subandean zones prior to 9–8 Ma, when the Tariquía Formation began to be deposited, then the isopach trends for the Yecua and Petaca Formations observed within the central Subandean zone would show maximum thicknesses greater than 825 m. Rock uplift rates prescribed for the Interandean and western Subandean zones in the gradual uplift model lead to maximum thicknesses for the Petaca-Yecua deposits that are on the order of 1 km. Although other studies have proposed that decreasing the flexural rigidity of the South American lithosphere can explain an increase in accommodation for the Tariquía Formation (Prezzi et al., 2009), we show that a constant flexural rigidity model can also fit the late Miocene isopachs well.

At 10 Ma in the models, both the Altiplano and Eastern Cordillera are sufficiently far from the central and eastern Subandean zones such that their contribution to the accommodation space is low compared to that of the Interandean and western Subandean zones. The deflection of an elastic beam in response to a point load decreases exponentially with distance from the

center of the load. As such, a comparison of the sediment thickness time-series results for the gradual and rapid uplift models shows little change in accommodation rates following 10 Ma in the eastern Subandean zone, even though the peak elevations of the Eastern Cordillera differ by 2 km (Fig. 7). Although the difference in peak elevation within the Eastern Cordillera does not strongly affect late Miocene foreland basin sediment accommodation, we can constrain the early surface uplift history of the Eastern Cordillera and Altiplano with the middle Cenozoic stratigraphy of the Eastern Cordillera and western Interandean zones because these depozones were located at least 100 km closer to the Eastern Cordillera than the central Subandean zone. DeCelles and Horton (2003) measured a thickness of over 2 km for the Camargo Formation within the Camargo syncline of the Eastern Cordillera. A comparison of sediment thicknesses deposited between 36 and 22 Ma predicted by the gradual and rapid uplift models shows that a sediment accommodation thickness of 2 km within the early foreland basin is not achievable unless rock uplift rates in the core of the Eastern Cordillera more closely resembled the rock uplift rates specified within the gradual uplift model (Fig. 7). Although there is uncertainty in the depositional age of the Camargo Formation, the rapid uplift model could not create sufficient sediment accommodation space if deposition began as early as 40 Ma. If the erosion rates in our model are close to the effective erosion rates during the early Miocene and late Oligocene, then the thick deposits of the Camargo Formation would imply that the peak elevation of the Eastern Cordillera was between 2 and 3 km by 22 Ma.

An Eastern Cordillera peak elevation of 2–3 km might have acted as an effective topographic barrier to moisture that was being transported west across the central Andes. Today, the topography of the eastern flank of the central Andean Plateau prevents a significant amount of moisture that originates from the Atlantic Ocean and Amazon Basin from being transported west into the Altiplano and Western Cordillera by the South American summer monsoon (Strecker et al., 2007). A similar behavior occurs in the southern Andes, where the moisture from the Pacific Ocean carried by the Southern Hemisphere westerlies rains out on the west coast of South America and the Patagonian Andes, leading to semiarid conditions on the eastern flank of the Andes. The southern Andes are an effective moisture barrier even though their mean elevation is around 2 km lower than the central Andes, with peak elevations ranging from 2 to 3 km between 38°S and 50°S. We envision the topography of the central Andes during the early to middle Miocene as being similar to the southern Andes today, which suggests that the orogen may have been an effective barrier to moisture being transported to the southwest from the Atlantic. If this is the case, then basins within the Altiplano and Western Cordillera regions of the central Andes should become increasingly arid around or soon after 22 Ma. Based on abrupt changes in lacustrine facies within basins, the onset of hyperaridity has been documented to have occurred between 10 and 6 Ma for basins within the Western Cordillera between 18°S and 22°S (Gaupp et al., 1999; Sáez et al., 1999). However, based on oxygen isotopes, soil morphological characteristics, and salt chemistry, Rech et al. (2006) and Rech et al. (2010) proposed that the onset of hyperaridity could have occurred in the Atacama Desert earlier, between 19 and 13 Ma. In northern Argentina between 25°S and 26°S, recent thermochronologic data from the Puna Plateau show that the Eastern Cordillera was deformed and exhumed at the same time as the Eastern Cordillera further to the north in southern Bolivia (Carrapa et al., 2011a, 2011b). Interestingly, Vandervoort et al. (1995) also documented an earlier shift from nonevaporitic to evaporitic sedimentary deposits within Puna Plateau basins located between 24°S and 26°S at ca. 15 Ma. A middle Miocene onset of aridity-hyperaridity within the Andean Plateau (perhaps as far south as the Puna) and basins on the western margin of the central Andes would be consistent with the peak elevations within the Eastern Cordillera that were necessary to generate sufficient accommodation space for the Camargo Formation.

Another output of our numerical model during the end-member uplift model experiment was a time series for sediment leaving the foreland basin (Figs. 5 and 6). Changes in the sediment bypass rates would not be directly recorded in the foreland basin stratigraphy, and, therefore, one must look at the stratigraphy of adjacent intracratonic basins or the continental shelf to directly sample this signal. Presently, a topographic divide exists in southern Bolivia that splits the flow of major rivers draining off the central Andes north into the Amazon and southeast into the Rio del La Plata cratonic basins. Based on the sediment bypass time series for each of the simulations, the sediment flux component from the central Andes should steadily increase from the late Oligocene into the Quaternary, with a local minimum between 8 and 3 Ma. Analysis of drill-core sediments from the Amazon fan and Ceara Rise show that Andean-derived sediments reached the continental margin between 16.5 and 11.3 Ma (Dobson et al., 2001; Figueiredo et al., 2009). Between 9 and 6.8 Ma, sedimentation rates increased for both the Amazon fan and Ceara Rise. This increase in sedimentation rates was interpreted as the establishment of a transcontinental Amazon drainage system that fully linked the Andean forelands to the Amazon fan. Sedimentation rates continued to increase into the Pliocene–Pleistocene during the period of most rapid uplift within the northern Andes (Hoorn et al., 1995). An overall increase in sedimentation rates on the Atlantic shelf is predicted by our model. The higher the mean elevation of the mountain belt, the more the sediment supply may exceed available accommodation space. Our model also predicts a factor of two increase in sediment flux leaving the foreland basin during the late Pliocene to Pleistocene as foredeep accommodation rates decrease in response to slower rock uplift rates in the Interandean and Subandean zones. One aspect of our model results that is not recorded in the Amazon fan is a late Miocene to early Pliocene decrease in sediment supply rate. The absence of a drop in sedimentation rates may be expected because a small portion of the Amazon Basin drains the central Andes, and, therefore, sediment supply to the Amazon fan is more predominantly influenced by the northern Andes. A stratigraphic data set that

would be more predominately influenced by the central Andes would be the deposits of deep-water fans offshore of the Rio del la Plata estuary, which samples the Andes between 18°S and 34°S. To our knowledge, no analysis has been conducted for the late Cenozoic sediments of the Rio del la Plata fans.

A hypothetical eclogite foundering event in the eastern Altiplano region during middle Miocene to Pliocene time was synchronous with increased grain size and depositional rate in the foreland basin stratigraphy. Our results show that the distance from the center of mass of the load is the most important parameter for determining how the growth and removal of a lower-crustal load would affect rock uplift and subsidence within the mountain belt and foreland basin (Fig. 8). An inflection point between subsidence and uplift occurs at a distance of $(\pi/2)\alpha$ (i.e., ~235 km for this study) from the growing eclogite root load. During the Oligocene to early Miocene, the foredeep depozone was located closer to the eclogite root than it would be for the remainder of the simulation (i.e., less than $[\pi/2]\alpha$ km). As a result, the accumulation of eclogite drove an additional 100 m of subsidence, which is more than a factor of 2 greater than subsidence added to the western and eastern Subandean depositional basins during the late Miocene. The greatest deflection occurred directly above the center of the eclogite root, which was located in the eastern Altiplano and westernmost Eastern Cordillera. This deflection led to ~0.5–1 km of rock uplift in the Altiplano. Therefore, an additional 2 km of surface uplift due to crustal thickening and sediment deposition are required to achieve the modern mean elevation of the Altiplano if it was located at 1 km around 10 Ma. Similar amounts of rock uplift due to crustal thickening (i.e., 1–2 km) are required in the Eastern Cordillera to reach its modern mean and spatially averaged maximum elevations. However, less rock uplift due to crustal thickening is necessary if the average thickness of the eclogite root was significantly larger than the value applied in this study. We infer that this is less likely because a thicker eclogite root would grow a significantly sized instability in less than a period of 3 m.y., which is the proposed length for the delamination period. Our results also show that <0.5 km of rock uplift is contributed to the eastern edge of the Eastern Cordillera and western Interandean zones by eclogite removal. Barke and Lamb (2006) calculated that the San Juan del Oro paleosurface, which overlies the eastern part of the Eastern Cordillera, was uniformly uplifted by ~1 km. Therefore, the eclogite root must be located closer to or beneath the forethrust of the Eastern Cordillera to uniformly uplift it by 1 km. However, tomography data suggest that foundering is more likely beneath the Altiplano and western part of the Eastern Cordillera (Beck and Zandt, 2002). Thus, part of the 1 km of rock uplift of the San Juan del Oro surface is likely the result of crustal thickening.

Climate change was the final process that we tested for the foreland basin of the central Andes of southern Bolivia. Erosion rates and transport coefficients between 43 and 9 Ma were a factor of 2 less than the values between 9 and 0 Ma. As a direct result, peak sediment flux into the foreland basin and sediment bypass rates were significantly less than the values resulting from the rapid uplift model (Fig. 10). At the onset of the South American monsoon, both sediment bypass rates and sediment flux rates into the foredeep increased significantly. The increase in sediment supply to the foredeep is predominantly controlled by the rapid uplift of the mountain belt instead of by the increase in mean annual precipitation. Increasing erosion rates and decreasing transport rates should lead to steeper slopes in the proximal foreland basin. Both sediment supply and transport capability increase as mean annual precipitation increases, which leads to minimal change in the regional slopes for the foreland basin, as these effects cancel each other. Regardless, regional slopes in the foreland basin component of the model appear to be predominately affected by rapid subsidence due to crustal thickening near the deformation front instead of by increasing precipitation. Another way to gauge the effect of climate change is to track the boundary between gravel and sand deposition. Paola et al. (1992) demonstrated that sinusoidal variations in both sediment supply and transport coefficients lead to migration of the gravel-sand interface, especially if the forcing period is small compared to the basin equilibrium time. Our results show that the first-order migration of the gravel-sand boundary appears to be unaffected by the factor of 2 increase in both of these parameters over the time scales that we are sampling (Fig. 11). Based on our results, the threshold slope term appears to be a more primary control on gravel progradation than the transport coefficient term. Rock uplift in the hinterland leads to steeper slopes that are capable of transporting gravel. As a result, gravel trapped near the mountain front can rapidly prograde toward the new deformation front when it is located in a wedge-top basin. Eventually, rock uplift leads to the exhumation of pre-Cenozoic units near the deformation front, which can produce gravel during bedrock incision. Conversely, the regional slope within the foreland is insufficient to transport gravels far into the foredeep. Therefore, a combination of processes causes progradation of the gravel-sand interface during forward propagation of the deformation front; climate change appears to have played a secondary role in controlling gravel progradation for the central Andes.

CONCLUSIONS

Based on our modeling results, we propose that the early surface uplift history (i.e., prior to 22 Ma) of the eastern margin of the central Andes more closely resembled the gradual uplift model. When the rigidity of the South American plate ranged between 1.5×10^{23} and 4.0×10^{24} Nm, surface uplift of the core of the Eastern Cordillera in the rapid uplift model produced grossly undermatched sediment accommodation during the deposition of the >2-km-thick Camargo Formation. The gradual uplift model more closely fits the observed sediment thicknesses for the Camargo Formation located in the Eastern Cordillera, and, thus, the core of the Eastern Cordillera experienced rock uplift rates that would have led to significant topography (i.e., peak elevations near 2–3 km as shown in Fig. 5A) by the early Miocene if basin-averaged erosion rates were on the order of 10^{-4} m/yr.

Further crustal thickening during the middle Miocene may have increased mean topography of the Eastern Cordillera to the point where it became an effective barrier to easterly moisture derived from the Atlantic (as shown in Fig. 5A); in turn, this initiated aridity in the Western Cordillera and Atacama Desert regions at that time. Our results suggest that peak topography of the Eastern Cordillera was well above 3 km prior to 10 Ma and, therefore, do not support the rapid uplift model, which predicts ~2 km of late Miocene rapid surface uplift within that region. However, our results do support rapid rock and surface uplift within the Interandean and western Subandean zones to produce the amount of accommodation required to store the thick Tariquía Formation, which was deposited within a period of 2 m.y.

Our results also support the hypothesis that the first-order trends in the Cenozoic foreland basin stratigraphy of the Subandean zone were predominantly influenced by its distance from the approaching mountain belt. An increase in topographic loads located less than $(3/4)\pi\alpha$ km from a depozone can cause deflections up to the order of a kilometer. Beyond this distance, deflection ranges up to hundreds of meters near the forebulge. During the late Miocene, the eclogite root was located more than a distance of $(3/4)\pi\alpha$ from the Subandean depozone. As such, the accommodation rates within the central-eastern Subandean zone foredeep were positive, but only on the order of tens of meters. Therefore, the deflection caused by an approaching mountain belt exceeded the deflection caused by eclogite delamination. Also based on our results, the factor-of-five increase in depositional rates was not likely the result of an increase in erosion rates and sediment transport rates caused by climate change. Increased erosion rates would lead to isostatic rebound of the mountain front, and thus decreased sediment accommodation within the foredeep, if crustal thickening does not increase as well. If crustal thickening does increase with erosion rates, as in our model, then the additional sediment supplied bypasses the foreland basin because it is already overfilled. In addition to accommodation rates, the location of the deformation front in time appears to exert a primary control on the gravel-sand interface. Increasing erosion rates and sediment transport rates by a factor of 2 due to the onset of the South American monsoon at 9 Ma did not generate a long-term progradation in the gravel-sand interface. Instead, the gravel front retreated toward the deformation front due to the rapid subsidence rates within the foredeep basin. Actively uplifting fold-and-thrust belts are able to achieve channel slopes that are above the threshold for gravel entrainment and produce gravel by bedrock incision. As a direct result, the gravel-sand interface rapidly prograde when the deformation front propagates into the foreland basin.

ACKNOWLEDGMENTS

Our research was supported in part by ExxonMobil funding. We thank Peter DeCelles and Joellen Russell for their helpful comments during the course of the work. We also thank Barbara Carrapa, Chris Poulsen, and Chris Beaumont for constructive reviews that improved the quality of the presentation significantly.

REFERENCES CITED

Barke, R., and Lamb, S., 2006, Late Cenozoic uplift of the Eastern Cordillera, Bolivian Andes: Earth and Planetary Science Letters, v. 249, p. 350–367, doi:10.1016/j.epsl.2006.07.012.

Barnes, J.B., Ehlers, T.A., McQuarrie, N., O'Sullivan, P.B., and Tawackoli, S., 2008, Thermochronometer record of the central Andean Plateau growth, Bolivia (19.5°S): Tectonics, v. 27, TC3003, doi:10.1029/2007TC002174.

Barnes, J.B., Ehlers, T.A., Insel, N., and McQuarrie, N., 2012, Linking orography, climate and exhumation across the central Andes: Geology, v. 40, p. 1135–1138, doi:10.1130/G33229.1.

Beck, S.L., and Zandt, G., 2002, The nature of orogenic crust in the central Andes: Journal of Geophysical Research, v. 107, no. 2230, doi:10.1029/2000JB000124.

Carrapa, B., Bywater-Reyes, S., DeCelles, P.G., Mortimer, E., and Gehrels, G.E., 2011a, Late-Pliocene basin evolution in the Eastern Cordillera of northwestern Argentina (25–26°S): Regional implications for Andean orogenic wedge development: Basin Research, v. 23, p. 1–20.

Carrapa, B., Trimble, J.D., and Stockli, D.F., 2011b, Patterns and timing of exhumation and deformation in the Eastern Cordillera of NW Argentina revealed by (U-Th)/He thermochronology: Tectonics, v. 30, TC3003, doi:10.1029/2010TC002707.

Chase, C.G., Sussman, A.J., and Coblentz, D.D., 2009, Curved Andes: Geoid, forebulge and flexure: Lithosphere, v. 1, p. 358–363, doi:10.1130/L67.1.

Coutand, I., Carrapa, B., Deeken, A., Schmitt, A.K., Sobel, E.R., and Strecker, M.R., 2006, Propagation of orographic barriers along an active range front: Insights from sandstone petrography and detrital apatite fission-track thermochronology in the intramontane Angastaco basin, NW Argentina: Basin Research, v. 18, p. 1–26, doi:10.1111/j.1365-2117.2006.00283.x.

DeCelles, P.G., and Giles, K.A., 1996, Foreland basin systems: Basin Research, v. 8, p. 105–123, doi:10.1046/j.1365-2117.1996.01491.x.

DeCelles, P.G., and Horton, B.K., 2003, Early to middle Tertiary foreland basin development and the history of Andean crustal shortening in Bolivia: Geological Society of America Bulletin, v. 115, p. 58–77, doi:10.1130/0016 -7606(2003)115<0058:ETMTFB>2.0.CO;2.

DeCelles, P.G., Gehrels, G.E., Quade, J., Ojha, T.P., Kapp, P.A., and Upreti, B.N., 1998, Neogene foreland basin deposits, erosional unroofing and the kinematic history of the Himalayan fold-thrust belt, western Nepal: Geological Society of America Bulletin, v. 110, p. 2–21, doi:10.1130/0016 -7606(1998)110<0002:NFBDEU>2.3.CO;2.

DeCelles, P.G., Ducea, M.N., Kapp, P., and Zandt, G., 2009, Cyclicity in Cordilleran orogenic systems: Nature Geoscience, v. 2, p. 251–257, doi:10.1038/ngeo469.

DeCelles, P.G., Carrapa, B., Horton, B.K., and Gehrels, G.E., 2011, Cenozoic foreland basin system in the central Andes of northwestern Argentina: Implications for Andean geodynamics and modes of deformation: Tectonics, v. 30, TC6013, doi:10.1029/2011TC002948.

Deeken, A., Sobel, E.R., Coutand, I., Haschke, M., Riller, U., and Strecker, M.R., 2006, Development of the southern Eastern Cordillera, NW Argentina, constrained by apatite fission track thermochronology: From Early Cretaceous extension to middle Miocene shortening: Tectonics, v. 25, TC6003, doi:10.1029/2005TC001894.

Dobson, D.M., Dickens, G.R., and Rea, D.K., 2001, Terrigenous sediment on Ceara Rise: A Cenozoic record of South American orogeny and erosion: Palaeogeography, Palaeoclimatology, Palaeoecology, v. 165, p. 215–229, doi:10.1016/S0031-0182(00)00161-9.

Echavarria, L., Hernandez, R., Allmendinger, R., and Reynolds, J., 2003, Subandean thrust and fold belt of northwestern Argentina: Geometry and timing of the Andean evolution: American Association of Petroleum Geologists Bulletin, v. 87, p. 965–985, doi:10.1306/01200300196.

Ege, H., Sobel, E.R., Scheuber, E., and Jacobshagen, V., 2007, Exhumation history of the southern Altiplano Plateau (southern Bolivia) constrained by apatite fission track thermochronology: Tectonics, v. 26, TC1004, doi:10.1029/2005TC001869.

Ehlers, T.A., and Poulsen, C.J., 2009, Influence of Andean uplift on climate and paleoaltimetry estimates: Earth and Planetary Science Letters, v. 281, p. 238–248, doi:10.1016/j.epsl.2009.02.026.

England, P., and Molnar, P., 1990, Surface uplift, uplift of rocks, and exhumation of rocks: Geology, v. 18, p. 1173–1177, doi:10.1130/0091-7613 (1990)018<1173:SUUORA>2.3.CO;2.

Figueiredo, J., Hoorn, C., van der Ven, P., and Soares, E., 2009, Late Miocene onset of the Amazon River and the Amazon deep-sea fan: Evidence from the Foz do Amazonas Basin: Geology, v. 37, p. 619–622, doi:10.1130/G25567A.1.

Flemings, P.B., and Jordan, T.E., 1989, A synthetic stratigraphic model for foreland basin development: Journal of Geophysical Research, v. 94, p. 3851–3866, doi:10.1029/JB094iB04p03851.

Ford, M., Lickorish, W.H., and Kusznir, N.J., 1999, Tertiary foreland sedimentation in the southern Subalpine chains, SE France: A geodynamic appraisal: Basin Research, v. 11, p. 315–336, doi:10.1046/j.1365 -2117.1999.00103.x.

Garcia-Castellanos, D., Fernandez, M., and Torne, M., 2002, Modeling the evolution of the Guadalquivir foreland basin (southern Spain): Tectonics, v. 21, 1018, doi:10.1029/2001TC001339.

Garzione, C.N., Hoke, G.D., Libarkin, J.C., Withers, S., MacFadden, B., Eiler, J., Ghosh, P., and Mulch, A., 2008, Rise of the Andes: Science, v. 320, p. 1304–1307, doi:10.1126/science.1148615.

Gaupp, R., Kott, A., and Worner, G., 1999, Palaeoclimatic implications of Mio-Pliocene sedimentation in the high altitude intra-arc Lauca Basin of northern Chile: Palaeogeography, Palaeoclimatology, Palaeoecology, v. 151, p. 79–100, doi:10.1016/S0031-0182(99)00017-6.

Ghosh, P., Garzione, C.N., and Eiler, J., 2006, Rapid uplift of the Altiplano revealed through $^{13}C-^{18}O$ bonds in paleosol carbonates: Science, v. 311, p. 511–515, doi:10.1126/science.1119365.

Göğüş, O.H., and Pysklywec, R.N., 2008, Near-surface diagnostics of dripping or delaminating lithosphere: Journal of Geophysical Research, v. 113, B11404, doi:10.1029/2007JB005123.

Gubbels, T.L., Isacks, B.L., and Farrar, E., 1993, High-level surfaces, plateau uplift, and foreland development, Bolivian central Andes: Geology, v. 21, p. 695–698, doi:10.1130/0091-7613(1993)021<0695:HLSPUA >2.3.CO;2.

Hack, J.T., 1957, Studies of Longitudinal Stream Profiles in Virginia and Maryland: U.S. Geological Survey Professional Paper 294-B, p. 45–97.

Hassan, M.A., Egozi, R., and Parker, G., 2006, Experiments on the effect of hydrograph characteristics on vertical grain sorting in gravel bed rivers: Water Resources Research, v. 42, W09408, doi:10.1029/2005WR004707.

Heller, P., Angevine, C.L., Winslow, N.S., and Paola, C., 1988, Two-phase stratigraphic model of foreland-basin sequences: Geology, v. 16, p. 501–504, doi:10.1130/0091-7613(1988)016<0501:TPSMOF>2.3.CO;2.

Hoke, G.D., Isacks, B.L., Jordan, T.E., Blanco, N., Tomlinson, A.J., and Ramezani, J., 2007, Geomorphic evidence for post–10 Ma uplift of the western flank of the central Andes 18°30′–22°S: Tectonics, v. 26, TC5021, doi:10.1029/2006TC002082.

Hoorn, C., Guerrero, J., Sarmiento, G.A., and Lorente, M.A., 1995, Andean tectonics as a cause for changing drainage patterns in Miocene northern South America: Geology, v. 23, p. 237–240, doi:10.1130/0091-7613 (1995)023<0237:ATAACF>2.3.CO;2.

Horton, B.K., 2005, Revised deformation history of the central Andes: Interferences from Cenozoic foredeep and intermontane basins of the Eastern Cordillera, Bolivia: Tectonics, v. 24, TC3011, doi:10.1029/2003TC001619.

Horton, B.K., and DeCelles, P.G., 1997, The modern foreland basin system adjacent to the central Andes: Geology, v. 25, p. 895–898, doi:10.1130/0091 -7613(1997)025<0895:TMFBSA>2.3.CO;2.

Horton, B.K., Hampton, B.A., LaReau, B.N., and Baldellon, E., 2002, Tertiary provenance history of the northern and central Altiplano (central Andes Bolivia): A detrital record of plateau-margin tectonics: Journal of Sedimentary Research, v. 72, p. 711–726, doi:10.1306/020702720711.

Howard, A.D., and Kerby, G., 1983, Channel changes in badlands: Geological Society of America Bulletin, v. 94, p. 739–752, doi:10.1130/0016-7606 (1983)94<739:CCIB>2.0.CO;2.

Insel, N., Poulsen, C.J., Ehlers, T.A., and Strum, C., 2012, Response of meteoric $\delta^{18}O$ to surface uplift—Implications for Cenozoic Andean Plateau growth: Earth and Planetary Science Letters, v. 317–318, p. 262–272, doi:10.1016/j.epsl.2011.11.039.

Isacks, B.L., 1988, Uplift of the central Andean Plateau and bending of the Bolivian orocline: Journal of Geophysical Research, v. 93, p. 3211–3231, doi:10.1029/JB093iB04p03211.

Kay, R.W., and Kay, S.M., 1993, Delamination and delamination magmatism: Tectonophysics, v. 219, p. 177–189, doi:10.1016/0040-1951(93)90295-U.

Kay, S.M., Coira, B., and Viramonte, J., 1994, Young mafic back-arc volcanic rocks as indicators of continental lithospheric delamination beneath the Argentine Puna Plateau, central Andes: Journal of Geophysical Research, v. 99, p. 24,323–24,339, doi:10.1029/94JB00896.

Kirby, E., and Whipple, K., 2001, Quantifying differential rock-uplift rates via stream profile analysis: Geology, v. 29, p. 415–418, doi:10.1130/0091-7613 (2001)029<0415:QDRURV>2.0.CO;2.

Kleinert, K., and Strecker, M.R., 2001, Climate change in response to orographic barrier uplift: Paleosol and stable isotope evidence from the late Neogene Santa Maria Basin, northwestern Argentina: Geological Society of America Bulletin, v. 113, p. 728–742, doi:10.1130/0016-7606 (2001)113<0728:CCIRTO>2.0.CO;2.

Marzo, M., and Steel, R.J., 2000, Unusual features of sediment supply-dominated, transgressive-regressive sequences: Paleogene clastic wedges, SE Pyrenean foreland basin, Spain: Sedimentary Geology, v. 138, p. 3–15, doi:10.1016/S0037-0738(00)00141-X.

McQuarrie, N., 2002, The kinematic history of the central Andean fold-thrust belt, Bolivia: Implications for building a high plateau: Geological Society of America Bulletin, v. 114, p. 950–963, doi:10.1130/0016-7606 (2002)114<0950:TKHOTC>2.0.CO;2.

McQuarrie, N., Horton, B.K., Zandt, G., Beck, S., and DeCelles, P.G., 2005, Lithospheric evolution of the Andean fold-thrust belt, Bolivia, and the origin of the central Andean Plateau: Tectonophysics, v. 399, p. 15–37, doi:10.1016/j.tecto.2004.12.013.

Molnar, P., and Garzione, C.N., 2007, Bounds on the viscosity coefficient of continental lithosphere from removal of mantle lithosphere beneath the Altiplano and Eastern Cordillera: Tectonics, v. 26, TC2013, doi:10.1029/2006TC001964.

Molnar, P., Anderson, R., Kier, G., and Rose, J., 2006, Relationships among probability distributions of stream discharges in floods, climate, bed load transport, and river incision: Journal of Geophysical Research, v. 111, F02001, doi:10.1029/2005JF000310.

Montgomery, D., and Brandon, M., 2002, Topographic controls on erosion rates in tectonically active mountain ranges: Earth and Planetary Science Letters, v. 201, p. 481–489, doi:10.1016/S0012-821X(02)00725-2.

Montgomery, D., Blanco, G., and Willett, S.D., 2001, Climate, tectonics and the morphology of the Andes: Geology, v. 29, p. 579–582, doi:10.1130/0091 -7613(2001)029<0579:CTATMO>2.0.CO;2.

Mugnier, J.L., Becel, D., and Granjeon, D., 2006, Active tectonics of the Subandean belt, *in* Willett, S., Hovius, N., Brandon, M., and Fisher, D., eds., Tectonics, Climate, and Landscape Evolution: Geological Society of America Special Paper 398, p. 352–369, doi:10.1130/2006.2398(04).

Muller, J.P., Kley, J., and Jacobshagen, V., 2002, Structure and Cenozoic kinematics of the Eastern Cordillera, southern Bolivia (21°S): Tectonics, v. 21, 1037, doi:10.1029/2001TC001340.

Oncken, O., Hindle, D., Kley, J., Elger, K., Victor, P., and Schemmann, K., 2006, Deformation of the central Andean upper plate system—Facts, fiction, and constraints for plateau models, *in* Oncken, O., Chong, G., Franz, G., Giese, P., Götze, H.J., Ramos, V., Strecker, M.R., and Wigger, P., eds., The Andes: Active Subduction Orogeny: Berlin, Springer, p. 3–27.

Paola, C., Heller, P.L., and Angevine, C.L., 1992, The large-scale dynamics of the grain-size variation in the alluvial basins: 1. Theory: Basin Research, v. 4, p. 73–90, doi:10.1111/j.1365-2117.1992.tb00145.x.

Pelletier, J.D., DeCelles, P.D., and Zandt, G., 2010, Relationships among climate, erosion, topography and delamination in the Andes: A numerical modeling investigation: Geology, v. 38, p. 259–262, doi:10.1130/G30755.1.

Poulsen, C.J., Ehlers, T.A., and Insel, N., 2010, Onset of convective rainfall during gradual late Miocene rise of the central Andes: Science, v. 328, p. 490–493, doi:10.1126/science.1185078.

Prezzi, C.B., Uba, C.E., and Gotze, H., 2009, Flexural isostasy in the Bolivian Andes: Chaco foreland basin development: Tectonophysics, v. 474, p. 526–543, doi:10.1016/j.tecto.2009.04.037.

Quade, J., Dettinger, M.P., Carrapa, B., DeCelles, P., Murray, K.E., Huntington, K.A., Cartwright, A., Canavan, R.R., Gehrels, G., and Clementz, M., 2015, this volume, The growth of the central Andes, 22°S–26°S, *in* DeCelles, P.G., Ducea, M.N., Carrapa, B., and Kapp, P.A., eds., Geodynamics of a Cordilleran Orogenic System: The Central Andes of Argentina and Northern Chile: Geological Society of America Memoir 212, doi:10.1130/2015.1212(15).

Quinteros, K., Ramos, V.A., and Jacovkis, P.M., 2008, Constraints on delamination from numerical models, *in* 7th International Symposium on Andean Geodynamics (ISAG 2008, Nice), Extended Abstracts, p. 417–420.

Rech, J.A., Currie, B.S., Michalski, G., and Cowan, A.M., 2006, Neogene climate change and uplift in the Atacama Desert, Chile: Geology, v. 34, p. 761–764, doi:10.1130/G22444.1.

Rech, J.A., Currie, B.S., Shullenberger, E.D., Dunagan, S.P., Jordan, T.E., Blanco, N., Tomlinson, A.J., Rowe, H.D., and Houston, J., 2010, Evidence for the development of the Andean rain shadow from a Neogene isotopic record in the Atacama Desert, Chile: Earth and Planetary Science Letters, v. 292, p. 371–382, doi:10.1016/j.epsl.2010.02.004.

Sáez, A., Cabrera, L., Jensen, A., and Chong, G., 1999, Late Neogene lacustrine record and palaeogeography in the Quillagua-Llamara basin, central Andean fore-arc (northern Chile): Palaeogeography, Palaeoclimatology, Palaeoecology, v. 151, p. 5–37, doi:10.1016/S0031-0182(99)00013-9.

Safran, E.B., Bierman, P.R., Aalto, R., Dunne, T., Whipple, K.X., and Caffee, M., 2005, Erosion rates driven by channel network incision in the Bolivian Andes: Earth Surface Processes and Landforms, v. 30, p. 1007–1024, doi:10.1002/esp.1259.

Schurr, B., Rietbrock, A., Asch, G., Kind, R., and Oncken, O., 2006, Evidence for lithospheric detachment in the central Andes from local earthquake tomography: Tectonophysics, v. 415, p. 203–223, doi:10.1016/j .tecto.2005.12.007.

Sempere, T., Butler, R.F., Richards, D.R., Marshall, L.G., Sharp, W., and Swisher, C.C., 1997, Stratigraphy and chronology of Late Cretaceous–early Paleocene strata in Bolivia and northwestern Argentina: Geological Society of America Bulletin, v. 109, p. 709–727, doi:10.1130/0016 -7606(1997)109<0709:SACOUC>2.3.CO;2.

Simpson, G., 2004, A dynamic model to investigate coupling between erosion, deposition and three-dimensional (thin-plate) deformation: Journal of Geophysical Research, v. 109, F03007, doi:10.1029/2003JF000078.

Snyder, N.P., Whipple, K.X., Tucker, G.E., and Merritts, D.J., 2000, Landscape response to tectonic forcing: Digital elevation model analysis of stream profiles in the Mendocino triple junction region, northern California: Geological Society of America Bulletin, v. 112, p. 1250–1263, doi:10.1130/0016-7606(2000)112<1250:LRTTFD>2.0.CO;2.

Stewart, J., and Watts, A.B., 1997, Gravity anomalies and spatial variations of flexural rigidity at mountain ranges: Journal of Geophysical Research, v. 102, p. 5327–5352, doi:10.1029/96JB03664.

Strecker, M.R., Alonso, R.N., Bookhagen, B., Carrapa, B., Hilley, G.E., Sobel, E.R., and Trauth, M.H., 2007, Tectonics and climate of southern central Andes: Annual Review of Earth and Planetary Sciences, v. 35, p. 747–787, doi:10.1146/annurev.earth.35.031306.140158.

Tassara, A., 2005, Interaction between the Nazca and South American plates and formation of the Altiplano-Puna Plateau: Review of a flexural analysis along the Andean margin (15°–34°S): Tectonophysics, v. 399, p. 39–57, doi:10.1016/j.tecto.2004.12.014.

Toth, J., Kusznir, N., and Flint, S., 1996, A flexural isostatic model of lithosphere shortening and foreland basin formation: Application to the Eastern Cordillera and Subandean belt of NW Argentina: Tectonics, v. 15, p. 213–223, doi:10.1029/95TC02291.

Turcotte, D.I., and Schubert, G., 2002, Geodynamics (2nd ed.): New York, Cambridge University Press, 472 p.

Uba, C.E., Heubeck, C., and Hulka, C., 2005, Facies analysis and basin architecture of the Neogene Subandean synorogenic wedge, southern Bolivia: Sedimentary Geology, v. 180, p. 91–123, doi:10.1016/j .sedgeo.2005.06.013.

Uba, C.E., Heubeck, C., and Hulka, C., 2006, Evolution of the late Cenozoic Chaco foreland basin, southern Bolivia: Basin Research, v. 18, p. 145–170, doi:10.1111/j.1365-2117.2006.00291.x.

Uba, C.E., Strecker, M.R., and Schmitt, A.K., 2007, Increased sediment accumulation rates and climatic forcing in the central Andes during the late Miocene: Geological Society of America Bulletin, v. 35, p. 979–982, doi:10.1130/G224025A.1.

Vandervoort, D.S., Jordan, T.E., Zeitler, P.K., and Alonso, R.N., 1995, Chronology of internal drainage development and uplift, southern Puna Plateau, Argentina central Andes: Geology, v. 23, p. 145–148, doi:10.1130/0091 -7613(1995)023<0145:COIDDA>2.3.CO;2.

Watts, A.B., 2001, Isostasy and Flexure of the Lithosphere: New York, Cambridge University Press, 480 p.

Whipple, K., and Tucker, G., 1999, Dynamics of the stream-power river incision model: Implications for height limits of mountain ranges, landscape response timescales, and research needs: Journal of Geophysical Research, v. 104, p. 17,661–17,674, doi:10.1029/1999JB900120.

MANUSCRIPT ACCEPTED BY THE SOCIETY 3 JUNE 2014
MANUSCRIPT PUBLISHED ONLINE 14 NOVEMBER 2014

The Geological Society of America
Memoir 212
2015

The Miocene Arizaro Basin, central Andean hinterland: Response to partial lithosphere removal?

P.G. DeCelles
B. Carrapa
Department of Geosciences, University of Arizona, Tucson, Arizona 85716, USA

B.K. Horton
Institute for Geophysics and Department of Geological Sciences, Jackson School of Geosciences,
University of Texas at Austin, Austin, Texas 78712, USA

J. McNabb
Department of Geological Sciences, University of Oregon, Eugene, Oregon 97403, USA

G.E. Gehrels
Department of Geosciences, University of Arizona, Tucson, Arizona 85716, USA

J. Boyd
Department of Geology and Geophysics, University of Wyoming, Laramie, Wyoming 82070, USA

ABSTRACT

The Arizaro Basin in northwestern Argentina sits today in the western Puna Plateau at elevations of 3800–4200 m along the eastern flank of the Miocene to modern magmatic arc. The basin is roughly circular in plan view and ~100 km in diameter, and it was filled during Miocene time (ca. 21–9 Ma) by >3.5 km of eolian, alluvial, fluvial, and lacustrine sediment in addition to ash-fall tuffs from the Andean magmatic arc. The basin fill was subsequently shortened in its central part, and it has been uplifted and topographically inverted. The Arizaro Basin is not obviously related to known faults, nor does it exhibit a peripheral belt of coarse-grained sedimentary rocks derived from flanking topographically higher regions. Sandstone modal framework compositions are arkosic, but not as rich in volcanic lithic fragments as typical intra-arc basins. Detrital zircon U-Pb age spectra implicate source terranes in locally exposed Ordovician granitoid rocks, more distal Upper Paleozoic–Mesozoic arc terranes in western Argentina and possibly northern Chile, and the local Miocene magmatic arc. Depositional-age zircons are present in most of the sandstones analyzed for detrital zircon U-Pb geochronology, and zircon U-Pb ages from volcanic tuff layers provide independent chronological control. The tectonic component of subsidence initiated at low rates, accelerated to ~0.6 mm/yr during the medial stage

DeCelles, P.G., Carrapa, B., Horton, B.K., McNabb, J., Gehrels, G.E., and Boyd, J., 2015, The Miocene Arizaro Basin, central Andean hinterland: Response to partial lithosphere removal? *in* DeCelles, P.G., Ducea, M.N., Carrapa, B., and Kapp, P.A., eds., Geodynamics of a Cordilleran Orogenic System: The Central Andes of Argentina and Northern Chile: Geological Society of America Memoir 212, p. 359–386, doi:10.1130/2015.1212(18). For permission to copy, contact editing@geosociety.org. © 2014 The Geological Society of America. All rights reserved.

of basin development, and tapered off to zero as the basin began to shorten internally and experience topographic inversion after ca. 10 Ma. Together, the data presented here suggest that the Arizaro Basin could have developed in response to the formation and gravitational foundering of a dense Rayleigh-Taylor–type instability in the lower crust and/or mantle lithosphere. Insofar as hinterland basins of uncertain tectonic affinity are widespread in the high central Andes, the model developed here may be relevant for other regions of enigmatic subsidence and sediment accumulation in the Andes and other cordilleran hinterland settings.

INTRODUCTION

Recent studies in cordilleran orogenic systems of the western Americas suggest that thickening of the lithosphere in the orogenic hinterland and magmatic arc may be accompanied or offset by partial removal of mantle lithosphere and/or lower crust. This view has evolved more or less independently from several different lines of evidence, including seismology, igneous and metamorphic petrology, regional structural geology, and geodynamic modeling. Kay and Kay (1993) and Kay et al. (1994) used the petrological compositions of Miocene lavas in the central Andean magmatic arc to infer that large segments of the Andean lower crust and mantle lithosphere had "delaminated" and foundered into the mantle. Schott and Schmelling (1998) studied the process of lithospheric removal using numerical models and showed kilometer-scale surface topographic effects using realistic model scenarios. Ducea and Saleeby (1998) and Ducea (2001) suggested that mantle xenoliths in Miocene volcanic rocks of the Sierra Nevada represent remnants of a former dense crustal root that was removed from beneath the southern Sierran arc in the North American Cordillera. Geophysical data in favor of this hypothesis show that the southern Sierra Nevada indeed does lack a root sufficient to support current elevations (e.g., Wernicke et al., 1996; Zandt et al., 2004; Jones et al., 2004; Frassetto et al., 2011). Beck and Zandt (2002), Yuan et al. (2002), Schurr et al. (2006), Heit et al. (2008), and Bianchi et al. (2013) reported seismic data from the central Andes showing large zones in the upper mantle and crust of anomalously slow and fast P-wave velocities that were interpreted as, respectively, regions from which chunks of Andean lithosphere had been gravitationally removed and the dense chunks themselves. McQuarrie et al. (2005) noted that the large amounts of upper-crustal shortening accommodated in the central Andean thrust belt require the disposal of large volumes of lithosphere beneath the Andean hinterland; the absence of such thick lithosphere in this region led them to propose significant amounts of lithospheric removal. Sobolev and Babeyko (2005) linked horizontal crustal shortening with production of metamorphic eclogite (in contrast to magmatic eclogitic restites) to drive episodic removal of lower crust and lithosphere beneath the central Andes. Garzione et al. (2006) documented evidence for an abrupt increase in surface paleoelevation during the late Miocene in the Altiplano of Bolivia and argued that the most feasible way to drive such a dramatic elevation gain was to remove

a regional-scale segment of Andean lithosphere by gravitational foundering of a Rayleigh-Taylor convective instability. Schoenbohm and Carrapa (2011) linked neotectonic extensional faults, young basalts, and local basin formation in the central Andes with the formation and gravitational foundering of local lithospheric roots. DeCelles et al. (2009) proposed a general model in which lithospheric thickening beneath Cordilleran hinterland regions results from horizontal shortening of the upper-plate lithosphere in oceanic-continental convergence zones; thickening is accompanied by melting to produce flare-ups in magmatic arcs, and dense residues accumulate in eclogitic roots that eventually drip into the asthenosphere when critical mass is achieved (Ducea, 2001). Although much of the work on this problem has been done in the actively shortening central Andes, workers in the North American Cordillera have begun to infer similar processes (e.g., Wells et al., 2012).

Local basin development in the hinterland is a predicted collateral result of the formation and removal of dense lower-crustal lithospheric roots (Molnar and Houseman, 2004). Pysklywec and Cruden (2004), Göğüş and Pysklywec (2008), and Krystopowicz and Currie (2013) constructed analog and finite-element models to illustrate the effects of formation and removal of lithospheric gravitational instabilities. These models predict that areas beneath which gravitational instabilities are forming and being removed should experience rapid surface subsidence that attenuates as the instability founders and completely detaches, followed by isostatic rebound as the detached root sinks into the asthenosphere (Göğüş and Pysklywec, 2008; Wang et al., this volume). The predicted magnitude of subsidence is kilometric in scale and depends in part on the volume, viscosity, and density of the instability relative to the ambient material (Currie et al., this volume; Wang et al., this volume). Such subsidence should be recorded in sedimentary basins that form above gravitational instabilities.

In this paper, we present data from a Miocene sedimentary basin located in the hinterland of the central Andean orogenic belt. This basin, which we refer to as the Arizaro Basin, is enigmatic insofar as it is not bounded by large faults and cannot be explained in terms of standard modes of basin subsidence (flexure, local stretching) in contractional orogenic belts. We propose that the Arizaro Basin formed and was later modified mainly in response to the development and subsequent removal of a lithospheric gravitational instability. This represents a novel way to interpret basins that are located deep within the interiors of cordilleran orogenic belts, including numerous other "hinterland" basins that are present in the modern central Andes (Horton, 2005, 2012; Leier et al., 2013).

GEOLOGICAL AND TECTONIC SETTING

The geology of the central Andes is dominated by the Puna-Altiplano high plateau and flanking Western and Eastern Cordilleras (Fig. 1). The Western Cordillera is the locus of Miocene to Holocene arc magmatism and is composed at the surface largely of ignimbrite sheets and young stratovolcanoes (e.g., De Silva et al., 2006; Kay and Coira, 2009). Western Cordillera pre-Cenozoic "basement" is exposed in the Cordillera de Domeyko and scattered outcrops south and east of the Salar de Atacama and along the western flank of the Salar de Arizaro, and consists of Upper Paleozoic and Mesozoic metamorphic and igneous rocks. Carboniferous, Permian, and Triassic plutonic and extrusive rocks are common. Cretaceous volcanogenic and siliciclastic strata are also exposed in the Cordillera de Domeyko. The Puna forms the southern half of the greater Central Andean Plateau (Isacks, 1988) and consists of a region of high average elevation with rugged basin-and-range topography (Fig. 1A). The ranges typically rise to elevations >5500 m, and average regional elevation is ~4200 m. Internally drained evaporative lakes (salars) are common between the numerous ranges. Puna basement consists of Paleozoic sedimentary and metasedimentary rocks, intruded by Ordovician granitoid rocks. The eastern margin of the Puna, defined as the divide that separates internally drained from externally drained topography, is marked by the Eastern Cordillera, a series of high ranges composed of Proterozoic and Paleozoic metasedimentary basement intruded locally by Ordovician and Cambrian granitoid rocks (Reutter et al., 1994; Kley and Monaldi, 2002; Hongn et al., 2010), Cretaceous to Paleocene rift-related deposits (Salfity and Marquillas, 1994; Marquillas et al., 2005), and Upper Paleocene through Miocene foreland basin strata related to the development of the Andean orogenic belt (Starck and Vergani, 1996; Reynolds et al., 2001; DeCelles et al., 2011). Pre-Cretaceous basement throughout the central Andes was deformed during both Paleozoic and Cenozoic orogenic events (e.g., Bahlburg and Hervé, 1997; Ramos, 2008).

The Arizaro Basin is a roughly 100-km-diameter accumulation of Miocene lacustrine, fluvial, and eolian strata that have been mapped variously as the Geste, Vizcachera, Pozuelos, Sijes, and Batín Formations by Donato (1987), Blasco et al. (1996), Zappettini and Blasco (2001), Coutand et al. (2001), Jordan and Mpodozis (2006), and Boyd (2010) (Figs. 2 and 3). In this paper, we employ the terminology of Zappettini and Blasco (2001), which attributes the bulk of the basin fill to the Vizcachera Formation (Fig. 3). New U-Pb geochronology reported herein justifies abandonment of the Geste Formation (which is Eocene in age; Pascual, 1983; Alonso, 1992; DeCelles et al., 2007) nomenclature for the lowermost part of the basin fill (Fig. 3). The western half of the basin is occupied mainly by the modern Salar de Arizaro at an elevation of ~3500 m, and the eastern half of the preserved basin fill is represented by extensive outcrops of Miocene strata locally interrupted by narrow north-south–trending outcrops of Ordovician granite (Fig. 2B). The modern Salar de Arizaro is filled mainly with halite (Alonso et al., 1991) and local borates (R.N. Alonso, 2013, personal commun.). Although subsurface data are not available for the region beneath the Salar de Arizaro, an ~1600-m-thick succession of Upper Miocene evaporitic strata (halite and gypsum) dips gently westward beneath its eastern fringe (Alonso et al., 1991; Zappettini and Blasco, 2001; Jordan and Mpodozis, 2006). Following Jordan and Mpodozis (2006), we infer that Miocene strata are also present beneath the salar (Fig. 2C). The eastward extent of the basin fill is also obscure; easternmost outcrops of Batín Formation and underlying Vizcachera Formation dip gently eastward toward the Salar de Pocitos, but these rocks apparently do not crop out east of the Salar de Pocitos.

Arizaro Basin is cradled between two fault-controlled southeastward protuberances of the Miocene–Holocene magmatic arc (Fig. 2A; Riller et al., 2001; Ramelow et al., 2006), and it is flanked by Lower Paleozoic metasedimentary and plutonic rocks on the east and by Ordovician and Permian plutonic rocks, Eocene volcanic rocks, and younger rocks of the Miocene to Holocene magmatic arc to the west. The Sierra de Macon (Fig. 2B; maximum elevation 5486 m) forms a narrow (up to 3 km wide) north-south–trending ridge that divides the northern half of the basin roughly in half; this ridge has up to ~1700 m of topographic relief above the surrounding Miocene outcrops and Quaternary deposits. The Sierra de Macon is composed mainly of Ordovician granitoid rocks (Poma et al., 2004) and has been raised above regional structural elevation by a steeply west-dipping reverse fault (Fig. 2C).

Previous work in the Salar de Arizaro region includes basic mapping and stratigraphic work by Donato (1987) and Zappettini and Blasco (2001), and regional structural and stratigraphic context provided by Coutand et al. (2001). Ages of Arizaro Basin strata have been reported by Alonso et al. (1991), Vandervoort (1993), Vandervoort et al. (1995), and Jordan and Mpodozis (2006). These ages range between 23.8 ± 0.4 Ma (^{40}Ar/^{39}Ar on hornblende) and 10.8 ± 2.0 Ma (zircon fission track) but are not located within the context of a detailed measured section. We incorporate these ages where possible into our measured stratigraphic framework. Based on these chronological constraints and lithofacies documentation, Alonso et al. (1991) and Vandervoort et al. (1995) recognized an early (before ca. 15 Ma) siliciclastic-dominated phase of basin evolution, followed by establishment of internal drainage and the onset of evaporitic sedimentation. Jordan and Mpodozis (2006) recognized a threefold stratigraphic division of the basin fill, which largely corresponds to that which we document in this study (Fig. 3).

Low-temperature thermochronology studies demonstrate that the rocks exposed at the surface of the Sierra de Macon have been cooler than the closure temperature for retention of apatite fission tracks (120–60 °C) since Cretaceous to early Eocene time

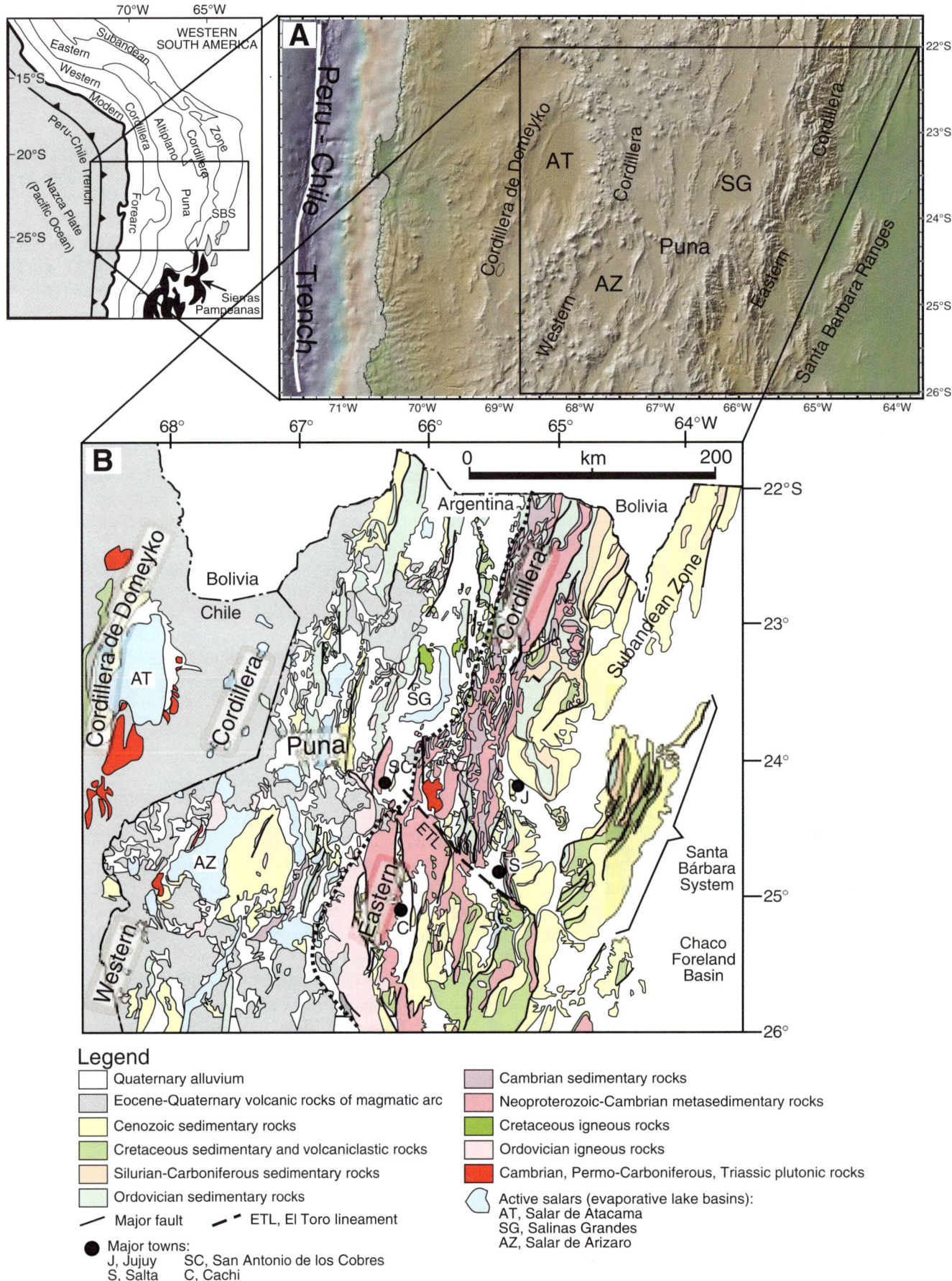

Legend

☐	Quaternary alluvium
☐	Eocene-Quaternary volcanic rocks of magmatic arc
☐	Cenozoic sedimentary rocks
☐	Cretaceous sedimentary and volcaniclastic rocks
☐	Silurian-Carboniferous sedimentary rocks
☐	Ordovician sedimentary rocks

— Major fault   ▰▰ ETL, El Toro lineament

● Major towns:
   J, Jujuy   SC, San Antonio de los Cobres
   S, Salta   C, Cachi

☐	Cambrian sedimentary rocks
☐	Neoproterozoic-Cambrian metasedimentary rocks
☐	Cretaceous igneous rocks
☐	Ordovician igneous rocks
☐	Cambrian, Permo-Carboniferous, Triassic plutonic rocks

⬡ Active salars (evaporative lake basins):
   AT, Salar de Atacama
   SG, Salinas Grandes
   AZ, Salar de Arizaro

Figure 1. (A) Digital elevation model (from GeoMapApp) of the central Andes showing major topographic features of the arc and retroarc region. A few of the many internally drained modern sedimentary basins are labeled: AT—Salar de Atacama; AZ—Salar de Arizaro; and SG—Salinas Grandes. Inset map (upper left) provides larger spatial context; SBS—Santa Bárbara system. (B) Geological map of northeastern Chile and northwestern Argentina (after Reutter et al., 1994).

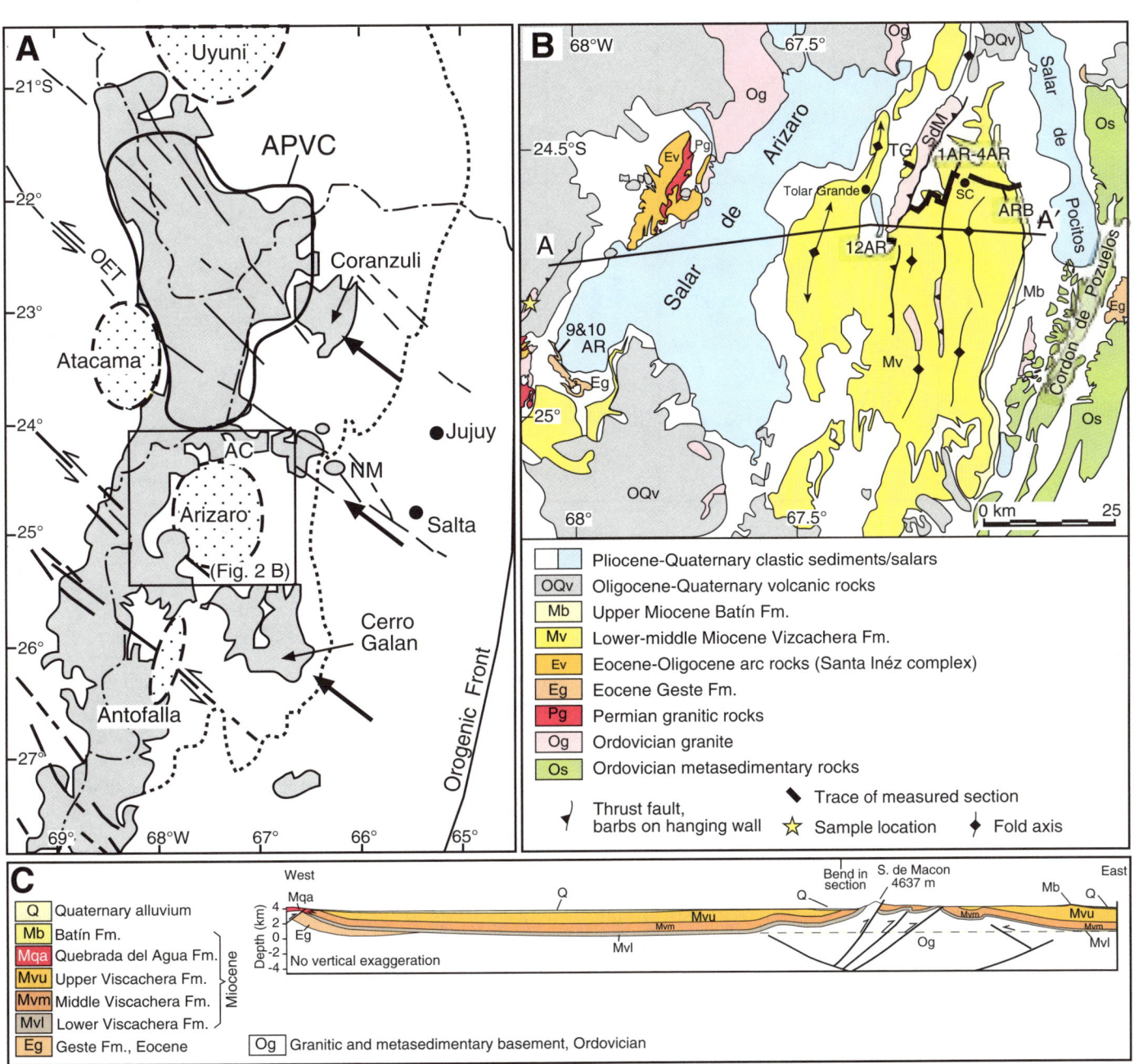

Figure 2. (A) Tectonic map of the Miocene–Holocene magmatic arc, showing locations of the Atacama, Arizaro, Antofalla, and Uyuni Basins (stippled ovals). Large black arrows indicate locations of southeastward protuberances in the arc (gray shaded area), possibly controlled by northwest-striking fault systems such as the Olacapato–El Toro fault zone (OET). Major Miocene–Pliocene calderas are highlighted at Cerro Galan, Negro Muerta (NM), Aguas Calientes (AC), Coranzuli, and the Altiplano-Puna volcanic complex (APVC). Dotted line highlights eastern margin of the central Andean Plateau. Figure is modified from Ramelow et al. (2006). (B) Geological map of the Arizaro Basin area, modified from Blasco et al. (1996), Salfity and Monaldi (1998), and Zappettini and Blasco (2001). Locations of measured sections are shown by black bars, labeled as follows: sections TG and 12AR are shown in Figure 5; sections labeled ARB and 1AR-4AR are a composite section that is illustrated in Figure 6; 9 and 10AR represent a section of rocks formerly considered to be Miocene Vizcachera Formation, but which our new geochronological results show to be of Eocene age. Table DR1 contains details of section locations (see text footnote 1). Yellow star near western map margin shows location of apatite fission-track samples discussed in text (see also Carrapa and DeCelles, this volume). SC—location of Siete Curvas; SdM—Sierra de Macon. Line labeled A–A′ is cross-section trace. (C) Cross section A–A′, based in part on Zappettini and Blasco (2001), Jordan and Mpodozis (2006), and detailed mapping by Boyd (2010).

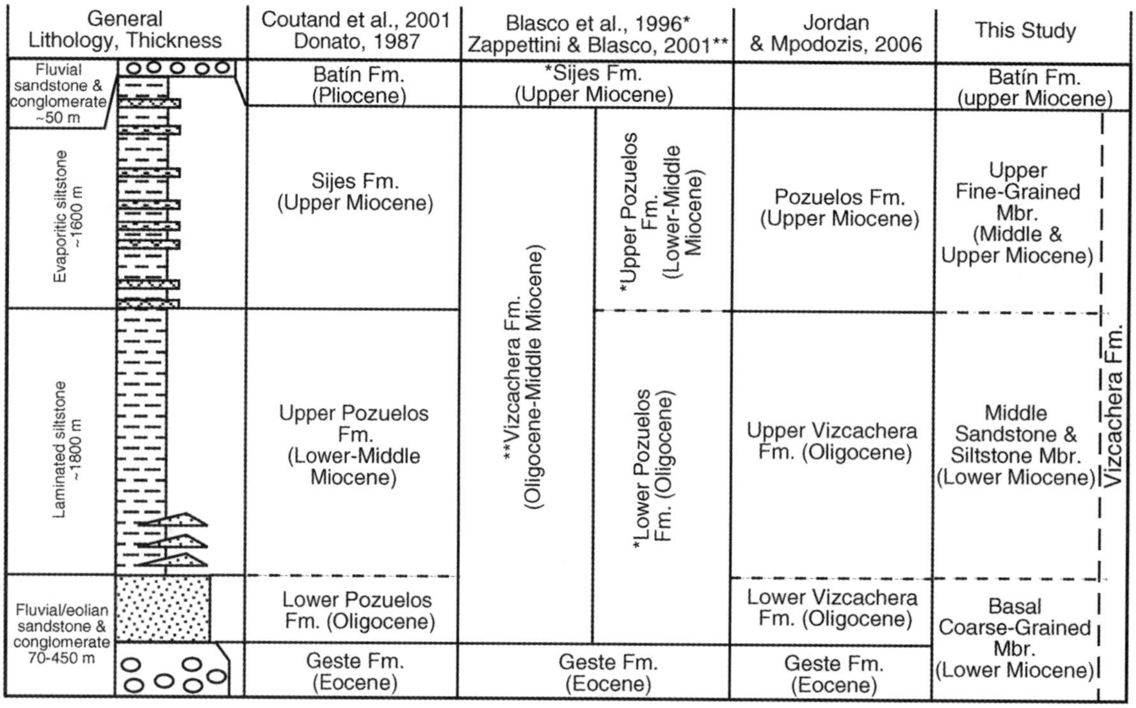

General Lithology, Thickness	Coutand et al., 2001 Donato, 1987	Blasco et al., 1996* Zappettini & Blasco, 2001**	Jordan & Mpodozis, 2006	This Study
Fluvial sandstone & conglomerate ~50 m	Batín Fm. (Pliocene)	*Sijes Fm. (Upper Miocene)		Batín Fm. (upper Miocene)
Evaporitic siltstone ~1600 m	Sijes Fm. (Upper Miocene)	*Upper Pozuelos Fm. (Lower-Middle Miocene)	Pozuelos Fm. (Upper Miocene)	Upper Fine-Grained Mbr. (Middle & Upper Miocene)
Laminated siltstone ~1800 m	Upper Pozuelos Fm. (Lower-Middle Miocene)	**Vizcachera Fm. (Oligocene-Middle Miocene) / *Lower Pozuelos Fm. (Oligocene)	Upper Vizcachera Fm. (Oligocene)	Middle Sandstone & Siltstone Mbr. (Lower Miocene)
Fluvial/eolian sandstone & conglomerate 70-450 m	Lower Pozuelos Fm. (Oligocene)		Lower Vizcachera Fm. (Oligocene)	Basal Coarse-Grained Mbr. (Lower Miocene)
	Geste Fm. (Eocene)	Geste Fm. (Eocene)	Geste Fm. (Eocene)	

Figure 3. Chart showing various stratigraphic nomenclatures that have been applied to the Arizaro Basin, and stratigraphic nomenclature employed in this paper.

(Carrapa et al., 2009; this paper). The absence of pre-Miocene sedimentary cover strata implies that the region was uplifted and shallowly eroded between Cretaceous and Eocene time. Regional studies of the history of foreland basin development and the timing of thrust-related exhumation show that by early to middle Miocene time, the front of the orogenic wedge had already migrated into the central part of the Eastern Cordillera, more than 100–150 km east of the Arizaro Basin (Coutand et al., 2001; Deeken et al., 2006; Carrapa et al., 2011a, 2011b; DeCelles et al., 2011; Pearson et al., 2013). Thus, the Arizaro Basin formed deep within the hinterland of the orogenic belt. Stable isotope paleoaltimetry data suggest that the basin formed and remained at high elevation throughout deposition of the basin fill (Canavan et al., 2014; J. Quade, 2013, personal commun.).

STRATIGRAPHY AND SEDIMENTOLOGY OF ARIZARO BASIN

The sedimentology of the Arizaro Basin was documented in stratigraphic sections measured on a bed-by-bed scale, down to individual layers as thin as a few centimeters. Samples were collected for sedimentary petrography, apatite fission-track (AFT) and (U-Th)/He thermochronology (Carrapa et al., 2009; Carrapa and DeCelles, this volume), and U-Pb geochronology.

The Arizaro Basin is filled with more than 3.5 km of nonmarine clastic sedimentary strata referred to as the Vizcachera Formation (Donato, 1987; Zappettini and Blasco, 2001) and a thin cap of conglomeratic sandstone, which we mapped as the Batín Formation (after Donato, 1987) along the eastern edge of the basin exposures (Figs. 2B and 3). Based on our observations, the Vizcachera Formation consists of three lithologically distinct, informal members: a lower coarse-grained sandstone and conglomerate member; a middle sandstone and siltstone interval; and an upper member composed mainly of siltstone and shale with abundant thin evaporite layers (Fig. 3). This stratigraphy is best seen in the central part of the basin, in the area between Siete Curvas and the Sierra de Macon. Previous maps and reports on the Arizaro Basin region consider the lower coarse-grained member of the Vizcachera Formation to be the Geste Formation of Eocene age (Figs. 3 and 4C; Donato, 1987; Blasco et al., 1996; Jordan and Mpodozis, 2006; Carrapa et al., 2009). As will be discussed herein, our U-Pb geochronological data show that all of the clastic strata in the Arizaro Basin near and east of the Sierra de Macon are of Miocene age. Additional exposures of Vizcachera Formation are mapped along the southwestern margin of the basin, but U-Pb ages on tuffs from these strata show them to be at least partly of Eocene age. Therefore, we map the section exposed along the southwestern edge of the modern salar as the Eocene Geste Formation (Fig. 2B). In the following description of the sedimentology of the Vizcachera Formation, standard lithofacies codes (defined in Table 1) are employed.

Figure 4. (A) Upright chevron folds in the middle siltstone and sandstone member in the center of Arizaro Basin. View toward north. Range in background is the Sierra de Macon. (B) Eolian dune deposits overlying desiccation-cracked siltstone in lower coarse-grained member. Hammer is 40 cm long. (C) Interbedded conglomerate (Gcm and Gcmi) and sandstone (Sh, St) cropping out along the western flank of Sierra de Macon. (See Table 1 for descriptions.) Person at lower right for scale.

Lower Coarse-Grained Member

The lower coarse-grained member is composed of sandstone and conglomerate up to ~500 m thick. The thickest occurrence is along the western slope of the Sierra de Macon (section TG; Figs. 2B and 5), where Ordovician granitic basement is overlain by steeply westward dipping beds of boulder to cobble conglomerate and intercalated sandstone. Coarse conglomeratic facies are also present along the southeastern flank of the Sierra de Macon (section 12AR; Fig. 5), in the high ridge to the south of the main road southeast of Tolar Grande, and in the area just west of Siete Curvas (Figs. 2B and 6).

Lithofacies in the lower member include clast-supported, massive or horizontally stratified conglomerate (Gcm, Gch; Fig. 4C; Table 1); horizontally laminated sandstone (Sh); trough cross-stratified sandstone (St); and large-scale (>2 m thick generally, locally >10 m thick) planar-tangential cross-stratified sandstone (Spl) (Fig. 4B). All of these lithofacies are well known and thoroughly documented in the sedimentological literature (for example, Reineck and Singh, 1975; Allen, 1984; Miall, 1996). We interpret the conglomeratic lithofacies as the deposits of shallow, laterally unstable, gravelly fluvial channels and intervening longitudinal bars (e.g., Hein and Walker, 1977; Church and Jones, 1982; Lunt and Bridge, 2004; Wooldridge and Hickin, 2005). Lithofacies Sh and St are common in fluvial channel deposits, the former developing as upper-flow-regime plane beds under very shallow flows, and the latter forming by the migration of subaqueous dunes (or large three-dimensional ripples) under unidirectional currents with velocities ranging from ~0.5 to 1.5 m/s (e.g., Cant, 1978; Cant and Walker, 1978). The large-scale cross-stratified sandstone lithofacies was deposited as grain flows (and reworked by wind-ripple migration) on slip faces in eolian dunes (Bagnold, 1954; Hunter, 1977; Kocurek and Dott, 1981). This lithofacies is abundant between the 140 m and 250 m levels of section TG (Fig. 5), as well as in the 0–300 m portion of section 3AR (Fig. 6). In the latter section, the dune deposits are amalgamated in the lowermost part of the section but become more isolated in lenticular sandstone bodies surrounded by laminated and rippled siltstone (~100–250 m levels). We interpret these isolated eolian sandstone bodies as starved dunes that migrated across sand/mudflats. Arid conditions are also indicated by the presence of desiccation cracks, often associated with the lower parts of dune foresets (Fig. 4B).

Middle Siltstone and Sandstone Member

Above the lower coarse-grained member, grain size abruptly diminishes, and the lithofacies of the Arizaro Basin become dominated by horizontally laminated very fine-grained, red sandstone and siltstone (Figs. 6 and 7A). Lithofacies include laminated siltstone (Fsl), rippled very fine-grained sandstone (Sr, Srw), and horizontally laminated sandstone (Sh). Lenticular bedding is rare. Thin layers of gypsum and zones of gypsiferous veins are locally present. Almost all sandstone beds are tabular over hundreds

TABLE 1. LITHOFACIES AND INTERPRETATIONS USED IN THIS STUDY, MODIFIED AFTER MIALL (1978) AND DECELLES ET AL. (1991)

Lithofacies code	Description	Interpretation
Fsl	Laminated red siltstone	Suspension settling in ponds and lakes
Sm	Massive medium- to fine-grained sandstone; bioturbated	Bioturbated or pedoturbated sand, penecontemporaneous deformation
Sr	Fine- to very fine-grained sandstone with small, asymmetric, 2-D and 3-D current ripples	Migration of small 2-D and 3-D ripples under weak (~20–40 cm/s), unidirectional flows in shallow channels
St	Medium-grained sandstone with trough cross-stratification	Migration of large 3-D ripples (dunes) under moderately powerful (40–100 cm/s), unidirectional flows in large channels
Sp	Fine- to medium-grained sandstone with very large-scale (>2 m) planar cross-stratification; locally associated with desiccation cracks	Migration of eolian dunes across dry playa surface
Sh	Fine- to medium-grained sandstone with plane-parallel lamination	Upper plane bed conditions under unidirectional flows, either strong (>100 cm/s) or very shallow
Srw	Fine-grained sandstone with symmetrical small ripples	Deposition of oscillatory current (orbital) ripples in shallow lakes and ponds
Gcm	Pebble to boulder conglomerate, poorly sorted, clast-supported, unstratified, poorly organized	Deposition from sheetfloods and clast-rich debris flows
Gcmi	Pebble to cobble conglomerate, moderately sorted, clast-supported, unstratified, imbricated (long-axis transverse to paleoflow)	Deposition by traction currents in unsteady fluvial flows
Gch, Gchi	Pebble to cobble conglomerate, well-sorted, clast-supported, horizontally stratified, imbricated (long-axis transverse to paleoflow)	Deposition from shallow traction currents in longitudinal bars and gravel sheets
Gmm	Massive, matrix-supported pebble to boulder conglomerate, poorly sorted, disorganized, unstratified, silty sandstone matrix	Deposition by semicohesive matrix-supported debris flows and hyperconcentrated flows

of meters laterally and are arranged in upward-thickening and upward-coarsening packages (Fig. 7B). A typical sequence consists of laminated red siltstone in its lower part; massive, laminated, or rippled very fine-grained sandstone beds that are intercalated with laminated red siltstone in the middle; and fine-grained sandstone beds containing cross-laminations produced by ripples, climbing ripples, and oscillatory current ripples at the top of the succession (Figs. 7C and 7D). Large-scale, low-angle planar-tangential cross-stratification is present in the uppermost beds of a few of these sequences. Thicknesses of sequences range between ~2 m and ~15 m. Bioturbation is extremely rare; desiccation cracks are present locally but not abundant. At least 45 of these packages are present in the central part of the basin fill.

We interpret the bulk of the middle member of the Vizcachera Formation to consist of lacustrine deposits. The laminated silty parts of the section represent relatively profundal parts of the lake, where sediment was deposited below wave base in stagnant water. The upward-coarsening and -thickening packages represent deposits of progradational lake-margin deltas. The abundance of ripples and climbing ripples at the tops of the progradational sequences is consistent with rapid influx of sediment-laden waters and high sediment fallout rates, perhaps during floods (e.g., Ashley et al., 1982).

Notably, the Arizaro lacustrine system preserved in the middle member of the Vizcachera Formation was seemingly devoid of macrobiological activity. Macrofossils are absent, and even the profundal facies contain virtually no organic material

or bioturbation. We did not observe bioturbation in any of the Vizcachera Formation, nor did we find evidence for pedogenesis. Moreover, the system contains practically no calcium carbonate. In the modern Puna and Atacama Deserts, carbonate is typically linked to plant cover, and hyperarid areas thus lack carbonate (Quade et al., 2013). Together, the absence of carbonate and paleosols, bioturbation, organic material, and macrofossils suggests that, like today, the Miocene Arizaro Basin presented a landscape devoid of plant cover and lacking in biological activity. Evidence for frequent desiccation of the lakes (evaporites, desiccation cracks, and disrupted layering) is also rare. Instead, the lakes of the Arizaro Basin during deposition of the middle siltstone and sandstone member appear to have been relatively fresh, but not biologically productive.

One possible cause of the peculiar sterility of the environment is acidification of Arizaro lake waters, which is common in volcanic settings such as that surrounding the Arizaro Basin (e.g., Varekamp et al., 2000). Acid saline lake waters characteristically

Figure 5. Stratigraphic logs of sections TG and 12AR, in the lower coarse-grained member of the Vizcachera Formation. Note that previously, these rocks have been considered to be the Eocene Geste Formation, but that new detrital U-Pb zircon ages require that these strata be no older than ca. 23 Ma. Conglomerate clast-count data are shown in pie charts. Locations of sections are provided in Table DR1 (see text footnote 1) and shown crudely on Figure 2B. For definitions of lithofacies codes, see Table 1.

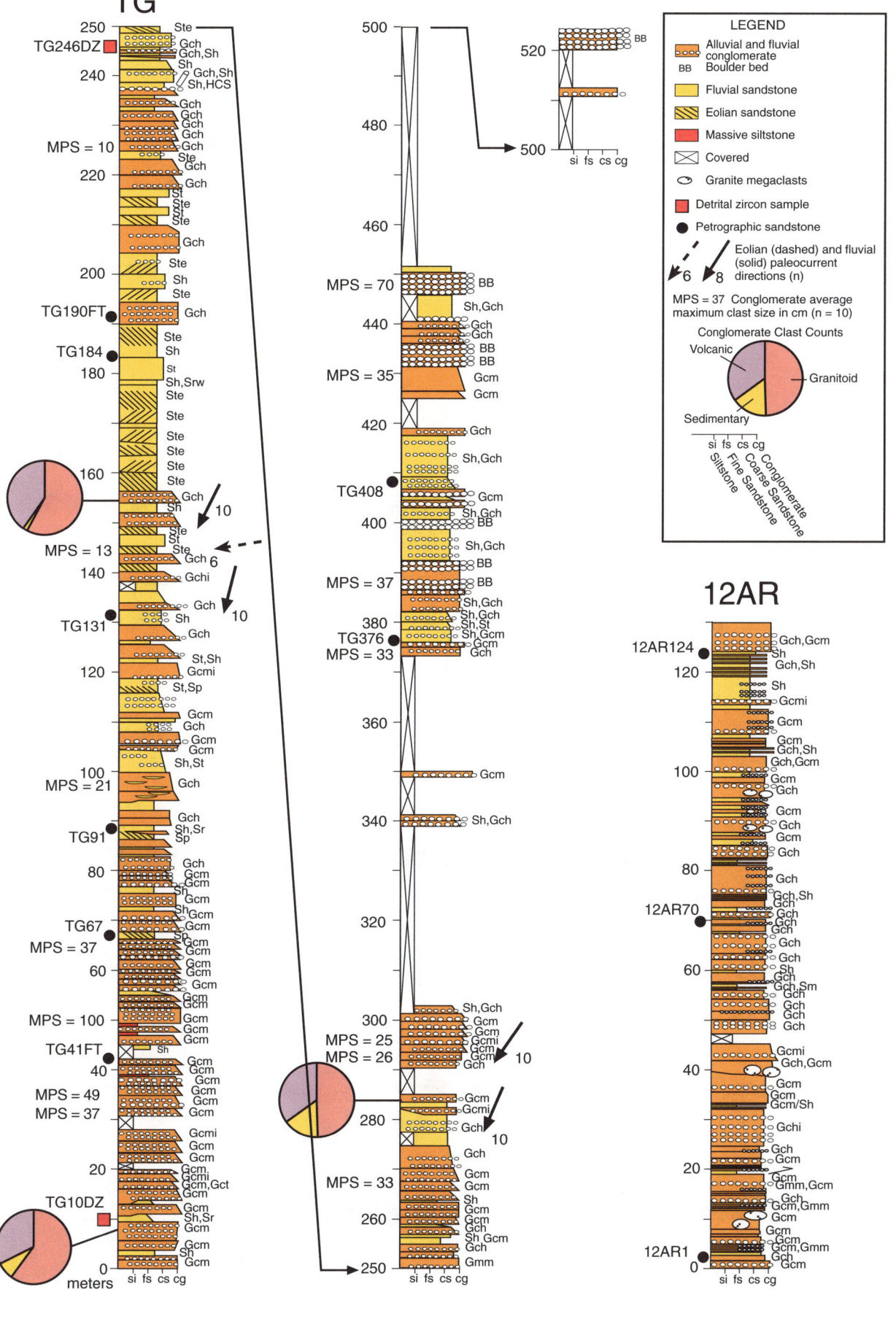

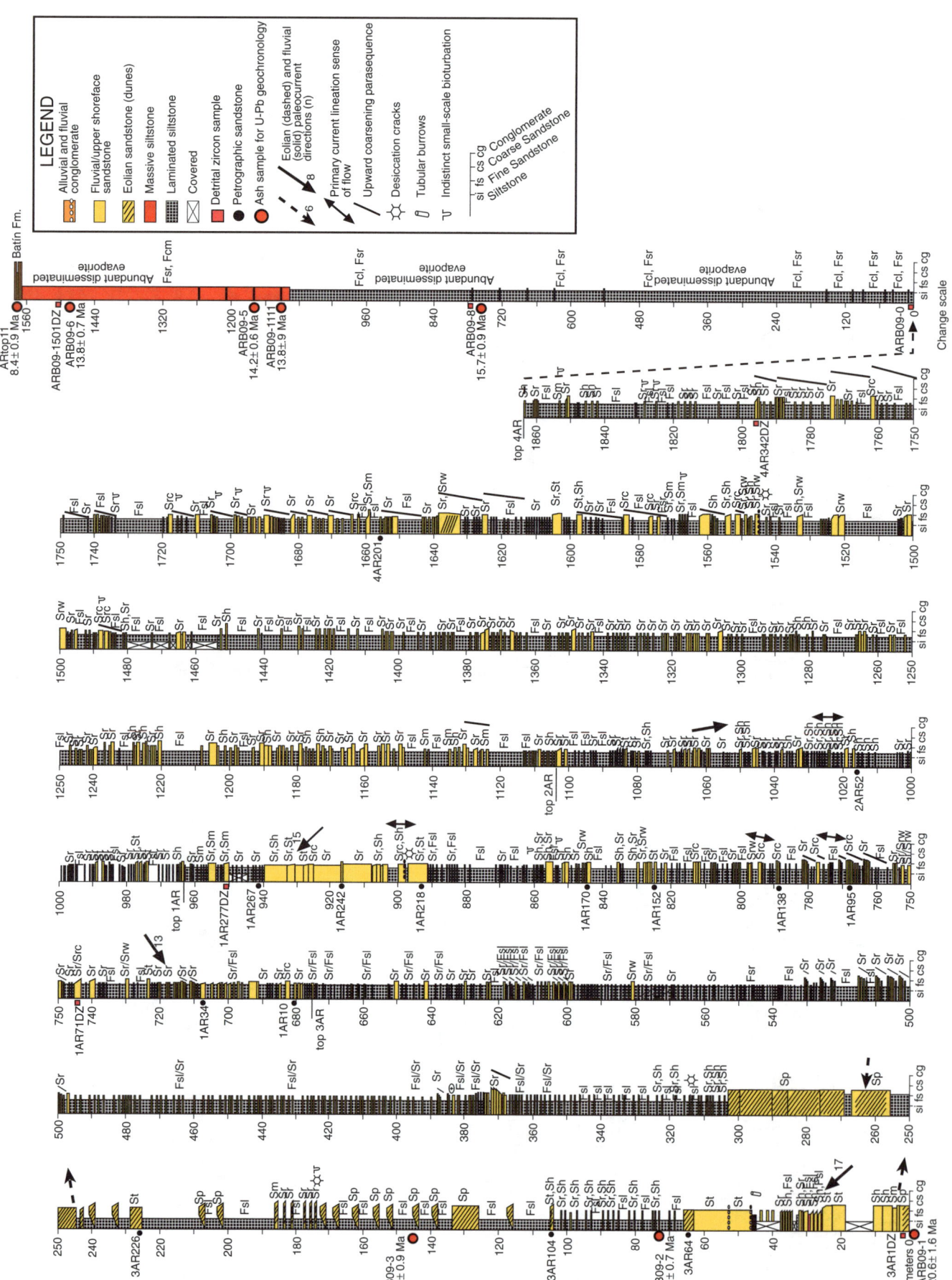

Figure 6. Stratigraphic log of sections 3AR, 1AR, 2AR, 4AR, and ARB09. Locations of sections are shown on Figure 2B and are provided in Table DR1 (see text footnote 1).

Figure 7. (A) Photograph of lower part of the middle sandstone and siltstone member (roughly the interval 700–800 m on section shown in Fig. 6), viewed toward the north. Range in background is Sierra de Macon. Note the remarkable lateral continuity of bedding. Three tents in left center for scale. (B) Four stacked progradational deltaic parasequences in the upper part of the middle sandstone and siltstone member (~45 m of section is visible). (C) Climbing ripple cross-lamination in upper part of progradational deltaic parasequence. (D) Symmetrical (oscillatory current) ripple cross-lamination in upper part of a shoaling-upward lacustrine parasequence in middle part of middle sandstone and siltstone member. Jacob staff is 1.5 m long, for scale.

are rich in dissolved Na-Cl-Mg-SO_4 and have high concentrations of Fe, Ca, Al, Si, Br, and Cu; however, they lack carbonate and bicarbonate (Risacher et al., 2002; Bowen and Benison, 2006). Acid lakes are highly toxic and commonly conducive to the deposition of hematitic jackets on sediment grains and the formation of red beds (Benison et al., 2007; Benison, 2008). In a synthesis study of >350 chemical analyses of acidic lakes in volcanic settings, Varekamp et al. (2000) found that none of the lakes studied had waters that are in equilibrium with secondary silicate minerals such as clays and zeolites. Petrographic analysis of Arizaro Basin sandstones (see following) documents abundant early (before deep burial and compaction) zeolite cements. Benison et al. (2007) documented small acid saline lakes in western Australia and showed that they produce sediments rich in evaporites and other evidence for complete desiccation, neither of which is present in the middle siltstone and sandstone member. In addition, distinctive early diagenetic minerals, including illite,

jarosite, and alunite, as well as halite and gypsum cements are characteristic of the Australian acid saline lake deposits, but they are absent from the middle member of the Vizcachera Formation. Thus, although the middle member has some attributes of acid saline lake deposits, such as hematitic red beds and practically no carbonate minerals, it lacks key features of extremely acidic lakes—the early diagenetic minerals in particular, as well as the textures documented by Benison et al. (2007). We therefore tentatively rule out low pH as the main reason for the apparent sterility of Arizaro lake waters.

A more plausible explanation is that the Arizaro Basin lakes were moderately to highly alkaline. Modern lakes in the Puna and neighboring Western Cordillera are almost all saline and neutral or alkaline (Risacher et al., 2002). Although Arizaro Basin lake-margin lithofacies do not contain chert (from magadiite), which is diagnostic of highly alkaline (pH > 9) lacustrine settings (Eugster, 1967), the abundance of early diagenetic zeolite cement is consistent with alkaline hydrology (Hay, 1977; Surdam, 1977; Eugster, 1986). The absence of Magadi-type chert in the Arizaro Basin lakes could be explained by the presence of boron-rich brines and/or minerals, which inhibit magadiite formation (Surdham and Sheppard, 1978; Eugster, 1986). Boron is abundant in felsic volcanic rocks such as those surrounding the Arizaro Basin, and late Neogene borates are present in economic quantities in salars of the eastern Puna (Alonso et al., 1991; Alonso, 1998). Moreover, the scarcity of gypsum and anhydrite in the middle siltstone and sandstone member is consistent with an alkaline setting (Eugster, 1986).

Upper Fine-Grained Member

The upper 1.5 km of the Vizcachera Formation consists of a monotonous succession of red and pink siltstone and shale, with common thin gypsum layers and veins. Alonso et al. (1991) reported ~1600 m of this unit in exposures directly east of the Salar de Arizaro, where it is dominated by halite and gypsum. Many layers are disrupted and massive. Rippled and horizontally laminated siltstone is common, but we did not observe organized progradational sequences as in the middle member. We sampled several silty ash layers that contain abundant zircon and biotite. These rocks crop out poorly and are deeply weathered.

We interpret the upper member of the Vizcachera Formation as the deposits of very shallow ephemeral saline lakes (see also Alonso et al., 1991). The ubiquitous presence of gypsum, as discrete layers, dense clusters of veins, or finely disseminated crystals, suggests that the depositional environment was highly evaporative. Abundant gypsum could reflect low pH conditions (Eugster, 1986), but other minerals diagnostic of acid waters (e.g., jarosite and alunite; Benison et al., 2007) are absent. Disrupted strata are also suggestive of desiccation. The absence of organized lacustrine parasequences such as those in the middle member suggests that water depth was highly transient, and lake-margin deltas had insufficient time to form and prograde. The lithofacies assemblage of the upper fine-grained member of the Vizcachera Formation bears many similarities to facies documented in modern evaporative saline lakes by Benison et al. (2007). Because of their ephemeral (highly seasonal) hydrology, these lakes deposit radically different sedimentary facies in close proximity, including evaporative halite and gypsum crusts and buckled layers, rippled subaqueous silt and very fine-grained sand, laminated mudflat deposits, and local eolian deposits. Red- and pink-colored sediments are abundant in these modern lakes.

Batín Formation

A narrow, continuous belt of conglomerate, sandstone, and minor siltstone crops out along the eastern edge of the Arizaro Basin. This package of lithofacies is no more than a few tens of meters thick, dips gently eastward, and overlies the Vizcachera Formation in a low-angle unconformity. Donato (1987) used the term Batín Formation for these rocks. Blasco et al. (1996) mapped this unit as the Sijes Formation. The main lithofacies include well-organized, horizontally stratified and imbricated pebble to cobble conglomerate, horizontally laminated sandstone, massive and laminated red and gray siltstone, and thin yellow tuffaceous layers. We interpret these deposits as low-sinuosity coarse-grained fluvial deposits.

STRUCTURE AND ARCHITECTURE OF ARIZARO BASIN

The three-dimensional shape of the Arizaro Basin is not readily apparent owing to the absence of subsurface data, relatively low topographic relief (a few hundred meters) in the exposed basin fill, and the generally low-angle dip of bedding. Nevertheless, it is possible to make some reasonable inferences about the basin's areal extent, three-dimensional shape, and relationship to key structures in the region. Cross sections through or near the basin have been published by Blasco et al. (1996), Zappettini and Blasco (2001), Coutand et al. (2001), and Jordan and Mpodozis (2006). Boyd (2010) produced a cross section of the eastern half of the basin. Coutand et al. (2001) depicted the basin fill as an abruptly westward-thickening wedge of sediment in the footwall of a major thrust fault along the western margin of the modern Salar de Arizaro. Although this is a plausible configuration for the basin fill, no direct observations can be made to test it because the basin fill does not resurface westward from beneath the modern salar. Jordan and Mpodozis (2006) depicted the basin fill beneath the modern salar as a broad, symmetrical synform truncated on its western side by a west-dipping, east-verging thrust fault with Lower Miocene volcanogenic rocks (ignimbrites and agglomerates of the Quebrada del Agua Formation; Zappettini and Blasco, 2001) in its hanging wall. Unpublished mapping and geochronology by L. Schoenbohm and B. Carrapa (2011) along the western flank of the modern salar document exposures of Quebrada del Agua Formation in the footwall beneath a west-dipping thrust fault that carries in its hanging wall Ordovician plutonic rocks overlain by Quebrada del Agua Formation. AFT data from the Ordovician rocks (sample locations indicated by yellow star on Fig. 2B) show that the most recent major cooling event was during the Eocene (Carrapa and DeCelles, this volume), suggesting that most fault displacement predated deposition of the Quebrada del Agua Formation. The eastward extent of the Quebrada del Agua Formation is not clear because it is buried by Quaternary alluvium.

Exposures along the western flank of the Sierra de Macon are generally poor, but satellite images and field observations suggest that the Vizcachera Formation rests unconformably on Ordovician granite and continues northward from our section TG as part of a several-kilometer-thick, westward-dipping succession. Farther west along the eastern margin of the Salar de Arizaro, the evaporitic upper fine-grained member of the Vizcachera Formation dips gently westward beneath the salar. Following Zappettini and Blasco (2001) and Jordan and Mpodozis

(2006), we presume that these strata, as well as underlying beds of the lower and middle Vizcachera Formation, continue westward beneath the modern salar (Fig. 2C).

Along the eastern flank of the Sierra de Macon, we observed a west-dipping fault zone with granitic rock in its hanging wall topographically above poorly exposed Vizcachera Formation in its footwall, suggesting the presence of a west-dipping thrust fault that was active after deposition of the Vizcachera Formation (Fig. 2C). In the central part of the basin, east of the Sierra de Macon, outcrops of the Vizcachera Formation expose numerous north-south–trending folds with vertical hinge surfaces and chevron geometries (Figs. 2C and 4A; Blasco et al., 1996; Boyd, 2010). Deformation dies out toward the perimeter of the basin. Along the eastern limit of Vizcachera Formation outcrops, the contact between the upper fine-grained member of the Vizcachera Formation and the overlying Batín Formation is a low-angle unconformity, and the Batín Formation dips eastward at an angle of <5°.

The eastern extent of the basin fill is unknown owing to burial beneath the Salar de Pocitos, but the presence of a thick succession of Ordovician low-grade metasedimentary rocks in the Cordon de Pozuelos (Fig. 2B) suggests that the Cenozoic section is truncated by a major west-verging thrust fault (Blasco et al., 1996; Coutand et al., 2001). Apatite fission-track and provenance data together with the presence of a syntectonic growth structure in the Eocene Geste Formation in the Cordon de Pozuelos demonstrate that the Ordovician rocks that constitute the range were exhumed and exposed at the surface by late Eocene time (Carrapa and DeCelles, 2008). Combined with similar evidence for Eocene exhumation along the western margin of the Arizaro Basin (discussed previously), this suggests that the faults along and near the basin margins were mainly active more than 15 m.y. before the Miocene onset of Arizaro Basin deposition, and that fault activity during the Miocene was minor.

Our cross section is restricted to the shallow crust because of the lack of subsurface data. It is clear, however, that deformation of the Arizaro Basin fill is concentrated in the center of the basin. The difference between deformed and undeformed bed lengths yields a shortening estimate of ~4.1 km. The Chamberlin (1919)–Dahlstrom (1990) method of calculating depth to detachment yields values of ~20–40 km, depending on subtle changes in shapes of the folds (e.g., Bulnes and Poblet, 1999). If the folds are cored by thrust faults branching upward from a regional detachment, the faults must be either improbably steep (for a bivergent pop-up geometry), or they must be part of an imbricate system rooted deep (>20 km) within the crust. Given the small displacements on the faults, the overall small amount of total shortening, and the relatively short wavelengths of many of the surface folds (e.g., Fig. 4A), we view neither of these possibilities as likely. A third possibility is that the basin fill is detached at a very shallow level, perhaps in evaporitic shales beneath the lower coarse-grained member. This is not consistent with the presence of crystalline basement rocks in the cores of several of the folds, including the Sierra de Macon.

An alternative explanation for the deformation of the basin fill is that it has been shortened by contractional fiber stresses that developed along the inner (upper) arc of a regionally downward-flexing panel of crust. The circular arc neutral surface subtending the angle corresponding to a 100 km chord length (the approximate east-west diameter of the basin) with a maximum 4 km amplitude of downward crustal flexure (an upper limit for basin subsidence) would be at a depth of ~11.3 km in order to produce 4 km of horizontal shortening at the surface. The other characteristic feature of this type of deformation would be that the displacements on the structures would increase upward, toward the center of curvature of the downward-flexing crustal arc. Although this explanation is somewhat speculative, it meets the conditions of the surface geology without calling for extraordinary structures and is consistent with our interpretation of the mode of basin formation, as discussed later herein.

PROVENANCE OF ARIZARO BASIN SANDSTONE AND DETRITAL ZIRCONS

Sandstone Petrography

Twenty-one samples of medium- to fine-grained sandstone were collected throughout the sandy portions of the Arizaro Basin fill (see Figs. 5 and 6 for sample locations) for petrographic analysis. Each thin section was stained for potassium feldspar and calcium plagioclase, and 450 grains were counted according to the Gazzi-Dickinson method (Ingersoll et al., 1984). Petrographic parameters identified in these point counts are listed in Table 2, and recalculated data are provided in Table 3.

The most abundant grain types in Arizaro Basin sandstones are, in order of decreasing abundance, monocrystalline quartz (Qm); plagioclase (P, predominantly calcium-rich, commonly zoned); potassium feldspar (K, mainly orthoclase, with a few microcline grains); and various types of volcanic lithic fragments. The volcanic grains are dominated by intermediate lathwork grains in which lath-shaped plagioclase crystals are set in a glassy matrix. Other volcanic grains have microlitic, glassy, and mafic textures/compositions. Sedimentary and metasedimentary lithic grains are relatively rare. The accessory mineral suite consists of epidote/zoisite/clinozoisite, tourmaline, muscovite, biotite, amphibole, and pyroxene. Cements are dominated by zeolites (mainly stilbite and heulandite), with some calcite and quartz. Reddish stain is common, mainly owing to the presence of hematite.

Figure 8 shows ternary diagrams for Arizaro Basin sandstones. These sandstones are arkosic to subarkosic in composition, and they plot within the basement uplift and dissected arc fields of Dickinson (1985). Average Qm, F, Lt = 41, 45, 14, and Qt, F, L = 46, 45, 9 (see Table 2 for definitions of petrographic parameters). The monomineralic fraction is dominated by quartz and plagioclase in subequal proportions (Qm, P, K = 49, 40, 11), and the lithic fraction consists mainly of volcanic grains with a few metasedimentary grains—mainly phyllite and quartz

TABLE 2. MODAL
PETROGRAPHIC POINT-COUNTING PARAMETERS

Symbol	Description
Qm	Monocrystalline quartz
Qp	Polycrystalline quartz
Qpt	Foliated polycrystalline quartz
Qms	Monocrystalline quartz in sandstone or quartzite lithic grain
C	Chert
S	Siltstone
Qt	Total quartzose grains (Qm + Qp + Qpt + Qms + C + S)
K	Potassium feldspar (including perthite, myrmekite, microcline)
P	Plagioclase feldspar (including Na and Ca varieties)
F	Total feldspar grains (K + P)
Lvm	Mafic volcanic grains
Lvf	Felsic volcanic grains
Lvv	Vitric volcanic grains
Lvx	Microlitic volcanic grains
Lvl	Lathwork volcanic grains
Lv	Total volcanic lithic grains (Lvm + Lvf + Lvv + Lvx + Lvl)
Lsh	Mudstone
Lph	Phyllite
Lsm	Schist (mica schist)
Lc	Carbonate lithic grains
Lm	Total metamorphic lithic grains (Lph + Lsm + Qpt)
Ls	Total sedimentary lithic grains (Lsh + Lc + C + S + Qms)
Lt	Total lithic grains (Ls + Lv + Lm + Qp)
L	Total nonquartzose lithic grains (Lv + Ls + Lph + Lsm + Lc)

Accessory minerals (in decreasing order from most abundant):

Biotite

Tourmaline

Amphibole

Muscovite

Epidote/zoisite

Chlorite

Magnetite

tectonite (average Lm, Lv, Ls = 25, 68, 07). The only significant trend in composition is an up-section increase in the amount of K-feldspar that is documented in the main part of the basin fill in sections east of the Sierra de Macon. Conglomerates in the lower member of the basin fill (formerly mapped as Eocene Geste Formation) along the western flank of the Sierra de Macon are dominated by granitoid and volcanic clasts, with small amounts of low-grade metasedimentary clasts (Fig. 5).

Detrital Zircon and Tephra U-Pb Ages

Eight samples of medium-grained sandstone were processed by standard methods for retrieving dense minerals, and detrital zircon grains were separated from these concentrates. Zircons were mounted in epoxy, polished, and analyzed for U-Pb ages by laser ablation–multicollector–inductively coupled plasma mass spectrometry (LA-MC-ICPMS) at the University of Arizona LaserChron Center. The methods employed are described in Gehrels et al. (2008). A total of 661 grains produced data of sufficient precision for geochronological interpretation. Analyses that yielded isotopic data of acceptable discordance, in-run fractionation, and precision are shown in Tables DR2 and DR3.[1] Because $^{206}Pb/^{238}U$ ages are generally more precise for younger ages, whereas $^{206}Pb/^{207}Pb$ ages are more precise for older ages, we report $^{206}Pb/^{238}U$ ages up to 1000 Ma and $^{206}Pb/^{207}Pb$ ages if the $^{206}Pb/^{238}U$ ages are older than 1000 Ma (Gehrels et al., 2008). These analyses are plotted on relative age-probability diagrams (Fig. 9), which represent a sum of the probability distributions of all analyses from a sample, normalized such that the areas beneath the probability curves are equal for all samples depicted in the figure. Age peaks on these diagrams are considered robust if defined by several analyses.

We also collected samples for U-Pb geochronology from ash beds throughout the stratigraphic section. These were processed similarly to the detrital zircon samples, and clear euhedral zircons were picked, mounted in epoxy, polished, and analyzed for U-Pb ages by LA-MC-ICPMS at the LaserChron Center. Mean age plots for these are provided in Figure 10. Minimum age clusters from the detrital age distributions and igneous zircon age clusters are shown together in stratigraphic order in Figure 11.

The detrital zircon ages for Arizaro Basin samples are dominated by clusters in the 20–40 Ma, 250–350 Ma, and 465–550 Ma ranges. Grains with 60–90 Ma ages as well as a broad scatter of ages >700 Ma are common but not abundant enough to define prominent age peaks (Fig. 9; Table DR2 [see footnote 1]). Zircons with latest Oligocene–early Miocene ages are present in all samples of the Vizcachera Formation. Sample 9AR386 from rocks formerly mapped as Vizcachera Formation along the southwestern margin of the Salar de Arizaro contains a much larger

---

[1]GSA Data Repository Item 2015009, Thermochronological and geochronological information from Arizaro Basin, Argentina, is available at www.geosociety.org/pubs/ft2015.htm, or on request from editing@geosociety.org or Documents Secretary, GSA, P.O. Box 9140, Boulder, CO 80301-9140, USA.

TABLE 3. RECALCULATED MODAL PETROGRAPHIC POINT-COUNT DATA

Sample	Qm (%)	F (%)	Lt (%)	Qt (%)	F (%)	L (%)	Qm (%)	P (%)	K (%)	K:K+P
TG67	34	21	45	67	21	12	61	29	10	0.26
TG91	47	40	13	49	39	11	55	44	01	0.02
TG131	44	50	06	47	50	03	47	53	00	0.00
TG184	35	58	07	37	57	05	38	45	18	0.28
TG376	41	55	04	42	55	03	43	51	06	0.10
TG408	36	55	08	37	55	07	40	56	04	0.07
3AR67	42	42	16	47	41	12	50	50	00	0.01
3AR104	42	44	14	44	43	12	49	51	00	0.01
3AR226	33	58	09	36	58	06	36	63	01	0.01
1AR10	35	60	05	37	60	04	37	37	26	0.41
1AR34	41	55	04	43	55	02	43	31	27	0.47
AR95	42	46	12	45	46	09	47	40	13	0.24
AR138	49	40	11	52	40	08	55	28	17	0.37
AR152	55	34	11	59	34	07	61	25	14	0.36
AR170	51	32	17	53	32	15	61	26	13	0.33
AR218	32	51	17	33	51	16	39	49	12	0.20
AR267	37	49	14	40	48	12	43	36	20	0.36
AR277DZ	50	35	15	56	35	09	59	27	14	0.35
2AR52	48	36	15	53	36	11	57	24	19	0.44
4AR201	39	45	16	43	45	12	47	46	07	0.13
12AR124	41	33	26	49	33	18	55	27	18	0.40
Averages	42	45	14	46	45	09	49	40	11	0.23

fraction of mid-Paleozoic grains and no Cenozoic-age grains younger than ca. 33 Ma.

Ash layers throughout the Vizcachera Formation produced well-defined zircon age clusters (Figs. 10 and 11). Sample ARB09-1 was collected from an ash bed, ~1 m thick, directly below the lowermost cliffs of fluvial and eolian sandstone in the lower coarse-grained member. This sample yielded eight zircon crystals with a mean age of 20.4 ± 1.6 Ma. Sample ARB09-2 was collected from a biotitic ash layer at a stratigraphic level a few meters above the top of the lower coarse-grained member. This sample yielded a mean U-Pb zircon age on 15 crystals of 18.8 ± 0.7 Ma. Sample ARB09-3 was collected from a 0.2-m-thick lenticular ash deposit within the starved dune/mudflat lithofacies assemblage at about the 145 m level of the stratigraphic section (Fig. 6). Fourteen crystals from this sample produced a mean U-Pb age of 17.9 ± 0.9 Ma. Sample ARB09-8 was collected from an ~5-cm-thick ash layer in the middle of the upper fine-grained member. This sample yielded 13 grains with a mean U-Pb age of 15.7 ± 0.9 Ma. Sample ARB09-1111 was collected from a sandy tuff layer and yielded a mean age of 13.8 ± 1.0 Ma from six crystals. Sample ARB09-5 produced 13 zircon crystals with a mean age of 14.1 ± 0.6 Ma from an ~10-cm-thick impure sandy biotitic ash layer. Sample ARB09-6, from an impure, biotitic, sandy ash layer, ~2–5 cm thick, in the upper part of the upper fine-grained member, produced a mean U-Pb age of 13.8 ± 0.7 Ma

from 14 crystals. The three samples ARB09-1111, ARB09-5, and ARB09-6 have statistically indistinguishable ages, despite the fact that they were collected over a stratigraphic interval of >1300 m. This suggests that the rate of sediment accumulation was rapid. Sample ARtop-11 was collected from an ~20-cm-thick sandy ash layer in the Batín Formation near the top of the upper fine-grained member and produced nine U-Pb ages with a mean of 8.4 ± 0.9 Ma. Alonso et al. (1991) reported a zircon fission-track age of 10.8 ± 2.0 Ma from an ash sample in the upper member of Vizcachera Formation that was collected ~50 m down section from our ARtop-11 collection locality.

Youngest age clusters in detrital zircon samples that represent plausible depositional ages were recovered from seven sandstone samples, and these are plotted together with tuff age populations in Figure 11. Samples TG10 and 12AR70 were collected from the lower part of the lower coarse-grained member on opposite sides of the Sierra de Macon, and they produced minimum age clusters in the 23–24 Ma range. Sample 3AR1 was collected from a medium- to coarse-grained sandstone just above the base of the lower coarse-grained member in the central part of the basin. This sample produced a minimum detrital zircon age cluster (six grains) with a mean of ca. 19 Ma. Sample TG246 was collected from the upper part of the lower coarse-grained member along the western flank of the Sierra de Macon and produced a minimum age cluster at 21 Ma. Sample 4AR342 was collected

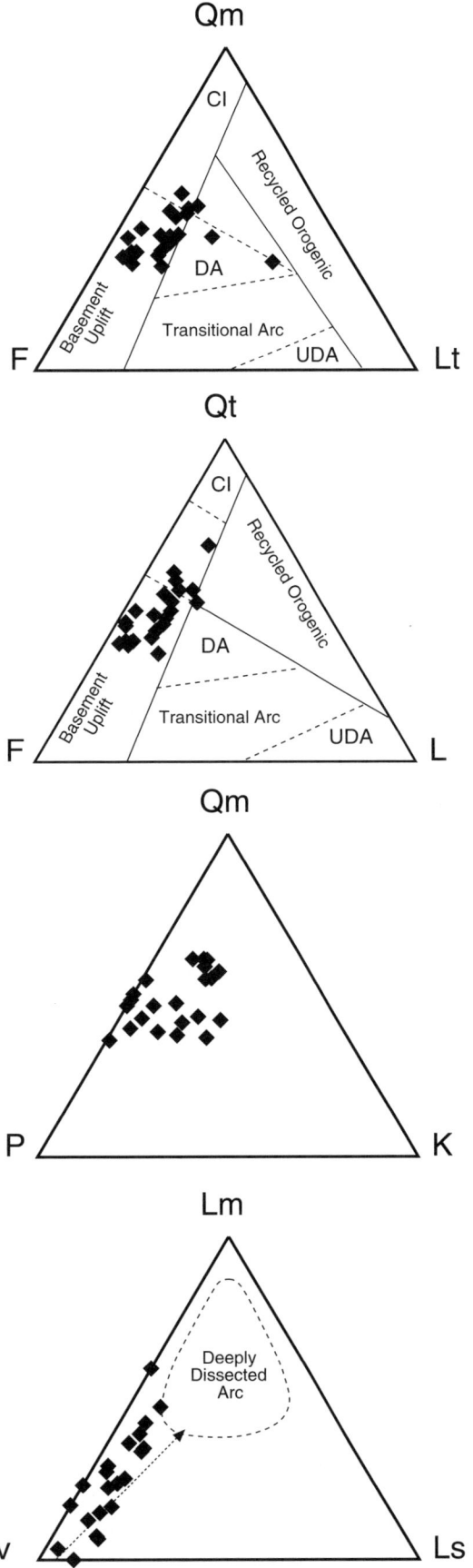

from the middle siltstone and sandstone member; the age peak of its minimum age cluster is ca. 20 Ma. Sample ARB09-1501 was collected from near the top of the upper fine-grained member. Two distinct early to middle Miocene age peaks were found in this sample, with the minimum age cluster at ca. 15 Ma.

Samples of ash and sandstone collected from section 9AR along the southwestern edge of the modern Salar de Arizaro demonstrate unequivocally that these rocks are much older than the Arizaro Basin fill. Two separate ash layers produced mean ages of 37–34 Ma (Fig. 10). Zappettini and Blasco (2001) reported a K/Ar age of 27 ± 1 Ma from a tuff in the upper part of this succession. Detrital zircons from sample 9AR386 define a minimum age cluster of ca. 37 Ma, and large clusters of ages in the 250–325 Ma and 480–517 Ma ranges (Fig. 9). The clusters of Eocene and late Paleozoic grains in this sample distinguish it from the Miocene Arizaro Basin samples and suggest that a large fraction of the zircons was derived from sources to the west and southwest. The Eocene age of these strata is similar to that of the Geste Formation of the western Eastern Cordillera (Pascual, 1983; Alonso et al., 1988; DeCelles et al., 2007), and the Quiñoas Formation in the Salar de Frailes area roughly 100 km to the south-southwest (Kraemer et al., 1999; Adelmann, 2001; Voss, 2002; Carrapa et al., 2005).

Provenance Interpretation

Modal petrographic point-count data from Arizaro Basin sandstones indicate continental block and dissected magmatic arc provenance (Fig. 8), consistent with derivation from granitic rocks of the Sierra de Macon and volcanic rocks of the surrounding Miocene magmatic arc, as well as igneous rocks farther west in northern Chile. The modal sandstone petrographic data are restricted to the lower two, relatively sandy members of the Vizcachera Formation. Together with abundant granitic clasts in the conglomerates of the lower coarse-grained member (Fig. 5), the petrographic data suggest that the main source for the basin center sections was the Sierra de Macon, which is composed of medium- to coarse-grained granodiorites, granites, and tonalites (Poma et al., 2004). Plagioclase, quartz, and alkali feldspar are the predominant minerals in these granitoid rocks, and biotite and amphibole are also present. Accessory phases include allanite, pyroxene, zircon, and tourmaline, and epidote is a common secondary replacement (Poma et al., 2004). All of these minerals are present in the thin sections that were point-counted.

The zircon age populations are consistent with sediment sources in the roughly coeval Miocene magmatic arc, Ordovician

Figure 8. Ternary diagrams illustrating recalculated modal petrographic data from Arizaro Basin sandstones. All parameters are defined in Table 2. Provenance fields are from Dickinson (1985), abbreviated as follows: CI—craton interior; DA—dissected arc; UDA—undissected arc.

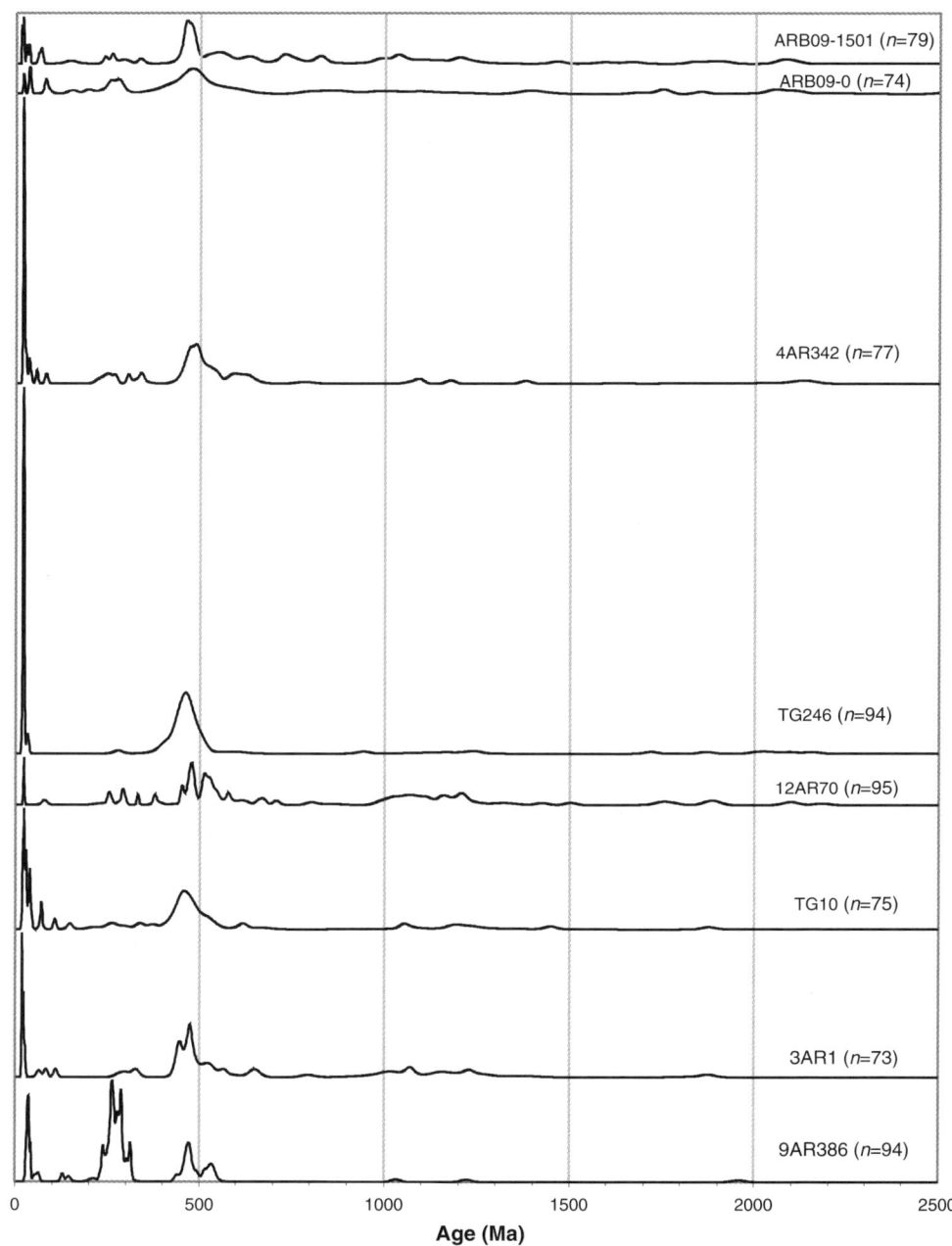

Figure 9. Relative probability plots for detrital zircon U-Pb ages from sandstones in Arizaro Basin. See Figures 5 and 6 for sample stratigraphic locations. *n*—number of analyses per sample. Plots are normalized such that the areas beneath each curve are equal. See Table DR2 for data (see text footnote 1).

granitic rocks of the Ocloyic orogenic event, and recycled sedimentary/metasedimentary sources in the Puna. Major sources in the Eastern Cordillera are unlikely because the latter are characterized by well-defined detrital zircon age peaks in the Ordovician and late Proterozoic, corresponding to the Ocloyic and Sunsas orogenic events (DeCelles et al., 2007, 2011; Einhorn et al., this volume; Bahlburg et al., 2009). Although grains of these ages are present, they do not define strong discrete populations in the age spectra (Fig. 9). The Sierra de Macon granitic rocks have been dated by ^{40}Ar/^{39}Ar at ca. 483 Ma (Koukharsky et al., 2002), and similar-aged rocks are present throughout the region (Poma et al., 2004). Sources of the Carboniferous- to

Permian-age zircons are available directly to the west of the modern Salar de Arizaro (Fig. 2B), north of the Sierra de Macon, and at more distal locales in northern Chile and westernmost Argentina (Poma et al., 2004). Cretaceous-age detrital zircons in the Arizaro Basin fill probably were derived from Cretaceous and Paleogene sedimentary and volcaniclastic rocks that crop out in the Cordillera de Domeyko in northern Chile. The absence of the Carboniferous–Permian and Cretaceous age populations in the samples from the western side of the Sierra de Macon, which instead are dominated by the Ordovician population, suggests that the Sierra de Macon area received sediment from local sources, while the rest of the basin received sediments from

DeCelles et al.

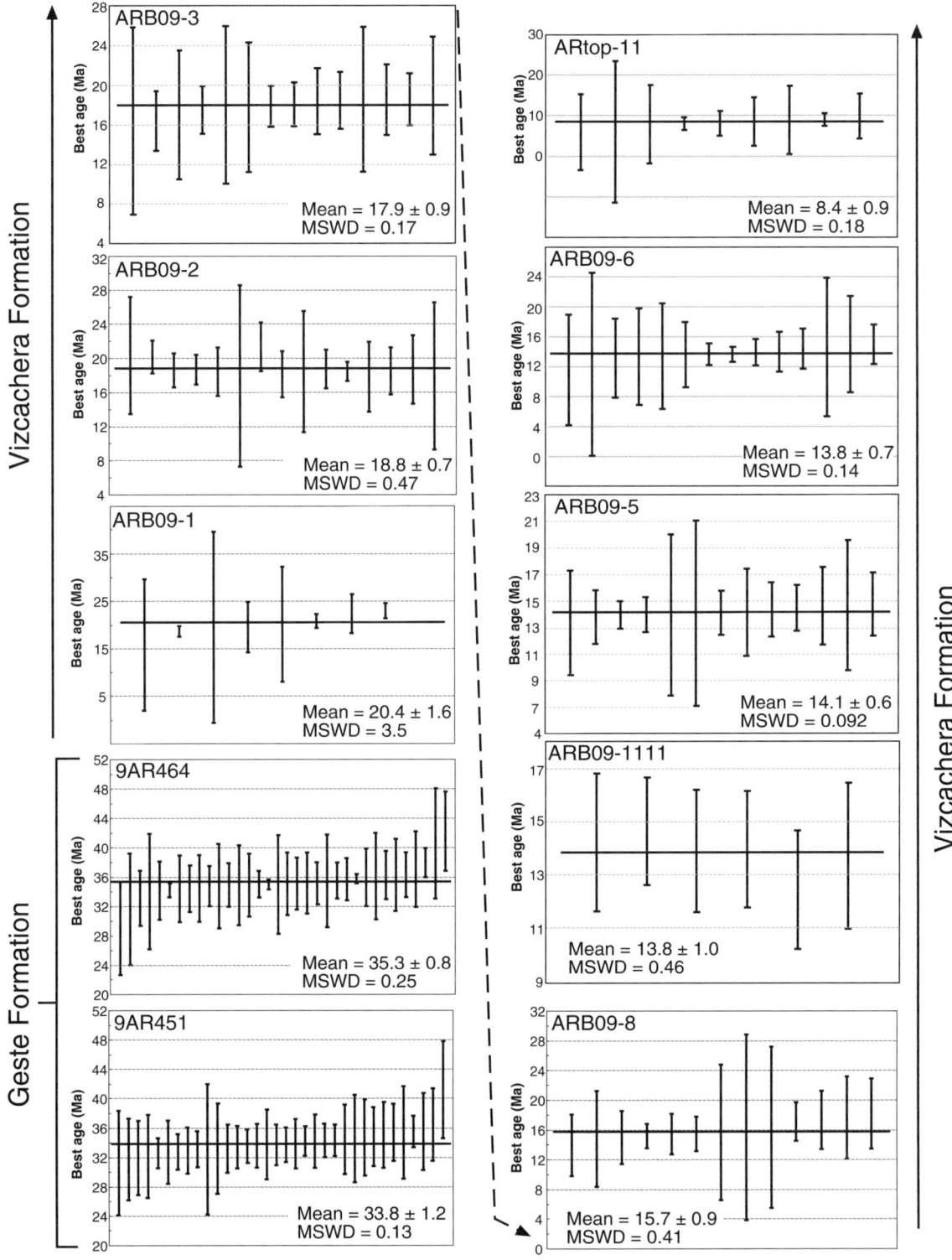

Figure 10. Mean U-Pb age (Ma) plots of zircons recovered from tuff layers in Arizaro Basin. Uncertainties include random and systematic errors. Error bars represent 2σ. See Table DR3 for data (see text footnote 1). MSWD—mean square of weighted deviates.

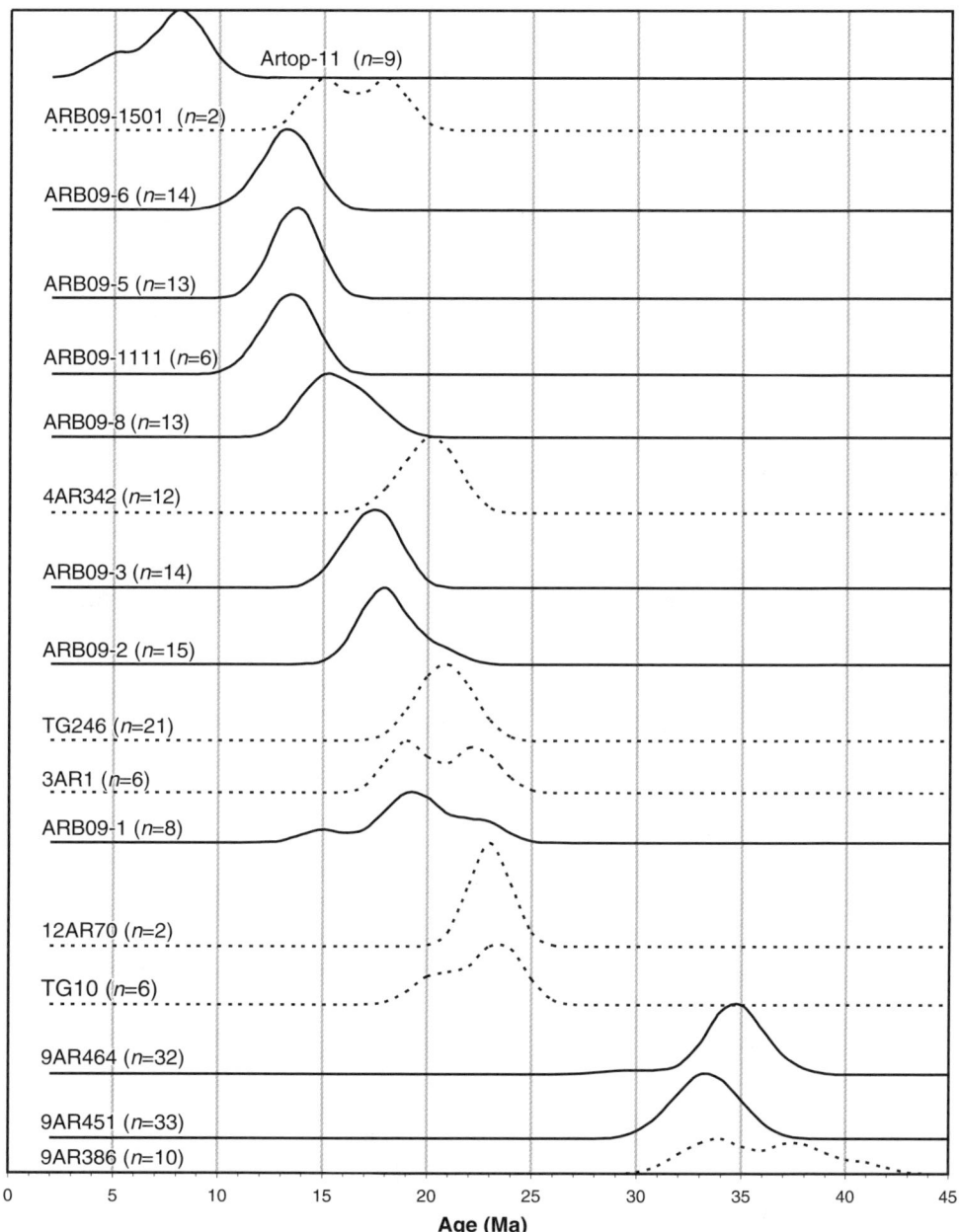

Figure 11. Relative probability plots for youngest detrital populations (dashed curves) and tuff samples (solid curves) arranged in stratigraphically upward-younging order. Not shown is sample ARB09-0, which lacks a young age population. *n*—number of grains/crystals used in each plot. See Tables DR2 and DR3 for data (see text footnote 1).

farther afield, possibly from sources as far west as the Cordillera de Domeyko. These distally derived grains could have been transported by rivers entering the Arizaro Basin, or perhaps by wind. If the former, then we may infer that the drainage catchment area that supplied water to the Arizaro lakes reached at least 100 km westward into the Cordillera de Domeyko.

Notably, the samples that were collected from the Eocene strata in section 9AR exhibit detrital zircon age spectra that are distinct from those of the Miocene Arizaro Basin fill, with much more abundant late Paleozoic zircons. This further highlights the fact that these strata, cropping out along the southwestern flank of the modern Salar de Arizaro, are not part of the Vizcachera Formation.

AGE, SUBSIDENCE, AND EXHUMATION HISTORY

Chronostratigraphy based on new U-Pb ages from sandstones and volcanic ash layers shows that the Arizaro Basin fill is early through middle Miocene in age—ca. 20.6–8.4 Ma. This age range is broadly consistent with previously reported geochronological data (Alonso et al., 1991; Vandervoort et al., 1995; Jordan and Mpodozis, 2006). A surprising result of the detrital and tephra geochronology is that the rocks previously mapped as Eocene Geste Formation in sections TG and 12AR and in the lower part of section 3AR contain abundant detrital and ash zircons of early Miocene age (Figs. 10 and 11). Thus, the previously mapped outcrops of "Geste Formation" (e.g., Blasco et al.,

1996) are actually part of the Vizcachera Formation. Conversely, outcrops along the southwestern flank of the modern Salar de Arizaro (sections 9AR and 10AR, not presented in this paper) contain ash layers that yield Eocene–Oligocene U-Pb zircon ages (ca. 37–34 Ma; Fig. 10), indicating that previously mapped "Vizcachera Formation" in that area (Zappettini and Blasco, 2001) is partly time equivalent to the Geste Formation.

The Miocene age data provide a basis for backstripping analysis of the Arizaro Basin fill. A potential problem with backstripping of the Arizaro Basin is that the exposed section dips at a generally very low angle (except in the relatively deformed central part of the basin); consequently, our data come from outcrops that are distributed along a nearly 20-km-long traverse of the eastern part of the basin. The stratigraphic zonation described earlier herein, however, is present throughout the basin, and the structural cross section suggests that our composite measured section is representative of the central part of the basin fill. Our analysis utilizes Nestor Cardoza's OSXBackStrip program, which assumes Airy isostatic compensation. Input parameters for the analysis, along with generally accepted values of porosity-depth coefficients that were used, are provided in Table 4. Any backstripping analysis is subject to problems in assessing the paleoelevation or paleobathymetry of the depositional surface through time; this is particularly problematic in nonmarine basins (e.g., Jordan et al., 1988). Errors associated with paleoelevation estimates are not systematic and reliably quantifiable but are probably in the range of 500–700 m (J. Quade, 2013, personal commun.). Stable isotope paleoaltimetry on ash beds in Arizaro Basin suggests that the elevation was on the order of 3–4 km throughout deposition (Canavan et al., 2014; Quade et al., this volume), so we assume that elevation did not change significantly during deposition. Bathymetry is not incorporated into the analysis because the basin remained above sea level and was probably never filled with more than ~20 m of standing water (based on thicknesses of lacustrine parasequences).

Total backstripped sediment accumulation and the tectonic component of this total are plotted in Figure 12. The tectonic component is approximately half of the total and follows a sigmoidal pattern; the tectonic subsidence rate gradually increased between ca. 20.6 and 18 Ma, reached maximum rates between ca. 17 Ma and 16 Ma, and subsequently decreased to near zero by

ca. 14 Ma. Between ca. 17.9 and 15.7 Ma, total sediment accumulation rate reached a maximum of nearly 1 mm/yr, which is comparable to the fastest rates of sediment accumulation in flexural, strike-slip, and extensional basins (Allen and Allen, 2005; Xie and Heller, 2009).

Apatite fission-track data from granitic rocks in the Sierra de Macon and from granitic clasts in the lower Vizcachera Formation show Jurassic (Deeken et al., 2006), Cretaceous, and Eocene ages (Carrapa et al., 2009). Apatite (U-Th)/He ages from the Sierra de Macon and pebbles derived from it are Eocene and Oligocene and indicate low-magnitude (<4 km) Cenozoic exhumation. Inverse thermal modeling of AFT data from a granitic clast in the Vizcachera Formation (derived from the Sierra de Macon) shows that basement rocks in the Arizaro Basin experienced slow cooling between ca. 60 and 20 Ma, moderate heating between ca. 20 and 10 Ma (Carrapa et al., 2009), and final cooling during exhumation to the surface after ca. 10 Ma (see Fig. DR1 [see footnote 1]). The early Miocene heating event is interpreted to represent burial by Arizaro Basin strata. Subsequent Miocene–Pliocene cooling and exhumation removed most of these sedimentary strata from the Sierra de Macon area. Apatite fission-track cooling ages from the southwestern Arizaro Basin (Fig. 2B) indicate that rocks exposed at the surface today have not experienced temperatures greater than 110 °C since ca. 37 Ma (Carrapa and DeCelles, this volume). Overall, these data indicate cooling associated with variable but limited exhumation (<3–4 km) since Cretaceous time.

TABLE 4. INPUT DATA FOR BACKSTRIPPING ANALYSIS

Unit	Base (m)	Age base (Ma)	Top (m)	Age top (Ma)	ρ	c	φ
1	3445	20.6	3372	18.8	2680	0.39	56
2	3372	18.8	3299	17.9	2680	0.39	56
3	3299	17.9	810	15.7	2720	0.51	63
4	810	15.7	410	14.2	2720	0.51	63
5	410	14.2	90	13.8	2680	0.51	63
6	90	13.8	0	8.4	2680	0.51	63

Note: ρ—density (kg/m^3); c—porosity-depth coefficient (km^{-1}); φ—initial porosity (%).

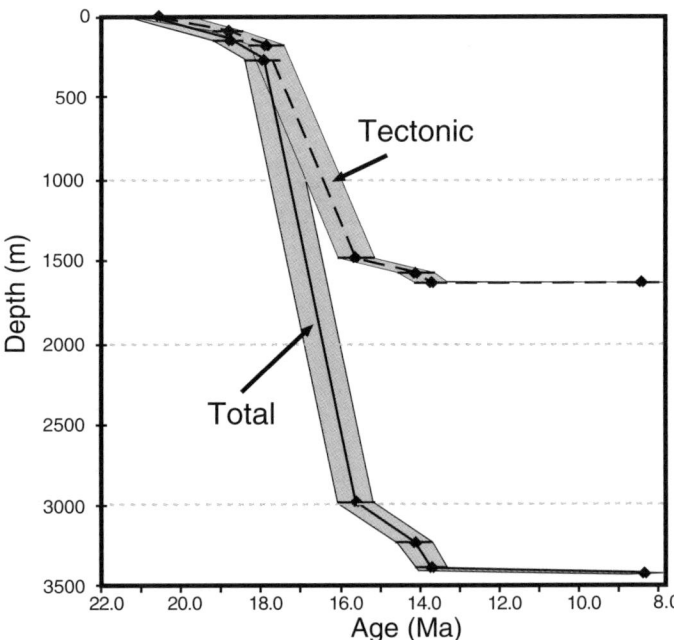

Figure 12. Geohistory (subsidence) curve for Arizaro Basin, showing total sediment accumulation and the tectonic component of subsidence. Error envelopes represent uncertainties of U-Pb zircon ages from tuffs (Fig. 10).

SUMMARY OF OBSERVATIONS AND INTERPRETATIONS

Key features in the history of the Miocene Arizaro Basin pertaining to possible geodynamic explanations are summarized as follows. The basin is a roughly circular depression, ~75–100 km in diameter, filled with ~3.5 km of nonmarine clastic sedimentary strata. The three-dimensional pattern of basin-fill thickness remains largely undocumented, but available information suggests that it is thickest in the interior part of the basin. Although thrust faults locally cut the basin fill, they had minor displacements during deposition of the Arizaro Basin fill, and thermochronological data suggest that the hanging walls of these thrust faults were rapidly exhumed during the Eocene, rather than the Miocene. From ca. 21 Ma to ca. 17.5 Ma, the basin was characterized by eolian and dry mudflat environments (Fig. 13A). Starved sandy dunes (probably barchans) migrated across the basin floor, which consisted of low-relief silty mudflats. The northern half of the basin was divided approximately in the middle by the narrow granitic ridge of the Sierra de Macon, which served as a source of coarse-grained alluvial fan and eolian deposits during the earliest part of the depositional record. From ca. 17.5 Ma to ca. 16 Ma, the basin was fed by small streams and rivers that were capable of maintaining an open-water lacustrine system. We find no evidence for large fluvial systems, but provenance data suggest that sand was transported into the main part of the Arizaro Basin from as far away as northern Chile. During this stage of basin development, the lake margin was an open sandy shoreface rimming a silty profundal depression with water generally <20 m deep (Fig. 13B). Small deltas prograded into the basin, depositing upward-coarsening deltaic parasequences. The lake waters were likely highly alkaline, inhibiting macrofaunal activity. Beginning ca. 16 Ma and recorded by the upper 1500 m of the basin fill, the basin became dominated by shallow, ephemeral, intensely evaporative playas (Fig. 13C). Volcanic ashes were deposited sporadically throughout the history of the basin. Stable isotope paleoaltimetry data suggest that the basin formed at elevations comparable to the present 3.5–4 km elevation (Canavan et al.,

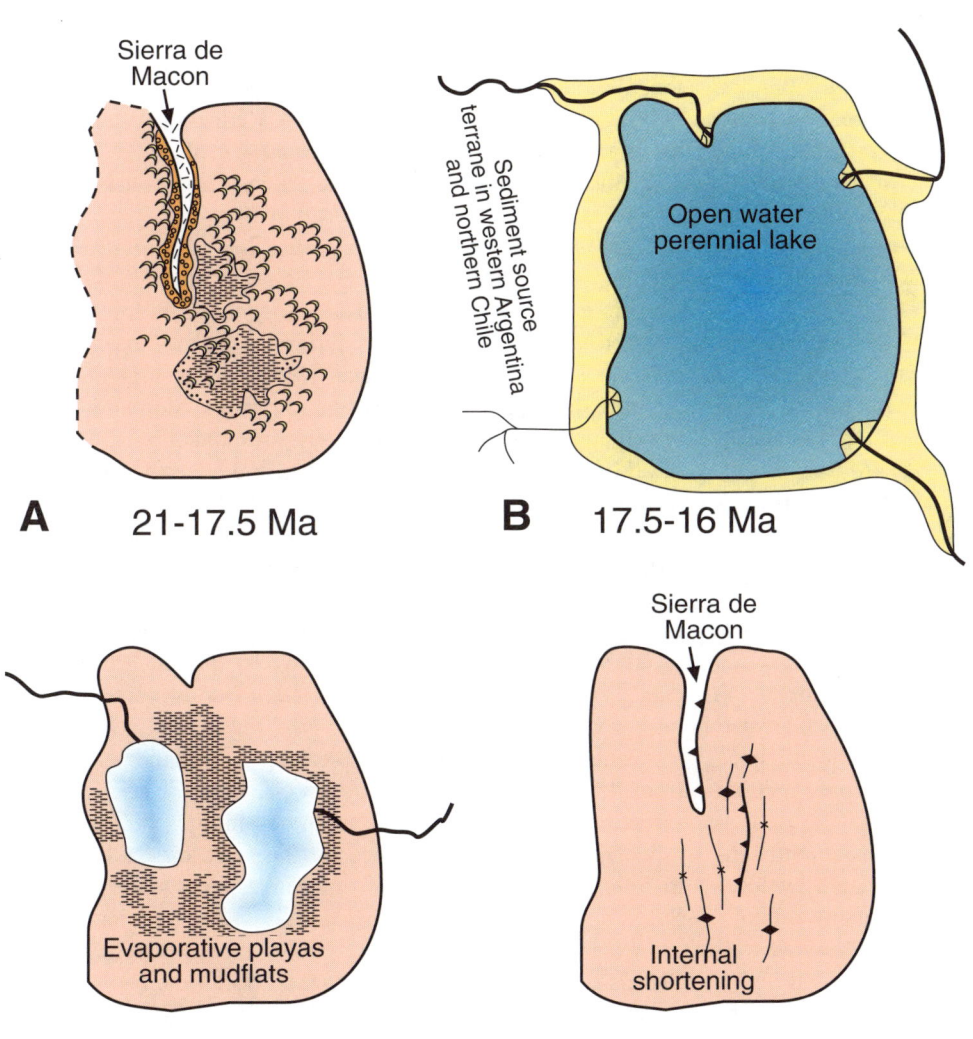

Figure 13. Sketch maps showing environmental changes that occurred during formation and filling of the Arizaro Basin. See text for discussion.

2014), and thermochronological data indicate that this region of the Puna was exhumed to shallow crustal levels during Cretaceous–Eocene time (Deeken et al., 2006; Carrapa et al., 2009). Soon after the basin was filled, its interior part was shortened, reactivating the Sierra de Macon ridge and forming trains of upright chevron-shaped folds (Fig. 13D). The central and eastern part of the Arizaro Basin fill today is uplifted (inverted) and being actively eroded. A significant proportion of basin inversion must postdate ca. 8.4 Ma, which is the age of the gently tilted Batín Formation along the eastern margin of the basin-fill outcrops.

GEODYNAMIC INTERPRETATION

The Arizaro Basin presents several features that are not explained by previously published models for basin evolution in contractional orogenic settings. Such basins are typically referred to as hinterland, wedge-top, extensional, or intra-arc basins, and substantial work has been done to document the typical features of each type of basin. Horton (2012) referred to basins isolated within the high central Andes as "hinterland basins" and noted that both the Altiplano basin in Bolivia and basins in northern Argentina initiated as low-elevation, regional foreland basins and were progressively incorporated into and uplifted within the interior of the orogenic belt as the strain front jumped eastward during late Eocene–early Oligocene time. The orogenic strain front in this part of the Argentine Andes also jumped eastward during late Eocene time (Carrapa and DeCelles, 2008; Carrapa et al., 2011a). However, this does not explain why a large saucer-shaped basin such as the Arizaro would form in the hinterland during the Miocene.

Extensional basins are associated with flanking normal faults and high-relief footwall topography, and many examples of extensional basins are present in the high Tibetan Plateau (Kapp et al., 2008; Taylor et al., 2012; Woodruff et al., 2013) and in the northern part of the Himalayan thrust belt (Murphy et al., 2002; Garzione et al., 2003; Saylor et al., 2010). The only significant extensional basin that has been documented in the Andes is the Callejon de Huaylas supradetachment basin in the Cordillera Blanca of Peru (McNulty and Farber, 2002; Giovanni et al., 2010; Horton, 2012). No large-displacement normal faults are documented in the Arizaro Basin region, and the basin fill does not thicken and coarsen systematically toward its margins.

The Arizaro Basin formed in a region that has been located directly adjacent to and within the Andean magmatic arc since middle to late Miocene time (e.g., Zappettini and Blasco, 2001; Kay and Coira, 2009; Fig. 2B); thus, it is reasonable to suggest that it could be an intra-arc basin (e.g., Smith and Landis, 1995). However, intra-arc basins are dominated by local extensional and transtensional structures and contain voluminous quantities of volcanogenic sediment and debris, as well as volcanic flows and hypabyssal intrusions (Smith and Landis, 1995; Busby, 2012). In contrast, the Arizaro Basin lacks extensional structural features, and Vizcachera Formation sandstone and conglomerate compositions are atypical of arc terranes.

Although volcanic grains dominate the lithic fraction, overall modal compositions are generally outside of the magmatic arc provenance field (Fig. 8). Detrital zircon ages implicate sources in a wide range of rocks throughout the Puna and the magmatic arc, but they are dominated by zircons derived from Paleozoic and Mesozoic rocks, rather than the coeval Miocene magmatic arc. Moreover, the Vizcachera Formation does not contain arc-proximal volcanogenic lithofacies (e.g., lava flows, pyroclastic flows). Instead, the main record of arc magmatism in the Arizaro Basin fill consists of distal air-fall tuffs. We therefore rule out an intra-arc basin interpretation.

Another possibility is that the Arizaro Basin formed on top of the active orogenic wedge in association with out-of-sequence thrusting. The term wedge-top basin refers to sediment accumulations that form above active thrust faults in the frontal part of the orogenic wedge (DeCelles and Giles, 1996; Ford, 2004; Sinclair, 2012), either as relatively isolated basins (e.g., piggyback basins; Ori and Friend, 1984) or as bodies of sediment that accumulate beneath a geomorphic surface that is connected to the open foreland basin (Horton and DeCelles, 1997). Several wedge-top accumulations have been documented in the Cenozoic stratigraphy of Bolivia and northern Argentina (e.g., Horton, 1998; Horton et al., 2002; Echavarria et al., 2003; Uba et al., 2006; Leier et al., 2010). By definition, wedge-top sediments are associated with growth structures (Anadón et al., 1986; DeCelles and Giles, 1996). In contrast, the Arizaro Basin lacks growth structures and began to form deep within the interior of the Andean orogenic belt, 15–20 m.y. after the orogenic strain front had migrated through the Puna and Eastern Cordillera. The basin was more than 200 km west of the orogenic strain front when it began to accumulate sediment during the early Miocene. Thus, it is not a typical foreland basin wedge-top accumulation. On the other hand, grain size coarsens toward the Sierra de Macon, which appears to be at least partially surrounded by a local halo of arkosic, conglomeratic alluvial-fan, and eolian facies derived from the Sierra de Macon granite (Fig. 4C). Together with the provenance data and detrital zircon ages, this information is consistent with the Sierra de Macon being a source for Arizaro Basin coarse-grained facies during the earliest Miocene. The range was buried, however, during deposition of the middle and upper parts of the Vizcachera Formation, as shown by thermochronological data discussed earlier herein (Fig. DR1 [see footnote 1]; Carrapa et al., 2009). Moreover, the size and lateral extent of the Miocene Arizaro Basin are much greater than would be expected from flexural loading by the narrow Sierra de Macon, which is restricted to the northern half of the basin and could not have influenced regional subsidence on the scale of the basin. We therefore rule out flexural subsidence in response to growth of the Sierra de Macon or other nearby ranges as the principal control on Arizaro Basin development. Rather, most exhumation of the Sierra de Macon took place during the Eocene and late Miocene; the range was buried during the early to middle Miocene, when the major pulse of basin subsidence was taking place (Carrapa et al., 2009; Fig. DR1).

Another possible explanation for the Arizaro Basin is that it developed in response to the growth of a dense, gravitationally unstable root beneath the basin, either in the lower crust or in the mantle lithosphere (Fig. 14; e.g., Houseman et al., 2000; Molnar and Houseman, 2004). Molnar and Houseman (2004) found that if the crust is weak and/or buoyant enough relative to the underlying mantle lithosphere, then paired Rayleigh-Taylor (RT) instabilities, or downwellings, develop on the flanks of the region of tectonic convergence. Conversely, where the crust is relatively strong, dense, or thin compared to the mantle lithosphere, a single RT instability will form beneath the center of the region of shortening. Pysklywec and Cruden (2004) and Göğüş and Pysklywec (2008) created analog and finite-element models to study the surface effects of dense root growth and removal. Both downwelling (or dripping) RT instabilities and delaminating slabs were modeled. Among the documented responses are a surface depression above the growing root and an annular peripheral bulge. The center of the depression experiences localized shortening at the acme of basin development, while the peripheral bulges are marked by local extensional faults. As the root begins to founder, the surface depression rebounds (Fig. 14D). The upper crust in the rebounding region experiences extension. Depression of the upper surface of the crust is symmetrical above a dripping instability but asymmetrical above a delaminated slab. As the delaminating slab pulls away from the upper crust, hot material wells upward to fill the growing gap, causing a large thermal perturbation. Deformation above the delaminating slab is mainly contractional above the part of the slab nearest the surface and extensional above the more distal region, where the slab has been completely removed. Recent work by Currie et al. (2011) suggests that a weak lithosphere allows for rapid inward flow to replace the material being removed in an RT instability; this suppresses the development of a surface depression, and in some cases actually thickens the crust. If the lithosphere is relatively strong and able to resist inward flow toward the region above the dripping instability, then a significant depression will form. These authors found that RT instabilities formed in the lower crust (as opposed to the mantle lithosphere) are more likely to produce surface depressions on the scale of the Arizaro Basin.

These modeling studies suggest that under certain circumstances, roughly circular surface depressions will form above gravitationally unstable roots. One common sequence of events is illustrated schematically (after Wang et al., this volume) in Figure 14. As the root grows and begins to pull down the overlying crust (Fig. 14A), the rate of subsidence increases up until the point at which the root begins to release from the lower crust and/or lithosphere. The basin fill will contract horizontally, causing internal shortening in the center of the basin (Fig. 14B), while the surrounding annular flexural bulge will experience minor extension (Pysklywec and Cruden, 2004). Rebound begins before the instability has completely pulled off of the crust or mantle lithosphere (Figs. 14C and 14D). This would predict that the basin fill should be at least partially inverted as it rises upward in response to removal of the dense root. The process is analogous to the

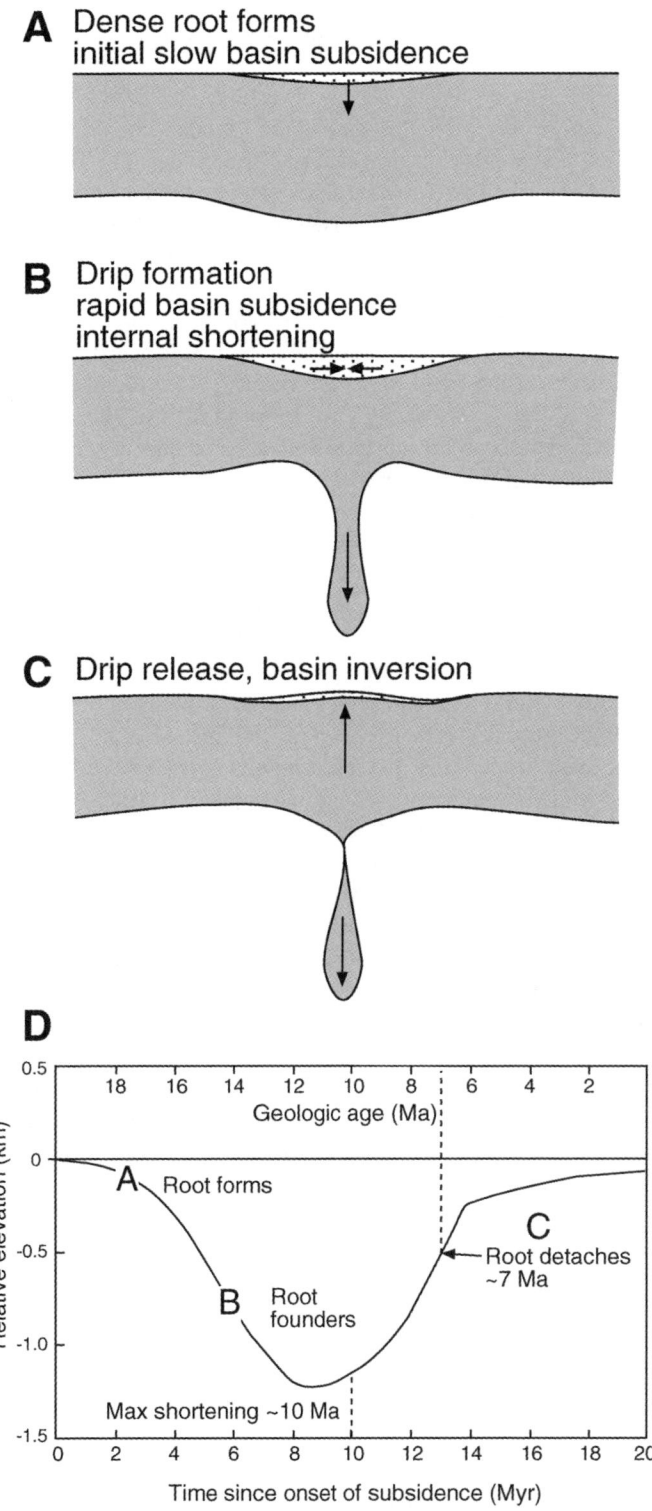

Figure 14. (A–C) Schematic, three-stage model for the formation of a sedimentary basin above a dense gravitational instability in the mantle lithosphere or lower crust (after Molnar and Houseman, 2004; Göğüş and Pysklywec, 2008). (D) Proposed three-stage trajectory of subsidence (and sediment accumulation) followed by uplift for a basin forming above the instability, abstracted from Wang et al. (this volume).

behavior of a "bobber" flotation device subjected to the activities of a curious fish, so we use the informal term "bobber basin" for this type of sedimentary basin.

The Arizaro Basin seems to exhibit several characteristics that are predicted by geodynamic models for depressions formed above growing gravitational instabilities: (1) The basin is generally circular to oval in shape, and its fill thickens inward. (2) The basin subsidence history is sigmoidal in shape (Fig. 12), as predicted for a basin in which subsidence is driven by downward viscous coupling between a Rayleigh-Taylor–type instability and the upper crust (Gögüs and Pysklywec, 2008; Currie et al., 2011). The relatively abrupt cessation of subsidence late in the history of the basin may have resulted from foundering and release of the RT instability. (3) The basin fill was shortened by internal folding and minor thrusting, accompanied by uplift of a narrow medial ridge of Ordovician basement. This is consistent with the acme of basin development roughly 10 m.y. after initial basin subsidence and prior to final removal of the root (Fig. 14D; Molnar and Houseman, 2004; Gögüs and Pysklywec, 2008). (4) Since late Miocene time, the central part of the basin fill has been uplifted and eroded, and the area of maximum Miocene subsidence is now almost completely inverted.

An evolutionary history of the Arizaro Basin in the context of models for convective removal of RT instabilities involves a three-stage history (Fig. 14). The first ~2 m.y. of basin filling took place while the dense root was growing toward a critical mass. As the root reached a size and mass sufficient to initiate foundering, the rate of basin subsidence accelerated (ca. 18–15 Ma; Figs. 12 and 14D). At this stage, the Arizaro Basin became dominated by open, relatively deep-water lacustrine environments, and the rate of sediment accumulation increased markedly. By ca. 15.5 Ma, the rate of subsidence had begun to decrease dramatically, perhaps in response to separation of the dense root from the overlying crust. Modeling by Wang et al. (this volume) suggests that shortening within the interior of the basin fill would have maximized ca. 10 Ma, and final root removal would have taken place by ca. 7 Ma (Fig. 14D). Topographic inversion must have begun prior to and continued after deposition of the Batín Formation (ca. 8.4 Ma), because it has been tilted and rests on slightly more steeply tilted beds of the upper Vizcachera Formation.

Gravity data suggest the presence of a dense anomaly extending from the Arizaro Basin region toward the northwest beneath the Salar de Atacama (Götze and Krause, 2002; Reutter et al., 2006). Crustal thickness estimates for the Arizaro region indicate that the Moho is shallower here than in most of the surrounding region, only ~42 km (Yuan et al., 2002; Bianchi et al., 2013). This suggests that thickened lower crust and lithosphere have been removed from beneath the Arizaro region. Ducea et al. (2013) argued that the chemistry of Pleistocene basalts and basaltic andesites in the Arizaro region suggests that they are products of pyroxenite melting at depths around 60 km, perhaps in a dripping RT instability body. Schoenbohm and Carrapa (2011) noted that the timing of local shortening, extension, sedimentation, and basaltic magmatism in the western part of the Arizaro Basin, the

Salar de Antofalla, and the Pasto Ventura region is evidence for diachronous, small-scale lithospheric foundering underneath the Puna since the Miocene. We suggest that basalt production and effusion mark just the final stages in a roughly 20 m.y. history of root growth and gravitational foundering. The bulk of this history is archived in the stratigraphic record of Arizaro Basin. Given the fact that numerous other anomalous hinterland basins are scattered throughout the Puna and Altiplano (Horton et al., 2002; Leier et al., 2010, 2013; Lamb, 2011; Horton, 2012), it is conceivable that at least some of these basins formed, or are presently forming, in response to gravitational removal of crust and mantle lithosphere. If this is the case, then it is to be expected that hinterland regions in cordilleran orogenic systems will exhibit complex temporal and spatial histories of paleoelevation and basin development.

CONCLUSIONS

The Miocene Arizaro Basin is situated in the high Puna portion of the Central Andean Plateau, at a present elevation of 3800–4200 m. The basin formed between ca. 21 Ma and 8.5 Ma, and filled with fluvial-deltaic, eolian and lacustrine sediments. Paleoaltimetry and low-temperature thermochronology indicate that the basin developed at high (approximately equal to modern) elevation in a region where exhumation and uplift had already taken place at least 20 m.y. before the onset of basin subsidence.

Arizaro Basin is not readily explained by previously published models for basin formation in contractional settings. Our analysis, coupled with geodynamic modeling studies, suggests that the basin formed in response to the development of a dense lower crustal/lithospheric root, which reached critical mass for foundering between ca. 16 and 10 Ma. Basin subsidence history, internal shortening, and stratigraphic relationships suggest that basin inversion has been active since ca. 10 Ma, and that the root was completely detached by late Miocene time (perhaps ca. 8–7 Ma).

Insofar as the Arizaro Basin is emblematic of basins formed above RT instabilities in the lower crust or upper mantle, potentially diagnostic basin characteristics might include: (1) oval to round, saucer-shape of the basin fill; (2) absence of bounding, coeval fault systems capable of producing the observed magnitude of subsidence; (3) location in regions of previously thickened, high-elevation crust; (4) low-magnitude contractional deformation in the basin interior; (5) slow initial subsidence, followed by extremely rapid subsidence, culminating in basin inversion; (6) predominance of lacustrine lithofacies (perennial and evaporative, probably depending strongly on climate); and (7) absence of features characteristic of wedge-top, rift, intra-arc, and strike-slip basins. Similar basins filled by lacustrine and eolian lithofacies are widely distributed throughout the Puna and Altiplano (e.g., Jordan and Alonso, 1987). Although some of these basins are attributable to more conventional mechanisms of basin formation (e.g., Horton, 1998; Leier et al., 2010), it is likely that basins formed by gravitational foundering of dense lithospheric roots are widespread in the high central Andes as well as other orogenic systems.

ACKNOWLEDGMENTS

Our work in the Arizaro Basin was supported by grants from Exxon-Mobil Corporation and the U.S. National Science Foundation. Ricardo Alonso provided logistical assistance and geological expertise. Discussions with Clare Currie, Huilin Wang, and Lindsay Schoenbohm provided geodynamic insights. We are grateful to Teresa Jordan and Timothy Lawton for insightful reviews that helped us to significantly improve the manuscript.

REFERENCES CITED

Adelmann, D., 2001, Känozoische Beckenentwicklung in der Südlichen Puna am Beispiel des Salar de Antofalla (NW-Argentinien) [Ph.D. thesis]: Berlin, Freie University Berlin, 180 p.

Allen, J.R.L., 1984, Sedimentary Structures, Their Character and Physical Basis, Unabridged (1 volume edition): Amsterdam, Elsevier, 593 p.

Allen, P.A., and Allen, J.R., 2005, Basin Analysis (2nd ed.): Malden, Massachusetts, Blackwell Publishing, 549 p.

Alonso, R.N., 1992, Estratigrafía del Cenozoico de la cuenca de Pastos Grandes (Puna Salteña) con énfasis en la Formación Sijes y sus boratos: Revista de la Asociación Geológica Argentina, v. 47, p. 189–199.

Alonso, R.N., 1998, Los Boratos de la Puna: Salta, Argentina, Camara de la Minerias de Salta, 196 p.

Alonso, R.N., Berman, W.D., Bond, M., Carlini, A.A., Pascual, R., and Reguero, M.A., 1988, Vertebrados Paleógenos de la Puna Austral: Sus aportes a la evolución biogeográfica: Jornadas Argentinas de Paleontología Vertebrados, Resúmenes, v. V, p. 38–39.

Alonso, R.N., Jordan, T.E., Tabbutt, K.T., and Vandervoort, D.S., 1991, Giant evaporite belts of the Neogene central Andes: Geology, v. 19, p. 401–404, doi:10.1130/0091-7613(1991)019<0401:GEBOTN>2.3.CO;2.

Anadón, P., Cabrera, L., Colombo, F., Marzo, M., and Riba, O., 1986, Syntectonic intraformational unconformities in alluvial fan deposits, eastern Ebro Basin margins (NE Spain), *in* Allen, P.A., and Homewood, P., eds., Foreland Basins: International Association of Sedimentologists Special Publication 8, p. 259–271.

Ashley, G.M., Southard, J.B., and Boothroyd, J.C., 1982, Deposition of climbing-ripple beds: A flume simulation: Sedimentology, v. 29, p. 67–79, doi:10.1111/j.1365-3091.1982.tb01709.x.

Bagnold, R.A., 1954, The Physics of Blown Sand and Desert Dunes: London, Methuen and Co., republished by Dover, 318 p.

Bahlburg, H., and Hervé, F., 1997, Geodynamic evolution and tectonostratigraphic terranes of northwestern Argentina and northern Chile: Geological Society of America Bulletin, v. 109, p. 869–884, doi:10.1130/0016-7606(1997)109<0869:GEATTO>2.3.CO;2.

Bahlburg, H., Vervoort, J.D., Du Frane, S.A., Barbara Bock, B., Augustsson, C., and Reimann, C., 2009, Timing of crust formation and recycling in accretionary orogens: Insights learned from the western margin of South America: Earth-Science Reviews, v. 97, p. 215–241, doi:10.1016/j.earscirev.2009.10.006.

Beck, S., and Zandt, G., 2002, The nature of orogenic crust in the central Andes: Journal of Geophysical Research, v. 107, no. B10, 2230, doi:10.1029/2000JB000124.

Benison, K.C., 2008, Life and death around acid saline lakes: Palaios, v. 23, p. 571–573, doi:10.2110/palo.2008.S05.

Benison, K.C., Bowen, B.B., Ikuenobe, F.E., Jagniecki, E.A., Laclair, D.A., Story, S.L., Mormile, M.R., and Hong, B., 2007, Sedimentary processes and products of ephemeral acid saline lakes in southern Western Australia: Journal of Sedimentary Research, v. 77, p. 366–388, doi:10.2110/jsr.2007.038.

Bianchi, M., Heit, B., Jakovlev, A., Yuan, X., Kay, S.M., Sandvol, E., Alonso, R.N., Coira, B., Brown, L., Kind, R., and Comte, D., 2013, Teleseismic tomography of the southern Puna plateau in Argentina and adjacent regions: Tectonophysics, v. 586, p. 65–83, doi:10.1016/j.tecto.2012.11.016.

Blasco, G., Zappettini, E.O., and Hongn, F., 1996, Hoja Geológica 2566-I, San Antonio de los Cobres, Provincias de Jujuy y Salta: Buenos Aires, Programa Nacional de Cartas Geológicas de la República Argentina (1:250,000), Secretaría Geológico Buenos Aires, Boletín (Instituto de Estudios de Poblacion y Desarrollo [Dominican Republic]) 217, 126 p.

Bowen, B.B., and Benison, K.C., 2006, Chemical diversity of natural waters in the acid saline systems of south Western Australia (abstract): Geological Society of America Abstracts with Programs, v. 38, no. 7, abstract 108332.

Boyd, J., 2010, Tectonic Evolution of the Arizaro Basin of the Puna Plateau, NW Argentina: Implications for Plateau-Scale Processes [M.S. thesis]: Laramie, Wyoming, University of Wyoming, 81 p.

Bulnes, M., and Poblet, J., 1999, Estimating the detachment depth in cross sections involving detachment folds: Geological Magazine, v. 136, p. 395–412, doi:10.1017/S0016756899002794.

Busby, C.J., 2012, Extensional and transtensional continental arc basins: Case studies from the southwestern United States, *in* Busby, C.J., and Azor, A., eds., Tectonics of Sedimentary Basins: Recent Advances (1st ed.): Oxford, UK, Blackwell Publishing, p. 382–404.

Canavan, R., Carrapa, B., Clementz, M.T., Quade, J., DeCelles, P.G., and Schoenbohm, L.M., 2014, Early Cenozoic uplift of the Puna Plateau, Central Andes, based on stable isotope paleoaltimetry of hydrated volcanic glass: Geology, v. 42, p. 447–450, doi:10.1130/G35239.1.

Cant, D.J., 1978, Development of a facies model for sandy braided river sedimentation: Comparison of the South Saskatchewan River and the Battery Point Formation, *in* Miall, A.D., ed., Fluvial Sedimentology: Canadian Society of Petroleum Geologists Memoir 5, p. 627–639.

Cant, D.J., and Walker, R.G., 1978, Fluvial processes and facies sequences in the sandy braided South Saskatchewan River, Canada: Sedimentology, v. 25, p. 625–648, doi:10.1111/j.1365-3091.1978.tb00323.x.

Carrapa, B., and DeCelles, P.G., 2008, Eocene exhumation and basin development in the Puna of northwestern Argentina: Tectonics, v. 27, TC1015, doi:10.1029/2007TC002127.

Carrapa, B., and DeCelles, P.G., 2015, this volume, Regional exhumation and kinematic history of the central Andes in response to cyclical orogenic processes, *in* DeCelles, P.G., Ducea, M.N., Carrapa, B., and Kapp, P.A., eds., Geodynamics of a Cordilleran Orogenic System: The Central Andes of Argentina and Northern Chile: Geological Society of America Memoir 212, doi:10.1130/2015.1212(11).

Carrapa, B., Adelmann, D., Hilley, G.E., Mortimer, E., Sobel, E.R., and Strecker, M.R., 2005, Oligocene uplift and development of plateau morphology in the southern central Andes: Tectonics, v. 24, TC401, doi:10.1029/2004TC001762.

Carrapa, B., DeCelles, P.G., Reiners, P.W., Gehrels, G.E., and Sudo, M., 2009, Apatite triple dating and white mica $^{40}Ar/^{39}Ar$ thermochronology of syntectonic detritus in the central Andes: A multiphase tectonothermal history: Geology, v. 37, p. 407–410, doi:10.1130/G25698A.1.

Carrapa, B., Trimble, J.D., and Stockli, D.F., 2011a, Patterns and timing of exhumation and deformation in the Eastern Cordillera of NW Argentina revealed by (U-Th)/He thermochronology: Tectonics, v. 30, TC3003, doi:10.1029/2010TC002707.

Carrapa, B., Reyes-Bywater, S., DeCelles, P.G., Mortimer, E., and Gehrels, G., 2011b, Cenozoic synorogenic basin evolution in the Eastern Cordillera of northwestern Argentina (25°–26°S): Regional implications for Andean orogenic wedge development: Basin Research, v. 23, p. 1–20, doi:10.1111/j.1365-2117.2011.00519.x.

Chamberlin, R.T., 1919, The building of the Colorado Rockies: The Journal of Geology, v. 27, p. 225–251, doi:10.1086/622658.

Church, M., and Jones, D., 1982, Channel bars in gravel bed rivers, *in* Hey, R.D., Bathurst, J.C., and Thorne, C.R., eds., Gravel-Bed Rivers: Chichester, UK, Wiley, p. 291–324.

Coutand, I., Cobbold, P.R., de Urreiztieta, M., Gautier, P., Chauvin, A., Gapais, D., Rossello, E.A., and Lòpez-Gamundí, O., 2001, Style and history of Andean deformation, Puna Plateau, northwestern Argentina: Tectonics, v. 20, p. 210–234, doi:10.1029/2000TC900031.

Currie, C.A., Ducea, M.N., DeCelles, P.G., Zandt, G., Gray, G., and Beaumont, C., 2011, Numerical models of the formation and removal of eclogite roots in Cordilleran volcanic arcs: San Francisco, California, American Geophysical Union, Fall Meeting 2011, abstract T13I-01.

Currie, C.A., Ducea, M.N., DeCelles, P.G., and Beaumont, C., 2015, this volume, Geodynamic models of Cordilleran orogens: Gravitational instability of magmatic arc roots, *in* DeCelles, P.G., Ducea, M.N., Carrapa, B., and Kapp, P.A., eds., Geodynamics of a Cordilleran Orogenic System: The Central Andes of Argentina and Northern Chile: Geological Society of America Memoir 212, doi:10.1130/2015.1212(01).

Dahlstrom, C.D.A., 1990, Geometric constraints derived from the law of conservation of volume and applied to evolutionary models for detachment folding: American Association of Petroleum Geologists Bulletin, v. 74, p. 336–344.

DeCelles, P.G., and Giles, K.N., 1996, Foreland basin systems: Basin Research, v. 8, p. 105–123, doi:10.1046/j.1365-2117.1996.01491.x.

DeCelles, P.G., Gray, M.B., Cole, R.B., Pequera, N., Pivnik, D.A., Ridgway, K.D., and Srivastava, P., 1991, Controls on synorogenic alluvial-fan architecture, Beartooth Conglomerate, Wyoming and Montana: Sedimentology, v. 38, p. 567–590.

DeCelles, P.G., Carrapa, B., and Gehrels, G.E., 2007, Detrital zircon U-Pb ages provide provenance and chronostratigraphic information from Eocene synorogenic deposits in northwestern Argentina: Geology, v. 35, p. 323–326, doi:10.1130/G23322A.1.

DeCelles, P.G., Ducea, M.N., Zandt, G., and Kapp, P., 2009, Cyclicity in Cordilleran orogenic systems: Nature Geoscience, v. 2, p. 251–257, doi:10.1038/ngeo469.

DeCelles, P.G., Carrapa, B., Horton, B.K., and Gehrels, G.E., 2011, Cenozoic foreland basin system in the central Andes of northwestern Argentina: Implications for Andean geodynamics and modes of deformation: Tectonics, v. 30, TC6013, doi:10.1029/2011TC002948.

Deeken, A., Sobel, E.R., Coutand, I., Haschke, M., Riller, U., and Strecker, M.R., 2006, Development of the southern Eastern Cordillera, NW Argentina, constrained by apatite fission track thermochronology: From Early Cretaceous extension to middle Miocene shortening: Tectonics, v. 25, TC6003, doi:10.1029/2005TC001894.

De Silva, S., Zandt, G., Trumbull, R., Viramonte, J.G., Salas, G., and Jimémenez, N., 2006, Large ignimbrite eruptions and volcano-tectonic depressions in the central Andes: A thermomechanical perspective, *in* Troise, C., De Natale, G., and Kilburn, C.R.J., eds., Mechanisms of Activity and Unrest at Large Calderas: Geological Society of London Special Publication 269, p. 47–63.

Dickinson, W.R., 1985, Interpreting provenance relations from detrital modes of sandstones, *in* Zuffa, G.G., ed., Provenance of Arenites: NATO Advanced Studies Institute Series 148: Dordrecht, Netherlands, Reidel, p. 333–361.

Donato, E., 1987, Características Estructurales del Sector Occidental de la Puna Saltena: Yacimientos Petroliferous Fiscales (YPF): Boletín de Informaciones Petroleras, v. 12, p. 89–97.

Ducea, M.N., 2001, The California arc: Thick granitic batholiths, eclogitic residues, lithospheric-scale thrusting, and magmatic flare-ups: GSA Today, v. 11, p. 4–10, doi:10.1130/1052-5173(2001)011<0004 :TCATGB>2.0.CO;2.

Ducea, M.N., and Saleeby, J., 1998, A case for delamination of the deep batholithic crust beneath the Sierra Nevada, California: International Geology Review, v. 40, p. 78–93, doi:10.1080/00206819809465199.

Ducea, M.N., Seclaman, A.C., Murray, K.E., and Jianu, D., 2013, Mantle-drip magmatism beneath the Altiplano-Puna Plateau, central Andes: Geology, v. 41, p. 915–918, doi: 10.1130/G34509.1.

Echavarria, L., Hernández, R., Allmendinger, R., and Reynolds, J., 2003, Subandean thrust and fold belt of northwestern Argentina: Geometry and timing of the Andean evolution: American Association of Petroleum Geologists Bulletin, v. 87, p. 965–985, doi:10.1306/01200300196.

Einhorn, J.C., Gehrels, G.E., Vernon, A., and DeCelles, P.G., 2015, this volume, U-Pb zircon geochronology of Neoproterozoic–Paleozoic sandstones and Paleozoic plutonic rocks in the central Andes (21°S–26°S), *in* DeCelles, P.G., Ducea, M.N., Carrapa, B., and Kapp, P.A., eds., Geodynamics of a Cordilleran Orogenic System: The Central Andes of Argentina and Northern Chile: Geological Society of America Memoir 212, doi:10.1130/2015.1212(06).

Eugster, H.P., 1967, Hydrous sodium silicates from Lake Magadi, Kenya: Precursors of bedded chert: Science, v. 157, p. 1177–1180, doi:10.1126/science.157.3793.1177.

Eugster, H.P., 1986, Lake Magadi, Kenya: A model for rift valley hydrochemistry and sedimentation?, *in* Frostick, L.E., et al., eds., Sedimentation in the African Rifts: Geological Society of London Special Publication 25, p. 177–189.

Ford, M., 2004, Depositional wedge tops: Interaction between low basal friction external orogenic wedges and flexural foreland basins: Basin Research, v. 16, p. 361–375, doi:10.1111/j.1365-2117.2004.00236.x.

Frassetto, A.M., Zandt, G., Gilbert, H., Owens, T.J., and Jones, C.H., 2011, Structure of the Sierra Nevada from receiver functions and implications for lithospheric foundering: Geosphere, v. 7, no. 4, p. 898–921, doi:10.1130/GES00570.1.

Garzione, C.N., DeCelles, P.G., Hodkinson, D.G., Ojha, T.P., and Upreti, B.N., 2003, East-west extension and Miocene environmental change in the southern Tibetan Plateau: Thakkhola graben, central Nepal: Geological Society of America Bulletin, v. 115, p. 3–20, doi:10.1130/0016 -7606(2003)115<0003:EWEAME>2.0.CO;2.

Garzione, C.N., Molnar, P., Libarkin, J.C., and MacFadden, B.J., 2006, Rapid late Miocene rise of the Bolivian Altiplano: Evidence for removal of mantle lithosphere: Earth and Planetary Science Letters, v. 241, p. 543–556, doi:10.1016/j.epsl.2005.11.026.

Gehrels, G.E., Valencia, V., and Ruiz, J., 2008, Enhanced precision, accuracy, efficiency, and spatial resolution of U-Pb ages by laser ablation–multicollector–inductively coupled plasma–mass spectrometry: Geochemistry Geophysics Geosystems, v. 9, Q03017, doi:10.1029/2007GC001805.

Giovanni, M.K., Horton, B.K., Garzione, C.N., McNulty, B., and Grove, M., 2010, Extensional basin evolution in the Cordillera Blanca, Peru: Stratigraphic and isotopic records of detachment faulting and orogenic collapse in the Andean hinterland: Tectonics, v. 29, TC6007, doi:10.1029/TC2010TC002666.

Göğüş, O.H., and Pysklywec, R.N., 2008, Near-surface diagnostics of dripping or delaminating lithosphere: Journal of Geophysical Research, v. 113, B11404, doi:10.1029/2007JB005123.

Götze, H.-J., and Krause, S., 2002, The central Andean gravity high, a relic of an old subduction complex?: Journal of South American Earth Sciences, v. 14, p. 799–811.

Hay, R.L., 1977, Geology of zeolites in sedimentary rocks, *in* Mumpton, F.A., ed., Mineralogy and Geology of Natural Zeolites: Mineralogical Society of America Short Course Notes, v. 4, p. 53–64.

Hein, F.J., and Walker, R.G., 1977, Bar evolution and development of stratification in the gravelly braided, Kicking Horse River, British Columbia: Canadian Journal of Earth Sciences, v. 14, p. 562–570, doi:10.1139/ e77-058.

Heit, B., Koulakov, I., Asch, G., Yuan, X., Kind, R., Alcozer, I., Tawackoli, S., and Wilke, H., 2008, More constraints to determine the seismic structure beneath the central Andes at 21°S using teleseismic tomography analysis: Journal of South American Earth Sciences, v. 25, p. 22–36, doi:10.1016/j .jsames.2007.08.009.

Hongn, F.D., Tubía, J.M., Aranguren, A., Vegas, N., Mon, R., and Dunning, G.R., 2010, Magmatism coeval with Lower Paleozoic shelf basins in NW-Argentina (Tastil batholith): Constraints on current stratigraphic and tectonic interpretations: Journal of South American Earth Sciences, v. 29, p. 289–305, doi:10.1016/j.jsames.2009.07.008.

Horton, B.K., 1998, Sediment accumulation on top of the Andean orogenic wedge: Oligocene to late Miocene basins of the Eastern Cordillera, southern Bolivia: Geological Society of America Bulletin, v. 110, p. 1174–1192, doi:10.1130/0016-7606(1998)110<1174:SAOTOT>2.3.CO;2.

Horton, B.K., 2005, Revised deformation history of the central Andes: Inferences from Cenozoic foredeep and intermontane basins of the Eastern Cordillera, Bolivia: Tectonics, v. 24, TC3011, doi:10.1029/2003TC001619.

Horton, B.K., 2012, Cenozoic evolution of hinterland basins in the Andes and Tibet, *in* Busby, C.J., and Azor, A., eds., Tectonics of Sedimentary Basins: Recent Advances (1st ed.): Oxford, UK, Blackwell Publishing Ltd., p. 427–444.

Horton, B.K., and DeCelles, P.G., 1997, The modern foreland basin system adjacent to the central Andes: Geology, v. 25, p. 895–898, doi:10.1130/0091 -7613(1997)025<0895:TMFBSA>2.3.CO;2.

Horton, B.K., Hampton, B.A., LaReau, B.N., and Baldellón, E., 2002, Tertiary provenance history of the northern and central Altiplano (central Andes, Bolivia): A detrital record of plateau-margin tectonics: Journal of Sedimentary Research, v. 72, p. 711–726, doi:10.1306/020702720711.

Houseman, G.A., Neil, E.A., and Kohler, M.D., 2000, Lithospheric instability beneath the Transverse Ranges of California: Journal of Geophysical Research, v. 105, p. 16,237–16,250, doi:10.1029/2000JB900118.

Hunter, R.E., 1977, Basic types of stratification in small eolian dunes: Sedimentology, v. 24, p. 361–387, doi:10.1111/j.1365-3091.1977.tb00128.x.

Ingersoll, R.V., Bullard, T.F., Ford, R.L., Grimm, J.P., Pickle, J.D., and Sares, S.W., 1984, The effect of grain size on detrital modes: A test of the Gazzi-Dickinson point-counting method: Journal of Sedimentary Petrology, v. 54, p. 103–116.

Isacks, B., 1988, Uplift of the central Andean Plateau and bending of the Bolivian orocline: Journal of Geophysical Research, v. 93, p. 3211–3231, doi:10.1029/JB093iB04p03211.

Jones, C.H., Farmer, G.L., and Unruh, J., 2004, Tectonics of Pliocene removal of lithosphere of the Sierra Nevada, California: Geological Society of America Bulletin, v. 116, p. 1408–1422, doi:10.1130/B25397.1.

Jordan, T.E., and Alonso, R.N., 1987, Cenozoic stratigraphy and basin tectonics of the Andes Mountains, 20°–28° South latitude: American Association of Petroleum Geologsts Bulletin, v. 71, p. 49–64.

Jordan, T.E., and Mpodozis, C., 2006, Estratigrafía y evolución tectónica de la cuenca Paleógena de Arizaro-Pocitos, Puna occidental (24°–25°S), *in* XI Congreso Geologico Chileno: Antofagasta, Chile, v. 2, p. 57–60.

Jordan, T.E., Flemings, P.B., and Beer, J.A., 1988, Dating thrust fault activity by use of foreland basin strata, *in* Kleinspehn, K., and Paolo, C., eds., New Perspectives in Basin Analysis: Berlin, Springer-Verlag, p. 307–330.

Kapp, P., Taylor, M., Stockli, D., and Ding, L., 2008, Development of active low-angle normal fault systems during orogenic collapse: Insight from Tibet: Geology, v. 36, p. 7–10, doi:10.1130/G24054A.1.

Kay, R.W., and Kay, S.M., 1993, Delamination and delamination magmatism: Tectonophysics, v. 219, p. 177–189, doi:10.1016/0040-1951(93)90295-U.

Kay, S.M., and Coira, B., 2009, Shallowing and steepening subduction zones, continental lithospheric loss, magmatism, and crustal flow under the central Andes, *in* Kay, S.M., Ramos, V.A., and Dickinson, W.R., eds., Backbone of the Americas: Shallow Subduction, Plateau Uplift, and Ridge and Terrane Collision: Geological Society of America Memoir 204, p. 229–259.

Kay, S.M., Coira, B., and Viramonte, J., 1994, Young mafic back arc volcanic rocks as indicators of continental lithospheric delamination beneath the Argentine Puna Plateau, central Andes: Journal of Geophysical Research–Solid Earth, v. 99, p. 24,323–24,339, doi:10.1029/94JB00896.

Ketcham, R.A., Carter, A., Donelick, R.A., Barbarand, J., and Hurford, A.J., 2007, Improved modeling of fission track annealing in apatite: American Mineralogist, v. 92, p. 799–810, doi:10.2138/am.2007.2281.

Kley J., and Monaldi, C.R., 2002, Tectonic inversion in the Santa Bárbara system of the central Andean foreland thrust belt, northwestern Argentina: Tectonics, v. 21, doi:10.1029/2002TC902003.

Kocurek, G., and Dott, R.H., 1981, Distinctions and uses of stratification types in the interpretation of eolian deposits: Journal of Sedimentary Petrology, v. 51, p. 579–595.

Koukharsky, M.L., Quenardelle, S., Litvak, V., Maisonnave, E.B., and Page, S., 2002, Plutonismo del Ordovícico inferior en el sector Norte de la sierra de Macon, provincia de Salta: Asociación Geológica Argentina: Revista, v. 57, p. 173–181.

Kraemer, B., Adelmann, D., Alten, M., Schnurr, W., Erpenstein, K., Kiefer, E., van den Bogaard, P., and Görler, K., 1999, Incorporation of the Paleogene foreland into the Neogene Puna Plateau: The Salar de Antofalla area, NW Argentina: Journal of South American Earth Sciences, v. 12, p. 157–182, doi:10.1016/S0895-9811(99)00012-7.

Krystopowicz, N.J., and Currie, C.A., 2013, Crustal eclogitization and lithosphere delamination in orogens: Earth and Planetary Science Letters, v. 361, p. 195–207, doi:10.1016/j.epsl.2012.09.056.

Lamb, S., 2011, Did shortening in thick crust cause rapid Late Cenozoic uplift in the northern Bolivian Andes?: Journal of the Geological Society of London, v. 168, p. 1079–1092, doi:10.1144/0016-76492011-008.

Leier, A.L., McQuarrie, N., Horton, B.K., and Gehrels, G.E., 2010, Upper Oligocene conglomerates of the Altiplano, central Andes: The record of deposition and deformation along the margin of a hinterland basin: Journal of Sedimentary Research, v. 80, p. 750–762, doi:10.2110/jsr.2010.064.

Leier, A.L., McQuarrie, N., Garzione, C.N., and Eiler, J., 2013, Stable isotope evidence for multiple pulses of rapid surface uplift in the central Andes, Bolivia: Earth and Planetary Science Letters, v. 371–372, p. 49–58, doi:10.1016/j.epsl.2013.04.025.

Lunt, I.A., and Bridge, J.S., 2004, Evolution and deposits of a gravelly braid bar, Sagavanirktok River, Alaska: Sedimentology, v. 51, p. 415–432, doi:10.1111/j.1365-3091.2004.00628.x.

Marquillas, R.A., del Papa, C., and Sabino, I.F., 2005, Sedimentary aspects and paleoenvironmental evolution of a rift basin: Salta Group (Cretaceous-Paleogene), northwestern Argentina: International Journal of Earth Sciences, v. 94, p. 94–113, doi:10.1007/s00531-004-0443-2.

McNulty, B., and Farber, D., 2002, Active detachment faulting above the Peruvian flat slab: Geology, v. 30, p. 567–570, doi:10.1130/0091-7613 (2002)030<0567:ADFATP>2.0.CO;2.

McQuarrie, N., Horton, B.K., Zandt, G., Beck, S., and DeCelles, P.G., 2005, Lithospheric evolution of the Andean fold-thrust belt, Bolivia, and the origin of the central Andean plateau: Tectonophysics, v. 399, p. 15–37, doi:10.1016/j.tecto.2004.12.013.

Miall, A.D., 1978, Lithofacies types and vertical profile models in braided river deposits: A summary, *in* Miall, A.D., ed., Fluvial Sedimentology: Memoirs of Canadian Society of Petroleum Geology, v. 5, p. 597–604.

Miall, A.D., 1996, The Geology of Fluvial Deposits: Berlin, Springer-Verlag, 582 p.

Molnar, P., and Houseman, G.A., 2004, The effects of buoyant crust on the gravitational instability of thickened mantle lithosphere at zones of intracontinental convergence: Geophysical Journal International, v. 158, p. 1134–1150, doi:10.1111/j.1365-246X.2004.02312.x.

Murphy, M.A., Yin, A., Kapp, P., Harrison, T.M., Manning, C.E., Ryerson, F.J., Lin, D., and Jinghui, G., 2002, Structural evolution of the Gurla Mandhata detachment system, southwest Tibet: Implications for the eastward extent of the Karakoram fault system: Geological Society of America Bulletin, v. 114, p. 428–447, doi:10.1130/0016-7606(2002)114<0428:SEOTGM>2.0.CO;2.

Ori, G.G., and Friend, P.G., 1984, Sedimentary basins, formed and carried piggyback on active thrust sheets: Geology, v. 12, p. 475–478, doi:10.1130/0091-7613(1984)12<475:SBFACP>2.0.CO;2.

Pascual, R., 1983, Novedosos marsupiales paleógenos de la Fm. Pozuelos de la Puna: Salta: Ameghiniana, v. 20, p. 265–280.

Pearson, D.M., Kapp, P., DeCelles, P.G., Reiners, P.W., Gehrels, G.E., Ducea, M.N., and Pullen, A., 2013, Influence of pre-Andean crustal structure on Cenozoic thrust belt kinematics and shortening magnitude: Northwestern Argentina: Geosphere, v. 9, no. 6, p. 1766–1782, doi:10.1130/GES00923.1.

Poma, S., Quenardelle, S., Litvak, B., Maisonnave, E.B., and Koukharsky, M., 2004, The Sierra de Macon, plutonic expression of the Ordovician magmatic arc, Salta Province, Argentina: Journal of South American Earth Sciences, v. 16, p. 587–597, doi:10.1016/j.jsames.2003.10.002.

Pysklywec, R.N., and Cruden, A.R., 2004, Coupled-crust mantle dynamics and intraplate tectonics: Two dimensional numerical and three dimensional analogue modeling: Geochemistry Geophysics Geosystems, v. 5, Q10003, doi:10.1029/2004GC000748.

Quade, J., Dettinger, M.P., Carrapa, B., DeCelles, P., Murray, K.E., Huntington, K.A., Cartwright, A., Canavan, R.R., Gehrels, G., and Clementz, M., 2015, this volume, The growth of the central Andes, 22°S–26°S, *in* DeCelles, P.G., Ducea, M.N., Carrapa, B., and Kapp, P.A., eds., Geodynamics of a Cordilleran Orogenic System: The Central Andes of Argentina and Northern Chile: Geological Society of America Memoir 212, doi:10.1130/2015.1212(15).

Ramelow, J., Riller, U., Romer, R.L., and Oncken, O., 2006, Kinematic link between episodic trapdoor collapse of the Negra Muerta caldera and motion on the Olacapato–El Toro fault zone, southern central Andes: International Journal of Earth Sciences, v. 95, p. 529–541, doi:10.1007/s00531-005-0042-x.

Ramos, V.A., 2008, The basement of the central Andes: The Arequipa and related terranes: Annual Reviews of Earth and Planetary Science, v. 36, p. 289–324, doi:10.1146/annurev.earth. 36.031207.124304.

Reineck, H.-E., and Singh, I.B., 1975, Depositional Sedimentary Environments: Berlin, Springer-Verlag, 439 p.

Reutter, K.-J., Döbel, R., Bogdanic, T., and Kley, J., 1994, Geological Map of the Central Andes between 20°S and 26°S, *in* Reutter, K.-J., et al., eds., Tectonics of the Southern Andes: Berlin, Heidelberg, Springer-Verlag, scale 1:1,000,000.

Reutter, K.-J., Charrier, R., Götze, H.-J., Churr, B., Wigger, P., Scheuber, E., Giese, P., Reuther, C.-D., Schmidt, S., Rietbrock, A., Chong, G., and Belmonte-Pool, A., 2006, The Salar de Atacama Basin: A subsiding block within the western edge of the Altiplano-Puna Plateau, *in* Oncken, O., et al., eds., The Andes—Active Subduction Orogeny: Berlin, Springer-Verlag, p. 303–326.

Reynolds, J.H., Hernández, R.M., Galli, C.I., and Idleman, B.D., 2001, Magnetostratigraphy of the Quebrada La Porcelana section, Sierra de Ramos, Salta Province, Argentina; age limits for the Neogene Orán Group and uplift of the southern Sierras Subandinas: Journal of South American Earth Sciences, v. 14, p. 681–692, doi:10.1016/S0895-9811(01)00069-4.

Riller, U., Petrinovic, I., Ramelow, J., Strecker, M., and Oncken, O., 2001, Late Cenozoic tectonism, collapse caldera and plateau formation in the central Andes: Earth and Planetary Science Letters, v. 188, p. 299–311, doi:10.1016/S0012-821X(01)00333-8.

Risacher, F., Alonso, H., and Salazar, C., 2002, Hydrochemistry of two adjacent acid saline lakes in the Andes of northern Chile: Chemical Geology, v. 187, p. 39–57, doi:10.1016/S0009-2541(02)00021-9.

Salfity, J.A., and Marquillas, R., 1994, Tectonic and sedimentary evolution of the Cretaceous–Eocene Salta Group, Argentina, *in* Salfity, J.A., ed., Cretaceous Tectonics of the Andes: Earth Evolution Sciences: Heidelberg, Vieweg, p. 266–315.

Salfity, J.A., and Monaldi, C.R., 1998, Mapa Geológico de la Provincia de Salta: Buenos Aires, Servicio Geológico Minero Argentino, scale 1:500,000.

Saylor, J., DeCelles, P.G., Gehrels, G.E., Murphy, M., Zhang, R., and Kapp, P., 2010, Basin formation in the High Himalaya by arc-parallel extension and tectonic damming: Zhada Basin, southwestern Tibet: Tectonics, v. 29, TC1004, doi:10.1029/2008TC002390.

Schoenbohm, L., and Carrapa, B., 2011, Evidence from the timing of contraction, extension, sedimentation and magmatism for small-scale lithospheric foundering in the Puna Plateau, NW Argentina: San Francisco, California, American Geophysical Union, Fall Meeting supplement, abstract T13I-07.

Schott, B., and Schmelling, H., 1998, Delamination and detachment of a lithospheric root: Tectonophysics, v. 296, p. 225–247, doi:10.1016/S0040-1951(98)00154-1.

Schurr, B., Rietbrock, A., Asch, G., Kind, R., and Oncken, O., 2006, Evidence for lithospheric detachment in the central Andes from local earthquake tomography: Tectonophysics, v. 415, p. 203–223, doi:10.1016/j.tecto.2005.12.007.

Sinclair, H., 2012, Thrust wedge/foreland basin systems, *in* Busby, C.J., and Azor, A., eds., Tectonics of Sedimentary Basins: Recent Advances (1st ed.): Oxford, UK, Blackwell Publishing Ltd., p. 522–537.

Smith, G.A., and Landis, C.A., 1995, Intra-arc basins, *in* Busby, C.J., and Ingersoll, R.V., eds., Tectonics of Sedimentary Basins: Oxford, Blackwell Science, p. 263–298.

Sobolev, S.V., and Babeyko, A.Y., 2005, What drives orogeny in the Andes?: Geology, v. 33, p. 617–620, doi:10.1130/G21557.1.

Stacey, J.S., and Kramers, J.D., 1975, Approximation of terrestrial lead isotope evolution by a two-stage model: Earth and Planetary Science Letters, v. 26, p. 207–221.

Starck, D., and Vergani, G., 1996, Desarrollo tecto-sedimentario del Cenozoico en el sur de la Provincia de Salta, Argentina, *in* XIII Congreso Geológico Argentino y III Congreso de Exploración de Hidrocarburos, Actas I: Buenos Aires, p. 433–452.

Surdam, R.C., 1977, Zeolites in closed hydrologic systems, *in* Mumpton, F.A., ed., Mineralogy and Geology of Natural Zeolites: Mineralogical Society of America Short Course Notes, v. 4, p. 65–91.

Surdham, R.C., and Sheppard, R.A., 1978, Zeolites in saline, alkaline-lake deposits, *in* Sand, L.B., and Mumpton, F.A., eds., Natural Zeolites: Occurrence, Properties, Use: Oxford, Pergamon Press, p. 145–74.

Taylor, M.H., Kapp, P.A., and Horton, B.K., 2012, Basin response to active extension and strike-slip deformation in the hinterland of the Tibetan Plateau, *in* Busby, C.J., and Azor, A., eds., Tectonics of Sedimentary Basins: Recent Advances (1st ed.): Oxford, UK, Blackwell Publishing Ltd., p. 445–460.

Uba, C.E., Heubeck, C., and Hulka, C., 2006, Evolution of the late Cenozoic Chaco foreland basin, southern Bolivia: Basin Research, v. 18, p. 145–170, doi:10.1111/j.1365-2117.2006.00291.x.

Vandervoort, D.S., 1993, Non-Marine Evaporite Basin Studies, Southern Puna Plateau, Central Andes [Ph.D. thesis]: Ithaca, New York, Cornell University, 177 p.

Vandervoort, D.S., Jordan, T.E., Zeitler, P.K., and Alonso, R.N., 1995, Chronology of internal drainage development and uplift, southern Puna Plateau, Argentine central Andes: Geology, v. 23, p. 145–148, doi:10.1130/0091-7613(1995)023<0145:COIDDA>2.3.CO;2.

Varekamp, J.C., Pasternack, G.B., and Rowe, G.L., Jr., 2000, Volcanic lake systematics: II. Chemical constraints: Journal of Volcanology and Geothermal Research, v. 97, p. 161–179, doi:10.1016/S0377-0273(99)00182-1.

Voss, R., 2002, Cenozoic stratigraphy of the southern Salar de Antofalla region, northwestern Argentina: Revista Geológica de Chile, v. 29, p. 167–189.

Wang, H., Currie, C.A., and DeCelles, P.G., 2015, this volume, Hinterland basin formation and gravitational instabilities in the central Andes: Constraints from gravity data and geodynamic models, *in* DeCelles, P.G., Ducea, M.N., Carrapa, B., and Kapp, P.A., eds., Geodynamics of a Cordilleran Orogenic System: The Central Andes of Argentina and Northern Chile: Geological Society of America Memoir 212, doi:10.1130/2015.1212(19).

Wells, M.L., Hoisch, T.D., Cruz-Uribe, A.M., and Vervoort, J.D., 2012, Geodynamics of synconvergent extension and tectonic mode switching: Constraints from the Sevier-Laramide orogen: Tectonics, v. 31, TC1002, doi:10.1029/2011TC002913.

Wernicke, B., Clayton, R., Ducea, M., Jones, C.H., Park, S., Ruppert, S., Saleeby, J., Snow, J.K., Squires, L., Fliedner, M., Jiracek, G., Keller, R., Klemperer, S., Luetgert, J., Malin, P., Miller, K., Mooney, W., Oliver, H., and Phinney, R., 1996, Origin of high mountains in the continents: The southern Sierra Nevada: Science, v. 271, p. 190–193, doi:10.1126/science.271.5246.190.

Woodruff, W.H., Jr., Horton, B.K., Kapp, P., and Stockli, D.F., 2013, Late Cenozoic evolution of the Lunggar extensional basin, Tibet: Implications for basin growth and exhumation in hinterland plateaus: Geological Society of America Bulletin, v. 125, p. 343–358, doi:10.1130/B30664.1.

Wooldridge, C.L., and Hickin, E.J., 2005, Radar architecture and evolution of channel bars in wandering gravel bed rivers: Fraser and Squamish Rivers, British Columbia, Canada: Journal of Sedimentary Research, v. 75, p. 844–860, doi:10.2110/jsr.2005.066.

Xie, X., and Heller, P.L., 2009, Plate tectonics and basin subsidence: Geological Society of America Bulletin, v. 121, p. 55–64.

Yuan, X., Sobolev, S.V., and Kind, R., 2002, Moho topography in the central Andes and its geodynamic implications: Earth and Planetary Science Letters, v. 199, p. 389–402, doi:10.1016/S0012-821X(02)00589-7.

Zandt, G., Gilbert, H., Owens, T.J., Ducea, M., Saleeby, J., and Jones, C.H., 2004, Active foundering of a continental arc root beneath the southern Sierra Nevada in California: Nature, v. 431, p. 41–46, doi:10.1038/nature02847.

Zappettini, E., and Blasco, G., 2001, Hoja Geológica 2569-II, Socompa, Provincia de Salta: Buenos Aires, Programa Nacional de Cartas Geológicas de la República Argentina (1:250,000), Instituto de Geología y Recursos Minerales, Servicio Geológico Minero Argentina, Boletín 260, 62 p.

MANUSCRIPT ACCEPTED BY THE SOCIETY 3 JUNE 2014
MANUSCRIPT PUBLISHED ONLINE 20 NOVEMBER 2014

The Geological Society of America
Memoir 212
2015

Hinterland basin formation and gravitational instabilities in the central Andes: Constraints from gravity data and geodynamic models

Huilin Wang*
Claire A. Currie
Department of Physics, University of Alberta, Edmonton, AB T6G 2E1, Canada

Peter G. DeCelles
Department of Geosciences, University of Arizona, Tucson, Arizona 85721, USA

ABSTRACT

The evolution of surface topography in an orogen provides information about the dynamics of the deep lithosphere. Within the high-elevation Altiplano-Puna Plateau of the central Andes, there are several local basins (~100 km wide) that sit >500 m lower than the surrounding plateau. These areas correspond to positive isostatic gravity anomalies, indicating high density in the lithosphere. There are also examples of former basins that are now at high elevation (e.g., the Miocene Arizaro Basin), suggesting that the basins are transient features that may be related to convective removal of lithosphere. Two-dimensional numerical models are used to investigate the topographic expression associated with removal of a high-density lithosphere root. A key result is that the presence of thick orogenic crust, as found in the Altiplano-Puna Plateau, can greatly affect the surface deflection above the detaching root. Three types of deflection are observed: (1) >500 m subsidence, followed by uplift, (2) little subsidence, and (3) uplift followed by collapse. The main control on the deflection is the viscous coupling between the root and surface, which decreases with increased root depth or weaker crust. If the crust is weak, the dense root induces crustal flow, resulting in thickened crust and either limited subsidence or uplift above the dripping lithosphere. Significant subsidence only occurs if the deep crust is relatively strong and the density anomaly is located within the crust. To produce surface deflection over a width of ~100 km, the near-surface rocks must be relatively weak.

*huilin1@ualberta.ca

Wang, H., Currie, C.A., and DeCelles, P.G., 2015, Hinterland basin formation and gravitational instabilities in the central Andes: Constraints from gravity data and geodynamic models, *in* DeCelles, P.G., Ducea, M.N., Carrapa, B., and Kapp, P.A., eds., Geodynamics of a Cordilleran Orogenic System: The Central Andes of Argentina and Northern Chile: Geological Society of America Memoir 212, p. 387–406, doi:10.1130/2015.1212(19). For permission to copy, contact editing@ geosociety.org.

INTRODUCTION

Subduction zones are regions of crustal deformation and continental growth through lateral lithosphere accretion and magmatism. They are also areas where removal of continental lithosphere may occur, for example, through subduction erosion or gravitational foundering. Here, we examine gravitational removal of lithosphere from the central Andes orogen of South America (Fig. 1A). This orogen formed in association with subduction of the Nazca plate below South America, with the majority of shortening in the last ~30 m.y. (Oncken et al., 2006, and references therein). This has produced the Altiplano-Puna Plateau, which has a 50–80-km-thick crust (e.g., Beck et al., 1996; Yuan et al., 2002; Oncken et al., 2006; Bianchi et al., 2013). In contrast, the mantle lithosphere below most of the plateau does not appear to be anomalously thick. Seismic tomography images of the shallow mantle show that much of the plateau is characterized by low seismic velocities and high attenuation (e.g., Myers et al., 1998; Beck and Zandt, 2002; Schurr et al., 2006; Bianchi et al., 2013; Beck et al., this volume). A seismic receiver function study indicated that the thickness of lithosphere is only ~100 km in the eastern Altiplano Plateau, which is ~30–70 km thinner than the surrounding areas (Heit et al., 2008). Magmatism throughout the plateau (e.g., Kay and Coira, 2009) and high surface heat flow (e.g., Springer and Förster, 1998) also suggest high mantle temperatures and a thin lithosphere. These observations have been interpreted to indicate the removal of mantle lithosphere, and possibly lower crust, from the orogen (e.g., Beck and Zandt, 2002; Schurr et al., 2006; Kay and Coira, 2009).

Mantle lithosphere is generally cooler and denser than the underlying mantle, and therefore it is gravitationally unstable. During orogenic shortening, lithospheric thickening may initiate convective removal of the deep lithosphere, through either a Rayleigh-Taylor–type instability (RT drip; e.g., Houseman and Molnar, 1997; Molnar and Houseman, 2004) or delamination (e.g., Bird, 1979). Foundering may also be triggered by the presence of high-density eclogitic rocks in the lower crust or mantle lithosphere, associated with metamorphic or magmatic processes (Kay and Kay, 1993; Leech, 2001; Ducea, 2002; Saleeby et al., 2003).

Lithosphere removal will modify the density and thermal structure of the orogen, and therefore removal events may be identified in the magmatic and elevation record. Localized

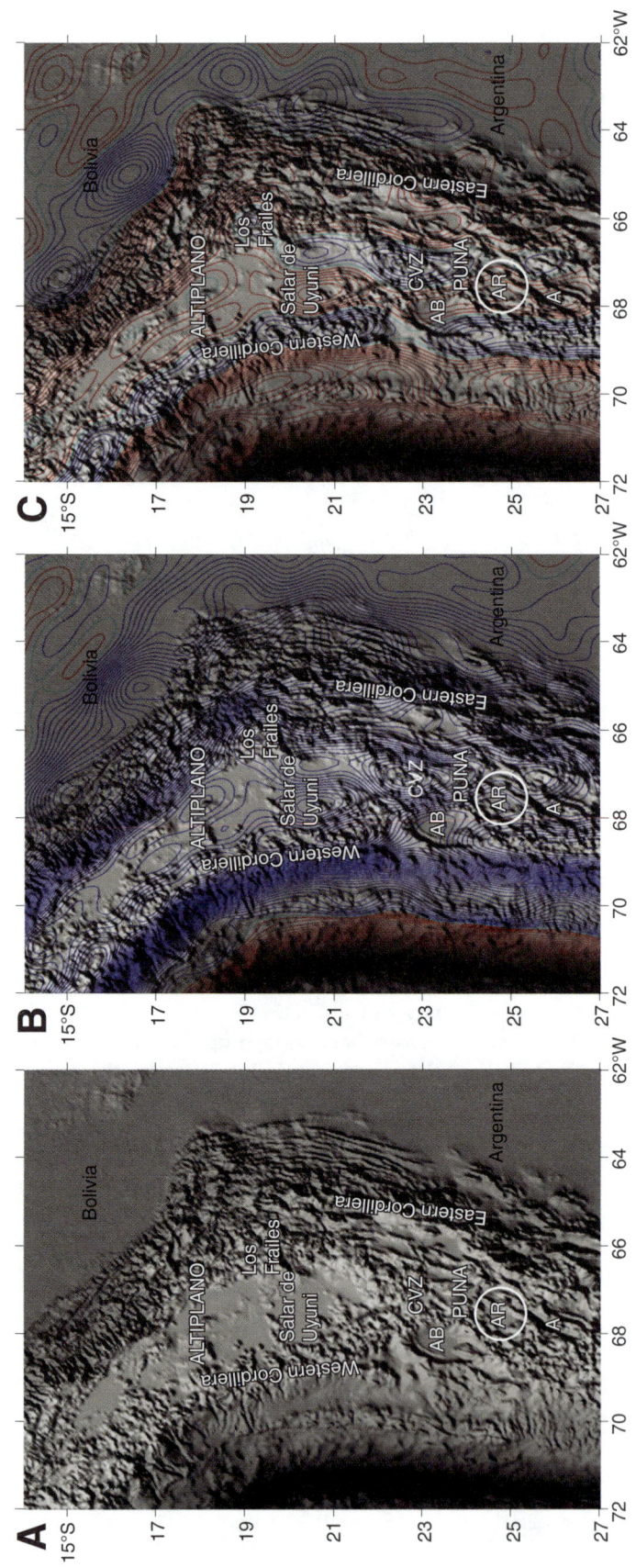

Figure 1. (A) Topography in the central Andes. CVZ—Central volcanic zone; AB—Atacama Basin; AR—Arizaro Basin; A—Antofalla Basin. The Arizaro Basin region consists of the Miocene basin in the east (now at high topography) and the modern Salar de Arizaro in the west. (B) Bouguer gravity map in the central Andes from gravity model Delft Gravity Model-1S (Delft Gravity Model, 2013). The red lines indicate positive gravity anomalies; the green line is zero anomaly; and the blue lines are negative anomalies. The contour interval is 10 mGal. (C) Isostatic gravity map in the central Andes (contours as in Fig. 1B).

magmatism in the Altiplano-Puna Plateau has been linked to convective removal of lithosphere (e.g., Kay et al., 1994; Kay and Coira, 2009). Ducea et al. (2013) showed that the composition of mafic magmas is consistent with small-scale lithosphere removal (<50 km diameter). As dense lithosphere is convectively removed, isostatic adjustment should produce uplift of Earth's surface. Garzione et al. (2006) and Leier et al. (2013) compiled paleoelevation data for the Altiplano and inferred a period of rapid >1 km uplift in the Miocene. The spatial extent of uplift appears to require large-scale foundering of the lithosphere across the width of the plateau (e.g., Hoke and Garzione, 2008).

In this study, we examine the topographic expression of lithosphere removal on a more local scale (100–200 km), similar to the scales inferred from magmatism (Ducea et al., 2013). Our work is motivated by observations of surface topography in the Altiplano-Puna Plateau. Regionally, the plateau has low relief and is internally drained, with an average elevation of ~4 km. On a smaller scale, there are several basins within the high-elevation hinterland (e.g., Horton, 2012). For example, the Atacama Basin and Salars de Arizaro and Uyuni stand out as areas of relatively low elevation within the modern high plateau (Fig. 1A). There are also examples of areas that were formerly basins but that now sit at high elevation. The most well studied of these is the Miocene Arizaro Basin in the Puna Plateau of northwest Argentina (DeCelles et al., this volume, and references therein). This basin is approximately circular, with a diameter of ~100 km. Basin subsidence initiated at ca. 21 Ma, at a time when paleoelevation data indicate that this part of the plateau had already undergone orogenic shortening and crustal thickening and was sitting at ~4 km elevation, close to its modern elevation (Carrapa et al., 2009, 2011; Quade et al., this volume). From 21 to ca. 8 Ma, the basin accumulated ~3400 m of lacustrine, fluvial, and eolian sedimentary rocks. Subsequently, the basin fill was internally shortened and then topographically inverted to its present ~4200 m elevation. It is now ~500 m higher than its surroundings and is located between the modern Salar de Arizaro to the west and the Salar de Pocitos to the east.

The formation of the Miocene Arizaro Basin within the high Puna Plateau is enigmatic. It does not appear to be controlled by neighboring thrust-faulted uplifts, loading associated with orogenesis, or extension (Schoenbohm and Carrapa, 2011; DeCelles et al., this volume). Given its roughly circular shape and transient nature, it has been proposed that the basin may be related to local removal of dense lithosphere as an RT drip (DeCelles et al., 2011, this volume; Schoenbohm and Carrapa, 2011). The goal of this study is to assess whether an RT drip can produce a surface deflection that is consistent with the observed characteristics of the Miocene Arizaro Basin. We first present gravity data for the central Andes. These provide information about the present-day density structure of the orogen and demonstrate that modern hinterland basins coincide with regions of high gravity, suggesting anomalously high density within the underlying lithosphere. We then use numerical models to examine the dynamics of RT drips

of either dense lower crust (model set A) or dense mantle lithosphere (model set B). The predicted surface deflection is compared to the geological record of the Arizaro Basin to constrain the relationship between RT drips and the temporal evolution of surface topography in an orogen.

ISOSTATIC GRAVITY DATA

Figure 1B shows the Bouguer gravity data for the central Andes, based on the satellite model Delft Gravity Model-1S (Delft Gravity Model, 2013) with spherical harmonic coefficients up to degree 250. The Altiplano-Puna Plateau appears as a negative Bouguer anomaly, with an average value of –300 mGal (Fig. 1B). In order to assess whether the present-day topography (Fig. 1A) is in isostatic equilibrium, we calculate the isostatic gravity anomaly. This calculation uses the present-day topography (Fig. 1A; ETOPO2v2 database with 2 min resolution) to determine the expected crustal root thickness ($w(x,y)$), based on Airy-Heiskanen isostasy, as given by Whitman (1999):

$$w(x,y) = - \frac{\rho_t}{\Delta\rho} \times h(x,y) \text{ (onshore)}, \quad (1)$$

$$w(x,y) = - \frac{\rho_t - \rho_w}{\Delta\rho} \times h(x,y) \text{ (offshore)}, \quad (2)$$

where ρ_t (2850 kg/m³) is the topographic (crustal) density, ρ_w (1030 kg/m³) is the water density, $\Delta\rho$ (450 kg/m³) is the density contrast across the Moho, and $h(x,y)$ is the topography or bathymetry, taken every 2 min. The zero-elevation crustal thickness is 40 km.

The expected crustal thickness is then gridded into rectangular prisms (x and y spacing of 0.25 degrees), and the associated gravity field at the surface is computed by combining the effects of all the prisms. Finally, the calculated gravity field is subtracted from the Bouguer gravity field to obtain the isostatic gravity field. We also subtract the gravitational effect of the Nazca plate (20 mGal; Whitman, 1999) from the entire region.

This approach provides a first-order view of the lithospheric density structure of the central Andes. Figure 1C shows the calculated isostatic gravity field, which is similar to the isostatic gravity field presented by Whitman (1999) and Götze and Krause (2002). Regions that have a nonzero gravity anomaly are areas that are not in isostatic equilibrium. On a regional scale, the Western Cordillera exhibits negative anomalies (~–30 mGal), which may be associated with the modern volcanic arc. In the Altiplano-Puna Plateau, the isostatic gravity anomaly is close to 0 mGal. The Eastern Cordillera has positive anomalies (>30 mGal), which have been related to plate flexure caused by underthrusting of the Brazilian Shield (Whitman, 1999).

On a smaller scale, areas with active volcanism, such as the Los Frailes volcanic field and Central volcanic zone, are characterized by negative residual anomalies of approximately

−30 mGal. This indicates low densities that may be associated with local thinning of the lithosphere and/or heating of the crust by volcanism. In contrast, modern salars and basins (e.g., Salar de Arizaro, Salar de Uyuni, and Atacama Basin) exhibit positive isostatic gravity anomalies, possibly related to anomalously high density in the lithosphere. In the Arizaro region, the largest positive anomaly is associated with the Salar de Arizaro in the west, and the gravity anomaly decreases to the east over the Miocene Arizaro Basin, which is now a topographic high. This suggests that the Salar de Arizaro is presently underlain by high-density material, whereas the Miocene basin is closer to isostatic equilibrium. For the Puna Plateau, seismic studies indicate low velocities in the shallow mantle, which have been interpreted to indicate that a thin mantle lithosphere in this region (Schurr et al., 2006). Therefore, it is unlikely that the inferred high density corresponds to unusually thick mantle lithosphere. The wavelength of the isostatic gravity anomalies is 100–200 km, which is consistent with anomalously dense material in either the deep crust or shallow mantle (<100 km depth). However, the wavelength of the gravity anomaly depends on both the depth and width of the density anomaly, and therefore the exact depth of the anomaly cannot be determined from the gravity field alone.

NUMERICAL MODELING METHODS

Model Geometry and Methods

The gravity data presented in the previous section are consistent with the idea that hinterland basins may form above areas with high-density material in the lithosphere. We now develop numerical models to address the dynamics of this material, in order to assess whether densification, followed by gravitational removal, is a viable mechanism to explain the Miocene Arizaro

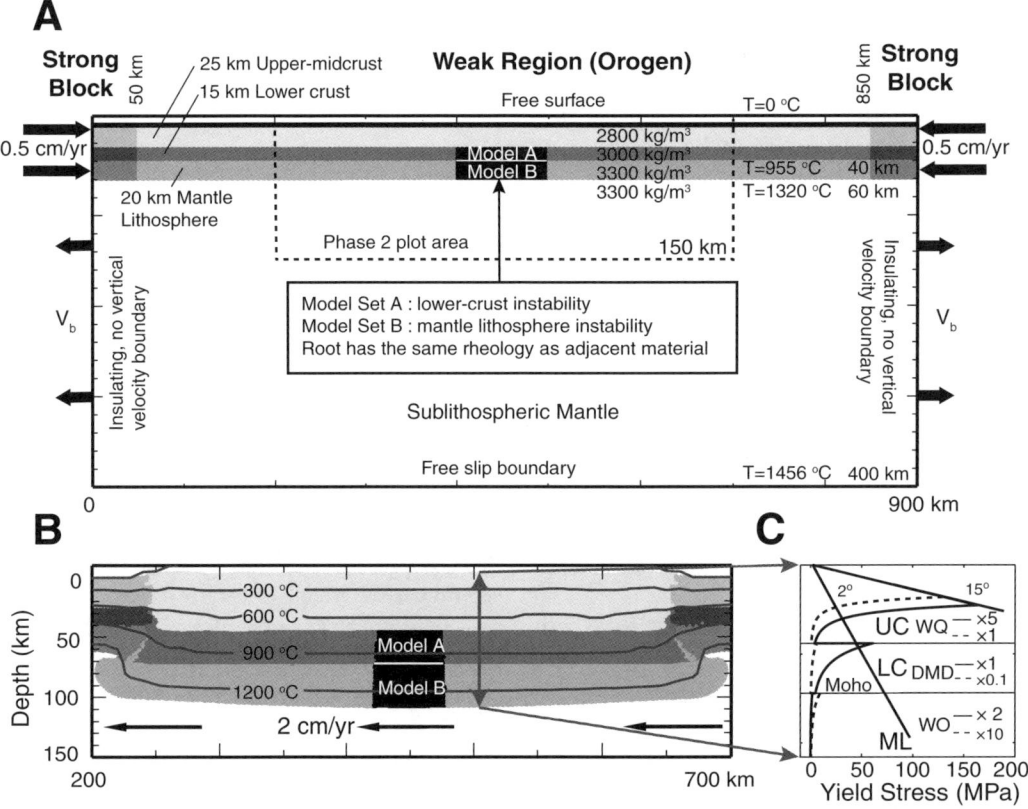

Figure 2. Numerical model setup. (A) Initial model geometry and thermal-mechanical boundary conditions for phase 1. Material parameters are given in Table 1. V_b is a small uniform outflux velocity assigned to the side boundaries of the sublithospheric mantle in order to maintain constant mass in the model domain. (B) Model geometry at the beginning of phase 2, after 380 km of shortening. In this phase, the lithosphere boundaries have zero velocity, and a 2 cm/yr flow to the left is assigned to the boundaries of the sublithospheric mantle to simulate mantle wedge flow. (C) Strength profiles for orogen lithosphere. The lines 15° and 2° show the frictional-plastic yield stress for unsoftened and softened materials, respectively. Solid lines show the viscous rheology used in the reference models; dashed lines are the strength variations that are tested in this study, based on variations in the scaling factor f (Eq. 4). The model thermal structure and a strain rate of 10^{-15} s^{-1} are used for the calculations. UC—upper-midcrust; LC—lower crust; ML—mantle lithosphere; WQ—wet quartzite; DMD—dry Maryland diabase; WO—wet olivine.

Basin. The initial geometry of the two-dimensional (2-D) thermal-mechanical numerical models is shown in Figure 2A. The model domain is 900 km wide and 400 km deep, with a 60-km-thick continental lithosphere overlying sublithospheric mantle. The lithosphere has an 800-km-wide weak zone (proto-orogen) between two stronger blocks. The initial (pre-orogenic) continental lithosphere consists of a 40-km-thick crust (25 km upper-midcrust and 15 km lower crust) and 20 km mantle lithosphere. In the models, we consider gravitational instability of anomalously dense material ("root") in either the orogen lower crust (model set A) or mantle lithosphere (model set B). The initial width of each of these zones is ~100 km (black material in Fig. 2A).

The models use the finite-element code SOPALE, in which arbitrary Eulerian-Lagrangian techniques are used to calculate the coupled thermal-mechanical evolution of the lithosphere–upper mantle system, under the assumptions of incompressibility, plane strain, and zero Reynolds number (Fullsack, 1995). Thermal-mechanical calculations are carried out on an Eulerian mesh, which has 180 elements in the horizontal direction (5 km width) and 82 elements in the vertical direction, with 56 elements in the upper 140 km (2.5 km height) and 26 elements below (10 km height). Benchmark tests show that this resolution is sufficient to obtain growth rates of Rayleigh-Taylor instabilities within 6% of expected values (Houseman and Molnar, 1997; Molnar et al., 1998). Material properties are tracked using Lagrangian particles that are advected with the model velocity field and are used to update the Eulerian mesh every time step. The Eulerian mesh is fixed horizontally but can stretch vertically to conform to the top boundary of the models (ground surface) as topography develops.

Material Properties

All materials have a temperature-dependent density and viscous-plastic rheology. At stresses above the frictional-plastic yield stress, material deformation follows the Drucker-Prager yield criterion:

$$J'_2 = P sin\phi_{eff} + c_0 cos\phi_{eff}, \qquad (3)$$

where J'_2 is the square root of the second invariant of the deviatoric stress σ'_{ij} ($J'^2_2 = \frac{1}{2}\sigma'_{ij}\sigma'_{ij}$), P is the pressure, ϕ_{eff} is the effective internal angle of friction, and c_0 is the cohesion. The frictional-plastic deformation is modeled as a viscous creeping flow by defining a viscosity that places the state of stress on yield (Fullsack, 1995; Willett, 1999). Crustal materials undergo frictional-plastic strain softening by decreasing ϕ_{eff} with increasing strain. This approximates rock weakening due to pore-fluid pressure variations, fault gouge formation, and mineral reactions during deformation (Huismans and Beaumont, 2002, 2003; Warren et al., 2008). In the models, ϕ_{eff} is linearly reduced from 15° to 2° over accumulated strain (I'_2) of 0.5 to 1.5 (Table 1; e.g., Huismans and Beaumont, 2003).

At stresses less than the frictional-plastic yield stress, deformation is viscous and follows a power-law creep rheology:

$$\eta^v_{eff} = f(B^*)(\dot{I}'_2)^{\frac{1-n}{n}} \exp\left(\frac{Q + PV^*}{nRT_K}\right), \qquad (4)$$

where η^v_{eff} is the effective viscosity, f is a scaling factor (see following), $(\dot{I}'^2_2)$ is the square root of the second invariant of the strain rate tensor $\dot{\varepsilon}_{ij}$ ($\dot{I}'^2_2 = \frac{1}{2}\dot{\varepsilon}_{ij}\dot{\varepsilon}_{ij}$), R is the gas constant, and T_K is absolute temperature. The pre-exponential factor (B^*), stress exponent (n), activation energy (Q), and activation volume (V^*) are rheological parameters derived from laboratory data.

Following previous studies (e.g., Beaumont et al., 2006), we chose rheological parameters from a few reliable laboratory studies. We used the parameters for wet quartzite (Gleason and Tullis, 1995), dry Maryland diabase (Mackwell et al., 1998), and wet olivine (Karato and Wu, 1993) for the upper-midcrust, lower crust, and mantle, respectively (Table 1). We then used the scaling parameter (f) to linearly scale the model viscosity relative to the laboratory data to approximate materials that are stronger or weaker than this base set. This provides a way to test reasonable variations in strength owing to moderate changes in composition or temperature, the effects of water fugacity, or uncertainties in the rheological parameters under the same ambient conditions (Beaumont et al., 2006). In the reference model, we use $f = 5$ for the upper-midcrust, $f = 1$ for the lower crust, and $f = 2$ for the mantle lithosphere in the orogen (Table 1). This is designed to approximate a strong, dry quartzo-feldspathic upper-midcrust, a strong lower crust with refractory, intermediate granulite rocks, and a relatively water-poor continental mantle lithosphere due to dehydration and melt depletion effects during lithosphere formation. The side blocks have values of f that are 5 times larger for each of the three units, so that the deformation is focused in the orogen region during shortening.

Initial Thermal Structure

The initial thermal structure of the model is calculated using a surface temperature of 0 °C, a temperature of 1456 °C at the base of the model ($z = 400$ km), and the thermal conductivity (k) and radiogenic heat production (A_r) values given in Table 1. A high thermal conductivity is assigned to the sublithospheric mantle to maintain a nearly constant heat flux to the base of the lithosphere and produce an adiabatic thermal gradient of 0.4 °C/km in the sublithospheric mantle, in order to simulate heat transfer by a convecting mantle (Pysklywec and Beaumont, 2004). The resulting thermal field has a surface heat flow of 72 mW/m², with temperatures of 955 °C at the Moho and 1320 °C at the base of the lithosphere (60 km depth). The assumption is that this lithosphere is in the backarc region of a subduction zone, and therefore the lithosphere is heated by convection associated with subduction (e.g., Currie and Hyndman, 2006).

TABLE 1. MATERIAL PARAMETERS IN THE REFERENCE NUMERICAL MODELS (A1 AND B1)

	Upper-midcrust	Lower crust	Mantle lithosphere	Sublithospheric mantle
Model geometry				
Initial thickness (km)	25	15	20	340
Thickness after phase 1 (km)	48	28	36	292
Plastic rheology				
c_0 (MPa)	2	2	0	0
ϕ_{eff}	15°–2°	15°–2°	15°	15°
Viscous rheology				
f	5	1	2	1
A (Pa^{-n} s^{-1})	1.10×10^{-28}	5.05×10^{-28}	3.91×10^{-15}	3.91×10^{-15}
B^* (Pa s$^{1/n}$)[†]	2.92×10^{6}	1.91×10^{5}	1.92×10^{4}	1.92×10^{4}
n	4.0	4.7	3.0	3.0
Q (kJ mol^{-1})	223	485	430	430
V^* (cm^3 mol^{-1})	0	0	10	10
Thermal parameters				
k (W m^{-1} K^{-1})	2.25	2.25	2.25	102.5
A_T (µW m^{-3})	1	0.4	0	0
c_p (J kg^{-1} K^{-1})	750	750	1250	1250
Density[§]				
ρ_0 (kg m^{-3})	2800	3000	3300	3300
T_0 (K)	900	900	900	900
α (K^{-1})	3.0×10^{-5}	3.0×10^{-5}	3.0×10^{-5}	3.0×10^{-5}

[†]$B^* = (2^{(1-n)/n}\, 3^{(n+1)/2n})A^{-1/n}$. The term in parentheses converts the pre-exponential viscosity parameter from uniaxial laboratory experiments (A) to the tensor invariant state of stress of the numerical models.

[§]All materials have a temperature-dependent density, given by $\rho(T) = \rho_0[1 - \alpha(T - T_0)]$, where ρ_0 is the reference density at temperature T_0, and α is the volumetric thermal expansion coefficient.

Modeling Approach and Boundary Conditions

For the Miocene Arizaro Basin, subsidence occurred after orogenic shortening, uplift, and shallow regional exhumation (Carrapa et al., 2009; Canavan et al., 2010, 2011; Quade et al., this volume). This is simulated by running the numerical models in two phases. During the first phase, the continental lithosphere is shortened to produce an orogen. Strong lithosphere is introduced through each of the side boundaries at 0.5 cm/yr (Fig. 2A), and shortening localizes within the region of weaker lithosphere. After 38 m.y., the orogen is ~400 km wide and has a ~76 km crust, similar to the geometry of the Puna Plateau (Fig. 2B). Note that we have not attempted to model the details of shortening but instead create the orogen through pure shear. One consequence is that at the end of shortening, the lithosphere thermal gradient has decreased, such that the model surface heat flow is somewhat lower than that observed (~50 mW/m² vs. >60 mW/m²; Springer and Förster, 1998), although the model Moho temperatures (~950 °C) are similar to those inferred in the central Andes (Babeyko et al., 2002).

In the second phase, shortening is stopped, and the density of the root region in the lower crust (model set A) or mantle litho-sphere (model set B) is progressively increased to initiate gravitational instability. As discussed later herein, the density change is attributed to either metamorphic eclogitization of the lower crust or the emplacement of dense, mafic magmas in the mantle lithosphere. During this phase, the mechanical boundary conditions are as follows: (1) a free surface, which allows topography to develop in response to the underlying dynamics, (2) a free slip basal boundary, (3) no vertical velocity on the side boundaries, (4) no horizontal velocity on the lithosphere side boundaries, and (5) a velocity of 2 cm/yr to the left through both side boundaries of the sublithospheric mantle. This flow simulates mantle wedge corner flow associated with Nazca plate subduction below the Puna Plateau. The thermal boundary conditions are: (1) insulating (no heat flux) side boundaries, (2) a constant temperature of 1456 °C for the basal boundary, and (3) a constant temperature of 0 °C for the top boundary.

We conducted a number of model experiments to examine the dynamics of gravitational removal of dense lower crust or mantle lithosphere, focusing on variations in the root density and rheological structure of the orogen (Table 2). The rheology variations in the lithosphere are shown in Figure 2C. All experiments were carried out in phase 2 of the models (after shortening has

TABLE 2. LIST OF MODELS SHOWING PARAMETER VARIATIONS TESTED IN THIS STUDY, FOCUSING ON
THE FRICTIONAL ANGLE (ϕ_{eff}) AND VISCOSITY SCALING FACTOR (f) FOR CRUST AND MANTLE MATERIALS

Model	Model type	Parameter	Upper-midcrust	Lower crust	Mantle lithosphere	Figure number
MODEL SET A: Lower-crust instability						
A1	FPSS	ϕ_{eff}	15°–2°	15°–2°	15°	3
(Ref.)	Strong crust & weak ML	f	5	1	2	
A2	High root densification rate (80 kg/m³/m.y.); other parameters as in A1					
A3	FPSS	ϕ_{eff}	15°–2°	15°–2°	15°	5
	Weak crust & weak ML	f	1	0.1	2	
A4	FPSS	ϕ_{eff}	15°–2°	15°–2°	15°	
	Strong crust & strong ML	f	5	1	10	
A5	No FPSS	ϕ_{eff}	15°	15°	15°	
	Strong crust & weak ML	f	5	1	2	
A6	No FPSS	ϕ_{eff}	2°	2°	15°	
	Strong crust & weak ML	f	5	1	2	
Lower-crust instability with sedimentation in basin						
A7	Sediment density = 2000 kg/m³; other parameters as in A1					8
A8	Sediment density = 2400 kg/m³; other parameters as in A1					
A9	Sediment density = 2000 kg/m³; other parameters as in A2					
MODEL SET B: Mantle lithosphere instability						
B1	FPSS	ϕ_{eff}	15°–2°	15°–2°	15°	10
(Ref.)	Strong crust & weak ML	f	5	1	2	
B2	FPSS	ϕ_{eff}	15°–2°	15°–2°	15°	
	Strong crust & strong ML	f	5	1	10	
B3	FPSS	ϕ_{eff}	15°–2°	15°–2°	15°	
	Weak crust & weak ML	f	1	0.1	2	
B4	FPSS	ϕ_{eff}	15°–2°	15°–2°	15°	
	Weak crust & strong ML	f	1	0.1	10	

Note: In all models, c_0 = 2 MPa in the upper-midcrust and lower crust, and c_0 = 0 MPa in the mantle lithosphere. Ref.—reference; FPSS—frictional-plastic strain softening; ML—mantle lithosphere.

stopped), and the times reported are the times in m.y. since the start of phase 2.

LOWER-CRUST INSTABILITY

Origin of the Crustal Density Anomaly

Model set A addresses gravitational removal of a local region of high-density material in the lower crust of the orogen (model A

block in Fig. 2). The high density may be related to metamorphic eclogitization of the lower crust. During orogenic shortening, the lower crust enters the eclogite stability field (generally temperatures >600 °C, pressures >1.2 GPa; Bousquet et al., 1997), but it may not immediately transform to eclogite. Eclogitization appears to be triggered by the presence of water, whereby localized shear zones or faults may enable the phase change in discrete regions (e.g., Austrheim, 1991; Leech, 2001; Jackson et al., 2004). Here, we test the idea that the granulitic lower crust remains metastable during orogenesis and then undergoes progressive eclogitization in the "root" region shown in Figure 2B, possibly associated with localized lithosphere hydration. This is simulated by gradually increasing the density of this zone, assuming that eclogitization occurs throughout the block. The material within the root initially has the density of lower crust (3000 kg/m^3), and its density is increased until it attains the density of mafic eclogite, 3550 kg/m^3 (Christensen and Mooney, 1995; Bousquet et al., 1997). We test densification rates of 40 kg/m^3/m.y. and 80 kg/m^3/m.y., corresponding to 7.2% and 14.5% eclogitization every million years

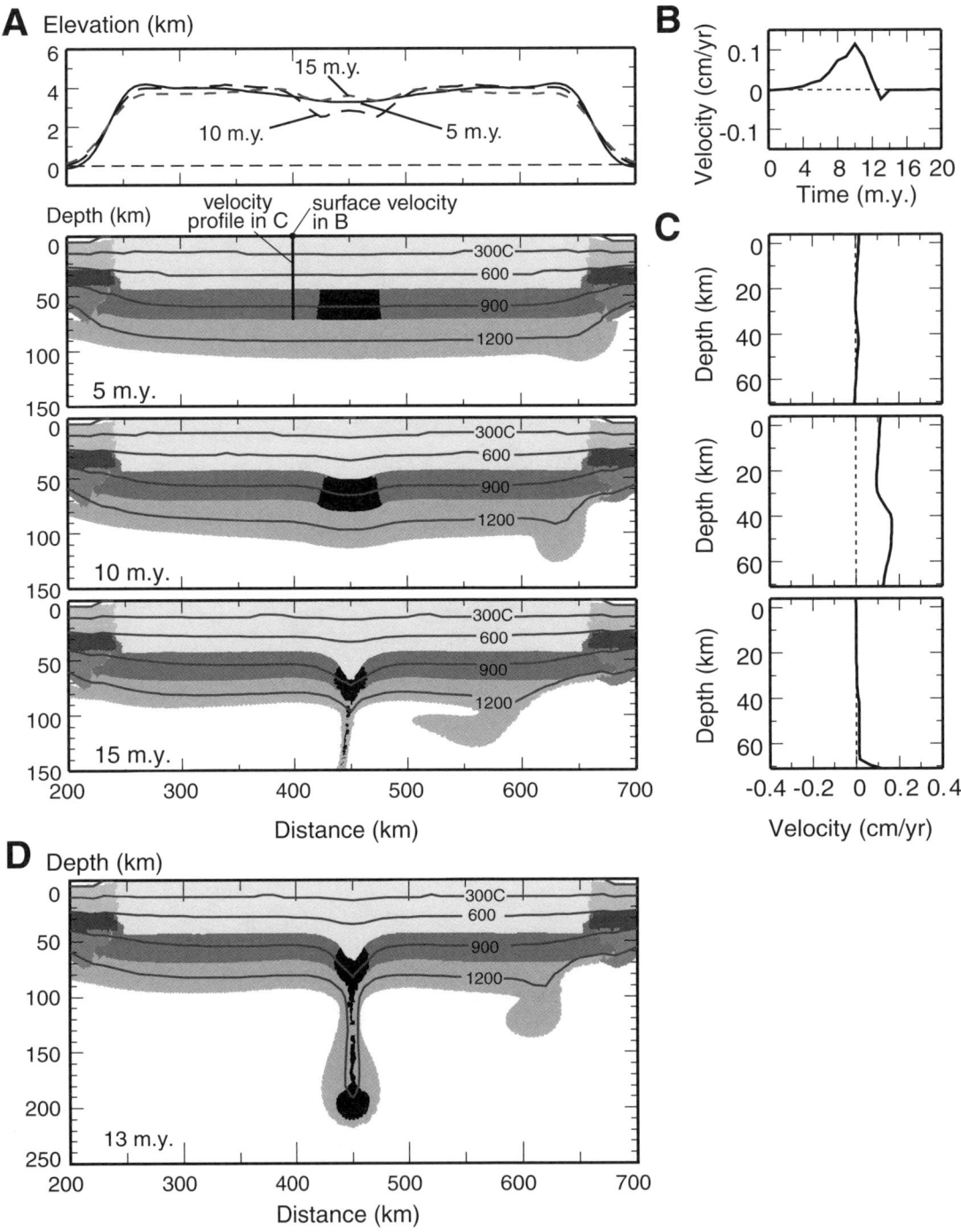

Figure 3. Evolution of reference model A1. (A) Surface topography and model geometry at the given times after the start of the model run. (B) Evolution of the horizontal surface velocity at $x = 400$ km. (C) Crustal flow velocity profile at $x = 400$ km at the times in A. (D) Model geometry at 13 m.y. when the drip is undergoing necking and detachment.

for the 1500 km³ (per unit kilometer along strike) root region. Another possibility is that the local region of high density is related to the emplacement of mafic magmas in the deep crust (Kay and Kay, 1993; Ducea, 2002; Kay and Coira, 2009).

In the models, only the root density is changed. Studies of the rheology of eclogite show that dry eclogite has a similar strength to the dry Maryland diabase used for the lower crust (Jin et al., 2001; Zhang and Green, 2007). Therefore, we assume that the root has the same rheological properties as its protolith. In the reference model, both the root and lower crust use $f = 1$ to represent dry, strong crust. Later, we test the case of weak, hydrated material ($f = 0.1$). Laboratory studies show that water may significantly weaken granulitic and eclogitic lower crust (Zhang and Green, 2007).

Effect of Root Densification Rate (Models A1 and A2)

Reference model A1 uses the model parameters given in Table 1, and the lower-crustal root undergoes densification at a rate of 40 kg/m³/m.y. Figure 3A shows the evolution of this model. Over time, the density of the root increases, and at ~7 m.y., it becomes more dense than the mantle, initiating gravitational instability. It starts to descend into the mantle, and the deep root detaches at ~13 m.y., leaving the uppermost part of the root intact. Figure 3D shows the time of root detachment. The deeper root entrains the surrounding weak mantle lithosphere, and thus there is no gap created in the deep lithosphere. The removed material is swept sideways by the 2 cm/yr mantle flow, as is the mantle lithosphere at the edges of the thickened orogen. The flow appears to only affect the root after detachment, and there is little effect on crustal dynamics or surface topography.

At the surface, the presence of the dense root results in the formation of a local basin with a width of ~100 km (top plot in Fig. 3A). Figure 4 shows the evolution of the depth of the basin. During the first 3 m.y., the subsidence rate is slow, as an isostatic response to root densification. It then increases to ~0.2 km/m.y. between 3 m.y. and 8 m.y. By 9 m.y., the subsidence rate decreases to zero, and the basin is at its maximum depth of ~1.2 km. Between 9 and 13 m.y. (the time of root detachment), the basin becomes shallower, with an uplift rate of ~0.2 km/m.y. The uplift rate then decreases, and by 20 m.y., the basin has uplifted ~1.1 km from its deepest point but is still ~0.1 km lower than the surroundings.

The development of the basin is also associated with contraction at the surface. Figure 3B shows the horizontal surface velocity at a fixed point 400 km from the left model boundary. Note that the model is symmetric about the center of the basin ($x = 450$ km). From 0 to 12 m.y., the velocity of the crustal flow is positive (toward the center of the basin), with a maximum velocity at 10 m.y.; in other words, the basin is experiencing contraction. After 10 m.y., the horizontal velocity decreases, and there is a small negative velocity as the root detaches at 13 m.y., indicating minor extension.

In this model, the maximum basin depth and surface contraction occur before the root is fully removed. This is a consequence of entrainment of the crust by the drip. Figure 3C shows a vertical velocity profile through the crust at $x = 400$ km. At 5 m.y., there is little crustal flow. At 10 m.y., the entire crust has a positive velocity as it undergoes contraction during growth of the root instability, with larger velocities in the deep crust. The enhanced velocities are associated with flow that appears to be driven by the negatively buoyant root, which induces a horizontal pressure gradient in the crust. This results in crustal thickening above the root, and by 10 m.y., crustal flow is sufficient to cause basin inversion and surface uplift prior to root detachment (Fig. 4). After the root detaches, the rate of surface uplift decreases and there is little flow within the crust (Fig. 3C).

Model A1 shows that the presence and removal of a dense crustal root can create a transient basin at surface. The density of the root is one important factor that controls its dynamics. The densification rate of the root depends on the efficiency of the eclogite phase change, which is not well constrained. Model A2 uses a densification rate of 80 kg/m³/m.y., twice that of reference model A1; all other parameters are unchanged. With the greater densification rate, the root undergoes instability and removal more rapidly, and the root detaches at ~8 m.y. At the surface, the dense root results in the formation of a basin with a maximum depth of ~1.4 km at 5 m.y. (Fig. 4), which produces a shorter-lived and slightly deeper basin than in reference model A1. Again, surface uplift is observed prior to root removal, owing to crustal thickening above the root.

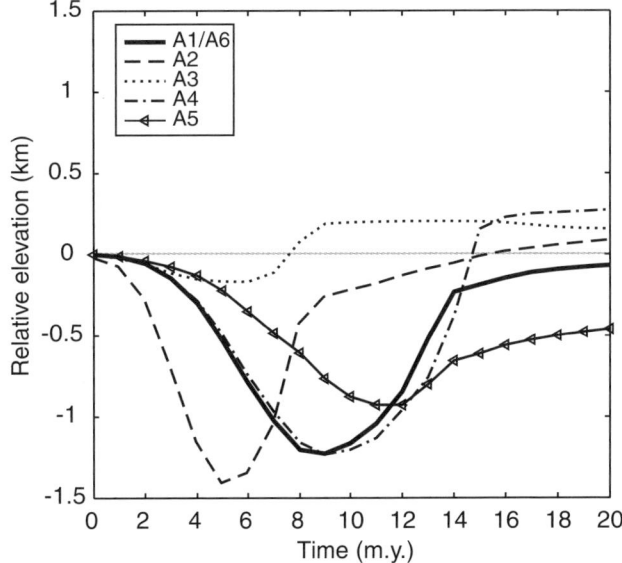

Figure 4. The evolution of the depth of the basin for a lower-crust instability (model set A). The basin depth is the difference between the average elevation of the plateau and the elevation at the center of the surface deflection.

Variations in Viscous Strength (Models A3 and A4)

The strength of the crust and strength of the mantle lithosphere are also important factors that control the dynamics of root removal. We first examine variations in crustal viscous strength, which primarily affects the deep regions of the crust (Fig. 2C). For crust and mantle, the viscous strength at a given temperature is a function of both composition and degree of hydration. Here, we use the scaling factor f (Eq. 4) to study the effect of variations in viscous strength. Reference model A1 uses a relatively strong crust (Table 2). Model A3 has a weaker crust, corresponding to more hydrated conditions or a more felsic composition. The strength is changed by decreasing f to 1 for the upper-midcrust (i.e., the base wet quartzite rheology of Gleason and Tullis, 1995) and 0.1 for the lower crust (10 times weaker than dry Maryland diabase rheology; Mackwell et al., 1998). The other parameters are the same as in model A1, with a root densification rate of 40 kg/m³/m.y.

With the weaker crust, the root detaches at ~9 m.y. (Fig. 5A), which is ~4 m.y. earlier than in the reference model. At the surface, there is little deflection in either the vertical (Fig. 4) or horizontal directions (Fig. 5B). The vertical velocity profile (Fig. 5C) shows that the weak crust is more easily entrained by the dripping root. As the root destabilizes, the flow velocity is ~0.3 cm/yr in the deep crust at 5 m.y., leading to thickening of the crust. Velocity decreases to ~0 cm/yr at 10 m.y. after the root is removed. At 15 m.y., there is negative velocity flow in the deep lower crust. This is caused by both relaxation of the thick, weak crust (i.e., flow driven by differential pressures associated with lateral variations in Moho depth) and shearing by flow in the underlying sublithospheric mantle.

Figure 6 compares the depth of the center of the basin and the cumulative crustal flux at $x = 400$ km (calculated by integrating the crustal flow velocities over the vertical direction and time; Figs. 3C and 5C). In model A1, most crustal flow occurs between 7 and 13 m.y., corresponding to gravitational removal of the root. At 9 m.y., when the basin has its maximum depth,

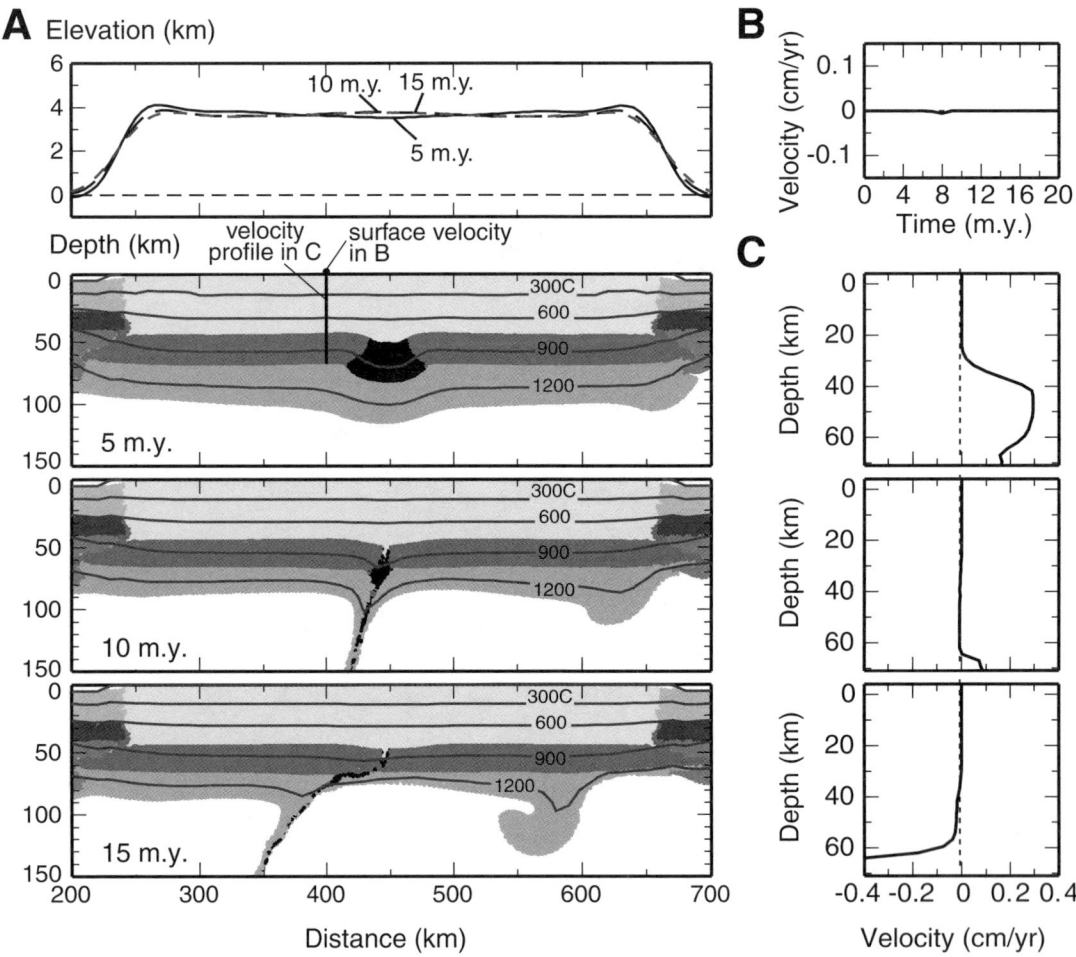

Figure 5. Evolution of model A3. (A) Surface topography and model geometry at the given times after the start of the model run. (B) Evolution of the horizontal surface velocity at $x = 400$ km. (C) Crustal flow velocity profile at $x = 400$ km at the times in A.

the accumulated crustal volume is >100 km³ per km along strike, and it continues to increase until the root detaches at 13 m.y. As a result, surface uplift associated with crustal thickening is greater than surface subsidence related to root removal, and basin inversion occurs prior to root detachment. Figure 6 also shows that the weaker crust of model A3 is more easily entrained by the detaching root. Crustal flow begins as early as 1 m.y., and by 6 m.y., the accumulated crustal volume is ~300 km³ per km along strike. With the enhanced crustal flow, there is little surface deflection associated with root removal.

In addition to crustal strength, root removal depends on mantle lithosphere strength because the root must descend through the mantle lithosphere. Model A4 has identical parameters to the reference model A1, but the mantle lithosphere is five times stronger ($f = 10$) (Tables 2 and 3) to test the effect of a more water-poor mantle lithosphere. In this model, the strong mantle lithosphere reduces the instability growth rate, and root detachment occurs at ~15 m.y., ~2 m.y. later than in the reference model. Figure 4 shows a similar subsidence history of the two models. The only difference is that model A4 has longer rebound time and greater uplift; at 20 m.y., the surface is at ~0.2 km higher than the surroundings. This may be due to a greater amount of crustal thickening above the root, owing to the stronger coupling between the crust and mantle lithosphere in this model.

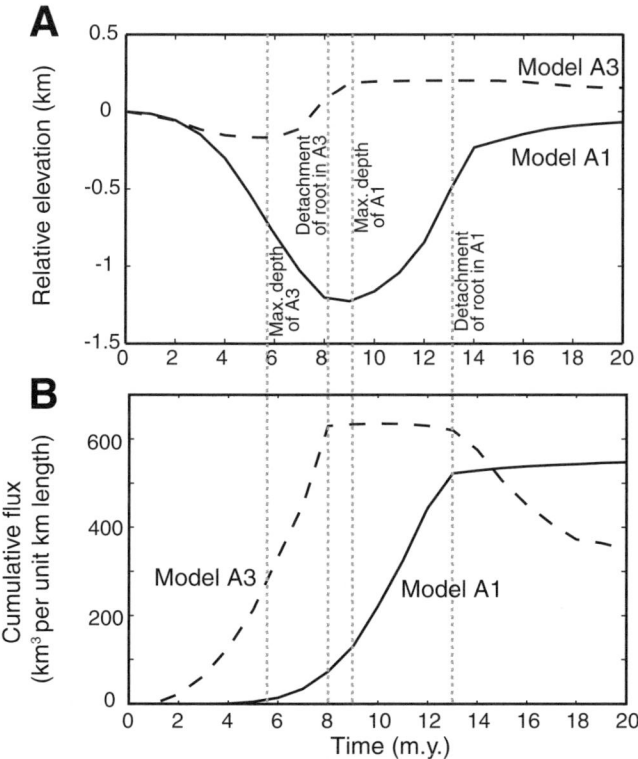

Figure 6. (A) The relative elevation of the center basin for models A1 and A3 (as in Fig. 4). (B) The cumulative crustal volume flowing through a vertical profile at $x = 400$ km per unit kilometer length perpendicular to the model plane.

Variations in Frictional-Plastic Strength (Models A5 and A6)

Deformation of the near-surface rocks (here called "upper crust") is controlled by their frictional-plastic strength (Eq. 3; Fig. 2C). In the previous models, frictional-plastic strain softening was included through a decrease in ϕ_{eff} with accumulated strain, consistent with laboratory and field observations that suggest that rocks may weaken during deformation (e.g., Huismans and Beaumont, 2003; Paterson and Wong, 2005). We now test models with no strain softening. Models A5 and A6 correspond to a strong upper crust ($\phi_{eff} = 15°$) and weak upper crust ($\phi_{eff} = 2°$), respectively (Fig. 2C); other parameters are the same as in model A1 (Tables 2 and 3). A lower ϕ_{eff} may be related to high pore-fluid pressure in the upper crust due to hydration (Babeyko and Sobolev, 2005, and references therein).

In models A5 and A6, the root removal process is similar to that in model A1. The main difference between the models is in the wavelength of surface deflection (Fig. 7). In model A5, the upper crust remains strong throughout model evolution, and the surface deflection has a long wavelength. The basin in model A5 is ~300 km wide, approximately three times that of model A1. In addition, the basin subsides more slowly and attains its maximum depth of ~1 km at 12 m.y. (Fig. 4). In contrast, there is little difference in the behavior of models A1 and A6. In model A6, the upper crust is weak throughout the model evolution, whereas the upper crust in model A1 weakens during deformation to the same strength. In both, the surface subsides to ~1.2 km at 9 m.y., and the width of the basin is ~100 km. These results demonstrate that the frictional-plastic strength of the crust strongly influences the geometry of the basin but has little effect on the dynamics of root removal; a narrower basin requires weak near-surface rocks.

Sedimentation Effects (Models A7, A8, and A9)

The previous models examine the development of a basin associated with lithosphere dynamics (i.e., tectonic subsidence). In nature, the basins are locations of sediment deposition. The introduction of sediments provides a surface load, which will increase basin depth. The thickness of the sediments depends on sedimentation rate, which is controlled by a number of factors, including erosion rate, the nature of the rocks in the source terranes, climate, and basin capacity. Another important factor is the sediment density, which will vary with composition and sediment porosity and may increase with depth due to compaction.

We examine the effect of sediment loading by adding sediments to the basin in model A1. For simplicity, we fill the basin to capacity at each time step, and we test two different sediment densities, 2000 kg/m³ (model A7) and 2400 kg/m³ (model A8; Table 2). The lower value represents more porous sediments that fill the basin to capacity or can be taken as a proxy for denser sediments that do not completely fill the basin. The latter value is comparable to the average value for 2–5-km-thick sediments in North America (Mooney and Kaban, 2010). Other properties in these models are as the same as those in model A1.

TABLE 3. EFFECTS OF KEY PARAMETERS ON SURFACE TOPOGRAPHY AND CRUSTAL FLOW IN NUMERICAL MODELS OF RAYLEIGH-TAYLOR (RT) DRIPS

Parameter change	Example of models	Surface elevation			Wavelength of basin	Duration of basin		Velocity of crustal flow	
		Greater subsidence	Limited subsidence	Uplift	Increase	Increase	Decrease	Increase	Decrease
Increase the depth of root	Set A vs. set B		√√	√√			√√		√√
Increase the densification rate of root	A1 vs. A2	√					√√	√	
Increase the width of root					√√				
Weaken the crust	A1 vs. A3; B1 vs. B3; B2 vs. B4		√√	√√			√	√√	
Strengthen the ML (weak crust)	B3 vs. B4			√		√		√	
Strengthen the ML (strong crust)	A1 vs. A4; B1 vs. B2	√				√		√	
Strengthen the near-surface rock	A1/A6 vs. A5		√		√√				
Add sedimentation	A1 vs. A7/A8; A2 vs. A9	√√				√			

Note: The number of check marks denotes the magnitude of each effect; more check marks indicate a greater effect. ML—mantle lithosphere.

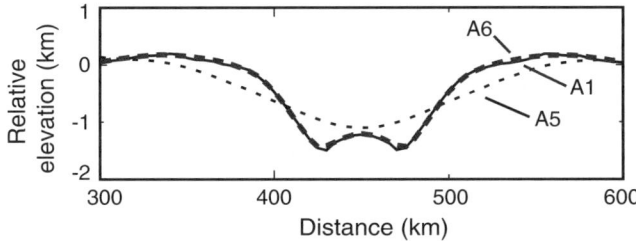

Figure 7. Surface topography for models A1, A5, and A6 at 9 m.y., 11 m.y., and 9 m.y. respectively, when the basin reaches its maximum depth in each model. Note the longer deflection wavelength for model A5, in which the frictional-plastic strength is highest.

Figure 8 shows the evolution of model A7, with sediment (black material at surface) deposited in the basin as it forms. The width of the sediment layer is ~100 km by ~10 m.y., which is consistent with the basin width in model A1. The root removal process is also similar to model A1, and there are only

minor differences in the surface contraction rate and crustal flow velocities. However, the gravitational load of the sediment increases the depth of the basin. The thickness of the sediment increases with time, with the maximum thickness of ~2.9 km at ~11 m.y. (Fig. 9). The change in subsidence rate at 4–5 m.y. appears to be caused by the weakening of the upper crust through frictional-plastic strain softening. A sediment density of 2400 kg/m^3 (model A8) results in a deeper basin (~4.1 km at ~11 m.y.) due to the greater surface load. The surface of the models undergoes isostatic rebound following root removal. Owing to the additional thickness of low-density sediments, the basin region becomes a topographic high, with elevations of ~1.2 km and ~0.8 km higher than the surroundings in models A7 and A8, respectively. Finally, model A9 tests the influence of sedimentation for a model with a greater root densification rate, based on the parameters in model A2 (Table 2). The overall removal time scale is similar to that in model A2, but the additional sediment load leads to a deeper basin, with a depth of ~3.7 km at 6 m.y. (Fig. 9).

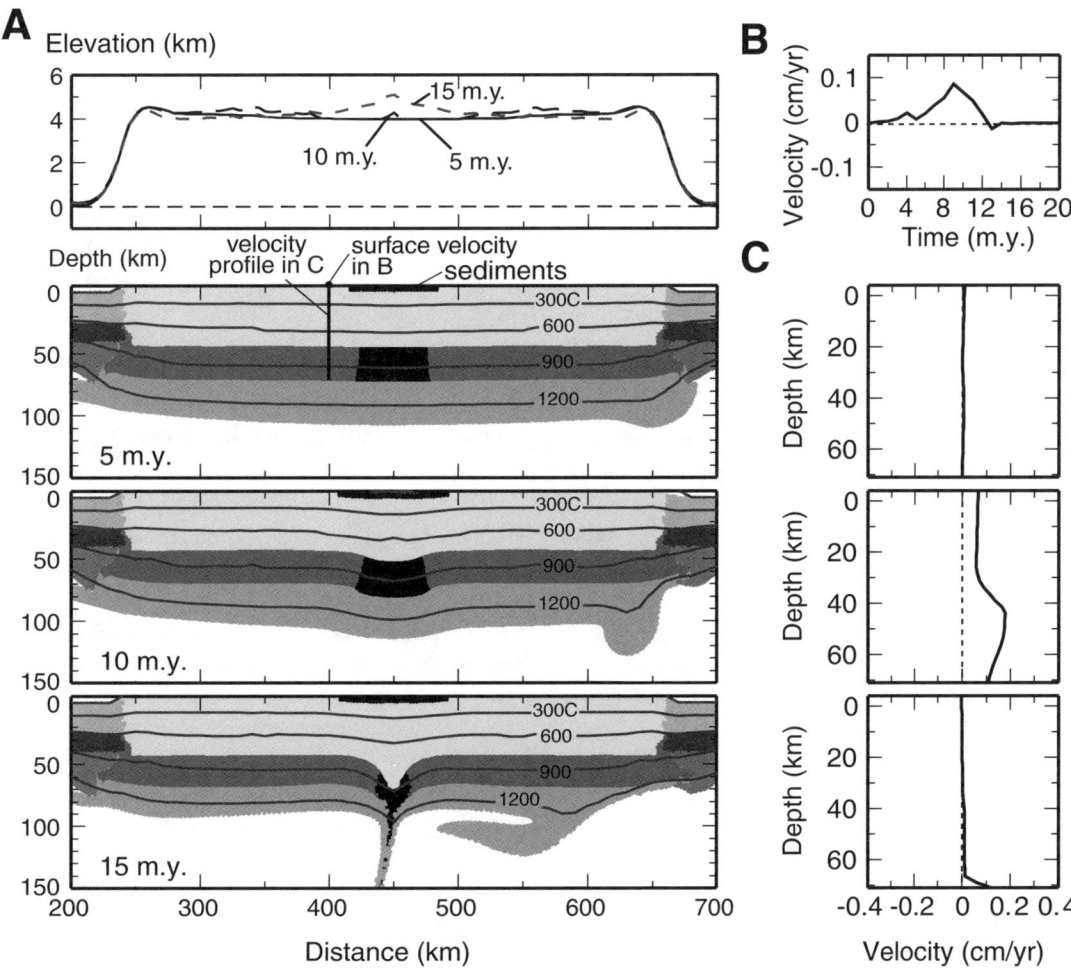

Figure 8. Evolution of model A7. (A) Surface topography and model geometry at the given times after the start of the model run. (B) Evolution of the horizontal surface velocity at $x = 400$ km. (C) Crustal flow velocity profile at $x = 400$ km at the times in A.

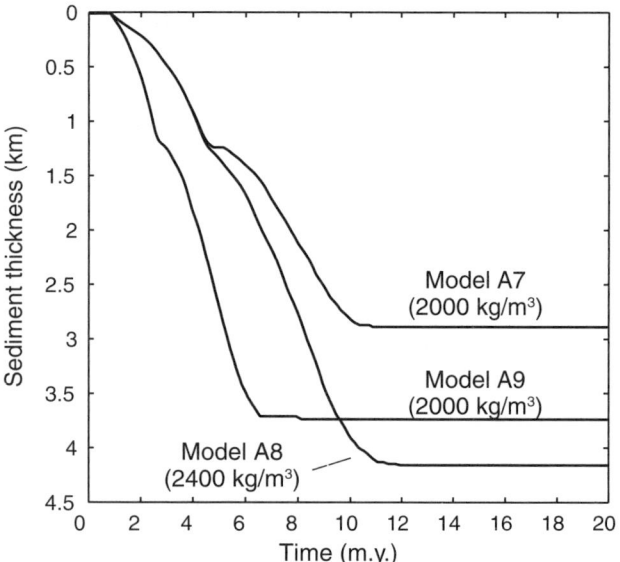

Figure 9. Accumulated sediment thickness over time at the center of the basin. Sediment density is 2000 kg/m³ in models A7 and A9, and 2400 kg/m³ in model A8.

MANTLE LITHOSPHERE INSTABILITY

Origin of the Mantle Lithosphere Instability

An alternate hypothesis is that high-density material is located in the mantle lithosphere of the orogen (block B in Fig. 2A), instead of the lower crust. In model set B, we consider a mantle lithosphere density anomaly associated with either magmatic eclogite or a lithospheric thickness perturbation. Magmatic eclogite can be produced from mantle-derived mafic magmas that pond near the base of the crust, as the crust acts as a density filter and prohibits ascent to the surface. Partial melting and differentiation lead to the development of a high-density eclogitic residue below the Moho, while the low-density, felsic melts rise up to shallower crustal levels (Richards, 2003; Saleeby et al., 2003). In our models, we assume that this process occurs in an ~50-km-wide region in the mantle lithosphere (block B in Fig. 2), and that the densification rate of this region depends on the rate of magma production and segregation. In Cordilleran volcanic arcs, 60–90 km³/m.y. of magma are erupted at the surface per kilometer along-strike arc length, and there may be 1–3 times more residual eclogite at depth (Ducea, 2002; Ducea and Barton, 2007). As magmatic eclogite (density 3550 kg/m³) replaces mantle lithosphere in the root region, the density of the root should increase at a rate of 22–99 kg/m³/m.y. In the models presented here, we use a densification rate of 40 kg/m³/m.y., which is consistent with the reference value used in model set A. We focus on the dynamics of high-density material in the mantle lithosphere, and therefore the models do not include the simultaneous formation of a low-density crustal batholith; inclusion of this is expected to enhance surface uplift of the models following root removal. It should be noted that this mechanism of root formation requires volcanism prior to basin development. There is evidence for arc-type magmatism across much of the Altiplano-Puna Plateau since 25 Ma (Trumbull et al., 2006), possibly associated with changes in dip of the subducting plate (Kay and Coira, 2009). The Arizaro Basin itself has been in close proximity to the volcanic arc since the Miocene (e.g., Kay and Coira, 2009).

Another possibility is that the high-density mantle lithosphere is a block of perturbed mantle lithosphere that is locally thicker than the surrounding lithosphere. A perturbation to the mantle lithosphere can become gravitationally unstable, leading to further thickening over time and then removal as an RT drip (Houseman and Molnar, 1997; Molnar and Houseman, 2004; Göğüş and Pysklywec, 2008). Our models approximate a growing instability by increasing the density in a small region of the mantle lithosphere (block B), instead of imposing variations in lithosphere thickness. The main purpose is to investigate the surface expression of the removal process, rather than a detailed assessment of the origin of the density anomaly.

Effects of a Local Mantle Lithosphere Instability (Model B1)

Reference model B1 has the same parameters as model A1 (Table 1), except that the high-density root (black material) is placed in the mantle lithosphere. Because the root region starts with the same density as the underlying mantle, it becomes gravitationally unstable as soon as its density is increased. The instability grows quickly, and the majority of the root detaches from the lithosphere at ~2 m.y. (Fig. 10A). During the removal process, the surface is hardly deflected and no basin forms (Figs. 10A and 11). In addition, the root has little effect on the crustal dynamics, with little surface contraction or crustal flow (Figs. 10B and 10C).

Variations in Viscous Strength (Models B2, B3, and B4)

We also test variations in the viscous strength of the mantle lithosphere and the crust. Model B2 has the same strong crust as model B1, but the viscosity of the mantle lithosphere is five times larger ($f = 10$; Table 2). The same rheology is used for both the root region and adjacent mantle lithosphere. With the stronger mantle lithosphere, root removal is slightly delayed, with detachment at ~4 m.y. The longer duration of the root in the lithosphere provides a greater negative buoyancy force on the base of the crust, causing the surface to be deflected downward and creating a basin with a maximum depth of ~0.3 km at 3 m.y. As seen in model set A, the basin starts to rebound prior to root removal, as a result of induced horizontal crustal flow and thickening.

We also consider the case in which the crust has a weaker rheology than the reference model, for both the reference mantle lithosphere (model B3) and the stronger mantle lithosphere (model B4; Table 2). The time of root removal in these models is the similar to that seen in models B1 and B2. As the root undergoes gravitational removal, the lateral pressure gradient

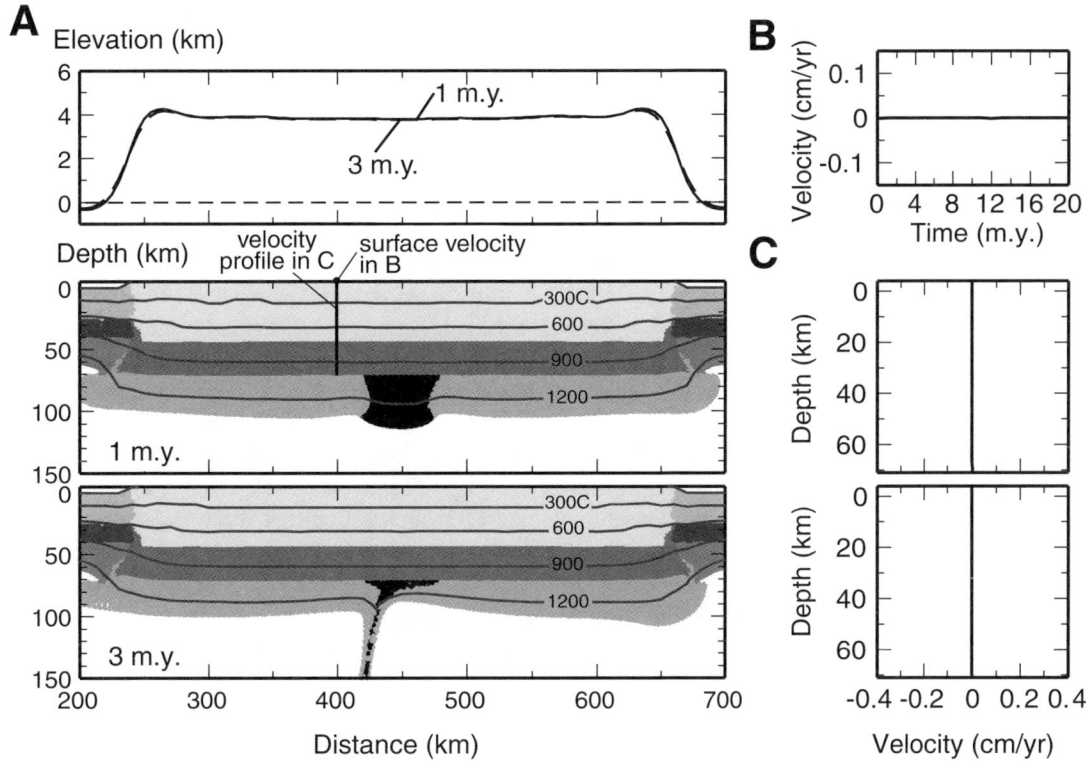

Figure 10. Evolution of reference model B1, in which the high-density (black) instability originates in the mantle lithosphere. (A) Surface topography and model geometry at the given times. (B) Evolution of the horizontal surface velocity at *x* = 400 km. (C) Crustal flow velocity profile at *x* = 400 km at the times in A.

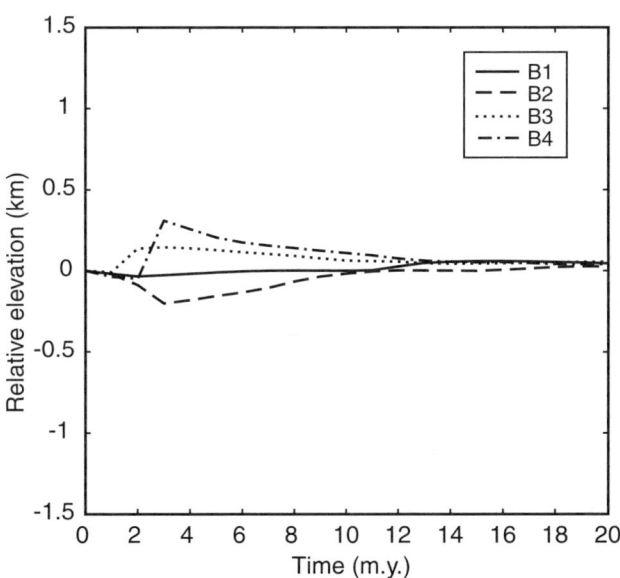

Figure 11. The evolution of the relative surface elevation above the Rayleigh-Taylor drip for a mantle lithosphere instability (model set B). The basin depth is the difference between the average elevation of the plateau and the elevation at the center of the surface deflection (at *x* = 450 km).

induces crustal flow in the weak crust. Flow velocities are sufficient to cause crustal thickening and surface uplift above the root (Fig. 11). In both cases, the surface is uplifted by ~0.3 km, with greater uplift in model B4. Following root removal, the surface subsides to its original elevation, as the weak crust flows outward from the topographic high (i.e., topography-induced crustal flow; Beaumont et al., 2006).

Overall, density anomalies located in the mantle lithosphere do not have a significant surface expression in our models. The maximum surface subsidence in mantle lithosphere instability models is ~0.3 km, which is much less than that of the Arizaro Basin (~1.6 km tectonic subsidence, DeCelles et al., this volume). If the crust is weak, the gravitational instability induces crustal flow and surface uplift above the root.

DISCUSSION AND CONCLUSIONS

Local hinterland basins of the Altiplano-Puna Plateau are enigmatic features that cannot be clearly linked to orogenic deformation or thrust loading (Garzione et al., 2006; Jordan et al., 2007; Horton, 2012; DeCelles et al., this volume). These basins appear to be transient features, as demonstrated by the sedimentary record of the Miocene Arizaro Basin. The present Puna

Plateau is characterized by several internally drained basins, such as the Salar de Arizaro, Salar de Pocitos, and Salar de Atacama basins. An analysis of the isostatic gravity field, derived from satellite gravity observations, shows that these basins are characterized by positive isostatic gravity anomalies (Fig. 1C), consistent with high-density material beneath the surface. The wavelength of the gravity anomaly is generally 100–200 km, indicating a lithosphere origin for the density anomaly. It is proposed that hinterland basins may be related to local gravitational removal of high-density lithosphere as an RT drip instability (e.g., DeCelles et al., this volume). This would explain both the gravity observations and the transient nature of the basins.

Summary of Modeling Results

The main goal of our study is to assess the dynamics of lithosphere removal and its surface expression, for cases in which the density anomaly ("root") is located in either the lower crust (model set A) or mantle lithosphere (model set B). In our models, the root density increases over time to simulate the formation of high-density material through metamorphic or magmatic processes. In all cases, the root region undergoes gravitational removal as an RT drip within 10 m.y. of the time at which its density exceeds that of the underlying mantle. Longer removal times are observed in models in which the root densification rate is low or the root and underlying mantle lithosphere have a high viscosity. These time scales are consistent with the observed time scales for lithospheric RT drips in other modeling studies (e.g., Houseman and Molnar, 1997; Jull and Kelemen, 2001; Elkins-Tanton, 2007; Göğüş and Pysklywec, 2008).

Table 3 summarizes the results of our models. To first order, the presence of the high-density root induces subsidence of the overlying surface, followed by uplift as the density anomaly is gravitationally removed. Initial subsidence occurs as an isostatic response to the development of the root, with more rapid subsidence occurring with a higher densification rate. However, with a high densification rate, the root becomes gravitationally unstable at an earlier time, and therefore the duration of subsidence is reduced. The geometry of the root also affects the surface expression. Subsidence is greater for a root located at a relatively shallow depth. In additional models not presented here, we found that an increase in the width of the root increases the magnitude of surface subsidence and the width over which subsidence occurs but slightly decreases the duration of subsidence, owing to the enhanced negative buoyancy of the larger root region.

An important result from the models is that the crust overlying the root may undergo deformation during root removal, which can significantly alter the surface expression. The strength of the near-surface rocks primarily affects the wavelength over which the surface is deflected. If the near-surface rocks are relatively weak, surface deflection is more localized and is primarily controlled by the width of the dense root. A weak upper crust may be the result of an intrinsically low frictional-plastic strength associated with a high pore-fluid pressure or weakness developed

during deformation (e.g., Huismans and Beaumont, 2003). The surface expression of an RT drip also depends on the strength of the deep crust, which is controlled by its viscous rheology. If the deep crust is relatively weak, the lateral pressure gradients associated with the presence of the high-density root can induce crustal flow. This is analogous to horizontal crustal flow that is driven by lateral pressure gradients associated with topographic variations (e.g., Beaumont et al., 2006, and references therein). Flow velocities increase for either a weaker crust or denser root, leading to crustal thickening above the root and a diminished surface deflection. Such flow has been observed in previous studies of RT drips (e.g., Neil and Houseman, 1999; Pysklywec and Beaumont, 2004; Elkins-Tanton, 2007).

Surface Expressions of RT Drips

Overall, we find that the two main controls on the surface expression of an RT drip are the depth of the high-density root and the strength of the crust overlying the root. Three main surface expressions are observed in the models: (1) surface subsidence of >500 m, followed by uplift (Fig. 12A), (2) little surface subsidence (<200 m; Figs. 12B and 12C), and (3) surface uplift followed by collapse (Fig. 12D).

The surface deflection depends on the degree of coupling between the root and the surface, which in part depends on the depth of the root. The viscous strength of the crust decreases with depth, owing to increasing temperatures (Fig. 2C), and therefore a shallower root is overlain by relatively strong material, allowing greater coupling between the root and near-surface material. Figure 13 compares the lithosphere strain rate field during root removal for models A1 and B1; the only difference between the models is the root depth. For the model times shown, the roots in models A1 and B1 have similar densities, 3360 kg/m^3 and 3420 kg/m^3, respectively. However, the crustal strain rate in model A1 is significantly higher (Fig. 13A), with a zone of high strain rate that originates from the dense crustal root and extends from the near-surface to the base of the lithosphere. The surface is depressed by stresses associated with root dynamics, and a basin forms (Fig. 12A). In contrast, a dense root located in the mantle lithosphere results in deformation localized in the deep crust and mantle lithosphere, with low strain rates in the shallower crust above the root (Fig. 13B). In this case, the root dynamics are decoupled from the shallower crust, and there is little deflection of the surface (Fig. 12B).

The other factor that determines coupling between the root and surface is the crustal viscosity. The reference models (A1 and B1) use rheological laws representative of dry, strong crustal rocks. If the crust is weaker than assumed, owing to a more felsic composition, hydration, or higher temperatures, the coupling stress between the root and surface decreases. For the weak crust end members examined in this study, horizontal flow is induced in the weak crust, leading to crustal thickening above the root. For a crustal root, the rate of crustal flow approximately balances the rate at which the root descends, and there is little surface deflection

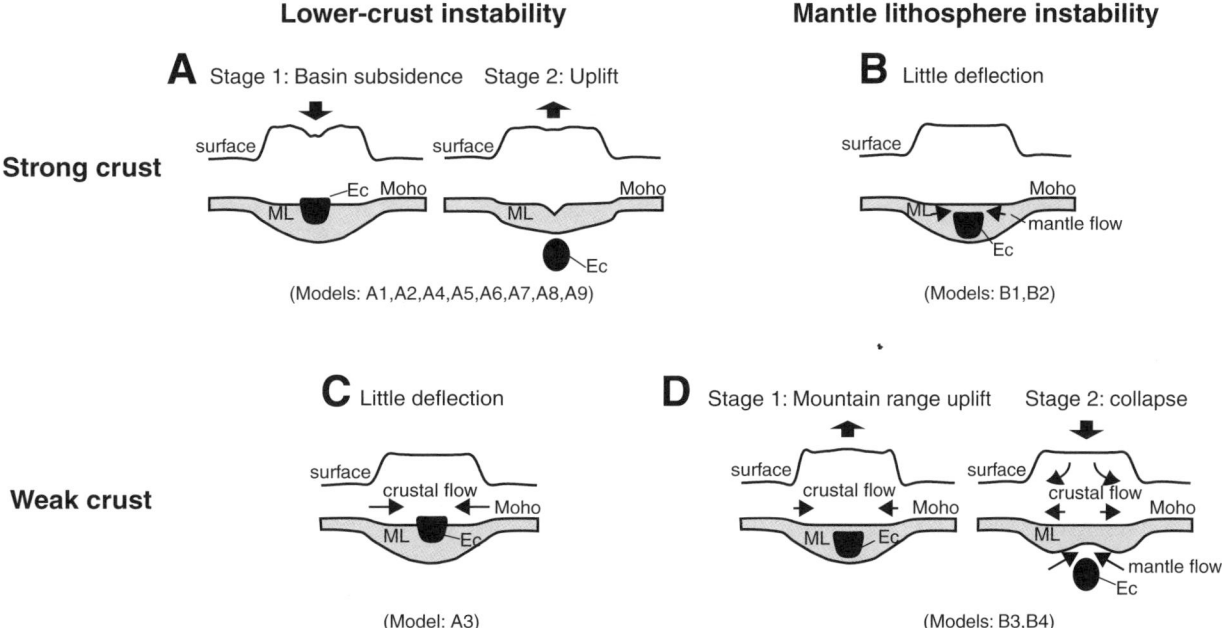

Figure 12. End-member models for the surface effects of Rayleigh-Taylor (RT) instabilities. (A) Lower-crust instability with a viscously strong crust. The high-density lower-crustal root is removed as an RT drip. The surface first subsides and then undergoes uplift as the crust thickens and the root is removed. (B) Mantle lithosphere (ML) instability with a viscously strong crust. Stresses associated with removal of the mantle lithosphere root are localized in the deep lithosphere, and there is little surface deflection. (C) Lower-crust instability with a viscously weak crust. Root removal induces crustal flow, and there is little surface deflection. (D) Mantle lithosphere instability with a viscously weak crust. Root removal induces crustal thickening, which uplifts the surface. After root removal, the topographic high collapses. Ec—eclogite.

(Fig. 12C). If the root is located in the mantle lithosphere, the rate of crustal thickening is sufficient to cause surface uplift of a few hundred meters (Fig. 12D). Once the root is removed, the crust relaxes, and the surface collapses to its original elevation.

Our models consider gravitational instabilities in a deformed and thickened orogen. In our models, the crust is thick (~70 km), and the Moho temperature is ~955 °C, and therefore the deep crust is relatively weak, even for the strongest rheology tested in our study. As a result, a mantle lithosphere instability is only weakly coupled to the shallower crust, leading to a minor surface deflection. Models with thin crust and relatively thick lithosphere may give different results. In this case, the Moho temperature

is low (<400 °C), and the crust is strong, which increases the coupling between the crust and mantle lithosphere and enhances surface subsidence (e.g., Göğüş and Pysklywec, 2008). Such models may be important for understanding the origin of basins in continental interiors.

Comparison with the Miocene Arizaro Basin

We now compare our model results with the geological record of the Miocene Arizaro Basin, which is taken as a representative example of a transient hinterland basin. DeCelles et al. (this volume) summarize the evolution of the Arizaro Basin.

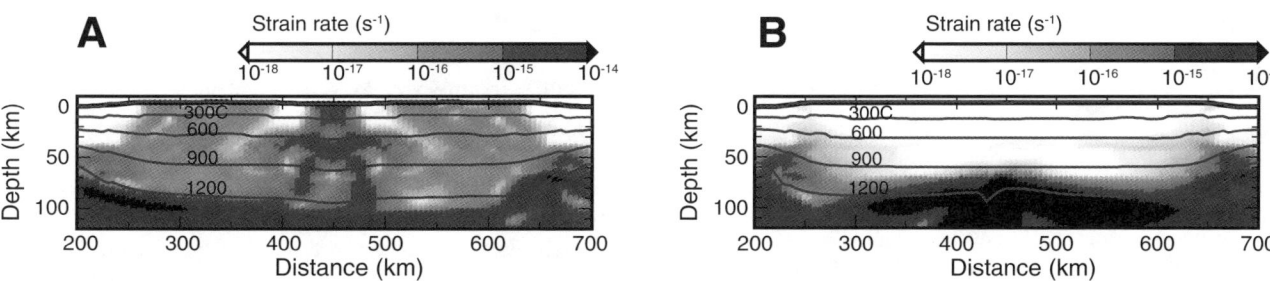

Figure 13. Strain rate field at the time of maximum basin depth for: (A) model A1 (lower-crust root) at 9 m.y. and (B) model B1 (mantle lithosphere root) at 3 m.y.

The subsidence history of the basin is given in Figure 14. Basin subsidence occurred from 21 to 8 Ma, to a maximum depth of ~3400 m, with an inferred ~1600 m of tectonic subsidence. The basin was subsequently exhumed, and today its surface is ~500 m higher than its surroundings. Also shown on Figure 14 is the surface elevation above the RT drip for models A2 and A9. Both models use the same parameters: a strong crustal rheology with frictional-plastic strain softening, and a crustal root with a densification rate of 80 kg/m^3/m.y.; the difference is that model A9 includes sedimentation (Table 2). These two models provide the best fit to the Arizaro Basin observations, including the following: (1) A symmetric basin of ~100 km width is formed, with greatest subsidence in the center. In three dimensions, this basin would exhibit a roughly circular shape. (2) A sigmoidal subsidence record is seen, with low subsidence during the first 1–3 m.y., followed by accelerating subsidence and then a decrease. Model A2 exhibits a maximum subsidence of ~1400 m. With the addition of sediments (model A9), the total subsidence is ~3600 m. Small differences between the observed and modeled subsidence histories can be reduced with minor changes in the densification rate of the root and in the sediment thickness/density. (3) The basin undergoes minor internal shortening (contraction) during formation. (4) The basin is then uplifted to a higher elevation than its surroundings. Uplift occurs as an isostatic response to removal of the dense root, low density associated with crustal thickening above the root, and the presence of low-density basin sediments.

Both the overall subsidence pattern of the Arizaro Basin and its localized distribution are consistent with the surface deformation associated with gravitational removal of anomalously dense lithosphere. The Arizaro Basin formed after the basement rocks of this part of the Puna Plateau had already experienced orogenic shortening and crustal thickening (Carrapa et al., 2009). Our models demonstrate that in regions of thick crust, gravitational removal of a dense root will only create a significant surface deflection if the root is located within the crust and the crust is relatively strong (Fig. 12A). Therefore, we favor a crustal origin for the root, possibly related to progressive metamorphic eclogitization of metastable granulitic crust or the emplacement of eclogitic magmatic restites (e.g., Kay and Kay, 1993; Leech, 2001; Kay and Coira, 2009). Even with a strong crust, gravitational instability of the root can induce flow in the deep crust if the temperature is relatively high. As a result, the crust will thicken as the root destabilizes, and the maximum subsidence and the onset of basin inversion may predate the time of root detachment.

Our study is primarily focused on the topographic expression associated with an RT drip. There are a number of other observations that may also be used to constrain the dynamics of this process. For example, after root detachment, the crust may be characterized by a V-shaped Moho with thickened crust in the former root region (e.g., Fig. 3). This geometry has been proposed to explain seismic data for the Sierra Nevada region of California, where root detachment has been inferred (Zandt et al., 2004). For the Arizaro region, ambient noise tomography images show that the lower crust has a low velocity and could be formed by downwelling of lower crust under the basin (Beck et al., this volume). Other seismic studies suggest that the crust may actually be thinner than surrounding regions (Yuan et al., 2002; Bianchi et al., 2013). However, these studies have only sparse data coverage of this area and may not be able to resolve Moho variations on lateral scales less than 50 km. Removal of a dense root may also be associated with magmatism, produced by melting of the descending root material, decompression melting of upwelling asthenosphere, and melting of the remaining continental lithosphere (e.g., Elkins-Tanton, 2007). Geochemical data for mafic magmatism in parts of the Altiplano-Puna Plateau are consistent with small-scale (<50 km width) convective instabilities over time scales of 1–5 m.y. (Ducea et al., 2013). This is consistent with the scale and timing of the instabilities in our models, but further work is needed to assess whether the material in our models will melt.

ACKNOWLEDGMENTS

We thank Jolante van Wijk and an anonymous reviewer for constructive comments that helped improve the manuscript. Numerical models in this study use the SOPALE numerical modeling code, developed under the direction of Christopher Beaumont (Dalhousie University, Halifax, Nova Scotia, Canada). Research was supported by grants from ExxonMobil Upstream Research Company and the Natural Sciences and Engineering Research Council (NSERC), and computational resources from Compute Canada (Westgrid).

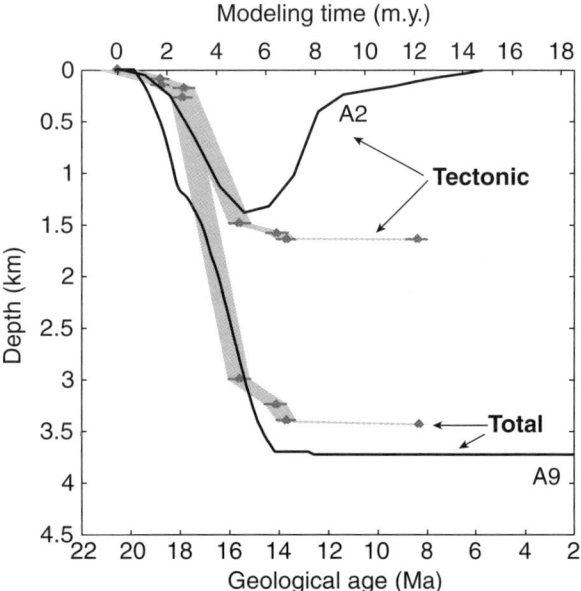

Figure 14. Comparison of the modeling results (models A2 and A9, solid lines) with the subsidence history of Miocene Arizaro Basin (gray circles). The gray envelopes and error bars are uncertainties in the tectonic component of subsidence and total sediment accumulation of Arizaro Basin (DeCelles et al., this volume).

REFERENCES CITED

Austrheim, H., 1991, Eclogite formation and dynamics of crustal roots under continental collision zones: Terra Nova, v. 3, p. 492–499, doi:10.1111/j .1365-3121.1991.tb00184.x.

Babeyko, A.Y., and Sobolev, S.V., 2005, Quantifying different modes of the late Cenozoic shortening in the central Andes: Geology, v. 33, p. 621–624, doi:10.1130/G21126.1.

Babeyko, A.Y., Sobolev, S.V., Trumbull, R.B., Oncken, O., and Lavier, L.L., 2002, Numerical models of crustal scale convection and partial melting beneath the Altiplano–Puna Plateau: Earth and Planetary Science Letters, v. 199, p. 373–388, doi:10.1016/S0012-821X(02)00597-6.

Beaumont, C., Nguyen, M.H., Jamieson, R.A., and Ellis, S., 2006, Crustal flow modes in large hot orogens, *in* Law, R.D., Searle, M.P., and Godin, L., eds., Channel Flow, Ductile Extrusion, and Exhumation in Continental Collision Zones: Geological Society, London, Special Publication 268, p. 91–145, doi:10.1144/GSL.SP.2006.268.01.05.

Beck, S.L., and Zandt, G., 2002, Nature of orogenic crust in the central Andes: Journal of Geophysical Research, v. 107, p. 2230, doi:10.1029/2000JB000124.

Beck, S.L., Zandt, G., Myers, S.C., Wallace, T.C., Silver, P.G., and Drake, L., 1996, Crustal-thickness variations in the central Andes: Geology, v. 24, p. 407–410, doi:10.1130/0091-7613(1996)024<0407:CTVITC> 2.3.CO;2.

Beck, S.L., Zandt, G., Ward, K.M., and Scire, A., 2015, this volume, Multiple styles and scales of lithospheric foundering beneath the Puna Plateau, central Andes, *in* DeCelles, P.G., Ducea, M.N., Carrapa, B., and Kapp, P.A., eds., Geodynamics of a Cordilleran Orogenic System: The Central Andes of Argentina and Northern Chile: Geological Society of America Memoir 212, doi:10.1130/2015.1212(03).

Bianchi, M., Heit, B., Jakovlev, A., Yuan, X., Kay, S.M., Sandvol, E., Alonso, R.N., Coira, B., Brown, L., Kind, R., and Comte, D., 2013, Teleseismic tomography of the southern Puna plateau in Argentina and adjacent regions: Tectonophysics, v. 586, p. 65–83, doi:10.1016/j.tecto.2012.11.016.

Bird, P., 1979, Continental delamination and the Colorado Plateau: Journal of Geophysical Research, v. 84, no. B13, p. 7561–7571, doi:10.1029/ JB084iB13p07561.

Bousquet, R., Goff, B., Henry, P., Le Pichon, X., and Chopin, C., 1997, Kinematic, thermal and petrological model of the central Alps: Lepontine metamorphism in the upper crust and eclogitisation of the lower crust: Tectonophysics, v. 273, p. 105–127, doi:10.1016/S0040-1951(96)00290-9.

Canavan, R., Clementz, M., Carrapa, B., Quade, J., DeCelles, P.G., Schoenbohm, L.M., and Boyd, J., 2010, Paleoelevation of the Puna Plateau, northwestern (NW) Argentina, inferred from deuterium isotopic analyses of volcanic glass: San Francisco, California, American Geophysical Union, Fall Meeting, abstract PP13B-1525.

Canavan, R., Clementz, M., Carrapa, B., Quade, J., DeCelles, P.G., and Schoenbohm, L.M., 2011, Paleoelevation of the Puna Plateau (northwest Argentina) inferred from geochemical analyses of volcanic glass: Geological Society of America Abstracts with Programs, v. 43, no. 5, p. 539.

Carrapa, B., DeCelles, P.G., Reiners, P.W., Gehrels, G.E., and Sudo, M., 2009, Apatite triple dating and white mica ^{40}Ar/^{39}Ar thermochronology of syntectonic detritus in the central Andes: A multiphase tectonothermal history: Geology, v. 37, no. 5, p. 407–410, doi:10.1130/G25698A.1.

Carrapa, B., Reyes-Bywater, S., DeCelles, P.G., Mortimer, E., and Gehrels, G., 2011, Cenozoic synorogenic basin evolution in the Eastern Cordillera of northwestern Argentina (25°–26°S): Regional implications for Andean orogenic wedge development: Basin Research, v. 23, p. 1–20, doi:10.1111/j.1365-2117.2011.00519.x.

Christensen, N.I., and Mooney, W.D., 1995, Seismic velocity structure and composition of the continental crust: A global view: Journal of Geophysical Research, v. 100, no. B6, p. 9761–9788, doi:10.1029/95JB00259.

Currie, C.A., and Hyndman, R.D., 2006, The thermal structure of subduction zone back arcs: Journal of Geophysical Research, v. 111, B08404, doi:10.1029/2005JB004024.

DeCelles, P.G., Carrapa, B., Horton, B.K., McNabb, J., and Boyd, J., 2011, Cordilleran hinterland basins as recorders of lithospheric removal in the central Andes: San Francisco, California, American Geophysical Union, Fall Meeting, abstract T13I-06.

DeCelles, P.G., Carrapa, B., Horton, B.K., McNabb, J., Gehrels, G.E., and Boyd, J., 2015, this volume, The Miocene Arizaro Basin, central Andean hinterland: Response to partial lithosphere removal? *in* DeCelles, P.G., Ducea, M.N., Carrapa, B., and Kapp, P.A., eds., Geodynamics of a Cordilleran Orogenic System: The Central Andes of Argentina and Northern Chile: Geological Society of America Memoir 212, doi:10.1130/2015.1212(18).

Delft Gravity Model, 2013, Delft Gravity Model release 1, Satellite-only (DGM-1S): www.citg.tudelft.nl/index.php?id=52752 (accessed April 2013).

Ducea, M.N., 2002, Constraints on the bulk composition and root foundering rates of continental arcs: A California arc perspective: Journal of Geophysical Research, v. 107, no. B11, 2304, doi:10.1029/2001JB000643.

Ducea, M.N., and Barton, M.D., 2007, Igniting flare-up events in Cordilleran arcs: Geology, v. 35, no. 11, p. 1047–1050, doi:10.1130/G23898A.1.

Ducea, M.N., Seclaman, A.C., Murray, K.E., Jianu, D., and Schoenbohm, L.M., 2013, Mantle-drip magmatism beneath the Altiplano-Puna Plateau, central Andes: Geology, v. 41, no. 8, p. 915–918, doi:10.1130/G34509.1.

Elkins-Tanton, L.T., 2007, Continental magmatism, volatile recycling, and a heterogeneous mantle caused by lithospheric gravitational instabilities: Journal of Geophysical Research–Solid Earth, v. 112, B03405, doi:10.1029/2005JB004072.

Fullsack, P., 1995, Arbitrary Lagrangian-Eulerian formulation for creeping flows and its application in tectonic models: Geophysical Journal International, v. 120, no. 1, p. 1–23, doi:10.1111/j.1365-246X.1995.tb05908.x.

Garzione, C.N., Molnar, P., Libarkin, J.C., and MacFadden, B.J., 2006, Rapid late Miocene rise of the Bolivian Altiplano: Evidence for removal of mantle lithosphere: Earth and Planetary Science Letters, v. 241, p. 543–556, doi:10.1016/j.epsl.2005.11.026.

Gleason, G.C., and Tullis, J., 1995, A flow law for dislocation creep of quartz aggregates determined with the molten salt cell: Tectonophysics, v. 247, no. 1–4, p. 1–23, doi:10.1016/0040-1951(95)00011-B.

Göğüş, O.H., and Pysklywec, R.N., 2008, Near-surface diagnostics of dripping or delaminating lithosphere: Journal of Geophysical Research, v. 113, B11404, doi:10.1029/2007JB005123.

Götze, H.J., and Krause, S., 2002, The central Andean gravity high, a relic of an old subduction complex?: Journal of South American Earth Sciences, v. 14, p. 799–811, doi:10.1016/S0895-9811(01)00077-3.

Heit, B., Koulakov, I., Asch, G., Yuan, X., Kind, R., Alcozer, I., Tawackoli, S., and Wilke, H., 2008, More constraints to determine the seismic structure beneath the central Andes at 21°S using teleseismic tomography analysis: Journal of South American Earth Sciences, v. 25, p. 22–36, doi:10.1016/j .jsames.2007.08.009.

Hoke, G.D., and Garzione, C.N., 2008, Paleosurfaces, paleoelevation, and the mechanisms for the late Miocene topographic development of the Altiplano Plateau: Earth and Planetary Science Letters, v. 271, p. 192–201, doi:10.1016/j.epsl.2008.04.008.

Horton, B.K., 2012, Cenozoic evolution of hinterland basins in the Andes and Tibet, *in* Busby, C.J., and Azor, A., eds., Tectonics of Sedimentary Basins: Recent Advances (1st ed.): Oxford, UK, Blackwell Publishing, p. 427–444.

Houseman, G.A., and Molnar, P., 1997, Gravitational (Rayleigh-Taylor) instability of a layer with non-linear viscosity and convective thinning of continental lithosphere: Geophysical Journal International, v. 128, no. 1, p. 125–150, doi:10.1111/j.1365-246X.1997.tb04075.x.

Huismans, R.S., and Beaumont, C., 2002, Asymmetric lithospheric extension: The role of frictional plastic strain softening inferred from numerical experiments: Geology, v. 30, no. 3, p. 211–214, doi:10.1130/0091-7613 (2002)030<0211:ALETRO>2.0.CO;2.

Huismans, R.S., and Beaumont, C., 2003, Symmetric and asymmetric lithospheric extension: Relative effects of frictional-plastic and viscous strain softening: Journal of Geophysical Research, v. 108, no. B10, 2496, doi:10.1029/2002JB002026.

Jackson, J.A., Austrheim, H., McKenzie, D., and Priestley, K., 2004, Metastability, mechanical strength, and the support of mountain belts: Geology, v. 32, no. 7, p. 625–628, doi:10.1130/G20397.1.

Jin, Z.M., Zhang, J., Green, H.W., and Jin, S., 2001, Eclogite rheology: Implications for subducted lithosphere: Geology, v. 29, no. 8, p. 667–670, doi:10.1130/0091-7613(2001)029<0667:ERIFSL>2.0.CO;2.

Jordan, T.E., Mpodozis, C., Munoz, N., Blanco, N., Pananont, P., and Gardeweg, M., 2007, Cenozoic subsurface stratigraphy and structure of the Salar de Atacama Basin, northern Chile: Journal of South American Earth Sciences, v. 23, p. 122–146, doi:10.1016/j.jsames.2006.09.024.

Jull, M., and Kelemen, P.B., 2001, On the conditions for lower crustal convective instability: Journal of Geophysical Research, v. 106, no. B4, p. 6423–6446, doi:10.1029/2000JB900357.

Karato, S.I., and Wu, P., 1993, Rheology of the upper mantle: A synthesis: Science, v. 260, no. 5109, p. 771–778, doi:10.1126/science.260.5109.771.

Kay, R.W., and Kay, S.M., 1993, Delamination and delamination magmatism: Tectonophysics, v. 219, p. 177–189, doi:10.1016/0040-1951(93)90295-U.

Kay, S.M., and Coira, B.L., 2009, Shallowing and steepening subduction zones, continental lithospheric loss, magmatism, and crustal flow under the central Andean Altiplano–Puna Plateau, *in* Kay, S.M., Ramos, V.A., and Dickinson, W.R., eds., Backbone of the Americas: Shallow Subduction, Plateau Uplift, and Ridge and Terrane Collision: Geological Society of America Memoir 204, p. 229–260, doi: 10.1130/2009.1204(11).

Kay, S.M., Coira, B., and Viramonte, J., 1994, Young mafic back-arc volcanic rocks as indicators of continental lithospheric delamination beneath the Argentine Puna Plateau, central Andes: Journal of Geophysical Research, v. 99, p. 24,323–24,339, doi:10.1029/94JB00896.

Leech, M.L., 2001, Arrested orogenic development: Eclogitization, delamination, and tectonic collapse: Earth and Planetary Science Letters, v. 185, no. 1–2, p. 149–159, doi:10.1016/S0012-821X(00)00374-5.

Leier, A., McQuarrie, N., Garzione, C., and Eiler, J., 2013, Stable isotope evidence for multiple pulses of rapid surface uplift in the central Andes, Bolivia: Earth and Planetary Science Letters, v. 371–372, p. 49–58, doi:10.1016/j.epsl.2013.04.025.

Mackwell, S.J., Zimmerman, M.E., and Kohlstedt, D.L., 1998, High-temperature deformation of dry diabase with application to tectonics on Venus: Journal of Geophysical Research, v. 103, no. B1, p. 975–984, doi:10.1029/97JB02671.

Molnar, P., and Houseman, G.A., 2004, The effects of buoyant crust on the gravitational instability of thickened mantle lithosphere at zones of intracontinental convergence: Geophysical Journal International, v. 158, no. 3, p. 1134–1150, doi:10.1111/j.1365-246X.2004.02312.x.

Molnar, P., Houseman, G.A., and Conrad, C.P., 1998, Rayleigh-Taylor instability and convective thinning of mechanically thickened lithosphere: Effects of non-linear viscosity decreasing exponentially with depth and of horizontal shortening of the layer: Geophysical Journal International, v. 133, no. 3, p. 568–584, doi:10.1046/j.1365-246X.1998.00510.x.

Mooney, W.D., and Kaban, M.K., 2010, The North American upper mantle: Density, composition, and evolution: Journal of Geophysical Research, v. 115, B12424, doi:10.1029/2010JB000866.

Myers, S., Beck, S., Zandt, G., and Wallace, T., 1998, Lithospheric-scale structure across the Bolivian Andes from tomographic images of velocity and attenuation for P and S waves: Journal of Geophysical Research, v. 103, p. 21,233–21,252, doi:10.1029/98JB00956.

Neil, E.A., and Houseman, G.A., 1999, Rayleigh-Taylor instability of the upper mantle and its role in intraplate orogeny: Geophysical Journal International, v. 138, no. 1, p. 89–107, doi:10.1046/j.1365-246x.1999.00841.x.

Oncken, O., Hindle, D., Kley, J., Elger, K., Victor, P., and Schemmann, K., 2006, Deformation of the central Andean upper plate system: Facts, fiction and constraints for plateau models, *in* Oncken, O., Chong, G., Franz, G., Giese, P., Götze, H.-J., Ramos, V.A., et al., eds., The Andes: Active Subduction Orogeny: Berlin, Springer-Verlag, Frontiers in Earth Sciences, v. 1, p. 3–28.

Paterson, M.S., and Wong, T.F., 2005, Experimental Rock Deformation—The Brittle Field (2nd ed.): Berlin, Springer, 348 p.

Pysklywec, R.N., and Beaumont, C., 2004, Intraplate tectonics: Feedback between radioactive thermal weakening and crustal deformation driven by mantle lithosphere instabilities: Earth and Planetary Science Letters, v. 221, no. 1–4, p. 275–292, doi:10.1016/S0012-821X(04)00098-6.

Quade, J., Dettinger, M.P., Carrapa, B., DeCelles, P., Murray, K.E., Huntington, K.A., Cartwright, A., Canavan, R.R., Gehrels, G., and Clementz, M., 2015, this volume, The growth of the central Andes, 22°S–26°S, *in* DeCelles, P.G., Ducea, M.N., Carrapa, B., and Kapp, P.A., eds., Geodynamics of a Cordilleran Orogenic System: The Central Andes of Argentina and Northern Chile: Geological Society of America Memoir 212, doi:10.1130/2015.1212(15).

Richards, J.P., 2003, Tectono-magmatic precursors for porphyry Cu-(Mo-Au) deposit formation: Economic Geology and the Bulletin of the Society of Economic Geologists, v. 98, no. 8, p. 1515–1533, doi:10.2113/gsecongeo.98.8.1515.

Saleeby, J., Ducea, M.N., and Clemens-Knott, D., 2003, Production and loss of high-density batholithic root, southern Sierra Nevada, California: Tectonics, v. 22, no. 6, 1064, doi:10.1029/2002TC001374.

Schoenbohm, L.M., and Carrapa, B., 2011, Evidence from the timing of contraction, extension, sedimentation and magmatism for small-scale lithospheric foundering in the Puna Plateau, NW Argentina: San Francisco, California, American Geophysical Union, Fall Meeting, abstract T13I-07.

Schurr, B., Rietbrock, A., Asch, G., Kind, R., and Oncken, O., 2006, Evidence for lithospheric detachment in the central Andes from local earthquake tomography: Tectonophysics, v. 415, no. 1–4, p. 203–223, doi:10.1016/j.tecto.2005.12.007.

Springer, M., and Förster, A., 1998, Heat-flow density across the central Andean subduction zone: Tectonophysics, v. 291, p. 123–139, doi:10.1016/S0040-1951(98)00035-3.

Trumbull, R.B., Riller, U., Oncken, O., Schueber, E., Munier, K., and Hongn, F., 2006, The time-space distribution of Cenozoic arc volcanism in the central Andes: A new data compilation and some tectonic considerations, *in* Oncken, O., Chong, G., Franz, G., Giese, P., Götze, H.-J., Ramos, V.A., et al., eds., The Andes: Active Subduction Orogeny: Berlin, Springer-Verlag, Frontiers in Earth Sciences, v. 1, p. 29–44.

Warren, C.J., Beaumont, C., and Jamieson, R.A., 2008, Deep subduction and rapid exhumation: Role of crustal strength and strain weakening in continental subduction and ultrahigh-pressure rock exhumation: Tectonics, v. 27, no. 6, TC6002, doi:10.1029/2008TC002292.

Whitman, D., 1999, The isostatic residual gravity anomaly of the central Andes, 12° to 29°S: A guide to interpreting crustal structure and deeper lithospheric processes: International Geology Review, v. 41, no. 5, p. 457–475, doi:10.1080/00206819909465152.

Willett, S.D., 1999, Rheological dependence of extension in wedge models of convergent orogens: Tectonophysics, v. 305, no. 4, p. 419–435, doi:10.1016/S0040-1951(99)00034-7.

Yuan, X., Sobolev, S.V., and Kind, R., 2002, Moho topography in the central Andes and its geodynamic implications: Earth and Planetary Science Letters, v. 199, no. 3–4, p. 389–402, doi:10.1016/S0012-821X(02)00589-7.

Zandt, G., Gilbert, H., Owens, T.J., Ducea, M., Saleeby, J., and Jones, C.H., 2004, Active foundering of a continental arc root beneath the southern Sierra Nevada in California: Nature, v. 431, p. 41–46, doi:10.1038/nature02847.

Zhang, J., and Green, H.W., 2007, Experimental investigation of eclogite rheology and its fabrics at high temperature and pressure: Journal of Metamorphic Geology, v. 25, p. 97–115, doi:10.1111/j.1525-1314.2006.00684.x.

MANUSCRIPT ACCEPTED BY THE SOCIETY 3 JUNE 2014
MANUSCRIPT PUBLISHED ONLINE 20 NOVEMBER 2014

The Geological Society of America
Memoir 212
2015

Temporal growth of the Puna Plateau and its bearing on the post–Salta Rift system subsidence of the Andean foreland basin at 25°30′S

Thomas P. Becker
Lori L. Summa
ExxonMobil Upstream Research Company, Houston, Texas 77027, USA

Mihai N. Ducea
Department of Geosciences, University of Arizona, Tucson, Arizona 85721, USA, and
Universitatea Bucuresti, Facultatea de Geologie Geofizica, Str. N. Balcescu Nr 1, Bucuresti 010041, Romania

Garry D. Karner
ExxonMobil Upstream Research Company, Houston, Texas 77027, USA

ABSTRACT

The Puna Plateau, a high-elevation portion of the central Andean Plateau, possesses some of the thickest crust on Earth, and its structural growth should be reflected in the adjacent foreland basin (present-day Eastern Cordillera and Santa Bárbara system) as a flexural response to crustal thickening via contractional deformation. The Cretaceous–Cenozoic stratigraphy preserved within the Eastern Cordillera and Santa Bárbara system also records the influence of the Cretaceous Salta Rift system, which heavily influenced depositional patterns in the region, particularly during postrift thermal subsidence. The Eastern Cordillera and Santa Bárbara system were significantly modified by Neogene inversion of Salta Rift basins, which subdivide the foreland basin and localize depocenters. Here, we examine results of two-dimensional kinematic models of basin formation and fill that proxy the thermal and mechanical behavior of the Salta rifting, and superimpose upon this rifting event two different scenarios for the temporal growth of the Puna Plateau—one with crustal thickening predominantly in the Eocene, and another with progressive crustal thickening beginning in the early Miocene. The two models attempt to forecast the combined effects of inherited rift history and growth of the Puna Plateau on the development of accommodation within the adjacent foreland basin. A Neogene (Miocene-age) Puna Plateau scenario creates a coeval foredeep within the Salta Rift system, but its magnitude and wavelength are influenced by crustal thickening in the Eastern Cordillera and Santa Bárbara system. In contrast, a Paleogene (Eocene-age) growth scenario for the Puna Plateau results in a substantial amount of coeval flexural accommodation in the

Becker, T.P., Summa, L.L., Ducea, M.N., and Karner, G.D., 2015, Temporal growth of the Puna Plateau and its bearing on the post–Salta Rift system subsidence of the Andean foreland basin at 25°30′S, *in* DeCelles, P.G., Ducea, M.N., Carrapa, B., and Kapp, P.A., eds., Geodynamics of a Cordilleran Orogenic System: The Central Andes of Argentina and Northern Chile: Geological Society of America Memoir 212, p. 407–433, doi:10.1130/2015.1212(20). For permission to copy, contact editing@geosociety.org. © 2015 The Geological Society of America. All rights reserved.

adjacent Eastern Cordillera that extends across most of the Salta Rift system, which is broken up by subsequent loading in the Eastern Cordillera and Santa Bárbara system. Thermal subsidence associated with thinned or delaminated mantle lithosphere in the Late Cretaceous also contributes to accommodation and is most prominent during periods of tectonic quiescence. Our modeling results show that: (1) Neogene topographic growth of the Puna Plateau produces a basin subsidence history that is consistent with the geologic record, (2) the Salta Rift system was not buried deeply prior to Neogene exhumation, (3) the eastward advance of the flexural foreland can be related to crustal thickening and elevation gain of the Puna Plateau and Eastern Cordillera at ca. 15 Ma, and (4) interpretations of foreland subsidence history across the Eastern Cordillera may need to consider the influence of thinned mantle lithosphere during Late Cretaceous Salta rifting, which continues to create some accommodation in the region through subtle thermal subsidence.

INTRODUCTION

The development and evolution of foreland basins on continental lithosphere reflect both the intrinsic rheological properties of the crust and the extrinsic processes that form and fill the basins. Intrinsic properties are typically a product of antecedent geologic history, which can create defects or buttresses in the crust that dramatically influence any subsequent accommodation for sediment and localization of deformation (e.g., Allmendinger et al., 1983; Allmendinger and Gubbels, 1996; Thomas, 1991, 2006; McGroder et al., this volume). Extrinsic processes, such as crustal shortening and the growth of continental arcs, create crustal masses that are accommodated isostatically via lithospheric flexure and create space for the accumulation of sediment (Beaumont, 1981; Jordan, 1981; Karner and Watts, 1983). At a first order, these properties and processes define the foreland basin architecture, regional stratigraphy, and timing. It follows, then, that the growth of the orogen is implicitly recorded within the foreland basin stratigraphy, and vice versa. The intent of this work is to evaluate the role of a localized antecedent tectonic event on the subsequent history of flexural accommodation and compare these scenarios with the stratigraphic record. The process provides context for evaluating a range of permissible basin formation scenarios and their potential influence on subsequent basin fill and thermal evolution. For this analysis, we have elected to evaluate the foreland basin evolution of northwestern Argentina (Fig. 1), which has been influenced by two principal basin-forming events: (1) the Cretaceous Salta Rift system, and (2) subsequent Andean convergence. Over the past several years, a collaborative research project on Cordilleran tectonic systems between ExxonMobil and the University of Arizona permitted a number of studies on the tectonic and topographic evolution of the central Andes in northwestern Argentina and central Chile. This combination of new data sets provides a useful framework for reevaluating the growth of the Andes in a region with marked contrast in tectonic style and shortening from the adjacent Altiplano–Tarijas Basin in Bolivia. The style of deformation within the internal to exter-

nal portions of the orogen (i.e., Eastern Cordillera and Santa Bárbara system) is characterized by contractional inversion of Cretaceous Salta Rift system structures, versus basement-detached shortening above a Paleozoic detachment in the Tarijas Basin of Bolivia (Allmendinger et al., 1983; Allmendinger and Gubbels, 1996; Kley and Monaldi, 1998; Echavarria et al., 2003; McGroder et al., this volume). In this paper, we apply numerical, kinematic models of lithospheric deformation to examine how alternative histories of Puna Plateau growth (both initial formation in Eocene and progressive Miocene growth), as well as deformation in the Eastern Cordillera and Santa Bárbara system (Fig. 1), might have modified the history of flexural subsidence in the adjacent foreland basin. A companion paper (Summa et al., this volume) discusses the influence of the different tectonic scenarios from the perspective of hydrocarbon systems modeling in the foreland basin region.

GEOLOGIC FRAMEWORK FOR THE HINTERLAND: CENTRAL ANDEAN PLATEAU

The central Andean Plateau, which extends from southern Peru to northwestern Argentina, is a large, topographically high (>4000 m above sea level), internally drained feature composed of two physiographic provinces: the Altiplano Plateau and the Puna Plateau (Fig. 1). The topographic evolution of the central Andean Plateau contributes to our understanding of tectonics along convergent margins, as well as several interrelated topics, including global climate, ecological evolution, and rates and magnitudes of surficial processes (e.g., Barnes et al., 2012). Estimates of orogenic paleotopography provide an important isostatic parameter with which to assess the development of an associated foreland basin through geodynamic modeling. Provided that the load is accommodated through a flexural response of the lithosphere (Karner and Watts, 1983), the wavelength and magnitude of flexural accommodation of the foreland basin can be estimated, which permit an even stronger linkage between the preserved stratigraphy and the growth of the orogen. Use of paleotopographic constraints provides a significant improvement over estimates of foreland basin flexural profiles that do not consider the magnitude of the crustal load responsible for the lithospheric deflection.

Research over the past 15 yr has provided several scenarios of temporal growth of the central Andean Plateau (e.g., Barnes and Ehlers, 2009). Allmendinger et al. (1997) reviewed geologic evidence that points to a Neogene age for the central Andean Plateau, as well as a diachronous topographic evolution of the Altiplano and slightly younger Puna Plateau. More recently, several paleotopographic proxies have been applied in an effort to determine the magnitude and timing of surface uplift of the central Andean Plateau. Within the Altiplano, paleoelevation proxies, such as stable isotopic compositions of pedogenic carbonates (Garzione et al., 2006; Quade et al., 2007), clumped isotope paleothermometry (Ghosh et al., 2006; Garzione et al., 2008), and paleofloral types (Gregory-Wodzicki, 2002), suggest that the central Andean Plateau region was rapidly uplifted over the past 10 m.y.

Evidence for a more progressive history of central Andean Plateau topographic growth comes from a number of field-based studies from the western forearc region. Charrier et al. (2007) and Hartley and Evenstar (2010) used deformed paleogeomorphic surfaces and facies distributions to suggest dominantly Neogene growth. Farías et al. (2005) estimated that the western Altiplano gained 1.7 km of structural relief during the early to middle Miocene, and Hoke et al. (2007) detailed field evidence of another 1.1 km of elevation growth since the late Miocene. Similarly, Jordan et al. (2010) used a combination of structural and geomorphic relationships along the western Puna Plateau to infer ~0.8 km of relief growth from 5 to 11 Ma and an additional 0.4 km since 5 Ma. There is ~5.1 km of structural relief along a margin-wide monocline between the foreland and Puna Plateau, which was formed since 17 Ma, with ~2.8 km created from 10 to 17 Ma and an additional ~2.3 km since 10 Ma (Jordan et al., 2010). In general, however, the structural relief points to a Neogene-age growth of the central Andean Plateau. Additional support for this idea comes from compilations of low-temperature thermochronometers (Reiners et al., this volume), which show that apatite U-Th/He ages from the present Puna surface are as young as 10 Ma.

In contrast to Neogene rise of the central Andean Plateau, new evidence suggests that the surface uplift was much earlier. As discussed in Quade et al. (this volume) and Canavan (2012), the topographic rise of the Andes is believed to be the causal mechanism for hyperaridity in the Atacama Desert, as it interrupted eastern moisture sources transported by easterly trade winds. Based on the age and isotopic compositions of volcanic glasses, and stable isotopic compositions of pedogenic soil carbonates in the Puna Plateau and Atacama Basin, Quade et al. (this volume) infer that the Puna Plateau attained its near-present elevation by the late Eocene. This inference is not without independent geologic support; Hongn et al. (2007) and Coutand et al. (2001) described field evidence of Andean deformation in the Puna–Eastern Cordillera in the middle Eocene.

GEOLOGIC FRAMEWORK FOR THE FORELAND BASIN REGION

Cretaceous Rifting

Basement rocks in the Eastern Cordillera and Santa Bárbara system foreland region (Fig. 1) consist of variably metamorphosed and deformed Paleozoic and Proterozoic strata, metasediments, and intrusive units, much of which is a vestige of the growth and quiescence of the Early Cambrian Pampean orogeny (Monaldi et al., 2001; Salfity and Monaldi, 2006). The Lower Cambrian Puncoviscana Formation is the most commonly exposed pre-Mesozoic unit. In places, the Puncoviscana is directly overlain by both Lower Paleozoic (Cambrian–Devonian) strata and/or by Cretaceous strata. Based on the spatial distribution of these relationships, it appears that the region was characterized by antecedent structural domes of thick crust, perhaps of late Paleozoic age, that generally plunge to the north beneath the Oŕan-Tarijas Basin in northern Argentina and southern Bolivia (e.g., Salfity et al., 1987; Comínguez and Ramos, 1995; Monaldi et al., 2008; McGroder et al., this volume).

Early Cretaceous rifting, associated with either the opening of the South Atlantic or back-arc extension due to Pacific subduction, resulted in the collapse and subsidence of these crustal domes into a landscape of horsts and graben, termed the Salta Rift system (Figs. 2 and 3); however, the rift system likely extended well south into the southern Sierras Pampeanas (Schmidt et al., 1995). The Salta Rift "event" was spatially and temporally diachronous, and it was quite protracted, with a history of almost continuous magmatism and rift-basin deposition throughout the Cretaceous (Viramonte et al., 1999; Marquillas et al., 2005). The earliest phase of extension-related magmatism (130–100 Ma) includes alkalic plutonic bodies, mostly granitoids (Viramonte et al., 1999), and was focused in the western portion of the Salta Rift system (Figs. 2 and 3; Brealito and Alemania subbasins; Marquillas et al., 2005) within the study area. The second stage (100–75 Ma) consists of alkali basalts that are spatially and temporally related to the main phase of rifting and are found in the Las Curtiembres and Los Blanquitos Formations of the Pirgua Subgroup (Fig. 4; Viramonte et al., 1999; Horton et al., 2012, personal commun.). The third and last phase (65–60 Ma) consists of lamproitic and basanitic sills and flows interbedded with sediments of the Balbuena and Santa Bárbara Subgroups.

From a basin-fill perspective, rift systems create accommodation by the thinning of the crust, rift-flank erosion, and magmatism and postrift subsidence following lithospheric thinning. Heat flow and associated subsidence can be heavily influenced by deformation of the mantle lithosphere, which is quite difficult to quantify or interrogate. In the Salta Rift system, however, some of the basalts and basanites host xenoliths that provide insight into mantle lithospheric processes, even though surface volcanism comprises only ~5% of the rift-related rocks.

There are three previously described xenolith locations within the Las Conchas basalts and basanites (ca. 75 Ma)

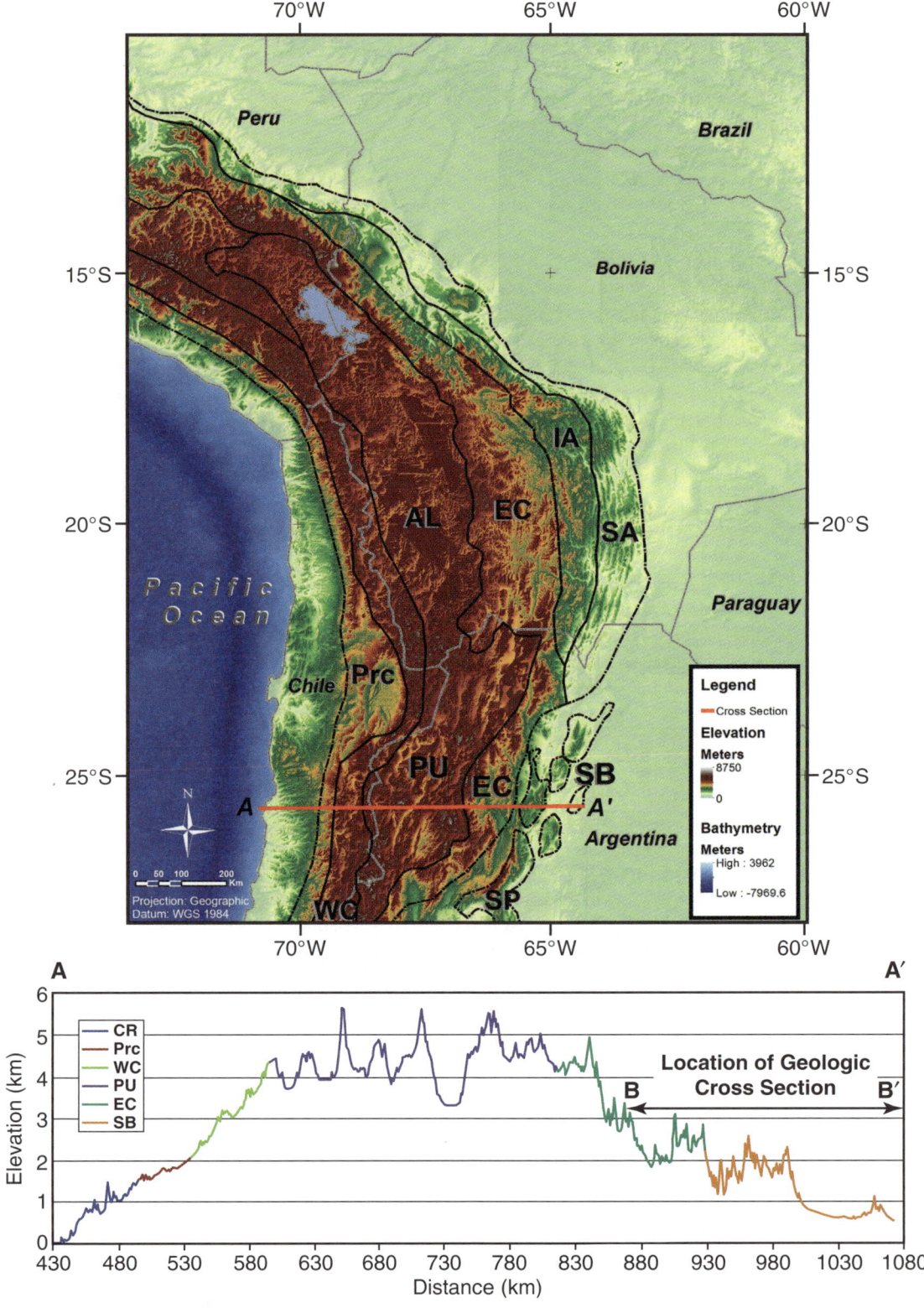

Figure 1. (Top) Digital elevation model of the central Andes with physiographic boundaries (modified from Barnes and Ehlers, 2009). The central Andean Plateau includes the Altiplano (AL) and Puna Plateau (PU). The Eastern Cordillera (EC) and Santa Bárbara system (SB) along cross section A–A′ are discussed in the text. Other regions labeled in the map are the Interandean zone (IA), the Subandean zone (SA), the Sierras Pampeanas (SP), the Western Cordillera (WC), the Precordillera (Prc), and Coast Ranges (CR). (Bottom) Topographic profile of A–A′, with colors depicting the different physiographic provinces, and the location of the geologic cross section (B–B′) referenced in text (Fig. 3).

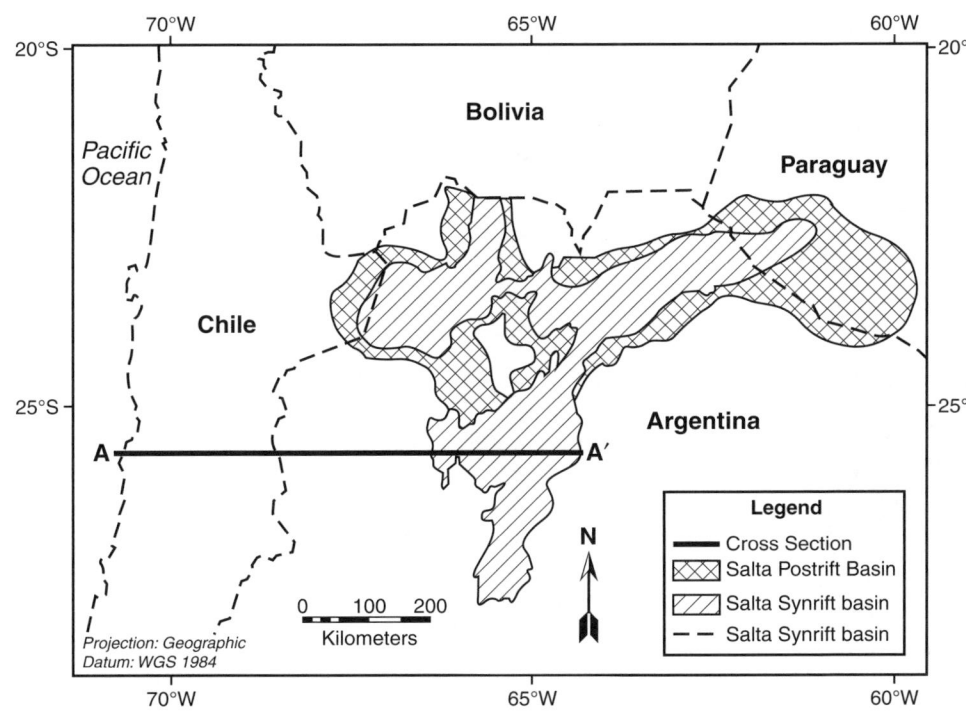

Figure 2. Location of the Salta Rift system in northern Argentina, based on Comínguez and Ramos (1995) and Monaldi et al. (2008). The rift system is superimposed on antecedent domes of thick crust (see McGroder et al., this volume). The three subbasins of the Salta Rift system discussed in the text are the Brealito, Alemania, and Metán. In the southern Salta Rift system, the synrift and postrift sections are generally superimposed on one another. The A–A' cross section from Figure 1 is also shown.

exposed in the Alemania region (Galliski and Viramonte, 1988). Xenoliths are of mantle (peridotites and pyroxenites) and crustal (metasediments, mafic and felsic granulites) origin (Lucassen et al., 1999). Most crustal xenoliths indicate equilibrium with high temperatures (>900 °C) and pressures around 10 kbar, which correspond to lowermost crustal depths beneath the extended crust during rifting (~35 km). Metamorphic (garnet Sm-Nd) ages of the mafic rocks are 95–110 ± 12 Ma, whereas felsic granulites have a younger age of 89–91.5 ± 3.5 Ma, indicating a period of at least 20 m.y. of high heat flux and metamorphism in the deep crust beneath the rift (Lucassen et al., 1999). New U-Pb data on felsic granulite SAR-04 from Alemania further constrain the age of magmatism in the lower crust beneath the Salta Rift. The xenolith is a large (~30 cm in diameter), S-type granitic bulk composition rock and consists of quartz–plagioclase–K-feldspar–garnet and accessory rutile. Although these distinctive xenoliths were referred to as felsic granulites by Lucassen et al. (1999), they could also be high-pressure granitic rocks. Thirty-four U-Pb analyses were performed on 28 zircon grains (some core and rim; most of them core only) separated from this sample (see analytical techniques in Appendix B). The crystallization age of the xenolith is 87 ± 0.7 Ma (Fig. 5; Data Repository[1]), which is identical to the garnet Sm-Nd age of similar rocks. A couple of inherited zircons of earlier Cretaceous age are synchronous with garnet Sm-Nd ages recorded in mafic

granulites from the same locations, whereas one grain has an age of 75 Ma, which is identical to the age of host basanite. These results indicate that the analyzed granitoid was emplaced as an igneous rock in the crust at 87 Ma, during the time of high heat flux characteristic for the Late Cretaceous extension, and it is not an older (Precambrian) rock that was metamorphosed during the Cretaceous. The source of this S-type granitoid is probably represented by high-pressure equivalents of the Lower Cambrian Puncoviscana Formation in the deep crust, because the bulk-rock initial Nd isotopes of the felsic granulites (granites) are similar ($\varepsilon_{Nd(0)}$ ~−10). Our results indicate that the lowermost crust of the Salta Rift partially melted during the Late Cretaceous extension due to lithospheric thinning and increased heat flow and that temperatures at ~35 km depth (close to the Cretaceous Moho) were >900 °C. The analyzed granite is the felsic end member of the "bimodal volcanism" often associated with continental extension; in this particular example, the felsic end member did not extrude as surface volcanic rock. Results are consistent with a lengthy period of high heat flux and mafic magmatic additions to the crust (110–75 Ma), which culminated with partial melting of the local basement. The paleothermometry and paleobarometry of xenoliths found within the Salta Rift system suggest that crustal thickness was no more than ~35 km at 110 Ma, and 30 km at 75 Ma, because the highest equilibration pressures on crustal rocks are ~10 kbar, and 8–9 kbar or less, respectively (Lucassen et al., 1999). Spinel peridotite xenoliths, representing the uppermost mantle from the same localities, are consistent with Moho depths of ~30 km during the Late Cretaceous and yield equilibration temperatures of ~1100 °C at the time of xenolith incorporation in the Late Cretaceous melts. These peridotites are

_____

[1]GSA Data Repository Item 2015012, Zircon U-Pb ages, is available at www.geosociety.org/pubs/ft2015.htm, or on request from editing@geosociety.org or Documents Secretary, GSA, P.O. Box 9140, Boulder, CO 80301-9140, USA.

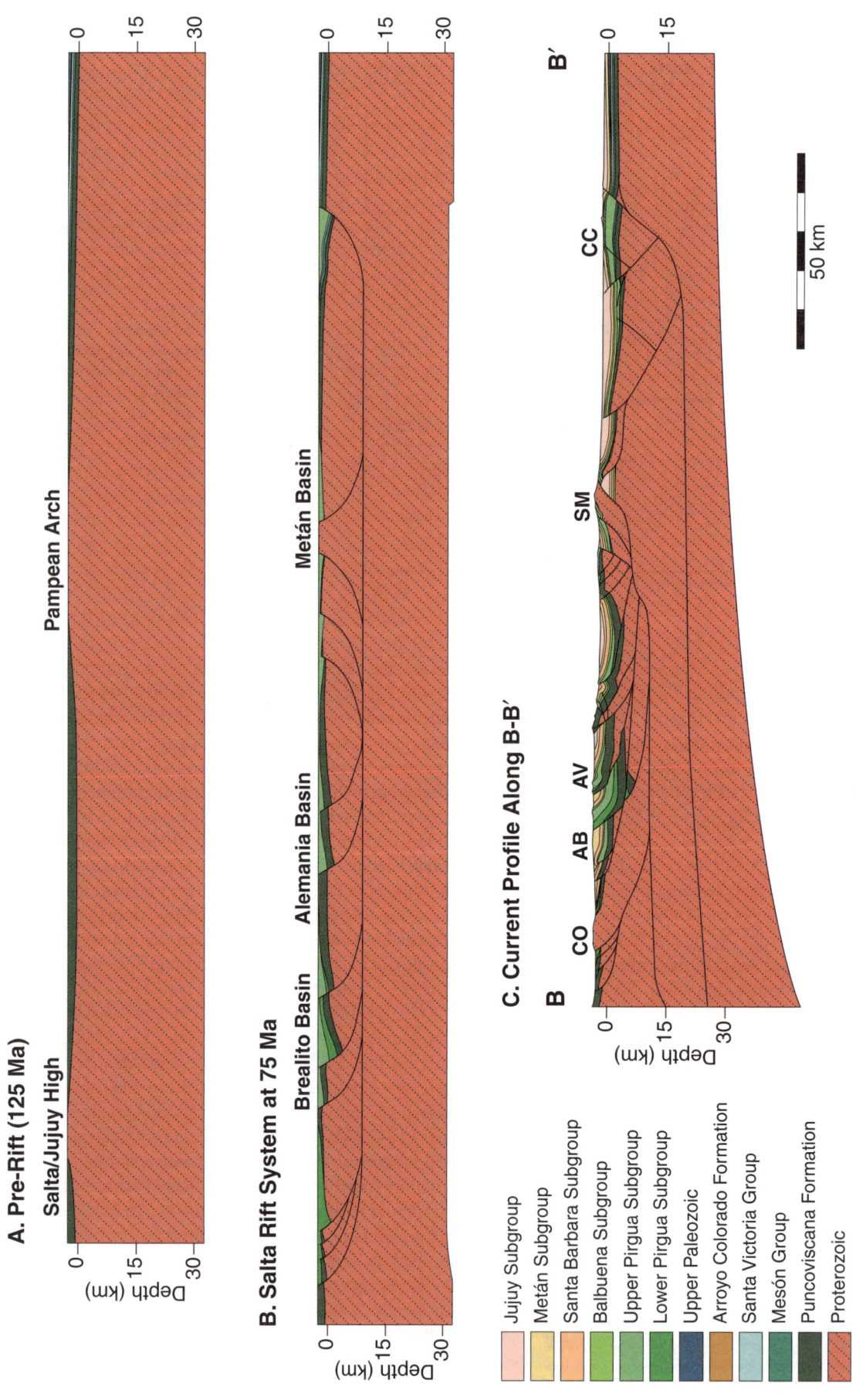

A. Pre-Rift (125 Ma)

Pampean Arch

Salta/Jujuy High

Depth (km)
0
15
30

B. Salta Rift System at 75 Ma

Brealito Basin   Alemania Basin   Metán Basin

Depth (km)
0
15
30

C. Current Profile Along B-B′

B   CO   AB   AV   SM   CC   B′

Depth (km)
0
15
30

50 km

Jujuy Subgroup
Metán Subgroup
Santa Bárbara Subgroup
Balbuena Subgroup
Upper Pirgua Subgroup
Lower Pirgua Subgroup
Upper Paleozoic
Arroyo Colorado Formation
Santa Victoria Group
Mesón Group
Puncoviscana Formation
Proterozoic

Figure 3. (A) A hypothetical cross section of the pre–Salta Rift system geology along B–B′ (location in Fig. 1, bottom graph) based on retrodeformed cross section 3C. The Salta/Jujuy High and Pampean Arch are two of the major antecedent domes that were extended in the Cretaceous Salta Rift system. (B) A cross section of the Salta Rift system at 75 Ma based on retrodeformation of Figure 3C, and stratal thicknesses reported in Marquillas et al. (2005, 2011). The locations of Salta Rift system subbasins (Brealito, Alemania, and Metán) are also labeled. (C) A cross section across the Salta Rift system at 25°30′S using the geologic maps of Monaldi et al. (2001) and Salfity and Monaldi (2006) that presently includes the Eastern Cordillera and Santa Bárbara system. The section strives to maintain material balance via plane strain; however, the section is undercorrected with contact and dip domain data. This prevents a rigorous balance, which is further complicated by a mix of detached and basement-involved deformation with presumably different deformation mechanisms. Note that the most of the rift stage faults have been removed in section C for simplicity—not all of them were reactivated in contraction. Two regional basal detachments are required to explain the depth of the active axial surface in the Santa Bárbara system, and observed deep earthquake hypocenters associated with contraction. Total shortening is ~50 km. CO—Cerro Overo, AB—Angastaco Basin, AV—Amblayo Valley, SM—Sierra de Metán, CC—Cerro Colorado.

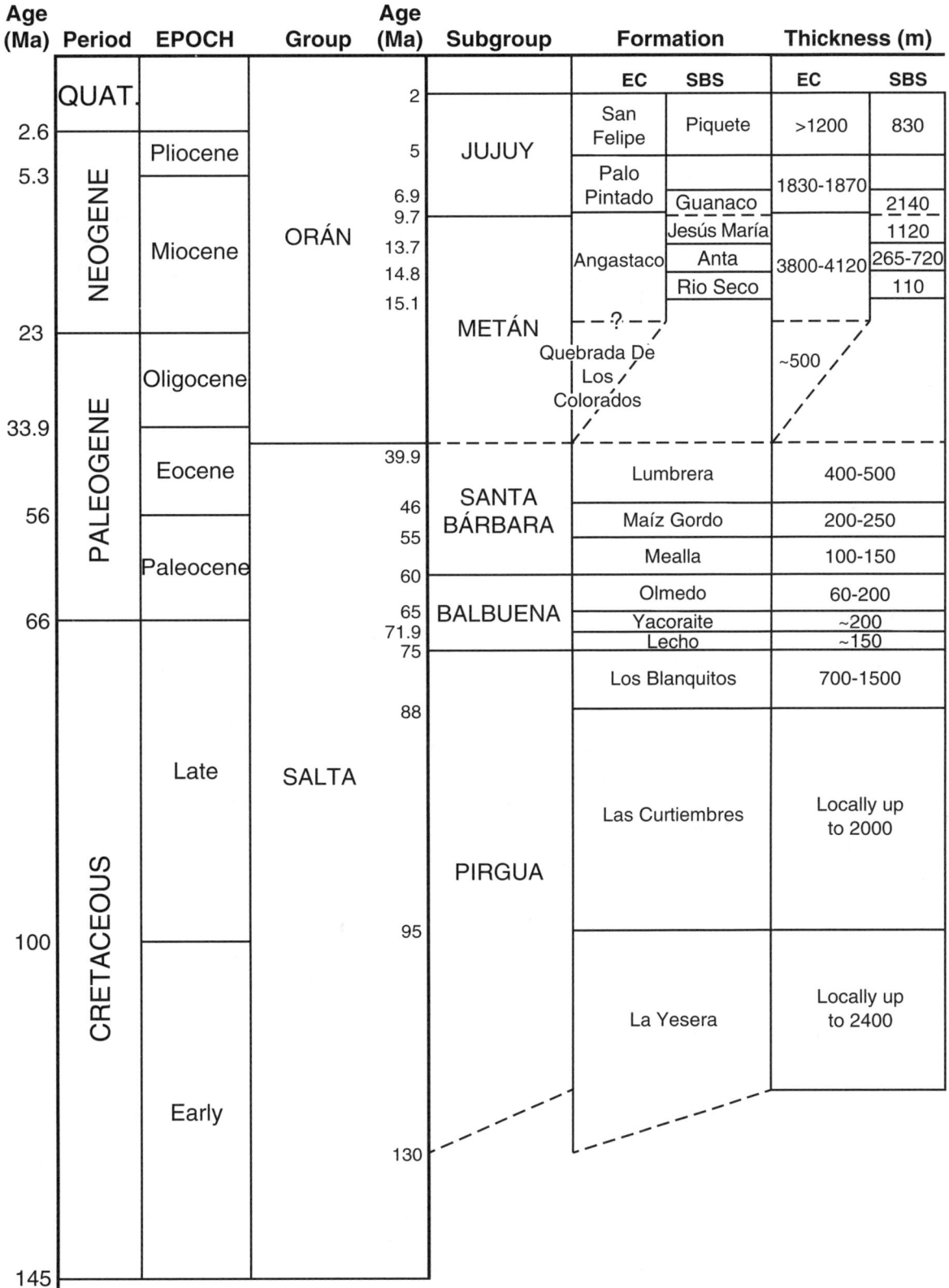

Figure 4. Cretaceous to present stratigraphic column of the Eastern Cordillera (EC) and Santa Bárbara system (SBS) region, compiled from stratigraphic descriptions and ages (Marquillas et al., 2005, 2011; del Papa et al., 2002; Carrapa et al., 2012; Reynolds et al., 2000). The time scale follows Walker et al. (2012).

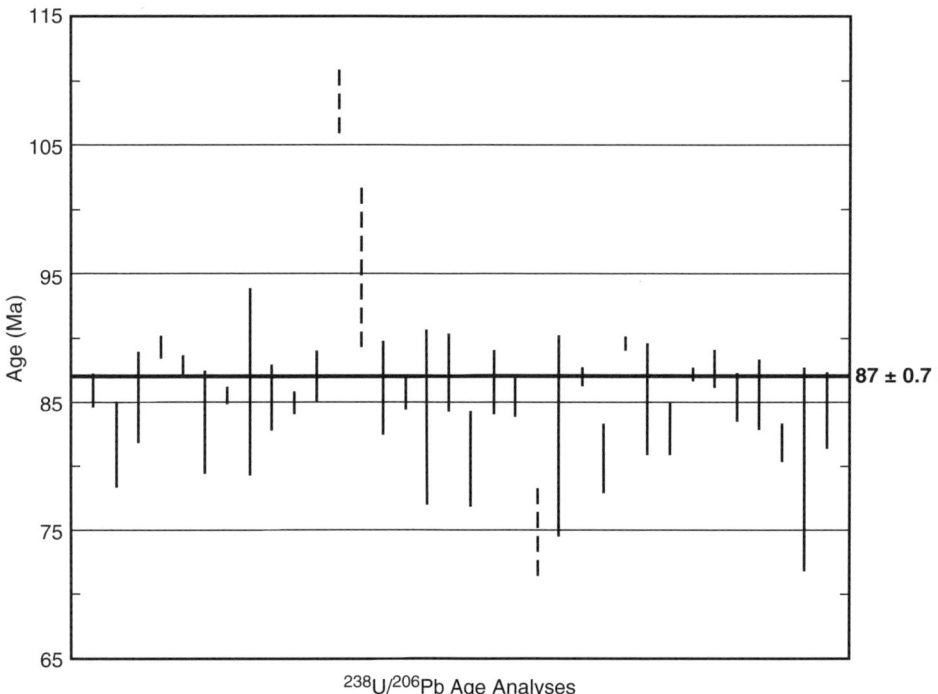

Figure 5. Weighted mean average U-Pb age of the SAR4 granulite xenolith derived from zircons within the sample. The analyses used in the determination are solid; those that may be inherited or crystallized during a later magmatic event are dashed. The regression yields an age of 87 ± 0.7 Ma (mean square of weighted deviates [MSWD] = 2.2).

clinopyroxene-poor harzburgites with isotopic characteristics of old South American lithosphere, and pressure-temperature data suggest that the lithosphere-asthenosphere boundary in the Campanian was at ~40–50 km beneath the surface. Assuming a starting pre-extension lithospheric thickness of ~125 km (Lucassen and Franz, 2005), the Cretaceous lithospheric mantle thinning factor (beta; Appendix A) was ~3. This is much higher than the observed crustal extension values (delta), which are estimated to be ~1.15 (Fig. 3). This disparity is indicative of either depth-dependent extension or possible mantle delamination in the Salta Rift during the Late Cretaceous.

ANDEAN CONVERGENCE

The tectonic history along a transect of the Andes at 25°30′S (Fig. 1) is potentially obscured by "tectonic erosion" or underplating of forearc crust (e.g., von Huene and Scholl, 1991; Scholl and von Huene, 2007), which may have removed a portion of the pre-Cretaceous crust along the Nazca Trench on the western margin of South America. Eastward-younging belts of Jurassic to Late Cretaceous magmatic arcs comprise the coastal Precordillera and western portion of the Western Cordillera (e.g., Viramonte et al., 1999). Evidence of Late Cretaceous deformation is preserved within the Western Cordillera and eastern Puna Plateau (Arriagada et al., 2006). The Puna Plateau lies west of the Eastern Cordillera and the Santa Bárbara system (Fig. 1). The Eastern Cordillera and Santa Bárbara system represent a combination of basement-detached and "thick-skinned" structural styles, with earthquake hypocenters that locate at 20–30 km depth (Cahill and Isacks, 1992; Rhea et al., 2008). The basement-involved structures are dominantly formed by inversion of the Salta Rift system basins and normal faults (Fig. 3; Allmendinger et al., 1983, 1997; Grier et al.., 1991; Salfity et al., 1993; Cristallini et al., 1997; Kley and Monaldi, 2002; Monaldi et al., 2008). Estimates of shortening vary some, but most are ~50–70 km (Fig. 3; Mingramm et al., 1979; Allmendinger et al., 1983; Grier et al., 1991), although Kley et al. (1999) suggested ~30 km. Within the Puna Plateau and Eastern Cordillera, there is evidence that contractional deformation initiated in the Paleocene (Hongn et al., 2007) and generally proceeded from the Eastern Cordillera to the Santa Bárbara system from the Miocene to present (Grier et al., 1991; Reynolds et al., 2000; Bywater-Reyes et al., 2010; Pearson et al., 2012, 2013). The foreland basin is located within the inverted basement-involved structures of the Eastern Cordillera and Santa Bárbara system ranges, and to their immediate east. The basin preserves a late Paleogene and Neogene depositional record that is thought to reflect Andean (Cenozoic) orogenic deformation (e.g., Jordan and Alonso, 1987; Reynolds et al., 2000; Fig. 4).

REGIONAL STRATIGRAPHY

Observed stratal thicknesses and lithology provide critical constraints needed to audit the forward models, and therefore they merit significant review. The following discussion focuses on key elements of the regional stratigraphy that have been used to appraise the modeled transect at 25°30′S, and this should not be considered a complete review.

The Salta Rift system encompasses (from west to east) the Brealito, Alemania, and Metán subbasins along the profile studied in this paper (Figs. 2 and 3B). These basins preserve the

Lower Cretaceous–Upper Eocene Salta Group, which includes the Pirqua, Balbuena, and Santa Bárbara Subgroups, which are generally thought to reflect the accommodation space created within the Salta Rift system. The Hauterivian–Maastrichtian Pirgua Subgroup (Fig. 4) is composed of the La Yesera, Las Curtiembres, and Los Blanquitos Formations, and it is interpreted to represent two stages of syn- to postrift deposition and locally exceeds 6 km thickness (Salfity and Marquillas, 1994; Marquillas et al., 2005). The units are lithostratigraphic, and the depositional age of the units is somewhat variable within the individual subbasins of the Salta Rift system (Marquillas et al., 2011). The La Yesera Formation is the oldest unit and is composed chiefly of conglomerate, sandstone, and siltstone deposited within the Brealito and Alemania rift subbasins. The thickness of the La Yesera is highly variable, ranging from 2400 m in the Brealito subbasin to 300 m in the Alemania subbasin. The La Yesera is capped by the Isonza basalt flows of 94–101 Ma. The Curtiembres Formation is thought to record the second (and most significant) stage of rifting, and it is found in all of the Salta Rift subbasins (Brealito, Alemania, and Metán) within the study area (Marquillas et al., 2005). In contrast to the La Yesera, the weakly laminated fine-grained clastic sediments of the Curtiembres Formation suggest the establishment of a perennial, underfilled lake (Marquillas et al., 2005). The Curtiembres Formation thickens to ~2000 m near graben-bounding faults. The Los Blanquitos exhibits coarsening-upward cycles of siltstone to coarse-grained sandstones and conglomerate, reaching thicknesses of 700–1500 m (Marquillas et al., 2005). The Los Blanquitos Formation contains evidence of paleosols and large terrestrial vertebrates, suggesting that the basin was, for a time, filled. The formation is capped by the Palmar Largo volcanics of ~70 ± 5 Ma.

The Balbuena Subgroup (Fig. 4) was deposited during the Maastrichtian to early Paleocene, and it includes the Lecho, Yacoraite, and Olmedo Formations. The Lecho Formation is mostly fine- to medium-grained, white calcareous sandstone, with varying quartz content, and has bedding consistent with both wind and shallow-water tractional shear. It is generally ~150 m thick (Marquillas et al., 2005). The Upper Cretaceous to Lower Paleocene Yacoraite Formation is a distinctive unit mostly composed of limestones and calcareous sandstones, but it also contains organic-rich intervals. Its thickness is variable and reaches a maximum of 200 m. The Lower Paleocene Olmedo-Tunal Formation is composed of black and gray shales, siltstones, dolomitic limestones, and evaporite deposits and ranges in thickness from 60 to 200 m (Marquillas et al., 2005). It is interpreted as an evaporative lacustrine system with relatively little clastic input.

The Paleocene–Eocene Santa Bárbara Subgroup (Fig. 4), composed of the Mealla, Maiz Gordo, and Lumbrera Formations, ranges in thickness from 700 to 900 m within the Santa Bárbara system and Eastern Cordillera and generally thickens toward the center of the Salta Rift basins (del Papa, 1999; Reynolds et al., 2000; del Papa et al., 2002; Marquillas et al., 2005). Although the Santa Bárbara Subgroup formations are distinctive, they are not very thick, and they possess evidence of a long-lived,

clastic-starved lake system within the general confines of the Salta Rift system (del Papa, 1999; del Papa et al., 2002; Marquillas et al., 2005). The Paleocene Mealla Formation, ranging in thickness from 100 to 150 m, is composed of fine- to medium-grained sandstones and siltstones with bedding features consistent with deposition in a shallow fluvial environment, interspersed with paleosols and small lakes with evaporative deposits (Marquillas et al., 2005). The Upper Paleocene–Lower Eocene Maiz Gordo Formation possesses beds of coarse- to fine-grained sandstone with erosive bases and pebble lags, unidirectional currents, and trough cross-beds, suggestive of a fluvial environment of deposition, although the chronostratigraphic equivalent units to the east are finer-grained and include limestones. It has been interpreted to represent progradation of a delta into a shallow, persistent lake (del Papa, 1999). The formation is ~200–250 m thick. The Lower–Upper Eocene Lumbrera Formation, the thickest within the Santa Bárbara Subgroup, is 400–500 m thick, composed of red sandstone and mudstone overlain by a distinctive green mudstone (Marquillas et al., 2005). These green beds of mudstone, referred to as the Faja Verde, preserve evidence of persistent lakes that existed during the Eocene (del Papa et al., 2002). Toward the top of the formation, the lakes appear to become more saline and evaporitic, suggesting a shift toward a drier climate.

Detrital zircon ages from selected locations in the Salta Group, reported in DeCelles et al. (2011), suggest that the source of sediment for the Pirgua and Balbuena Subgroups is of lower Paleozoic age, which is presumably derived from local Salta footwall uplifts. Only the Lumbrera Formation, which caps the Santa Bárbara Subgroup, contains detrital zircon ages younger than early Paleozoic. The reported younger age probability peaks within these two samples of Lumbrera Formation are 133 and 167 Ma, which are also coeval with pre–Salta Rift magmatism (e.g., Viramonte et al., 1999) and may not represent an extrabasinal source. Alternatively, the sediment could have been derived from the Puna Plateau, where rocks of this age also crop out.

The Oligocene–Quaternary Orán Group (Fig. 4) represents the depositional record of proximal, eastward-encroaching Andean deformation and contains the Upper Oligocene–Middle Miocene Metán Subgroup and the late Miocene–Quaternary Jujuy Subgroup (Reynolds et al., 2000). Because of the progressive tectonic deformation across the Andean foreland during the Neogene, the stratigraphic units vary in character and thickness from the Eastern Cordillera to the Santa Bárbara system, and the following is an attempt to correlate those units chronostratigraphically, based on work by Reynolds et al. (2000), Coutand et al. (2006), Carrapa et al. (2012), and DeCelles et al. (2011). The Metán Subgroup within the Eastern Cordillera includes the Oligocene–Lower(?) Miocene Quebrada de los Colorados and the middle Miocene Angastaco Formation, which has partial equivalents in the middle Miocene Rio Seco, Anta, and Jesus Maria Formations in the Santa Bárbara system. The Jujuy Subgroup encompasses the Upper Miocene to Pliocene Palo Pintado and Pliocene–Pleistocene San Felipe Formations in the Eastern Cordillera. In the Santa Bárbara system, the Jujuy Subgroup includes the Upper Miocene Guanaco and Pliocene–Pleistocene Piquete Formations (see Fig. 4).

The Metán Subgroup (Fig. 4) is often interpreted as the depositional record of the transition to Andean deformation into the Eastern Cordillera that built the present orogen. The thickness of the Metán Subgroup appears to vary in general correspondence to the Salta Rift system (Galli, 1995; Reynolds et al., 2000). In the Eastern Cordillera, the base of the Metán Subgroup includes the Priabonian–Langhian Quebrada de los Colorados Formation. The Quebrada de los Colorados is a succession of interbedded red sandstones (conglomeratic in places) and siltstones, often exhibiting trough cross-stratification and lenticular geometries (DeCelles et al., 2011). It thins dramatically from west to east, reaching thicknesses of at least ~400 m in the Eastern Cordillera (DeCelles et al., 2011) and tapering to zero thickness at Alemania, just ~125 km from the eastern margin of the Puna Plateau. It is possibly conformable with the underlying Santa Bárbara Subgroup, but it onlaps a progressive unconformity migrating to the east. Conglomerate clasts appear to be from Lower Paleozoic formations and lithologies, not Salta Group units (DeCelles et al., 2011; Carrapa et al., 2012), suggesting that the Puna and Eastern Cordillera region (outside of the Salta Rift) was the source and that the Puna Plateau may not have been a hydraulically closed basin. Overlying the Quebrada de los Colorados is the Lower(?) Miocene Angastaco Formation, preserved within the Angastaco Basin, which is apparently partially localized over a Salta Rift horst or structural high that was buried during the Neogene (Fig. 4). The 23(?)–9.7 Ma Angastaco Formation is comparatively thick, approaching 3300–3800 m of sandstone and conglomerate (Coutand et al., 2006; Carrapa et al., 2012). The base of the Angastaco Formation is characterized by bed forms characteristic of eolian deposition that grade upward to fluvial and alluvial facies (Carrapa et al., 2012). In the Santa Bárbara system, the Metán Subgroup is a succession of lacustrine and fluvial sediments that predate the local inversion of Salta Rift structures at ca. 9.7 Ma (Reynolds et al., 2000; Fig. 4). The basal formation of the Metán Subgroup is the Langhian Rio Seco Formation, which is separated from the underlying Quebrada de los Colorados Formation by a 22 m.y. disconformity (Reynolds et al., 2000; Fig. 4). The Rio Seco reaches a maximum thickness of ~110 m and is composed of reddish sandstones and siltstones with lenses of conglomerate (Reynolds et al., 2000). Sandstone petrology and paleocurrent measurements from the Rio Seco indicate a dramatic change in sediment source from east-derived (cratonic) metamorphic rocks to west-derived, Andean sources (Reynolds et al., 2000). Detrital zircon age populations from the Orán Group, although generally similar to the Salta Group, commonly possess Cenozoic-age populations that are nearly coeval with the depositional age of the unit (DeCelles et al., 2011; Carrapa et al., 2011), also pointing to shift to an Andean-orogen source. The Rio Seco is overlain by the Langhian–Lower Seravallian Anta Formation, which ranges in thickness from 265 to 720 m and is characterized by very fine-grained sandstone, siltstone, mudstone, and thin carbonate shoals (Reynolds et al., 2000). The Seravallian–mid-Tortonian Jesus Maria Formation is the stratigraphically highest unit in the Metán Subgroup. Its thickness is variable, but it reaches a maximum of 1120 m and is distinguished by beds of red-gray siltstone, sandstone, and conglomerate (Reynolds et al., 2000). Sedimentary petrography of the Jesus Maria indicates cyclic changes in provenance lithotype from metamorphic rock fragments to lithic fragments (Reynolds et al., 2000), suggestive of unroofing of inverted Salta Rift hanging walls.

The middle Miocene to Pliocene Jujuy Subgroup (Fig. 4) unconformably overlies the Metán Subgroup and generally represents deposition during more proximal deformation and exhumation in the Eastern Cordillera and within the Santa Bárbara system. In the Eastern Cordillera, the middle Tortonian–Messinian Palo Pintado Formation is an 1800-m-thick succession of dominantly mudstone with interbeds of sandstone, interpreted to be deposited in a low-gradient fluvial environment (Coutand et al., 2006; Carrapa et al., 2012). Clasts within the Palo Pintado conglomerates are the oldest to have lithologies distinctive of the Salta Group (e.g., Late Cretaceous Yacoraite Formation), suggesting a provenance from east of Angastaco. The Pliocene to Pleistocene San Felipe Formation caps the Jujuy Subgroup in this region of the Eastern Cordillera, which is locally at least 800 m thick (Carrapa et al., 2012). The cross section and geologic map by Coutand et al. (2006) suggest that most of the Angastaco Basin deformation occurred in the late Pliocene, although the surrounding Salta Rift basins were likely inverted in the middle Miocene through early Pliocene (Reynolds et al., 2000; Coutand et al., 2006; Carrapa et al., 2011). Within the Santa Bárbara system, the Tortonian Guanaco Formation marks the base of the Jujuy Subgroup and may be separated from the underlying Jesus Maria Formation by an angular unconformity. The Guanaco Formation is as much as 2140 m thick and is composed chiefly of red-gray, medium- to coarse-grained sandstone and conglomerate (Reynolds et al., 2000). Petrographic analysis suggests that until ca. 8.7 Ma, the majority of the detritus was derived from metamorphic rocks, likely from the Puna Plateau or Eastern Cordillera, followed by a switch to more lithic fragments suggestive of unroofing and recycling of Salta Group sediments (Reynolds et al., 2000). The Pliocene–Pleistocene Piquete Formation, with thicknesses measured to 820 m, is dominantly a succession of conglomerates and sandstones. Limestone clasts within the Piquete Formation conglomerate beds are quite common and are believed to be derived from the Salta Group.

At present, the Bermejo, Juramento/Salado, and Dulce Rivers drain most of the bordering Eastern Cordillera along the Puna Plateau. The Bermejo River fills and overruns the limits of the modern flexural foreland basin (Chase et al., 2009; Hartley et al., 2010; Cohen et al., this volume). The Juramento/Salado River drains much of the modern foreland region described herein, depositing sediment via a southeast-trending distributive fan in the southern Argentine provinces of Salta and Santiago del Estero.

MODELING METHODOLOGY

The evolution of the Salta Rift system, and its effect on the Metán foreland during subsequent Andean deformation, was modeled using a kinematic approach that determines the accommodation and uplift caused by the repeated extensional and contractional deformation of the lithosphere, termed quantitative basin analysis (QBA; Driscoll and Karner, 1998; Karner et al., 2003). The program accounts for the fundamental processes responsible for both the generation and destruction of accommodation, such as flexural isostasy, the vertical and lateral flow of lithospheric heat, sediment compaction, eustasy, crustal thinning and thickening, and lithospheric mantle thickening and thinning. The modeling is constrained using empirical data sets, such as seismic reflection (stratal geometries) and refraction data (crustal velocities and basement and Moho topographies), measured stratigraphic sections, well data, gravity, and (paleo-) water depths. The effective elastic thickness, which controls the degree of lithospheric flexural support for applied geological loads, is estimated using the depth to a controlling isotherm (e.g., the 450° isotherm; Appendix A; Karner and Watts, 1983). The position of this isotherm within the lithosphere varies due to the advection of heat during deformation and its consequent conductive cooling by vertical and lateral heat flow following deformation. Whereas QBA is particularly effective in extensional systems that thin the lithosphere, in this analysis, it is mainly being used to model lithospheric thickening and its attendant thermal and isostatic effects. While the use of the program provides insights into major processes that influence hydrocarbon systems, there are some limitations:

1. Mass is not conserved. It is added to the system to approximate load geometries and observed basin accommodation. The foreland is filled by defining a datum below which sediment is deposited. The addition of sediment adds mass to the system and modifies basin sediment density and topography, but more importantly, its accumulation tracks lithospheric deformation. For convenience, erosion is not included in the models. Because the load of the sediment in the basin is not subtracted from the orogenic load, the topographic loads are at a maximum, which tend to enhance basin accommodation. This is considered a minor problem.

2. One-dimensional profiles of the lithospheric thickness are used for the QBA calculations. In convergent systems, crust is often thickened by imbrication or duplication of the crust where strata (and their basins) are translated over an adjacent footwall made up of similar strata. In the present study area, the orogen has been translated ~50 km over the past 15 m.y. (Fig. 3). While this complexity is not captured within the modeled orogen, the evolving load represented by this convergence is included, as is the corresponding flexural deformation of the adjacent foreland region.

3. The central Andean Plateau resides above an active subduction zone on the edge of a continental lithospheric plate, meaning that a portion of the mantle lithosphere beneath it is missing. This condition is not accounted for in the model, per se. However, the motivation for the modeling is an examination of the retroarc foreland basin, which is underlain by a mantle lithosphere, and the general patterns of flexural and thermal subsidence over time are likely captured, perhaps rendering this a relatively minor problem.

Our models also do not account for the influence of extrinsic processes such as denudation and asthenospheric convection (i.e., dynamic effects), nor do they include the effects of crustal radioactivity to heat flow. The effects of in-plane stress were not applied to these models in the interest of reducing the number of variables, and the recognition that such effects are likely second order. Dynamic subsidence created by asthenospheric convection (i.e., dynamic topography; Mitrovica et al., 1989; Gurnis, 1990; Shephard et al., 2012) remains controversial (e.g., Wheeler and White, 2000) but could also be an important consideration. The crust of the Puna Plateau is some of the thickest on the planet (e.g., Mooney et al., 1998; Chulick et al., 2013), which puts the surface elevation well above the meteoric dry line. As a result, the Puna Plateau is internally drained, and almost none of the crustal mass is eroded and transported to the modern foreland basin. Back-arc rifting may create complications in the thermal state and composition of the lithosphere, leading to localized uplift and subsidence, but this is unrelated to the flexural response to crustal thickening.

Modeling Constraints

The crustal thickening (or thinning) parameters required by QBA can be obtained from a number of data sets, including topographic profiles or geologic cross sections. In the models presented here, a combination of the two was employed. Within the Eastern Cordillera and Santa Bárbara system, series of cross sections (Fig. 3) were used to calibrate the appropriate distribution and magnitude of crustal thinning and thickening. A modern topographic profile extracted from a 90 m Shuttle Radar Topography Mission (SRTM) digital elevation model trending due west from the end of the cross section was used to calibrate total crustal thickening values within the orogen. The model outputs were audited by a comparison to overall structural geometry of the cross sections and comparison with measured sections and well data from within the Eastern Cordillera and Santa Bárbara system (Reynolds et al., 2000; DeCelles et al., 2011; Carrapa et al., 2012). In addition, sedimentary provenance data were used as a check on the paleogeographic interpretation.

Salta Rift (Neocomian and Campanian)

The Salta Rift region was modeled with an initial crustal thickness of 35 km, an initial elastic thickness of ~42 km (see Appendix A), and initial background surface heat flow of 60 mW/m², estimated using nearby vitrinite data. This heat flow may be somewhat conservative, based on compilations of nearby and global retroarc surface heat-flow values (e.g., Currie and

Hyndman, 2006), but were reasonably close. Neocomian extension was described using depth-independent stretching for the crust and mantle lithosphere (i.e., McKenzie, 1978), and it was allowed to thermally decay. As described already, depths below a prescribed datum and above the initial "basement" are allowed to fill with sediment. Throughout the Salta Rift models, the datum was set at 500 m above modern sea level. As many modern rift basins are hydraulically separated from global oceans, this is not an unreasonable assumption. One potential issue with this datum, versus one below sea level, is that it may initially be allowing too much sediment into the basins. In the nearby Lomas de Olmedo Basin, seismic profiles show the amount of extension, and available accommodation, greatly exceeds the quantity of sediment (e.g., Comínguez and Ramos, 1995; Starck, 2011), even after correcting for the effects of compaction, implying the basin was "starved" of clastic sediment. The persistence of lacustrine facies in the Salta Rift basins throughout the Cretaceous and Paleogene (del Papa et al., 2002; Marquillas et al., 2005) also suggests that these basins were chronically starved of sediment; i.e., subsidence outpaced sedimentation. Varying the sedimentary datum to account for these variations in mass, however, results in expanding/condensing the "stratigraphic record" in these models somewhat arbitrarily. To remove this potential bias, we chose to maintain a constant datum, despite its potential inaccuracy. A retrodeformed cross section (Fig. 3) was used as a basis for modeling the magnitude of extension (both crustal and mantle lithospheric; Appendix A) to simulate the appropriate amount of accommodation in the basins. This was solved iteratively until a sufficient match between the modeled and interpreted rift fill was achieved.

In contrast to Neocomian extension, depth-dependent stretching was associated with the Salta extension during the Cenomanian–Campanian at 100–75 Ma, reflecting petrologic information derived from xenoliths hosted in Late Cretaceous basalts and basanites. Crustal stretching was of similar magnitude to the initial Neocomian rifting, but a much higher degree of lithospheric mantle thinning was required to approximate highly attenuated mantle lithosphere, which has a significant impact on heat flow, synrift uplift, and postrift subsidence. Differences in accommodation created by depth-independent and depth-dependent stretching/mantle delamination are illustrated in Figure 6 and principally reflect the contrast in the thermal state of the lithosphere. The associated postrift thermal uplift and subsequent relaxation are dramatic and result in a progressively thicker package of sediment in the center of the rift basins than expected for moderate mantle lithospheric strains and less thermal advection (Fig. 6). The scale of the mantle lithosphere delamination remains poorly defined, but it is worth noting that the uninverted Lomas de Olmedo Basin shows similar patterns of protracted subsidence (Comínguez and Ramos, 1995; Starck, 2011).

Next, we summarize the results of both the extensional and contractional evolution of the Andean foreland region at 25°30′S. The specific parameters and timing of episodes that were used in the QBA modeling are listed in Appendix A.

Andean Deformation (Eocene to Present)

As stated in the introduction, two contrasting phases of Cenozoic crustal thickening were modeled, based on alternative assumptions of the eastward growth of the Puna Plateau, the Eastern Cordillera, and the Santa Bárbara system. The modeling approach was similar to that used for the Salta Rift, although the geologic constraints are more limited. As discussed earlier, QBA does not transfer mass laterally, and we have not included erosion in the model. In contractional environments (i.e., within the orogen), this is obviously a limitation (see Appendix A). Recognizing this, the goal of the study was to evaluate basin subsidence patterns in relatively unstructured areas to try to gain insight, not necessarily to be predictive (e.g., Oreskes et al., 1994). Instead of attempting to match the age of outcrops along the Puna-Metán transect, we opted to use the modern topographic profile (from 90 m elevation grids) and progressively thickened the crust in the regions of the Puna Plateau, Eastern Cordillera, and Santa Bárbara system until a reasonable match was achieved (Figs. 3 and 7). The QBA model ensures that the imposed thickening retains flexural isostatic balance. The present-day cross section (Fig. 3) was used to honor regions of localized inversion. Because erosion is not considered, the modern topographic profile presents a minimum case for flexural subsidence. We view this as a minor problem, given the assumption that there is a relatively small degree of erosion along the modeled transect. Surface thermochronometry from the Puna region (Insel et al., 2012) and the Eastern Cordillera (Pearson et al., 2012, 2013; Löbens et al., 2013; Carrapa

---

Figure 6. Model output showing the Salta Rift system with two different modes of lithospheric extension during the second stage of rifting at 100–75 Ma (Las Curtiembres and Los Blanquitos) at the end of Pirgua Subgroup deposition (ca. 75 Ma): (A) depth-independent lithospheric extension (mantle and crustal lithosphere thinned by same magnitudes; beta = delta; Appendix A) and (B) depth-dependent lithospheric extension (mantle and crustal lithosphere thinned independently; beta ≠ delta; Appendix A). Note sections are vertically exaggerated ~2:1. The magnitudes and locations of extension are based on the cross section in Figure 3B. The two stages of rifting (crustal extension manifest through faulting) and accommodation/fill at 125 Ma and 100 Ma are colored different shades of green. In section B, the beta values are ~3 in the core of the rift system (see text). (C) Modeled effective elastic thickness (*Te*) of depth-independent stretching in the foreland region over time (profile A). The small degree of extension results in little displacement of the effective elastic isotherm, so the rift has little influence on the strength of the lithosphere and little thermal subsidence. (D) Modeled excess heat flow at the top of basement from the modeled depth-dependent extension (profile B) with disproportionate mantle lithosphere thinning that results in thermal uplift of the region. The advected asthenosphere results in excess heat-flow values that persist to present in the eastern portion of the cross section. (E) The modeled effective elastic thickness at 125–0 Ma for depth-dependent stretching, which is significantly reduced in the Salta Rift system due to advection of the effective elastic isotherm. Note that depth-independent stretching of the Early Cretaceous (125–100 Ma) rifting did not significantly reduce *Te*. The Cenozoic histories of *Te* and heat flow for parts C, D, and E are from the Neogene Puna Plateau growth scenario.

419

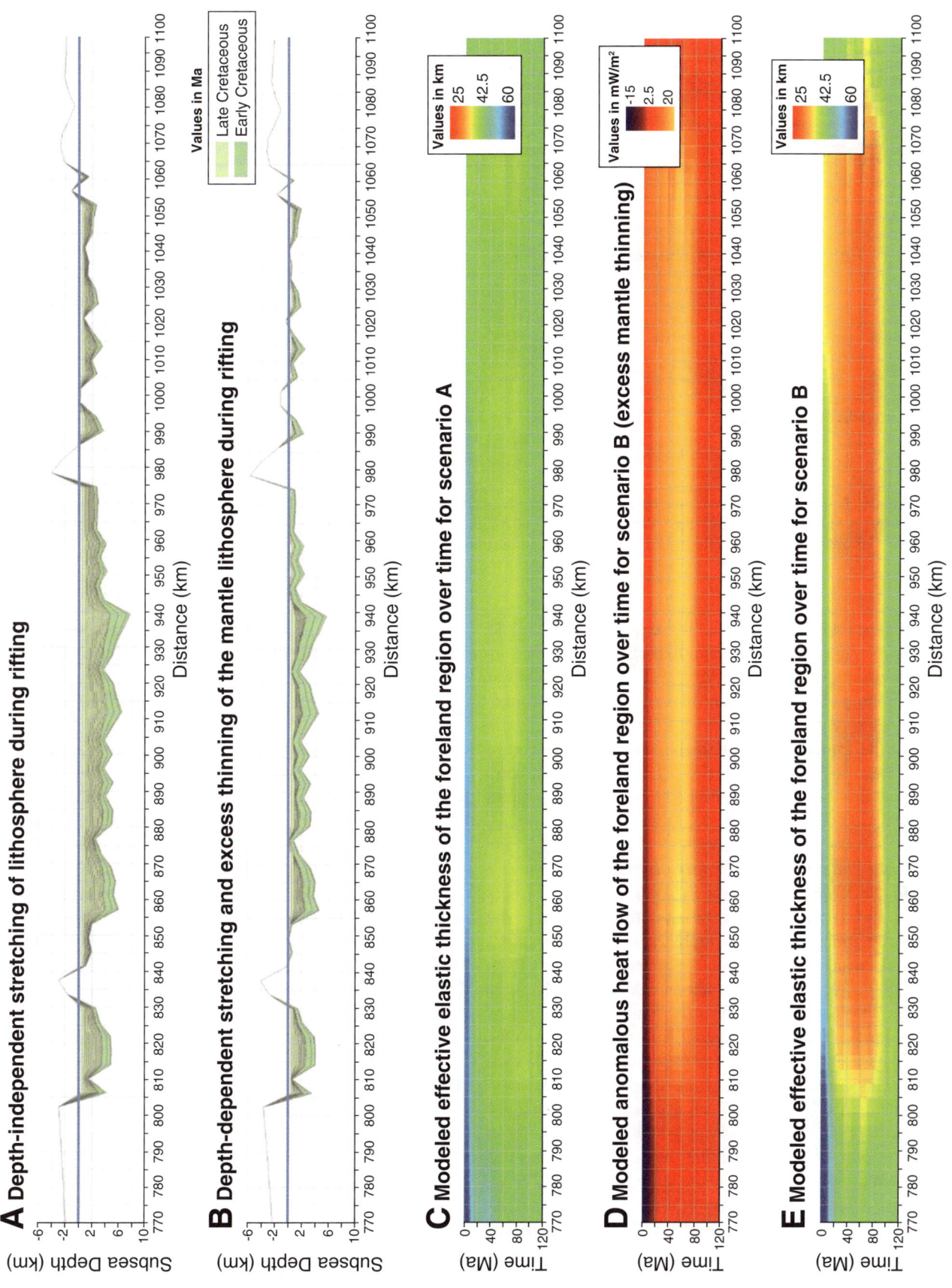

A Depth-independent stretching of lithosphere during rifting

Subsea Depth (km)

Distance (km)

Values in Ma
Late Cretaceous
Early Cretaceous

B Depth-dependent stretching and excess thinning of the mantle lithosphere during rifting

Subsea Depth (km)

Distance (km)

C Modeled effective elastic thickness of the foreland region over time for scenario A

Time (Ma)

Distance (km)

Values in km
25
42.5
60

D Modeled anomalous heat flow of the foreland region over time for scenario B (excess mantle thinning)

Time (Ma)

Distance (km)

Values in mW/m²
-15
2.5
20

E Modeled effective elastic thickness of the foreland region over time for scenario B

Time (Ma)

Distance (km)

Values in km
25
42.5
60

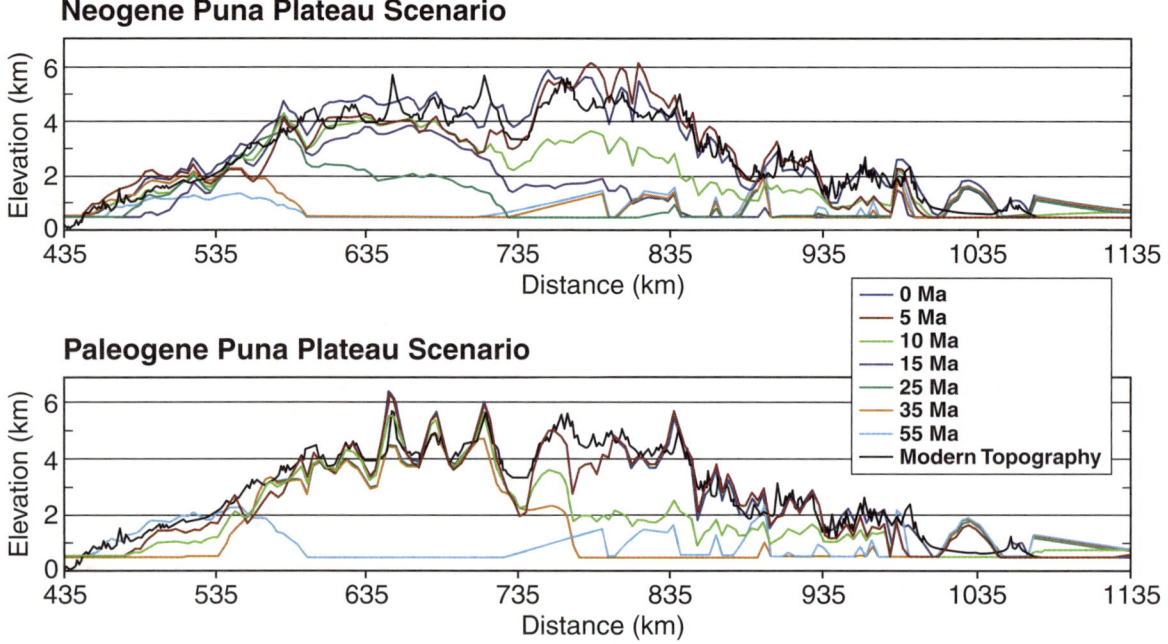

Figure 7. Topographic evolution of the progressive Neogene Puna Plateau and Paleogene growth scenario for the A–A′ profile (from Fig. 1). The pre-Puna topography is dominated by the Salta Rift footwall uplifts, which subside during the Cenozoic as a result of flexural downwarping and thermal subsidence.

et al., 2013) support this assumption. These regions preserve old ages, indicating that exhumation magnitudes have been low during the Cenozoic (<3 km). In addition, Late Cretaceous crustal thickening in the modern-day Precordillera and Western Cordillera was modeled (e.g., Arriagada et al., 2006; Quade et al., this volume), but the foredeep was relatively small and localized to the Puna Plateau region and well outside of the Salta Rift system. As the focus of this paper is in the modern Eastern Cordillera and Santa Bárbara system, it will not be further considered.

Neogene Puna Plateau Model: Progressive Growth Initiating in the Miocene

The first model scenario creates major crustal growth of the Puna Plateau beginning at 35 Ma to represent Incaic deformation, progressing eastward at 25–16 Ma in the central and eastern Puna region. Major crustal thickening events in Eastern Cordillera are prescribed at 15 Ma to 6 Ma and in the Santa Bárbara system from 5 Ma to present (Fig. 7). In detail, crustal thickening in the Puna and Eastern Cordillera continue to present, as can be seen in Figure 7. This scenario follows observations from Hoke et al. (2007), Jordan et al. (2010), and Rech et al. (2010) that suggest Puna Plateau topographic growth was continuous beginning in the early to middle Miocene, and that the deformation stepped out progressively, mostly in the Neogene (e.g., Reynolds et al., 2000; Carrapa et al., 2011; DeCelles et al., 2011; Pearson et al., 2012, 2013). From the perspective of the foreland basin, this would suggest that any major flexural foreland basins should be of similar age (<35 Ma) age. The structural relief and elevation

gains summarized in Jordan et al. (2010) are generally followed. The separate growth of the Eastern Cordillera and Santa Bárbara system creates series of eastward-stepping foreland basins, which were sequentially exhumed and deformed (Fig. 7).

Paleogene Puna Plateau Model: Mostly Eocene Growth

In the second model scenario, the Puna Plateau attains ~90% of its present elevation through crustal thickening from 55 Ma to 40 Ma (Fig. 7), as suggested by stable isotope paleoaltimetric data sets (Canavan, 2012; Quade et al., this volume). Following a 25 m.y. hiatus in crustal thickening, the Eastern Cordillera grew from 15 Ma to 6 Ma and the Santa Bárbara system rose from 5 Ma to present, using a similar kinematic history to the Neogene Puna scenario, based on several studies (Reynolds et al., 2000; Carrapa et al., 2011; DeCelles et al., 2011; Pearson et al., 2012, 2013). The spatial distribution of topographic growth in the Paleogene Puna Plateau model is generally captured following Quade et al.'s (this volume) conceptualization (Fig. 7).

DISCUSSION

Modeling Results

Although the purpose of the modeling in this region is an attempt to evaluate the temporal and spatial history of flexural accommodation of the Puna foreland region based on different models of topographic evolution of the Puna Plateau, the Salta Rift system figures prominently into the history of subsidence (thermal and elastic). One of the major conclusions from the modeling is that the Salta Rift system influenced sediment distribution patterns all the way through the Cenozoic. Figure 8 shows the results from the Paleogene Puna Plateau development with and without Cretaceous Salta Rift system extension. It is evident that thermal subsidence associated with the rift enhanced accommodation during the isostatic response to crustal thickening in the Cenozoic, particularly with respect to the thickness and distribution of the Santa Bárbara and Metán Subgroups. In these models, additional thermal subsidence was ~500 m.

The models of Paleogene and Neogene Puna Plateau evolution begin to diverge after 55 Ma, as the two different models of growth begin to exert major influence on creation and destruction of accommodation in the adjacent regions. The results from the models are shown in Figures 9 and 10 and show differing total thickness of the Cretaceous and Cenozoic fill. Time slices of 25 Ma and present are shown (Figs. 9 and 10) because they depict the divergence in tectonic growth of the Puna Plateau.

In the model of progressive Neogene growth of the Puna Plateau, the accommodation over the Salta Rift basin at 25 Ma (Fig. 9A) reflects the long-term influence of thermal subsidence from mantle delamination. The western third of the basin (770–880 km) exhibits a westerly thickening wedge of sediment with associated with modeled Incaic deformation that encroached on the western Puna Plateau (e.g., Carrapa et al., 2005). The rest of the profile (900–1070 km) is located east of the flexural forebulge, but continued thermal subsidence following Late Cretaceous thinning of the mantle lithosphere results in continuous accommodation. This is represented as a uniform expansion of the Paleogene section in Figure 9, which creates ~1 km of strata at 25 Ma. Rift shoulders remain as topographic highs within the models, due to the lack of erosion and the thick, unrifted crust that buoys it above the basin floor. The forebulge associated with the modeled Incaic load is isolated within the Eastern Cordillera and would imply that observed exposure surfaces (paleosols) within the Eastern Cordillera and Santa Bárbara system (e.g., Reynolds et al., 2000; Starck, 2011; DeCelles et al., 2011) could also represent features associated with early inversion of Salta Rift system hanging walls. Until ca. 10 Ma, the eastern margin of the flexural foreland was confined within the limits of the Salta Rift system (950–1085 km; Fig. 9B). To a first order, this is because the eastern edge of the rift basin is buoyed by thick (35 km), unextended continental crust that remains above the depositional datum, despite flexural influence. Secondarily, the Salta Rift is still subsiding thermally, which continues to enhance the foredeep. The eastward growth of the Eastern Cordillera and the Santa Bárbara system provided enough of a tectonic load to bury the bounding rift shoulder. Overall, this model suggests relatively little accommodation throughout the Cretaceous and Cenozoic, with a total stratal thickness generally not exceeding 3.5 km thickness.

In contrast, the Paleogene Puna Plateau loading modeled in the Paleogene superimposes a large flexural profile on the thermally subsiding Salta Rift basin (Fig. 10). The load from the early Puna Plateau growth, coupled with the reduced effec-

tive elastic thickness, produces a significant flexural foredeep that creates ~5.5 km of flexural subsidence (Fig. 10). The wavelength of the Paleogene foredeep extends across two thirds of the Salta Rift system (1010 km; Fig. 10A). Superimposed on this flexural profile is additional thermal subsidence, the effects of which can be seen by several closely spaced "stratal" chrons that extend across the Salta Rift basins (Fig. 10) from 40 Ma to 15 Ma. From 15 Ma to 5 Ma, the Eastern Cordillera region was exhumed, inverting what was a foreland basin into a high-standing range that produces its own foredeep in the Santa Bárbara region (Fig. 10B). Like the Eocene depocenter, this Miocene flexural depression is mostly confined to the Salta Rift system until 7 Ma, and it creates as much as 3 km of accommodation. At the final stage of the model, the only active accommodation is being created as a "wedge-top" basin between the Santa Bárbara system uplifts (Fig. 10B).

The two scenarios for Puna Plateau topographic evolution result in very different foreland subsidence histories. Perhaps the most dramatic is the formation of a thick Paleogene foredeep with the Paleogene Puna Plateau model that buries the Salta Rift assemblage in the Eastern Cordillera to depths of 4–5 km. In contrast, the Neogene Puna Plateau model shows much less burial in the Eastern Cordillera, which is dominated by thermal subsidence and essentially uniform stratal geometries through the Paleogene, similar to hypotheses advanced by Cristallini et al. (1991), Salfity and Marquillas (1994), del Papa (1999), del Papa and Salfity (1999), del Papa et al. (2002), and Marquillas et al. (2005) to explain stratal thicknesses and facies relationships in the Salta Rift basins. The thermal consequences of these two scenarios on a potential petroleum system are discussed in a companion paper (Summa et al., this volume).

Geologic Implications

The cross sections from the basin model deviate in detail from the geologic observations, due to reasons that have been mentioned previously (i.e., the lack of erosion in the models, and the lack of lateral translation as a means of thickening crust). Nonetheless, several observations can be made regarding the development of the Andean foreland. In both scenarios, Salta Rift system horsts generally tend to remain as topographic highs (although this is also partially conditioned by the lack of erosion in the model). Field relationships and thermochronometric data suggest these features were indeed beveled and buried (Kley et al., 2005; Starck, 2011; Carrapa et al., 2013), although the structures exhibiting this relationship are inverted rift hanging walls (basins), not potentially high-standing Cretaceous rift flanks. The magnitude and distribution of subsequent Neogene Andean crustal thickening were never great enough to deeply bury these rift features by a flexural response. It is not believed that this is solely an artifact of the boundary conditions of the modeling, as Insel et al. (2012), Löbens et al. (2013), Carrapa et al. (2013), and Pearson et al. (2013) report Cretaceous thermochronometric ages within the Eastern Cordillera. The influence of the Salta Rift subsidence on

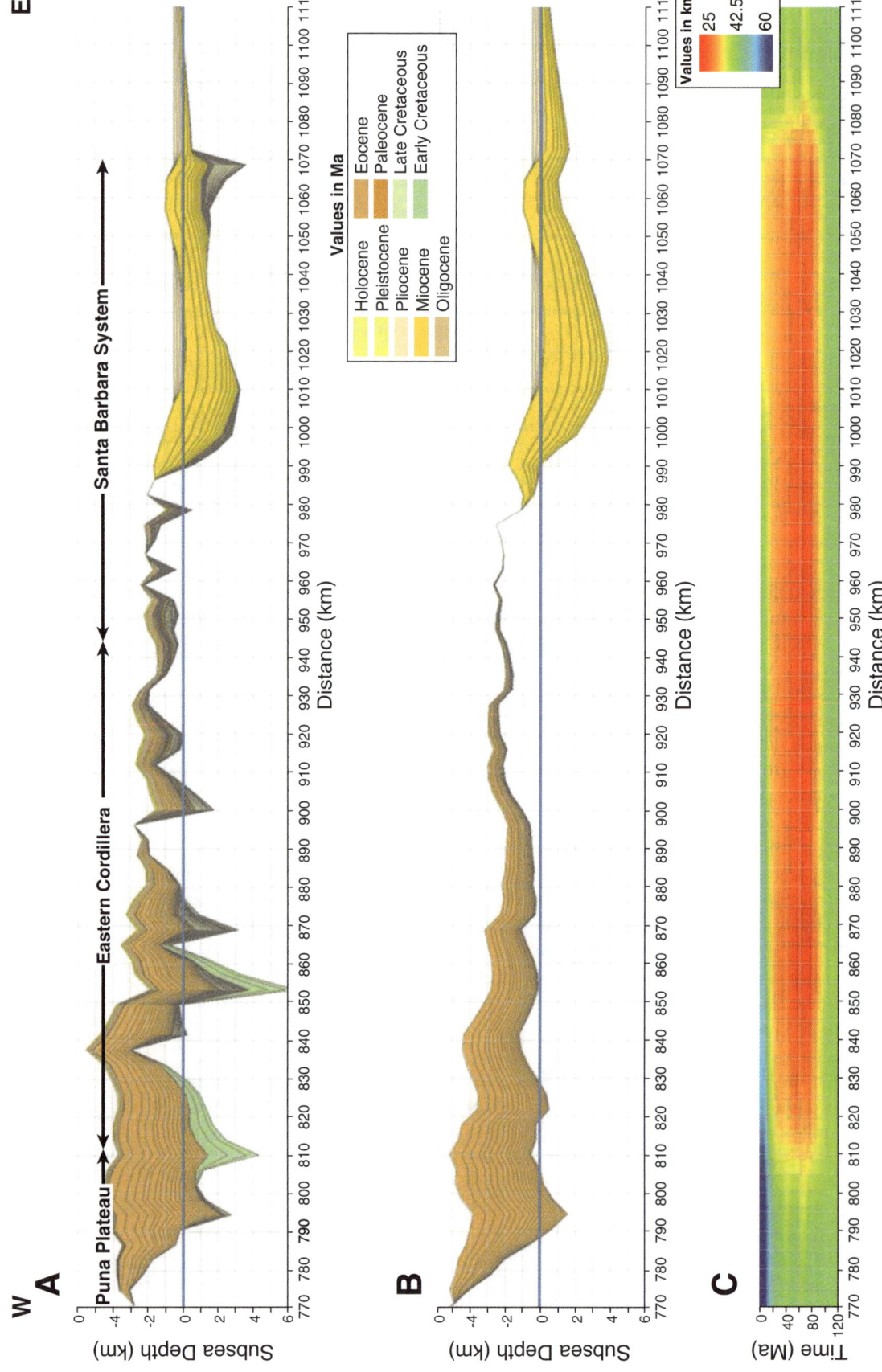

Figure 8. A comparison of quantitative basin analysis (QBA) models of Paleogene Puna development with (A) and without (B) the Salta Rift system. (A) The Salta Rift system experiences two stages of rifting and severe mantle thinning (akin to Fig. 6B). (B) The same Cenozoic tectonic history is superimposed on unrifted lithosphere (35 km thick crust). (C) The effective elastic thickness of section A as a function of space/time from profile A. The Salta Rift thermal subsidence creates accommodation irrespective of position within the flexural profile, which creates more continuous deposition across the basin. The modeled Cenozoic section is also incrementally thicker in A vs. B. Note that the profiles are vertically exaggerated (~4:1).

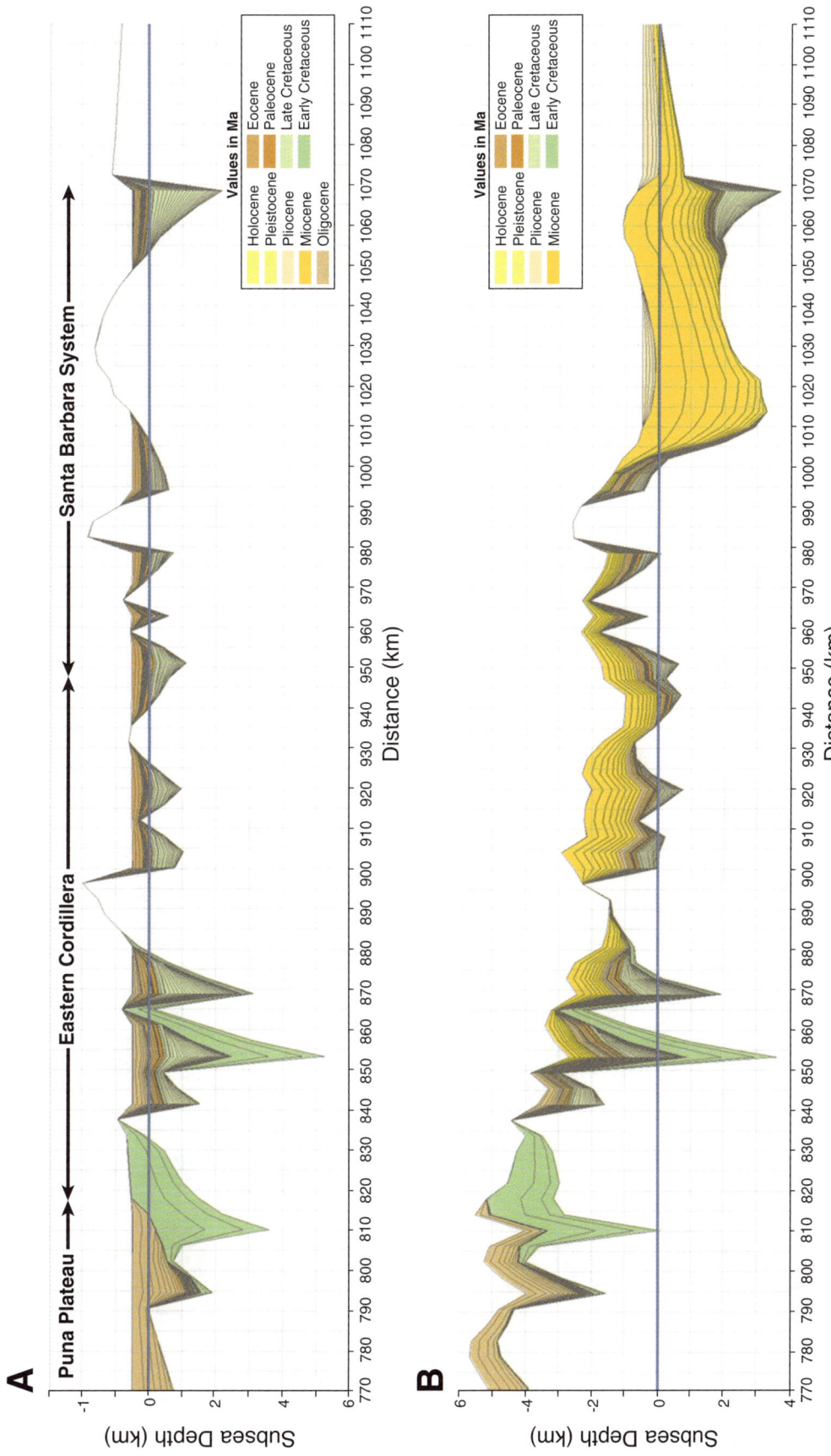

Figure 9. Two sequential cross sections of the foreland basin region using the progressive Neogene Puna Plateau modeling scenario at the (A) 25 Ma and (B) present time steps. As can be seen in A, thermal subsidence dominates basin accommodation until 25 Ma, although flexural subsidence associated with the Incaic deformational event is evident in the Eastern Cordillera. Cross section B shows the combined flexural and thermal subsidence associated with Incaic, Puna Plateau, Eastern Cordillera, and Santa Bárbara system crustal thickening, the majority of which occurs in the Neogene. These profiles are vertically exaggerated (~10:1) to exhibit the history of accommodation.

424

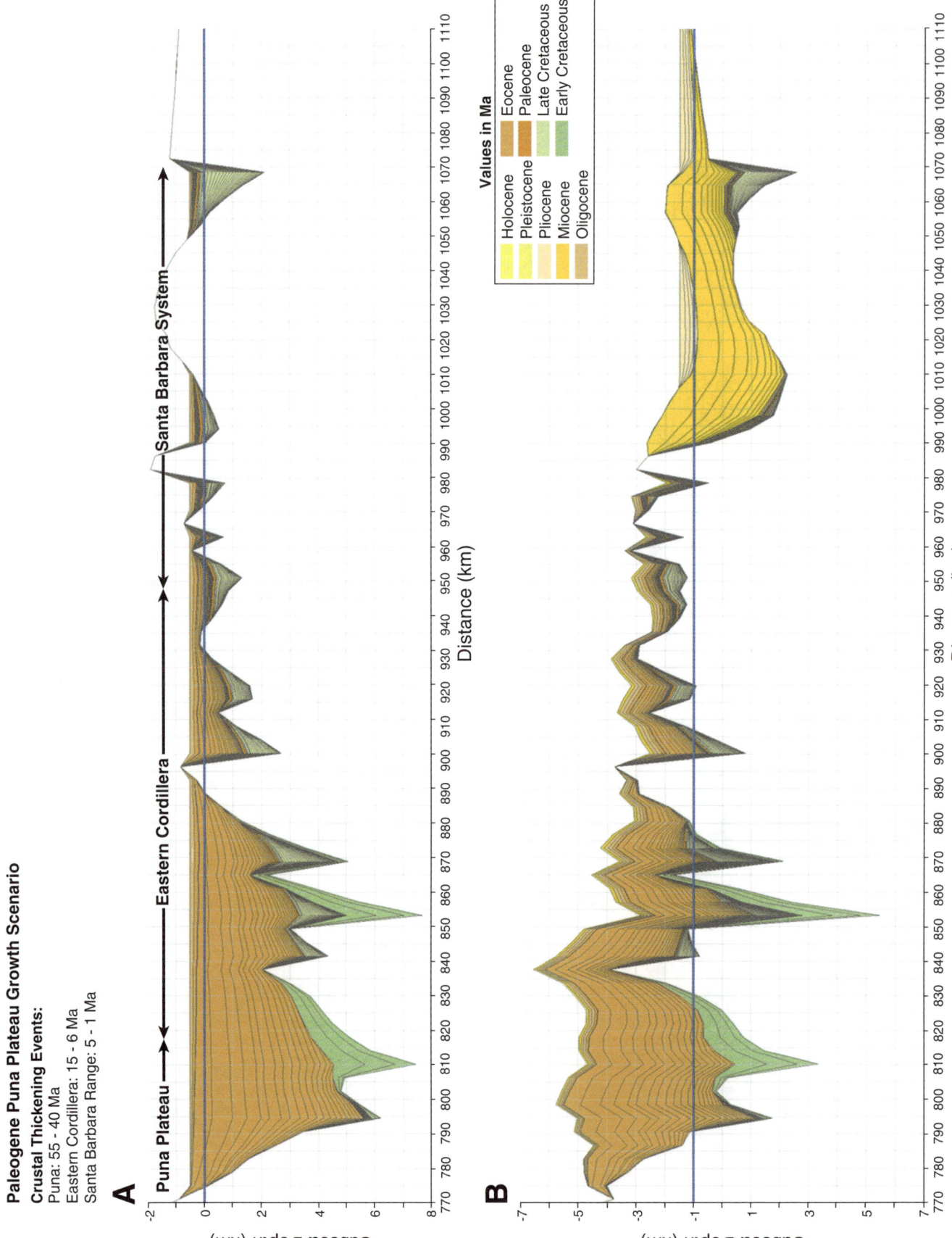

Paleogene Puna Plateau Growth Scenario

Crustal Thickening Events:
Puna: 55 - 40 Ma
Eastern Cordillera: 15 - 6 Ma
Santa Barbara Range: 5 - 1 Ma

potential facies distributions and paleoenvironments should also be considered in interpretations of subsidence and accommodation based on the stratigraphic record. For example, the isopach patterns of the Miocene Metán Subgroup generally follow the boundaries of the Salta Rift system (Galli, 1995; Reynolds et al., 2000). The thermal subsidence becomes increasingly subtle over time, yet it likely remains an active process to the present.

Perhaps the most interesting aspect of the models, and the one that motivated this study, is the influence of differing hypotheses of timing of topographic growth of the Puna Plateau on accommodation and sediment fill. The progressive Neogene Puna Plateau model creates much less Cenozoic burial in the Eastern Cordillera. During Paleogene time in this model, the Eastern Cordillera is dominated by thermal subsidence to produce essentially uniform stratal geometries. In contrast, the model of the Paleogene Puna Plateau scenario produces a thick Paleogene foredeep that buries the Salta Rift assemblage in the Eastern Cordillera to depths of 4–5 km. Some of the implicit and explicit assumptions embedded within the modeling may have influenced these results, and as such, are worthy of further discussion, including: (1) thinning of the mantle lithosphere, (2) the elastic strength of the lithosphere, and (3) whether the Andean orogen was providing sediment to the foredeep. In the following paragraphs, we discuss some of the more relevant geologic observations that bear on these assumptions.

Although there is good petrologic evidence of mantle lithospheric thinning beneath the Salta Rift system, the timing and extent of it are not well understood. In addition, there remains a temporal disconnect between the surface manifestation of Salta rifting as viewed by igneous petrology, which suggests more or less continuous lithospheric extension/thinning from ca. 125 to 60 Ma (reviewed earlier herein), and regional stratigraphic relationships that suggest two episodes of extension, the first from ca. 125 to 100 Ma and the second from ca. 100 to 70 Ma (e.g., Marquillas et al., 2005). The actual thinning of the mantle lithosphere beneath the Salta Rift system is constrained between 110 and 75 Ma but is not well known. Additional work may help to better resolve the timing of the mantle lithospheric extension, or if the thinning of the mantle lithosphere is associated with crustal extension.

The modern effective elastic thickness of the lithosphere in this region of South America is a topic of active research and debate, and several proposed solutions have recently been published, for example: 30–50 km (Stewart and Watts, 1997), 5–15 km (Tassara et al., 2007), 10–30 km (Perez-Gussinye et al., 2008, 2009), and 50 km (Chase et al., 2009). Given the uncertainty in calculating a "static" modern system, estimating the paleoelastic properties of the lithosphere may be problematic. As discussed in Appendix A and Roberts et al. (1998), the modeling efforts serve to test whether the lithosphere did or did not have elastic strength, not necessarily to produce precise values of effective elastic thickness. Although we are unable to produce exact matches of the geologic record for a variety of reasons, the choice of a 450° elastic isotherm (initial effective elastic thickness of ~42 km) creates reasonably similar magnitudes and distributions of accommodation. Models (not shown) of the foreland simulating local loading (no effective elastic thickness) can produce an Eocene-age Puna Plateau without a foredeep; however, there is a poor match to the observed geology in region prior to, and following, the Paleogene.

Internal drainage appears to be a necessary property of orogenic plateau development, if the Tibetan Plateau and central Andean Plateau are characteristic examples, and this naturally has implications for sources of sediment to the foreland basin. For the study area, this would imply that (1) only the eastern margin of the Paleogene central Andean Plateau would be a source of sediment or (2) the central Andean Plateau was elevated yet integrated into drainage networks likely flowing to the east (given the prevailing wind direction and moisture source). Alonso et al. (1991) and Vandervoort et al. (1995) suggested that the Puna Plateau became hydraulically isolated between 14.1 and 24.2 Ma, most likely at ca. 15 Ma, based on the age of evaporite deposits. This is consistent with the Neogene formation of the Puna Plateau (Figs. 7 and 9) and would permit Puna-derived sediment to fill only the early Miocene foredeep in the Eastern Cordillera. Miocene–present growth of the Puna Plateau would also create a significant foredeep in the present-day Eastern Cordillera, prior to its exhumation in the middle Miocene. The ~4-km-thick clastic deposits within the Angastaco Basin are of the appropriate age and thickness to represent that foredeep, and paleocurrent, petrologic, and detrital zircon age data also hint at a possible sediment source from the Puna Plateau (Coutand et al., 2006; DeCelles et al., 2011; Carrapa et al., 2012). Sedimentary petrography of the middle Miocene Rio Seco and Anta Formations within the Metán Basin also suggests a source from the Puna and/or Eastern Cordillera (Reynolds et al., 2000). The Rio Seco and Anta Formations, as well as a significant portion of the Angastaco Formation, are younger than 15 Ma, which either implies that the Puna Plateau to the west of Angastaco Basin was not internally drained and did provide sediment to the region, or that the sediment is from more proximal sources in the surrounding Eastern Cordillera. Assuming the latter, the growth of the Puna Plateau would only contribute to the subsidence, not the filling, of the Angastaco Basin. The coincidence in timing between the interpreted isolation of the Puna Plateau (Vandervoort et al., 1995) and the resumption of subsidence in the Metán Basin (Reynolds et al., 2000) within the Santa Bárbara system at ca. 15 Ma is also noteworthy.

Figure 10. Time-sequential cross sections of the foreland basin region using the Paleogene Puna Plateau growth scenario at the (A) 25 Ma and (B) present time steps. The Eocene flexural basin is very prominent and creates as much as 6 km of subsidence in the eastern Puna Plateau region. The flexural forebulge for this load, at around 960–1050 km, prevents the formation of accommodation, despite ubiquitous thermal subsidence. As a result, the model predicts there should be little to no Eocene–middle Miocene Metán Subgroup present in the Santa Bárbara system. These profiles are vertically exaggerated (~9:1) to better show accommodation distribution over time.

In the Paleogene Puna scenario, the flexural foredeep accommodation created by the development of the Puna Plateau is filled with ~5 km of sediment at its maximum, and it would be inferred that this is coming from an adjacent uplift. Although this outcome is a construct of a model that fills all accommodation below 500 m above sea level with sediment, it would realistically require that the Puna be hydraulically connected. Since a 6-km-thick Eocene deposit has not been described in the Eastern Cordillera, this foreland might have remained starved of sediment, although that scenario might also have resulted in a large lake or marine-connected basin filled with condensed deep-water facies. The 400–500-m-thick Eocene Lumbrera Formation (Fig. 4) represents deposition in the Eastern Cordillera during that time and is interpreted as a lacustrine system with marginal deltas filling it (del Papa et al., 2002). However, the unit thickens toward the interior of the Salta Rift system basins, not the orogen (del Papa et al., 2002; Marquillas et al., 2005), as would be expected with an Eocene foredeep in this region according to the flexural model. The overlying Eocene–Oligocene Quebrada de los Colorados Formation (Fig. 4), found within the Eastern Cordillera, has been interpreted as evidence of an eastward-propagating orogen producing a foredeep (DeCelles et al., 2011), which would actually slightly postdate the proposed early uplift of the Puna Plateau (Quade et al., this volume). Eocene–Oligocene contractional deformation, associated with the Incaic phase of Andean tectonism and foredeep deposition, is generally restricted to the regions that presently comprise the central Andean Plateau (Jordan and Alonso, 1987; Allmendinger et al., 1997; Horton et al., 2002; Carrapa et al., 2005). Incaic deformation is a plausible mechanism for the Eocene–Oligocene flexural subsidence in the Eastern Cordillera, if the Puna Plateau remained under ~2 km elevation (similar to the Neogene Puna Plateau growth model; Fig. 9).

The modeling results (Figs. 9 and 10) generally represent the geology along the profile, but the output does not lend itself well to direct comparison with a cross section (Fig. 3C). In an attempt to represent the modeling results within a geologically amenable context, the location and thickness of modeled Cenozoic basins and loads from the Paleogene and Neogene Puna scenarios were projected onto the modern cross section layer by layer in the appropriate locations (Fig. 11) using a restoration. Although it is seemingly straightforward to compare the model results in this manner, implicitly the differences have significant consequences. For example, the thicker Paleogene foredeep modeled in the Eastern Cordillera of the Paleogene Puna scenario (Fig. 11C) would require more erosion in the Eastern Cordillera and deposition in the Santa Barbara system to create the synthetic cross section and maintain isostatic balance. Variations in the depth and magnitude of shortening on faults within the Eastern Cordillera and Santa Barbara system influence the degree of crustal thickening. The isostatic response to this crustal thickening, which is subsequently modified by erosion, determines what stratigraphy is exposed at the surface (this isostatic interplay among crustal thickness, detachment depth, and fault displacement is discussed briefly in Appendix A). With this in mind, the thermal history in the Eastern

Cordillera for the two different Puna Plateau growth scenarios is predicted to be very different, as described in Summa et al. (this volume). Low-temperature thermochronometric ages (apatite fission-track ages) from the Eastern Cordillera are unlikely to have been completely annealed prior to exhumation if the Puna Plateau grew during the Miocene, unlike the subsidence and exhumation predicted for Paleogene Puna Plateau elevation growth, which would have completely reset Miocene–present ages.

CONCLUSIONS

Two-dimensional kinematic forward models that adhere to the principles of flexural isostasy and a two-layer lithosphere (Appendix A) offer a means of testing alternate tectonic histories as preserved in potential foreland basin deposits. Proxies for paleorelief and paleoaltimetry (e.g., Jordan et al., 2010; Canavan, 2012; Quade et al., this volume) also provide important constraints on the mass of the orogen responsible for the foreland flexural response, which is otherwise difficult to constrain. As demonstrated herein, the preexisting Salta Rift system creates a complex mechanical and thermal response that influences the flexural response to orogenic growth and the subsidence history.

The Cretaceous and Paleogene stratigraphy within the foreland basin adjacent to the Puna Plateau appears to be localized by the Salta Rift system and not influenced by major early Andean crustal thickening. The modeling results presented in this paper illustrate the potential ramifications of continental mantle lithospheric thinning and delamination on the evolution of the Salta Rift basins (and the later Puna Plateau foredeep), an inference based on the study of lower-crustal and mantle xenoliths. The thermal pulse associated with advected asthenosphere transiently reduces the effective elastic thickness of the foreland lithosphere and results in a long-lived thermal spike that requires tens of millions of years to dissipate. The observed accommodation created by this mantle lithospheric thinning event is not disproved by observed Cretaceous stratal isopachs. If correctly identified, the thinned mantle lithosphere created during the Late Cretaceous phase of rifting had long-lasting effects on accommodation that are still operative today. Given the important influence on the subsidence history imparted by the Late Cretaceous mantle lithospheric thinning, future work should focus on constraining the timing, magnitude, and duration of that event.

Neogene growth of the Puna Plateau radically altered subsidence and accommodation within the adjacent foreland basin

Figure 11. Comparison of the cross section of Figure 3C through the Eastern Cordillera and Santa Bárbara system (A) with the modeling results of the Neogene Puna Plateau growth (B) and the Paleogene Puna Plateau growth scenarios (C) at the same scale. The stratal thicknesses modeled for each were superimposed on retrodeformed sections to locate them on appropriate fault blocks. Section B (progressive Neogene Puna Plateau scenario) is more consistent with the preserved geologic record.

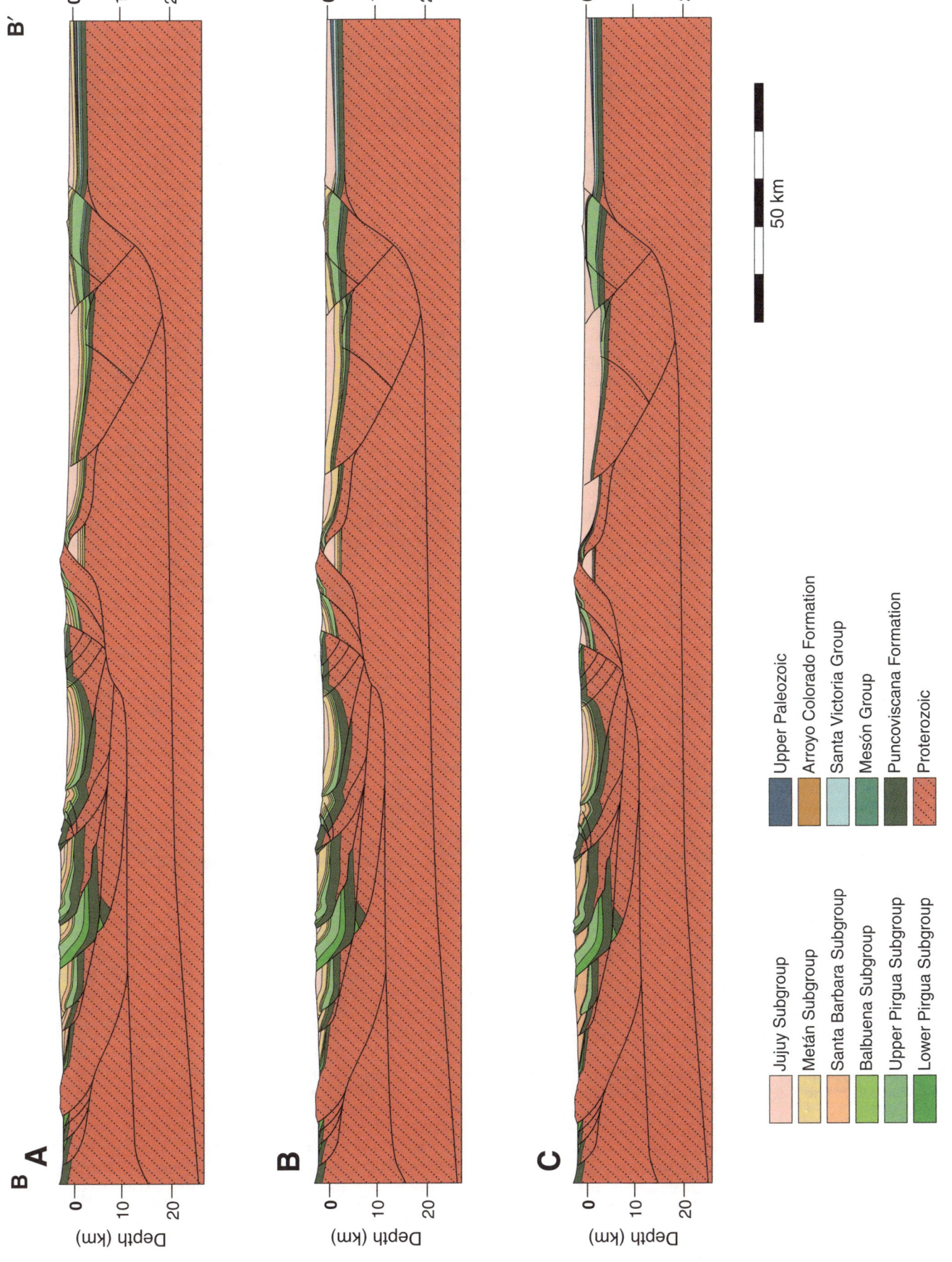

B' B

A

B

C

Depth (km)
0
10
20

Depth (km)
0
10
20

Depth (km)
0
10
20

0
10
20

0
10
20

0
10
20

50 km

Upper Paleozoic

Arroyo Colorado Formation

Santa Victoria Group

Mesón Group

Puncoviscana Formation

Proterozoic

Jujuy Subgroup

Metán Subgroup

Santa Barbara Subgroup

Balbuena Subgroup

Upper Pirgua Subgroup

Lower Pirgua Subgroup

that now forms the Eastern Cordillera and Santa Bárbara system. The profound influence of the Salta Rift system is evident in the foreland basin profile, irrespective of the timing of Puna Plateau growth (Figs. 9–10). The contrasting models of the evolution of the Puna Plateau, Eastern Cordillera, and Santa Bárbara system are conceptual in nature but can be used in conjunction with field-based interpretations (e.g., Reynolds et al., 2000; DeCelles et al., 2011; Carrapa et al., 2011) to evaluate alternative concepts about foreland basin subsidence and accommodation. The modeling suggests that a topographically high Puna Plateau should have created a significant foreland in the region of the present-day Eastern Cordillera either during the Miocene or Eocene. Stratigraphic evidence of an extensive Eocene foreland deposit (Paleogene Puna Plateau scenario) is lacking, as the corresponding stratigraphic unit in the Eastern Cordillera (Quebrada de los Colorados Formation) is only ~400 m thick (Carrapa et al., 2011) in a location where it is predicted to be several kilometers. One possibility is that the Eastern Cordillera began inverting earlier than modeled, although this merely displaces the location of the foredeep to the east, where support for a thick, extensive foreland deposit is also lacking. In addition, modeling demonstrates a probable linkage between hydrologic isolation of the Puna Plateau at ca. 15 Ma and the eastward expansion of the flexural foredeep in the Metán Basin region, adding evidence that the Puna Plateau did not grow in any wholesale way as previously implied (Garzione et al., 2006, 2008).

Future work refining the structural and petrologic relationships in the region will provide a better basis for understanding the complex tectonic history of the region. In addition, application of paleothermometric tools, such as low-temperature thermochronometry, illite age analysis, vitrinite reflectance, optical fluid-inclusion thermometry, etc., can provide a more complete picture of the timing and extent of the potential Cretaceous mantle lithosphere delamination event beneath the Salta Rift system. Development of criteria for recognizing the potential role of the mantle lithosphere within convergent systems, as well as its ramifications, remains an important goal for understanding plate tectonics.

APPENDIX A: QUANTITATIVE BASIN MODELING

Kinematic Modeling of Lithospheric Deformation

A kinematic model was used to map the repeated extensional and contractional deformation of the lithosphere. Extension of the lithosphere, with prerift thickness, and crustal thickness, etc., is accommodated by brittle failure and hanging-wall collapse in the upper crust and an equal, although not necessarily spatially coincident, amount of ductile extension in the lower crust and upper-mantle lithosphere. Brittle faults often detach into a region of plasticity, termed the depth of necking. We assume that the form of the necking depth relates to a depth-dependent crustal decoupling zone as described by Driscoll and Karner (1998) and Karner and Driscoll (1999)—the upper plate is the zone above the decoupling zone, and the lower plate is the crustal and lithospheric mantle below the decoupling zone. Upper- and lower-crustal thinning is measured relative to the decoupling zone. For example, if the decoupling zone is the Moho, then extension generates a kinematic hole that is flexurally uplifted and placed back into equilibrium. Foot-

wall uplift is a consequence of the flexural rebound of the lithosphere due to unloading of the upper plate (Weissel and Karner, 1989).

The ratios of pre- and postrift thickness of the hanging-wall and footwall blocks (upper and lower plates) are $\delta(x)$ and $\beta(x)$, respectively. The values of $\delta(x)$ and $\beta(x)$ across the margin determine the magnitude and distribution of rift-induced subsidence, while $\beta(x)$ is the primary control on postrift thermal subsidence. Extension is represented by $\delta(x)$ and $\beta(x) > 1$; inversion is represented by $\delta(x)$ and $\beta(x) < 1$. Extension perturbs the thermal structure of the lithosphere and leads to passive shallowing of the lithosphere-asthenosphere boundary. Flexural uplift of the rift flanks occurs in response to mechanical unloading of the lithosphere during extension (e.g., Weissel and Karner, 1989). In the modeling by Driscoll and Karner (1996, 1998), finite rifting is treated as a series of instantaneous events, rather than a single large event with subsequent cooling (cf. McKenzie, 1978). This allows heat to dissipate during rifting, which may be significant if the rate of extension is slow relative to the thermal half-life of the lithosphere. Finite rifting results in increased subsidence during rifting, and a corresponding decrease in the postrift subsidence (e.g., Cochran, 1983; Karner et al., 1997).

The computer code used to investigate margin evolution is a 2-D kinematic forward modeling program that tracks the thermal and mechanical response of the lithosphere both during and after rifting and, thus, the history of rift and postrift basin subsidence. The premise of the modeling technique is that the distribution and thickness of the preserved synrift and postrift sediment relate to the distribution and amplitude of lithospheric extension. Features of our model include:

1. Basin subsidence and rift-flank uplift are calculated as a function of space and time. Lithospheric cooling is by both vertical and lateral heat flow. The proportion of space filled by sediment is controlled by paleo–water depth profiles, generated to simulate the preserved sediment thickness and constrained, where possible, using biostratigraphic data.

2. A kinematic description of the amount and distribution of lithospheric extension and lithospheric mantle thinning (termed delta and beta functions, respectively) is incorporated.

3. A kinematic description of the amount and distribution of crustal convergence (0 < delta and beta functions < 1) is incorporated to simulate basin inversion.

4. Crustal extension across faults in the hanging-wall blocks can be incorporated.

5. The extension computer code incorporates finite rifting, re-rifting, and inversion events.

6. Basin fill with compacting sediments is included. The compaction characteristics are spatially variable and parameterized using Athy's law. The distribution and the amount of basin fill are regulated by the paleo–water depth profiles. As such, the sediment transport and depositional systems are not process or physics based.

7. Erosion of emergent and submarine topography producing peneplains and unconformities is incorporated.

8. Eustatic variations can be included (both first and second order).

9. Flexural isostasy is employed throughout the calculations, during both rift and postrift development, and during erosion, inversion, and magmatic underplating. The effective elastic thickness of the lithosphere changes as a function of space and time and as tracked by the depth to a particular isotherm (here assumed to be approximated by the 450 °C; e.g., Karner et al., 1997; Pérez-Gussinye and Watts, 2005).

10. Display of the time-line stratigraphy and the basin crustal geometry at any time during evolution of the system, i.e., both during and after rifting and inversion events, is included.

11. The process of erosion is also treated as an exponential function that defines the destruction of topography in terms of a "half-life." Erosion is regarded as a function of height (i.e., the highest topography is eroded fastest), where height is measured relative to the input paleo–water depth for that time interval. The assumptions dictating subaerial and submarine erosion serve only to approximate the process.

12. Likewise, the concept of effective elastic thickness (*Te*) requires some discussion. The spatial variation of continental *Te* appears to be generally controlled by the thermal structure of the lithosphere at the time of loading. Thus, for example, the Basin and Range region of the western United States is characterized by low but finite *Te*, while the Canadian Shield is associated with significantly larger values (e.g., Lowry and Smith, 1995; Audet and Mareschal, 2004; Pérez-Gussinye and Watts, 2005). The strength of extended lithosphere appears to be maintained during rifting despite fracturing of the crust by normal faulting and relatively high heat flow, as demonstrated by the existence of rift-flank topography and large-amplitude gravity anomalies (a proxy for lithospheric stress difference) over rift basins (e.g., Karner et al., 2000). In this paper, we track the depth to the 450 °C isotherm as a function of space and time as a proxy for the effective elastic thickness of the lithosphere. The exact isotherm and, thus, *Te* is not particularly critical—as noted by Roberts et al. (1998), what is important is not the exact value of *Te* but whether it is zero or finite.

Parameters for Models

Initial Conditions

Profile length: 2000 km
Initial crustal thickness: 35 km
Controlling elastic isotherm: 450 °C
Asthenosphere temperature: 1333 °C
Sediment grain density: 2650 kg/m^3
Surface porosity: 60%
Porosity characteristic depth: 1.5 km
Initial surface heat flow (at basement interface): 55 mW/m^2
Sedimentation datum: 500 m

Salta Rift Modeling Parameters

Neocomian Rift Event (125–110 Ma):
Cooling interval: 10 m.y.
Beta: 1.15
Delta: 1.15
Campanian Rift Event (100–70 Ma):
Cooling interval: 5 m.y.
Beta: variable; maximum of 3 throughout Salta Rift system
Delta: 1.15

TABLE A1. HEAT FLOW FOR
LOCATION IN CORE OF SALTA RIFT

Time (Ma)	Basal heat flow (mW/m^2)
130	55
120	57
115	59
110	62
100	62
90	67
85	72
80	77
70	82
60	81
50	77
40	74
30	72
20	69
10	67
0	65

Neogene Puna Plateau Scenario

Domeyko Shortening (60–56 Ma)

Cooling interval: 1 m.y.
Profile position: 435–600 km

Incaic Shortening (35–26 Ma)

Cooling interval: 1 m.y.
Profile position: 440–759 km

Puna Plateau Growth (25–16 Ma)

Cooling interval: 1 m.y.
Profile position: 440–911 km

Puna Plateau–Eastern Cordillera Growth (15–6 Ma)

Cooling interval: 1 m.y.
Profile position: 440–1020 km

Santa Bárbara System Growth (5–1 Ma)

Cooling interval: 1 m.y.
Profile position: 434–790; 935–1072 km

Paleogene Puna Plateau Scenario

Domeyko Shortening (60–56 Ma)

Cooling interval: 1 m.y.
Profile position: 435–600 km

Puna Plateau Growth (55–41 Ma)

Cooling interval: 20 m.y.
Profile position: 493–794 km

Eastern Cordillera Growth (15–6 Ma)

Cooling interval: 1 m.y.
Profile position: 434–1021 km

Santa Bárbara System Growth: (5–1 Ma)

Cooling interval: 1 m.y.
Profile position: 888–1072 km

Relationship between Crustal Shortening and Thickening

Kinematic restorations of structurally deformed regions typically strive to achieve material balance using a combination of geometric relationships and internal strain algorithms (e.g., Dahlstrom, 1969; Elliott, 1983; Suppe, 1983; Jamieson, 1987; Allmendinger, 1998). A complementary approach to material balance, at perhaps a larger scale, includes flexural isostatic balance. In the QBA code, shortening of the upper plate, delta, where 0 < delta < 1, can be expressed as:

$$\delta = l_c - l_o / l_o,$$

where l_c is the finite length of the shortened crust, and l_o is the original length of crust.

A reciprocal of delta:

$$Tc_f = Tc_o / \delta,$$

can be used to calculate the finite crustal thickness (Tc_f) from the original thickness (Tc_o). In most convergent systems, crustal thickening does not involve wholesale pure shear shortening of the crust, but employment of a weak detachment. Because of this, a delta value cannot be chosen arbitrarily because it carries implications for a corresponding depth of detachment and magnitude of shortening to create the estimated crustal thickness. In the simplest arrangement of pure shear above a flat detachment, one can estimate the corresponding crustal thickness (and delta):

$$Tc_f = (D_d \times l_o)/(l_o - S) + (Tc_o - D_d),$$

where D_d is the depth of the detachment, and S is the magnitude of shortening. Balanced cross sections provide a more sophisticated means of determining the magnitude of shortening and the depth of detachment, but this equation provides a quick check on the validation of a one-dimensional parameter in geodynamic models.

APPENDIX B: U-Pb GEOCHRONOLOGIC METHODS FOR IGNEOUS ZIRCON ANALYSIS

The sample was prepared and run at the University of Arizona. The xenolith was crushed and pulverized using a jaw crusher and roller mill. Density separation and sieving to collect grains <350 μm were done using a Wilfley table. Additional density separation was done with methylene iodide and magnetic separation using a Franz magnetic separator. The sample with >90% purity was mounted in epoxy within 1-inch-diameter (2.54-cm-diameter) rings and sanded down to ~40 μm depth to expose zircon interiors. Analysis followed procedures described in detail by Gehrels et al. (2008). General information about data acquisition and reduction are presented next.

U-Pb geochronology of zircons was conducted by laser ablation–multicollector–inductively coupled plasma–mass spectrometry (LA-MC-ICP-MS) at the Arizona LaserChron Center. The analyses involve ablation of zircon with a New Wave/Lambda Physik DUV193 Excimer laser (operating at a wavelength of 193 nm) using a spot diameter of 25 or 35 μm. The ablated material is carried with helium gas into the plasma source of a GV Instruments Isoprobe, which is equipped with a flight tube of sufficient width that U, Th, and Pb isotopes are measured simultaneously. All measurements are made in static mode, using Faraday detectors for ^{238}U and ^{232}Th, an ion-counting channel for ^{204}Pb, and either faraday collectors or ion-counting channels for $^{208}Pb-^{206}Pb$. Ion yields are ~1 mv per ppm. Each analysis consists of one 20 s integration on peaks with the laser off (for backgrounds), twenty 1 s integrations with the laser firing, and a 30 s delay to purge the previous sample and prepare for the next analysis. The ablation pit is ~15 μm in depth.

For each analysis, the errors in determining $^{206}Pb/^{238}U$ and $^{206}Pb/^{204}Pb$ result in a measurement error of ~1% (at 2σ level) in the $^{206}Pb/^{238}U$ age. The errors in measurement of $^{206}Pb/^{207}Pb$ and $^{206}Pb/^{204}Pb$ also result in ~1% (2σ) uncertainty in age for grains that are older than 1.0 Ga, but errors are substantially larger for younger grains due to low intensity of the ^{207}Pb signal. For most analyses, the crossover in precision of $^{206}Pb/^{238}U$ and $^{206}Pb/^{207}Pb$ ages occurs at ca. 1.0 Ga.

Common Pb correction is accomplished by using the measured ^{204}Pb and assuming an initial Pb composition from Stacey and Kramers (1975; with uncertainties of 1.0 for $^{206}Pb/^{204}Pb$ and 0.3 for $^{207}Pb/^{204}Pb$). Our measurement of ^{204}Pb is unaffected by the presence of ^{204}Hg because backgrounds are measured on peaks (thereby subtracting any background ^{204}Hg and ^{204}Pb), and because very little Hg is present in the argon gas. Interelement fractionation of Pb/U is generally ~20%, whereas fractionation of Pb isotopes is generally <2%. In-run analysis of fragments of a large Sri Lanka zircon crystal (generally every fifth measurement) with known age of 564 ± 4 Ma (2σ error) is used to correct for this fractionation (see Gehrels et al., 2008). The uncertainty resulting from the calibration correction is generally ~1% (2σ) for both $^{206}Pb/^{207}Pb$ and $^{206}Pb/^{238}U$ ages.

The reported ages are determined from the weighted mean of the $^{206}Pb/^{238}U$ ages of the concordant and overlapping analyses (Ludwig, 2003). The reported uncertainty (labeled "mean") is based on the scatter and precision of the set of $^{206}Pb/^{238}U$ or $^{206}Pb/^{207}Pb$ ages, weighted according to their measurement errors (shown at 1σ). The systematic error, which includes contributions from the standard calibration, age of the calibration standard, and composition of common Pb and U decay constants, is generally ~1%–2% (2σ).

ACKNOWLEDGMENTS

Discussions in the field with Gary Gray, Peter DeCelles, Barbara Carrapa, Patricio Figueredo, and Dave Pearson helped introduce the first author to the interesting geology in the Salta and Jujuy provinces. This study was an outgrowth of the Convergent Orogenic Systems Analysis Collaborative between ExxonMobil and the University of Arizona. Lee Nachtegaele and Becky Miller helped create some of the figures in the paper. Constructive reviews by Mike McGroder, James Reynolds, Barbara Carrapa, Marlies ter Voorde, and an anonymous reviewer helped clarify and improve the manuscript.

REFERENCES CITED

Allmendinger, R.W., 1998, Inverse and forward numerical modeling of trishear fault-propagation folds: Tectonics, v. 17, p. 640–656, doi:10.1029/98TC01907.

Allmendinger, R.W., and Gubbels, T., 1996, Pure and simple shear plateau uplift, Altiplano-Puna, Argentina and Bolivia: Tectonophysics, v. 259, p. 1–13, doi:10.1016/0040-1951(96)00024-8.

Allmendinger, R.W., Ramos, V.A., Jordan, T.E., Palma, M., and Isacks, B.L., 1983, Paleogeography and Andean structural geometry, northwest Argentina: Tectonics, v. 2, p. 1–16, doi:10.1029/TC002i001p00001.

Allmendinger, R.W., Isacks, B.L., Jordan, T.E., and Kay, S.M., 1997, The evolution of the Altiplano-Puna Plateau of the central Andes: Annual Review of Earth and Planetary Sciences, v. 25, p. 139–174, doi:10.1146/annurev.earth.25.1.139.

Alonso, R.N., Jordan, T.E., Tabbutt, K.T., and Vandervoort, D.S., 1991, Giant evaporite belts of the Neogene central Andes: Geology, v. 19, p. 401–404, doi:10.1130/0091-7613(1991)019<0401:GEBOTN>2.3.CO;2.

Arriagada, C., Cobbold, P.R., and Roperch, P., 2006, Salar de Atacama Basin: A record of compressional tectonics in the central Andes since the mid-Cretaceous: Tectonics, v. 25, TC1008, doi:10.1029/2004TC001770.

Audet, P., and Mareschal, J.-C., 2004, Variations in elastic thickness in the Canadian Shield: Earth and Planetary Science Letters, v. 226, p. 17–31, doi:10.1016/j.epsl.2004.07.035.

Barnes, J., and Ehlers, T.A., 2009, End member models for Andean Plateau uplift: Earth-Science Reviews, v. 97, p. 105–132, doi:10.1016/j.earscirev.2009.08.003.

Barnes, J.B., Ehlers, T.A., Insel, N., McQuarrie, N., and Poulson, C.J., 2012, Linking orography, climate, and exhumation across the central Andes: Geology, v. 40, p. 1135–1138, doi:10.1130/G33229.1.

Beaumont, C., 1981, Foreland basins: Geophysical Journal of the Royal Astronomical Society, v. 65, p. 291–329, doi:10.1111/j.1365-246X.1981.tb02715.x.

Bywater-Reyes, S., Carrapa, B., Clementz, M., and Schoenbohm, L., 2010, Effect of late Cenozoic aridification on sedimentation in the Eastern Cordillera of northwest Argentina (Angastaco basin): Geology, v. 38, p. 235–238, doi:10.1130/G30532.1.

Cahill, T., and Isacks, B.L., 1992, Seismicity and shape of the subducted Nazca plate: Journal of Geophysical Research, v. 97, p. 17,503–17,529, doi:10.1029/92JB00493.

Canavan, R., 2012, Cenozoic Paleoelevation Reconstructions of the Puna Plateau, NW Argentina [M.S. thesis]: Laramie, Wyoming, University of Wyoming, 104 p.

Carrapa, B., Adelmann, D., Hilley, G.E., Mortimer, E., Sobel, E.R., and Strecker, M.R., 2005, Oligocene range uplift and development of plateau morphology in the southern central Andes: Tectonics, v. 24, TC4011, doi:10.1029/2004TC001762.

Carrapa, B., Trimble, J.D., and Stockli, D.F., 2011, Patterns and timing of exhumation and deformation in the Eastern Cordillera of NW Argentina revealed by (U-Th)/He thermochronology: Tectonics, v. 30, TC3003, doi:10.1029/2010TC002707.

Carrapa, B., Bywater-Reyes, S., DeCelles, P.G., Mortimer, E., and Gehrels, G.E., 2012, Late Eocene–Pliocene basin evolution in the Eastern Cordillera of northwestern Argentina (25°–26°S): Regional implications for Andean orogenic wedge development: Basin Research, v. 24, p. 249–268, doi:10.1111/j.1365-2117.2011.00519.x.

Carrapa, B., Bywater-Reyes, S., Safipour, R., Sobel, E.R., Schoenbohm, L.M., DeCelles, P.G., Reiners, P.W., and Stockli, D., 2013, The effect of inherited paleotopography on exhumation of the central Andes of NW Argentina: Geological Society of America Bulletin, v. 126, p. 66–77, doi:10.1130/B30844.1.

Charrier, R., Pinto, L., and Rodriguez, M.P., 2007, Tectonostratigraphic evolution of the Andean orogen in Chile, *in* Moreno, T., and Gibbons, W., eds., The Geology of Chile: London, Geological Society, p. 21–114.

Chase, C.G., Sussman, A.J., and Coblentz, D.D., 2009, Curved Andes: Geoid, forebulge, and flexure: Lithosphere, v. 1, p. 358–363, doi:10.1130/L67.1.

Chulick, G.S., Detweiler, S., and Mooney, W.D., 2013, Seismic structure of the crust and uppermost mantle of South America and surrounding oceanic basins: Journal of South American Earth Sciences, v. 42, p. 260–276, doi:10.1016/j.jsames.2012.06.002.

Cochran, J.R., 1983, Effects of finite extension times on the development of sedimentary basins: Earth and Planetary Science Letters, v. 66, p. 289–302, doi:10.1016/0012-821X(83)90142-5.

Cohen, A., McGlue, M.M., Ellis, G.S., Zani, H., Swarzenski, P.W., Assine, M.L., and Silva, A., 2015, this volume, Lake formation, characteristics, and evolution in retroarc deposystems: A synthesis of the modern Andean orogen and its associated basins, *in* DeCelles, P.G., Ducea, M.N., Carrapa, B., and Kapp, P.A., eds., Geodynamics of a Cordilleran Orogenic System: The Central Andes of Argentina and Northern Chile: Geological Society of America Memoir 212, doi:10.1130/2015.1212(16).

Comínguez, A.H., and Ramos, V.A., 1995, Geometry and seismic expression of the Cretaceous Salta Rift of northwestern Argentina, *in* Tankard, A.J., Suarez Soruco, R., and Welsink, H.J., eds., Petroleum Basins of South America: American Association of Petroleum Geologists Memoir 62, p. 325–334.

Coutand, I., Cobbold, P.R., de Urreiztieta, M., Gautier, P., Chauvin, A., Gapais, D., Rossello, E.A., and López-Gamundí, O., 2001, Style and history of Andean deformation, Puna Plateau, northwestern Argentina: Tectonics, v. 20, p. 210–234, doi:10.1029/2000TC900031.

Coutand, I., Carrapa, B., Deeken, A., Schmitt, A.K., Sobel, E.R., and Strecker, M.R., 2006, Propagation of orographic barriers along an active range front: Insights from sandstone petrography and detrital apatite fission-track thermochronology in the intermontane Angastaco basin, NW Argentina: Basin Research, v. 18, p. 1–26.

Cristallini, E., Comínguez, A.H., and Ramos, V.A., 1997, Deep structure of the Metán-Guachipas region: Tectonic inversion in northwestern Argentina: Journal of South American Earth Sciences, v. 10, p. 403–421, doi:10.1016/S0895-9811(97)00026-6.

Currie, C.A., and Hyndman, R.D., 2006, The thermal structure of subduction zone back arcs: Journal of Geophysical Research, v. 111, B08404, doi:10.1029/2005JB004024.

Dahlstrom, C.D.A., 1969, Balanced cross sections: Canadian Journal of Earth Sciences, v. 6, p. 743–757, doi:10.1139/e69-069.

DeCelles, P.G., Carrapa, B., Horton, B.K., and Gehrels, G.E., 2011, Cenozoic foreland basin system in the central Andes of northwestern Argentina: Implications for Andean geodynamics and modes of deformation: Tectonics, v. 30, TC6013, doi:10.1029/2011TC002948.

del Papa, C.E., 1999, Sedimentation on a ramp type lake margin: Paleocene–Eocene Maiz Gordo Formation, northwestern Argentina: Journal of South American Earth Sciences, v. 12, p. 389–400, doi:10.1016/S0895-9811(99)00025-5.

del Papa, C.E., and Salfity, J., 1999, Non-marine Paleogene sequences, northwest Argentina: Acta Geologica Hispanica, v. 34, no. 2–3, p. 105–122.

del Papa, C., Garcia, V., and Quattrocchio, M., 2002, Sedimentary facies and palynofacies assemblages in an Eocene perennial lake, Lumbrera Formation, northwest Argentina: Journal of South American Earth Sciences, v. 15, p. 553–569, doi:10.1016/S0895-9811(02)00081-0.

Driscoll, N.W., and Karner, G.D., 1996, Tectonic and Stratigraphic Evolution of the Carnarvon Basin, Northwest Australia: Report of the Minerals and Energy Research Institute of Western Australia 170.

Driscoll, N.W., and Karner, G.D., 1998, Lower crustal extension across the northern Carnarvon Basin, Australia: Evidence for an eastward dipping detachment: Journal of Geophysical Research, v. 103, p. 4975–4992, doi:10.1029/97JB03295.

Echavarria, L., Hernandez, R., Allmendinger, R., and Reynolds, J., 2003, Subandean thrust and fold belt of northwestern Argentina: Geometry and timing of the Andean evolution: American Association of Petroleum Geologists Bulletin, v. 87, p. 965–985, doi:10.1306/01200300196.

Elliott, D., 1983, The construction of balanced cross-sections: Journal of Structural Geology, v. 5, p. 101, doi:10.1016/0191-8141(83)90035-4.

Farías, M.R., Charrier, R., Comte, D., Martinod, J., and Herail, G., 2005, Late Cenozoic deformation and uplift of the western flank of the Altiplano: Evidence from the depositional, tectonic, and geomorphologic evolution and shallow seismic activity (northern Chile at 19°30′S): Tectonics, v. 24, TC4001, doi:10.1029/2004TC001667.

Galli, C.I., 1995, Estratigrafía y Sedimentologa de Subgrupo Metán (Grupo Orán-Terciaro), Provincial de Salta, Argentina [Ph.D. thesis]: Salta, Argentina, Universidad Nacional de Salta, 109 p.

Galliski, M.A., and Viramonte, J.G., 1988, The Cretaceous paleorift in northwestern Argentina: A petrologic approach: Journal of South American Earth Sciences, v. 1, p. 329–342, doi:10.1016/0895-9811(88)90021-1.

Garzione, C.N., Hoke, G.D., Libarkin, J.C., and MacFadden, B.J., 2006, Rapid late Miocene rise of the Bolivian Altiplano: Evidence for removal of mantle lithosphere: Earth and Planetary Science Letters, v. 241, p. 543–556, doi:10.1016/j.epsl.2005.11.026.

Garzione, C.N., Hoke, G.D., Libarkin, J.C., Withers, S., MacFadden, B., Eiler, J., Ghosh, P., and Mulch, A., 2008, Rise of the Andes: Science, v. 320, p. 1304–1307, doi:10.1126/science.1148615.

Gehrels, G.E., Valencia, V., and Ruiz, J., 2008, Enhanced precision, accuracy, efficiency, and spatial resolution of U-Pb analysis by laser-ablation–multicollector–inductively-coupled-plasma–mass spectrometry: Geochemistry Geophysics Geosystems, v. 9, Q03017, doi:10.1029/2007GC001805.

Ghosh, P., Garzione, C.N., and Eiler, J.M., 2006, Rapid uplift of the Altiplano revealed through ^{13}C–^{18}O bonds in paleosol carbonates: Science, v. 311, p. 511–515, doi:10.1126/science.1119365.

Gregory-Wodzicki, K.M., 2002, A late Miocene subtropical-dry flora from the northern Altiplano, Bolivia: Palaeogeography, Palaeoclimatology, Palaeoecology, v. 180, p. 331–348, doi:10.1016/S0031-0182(01)00434-5.

Grier, M.E., Salfity, J.A., and Allmendinger, R.W., 1991, Andean reactivation of the Cretaceous Salta Rift, northwestern Argentina: Journal of South American Earth Sciences, v. 4, p. 351–372, doi:10.1016/0895-9811(91)90007-8.

Gurnis, M., 1990, Bounds on global dynamic topography from Phanerozoic flooding of continental platforms: Nature, v. 344, p. 754–756, doi:10.1038/344754a0.

Hartley, A.J., and Evenstar, L., 2010, Cenozoic stratigraphic development in the north Chilean forearc: Implications for basin development and uplift history of the central Andean margin: Tectonophysics, v. 495, p. 67–77, doi:10.1016/j.tecto.2009.05.013.

Hartley, A.J., Weissmann, G., Nichols, G.J., and Warwick, G.L., 2010, Large distributive fluvial systems: Characteristics, distribution, and controls on development: Journal of Sedimentary Research, v. 80, p. 167–183, doi:10.2110/jsr.2010.016.

Hoke, G.D., Isacks, B.L., Jordan, T.E., Blanco, N., Tomlinson, A.J., and Ramezani, J., 2007, Geomorphic evidence for post–10 Ma uplift of the western flank of the central Andes 18°30′–22°S: Tectonics, v. 26, TC5021, doi:10.1029/2006TC002082.

Hongn, F., del Papa, C., Powell, J., Petrinovic, I., Mon, R., and Deraco, V., 2007, Middle Eocene deformation and sedimentation in the Puna–Eastern Cordillera transition (23°–26°S): Control by preexisting heterogeneities on the pattern of initial Andean shortening: Geology, v. 35, p. 271–274, doi:10.1130/G23189A.1.

Horton, B.K., Hampton, B.A., LaReau, B.N., and Baldellon, E., 2002, Cenozoic provenance history of the northern and central Altiplano (central Andes, Bolivia): A detailed record of plateau-margin tectonics: Journal of Sedimentary Research, v. 72, p. 711–726, doi:10.1306/020702720711.

Insel, N., Grove, M., Haschke, M., Barnes, J.B., Schmitt, A.K., and Strecker, M.R., 2012, Paleozoic to early Cenozoic cooling and exhumation of the basement underlying the eastern Puna Plateau margin prior to plateau growth: Tectonics, v. 31, TC6006, doi:10.1029/2012TC003168.

Jamison, W.R., 1987, Geometric analysis of fold development in overthrust terranes: Journal of Structural Geology, v. 9, p. 207–219.

Jordan, T.E., 1981, Thrust loads and foreland basin evolution, Cretaceous, western United States: American Association of Petroleum Geologists Bulletin, v. 65, p. 2506–2520.

Jordan, T.E., and Alonso, R.N., 1987, Cenozoic stratigraphy and basin tectonics of the Andes mountains, 20–28° south latitude: American Association of Petroleum Geologists Bulletin, v. 71, p. 49–64.

Jordan, T.E., Nester, P.L., Blanco, N., Hoke, G.D., Davila, F., and Tomlinson, A.J., 2010, Uplift of the Altiplano-Puna Plateau: A view from the west: Tectonics, v. 29, TC5007, doi:10.1029/2010TC002661.

Karner, G.D., and Driscoll, N.W., 1999, Style, timing, and distribution of tectonic deformation across the Exmouth Plateau, northwest Australia, determined from stratal architecture and kinematic basin modeling, *in* MacNiocaill, C., and Ryan, P.D., eds., Continental Tectonics: Geological Society, London, Special Publication 164, p. 287–323.

Karner, G.D., and Watts, A.B., 1983, Gravity anomalies and flexure of the lithosphere at mountain ranges: Journal of Geophysical Research, v. 88, p. 10,449–10,477, doi:10.1029/JB088iB12p10449.

Karner, G.D., Driscoll, N.W., McGinnis, J.P., Brumbaugh, W.D., and Cameron, N., 1997, Tectonic significance of syn-rift sedimentary packages across the Gabon-Cabinda continental margin: Marine and Petroleum Geology, v. 14, p. 973–1000, doi:10.1016/S0264-8172(97)00040-8.

Karner, G.D., Byamungu, B.R., Ebinger, C.J., Kampunzu, A.B., Mukasa, R.K., Nyakaana, J., Rubondo, E.N.T., and Upcott, N.M., 2000, Distribution of crustal extension and regional basin architecture of the Albertine rift system, East Africa: Marine and Petroleum Geology, v. 17, p. 1131–1150, doi:10.1016/S0264-8172(00)00058-1.

Karner, G.D., Driscoll, N.W., and Barker, D.H.N., 2003, Synrift subsidence across the West African continental margin: The role of lower plate ductile extension, *in* Arthur, T.J., MacGregor, D.S., and Cameron, N.R., eds., Petroleum Geology of Africa: New Themes and Developing Technologies: Geological Society, London, Special Publication 207, p. 105–125.

Kley, J., and Monaldi, C.R., 1998, Tectonic shortening and crustal thickness in the central Andes; how good is the correlation?: Geology, v. 26, p. 723–726, doi:10.1130/0091-7613(1998)026<0723:TSACTI>2.3.CO;2.

Kley, J., and Monaldi, C.R., 2002, Tectonic inversion in the Santa Bárbara system of the central Andean foreland thrust belt, northwestern Argentina: Tectonics, v. 21, no. 6, 1061, doi:10.1029/2002TC902003.

Kley, J., Monaldi, C.R., and Salfity, J.A., 1999, Along-strike segmentation of the Andean foreland: Causes and consequences: Tectonophysics, v. 301, p. 75–94, doi:10.1016/S0040-1951(98)90223-2.

Kley, J., Rossello, E.A., Monaldi, C.R., and Habighorst, B., 2005, Seismic and field evidence of Cretaceous normal faults and selective inversion, Salta rift, Northwest Argentina: Tectonophysics, v. 399, p. 155–172, doi:10.1016/j.tecto.2004.12.020.

Löbens, S., Sobel, E.R., Bense, F.A., Wemmer, K., Dunkl, I., and Siegesmund, S., 2013, Refined exhumation history of the northern Sierras Pampeanas, Argentina: Tectonics, v. 32, p. 453–472, doi:10.1002/tect.20038.

Lowry, A.R., and Smith, R.B., 1995, Strength and rheology of the western U.S. Cordillera: Journal of Geophysical Research, v. 100, p. 17,947–17,963, doi:10.1029/95JB00747.

Lucassen, F., and Franz, G., 2005, The early Paleozoic orogeny in the central Andes; a non-collisional orogeny comparable to the Cenozoic high plateau, *in* Vaughan, A.P.M., Leat, P.T., and Pankhurst, R.J., eds., Terrane Processes at the Margins of Gondwana: Geological Society, London, Special Publication 246, p. 257–273, doi:10.1144/GSL.SP.2005.246.01.09.

Lucassen, F., Leverenz, S., Franz, G., Viramonte, J., and Mezger, K., 1999, Metamorphism, isotopic ages and composition of lower crustal granulite xenoliths from the Cretaceous Salta Rift, Argentina: Contributions to Mineralogy and Petrology, v. 134, p. 325–341, doi:10.1007/s004100050488.

Ludwig, K., 2003, Mathematical-statistical treatment of data and errors for ^{230}Th/U geochronology, *in* Bourdon, B., Henderson, G.M., Lundstrom, C.C., and Turner, S.P., eds., Uranium-Series Geochemistry: Reviews in Mineralogy and Geochemistry, v. 52, p. 631–656, doi:10.2113/0520631.

Marquillas, R.A., del Papa, C., and Sabino, I.F., 2005, Sedimentary aspects and paleoenvironmental evolution of a rift basin: Salta Group (Cretaceous-Paleogene), northwestern Argentina: International Journal of Earth Sciences, v. 94, p. 94–113, doi:10.1007/s00531-004-0443-2.

Marquillas, R.A., Salfity, J.A., Matthews, S.J., Matteini, M., and Dantas, E., 2011, U-Pb zircon age of the Yacoraite Formation and its significance to the Cretaceous-Cenozoic boundary in the Salta Basin, Argentina, *in* Salfity J.A., and Marquillas, R.A., eds., Cenozoic Geology of the Central Andes of Argentina: Salta, Argentina, SCS Publisher, p. 227–246.

McGroder, M.F., Lease, R.O., and Pearson, D.M., 2015, this volume, Along-strike variation in structural styles and hydrocarbon occurrences, Subandean fold-and-thrust belt and inner foreland, Colombia to Argentina, *in* DeCelles, P.G., Ducea, M.N., Carrapa, B., and Kapp, P.A., eds., Geodynamics of a Cordilleran Orogenic System: The Central Andes of Argentina and Northern Chile: Geological Society of America Memoir 212, doi:10.1130/2015.1212(05).

McKenzie, D., 1978, Some remarks on the development of sedimentary basins: Earth and Planetary Science Letters, v. 40, p. 25–32, doi:10.1016/0012-821X(78)90071-7.

Mingramm, A., Russo, A., Pozzo, A., and Casau, L., 1979, Sierras Subandinas, *in* Segundo Simposiode Geologica Regional Argentina, Cordoba: Cordoba, Argentina, Academi Nacional Ciencias p. 95–138.

Mitrovica, J., Beaumont, C., and Jarvis, G., 1989, Tilting of continental interiors by the dynamical effects of subduction: Tectonics, v. 8, p. 1079–1094, doi:10.1029/TC008i005p01079.

Monaldi, C.R., Alonso, R.N., Gonzalez, R.E., Igarzabal, A.P., Ramallo, E., Godeas, M., Fuertes, A., Garcia, R., and Moya, F., 2001, Programa Nacional de Cartas Geologicas de la Republica Argentina, Cachi, Provincias de Salta y Catamarca: Instituto de Geologica y Recursos Minerales Hoja Geologica 2566-III, scale 1:250,000, 1 sheet.

Monaldi, C.R., Salfity, J.A., and Kley, J., 2008, Preserved extensional structures in an inverted Cretaceous rift basin, northwestern Argentina: Outcrop examples and implications for fault reactivation: Tectonics, v. 27, TC1001, doi:10.1029/2006TC001993.

Mooney, W.D., Laske, G., and Masters, T.G., 1998, CRUST 5.1: A global crustal model at 5° × 5°: Journal of Geophysical Research, v. 103, p. 727–747, doi:10.1029/97JB02122.

Oreskes, N., Shrader-Frechette, K., and Belitz, K., 1994, Verification, validation, and confirmation of numerical models in the earth sciences: Science, v. 263, p. 641–646, doi:10.1126/science.263.5147.641.

Pearson, D.M., Kapp, P., Reiners, P.W., Gehrels, G.E., Ducea, M.N., Pullen, A., Otamendi, J.E., and Alonso, R.N., 2012, Major Miocene exhumation by fault-propagation folding within a metamorphosed, early Paleozoic thrust belt: Northwestern Argentina: Tectonics, v. 31, TC4023, doi:10.1029/2011TC003043.

Pearson, D.M., Kapp, P., DeCelles, P.G., Reiners, P.W., Gehrels, G.E., Ducea, M.N., and Pullen, A., 2013, Influence of pre-Andean crustal structure on Cenozoic thrust belt kinematics and shortening magnitude: Northwestern Argentina: Geosphere, v. 9, p. 1766–1782, doi:10.1130/GES00923.1.

Pérez-Gussinye, M., and Watts, A.B., 2005, The long-term strength of Europe and its implications for plate-forming processes: Nature, v. 436, p. 381–384, doi:10.1038/nature03854.

Perez-Gussinye, M., Lowry, A.R., Phipps Morgan, J., and Tassara, A., 2008, Effective elastic thickness along the Andean margin and their relationship to subduction geometry: Geochemistry Geophysics Geosystems, v. 9, doi:10.1029/2007GC001786.

Quade, J., Rech, J.A., Latorre, C., Betancourt, J.L., Gleeson, E., and Kalin, M.T.K., 2007, Soils at the hyperarid margin: The isotopic composition of soil carbonate from the Atacama Desert, northern Chile: Geochimica et Cosmochimica Acta, v. 71, p. 3772–3795, doi:10.1016/j.gca.2007.02.016.

Quade, J., Dettinger, M.P., Carrapa, B., DeCelles, P., Murray, K.E., Huntington, K.A., Cartwright, A., Canavan, R.R., Gehrels, G., and Clementz, M., 2015, this volume, The growth of the central Andes, 22°S –26°S, *in* DeCelles, P.G., Ducea, M.N., Carrapa, B., and Kapp, P.A., eds., Geodynamics of a Cordilleran Orogenic System: The Central Andes of Argentina and Northern Chile: Geological Society of America Memoir 212, doi:10.1130/2015.1212(15).

Rech, J.A., Currie, B.S., Shullenberger, E.D., Dunagan, S.P., Jordan, T.E., Blanco, N., Tomlinson, A.J., Rowe, H.D., and Houston, J., 2010, Evidence for the development of the Andean rain shadow from a Neogene isotopic record in the Atacama Desert, Chile: Earth and Planetary Science Letters, v. 292, p. 371–382, doi:10.1016/j.epsl.2010.02.004.

Reiners, P.W., Vernon, A., Thomson, S.N., Zattin, M., Einhorn, J., Gehrels, G., Quade, J., Pearson, D., Murray, K.E., and Cavazza, W., 2015, this volume, Low-temperature thermochronologic trends across the central Andes, 21°S–28°S, *in* DeCelles, P.G., Ducea, M.N., Carrapa, B., and Kapp, P., eds., Geodynamics of a Cordilleran Orogenic System: The Central Andes of Argentina and Northern Chile: Geological Society of America Memoir 212, doi:10.1130/2015.1212(12).

Reynolds, J.H., Galli, C.I., Hernandez, R.M., Idleman, B.D., Kotila, J.M., Hilliard, R.V., and Naeser, C.W., 2000, Middle Miocene tectonic development of the Transition zone, Salta Province, northwest Argentina: Magnetic stratigraphy from the Metán Subgroup, Sierra de Gonzalez: Geological Society of America Bulletin, v. 112, p. 1736–1751, doi:10.1130/0016-7606(2000)112<1736:MMTDOT>2.0.CO;2.

Rhea, S., Hayes, G., Villasenor, A., Furlong, K.P., Tarr, A.C., and Benz, H., 2008, Seismicity of the Earth: Nazca Plate and South America: U.S. Geological Survey Open-File Report 2010-1083-E, scale 1:12,000,000, 1 sheet.

Roberts, A.M., Kuznir, N.J., Yielding, G., and Styles, P., 1998, 2D flexural backstripping of extensional basins: The need for a sideways glance: Petroleum Geoscience, v. 4, p. 327–338, doi:10.1144/petgeo.4.4.327.

Salfity, J.A., and Marquillas, R., 1994, Tectonic and sedimentary evolution of the Cretaceous–Eocene Salta Group, Argentina, in Salfity, J.A., ed., Cretaceous Tectonics of the Andes; Earth Evolution Sciences: Heidelberg, Viewag, p. 266–315.

Salfity, J.A., and Monaldi, C.R., 2006, Programa Nacional de Cartas Geologicas de la Republica Argentina, Metán, Provincia de Salta: Instituto de Geologica y Recursos Minerales Hoja Geologica 2566-IV, scale 1:250,000, 1 sheet.

Salfity, J.A., Azcuy, C.L., Lopez, O., Valencio, D.A., Vilas, J.F., Cuerda, A., and Laffitte, G., 1987, Cuenca Tarija, in Archangelsky, S., ed., El Sistema Carbonifero en la Republica Argentina: SCCS-Project PICG 211: Cordoba, Argentina, Academia Nacional de Ciencias de Cordoba, p. 15–39.

Salfity, J.A., Monaldi, R., Marquillas, R., and Gonzalez, R., 1993, La inversion tectonica del Umbral de los Gallos en la Cuenca del Grupo Salta durante la fase Incaica, in XII Congreso Geologico Argentino: Mendoza, Argentina, Association Geologico Argentina, p. 200–210.

Schmidt, C.J., Astini, R.A., Costa, C.H., Gardini, C.E., and Kraemer, P.E., 1995, Cretaceous rifting, alluvial fan sedimentation and Neogene inversion, southern Sierras Pampeanas, Argentina, in Tankard, A.J., Suarez, R., and Welsink, H.J., eds., Petroleum Basins of South America: American Association of Petroleum Geologists Memoir 62, p. 341–358.

Scholl, D.W., and von Huene, R., 2007, Crustal recycling at modern subduction zones applied to the past-issues of growth and preservation of continental basement crust, mantle geochemistry, and supercontinent reconstruction, in Hatcher, R.D., Jr., Carlson, M.P., McBride, J.H., and Martinez Catalán, J.R., eds., 4-D Framework of Continental Crust: Geological Society of America Memoir 200, p. 9–32.

Shephard, G.E., Liu, L., Muller, R.D., and Gurnis, M., 2012, Dynamic topography and anomalous negative residual depth of the Argentine Basin: Gondwana Research, v. 22, p. 658–663, doi:10.1016/j.gr.2011.12.005.

Starck, D., 2011, Cuenca Cretacica–Paleogena del Noroeste Argentino, in IAPG VIII Congreso de Exploracion y Desarrollo de Hidrocarburos, Sim-posio Cuencas Argentinas: Mar del Plata, Argentina, vision actual, IAPG, p. 407–453.

Stewart, J., and Watts, A.B., 1997, Gravity anomalies and spatial variations of flexural rigidity at mountain ranges: Journal of Geophysical Research, v. 102, p. 5327–5352.

Summa, L., Becker, T., Gray, G., and Awwiller, D., 2015, this volume, Evolving genetic concepts and their influence on hydrocarbon systems predictions, Subandean fold belt and deformed foreland, Argentina, in DeCelles, P.G., Ducea, M.N., Carrapa, B., and Kapp, P.A., eds., Geodynamics of a Cordilleran Orogenic System: The Central Andes of Argentina and Northern Chile: Geological Society of America Memoir 212, doi:10.1130/2015.1212(21).

Suppe, J., 1983, Geometry and kinematics of fault-bend folding: American Journal of Science, v. 283, p. 684–721, doi:10.2475/ajs.283.7.684.

Tassara, A., Swain, C., Hackney, R., and Kirby, J., 2007, Elastic thickness structure of South America estimated using wavelets and satellite-derived gravity data: Earth and Planetary Science Letters, v. 253, p. 17–36.

Thomas, W.A., 1991, The Appalachian-Ouachita rifted margin of southeastern North America: Geological Society of America Bulletin, v. 103, p. 415–431, doi:10.1130/0016-7606(1991)103<0415:TAORMO>2.3.CO;2.

Thomas, W.A., 2006, Tectonic inheritance at a continental margin: GSA Today, v. 16, no. 2, p. 4–11, doi:10.1130/1052-5173(2006)016[4:TIAACM]2.0.CO;2.

Vandervoort, D.S., Jordan, T.E., Zeitler, P.K., and Alonso, R.N., 1995, Chronology of internal drainage development and uplift, southern Puna Plateau, Argentine central Andes: Geology, v. 23, p. 145–148, doi:10.1130/0091-7613(1995)023<0145:COIDDA>2.3.CO;2.

Viramonte, J.G., Kay, S.M., Becchio, R., Escayola, M., and Novitski, I., 1999, Cretaceous rift related magmatism in central-western South America: Journal of South American Earth Sciences, v. 12, p. 109–121, doi:10.1016/S0895-9811(99)00009-7.

von Huene, R., and Scholl, D.W., 1991, Observations at convergent margins concerning sediment subduction, subduction erosion, and the growth of continental crust: Reviews of Geophysics, v. 29, p. 279–316, doi:10.1029/91RG00969.

Walker, J.D., Geissman, J.W., Bowring, S.A., and Babcock, L.E., compilers, 2012, Geologic Time Scale Version 4.0: Boulder, Colorado, Geological Society of America, doi:10.1130/2012.CTS004R3C.

Weissel, J.K., and Karner, G.D., 1989, Flexural uplift of rift flanks due to mechanical unloading of the lithosphere during extension: Journal of Geophysical Research, v. 94, p. 13,919–13,950, doi:10.1029/JB094iB10p13919.

Wheeler, P., and White, N., 2000, Quest for dynamic topography: Observations from Southeast Asia: Geology, v. 28, p. 963–966, doi:10.1130/0091-7613(2000)28<963:QFDTOF>2.0.CO;2.

MANUSCRIPT ACCEPTED BY THE SOCIETY 3 JUNE 2014
MANUSCRIPT PUBLISHED ONLINE 20 NOVEMBER 2014

The Geological Society of America
Memoir 212
2015

Evolving genetic concepts and their influence on hydrocarbon systems predictions, Subandean fold belt and deformed foreland, Argentina

L. Summa
T. Becker
G. Gray
D. Awwiller

ExxonMobil Upstream Research Company, 22777 Springwoods Village Parkway, Spring, Texas 77210, USA

ABSTRACT

Convergent orogenic systems pose challenges for the prospector seeking to predict oil and gas reservoirs, as these regions are often data poor. Explorationists, therefore, must often rely on conceptual models to predict the occurrence of oil and gas, and to choose the most favorable exploration areas. This paper examines how different conceptual models of hinterland evolution can influence predictions of hydrocarbon systems in an adjacent foreland. A one-dimensional (1-D) method was developed for assessing incremental hydrocarbon yield during progressive fold-belt deformation and source rock burial and maturation. This method was employed to evaluate the way in which hydrocarbon yield is affected by two different scenarios for the timing of uplift of the Puna Plateau and subsequent burial of the adjacent Metán region of northwestern Argentina. In the first scenario, plateau growth and burial of the Metán region begin in the early Miocene and progress to the present day. In the second scenario, plateau growth and basin formation occur predominantly in the Eocene, with minor deformation from middle Miocene to present. The later load timing decreases the relative volume of liquid hydrocarbons available to fill structural traps. The differences in charge volume and timing between the two scenarios are enhanced in this region as a result of thinned crust and relatively high heat flow that linger from a Cretaceous rifting event. The results of the study provide an example of how boundary conditions obtained from studies of an orogenic hinterland can be used to calculate risk for exploring a potential play within a basin where data are incomplete or inconclusive. They also yield general insights on the use of regional genetic concepts to predict thermal history and source rock maturation and yield in data-poor areas.

Summa, L., Becker, T., Gray, G., and Awwiller, D., 2015, Evolving genetic concepts and their influence on hydrocarbon systems predictions, Subandean fold belt and deformed foreland, Argentina, *in* DeCelles, P.G., Ducea, M.N., Carrapa, B., and Kapp, P.A., eds., Geodynamics of a Cordilleran Orogenic System: The Central Andes of Argentina and Northern Chile: Geological Society of America Memoir 212, p. 435–457, doi:10.1130/2015.1212(21). For permission to copy, contact editing@geosociety.org. © 2014 The Geological Society of America. All rights reserved.

INTRODUCTION

Convergent systems have been attractive regions to explore for oil and gas since the mid-late 1800s, due in large part to the prevailing genetic basin concept of that period: the anticlinal theory of oil and gas accumulation. Among other features, convergent systems are effective at forming anticlinal traps, and many early petroleum discoveries occurred within these systems and on their peripheries. Today, they remain an important regime for conventional exploration. A statistical assessment of global hydrocarbon reserves suggests that the world's fold-and-thrust belts contain nearly 15% of the remaining undiscovered conventional hydrocarbon potential (Cooper, 2007). Many of the world's emerging "unconventional" plays are also located in foreland basins.

Many prolific fold belts have significant remaining petroleum potential, but the volume and type of hydrocarbon can be highly variable along strike (e.g., McGroder et al., this volume). Accurate prediction of hydrocarbon charge, including the relative proportion of oil versus gas, requires a good understanding of the type of source rocks present in the basin, their position in the stratigraphic column, their maximum depth of burial, and the timing of trap development versus hydrocarbon charge. In many fold belts, these elements can be difficult to predict due to the absence of basic data. As a consequence, play element predictions are often model-driven, relying on fundamental "genetic" analysis of how fold belts evolve, supplemented with loosely constrained forward models of thermal history.

We use the term "genetic" to describe a full-systems approach that addresses how basins form, fill, and evolve, in order to identify fundamental relationships that control hydrocarbon play elements. All sedimentary basins are unique, but the processes that control basin formation, fill, and evolution are common and predictable. Plate interaction and interplate events drive intraplate deformation, which in turn generates subsidence and uplift. Subsidence and uplift, convolved with climate and eustasy, control accommodation history, which in turn strongly influences depositional systems and local paleogeography. Postdepositional processes modify the previously developed elements and control fluid migration. This hierarchy of processes provides the fundamental underpinning for petroleum systems predictions in complex settings with limited data, and it is a valuable tool for basins with multiple episodes of deformation and fill. Foreland fold-belt regions are particularly amenable to regional genetic analysis, since orogenic and basin formation processes are coupled, and there is typically information from the orogenic belt (e.g., source and reservoir presence) that relates directly to play elements in the adjacent foreland basin. In addition, many current deformed foreland basins inherit the characteristics or "genes" from earlier tectonic events, and regional genetic analysis provides a systematic approach to unraveling the influence of those "genes."

New concepts of how the South American Andes have evolved, described later herein, provide an opportunity to examine how different regional assumptions influence hydrocarbon systems predictions. This discussion begins with an illustration of basic concepts for source rock maturation and yield, using results from a traditional exploration study of Subandean basins. This is contrasted with a recent analysis in Argentina that compares hydrocarbon systems predictions for two different evolutionary scenarios. The first scenario examines the "conventional" model of a long-lived, continuously growing and migrating fold belt with constant sediment supply to an adjacent foreland basin. The second scenario examines a system with a more punctuated migration of the orogenic wedge and sediment supply to the adjacent foreland. These differing concepts yield different estimates of regional subsidence and exhumation, and they are examined regarding the predicted variations in hydrocarbon maturation timing and fluid type.

GEOLOGIC FRAMEWORK

The Andean margin of western South America is the prototypical arc-continent convergent margin involving the subduction of oceanic crust beneath a continent. It is believed that western South America has generally maintained its present convergent plate geometry for most of the Phanerozoic (e.g., Allmendinger et al., 1997; Lucassen et al., 2000; Ramos, 2009; McGroder et al., this volume, and references therein). This geometry was disrupted by the accretion of exotic terranes during the Cambrian, Ordovician, Devonian, and Permian (e.g., Ramos et al., 1986, 2008; Astini et al., 1995; Rapela et al., 1998; Cawood, 2005). The most recent episode of Andean-style mountain building began in earnest in the Western Cordillera and eastern Puna Plateau during the Paleogene (e.g., Isacks, 1988; Allmendinger et al., 1997; Davila et al., 2003; Arriagada et al., 2006). Despite the relative consistency in tectonic geometry over the past 550 m.y., there are dramatic along-strike variations in tectonic architecture, owing to a combination of heterogeneous continental crust and inherited tectonic events (e.g., Allmendinger et al., 1997; Kley et al., 1999; McGroder et al., this volume, and references therein).

Over the past several years, various new data sets and observations from the central Andes have spawned a number of intriguing hypotheses about the evolution of orogenic systems. Processes such as the formation and removal of gravitationally unstable convective instabilities in the mantle lithosphere (Houseman et al., 1981; Molnar and Houseman, 2004; Garzione et al., 2008), lower crust and lithosphere delamination (Kay and Kay, 1993; Kay et al., 1994; Schurr et al., 2006), the formation and removal of dense eclogitic restites beneath the magmatic arc (Ducea, 2001; Ducea and Barton, 2007; DeCelles et al., 2009), and subduction erosion (von Huene and Scholl, 1991; Scholl and von Huene, 2007) are potentially reshaping the view of how these convergent margins evolve, and providing a more holistic view of plate convergence that includes the roles of mantle and surface processes (e.g., Pelletier et al., 2010).

The central Andean Plateau, which extends from Peru to northwestern Argentina, is a large, topographically high feature composed of two physiographic provinces: the Altiplano and the Puna Plateaus (e.g., Allmendinger et al., 1997). Several mechanisms have been proposed to explain the creation of the plateau, including erosion of the sublithospheric mantle and underthrusting of the Brazilian craton (Barke and Lamb, 2006), delamination of the subcrustal lithosphere (Garzione et al., 2008), increased basal traction from subduction of old, thickened oceanic lithosphere (Capitanio et al., 2011), and climate-linked enhancement of frictional stress along the Nazca–South America plate boundary downdip from the Nazca Trench (Lamb and Davis, 2003). The formation of the orogenic plateau (and thickened crust) may be important for initiating the stable formation of dense "arclogitic" roots (Kay and Kay, 1993; Leech, 2001). Creation and loss of these dense roots have been postulated as mechanisms for modulating deformation in the adjacent foreland (DeCelles et al., 2009).

The high plateaus are flanked by basins that formed as the South American lithosphere was loaded by plateau growth (Salfity and Marquillas, 1994; Cominguez and Ramos, 1995; Starck and Vergani, 1996; Hernández et al., 1999; Horton, 2005; Uba et al., 2006; Carrera and Muñoz, 2008; Carrapa et al., 2011; DeCelles et al., 2011). Many of these basins are typical of classic foreland petroleum provinces. Source rocks were buried into the oil- and gas-generation windows during late Cenozoic deformation, and the relative timing of source rock yield and trap formation defines the success or failure of the petroleum system. Variability of petroleum systems between basins can be correlated to changes in basement framework and inherited Mesozoic tectonic events that control how the crust behaves within individual basins (e.g., McGroder et al., this volume).

This discussion focuses on the depositional record of the Metán region of the Cenozoic foreland, located east of the Puna Plateau (Fig. 1). The history of the Metán region is characterized by two distinct stages: (1) Neocomian–Campanian rifting, driven by back-arc extension and/or the opening of the South Atlantic, which just predated (2) Cenozoic subsidence and deformation, driven by Andean tectonics. Strata of the Pirgua and Balbuena Subgroups record the Cretaceous rifting and subsequent early Cenozoic thermal relaxation and contain organic-rich lacustrine facies (Yacoraite Formation) that may have acted as petroleum source rocks (Fig. 1). Lacustrine facies of the Lumbrera Group are also reported to have source potential (del Papa et al., 2002). Andean contraction is generally assumed to have driven trap formation and source rock heating.

PETROLEUM SYSTEMS CONCEPTS AND METHODS

As noted already, the relative timing of trap formation versus hydrocarbon yield is a constant concern in fold-belt exploration. This is because maturation and trap formation are frequently driven by the same tectonic event, and if hydrocarbon yield slightly predates trap formation, the result can be dry, underfilled, or gas-filled traps. Evaluation of alternate scenarios for hydrocarbon maturation and yield is thus a standard element of petroleum systems studies. The following discussion provides an introduc-

tion to these concepts and background for subsequent analysis of the Metán region.

The first step in hydrocarbon maturation and yield evaluation is to construct one-dimensional (1-D) subsidence and thermal history models for prospective play areas. These calculations provide a means for understanding the basic constraints on source rock burial and maturation. An example of this type of analysis is shown for representative wells from the central Andean foreland system that straddles the Bolivia-Argentina border (Fig. 2). The models are typical of frontier exploration areas in the broader Andean basins and were chosen because the wells had good calibration data. They show the variations that can be expected relatively close to the region of interest. Calibration data include temperatures measured in boreholes, thermal maturities measured on cuttings samples, and literature-based regional heat-flow data. Traditional indicators of organic maturity such as vitrinite reflectance (Ro), pyrolysis (T_{max}), and spore color or thermal alteration index (TAI) were supplemented where possible with inorganic thermal history indicators, including apatite fission tracks, fluid inclusion homogenization temperatures, and quartz cement abundance. Typical parameters that are calculated in order to evaluate the effect of different scenarios for basin formation and fill include present-day thermal maturity versus depth, and the thermal history of individual source rock units. For the wells in this example, the depth to the onset of oil generation, also called "the top of the oil window," varies from ~1 to 2 km, and the depth to the onset of dry gas generation, or the "top of the gas window," varies from 4.5 to 5.5 km (Fig. 2A). Oil generation from Silurian source rocks began during the Paleozoic and continued through the latest phase of Andean deformation, beginning in the middle Miocene (Fig. 2B). At 1–2 km, the present-day top of the oil window is relatively shallow in these wells, which begs the question of whether this surface varies regionally in any significant way.

The regional variation in the oil and gas windows was investigated by extending the analysis described earlier to multiple wells with good calibration data located in all of the major basins between central Colombia and southern Bolivia (Fig. 3). This analysis was undertaken as part of a regional exploration study. It allows at least a qualitative comparison of variability within and between basins. As shown in Figure 3, the depth to the top of the oil and gas windows varies significantly along the length of the Andes. There are multiple possible explanations for the variation in oil and gas windows, including changes in basal heat flow, and recent exhumation history. The importance of the variability is that it implies significant differences in hydrocarbon systems predictions from north to south, and it reinforces the need for well-founded genetic concepts that allow reasonable predictions to be made in data-poor areas.

The plots shown in Figures 2 and 3 provide general insight on source rock burial history and the present-day thermal state of these basins along the Andean fold belt, but they provide little information about the type and volume of hydrocarbons available to fill a specific trap at any given time. Such predictions require

Summa et al.

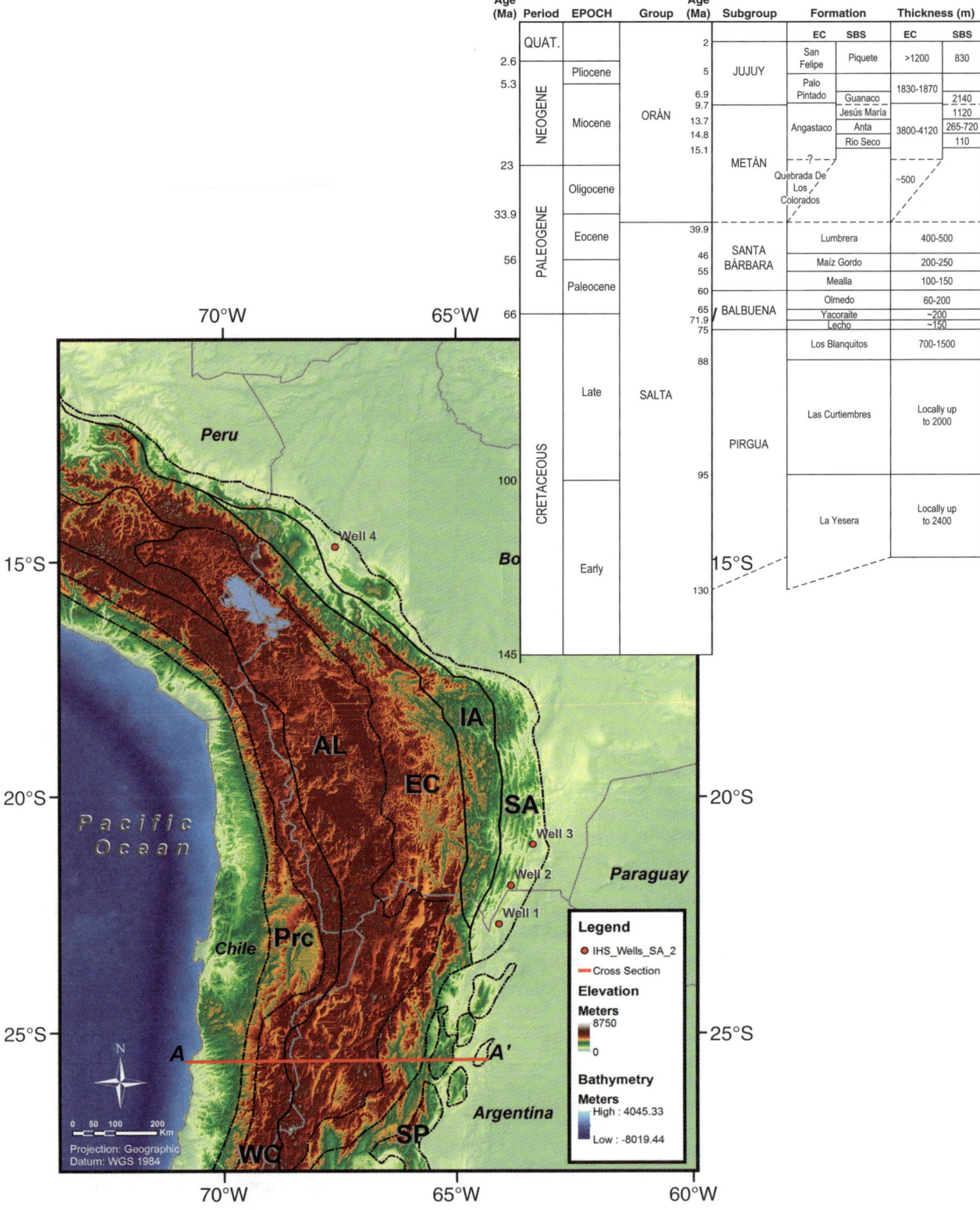

Figure 1. Major tectonic provinces of the central Andes. Abbreviations indicate the following provinces: PrC—Precordillera; PU—Puna Plateau; WC—Western Cordillera; SP—Sierras Pampeanas; SA—Subandes; EC—Eastern Cordillera; AL—Altiplano; IA—Interandean; SBS—Santa Bárbara system. The Metán region is located along the eastern quarter of the cross section. A–A′ shows the location of the modeled cross section discussed in the following sections. Wells 1–4 show the locations of wells referenced in Figure 2. The stratigraphic column on the right introduces the key units and sediment thicknesses referred to in the discussion.

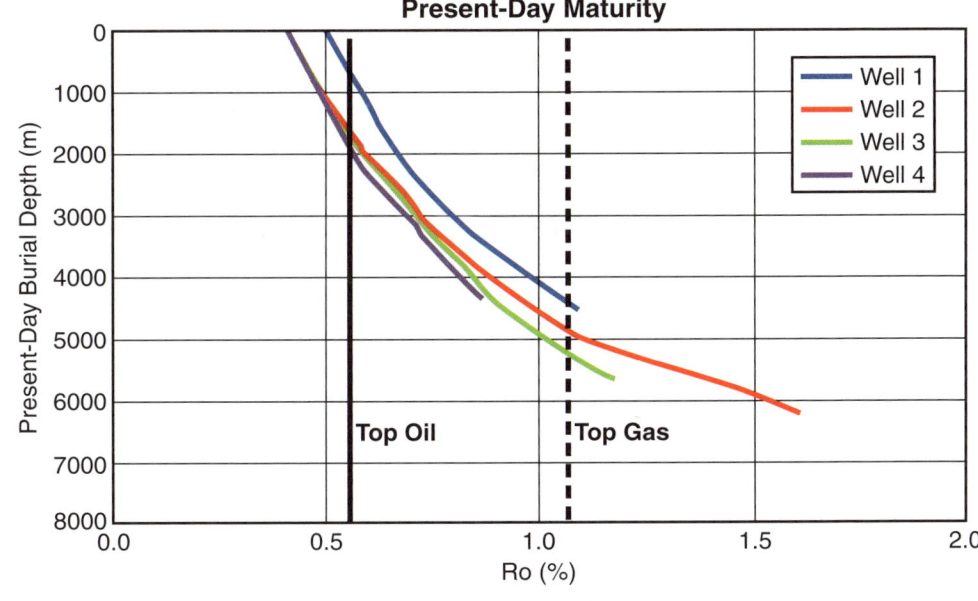

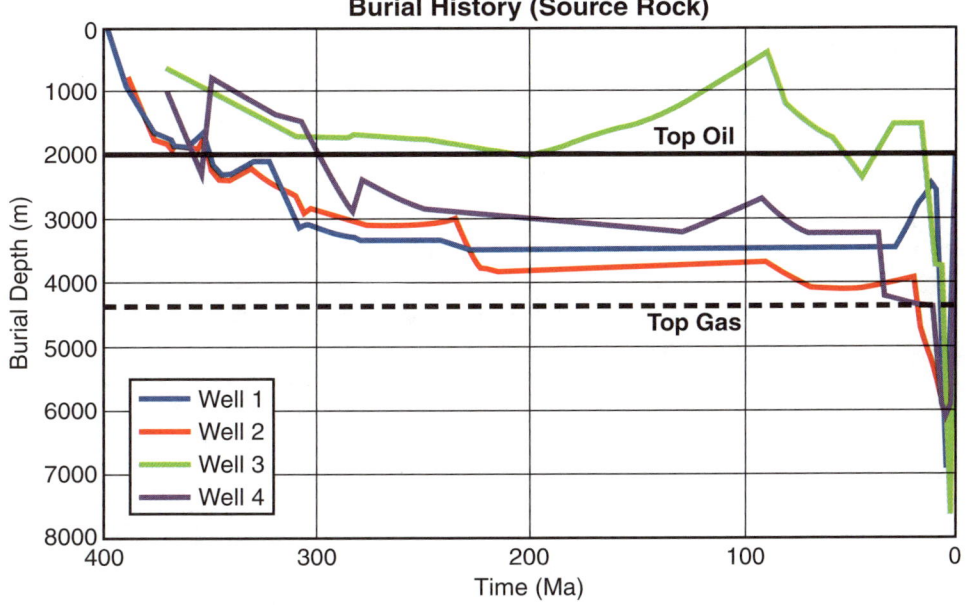

Figure 2. (A) Present-day maturity/depth plot and (B) burial history plot for four representative wells from Bolivia to the northern Argentina foreland region. The burial history is for a Silurian source rock horizon. The depth to the onset of oil generation, also called the top of the oil window, is shown by the solid black lines. The depth to the onset of dry gas generation, also called the top of the gas window, is shown by the dashed black lines. Ro is vitrinite reflectance (in %).

an additional calculation of incremental source rock yield, in the context of a specific model for the timing of trap formation. A proven method of addressing this issue is to build quantitative two-dimensional (2-D) deformation and thermal models for individual fold-belt transects (e.g., Roure et al., 2003; Schneider et al., 2004). Because the objective of this study was to develop broad predictive models in the absence of data, we constructed a set of theoretical calculations using a generalized scenario in which traps develop as the fold belt and its associated foreland basin migrate continuously toward the craton (Fig. 4). This scenario is based on the conventional model of a continually migrating fold belt, and it corresponds to one of the end-member scenarios for Metán region example that follows. The scenario assumes that source rocks at

the frontal part of the fold belt initially undergo relatively slow burial. Burial rate increases as the thrust front advances.

Incremental hydrocarbon yields were calculated using 1-D thermal histories modeled with Stellar™, ExxonMobil's proprietary 1-D, 2-D, and three-dimensional (3-D) basin modeling software. This modeling software incorporates an extensive geological parameterization, which includes heat capacity and conductivity variation as a function of lithology and compaction, surface temperature as a function of water depth, paleolatitude, and geologic time, and basal heat flow as a function of crustal composition and tectonic events.

Figure 5 shows source rock yields (i.e., transformation ratios) over the period of rapid burial and deformation shown

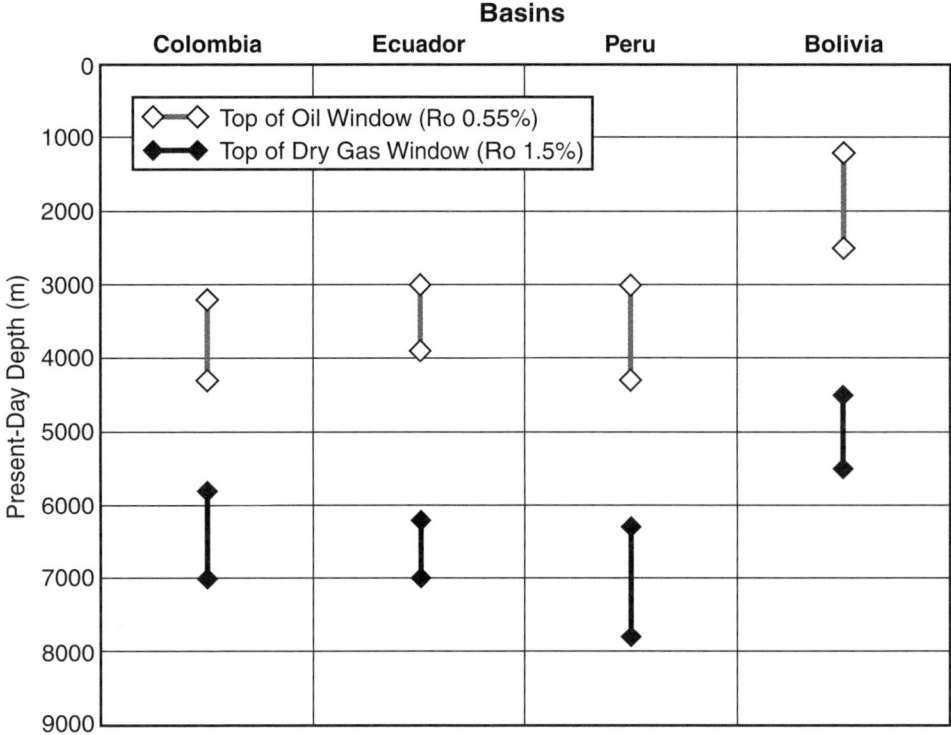

Figure 3. Depths to the onset of oil and dry gas maturity windows for foreland basins along the Andean margin, from central Colombia through southern Bolivia. Basins have been grouped by country to show north-south variability. Calculation is based on one-dimensional thermal history models from nearly 30 representative wells with high-quality calibration data. Note variability in top oil and gas windows along the margin. Ro is vitrinite reflectance (in %).

in Figure 4. Transformation ratios were calculated for a typical rapid heating rate of 5 °C/m.y. The calculations used ExxonMobil proprietary kinetics for type II marine source rocks:

$$dK/dt = -A \cdot K,\qquad(1)$$

where K is the amount of unreacted kerogen at time t, A is the extent of kerogen reaction to hydrocarbon in a given time step (fraction), and dK/dt is the instantaneous reaction rate of kerogen reaction to hydrocarbon (m.y.$^{-1}$).

A in Equation 1 is determined using a modified first-order Arrhenius kinetic expression:

$$A = ff \cdot e^{\left(\frac{-E_a}{RT}\right)},\qquad(2)$$

where ff is a pre-exponential constant (1/time), E_a is an array of activation energies (cal/mole), R is the universal gas constant = 1.987 (cal/mole K), and T is temperature (K).

Source rock yield initiates at ~90 °C (Fig. 5B). From the time of initial yield, total transformation of kerogen to liquids plus gas requires 30–35 m.y. for source rocks that are only buried to temperatures of 40 °C prior to rapid burial, but <20 m.y. for source rocks that are buried to 90 °C or greater. Figure 6 compares yield timing for variable heating rates and integrates the yields over the assumed time of trap formation to estimate the total fraction of hydrocarbon yield available to fill a given trap (Fig. 6D). The calculations lead to some interesting, and

perhaps nonintuitive results. Rapid heating actually allows for higher fractional hydrocarbon yield at a given time. This is because liquid hydrocarbon generation follows a first-order Arrhenius kinetic equation. Thus, yield increases proportionally to the change in ΔT/m.y. (Figs. 6A–6C). In other words, rapid heating is more "efficient," in that a greater fraction of the hydrocarbon yield is available to fill a trap (Fig. 6D). The results also show that the starting temperature of the source rocks at the initiation of rapid heating is equally as important as the heating rate. If the system has undergone some degree of burial and maturation during a previous tectonic episode, then traps formed during the latest tectonic event will receive only the final increments of yield and will likely contain gas. The implication is that for orogenic systems similar to the Andes, the most "forgiving" hydrocarbon system is one with high-quality source rocks, and (1) a low–moderate source rock temperature at the initiation of rapid burial (heating) and thrust emplacement, and (2) a high heating rate during the time of thrust emplacement.

These numerical experiments used heating rates empirically derived from basins with abundant data. However, as noted earlier herein, large areas of the Andean fold belt and its associated foreland have limited calibration data, and there are clear along-strike variations in thermal state, deformational style, and tectonic history. Thus, the aim is to determine whether regional genetic concepts can be used in the absence of calibration data to infer heating rate and the starting temperature of source rocks

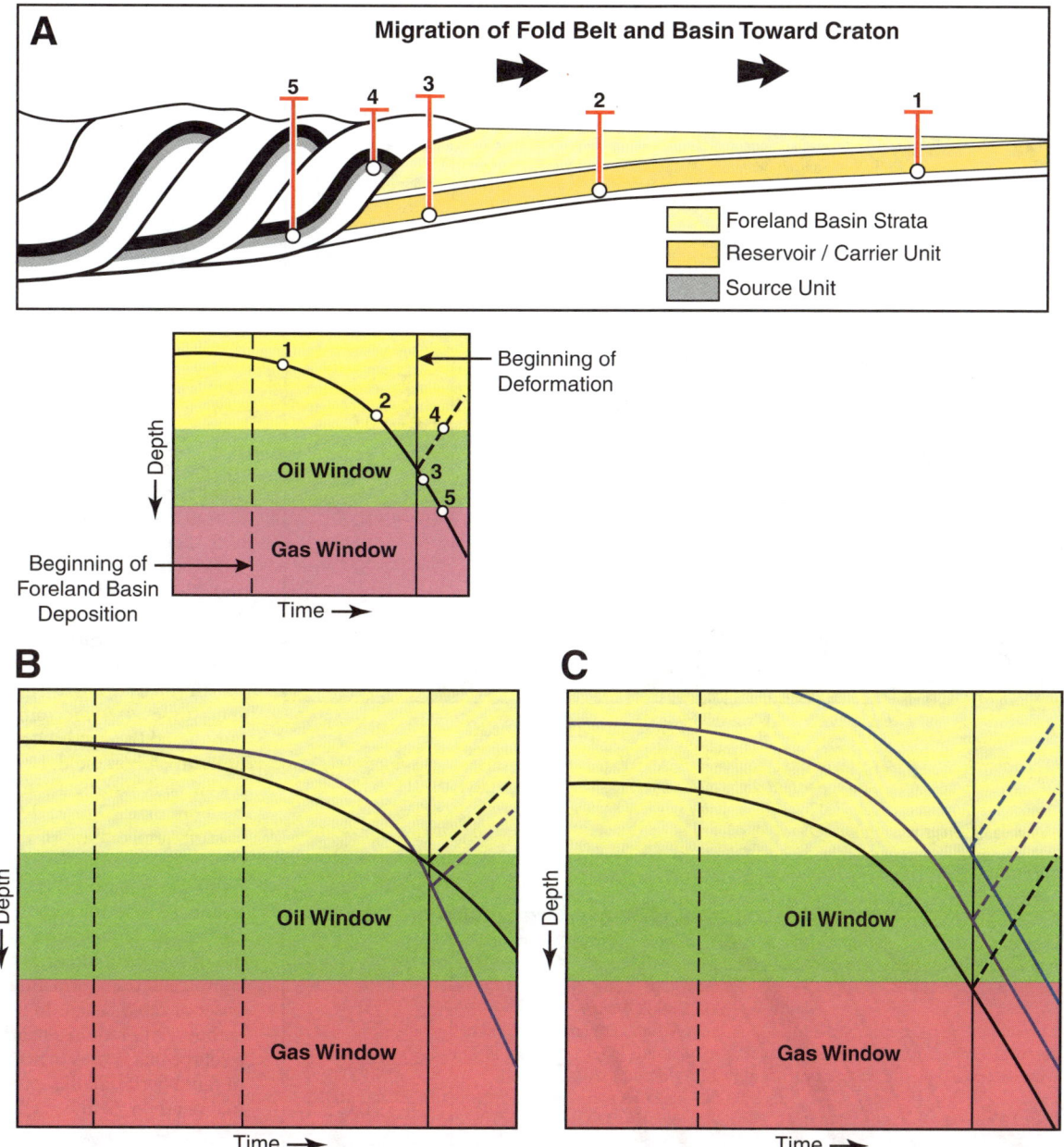

Figure 4. Trap timing and maturation scenarios typical of a coupled fold-belt–foreland system. (A) A generalized burial history curve and cross section, constructed to represent multiple well locations between the fold belt and foreland. As the fold belt migrates toward the craton, the current position of well 2 on the cross section becomes equivalent to the current position of well 3, etc. Trap formation timing is represented by points 3–5, to the right of the vertical black line indicating the beginning of local deformation. On the burial history curve, each numbered point corresponds to the final burial depth of the basal unit in the similarly numbered well on the cross section. (B–C) Variations of the burial history for differing crustal and/or lithosphere strengths (B), and different pre-foreland heat flow (C). In B, weaker lithosphere is shown in purple. In C, the lowest heat flow is shown in blue, and the highest heat flow is in black.

at the time of maturation, and thereby obtain insight into relative abundance of trapped oil and gas.

The Metán region of the Salta Province provides the opportunity to test this idea. The combined Oran-Metan petroleum province was discovered in the 1930s. The proven and probable reserves in the Orán basin are a modest 116 million barrels of oil and 220 billion cubic feet of gas (Urien et al., 1995). As such, there is some subsurface stratigraphic information that can be used to constrain thermal history and yield models, but there are also underexplored areas in which regional analysis might provide new exploration ideas. In the next section, we will discuss detailed scenarios for source rock maturation and

Summa et al.

Heating Rate = 5 °C/my

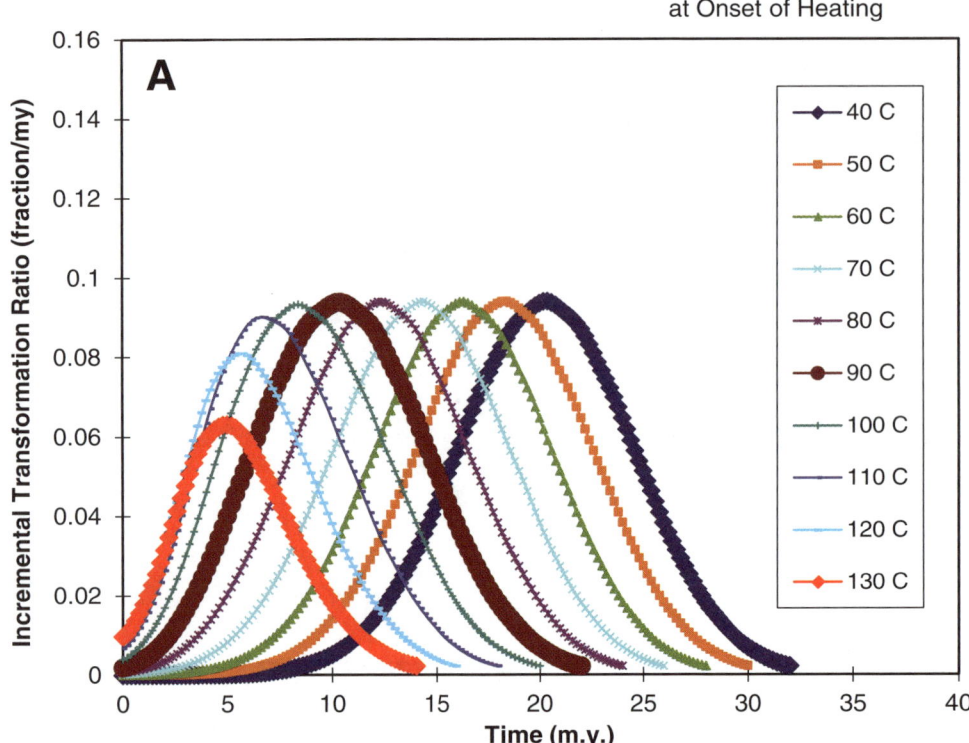

Figure 5. Quantitative source rock yield for the hypothetical scenarios illustrated in Figure 4. The heating rate is 5 °C/m.y., and the starting temperatures of source rocks vary between 40 and 130 °C at the onset of rapid burial associated with thrust loading. The calculations used kinetics for typical marine source rocks. (A) The incremental transformation ratio. The *x* axis represents time after the initiation of rapid burial, and the *y* axis shows the fraction of hydrocarbon yield. Note that the timing of peak hydrocarbon yield can vary by more than 20 m.y., depending on the starting temperature of the source rock. (B) The cumulative transformation ratio. Hydrocarbon yield is complete within 10 m.y. for source rocks that are already at temperatures of 130 °C at the onset of rapid burial. This assumes significant transformation prior to the onset of rapid burial. In contrast, hydrocarbon yield requires nearly 25 m.y. for a source that is only at 40 °C at the onset of rapid burial. In this case, 100% of the hydrocarbon yield occurs after the onset of rapid burial.

Heating Rate = 5 °C/my

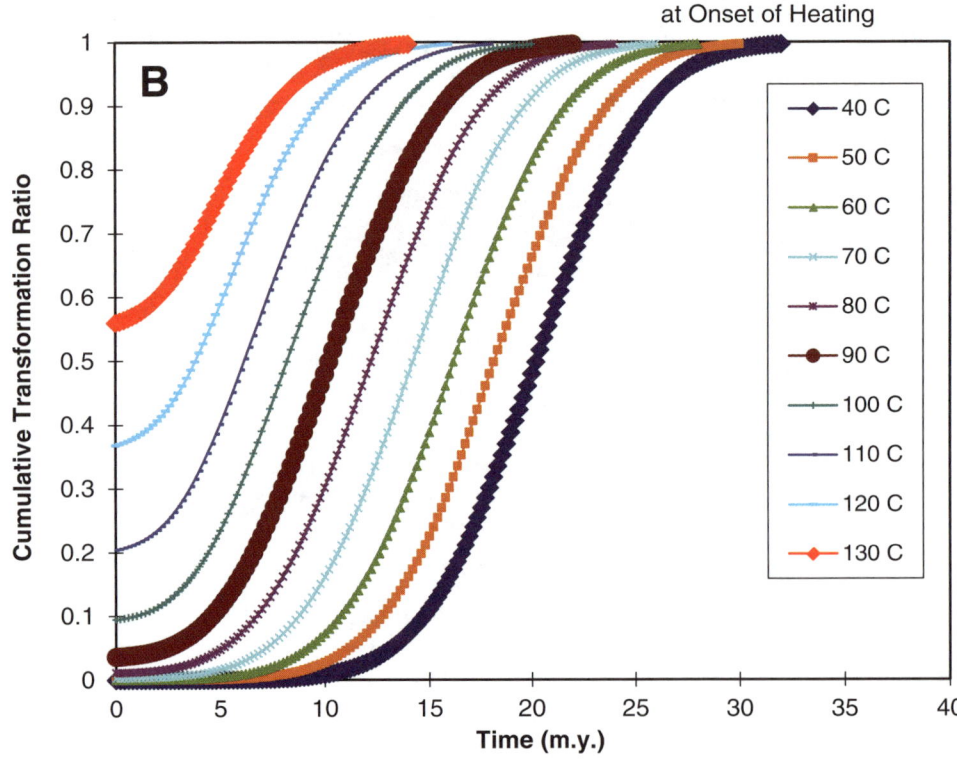

Heating Rate during Rapid Burial = 2 deg C per m.y.

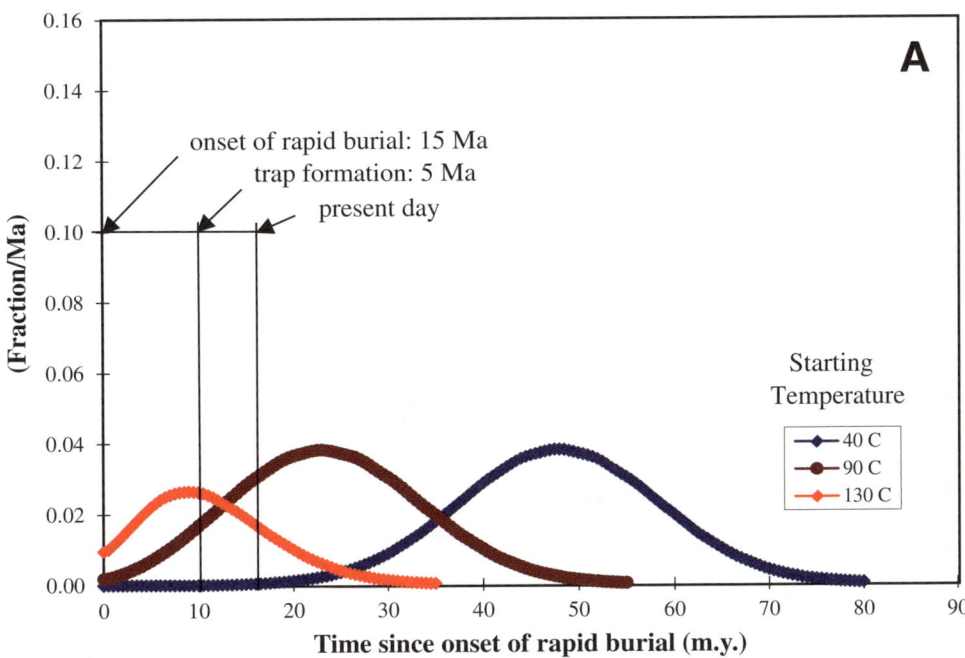

Heating Rate during Rapid Burial = 5 deg C per m.y.

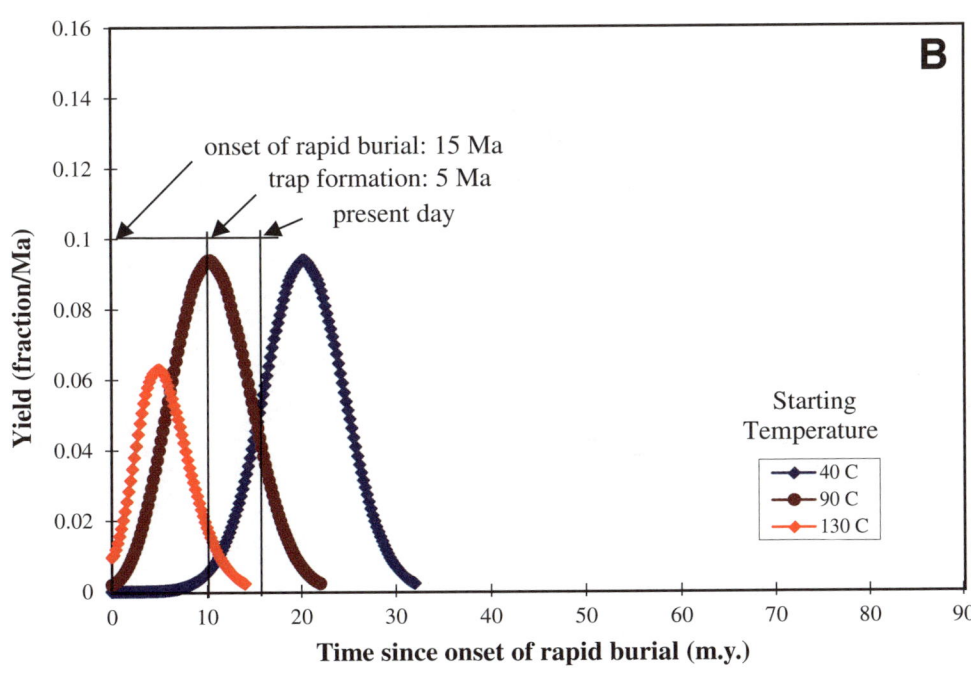

Figure 6 (*Continued on following page*). Quantitative source rock yields for hypothetical scenarios illustrated in Figure 4B. Calculations used end-member heating rates typical of the Andean basins. In this case, hypothetical timing of trap development has been superimposed on the calculations of transformation ratio. Assumed times for the calculation include: (1) onset of rapid loading and burial, (2) top formation: 5 Ma, and (3) present day. Only hydrocarbons generated between 5 Ma and present are available to fill a trap. Individual curves are for starting temperatures of 40–130 °C, which represent end-member scenarios for source rock burial depth prior to rapid loading. For the low-heating-rate case, it takes ~50 m.y. to generate all of the hydrocarbons from the source rock. Most of the hydrocarbons are generated well after the assumed time of trap formation, except in the highest starting temperature calculation (130 °C). (B–C) Same plots as in A but for heating rates of 5 °C/m.y. and 8 °C/m.y. In these cases, maximum hydrocarbon yield predates trap formation, except for the lowest-temperature source rocks. (D) Total fractional yield available to fill traps, derived from integration of incremental yields over the hypothetical trap formation time. A greater fraction of hydrocarbon yield is available to fill traps in the rapid heating case relative to the slower heating case, due to higher yields for larger changes in temperature. In the high-heating-rate case, the viability of the hydrocarbon system may be enhanced, because the volume of hydrocarbon available during the critical time of trap formation is higher than in the slow burial case. For lean source rocks, there is a trade-off between enhanced yield efficiency and total hydrocarbon yield. Slower burial and trap formation may be preferable in order to allow sufficient expelled product to fill a trap.

Heating Rate during Rapid Burial = 8 deg C per m.y.

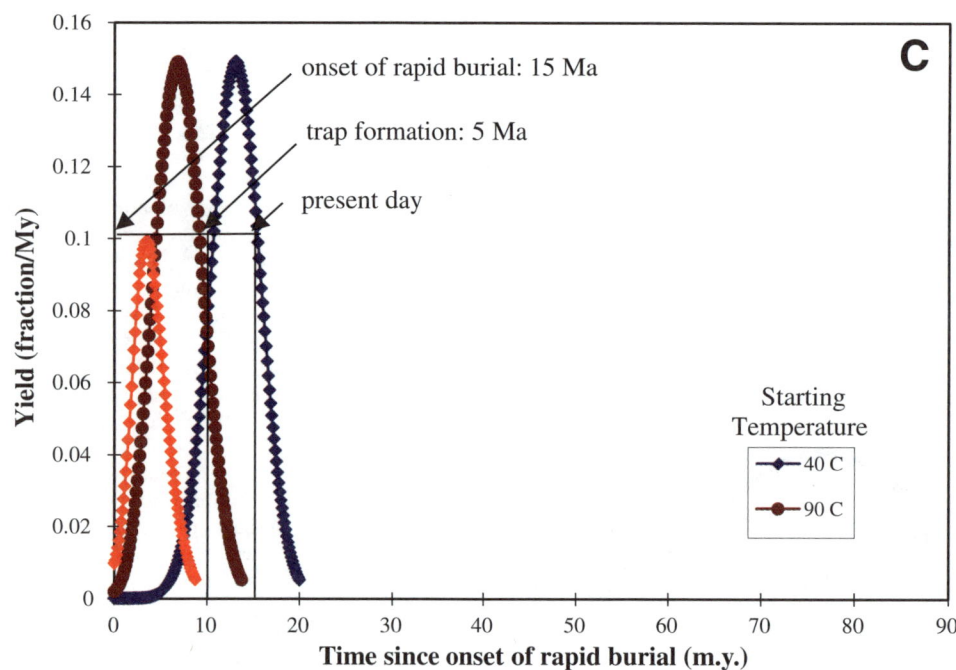

Figure 6 (*Continued*).

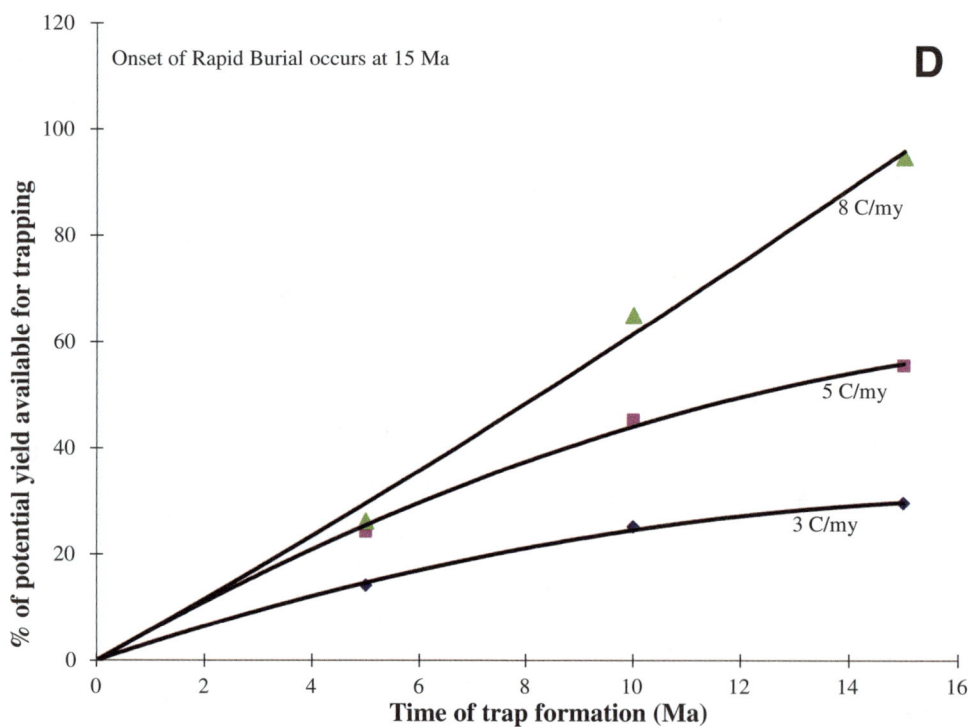

yield in this area. However, prior to developing those scenarios, we can use basic genetic concepts, coupled with the generalized calculations, to make predictions for hydrocarbon yield.

As summarized in the introduction, the history of the Metán region is characterized by two distinct stages: (1) Neocomian–Campanian rifting, and (2) Cenozoic contractional deformation and subsidence. If the starting temperature and heating rate are the principal controls on source yield for rocks undergoing rapid burial, then key controls on starting temperature and heating rate in this system should include (1) the distribution and magnitude of transient heat flow and thermal subsidence associated with the Cretaceous rift event, and (2) the timing and proximity of Cenozoic crustal loads that drive contractional deformation. In regions distal to the load (e.g., similar to site 2 in Fig. 4; or the SP province in Fig. 1), alternate scenarios for rifting and load timing are unlikely to influence local source rock yield. For a typical flexural half width of 300 km, and topographic elevation of 2.5 km, source rocks more than 150 km from the load are unlikely to experience burial of more than ~1 km due to that load (e.g., DeCelles et al., 2011). One kilometer of burial would lead to a maximum temperature of ~35–40 °C for a normal geothermal gradient in a temperate region. Given an average heating rate of 5 °C/m.y., source rocks must be at a temperature of at least 100 °C for hydrocarbon generation (Fig. 5). This is ~60 °C higher than the predicted temperature, which implies nearly 2 km of additional burial, and it is difficult to imagine the mechanism that would lead to that much accommodation far from the crustal load. Elevated heat flow associated with thermal decay of a Cretaceous rift event would not raise temperatures of near-surface source rocks enough to make a difference. In contrast, source rocks located closer to the Cenozoic load (e.g., similar to site 5 in Fig. 4; or the eastern portion of the EC region in Fig. 1) should have significantly different yield histories. In these regions, 3–4 km of burial would be typical for the flexural system described above (DeCelles et al., 2011). For potential source rocks near the base of the sedimentary section, this amount of burial might be sufficient to raise the source rock temperature to 100 °C, and it could lead to nearly 100% source rock yield over 20 m.y. for a heating rate of 5 °C/m.y. Heating rates and starting temperatures could be even higher if local heat flow was elevated due to inherited extensional events. If the crustal load was emplaced early, then oil would be found only in early-formed traps that were not subsequently breached. Later-formed traps would be either dry or gas-filled.

SUBSIDENCE MODELING

These predictions regarding the influence of contrasting scenarios for timing and magnitude of Cretaceous rifting and Andean deformation on hydrocarbon systems were examined using software that follows a Quantitative Basin Analysis (QBA) approach (e.g., Karner and Driscoll, 1999; Karner and Watts, 1983; Karner et al., 2004). The modeling approach is described in detail in Becker et al. (this volume), and it is summarized briefly here. The software uses a 2-D kinematic model to calculate elastic and thermal response of the lithosphere to thinning and thickening. It extends the 1-D modeling approach described previously, in that it captures lithospheric-scale processes responsible for maintaining isostatic balance, and it uses empirical data sets, such as seismic-reflection and seismic-refraction data, well data, gravity, (paleo-) water depths, etc., to obtain insight into the governing processes that affect basin formation. The lithospheric thickening (or thinning) parameters required by the model can be obtained from a number of data sets, including topographic profiles and geologic cross sections. In the model output described later herein (from Becker et al., this volume), a combination of two inputs was employed. A modern topographic profile extracted from a 90 m Shuttle Radar Topography Mission (SRTM) digital elevation model was used as a bounding condition in the forward models for crustal thickening across the Andean orogen. The model outputs were audited by a comparison to overall structural geometry of the cross sections and comparison with measured sections and well data from within the Eastern Cordillera and Santa Barbara system (Reynolds et al., 2000; DeCelles et al., 2011; Carrapa et al., 2011). In addition, sedimentary provenance data and low-temperature thermochronometry were used as a check on the paleogeographic interpretation. The model does not capture all of the geologic complexities of the system, such as erosion of uplifted sediments, but it does provide a 2-D snapshot in time to track changes associated with basin dynamics.

Published geologic maps of Monaldi et al. (2001) and Salfity and Monaldi (2006), and limited seismic and well data were used to construct a cross section through the Eastern Cordillera and Santa Barbara system at 25°30′S. This cross section was retrodeformed to constrain the locations of Cretaceous–Early Tertiary Salta Rift system basins, and to estimate the magnitude of shortening (and crustal thickening) along the profile (Fig. 7). Major basin-forming events that were incorporated into the model include (1) extension/heating associated with Cretaceous Salta rifting, and (2) contraction during Cenozoic Andean deformation. Four specific scenarios were evaluated, including two scenarios for Cretaceous rifting, and two scenarios for Cenozoic contraction (Table 1). Alternate scenarios for Cretaceous lithospheric delamination (depth-independent vs. depth-dependent stretching) are based on sparse data on composition and age of mantle xenoliths preserved in basaltic sills in the Pirgua Subgroup (Becker et al., this volume). The Neogene scenario for Puna Plateau shortening and structural growth follows the interpretations of Garzione et al. (2006), Ghosh et al. (2006), Hoke et al. (2007), Jordan et al. (2010), Rech et al. (2010), and Echavarria et al. (2003). The Paleogene scenario for Puna Plateau shortening and structural growth follows the interpretation of Jordan et al. (1997), Coutand et al. (2001), Deeken et al. (2006), Hogn et al. (2007), Carrapa and DeCelles (2008), Carrapa et al. (2009, 2011), and DeCelles et al. (2011). In addition, Quade et al. (this volume) reported δD values of modern and ancient volcanic glasses back to 34 Ma to estimate paleoaltitude and found little change in elevation in the Puna and western Eastern Cordillera since that time.

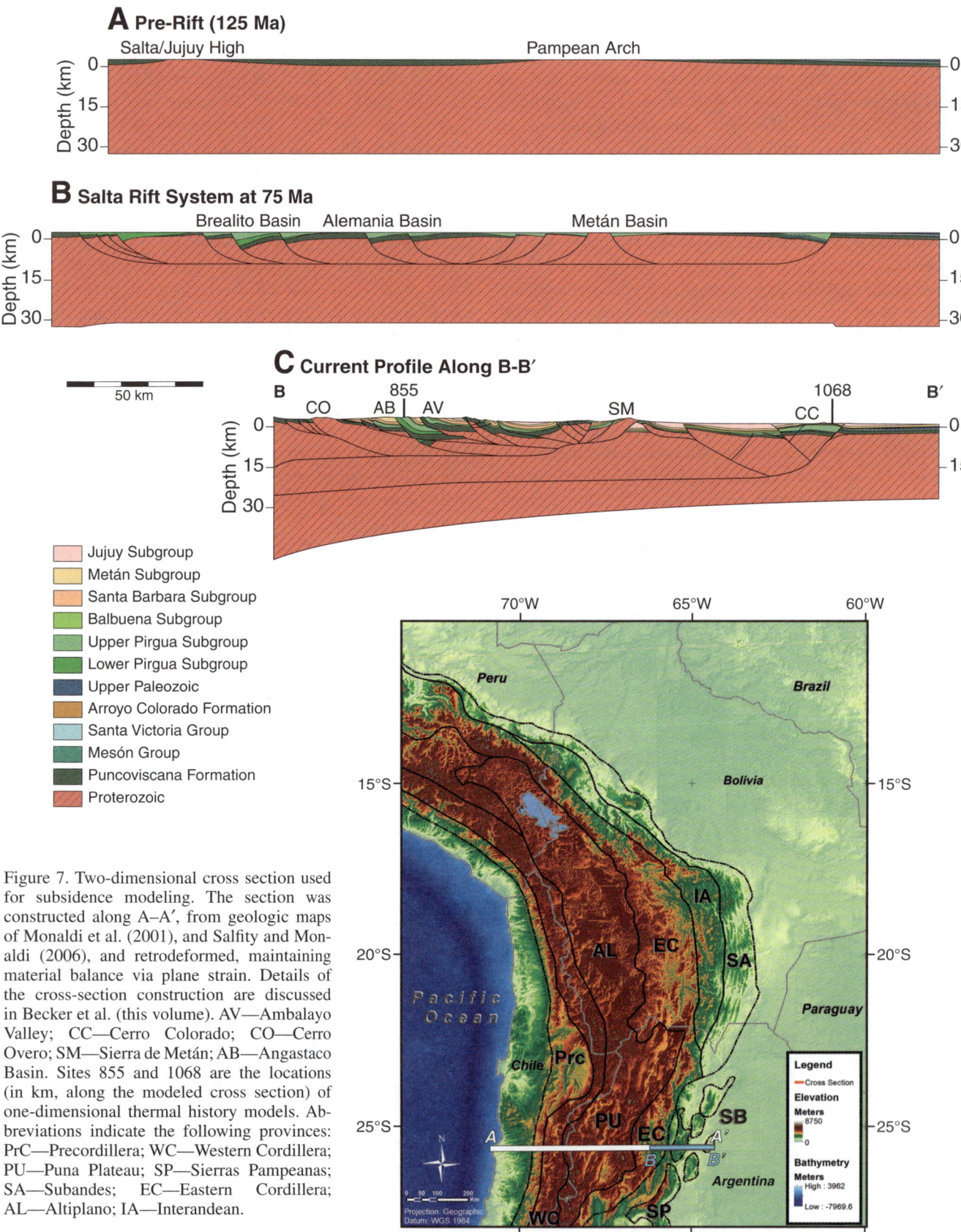

Figure 7. Two-dimensional cross section used for subsidence modeling. The section was constructed along A–A', from geologic maps of Monaldi et al. (2001), and Salfity and Monaldi (2006), and retrodeformed, maintaining material balance via plane strain. Details of the cross-section construction are discussed in Becker et al. (this volume). AV—Ambalayo Valley; CC—Cerro Colorado; CO—Cerro Overo; SM—Sierra de Metán; AB—Angastaco Basin. Sites 855 and 1068 are the locations (in km, along the modeled cross section) of one-dimensional thermal history models. Abbreviations indicate the following provinces: PrC—Precordillera; WC—Western Cordillera; PU—Puna Plateau; SP—Sierras Pampeanas; SA—Subandes; EC—Eastern Cordillera; AL—Altiplano; IA—Interandean.

TABLE 1. MODELED TECTONIC SCENARIOS

Scenario name	Cretaceous events	Tertiary events
Neogene—delamination (NCD)	Campanian rift event associated with lithospheric delamination, i.e., depth-dependent (nonuniform) stretching associated with highly attenuated mantle lithosphere	Puna Plateau emplaced in early Miocene, with continuous growth to present
Neogene—no delamination (NND)	*No* lithospheric delamination during the Campanian rift event; i.e., depth-independent (uniform) stretching (McKenzie, 1978)	Same as above
Paleogene—delamination (PCD)	Campanian rift event associated with lithospheric delamination	Puna Plateau emplaced during late Eocene; shortening slows until late Miocene; and increases again from 10 Ma to present
Paleogene—no delamination (PND)	*No* lithospheric delamination during the Campanian rift event	Same as above

Forward models of subsidence and sediment fill for the two different Cretaceous extension scenarios are shown in Figures 8A and 8B, using only the Paleogene scenario for Puna Plateau emplacement. In the depth-independent stretching scenario, the flexural response associated with the emplacement of the adjacent Puna Plateau is manifest as a broader and deeper foredeep, relative to the depth-dependent, mantle thinning scenario (Fig. 8A). This is because the thermally weakened lithosphere accommodates less of the Puna crustal load by flexure and more by local compensation (i.e., Airy isostasy). The effect is most evident away from the crustal load, where subsidence is dominated by flexure (Fig. 8A). The 1-D bed history models for potential Balbuena Subgroup source rocks adjacent to, and distant from, the load illustrate the differences in burial history across the section (sites 855 and 1068). At site 1068, far from the load, absence of crustal delamination leads to ~500 m of additional sediment load, or ~15 °C for a normal geothermal gradient. The additional load of 200–300 m at site 855 (eastern edge of the present-day Eastern Cordillera) is within model uncertainty.

Sediment fill for the two different scenarios of Puna Plateau emplacement (and depth-dependent Cretaceous rifting) is shown in Figures 9A and 9B. Close to the crustal load (Fig. 9B, site 855), the two scenarios predict very different sediment accumulation rates and timing (Fig. 9B). In the Paleogene Puna Plateau emplacement scenario, maximum postrift sediment thickness reaches nearly 4 km, and sediments at site 855 were largely deposited by 38 Ma. In the Neogene Puna Plateau emplacement scenario, maximum postrift sediment thickness reaches ~2.5 km, and sediments at site 855 were largely deposited between 38 and 10 Ma. In both scenarios, sediment accumulation in the present-day Santa Bárbara system is 2–3.5 km, and sediments were deposited mainly in the last 10 m.y., when the Andean deformation front reached the Eastern Cordillera and exhumed it above the prescribed model depositional datum of 500 m (Figs. 9A and 9B, site 1068). The thicker, stronger lithosphere of the depth-independent rifting case would accommodate more of the crustal load, resulting in greater space for Neogene sediment, compared with the thinned mantle lithosphere of Figure 9B. These end-member scenarios provide a useful framework within which to examine the effect of an earlier crustal load on sediment fill and source rock maturation, and, in general, the calculated depositional rates and timing do not fall far outside of observations from the rock record. However, there are significant local differences in the age and thickness of specific stratigraphic units that would need to be reconciled if these models were to be used for exploration purposes (Becker et al., this volume; Reynolds et al., 2001; DeCelles et al., 2011).

PETROLEUM SYSTEMS IMPLICATIONS

Sites 855 and 1068 were extracted from the 2-D subsidence models for quantitative comparison of hydrocarbon maturation and yield, and to test the effect of different subsidence models on hydrocarbon systems. These two sites effectively represent the likely range of thermal histories for potential source rocks in the region and, as such, allow easy extrapolation to other sites along the cross section. Thermal boundary conditions for the models are generally equivalent for both Paleogene and Neogene loading scenarios (Fig. 10). Sediment type and conductivity, and source rock TOC (total organic carbon) and kinetics were held constant between the two scenarios. Two different heat-flow histories were tested, along with the two different scenarios for sediment load timing. The Cretaceous rift event was modeled both with and without lithospheric delamination (i.e., depth-dependent and depth-independent stretching). Initial basal heat flow for both models is 55 mW/m², which is typical for the southern Subandean region. Lithospheric delamination leads to elevated basal heat flow of 80–85 mW/m² at 65 Ma, whereas no delamination results in basal heat flow of only 65–70 mW/m² at 65 Ma. Thermal decay continues to present day in both scenarios.

Burial-thermal history and yield calculations are shown in Figures 11 and 12 for presumed Yacoraite Formation source

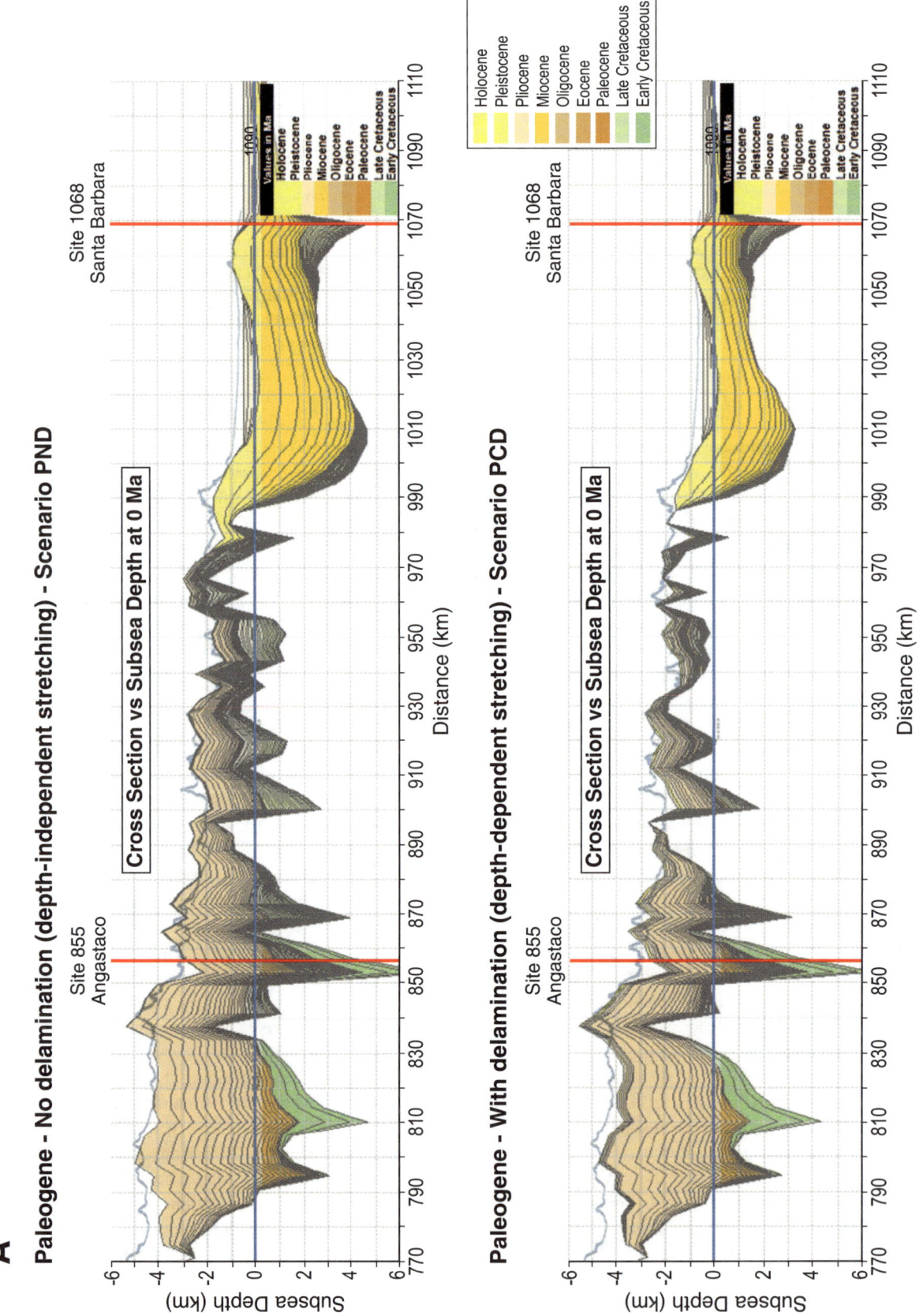

Figure 8 (*Continued on facing page*). Output from Quantitative Basin Analysis (QBA)–based two-dimensional subsidence calculations. (A) In both cross sections, the Puna Plateau was emplaced during the Paleogene, and the models show alternate scenarios for lithospheric delamination during the Cretaceous heating event. Cretaceous strata are green, Paleogene strata are buff to tan, and Neogene strata are yellow. Depth-dependent stretching (i.e., lithospheric delamination) leads to greater localized sediment fill (e.g., site 855). Depth-independent stretching (i.e., no delamination) leads to more distributed sediment fill. The differences in distribution of sediment fill are most evident from 870 to 970 km, where the no-delamination scenario has ~1 km more sediment than the delamination scenario.

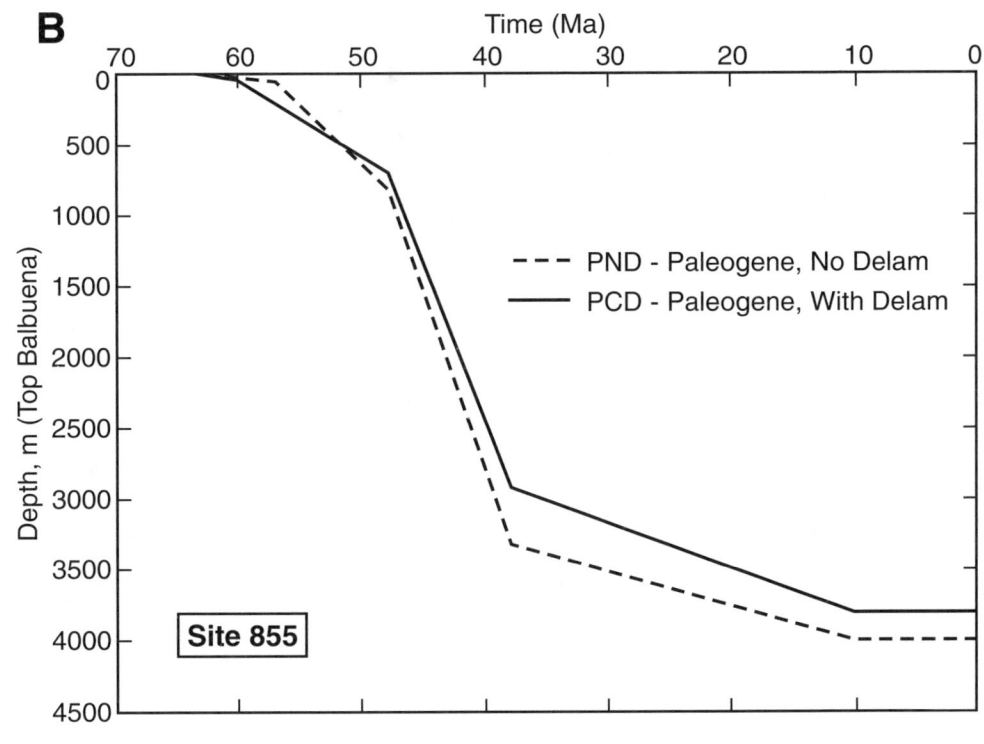

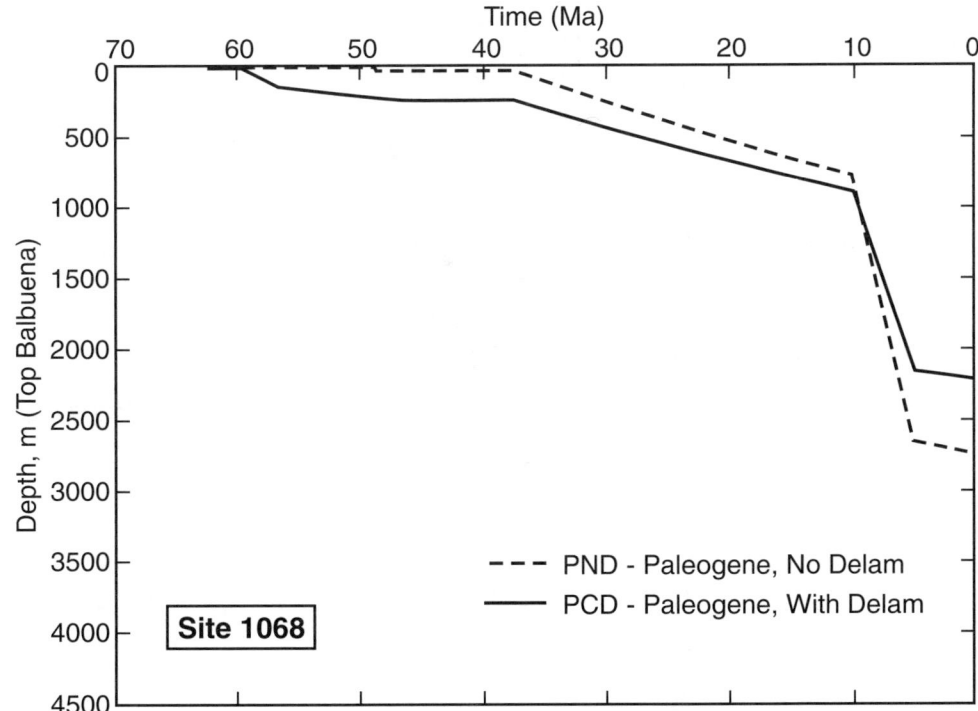

Figure 8 (*Continued*). (B) The sediment fill history for potential source rocks near the top of Yacoraite Formation at two representative sites proximal and distal to the plateau. At site 855, the differences in sediment fill history are within the uncertainty of the models. At site 1068, where flexural effects are more pronounced, the difference in present-day burial is ~500 m, which will have at least minor consequences for the thermal history.

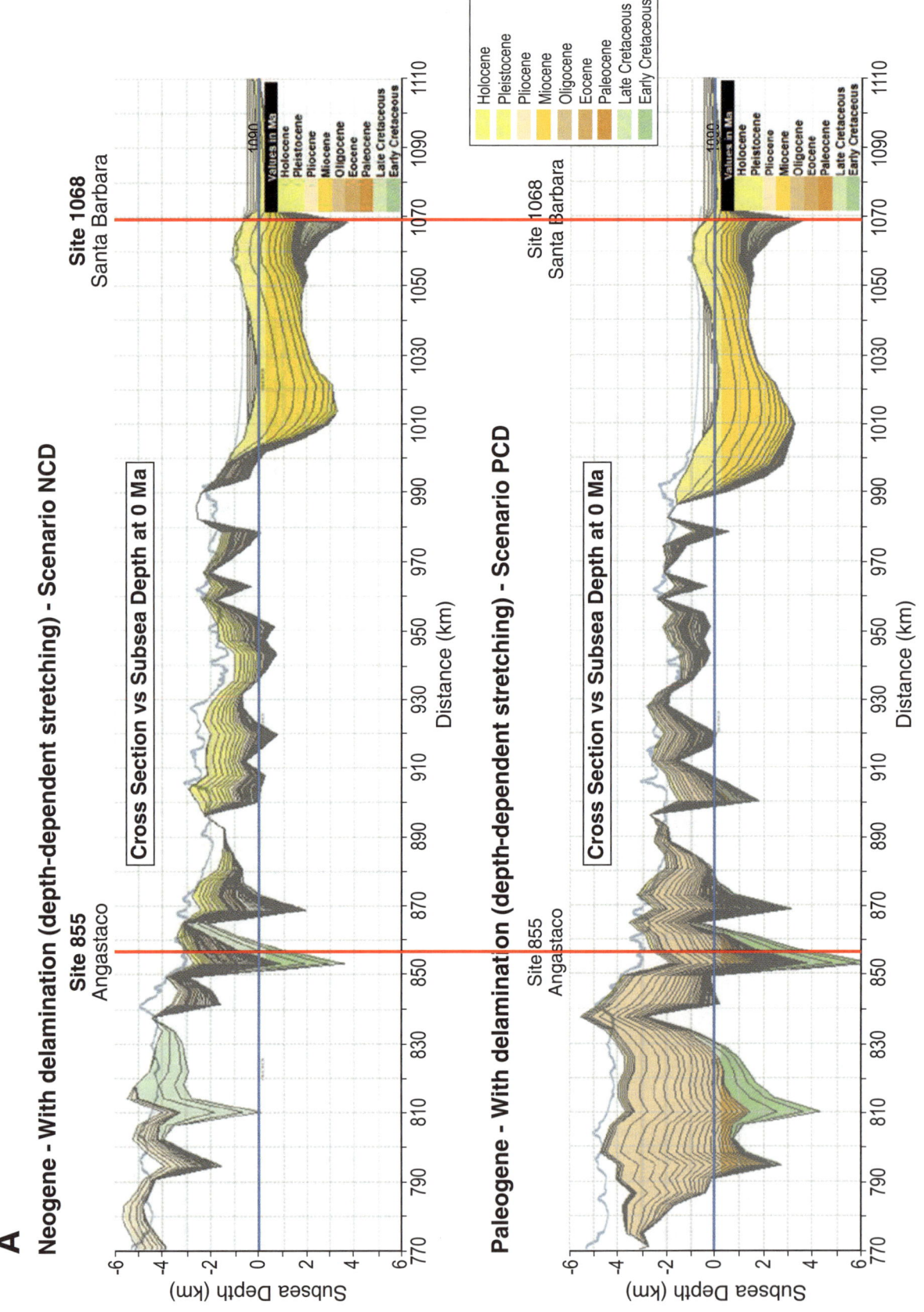

Figure 9 (*Continued on facing page*). Output from Quantitative Basin Analysis (QBA)–based two-dimensional subsidence calculations showing results of alternate scenarios for timing of Puna Plateau emplacement. (A) Both scenarios use depth-dependent stretching (lithospheric delamination). Maximum sediment fill in the present day is ~8 km for the model of Neogene Puna emplacement, but only ~6 km for the model of Paleogene Puna emplacement.

B

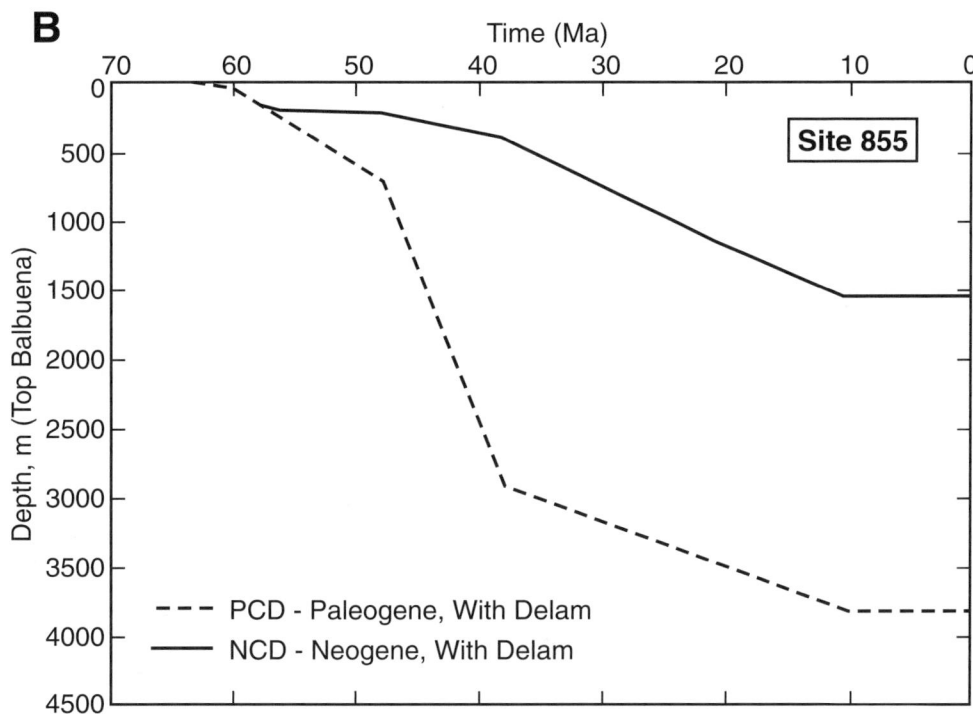

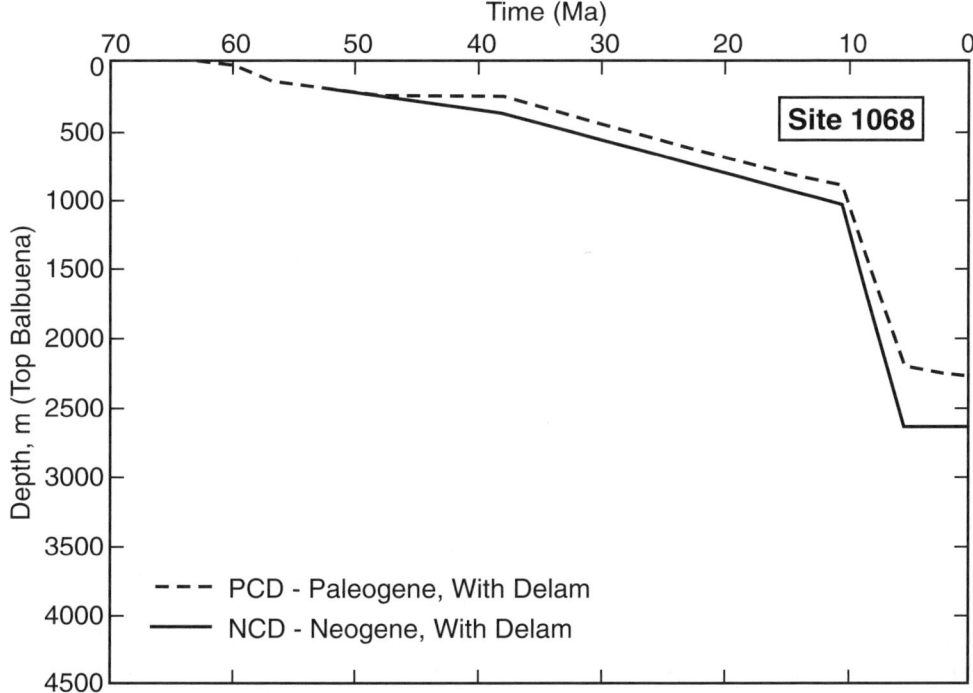

Figure 9 (*Continued*). (B) The one-dimensional (1-D) burial history model for site 855 illustrates the differences in sediment fill timing associated with different load timing (Fig. 9B). Sediment accumulation for the Paleogene emplacement model largely predates 40 Ma, whereas sediment accumulation for the Neogene emplacement model largely postdates 40 Ma. At distal site 1068, in the present-day Santa Bárbara ranges, sediment accumulation sufficient to drive source rock maturation occurs in the last 10 m.y. in both scenarios.

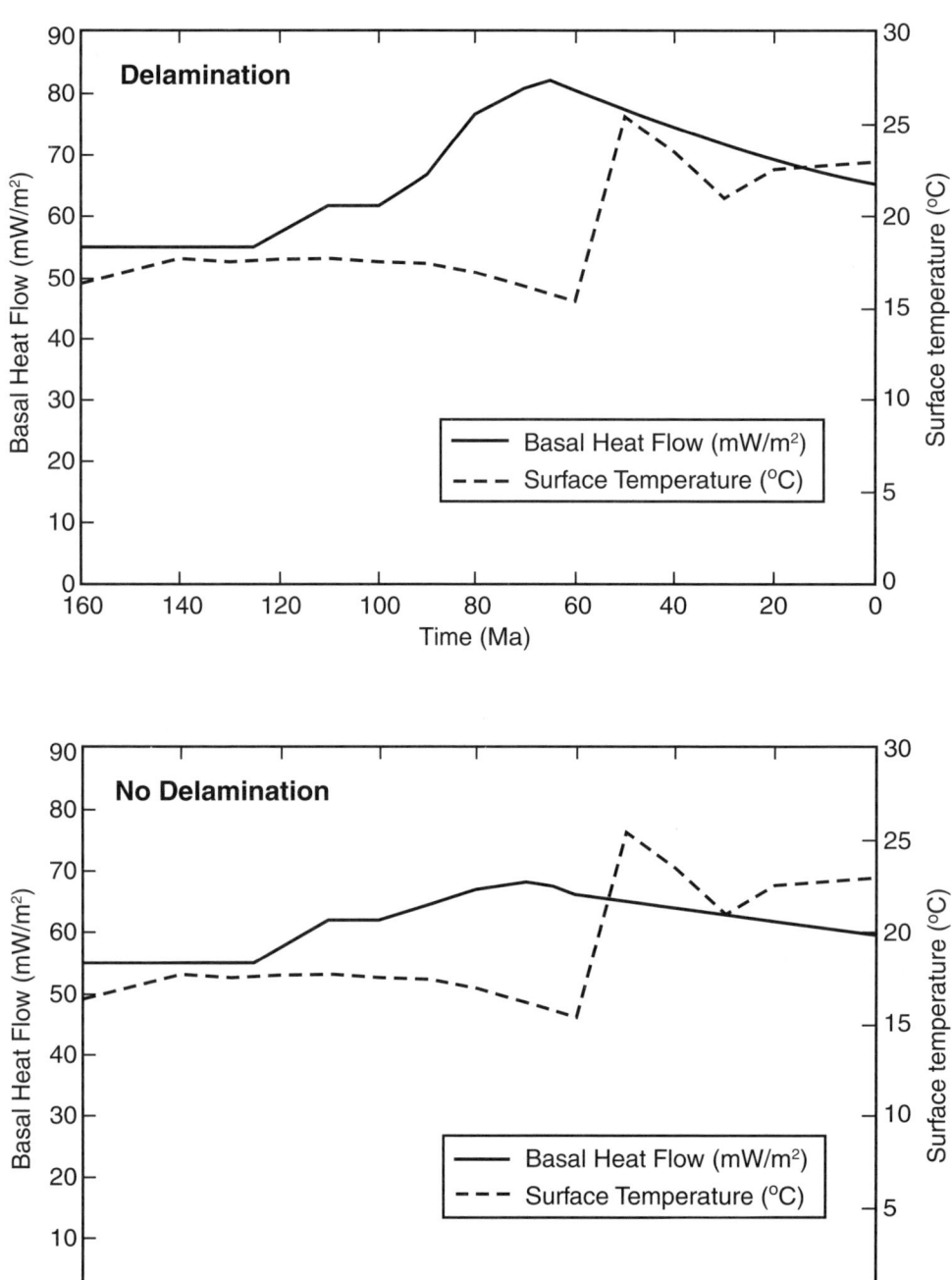

Figure 10. Thermal boundary conditions for maturation and yield models. Surface temperature and basal heat flow are identical for both scenarios of Puna emplacement timing. Lithospheric delamination during the Cretaceous heating event leads to a larger thermal perturbation, ~15 mW/m² greater than the scenario with no delamination. For both scenarios of Puna Plateau emplacement, the thermal pulse associated with the Cretaceous event is still decaying at the time the crustal load is emplaced. However, for either Paleogene or Neogene Puna emplacement timing, the difference in basal heat flow between the two scenarios is only ~10 mW/m².

rocks within the Balbuena Subgroup. Only the models that include lithospheric delamination are shown, as they generally represent maximum thermal maturity and yield. There is no added thermal effect in the absence of lithospheric delamination, because the burial depths for the two rifting scenarios are essentially identical (Fig. 8B). In the region of the Santa Bárbara system (site 1068), present-day maturities for this interval are nearly identical for both Puna emplacement scenarios (Fig. 11). This is not a surprise, given that basal heat flow and sediment con-

ductivity were consistent for both models, and total burial depth is similar. Source rocks just reach the top of the oil window in both scenarios and, as such, are not expected to yield significant hydrocarbons. Only in the case where there was no lithospheric delamination, and flexural loading might have allowed ~500 m of additional sediment load (Fig. 8B), would any significant hydrocarbon yield be expected. This may explain why the estimates of remaining oil reserves for the Oran basin are extremely low (<1 million barrels) (Yrigoyen, 1990). In contrast, with 2 km of

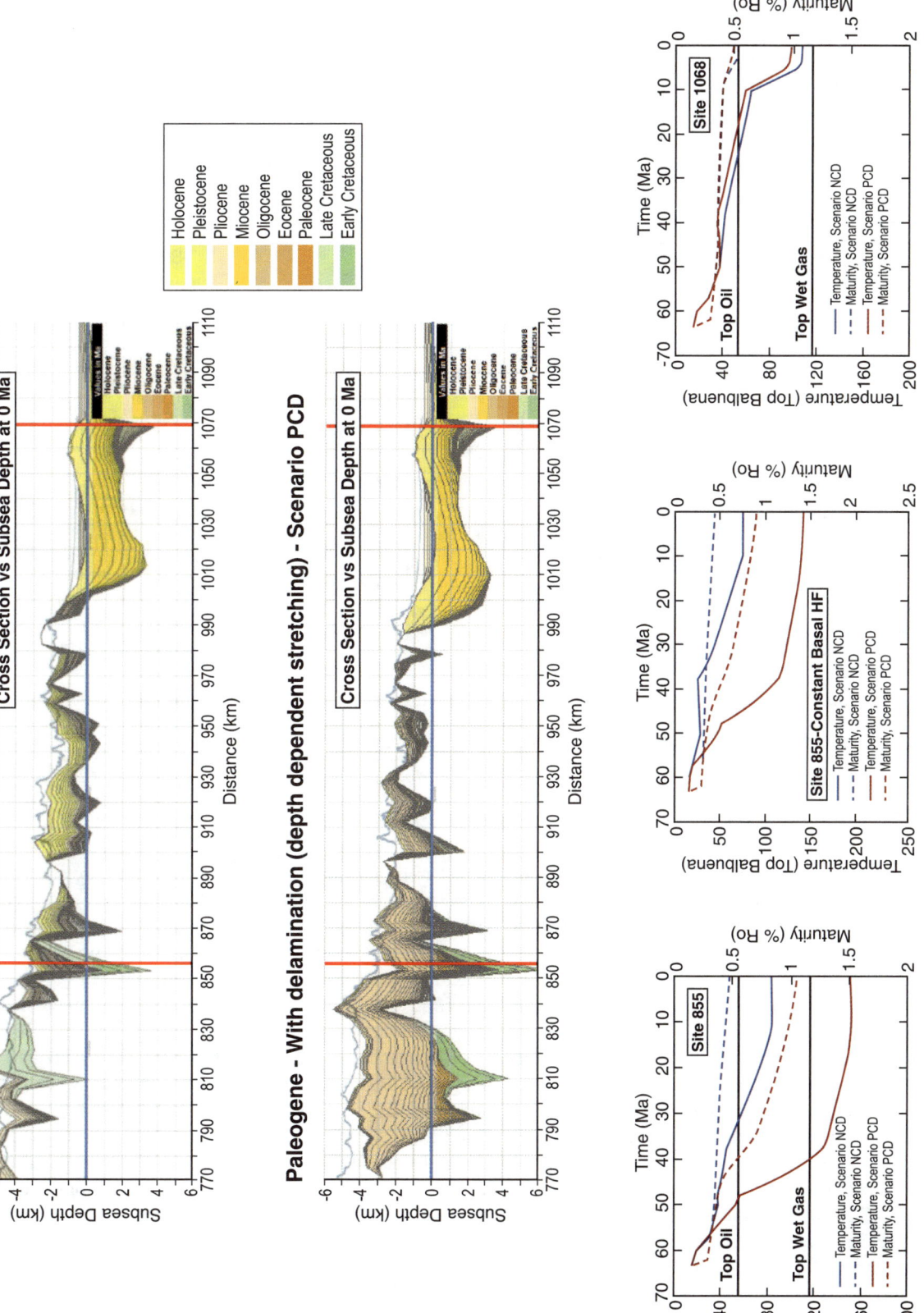

Figure 11. Modeled depth and thermal history for potential Yacoraite source rocks within the Metán region. Both early and late growth cases of the Puna Plateau are shown for sites 855 and 1068. At site 1068, the source rocks just reach the top of the oil window in the Eocene emplacement scenario, which had ~500 m of additional burial. The two scenarios for timing of plateau emplacement have a significant effect on source rock maturation at site 855. For the Eocene emplacement scenario, source rocks enter the oil window at ca. 40 Ma and remain within the window until present. For the Miocene emplacement scenario, source rocks remain immature. HF—heat flow. Ro is vitrinite reflectance in %; scenarios are NCD and PCD (Neogene—with delamination and Paleogene—with delamination).

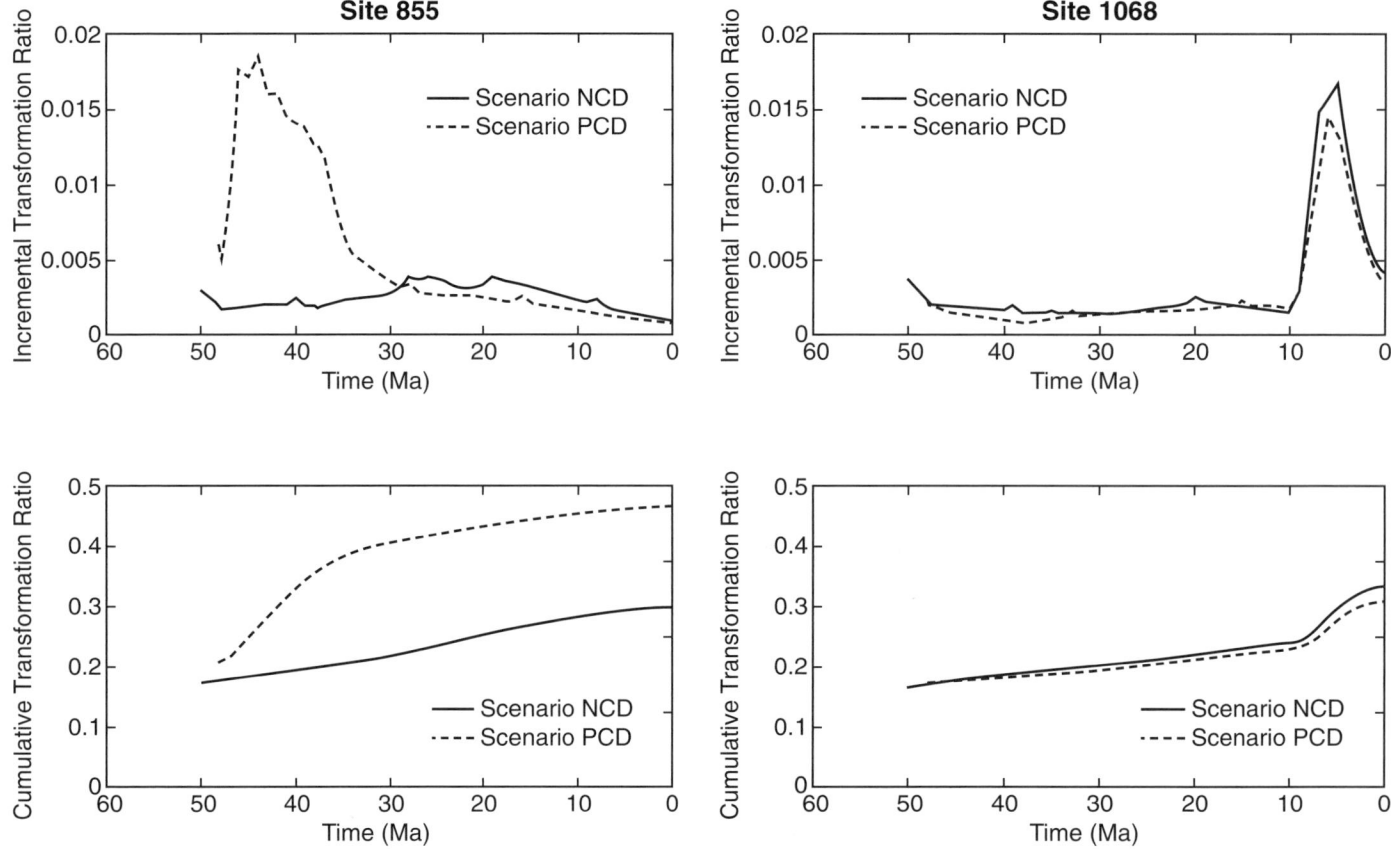

Figure 12. Comparison of modeled kerogen transformation ratios within the Metán region for the Eocene versus Miocene Puna growth cases. Site 855 is on the left, and site 1068 is on the right. Rapid transformation is associated with periods of rapid burial. In all cases, except the early Puna emplacement case at site 855, cumulative yield is only 30%. Scenarios are NCD and PCD (Neogene—with delamination and Paleogene—with delamination).

additional burial in the eastern portion of the Eastern Cordillera (site 855) in the Paleogene Puna scenario, source rocks in the Yacoraite interval entered the oil window ca. 40 Ma and would have continued to generate oil to present day in the absence of exhumation. In the Neogene Puna scenario, source rocks at this particular site remain immature, although they likely reached the top of the oil window in the small kitchens to the east where Miocene and younger sediment load led to close to 3 km of source rock burial.

A third model was run using a constant basal heat flow of 60 mW/m² (i.e., no basal heat flow spike at 65 Ma), as a means of testing the relative influence of heating due to the Cretaceous rift event, relative to heating associated with the emplacement of the Puna Plateau. The results of this model were calculated only for site 855, since source rocks did not reach maturity even with the additional thermal input at site 1068. Thermal history and maturation histories of Yacoraite source rocks are essentially identical with and without Cretaceous heating. As expected, source rock temperatures and maturities are slightly lower during initial burial without additional Cretaceous heating, but the

effect is minor, and the final maturities are determined by the subsequent emplacement of the Puna Plateau. This is because the Yacoraite source rocks had experienced very little burial at the time of the Cretaceous heating event. If the source rocks had been slightly older, and/or the heating event had been slightly younger, the effect of that event on thermal history and maturation would have been more dramatic.

Thermochronology data from the eastern portion of the Eastern Cordillera can be used to audit the modeled thermal maturity values. Apatite fission-track (AFT) ages were calculated from the upper part of the Pirgua Subgroup and the base of the Miocene at outcrops located west of Angastaco (Fig. 7; B. Carrapa, 2014, personal commun.). These outcrops are relatively close to site 855, although they almost certainly experienced slightly less total burial. The Pirgua samples from these outcrops show evidence of partial annealing: One of the two detrital AFT populations is 69.6 Ma, i.e., younger than the depositional age (Fig. 13). This suggests that the temperature for the top of the Pirgua Subgroup reached 80–120 °C. The second sample from Miocene rocks at the same location has detrital populations that are all older than the depositional age, suggesting the maximum temperature for the base of the Miocene was <110 °C (Fig. 11 *in* Carrapa et al., 2011). A second set of samples that was collected from early Cenozoic outcrops north of Angastaco all have ages younger than the depositional age of the strata, suggesting that temperatures reached >80 °C (Carrapa et al., 2011). Although not a comprehensive study, these data do provide supporting evidence that the temperatures calculated in the thermal models are not unrealistic, and they provide the foundation for possible future investigations.

Hydrocarbon yield comparisons (Fig. 12) confirm that incremental and cumulative transformation ratios for the two scenarios are similar at site 1068. However, the two scenarios produce very different predictions for hydrocarbon charge at site 855. In the Neogene Puna emplacement scenario, source rocks did not reach the oil window, resulting in essentially no hydrocarbon yield. In the Paleogene Puna Plateau emplacement scenario, the majority of the hydrocarbon yield occurs rapidly, between 45 and 35 Ma. This is consistent with heating rates of nearly 8 °C/m.y. (Fig. 6), leading to extraordinarily efficient yield. Assuming that unbreached traps in this region formed later than the yield, most of the early-generated hydrocarbons would have been lost to the surface along migration pathways. However, as noted in the preceding section, the timing of subsidence and sediment fill calculated by the forward model varies relative to the local stratigraphy. As such, the next step in this analysis would be to determine

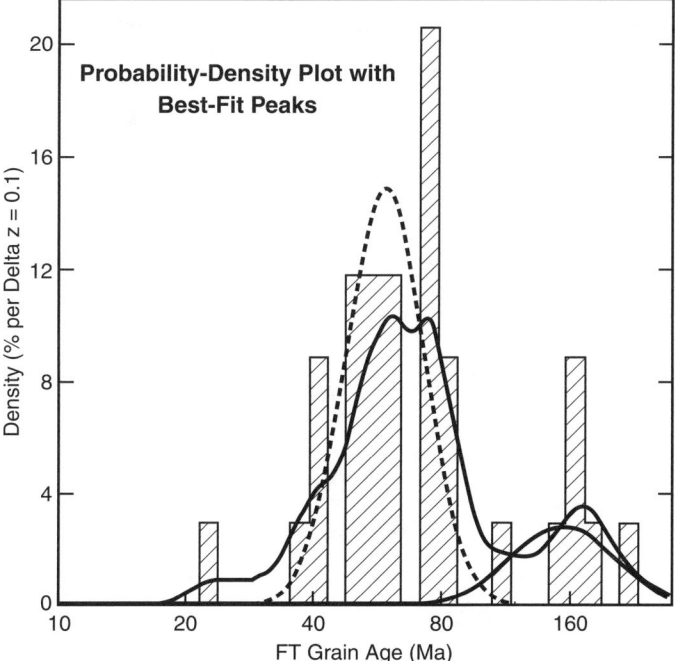

Figure 13. Probability-density plot of fission-track ages for outcrop samples near the top of the Pirgua Group at Angastaco (AB along profile C in Fig. 7). Ages range from ca. 25 to 170 Ma, both younger and older than the depositional age. This partial resetting of fission-track ages suggests maximum temperatures of 80–120 °C for rocks in this section.

whether the latest possible yield timing might overlap with the earliest possible trap timing.

CONCLUSIONS

Regional tectonic and petroleum systems analyses demonstrate significant along-strike variations in crustal structure and coincident variability in the present-day thermal state of Andean foreland basins from Colombia through Argentina (McGroder et al., this volume). Variable thermal histories associated with these along-strike changes can lead to very different scenarios for trap fill and hydrocarbon type. The critical parameters that control these variations include the starting temperature and heating rate of source rocks during the period of trap development, which implies that subtle variations in heat flow and burial rate may greatly influence the potential outcome of hydrocarbon resource type for conventional accumulations. Unfortunately, many active exploration areas lack the data needed to develop well-calibrated models for burial rate and heat flow. In these settings, models based on genetic concepts are required to predict trap fill type and volume. This example from the Metán region of northwestern Argentina demonstrates the extent to which evolving concepts of hinterland evolution can influence predictions of the hydrocarbon system behavior in the associated foreland, assuming no lateral migration of hydrocarbons from within the fold belt. Qualitative predictions for the Metán region based on simple genetic concepts and theoretical hydrocarbon yield models suggest that variable timing of Puna Plateau emplacement would only affect thermal and maturation history of source rocks located near the load. These qualitative predictions are consistent with calculations based on quantitative 2-D subsidence models.

Neither scenario for Puna Plateau emplacement is particularly optimistic for regional hydrocarbon systems. Paleogene emplacement of the Puna Plateau likely results in hydrocarbon yield from Upper Cretaceous source rocks that predates trap formation. Neogene Puna Plateau emplacement may not allow sufficient burial of Upper Cretaceous source rocks for maturation and yield to occur. In this system, the source rocks are sufficiently young that they were only subjected to minor burial at the time of Cretaceous heating, and thus they remained relatively unaffected by that thermal event. Had the source rocks been deposited earlier, or had the heating event been slightly younger, the outcome could have been very different.

Despite uncertainties in the models, the analyses suggest that the most robust predictions of hydrocarbon type and volume for data-poor areas should focus on ambiguities in the inherited characteristics of the deformed foreland. These studies also illustrate that genetic concepts for the style and exhumation history of an orogenic hinterland can be used to gain insight into hydrocarbon prospectivity within an adjacent foreland basin. In addition, emerging tools such as low-temperature thermochronometry, xenolith petrology, and paleoaltimetry can provide boundary conditions to test scenarios in areas with limited foreland data.

ACKNOWLEDGMENTS

This study was an outgrowth of the Convergent Orogenic Systems Analysis (COSA) collaborative project between Exxon-Mobil and the University of Arizona. We gratefully acknowledge the thermochronology data provided by Barbara Carrapa. Constructive reviews by Mike McGroder, Peter DeCelles, Barbara Carrapa, Francois Roure, and Xiaoli Liu led to significant improvements in the manuscript. Lee Nachtegaele and Rebecca Miller helped create the figures.

REFERENCES CITED

Allmendinger, R.W., Isacks, B.L., Jordan, T.E., and Kay, S.M., 1997, The evolution of the Altiplano-Puna Plateau of the central Andes: Annual Review of Earth and Planetary Sciences, v. 25, p. 139–174, doi:10.1146/annurev.earth.25.1.139.

Arriagada, C., Cobbold, P.R., and Roperch, P., 2006, Salar de Atacama Basin: A record of compressional tectonics in the central Andes since the mid-Cretaceous: Tectonics, v. 25, no. 1, doi:10.1029/2004TC001770.

Astini, R.A., Benedetto, J.L., and Vaccari, N.E., 1995, The early Paleozoic evolution of the Argentine Precordillera as a Laurentian rifted, drifted, and collided terrane: A geodynamic model: Geological Society of America Bulletin, v. 107, p. 253–273, doi:10.1130/0016-7606(1995)107<0253:TEPEOT>2.3.CO;2.

Barke, R., and Lamb, S., 2006, Late Cenozoic uplift of the Eastern Cordillera, Bolivian Andes: Earth and Planetary Science Letters, v. 249, p. 350–367, doi:10.1016/j.epsl.2006.07.012.

Becker, T.P., Summa, L.L., Ducea, M., and Karner, G.D., 2015, this volume, Temporal growth of the Puna Plateau and its bearing on the post–Salta Rift system subsidence of the Andean foreland basin at 25°30′S, in DeCelles, P.G., Ducea, M.N., Carrapa, B., and Kapp, P.A., eds., Geodynamics of a Cordilleran Orogenic System: The Central Andes of Argentina and Northern Chile: Geological Society of America Memoir 212, doi:10.1130/2015.1212(20).

Capitanio, F.A., Faccenna, C., Zlotnik, S., and Stegman, D.R., 2011, Subduction dynamics and the origin of Andean orogeny and the Bolivian orocline: Nature, v. 480, p. 83–86, doi:10.1038/nature10596.

Cawood, P.A., 2005, The Terra Australis orogen: Rodinia break-up and development of the Pacific and Iapetus margins of Gondwana during the Neoproterozoic and Paleozoic: Earth-Science Reviews, v. 69, p. 249–279, doi:10.1016/j.earscirev.2004.09.001.

Carrapa, B., and DeCelles, P.G., 2008, Eocene exhumation and basin development in the Puna of northwestern Argentina: Tectonics, v. 27, TC1015, doi:10.1029/2007TC002127.

Carrapa, B., DeCelles, P.G., Reiners, P.W., Gehrels, G.E., and Sudo, M., 2009, Apatite triple dating and white mica ^{40}Ar/^{39}Ar thermochronology of syntectonic detritus in the central Andes: A multiphase tectonothermal history: Geology, v. 37, no. 5, p. 407–410, doi:10.1130/G25698A.1.

Carrapa, B., Mortimer, E., DeCelles, P.G., Bywater, S., Trimble, J., and Gehrels, G.E., 2011, Eocene–Miocene synorogenic basin evolution in the Eastern Cordillera of northwestern Argentina (25°–26°S): Regional implications for Andean orogenic wedge development: Basin Research, v. 23, p. 1–20.

Carrera, N., and Muñoz, J.A., 2008, Thrusting evolution in the southern Cordillera Oriental (northern Argentine Andes): Constraints from growth strata: Tectonophysics, v. 459, p. 107–122.

Cominguez, A.H., and Ramos, V.A., 1995, Geometry and seismic expression of the Cretaceous Salta Rift of northwestern Argentina, in Tankard, A.J., Suarez, R., and Welsink, H.J., eds., Petroleum Basins of South America: American Association of Petroleum Geologists Memoir 62, p. 325–339.

Cooper, M., 2007, Structural style and hydrocarbon prospectivity in fold and thrust belts: A global review, in Ries, A.C., Butler, R.W.H., and Graham, R.H., eds., Deformation of the Continental Crust: The Legacy of Mike Coward: Geological Society, London, Special Publication 272, p. 447–472.

Coutand, I., Cobbold, P.R., de Urreiztieta, M., Gautier, P., Chauvin, A., Gapais, D., Rossello, E.A., and López-Gamundí, O., 2001, Style and history of Andean deformation, Puna Plateau, northwestern Argentina: Tectonics, v. 20, no. 2, p. 210–234.

Dávila, F.M., Astini, R.A., and Schmidt, C., 2003, Unraveling 470 my of shortening in the central Andes and documentation of type 0 super-

posed folding: Geology, v. 31, p. 275–278, doi:10.1130/0091-7613(2003)031<0275:UMYOSI>2.0.CO;2.

DeCelles, P.G., Ducea, M.N., Kapp, P., and Zandt, G., 2009, Cyclicity in Cordilleran orogenic systems: Nature Geoscience, v. 2, p. 251–257, doi:10.1038/ngeo469.

DeCelles, P.G., Carrapa, B., Horton, B.K., and Gehrels, G.E., 2011, Cenozoic foreland basin system in the central Andes of northwestern Argentina: Implications for Andean geodynamics and modes of deformation: Tectonics, v. 30, no. 6, doi:10.1029/2011TC002948.

Deeken, A., Sobel, E.R., Coutand, I., Haschke, M., Riller, U., and Strecker, M.R., 2006, Development of the southern Eastern Cordillera, NW Argentina, constrained by apatite fission track thermochronology: From Early Cretaceous extension to middle Miocene shortening: Tectonics, v. 25, no. 6, TC6003, doi:10.1029/2005TC001894.

del Papa, C., Garcia, V., and Quattrocchio, M., 2002, Sedimentary facies and palynofacies assemblages in an Eocene perennial lake, Lumbrera Formation, northwest Argentina: Journal of South American Earth Sciences, v. 15, p. 553–569, doi:10.1016/S0895-9811(02)00081-0.

Ducea, M.N., 2001, Constraints on the bulk composition and root foundering rates of continental arcs: A California arc perspective: Journal of Geophysical Research, v. 107, no. B11, p. ECV 15.1–ECV 15.13, doi:10.1029/2001JB000643.

Ducea, M.N., and Barton, M.D., 2007, Igniting flare-up events in Cordilleran arcs: Geology, v. 35, no. 11, p. 1047–1050, doi:10.1130/G23898A.1.

Echavarria, L., Hernandez, R., Allmendinger, R., and Reynolds, J., 2003, Subandean thrust and fold belt of northwestern Argentina: Geometry and timing of the Andean evolution: American Association of Petroleum Geologists Bulletin, v. 87, p. 965–985, doi:10.1306/01200300196.

Garzione, C.N., Hoke, G.D., Libarkin, J.C., and MacFadden, B.J., 2006, Rapid late Miocene rise of the Bolivian Altiplano: Evidence for removal of mantle lithosphere: Earth and Planetary Science Letters, v. 241, p. 543–556, doi:10.1016/j.epsl.2005.11.026.

Garzione, C.N., Hoke, G.D., Libarkin, J.C., Withers, S., MacFadden, B., Eiler, J., Ghosh, P., and Mulch, A., 2008, Rise of the Andes: Science, v. 320, p. 1304–1307, doi:10.1126/science.1148615.

Ghosh, P., Garzione, C.N., and Eiler, J.M., 2006, Rapid uplift of the Altiplano revealed through ^{13}C–^{18}O bonds in paleosol carbonates: Science, v. 311, p. 511–515, doi:10.1126/science.1119365.

Hernández, R.M., Galli, C.I., and Reynolds, J., 1999, Estratigrafía del Terciario en el noroeste Argentino, in Gonzalez, B., Bonorino, R., and Viramonte, J., eds., Geologia del Noroeste Argentino: Relatorio XIV Congreso Geologico Argentino, v. 1, p. 316–328.

Hoke, G.D., Isacks, B.L., Jordan, T.E., Blanco, N., Tomlinson, A.J., and Ramezani, J., 2007, Geomorphic evidence for post–10 Ma uplift of the western flank of the central Andes 18°30′–22°S: Tectonics, v. 26, TC5021, doi:10.1029/2006TC002082.

Hongn, F., del Papa, C., Powell, J., Petrinovic, I., Mon, R., and Deraco, V., 2007, Middle Eocene deformation and sedimentation in the Puna–Eastern Cordillera transition (23°–26°S): Control by preexisting heterogeneities on the pattern of initial Andean shortening: Geology, v. 35, p. 271–274, doi:10.1130/G23189A.1.

Horton, B.K., 2005, Revised deformation history of the central Andes: Inferences from Cenozoic foredeep and intermontane basins of the Eastern Cordillera, Bolivia: Tectonics, v. 24, TC3011, doi:10.1029/2003TC001619.

Houseman, G.A., McKenzie, D.P., and Molnar, P., 1981, Convective instability of a thickened boundary-layer and its relevance for the thermal evolution of continental convergent belts: Journal of Geophysical Research, v. 86, p. 6115–6132, doi:10.1029/JB086iB07p06115.

Isacks, B., 1988, Uplift of the central Andean Plateau and bending of the Bolivian orocline: Journal of Geophysical Research, v. 93, p. 3211–3231, doi:10.1029/JB093iB04p03211.

Jordan, T.E., Reynolds, J.H., and Erikson, J.P., 1997, Variability in age of initial shortening and uplift in the central Andes, 16–33°30′S, in Ruddiman, W.F., ed., Tectonic Uplift and Climate Change: New York, Plenum Press, p. 41–61, doi:10.1007/978-1-4615-5935-1_3.

Jordan, T.E., Nester, P.L., Blanco, N., Hoke, G.D., Davila, F., and Tomlinson, A.J., 2010, Uplift of the Altiplano-Puna Plateau: A view from the west: Tectonics, v. 29, TC5007, doi:10.1029/2010TC002661.

Karner, G.D., and Driscoll, N.W., 1999, Tectonic and stratigraphic development of the West African and eastern Brazilian margins: Insights from quantitative basin modeling, in Cameron, N.R., Bate, R.H., and Clure, V.S., eds., The Oil and Gas Habitats of the South Atlantic: Geological

Society, London, Special Publication 153, p. 11–40, doi:10.1144/GSL.SP.1999.153.01.02.

Karner, G.D., and Watts, A.B., 1983, Gravity anomalies and flexure of the lithosphere at mountain ranges: Journal of Geophysical Research, v. 88, p. 10,449–10,477, doi:10.1029/JB088iB12p10449.

Karner, G.D., Taylor, B., Driscoll, N.W., and Kohlstedt, D.L., eds., 2004, Rheology and Deformation of the Lithosphere at Continental Margins: Cambridge, UK, Cambridge University Press, 408 p.

Kay, R.W., and Kay, S.M., 1993, Delamination and delamination magmatism: Tectonophysics, v. 219, p. 177–189, doi:10.1016/0040-1951(93)90295-U.

Kay, S.M., Coira, B., and Viramonte, J., 1994, Young mafic back arc volcanic rocks as indicators of continental lithospheric delamination beneath the Argentine Puna Plateau, central Andes: Journal of Geophysical Research, v. 99, no. B12, p. 24,323–24,339, doi:10.1029/94JB00896.

Kley, J., Monaldi, C.R., and Salfity, J.A., 1999, Along-strike segmentation of the Andean foreland: Causes and consequences: Tectonophysics, v. 301, p. 75–94, doi:10.1016/S0040-1951(98)90223-2.

Lamb, S., and Davis, P., 2003, Cenozoic climate change as a possible cause for the rise of the Andes: Nature, v. 425, p. 792–797, doi:10.1038/nature02049.

Leech, M.L., 2001, Arrested orogenic development: Eclogitization, delamination, and tectonic collapse: Earth and Planetary Science Letters, v. 185, p. 149–159, doi:10.1016/S0012-821X(00)00374-5.

Lucassen, F., Becchio, R., Wilke, H.G., Franz, G., Thirlwall, M.F., Viramonte, J., and Wemmer, K., 2000, Proterozoic–Paleozoic development of the basement of the central Andes (18–26°S)—A mobile belt of the South American craton: Journal of South American Earth Sciences, v. 13, p. 697–715, doi:10.1016/S0895-9811(00)00057-2.

McGroder, M.F., Lease, R.O., and Pearson, D.M., 2015, this volume, Along-strike variation in structural styles and hydrocarbon occurrences, Subandean fold-and-thrust belt and inner foreland, Colombia to Argentina, *in* DeCelles, P.G., Ducea, M.N., Carrapa, B., and Kapp, P.A., eds., Geodynamics of a Cordilleran Orogenic System: The Central Andes of Argentina and Northern Chile: Geological Society of America Memoir 212, doi:10.1130/2015.1212(05).

McKenzie, D., 1978, Some remarks on the development of sedimentary basins: Earth and Planetary Science Letters, v. 40, p. 25–32, doi:10.1016/0012-821X(78)90071-7.

Molnar, P., and Houseman, G.A., 2004, The effects of buoyant crust on the gravitational instability of thickened mantle lithosphere at zones of intracontinental convergence: Geophysical Journal International, v. 158, p. 1134–1150, doi:10.1111/j.1365-246X.2004.02312.x.

Monaldi, C.R., Alonso, R.N., Gonzalez, R.E., Igarzabal, A.P., Ramallo, E., Godeas, M., Fuertes, A., Garcia, R., and Moya, F., 2001, Programa Nacional de Cartas Geologicas de la Republica Argentina, Cachi, Provincias de Salta y Catamarca: Instituto de Geologica y Recursos Minerales Hoja Geologica 2566-III, scale 1:250,000, 1 sheet.

Pelletier, J.D., DeCelles, P.G., and Zandt, G., 2010, Relationships among climate, erosion, topography, and delamination in the Andes: A numerical modeling investigation: Geology, v. 38, p. 259–262, doi:10.1130/G30755.1.

Quade, J., Dettinger, M.P., Carrapa, B., DeCelles, P., Murray, K.E., Huntington, K.A., Cartwright, A., Canavan, R.R., Gehrels, G., and Clementz, M., 2015, this volume, The growth of the central Andes, 22°S–26°S, *in* DeCelles, P.G., Ducea, M.N., Carrapa, B., and Kapp, P.A., eds., Geodynamics of a Cordilleran Orogenic System: The Central Andes of Argentina and Northern Chile: Geological Society of America Memoir 212, doi:10.1130/2015.1212(15).

Ramos, V.A., 2009, Anatomy and global context of the Andes: Main geologic features and the Andean orogenic cycle, *in* Kay, S.M., Ramos, V.A., and Dickinson, W.R., eds., Backbone of the Americas: Shallow Subduction, Plateau Uplift, and Ridge and Terrane Collision: Geological Society of America Memoir 204, p. 31–65, doi:10.1130/2009.1204(02).

Ramos, V.A., Jordan, T.E., Allmendinger, R.W., Mpodozis, C., Kay, S.M., Cortés, J.M., and Palma, M.A., 1986, Paleozoic terranes of the central Argentine-Chilean Andes: Tectonics, v. 5, p. 855–880, doi:10.1029/TC005i006p00855.

Rapela, C.W., Pankhurst, R.J., Casquet, C., Baldo, E., Saavedra, J., Galindo, C., and Fanning, C.M., 1998, The Pampean orogeny of the southern proto-Andes: Cambrian continental collision in the Sierras de Cordoba, *in* Pankhurst, R.J., and Rapela, C.W., eds., The Proto-Andean Margin of Gondwana: Geological Society, London, Special Publication 142, p. 181–217, doi:10.1144/GSL.SP.1998.142.01.10.

Rech, J.A., Currie, B.S., Shullenberger, E.D., Dunagan, S.P., Jordan, T.E., Blanco, N., Tomlinson, A.J., Rowe, H.D., and Houston, J., 2010, Evidence for the development of the Andean rain shadow from a Neogene isotopic record in the Atacama Desert, Chile: Earth and Planetary Science Letters, v. 292, p. 371–381, doi:10.1016/j.epsl.2010.02.004.

Reynolds, J.H., Galli, C.I., Hernández, R.M., Idleman, B.D., Kotila, J.M., Hilliard, R.V., and Naeser, C.W., 2000, Middle Miocene tectonic development of the Transition zone, Salta Province, northwest Argentina: Magnetic stratigraphy from the Metán Subgroup, Sierra de González: Geological Society of America Bulletin, v. 112, no. 11, p. 1736–1751, doi:10.1130/0016-7606(2000)112<1736:MMTDOT>2.0.CO;2.

Reynolds, J.H., Hernández, R.M., Galli, C.I., and Idleman, B.D., 2001, Magnetostratigraphy of the Quebrada La Porcelana section, Sierra de Ramos, Salta Province, Argentina; age limits for the Neogene Orán Group and uplift of the southern Sierras Subandinas: Journal of South American Earth Sciences, v. 14, p. 681–692, doi:10.1016/S0895-9811(01)00069-4.

Rhea, S., Hayes, G., Villasenor, A., Furlong, K.P., Tarr, A.C., and Benz, H., 2008, Seismicity of the Earth: Nazca Plate and South America: U.S. Geological Survey Open-File Report 2010-1083-E, 1 map sheet, scale 1:12,000,000.

Roure, F., Bordas-Lefloch, N., Toro, J., Aubourg, C., Guilhaumou, N., Hernandez, E., Lecornec-Lance, S., Rivero, C., Robion, P., and Sassi, W., 2003 Petroleum systems and reservoir appraisal in the Subandean basins (eastern Venezuela and eastern Colombian foothills), *in* Bartolini, C., Burke, K., Buffler, R., Blickwede, J., and Burkart, B., eds., Mexico and the Caribbean Region: Plate Tectonics, Basin Formation and Hydrocarbon Habitats: American Association of Petroleum Geologists Memoir 79, p. 750–775.

Salfity, J.A., and Marquillas, R., 1994, Tectonic and sedimentary evolution of the Cretaceous–Eocene Salta Group, Argentina, *in* Salfity, J.A., ed., Cretaceous Tectonics of the Andes: Heidelberg, Vieweg, Earth Evolution Sciences, p. 266–315.

Salfity, J.A., and Monaldi, C.R., 2006, Programa Nacional de Cartas Geologicas de la Republica Argentina, Metán, Provincia de Salta: Instituto de Geologica y Recursos Minerales Hoja Geologica 2566-IV, scale 1:250,000, 1 sheet.

Schneider, F., Pagel, M., and Hernandez, E., 2004, Basin modeling in complex area: Example from the eastern Venezuelan Foothills, *in* Swennen, R., Roure, F., and Granath, J., eds., Deformation, Fluid Flow and Reservoir Appraisal in Foreland Fold-and-Thrust Belts: American Association of Petroleum Geologists Hedberg Series 1, p. 357–369.

Scholl, D.W., and von Huene, R., 2007, Crustal recycling at modern subduction zones applied to the past—Issues of growth and preservation of continental basement crust, mantle geochemistry, and supercontinent reconstruction, *in* Hatcher, R.D., Jr., Carlson, M.P., McBride, J.H., and Martínez Catalán, J.R., eds., 4-D Framework of Continental Crust: Geological Society of America Memoir 200, p. 9–32, doi:10.1130/2007.1200(02).

Schurr, B., Rietbrock, A., Asch, G., Kind, R., and Oncken, O., 2006, Evidence for lithospheric detachment in the central Andes from local earthquake tomography: Tectonophysics, v. 415, p. 203–223, doi:10.1016/j.tecto.2005.12.007.

Starck, D., and Vergani, G., 1996, Desarrollo tecto-sedimentario del Cenozoico en el sur de la Provincia de Salta—Argentina, *in* XIII Congreso Geológico Argentino y III Congreso de Exploración de Hidrocarburos, Actas I: Buenos Aires, p. 433–452.

Uba, C.E., Heubeck, C., and Hulka, C., 2006, Evolution of the late Cenozoic Chaco foreland basin, southern Bolivia: Basin Research, v. 18, p. 145–170, doi:10.1111/j.1365-2117.2006.00291.x.

Urien, C.M., Zambrano, J.J., and Yrigoyen, M.R., 1995, Petroleum Basins of South America, an Overview, *in* Tankard, A.J., Suarez-Soruco, R., and Welsink, H.J., eds., Petroleum Basins of South America: American Association of Petroleum Geologists Memoir 62, p. 63–77.

von Huene, R., and Scholl, D.W., 1991, Observations at convergent margins concerning sediment subduction, subduction erosion, and the growth of continental crust: Reviews of Geophysics, v. 29, p. 279–316, doi:10.1029/91RG00969.

Yrigoyen, M.R., 1990, Sub-andean hydrocarbon resources of Argentina, *in* Ericksen, G.E., Canas Pinochet, M.T., and Reinemund, J.A., eds., 1989, Geology of the Andes and Its Relation to Hydrocarbon and Mineral Resources: Houston, Texas, Circum-Pacific Council for Energy and Mineral Resources Earth Science Series, v. 11, p. 439–452.

MANUSCRIPT ACCEPTED BY THE SOCIETY 3 JUNE 2014
MANUSCRIPT PUBLISHED ONLINE 20 NOVEMBER 2014

The Geological Society of America
Memoir 212
2015

Cyclical orogenic processes in the Cenozoic central Andes

P.G. DeCelles
G. Zandt
S.L. Beck
Department of Geosciences, University of Arizona, Tucson, Arizona 85716, USA

C.A. Currie
Department of Physics, University of Alberta, Edmonton, Alberta T6G 2E1, Canada

M.N. Ducea
Department of Geosciences, University of Arizona, Tucson, Arizona 85716, USA, and
Facultatea de Geologie Geofizica, Universitatea Bucuresti, Strada N. Balcescu Nr 1, Bucuresti 010041, Romania

P. Kapp
G.E. Gehrels
B. Carrapa
J. Quade
Department of Geosciences, University of Arizona, Tucson, Arizona 85716, USA

L.M. Schoenbohm
Department of Chemical and Physical Sciences, University of Toronto–Mississauga, Mississauga, Ontario L5L 1C6, Canada

INTRODUCTION

The central Andes of Bolivia, northern Chile, and northwestern Argentina provide a unique natural laboratory for understanding cordilleran mountain belts. Rapid convergence between the oceanic Nazca and continental South American plates is accommodated by subduction of the former along the offshore Peru-Chile Trench, as well as crustal shortening and thickening in the Andes of the overriding South American plate (James, 1971; Isacks, 1988). The central sector of the Andean orogenic system contains an erosional forearc, an active magmatic arc, a high-elevation plateau, an active retroarc fold-and-thrust belt, and a subcontinental-scale retroarc foreland basin system (Fig. 1). Numerous studies have documented widely variable (in time and space) tectonic, magmatic, and surficial processes in this region (e.g., Oncken et al., 2006; Strecker et al., 2007; McQuarrie et al., 2008). For example, many workers have studied the obvious con-

tractional structural belts along the eastern side of the orogenic system, but extensional and strike-slip faults are also widely distributed within the forearc, hinterland, and in areas marginal to the high plateau (e.g., Marrett et al., 1994; Delouis et al., 1998; Jordan et al., 2007; Schoenbohm and Strecker, 2009). Magmatism in the region varies widely in location, composition, and character, from linear arc trends to back-arc positions, from mafic to ultrapotassic to felsic, and from monogenetic flows to stratovolcanoes to vast ignimbrite sheets (de Silva et al., 2006; Mazzuoli et al., 2008; Kay and Coira, 2009). The lower crust and upper mantle beneath the high plateau exhibit extreme spatial complexity in seismic wave speeds, interpreted to be a result of dynamic processes of lithospheric removal from the South American plate (Myers et al., 1998; Beck and Zandt, 2002; Schurr et al., 2006; Heit et al., 2008). Sedimentary basins—which record changing surficial environments, paleoelevations, and tectonic boundary conditions—are preserved throughout the Andean orogenic

DeCelles, P.G., Zandt, G., Beck, S.L., Currie, C.A., Ducea, M.N., Kapp, P., Gehrels, G.E., Carrapa, B., Quade, J., and Schoenbohm, L.M., 2015, Cyclical orogenic processes in the Cenozoic central Andes, *in* DeCelles, P.G., Ducea, M.N., Carrapa, B., and Kapp, P.A., eds., Geodynamics of a Cordilleran Orogenic System: The Central Andes of Argentina and Northern Chile: Geological Society of America Memoir 212, p. 459–490, doi:10.1130/2015.1212(22). For permission to copy, contact editing@geosociety.org. © 2015 The Geological Society of America. All rights reserved.

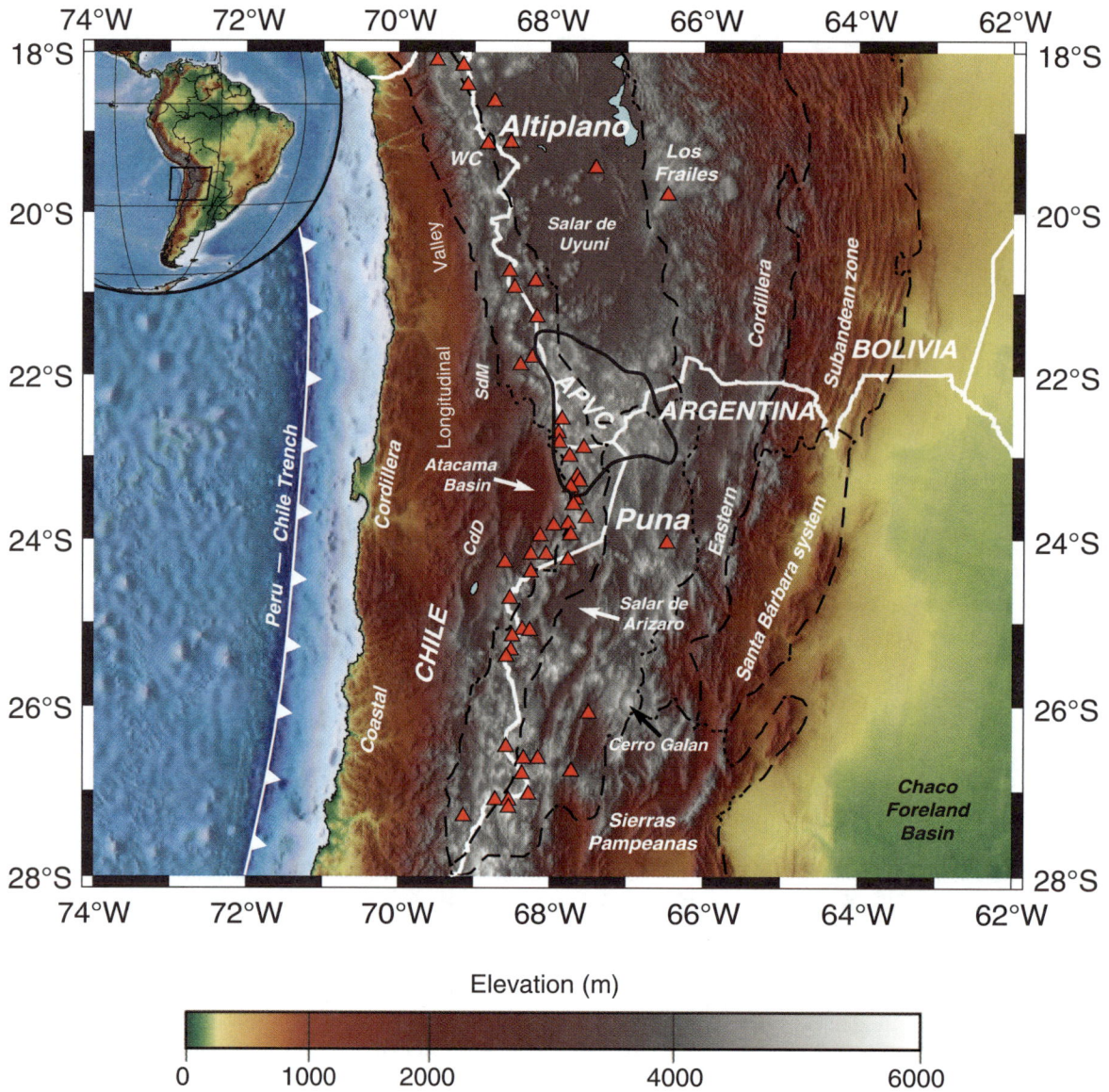

Figure 1. Digital elevation model of the study region in the central Andes. Red triangles represent Quaternary to active volcanoes. Chilean Precordillera is represented by the Sierra de Moreno (SdM) and Cordillera de Domeyko (CdD). APVC—Altiplano-Puna volcanic complex; WC—Western Cordillera.

system, from the forearc to the distal retroarc (e.g., Jordan, 1995; Hartley, 2003; Jordan et al., 2007; Horton, 2012). In many cases, the various geological records preserved in the central Andes (as well as other cordilleran orogenic systems) present a seemingly confounding array of sometimes antithetical processes: Studies in the hinterland indicate extension at certain times (Jordan et al., 2007), whereas evidence from the coeval retroarc thrust belt and foreland basin system indicates shortening (Reynolds et al., 2000; Carrapa et al., 2011a; DeCelles et al., 2011); magmatic activity may appear to be dominated by processes in the subducting Nazca plate (e.g., changes in slab dip; Kay and Coira, 2009), whereas alternative processes confined to the upper South Ameri-

can plate may be equally well invoked (Ducea et al., 2013); and stable isotopic records of paleoelevation in different districts of the central Andes may in some cases appear to be at odds with each other (Garzione et al., 2006, 2014; Insel et al., 2012; Canavan et al., 2014; Quade et al., this volume).

In this paper, we attempt to weave together multifarious strands of recent research in the region to form a coherent model for the way in which cordilleran orogenic systems operate at 25–100 m.y. time scales. Our initial hypothesis was that the central Andes are the product of systematic, predictable, interconnected processes that operate on a cyclical schedule over time scales of ~25 m.y. We refer informally to this hypothesis as the cordilleran cycle (Fig. 2). As originally conceived (DeCelles et al., 2009), the cycle is characterized by periodic episodes of high-flux arc magmatism, which are fueled by underthrusting foreland lithosphere. Arc magmatism produces dense "arclogitic" residues (Ducea, 2001), and metamorphic eclogite forms beneath the hinterland as foreland lithosphere is buried beneath thickened crust (Sobolev et al., 2006). The dense bodies beneath the hinterland and arc founder into the mantle, relieving the space problem caused by continued lithospheric underthrusting (McQuarrie et al., 2005). Isostatic adjustment of the topographic surface in response to removal of the dense instabilities causes rapid regional elevation gain and drives the forearc and retroarc orogenic wedges into a supercritical state such that orogenic strain fronts propagate outward toward the craton and the trench (Garzione et al., 2006; Pelletier et al., 2010). Renewed rapid foreland underthrusting recharges the magmatic system, and the cycle is rejuvenated. Thus, retroarc underthrusting (complemented by thrust belt shortening), arc magmatism, and gravitational foundering of dense instabilities beneath the orogenic hinterland are linked in process-feedback loops. The overarching controls on the system of processes involved in the cordilleran cycle are the rate and direction of convergence between the two plates involved and the rate and direction of migration of the subduction trench (e.g., Schellart, 2008). Details of the numerous individual processes involved are modulated by rheology, temperature distribution, lithological composition, and climate, among other variables. This model incorporates numerous processes that have been documented individually in different orogenic belts, but no attempt has been made to test the model on an active cordilleran orogenic system.

What makes the central Andes particularly attractive for such a study is the well-documented record of post-Triassic magmatic activity, which exhibits a cyclical record of changing composition and flux (Allmendinger et al., 1997; Haschke et al., 2002, 2006; Trumbull et al., 2006; Schnurr et al., 2007; Kay and Coira, 2009). Using the Cenozoic part of this record (Fig. 3A) as a starting point, we explore the temporal and spatial distributions of many other tectonic and surficial processes in the region in order to assess potential linkages among them. As we discuss in this chapter, and as documented in the other chapters of this volume, the original cordilleran cycle model requires some significant modifications (Fig. 2) in order to satisfy the observations. Nevertheless, we conclude that the central Andes exhibit a clear cycle of cordilleran-style orogenic activity.

This is not the first attempt to produce a holistic interpretation that integrates multiple processes in the central Andes through Cenozoic time. Previous efforts have focused on inherent properties of the Andean lithosphere (Allmendinger et al., 1997; Kley and Monaldi, 1998; Mamani et al., 2008); the role of plate-scale variables (e.g., rates of trench retreat and upper-plate migration toward the trench, variations in the age of the subducting plate and thickness of the overriding plate; Schellart, 2008; Capitanio et al., 2011; O'Driscoll et al., 2012; Maloney et al., 2013); aspects of the force balance between the converg-

ing South American and Nazca plates (Richardson and Coblentz, 1994; Iaffaldano and Bunge, 2008; Meade and Conrad, 2008; Husson et al., 2012); the roles of lithospheric underthrusting and gravitational removal beneath the hinterland (Schmitz, 1994; McQuarrie et al., 2005); the importance of flat-slab subduction in potentially controlling rates of plate convergence (Martinod et al., 2010); interactions among magmatic and tectonic processes, with emphasis on delamination of the mantle lithosphere (Kay and Kay, 1993; Schnurr et al., 2007; Mazzuoli et al., 2008; Kay and Coira, 2009) or break-off of the subducting slab (Haschke et al., 2006); linkages among tectonic, metamorphic, and erosional processes (Sobel et al., 2003; Pelletier et al., 2010); the role of climate in controlling flux of sediment into the Peru-Chile Trench, with implications for the strength of coupling between the two converging plates (Lamb and Davis, 2003; Meade and Conrad, 2008); and relationships among deformation, magmatism, plate kinematics, and climate (Oncken et al., 2006; McQuarrie et al., 2008; Barnes and Ehlers, 2009). The importance of low-angle (or "flat-slab") subduction of the Nazca plate is highlighted in many previous syntheses (Isacks, 1988; James and Sacks, 1999; Haschke et al., 2002, 2006; Kay and Coira, 2009; Ramos and Folguera, 2009; Martinod et al., 2010; O'Driscoll et al., 2012). In this synthesis, we borrow directly from the observations and insights provided by these and many other previous efforts. Our intention is not to produce a "better" synthesis or to reveal a previously unknown process; most of the processes involved in construction of the central Andean orogenic system are well known in and of themselves. What we introduce are some novel relationships among these processes, and a general model for how Andean-type (or cordilleran) orogenic systems operate.

CENOZOIC TECTONIC AND GEOLOGICAL SETTING OF THE CENTRAL ANDES

Pardo-Casas and Molnar (1987), Somoza (1998), and Sdrolias and Müller (2006) documented the post-Cretaceous rate of convergence between the Nazca and South American plates (Fig. 3H). Although the results from these studies are strictly representative of various latitudinal zones of the Andean margin, and minor differences exist among the three reconstructions, Pardo-Casas and Molnar (1987) and Sdrolias and Müller (2006) documented high rates (>150 mm/yr) of convergence during the late Eocene, and all three studies show rapid convergence ca. 25–10 Ma (>110 mm/yr). Although minor local variations are present, the changes between slow and fast convergence, in the words of Pardo-Casas and Molnar (1987), "are inescapable." Schellart (2008) proposed that the most important variables leading to increased shortening and crustal thickening in the central Andes are the rapid absolute rate of plate convergence and the great distance of the central Andean region from the lateral edges of the subducting Nazca plate. The latter is important because westward rollback of the Nazca plate is impeded by a stagnation point in the mantle beneath the central part of the plate (e.g., Russo and Silver, 1996). Norabuena et

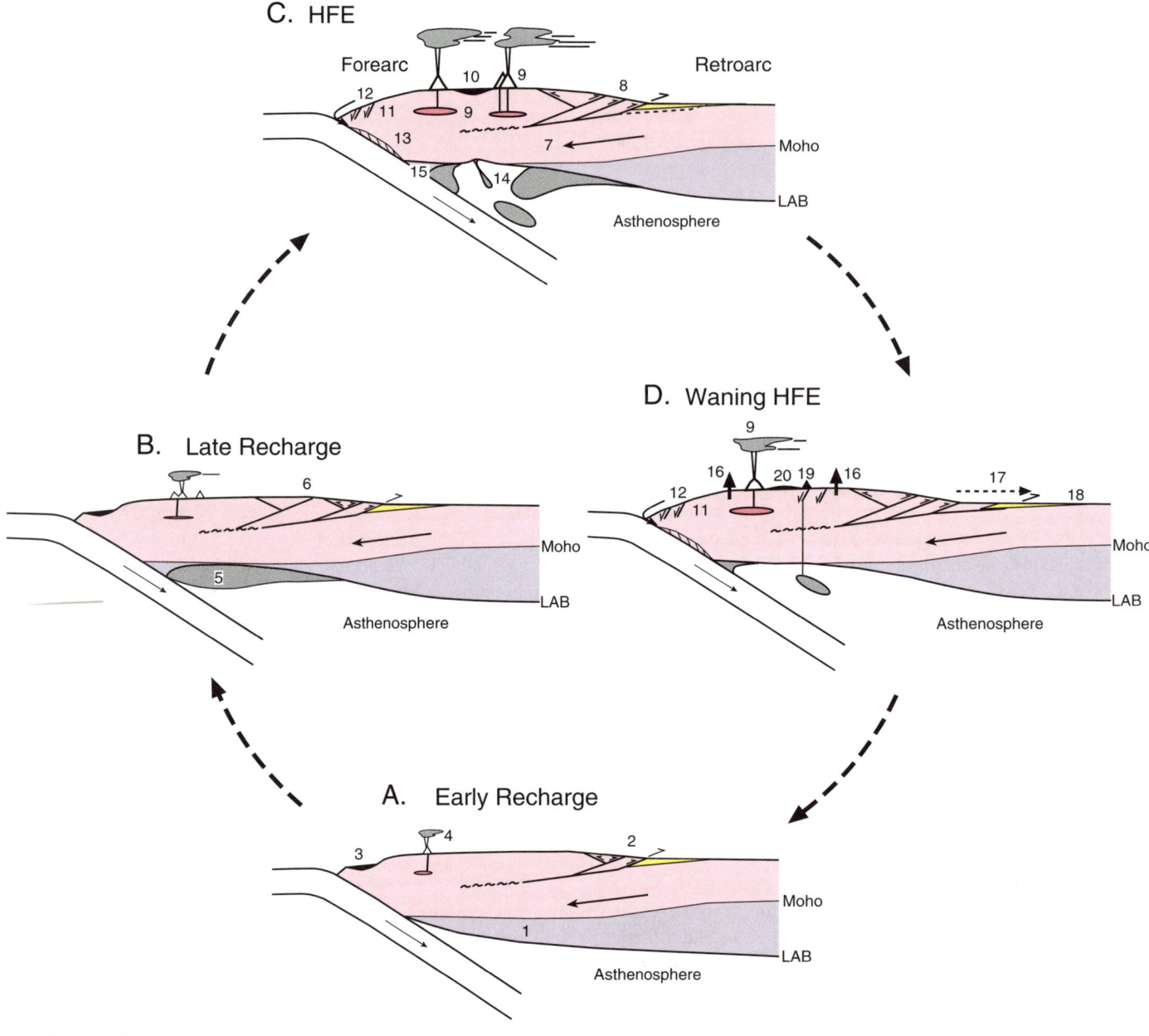

Legend
1. Thickening of hinterland lithosphere
2. Internal shortening, slow forward propagation of thrust belt
3. Forearc subsidence, sediment accumulation
4. Magmatic arc lull
5. Magmatic recharge & metamorphism, eclogite development
6. Internal shortening in thrust belt
7. Underthrusting of craton
8. Upper-crustal shortening
9. High-flux arc magmatism
10. Subsidence of hinterland bobber basins

11. Forearc normal faulting and collapse
12. Forearc tectonic erosion
13. Forearc underplating
14. Eclogite removal by delamination and RTI
15. Increased interplate coupling
16. Regional isostatic uplift of hinterland (<1 km)
17. Rapid forward propagation of retroarc thrust belt
18. Rapid forward migration of foreland flexural wave
19. Mafic magmatism associated with extension
20. Inversion of bobber basins

Figure 2. Diagram illustrating the cordilleran cycle hypothesis, revised from DeCelles et al. (2009). No time scale is implied, but studies in cordilleran magmatic arcs suggest a periodicity of 25–50 m.y. Numbers indicate individual processes listed at bottom. HFE—high-flux event; LAB—lithosphere-asthenosphere boundary; RTI—Rayleigh-Taylor–type instability.

Data sources
1. Trumbull et al., 2006
2. Kay & Coira, 2009
3. Haschke et al., 2002
4. Schoenbohm & Carrapa, this volume
5. Echavarria et al., 2003
6. Carrapa et al., 2011b
7. DeCelles et al., 2011
8. Oncken et al., 2006
9. Schoenbohm & Strecker, 2009
10. Carrapa & DeCelles, this vol.
11. Quade et al., this vol.
12. Canavan et al., 2014
13. DeCelles et al., this vol.
14. Roperch et al., 1999
15. Hartley et al., 2000; Clift &
    Hartley 2007
16. Arriagada et al., 2003
17. Pananont et al., 2004; Jordan
    et al., 2007
18. Jordan et al., 2010
19. Metcalf & Kapp, this vol.
20. Pardo-Casas & Molnar, 1987
21. Somoza, 1998
22. Sdrolias & Muller, 2006
23. Yanez et al., 2001

Arc high-flux event

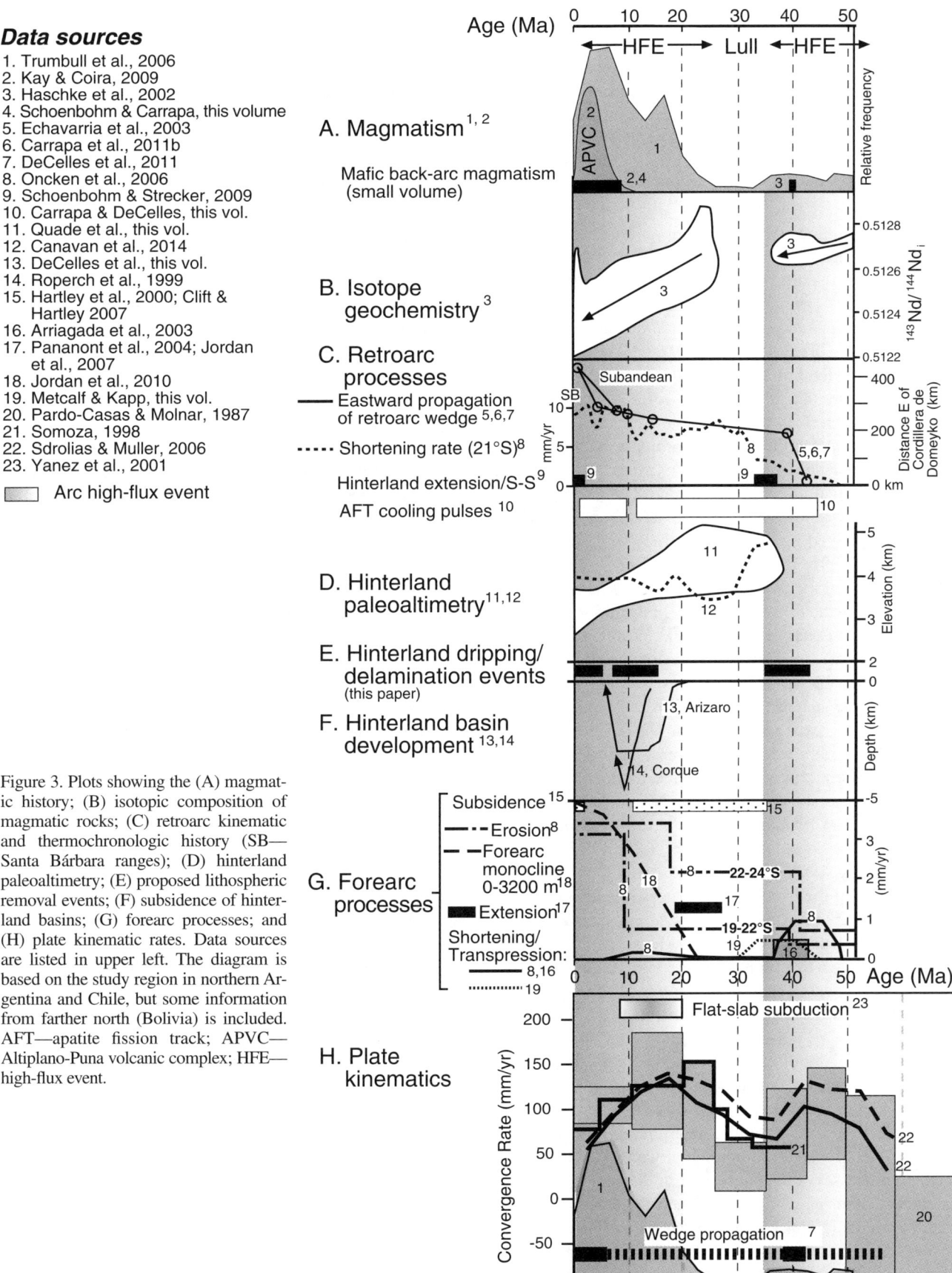

Figure 3. Plots showing the (A) magmatic history; (B) isotopic composition of magmatic rocks; (C) retroarc kinematic and thermochronologic history (SB—Santa Bárbara ranges); (D) hinterland paleoaltimetry; (E) proposed lithospheric removal events; (F) subsidence of hinterland basins; (G) forearc processes; and (H) plate kinematic rates. Data sources are listed in upper left. The diagram is based on the study region in northern Argentina and Chile, but some information from farther north (Bolivia) is included. AFT—apatite fission track; APVC—Altiplano-Puna volcanic complex; HFE—high-flux event.

al. (1999), Iaffaldano and Bunge (2008), and Meade and Conrad (2008) also suggested that a negative feedback relationship exists between growth of the Andean topographic edifice and plate convergence rate, with the Miocene deceleration of plate convergence owing partly to increased interplate coupling in response to stresses produced by the growing Andes (e.g., Richardson and Coblentz, 1994).

Isacks (1988) defined the region with >3 km average elevation between latitudes 15°S and 27°S as the central Andean Plateau, composed of the northern Altiplano and the southern Puna. The regional tectonic geology of the central Andes is clearly explained by numerous recent papers (e.g., Allmendinger et al., 1997; Oncken et al., 2006; Ramos, 2009), which together provide a basis for subdividing the central Andes into longitudinal tectonomorphic zones that include, from west to east: the forearc belt, consisting of the Coastal Cordillera, Longitudinal Valley, and Chilean Precordillera; the Western Cordillera, which consists of the Neogene to modern magmatic arc (Kay and Coira, 2009); the high Puna-Altiplano Plateau; the Eastern Cordillera, mainly consisting of Proterozoic–Paleozoic igneous and metasedimentary rocks incorporated into a bivergent (east and west) thrust belt (Grier et al., 1991; McQuarrie and DeCelles, 2001; Pearson et al., 2013); and the frontal Subandean and Santa Bárbara Ranges, which form the active frontal part of the thrust belt (Kley and Monaldi, 2002; Echavarria et al., 2003). To the east lies the Andean foreland basin system, which is dominated by fluvial-lacustrine deposystems (Cohen et al., this volume). Significant differences in plate subduction angle, magmatism, structural style, and shortening exist along strike in the Andes (Kley and Monaldi, 1998; Gutscher et al., 2000; Ramos, 2009; McGroder et al., this volume). In northwestern Argentina, an important inherited feature of the geology is the Cretaceous Salta Rift (Salfity and Marquillas, 1994), bounding faults of which were reactivated (inverted) during the development of the Eastern Cordillera and Santa Bárbara Ranges (Kley et al., 2005).

Early studies of the history of the Andes defined the Incaic and Quechua orogenic phases, during the late Eocene and Miocene, respectively (Steinmann, 1929; Mégard, 1984). More recent work has shown that orogenic activity in the central Andes during the Cenozoic has been more or less continuous in time but variable in spatial location and structural style (e.g., McQuarrie et al., 2005, 2008; Oncken et al., 2006; Pearson et al., 2013). Nevertheless, the record of deformation does show abrupt relocations of the orogenic strain front (approximated by the front of the thrust belt, with the understanding that minor layer-parallel shortening may extend beyond the actual limit of fault displacement; Mitra, 1994), which are correlated with other magmatic and tectonic processes. Total shortening in the central Andes is greater than in other parts of the Andes to the north and south (Kley and Monaldi, 1998; Oncken et al., 2006), but it remains poorly constrained at the upper ends of available estimates, mainly because shortened rocks in the western part of the central Andean Plateau are obscured by young volcanic and sedimentary cover. Kley and Monaldi (1998) and Oncken et al. (2006) com-

piled shortening estimates based on restored balanced regional cross sections. In Bolivia (north of ~21°S), estimates range up to 300–330 km (McQuarrie, 2002), whereas south of this latitude, in northwestern Argentina, shortening is less well constrained but generally agreed to be much less, between ~60 and 200 km (e.g., Grier et al., 1991; Pearson et al., 2013).

UPPER-MANTLE STRUCTURE AND PROCESSES

Broadband seismological studies over the past 15 yr have developed a complex picture of the upper mantle beneath the central Andes. Although details vary among the various groups of workers, geometric complexity and strong evidence for piecemeal removal of the lithosphere and lower crust are consistently imaged (e.g., Beck and Zandt, 2002; Schurr et al., 2006; Heit et al., 2008; Bianchi et al., 2013).

In this section, we review seismological constraints related to lithospheric foundering in the Puna region between 21°S and 27°S. In the southern Altiplano and northern Puna, several portable seismic deployments between 1994 and 1997 by the German GeoForschungsZentrum (GFZ) group imaged the lithosphere between 21°S and 24°S and were reviewed in Kay and Coira (2009) and Beck et al. (this volume). A large part of the upper mantle beneath the Puna shows low Vp, consistent with the removal of much of the lithospheric mantle. The tomography studies of the Altiplano and Puna show significant heterogeneities in the uppermost mantle but in general suggest that the upper mantle under the central part of the Altiplano is seismically faster compared to that beneath the Puna. Between 22°S and 24°S, Schurr et al. (2006) and Asch et al. (2006) identified ~100-km-scale high-velocity anomalies in the upper mantle close to the subducting Nazca slab, which they interpreted as delaminated blocks of continental lithosphere.

Two new tomographic studies from the recently completed U.S.-Argentine-German PUNA-PUDEL experiment (~25°S–28°S) provided some of the first images for this region. Bianchi et al. (2013) combined teleseismic traveltime data (International Seismological Center data) and regional earthquake data to image the upper mantle in the southern Puna region (25°S–28°S). They found overall low Vp values between 25°S and 27°S in the southeastern part of the Puna above the Nazca slab. Below Salar de Hombre Muerto (~25.5°S, 66.5°W), just north of Cerro Galán, Bianchi et al. (2013) identified an ~50-km-diameter fast anomaly at ~100 km depth that they interpreted as a delaminated block. Using surface wave tomography, Calixto et al. (2013) found a similar fast anomaly just north of Cerro Galán at 67°W at a depth of ~150 km.

Beck et al. (this volume) used two recently completed large-scale studies to document two different styles and scales of lithospheric removal in the Puna region of the central Andes. One is an ambient noise tomography (ANT) study that combined all the available broadband data in the central Andes producing a shear-wave (V_{sv}) velocity model for the crust down to ~50 km (Ward et al., 2013). The second is a teleseismic P-wave tomography study that combined data from many of the portable seismic deployments in the region to image the upper mantle (Scire et al., this volume).

Ward et al. (2013) used Rayleigh waves, so their study is sensitive to the absolute velocities of the vertically polarized S-wave, or V_{sv}. Results for three depth layers are shown in Figures 4A–4C for our study area. The ANT results show large variations in shear-wave velocities beneath the Puna. Across the entire region, calderas, ignimbrite volcanic fields (e.g., Kay and Coira, 2009), and surface deformation locations (Pritchard and Simons, 2004) (Fig. 4A) are underlain by a low-velocity zone at ~15 km depth. However, the crust at ~24°S associated with the Central Andean gravity high (Götze and Krause, 2002) is seismically fast, with shear-wave velocities of >3.6 km/s throughout the upper crust, suggesting that it is different than the crust to the north and south.

The largest anomaly in the shear-wave velocity model is a low-velocity body that is visible throughout the crust and coincides with the area of the Altiplano-Puna volcanic complex in the northern Puna region (Figs. 4A–4C). The Altiplano-Puna volcanic complex is the largest Neogene ignimbrite center in the Andes (de Silva et al., 2006). Based on new results in the upper crust (<30 km), the anomaly has a slightly elliptical plan-view shape ~200 km in diameter. In the lower crust (>30 km), the anomaly continues downward as a relative low, although the absolute velocity increases, perhaps reflecting increasingly mafic composition of the crust. By 40–45 km depth, the central portion of the anomaly is relatively fast, indicating that the main portion of the thermal anomaly in this region is located in the upper crust (Fig. 4C). Ward et al. (2013) suggested that the upper-crustal low-velocity anomaly is a magma mush zone related to ongoing silicic pluton formation, perhaps triggered by a lithospheric delamination event.

In the southern Puna region (~24°S–27°S), the crustal shear-wave velocities are generally higher and more variable with depth (Figs. 4A–4C). In the northern part of this region, a NW-SE alignment of relatively high crustal velocities underlies the Salar de Atacama Basin in the forearc and connects to a zone of high velocities centered under the Salar de Arizaro Basin. This high-velocity zone correlates closely to the Central Andean gravity high as first defined by Götze and Krause (2002) and is bounded on its northeastern edge by the El Toro fault zone (Riller et al., 2001). South of, and roughly parallel to, this high-velocity zone, there is a low-velocity zone that extends from the arc southeastward to the Cerro Galán volcanic field and beyond (Figs. 4A and 4B). In the upper crust, this low-velocity zone is significantly smaller compared to the one beneath the Altiplano-Puna volcanic complex. These velocity patterns are most prominent in the upper crust but extend into the lower crust with diminished amplitudes and changing shapes. In the lower crust (35–45 km depth), the low-velocity anomaly has a circular shape, centered between the Arizaro Basin and Cerro Galán (Fig. 4C). Beck et al. (this volume) suggest that this lower-crustal anomaly is the thermal and compositional signature of a recent lithospheric foundering event.

Scire et al. (this volume) used data collected from 284 seismic stations from 11 different networks operating between 1994 and 2009 for teleseismic P-wave tomography. P and PKIKP arrivals were picked (in four frequency bands) using a multichannel cross-correlation technique. A finite-frequency P-wave teleseismic inversion scheme was used to image the mantle from the base of the crust to 700 km depth. Despite the fact that not all the stations were deployed at the same time, overlapping coverage and the use of permanent stations provide sufficient corrections for datum offsets as shown by resolution tests and the continuity of the imaged subducted slab.

The resolution tests show that the slab and surrounding mantle are well imaged from ~150 km to 700 km, with some vertical smearing in the upper 100–150 km. Hence, a 4% high-velocity slab based on the geometry of the Slab1.0 model (Hayes et al., 2012) was prescribed for the slab in the starting model in order to sharpen the images in the upper 150 km. However, the major interpreted anomalies are present in both the "constrained slab" and "unconstrained slab" inversions. Figures 4D–4F show depth slices at 95 km, 130 km, and 165 km for the P-wave velocity model as percent perturbations from a starting model with the prescribed slab. The results resolve considerable P-wave velocity variations in the upper mantle above the slab beneath the Altiplano and Puna regions. Aside from the slab, the most prominent anomaly is an ~100-km-diameter, roughly circular high-velocity anomaly at ~25°S and 65.5°W that is surrounded by a broad halo of lower velocities. This anomaly is approximately cylindrical and similar in size and shape to the "Isabella anomaly" associated with the purported southern Sierra Nevada "delamination" (Zandt et al., 2004). Beck et al. (this volume) suggest that this anomaly represents mantle lithosphere recently delaminated from beneath the southern Puna and displaced eastward. Although parts of this anomaly were detected by investigators of the PUNA-PUDEL experiment (Bianchi et al., 2013; Calixto et al., 2013), much of the anomaly was on the edge of that deployment, and its precise lateral location and full size and shape were only resolved after combining data from previous deployments. Another high-velocity anomaly farther south (27.5°S) underlies the Sierras Pampeanas (also detected by Bianchi et al., 2013), where we might expect the presence of a thicker mantle lithosphere.

The shear-wave velocities in the crust from ANT, and P-wave velocity perturbations in the mantle reveal very different lithospheric structures along strike. At 22°S, there are very low shear-wave velocities in the upper and middle crust, consistent with the previously identified crustal magma/mush body. At this latitude, the upper mantle has generally low velocities and smaller variations in P-wave velocities across the mantle wedge. The seismically slowest mantle is located directly beneath the slowest crust. The area near 25°S that crosses the Arizaro Basin and the back-arc near Cerro Galán is different, with an upper-mantle, very low-velocity anomaly centered under the Puna region and a high-velocity anomaly to the east, as discussed already. Directly above the lowest velocity zone in the mantle, there is a circular low-velocity anomaly that lies between the arc to the west and

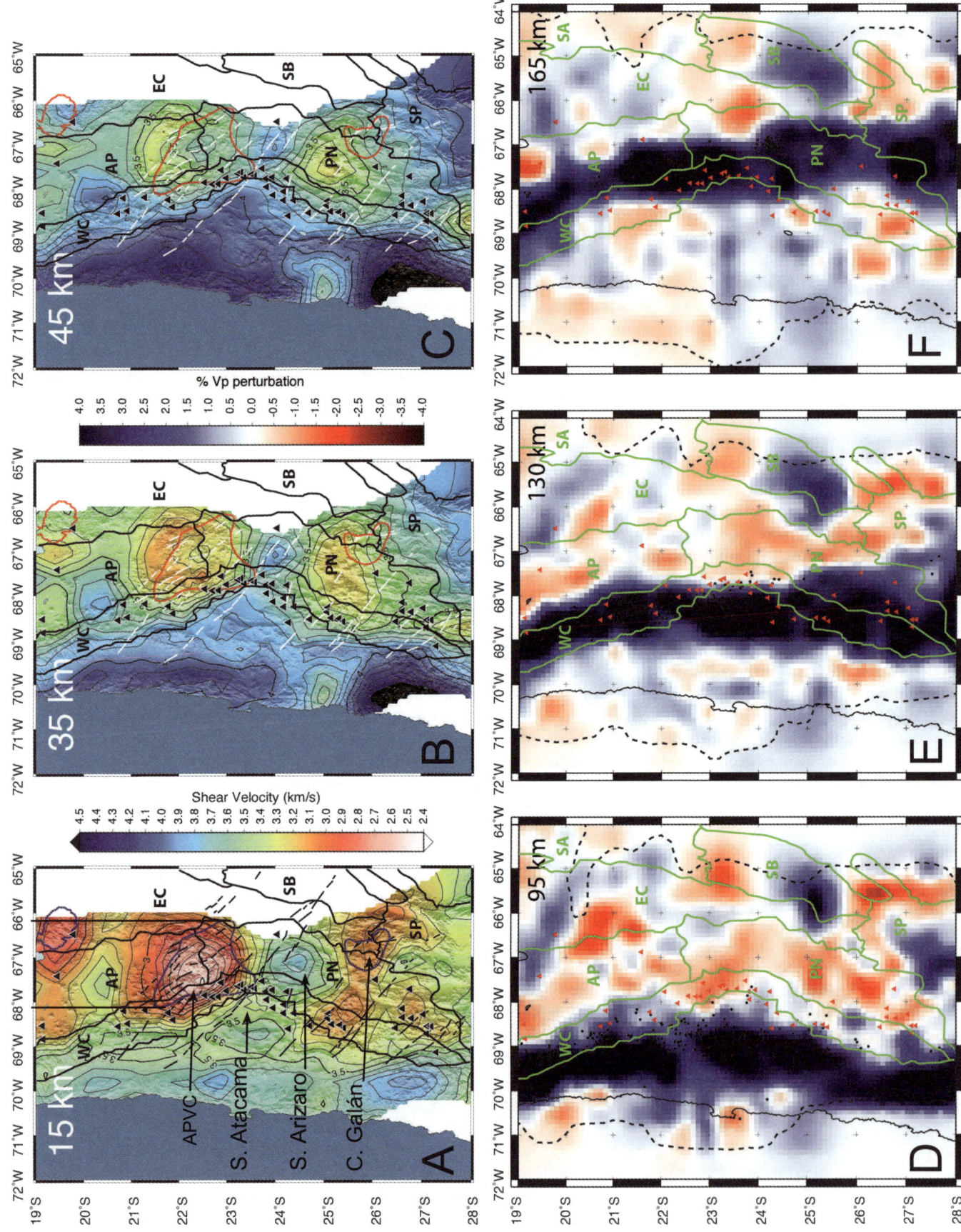

Cerro Galán to the east. The southernmost area at 27.5°S crosses the cordillera near the southern termination of the Central volcanic zone and the Puna and passes into the northern Sierras Pampeanas. The dip of the subducting slab is starting to flatten southward into the well-known Sierras Pampeanas flat-slab region. Here, we observe a persistently thick, low-velocity crust under the high topography, but a more "normal" higher-velocity crust under the lower average elevation of the block-faulted foreland. In the mantle, we observe low velocities under the arc and the previously described high-velocity anomaly directly beneath the crust of the Sierras Pampeanas.

Although seismic imaging can only delineate present-day structure, Beck et al. (this volume) hypothesize that the images in Figure 4 show the lithospheric structure at different stages of a piecemeal (~200-km-scale areas) delamination process that progressed in a nonuniform fashion from north to south. They suggest that the two low-velocity zones in the lower crust mark locations of two main delamination centers under the Puna: one under the Altiplano-Puna volcanic complex and the other between the Arizaro Basin and Cerro Galán. The southernmost area (south of ~27°S) shows the structure prior to or at an incipient delamination time (pre-delamination stage); the southern Puna area (24°S–27°S) shows the structure associated with recent or ongoing removal; and the northern Puna area (21°S–24°S) shows the evolved structure after some time has passed since delamination. A similar argument, based on the timing and composition of volcanic rocks, was presented by Schnurr et al. (2007). At face value, a comparison of the width of the high cordillera and the elevation difference between 27.5°S and 25°S suggests that surface uplift associated with removal of lithospheric mantle combined with any crustal thickening that occurred during or prior to the removal event can account for an ~250 km increase in the width of the high plateau and ~2 km of surface uplift, roughly consistent with modeling predictions (Krystopowicz and Currie,

---

Figure 4. Map views of shear-wave velocity for the crust and P-wave velocity perturbations for the upper mantle beneath the study area. The crustal images are from the ambient noise tomography study (Ward et al., 2013) and show absolute Vsv velocity values using the shear-wave color palette and velocity contours in thin black lines for depths of 15 km (A), 35 km (B), and 45 km (C). Also plotted on these panels are: NW-SE strike-slip faults (black lines in A, and white lines in B and C); Holocene volcanic centers (black triangles), and major ignimbrite regions (purple lines on A and red lines in B and C; Altiplano-Puna volcanic complex and Cerro Galán); and physiographic provinces in thicker black lines with labels (WC—Western Cordillera; AP—Altiplano; EC—Eastern Cordillera; PN—Puna Plateau; SB—Santa Bárbara Ranges; SP—Sierra Pampeanas). Teleseismic P-wave velocity model (from Scire et al., this volume) is plotted as % Vp perturbations from a starting model with a prescribed slab shown for depth slices of 95 km (D), 130 km (E), and 165 km (F). Dashed black lines enclose region of good resolution (for details, see Scire et al., this volume). Holocene volcanoes are plotted as red triangles. Physiographic provinces are plotted in green lines with labels same as above (SA—Subandes). APVC—Altiplano-Puna volcanic complex.

2013). In the northernmost area where the Altiplano-Puna volcanic complex is located, the region of high elevations is somewhat wider, and relief is more subdued. We attribute this to thermal relaxation of the heat input following the delamination process. Heat is progressively rising into the crust from the mantle, causing crustal melting and crustal flow, widening and flattening the high topography and leaving a cooling mantle (figure 7 *in* Beck et al., this volume). However, we caution that the link between lithospheric foundering and surface topography and uplift is not always straightforward because it depends on the density and rheology of the crust and mantle lithosphere. The pattern of surface uplift varies with the type of lithospheric delamination and changes through time as the delamination event progresses. Hence, surface uplift is affected by many factors that can lead to a wide range of uplift magnitudes and histories (Krystopowicz and Currie, 2013; Wang et al., this volume).

The other key observation from the seismic data is the scale of the removal process. The data suggest that the crust and upper mantle beneath the hinterland of the central Andes have experienced and continue to experience delamination/dripping events that are on at least two distinct spatial scales: >200 km in diameter and ≤100 km. This is supported by independent geological data that indicate variable scales, spatial heterogeneity, and temporal variability of dripping events beneath the northern and southern Puna (Schnurr et al., 2007; Schoenbohm and Carrapa, this volume; DeCelles et al., this volume). This in turn suggests that geodynamic processes in the crust that are related to the removal events (e.g., thrust wedge kinematics and surface elevation changes) should be manifest at similar scales. For example, the kinematic history of the frontal thrust belt might be expected to vary along strike at similar scales. Unfortunately, the seismic results do not provide a time scale for the processes involved in delamination/dripping. Nevertheless, the magmatic history of the region, including the Altiplano-Puna volcanic complex and Cerro Galán, suggests that the time scale is on the order of 5–10 m.y. Geodynamic modeling (Göğüş and Pysklywec, 2008; Krystopowicz and Currie, 2013; Currie et al., this volume; Wang et al., this volume) supports this time scale, and together with the spatial scales indicated by the seismology, implies that the granularity of processes related to mass removal in the orogenic belt is on the scale of 100–200 km and 5–10 m.y.

MAGMATIC HISTORY

The western margin of South America contains a record of Neoproterozoic rifting that marks the breakup of Rodinia, followed by a passive-margin history until subduction initiation (Bahlburg and Hervé, 1997; Ramos, 2008, and references therein). Ancient arc-related magmatic products have been present within the modern central Andes since the Cambrian; the Pampean-Famatinian arcs are regionally extensive belts that developed on the former passive margin and crop out within the Altiplano-Puna Plateau as well as basement uplifts in the Sierras Pampeanas to the south (Lucassen et al., 2000, 2011; Ducea et al., this volume). Magmatic episodes of lesser significance are documented intermittently throughout the Paleozoic (Bahlburg and Hervé, 1997; Lucassen et al., 2000, 2011; Einhorn et al., this volume).

However, Andean subduction magmatism sensu lato is defined as Mesozoic and younger in age (Haschke et al., 2002). The majority of ancient arc-related products in the central Andes are located within the modern forearc and as framework rocks to the modern arc of the Western Cordillera. Older arcs occupy lithologic domains characterized by steep to vertical foliations, and, as such, they represent a deformed but comprehensive archive of arc magmatism along the South American margin. They have been brought into their present position structurally via convective overturn and reworking within the younger arc (Babeyko et al., 2002, 2006), arc-continent collision or accretion (Ramos, 2009), or by thrust faulting (DeCelles et al., 2009). The majority of earlier magmatic products (Triassic and Jurassic) are mafic arcs and back-arc centers, which are located within the Western Cordillera in Chile and westernmost Argentina (Oliveros et al., 2007; Rossel et al., 2013) and may have originated west of South America only to be subsequently accreted to the continent.

In addition to magmatic products described previously, arc-related magmatism extended into the Andean interior, within the modern Altiplano-Puna Plateau and, rarely, even farther eastward since the Late Cretaceous (Galliski and Viramonte, 1988), in a manner similar to North American Cordilleran interior magmatism (Barton, 1996). Magmatism was particularly widespread in the Andean interior during the late Miocene and Pliocene, leading to the development of some of the largest intermediate composition volcanoes on the planet (de Silva, 1989). Moreover, the chemical compositions of Andean interior magmas are very much like all Andean subduction-related magmatic rocks. This leads to some confusion as to where the arc is located or was located in recent times—Is it just the Western Cordillera, thus making all Altiplano-Puna magmatism of the back-arc type (Kay et al., 1994), or is the Western Cordillera the location of just the "frontal" arc, the first line of calc-alkaline volcanic rocks of a wider arc? We use the second definition here, where all Cenozoic calc-alkaline rocks in the central Andes represent collectively the archive of arc magmatism.

Trumbull et al. (2006) and Haschke et al. (2002, 2006) compiled data for the timing and composition of volcanic rocks in the central Andes (Fig. 3A). Together these data sets show spatial and temporal cyclicity and provide the strongest evidence that tectonic and magmatic processes in the central Andes are operating on a predictable schedule (e.g., Ducea, 2001; DeCelles et al., 2009). Post–200 Ma magmatic activity has occurred during four major episodes in progressively eastward-migrating arc trends. We suggest that these four magmatic episodes represent high-flux events in the Andean magmatic arc, as defined in Ducea (2001) and Ducea and Barton (2007). As noted by Ducea (2001), high-flux events in continental arcs are characterized by relatively brief periods (10–15 m.y.) during which magmatic flux rises up to about an order of magnitude greater than normal

background levels (e.g., Armstrong, 1988; Ghosh, 1995; Barton, 1996; Gaschnig et al., 2010). The central Andean record is also punctuated by short-lived, low-volume mafic magmatic activity during the latter stages of and immediately following the high-flux events (e.g., Lamb and Hoke, 1997; Haschke et al., 2006; Kay and Coira, 2009; Schoenbohm and Carrapa, this volume; Murray et al., this volume). Magmatic arc high-flux events characteristically exhibit increased crustal and upper-plate lithospheric contributions relative to the lulls in magmatism between the high-flux events. Haschke et al. (2002) documented ratios of La/Yb, $^{87}Sr/^{86}Sr_i$, and $^{143}Nd/^{144}Nd_i$ that show increasingly evolved compositions during each of the four high-flux events, and returns to more primitive compositions between high-flux events (Figs. 3A and 3B; e.g., Hoke and Lamb, 2007). Moreover, the long-term (since ca. 200 Ma) isotopic and trace-element trends suggest that magma sources have become more evolved through time in the central Andes, perhaps in response to melting of older, more inboard South American crust through time. These types of trends also are evident in all major North American batholiths over the course of their >100 m.y. lifetime (Ducea, 2001; Gehrels et al., 2009; Paterson et al., 2011; Girardi et al., 2012; Gaschnig et al., 2013; Barth et al., 2013). We focus here on the two most recent high-flux events, which occurred during latest Cretaceous–Eocene and Miocene–Pliocene times (Fig. 3A), mainly because this is the time frame during which most of the Andean orogenic edifice was constructed, and also because the regional tectonic and kinematic picture becomes increasingly less clear prior to the Cenozoic record.

Since the late Oligocene, magmatism has been active both along the Western Cordillera frontal arc and within the Altiplano-Puna Plateau (Allmendinger et al., 1997; Trumbull et al., 2006; Hoke and Lamb, 2007; Schnurr et al., 2007; Kay and Coira, 2009). Miocene to Holocene volcanic rocks ranging from basalts (Kay et al., 1994; Drew et al., 2009; Ducea et al., 2013) to dacites and rhyolites are preserved within the Altiplano-Puna Plateau. The Miocene–Pliocene Altiplano-Puna volcanic complex (Fig. 1; de Silva, 1989) is the largest and best-known example of the extension of arc magmatism into the Andean interior. The Altiplano-Puna volcanic complex became active ca. 11–10 Ma, with greatest eruptive activity concentrated in the interval 8.4–4 Ma and tapering off until <1 Ma (e.g., de Silva et al., 2006; Kay and Coira, 2009; Kay et al., 2010; Salisbury et al., 2011). No obvious spatio-temporal patterns exist in the Altiplano-Puna volcanic complex record (Kay and Coira, 2009; Salisbury et al., 2011), nor are there clear relationships with changes in plate kinematic parameters (e.g., Fig. 3H). Flat-slab subduction under the central Andes, in part due to the subduction of the Juan Fernandez Ridge (Yáñez et al., 2001), could have widened the area of slab dehydration and dehydration melting in the mantle above the slab, thus widening the arc beneath the orogen (Kay and Coira, 2009). The Altiplano-Puna volcanic complex consists of a calc-alkaline suite of volcanic rocks that are scattered on the Altiplano-Puna Plateau instead of being focused like the products of the frontal arc; they represent mixtures of mantle wedge–derived magmas and local crustal materials, typical of cordilleran arcs (Ort et al., 1996; Lindsay et al., 2001; Kay et al., 2010). Because the Altiplano-Puna volcanic complex and related rocks are scattered and not focused under stratovolcanic edifices, wider sets of compositions are exposed.

It is also clear that the development of the Altiplano-Puna volcanic complex and other magmatic products on the Altiplano-Puna Plateau since the latest Oligocene coincides with overall widening of the Andean orogen (see following sections). The Altiplano-Puna volcanic complex was produced by a high-flux event that peaked during latest Miocene–Pliocene time. The trigger for this high-flux event is unclear: Some argue that it resulted from a lithosphere delamination or dripping event (de Silva et al., 2006; Schnurr et al., 2007; Kay and Coira, 2009; Kay et al., 2010), but it is also plausible that it simply represents a flare-up driven by crustal thickening and partial melting of lithosphere that was heated by the influx of basaltic melts from the underlying mantle. Seismological considerations discussed earlier herein and in Beck et al. (this volume) suggest that the Altiplano-Puna volcanic complex region has experienced recent lower-crust and upper-mantle delamination/dripping.

A similar feature to the Altiplano-Puna volcanic complex is the volcanic complex centered around Cerro Galán in the southern Puna region (Kay and Coira, 2009). Here, magmatism appears to have peaked during the Pleistocene (ca. 2 Ma). The significance of the mid-Miocene and Pleistocene high-flux events is unresolved. However, all magmatism on the Altiplano-Puna Plateau appears to be confined temporally to one cordilleran tectono-magmatic cycle. At this temporal scale, no clear patterns of magmatic spatial migration and/or changes in compositions are evident.

Mafic low-volume eruptions are noteworthy during the late Eocene, latest Oligocene–early Miocene, and again after ca. 8 Ma. Kay and Coira (2009) also highlighted the potential importance of partial lithosphere/crust removal by delamination and the possible effect of periods of flat-slab subduction, and Hoke and Lamb (2007) presented He isotope data that indicate widespread mantle contributions to the behind-arc magmatism in the mafic lavas of the Altiplano. Schoenbohm and Carrapa (this volume) document the close association between small-slip normal faults and mafic magmatism and present arguments that support the idea that both are related to local removal of lithosphere.

The role of trench and forearc rocks in influencing arc magmatism is unresolved. This is a general problem for cordilleran magmatic arc systems because trench and forearc materials are derived from older arcs. Two mechanisms may be responsible for delivering materials from the forearc side of the orogen into the source region of magmatism: (1) steady-state sediment subduction (Cloos and Shreve, 1988), followed by upwelling into the mantle wedge during dehydration reactions (Plank, 2005), diapiric uprising, or "relamination" (Behn et al., 2011); or (2) structural burial of forearc material via tectonic underplating (Platt, 1986; Stern, 2004). The first set of mechanisms cannot supply more than ~5% of the required "crustal" signature in terms of

major elements during high-flux events, even if all available sediment is transferred into arc magmas (Ducea and Barton, 2007). Because of this, "normal" steady-state subducted sediment has no significance for triggering and sustaining high-flux event magmatism. Thickening of arc crust by tectonic underplating and/or addition of upper-crustal materials to the bottom of arc crust (which takes place during shallow/flat subduction events) can, in principle, be responsible for generating intermediate to silicic arc magmas during high-flux events in conjunction with baseline melting of the mantle wedge. However, observations in the North and South American Cordilleras show that these flat-slab events typically shut down magmatism in the main arc, and the majority of underplated sections unroofed by subsequent tectonic exhumation are essentially melt-free (e.g., Kidder and Ducea, 2006). Following the argument presented in Ducea and Barton (2007) for a positive shift in ε_{Nd} values during episodes of subduction erosion and underplating, the majority of the Altiplano-Puna rocks lack the isotopic characteristics required by a forearc or trench origin. Trench-side rocks are a mixture of a primitive mid-ocean-ridge basalt (MORB)–like component with a component as "evolved" as the nearby baseline arc. Such a mixture cannot be more "evolved" than the baseline arc. Moreover, the arc constitutes a topographic barrier to transport of sediments from the continental interior to the trench. Forearc sediments are typically dominated by arc provenance, with lesser additions from the accretionary prism (Ingersoll, 1983). These basic observations and inferences from modern and ancient cordilleran arc regions suggest that although subduction erosion is important along many cordilleran margins, trench/forearc rocks do not play a significant role in high-flux event magmatism.

CENOZOIC KINEMATIC HISTORY OF THE FOREARC OROGENIC WEDGE

From west to east, the modern northern Chilean forearc consists of the trench, Coastal Cordillera, Longitudinal Valley (in the north)–Central Depression (in the south), and the Precordillera, which rises eastward to the axis of the modern arc front in the Western Cordillera (Fig. 1). North of 21°S, the Precordillera corresponds to the western topographic escarpment of the Andes (or Western Andean slope). Between 21°S and 25°S, the physiography is more complex. Here, the Precordillera consists of series of north-south–trending ranges (Sierra de Moreno in the north and the Cordillera de Domeyko in the south) that are separated from the Western Cordillera by intermontane basins. The largest of these basins is occupied by the Salar de Atacama. Its elevation (~2.3 km) is anomalously low, given that it is underlain by some of the thickest crust in the Andes (>60 km; Yuan et al., 2002; Schurr and Rietbrock, 2004). The axes of the Western Andean slope and Western Cordillera are deflected inboard around the Salar de Atacama, such that the distance between the trench and arc front at this latitude is anomalously large, compared to the typical arc-trench distance of ~280 km elsewhere along strike in the central Andes.

As a consequence of long-lived subduction erosion along the western margin of the central Andes, the longitudinal position of Jurassic to Cenozoic arc magmatism has shifted hundreds of kilometers inboard (Scheuber and Reutter, 1992; von Heune et al., 1999; Stern, 2004). The result is that the inner parts of the forearc underwent a transition from an initial back-arc setting to an intra-arc setting prior to final incorporation into a forearc setting with the initiation of magmatism in the modern Western Cordillera at ca. 26 Ma. The history of this inboard sweep in magmatism, and associated deformation and basin development, is briefly summarized here.

During Jurassic–Early Cretaceous time, a continental magmatic arc was established in the modern position of the Coastal Cordillera. Subduction of the Phoenix oceanic plate at this time was highly oblique (Jaillard et al., 1990; Seton et al., 2012) and resulted in intra-arc sinistral transtension within the >1000-km-long, arc-parallel Atacama fault zone (Buddin et al., 1993; Taylor et al., 1998; Scheuber and Gonzalez, 1999). The arc was associated with a generally shallow-marine, back-arc transtensional basin, which is now exposed along the modern Central Depression.

During Early Cretaceous time, magmatism migrated ~100 km inboard such that the main axis of mid-Cretaceous magmatism overprinted the former back-arc basin within the Central Depression. This time marks the establishment of a highly erosive (tectonically) margin, concomitant with the opening of the South Atlantic Ocean (Coney and Evenchick, 1994). Intra- to back-arc strata of Early to mid-Cretaceous age are scarce, and the structural setting (contractional vs. extensional) is poorly constrained. The arc continued to migrate inboard during the Late Cretaceous at a rate of ~1.5 mm/yr. The modern Coastal Cordillera was placed into a forearc position, and the Atacama fault zone was reactivated by sinistral transpression (Taylor et al., 1998). Inboard of the Late Cretaceous arc, nonmarine strata of the Upper Cretaceous Purilactis Formation were deposited. It is debated whether the Purilactis Formation represents early contractional retroarc foreland basin deposits (Mpodozis et al., 2005; Arriagada et al., 2006) or extensional back-arc basin deposits (Hartley et al., 1992). Farther east, the Salta Rift developed during Early–Late Cretaceous time, either as part of the back-arc extensional system or as part of the opening South Atlantic Ocean system (Salfity and Marquillas, 1994).

During the Eocene–Oligocene, arc magmatism increased. The detrital record in the Sierra de Moreno (Umlauf, 2011) reveals a peak in Cenozoic zircon ages between 50 and 40 Ma, which we interpret as the detrital signal of a high-flux event (Fig. 3A). At this time, the locus of magmatism jumped inboard into the modern position of the Precordillera, which could be explained by a phase of accelerated subduction erosion (Fig. 3G). The intra-arc Precordillera experienced transpression, crustal thickening, rapid exhumation of 4–5 km, and syncontractional intramontane basin development (Maksaev and Zentilli, 1999; Mpodozis et al., 2005; Arriagada et al., 2006). Oncken et al. (2006) noted increased rates of forearc erosion in regions north of our study area at 40 ± 2 Ma,

and increased forearc shortening between ca. 48 and 37 Ma (Fig. 3G). Thermochronologic studies in the Sierra de Moreno reveal an episode of rapid exhumation between 35 and 30 Ma (Maksaev and Zentilli, 1999; Reiners et al., this volume), which we interpret to be the consequence of accelerated transpressional deformation. The Cordillera de Domeyko to the south underwent rapid exhumation slightly earlier, at ca. 45–40 Ma (Reiners et al., this volume). Overall, the resolved history of a high-flux event at 50–40 Ma, accelerated subduction erosion at ca. 40–36 Ma, and accelerated deformation and exhumation in the 45–30 Ma time interval are broadly consistent with predictions of the cordilleran cycle model (DeCelles et al., 2009).

By late Oligocene to early Miocene time, the arc axis had migrated inboard to near its modern position in the Western Cordillera, leaving the Precordillera in a forearc position. An exception is inboard of the Salar de Atacama, where there is an apparent gap in arc volcanism until ca. 15 Ma (Kay et al., 1999; Victor et al., 2004). We speculate that the Precordillera between 21°S and 25°S may have been at a similar or even higher elevation than regions to the east during Eocene–Oligocene time, when it was in an intra-arc position (in a purely relative sense; we refrain from guessing the magnitude of elevation). This is based on the observation that the highest elevations in modern ocean-continent convergent margins are found along the axes of their associated active arc fronts. It follows then, that the surface elevation of the Precordillera may have decreased in response to subduction refrigeration and thermal subsidence as it was transferred into the forearc. This could explain widespread subsidence and accumulation of eastward-derived basin fill above an unconformity within the Precordillera beginning at 35–25 Ma (Hartley et al., 2000; Hartley and Evenstar, 2010), coeval with early activity on west-vergent reverse faults exposed in incised canyons along the Western Andean slope (north of 21°S; Wörner et al., 2002; Victor et al., 2004). Deformation within the Precordillera during late Oligocene–early Miocene time is also enigmatic. Interpretations of seismic-reflection profiles suggest that the Salar de Atacama was downdropped relative to the Precordillera along an Oligocene–Miocene east-dipping normal fault (Pananont et al., 2004; Jordan et al., 2007). Coeval transtensional faults and associated basins have also been documented to the north near Calama (Tomlinson et al., 2001; Jordan et al., 2006). A speculative explanation for this extension within the inner forearc is that it marks gravity-driven topographic collapse of the Precordillera as it was transferred into the forearc.

The structural and surface uplift history of the Western Andean slope has been a major focus of research. Slip on west-vergent thrust faults is documented within the Precordillera north of 21°S between 30 and 5 Ma (Muñoz and Charrier, 1996; Wörner et al., 2002; Victor et al., 2004; Farías et al., 2005), but there is compelling evidence that these faults did not generate all of the presently observed relief across this topographic escarpment. Analysis of stream profiles and stratigraphic surfaces of known age suggest that ~1 km of long-wavelength monoclinal relief has been generated across the Western Andean slope north of ~21°S and in the south between 26°S and 28°S since 10–11 Ma (Hoke et al., 2007; Schildgen et al., 2007; Riquelme et al., 2007; Jordan et al., 2010; Fig. 3G). Interestingly, the magnitude of relief generation along the Western Andean slope separating the northern Puna and the Salar de Atacama is estimated to be twice as much over the same time interval (Jordan et al., 2010). This larger magnitude of relief generation may be related to enhanced magmatic activity and addition to the crust within the southern Altiplano and northern Puna since ca. 10 Ma (as represented by ignimbrites of the Altiplano-Puna volcanic complex and the seismically imaged midcrustal Altiplano-Puna magma body; Fig. 4A) and associated heating of the eastern Atacama block as suggested by Jordan et al. (2010). We add that lithospheric removal beneath the Puna (and possibly concomitant lithospheric root development beneath the Salar de Atacama) may also have contributed to the enhanced generation of relief along this latitude of the Western Andean slope through their associated isostatic and thermal consequences.

The central Andean forearc between 21°S and 25°S is anomalous in other regards. Surface elevations of the Coastal Cordillera and Central Depression are greatest at this latitude. Whereas no thrust faults younger than 5 Ma have been documented within the Precordillera to the north, active north-south–striking thrust faults bound mountain ranges of the Sierra de Moreno and Cordillera de Domeyko (Fig. 1; Jolley et al., 1990; Buddin et al., 1993). Pliocene–Quaternary north-south–striking normal faults and surface uplift are also most pronounced in the Coastal Cordillera at this latitude (e.g., Armijo and Thiele, 1990; Riquelme et al., 2003). Geodynamically, the forearc can be assessed as a single or composite (inner and outer), critically tapered frictional wedge(s), and the generation of coeval inner forearc compression and outer forearc extension can be explained in this context (Lallemand et al., 1994; Adam and Reuther, 2000). In particular, extension and uplift within the Coastal Cordillera are well explained by frontal erosion of the outermost forearc concomitant with underplating beneath the coastal portion of the outer forearc. Outer forearc uplift and extension may also be amplified, however, by other factors (e.g., Delouis et al., 1998; Loveless et al., 2010; Metcalf and Kapp, this volume), such as coupling at the subduction interface and resultant bending of the upper plate, elastic rebound during the earthquake cycle, the presence/absence of preexisting structures favorable for reactivation, negative buoyancy beneath the Salar de Atacama, lithospheric root removal beneath the hinterland, and even subtle changes in slab dip angle through time. From our perspective, future work (ideally a combination of geodynamic modeling and high-resolution geological work) is needed to better understand the relative roles of the various mechanisms that may be driving outer forearc extension and inner forearc contraction in northern Chile.

Returning to the issue of how the orogenic cycle may modulate the dynamics of the central Andean forearc wedge, we are left with more questions than answers. An acceleration of inboard arc migration at ca. 40 Ma corresponds to the waning stage of an apparent magmatic high-flux event. This could be interpreted to mark an accelerated phase of subduction erosion driven by root removal in the hinterland (Fig. 3G) or flattening of the subducting slab. It is tempting to attribute the anomalous Miocene to Holocene tilting of the Western Andean slope and prominent inner forearc shortening and outer forearc extension at the latitude of the Salar de Atacama to the most recent phase of root removal beneath the Puna. However, the preexisting heterogeneities of the Atacama fault zone and cold, dense lithosphere of the Atacama block at this latitude preclude an unambiguous interpretation (Metcalf and Kapp, this volume). The central Andean forearc shows distinct phases of deformation, subsidence, and uplift since the Oligocene. Available evidence, however, suggests that this history was relatively consistent regionally along strike in the central Andean forearc (e.g., Clift and Hartley, 2007; Hartley and Evenstar, 2010), challenging the notion that forearc dynamics are modulated by a process like lithospheric root removal that operates at the scale of ~200 km. At the same time, it is arguable that our understanding of the geological evolution of the Andean forearc is still in its infancy, and we anticipate that the cordilleran cycle concept will stimulate additional research and add a hinterland perspective to the subduction-interface–dominated view about forearc dynamics. Perhaps a better setting to test for cyclicity is in accretionary forearcs, where sedimentary archives of forearc wedge behavior are more continuous in the stratigraphic record.

CENOZOIC KINEMATIC HISTORY OF THE RETROARC OROGENIC WEDGE

The central Andean retroarc region contains a major thrust belt that continues more than 2000 km along strike from Peru to west-central Argentina (Allmendinger et al., 1997; Kley and Monaldi, 1998; Ramos, 2009; McGroder et al., this volume). In northwestern Argentina (Fig. 1), this thrust belt consists of three domains characterized by distinct structural styles: north of ~23°S, the Subandean fold-and-thrust belt consists of eastward-verging thin-skinned thrust sheets that continue northward into the Bolivian Subandean zone (Echavarria et al., 2003). South of this thin-skinned belt, the Paleozoic stratigraphic section thins abruptly, and the structural style in the frontal part of the thrust belt changes to thick-skinned thrusting in the Santa Bárbara belt, with deeper-level detachments feeding slip into eastward-dipping back thrusts at the surface (Grier et al., 1991; Cristallini et al., 1997; Kley and Monaldi, 2002; Pearson et al., 2013; McGroder et al., this volume). West of the frontal Subandean–Santa Bárbara belts lies the Eastern Cordillera, which consists of Paleozoic through Cretaceous sedimentary rocks and Upper Proterozoic low-grade metasedimentary rocks, along with early Paleozoic granitoid rocks. Thrust faults in the Eastern Cordillera tend to be somewhat steeper than those in the frontal thrust belt, and both east- and west-verging faults are present (e.g., Blasco et al., 1996). Most interpretations of the structure beneath the Eastern Cordillera call for a regional detachment between 20 and

10 km depth (Grier et al., 1991; Cristallini et al., 1997; Kley and Monaldi, 2002; Pearson et al., 2013; Kortyna, 2013), and many interpretations attribute the abundance of west-verging thrusts to reactivation of Salta Rift–related faults (Cristallini et al., 1997; Kley et al., 2005; Carrera and Muñoz, 2008; Kortyna, 2013).

The kinematic history of the central Andean retroarc region is based on thermochronological, structural, and basin evolution studies. In select cases, the thermochronological data may be reasonably inferred to record the timing of exhumation of structures during episodes of thrust-driven rock uplift (Coutand et al., 2006; Barnes et al., 2008; McQuarrie et al., 2008; Carrapa et al., 2011a). Accurately dated growth structures formed in proximal alluvial deposits associated with coeval-growing contractional structures also provide information about the location of the orogenic strain front through time. Finally, the passage of the regional-scale retroarc foreland basin system through the central Andes provides an independent means of tracking the strain front through time (Starck and Vergani, 1996; Horton and DeCelles, 1997; DeCelles and Horton, 2003; DeCelles et al., 2011; Carlotto, 2013; Engelder and Pelletier, this volume).

Drawing upon previously published and new data sets from northwestern Argentina, Carrapa and DeCelles (this volume) and Reiners et al. (this volume) synthesize the history of exhumation based on low-temperature thermochronology, mainly apatite fission-track (AFT) and apatite (U-Th)/He dating. Together, these thermochronometers document the cooling history of rocks over a temperature range of ~120–40 °C (Gleadow and Fitzgerald, 1987; Wolf et al., 1998). Given a normal to slightly elevated geothermal gradient, these data record the exhumation history of rocks within ~6 km of the surface. In the case of the central Andes, the level of exhumation is shallow enough that higher-temperature thermochronometers were generally not reset during the Cenozoic (e.g., Carrapa et al., 2009). The shallow level of exhumation in the central Andes is characteristic of an orogenic system that has not been subjected to high rates of erosion, probably because easterly moisture sources have been blocked by the high frontal ranges, effectively starving the hinterland region of moisture and inhibiting erosion and transport of sediment out of the highlands (e.g., Sobel et al., 2003; Strecker et al., 2007; Quade et al., this volume).

Carrapa et al. (2011a) and Carrapa and DeCelles (this volume) demonstrate that AFT data show an unsteady eastward progression of syntectonic exhumation, which they interpret to reflect the migration of the orogenic front. The analyses of Strecker et al. (2007) and Reiners et al. (this volume) support this concept, insofar as almost all of the rapid exhumation in the central Andes is localized to the frontal part of the orogenic wedge owing to rain out of moisture borne on easterly winds from the Atlantic Ocean. The locus of rapid exhumation shows two major eastward relocations: from the Chilean Precordillera (Cordillera de Domeyko) into the eastern Puna region during the late Eocene, and thence into the frontal Santa Bárbara Ranges during the latest Miocene–Pliocene (Fig. 3C). These events are traceable northward along strike into Bolivia and represent episodes of sig-

nificant eastward propagation of the Andean orogenic wedge (see following). Rapid exhumation also took place during the interval ca. 25–10 Ma but was confined to a relatively narrow region in the Eastern Cordillera. Carrapa et al. (2011a) interpreted this pattern to represent a period of slow forward (eastward) propagation of the orogenic strain front. Pearson et al. (2013) reported apatite (U-Th)/He thermochronological dates from rocks in the Eastern Cordillera along a profile at latitude 24.5°S–25°S, which they interpreted to indicate an unsteady forward and then backward breaking succession of thrust faulting between ca. 15 and 4 Ma (Fig. 5). On the other hand, Kortyna (2013) reported apatite (U-Th)/He data from along strike to the south, in the Tonco-Amblayo part of the Eastern Cordillera (25°S–26°S), that suggest an eastward sweep of thrust activity from ca. 12 Ma to 4 Ma. In any case, although the sequence of thrusting in the Eastern Cordillera is demonstrably complex at spatial scales of a few tens of kilometers, the overall long-term pattern was an eastward migration of deformation from the eastern Puna, through the Eastern Cordillera, into the Santa Bárbara Ranges from late Eocene to Pliocene time (Fig. 5).

Another source of information regarding the regional kinematic history of the central Andean orogenic wedge is the age and location of coarse-grained proximal deposits that record growth of nearby contractional structures (e.g., Vergani and Stark, 1989; Monaldi et al., 1996). Such growth structures have been reported in Upper Eocene alluvial-fan and fluvial deposits of the eastern Puna region (Carrapa and DeCelles, 2008) and Eastern Cordillera (Hongn et al., 2007; Bosio et al., 2009); the Middle to Upper Miocene Angastaco Formation (Carrera and Muñoz, 2008; Carrapa et al., 2011b), Upper Miocene Agujas Conglomerate (Mazzuoli et al., 2008; DeCelles et al., 2011), and the Pliocene–Pleistocene Piquete Formation (Carrera and Muñoz, 2008) of the central Eastern Cordillera; the Upper Miocene Pisungo Formation in the eastern part of the Eastern Cordillera (Siks and Horton, 2011); in Pleistocene conglomeratic facies in the Eastern Cordillera and near the western boundary of the Santa Bárbara system (Monaldi et al., 1996); and in uppermost Miocene–Pliocene deposits in the frontal Subandes of northernmost Argentina (Echavarria et al., 2003) and adjacent Bolivia (Uba et al., 2009).

Most of the growth structures mentioned here are in the upper parts of thick (3–5 km) successions of generally upward-coarsening fluvial and alluvial-fan deposits that record development and eastward migration of a regional foreland basin system (DeCelles et al., 2011; Siks and Horton, 2011; Carrapa et al., 2011b). Most of these structures can be directly attributed to mapped thrust faults, such as the Solá thrust (DeCelles et al., 2011), the Calchaqui thrust (Carrera and Muñoz, 2008), and thrust-related antiforms in the Subandean belt (Echavarria et al., 2003; Hernández and Echavarria, 2009). Palinspastic reconstruction of this foreland basin system from 60 Ma to the present demonstrates a highly unsteady pace of flexural wave migration. Two episodes of rapid (~20–50 mm/yr) eastward flexural wave migration are recorded during the intervals ca. 50–40 Ma and <10 Ma (DeCelles et al., 2011). The intervening period was characterized

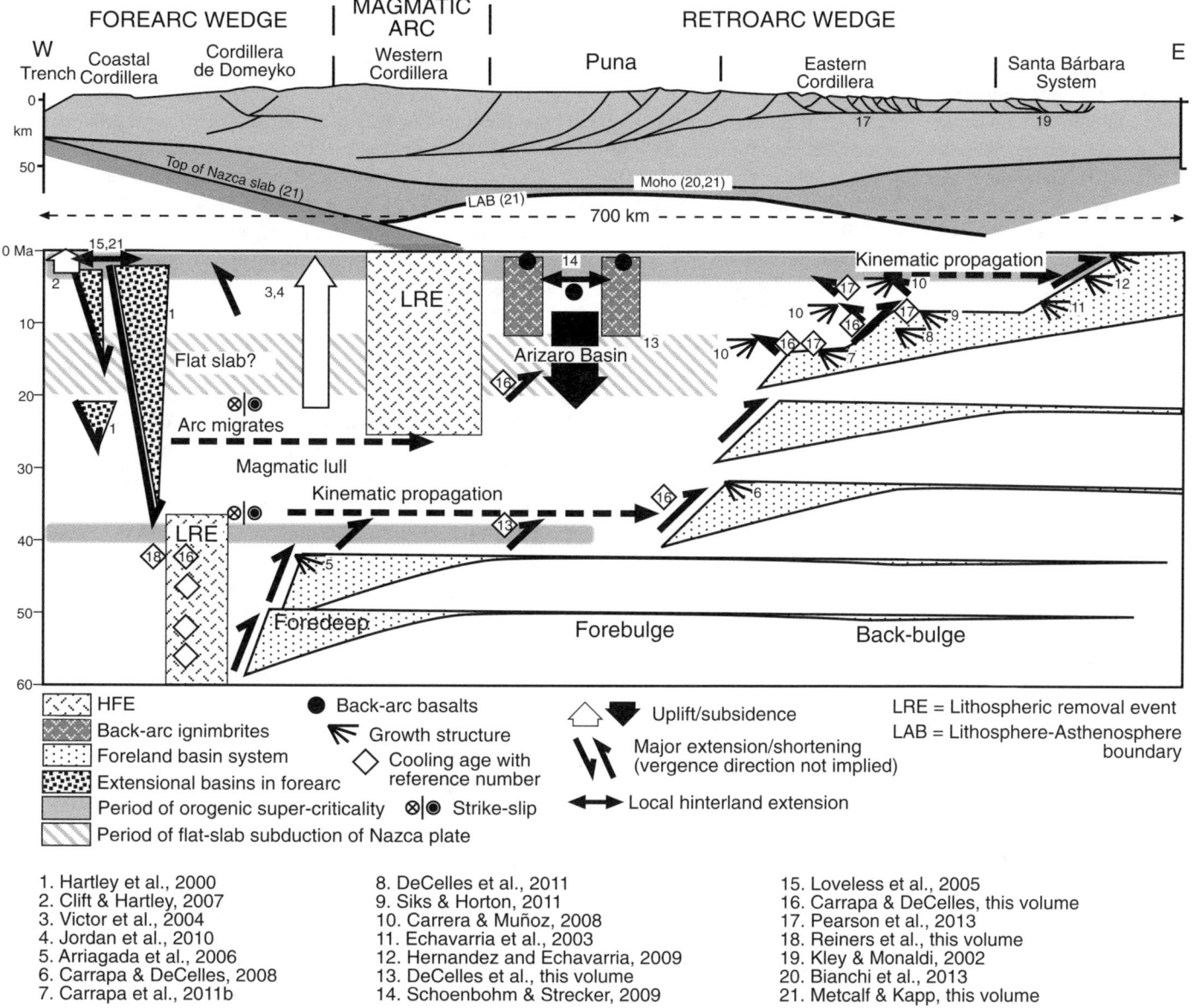

Figure 5. Time-space history diagram of the central Andes, based on the sources listed. LRE—lithospheric removal event (including dripping or delamination); LAB—lithosphere-asthenosphere boundary; HFE—high-flux event. Generalized magmatic history is taken from Trumbull et al. (2006), de Silva et al. (2006), Kay and Coira (2009, 2010), Salisbury et al. (2011), and references therein.

by very slow (~4–5 mm/yr) eastward migration and local out-of-sequence or break-back thrusting (Figs. 3C and 5).

Together, the thermochronological data, growth structures, and reconstructed pattern of flexural wave migration provide a means of tracking the orogenic strain front through time. The three independent data sets indicate rapid eastward propagation events in the central Andean orogenic wedge during late Eocene and late Miocene–Pliocene time (Figs. 3C and 5). During the intervening time (ca. 35–8 Ma), the strain front migrated relatively slowly from the eastern Puna through the Eastern Cordillera.

In regional assessments of the timing of shortening along west-to-east transects in Bolivia, Elger et al. (2005), Oncken et al.

(2006), McQuarrie et al. (2008), and Barnes et al. (2012) showed regional patterns that are similar to the one we find in northwestern Argentina: The locus of shortening migrated abruptly from the Chilean Precordillera into the Eastern Cordillera at ca. 45–40 Ma (see also DeCelles and Horton, 2003; Horton, 2005; Gillis et al., 2006; Ege et al., 2007), marched steadily eastward through the Eastern Cordillera and Interandean zone from ca. 35 to 15 Ma, and then migrated relatively rapidly eastward once again, into the Subandean frontal thrust belt during late Miocene time (Gubbels et al., 1993; Barnes et al., 2012). During the interval ca. 30–25 Ma, and again ca. 20–10 Ma, out-of-sequence shortening took place within the Altiplano and western Eastern Cordillera (Elger et al.,

2005; Lamb, 2011; Leier et al., 2010). Oncken et al. (2006) noted that the overall regional strain history is roughly consistent along strike between 15°S and 23°S.

Evidence for widespread late Miocene and younger normal and strike-slip faulting in the Puna and Altiplano was summarized by Liu et al. (2002), Schoenbohm and Strecker (2009), and Zhou et al. (2013). These data (Mercier, 1981; Sébrier et al., 1985; Allmendinger, 1986; Cabrera et al., 1987; Allmendinger et al., 1989; Mercier et al., 1992; Marrett et al., 1994; Cladouhos et al., 1994; Marrett and Strecker, 2000; Lamb, 2000; Santimano and Riller, 2012) show that since ca. 7.3 Ma, much of the central Andean hinterland region (including the Puna) has been under extension, resulting in small-slip normal and strike-slip faulting. Mafic to intermediate magmatic activity is commonly associated with the faulting (e.g., Mazzuoli et al., 2008; Schoenbohm and Carrapa, this volume). Schoenbohm and Strecker (2009) and Pearson et al. (2013) attributed this phenomenon to a change in the state of stress, with the maximum principal stress rotating vertically away from horizontal, associated with increased gravitation potential energy induced by isostatic rebound during removal of mantle lithosphere and lower crust beneath the southern Puna region. These results are also consistent with geodynamic models that predict extension in the central Andean Plateau coeval with shortening in the frontal parts of the retroarc thrust belt (e.g., Liu et al., 2002).

PALEOELEVATION HISTORY

Paleoaltimetry studies in the Altiplano and Eastern Cordillera of Bolivia suggest pulses of uplift during the Miocene (Gregory-Wodzicki, 2000; Ghosh et al., 2006; Garzione et al., 2008, 2014; Leier et al., 2013). Isotopic analysis of pedogenic carbonate nodules coupled with paleobotanical data from the Altiplano suggest up to 3 km of surface uplift between ca. 11 and 6 Ma (Garzione et al., 2006, 2008), which the authors attributed to isostatic rebound following lithospheric removal (Molnar and Garzione, 2007). Alternatively, climate models suggest a more gradual uplift of the Bolivian Andes starting at ca. 25 Ma (Poulsen et al., 2010). A recent study from the Eastern Cordillera of Bolivia shows >1 km of uplift between 24 and 15 Ma, which the authors also interpreted as isostatic rebound following lithospheric removal (Leier et al., 2013).

In Argentina, deuterium isotopic analyses (δD) of volcanic glass from tuffs intercalated within Eocene to modern deposits indicate values similar to present-day values at Salares de Antofalla, Arizaro, and Fraile by ca. 34 Ma (Canavan et al., 2014; Quade et al., this volume). Given that the deformation front was located east of the Puna Plateau, in the Salar de Pastos Grande, by ca. 38 Ma (Carrapa and DeCelles, 2008), this suggests that high elevation was achieved in the Puna during shortening and crustal thickening. Multiple-proxy isotopic analyses of pedogenic carbonate nodules, marls, and δD of volcanic glasses of Miocene to Holocene deposits in the Eastern Cordillera and southern Puna (Salar de Antofalla, Pasto Ventura, Salar de Pocitos, Cerro Galán,

and San Antonio de los Cobres), combined with existing stratigraphic and sedimentological data, indicate that modern elevations were achieved between ca. 21 and 14 Ma in the Eastern Cordillera and by 11 Ma in the southern Puna Plateau (Quade et al., this volume; Canavan et al., 2014). The time of attainment of high elevation in the Eastern Cordillera coincides with the time of exhumation of the Eastern Cordillera (Deeken et al., 2006; Coutand et al., 2006; Carrapa et al., 2011b). Growth structures in the Angastaco Basin and Quebrada del Toro in 14–11 Ma strata (DeCelles et al., 2011; Carrapa et al., 2011a) indicate active shortening at this time.

The modern and ancient distribution of evaporites strongly supports the notion of eastward progression of uplift of the Andes to >3000 m, starting by at least the late Eocene in the Puna region. Evaporites have been used to indicate the development of aridity and internal drainage in the Puna (Alonso et al., 1991; Vandervoort et al., 1995). Quade et al. (this volume) compiled the elevations of modern evaporites at latitudes <26°S in South America, showing that they are confined to the arid high Andes at elevations >3000 m. The presence of evaporites as far back as the late Eocene in the Puna suggests that this area must have been elevated >3000 m, in agreement with interpretations of isotopic analysis of volcanic glass (Quade et al., this volume). The maximum ages of evaporites decrease eastward, first appearing by ca. 15 Ma in the Salar de Hombre Muerto (Vandervoort et al., 1995). This also matches isotopic evidence for uplift and hydrographic isolation of this region of the Eastern Cordillera during the middle to late Miocene.

Overall data from the Puna Plateau and Eastern Cordillera of northwestern Argentina show that high elevations were produced and maintained by crustal thickening, which was ongoing throughout the Cenozoic. Elevation increases in response to other processes in the cordilleran cycle are not obvious, although one could argue that elevation was highest during the magmatic lull and has declined by ~1 km since the late Oligocene (Fig. 3D). Modeling studies (Pelletier et al., 2010; Currie et al., this volume; Wang et al., this volume) suggest that episodes of gravitational foundering beneath the hinterland are not likely to produce more than a few hundred meters of isostatic elevation gain, which is probably not resolvable given the present level of uncertainty attributed to paleoaltimetry data sets.

HINTERLAND BASIN DEVELOPMENT

Following Ducea (2001), the cordilleran cycle model suggests that dense garnet pyroxenite eclogites ("arclogites") will form in the lower crust during high-flux events. Eclogites may also form beneath the Andean hinterland in response to metamorphic processes (e.g., Beck and Zandt, 2002; Sobolev et al., 2006; Krystopowicz and Currie, 2013). Geodynamic modeling studies show that once such density anomalies reach critical mass, they will be gravitationally removed by delamination or by dripping of Rayleigh-Taylor–type convective instabilities (Molnar and Houseman, 2004; Sobolev and Babeyko, 2005; Göğüş and Pysklywec, 2008; Wang et al., this volume). Wang et al. (this volume) show that for rheological and thermal characteristics typical of the central Andes, dense anomalies that form in the lower crust (rather than in the mantle lithosphere) are most likely to drive significant surface subsidence.

Isolated basins characterized by internal drainage and lacustrine depositional systems with sporadic evaporite deposition are widespread in the Puna and Altiplano (Alonso et al., 1991; Vandervoort et al., 1995). These basins have formed by a variety of processes, and many of them exhibit superimposed mechanisms of subsidence through time (Horton, 2012). Modern examples include the Atacama, Arizaro, and Uyuni topographic basins, all of which are occupied by saline lakes and salt pans (salars). Miocene examples include the Arizaro and Antofalla Basins in northwestern Argentina (DeCelles et al., this volume; Quade et al., this volume), and perhaps the Miocene phase of subsidence in the Corque Basin in Bolivia (Roperch et al., 1999; Garzione et al., 2008).

Geophysical data show that the modern Atacama, Arizaro, and Uyuni Basins are underlain by dense, seismically fast regions in the lower crust and upper mantle (Myers et al., 1998; Beck and Zandt, 2002; Götze and Krause, 2002; Reutter et al., 2006; Beck et al., this volume; Metcalf and Kapp, this volume). The Central Andean gravity high extends from the modern Arizaro Basin to the Atacama Basin (Götze and Krause, 2002), and positive isostatic gravity anomalies exist beneath the Arizaro, Atacama, and Altiplano (including the Uyuni) Basins (Wang et al., this volume). The Atacama Basin is underlain by anomalously thick crust (~65 km) for its relatively low elevation (2300 m; Reutter et al., 2006), whereas the Arizaro Basin, at ~4000 m elevation, sits atop anomalously thin crust (~42 km; Yuan et al., 2002; Bianchi et al., 2013). These observations suggest that crustal subsidence in the central Andean hinterland may be coupled with processes in the upper mantle. Modeling studies demonstrate that basins produced by the formation and removal of dense gravitationally unstable bodies in the lower crust exhibit a sigmoidal (slow, then rapid, then slow) subsidence pattern, followed by rebound and possibly even uplift to elevations greater than the original elevation of the surface (Göğüş and Pysklywec, 2008; Wang et al., this volume; DeCelles et al., this volume).

A detailed study of the Miocene fill of Arizaro Basin highlights the fact that traditional mechanisms for driving subsidence are inadequate to explain the origin of the basin, and that the most likely cause of Miocene subsidence was the removal of a dense convective instability from the lower crust and/or mantle lithosphere (DeCelles et al., this volume). Similar studies are needed in other hinterland basins in the central Andes in order to demonstrate the extent to which this process may be acting. In the context of the cordilleran cycle model, hinterland basin development owing to growth and removal of dense roots is expected to coincide with periods of high-flux magmatism during which dense roots are formed and begin to founder beneath the arc and thrust belt hinterland. Wang et al. (this volume) show that such basins scale spatially with the size of the foundering body and

have tectonic subsidence rates of up to ~200 m/m.y. Addition of sedimentary fill roughly doubles the total subsidence. Basin inversion commences before the foundering root is completely detached, and basins that are filled with sediment will rebound to elevations higher than the original elevation. An important result of the modeling (Wang et al., this volume) is that dense instabilities that form in the mantle lithosphere have relatively little effect on the surface, whereas those that develop in the lower crust have a strong effect. Other enigmatic Miocene–Holocene basins in the central Andean hinterland region are likely candidates for this mechanism of subsidence and rebound, including the Uyuni, Corque, and Titicaca Basins in Bolivia and Peru, the Atacama Basin in northern Chile, and the Pasto Ventura Basin in Argentina.

GEODYNAMIC MODELS

Several geodynamic studies have investigated the evolution of the central Andes. Many of these studies focused on the relationship between subduction and continental shortening, concluding that high rates of shortening result from the combination of a strong subduction interface, weak continental lithosphere, and westward motion of South America (e.g., Wdowinski and Bock, 1994; Sobolev and Babeyko, 2005; Sobolev et al., 2006; Luo and Liu, 2009). Other important factors may be subduction-related basal shearing of the continental lithosphere (Pope and Willett, 1998; Capitanio et al., 2011) and interactions between a low-angle slab and thick cratonic lithosphere (O'Driscoll et al., 2012). Some authors suggest that growth of the high Andes created greater coupling with the subducting plate, thereby decreasing the rate of plate convergence (Iaffaldano and Bunge, 2008, 2009) and that coupling may have been enhanced by a reduction in trench sediments during climate aridification (Lamb and Davis, 2003). Numerical models have also highlighted additional processes that may affect the orogen crust, including intracrustal convection (Babeyko et al., 2002, 2006), topographically driven channel flow (Husson and Sempere, 2003; Gerbault et al., 2005), and failure of weak foreland sediments (Babeyko and Sobolev, 2005).

Here, we focus on gravitational removal of negatively buoyant lithosphere. As discussed earlier herein, this appears to be a key process in the evolution of the central Andes. Foundering may involve only the mantle lithosphere, which is cooler and therefore denser than the underlying mantle. In this case, removal must occur within 10–20 m.y., before thermal diffusion smooths the density perturbation (Conrad and Molnar, 1999). However, only the lower 30%–60% of mantle lithosphere may be removed, owing to its strongly temperature-dependent rheology (Buck and Toksöz, 1983; Conrad and Molnar, 1999; Conrad, 2000). To remove a greater thickness of lithosphere, additional negative buoyancy is required. This may arise from either metamorphic eclogitization of thickened crust (e.g., Bousquet et al., 1997; Leech, 2001; Krystopowicz and Currie, 2013), or the presence of a magmatically derived garnet pyroxenite ("arclogite") residue in the deep crust or shallow mantle (e.g., Kay and Kay, 1993; Jull and Kelemen, 2001; Ducea, 2002; Kay and Coira, 2009; Currie et al., this volume).

Geodynamic studies differentiate between two modes of lithosphere removal (Fig. 6):

1. In a Rayleigh-Taylor–type gravitational instability (Fig. 6A), deep lithosphere is removed through subvertical dripping (e.g., Houseman et al., 1981). Jull and Kelemen (2001) showed that dense eclogite induces a Rayleigh-Taylor–type gravitational instability that removes both the eclogite and underlying lithosphere within 10 m.y. However, for many rheologies, the adjacent mantle lithosphere is entrained and fills the gap created by foundering of the Rayleigh-Taylor–type gravitational instability, so no significant thinning occurs (Wang et al., this volume).

2. Delamination involves lateral peeling of the deep lithosphere along a shallow detachment layer (Fig. 6B; e.g., Bird, 1979). This removes the entire thickness of the dense lithosphere, leaving a wide area of thin lithosphere (e.g., Bird, 1979; Göğüş and Pysklywec, 2008). Once initiated, delamination proceeds quickly, with removal over a few million years (e.g., Göğüş and Pysklywec, 2008; Morency and Doin, 2004; Ueda et al., 2012; Krystopowicz and Currie, 2013). Delamination requires a weak layer in the deep crust, possibly associated with a weak rheology and high temperatures (Meissner and Mooney, 1998; Schott and Schmeling, 1998; Morency and Doin, 2004; Krystopowicz and Currie, 2013). If the deep crust is too strong, removal occurs through a Rayleigh-Taylor–type gravitational instability.

Rayleigh-Taylor–type gravitational instabilities and delamination can be differentiated through surface observations, such as the area affected by removal, crustal deformation, surface uplift, and magmatism (Fig. 6; Göğüş and Pysklywec, 2008). In general, Rayleigh-Taylor–type gravitational instabilities tend to be symmetric and more spatially confined (~50–200 km width). For eclogite-driven Rayleigh-Taylor–type gravitational instabilities, the width is determined by the size of the eclogite body; this may be <100 km for eclogite produced through subduction-related magmatism (Elkins-Tanton, 2007; Wang et al., this volume). Rayleigh-Taylor–type gravitational instabilities produce contraction of the crust above the drip (Göğüş and Pysklywec, 2008). If the crust is strong, removal is accompanied by surface subsidence, whereas weak crust experiences viscous thickening and surface uplift (Neil and Houseman, 1999; Molnar and Houseman, 2004; Göğüş and Pysklywec, 2008; Molnar and Houseman, 2013; Wang et al., this volume). Localized magmatism may arise from conductive heating and melting of the descending lithosphere, decompression melting of the upwelling asthenosphere, and heating and melting of the remaining lithosphere following drip removal (e.g., Elkins-Tanton, 2007; Ducea et al., 2013).

In contrast, delamination causes asymmetric removal over a wider area (>200 km), with the spatial extent governed by the region over which the density structure and crustal strength promote detachment. Two styles of delamination are observed in numerical models, depending on the strength of the mantle

A RAYLEIGH-TAYLOR INSTABILITY

Syn-drip Subsidence

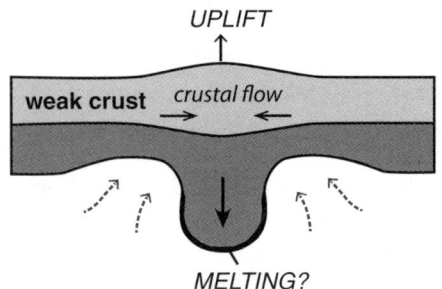

Syn-drip Uplift

B DELAMINATION

Retreating Delamination

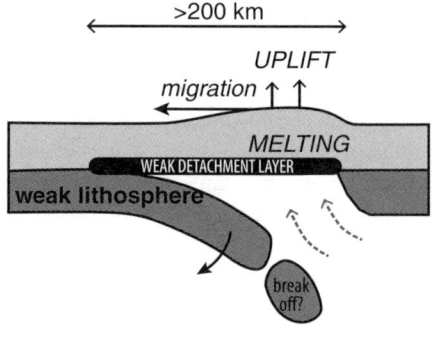

Stationary Delamination

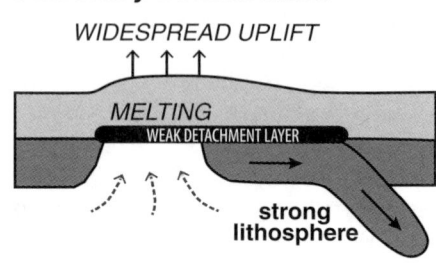

Figure 6. Lithosphere removal and associated surface expressions for (A) Rayleigh-Taylor–type instability (RTI drip) and (B) delamination. Gravitational removal of the dark-gray region is driven by cool, dense mantle lithosphere ± high-density eclogite associated with magmatic or metamorphic processes.

lithosphere (Fig. 6B; Krystopowicz and Currie, 2013). In retreating delamination, lateral migration of the detachment line creates a contemporaneous wave of crustal deformation and heating, surface uplift, and magmatism. In stationary delamination, the deep lithosphere slides into the mantle at a fixed detachment line, producing rapid removal and widespread uplift, crustal heating, and magmatism.

Most numerical models predict lithosphere removal within 10 m.y. of the onset of instability for both Rayleigh-Taylor–type gravitational instability and delamination, with shorter times where lithosphere is rheologically weak or high-density eclogite is present. Removal may be delayed if the lithosphere is relatively strong or if it is held in place by an episode of flat subduction. The temporal and spatial scales predicted by numerical models are consistent with observational data for the central Andes. For example, removal over a width of several hundred kilometers has been proposed based on paleoelevation data for the Altiplano Plateau (Garzione et al., 2008), and magmatic records and seismic tomography for the Puna Plateau (Beck et al., this volume); such large-scale removal may reflect wholesale delamination. There are also areas, especially in the Puna, where localized removal (~100 km width) has been inferred from magmatic records (Kay et al., 1994; Kay and Coira, 2009; Ducea et al., 2013) and the presence of transient basins (DeCelles et al., this volume). These observations are more consistent with Rayleigh-Taylor–type gravitational instability style of removal (e.g., Wang et al., this volume). This suggests that multiple modes of lithosphere removal may operate within the central Andes, which likely reflect the heterogeneous strength and density structure of the orogen.

SYNTHESIS

Spatial and temporal (Figs. 3 and 5) context for significant tectonic and magmatic processes in the central Andes (21°S–25°S) provides a test of the cordilleran cycle model (Fig. 2). We compile observations from our own work as well as those previously derived from an extensive study of the Cenozoic geological history of the Bolivian and northern Chilean central Andes (Oncken et al., 2006) and other regional studies (e.g., Kay and Coira, 2009; Jordan et al., 2010).

As discussed earlier herein, the magmatic history of the region provides the baseline frequency for the cordilleran cycle. Predominantly calc-alkaline high-flux events in the central Andean magmatic arc took place during the late Eocene and the middle to late Miocene, peaking ~30 m.y. apart (Haschke et al., 2002; Trumbull et al., 2006). The interval ca. 36–26 Ma was a relative lull, during which magmatic activity decreased markedly but did not completely terminate (Trumbull et al., 2006), and the time since ca. 2 Ma may be the beginning of another magmatic lull (Trumbull et al., 2006; Kay and Coira, 2009). Small-volume basaltic magmatism was active mainly during the late Eocene (at the latitude of the Puna) and during the latest Miocene–Pleistocene. We suggest that the large-volume calc-

alkaline magmatism during the Eocene and Miocene–Pliocene represents typical high-flux events, whereas the mafic magmatism that occurred during the waning phases of these high-flux events represents episodes of postdrip/delamination magmatism (Ducea et al., 2013; Schoenbohm and Carrapa, this volume; Murray et al., this volume). Kay and Kay (1993) proposed that recent mafic magmatism in the Puna Plateau could be related to removal of lithosphere beneath the region. In addition, Kay and Coira (2009) suggested that the major back-arc intermediate to felsic magmatism represented by the Altiplano-Puna volcanic complex and other massive back-arc ignimbrite eruptions may be a consequence of lithosphere removal. Other workers have suggested that large Miocene–Pliocene dacitic ignimbrites in the Altiplano-Puna volcanic complex require some component of mantle contribution (Ort et al., 1996; Lindsay et al., 2001). If this is the case, then the Neogene high-flux event may have climaxed during the late Miocene, and younger magmatism may represent melting and eruption in response to removal of dense instabilities beneath the Puna region. In this view, the basalts, which exhibit garnet-pyroxenite melt compositions (Ducea et al., 2013), would represent just the last stages of "drip-related" magmatism. Recent geochemical and structural data from the western Salar de Arizaro and Pasto Ventura area in the southern Puna Plateau indicate that mafic volcanism is associated with extension and was preceded by shortening, as predicted by analog modeling for lithospheric removal (Schoenbohm and Carrapa, this volume). These data suggest at least two episodes of dripping: a 100-km-scale event in the vicinity of Salar de Arizaro during the early Miocene, and a smaller "driplet" beneath the southern Puna at 10–5 Ma (Schoenbohm and Carrapa, this volume).

Haschke et al. (2002) provided trace-element and isotopic data that reinforce the idea that the late Eocene and middle Miocene high-flux events resulted mainly from melting of underthrusting South American lithosphere (Fig. 3B). We infer that the magmatic history of the region entailed the production of eclogitic arc residues in the lower crust and upper mantle during late Eocene and middle Miocene times, and that these residues began to founder into the mantle during the waning phases of the high-flux events (Fig. 3E).

The isotopic data from arc magmatic rocks and mass balance considerations (Ducea, 2001; DeCelles et al., 2009) suggest that the high-flux events are primarily the result of melting of continental crust that is underthrust beneath the magmatic arc (Ducea and Barton, 2007; Girardi et al., 2012). In the case of the central Andes, underthrusting South American crust is proposed to have fueled the two high-flux events shown in Figure 3A. Thermal relaxation times on the order of 10–25 m.y. are required before the underthrusted crust begins to melt (Ducea, 2001), probably depending on composition and volatile content of the crust and the underthrusting rate. In Figure 3C, we show the time periods ca. 42–39 Ma and ca. 8–0 Ma as times when the rate of eastward propagation of the strain front was relatively rapid. During the Eocene propagation event, the orogenic strain front migrated rapidly from the Cordillera de Domeyko in Chile into the eastern part of the Puna at ca. 39 Ma. During the second propagation event, the strain front migrated from the Eastern Cordillera into the Santa Bárbara and Subandean Ranges at ca. 4–2 Ma (Carrapa et al., 2011b) and ca. 8 Ma (Echavarria et al., 2003), respectively. It is important to distinguish between these thrust belt propagation events and changes in shortening rate. Shortening and crustal thickening were probably ongoing throughout Cenozoic time in the central Andes, whereas the eastern limit of this deformation migrated eastward at an unsteady pace that was in phase with the latter stages of magmatic high-flux events (Figs. 3A and 3C). Immediately following these episodes of rapid eastward propagation, the hinterland region of the Puna experienced minor extension and/or strike-slip faulting (Fig. 3C). Low-temperature thermochronological data demonstrate that the major periods of erosional exhumation in the central Andes occurred immediately after these eastward jumps in the strain front, as newly activated thrust faults began to structurally elevate the rocks in their hanging walls, and erosion was invigorated by increased slopes (Fig. 3C; Carrapa and DeCelles, this volume). Together, these results suggest that the retroarc orogenic wedge became supercritically tapered during the Eocene and Miocene–Pliocene; the eastward propagation events in the strain front and local hinterland extension and strike-slip would represent responses of the orogenic wedge to increased gravitational potential energy in the hinterland (e.g., England and Molnar, 1993; Garzione et al., 2006). A similar process may be invoked to explain aspects of the kinematic and tectonic erosional history of the forearc region (Fig. 3G).

Hinterland elevation remains one of the least-documented parameters of the cordilleran cycle model in the central Andes. Nevertheless, preliminary stable isotope data from volcanic glass samples (Canavan et al., 2014; Quade et al., this volume) suggest that elevation in the Arizaro Basin and Salar de Frailes regions, deep within the Puna, was already >4000 m by late Eocene time. The data hint at temporal variations in elevation, on the order of 0.5–1.0 km, but uncertainties in the methods of converting isotopic values to paleoelevation remain too high to correlate documented changes in paleoelevation with particular events in the proposed cordilleran cycle. Most significantly, the attainment of high elevation in the western Puna region during late Eocene time occurred during the latter part of the high-flux event, suggesting that elevation gain was mainly related to crustal thickening and eastward migration of the orogenic strain front from the western Puna and Cordillera de Domeyko into the eastern Puna region (Figs. 3C and 3D). This suggests that the construction of persistent high-elevation topography in the central Andes depends primarily upon crustal thickening, rather than transient effects owing to lithosphere or lower-crust removal, as originally suggested by DeCelles et al. (2009). In turn, the paleoaltimetry data raise the question of how Puna crust was thickened, because existing cross sections through the study area do not suggest horizontal shortening sufficient to raise surface elevation to 4 km by Eocene time (e.g., Pearson et al., 2013). Material removed from the forearc region and "relaminated" beneath the retroarc region

may have contributed to crustal thickening, but such a process has not been investigated in the Andes.

Another key implication of the cordilleran cycle model for the central Andes is that hinterland basin subsidence should coincide with development of dense convective instabilities in the lower crust or mantle lithosphere. The Arizaro Basin subsided, filled with sediment, and shortened internally during the period 20–8 Ma, coeval with the proposed early-middle Miocene high-flux event (Figs. 3A, 3E, and 3F). Uplift and incision of the basin fill commenced soon thereafter (Carrapa et al., 2009; DeCelles et al., this volume), possibly in response to the onset of foundering of dense eclogitic roots. Although the Arizaro Basin is a relatively local feature and is probably not representative of the dynamics of the entire region, its timing and character are consistent with the regional history of magmatism in the context of the cordilleran cycle. Other Miocene basins in the region may ultimately provide additional tests of the model. For example, the Corque Basin in the northern Altiplano Plateau has a sediment accumulation and basin inversion history similar to that of the Arizaro Basin (Fig. 3F; Roperch et al., 1999).

In the forearc region (Fig. 3G), Eocene increases in shortening and subduction erosion were coeval with major eastward migration of the retroarc strain front (Fig. 3C). Miocene forearc events were less clustered in time and seem to have slightly predated the Miocene–Pliocene strain-front migration event in the retroarc thrust belt. The Miocene events were also broadly coeval with a roughly 3 km increase in the structural and topographic relief along the western slope of the Western Cordillera (Jordan et al., 2010). Insofar as the trench outboard of the central Andes has been starved of sediment during much of the Cenozoic (Lamb and Davis, 2003; Oncken et al., 2006), periods of orogenic supercriticality should be marked by onset of rapid subduction erosion as the surface slope of the forearc increases in response to hinterland elevation gain (Lallemand et al., 1994). If the subduction erosion rate calculated by Oncken et al. (2006) for latitudes 19°S–22°S is representative, then the patterns shown by this rate, the shortening rate, and magmatic high-flux events are internally consistent (Figs. 3C and 3G). Shortening events in the forearc took place just before and coeval with major subduction erosion events and were simultaneous with high-flux events. Internal shortening of the forearc orogenic wedge may have been a response to decreased orogenic taper during growth of dense eclogitic residues beneath the arc during high-flux events. Alternatively, forearc shortening may be controlled by increased interplate coupling and/or decreased trench rollback, independent of the cordilleran cycle.

In summary, periods of retroarc thrust belt migration, forearc shortening, and subduction erosion appear to have been more or less in phase with the peaks in magmatic high-flux events in the arc (Fig. 3). The Eocene episode was relatively tightly clustered over a period of ~5–10 m.y., whereas the Miocene event was more diffuse, possibly continuing for a period of >10 m.y. The onset of high-flux event flare-ups generally predated the major kinematic events by 10–15 m.y. Episodes of minor extension in the hinterland region took place concurrent with or immediately after the Eocene and Miocene events, in spatial and temporal association with local mafic magmatism. Hinterland basin development took place during the rising part of the Miocene magmatic high-flux event. High elevation in the Puna was achieved by late Eocene time and shows little correspondence to tectonic and magmatic events at the present level of resolution. However, the gain of >3 km of structural and topographic relief on the western flank of the high Andes was almost exactly coeval with the Miocene high-flux event. Insofar as much of this relief gain took place to the east of the modern Salar de Atacama and Cordillera de Domeyko thrust belt, this observation seems to conflict with the stable isotope paleoaltimetry data, which indicate the Puna region was at modern elevations by ca. 36 Ma (Canavan et al., 2014; Quade et al., this volume). However, the relief gain in the forearc is not referenced to sea level because of the methods used in the analysis (Jordan et al., 2010), and it is equally plausible that the development of such high relief owes to collapse of the forearc as the magmatic arc migrated through the region from Paleogene to Neogene time. Conceivably, the present forearc region lay at much higher elevation during the Paleogene, and subsequently collapsed to lower elevations after the magmatic arc moved inboard to its present location in the Western Cordillera (Quade et al., this volume).

We propose that this assemblage of disparate processes and events over the past 50 m.y. in the central Andes may be explained as the integrated response to alternating periods of rapid advance of the orogenic wedge (ca. 40–38 Ma and 8–0 Ma), high-flux magmatism (ca. 50–40 Ma and ca. 25–2 Ma), and removal of dense arc roots and overthickened eclogitic lower crust by foundering of convective instabilities (ca. 42–36 Ma and post–15 Ma) (Figs. 2 and 3). Shortening thickened the crust beneath the hinterland and introduced melt-fertile lower crust and possibly mantle lithosphere into the region of high temperature beneath the arc. After lag times of ~10–15 m.y., these rocks began to melt, initiating a high-flux event. Eclogitic roots could have developed in response to both metamorphic reactions and magmatic differentiation (Ducea, 2001). Although the post–50 Ma record of the central Andes contains only two high-flux events and one obvious recharge interval (ca. 36–26 Ma), the data presented by Haschke et al. (2002) suggest a second recharge interval during the period ca. 80–75 Ma. In addition, the last few million years may have witnessed a shift to recharge mode as the Miocene high-flux event appears to be ending (since Pliocene–Pleistocene time; Trumbull et al., 2006; Kay and Coira, 2009). High elevation was attained mainly in response to crustal thickening associated with shortening. The major eastward relocations of the strain front in the retro-arc orogenic wedge took place at the tail ends of the two high-flux events, as did the small-volume mafic eruptions, suggesting that these events may be related to the process of convective instability foundering and delamination. Local hinterland extension also took place at nearly the same times, again suggesting that the maximum principal stress direction in the hinterland may have rotated off of horizontal in response to increased elevation. The magnitudes of these elevation changes are unlikely to have been

more than a few hundred meters, however (Pelletier et al., 2010; Ducea et al., 2013; Currie et al., this volume; Wang et al., this volume). We envision a process in which a number of dense bodies were removed over a several-million-year period of time. The cumulative effect of these removal events was to add incremental elevation to a region that already had considerably high elevation owing to crustal thickening (Quade et al., this volume). Subsidence of the Arizaro hinterland basin took place during the Miocene high-flux event, and basin inversion began after 8 Ma, consistent with a component of the inversion being driven by isostatic rebound following removal of a dense instability from beneath the basin. In the forearc, shortening and subduction erosion during the late Eocene correlate with the proposed foundering event centered roughly at ca. 40 Ma. Correlation of forearc events during the Miocene high-flux event is less obvious but generally follows the same, albeit more diffuse, pattern as the Eocene event, with shortening and subduction erosion increasing during the peak of the Miocene high-flux event. Growth of relief on the western Andean flank monocline was coeval with these events.

DISCUSSION

Period of Cyclicity and the Overarching Control

Magmatic high-flux events in the central Andes have occurred four times since ca. 200 Ma, peaking with a frequency of ~25–45 m.y. (Haschke et al., 2006). The most recent pair of high-flux events peaked during the Eocene and late Miocene, ~25–30 m.y. apart. One obvious possible control on the timing of high-flux events is the rate of plate convergence (DeCelles et al., 2009); faster convergence might produce higher-frequency high-flux events. Comparison of the frequencies of high-flux events in the North American and South American Cordilleras does not support this idea: High flux events in the North American Cordilleran magmatic arc are spaced at time intervals of ~50 m.y. (Barton, 1996; Ducea, 2001; Gehrels et al., 2009; Mahoney et al., 2009; Paterson et al., 2011; Barth et al., 2013), against a backdrop of plate convergence rates (Engebretson et al., 1985) that were comparable to the Nazca–South American convergence rates over the past 60 m.y. Plate kinematic studies (Pardo-Casas and Molnar, 1987; Somoza, 1998; Sdrolias and Müller, 2006) show that Nazca–South America convergence velocity along the trench outboard of the central Andean orogen ranged between ~40 and 150 mm/yr during the Cenozoic, with peak rates at ca. 50–40 Ma and ca. 20–15 Ma (Fig. 3H). Plate convergence rate is not in phase with the high-flux events in the magmatic arc, but the time difference between the maxima in plate kinematic rates is ~25–30 m.y. The closeness of this time span to the time elapsed between the peaks of the two most recent high-flux events suggests some level of control, but the nature of this control is not obvious. It appears that the peaks in plate rates precede peaks in high-flux events by ~5–10 m.y., hinting at the possibility of a lag time between peak plate rates and magmatic flux rates. The major eastward jumps in the location of the retroarc strain front took place during periods of decelerating plate convergence and were coeval with the peaks in the high-flux events. This suggests that rapid plate convergence is not converted directly into rapid orogenic wedge propagation, nor is the associated crustal shortening converted directly into a magmatic high-flux event; rather, the kinematic propagation events seem to take place in response to magmatic and isostatic processes during the latter stages of high-flux events. It is also conceivable that increased interplate coupling owing to growth of the orogenic belt causes decreased plate convergence rate (Richardson and Coblentz, 1994; Norabuena et al., 1999; Iaffaldano and Bunge, 2008; Meade and Conrad, 2008), and the data from the central Andes are consistent with this view (Figs. 2C and 3).

Another mechanism by which plate convergence rate could be controlling the cordilleran cycle is by the vigor of mantle hydration by the subducting plate, which will affect the rate of mantle wedge melting and heat advection by mantle melts into the upper plate. Greater convergence velocity should increase the rate and extent of mantle wedge hydration, as well as the rate at which melt-fertile continental lithosphere is fed into the zone of high heat flux and melting. If this is the case, then the cordilleran cycle can be understood in terms of a cascading series of processes, beginning with the flux of heat into the overriding plate. This heat flux is controlled over time spans of ~50 m.y. by the rate of partial melting in the mantle wedge and transport of hot magmas into the lower crust and mantle lithosphere. Shortening in the upper plate provides melt-fertile lower crust to the region of high heat flux. High-flux magmatism initiates after a lag time of ~10 m.y. (in the central Andean case), and dense garnet-pyroxenite residues accumulate in the lower crust while batholith-forming felsic magmas ascend into the middle and upper crust. Metamorphic eclogite may also form during thickening of the crust caused by shortening of the retroarc orogenic wedge. As the dense arc roots become gravitationally unstable, they begin to founder as Rayleigh-Talyor drips or delaminating slabs. We suggest that in the central Andes, this process removed at least several drips and delaminating slabs over a several-million-year period (Fig. 2C). Removal of the dense arc roots allowed the remaining crust to isostatically rebound, increasing gravitational potential energy and causing the maximum principal stress to rotate off of the horizontal in the orogenic hinterland. The results of this change in the hinterland state of stress were threefold: Local strike-slip and normal faults developed in the hinterland; some of these faults served as conduits or triggers for caldera eruptions (de Silva et al., 2006; Mazzuoli et al., 2008) and, especially later during episodes of arc root removal, as conduits for low-volume basaltic eruptions (Schoenbohm and Carrapa, this volume); and the contractional strain front advanced rapidly eastward, driving with it the foreland basin flexural wave (DeCelles et al., 2011). In the forearc, the rate of shortening increased, and subduction erosion may have been promoted by increased mass wasting toward the trench (Metcalf and Kapp, this volume). Removal of arc roots created new space for additional underthrusting from the foreland side of the system, which in turn recharged melt-fertile material

into the hot region beneath the arc (Figs. 2D and 2A). After passage of sufficient time for melting to occur, the next high-flux event was "ignited."

Cordilleran orogenic belts, given sufficient rates of crustal thickening owing to generally rapid convergence and high resistance to subduction zone rollback (Schellart, 2008), are inherently prone to cyclic behavior owing to the lag time between heating and melting in the deep continental crust. If this is the case, then cause-and-effect linkages between changes in plate convergence velocity and the cordilleran cycle are unnecessary, and upper-plate processes may in fact control rates of convergence, rather than vice versa (Iaffaldano and Bunge, 2008; Meade and Conrad, 2008).

Along-Strike Differences

The central Andes exhibit marked differences in structure and topography along strike from southern Peru, through Bolivia, and into northern Argentina (e.g., Allmendinger et al., 1997; Kley and Monaldi, 1998; McGroder et al., this volume). One of the most striking differences is the topographic dissimilarity between the Altiplano and Puna sectors of the central Andean Plateau (Isacks, 1988). The Altiplano is mainly occupied by a vast Miocene to Holocene sedimentary basin (Horton, 2012), whereas the Puna has more rugged topography, higher elevation, and local topographic depressions. Insofar as at least some of the modern and late Cenozoic basins in the Puna may be related to stresses exerted by dripping or delaminating parcels of lower crust and mantle lithosphere, it is tempting to speculate that the Altiplano basin (which is a composite late Miocene to Holocene feature; Oncken et al., 2006) is the surface manifestation of ongoing densification of the lithosphere beneath this portion of the central Andean hinterland. Conceivably, the Altiplano region is in the recharge portion of the cordilleran cycle (Figs. 2A and 2B), while the Puna is in the late high-flux event part of the cycle (Fig. 2D). This would suggest that the Altiplano is several million years "ahead" of the Puna in terms of cordilleran cyclicity, consistent with somewhat earlier (ca. 12–10 Ma) eastward propagation of the orogenic strain front into the Subandean zone in Bolivia (Barnes et al., 2012) as compared to the post–8 Ma timing of strain propagation into the Argentine Subandes (Echavarria et al., 2003) and Santa Bárbara Ranges (Reynolds et al., 2000). This might also suggest that the Altiplano and Puna regions represent nearly opposite end-member states of the cycle as manifested in a mature, fully developed cordilleran orogenic system. The Puna versus Altiplano comparison also highlights the potential effects of differing scales of lithospheric removal: Garzione et al. (2008) argued for regional-scale, wholesale delamination during the late Miocene beneath the northern Altiplano, whereas various petrologic, seismic, and sedimentologic studies suggest smaller-scale (~100–200 diameter) lithospheric removal events beneath the Puna region (e.g., Beck et al., this volume). In general, the central Andean system suggests that large variations in structure, kinematic history, magmatism, elevation, and upper-mantle structure and composition are to be expected along strike in cordilleran orogenic belts, at spatial scales on the order of 100–200 km and time scales of 5–20 m.y.

Significance of Flat-Slab Subduction

Many previous treatments of the tectonic and magmatic history of the central Andes emphasized the importance of low-angle or flat subduction of the Nazca oceanic plate (e.g., Gutscher et al., 2000; Yáñez et al., 2001; Haschke et al., 2006; Kay and Coira, 2009; Ramos and Folguera, 2009; Martinod et al., 2010; O'Driscoll et al., 2012). Although we do not discount flat-slab subduction as an important process in the creation of the central Andes, we note that several aspects of models that call for flat-slab subduction lead us to question the interpretation that this is an all-important process in the central Andes.

First, episodes of flat-slab subduction in the Andes are generally interpreted on the basis of the distribution of magmatic activity in time and space. In particular, episodes of eastward expansion or migration of the arc are inferred to indicate decreasing slab descent angle, and gaps in the record of arc-proper magmatic activity are commonly interpreted to represent periods during which the trajectory of subduction is essentially horizontal beneath the retroarc region. However, it is not clear how reliable this signal is for reconstructing the details of flat-slab events. For example, Ramos et al. (2002) depicted the depth of magma origination from the subducting slab as anywhere between 120 and 220 km, and the history of magmatism they depict is not consistent with a simple eastward flattening of the Nazca slab (their figure 10). Similarly, Kay and Coira (2009) depicted magma generation over a roughly 100–200-km-deep zone along the subducting slab, and the flat-slab events they illustrate (their figure 6) are geometrically subtle and seldom more than ~100 km wide perpendicular to the trench. Several authors have interpreted eastward migration of the Andean arc front as a result of inboard migration of the trench wall, as subduction erosion has trimmed the western lip of the South American plate (Trumbull et al., 2006; Kukowski and Oncken, 2006; Haschke et al., 2006). This would negate the need to call upon flat-slab subduction to explain *inboard* migration of the magmatic arc.

Second, the magmatic lulls are too brief to be explained as a result primarily of flat-slab subduction. Geochronological data sets presented by Allmendinger et al. (1997), Haschke et al. (2006), Trumbull et al. (2006), and Kay and Coira (2009) all show that the magmatic lulls were not devoid of arc magmatism, but rather that the intensity of magmatism was simply lower for a few million years. The lulls are also very brief, on the order of only a few million years. The two late Miocene to Holocene bona fide flat-slab events in Peru and central Argentina are associated with magmatic lulls in the arc proper that commenced ca. 4 Ma and ca. 9 Ma, respectively (Ramos et al., 2002; Rosenbaum et al., 2005), and the associated flat slabs show no signs of rolling back at this point in time (Wagner et al., 2006). The magmatic gap in the main arc trend associated with the Laramide flat-slab

event in North America persisted for several tens of millions of years (Coney and Reynolds, 1977; Constenius, 1996; Dickinson, 2013). The amount of time required to emplace a flat slab eastward beneath the central Andes at convergence rates reconstructed for the Cenozoic (as high as ~160 km/m.y. to as low as 60 km/m.y.; Pardo-Casas and Molnar, 1987; Somoza, 1998) suggests that at least 5–10 m.y. would be required before the slab was flat over a horizontal district as large as the present Chile-Argentina flat-slab segment (stretching as much as 800 km from the trench).

Third, the Eocene flat-slab event postulated by Martinod et al. (2010) is based on an uncertain reconstruction of the Juan Fernandez Ridge and lacks support in the geological record of the upper plate. Although Martinod et al. (2010) called attention to a plausible overlap of deformation timing in the Eastern Cordillera with their Eocene flat-slab event, the deformation is well documented to have begun earlier than the proposed flat-slab event.

Fourth, Yáñez et al. (2001) proposed that the Juan Fernandez Ridge, which currently appears to prop up the Chile-Argentina flat slab, swept obliquely through our study region over the period ca. 20–8 Ma, and Kay and Mpodozis (2002) and Kay and Coira (2009) amplified this model by suggesting correlations among tectonic and magmatic events and the proposed migration of the ridge. However, we note that although magmatism did spread eastward beginning ca. 25 Ma, a magmatic gap in the main arc is absent over this time interval (cf. Allmendinger et al., 1997). The Miocene flat-slab event postulated by Kay and Coira (2009) and Ramos and Folguera (2009) would seem to spatially interfere with the proposed lithospheric delamination events of Kay et al. (1994) and Kay and Coira (2009). Moreover, the proposed early to middle Miocene flat-slab event in the central Andes was not accompanied by the development of isolated, major basement block uplifts and intervening basins, as are the present Peru and Chile-Argentina flat slabs. The part of the Juan Fernandez Ridge that is still visible on the seafloor consists of a chain of seamounts with only moderately thickened crust (Moho depth of −12 km; Kopp et al., 2004); thus its ability to create a significant buoyant force capable of disrupting the upper plate several hundred kilometers inboard from the trench is limited. Trumbull et al. (2006) arrived at a similarly skeptical conclusion regarding the importance of flat-slab subduction of the Juan Fernandez Ridge based on the magmatic record.

Although we view the importance of flat-slab subduction to be somewhat overstated in some of the literature, the compilation of events in Figure 5 suggests that some of the oddities of the Miocene record, particularly in the forearc region, may in fact be explained by a minor flat-slab event as postulated by Yáñez et al. (2001). Large-magnitude subduction erosion and elevated shortening rate (Oncken et al., 2006), as well as the development of several kilometers of relief along the western flank monocline (Jordan et al., 2010), correlate in time with the interval during which the Juan Fernandez aseismic ridge is considered to have passed beneath the study region (e.g., Kay and Coira, 2009). We suggest that, at a minimum, subduction of seamounts along the ridge would have disrupted the forearc region. It is unlikely that this would have had an effect at the scale of the western flank monocline (which extends for many hundreds of kilometers along strike). Flat-slab subduction events may vary significantly in terms of the impact they have on the upper plate, mainly owing to differences in the thickness and buoyancy of the subducting plate. Subduction of large oceanic plateaux is likely to cause considerable damage to the upper plate (Saleeby, 2003), whereas smaller features may leave little in the geological record to mark their passage. In the case of the proposed Miocene flat-slab event owing to passage of the Juan Fernandez Ridge beneath the central Andes (Yáñez et al., 2001), the muted expression of this event may testify to the narrowness of the ridge itself (Kopp et al., 2004) and the sideways oblique trajectory that it followed beneath the Puna (Yáñez et al., 2001).

Relevance to the North American Cordillera

An obvious question is the extent to which the cordilleran cycle model may be applied to the North American Cordillera. Although this type of synthesis has yet to be undertaken in the North American Cordillera, all of the processes operating during the Cenozoic in the central Andes also were active in the North American system from Late Jurassic time until the mid-Cenozoic. A retroarc thrust belt with several hundred kilometers of shortening developed to the east of a hinterland plateau underlain by roughly double-thick crust (Coney and Harms, 1984; DeCelles and Coogan, 2006; Evenchick et al., 2007). Cyclical patterns of magmatic activity in batholiths of the North American Cordillera have been recognized for many years (e.g., Armstrong, 1988; Barton, 1996), and recent work in the California (Sierra Nevada) and Coast Mountains batholiths, as well as the detrital records of their erosion, further attests to the cyclical frequency and composition of arc magmatism (Ducea, 2001;

Gehrels et al., 2009; Girardi et al., 2012; Barth et al., 2013). The Coast Mountains batholith exhibits a roughly 50 m.y. period for high-flux events, similar to that documented in the California batholith (Barton, 1996; Ducea, 2001; Barth et al., 2013) and the central Andes (Fig. 7). Of particular interest are the ca. 150 ± 10 and 100 ± 10 Ma high-flux events exhibited by the Coast Mountains and California batholiths. In the western United States, each of these high-flux events was associated with an episode of major crustal shortening and eastward propagation of the orogenic strain front in the retroarc thrust belt (Yonkee et al., 1997; DeCelles, 2004). Wells et al. (2012) explicitly invoked delamination/dripping of dense instabilities, isostatic uplift, and decompression melting to explain the history of magmatism and metamorphism in the hinterland during the Late Cretaceous. Laskowski et al. (2013) and Barth et al. (2013) showed that the cyclical magmatic history of the North American Cordilleran arc is also reflected in the detrital zircon U-Pb age record of sandstones from the Cordilleran foreland basin. Whereas testing the cordilleran cycle model in the North American Cordillera is challenging because of the post–early Cenozoic transition from convergent to transform tectonics along the western plate margin, the ensuing extensional collapse of the orogenic belt provides opportunities to assess processes in the middle crust that are inaccessible in the modern central Andes.

CONCLUSIONS

Our analysis of the central Andean orogenic belt demonstrates that the cordilleran cycle model provides a long-term (~40 m.y.) context for understanding the linkages among lithospheric shortening, magmatism, kinematic history, paleoelevation, and upper-mantle geodynamics. However, some important modifications of the original model (DeCelles et al., 2009) are required to better satisfy available data sets from the central

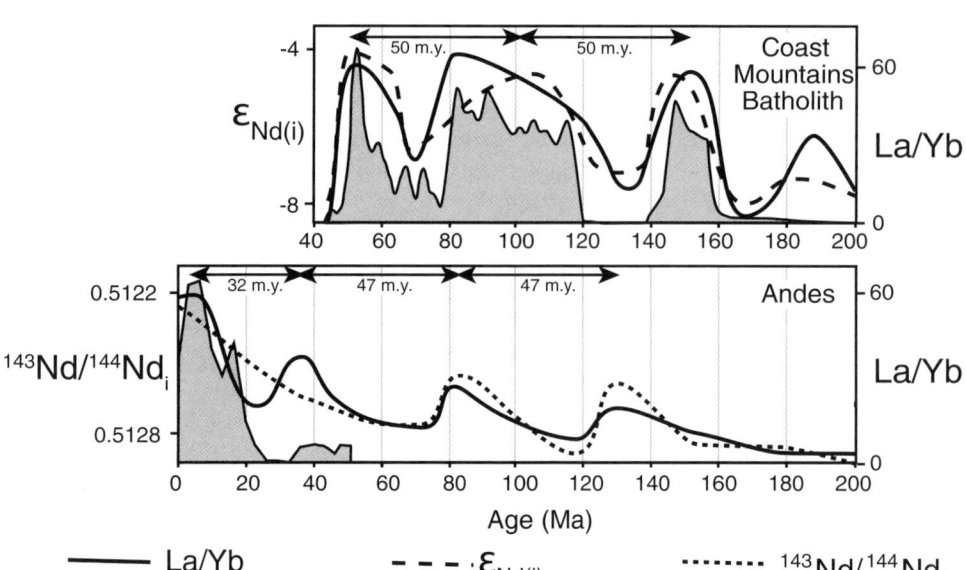

Figure 7. Comparison of magmatic flux, Nd isotopic, and La/Yb geochemical patterns of the Andes and the Coast Mountains batholith of British Columbia. Double-headed arrows show approximate durations of each cycle. Data from Coast Mountains batholith are from Gehrels et al. (2009) and Girardi et al. (2012). Andean data are from Haschke et al. (2002).

Andes. In particular, the original suggestion that high-flux events are terminated by lithospheric removal events is not supported by the central Andean magmatic record, which shows instead that lithospheric removal events overlapped in time with the two most recent high-flux events. Similarly, the major eastward propagation events in the retroarc strain front occurred not after the high-flux events, but during their peak and waning phases, and high elevation developed during protracted crustal shortening, not simply in response to isostatic adjustments owing to lithospheric removal events. Our analysis suggests that the central Andean orogenic system oscillates between two end-member conditions. Intense arc magmatism with evolved compositions, eclogite production and removal by delamination and dripping, crustal shortening and rapid outward strain propagation, rapid tectonic erosion of the forearc, hinterland basin development, and attainment of high regional elevation characterize the high-flux state. Reduced magmatism, stagnation of the orogenic strain front, and internal shortening are typical of the recharge state. Regional-scale differences in topography and orogenic history in the central Andes (e.g., the Altiplano vs. Puna) may be explained in part as the results of variations in the pace of the cordilleran cycle along strike, and the volumetric scale of lithospheric removal events. Numerical models have provided insight into the dynamics of key aspects of the orogenic system, including subduction-induced upper-plate shortening and lithosphere removal, but they have primarily been carried out as studies of isolated processes. Future models that incorporate subduction, upper-plate shortening, magmatic and metamorphic processes, crustal deformation, and surface evolution are needed to fully assess the linkages outlined here.

ACKNOWLEDGMENTS

We are grateful to all of the participants in the Convergent Orogenic Systems Analysis (COSA) project, and to ExxonMobil for funding the bulk of the work summarized in this chapter. Two anonymous reviewers provided valuable advice on how to improve this chapter, and we are thankful for their willingness to read such a long manuscript.

REFERENCES CITED

Adam, J., and Reuther, C.D., 2000, Crustal dynamics and active fault mechanics during subduction erosion. Application of frictional wedge analysis on to the north Chilean forearc: Tectonophysics, v. 321, p. 297–325, doi:10.1016/S0040-1951(00)00074-3.
Allmendinger, R.W., 1986, Tectonic development, southeastern border of the Puna Plateau, northwestern Argentine Andes: Geological Society of America Bulletin, v. 97, p. 1070–1082, doi:10.1130/0016-7606(1986)97<1070:TDSBOT>2.0.CO;2.
Allmendinger, R.W., Strecker, M.R., Eremchuk, J.E., and Francis, P.W., 1989, Neotectonic deformation of the southern Puna Plateau, northwestern Argentina: Journal of South American Earth Sciences, v. 2, p. 111–130, doi:10.1016/0895-9811(89)90040-0.
Allmendinger, R., Jordan, T., Kay, S., and Isacks, B., 1997, The evolution of the Altiplano-Puna Plateau of the central Andes: Annual Review of Earth and Planetary Sciences, v. 25, p. 139–174, doi:10.1146/annurev.earth.25.1.139.

Alonso, R.N., Jordan, T.E., Tabutt, K.T., and Vandervoort, D.S., 1991, Giant evaporite beds of the Neogene central Andes: Geology, v. 19, p. 401–404, doi:10.1130/0091-7613(1991)019<0401:GEBOTN>2.3.CO;2.
Armijo, R., and Thiele, R., 1990, Active faulting in northern Chile: Ramp stacking and lateral decoupling along a subduction plate boundary?: Earth and Planetary Science Letters, v. 98, p. 40–61, doi:10.1016/0012-821X(90)90087-E.
Armstrong, R.L., 1988, Mesozoic and early Cenozoic magmatic evolution of the Canadian Cordillera, in Clark, S.P., Burchfiel, B.C., and Suppe, J., eds., Processes in Continental Lithospheric Deformation: A Symposium to Honor John Rodgers: Geological Society of America Special Paper 218, p. 55–91.
Arriagada, C., Cobbold, P.R., and Roperch, P., 2006, Salar de Atacama Basin: A record of compressional tectonics in the central Andes since the mid-Cretaceous: Tectonics, v. 25, TC1008, doi:10.1029/2004TC001770.
Asch, G., Schurr, B., Bohm, M., C., Haberland, C., Heit, B., Kind, R., Wolbern, I., Yuan, X., Bataille, K., Comte, D., Pardo, M., Viramonte, J., Rietbrock, A., and Giese, P., 2006, Seismological studies of the central and southern Andes, in Oncken, O., Chong, G., Franz, G., Giese, P., Götze, H.-J., Ramos, H.-J., Strecker, M.R., and Wigger, P., eds., Active Subduction Orogeny: Berlin, Springer-Verlag, p. 443–457.
Babeyko, A.Y., and Sobolev, S.V., 2005, Quantifying different modes of the late Cenozoic shortening in the central Andes: Geology, v. 33, p. 621–624, doi:10.1130/G21126.1.
Babeyko, A.Y., Sobolev, S.V., Trumbull, R.B., Oncken, O., and Lavier, L.L., 2002, Numerical models of crustal scale convection and partial melting beneath the Altiplano-Puna Plateau: Earth and Planetary Science Letters, v. 199, p. 373–388, doi:10.1016/S0012-821X(02)00597-6.
Babeyko, A.Y., Sobolev, S.V., Victor, T., Oncken, O., and Trumbull, R.B., 2006, Numerical study of weakening processes in the central Andean back arc, in Oncken, O., Chong, G., Franz, G., Giese, P., Götze, H.-J., Ramos, H.-J., Strecker, M.R., and Wigger, P., eds., The Andes: Active Subduction Orogeny: Berlin, Springer-Verlag, p. 495–512.
Bahlburg, H., and Hervé, F., 1997, Geodynamic evolution and tectonostratigraphic terranes of northwestern Argentina and northern Chile: Geological Society of America Bulletin, v. 109, p. 869–884, doi:10.1130/0016-7606(1997)109<0869:GEATTO>2.3.CO;2.
Barnes, J.B., and Ehlers, T.A., 2009, End member models for Andean Plateau uplift: Earth-Science Reviews, v. 97, p. 105–132, doi:10.1016/j.earscirev.2009.08.003.
Barnes, J.B., Ehlers, T.A., McQuarrie, N., O'Sullivan, P.B., and Tawackoli, S., 2008, Thermochronometer record of central Andean Plateau growth, Bolivia (19.5°S): Tectonics, v. 27, TC3003, doi:10.1029/2007TC002174.
Barnes, J.B., Ehlers, T.A., Insel, N., McQuarrie, N., and Poulsen, C.J., 2012, Linking orography, climate, and exhumation across the central Andes: Geology, v. 40, no. 12, p. 1135–1138, doi:10.1130/G33229.1.
Barth, A.P., Wooden, J.L., Jacobson, C.E., and Economos, R.C., 2013, Detrital zircon as a proxy for tracking the magmatic arc system: The California arc example: Geology, v. 41, p. 223–226, doi:10.1130/G33619.1.
Barton, M.D., 1996, Granitic magmatism and metallogeny of southwestern North America: Transactions of the Royal Society of Edinburgh–Earth Sciences, v. 87, p. 261–280, doi:10.1017/S0263593300006672.
Beck, S., and Zandt, G., 2002, The nature of orogenic crust in the central Andes: Journal of Geophysical Research, v. 107, no. B10, doi:10.1029/2000JB000124.
Beck, S.L., Zandt, G., Ward, K.M., and Scire, A., 2015, this volume, Multiple styles and scales of lithospheric foundering beneath the Puna Plateau, central Andes, in DeCelles, P.G., Ducea, M.N., Carrapa, B., and Kapp, P., eds., Geodynamics of a Cordilleran Orogenic System: The Central Andes of Argentina and Northern Chile: Geological Society of America Memoir 212, doi:10.1130/2015.1212(03).
Behn, M.D., Kelemen, P.B., Hirth, G., Hacker, B.R., and Massonne, H.J., 2011, Diapirs as the source of the sediment signature in arc lavas: Nature Geoscience, v. 4, no. 9, p. 641–646, doi:10.1038/ngeo1214.
Bianchi, M., Heit, B., Jakovlev, A., Yuan, X., Kay, S.M., Sandvol, E., Alonso, R.N., Coira, B., Brown, L., Kind, R., and Comte, D., 2013, Teleseismic tomography of the southern Puna Plateau in Argentina and adjacent regions: Tectonophysics, v. 586, p. 65–83.
Bird, P., 1979, Continental delamination and the Colorado Plateau: Journal of Geophysical Research, v. 84, p. 7561–7571, doi:10.1029/JB084iB13p07561.
Blasco, G., Zappettini, E.O., and Hongn, F., 1996, Hoja Geológica 2566-I, San Antonio de los Cobres, Provincias de Jujuy y Salta, Programa Nacional

de Cartas Geológicas de la República Argentina (1:250,000), Secretaría Geológico Buenos Aires: Boletin (Instituto de Estudios de Poblacion y Desarrollo [Dominican Republic]), v. 217, p. 1–126.

Bosio, P.P., Powell, J., del Papa, C., and Hongn, F., 2009, Middle Eocene deformation–sedimentation in the Luracatao Valley: Tracking the beginning of the foreland basin of northwestern Argentina: Journal of South American Earth Sciences, v. 28, p. 142–154, doi:10.1016/j.jsames.2009.06.002.

Bousquet, R., Goff, B., Henry, P., Le Pichon, X., and Chopin, C., 1997, Kinematic, thermal and petrological model of the central Alps: Lepontine metamorphism in the upper crust and eclogitisation of the lower crust: Tectonophysics, v. 273, p. 105–127, doi:10.1016/S0040-1951(96)00290-9.

Buck, W.R., and Toksoz, M.N., 1983, Thermal effects of continental collisions: Thickening a variable viscosity lithosphere: Tectonophysics, v. 100, p. 53–69, doi:10.1016/0040-1951(83)90178-6.

Buddin, T.S., Stimpson, I.G., and Williams, G.D., 1993, North Chilean forearc tectonics and Cenozoic plate kinematics: Tectonophysics, v. 220, p. 193–203, doi:10.1016/0040-1951(93)90231-8.

Cabrera, J., Sébrier, M., and Mercier, J.L., 1987, Active normal faulting in high plateaus of central Andes: The Cuzco region (Peru): Annales Tectonicae, v. 1, p. 116–138.

Calixto, F.J., Sandvol, E., Kay, S., Mulcahy, P., Heit, B., Yuan, X., Coira, B., Comte, D., and Alvarado, P., 2013, Velocity structure beneath the southern Puna Plateau: Evidence for delamination: Geochemistry Geophysics Geosystems, v. 14, p. 4292–4305, doi:10.1002/ggge.20266.

Canavan, R., Carrapa, B., Clementz, M.T., Quade, J., DeCelles, P.G., and Schoenbohm, L.M., 2014, Early Cenozoic uplift of the Puna Plateau, Central Andes, based on stable isotope paleoaltimetry of hydrated volcanic glass: Geology, v. 42, p. 447–450, doi:10.1130/G35239.1.

Capitanio, F.A., Faccenna, C., Zlotnik, S., and Stegman, D.R., 2011, Subduction dynamics and the origin of the Andean orogeny and the Bolivian orocline: Nature, v. 480, p. 83–86, doi:10.1038/nature10596.

Carlotto, V., 2013, Paleogeographic and tectonic controls on the evolution of Cenozoic basins in the Altiplano and Western Cordillera of southern Peru: Tectonophysics, v. 589, p. 195–219, doi:10.1016/j.tecto.2013.01.002.

Carrapa, B., and DeCelles, P.G., 2008, Eocene exhumation and basin development in the Puna of northwestern Argentina: Tectonics, v. 27, TC1015, doi:10.1029/2007TC002127.

Carrapa, B., and DeCelles, P.G., 2015, this volume, Regional exhumation and kinematic history of the central Andes in response to cyclical orogenic processes, *in* DeCelles, P.G., Ducea, M.N., Carrapa, B., and Kapp, P., eds., Geodynamics of a Cordilleran Orogenic System: The Central Andes of Argentina and Northern Chile: Geological Society of America Memoir 212, doi:10.1130/2015.1212(11).

Carrapa, B., DeCelles, P.G., Reiners, P., and Gehrels, G.E., 2009, Apatite triple dating and white mica $^{40}Ar/^{39}Ar$ thermochronology of syn-tectonic detritus in the central Andes: A multi-phase tectono-thermal history: Geology, v. 37, p. 407–410, doi:10.1130/G25698A.1.

Carrapa, B., Bywater-Reyes, S., DeCelles, P.G., Mortimer, E., and Gehrels, G.E., 2011a, Eocene–Miocene synorogenic basin evolution in the Eastern Cordillera of northwestern Argentina (25°–26°S): Regional implications for Andean orogenic wedge development: Basin Research, v. 23, p. 1–20.

Carrapa, B., Trimble, J.D., and Stockli, D.F., 2011b, Patterns and timing of exhumation and deformation in the Eastern Cordillera of NW Argentina revealed by (U-Th)/He thermochronology: Tectonics, v. 30, TC3003, doi:10.1029/2010TC002707.

Carrera, N., and Muñoz, J.A., 2008, Thrusting evolution in the southern Cordillera Oriental (northern Argentine Andes): Constraints from growth strata: Tectonophysics, v. 459, p. 107–122, doi:10.1016/j.tecto.2007.11.068.

Cladouhos, T.T., Allmendinger, R.W., Coira, B., and Farrar, E., 1994, Late Cenozoic deformation in the central Andes; fault kinematics from the northern Puna, northwestern Argentina and southwestern Bolivia: Journal of South American Earth Sciences, v. 7, p. 209–228, doi:10.1016/0895-9811(94)90008-6.

Clift, P.D., and Hartley, A.J., 2007, Slow rates of subduction erosion and coastal underplating along the Andean margin of Chile and Peru: Geology, v. 35, p. 503–506, doi:10.1130/G23584A.1.

Cloos, M., and Shreve, R.L., 1988, Subduction-channel model of prism accretion, mélange formation, sediment subduction, and subduction erosion at convergent plate margins: Part I. Background and description: Pure and Applied Geophysics, v. 128, no. 3–4, p. 455–500, doi:10.1007/BF00874548.

Cohen, A., McGlue, M.M., Ellis, G.S., Zani, H., Swarzenski, P.W., Assine, M.L., and Silva, A., 2015, this volume, Lake formation, characteristics, and evolution in retroarc deposystems: A synthesis of the modern Andean orogen and its associated basins, *in* DeCelles, P.G., Ducea, M.N., Carrapa, B., and Kapp, P.A., eds., Geodynamics of a Cordilleran Orogenic System: The Central Andes of Argentina and Northern Chile: Geological Society of America Memoir 212, doi:10.1130/2015.1212(16).

Coney, P.J., and Evenchick, C.A., 1994, Consolidation of the American Cordilleras: Journal of South American Earth Sciences, v. 7, p. 241–262, doi:10.1016/0895-9811(94)90011-6.

Coney, P.J., and Harms, T., 1984, Cordilleran metamorphic core complexes: Cenozoic extensional relics of Mesozoic compression: Geology, v. 12, p. 550–554, doi:10.1130/0091-7613(1984)12<550:CMCCCE>2.0.CO;2.

Coney, P.J., and Reynolds, S.J., 1977, Cordilleran Benioff zones: Nature, v. 270, p. 403–406, doi:10.1038/270403a0.

Conrad, C.P., 2000, Convective instability of thickening mantle lithosphere: Geophysical Journal International, v. 143, p. 52–70, doi:10.1046/j.1365-246x.2000.00214.x.

Conrad, C.P., and Molnar, P., 1999, Convective instability of a boundary layer with temperature- and strain-dependent viscosity in terms of 'available buoyancy': Geophysical Journal International, v. 139, p. 51–68, doi:10.1046/j.1365-246X.1999.00896.x.

Constenius, K.N., 1996, Late Paleogene extensional collapse of the Cordilleran foreland fold and thrust belt: Geological Society of America Bulletin, v. 108, p. 20–39, doi:10.1130/0016-7606(1996)108<0020:LPECOT>2.3.CO;2.

Coutand, I., Carrapa, B., Deeken, A., Schmitt, A.K., Sobel, E., and Strecker, M.R., 2006, Orogenic plateau formation and lateral growth of compressional basins and ranges: Insights from sandstone petrography and detrital apatite fission-track thermochronology in the Angastaco Basin, NW Argentina: Basin Research, v. 18, p. 1–26, doi:10.1111/j.1365-2117.2006.00283.x.

Cristallini, E., Cominguez, A.H., and Ramos, V.A., 1997, Deep structure of the Metan-Guachipas region: Tectonic inversion in northwestern Argentina: Journal of South American Earth Sciences, v. 10, p. 403–421, doi:10.1016/S0895-9811(97)00026-6.

Currie, C.A., Ducea, M.N., DeCelles, P.G., and Beaumont, C., 2015, this volume, Geodynamic models of Cordilleran orogens: Gravitational instability of magmatic arc roots, *in* DeCelles, P.G., Ducea, M.N., Carrapa, B., and Kapp, P.A., eds., Geodynamics of a Cordilleran Orogenic System: The Central Andes of Argentina and Northern Chile: Geological Society of America Memoir 212, doi:10.1130/2015.1212(01).

de Silva, S., 1989, Altiplano-Puna volcanic complex of the central Andes: Geology, v. 17, p. 1102–1106, doi:10.1130/0091-7613(1989)017<1102:APVCOT>2.3.CO;2.

de Silva, S., Zandt, G., Trumbull, R., Viramonte, J.G., Salas, G., and Jiménez, N., 2006, Large ignimbrite eruptions and volcano-tectonic depressions in the central Andes: A thermomechanical perspective, *in* Troise, C., De Natale, G., and Kilburn, C.R.J., eds., Mechanisms of Activity and Unrest at Large Calderas: Geological Society, London, Special Publication 269, p. 47–63.

DeCelles, P.G., 2004, Late Jurassic to Eocene evolution of the Cordilleran thrust belt and foreland basin system, western USA: American Journal of Science, v. 304, p. 105–168.

DeCelles, P.G., and Coogan, J.C., 2006, Regional structure and kinematic history of the Sevier fold-thrust belt, central Utah: Implications for the Cordilleran magmatic arc and foreland basin system: Geological Society of America Bulletin, v. 118, p. 841–864, doi:10.1130/B25759.1.

DeCelles, P.G., and Horton, B.K., 2003, Early to middle Tertiary foreland basin development and the history of Andean crustal shortening in Bolivia: Geological Society of America Bulletin, v. 115, p. 58–77, doi:10.1130/0016-7606(2003)115<0058:ETMTFB>2.0.CO;2.

DeCelles, P.G., Ducea, M.N., Kapp, P., and Zandt, G., 2009, Cyclicity in Cordilleran orogenic systems: Nature Geoscience, v. 2, p. 251–257, doi:10.1038/ngeo469.

DeCelles, P.G., Carrapa, B., Horton, B.K., and Gehrels, G.E., 2011, Cenozoic foreland basin system in the central Andes of northwestern Argentina: Implications for Andean geodynamics and modes of deformation: Tectonics, v. 30, TC6013, doi:10.1029/2011TC002948.

DeCelles, P.G., Carrapa, B., Horton, B.K., McNabb, J., Gehrels, G.E., and Boyd, J., 2015, this volume, The Miocene Arizaro Basin, central Andean hinterland: Response to partial lithosphere removal?, *in* DeCelles, P.G., Ducea, M.N., Carrapa, B., and Kapp, P.A., eds., Geodynamics of a Cordilleran Orogenic System: The Central Andes of Argentina and Northern Chile: Geological Society of America Memoir 212, doi:10.1130/2015.1212(18).

Deeken, A., Sobel, E.R., Coutand, I., Haschke, M., Riller, U., and Strecker, M.R., 2006, Development of the southern Eastern Cordillera, NW Argentina, constrained by apatite fission track thermochronology: From Early Cretaceous extension to middle Miocene shortening: Tectonics, v. 25, TC6003, doi:10.1029/2005TC001894.

Delouis, B., Philip, H., Dorbath, L., and Cisternas, A., 1998, Recent crustal deformation in the Antofagasta region (northern Chile) and the subduction process: Geophysical Journal International, v. 132, p. 302–338, doi:10.1046/j.1365-246x.1998.00439.x.

Dickinson, W.R., 2013, Phanerozoic palinspastic reconstructions of Great Basin geotectonics (Nevada-Utah, USA): Geosphere, v. 9, no. 5, p. 1384–1396, doi:10.1130/GES00888.1.

Drew, S., Ducea, M.N., and Schoenbohm, L.M., 2009, Mafic volcanism on the Puna Plateau, NW Argentina: Implications for lithospheric composition and evolution with an emphasis on lithospheric foundering: Lithosphere, v. 1, p. 305–318, doi:10.1130/L54.1.

Ducea, M., 2001, The California arc: Thick granitic batholiths, eclogitic residues, lithospheric-scale thrusting, and magmatic flare-ups: GSA Today, v. 11, no. 11, p. 4–10, doi:10.1130/1052-5173(2001)011<0004:TCATGB>2.0.CO;2.

Ducea, M.N., 2002, Constraints on the bulk composition and root foundering rates of continental arcs: A California arc perspective: Journal of Geophysical Research, v. 107, no. B11, 2304, doi:10.1029/2001JB000643.

Ducea, M.N., and Barton, M.D., 2007, Igniting flare-up events in Cordilleran arcs: Geology, v. 35, p. 1047–1050, doi:10.1130/G23898A.1.

Ducea, M.N., Seclaman, A.C., Murray, K.E., Jianu, D., and Schoenbohm, L.M., 2013, Mantle-drip magmatism beneath the Altiplano-Puna Plateau, central Andes: Geology, v. 41, p. 915–918, doi:10.1130/G34509.1.

Ducea, M.N., Otamendi, J.E., Bergantz, G.W., Jianu, D., and Petrescu, L., 2015, this volume, The origin and petrologic evolution of the Ordovician Famatinian-Puna arc, *in* DeCelles, P.G., Ducea, M.N., Carrapa, B., and Kapp, P.A., eds., Geodynamics of a Cordilleran Orogenic System: The Central Andes of Argentina and Northern Chile: Geological Society of America Memoir 212, doi:10.1130/2015.1212(07).

Echavarria, R., Hernández, R., Allmendinger, R.W., and Reynolds, J.H., 2003, Subandean thrust and fold belt of northwest Argentina: Geometry and timing of the Andean evolution: American Association of Petroleum Geologists Bulletin, v. 87, p. 965–985, doi:10.1306/01200300196.

Ege, H., Sobel, E.R., Scheuber, E., and Jacobshagen, V., 2007, Exhumation history of the central Andean Plateau (southern Bolivia) constrained by apatite fission track thermochronology: Tectonics, v. 26, TC1004, doi:10.1029/2005TC001869.

Einhorn, J.C., Gehrels, G.E., Vernon, A., and DeCelles, P.G., 2015, this volume, U-Pb zircon geochronology of Neoproterozoic–Paleozoic sandstones and Paleozoic plutonic rocks in the Central Andes (21°S–26°S), *in* DeCelles, P.G., Ducea, M.N., Carrapa, B., and Kapp, P.A., eds., Geodynamics of a Cordilleran Orogenic System: The Central Andes of Argentina and Northern Chile: Geological Society of America Memoir 212, doi:10.1130/2015.1212(06).

Elger, K., Oncken, O., and Glodny, J., 2005, Plateau-style accumulation of deformation: Southern Altiplano: Tectonics, v. 24, TC4020, doi:10.1029/2004TC001675.

Elkins-Tanton, L.T., 2007, Continental magmatism, volatile recycling, and a heterogeneous mantle caused by lithospheric gravitational instabilities: Journal of Geophysical Research, v. 112, B03405, doi:10.1029/2005JB004072.

Engebretson, D.C., Cox, A., and Gordon, R.G., 1985, Relative Motions between Oceanic and Continental Plates in the Pacific Basin: Geological Society of America Special Paper 206, 59 p.

Engelder, T.M., and Pelletier, J.D., 2015, this volume, Simulating foreland basin response to mountain belt kinematics and climate change in the Eastern Cordillera and Subandes: An analysis of the Chaco foreland basin in southern Bolivia, *in* DeCelles, P.G., Ducea, M.N., Carrapa, B., and Kapp, P.A., eds., Geodynamics of a Cordilleran Orogenic System: The Central Andes of Argentina and Northern Chile: Geological Society of America Memoir 212, doi:10.1130/2015.1212(17).

England, P., and Molnar, P., 1993, Cause and effect among thrust and normal faulting, anatectic melting and exhumation in the Himalaya, *in* Treloar, P.J., and Searle, M.P., eds., Himalayan Tectonics: Geological Society, London, Special Publication 74, p. 401–411.

Evenchick, C.A., McMechan, M.E., McNicoll, V.J., and Carr, S.D., 2007, A synthesis of the Jurassic–Cretaceous tectonic evolution of the central and southeastern Canadian Cordillera: Exploring links across the oro-gen, *in* Sears, J.W., and Evenchick, C.A., eds., Whence the Mountains? Inquiries into the Evolution of Orogenic Systems: A Volume in Honor of Raymond A. Price: Geological Society of American Special Paper 433, p. 117–145.

Farías, M., Charrier, R., Comte, D., Martinod, J., and Herail, G., 2005, Late Cenozoic deformation and uplift of the western flank of the Altiplano: Evidence from the depositional, tectonic, and geomorphologic evolution and shallow seismic activity (northern Chile at 19°30′S): Tectonics, v. 24, TC4001, doi:10.1029/2004TC001667.

Galliski, M.A., and Viramonte, J.G., 1988, The Cretaceous paleorift in northwestern Argentina: A petrological approach: Journal of South American Earth Sciences, v. 1, p. 329–342, doi:10.1016/0895-9811(88)90021-1.

Garzione, C.N., Molnar, P., Libarkin, J.C., and MacFadden, B.J., 2006, Rapid late Miocene rise of the Bolivian Altiplano: Evidence for removal of the mantle lithosphere: Earth and Planetary Science Letters, v. 241, p. 543–556, doi:10.1016/j.epsl.2005.11.026.

Garzione, C., Hoke, G., Libarkin, J., and Withers, S., 2008, Rise of the Andes: Science, v. 320, p. 1304–1307, doi:10.1126/science.1148615.

Garzione, C., Auerbach, D.J., Smith, J.J.-S., Rosario, J.J., Passey, B.H., Jordan, T.E., and Eiler, J.M., 2014, Clumped isotope evidence for diachronous surface cooling of the Altiplano and pulsed surface uplift of the central Andes: Earth and Planetary Science Letters, v. 393, p. 173–181, doi:10.1016/j.epsl.2014.02.029.

Gaschnig, R.M., Vervoort, J.D., Lewis, R.S., and McClelland, W.C., 2010, Migrating magmatism in the northern US Cordillera: In situ U-Pb geochronology of the Idaho batholith: Contributions to Mineralogy and Petrology, v. 159, p. 863–883, doi:10.1007/s00410-009-0459-5.

Gaschnig, R.M., Vervoort, J.D., Lewis, R.S., and Tikoff, B., 2013, Probing for Proterozoic and Archean crust in the northern U.S. Cordillera with inherited zircon from the Idaho batholith: Geological Society of America Bulletin, v. 125, p. 73–88, doi:10.1130/B30583.1.

Gehrels, G., Rusmore, M., Woodsworth, G., Crawford, M., Andronicos, C., Hollister, L., Patchett, J., Ducea, M., Butler, R., Klepeis, K., Davidson, C., Friedman, R., Haggart, J., Mahoney, J., Crawford, W., Pearson, D., and Girardi, J., 2009, U-Th-Pb geochronology of the Coast Mountains batholith in north-coastal British Columbia: Constraints on age, petrogenesis, and tectonic evolution: Geological Society of America Bulletin, v. 121, p. 1341–1361, doi:10.1130/B26404.1.

Gerbault, M., Martinod, J., and Herail, G., 2005, Possible orogeny-parallel lower crustal flow and thickening in the central Andes: Tectonophysics, v. 399, p. 59–72, doi:10.1016/j.tecto.2004.12.015.

Ghosh, D.K., 1995, Nd-Sr isotopic constraints on the interactions of the Intermontane superterrane with the western edge of North America in the southern Canadian Cordillera: Canadian Journal of Earth Sciences, v. 32, p. 1740–1758, doi:10.1139/e95-136.

Ghosh, P., Eiler, J.M., and Garzione, C., 2006, Rapid uplift of the Altiplano revealed in abundances of ^{13}C–^{18}O bonds in paleosol carbonate: Science, v. 311, p. 511–515, doi:10.1126/science.1119365.

Gillis, R.J., Horton, B.K., and Grove, M., 2006, Thermochronology, geochronology, and upper crustal structure of the Cordillera Real: Implications for Cenozoic exhumation of the central Andean Plateau: Tectonics, v. 25, TC6007, doi:10.1029/2005TC001887.

Girardi, J.D., Patchett, P.J., Ducea, M.N., Gehrels, G.E., Cecil, M.R., Rusmore, M.E., Woodsworth, G.J., Pearson, D.M., Manthei, C., and Wetmore, P., 2012, Elemental and isotopic evidence for granitoid genesis from deep-seated sources in the Coast Mountains batholith, British Columbia: Journal of Petrology, v. 53, p. 1505–1536, doi:10.1093/petrology/egs024.

Gleadow, A.J.W., and Fitzgerald, P.G., 1987, Tectonic history and structure of the Transantarctic Mountains: New evidence from fission track dating in the Dry Valleys area of southern Victoria Land: Earth and Planetary Science Letters, v. 82, p. 1–14, doi:10.1016/0012-821X(87)90102-6.

Göğüş, O.H., and Pysklywec, R.N., 2008, Near surface diagnostics of dripping and delaminating lithosphere: Journal of Geophysical Research, v. 113, B11404, doi:10.1029/2007JB005123.

Götze, H.-J., and Krause, S., 2002, The Central Andean gravity high, a relic of an old subduction complex?: Journal of South American Earth Sciences, v. 14, p. 799–811, doi:10.1016/S0895-9811(01)00077-3.

Gregory-Wodzicki, K.M., 2000, Uplift history of the central northern Andes: Geological Society of America Bulletin, v. 112, no. 7, p. 1091–1105, doi:10.1130/0016-7606(2000)112<1091:UHOTCA>2.0.CO;2.

Grier, M.E., Salfity, J.A., and Allmendinger, R.W., 1991, Andean reactivation of the Cretaceous Salta Rift, northwestern Argentina: Journal of

South American Earth Sciences, v. 4, p. 351–372, doi:10.1016/0895 -9811(91)90007-8.

Gubbels, T.L., Isacks, B.L., and Farrar, E., 1993, High-level surfaces, plateau uplift, and foreland development, Bolivian central Andes: Geology, v. 21, p. 695–698, doi:10.1130/0091-7613(1993)021<0695:HLSPUA>2.3.CO;2.

Gutscher, M.-A., Spakman, W., Bijwaard, H., and Engdahl, E.R., 2000, Geodynamics of flat subduction: Seismicity and tomographic constraints from the Andean margin: Tectonics, v. 19, p. 814–833, doi:10 .1029/1999TC001152.

Hartley, A.J., 2003, Andean uplift and climate change: Journal of the Geological Society, London, v. 160, p. 7–10, doi:10.1144/0016-764902-083.

Hartley, A.J., and Evenstar, L., 2010, Cenozoic stratigraphic development in the north Chilean forearc: Implications for basin development and uplift history of the central Andean margin: Tectonophysics, v. 495, p. 67–77, doi:10.1016/j.tecto.2009.05.013.

Hartley, A.J., Flint, S., Turner, P., and Jolley, E.J., 1992, Tectonic controls on the development of a semi-arid, alluvial basin as reflected in the stratigraphy of the Purilactis Group (Upper Cretaceous–Eocene), northern Chile: Journal of South American Earth Sciences, v. 5, p. 275–296, doi:10.1016/0895-9811(92)90026-U.

Hartley, A.J., May, G., Chong, G., Turner, P., Kape, S.J., and Jolley, E.J., 2000, Development of a continental forearc: A Cenozoic example from the central Andes, northern Chile: Geology, v. 28, no. 4, p. 331–334, doi:10.1130/0091-7613(2000)28<331:DOACFA>2.0.CO;2.

Haschke, M., Siebel, W., Gunther, A., and Scheuber, E., 2002, Repeated crustal thickening and recycling during the Andean orogeny in north Chile (21°–26° S): Journal of Geophysical Research–Solid Earth, v. 107, no. B1, p. 1–18, doi:10.1029/2001jb000328.

Haschke, M., Gunther, A., Melnick, D., Echtler, H., Reutter, K.J., Scheuber, E., and Oncken, O., 2006, Central and southern Andean tectonic evolution inferred from arc magmatism, *in* Oncken, O., Chong, G., Franz, G., Giese, P., Götze, H.-J., Ramos, H.-J., Strecker, M.R., and Wigger, P., eds., The Andes—Active Subduction Orogeny: Berlin, Springer-Verlag, Frontiers in Earth Sciences, v. 1, p. 337–353.

Hayes, G.P., Wald, D.J., and Johnson, R.J., 2012, Slab1.0: A three-dimensional model of global subduction zone geometries: Journal of Geophysical Research, v. 117, B01302, doi:10.1029/2011JB008524.

Heit, B., Koulakov, I., Asch, G., Yuan, X., Kind, R., Alcozer, I., Tawackoli, S., and Wilke, H., 2008, More constraints to determine the seismic structure beneath the central Andes at 21°S using teleseismic tomography analysis: Journal of South American Earth Sciences, v. 25, p. 22–36, doi:10.1016/j .jsames.2007.08.009.

Hernández, R.M., and Echavarria, L., 2009, Faja plegada y corrida Subandina del noroeste Argentina: Estratigrafía, geometría y cronología de la deformación: Asociacion Geologic Argentina Revista, v. 65, p. 68–80.

Hoke, G.D., Isacks, B.L., Jordan, T.E., Blanco, N., Tomlinson, A.J., and Ramezani, J., 2007, Geomorphic evidence for post–10 Ma uplift of the western flank of the central Andes (18°30′–22°S): Tectonics, v. 26, TC5021, doi:10.1029/2006TC002082.

Hoke, L., and Lamb, S., 2007, Cenozoic behind-arc volcanism in the Bolivian Andes, South America: Implications for mantle melt generation and lithospheric structure: Journal of the Geological Society, London, v. 164, p. 795–814, doi:10.1144/0016-76492006-092.

Hongn, F., del Papa, C., Powell, J., Petrinovic, I., Mon, R., and Deraco, V., 2007, Middle Eocene deformation and sedimentation in the Puna–Eastern Cordillera transition (23°–26°S): Control by preexisting heterogeneities on the pattern of initial Andean shortening: Geology, v. 35, p. 271–274, doi:10.1130/G23189A.1.

Horton, B.K., 2005, Revised deformation history of the central Andes: Inferences from Cenozoic foredeep and intermontane basins of the Eastern Cordillera, Bolivia: Tectonics, v. 24, TC3011, doi:10.1029/2003TC001619.

Horton, B.K., 2012, Cenozoic evolution of hinterland basins in the Andes and Tibet, *in* Busby, C.J., and Azor, A., eds., Tectonics of Sedimentary Basins: Recent Advances (1st ed.): Hoboken, New Jersey, Blackwell Publishing, p. 427–444.

Horton, B.K., and DeCelles, P.G., 1997, The modern foreland basin system adjacent to the central Andes: Geology, v. 25, p. 895–898, doi:10.1130/0091 -7613(1997)025<0895:TMFBSA>2.3.CO;2.

Houseman, G.A., McKenzie, D.P., and Molnar, P., 1981, Convective instability of a thickened boundary layer and its relevance for the thermal evolution of continental convergence belts: Journal of Geophysical Research, v. 86, p. 6115–6132, doi:10.1029/JB086iB07p06115.

Husson, L., and Sempere, T., 2003, Thickening the Altiplano crust by gravity-driven crustal channel flow: Geophysical Research Letters, v. 30, 1243, doi:10.1029/2002GL016877.

Husson, L., Conrad, C.P., and Faccenna, C., 2012, Plate motions, Andean orogeny, and volcanism above the South Atlantic convection cell: Earth and Planetary Science Letters, v. 317–318, p. 126–135, doi:10.1016/j .epsl.2011.11.040.

Iaffaldano, G., and Bunge, H.-P., 2008, Strong plate coupling along the Nazca–South America convergent margin: Geology, v. 36, p. 443–446, doi:10.1130/G24489A.1.

Iaffaldano, G., and Bunge, H.-P., 2009, Relating rapid plate-motion variations to plate-boundary forces in global coupled models of the mantle/ lithosphere system: Effects of topography and friction: Tectonophysics, v. 474, p. 393–404, doi:10.1016/j.tecto.2008.10.035.

Ingersoll, R.V., 1983, Petrofacies and provenance of late Mesozoic forearc basin, northern and central California: American Association of Petroleum Geologists Bulletin, v. 67, p. 1125–1142.

Insel, N., Christopher J. Poulsen, C.J., Ehlers, T.A., and Sturm, C., 2012, Response of meteoric $\delta^{18}O$ to surface uplift—Implications for Cenozoic Andean Plateau growth: Earth and Planetary Science Letters, v. 317–318, p. 262–272.

Isacks, B., 1988, Uplift of the central Andean Plateau and bending of the Bolivian orocline: Journal of Geophysical Research, v. 93, p. 3211–3231, doi:10.1029/JB093iB04p03211.

Jaillard, E., Soler, P., Carlier, C., and Mourier, T., 1990, Geodynamic evolution of the northern and central Andes during the middle Mesozoic times: A Tethyan model: Journal of the Geological Society, London, v. 147, p. 1009–1022, doi:10.1144/gsjgs.147.6.1009.

James, D.E., 1971, Plate tectonic model for the evolution of the central Andes: Geological Society of America Bulletin, v. 82, p. 3325–3346, doi:10.1130/0016-7606(1971)82[3325:PTMFTE]2.0.CO;2.

James, D.E., and Sacks, I.W., 1999, Cenozoic formation of the central Andes: A geophysical perspective, *in* Skinner, B., ed., Geology and Ore Deposits of the Central Andes: Society of Economic Geologists Special Publication 7, p. 1–25.

Jolley, E., Turner, P., Williams, G., Hartley, A., and Flint, S., 1990, Sedimentologic responses of an alluvial system to Neogene thrust tectonics: Journal of the Geological Society, London, v. 147, p. 769–784, doi:10.1144/ gsjgs.147.5.0769.

Jordan, T.E., 1995, Retroarc foreland and related basins, *in* Busby, C.J., and Ingersoll, R.V., eds., Tectonics of Sedimentary Basins: Oxford, UK, Blackwell Science, p. 331–362.

Jordan, T.E., Blanco, N., Dávila, F.M., and Tomlinson, A., 2006, Sismoestratigrafía de la Cuenca Calama (22°–23°S), Chile, *in* XI Congreso Geologico Chileno: Antofagasta, Chile, v. 2, p. 53–56.

Jordan, T.E., Mpodozis, C., Muñoz, N., Blanco, N., Pananont, P., and Gardeweg, M., 2007, Cenozoic subsurface stratigraphy and structure of the Salar de Atacama Basin, northern Chile: Journal of South American Earth Sciences, v. 23, p. 122–146, doi:10.1016/j.jsames.2006.09.024.

Jordan, T.E., Nester, P.L., Blanco, N., Hoke, G.D., Dávila, F., and Tomlinson, A.J., 2010, Uplift of the Altiplano-Puna Plateau: A view from the west: Tectonics, v. 29, TC5007, doi:10.1029/2010TC002661.

Jull, M., and Kelemen, P.B., 2001, On the conditions for lower crustal convective instability: Journal of Geophysical Research, v. 106, p. 6423–6446, doi:10.1029/2000JB900357.

Kay, R.W., and Kay, S.M., 1993, Delamination and delamination magmatism: Tectonophysics, v. 219, p. 177–189, doi:10.1016/0040-1951(93)90295-U.

Kay, S.M., and Coira, B.L., 2009, Shallowing and steepening subduction zones, continental lithospheric loss, magmatism, and crustal flow under the central Andean Altiplano–Puna Plateau, *in* Kay, S.M., Ramos, V.A., and Dickinson, W.R., eds., Backbone of the Americas: Shallow Subduction, Plateau Uplift, and Ridge and Terrane Collision: Geological Society of America Memoir 204, p. 229–259.

Kay, S.M., and Mpodozis, C., 2002, Magmatism as a probe to the Neogene shallowing of the Nazca plate beneath the modern Chilean flat-slab: Journal of South American Earth Sciences, v. 15, p. 39–57, doi:10.1016/ S0895-9811(02)00005-6.

Kay, S.M., Coira, B., and Viramonte, J., 1994, Young mafic back arc volcanic rocks as indicators of continental lithospheric delamination beneath the Argentine Puna Plateau, central Andes: Journal of Geophysical Research, v. 99, p. 24,323–24,339, doi:10.1029/94JB00896.

Kay, S.M., Mpodozis, C., and Coira, B., 1999, Neogene magmatism, tectonism, and mineral deposits of the central Andes (22°S to 33°S latitude), *in*

Skinner, B.J., ed., Geology and Ore Deposits of the Central Andes: Society of Economic Geologists Special Publication 7, p. 27–59.

Kay, S.M., Coira, B.L., Caffe, P.J., and Chen, C.-H., 2010, Regional chemical diversity, crustal and mantle sources and evolution of central Andean Puna Plateau ignimbrites: Journal of Volcanology and Geothermal Research, v. 198, p. 81–111, doi:10.1016/j.jvolgeores.2010.08.013.

Kidder, S., and Ducea, M.N., 2006, High temperatures and inverted metamorphism in the schist of Sierra de Salinas, California: Earth and Planetary Science Letters, v. 241, p. 422–437, doi:10.1016/j.epsl.2005.11.037.

Kley, J., and Monaldi, C.R., 1998, Tectonic shortening and crustal thickness in the central Andes: How good is the correlation?: Geology, v. 26, p. 723–726, doi:10.1130/0091-7613(1998)026<0723:TSACTI>2.3.CO;2.

Kley, J., and Monaldi, C.R., 2002, Tectonic inversion in the Santa Bárbara system of the central Andean foreland thrust belt, northwestern Argentina: Tectonics, v. 21, 1061, doi:10.1029/2002TC902003.

Kley, J., Rossello, E.A., Monaldi, C.R., and Habighorst, B., 2005, Seismic and field evidence for selective inversion of Cretaceous normal faults, Salta Rift, northwest Argentina: Tectonophysics, v. 399, p. 155–172, doi:10.1016/j.tecto.2004.12.020.

Kopp, H., Flueh, E.R., Papenberg, C., and Klaeschen, D., 2004, Seismic investigations of the O'Higgins Seamount Group and Juan Fernandez Ridge: Aseismic ridge emplacement and lithosphere hydration: Tectonics, v. 23, TC2009, doi:10.1029/2003TC001590.

Kortyna, C., 2013, Structural and Thermochronologic Constraints on Kinematics, Timing and Shortening during Inversion of the Salta Rift into the Andean Fold-Thrust Belt, Northwest Argentina [M.S. thesis]: Tucson, Arizona, University of Arizona, 98 p.

Krystopowicz, N.J., and Currie, C.A., 2013, Crustal eclogitization and lithosphere delamination in orogens: Earth and Planetary Science Letters, v. 361, p. 195–207, doi:10.1016/j.epsl.2012.09.056.

Kukowski, N., and Oncken, O., 2006, Subduction erosion—The "normal" mode of fore-arc material transfer along the Chilean margin?, *in* Oncken, O., Chong, G., Franz, G., Giese, P., Götze, H.-J., Ramos, H.-J., Strecker, M.R., and Wigger, P., eds., Active Subduction Orogeny: Berlin, Springer-Verlag, p. 217–236.

Lallemand, S.E., Schnürle, P., and Malavieille, J., 1994, Coulomb theory applied to accretionary and nonaccretionary wedges: Possible causes for tectonic erosion and/or frontal accretion: Journal of Geophysical Research, v. 99, p. 12,033–12,055, doi:10.1029/94JB00124.

Lamb, S., 2000, Active deformation in the Bolivian Andes, South America: Journal of Geophysical Research, v. 105, p. 25,627–25,653, doi:10.1029/2000JB900187.

Lamb, S., 2011, Did shortening in thick crust cause rapid late Cenozoic uplift in the northern Bolivian Andes?: Journal of the Geological Society, London, v. 168, p. 1079–1092, doi:10.1144/0016-76492011-008.

Lamb, S., and Davis, S.D., 2003, Cenozoic climate change as a possible cause for the rise of the Andes: Nature, v. 425, p. 792–797, doi:10.1038/nature02049.

Lamb, S., and Hoke, L., 1997, Origin of the high plateau in the central Andes, Bolivia, South America: Tectonics, v. 16, p. 623–649, doi:10.1029/97TC00495.

Laskowski, A.K., DeCelles, P.G., and Gehrels, G.E., 2013, Detrital zircon geochronology of Cordilleran retroarc foreland basin strata, western North America: Tectonics, v. 32, p. 1027–1048, doi:10.1002/tect.20065.

Leech, M.L., 2001, Arrested orogenic development: Eclogitization, delamination, and tectonic collapse: Earth and Planetary Science Letters, v. 185, no. 1–2, p. 149–159, doi:10.1016/S0012-821X(00)00374-5.

Leier, A.L., McQuarrie, N., Horton, B.K., and Gehrels, G.E., 2010, Upper Oligocene conglomerates of the Altiplano, central Andes: The record of deposition and deformation along the margin of a hinterland basin: Journal of Sedimentary Research, v. 80, p. 750–762, doi:10.2110/jsr.2010.064.

Leier, A., McQuarrie, N., Garzione, C., and Eiler, J., 2013, Stable isotope evidence for multiple pulses of rapid uplift in the central Andes: Earth and Planetary Science Letters, v. 371–372, p. 49–58, doi:10.1016/j.epsl.2013.04.025.

Lindsay, J., Schmitt, A., Trumbull, R., de Silva, S.L., Siebel, W., and Emmermann, R., 2001, Magmatic evolution of the La Pacana caldera system, central Andes, Chile: Compositional variation of two cogenetic, large-volume felsic ignimbrites: Journal of Petrology, v. 42, p. 459–486, doi:10.1093/petrology/42.3.459.

Liu, M., Yang, Y., Stein, S., and Klosko, E., 2002, Crustal shortening and extension in the central Andes: Insights from a viscoelastic model, *in* Stein, S., and Freymueller, J.T., eds., Plate Boundary Zones: Washington, D.C., American Geophysical Union, p. 325–339, doi:10.1029/GD030p0325.

Loveless, J.P., Hoke, G.D., Allmendinger, R.W., González, G., Isacks, B.L., and Carrizo, D.A., 2005, Pervasive cracking of the northern Chilean Coastal Cordillera: New evidence for forearc extension: Geology, v. 33, p. 973–976, doi:10.1130/G22004.1.

Loveless, J.P., Allmendinger, R.W., Pritchard, M.E., and Gonzalez, G., 2010, Normal and reverse faulting driven by the subduction zone earthquake cycle in the northern Chilean forearc: Tectonics, v. 29, TC2001, doi:10.1029/2009TC002465.

Lucassen, F., Becchio, R., Wilke, H.G., Franz, G., Thirlwall, M.F., Viramonte, J., and Wemmer, K., 2000, Proterozoic–Paleozoic development of the basement of the central Andes (18–26°S)—A mobile belt of the South American craton: Journal of South American Earth Sciences, v. 13, p. 697–715, doi:10.1016/S0895-9811(00)00057-2.

Lucassen, F., Becchio, R., and Franz, G., 2011, The early Palaeozoic high-grade metamorphism at the active continental margin of West Gondwana in the Andes (NW Argentina/N Chile): International Journal of Earth Sciences, v. 100, p. 445–463, doi:10.1007/s00531-010-0585-3.

Luo, G., and Liu, M., 2009, How does trench coupling lead to mountain building in the Subandes? A viscoelastoplastic finite element model: Journal of Geophysical Research, v. 114, B03409, doi:10.1029/2008JB005861.

Mahoney, J.B., Gordee, S.M., Haggart, J.W., Friedman, R.M., Diakow, L.J., and Woodsworth, G.J., 2009, Magmatic evolution of the eastern Coast plutonic complex, Bella Coola region, west-central British Columbia: Geological Society of America Bulletin, v. 121, p. 1362–1380, doi:10.1130/B26325.1.

Maksaev, V., and Zentilli, M., 1999, Fission track thermochronology of the Domeyko Cordillera, northern Chile: Implications for Andean tectonics and porphyry copper metallogenesis: Exploration and Mining Geology, v. 8, p. 65–89.

Maloney, K.T., Clarke, G.L., Klepeis, K.A., and Quevedo, L., 2013, The Late Jurassic to present evolution of the Andean margin: Drivers and the geological record: Tectonics, v. 32, p. 1049–1065, doi:10.1002/tect.20067.

Mamani, M., Tassara, A., and Woerner, G., 2008, Composition and structural control of crustal domains in the central Andes: Geochemistry Geophysics Geosystems, v. 9, Q03006, doi:10.1029/2007GC001925.

Marrett, R.A., and Strecker, M.R., 2000, Response of intracontinental deformation in the central Andes to late Cenozoic reorganization of South American plate motions: Tectonics, v. 19, p. 452–467, doi:10.1029/1999TC001102.

Marrett, R.A., Allmendinger, R.W., Alonso, R.N., and Drake, R.E., 1994, Late Cenozoic tectonic evolution of the Puna Plateau and adjacent foreland, northwestern Argentine Andes: Journal of South American Earth Sciences, v. 7, p. 179–207, doi:10.1016/0895-9811(94)90007-8.

Martinod, J., Husson, L., Roperch, P., Guillaume, B., and Espurt, N., 2010, Horizontal subduction zones, convergence velocity and the building of the Andes: Earth and Planetary Science Letters, v. 299, p. 299–309, doi:10.1016/j.epsl.2010.09.010.

Mazzuoli, R., Vezzoli, L., Omarini, R., Acocella, V., Gioncada, A., Matteini, M., Dini, A., Guillou, H., Hauser, N., Uttini, A., and Scaillet, S., 2008, Miocene magmatism and tectonics of the easternmost sector of the Calama–Olacapato–El Toro fault system in central Andes at ~24°S: Insights into the evolution of the Eastern Cordillera: Geological Society of America Bulletin, v. 120, p. 1493–1517, doi:10.1130/B26109.1.

McGroder, M.F., Lease, R.O., and Pearson, D.M., 2015, this volume, Along-strike variation in structural styles and hydrocarbon occurrences, Subandean fold-and-thrust belt and inner foreland, Colombia to Argentina, *in* DeCelles, P.G., Ducea, M.N., Carrapa, B., and Kapp, P.A., eds., Geodynamics of a Cordilleran Orogenic System: The Central Andes of Argentina and Northern Chile: Geological Society of America Memoir 212, doi:10.1130/2015.1212(05).

McQuarrie, N., 2002, The kinematic history of the central Andean fold-thrust belt, Bolivia: Implications for building a high plateau: Geological Society of America Bulletin, v. 114, p. 950–963, doi:10.1130/0016-7606(2002)114<0950:TKHOTC>2.0.CO;2.

McQuarrie, N., and DeCelles, P.G., 2001, Geometry and structural evolution of the Central Andean backthrust belt, Bolivia: Tectonics, v. 20, p. 669–692.

McQuarrie, N., Horton, B.K., Zandt, G., Beck, S., and DeCelles, P.G., 2005, Lithospheric evolution of the Andean fold–thrust belt, Bolivia, and the origin of the central Andean Plateau: Tectonophysics, v. 399, p. 15–37, doi:10.1016/j.tecto.2004.12.013.

McQuarrie, N., Barnes, J.B., and Ehlers, T.A., 2008, Geometric, kinematic, and erosional history of the central Andean Plateau, Bolivia (15–17°S): Tectonics, v. 27, TC3007, doi:10.1029/2006TC002054.

Meade, B.J., and Conrad, C.P., 2008, Andean growth and the deceleration of South American subduction: Time evolution of a coupled orogen-subduction system: Earth and Planetary Science Letters, v. 275, p. 93–101, doi:10.1016/j.epsl.2008.08.007.

Mégard, F., 1984, The Andean orogenic period and its major structures in central and northern Peru: Journal of the Geological Society, London, v. 141, p. 893–900, doi:10.1144/gsjgs.141.5.0893.

Meissner, R.O., and Mooney, W.D., 1998, Weakness of the lower continental crust: A condition for delamination, uplift, and escape: Tectonophysics, v. 296, p. 47–60, doi:10.1016/S0040-1951(98)00136-X.

Mercier, J.L., 1981, Extensional compressional tectonics associate with the Aegean Arc: Comparison with the Andean Cordillera of south Peru–north Bolivia: Philosophical Transactions of the Royal Society of London, ser. A, v. 300, p. 337–355, doi:10.1098/rsta.1981.0068.

Mercier, J.L., Sebrier, M., Lavenu, A., Cabrera, J., Bellier, O., Dumont, J.-F., and Machrare, J., 1992, Changes in the tectonic regime above a subduction zone of Andean type: The Andes of Peru and Bolivia during the Pliocene–Pleistocene: Journal of Geophysical Research, v. 97, p. 11,945–11,982, doi:10.1029/90JB02473.

Metcalf, K., and Kapp, P., 2015, this volume, Along-strike variations in crustal seismicity and modern lithospheric structure of the central Andean forearc, in DeCelles, P.G., Ducea, M.N., Carrapa, B., and Kapp, P.A., eds., Geodynamics of a Cordilleran Orogenic System: The Central Andes of Argentina and Northern Chile: Geological Society of America Memoir 212, doi:10.1130/2015.1212(04).

Mitra, G., 1994, Strain variation in thrust sheets across the Sevier fold-and-thrust belt (Idaho-Utah-Wyoming): Implications for section restoration and wedge taper evolution: Journal of Structural Geology, v. 16, p. 585–602, doi:10.1016/0191-8141(94)90099-X.

Molnar, P., and Garzione, C.N., 2007, Bounds on the viscosity coefficient of continental lithosphere from removal of mantle lithosphere beneath the Altiplano and Eastern Cordillera: Tectonics, v. 26, TC2013, doi:10.1029/2006TC001964.

Molnar, P., and Houseman, G.A., 2004, The effects of buoyant crust on the gravitational instability of thickened mantle lithosphere at zones of intra-continental convergence: Geophysical Journal International, v. 158, no. 3, p. 1134–1150, doi:10.1111/j.1365-246X.2004.02312.x.

Molnar, P., and Houseman, G.A., 2013, Rayleigh-Taylor instability, lithospheric dynamics, surface topography at convergent mountain belts, and gravity anomalies: Journal of Geophysical Research, v. 118, p. 2544–2557, doi:10.1002/jgrb.50203.

Monaldi, C.R., Gonzalez, R.E., and Salfity, J.A., 1996, Thrust fronts in the Lerma Valley (Salta, Argentina) during the Piquete Formation deposition (Pliocene-Pleistocene), in Third ISAG (International Symposium of Andean Geodynamics) Abstracts: St. Malo, France, p. 447–450.

Morency, C., and Doin, M.-P., 2004, Numerical simulations of the mantle lithosphere delamination: Journal of Geophysical Research, v. 109, B03410, doi:10.1029/2003JB002414.

Mpodozis, C., Arriagada, C., Basso, M., Roperch, P., Cobbold, P., and Reich, M., 2005, Late Mesozoic to Paleogene stratigraphy of the Salar de Atacama Basin, Antofagasta, northern Chile: Implications for the tectonic evolution of the central Andes: Tectonophysics, v. 399, p. 125–154, doi:10.1016/j.tecto.2004.12.019.

Muñoz, N., and Charrier, R., 1996, Uplift of the western border of the Altiplano on a west vergent thrust system, northern Chile: Journal of South American Earth Sciences, v. 9, p. 171–181, doi:10.1016/0895-9811(96)00004-1.

Murray, K.E., Ducea, M.N., and Schoenbohm, L., 2015, this volume, Foundering-driven lithospheric melting: The source of central Andean mafic lavas on the Puna Plateau (22°S–27°S), in DeCelles, P.G., Ducea, M.N., Carrapa, B., and Kapp, P., eds., Geodynamics of a Cordilleran Orogenic System: The Central Andes of Argentina and Northern Chile: Geological Society of America Memoir 212, doi:10.1130/2015.1212(08).

Myers, S.C., Beck, S., Zandt, G., and Wallace, T., 1998, Lithospheric-scale structure across the Bolivian Andes from tomographic images of velocity and attenuation for P and S waves: Journal of Geophysical Research, v. 103, p. 21,233–21,252, doi:10.1029/98JB00956.

Neil, E.A., and Houseman, G.A., 1999, Rayleigh-Taylor instability of the upper mantle and its role in intraplate orogeny: Geophysical Journal International, v. 138, no. 1, p. 89–107, doi:10.1046/j.1365-246x.1999.00841.x.

Norabuena, E.O., Dixon, T.H., Stein, S., and Harrison, C.G.A., 1999, Decelerating Nazca–South America and Nazca-Pacific plate motions: Geophysical Research Letters, v. 26, p. 3405–3408, doi:10.1029/1999GL005394.

O'Driscoll, L.J., Richards, M.A., and Humphreys, E.D., 2012, Nazca–South America interactions and the late Eocene–late Oligocene flat-slab episode in the central Andes: Tectonics, v. 31, TC2013, doi:10.1029/2011TC003036.

Oliveros, V., Morata, D., Aguirre, L., Féraud, G., and Fornari, M., 2007, Jurassic to Early Cretaceous subduction-related magmatism in the Coastal Cordillera of northern Chile (18°30′–24°S): Geochemistry and petrogenesis: Revista Geológica de Chile, v. 34, p. 209–232.

Oncken, O., Hindle, D., Kley, J., Elger, K., Victor, P., and Schemmann, K., 2006, Deformation of the central Andean upper plate system—Facts, fiction, and constraints for plateau models, in Oncken, O., Chong, G., Franz, G., Giese, P., Götze, H.-J., Ramos, H.-J., Strecker, M.R., and Wigger, P., eds., The Andes—Active Subduction Orogeny: Berlin, Springer-Verlag, p. 3–27.

Ort, M., Coira, B., and Mazzoni, M., 1996, Generation of a crust-mantle magma mixture: Magma sources and contamination at Cerro Panizos, central Andes: Contributions to Mineralogy and Petrology, v. 123, p. 308–322, doi:10.1007/s004100050158.

Pananont, P., Mpodozis, C., Blanco, N., Jordan, T.E., and Brown, L.D., 2004, Cenozoic evolution of the northwestern Salar de Atacama Basin, northern Chile: Tectonics, v. 23, TC6007, doi:10.1029/2003TC001595.

Pardo-Casas, F., and Molnar, P., 1987, Relative motion of the Nazca (Farallon) plate since Late Cretaceous time: Tectonics, v. 6, p. 233–248, doi:10.1029/TC006i003p00233.

Paterson, S.R., Okaya, D., Memeti, V., Economos, R., and Miller, R.B., 2011, Magma addition and flux calculations of incrementally constructed magma chambers in continental margin arcs: Combined field, geochronologic, and thermal modeling studies: Geosphere, v. 7, p. 1439–1468, doi:10.1130/GES00696.1.

Pearson, D.M., Kapp, P., DeCelles, P.G., Reiners, P.W., Gehrels, G.E., Ducea, M.N., and Pullen, A., 2013, Influence of pre-Andean crustal structure on Cenozoic thrust belt kinematics and shortening magnitude: Northwestern Argentina: Geosphere, v. 9, no. 6, p. 1766–1782, doi:10.1130/GES00923.1.

Pelletier, J.D., DeCelles, P.G., and Zandt, G., 2010, Relationships among climate, erosion, topography, and delamination in the Andes: A numerical modeling investigation: Geology, v. 38, p. 259–262, doi:10.1130/G30755.1.

Plank, T., 2005, Constraints from Th/La on sediment recycling at subduction zones and the evolution of the continents: Journal of Petrology, v. 46, p. 921–944, doi:10.1093/petrology/egi005.

Platt, J.P., 1986, Dynamics of orogenic wedges and the uplift of high-pressure metamorphic rocks: Geological Society of America Bulletin, v. 97, p. 1037–1053, doi:10.1130/0016-7606(1986)97<1037:DOOWAT>2.0.CO;2.

Pope, D.C., and Willett, S.D., 1998, Thermal-mechanical model for crustal thickening in the central Andes driven by ablative subduction: Geology, v. 26, p. 511–514, doi:10.1130/0091-7613(1998)026<0511:TMMFCT>2.3.CO;2.

Poulsen, C.J., Ehlers, T.A., and Insel, N., 2010, Onset of convective rainfall during gradual late Miocene rise of the central Andes: Science, v. 328, p. 490–493, doi:10.1126/science.1185078.

Pritchard, M.E., and Simons, M., 2004, Surveying volcanic arcs with satellite radar interferometry: The central Andes, Kamchatka and beyond: GSA Today, v. 14, no. 8, p. 4–11, doi:10.1130/1052-5173(2004)014<4:SVAWSR>2.0.CO;2.

Quade, J., Dettinger, M.P., Carrapa, B., DeCelles, P., Murray, K.E., Huntington, K.A., Cartwright, A., Canavan, R.R., Gehrels, G., and Clementz, M., 2015, this volume, The growth of the central Andes, 22°S–26°S, in DeCelles, P.G., Ducea, M.N., Carrapa, B., and Kapp, P.A., eds., Geodynamics of a Cordilleran Orogenic System: The Central Andes of Argentina and Northern Chile: Geological Society of America Memoir 212, doi:10.1130/2015.1212(15).

Ramos, V.A., 2008, The basement of the central Andes: The Arequipa and related terranes: Annual Reviews of Earth and Planetary Science, v. 36, p. 289–324, doi:10.1146/annurev.earth.36.031207.124304.

Ramos, V.A., 2009, Anatomy and global context of the Andes: Main geologic features and the Andean orogenic cycle, in Kay, S.M., Ramos, V.A., and Dickinson, W.R., eds., Backbone of the Americas: Shallow Subduction, Plateau Uplift, and Ridge and Terrane Collision: Geological Society of America Memoir 204, p. 31–65, doi:10.1130/2009.1204(02).

Ramos, V.A., and Folguera, A., 2009, Andean flat-slab subduction through time, in Murphy, J.B., Keppie, J.D., and Hynes, A.J., eds., Ancient Orogens and Modern Analogues: Geological Society, London, Special Publication 327, p. 31–54, doi:10.1144/SP327.3.

Ramos, V.A., Cristallini, E.O., and Pérez, D.J., 2002, The Pampean flat-slab of the Central Andes: Journal of South American Earth Sciences, v. 15, p. 59–78.

Reiners, P.W., Thomson, S.N., Vernon, A., Willett, S.D., Zattin, M., Einhorn, J., Gehrels, G., Quade, J., Pearson, D., Murray, K.E., and Cavazza, W., 2015, this volume, Low-temperature thermochronologic trends across the central Andes, 21°S–28° S, *in* DeCelles, P.G., Ducea, M.N., Carrapa, B., and Kapp, P.A., eds., Geodynamics of a Cordilleran Orogenic System: The Central Andes of Argentina and Northern Chile: Geological Society of America Memoir 212, doi:10.1130/2015.1212(12).

Reutter, K.J., Charrier, R., Götze, H.J., Schurr, B., Wigger, P., Scheuber, E., Giese, P., Reuther, C.-D., Schmidt, S., Rietbrock, A., Chong, G., and Belmonte-Pool, A., 2006, The Salar de Atacama Basin: A subsiding block within the western edge of the Altiplano-Puna Plateau, *in* Oncken, O., Chong, G., Franz, G., Giese, P., Götze, H.J., Ramos, V.A., Strecker, M.R., and Wigger, P., eds., The Andes—Active Subduction Orogeny: Berlin, Springer, p. 303–325.

Reynolds, J., Galli, C., Hernández, R., Idleman, B., Kotila, J., Hilliard, R., and Naeser, C., 2000, Middle Miocene tectonic development of the Transition zone, Salta Province, northwest Argentina: Magnetic stratigraphy from the Metán Subgroup, Sierra de González: Geological Society of America Bulletin, v. 112, p. 1736–1751, doi:10.1130/0016 -7606(2000)112<1736:MMTDOT>2.0.CO;2.

Richardson, R.M., and Coblentz, D.D., 1994, Stress modeling in the Andes: Constraints on the South American intraplate stress magnitudes: Journal of Geophysical Research, v. 99, p. 22,015–22,025, doi:10.1029/94JB01751.

Riller, U., Petrinovic, I., Ramelow, J., Strecker, M., and Oncken, O., 2001, Late Cenozoic tectonism, collapse caldera and plateau formation in the central Andes: Earth and Planetary Science Letters, v. 188, p. 299–311, doi:10.1016/S0012-821X(01)00333-8.

Riquelme, R., Martinod, J., Hérail, G., Darrozes, J., and Charrier, R., 2003, A geomorphological approach to determining the Neogene to Recent tectonic deformation in the Coastal Cordillera of northern Chile (Atacama): Tectonophysics, v. 361, p. 255–275, doi:10.1016/S0040-1951(02)00649-2.

Riquelme, R., Hérail, G., Martinod, J., Charrier, R., and Darrozes, J., 2007, Late Cenozoic geomorphologic signal of forearc deformation and tilting associated with the uplift and climate changes of the Andes, southern Atacama Desert (26°S–28°S): Geomorphology, v. 86, p. 283–306, doi:10.1016/j .geomorph.2006.09.004.

Roperch, P., Herail, G., and Fornari, M., 1999, Magnetostratigraphy of the Miocene Corque Basin, Bolivia; implications for the geodynamic evolution of the Altiplano during the Late Tertiary: Journal of Geophysical Research, v. 104, p. 20,415–20,429, doi:10.1029/1999JB900174.

Rosenbaum, G., Giles, D., Saxon, M., Betts, P.G., Weinberg, R.F., and Duboz, C., 2005, Subduction of the Nazca Ridge and the Inca Plateau: Insights into the formation of ore deposits in Peru: Earth and Planetary Science Letters, v. 239, p. 18–32, doi:10.1016/j.epsl.2005.08.003.

Rossel, P., Oliveros, V., Ducea, M.N., Reynaldo Charrier, R., Scaillet, S., Retamal, L., and Figueroa, O., 2013, The Early Andean subduction system as an analog to island arcs: Evidence from across-arc geochemical variations in northern Chile: Lithos, v. 179, p. 211–230, doi:10.1016/j .lithos.2013.08.014.

Russo, R.M., and Silver, P.G., 1996, Cordillera formation, mantle dynamics, and the Wilson cycle: Geology, v. 24, p. 511–514, doi:10.1130/0091 -7613(1996)024<0511:CFMDAT>2.3.CO;2.

Saleeby, J., 2003, Segmentation of the Laramide slab—Evidence from the southern Sierra Nevada region: Geological Society of America Bulletin, v. 115, p. 655–668, doi:10.1130/0016-7606(2003)115<0655:SOTLSF >2.0.CO;2.

Salfity, J.A., and Marquillas, R., 1994, Tectonic and sedimentary evolution of the Cretaceous–Eocene Salta Group, Argentina, *in* Salfity, J.A., ed., Cretaceous Tectonics of the Andes: Heidelberg, Vieweg, Earth Evolution Sciences, p. 266–315.

Salisbury, M.J., Jicha, B.R., de Silva, S.L., Singer, B.S., Jiménez, N.C., and Ort, M.H., 2011, ^{40}Ar/^{39}Ar chronostratigraphy of Altiplano-Puna volcanic complex ignimbrites reveals the development of a major magmatic province: Geological Society of America Bulletin, v. 123, p. 821–840, doi:10.1130/B30280.1.

Santimano, T., and Riller, U., 2012, Kinematics of Tertiary to Quaternary intracontinental deformation of upper crust in the Eastern Cordillera, southern central Andes, NW Argentina: Tectonics, v. 31, TC4002, doi:10.1029/2011TC003068.

Schellart, W.P., 2008, Subduction zone trench migration: Slab driven or overriding-plate driven?: Physics of the Earth and Planetary Interiors, v. 170, p. 73–88, doi:10.1016/j.pepi.2008.07.040.

Scheuber, E., and Gonzalez, G., 1999, Tectonics of the Jurassic–Early Cretaceous magmatic arc of the north Chilean Coastal Cordillera (22°–26°S): A story of crustal deformation along a convergent plate boundary: Tectonics, v. 18, no. 5, p. 895–910, doi:10.1029/1999TC900024.

Scheuber, E., and Reutter, K.J., 1992, Magmatic arc tectonics in the central Andes between 21° and 25°S: Tectonophysics, v. 205, no. 1–3, p. 127–140, doi:10.1016/0040-1951(92)90422-3.

Schildgen, T., Hodges, K.V., Whipple, K.X., Reiners, P.W., and Pringle, M.S., 2007, Uplift of the western margin of the Andean Plateau revealed from canyon incision history, southern Peru: Geology, v. 35, p. 523–526, doi:10.1130/G23532A.1.

Schmitz, M., 1994, A balanced model of the southern central Andes: Tectonics, v. 13, p. 484–492, doi:10.1029/93TC02232.

Schnurr, W.B.W., Trumbull, R.B., Clavero, J., Hahne, K., Siebel, W., and Gardeweg, M., 2007, Twenty-five million years of silicic volcanism in the southern central volcanic zone of the Andes: Geochemistry and magma genesis of ignimbrites from 25 to 27°S, 67 to 72°W: Journal of Volcanology and Geothermal Research, v. 166, p. 17–46, doi:10.1016/j .jvolgeores.2007.06.005.

Schoenbohm, L.M., and Carrapa, B., 2015, this volume, Miocene–Pliocene shortening, extension, and mafic magmatism support small-scale lithospheric foundering in the central Andes, NW Argentina, *in* DeCelles, P.G., Ducea, M.N., Carrapa, B., and Kapp, P., eds., Geodynamics of a Cordilleran Orogenic System: The Central Andes of Argentina and Northern Chile: Geological Society of America Memoir 212, doi:10.1130/2015.1212(09).

Schoenbohm, L.M., and Strecker, M.R., 2009, Normal faulting along the southern margin of the Puna Plateau, northwest Argentina: Tectonics, v. 28, TC5008, doi:10.1029/2008TC002341.

Schott, B., and Schmeling, H., 1998, Delamination and detachment of a lithospheric root: Tectonophysics, v. 296, p. 225–247, doi:10.1016/S0040 -1951(98)00154-1.

Schurr, B., and Rietbrock, A., 2004, Deep seismic structure of the Atacama Basin, northern Chile: Geophysical Research Letters, v. 31, L12601, doi:10.1029/2004GL019796.

Schurr, B., Rietbrock, A., Asch, G., Kind, R., and Oncken, O., 2006, Evidence for lithospheric detachment in the central Andes from local earthquake tomography: Tectonophysics, v. 415, p. 203–223, doi:10.1016/j .tecto.2005.12.007.

Scire, A., Biryol, C.B., Zandt, G., and Beck, S.L., 2015, this volume, Imaging the Nazca slab and surrounding mantle to 700 km depth beneath the central Andes (18°S to 28°S), *in* DeCelles, P.G., Ducea, M.N., Carrapa, B., and Kapp, P., eds., Geodynamics of a Cordilleran Orogenic System: The Central Andes of Argentina and Northern Chile: Geological Society of America Memoir 212, doi:10.1130/2014.1212(02).

Sdrolias, M., and Müller, R.D., 2006, Controls on back-arc basin formation: Geochemistry Geophysics Geosystems, v. 7, Q04016, doi:10 .1029/2005GC001090.

Sébrier, M., Mercier, J.L., Megard, F., Laubacher, G., and Garey-Gailhardies, E., 1985, Quaternary normal and reverse faulting and the state of stress in the central Andes of south Peru: Tectonics, v. 4, p. 739–780, doi:10.1029/ TC004i007p00739.

Seton, M., Müller, R.D., Zahirovic, S., Gaina, C., Torsvik, T., Shephard, G., Talsma, A., Gurnis, M., Turner, M., Maus, S., and Chandler, M., 2012, Global continental and ocean basin reconstructions since 200 Ma: Earth-Science Reviews, v. 113, p. 212–270, doi:10.1016/j.earscirev.2012.03.002.

Siks, B.C., and Horton, B.K., 2011, Growth and fragmentation of the Andean foreland basin during eastward advance of fold-thrust deformation, Puna Plateau and Eastern Cordillera, northern Argentina: Tectonics, v. 30, TC6017, doi:10.1029/2011TC002944.

Sobel, E.R., Hilley, G.E., and Strecker, M.R., 2003, Formation of internally drained contractional basins by aridity-limited bedrock incision: Journal of Geophysical Research, v. 108, 2344, doi:10.1029/2002JB001883.

Sobolev, S.V., and Babeyko, A.Y., 2005, What drives orogeny in the Andes?: Geology, v. 33, p. 617–620, doi:10.1130/G21557.1.

Sobolev, S.V., Babeyko, A.Y., Koulakov, I., and Oncken, O., 2006, Mechanism of the Andean orogeny: Insight from numerical modeling, *in* Oncken, O., Chong, G., Franz, G., Giese, P., Götze, H.J., Ramos, V.A., Strecker, M.R., and Wigger, P., eds., The Andes—Active Subduction Orogeny: Berlin, Springer-Verlag, Frontiers in Earth Sciences, v. 1, p. 513–535.

Somoza, R., 1998, Updated Nazca (Farallon)–South America relative motions during the last 40 Ma: Implications for mountain building in the central Andean region: Journal of South American Earth Sciences, v. 11, p. 211–215, doi:10.1016/S0895-9811(98)00012-1.

Starck, D., and Vergani, G., 1996, Desarrollo tecto-sedimentario del Cenozoico en el sur de la Provincia de Salta, Argentina, *in* XIII Congreso Geológico Argentino y III Congreso de Exploración de Hidrocarburos, Actas I: Buenos Aires, p. 433–452.

Steinmann, G., 1929, Geologie von Peru: Heidelberg, Karl Winter, 448 p.

Stern, C.R., 2004, Active Andean volcanism: Its geologic and tectonic setting: Revista Geologica de Chile, v. 31, p. 161–206.

Strecker, M.R., Alonso, R.N., Bookhagen, B., Carrapa, B., Hilley, G.E., Sobel, E.R., and Trauth, M.H., 2007, Tectonics and climate of the southern central Andes: Annual Review of Earth and Planetary Sciences, v. 35, p. 747–787, doi:10.1146/annurev.earth.35.031306.140158.

Taylor, G.K., Grocott, J., Pope, A., and Randall, D.E., 1998, Mesozoic fault systems, deformation and fault block rotation in the Andean forearc: A crustal scale strike-slip duplex in the Coastal Cordillera of northern Chile: Tectonophysics, v. 299, p. 93–109, doi:10.1016/S0040-1951(98)00200-5.

Tomlinson, A.J., Blanco, N., Maksaev, V., Dilles, J.H., Grunder, A., and Ladino, M., 2001, Geología de la Precordillera Andina de Quebrada Blanca Chuquicamata, Regiones I y II [20300–22300]: Santiago, Chile, Servicio Nacional de Geología y Minería, 444 p.

Trumbull, R.B., Riller, U., Oncken, O., Scheuber, E., Munier, K., and Hongn, F., 2006, The time-space distribution of Cenozoic volcanism in the south-central Andes: A new data compilation and some tectonic implications, *in* Oncken, O., Chong, G., Franz, G., Giese, P., Götze, H., Ramos, V.A., Strecker, M.R., and Wigger, P., eds., The Andes—Active Subduction Orogeny: Berlin, Springer-Verlag, p. 29–43.

Uba, C.E., Kley, J., Strecker, M.R., and Schmitt, A.K., 2009, Unsteady evolution of the Bolivian Subandean thrust belt; the role of enhanced erosion and clastic wedge progradation: Earth and Planetary Science Letters, v. 281, p. 134–146, doi:10.1016/j.epsl.2009.02.010.

Ueda, K., Gerya, T.V., and Burg, J.-P., 2012, Delamination in collisional orogens: Thermomechanical modeling: Journal of Geophysical Research, v. 117, B08202, doi:10.1029/2012JB009144.

Umlauf, K., 2011, Insights into the Timing of Uplift along the Western Edge of the Central Andes, Northern Chile [M.S. thesis]: Tucson, Arizona, University of Arizona, 77 p.

Vandervoort, D.S., Jordan, T.E., Zeitler, P.K., and Alonso, R.N., 1995, Chronology of internal drainage development and uplift, southern Puna Plateau, Argentine central Andes: Geology, v. 23, p. 145–148, doi:10.1130/0091-7613(1995)023<0145:COIDDA>2.3.CO;2.

Vergani, G., and Stark, D., 1989, Aspectos estructurales del valle de Lerma, al sur de la ciudad de Salta Boletin de Informaciones Petroleras: Yacimientos Petrolíferos Fiscales, v. 20, p. 2–9.

Victor, P., Oncken, O., and Glodny, J., 2004, Uplift of the western Altiplano Plateau: Evidence from the Precordillera between 20 degrees and 21 degrees S (northern Chile): Tectonics, v. 23, TC4004, doi:10.1029/2003TC001519.

von Heune, R., Weinrebe, W., and Heeren, F., 1999, Subduction erosion along the north Chile margin: Journal of Geodynamics, v. 27, p. 345–358, doi:10.1016/S0264-3707(98)00002-7.

Wagner, L.S., Beck, S.L., Zandt, G., and Ducea, M., 2006, Depleted lithosphere, cold, trapped asthenosphere, and frozen melt puddles above the flat slab in central Chile and Argentina: Earth and Planetary Science Letters, v. 245, p. 289–301, doi:10.1016/j.epsl.2006.02.014.

Wang, H., Currie, C.A., and DeCelles, P.G., 2015, this volume, Hinterland basin formation and gravitational instabilities in the central Andes: Constraints from gravity data and geodynamic models, *in* DeCelles, P.G., Ducea, M.N., Carrapa, B., and Kapp, P.A., eds., Geodynamics of a Cordilleran Orogenic System: The Central Andes of Argentina and Northern Chile: Geological Society of America Memoir 212, doi:10.1130/2015.1212(19).

Ward, K., Porter, R., Zandt, G., Beck, S., Wagner, L., Minaya, E., and Tavera, H., 2013, Ambient noise tomography across the central Andes: Geophysical Journal International, v. 194, p. 1559–1573, doi:10.1093/gji/ggt166.

Wdowinski, S., and Bock, Y., 1994, The evolution of deformation and topography of high elevated plateaus: 2. Application to the central Andes: Journal of Geophysical Research, v. 99, p. 7121–7130, doi:10.1029/93JB02396.

Wells, M.L., Hoisch, T.D., Cruz-Uribe, A.M., and Vervoort, J.D., 2012, Geodynamics of synconvergent extension and tectonic mode switching: Constraints from the Sevier-Laramide orogen: Tectonics, v. 31, TC1002, doi:10.1029/2011TC002913.

Wolf, R.A., Farley, K.A., and Kass, D.M., 1998, Modeling of the temperature sensitivity of the apatite (U-Th)/He thermochronometer: Chemical Geology, v. 148, p. 105–114, doi:10.1016/S0009-2541(98)00024-2.

Wörner, G., Uhlig, D., Kohler, I., and Seyfried, H., 2002, Evolution of the West Andean Escarpment at 18°S (N. Chile) during the last 25 Ma: Uplift, erosion and collapse through time: Tectonophysics, v. 345, p. 183–198.

Yáñez, G.A., Ranero, C.R., von Huene, R., and Diaz, J., 2001, Magnetic anomaly interpretation across the southern central Andes (32°–34°S): The role of the Juan Fernández Ridge in the Late Tertiary evolution of the margin: Journal of Geophysical Research, v. 106, p. 6325–6345, doi:10.1029/2000JB900337.

Yonkee, W.A., DeCelles, P.G., and Coogan, J.C., 1997, Kinematics and synorogenic sedimentation of the eastern frontal part of the Sevier orogenic wedge, northern Utah: *in* Link, P.K., and Kowallis, B.J., eds., Brigham Young University Geology Studies, v. 42, part 1, p. 355–380.

Yuan, X., Sobolev, S.V., and Kind, R., 2002, Moho topography in the central Andes and its geodynamic implications: Earth and Planetary Science Letters, v. 199, p. 389–402, doi:10.1016/S0012-821X(02)00589-7.

Zandt, G., Gilbert, H., Owens, T.J., Ducea, M., Saleeby, J., and Jones, C.H., 2004, Active foundering of a continental arc root beneath the southern Sierra Nevada, California: Nature, v. 431, p. 41–46, doi:10.1038/nature02847.

Zhou, R., Schoenbohm, L.M., and Cosca, M., 2013, Recent, slow normal and strike-slip faulting in the Pasto Venture region of the southern Puna Plateau, NW Argentina: Tectonics, v. 32, p. 19–33, doi:10.1029/2012TC003189.

MANUSCRIPT ACCEPTED BY THE SOCIETY 16 JULY 2014
MANUSCRIPT PUBLISHED ONLINE 20 NOVEMBER 2014